# Toxicology of Fishes

This up-to-date, comprehensive toxicology handbook is devoted to the effects of environmental pollution on fish. Fish species represent nearly half of all vertebrates and have become important sentinels for environmental contamination and model organisms for understanding adverse outcomes from exposures. This new edition is written by recognized experts, and it highlights the significant research progress in fish toxicology that has resulted from rapid technological developments in analytical, biochemical, and genomic sciences. The book:

- Discusses fundamental topics such as toxicokinetics in fishes, processes governing biotransformation within these organisms, and reactive oxygen species and oxidative stress.
- Explains key target organ systems for chemical impacts in fish, such as the nervous and immune systems, and how fishes can develop resistance to chemical toxicity.
- Covers multi-transgenerational effects on fishes, epigenetics, proteomics and metabolomics, and adverse outcome pathways.

Replacing the case studies in the first edition, this update delves into the impacts of microplastics, pharmaceuticals, and oil spills in dedicated final chapters. With nearly 200 illustrations and tables, this comprehensive reference work presents concepts in a way that is useful for both novices to and experts in the field of fish toxicology.

# Toxicology of Fishes

Edited By
Kristine L. Willett
Neelakanteswar Aluru

CRC Press
Taylor & Francis Group
Boca Raton London New York

CRC Press is an imprint of the
Taylor & Francis Group, an **informa** business

Cover © shutterstock

Second edition published 2024
by CRC Press
2385 NW Executive Center Drive, Suite 320, Boca Raton FL 33431

and by CRC Press
4 Park Square, Milton Park, Abingdon, Oxon, OX14 4RN

*CRC Press is an imprint of Taylor & Francis Group, LLC*



ISBN: 978-0-367-74997-2 (hbk)
ISBN: 978-0-367-75000-8 (pbk)
ISBN: 978-1-003-16069-4 (ebk)

DOI: 10.1201/9781003160694

Typeset in Times
by codeMantra

# Contents

# The Editors

**Kristine L. Willett, PhD** is Chair of the Department of BioMolecular Sciences in the School of Pharmacy at the University of Mississippi (UM). A professor of Pharmacology and Environmental Toxicology, she has taught both graduate and undergraduate courses in toxicology and environmental toxicology at UM since 2000. Dr. Willett earned her BA in Chemistry at the University of North Carolina and a PhD in Toxicology from Texas A&M University. She was a Dreyfus postdoctoral fellow in environmental chemistry at Indiana University followed by an RJR Leon Golberg postdoctoral fellow in toxicology at Duke University. Dr. Willett's research has been funded over the years by NIDA, NIEHS, NOAA, USGS, and the Army Corps of Engineers and resulted in 80+ manuscripts and book chapters. Throughout her career, she has led research projects which were designed to fundamentally understand the molecular mechanisms underlying toxicity and/or shed light on the potential adverse outcomes due to relevant anthropogenic contamination. Her laboratory has studied the developmental, reproductive and multi-generational impacts of nanosilver, cannabinoid, and benzo[a]pyrene exposures using fish models. She also studies consequences of environmental stressors on oysters. She serves as a Deputy Editor of *Toxicological Sciences*. She has been elected to leadership positions in the Society of Environmental Toxicology and Chemistry and the Society of Toxicology (SOT). SOT and American Association of Colleges of Pharmacy (AACP) have also recognized her with undergraduate and graduate educator awards, respectively. At UM, she earned the 2022 Research and Distinguished Achievement Award and the 2017 Faculty Achievement Award.

**Neelakanteswar Aluru, PhD** is an Associate Scientist in the Biology Department at Woods Hole Oceanographic Institution (WHOI). Dr. Aluru earned his Bachelor of Fisheries Science at the Orissa University of Agriculture and Technology in Berhampur, India, followed by a Master of Fisheries Science (Aquaculture) from Kerala Agricultural University Cochin, India. He earned a Master of Science from the College of Fishery Science, University of Tromsø, Norway. He received his PhD in Biology from the University of Waterloo, Ontario, Canada, where he studied influence of persistent organic pollutants on the endocrine stress axis in rainbow trout. He has held postdoctoral research appointments at the University of Guelph, Ontario, Canada, and at WHOI. Since 2012, he has been conducting research as an independent scientist at WHOI. His research interests are mainly focused in the field of environmental epigenetics, particularly the epigenetic processes involved in determining phenotypic (and/or developmental) plasticity in response to environmental cues/stressor exposures. His laboratory uses a variety of vertebrate and invertebrate model systems and employs a number of different molecular biology methods – gene-specific to high-throughput sequencing to study epigenetic modes of action. He is part of the Woods Hole Center for Oceans and Human Health.

# Contributors

**Bethanie Carney Almroth**
Department of Biology and Environmental Sciences, University of Gothenburg, Göteborg, Sweden

**Neelakanteswar Aluru**
Biology department, Woods Hole Oceanographic Institution, Woods Hole, Massachusetts, USA

**Gerald T. Ankley**
U.S. Environmental Protection Agency, Center for Computational Toxicology and Exposure, Duluth, Minnesota, USA

**Jon A. Arnot**
ARC Arnot Research and Consulting, Toronto, Ontario, Canada

**Tracie R. Baker**
Department of Environmental and Global Health, College of Public Health and Health Professions, University of Florida, Gainesville, Florida, USA

**Mace G. Barron**
U.S. Environmental Protection Agency (retired), Office of Research and Development, Environmental Effects Research Laboratory, Gulf Breeze, Florida, USA

**Steven P. Bradbury**
Natural Resource Ecology and Management, Iowa State University, Ames Iowa, USA

**Susanne M. Brander**
Department of Fisheries, Wildlife, and Conservation Sciences, Coastal Oregon Marine Experiment Station, Oregon State University, Corvallis, Washington USA

**Nadja R. Brun**
Department of Biological Sciences, University of Bergen, 5020 Bergen, Norway

**Michael J. Carvan III**
School of Freshwater Sciences, University of Wisconsin-Milwaukee, Milwaukee, Wisconsin, USA

**Bryan W. Clark**
U.S. Environmental Protection Agency, Atlantic Ecology Division, Narragansett, Rhode Island, USA

**Nancy D. Denslow**
Center for Environmental and Human Toxicology, College of Veterinary Medicine, University of Florida, Gainesville, Florida, USA

**Richard T. Di Giulio**
Environmental Sciences and Policy, Nicholas School of the Environment, Duke University, Durham, North Carolina, USA

**Karl Fent**
ETH Zurich, Institute of Biogeochemistry and Pollutant Dynamics, 8092 Zürich, Switzerland

**Fernando Galvez**
Department of Biological Sciences, Louisiana State University, Baton Rouge, Louisiana, USA

**Jed Goldstone**
Department of Biology, Woods Hole Oceanographic Institution, Woods Hole, Massachusetts, USA

**Leah Thornton Hampton**
Toxicology Department, Southern California Coastal Water Research Project, Costa Mesa, California, USA

**Margaret O. James**
Department of Medicinal Chemistry, College of Pharmacy, University of Florida, Gainesville, Florida, USA

**Azora König**
Department of Biology and Environmental Sciences, University of Gothenburg, Göteborg, Sweden

**Ramon Lavado**
Department of Environmental Science, Baylor University, Waco, Texas USA

**Christopher J. Martyniuk**
Department of Physiological Sciences, Center for Environmental and Human Toxicology, University of Florida, Gainesville, Florida USA

**Joel N. Meyer**
Environmental Sciences and Policy, Nicholas School of the Environment, Duke University, Durham, North Carolina, USA

**Danielle Meyer**
Department of Environmental and Global Health, College of Public Health and Health Professions, University of Florida, Gainesville, Florida, USA

**Diane E. Nacci**
U.S. Environmental Protection Agency (Retired), Office of Research and Development, Center for Environmental Measurement and Modeling, Atlantic Coastal Environmental Sciences Division, Narragansett, Rhode Island, USA

**John W. Nichols**
U.S. Environmental Protection Agency (retired), Office of Research and Development, Center for Computational Toxicology and Exposure, Great Lakes Toxicology and Ecology Division, Duluth, Minnesota, USA

**Charles D. Rice**
Department of Biological Sciences, College of Science, Clemson University, Clemson, South Carolina, USA

**Daniel Schlenk**
Environmental Sciences Department, University of California Riverside, Riverside, California, USA

**Helmut Segner**
Centre for Fish and Wildlife Health, University of Bern, Bern, Switzerland

**Megan E. Solan**
Department of Environmental Science, Baylor University, Waco, Texas USA

**Alicia R. Timme-Laragy**
Department of Environmental Health Science, School of Public Health and Health Sciences, University of Massachusetts Amherst, Amherst, Massachusetts, USA

**Peter van den Hurk**
Department of Biological Sciences, College of Science, Clemson University, Clemson, South Carolina, USA

**Daniel L. Villeneuve**
U.S. Environmental Protection Agency, Great Lakes Toxicology and Ecology Division, Duluth, Minnesota, USA

**Andrew Whitehead**
Environmental Toxicology, University of California Davis, Davis California, USA

**Kristine L. Willett**
Department of BioMolecular Sciences, School of Pharmacy, University of Mississippi, Mississippi, USA

**Judy Zelikoff**
Department of Medicine, New York University School of Medicine, New York, USA

**Kun Zhang**
State Environmental Protection Key Laboratory of Environmental Health Impact Assessment of Emerging Contaminants, School of Environmental Science and Engineering, Shanghai Jiao Tong University, Shanghai 200240, China

**Yanbin Zhao**
State Environmental Protection Key Laboratory of Environmental Health Impact Assessment of Emerging Contaminants, School of Environmental Science and Engineering, Shanghai Jiao Tong University, Shanghai 200240, China

# 1 Introduction

*Neelakanteswar Aluru*
Woods Hole Oceanographic Institution

*Kristine L. Willett*
University of Mississippi

Toxicology is an applied science focused on understanding the adverse outcomes associated with exposure to stressors that may range from consequences of climatic change, to toxins, metals, pharmaceuticals, and a wide range of environmental contaminants. The history of toxicology dates back to ancient times as humans began to first appreciate consequences of natural poisons, followed by recognition of forensic and occupational aspects of toxicology. As pharmacology advanced, toxicology became a fundamental aspect of drug development. In the last 50–60 years, the implications of wildlife exposures spawned a subdiscipline of toxicology coined ecotoxicology. The influence of human activities in altering aquatic ecosystems in the past few decades has increased in an unprecedented way. Industrialization, agricultural runoff, and urbanization have resulted in the degradation of the water quality in rivers and coastal waters adversely impacting health of aquatic organisms as well as ecosystem health. These changes are also directly influencing human health by threatening water and food security. With the appreciation that effects in wildlife could both be an indicator of potential human effects, and reflect extreme ecological consequences of toxic exposures, fish became a key species of toxicological interest.

Fish species account for nearly half of all vertebrate species. Of the nearly 30,500 species of ray-finned fish (Actinopterygii), teleosts make up 98% of the species (Ravi & Venkatesh, 2018). Fish species have evolved with distinctive morphologies and physiologies to survive in diverse aquatic and marine habitats including those (e.g., Salmonids) that live in both at various life stages. Despite the fact that saltwater covers nearly 70% of the planet, there are more species of freshwater (~15,150) versus marine fish (~14,740) (Carrete Vega & Wiens, 2012). According to the latest assessment report of the Intergovernmental Panel on Climate Change (IPCC, 2022), projected changes in the amount and variability of precipitation and increasing temperature could affect the river flows and hence the mobility and dilution of contaminants and surface water quality. Because of the rich genetic and species diversity, fish biodiversity can be used to both track ecosystem stress and conservation efficacy (Manel et al., 2020). In ecotoxicology, fish are often sentinels of contamination and particular physiological changes have become recognized biomarkers of exposure. For example, exposure to estrogenic endocrine disruptors causes expression of the egg yolk protein vitellogenin in male fish and feminization of testis (Delbes et al., 2022). Induction of liver CYP1A activity is used to assess exposure to polycyclic aromatic hydrocarbons (PAHs) and polychlorinated biphenyls (PCBs) (Oris & Roberts, 2007). Adverse consequences of oil spills on wild fish species are becoming well established and include cardiac toxicity, immune suppression, gill toxicity, and reduced reproductive success (Takeshita et al., 2021). As the diversity of anthropogenic chemicals impacting aquatic organisms has grown in the past few decades, the field of fish toxicology has been in the forefront of research in assessing the risk of exposures to these chemicals. In 2008, the 1st edition of *The Toxicology of Fishes* was edited by Drs. Richard Di Giulio and David Hinton. For years, the first edition has provided a comprehensive and foundational review of the impacts of various chemicals on different organ systems in fish.

As the human population grows, fish are also becoming an ever more important source of dietary protein. While conservation measures have been enacted to preserve some wild fisheries, aquaculture of fish has grown exponentially. The United Nations Food and Agriculture Organization (FAO) 2018 State of the World Fisheries indicates that finfish aquaculture in 2016 generated more than 54 million tons live weight primarily driven by ~44 million tons of inland aquaculture in Asia. Marine and coastal aquaculture represented ~6.5 million tons (FAO, 2018). Nile tilapia, carp species, and salmon are the most commonly cultured. While finfish aquaculture may buffer food insecurity, it is not without criticisms including ecological damage, reduced diversity, pharmaceutical overuse, and animal welfare concerns (Franks et al., 2021). Understanding the adverse effects of ecological damage due to anthropogenic activities, often using specific fish species as model organisms, is critical for both aquatic organismal and human health.

In addition to the aquatic and ecotoxicology studies using fish, fish models are also becoming highly utilized in biomedical studies because of the many aspects of conserved physiology. Whole genome sequencing of 60+ teleosts has allowed for comparisons across vertebrate genomes (Ravi & Venkatesh, 2018). Zebrafish particularly have become highly utilized in laboratories across the world because of their small size, high fecundity, and relatively easy and cost-effective

DOI: 10.1201/9781003160694-1

culture needs. A Zebrafish International Resource Center (ZIRC) and the Zebrafish Information Network (zfin.org) support the species as a research organism serving as a strain and data repository, respectively (Bradford et al., 2022).

Since the publishing of the first edition of this book (Di Giulio & Hinton, 2008), there has been significant research progress in fish toxicology. This stems from the rapid technological developments in analytical, biochemical, and genomic sciences. Mass spectrometry has become a powerful analytical technique in the field of environmental toxicology and its sensitivity, precision, and dynamic range has enabled identification and quantification of trace levels of chemicals in different matrices. Similarly, advancements in biochemical methods and genomics have led to significant process in understanding the effects of exposure to toxicants. Over the last 30–40 years, biomarkers have greatly expanded as early and specific endpoints for monitoring cellular responses to exposures and as predictors of health outcomes. Currently, genomic, proteomic, and metabolomic biomarkers represent a continuum of cellular responses to chemical exposures and provide linkages to mechanisms of action of chemicals. Until recently, application of these approaches was mainly restricted to model systems such as zebrafish. In the past 15–20 years, their application has greatly expanded to any species of interest, and this has revolutionized aquatic toxicology.

In this revised edition of the book, we have updated the scientific discoveries that have taken place in both fundamental and new or emerging subfields within toxicology. Here, we have merged the prior chapters on bioavailability and toxicokinetics into a single comprehensive chapter. The fish biotransformation chapter includes latest developments in P450 diversity. Because receptor-mediated mechanisms of action permeate much of contaminant toxicology, those aspects are integrated across most of the chapters. Please refer to the first edition for the topics of liver and osmoregulatory toxicology, cancer, and development. In contrast, in this edition Chapters 9–15 are all entirely new and highlight the recent molecular advances and new emerging contaminants in the field of fish toxicology. In addition, we have expanded the chapters in oxidative stress, neurological and immune effects, toxicity resistance, and oil spills. The new chapters include epigenetic and multigenerational effects of toxicants, proteomic, and metabolomic responses, endocrine disruptors and emerging contaminants of concern (e.g., microplastics, pharmaceuticals). This addition includes a chapter on Adverse Outcome Pathways, which provide a risk assessment framework for integrating all the new molecular initiating events gleaned from the new molecular and analytical technologies with population-level adverse impacts.

Our goal with this book was to highlight all of the important advances taking place in the area of fish toxicology. We would like to thank all the authors who participated in this edition of the book and acknowledge the previous editors who provided the framework for us to expand and update. Our hope is that this text will provide a comprehensive resource to both those coming to the field of fish toxicology as a novice as well as those who have been researching fish throughout their careers.

## REFERENCES

Bradford, Y. M., Van Slyke, C. E., Ruzicka, L., Singer, A., Eagle, A., Fashena, D., Howe, D. G., Frazer, K., Martin, R., Paddock, H., Pich, C., Ramachandran, S., & Westerfield, M. (2022). Zebrafish information network, the knowledgebase for Danio rerio research. *Genetics*, *220*(4). https://doi.org/10.1093/genetics/iyac016

Carrete Vega, G., & Wiens, J. J. (2012). Why are there so few fish in the sea? *Proceedings of the Royal Society B: Biological Sciences*, *279*(1737), 2323–2329. https://doi.org/10.1098/rspb.2012.0075

Delbes, G., Blázquez, M., Fernandino, J. I., Grigorova, P., Hales, B. F., Metcalfe, C., Navarro-Martín, L., Parent, L., Robaire, B., Rwigemera, A., Van Der Kraak, G., Wade, M., & Marlatt, V. (2022). Effects of endocrine disrupting chemicals on gonad development: Mechanistic insights from fish and mammals. *Environmental Research*, *204*, 112040. https://doi.org/10.1016/j.envres.2021.112040

Di Giulio, Richard T., & Hinton, D. E. (Ed.). (2008). *The Toxicology of Fishes* (1st ed.). CRC Press, Boca Raton, FL.

FAO. (2018). *The State of World Fisheries and Aquaculture 2018- Meeting the Sustainable Development Goals*, Rome. License: CC BY-NC-SA 3.0 IGO. https://www.fao.org/state-of-fisheries-aquaculture/2018/en

Franks, B., Ewell, C., & Jacquet, J. (2021). Animal welfare risks of global aquaculture. *Science Advances*, *7*(14), 1–8. https://doi.org/10.1126/sciadv.abg0677

IPCC, 2022: Climate change 2022: Impacts, adaptation and vulnerability. Contribution of Working Group II to the Sixth Assessment Report of the Intergovernmental Panel on Climate Change [H.-O. Pörtner, D. C. Roberts, M. Tignor, E. S. Poloczanska, K. Mintenbeck, A. Alegría, M. Craig, S. Langsdorf, S. Löschke, V. Möller, A. Okem, B. Rama (eds.)]. Cambridge University Press, Cambridge and New York, 3056 pp., doi:10.1017/9781009325844.

Manel, S., Guerin, P. E., Mouillot, D., Blanchet, S., Velez, L., Albouy, C., & Pellissier, L. (2020). Global determinants of freshwater and marine fish genetic diversity. *Nature Communications*, *11*(1), 1–9. https://doi.org/10.1038/s41467-020-14409-7

Oris, J. T., & Roberts, A. P. (2007). Statistical analysis of cytochrome P4501A biomarker measurements in fish. *Environmental Toxicology and Chemistry*, *26*(8), 1742. https://doi.org/10.1897/07-039R.1

Ravi, V., & Venkatesh, B. (2018). The divergent genomes of teleosts. *Annual Review of Animal Biosciences*, *6*, 47–68. https://doi.org/10.1146/annurev-animal-030117-014821

Takeshita, R., Bursian, S. J., Colegrove, K. M., Collier, T. K., Deak, K., Dean, K. M., De Guise, S., DiPinto, L. M., Elferink, C. J., Esbaugh, A. J., Griffitt, R. J., Grosell, M., Harr, K. E., Incardona, J. P., Kwok, R. K., Lipton, J., Mitchelmore, C. L., Morris, J. M., Peters, E. S., ... Hall, A. J. (2021). A review of the toxicology of oil in vertebrates: what we have learned following the Deepwater Horizon oil spill. *Journal of Toxicology and Environmental Health - Part B: Critical Reviews*, *24*(8), 355–394. https://doi.org/10.1080/10937404.2021.1975182

# 2 Toxicokinetics in Fishes

*John W. Nichols*
U.S. Environmental Protection Agency (retired)

*Jon A. Arnot*
ARC Arnot Research and Consulting

*Mace G. Barron*
U.S. Environmental Protection Agency (retired)

**Abstract**

The term *toxicokinetics* refers to the study of absorption, distribution, metabolism (biotransformation), and excretion (ADME) of toxicants in living organisms in relation to time. In this chapter, we employ a physiological perspective to address the topic of chemical toxicokinetics in fish. Introductory sections provide an overview of chemical absorption, distribution, and excretion, and highlight how adaptations of fish to their aquatic environment shape these processes. An extended section on toxicokinetic modeling describes how ADME processes may be represented mathematically to develop computational models of chemical uptake and disposition in fish. Included are descriptions of different modeling approaches, methods used to estimate model parameter values, and the scientific and regulatory questions to which these models can be applied. Special emphasis is given to the topic of chemical bioaccumulation assessment as this historically represents the most important application of toxicokinetic models for fish. Additional sections address the need to quantify chemical biotransformation as a means of representing this activity in predictive computational models. Recognizing the desire to reduce the use of live animals in chemical testing, a concluding section describes how toxicokinetic models can be used to relate observed *in vitro* effects to environmental exposures that could be expected to elicit similar effects *in vivo*.

## INTRODUCTION

Fish are intimately linked to their aqueous habitat. They respire, osmoregulate, achieve acid–base balance, and obtain their thermal character in relation to the surrounding water. In addition, water serves as a common conduit for many essential life processes. Most fish eat, drink, urinate, defecate, swim, obtain sensory information, reproduce, and spatially orient within a single surrounding and contiguous medium. The structural and physiological adaptations that allow fish to thrive in an aqueous environment strongly impact their interactions with xenobiotic chemicals. The solvent properties of water and its contrasting character relative to that of many environmental contaminants further shape these interactions.

The toxicity of a xenobiotic chemical to fish is substantially determined by the processes of absorption, distribution, biotransformation, and excretion (ADME[1]). These processes influence whole-organism dose–response relationships by altering the effective chemical concentration over time at a target site. The degree to which any individual process impacts the concentration time course depends on the character of the chemical and attributes of the exposed organism. The role that individual organs play in each process and the molecular susceptibility of specific tissues and organs provide the basis for specific toxicities. Most of the literature dealing with the ADME of chemicals in living organisms has been obtained in studies with drugs. The term *pharmacokinetics* is used to describe the quantitative study of these ADME processes in relation to time. The term *toxicokinetics* represents an extension of pharmacokinetic principles to chemicals (toxicants) that cause adverse effects in exposed organisms. This distinction aside, the two terms are synonymous.

For decades, aquatic toxicologists have used mathematical models to better understand adverse effects expressed at the whole-organism level. The goal in such cases is to more directly relate an external exposure to the observed effects by describing the corresponding internal exposure. Models developed to describe the toxicokinetic behavior of chemicals in fish vary substantially with respect to their complexity and data requirements. This variability reflects, in turn, the different purposes for which the models are developed, the time scale of interest, the complexity of ADME processes relevant to a particular chemical and/or dosing scenario, and the availability of data for model calibration and evaluation.

ADME processes also determine the extent to which xenobiotics accumulate in fish. The term *bioconcentration* refers to chemical accumulation that occurs when an aquatic organism is exposed to a chemical in water only. *Bioaccumulation* is a more inclusive term that describes chemical accumulation that occurs from all possible route of exposure, including water, diet, and sediments. *Biomagnification*, which may occur as an outcome of dietary uptake, refers to an increase in chemical concentrations in organisms that occupy successively higher trophic levels within an aquatic food web.[2] Bioconcentration, bioaccumulation, and biomagnification, particularly of persistent chemicals, serve as indicators of past exposure for the animal; as indicators of potential exposure for higher trophic

DOI: 10.1201/9781003160694-2

level consumers; and as markers of potential toxicity. As such, bioaccumulation assessment is a critical component of most chemical risk assessment approaches.

This chapter builds on a contribution given previously by Kleinow et al. (2008). As in the original text, our goal was to link anatomical, physiological, and environmental determinants of chemical uptake, accumulation, and elimination in fish with kinetic modeling approaches that describe this behavior. Introductory sections review physicochemical concepts relevant to the description of chemical kinetics in natural systems, highlight physiological differences between fish and mammals, and describe mechanisms of xenobiotic transport across biological membranes. The sections that follow describe the processes of absorption, distribution, and excretion in fish one by one. An extended section on toxicokinetic modeling begins with a discussion of basic clearance concepts. This text is followed by a description of empirical and physiologically based modeling approaches. In comparison with the chapter provided by Kleinow et al. (2008), the modeling text has been amended to deemphasize the presentation of mathematical equations while placing relatively greater emphasis on model evaluation and use.

Bioaccumulation modeling is addressed as a special topic due to the large amount of focused research in this area and because prediction of various bioaccumulation metrics represents the most common application of toxicokinetic models for fish. A critical input to these models is the rate of chemical biotransformation, typically expressed as an apparent whole-body elimination rate constant ($k_B$; 1/day). *In silico* and *in vitro* methods for the prediction of $k_B$ are described, and the domain of applicability of such methods is discussed. A concluding section describes the use of *reverse toxicokinetic* modeling approaches to extrapolate *in vitro* effects information to a corresponding whole-organism exposure.

The principal focus of this chapter is on the toxicokinetics of organic xenobiotic chemicals. In a section on *semi-physiological* modeling approaches, we discuss models that have been developed to describe the kinetics of several metals and metalloids in fish. No attempt is made, however, to review the physiological and biochemical determinants of this behavior (e.g., the presence/absence and activities of specific metal binding proteins) as these determinants tend to be highly substance specific (Olsson et al., 1998; Roesijadi and Robinson, 1994). Similarly, this chapter does not address the behavior of engineered nanomaterials or naturally derived toxins.

## BASIC CONCEPTS

### Fish as Wet Vertebrates

The basic requirements for life are the same for fish as other organisms. Life in water, however, presents several challenges to fish that are not equitably shared by most terrestrial animals. Water has a high specific heat capacity, which makes it impossible for all but the largest fish to maintain internal temperatures different from those of the immediate surroundings. Compared with air, water contains low levels of dissolved gases. The solubilities of physiologically important gases in water also differ. These facts necessitate both specialized gas-exchange surfaces and a different approach to acid–base balance. The viscosity of water requires that fish employ specialized approaches to locomotion and expend considerable energy to support respiration.

Although the physiological and biochemical mechanisms available to fish for homeostasis are like those of other vertebrates, the ways they are used may or may not be the same. This is important to keep in mind when comparing the kinetics of xenobiotics among fish and mammals, among fishes of diverse groups, and even among individuals of the same species held under different environmental conditions. Table 2.1 lists some generalized structural and functional differences between fish and mammals (Cunningham, 1997; Evans, 1993; Ruckesbusch et al., 1991). Many of these differences influence xenobiotic absorption, distribution, and elimination in fishes. As an example, poikilothermy influences membrane structure and function, enzymatic isozyme profiles, cardiac output, bile flow rate, and gut transit time.

Uptake of xenobiotics by fish and mammals may occur by inhalation, ingestion, or dermal exposure (Figure 2.1). Likewise, the excretion of xenobiotics by fish and mammals occurs at respiratory and skin surfaces, as well as by urinary, biliary, and fecal routes. Even with these commonalities, there are some obvious and not so obvious differences between fish and mammals regarding the nature of these pathways and their relative importance for different environmental contaminants. For example, the mammalian lung is exposed primarily to volatile chemicals and chemicals associated with or existing as particulates. In contrast, the fish gill is commonly exposed to water-soluble chemicals, as well as nonpolar chemicals bound to dissolved and particulate organic matter (DOM and POM, respectively), which are typically quantified as dissolved and particulate organic carbon (DOC and POC, respectively). Branchial uptake requires diffusion across the gill membranes, which possess a hydrophobic core. For nonpolar chemicals bound to organic matter in water, uptake requires first that they dissociate from the organic material and traverse a polar water and mucous interface before diffusing across a compatible hydrophobic membrane. The same phase transitions exist for chemical excretion across the gills. In general, the gills play a much greater role in the absorption and excretion of xenobiotics by fish than the lung plays in mammals.

The skin of mammals is dry, dead, and keratinized, and does not possess a circulatory component. When present, fur may further limit dermal uptake of xenobiotics in mammals. In contrast, the skin of fishes is a living epidermis that is perfused by an underlying secondary circulation. In very small fish or benthic species that are in contact with contaminated sediment, the skin becomes an important exchange surface for xenobiotic chemicals.

The gastrointestinal tract plays an important role in the absorption of xenobiotics in both fish and mammals. The structure and function of the gastrointestinal tract in both taxa are similar in many respects. In fish, however, the absence of a lymphatic system and lack of classical mammalian villi may influence the absorptive process. The feces

**TABLE 2.1**
**Generalized Structural, Functional, and Environmental Differences between Fish and Mammals**

| Characteristic | Fish | Mammals |
|---|---|---|
| | **Media** | |
| Surrounding media | Water | Air |
| Specific heat | High | Low |
| Solvent properties | Universal solvent | Nonsolvent |
| Dissolved gases | Low levels | High levels |
| Viscosity | High | Low |
| | **Temperature** | |
| Modality | Poikilotherms | Homeotherms |
| Metabolic rate | Slow | Fast |
| Primary energy currency | Amino acids and lipid | Carbohydrates |
| Feed/weight conversion | Very efficient | Less efficient |
| Membrane composition | Homeoviscous adaptation | No adaptation necessary |
| Body temperature regulation | Behavioral | Internal set point |
| | **Respiratory surface** | |
| Primary organ | Gill | Lung |
| Blood/gas flows | Counter current | Directional/tidal flow |
| Gas exchange | Yes | Yes |
| Acid–base balance | Yes | Yes |
| Nitrogen excretion | Yes | No |
| Osmoregulation | Yes | No |
| Primary driver of respiration | Low oxygen | High carbon dioxide |
| | **Acid–base balance** | |
| Primary organ | Gill | Lung and kidney |
| Primary form | Ion based | Gas (lung) and ion (kidney) based |
| | **Nitrogenous waste elimination** | |
| Primary organ | Gill | Kidney |
| Primary form | Ammonia | Urea |
| | **Circulation** | |
| Heart | Two chambers | Four chambers |
| Arrangement | Heart and gill in series | Heart and lung in parallel |
| Vessels | Arteries and veins less distinct | Distinct arteries and veins |
| Lymphatics | No | Yes |
| Secondary circulation | Yes | No |
| Blood pressure and flow | Low | High |
| Red blood cells | Nucleated | Non-nucleated |
| Plasma proteins | Lower levels | Higher levels |
| Blood pH | Higher than mammals | Lower than fish |
| | **Kidney** | |
| Nephron | Less complex. SW – no distal tubule; aglomerular. FW, SW – modified loop of Henle | More complex, including glomerulus, distal tubule, and loop of Henle |
| Blood supply | Renal artery and renal portal | Renal artery |
| Hematopoiesis | Many fishes | NA |
| Waste excretion | Minor site | Primary site |
| Fluid balance | Primary site | Primary site |
| Ion balance | Secondary site | Primary site |
| Acid–base balance | Secondary site | Primary site |
| | **Liver** | |
| Functional unit | Tubule, with no triads or zonal relationships | Lobule, with triads and zonal arrangement |
| Gall bladder | May or may not be present | May or may not be present |
| Bile flow | Slower than mammals | Faster than fish |
| Biotransformation rates | Slower than mammals | Faster than fish |

*(Continued)*

**TABLE 2.1** (*Continued*)
**Generalized Structural, Functional, and Environmental Differences between Fish and Mammals**

| Characteristic | Fish | Mammals |
|---|---|---|
| | **Gastrointestinal tract** | |
| Structural divisions | Gradual transitions | Clear delineation |
| Gastric stomach | Absent in 15% of species | Present |
| Absorptive surface | Longitudinal folds | Villi |
| Lipid absorption | Primarily into blood | Size dependent into lymph |
| Change in temperature | Change in length/morphometry | NA |
| | **Skin** | |
| Epidermis | Living, moist | Dead, dry, keratinized |
| Mucous | Present | Absent |

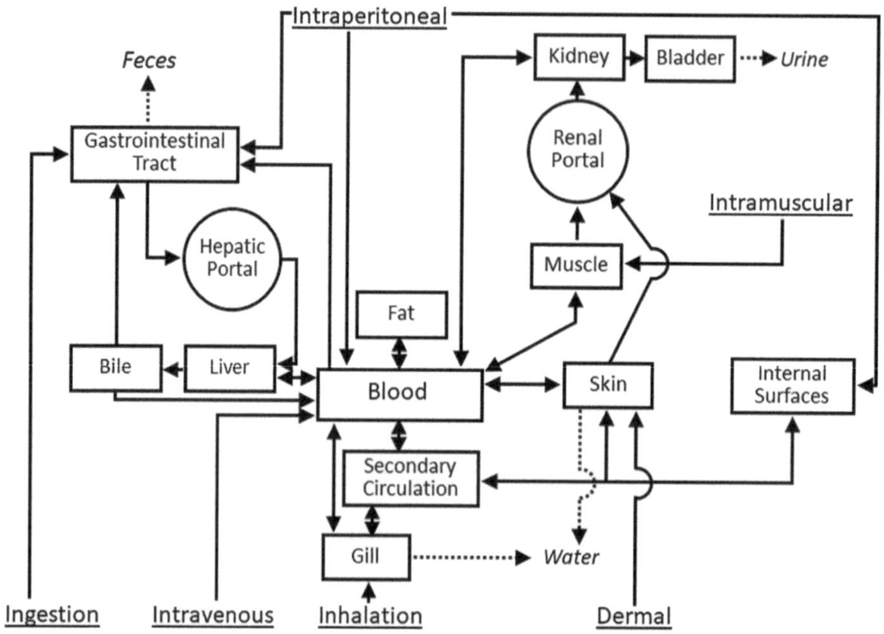


**FIGURE 2.1** Pathways for absorption, distribution, and excretion of xenobiotic chemicals in fish. Potential routes of exposure are underlined. Solid lines represent chemicals movements into and within the animal. Circled portal blood flows represent venous blood originating from the identified tissues. Dashed lines show pathways by which chemicals are lost from fish to the environment.

represent an important route of elimination in fish, both as a purveyor of bile and as a matrix with affinity for chemicals in the organism. Although bile formation is slow in fishes relative to mammals, many xenobiotics are excreted in bile.

In most cases, urinary elimination of xenobiotics is less important in fish than it is for mammals. This is due in part to the contribution of the gills to chemical elimination and to the modified role of the kidneys in comparison with mammals. In some cases, however, the urine represents the primary route of elimination. Urinary elimination may be particularly important for ionized chemicals that are substrates for membrane transporters within the renal tubular epithelium.

## The Environment

Large differences may exist between the concentration of a xenobiotic chemical in the environment, the concentration in the systemic circulation of a fish, and the concentration involved in specific target organ toxicity. These differences are due to both environmental factors and factors associated with the fish. The amount of chemical released to the environment, its movement among compartments (e.g., air, water, sediment), and susceptibility to biotic and abiotic transformation processes are primary determinants of chemical exposure to the exchange surfaces of fish. Conceptually similar processes operate within the fish, modulating exposure to the sites of toxicity.

Water dissolves more substances and dissolves them more completely than any other liquid; nevertheless, some nonpolar substances are relatively insoluble in water. In general, polar chemicals that are nonvolatile will remain unbound in the water column. Alternatively, if the chemical is nonpolar and nonvolatile, it will bind to DOC and POC in water, reducing the amount that is unbound and available for uptake. These characteristics may influence both chemical bioavailability to fish and the primary route of exposure.

Water in natural systems is a composite of $H_2O$ and the materials dissolved and suspended in it. Because these materials influence the complexation, binding, precipitation, and form of many chemicals, water cannot be viewed as a uniform or generic medium. Water quality as it relates to aquatic toxicology includes both biotic and abiotic components. Living organisms and organically derived substances contribute to the biotic character of water. Abiotic factors include pH, hardness, alkalinity, salinity, dissolved solids, and temperature. Often, the influence of water quality on chemical bioavailability and toxicity is a collective action, with each characteristic exerting independent as well as interdependent effects.

### Membrane Transport

Xenobiotic absorption, distribution, and elimination are dependent on transport across one or more biological membranes. All known cellular membranes are composed of lipid bilayers arranged with hydrophilic polar regions facing the outer surfaces and hydrophobic regions oriented toward the interior. On either surface or traversing the entire width of the membrane are globular proteins. Hydrophobic and electrostatic molecular interactions, along with the cytoskeleton, maintain the association of components, regionalization, and structural integrity of the membrane. Throughout the fish, and even within each cell, membranes may possess different morphological and biochemical attributes. External influences such as diet and temperature may alter membrane composition and character.

### Passive Diffusion

The primary mechanism of xenobiotic transport across lipid membranes is passive diffusion. Moderately nonpolar, lipid-soluble chemicals diffuse easily across biological membranes. Polar and very nonpolar chemicals diffuse across membranes less easily, although for different reasons. Polar chemicals have limited access to the nonpolar portion of the membrane, due to their low lipid solubility. In contrast, slow diffusion of nonpolar chemicals across hydrophilic regions may limit flux.

Diffusion across a membrane bathed on either side by an aqueous solution is often described using Fick's law of diffusion:

$$\text{Flux} = (D\ P_{MW}/h)(C_1 - C_2) \tag{2.1}$$

where $D$ is the diffusion coefficient (cm$^2$/s), $P_{MW}$ is the membrane–water partition coefficient (unitless), $h$ is the thickness of the diffusion path or membrane (cm), and $C_1$ and $C_2$ are chemical concentrations on each side of the membrane (mg/cm$^3$). When aggregated, $D$, $P_{MW}$, and $h$ define a permeability coefficient with units of cm/sec. For a unit area membrane, the rate of chemical transfer across the membrane (flux) has units of mg/s/cm$^2$. Multiplication by the membrane surface area provides a measure of total flux with units of mg/sec. This equation indicates that diffusional flux is first order and linear with respect to the chemical concentration gradient and progresses by a fractional rate constant.

Fick's law is a useful construct insofar as it highlights factors that promote or inhibit membrane flux. For example, the diffusion coefficient is a function of molecular size, molecular conformation, and the presence or absence of specific functional groups. In general, diffusion is most rapid for small, moderately hydrophobic molecules that possess little or no charge character. The chemical concentration gradient refers to the concentrations of diffusing chemical species on either side of the membrane. Generally, it is assumed that the principle diffusing species are those that are unbound in solution. Factors that reduce this unbound concentration, such as plasma binding or binding to DOC in water, can be expected, therefore, to limit flux. The membrane–water partition coefficient is a measure of the relative solubility of the chemical in the membrane and is generally estimated from measured lipid–water partitioning or a surrogate measurement such as the octanol–water partition coefficient (see *Chemical Hydrophobicity and the Octanol–Water Partition Coefficient* ($K_{OW}$)).

Empirical evidence suggests that nonionized chemicals diffuse much more readily across membranes than ionized forms. These observations have important implications for membrane flux of weak organic acids and bases. If the nonionized moiety has a lipid–water partition coefficient that favors membrane penetration, it will tend to reach an equilibrium concentration on both sides of the membrane, while the ionized form may be much more limited in its movements. As an approximation, therefore, the equilibrium concentration across the membrane will be based on the concentration of the nonionized form, with Fick's law of diffusion applying only to this nonionized form. The ratio of the two ionization states is dependent on the dissociation constant (p$K_a$) of the chemical and the pH of the surrounding media. This relationship is described by the Henderson–Hasselbalch equation:

For weak organic acids:

$$\text{pH} = \text{p}K_a + \log\ (\text{ionized/nonionized}) \tag{2.2}$$

For weak organic bases:

$$\text{pH} = \text{p}K_a + \log\ (\text{nonionized/ionized}) \tag{2.3}$$

For a weak organic acid, a decrease of one pH unit results in a tenfold increase in the concentration of the nonionized form. Conversely, an increase of one pH unit results in a tenfold increase in the concentration of the ionized form. Weak organic bases behave in the opposite manner; for example, a weak organic acid (p$K_a$ of 7) in moderately acidic water (pH of 5) would exist in a nonionized/ionized ratio of 100/1. For a weak organic base, the ratio would be 1/100. The effect of pH on the nonionized/ionized ratio can be extreme; thus, a pH change from 5 to 1 would change the nonionized/ionized ratio of a p$K_a$ 7 organic acid from 100/1 to 1,000,000/1.

Differences in pH across biological membranes will cause the degree of ionization to differ on either side. The side with the greatest degree of ionization may have a much higher total xenobiotic concentration because the equilibrium distribution of chemical is based on concentrations of the diffusing nonionized form. This phenomenon is referred to as *ion trapping*. The theoretical equilibrium concentration ratio ($R_{xy}$) of a xenobiotic across a membrane with two sides, $x$ and $y$, can be calculated as follows:

For weak organic acids:

$$R_{xy} = \left[1 + \text{antilog}\left(\text{pH}_x - \text{p}K_a\right)\right] / \left[1 + \text{antilog}\left(\text{pH}_y - \text{p}K_a\right)\right] \quad (2.4)$$

For weak organic bases:

$$R_{xy} = \left[1 + \text{antilog}\left(\text{p}K_a - \text{pH}_x\right)\right] / \left[1 + \text{antilog}\left(\text{p}K_a - \text{pH}_y\right)\right] \quad (2.5)$$

With respect to these relationships, fish present several interesting nuances that are not typically encountered in mammals. While ion trapping in mammals is generally limited to internal membranes (e.g., blood to gastrointestinal tract, blood to milk, blood to urine), ion trapping in fish may occur across external surfaces such as the gills and skin (blood to water). The range of environmental pH values that will support fish populations is about 5–10. For a given chemical, therefore, diffusion across the gills and skin may occur rapidly or very slowly, depending on the pH.

Physiological differences between fish and mammals also exist. Most fish species have blood pH values that range from 7.7 to 8.0. For most mammals, the blood pH is ~7.4. Finally, the p$K_a$ for a particular chemical or functional group will vary somewhat with temperature. Because fish are poikilotherms, changes in chemical dissociation may occur as they encounter different temperature environments.

Limited diffusion of ions across biological membranes also occurs. In such cases, the driving force for diffusion is the electrochemical gradient, which is a function of the chemical concentration gradient across the membrane, the charge distribution across the membrane (i.e., the electrical potential), and the charge on the diffusing species. The transmembrane electrical potential for a typical animal cell is −50 to −70 mV (inside relative to the outside). This electrical potential will tend to favor the inward diffusion of cations while promoting the outward diffusion of anions.

### Carrier-Mediated Transport

Biological membranes allow for compartmentalization and control of biological function by acting as diffusion barriers for inorganic ions and polar and/or or charged organic molecules, including energy substrates and structural components (e.g., sugars, fatty acids, and amino acids), signaling molecules (e.g., hormones and neurotransmitters), and substances that regulate osmotic pressure and pH (various organic and inorganic ions). The transport of these ions and organic molecules across membranes is largely controlled by membrane-bound channel and transporter proteins. These proteins may facilitate the diffusion of a substance down its electrochemical gradient or transport it against its electrochemical gradient via an active mechanism. The energy required for active transport can be provided by hydrolysis of adenosine 5′-triphosphate (ATP; primary active mechanism) or by coupling transport to the electrochemical potential of a second solute (secondary active mechanism).

Transporters responsible for transmembrane flux of drugs and toxicologically relevant substrates fall into two superfamilies: the ATP-binding cassette (ABC) transporter superfamily and the solute carrier (SLC) superfamily. ABC transporters act via the primary active mechanism and generally operate as efflux pumps, moving substrates out of cells or into cell organelles. Included in this family are the well-known multidrug resistance-related proteins (Deeley et al., 2006). SLC transporters that act by transporting substrates against their electrochemical gradient do so via the secondary active mechanism (Nigam et al., 2015; Roth et al., 2012). Other SLC transporters operate by facilitating membrane diffusion. The tissue distribution, regulation, and physiological functions of ABC transporters in fish are relatively well known (Ferreira et al., 2014; Luckenbach et al., 2014). By comparison, the SLC transporters in fish are less well described. Nevertheless, functional studies performed using isolated renal proximal tubules (Aslamkhan et al., 2006) and *in vitro* cellular expression systems (Aslamkhan et al., 2006; Popovic et al., 2014) have shown that teleost SLC transporters can transport xenobiotic environmental contaminants against their concentration gradient.

The different roles that membrane transporters play in supporting biliary and urinary elimination of xenobiotics in fish are described in greater detail below. In mammals, interactions of xenobiotics with specific nuclear receptors can result in coordinated expression of biotransformation enzymes that metabolize these chemicals as well as membrane transporters that promote the elimination of resulting metabolic products. Similar processes are thought to operate in fish (Bard, 2000; Ferreira et al., 2014). Because the total number of membrane transporters is finite, there is a potential for the kinetics of transport to saturate at high substrate concentrations. Similarly, there is a potential for one substrate to competitively inhibit the transport of other, similar chemicals. This type of inhibition is an important cause of drug–drug interactions. Its relevance for chemical mixture interactions in an ecotoxicological context is largely unknown.

### Pinocytosis

Pinocytosis is a process whereby the cell membrane invaginates, capturing external material within the lumen of a vesicle. The vesicle may empty its contents within the cell or fuse with another portion of the cell membrane, releasing its contents to the intercellular space. Uptake of macromolecular proteins by pinocytosis is well documented in the fish intestine (Rombout et al., 1985; Stroband et al., 1979; Stroband and Kroon, 1981).

## Chemical Hydrophobicity and the Octanol–Water Partition Coefficient ($K_{OW}$)

While there are several physicochemical properties that influence chemical toxicokinetics in fish, historically the most relevant is the octanol–water partition coefficient ($K_{OW}$, sometimes referred to as $P$ or $P_{OW}$). The $K_{OW}$ is defined as the ratio of the equilibrium concentrations of a dissolved substance in a system consisting of *n*-octanol and water. Numerous published relationships exist between $K_{OW}$ and acute toxicity, bioaccumulation, and other parameters relating to environmental toxicology and chemistry (Boethling and Mackay, 2000; Schwarzenbach et al., 2016). Similar relationships have been developed and applied in the human health sciences (Hansch, 1972; Hansch et al., 1995). The primary reason that $K_{OW}$ has proven to be a useful property is that *n*-octanol is a good surrogate for many organic biological and environmental phases including neutral storage lipids, membrane phospholipids, proteins, POC, DOC, and sediment organic carbon. For neutral organic chemicals, solubility in *n*-octanol tends to be relatively constant while solubility in water can vary enormously (Admire and Yalkowsky, 2013; Leo, 2000; Mackay et al., 1980; Meylan and Howard, 1995; Pinsuwan et al., 1995; Ran et al., 2001; Sangster, 1989). For this reason, the term "hydrophobic" ("water-hating") should be used preferentially to "lipophilic" ("lipid loving") when characterizing a chemical's relative tendency to partition out of water and into organic phases such as lipids.

The $K_{OW}$ may be measured using a variety of standardized methods (OECD 1995, 2004, 2006). In a system containing *n*-octanol and water, hydrophobic chemicals are thermodynamically driven from water into *n*-octanol because of their low aqueous solubility (Admire and Yalkowsky, 2013; Mackay et al., 1980; Meylan and Howard, 1995; Ran et al., 2001). At equilibrium, the chemical activity (or fugacity) in each phase is the same but measured chemical concentrations may differ by many orders of magnitude. Because of the large range of $K_{OW}$ values, $K_{OW}$ is often expressed as log $K_{OW}$ (base 10 logarithm). For chemicals that are very poorly soluble in water, reliable measurements of $K_{OW}$ may be difficult to obtain. In such cases, it may be necessary to predict $K_{OW}$ using a quantitative structure–activity relationship (QSAR; Cappelli et al., 2015) or polyparameter linear free energy relationship (pp-LFER; Stenzel et al., 2013a, 2013b). Specific chemical classes may present technical challenges that preclude the direct determination of reliable $K_{OW}$ values. For example, surfactants may form aggregates in aqueous solution and/or accumulate at the interface between *n*-octanol and water (Hodges et al., 2019), while some pigments and dyes are essentially insoluble in water (Anliker and Moser, 1987).

Chemical hydrophobicity is an important determinant of chemical distribution within the environment, between the environment and biota, and within an organism. Chemicals bound to POC and DOC are not available for uptake by fish across the gills (Black and McCarthy, 1988; Freidig et al., 1998; Johnsen et al., 1989; McCarthy and Jimenez, 1985; Schrap and Opperhuizen, 1990). Thus, consideration of bound and unbound fractions in water becomes important when quantifying chemical uptake from water or expressing the total concentration in a fish relative to that in water. For neutral organic chemicals, the issue of bioavailability in the water is most relevant chemicals with log $K_{OW}$>~5 (Arnot et al., 2010; Gobas and Morrison, 2000; Parkerton et al., 2008).

When a chemical is taken up by a fish, it distributes between water and various organic phases (Bertelsen et al., 1998). For many chemicals, the internal distribution at equilibrium can be predicted from $K_{OW}$ and the proximate composition of blood and individual tissues (e.g., the fractional content of water, lipid, and protein). A similar approach can be used to predict a partitioning-based equilibrium distribution between the fish and water. In such cases, a proportionality constant may be used to relate chemical solubility in *n*-octanol to that in different organic phases. A proportionality constant of 1.0 generally provides a good prediction of chemical partitioning to neutral and membrane lipids (Mackay, 1982; Chiou, 1985; Vaes et al., 1998). A proportionality constant of 0.05 has been determined for proteins (i.e., the solubility of neutral organic chemicals in protein is about 5% of that in octanol; deBruyn and Gobas, 2007).

The $K_{OW}$ is not a suitable parameter for predicting the partitioning behavior of fully ionized substances. For weak acids and bases, however, it may be useful to characterize partitioning between *n*-octanol and water using a distribution coefficient (*D* or log *D*), which is equal to the ratio of the sum of all forms of the chemical (neutral and ionized) in both phases. The ionization of a weak acid or base depends on its dissociation constant and the pH of the local environment. The interpretation of a stated *D* value requires, therefore, that the pH be identified (typically as a subscript, e.g., log $D_{7.4}$).

Although the $K_{OW}$ has proven to be a useful predictor of chemical partitioning to lipids and other organic phases, more advanced methods for the prediction of chemical partitioning are being developed and tested. In particular, the pioneering work of Abraham and colleagues (Abraham et al., 1999, 2004), as well as Goss and co-workers (Goss and Schwarzenbach, 2001; Endo and Goss, 2014) has led to the development and validation of pp-LFERs for predicting partition coefficients between water and storage lipids (Geisler et al., 2012), water and membrane lipids (Endo et al., 2011), and water and various proteins (Endo and Goss, 2011; Endo et al., 2012). These pp-LFERs are becoming more accessible to the scientific community (Ulrich et al., 2017) and may in time supplant simple $K_{OW}$-based prediction approaches.

## ABSORPTION, DISTRIBUTION, AND EXCRETION

### Xenobiotic Absorption

Xenobiotics are absorbed by fish across the gills, skin, and gut. In terms of gross morphology, these structures differ greatly from one another. All three, however, possess two basic features that contribute to their role as chemical exchange surfaces: (1) large surface area; and (2) separation of the environment, or an extension thereof, from the circulatory system by a membrane consisting of one, two, or a few cell layers. The path a chemical takes when it is absorbed at one of these surfaces involves: (1)

presentation to the absorbing epithelium in water or gut contents; (2) transport across the epithelium into blood; (3) incorporation into blood, including binding to plasma lipid and proteins; (4) transport via the systemic circulation to various tissues; and (5) transport from blood into tissues. From mass-balance considerations, the rate of uptake cannot exceed the rate at which chemical is presented to the exchange surface. Rates of diffusion across the membrane barrier and removal by the circulatory system also have the potential to limit the overall rate of chemical uptake.

## The Gills

### *Branchial Absorption of Xenobiotics*

The anatomical and physiological features of fish gills that promote efficient exchange of respiratory gases also contribute to uptake of xenobiotic chemicals from water; namely, a thin membrane separating blood and water, large surface area, and high rates of counter-current blood (perfusion) and water (ventilation) flow. Direct measurements of chemical uptake across fish gills have been obtained using fish respirometer-metabolism chambers, which separate inspired and expired water flows (McKim and Goeden, 1982; McKim and Heath, 1983). Using this model system, McKim and coworkers (Bradbury et al., 1986; McKim et al. 1985, 1986, 1987a, 1987b) measured branchial uptake rates in adult rainbow trout (*Oncorhynchus mykiss*) for a heterogeneous group of organic chemicals. These measurements suggested a consistent relationship between a chemical's uptake rate and its relative hydrophobicity, as indicated by the log of its octanol–water partition coefficient (log $K_{OW}$; Figure 2.2A). Uptake rates were low for chemicals with log $K_{OW}$ values less than 1, increased about four-fold between log $K_{OW}$ 1 and 3, leveled off between log $K_{OW}$ 3 and 6, and declined when log $K_{OW}$ exceeded 6. Working with a series of phenols, anisoles, and carboxylic acids, Saarikoski et al. (1986) found a similar relationship between absorption rate and log $K_{OW}$ in the guppy (*Poecilia reticulata*) (Figure 2.2B).

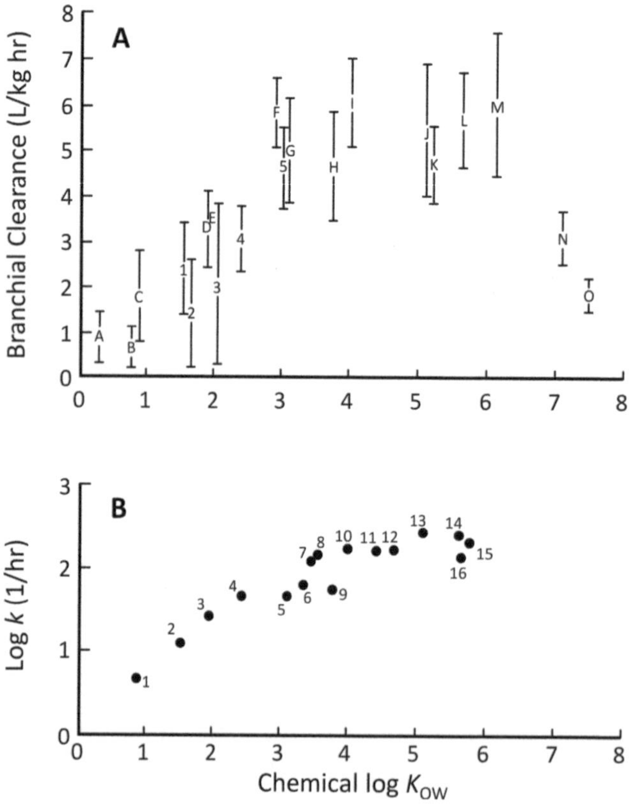


**FIGURE 2.2** Relationship between branchial uptake of chemicals by rainbow trout and guppies as a function of chemical hydrophobicity, expressed as the log of the octanol–water partition coefficient (log $K_{OW}$). For ionizable chemicals, the log $K_{OW}$ value is that of the neutral form. The uptake rate for rainbow trout is expressed as a clearance constant; that for guppies is given as the log of the uptake rate constant from water when pH < chemical $pK_a$. Panel A, *Rainbow trout sublethal exposures*: A, ethyl formate; B, ethyl acetate; C, 1-butanol; D, nitrobenzene; E, p-cresol; F, chlorobenzene; G, 2,3-dichlorophenol; H, 2,4,5-trichlorophenol; I, 1-decanol; J, 1-dodecanol; K, pentachlorophenol; L, hexachlorobenzene; M, 2,2′,5,5′-tetrachlorobiphenyl; N, dinitrophenol; O, Mirex. *Rainbow trout lethal exposures*: 1, benzaldehyde; 2, 2,4-dinitrophenol; 3, MS-222; 4, malathion; 5, 1-octanol (Data from Erickson and McKim, 1990a, 1990b). Panel B, *Guppy sublethal exposures*: 1, butyric acid; 2, phenol; 3, benzoic acid; 4, 4-phenylbutyric acid; 5, 2,4-dichlorophenol; 6, 2-sec butyl-4,6-dinitrophenol; 7, 3,4-dichlorobenzoic acid; 8, 2,6-dibromo-4-nitrophenol; 9, 2,4,5-trichlorophenol; 10, 2,4,6-trichlorophenol; 11, 2,3,4,6-tetrachlorophenol; 12, tetrachloroverathrol; 13, pentachlorophenol; 14, pentachloroanisole; 15, 2,4,6-trichloro-5-phenylphenol; 16, DDT; 17, 2,3,6-trichloro-4-nitrophenol (Adapted from Saarikoski et al., 1986).

Several physiological and chemical factors may account for these trends in branchial uptake with chemical hydrophobicity. Both McKim et al. (1985) and Saarikoski et al. (1986) attributed increases in absorption between log $K_{OW}$ 1 and 3 to greater membrane permeability for the more hydrophobic chemicals. Drawing upon the two-phase resistance model given by Flynn and Yalkowsky (1972), Gobas and co-workers (Gobas et al., 1986; Gobas and Mackay, 1987) proposed that branchial uptake was regulated by chemical partitioning into the lipid portion of the gill epithelium and diffusion through separate aqueous and organic (gill membrane) layers. Operationally, this description assumes that Fick's law of diffusion applies to two adjacent membranes possessing different molecular properties. Several authors have proposed molecular size cutoffs (e.g., effective cross-sectional diameter) for chemical diffusion across gill membranes (Dimitrov et al., 2003; Opperhuizen et al., 1985) as a means of explaining patterns in reported fish bioconcentration factors (BCFs; defined as the steady-state ratio of the chemical concentration in fish to that in water resulting from a water-only exposure). Arnot et al. (2010) cautioned against the use of strict molecular size criteria to predict chemical bioaccumulation potential, arguing that such cutoffs reflect errors in the interpretation measured BCFs. Other researchers, noting the downward trend in BCFs for chemicals with log $K_{OW}$ values greater than 6, have proposed that structural attributes of these chemicals may limit branchial uptake (Bruggeman et al., 1984). However, the principal limitation on branchial uptake of very hydrophobic chemicals is thought to be chemical binding to DOC and POC, which reduces the unbound or bioavailable concentration of the chemical in respired water (Black and McCarthy, 1988; Freidig et al., 1998; Johnsen et al., 1989; McCarthy and Jimenez, 1985; Schrap and Opperhuizen, 1990).

Water and blood flows through the gills maintain xenobiotic diffusion gradients across the gill epithelium. Norstrom et al. (1976), Neely (1979), and Bruggeman et al. (1981) suggested that branchial uptake should be correlated to the rate of water flow across the gills and thus to respiration rate. Barber et al. (1988) presented a hydrodynamic-based model for chemical bioconcentration in fish that incorporates effects of advection and diffusion in water flowing through the gills. Hayton and Barron (1990) suggested that blood flow, water flow, and diffusional barriers all have the potential to influence branchial flux, and proposed a model based on the concept of serial resistance. Erickson and McKim (1990a, 1990b) developed counter-current models for branchial flux of nonionized chemicals that account for both flow and diffusion limitations, as well as binding to DOC.

### *Physiological Impacts on Branchial Absorption*

Gill ventilation and blood perfusion rates vary with activity level, temperature, and environmental $O_2$ tension. Differences in environmental conditions and activity level also impact the toxicity and bioconcentration of chemicals in fish. These observations are now understood to be linked; thus, changes in environment and activity increase or decrease branchial absorption of xenobiotics by impacting the rate-limiting processes that control this uptake. For example, Lloyd (1961) observed increases in the toxicity of several chemicals to rainbow trout as ambient $O_2$ concentrations were reduced from 100% to 30% of saturation. Based on these observations, he suggested that the increase in toxicity was caused by an increase in ventilation volume that occurred at reduced $O_2$ concentrations, resulting in a larger quantity of chemical being brought in close contact with the gills.

McKim and Goeden (1982) measured changes in branchial uptake of endrin in response to a stepwise decrease in afferent $O_2$ concentration. Initial oxygen and endrin extraction efficiencies were nearly constant at $O_2$ saturation values of 100% and 80% but dropped progressively at $O_2$ saturation values of 50% and 30%. These observed declines in extraction efficiency may have been due to an increase in ventilatory stroke volume, resulting in greater bypass of water around the gill lamellar channels, or an increase in physiological dead space associated with increased water velocities. The primary effect of decreasing $O_2$ content, however, was to cause an increase in ventilation volume. The net result of these changes was that the rate of endrin uptake at 50% of $O_2$ saturation was approximately double that at 100%.

Environmental temperature changes can cause dramatic changes in the metabolic rates of poikilothermic animals, which affects their demand for $O_2$. An increase in $O_2$ demand due to increased temperature is especially problematic for fish because the solubility of $O_2$ in water is inversely related to temperature. To obtain more $O_2$, a fish in most cases adjusts its ventilation volume and the number of perfused lamellae in contact with respiratory water (Randall, 1982). This suggests that, if chemical uptake is controlled by water flow across the gills, an increase in temperature will result in an increase in branchial uptake rate like that caused by a decrease in $O_2$ content. Changes in temperature also affect cardiac output and could potentially impact branchial absorption if blood flow to the gills was a limiting factor. Several authors have shown that chemical uptake correlates with $O_2$ consumption ($VO_2$) as temperature changes (Black et al., 1991; Murphy and Murphy, 1971; Rodgers and Beamish, 1981). For example, Black et al. (1991) found that the branchial uptake of three moderately hydrophobic chemicals – benzo[a]pyrene, 2,2′,5,5′-tetrachlorobiphenyl, and naphthalene – changed in direct proportion to $VO_2$ when rainbow trout were subjected to an acute reduction in temperature from 17°C to 8°C. Finally, it is well known that chemical diffusion rates in solution vary with temperature. Similar changes are expected to occur in biological membranes and could be important if diffusion across the gill epithelium limits the rate of branchial flux. Fish may also respond to prolonged changes in temperature by altering the molecular structure of the gill epithelium, possibly changing its permeability to xenobiotic chemicals.

### The Gastrointestinal Tract

#### *Gastrointestinal Absorption of Xenobiotics*

The gastrointestinal tract (GIT) of fishes functions in digestion, nutrient absorption, excretion, and as a barrier to the external environment. For some groups of fishes, the GIT also plays a role in osmoregulation, buoyancy, ion regulation, and placental nutrition. Diverse structural adaptations in different species reflect a wide array of digestive strategies. Gross differences among species include the presence or absence of an acid-secreting stomach, the presence or absence of a gizzard, large differences in GIT length, and an absence or varying number of blind diverticula called *pyloric ceca* that project from the intestine near the pylorus of the stomach. In general, nutrient uptake and blood flows are highly regionalized and respond to local stimuli initiated by the food bolus. As the food bolus moves down the GIT, the character of the ingesta changes, as does its volume. Nutrients are absorbed and mucus, bacteria, and epithelial cells are added. As a result, different regions of the absorptive surface are presented with ingesta that differs markedly in composition. Reviews of fish GIT structure and function have been provided by several authors (Bucking and Anderson, 2020; Kapoor et al., 1975; Kleinow and James, 2001; Smith, 1989).

Because of its high surface area and role in nutrient uptake, the proximal intestine is the most important site for xenobiotic absorption. Digestive processes liberate xenobiotics from the food matrix, transport them to the mucosal epithelium, and control their presentation to the absorptive surface. Gut uptake of hydrophobic contaminants has been strongly associated with digestion and absorption of dietary lipid. Mechanical mixing, pancreatic lipases, and bile salts disperse dietary lipid and contribute to the formation of an equilibrium phase consisting of small mixed micelles (Kleinow and James, 2001). Hydrophobic contaminants reside in the interior of these micelles, while the hydrophilic exterior allows the micelles to remain soluble in the aqueous milieu of the luminal contents. Upon reaching the enterocytes, the micelles are dissociated by the combined action of pH changes and mucosal lipases, liberating stored xenobiotic. This process creates a highly localized concentration gradient for uptake of both xenobiotic and fatty acids across the absorptive epithelium. Uptake of hydrophobic organic chemicals by fish in an environmental setting may be dominated by the dietary route of exposure (Bruggeman et al., 1984; Clark et al., 1990; Qiao et al., 2000). This outcome is a product of limitations on branchial uptake that occur because of chemical binding to DOC and POC, as well as chemical accumulation in prey items (e.g., plankton and benthos) that occupy lower levels of aquatic food webs.

An examination of literature of the topic of gastrointestinal absorption in fish reveals some inconsistencies in the use of terminology. Here, we adopt terminology suggested by Arnot and Mackay (2018), which is consistent with that used in mammalian toxicokinetic studies. *Dietary absorption efficiency* ($E_D$) is defined as the fractional amount of chemical consumed by a fish that is absorbed across the gastrointestinal mucosa, while *oral bioavailability* ($F$) refers to the fraction amount of chemical consumed that enters the systemic circulation. The $E_D$ for a persistent, poorly metabolized chemical may be estimated from data collected in a standardized feeding experiment that provides information needed to account for the effect of chemical elimination (OECD, 2012). For chemicals that are subject to biotransformation before they enter the systemic circulation, an "$E_D$" value determined in this manner is more appropriately viewed as an estimate of $F$ (Arnot and Mackay, 2018). An estimate of $F$ may also be obtained by dividing the area under the blood concentration–time curve (AUC) that results from a dietary exposure by that which results from an equivalent intravascular dose (Barron et al., 1990). If an ingested chemical undergoes no biotransformation, $E_D$ and $F$ will be equal. If a chemical undergoes biotransformation in the liver and/or gastrointestinal mucosa, $E_D$ will be greater than $F$. Biotransformation in the gastrointestinal lumen would tend to reduce $E_D$ and $F$ in a similar manner.

Oral bioavailability ($F$) values determined using the AUC method have been reported for several chemicals in fish (Kleinow et al., 2008), but this limited dataset provides little information regarding the physicochemical and biological factors (e.g., hydrophobicity, metabolic stability) that control this process. In contrast, $E_D$ values have been estimated for a substantial number of chemicals (Gobas et al., 1988; Arnot and Quinn, 2015; Figure 2.3). Previously, it was suggested that $E_D$ values are relatively constant for chemicals with log $K_{OW}$ values between 4 and 6 (averaging about 50% across studies), then decrease progressively with log $K_{OW}$ at log $K_{OW}$ values greater than 6 (Gobas et al., 1988; Figure 2.3A). A more recent review of $E_D$ data casts some doubt on these generalizations (Arnot and Quinn 2015; Figure 2.3B). Complicating this assessment is that fact that there is considerable scatter in existing $E_D$ data, even for the same chemical. The cause of this scatter is not entirely known but likely reflects true biological variability among species, the variable and largely unknown impact of biotransformation (i.e., whether reported $E_D$ values are in fact estimates of $F$), and differences in experimental design (e.g., fish size, temperature, diet type, and number and size of individual feedings). It is possible that trends with respect to the log $K_{OW}$ dependence of $E_D$ are obscured by variability in existing datasets. Additional research involving a substantial number of chemicals and a highly standardized dietary dosing protocol would be required to address this question.

Empirically based algorithms for the prediction of $E_D$ have been developed using data for chemicals with log $K_{OW}$ values <9 (Gobas et al., 1988; Figure 2.3A; also see *Bioaccumulation Assessment and Modeling*). Presently, there are very few $E_D$ measurements for chemicals with log $K_{OW}$ values >9. When existing algorithms are applied to a chemical with a log $K_{OW}$ >9, dietary uptake is predicted to be negligible. Accumulating field data indicates, however,

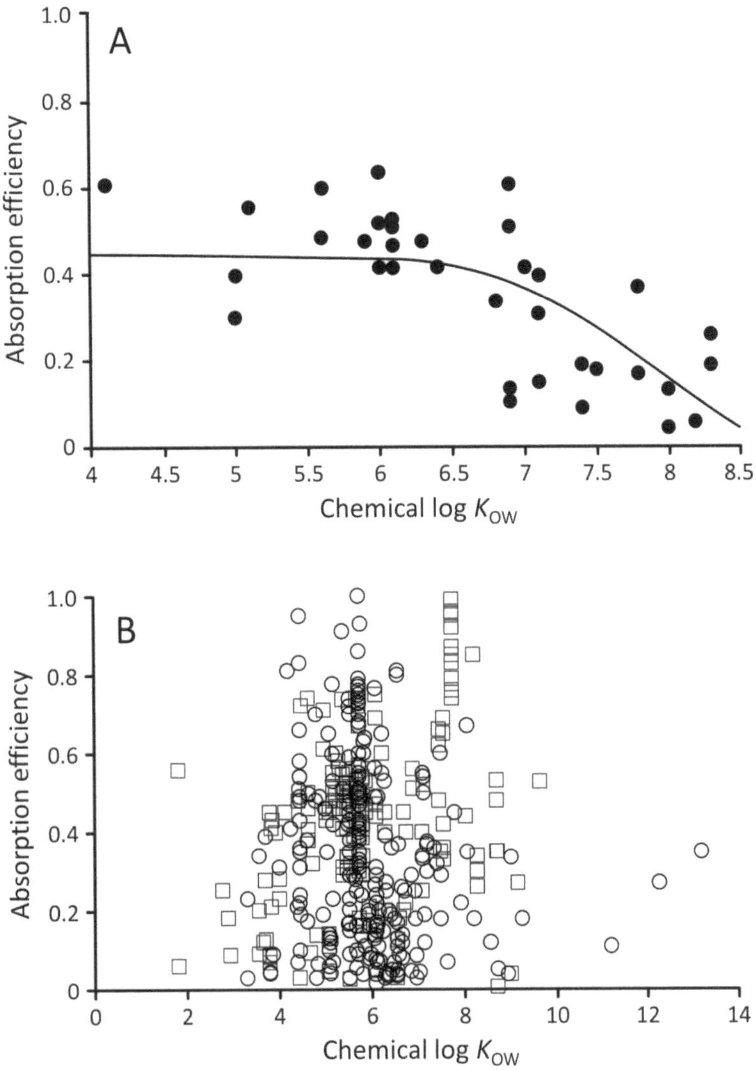


**FIGURE 2.3** Dependence of dietary absorption efficiency ($E_D$) on chemical log $K_{OW}$. Panel A: $E_D$ data presented by Gobas et al. (1988). The solid line was generated using the fugacity-based model given by the authors. Panel B: $E_D$ data summarized by Arnot and Quinn (2015). The data shown are "high" (open circles) and "medium" (open squares) confidence values from longer term (≥7 day) exposure tests.

that dietary uptake of such chemicals may contribute to substantial bioaccumulation (He et al., 2012). It is possible that diffusive uptake of extremely hydrophobic chemicals, while limited, still occurs. Alternatively, chemical uptake via pinocytosis in the GIT may provide a "baseline" level of $E_D$ regardless of log $K_{OW}$.

*Physiological Impacts on Gastrointestinal Absorption*

Little is known regarding the direct effects of GIT structure and function on dietary uptake of xenobiotics by fish. Numerous studies indicate, however, that pH gradients, gastrointestinal passage times, blood flow patterns, and other aspects of digestive physiology vary with species, feeding frequency, meal composition and size, and environmental influences such as temperature. For fish that possess a stomach, gastric evacuation generally proceeds according to an exponential decay model after an initial lag phase (Persson, 1986). The rate constant for gastric emptying has been shown to increase with acclimation temperature (Windell et al., 1976) and decrease with food particle size (Jobling, 1987). Gut transit times tend to decrease with increased acclimation temperature (Hofer et al., 1982; Shrable et al., 1969).

Changes in the character of the ingesta can facilitate or hinder the absorption of xenobiotics; for example, moderate levels of dietary lipid appear to facilitate the absorption of hydrophobic contaminants, but low or very high lipid diets are associated with reduced absorption efficiency (Van Veld, 1990). High levels of dietary fat may overwhelm the ability of the system to digest and remove this material. Unprocessed lipid would then provide a partitioning phase for hydrophobic chemicals within the gut lumen. Qualitative features of dietary lipid such as fatty acid chain length and composition may also determine xenobiotic solubility in micelles and subsequent systemic availability (Doi et al., 2000).

### The Skin

The skin of a fish functions as a physical barrier to separate the internal and external environments. As such, it maintains the ionic and osmotic integrity of the internal environment and provides protection from abrasion and disease organisms. Physiological functions of fish skin include respiration, ion exchange, and acid–base regulation. Based on morphometric and anatomical information, researchers have long suggested that cutaneous $O_2$ flux contributes significantly to total respiration in larval fishes (McDonald and McMahon, 1977; McElman and Balon, 1980; Oikawa and Itazawa, 1985; Rombough and Moroz, 1990). Because the gill epithelium of both small and large fish consists of one or a few cell layers, its thickness does not change much with fish size. In contrast, skin thickness tends to decrease with decreasing fish size and in small fish may approach the thickness of the gill epithelium. In larval Chinook salmon (*Oncorhynchus kisutch*), as much as 80% of $O_2$ uptake takes place across the skin (Rombough and Moroz, 1990).

Direct measurements of dermal uptake in rainbow trout and channel catfish (*Ictalurus punctatus*) were obtained by exposing fish confined to respirometer-metabolism chambers to a mixture of three chloroethanes (McKim et al., 1996). Although the rates at which both species approached steady state were close to those observed in inhalation exposures to the same chemicals, steady-state chemical concentrations in blood were much lower. An analysis of the trout dataset indicated that dermal absorption would contribute 2%–4% of initial uptake (dermal plus branchial) in a hypothetical waterborne exposure. A similar analysis for channel catfish suggested that dermal absorption would contribute 7%–9% of initial uptake. Based on these findings, it was concluded that dermal uptake is a minor route of exposure in large fish, except perhaps for species that live in intimate contact with contaminated sediments.

In contrast, several studies have suggested that dermal uptake contributes substantially to total uptake of waterborne chemicals by small fish and juveniles of larger species. Tovell et al. (1975) measured anionic detergent uptake across the skin of small goldfish (*Carassius auratus*) and found that 20% of total uptake occurred by this route. Saarikoski et al. (1986) obtained dermal uptake values for guppies exposed to a series of phenols, anisoles, and carboxylic acids. To distinguish between gill and skin absorption, fish were positioned into a hole cut in a rubber membrane separating two exposure chambers. Estimates of dermal absorption ranged from 25% to 40% of total absorption. Japanese medaka (*Oryzias latipes*) exposed to 2,2′,5,5′-tetrachlorobiphenyl in water accumulated considerably more chemical than could be explained by inhalation uptake (Lien and McKim, 1993). Similar results were reported for fathead minnows (*Pimephales promelas*) exposed to three chloroethanes (Lien et al., 1994). Lien and McKim (1993) suggested that dermal uptake should increase in relative importance in small fish because of the way that gill and skin surface areas scale to fish body weight. In small-mouth bass, gill surface area scales to body weight by a fractional exponent of about 0.78, consistent with the role of the gill in supporting metabolic activity (Price, 1931). Skin surface area scales instead to a body weight exponent of about 0.67 (Schmidt-Nielsen, 1984). Using these values, Lien and McKim (1993) predicted that skin surface area approaches and may even exceed that of the gills in fish weighing less than 5 g.

## Xenobiotic Distribution

### Primary Determinants

A variety of processes act to distribute absorbed xenobiotic chemicals within fish. A portion of the absorbed chemical may distribute to a site of action where toxic effects are expressed. Other sites serve as repositories for the chemical over short or long periods of time, while others are mobile and provide a means of transport within and out of the animal. During its residence in the body, a xenobiotic will redistribute continuously due to changes in concentration gradients that drive chemical movement.

Five primary factors control the distribution of xenobiotics from blood to peripheral tissues: (1) the physicochemical characteristics of the chemical (e.g., $pK_a$, lipid solubility, molecular volume); (2) the concentration gradient between blood and tissues; (3) the ratio of blood flow to tissue mass; (4) the chemical sorptive capacity of blood and tissue constituents; and (5) the activity of membrane transport proteins. For many xenobiotics, transport from blood to tissues occurs by simple diffusion following a concentration gradient. When a chemical diffuses rapidly across biological membranes, an equilibrium may become established between a tissue and venous blood exiting the tissue. In other cases, the rate of membrane diffusion limits chemical flux between blood and tissues with the result that these phases do not achieve equilibrium at early time points in an exposure. Chemical binding to blood and tissue constituents influences the distribution process by defining the concentration gradient that drives diffusion. Distribution processes controlled by blood flow or simple diffusion generally exhibit first-order kinetics. In contrast, distribution controlled by the activity of a membrane transport protein could exhibit nonlinear (saturable) kinetics.

At early time points in an exposure, chemical distribution to tissues may be highly influenced by the route of administration. High concentrations in the skin following exposure to contaminated sediments, the gastrointestinal mucosa following dietary exposure, or muscle following an intramuscular injection are obvious examples. Anatomical and physiological peculiarities of fish may also result in characteristic distribution patterns; for example, xenobiotics administered by intramuscular injection in the trunk muscle are initially transported to the kidney because venous blood from the trunk muscle collects in the caudal vein, which is part of the renal portal circulation in fish (Figure 2.1). Similarly, chemicals taken up from the GIT or intraperitoneal cavity are transported first to the liver by the hepatic portal vein and then to the gills via the ventral aorta before distributing into the general circulation. The systemic availability of a chemicals taken up by this route may be influenced, therefore, by elimination pathways operating in the gut, liver, and gills. For example, the fish anesthetic tricaine methane sulfonate (MS 222) will not produce anesthesia when given by intraperitoneal injection anesthesia because of its rapid metabolism in the liver and the ease with which both metabolites and parent chemical diffuse out across the gills (Hunn and Allen, 1974). Anesthetic concentrations of MS 222 in arterial blood can only be achieved by maintaining high MS 222 concentrations in water.

### Xenobiotic Transport and Binding in Plasma

Tissues with the highest blood perfusion rates in fish include the kidney, red muscle, pyloric caeca, intestine, spleen, and liver (Barron et al., 1987a). Tissues receiving an intermediate level of blood perfusion include the gonads, skin, and white muscle, while adipose tissue and bone are poorly perfused. The influence of blood flow on chemical distribution was demonstrated in rainbow trout exposed to linear alkylbenzene sulfonate (LAS) in water (Tolls et al., 2000a). Concentrations of LAS in the internal organs (primarily kidney and GIT) and liver exceeded 80% of their steady-state values after 8 hours of exposure, while concentrations in the muscle and skin continued to increase until 78 hours.

Blood flow to some tissues may change substantially in response to physiological stimuli. An example of this phenomena is provided by the pre- and post-prandial intestine (Axelsson and Fritsche, 1991; Axelsson et al., 1989, 2000). Exercise (Neumann et al., 1983) and hypoxia (Cameron, 1975) were shown to alter blood flow patterns in rainbow trout and arctic grayling (*Thymallus arcticus*), respectively. In rainbow trout, an increase in cardiac output ($Q_C$) associated with increasing ambient temperature was accompanied by a decrease in relative blood perfusion to most tissues (expressed as a percent of $Q_C$), resulting in maintenance of tissue blood flow rates (Barron et al., 1987a). These changes were offset, however, by a redistribution of $Q_C$ to white muscle, substantially increasing the perfusion rate of this tissue.

Another circulatory feature of fish with unknown but potential significance to xenobiotic distribution is the secondary circulation. This network of anastomosing vessels arises from the walls of the primary arteries and parallels the primary circulation of arteries, capillaries, and veins (Olson, 1996). The secondary circulation shares some characteristics with the mammalian lymphatic system, such as structural attributes of the vessels and restricted access for formed elements but is more limited in its distribution. Tissues perfused by the secondary circulation include the gill filaments, skin, peritoneal lining, and oral mucosa. Limited studies suggest that the secondary circulation has a volume greater than that of the primary circulation and a turnover time of several hours (Steffensen and Lomholt, 1992). When the primary and secondary circulations are combined and expressed as a percentage of body weight, the total circulatory volume places teleost fishes well into the upper range for vertebrates.

Although many chemicals are transported in blood as unbound forms, others are transported primarily in association with proteins, lipoproteins, lipids, or cellular components. Xenobiotics interact with these components by several mechanisms. Covalent binding usually restricts further distribution of the chemical. In contrast, specific, noncovalent ligand–protein interactions with binding proteins such as albumin may result in reversible binding that follows the law of mass action. When more than one specific binding interaction is possible, the distribution of a chemical among binding proteins depends on both the binding affinity of each protein and their relative concentrations. These binding affinities may change with changes in ionic strength, pH, temperature, and protein conformation. If this binding is reversible, however, an equilibrium will tend to be reestablished, providing an efficient means by which xenobiotics are transported and redistributed. Hydrophobic organic chemicals will partition to both lipids and proteins in blood. This type of binding is reversible, nonspecific and nonsaturable.

Total plasma protein concentrations in fish are generally lower than those in mammals, possibly resulting in a reduced number of binding sites. In mammals, plasma albumin plays an important role in binding weak organic acids, while weak organic bases bind to plasma glycoproteins. The structural diversity of plasma albumins in fish exceeds that of albumins from higher vertebrates (Andreeva, 2010). Some species do not possess plasma albumin at all, including some sharks and rays (Metcalf et al., 1999; Weisiger et al., 1984), or exhibit only very low concentrations (De Smet et al., 1998). Others possess relatively high concentrations, including some phylogenetically ancient species (Gray and Doolittle, 1992; Metcalf et al., 2007). The plasma albumin of rainbow trout has been described as *para-albumin* because of significant functional differences from the albumin of mammals (Perrier et al., 1977). Thus, the characteristic binding of xenobiotics to mammalian albumin and other plasma proteins may not directly extrapolate to fish. For example, low concentrations of the antibiotic sulfadimethoxine become highly (>90%) bound when added to rat plasma. In rainbow trout, sulfadimethoxine binding in plasma ranges from 13% to 17% over the same concentration range, suggesting a nonsaturable, nonspecific interaction. This low degree of protein binding in trout facilitates the elimination of sulfadimethoxine and may result in a relatively larger apparent volume of distribution when compared with mammals (Kleinow and Lech, 1988).

The binding of several neutral and basic drugs was shown to be similar in rainbow trout and human plasma (Henneberger et al., 2022). In contrast, acidic drugs exhibited up to 70-fold higher binding in human plasma than in trout plasma. Similar binding was observed for 1-butanol, phenol, nitrobenzene, and pentachlorophenol in plasma from rainbow trout and rats (Schmieder and Henry, 1988). This binding was shown to correlate positively with chemical log $K_{OW}$. An empirically based algorithm (Fitzsimmons et al., 2001) that predicts binding in trout plasma from chemical log $K_{OW}$ provided a good fit to measured values for 20 organic chemicals, most of which are neutral at physiological pH (Nichols et al., 2013).

### Chemical Partitioning to Neutral Lipid

Although chemical distribution at the beginning of an exposure is often determined by relative tissue blood flows, many chemicals redistribute over time in accordance with their relative affinity for various tissue macromolecules. Hydrophobic xenobiotics, such as polychlorinated biphenyls (PCBs), tend to partition into tissues that have a high neutral ("storage") lipid content. Distributional differences among species may therefore occur due to different patterns of lipid deposition in organs and tissues. For example, Guiney and Peterson (1980) found that 60% of a dose of 2,2′,5,5′-tetrachlorobiphenyl in rainbow trout was contained in skeletal muscle and carcass, but in yellow perch (*Perca flavescens*) 70% was contained in viscera and carcass (excluding skeletal muscle). These patterns were correlated with differences in the lipid content of each tissue, relative to that of other tissues.

Tissues that contain a large amount neutral lipid may act as storage depots for hydrophobic xenobiotics. This storage may protect against adverse effects by isolating a chemical away from its sites of toxic action. Conversely, storage in these sites may prolong the overall residence time of a chemical in the body and promote accumulation during chronic exposures. By reducing adipose lipid stores, starvation may result in a relatively rapid mobilization of accumulated chemical. When the rate of elimination from the whole animal is low, this starvation-induced decrease in whole-body lipid content may cause chemical concentrations in relatively lean tissues to increase substantially, even as whole-body chemical concentrations remain relatively constant (Gruger et al., 1975; Lieb et al., 1974).

### Lipid Mobilization and Xenobiotic Redistribution in Reproducing Fish

Female fish support egg development by mobilizing lipids from body stores. This lipid may be transferred to the growing egg mass or incorporated into larger energy storage molecules such as vitellogenin. Additional lipid may be mobilized in males and females to provide energy for spawning behaviors such as migration and nest defense. In either case, xenobiotics may redistribute from fat storage depots to the developing gonads. To investigate this phenomenon, Vodicnik and Peterson (1985) exposed female yellow perch to [$^{14}$C] 2,2′,5,5′-tetrachlorobiphenyl in water for 24 hours and then monitored tissue distribution and elimination for 5 months. Two weeks after exposure, 30% of chemical retained by fish was present in the developing ovaries. This value increased to 50% just prior to spawning, which occurred 4 months into the study. Similar studies with rainbow trout demonstrated the redistribution of [$^{14}$C] 2,2′,5,5′-tetrachlorobiphenyl to eggs and sperm (Guiney et al., 1979).

The percentage of an accumulated hydrophobic contaminant that is redistributed to the developing gonads depends on the lipid content of the fish prior to gonad development, the fraction of whole-body lipid content that is mobilized and incorporated into the gonads, and the size of the gonads relative to total body weight. Niimi (1983) examined these features in five fish species: rainbow trout, yellow perch, smallmouth bass (*Micropterus dolomieu*), white bass (*Morone chrysops*), and white sucker (*Catostomus commersonii*). Yellow perch were the leanest of these five species (5.1%) and transferred the greatest percentage of accumulated contaminants (25.5%) to their eggs. In contrast, rainbow trout had the highest starting lipid content (11.4%) and transferred the lowest percentage of contaminants to their eggs (5.5%). The high percentage of contaminant transfer in spawning yellow perch was due to the large size of its egg mass (22.3% of body weight) and the fact that it transfers a high percentage of its limited whole-body lipid stores to the developing ovaries (27.1%).

Reproductive life history may also influence the redistribution of hydrophobic contaminants in fish. The Chinook salmon, a semelparous (once-bearing) species, transfers most of its stored lipid into a single spawn of eggs. By comparison, iteroparous (multiple bearing) lake trout (*Salvelinus namaycush*) invest a lower proportion of stored lipid into each reproductive effort. In a study of these two species in Lake Michigan, Miller (1993) found that female salmon transferred 28%–39% of their accumulated PCB body burden to the eggs during a single spawn. In contrast, female lake trout eliminated 3%–5% of whole-body PCBs during each of what may be several spawns.

## Xenobiotic Excretion

### Branchial Excretion

In teleosts, the most important route of excretion for neutral, water-soluble, low-molecular-weight chemicals is across the gills. Working with the Dolly Varden char (*Salvelinus malma*) in a split chamber system, Thomas and Rice (1981) showed that aromatic hydrocarbons which exhibit low to moderate hydrophobicity (log $K_{OW}$ 1–4) are excreted across the gills at a greater rate than those which are very hydrophobic (log $K_{OW}$ 4–7). A similar finding was reported for goldfish exposed to a series of substituted phenols with log $K_{OW}$ values ranging from 1 to 5 (Nagel and Urich, 1980). Erickson and McKim (1990b) compiled data from several studies involving guppies and found that excretion rates declined linearly with chemical log $K_{OW}$ across a wide range (3–8) of values. This dependence of branchial excretion on chemical hydrophobicity is due largely to nonspecific binding of chemicals in blood. This binding lowers the diffusion gradient across the gill epithelium by reducing the concentration of chemical in blood that is unbound.

### Biliary Excretion

The liver plays a central role in the elimination of xenobiotic chemicals, both as the primary site of enzymatic biotransformation and via biliary excretion. Here, we consider biliary excretion as a process distinct from biotransformation. It should be noted, however, that biotransformation may play a pivotal role in biliary excretion of xenobiotics by creating metabolites (e.g., products of phase II conjugation pathways) that are substrates for active membrane transporters within the canalicular membrane.

Biliary excretion of xenobiotics and their metabolites depends on the formation and secretion of bile, chemical uptake by hepatocytes, and the transport of these chemicals into bile. A variety of structural arrangements have been described for the liver and biliary tree of fish. In general, however, fish do not possess the lobular structure and zonally oriented portal tracts observed in mammalian livers (Eurell and Haensly, 1982; Hampton et al., 1989; Hinton and Couch, 1998; Robertson and Bradley, 1992; Schar et al., 1985). Instead, the hepatocytes are commonly arranged into tubular elements. In rainbow trout, hepatocyte apices are directed toward the center of the tubule. At the center of the tubule, plasma membranes of adjacent cells form bile canaliculi. Where a biliary space exists, the biliary epithelial cells form junctional complexes with each other, and the hepatocytes on either side form a wall of the biliary passageway. These biliary passageways lead in turn to the cholangioles, small bile ducts, and large bile ducts. The contralateral or basal side of each hepatocyte faces a blood-filled sinusoid surrounding the tubule. Individual tubules curve, branch, and anastomose to form a complex network of parenchyma.

In salmon, multiple hepatic bile ducts empty into an elongated major duct that runs along the posterior ventral margin of the liver. The cystic duct branches from the major bile duct and serves to divert bile to the gall bladder. Bile originating from the major bile duct and gall bladder is collected in the common bile duct and empties into the proximal intestine just posterior to the pyloric curvature of the stomach. Some fish species, including members of the cod family, do not possess a gall bladder. For these species, it is thought that bile is secreted continuously into the intestine.

Biliary excretion may be facilitated by the activity of ABC transporters and transporters of the SLC family. ABC transporters move substrates out of the hepatocyte and into bile or blood. Substrates for these transporters include ionized xenobiotics and xenobiotics conjugated with sulfate, glutathione (GSH) and glucuronic acid. Additional substrates include a wide range of endogenous molecules including steroids, lipids, and peptides. SLC transporters residing within the sinusoidal (basolateral) membrane transport organic anions from blood to the interior of the hepatocyte.

Fish excrete a wide variety of xenobiotics in the bile. A partial listing of these chemicals includes antibiotics (Samuelsen et al., 1995), insecticides (Lech et al., 1973), dyes (Plakas et al., 1992a), herbicides (Schlenk and Moore, 1993), polycyclic aromatic hydrocarbons (Collier and Varanasi, 1991), metals (Grosell et al., 2001), and natural products (Sahin et al., 1996). Early studies with rainbow trout showed that chemical concentrations in bile may be much higher than those in surrounding water (Lech et al., 1973; Statham et al., 1976). Guarino and Lech (1986) summarized the results of studies with fish exposed to more than 40 drugs and other xenobiotics of varying molecular weight. The data were examined to determine bile/plasma concentration ratios and the percentage of chemical eliminated in bile or urine. Bile/plasma concentration ratios were greater than 1 for over 85% of chemicals examined, suggesting that fish were capable of concentrating xenobiotics in bile relative to plasma. The relative importance of biliary secretion as a route of excretion increased with molecular weight, although other factors such as the charge on the molecule were important. Generally, chemicals that are highly polar and have molecular weights greater than 600 are excreted largely in bile (Gingerich et al., 1977; Schmidt and Weber, 1973). Xenobiotics of intermediate molecular weight (300–600) and polarity may be excreted in bile as well as by other routes (Allen et al., 1979; Hunn and Allen, 1975).

Secretion of bile into the intestinal tract is stimulated by nutritional and digestive signals. Due to the episodic nature of this process and potential for xenobiotic reabsorption within the intestine, it is difficult to assess the contribution of biliary secretion to the overall process of chemical elimination. The gall bladder of a recently fed fish is generally void of bile. By necessity, therefore, chemical concentrations in bile are generally determined using fasted animals.

Techniques for continuous collection of bile in unanesthetized rainbow trout have been described by several authors (Gingerich et al., 1977; Sanz et al., 1993; Schmidt and Weber, 1973). Similar methods have been developed for the skate (*Raja erinacea*) and dogfish shark (*Squalus acanthias*) (Boyer et al., 1976a, 1976b, 1976c). Basal bile flows measured using these techniques range from 50 to 200 μL/h/kg in rainbow trout, 30–70 μL/h/kg in the skate, and 80–110 μL/h/kg in the dogfish. Bile flow rates in mammals may be 50 times greater than these values (Klaassen and Plaa, 1967). Differences in the rates of biliary excretion of bromosulfophthalein between rainbow trout and rats were found to be well correlated with differences in bile flow rate (Gingerich et al., 1977), suggesting that bile formation rate limits the rate of bromosulfophthalein excretion in bile.

### Renal Excretion

Fish kidneys perform two major functions: (1) ion and water balance; and, (2) excretion of endogenous and exogenous solutes. The diversity of kidney structure–function across species is enormous, but basic features are shared by both freshwater and saltwater teleosts (Hickman and Trump, 1969; Nishimura and Imai, 1982; Pritchard, 1981; Pritchard and Miller, 1993; Pritchard and Renfro, 1984). The functional unit of the freshwater fish kidney consists of an encapsulated capillary network called the *glomerulus*, a tubule lined with epithelial cells, and bladder. The kidney tubule can be subdivided further into two proximal segments (I and II), a distal tubule and a collecting tubule. The functional kidney unit of saltwater fish is like that of freshwater fish but lacks the distal tubule. In some saltwater species, the glomerulus may also be absent.

Because they are hyperosmotic with respect to their environment, freshwater fish must eliminate large volumes of dilute urine and retain salt (principally NaCl). High urine flow rates are supported by a high glomerular filtration rate (GFR; L/h), where the GFR defined as the volume of fluid filtered from the glomerular capillaries per unit of time. In contrast, saltwater teleosts are hypoosmotic with respect to their environment and must overcome dehydration by drinking seawater and absorbing water and minerals (primarily monovalent ions) through the gut. Excess salts are eliminated across the gills (both mono- and divalent ions) and kidney (primarily divalent ions), and water loss is minimized by producing small quantities of relatively concentrated urine. The GFR in saltwater fish is much lower than that of freshwater fish. Urine osmolarity is maintained by an exchange of divalent and monovalent ions within the proximal tubule. Water is then reabsorbed with NaCl in the bladder. Because it comprises up to 40% of total tubular surface, the proximal tubule of saltwater fish has been used extensively for both *in vitro* and *in vivo* studies of kidney function (Kinter, 1966, 1975; Miller and Pritchard, 1997; Pritchard and Miller, 1993; Pritchard and Renfro, 1984).

The mechanisms that have evolved for handling ions and water in the fish kidney also contribute to the excretion of endogenous and exogenous organic solutes. The first step in urine formation is ultrafiltration of plasma through the capillary network of the glomerulus. Solutes within this glomerular filtrate may be reabsorbed back into plasma. Alternatively, solutes may be added to the filtrate by transport of chemical from the plasma across the renal tubular epithelium. The final composition of urine is determined by characteristics of the epithelium in both the tubule and bladder (Miller, 1987).

Molecules with a radius of less than about 20 Å are completely filtered at the glomerulus, those with a radius between 20 and 40 Å are partially filtered, and those with a radius greater than 40 Å tend to be retained in plasma (Miller, 1987). In fish as in mammals, chemicals with molecular weights greater than about 500 tend to be excreted in the liver rather than the kidney (Pritchard and Renfro, 1984). Plasma binding is an important determinant of urinary excretion because glomerular filtration and tubular secretion act primarily on unbound chemicals in plasma. Because they bind to plasma proteins, hydrophobic chemicals are generally retained by plasma as it flows though the glomerulus. Hydrophobic chemicals contained within the glomerular filtrate may be reabsorbed from the tubule. This is particularly true of saltwater fish due to the concentrating effect of water reabsorption. The caudal vein in fish drains into the kidney, providing a substantial supply of portal blood. Most of this portal blood bathes the kidney tubules and contributes to the renal secretory system (Pritchard and Renfro, 1984). Overall, up to 80% of $Q_C$ flows through the fish kidney, as compared with about 20% in mammals.

Some of the xenobiotics that fish are exposed to exist as organic anions or cations, while others are taken up as neutral species and converted to anions or cations by various biotransformation pathways. In all vertebrates studied to date, anionic and cationic secretory systems located in the proximal tubule transport charged chemicals from plasma into the tubule urine (Pritchard and Miller, 1993). The organic anion secretory system uses metabolic energy to move substrate molecules into tubule epithelial cells against their electrochemical gradient. This mechanism is carrier-mediated, saturable, and inhibited competitively by other substrates (Pritchard, 2001). Organic anion secretion is a three-step process: (1) ATP is hydrolyzed to drive the $Na^+$ pump and create an inward $Na^+$ gradient; (2) the $Na^+$ gradient into the cell drives the uptake of α-ketoglutarate ($\alpha KG^{-2}$), thereby creating an outward $\alpha KG^{-2}$ gradient; and (3) the organic anion moves into the cell in exchange for the outward movement of $\alpha KG^{-2}$. Efflux of the organic anion into the tubule may occur by facilitated diffusion or through the action of an ABC transporter (Miller and Pritchard, 1997).

Organic cation secretion is also a multistep process. In contrast to anion secretion, however, chemical movement into the epithelial cell (by facilitated diffusion) is energetically favored because of the strong negative charge on the cell interior. Cation transport into the lumen of the tubule is opposed by the membrane potential and must be tied to the hydrolysis of ATP (Pritchard, 2001). As with anion secretion, the cationic secretory system is susceptible to substrate competition.

The mechanisms responsible for renal excretion of chemicals can be distinguished by dividing measured renal clearance rates ($CL_R$) by the GFR (McKim et al., 1999; Pritchard and Renfro, 1984). The $CL_R$ is the volume of plasma completely cleared of chemical by the kidney in a fixed period and may be calculated by dividing the excretion rate of chemical in urine by the chemical concentration in plasma at the midpoint of a sampling period. The GFR is typically estimated using an *in vivo* [$^{14}C$]-polyethylene glycol method (Beyenbach and Kirschner, 1976). Figure 2.4 shows $CL_R$/GFR ratios for several chemicals in both freshwater and saltwater fish. In the absence of substantial plasma binding, a $CL_R$/GFR ratio <1 suggests that a chemical is reabsorbed within the kidney tubule, a ratio of 1 indicates no net tubular transport, and a ratio >1 signifies active secretion into the tubular fluid. Chemical binding in plasma can reduce renal tubular secretion as well as chemical filtration at the glomerulus. In the absence of active section to urine, a high degree of plasma binding may result in a $CL_R$/GFR ratio close to 0. For this type of chemical, a $CL_R$/GFR ratio between 0 and 1 can provide evidence for active secretion (e.g., PFOA; no. 15 in Figure 2.4). For a given mechanism of renal excretion, $CL_R$/GFR ratios in saltwater fish tend to

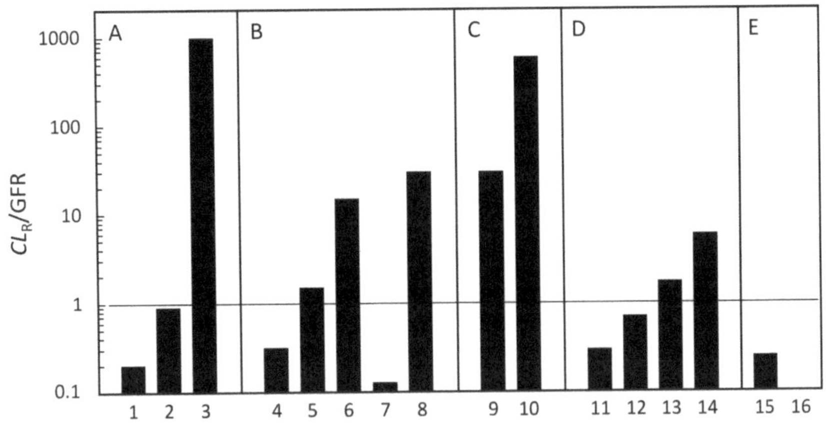


**FIGURE 2.4** Influence of tubular transport, biotransformation, and plasma binding on renal excretion in marine teleosts (A–C; Pritchard and Renfro, 1984) and freshwater teleosts (D, E; McKim et al., 1999; Consoer et al., 2014, 2016). Data for each chemical are expressed as the ratio of renal clearance ($CL_R$) to glomerular filtration rate (GFR). (A) *Reference chemicals*: (1) glucose, reabsorbed by active transport; (2) 2-deoxyglucose, with no net tubular transport; (3) *p*-aminohippuric acid, a prototypical substrate for active anion secretion pathway. (B) *Effect of biotransformation*: (4) benzo[a]pyrene (BaP), reabsorbed passively; (5) BaP-7-phenol conjugates (largely glucuronide), limited anion secretion; (6) BaP-7,8-dihydroxydiol conjugates (largely sulfate), a better substrate for anion secretion; (7) DDT, passive reabsorption; (8) DDA, effective tubular secretion. (C) *Effect of plasma binding*: (9) DDA, plasma binding of 97%; (10) 2,4-D, plasma binding of 70%. DDA and 2,4-D are transported similarly *in vitro* (without binding). (D) *Effects of reabsorption and biotransformation*: (11) hydroxyquinone, strong passive reabsorption; (12) phenol, slight passive reabsorption; (13) phenyl glucuronide, limited anion secretion; (14) phenyl sulfate, better substrate of anion secretion. (E) *Renal handling of PFAS*: (15) perfluorooctanoate (PFOA), substantial clearance despite ~100% plasma binding suggests active anion secretion; (16) perfluorooctane sulfonate (PFOS), binding also close to 100% but no evidence for active anion secretion.

exceed those of freshwater fish. This pattern appears to be due primarily to higher GFR values in freshwater fish.

## Fecal Excretion

Xenobiotics incompletely absorbed from the diet will be excreted in feces while the absorbed fraction will be transferred to the liver and subjected to biotransformation and biliary excretion. Parent chemicals and their metabolites excreted in bile may be expelled with the feces (Statham et al., 1976); however, hydrolysis of a biliary conjugate by enzymes present in gut microflora may liberate the parent chemical for reabsorption and enterohepatic recirculation (Collicutt and Eales, 1974; Fricker et al., 1997; James, 1987; Layiwola et al., 1983; Schultz et al., 2001). Hydrophobic xenobiotics that are poorly cleared by hepatic, renal, and branchial routes may be excreted by passive diffusion from blood to contents of the intestinal lumen (Ingebrigtsen and Solbakken, 1985; Ingebrigtsen et al., 1988; Kleinow et al., 1996).

## Dermal Excretion

The extent to which chemicals and their metabolites are excreted across the skin is poorly known; however, anatomical and physiological factors that promote chemical uptake across the skin of some fish species (e.g., high degree of vascularization and large ratio of skin surface area to gill surface area) also might be expected to facilitate chemical excretion. Naphthalene was recovered from the mucus of rainbow trout soon after it was administered by intraperitoneal injection or dietary exposure (Varanasi et al., 1978). At later time points, only naphthalene metabolites were found. Whether or not these metabolites were formed in skin prior to their diffusion into mucus was not investigated.

## Excretion via the Gametes

The redistribution of hydrophobic chemicals from somatic lipid stores to developing gametes was described earlier in this chapter. Although spawning eliminates xenobiotics from the animal, reducing the whole-animal contaminant load, whole-body chemical concentrations may go up (Tietge et al., 1998), down (Guiney et al., 1979; Niimi, 1983; Vodicnik and Peterson, 1985), or stay the same (Niimi, 1983), depending on the chemical concentration in gametes relative to that in the rest of the body. If the contaminant concentration in the gametes is lower than that of the rest of the fish, gamete release increases the chemical concentration expressed on a whole-body basis. The reverse is true if the contaminant becomes highly concentrated in the gametes, relative to the rest of the animal.

The importance of maternal transfer in viviparous species is unknown. Wourms et al. (1988) estimated that, among teleosts, live-bearing strategies have developed in only 2%–3% of species; however, two of these species, the mosquitofish (*Gambusia affinis*) and guppy, have been used extensively in aquatic toxicology research. To the extent that an exchange of blood occurs between female fish and the developing young (as in Poeciliidae), it can be speculated that maternal transfer of chemicals to offspring occurs in some live-bearing species. In species for which most early development takes place after the egg has left the follicle (Embiotocidae and live-bearing sharks), the opportunity for maternal transfer may be reduced.

## TOXICOKINETIC MODELING

The kinetic behavior of xenobiotic chemicals in fish has been described using a variety of modeling approaches. Previous reviews of this topic have distinguished between *empirical* (a.k.a. *data-based* or *compartmental*) and *physiologically based* models (Barron et al., 1990; Gretch et al., 2016; Kleinow et al., 2008; Landrum et al., 1992). Empirical models are generally developed by fitting modeled simulations to concentration time course data from controlled toxicokinetic studies. Common types include *rate constant-based* models that describe chemical movements within a system using first-order rate constants and *clearance-volume* models that describe these fluxes using first-order clearance constants. Physiologically based toxicokinetic (PBTK) models are based on the premise that chemical uptake and disposition can be described from anatomical, physiological, and biochemical attributes of the exposed organism and physicochemical characteristics of the chemical. The compartments in a PBTK model correspond to actual tissues and organs, and chemical transport within the animal is defined by blood flow relationships. In principle, it is possible to develop a PBTK model in the absence of measured toxicokinetic data.

Here, we retain these distinctions, primarily as an aid to describing modeling concepts and equations. In practice, however, these distinctions have become increasingly blurred. Thus, many recent PBTK models include parameters determined by fitting modeled simulations to measured kinetic data. Similarly, some older empirical models have been amended to include a substantial amount of physiological information, becoming in the process simple PBTK models (see *Bioaccumulation Assessment and Modeling*). Finally, there has been a proliferation of models that include compartments corresponding to specific tissues and organs; however, most or all intercompartmental rate constants are obtained by a process of data fitting. Such models share attributes of traditional clearance-volume models while possessing traits more commonly associated with PBTK models. Models of this type are described in a section on *Semi-physiological Modeling Approaches*.

Most toxicokinetic models consist of one or more compartments, each representing tissues and organs with similar kinetic properties. Differential equations are then developed to describe the rate of change of the mass of chemical in each compartment ($dX/dt$; mg/h) as the difference between competing rates of input and output. A listing of commercial software used to develop and run toxicokinetic models is provided on the website for Pharmacokinetics and Pharmacodynamic Resources (Bourne, 2021). A detailed discussion of software tools used to perform PBTK modeling was given by Hack et al. (2020). It may be noted, however, that while commercial modeling software has been provided by numerous vendors, there is a trend toward increased use of open source modeling tools. In comparison with proprietary software, open source tools provide greater transparency and may make it easier to replicate of other groups' modeled findings.

The following text provides an overview of clearance concepts common to both empirical and PBTK models. These two modeling approaches are then taken up individually. A brief section on noncompartmental modeling is also provided. Although seldom employed in aquatic toxicology, noncompartmental models offer important advantages when dealing with sparse datasets. Model inputs and calculated terms are accompanied upon first usage by typical units (e.g., L/h for $Q_C$). Unless otherwise noted, the listed units are for values that have not been normalized to fish body weight.

### Clearance Concepts

Xenobiotic chemicals taken up by fish may be eliminated by several pathways, including biotransformation and excretion in bile, urine, and respiratory water. In some cases, elimination rates are extremely slow, and chemicals are retained for days, weeks, or even months after fish are transferred to a clean environment (Niimi and Oliver, 1983). Alternatively, elimination pathways may clear >99% of an absorbed dose within a few hours. In either case, elimination may be characterized by an *elimination clearance* ($CL$; L/h). The whole-body clearance ($CL_{wb}$) is defined as the rate of chemical elimination ($dX/dt$) from the body divided by the chemical concentration in a reference region (here assumed to be blood plasma) that is taken to represent the body ($C_P$; mg/L):

$$CL_{WB} = (dX/dt)/C_P \tag{2.6}$$

Normalization to body size (weight or surface area) may also be appropriate, particularly when averaging values from fish of different sizes (Hendriks et al., 2001). The $CL_{WB}$ is the sum of the individual clearances for all relevant elimination pathways; for example, the chemical may be simultaneously metabolized in the liver and excreted by the kidney and at the gills:

$$CL_{WB} = CL_H + CL_R + CL_B \tag{2.7}$$

where the subscripts refer to hepatic (H), renal (R), and branchial (B) clearance.

#### Hepatic Clearance

The $CL_H$ is limited by the rate of blood flow to the liver ($Q_L$; L/h). If blood perfusing the liver is completely cleared of chemical, then $CL_H$ is equal to $Q_L$. However, $CL_H$ may be less than $Q_L$ if a chemical is poorly metabolized or if it is reversibly bound to plasma proteins, limiting the unbound fraction that is available to biotransformation enzymes. A physiologically based model of hepatic clearance may be developed by assuming that the liver is a well-mixed compartment (Rowland et al., 1973; Wilkinson and Shand, 1975). Chemical concentrations in afferent blood entering the liver and venous blood draining the liver are denoted as $C_{AFF}$ (mg/L) and $C_{VL}$ (mg/L), respectively. Here the term $-dX/dt$ is the rate of chemical elimination by the liver (the minus sign indicates that chemical is being lost from the system). For this well-mixed liver model, the rate of elimination is:

$$-dX/dt = Q_L(C_{AFF} - C_{VL}) \quad (2.8)$$

If the right-hand side of this equation is multiplied by $C_{AFF}/C_{AFF}$ one obtains:

$$-dX/dt = Q_L(C_{AFF} - C_{VL})C_{AFF}/C_{AFF} \quad (2.9)$$

The ratio $(C_{AFF} - C_{VL})/C_{AFF}$ is termed the *hepatic extraction ratio* ($E$; unitless). Substituting, the rate of elimination can be written as:

$$-dX/dt = Q_L E C_{AFF} \quad (2.10)$$

Because $(-dX/dt)/C_{AFF}$ is equal to $CL_H$, it is apparent that $CL_H = Q_L E$.

The $CL_H$ is controlled by both $Q_L$ and the capacity of hepatic biotransformation enzymes. It is useful, therefore, to separate these two influences. This can be accomplished by defining the *intrinsic clearance* ($CL_{INT}$; L/h). The $CL_{INT}$ is the rate of chemical elimination divided by the chemical concentration in liver tissue ($C_L$; mg/L); that is, the concentration in contact with biotransformation enzymes:

$$-dX/dt = CL_H C_{AFF} = CL_{INT} C_L \quad (2.11)$$

and

$$CL_{INT} = CL_H C_{AFF}/C_L = CL_H/(1-E) \quad (2.12)$$

The rate of chemical elimination by the liver is therefore proportional via the hepatic clearance to the chemical concentration in blood entering the liver or via the intrinsic hepatic clearance to the chemical concentration in liver tissue.

A more general model for hepatic clearance that accounts for the effect of chemical binding in blood may be written by defining a new reference concentration, the unbound chemical concentration in the liver ($C_{L,U}$; mg/L), and a new clearance parameter, the *free (unbound) intrinsic clearance* ($CL_{INT,U}$; L/h). From the well-mixed tissue assumption, $C_{L,U}$ equals the unbound concentration in venous blood draining the liver ($C_{VL,U}$), which is equal to the product term $f_{U,B}$ $C_{V,L}$, where $f_{U,B}$ (unitless) is the unbound fraction of chemical in blood. Accounting for effect of blood binding, the rate of elimination may therefore be written as:

$$-dX/dt = CL_{INT,U} C_{L,U} = CL_{INT,U} f_{U,B} C_{VL} \quad (2.13)$$

The $CL_{INT,U}$ represents the activity of biotransformation enzymes toward the chemical under first-order conditions and can be understood in terms of the Michaelis–Menten relationship: $V_{MAX}/(K_M + C_{L,U})$. When $C_{L,U}$ is low relative to $K_M$ (<10%), the relationship simplifies to $V_{MAX}/K_M$, and the rate of elimination becomes first order with respect to $C_{L,U}$ and by extension $f_{U,B}$ $C_{VL}$:

$$-dX/dt = (V_{MAX}/K_M)C_{L,U} = (V_{MAX}/K_M) f_{U,B} C_{VL} \quad (2.14)$$

Given values for $CL_{INT,U}$, $f_{U,B}$, and $Q_L$, questions can be answered about how $CL_H$, $E$, and $F$ (oral bioavailability) will change if there is a change in hepatic blood flow, plasma protein binding, or hepatic biotransformation. The useful working equations for this model are:

$$E = f_{U,B} CL_{INT,U} / (Q_L + f_{U,B} CL_{INT,U}) \quad (2.15)$$

$$CL_H = Q_L f_{U,B} CL_{INT,U} / (Q_L + f_{U,B} CL_{INT,U}) \quad (2.16)$$

The letter $F$ is commonly used to represent the fraction of an oral dose that enters the systemic circulation and has two contributing parts: $E_D$, which represents the fraction of the dose that is absorbed from the GIT into portal blood (abbreviated $F_{AB}$ by Kleinow et al., 2008), and $F_{FP}$, which represents the fraction of an absorbed dose that escapes elimination by first-pass clearance in the liver (see also *The Gastrointestinal Tract, Gastrointestinal Absorption of Xenobiotics*). The overall bioavailability of an orally administered chemical is the product of these two fractions:

$$F = E_D F_{FP} \quad (2.17)$$

$F_{FP}$ can also be calculated from $CL_{INT}$, $f_U$, and $Q_L$ using the following relationship:

$$F_{FP} = 1 - E = Q_L/(Q_L + f_{U,B} CL_{INT,U}) \quad (2.18)$$

The independent variables in Equations 2.15, 2.16, and 2.18 are $CL_{INT,U}$, $f_{U,B}$, and $Q_L$. The dependent variables are $E$, $CL_H$, and $F_{FP}$. The relationships among these independent and dependent variables are simplified when the value of $E$ is small (<0.25) or large (>0.75); for example, for a large $E$, $CL_{INT,U} >> Q_L$, and Equation 2.16 becomes $CL_H = Q_L$. For some toxicants the value of $E$ lies between 0.25 and 0.75, and the full equations for $E$ and $CL_H$ must be used. In such cases, there are no direct proportionalities: $CL_{INT,U}$, $f_U$, and $Q_L$ all determine $E$, $CL_H$, and $F_{FP}$, and a change in any one of the former variables produces a less than proportional change in the latter.

## Renal Clearance

The kidney clears blood of xenobiotic chemicals by filtration and extraction. Filtration involves the passage of plasma water across the glomerular capillary membrane. Chemicals dissolved in this water are transported across the membrane and end up in the pre-urine. When protein binding occurs, protein-bound chemicals remain with the protein in plasma. The plasma concentration of unbound chemical remains constant during filtration, as both water and unbound chemical are removed by the filtration process. The equilibrium between bound and unbound chemical is therefore not disturbed; consequently, the maximum possible clearance via filtration is the GFR multiplied by the fraction of chemical that is unbound in plasma.

The extraction mechanism only applies when a xenobiotic is a substrate for an active tubular secretory pathway. Unbound chemical contained in post-glomerular and renal portal blood is transported across the proximal tubule into pre-urine. As in the liver, this may be a high $E$ or low $E$ process. The same extraction model described for the liver can be used for active tubular secretion. In this case, $CL_{INT}$ is the activity of the secretory mechanism.

Total clearance by the kidney reflects the net result of filtration, active secretion, and passive reabsorption:

$$CL_R = \text{filtration clearance} + \text{secretion clearance} - \text{reabsorption} \quad (2.19)$$

Passive reabsorption is expected to be minimal in freshwater fish because the kidney does not reabsorb much water from urine and thereby concentrate chemicals contained therein. In marine fish, the kidney conserves water by reabsorption and a greater potential exists for chemical reabsorption from urine.

## Branchial Clearance

Fish gills are an important site for xenobiotic uptake and elimination, and both processes can be described using clearance concepts. Unlike hepatic and renal clearance, however, the mathematical description of branchial clearance must be bidirectional to reflect the fact that chemical flux occurs in both directions. Depending on the concentration gradient across the gills, fish may clear chemical from inspired water or from blood flowing through the gills. The principal limitations on chemical uptake at fish gills were identified by Hayton and Barron (1990). If the permeability of the gill epithelium to a chemical is low, diffusion can limit exchange, and branchial clearance ($CL_B$; L/h) is controlled by the product of a gill permeability coefficient ($k_D$; m/h) and the surface area for diffusion ($A$; m$^2$):

$$CL_B = k_D A \quad (2.20)$$

The permeability coefficient is proportional to chemical diffusivity in the gill epithelium ($D$; m$^2$/h) and inversely related to the effective thickness of the diffusion barrier ($h$; m):

$$k_D = D/h \quad (2.21)$$

To obtain units of L/h, the $CL_B$ calculated by Equation 2.20 must be multiplied by 1,000.

If diffusive flux is limited by chemical diffusion within nonaqueous portions of the gill epithelium, it is appropriate to incorporate a membrane–water partition coefficient ($P_{MW}$), which would be expected to correlate positively with chemical hydrophobicity:

$$k_D = DP_{MW}/h \quad (2.22)$$

The resulting permeability coefficient is identical to that given previously in Equation 2.1. Alternatively, if chemical diffusion in the aqueous phase of the membrane limits its flux, Equation 2.21 is appropriate (see *Physiologically based Toxicokinetic Models, Route of Exposure, Branchial Uptake*). In either case, the value of $D$ applies to the phase that constitutes the principal barrier to diffusive flux. Chemical diffusivity in aqueous solution may be calculated from equations given in standard physical chemistry texts and is expected to decline with increasing molecular volume.

If the permeability of the gill epithelium to a chemical is sufficiently high, branchial uptake may be limited by the rate of blood flow through the gill lamellae. Chemical binding to plasma proteins or to cells in the blood would increase the capacity of blood to carry chemical away from the gills. The uptake clearance when blood flow is the rate-limiting factor may be expressed as:

$$CL_B = Q_C P_{BW} \quad (2.23)$$

where $Q_C$ (L/h) is the total cardiac output, and $P_{BW}$ is an equilibrium blood–water partition coefficient.

Because the value of $P_{BW}$ can be very large, it is possible for water flow across the absorbing surface of the gills to control the rate of chemical uptake. In this case, $CL_B$ is equal to the effective respiratory volume ($Q_W$; L/h), which is the flow rate of inspired water that exchanges with blood:

$$CL_B = Q_W \quad (2.24)$$

To a first approximation, the potential rate limitations on branchial flux can be treated as if they were resistances in series in an electrical circuit (Hayton and Barron, 1990):

$$CL_B = 1/\left(h/DP_{MW}A + 1/Q_C P_{BW} + 1/Q_W\right) \quad (2.25)$$

A more complex model based on the counter-current structure of fish gills (Erickson and McKim, 1990a, 1990b) is described below in the section on physiologically based toxicokinetic models; however, in the case where one resistance is the principal determinant of chemical flux (by virtue of being much greater than the other two), the two models reduce to a common description.

### Compartmental Models for Fish

The use of compartmental models to characterize the kinetics of absorption, distribution, and elimination of exogenous and endogenous substances by a variety of species is well established (e.g., Derendorf and Schmidt, 2020; Gibaldi and Perrier, 1982; Wagner, 1993). Riggs (1963) defined the term *compartment* as follows:

> If a substance, S, is present in a biological system in several distinguishable forms or locations, and if S passes from one form or location to another form or location at a measurable rate, then each form or location constitutes a separate compartment for S.

In the context of toxicokinetics, a xenobiotic and its metabolites would be forms of a substance S, and a location would be one or more tissues in which the forms had similar kinetic behavior. As an example, a tank of exposure water might be a compartment from which chemical absorption occurs, the blood and highly perfused tissues might constitute a second compartment, the poorly perfused tissues would be a third compartment, and a metabolite might require specification of a fourth compartment (Figure 2.5A).

Compartments are assumed to behave like a tank that contains a well-mixed fluid; that is, chemical that enters a compartment is assumed to distribute instantaneously and linearly, such that the concentration is everywhere proportional to the amount of chemical in the compartment. Body compartments are usually arranged so chemical enters and exits from only one of the compartments, the so-called central compartment. With multiple-compartment models, the added peripheral compartments are usually attached to the central compartment so that chemical exchanges between the central and each peripheral compartment; this configuration is termed *mammillary*, as opposed to the *catenary* configuration, which refers to a chain-like series of compartments. Typical mammillary configurations are shown in Figure 2.5B. Some considerations in the selection of the number of compartments include the purpose of the model, the number of sites from which samples are taken for determination of parent chemical and metabolite concentrations, the number and spacing of individual sampling times, and attributes of the resulting dataset including apparent mono- or multi-exponential behavior.

## Volume of Distribution

The size of a compartment is characterized by an apparent volume of distribution ($V$) that has units of volume (e.g., L) or volume normalized to body size (weight or surface area; L/kg or L/m$^2$). The apparent volume of distribution of the central compartment (compartment 1 in Figure 2.5B) is the amount of chemical in the compartment divided by the chemical concentration in a reference region. The reference region is the fluid or tissue in which the chemical concentration is measured, commonly blood or plasma. In the case of plasma, for example, the value of $V_1$ would represent the volume of plasma that would be required to account for all chemical in the central compartment. In studies with small fish, the exposure water is commonly used as the reference region and the value of $V_1$ represents the volume of exposure water that would be required to account for all chemical in the central compartment under equilibrium conditions, absent clearance from the central compartment. Under these conditions, the concentration in the exposure water equals the concentration of unbound chemical in plasma water.

The apparent volumes of distribution of the peripheral compartments are based on the same reference region as that of the central compartment, and the summed volume of all compartments is the apparent steady-state volume of distribution ($V_{SS}$). The magnitude of the $V_{SS}$ is determined by the chemical sorptive capacity of the reference region relative to that of other tissues and fluids. The fractional water content of most fish tissues is about 65%–80% (Bertelsen et al., 1998). If plasma is used as the reference region, a $V_{SS}$ of about 0.65–0.8 L plasma/kg would be expected for a chemical that distributes only into body water and does not bind to plasma proteins. Alternatively, if a chemical is bound extensively to plasma proteins and has relatively little affinity for other tissues, the $V_{SS}$ would

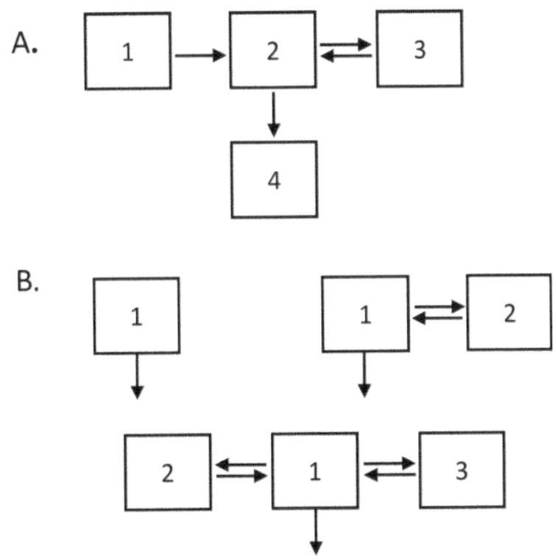


**FIGURE 2.5** Examples of compartmental models. (A) A model appropriate for xenobiotic absorbed from the gastrointestinal tract (GIT) and eliminated by formation of a metabolite. Compartment 1, GIT; compartments 2 and 3, the body lumped into rapidly and slowly equilibrating tissues; compartment 4, the metabolite. Arrows represent first-order kinetic processes. (B) Mammillary models that represent the body as one, two, or three well-mixed compartments. The depicted mammillary models indicate that a chemical taken up by the animal (e.g., by bolus injection) can be eliminated from the body. Such models are termed *open* models.

be less than 0.65 L plasma/kg. Conversely, if a chemical has high affinity for one or more tissues (e.g., high lipid solubility resulting in high concentrations in adipose fat), the $V_{SS}$ could be substantially greater than 1 L plasma/kg. Reported $V_{SS}$ values for fish determined using plasma as the reference region range from 0.08 L plasma/kg for the urinary tract antibiotic nitrofurantoin in channel catfish (Stehly and Plakas, 1993) to 6.1 L plasma/kg for the hydrophobic steroid methyltestosterone in rainbow trout (Vick and Hayton 2001). Reported $V_{SS}$ values determined using the exposure water as the reference region range from 0.55 L water/kg for aminoantipyrine in goldfish (Kaka and Hayton, 1978) to 19,000 L water/kg for di-2-ethylhexyphthalate in sheepshead minnows (Karara and Hayton, 1984).

## One-Compartment Empirical Model

### *Intravascular Administration*

The simplest empirical model assumes that the fish body behaves like a single, well-mixed compartment. In this section, the one-compartment model is developed for a xenobiotic that is administered by intravascular injection. Intravascular administration is usually accomplished by injecting a solution of the chemical via an indwelling catheter placed in the dorsal aorta. The dose may be administered as a rapid injection (bolus) or as a constant rate infusion using an infusion pump. After the dose is administered, samples of plasma are removed at various times to measure the chemical concentration ($C_P$; mg/L).

#### *Bolus Dose*

For a chemical administered as a bolus dose, its rate of elimination ($dX/dt$) is:

$$-dX/dt = CL_{WB}C_P \tag{2.26}$$

Division of both sides by $V$, followed by integration, yields an equation that predicts the time course of $C_P$:

$$C_P = C_0 e^{-(CL_{WB}/V)t} \tag{2.27}$$

where $t$ (h) is time and $C_0$ (mg/L) is the plasma concentration at $t=0$ (i.e., the dose/$V$). This equation predicts that the plasma concentration will decline exponentially, with the result that a graph of log $C_P$ vs. time gives a straight line (Figure 2.6A). From the slope of this line, the ratio $CL_{WB}/V$ may be obtained, which is commonly referred to as the elimination rate constant ($k_{EL}$; 1/h):

$$k_{EL} = -2.3 \cdot \text{slope} = CL_{WB}/V \tag{2.28}$$

The elimination half-life ($t_{1/2}$; h) is calculated from $k_{EL}$ as:

$$t_{1/2} = 0.693/k_{EL} = 0.693V/CL_{WB} \tag{2.29}$$

This relationship shows that a change in $t_{1/2}$ may result from a change in $CL_{WB}$ or $V$. The $y$-axis intercept of the semi-log plot gives an estimate of $C_0$ that can be used to calculate a value for $V$:

$$V = \text{dose}/C_0 \tag{2.30}$$

Ideally, the number of plasma samples should be three times the number of parameters to be estimated (in this case, six samples to estimate the two parameters $CL_{WB}$ and $V$). These samples should be uniformly spaced and taken over a time span at least three times the $t_{1/2}$.

The value of a TK study performed in this manner is difficult to overstate. The resulting $k_{EL}$ estimate is equivalent to that obtained from an aqueous uptake–depuration study protocol, where measured chemical concentrations are expressed on a whole-fish basis (e.g., OECD Test Guideline 305; OECD, 2012). However, because all data are obtained from a single animal (eliminating the confounding effect of individual variability), it is possible to obtain estimates of $CL_{WB}$ and $V$ that, when evaluated across multiple animals, can be used to characterize the population variance in these parameters.

#### *Infusion Dose*

In some cases, it may be advantageous to administer the dose at a constant rate ($R_0$; mg/h). This approach avoids the high transient plasma concentration associated with a bolus dose and provides a larger volume of solvent to dissolve the chemical. The plasma concentration will increase exponentially during the infusion, followed by an exponential decline after the infusion is stopped. During the infusion, the predicted plasma concentration is:

$$C_P = (R_0/CL_{WB})(1 - e^{-k_{EL}t}) \tag{2.31}$$

After three to four half-lives, the exponential approaches a value of zero, and a steady-state condition is achieved in which the rate of infusion equals the rate of chemical elimination.

Plasma samples are generally obtained after the infusion has been stopped using the same cannula employed to infuse the chemical. The value of $k_{EL}$ is obtained from the slope of a semi-log plot of the post-infusion plasma data (Equation 2.31). If the infusion is sufficient to achieve a steady-state condition, $CL_{WB}$ can be calculated from:

$$CL_{WB} = R_0/C_{SS} \tag{2.32}$$

where $C_{SS}$ (mg/L) is the steady-state plasma concentration, estimated from the $y$-axis intercept of the semi-log plot. The volume of distribution ($V$) can then be calculated as:

$$V = CL_{WB}/k_{EL} \tag{2.33}$$

If the infusion was stopped prior to steady state, $k_{EL}$ may still be estimated from the slope of the semi-log plot, but estimation of $CL_{WB}$ and $V$ is less straightforward because the amount of chemical in the fish at the end of the infusion is unknown. Equation 2.31 may be rearranged to give the following expression for estimation of $CL_{WB}$, which can then be used along with $k_{EL}$ to estimate $V$:

$$CL_{WB} = (R_0/C_{END})\left(1 - e^{-k_{EL} t_{INF}}\right) \quad (2.34)$$

where $C_{END}$ (mg/L) is the concentration in plasma at the end of the infusion and $t_{INF}$ (h) is the duration of the infusion.

*Waterborne Exposure*

Kinetic studies with fish may be conducted by exposing small fish or juveniles of larger species to a chemical in water. Typically, the exposure is initiated with a fixed number of fish, and at each sampling time a subset of animals is collected to determine whole-body chemical concentrations. The exposure may be conducted using a flow-through system that maintains a constant concentration of chemical in the exposure water. Alternatively, fish may be exposed in a static system, which may or may not be renewed during the progress of a test. The aqueous concentration of chemical in a static exposure system may decline over time due to absorption by the test organisms. Additional losses may occur due to adsorption to the test container, volatilization, photodegradation, and microbial biotransformation.

The rate of chemical accumulation in fish dosed in a flow-through exposure is proportional to the difference between the concentration of chemical in the exposure water ($C_W$; mg/L) and the concentration unbound in plasma water ($C_{P,U}$; mg/L):

$$dX/dt = \rho\left(C_w - C_{P,U}\right) \quad (2.35)$$

The $C_W$ is constant during the exposure, and the $C_{P,U}$ can be replaced with $X/V$, where $V$ is the apparent volume of distribution of the chemical referenced to the exposure water. The units of $V$ and $\rho$ are usually normalized to body weight; for example, if $X$ had units of mg/kg, and $C_W$ and $C_{P,U}$ had units of mg/L, then $V$ would have units of L/kg and $\rho$ would have units of L/h/kg. Integration of Equation 2.35 gives an equation for $X$ as a function of time:

$$X = VC_W\left(1 - e^{-(\rho/V)t}\right) \quad (2.36)$$

This equation is fit to experimentally determined values of $X$ at various times to determine values for $\rho$ and $V$. The elimination half-life can then be calculated as:

$$t_{1/2} = \ln 2V/\rho = 0.693V/\rho \quad (2.37)$$

Equation 2.36 predicts that $X$ will exponentially approach a limiting value equal to $V\,C_W$. In this case, the value of $V$ is equivalent to the fish BCF, as defined earlier in the chapter.

When a metabolite of the test chemical is formed during a flow-through exposure, the model given above may be extended to permit simultaneous characterization of its toxicokinetics. A second modification of the flow-through model that accounts for chemical accumulation by fish can be developed to describe uptake in a static exposure scenario. Derivations and equations for the flow-through metabolite and static exposure models were given by Kleinow et al. (2008).

*Oral Administration*

A single oral dose may be administered to a fish that has been prepared with an indwelling vascular catheter. If the absorption kinetics are first order and monoexponential, the rate of appearance of chemical in the fish is:

$$dX/dt = k_{AB}X_G - CL_{WB}C_P \quad (2.38)$$

where $k_{AB}$ (1/h) is the absorption rate constant, and $X_G$ (mg) is the amount of chemical remaining in the GIT. From this equation, it is possible to derive the following relationship:

$$C_P = k_{AB}F \cdot \text{dose}\left(e^{-(CL_{WB}/V)t} - e^{-k_{AB}t}\right)/V\left(k_{AB} - (CL_{WB}/V)\right) \quad (2.39)$$

where $F$ (unitless) is the fraction of the dose that reaches the systemic circulation of the fish (i.e., oral bioavailability).

When the dietary exposure is continuous, the rate of change of the amount of chemical in the fish is:

$$dX/dt = FR_{IN} - CL_{WB}C_P \quad (2.40)$$

where $R_{IN}$ (mg/h) is the rate of chemical ingestion. In this case, the mass of chemical exponentially approaches a constant or steady-state value:

$$X_{SS} = (F\ R_{IN}V\ /\ CL_{WB})\left(1 - e^{-k_{EL}t}\right) \quad (2.41)$$

## Two-Compartment Empirical Model

A one-compartment model assumes that a chemical distributes instantaneously throughout all tissues of the exposed animal. In principle, this is impossible because chemical distribution to each tissue is linked to blood perfusion rate, and tissue-specific perfusion rates vary considerably. However, if an internal distribution equilibrium is rapidly established, errors associated with application of a one-compartment modeling assumption are small and can be ignored. Alternatively, if the time required for establishment of an internal equilibrium is a significant fraction of total sampling time, this will impact the shape of the plasma concentration–time profile and a higher order compartment model may be required. A common example is when log-transformed plasma concentrations from an intravascular bolus dosing study decrease in a *biphasic* manner (Figure 2.6B). The early decline in plasma concentration (α phase) is due primarily to redistribution of the injected chemical, while the later elimination or β phase reflects rate limitations on distribution and elimination. A two-compartment model for intravascular administration of a bolus dose is described in the following section. Space does not permit the presentation of other multiple-compartment models; however, the principles illustrated by this relatively simple model apply generally to more complex models.

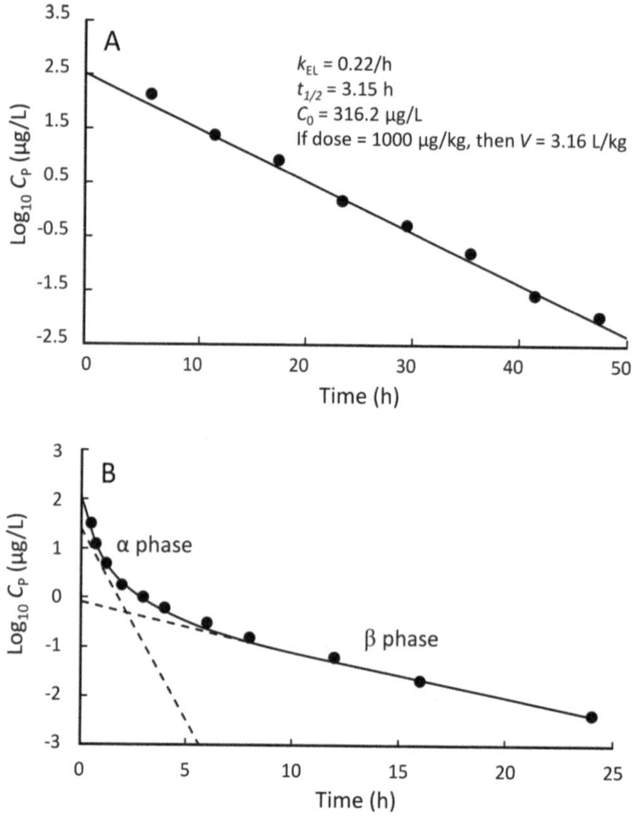


**FIGURE 2.6** Hypothetical plasma concentration–time profiles for chemicals exhibiting monoexponential (A) or biexponential (B) elimination kinetics following intravascular injection. Solid lines represent the least-square fit of a one-compartment (A) or two-compartment (B) model. The two-compartment model is typically fit by nonlinear regression analysis. A dashed line in panel B shows the slope of the terminal (elimination or $\beta$) phase extrapolated to $t=0$. A second dashed line shows the slope of the distribution ($\alpha$) phase residuals (not shown) obtained by subtracting extrapolated $C_P$ values (from the $\beta$ phase equation) from measured $C_P$ values (Gibaldi and Perrier, 1982). Symbols and abbreviations are defined in the text.

*Intravascular Bolus Administration*

This widely used model represents the body as two compartments (Figure 2.5B). Chemical enters the central compartment, distributing instantaneously. From there, it moves into a peripheral compartment. Once it is in the peripheral compartment, the chemical is again assumed to distribute instantaneously. The central compartment generally represents the blood and highly perfused tissues such as the liver, kidney, and GIT. The peripheral compartment usually represents poorly perfused tissues, including bone, connective tissue, white muscle, and fat. The brain may be associated with either compartment, depending on whether the blood–brain barrier limits the rate of chemical exchange between the blood and brain tissues.

Following the intravascular administration of a bolus dose ($X_0$; mg), mass-balance equations that describe the rate of change of the amount of chemical in compartments 1 and 2 may be written as follows:

$$dX_1/dt = k_{21}X_2 - k_{10}X_1 - k_{12}X_1 \quad (2.42)$$

$$dX_2/dt = k_{12}X_1 - k_{21}X_2 \quad (2.43)$$

where $X_1$ and $X_2$ (mg) are the amounts of chemical in compartments 1 and 2, $k_{12}$ and $k_{21}$ (1/h) are first-order intercompartmental transfer rate constants, and $k_{10}$ (1/h) is a first-order rate constant for chemical elimination from the central compartment.

If plasma is used as the reference region, an equation may be derived that predicts the decrease in plasma concentration over time (Kleinow et al., 2008):

$$C_P = Ae^{-\alpha t} + Be^{-\beta t} \quad (2.44)$$

This equation can be fit to the $C_P$–time profile by nonlinear regression analysis to obtain estimates of $A$ (mg/L), $\alpha$ (1/h), $B$ (mg/L), and $\beta$ (1/h), where $\alpha$ and $\beta$ are hybrid rate constants that are comprised of the model rate constants. These estimated terms may then be used to calculate model parameters and other parameters of interest (Gibaldi and Perrier, 1982):

$$V_1 = X_0/(A+B) \quad (2.45)$$

$$k_{21} = (A\alpha + B\beta)/(A+B) \quad (2.46)$$

$$k_{10} = \alpha\beta / k_{21} \tag{2.47}$$

$$k_{12} = \alpha + \beta - k_{21} - k_{10} \tag{2.48}$$

$$V_2 = V_1\left(k_{12}/k_{21}\right) \tag{2.49}$$

$$V_{SS} = V_1 + V_2 \tag{2.50}$$

$$V_\beta = V_1\left(k_{12} + k_{21} - B\right)/\left(k_{21} - \beta\right) = V_1 k_{10}/\beta \tag{2.51}$$

$$CL = k_{10}V_1 = \beta V_\beta \tag{2.52}$$

$$t_{1/2,\alpha} = \ln 2/\alpha \tag{2.53}$$

$$t_{1/2,\beta} = \ln 2/\beta \tag{2.54}$$

The $t_{1/2,\alpha}$ term is characteristic of the time required for distribution, while $t_{1/2,\beta}$ is the elimination half-life of the chemical. Unlike the half-life for a one-compartment model, the value of $t_{1/2,\beta}$ is a function of both the rate constant for elimination from the central compartment and rate constants that describe chemical flux between the two compartments. Here, the term $V_{SS}$ refers to the apparent volume of distribution at steady state, which is equal to the sum of the apparent volumes of the central and peripheral compartments. In this model, clearance is assumed to be limited to the central compartment.

### Utility of Compartmental Models for Fish

Compartmental models have played a central role in the evolution of thought regarding factors that control uptake, accumulation, and distribution of organic chemicals in fish. One-compartment models with fitted first-order uptake and elimination rate constants were used extensively in the 1970s and 1980s to evaluate uptake and depuration data from controlled waterborne exposures (Neely et al., 1974; Branson et al., 1975). Generally, these models were structured such that the uptake term had units of clearance (e.g., L of water cleared of chemical per unit of time, normalized to body weight), while the elimination term was a first-order rate constant with units of inverse time. Derived rate constants were correlated with chemical log $K_{OW}$ to develop predictive empirical relationships. Additional studies were performed to determine how uptake and elimination rate constants change with physiological factors such as fish size and temperature (Neely, 1979), and the degree of $O_2$ saturation (Opperhuizen and Schrap, 1987). In time, it was recognized that chemical uptake from water could be predicted from the rate at which water is delivered to the gills and an extraction efficiency term, which could itself be related to log $K_{OW}$ (Neely, 1979; Norstrom et al., 1976). These efforts led to the development of mechanistic descriptions of chemical flux across fish gills described elsewhere in this text.

The one compartment model for aqueous exposure was subsequently expanded to include uptake from dietary exposures (Bruggeman et al., 1981, 1984). The integrated form of this model for a diet-only exposure has been used extensively for the estimation of $E_D$ values in feeding studies with fish, including values summarized in Figure 2.3 (also see *The Gastrointestinal Tract, Gastrointestinal Absorption of Xenobiotics*).

In some cases, the observed kinetics of a chemical are best described by a two- or three-compartment model. In many such instances, the chemical is administered as a single intravascular or oral dose and the sampling period is relatively short. Experiments of this type tend to reveal the distributional behavior of a chemical. Multiexponential kinetic behavior may also be evident in whole-body depuration data collected following a long-term aqueous or dietary exposure (Könemann and van Leeuwen, 1980; Fisk et al., 1998).

Multicompartment models developed from measured chemical concentrations in plasma provide information regarding rates of chemical distribution within the animal as well as the sorptive capacity of tissues and organs represented by each compartment. Modeled terms derived for a particular chemical may be compared to those developed in modeling efforts with mammals to evaluate species differences in chemical disposition and toxicity. For example, Abbas and Hayton (1996) used a three-compartment model to evaluate the kinetics of parathion and its metabolite paraoxon in rainbow trout after intravascular administration. Parathion was shown to have greater persistence in trout than in rats, suggesting that observed differences in sensitivity to this chemical (rats>trout) are not due to toxicokinetic difference between the species. Fitted model terms also may be interpreted in the context of other biological information to infer the importance of specific elimination processes. For example, if mammals are known to eliminate a chemical primarily into urine, a fitted $CL_{WB}$ comparable to that of estimated renal blood flow could be viewed as evidence for the same elimination process in fish (Schultz et al., 1995).

Clearance-volume models have been used to investigate a variety of factors that impact chemical kinetics in fish. For example, Barron et al. (1987b) used a three-compartment model to investigate the effect of acclimation temperature on the kinetics of di-2-ethylhexylphthalate (DEHP) administered as an intraarterial dose to rainbow trout. The $t_{1/2,\beta}$ for DEHP increased with acclimation temperature, primarily due to changes in $V_{SS}$ and not $CL_{WB}$. A three-compartment model with elimination from both the central compartment and a separate gill compartment was used to investigate the uptake and accumulation of DEHP in trout exposed via the water (Barron et al., 1989). Modeled findings suggested that pre-systemic branchial biotransformation limits uptake of DEHP. Subsequent studies performed using isolated perfused gill arches showed that trout metabolize DEHP to monoethylhexylphthalate. Several authors have combined multicompartment modeling with a detailed study of metabolite concentrations in plasma, urine, and bile to identify biotransformation pathways and characterize routes and rates of excretion for parent chemicals and their metabolic products (Bradbury et al., 1993; Kleinow et al., 1992; Plakas et al. 1992b; Stehly and Hayton, 1989; Stehly and Plakas, 1992, 1993).

A two-compartment model was used by Schultz and Hayton (1994) to investigate the effect of body size on uptake and accumulation of trifluralin in rainbow trout. The uptake clearance from water decreased with increasing body weight. This change was accompanied by size-dependent increases in $V_{SS}$ and $t_{1/2,\beta}$. The kinetics of trifluralin following intravascular injection was evaluated in hatchery-reared and wild steelhead trout using a three-compartment model (Schultz and Hayton, 1997). Derived $V_{SS}$ and $t_{1/2,}\beta$ values for hatchery trout were substantially greater than those determined for wild trout. These differences were consistent with differences in whole-body lipid content, illustrating the effect of lipid content on elimination of hydrophobic chemicals.

## Physiologically based Toxicokinetic Models

Mass-balance differential equations that describe chemical kinetics in a tissue exchanging with blood were developed in the 1930s (Teorell, 1937), but the computing power required to simultaneously solve these equations for several tissues was not widely available until the 1960s. Early use of the physiological modeling approach consisted primarily of describing the kinetics of chemotherapeutic agents (Gerlowski and Jain, 1983). This approach was subsequently adopted by toxicologists interested in extrapolating data from laboratory test animals to humans (Ramsey and Andersen, 1984). One early application of the PBTK modeling approach was to compare chemical distribution and elimination in fish and rodents. Models for methotrexate in stingrays (*Dasyatidae sabina* and *D. sayi*; Zaharko et al., 1972) and phenol red in the dogfish shark (Bungay et al., 1976) were designed to simulate experiments in which the chemicals were injected intravascularly, and elimination was limited to urinary and fecal routes. The first PBTK model for fish that incorporated an environmentally relevant route of exposure (branchial) was developed by Nichols et al. (1990). Later models have incorporated both dermal (Lien and McKim, 1993; Lien et al., 1994; Nichols et al., 1996) and dietary routes of exposure (Nichols et al., 1998, 2004).

In this section, we review the principles of PBTK modeling with fish. Mathematical details are deemphasized, as this information is available is several excellent texts (Clewell et al., 2014; Fisher et al., 2020). This section focuses instead on route of exposure considerations and specific applications as they relate to fish. An emphasis is placed on factors that distinguish fish and mammalian models, either in terms of their structure or the ways in which the models have been used.

### Model Structure

A PBTK model consists of one or more tissue compartments. For multicompartmental descriptions, these compartments are connected in a manner consistent with the cardiovascular system of the exposed organism (Figure 2.7). Additional details follow from the type of exposure, characteristics of the test chemical, and the goals of the modeling exercise. Often, it is necessary to ignore known or suspected complexity to simplify the task of parameter estimation. In general, a modeler strives to create the simplest structure that will satisfactorily represent the behavior of the system under study. The number and identity of tissue compartments can vary greatly. An emphasis is usually placed on tissues that control the kinetics of uptake and elimination, as well as those that are important toxicologically.

Physiological and anatomical attributes of fish that influence chemical uptake and disposition were reviewed earlier in this chapter. Several are singled out here because they tend to distinguish fish and mammalian PBTK models. First, chemical uptake by inhalation exposure can be adequately simulated in mammals by modeling the lungs as a mixed chemical reactor. Chemicals taken up across the lung travel first to the heart and then to the rest of the body. In contrast, fish gills are more appropriately modeled as a counter-current exchange system, and chemicals taken up across the gills travel throughout the body before reaching the heart. Second, the kidney in many fish is supplied by venous blood draining both the trunk musculature and skin. Although this portal blood does not appear to be available for glomerular filtration, it may contribute to urinary elimination of chemicals that are actively secreted in the proximal tubule. Because fish are poikilothermic, physiological parameters such as ventilation volume and $Q_C$ can vary greatly with ambient temperature, impacting chemical kinetics. Finally, it is important to recognize that physiological differences may exist among fish species because of differences in life history and reproductive strategy, dietary preferences, and the need to adapt to different environmental conditions. Incorporation of these differences into any given model depends on whether they influence the kinetics of the chemical.

The compartments in a PBTK model correspond to tissues and organs with similar kinetic characteristics. For a well-mixed compartment (I) that receives only arterial blood, a mass-balance equation may be written as:

$$dX_I/dt = Q_I\left(C_A - C_{VI}\right) \qquad (2.55)$$

where $X_I$ (µg) is the mass of chemical in the compartment, $Q_I$ (L/h) is the rate of arterial blood flow to the compartment, $C_A$ (µg/L) is the chemical concentration in arterial blood, and $C_{VI}$ (µg/L) is the chemical concentration in venous blood draining the compartment. The rate of chemical accumulation by a tissue cannot exceed the product term $Q_IC_A$. The actual rate at any point in time depends on the prior exposure history (i.e., the extent to which chemical has already accumulated) and the rate of chemical exchange between blood and tissues, both of which have the potential to impact $C_{VI}$. It may be noted from this description that $Q_I$ has units flow, as is the case for clearance constants used in an empirical clearance-volume model. In fact, PBTK

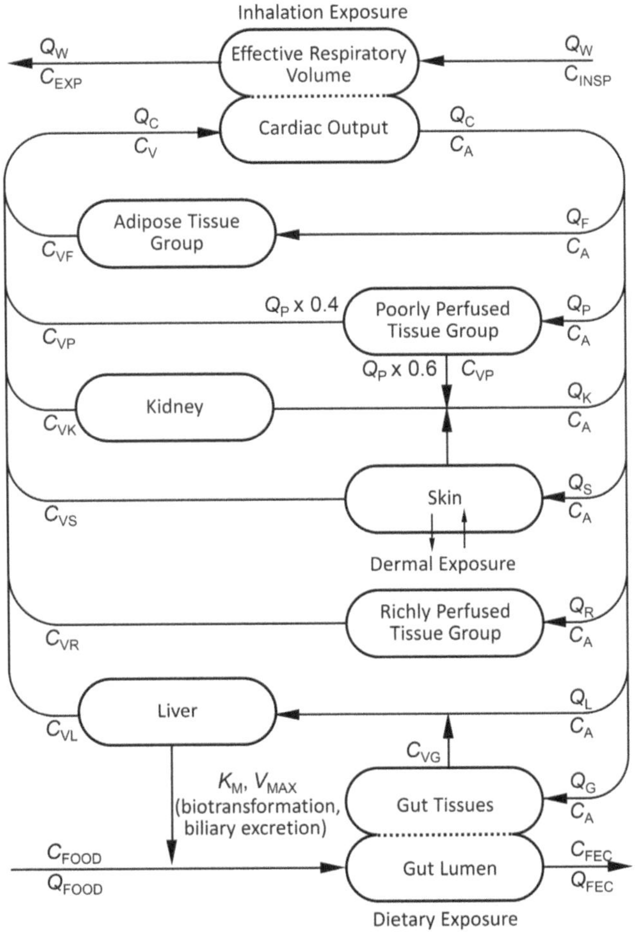


**FIGURE 2.7** Schematic representation of a physiologically based toxicokinetic model for fish. $Q_F$, $Q_P$, $Q_K$, $Q_S$, $Q_R$, $Q_L$, and $Q_{GT}$ refer to adipose tissue, poorly perfused tissue (white muscle, bone, and connective tissue), kidney, skin, richly perfused tissue (stomach, upper intestine, spleen, and gonads), liver, and lower intestine blood flow as proportions of $Q_C$. $Q_W$ is the flow rate of inspired water that is available to exchange with blood at the gills, and $C_{INSP}$ is the chemical concentration in this water. $Q_{FOOD}$ and $Q_{FEC}$ are the bulk flow rates of food and feces. $C_{FOOD}$ and $C_{FEC}$ are chemical concentrations in food and feces. $C_{VF}$, $C_{VP}$, $C_{VK}$, $C_{VS}$, $C_{VR}$, $C_{VL}$, and $C_{VGT}$ refer to chemical concentrations in venous blood draining these tissues, while $C_{VEN}$ is the chemical concentration in mixed venous blood. $K_M$ and $V_{MAX}$ are kinetic rate constants for saturable biotransformation in the liver.

models are properly viewed as a type of clearance-volume model. In a PBTK model, however, the rate constants that describe chemical flux into and out of modeled compartments are developed from physiological and biochemical process information.

A relatively simple tissue description can be developed if it is assumed that the rate of chemical exchange between blood and tissue is fast relative to the blood perfusion rate. Under these circumstances, chemical partitioning proceeds to equilibrium (consistent with the well-mixed tank assumption) with the result that $C_{VI}$ is equal to the chemical concentration in the tissue ($C_I$; μg/L) divided by an equilibrium tissue–blood partition coefficient $P_I$:

$$C_{VI} = C_I / P_I \quad (2.56)$$

To further simplify the model, it is generally assumed that blood and tissues have a specific gravity of 1 with the result that all partition coefficients are unitless and concentrations expressed on a volume basis can be equated to those expressed on a wet weight basis. Substituting Equation 2.56 into Equation 2.55, one obtains:

$$dX_I/dt = Q_I\left(C_{ART} - C_I/P_I\right) \quad (2.57)$$

or equivalently:

$$V_I\, dC_I/dt = Q_I\left(C_{ART} - C_I/P_I\right) \quad (2.58)$$

where $V_I$ (L) is the tissue volume. This type of chemical distribution is said to be *flow limited* because the rate of chemical accumulation is controlled by the blood flow rate.

Chemical distribution is said to be *diffusion limited* if the rate of diffusion across a biological membrane limits chemical flux between blood and tissues. Under these circumstances, the tissue must be divided into two or more subcompartments. Chemical flux between subcompartments is then modeled using the Fick relationship (Equation 2.1).

In practice, chemical permeability (i.e., $DP_{MW}$/h) and the surface area for diffusion are difficult to estimate. The permeability-area (PA) product is therefore frequently treated as a first-order clearance constant ($k_I$; L/h), the value of which is determined from exposure data. Mathematic equations that describe diffusion limited chemical flux between blood and tissues were provided by Kleinow et al. (2008).

Some chemicals exhibit high affinity binding to proteins in blood. This binding may follow a Langmuir-type isotherm, saturating at high substrate concentrations. Other chemicals form stable complexes with intracellular proteins in blood (e.g., hemoglobin) and/or tissues. In either case, it may be necessary to model the effect of this binding to determine unbound chemical concentrations (Gerlowski and Jain, 1983).

In view of these potential complexities, it is perhaps remarkable that the assumption of flow-limited distribution has been shown to approximate the kinetic behavior of many chemicals in both fish and mammals. The principal advantage of the flow-limited tissue description is that it does not require the fitting of any kinetic parameters. For this reason, it is often used as a default assumption in model formulation. Exposure data can then be examined for evidence of a diffusion limitation on chemical flux, typically presented as a slower-than-anticipated rate of accumulation in one or more tissues. This type of evidence would be insufficient to determine where the diffusion limitation exists (i.e., which membrane(s)) and to what extent, if any, strong binding plays a role in retarding flux. One might conclude, however, that additional studies are needed to guide model development.

The kidney and liver of fish receive both arterial and portal blood. Generally, it is assumed that arterial and portal blood mix before exchanging with the tissue. With this simplification, total tissue blood flow ($Q_{MI}$; L/h) equals the sum of arterial ($Q_I$) and portal inputs ($Q_{PI}$; L/h):

$$Q_{MI} = Q_I + Q_{PI} \tag{2.59}$$

where $Q_{PI}$ is equal to all or a fraction of arterial flow to the upstream compartment. The chemical concentration in mixed blood ($C_{MI}$) is then calculated by summing the flow-weighted contributions of arterial ($C_{ART}$) and portal blood ($C_{PI}$):

$$C_{MI} = \left(Q_I C_{ART} + Q_{PI} C_{PI}\right)/Q_{MI} \tag{2.60}$$

where $C_{PI}$ is equal to the concentration in venous blood exiting the upstream compartment. Assuming further that chemical uptake by the tissue is flow limited, a mass balance for the compartment may be written as:

$$dX_i/dt = Q_{MI}\left(C_{MI} - C_{VI}\right) \tag{2.61}$$

When the kinetic descriptions for all tissues have been defined, the chemical concentration in mixed venous blood is calculated from the flow-weighted contributions of venous blood draining each tissue compartment. In a model with chemical uptake at the gills, this mixed venous concentration provides one of the inputs to the branchial exchange description (see below).

## Routes of Exposure

A critical part of any PBTK model is the route of exposure description. In experimental dosing studies, the investigator controls the route and timing of the exposure. Natural exposures are more likely to be of a mixed type, involving chemical flux at more than one exchange surface. Environmental routes of uptake (branchial, dermal, dietary) can also function as routes of elimination. In contrast, some experimental routes of exposure, such as intravascular infusion, function only as routes of uptake.

### *Branchial uptake*

The structure of fish gills reflects their primary function as a gas-exchange and osmoregulatory organ. A large surface area for exchange is achieved by an elaboration of plate-like structures called *secondary lamellae*, which form narrow channels above and below each gill filament. In most species, blood flows through the lamellae in a posterior-to-anterior direction, providing for highly efficient counter-current exchange with inspired water.

The principal limitations on chemical uptake at fish gills were outlined by Hayton and Barron (1990). If the permeability of the gill epithelium to a chemical is low, diffusion can limit exchange, and uptake flux is controlled by the product of gill permeability and the surface area for diffusion. If, on the other hand, the permeability of the epithelium is high, uptake may be limited by the chemical capacities of blood and water flowing to the gills. A blood flow limitation exists if more chemical is delivered to the gills in respiratory water than can be carried away in blood. A water flow limitation exists if the capacity of blood to carry chemical away from the gills exceeds the chemical capacity of respiratory water.

Ignoring diffusion limitations on uptake, Erickson and McKim (1990a) developed a model for the branchial flux ($F_G$; µg/h) of organic chemicals based on flow limitations and the behavior of counter-current exchange systems. The complete flow-limited model may be stated as:

$$F_G = \text{Min}\left[Q_W, Q_C P_{BW}\right]\left[C_W - \left(C_V/P_{BW}\right)\right] \tag{2.62}$$

where $Q_W$ (L/h) is the flow of inspired water that exchanges with blood, $P_{BW}$ (unitless if blood is assumed to have a specific gravity of 1) is an equilibrium blood–water partition coefficient, $C_W$ (µg/L) is the total chemical concentration in water, and $C_V$ (µg/L) is the chemical concentration in venous blood entering the gills. In this equation, $F_G$ is calculated as the product of the concentration gradient between inspired water and venous blood entering the gills (second term in brackets) and the chemical capacity of blood or respiratory water (first term), whichever is less. When the flow-limited model was parameterized using data for rainbow trout, it reproduced the well-known dependence of branchial uptake efficiency on chemical log $K_{OW}$ for chemicals of low to intermediate hydrophobicity but did not predict the observed decline in uptake for high log $K_{OW}$ chemicals (Figure 2.2A). In addition, the model tended to overestimate the maximal rate of uptake by about 20%.

Subsequently, Erickson and McKim (1990b) published a more comprehensive gill model that includes both flow and diffusion limitations, and accounts for binding of high log $K_{OW}$ chemicals to DOC and POC. The full model, which accounts for advective (direction of fluid flow) and diffusive (perpendicular to fluid flow) chemical transport in both blood and water, did not yield an analytical solution. An approximate analytical solution was obtained, however, by segregating the lamellar unit into layers in which only advective or diffusive transport predominates. With these simplifications, branchial flux can be calculated as the product of the concentration gradient between inspired water and venous blood and an exchange coefficient $k_X$ (L/h):

$$F_G = k_X\left(C_W - C_V P_{BW}\right) \tag{2.63}$$

where

$$k_X = \left(e^{-k_D/k_B} - e^{-k_D/k_W}\right) \Big/ \left[\left(e^{-k_D/k_B}/k_W\right) - \left(e^{-k_D/k_W}/k_B\right)\right] \tag{2.64}$$

For this model, $C_W$ refers to the unbound chemical concentration and the terms $k_W$ (L/h), $k_B$ (L/h), and $k_D$ (L/h) represents the capacity for respiratory water, blood flowing through the gills, and chemical diffusion across the gills, respectively, to support branchial flux.

$$k_W = Q_W \tag{2.65}$$

$$k_B = Q_C P_{BW} \tag{2.66}$$

$$k_D = DA/h \tag{2.67}$$

Here, $D$ (m$^2$/h), $A$ (m$^2$), and $h$ (μm) refer to the chemical diffusivity in, the total area of, and the effective thickness of the gill diffusion barrier, respectively. Importantly, this model assumes that chemical diffusion in the aqueous phase of the gill epithelium limits flux. Simulations for rainbow trout generated using the complete model accurately describe all the observed trends in branchial uptake data (Figure 2.2A), including the reduction in uptake of high log $K_{OW}$ chemicals.

The complete gill model has been used to describe branchial uptake of waterborne chemicals by several fish species, requiring the collection of necessary physiological and gill morphometric information (Figure 2.8). For many chemicals, however, a good approximation of branchial flux can be obtained using the simpler flow-limited model. Direct experimental evidence for the predominant influence of flow limitations on branchial uptake has been obtained by independently varying $Q_W$ and $Q_C$ for chemicals with differing hydrophobicity (Schmieder and Weber, 1992). Finally, it is important to note that both the flow-limited and complete gill models separate the blood circulation into arterial and venous sides. In a PBTK model, chemical taken up across the gills is added to that already present in venous blood to provide a new value for the arterial blood concentration. The arterial blood concentration is then used as an input to each of the tissue descriptions (Figure 2.7).

### Dermal Uptake

Dermal uptake may be important for small fish, juveniles of larger species, and fish that live in direct contact with contaminated sediments. The general architecture of fish skin (epidermis, dermis, and hypodermis) is similar in most species, but the thickness of these layers varies widely. Additional differences exist with respect to the presence or absence of scales and the number and location of sensory organs, mucous glands, and other specialized structures.

A PBTK model for dermal uptake in small fish was developed by Lien and co-workers (Lien and McKim, 1993; Lien et al., 1994) from the gill model given in Equation 2.64 by assuming that water flow does not limit chemical transport across the skin. This model can be visualized as a parallel network of capillaries with chemical diffusion from water to blood perpendicular to the capillary axis. With these assumptions, Equation 2.64 simplifies to:

$$k_X = k_B\left(1 - e^{-k_D/k_B}\right) \tag{2.68}$$

This description suggests that dermal flux is limited by the capacity of the skin surface to support chemical diffusion ($k_D$) and the capacity of blood flowing to the skin ($k_B$) to remove chemical once it has been absorbed. The terms $k_B$ and $k_D$ are analogous to those given in Equations 2.66 and 2.67, except that in this case diffusivity, area, and thickness refer to the skin diffusion barrier, and the relevant blood flow rate is that to skin. As in the gill model given by Erickson and McKim (1990b), diffusive flux was assumed to be limited by chemical diffusion in the aqueous phase of the diffusion barrier. Using this model, dermal flux ($F_S$; μg/h) may be calculated as the product of $k_X$ and the concentration gradient for uptake:

$$F_S = k_X\left(C_W - C_{ART}/P_{BW}\right) \tag{2.69}$$

The chemical concentration in venous blood draining the skin is then calculated by adding this flux to the amount of chemical already present in arterial blood perfusing the skin.

A more complex description of dermal uptake is required if chemical accumulation by skin tissue impacts chemical flux between blood and water. Following the approach used by McDougal et al. (1990) to describe dermal uptake of chemical vapors by rats, Nichols et al. (1996) developed a dermal uptake model for fish that incorporates a discrete skin compartment. This model was then evaluated by

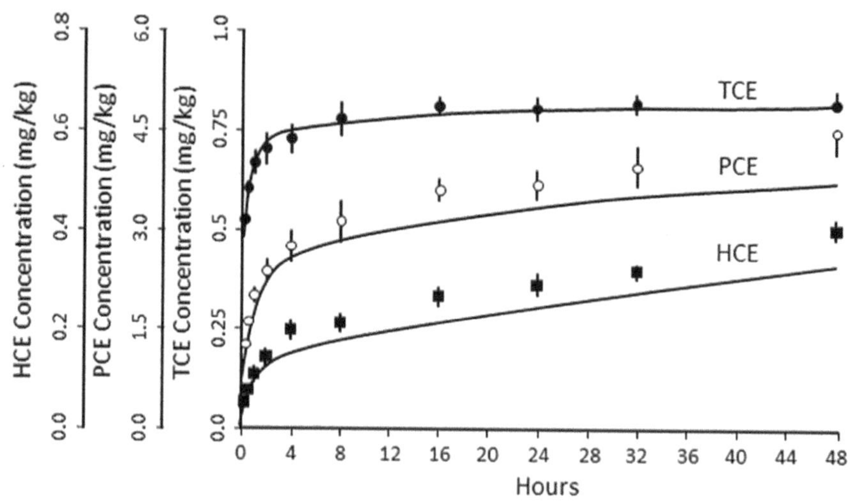


**FIGURE 2.8** Concentration time course of tetrachloroethane (TCE), pentachloroethane (PCE), and hexachloroethane (HCE) in arterial blood of adult channel catfish (adapted from Nichols et al., 1993). Catfish were simultaneously exposed to all three chloroethanes in water. Solid lines represent simulations generated using a PBTK model incorporating the branchial uptake description given by Erickson and McKim (1990b). Measured values are shown as individual points. No effort was made to calibrate model simulations to measured data. The model was run using tissue–blood partition coefficients determined using an *in vitro* vial equilibration method (Hoffman et al., 1992).

exposing large trout and channel catfish to a series of chlorinated ethanes in a chambered exposure system (McKim et al., 1996). For these studies, only the trunk region of the fish was exposed. By using chemicals for which validated inhalation models already existed, it was possible to fit a set of apparent skin permeability coefficients. These fitted permeability coefficients increased little if at all with increasing chemical affinity for skin, suggesting that diffusion through tissue water limits dermal flux of moderately hydrophobic chemicals (Nichols et al., 1996).

### Dietary Uptake

Chemical and physiological factors that control xenobiotic uptake from the diet were described earlier in this chapter. From a modeling perspective, the challenge presented by these findings is to describe a highly dynamic process impacted by a multiplicity of factors using a relatively simple mathematical formulation. Detailed PBTK models that include multiple gut sub-compartments have been developed to simulate dietary uptake of methotrexate in sting rays (Zaharko et al., 1972) and 2,2′,5,5′-tetrachlorobiphenyl in rainbow trout (Nichols et al. 2004); however, routine application of such models is impractical. An important simplifying principle that can be applied to this challenge relates to the impact of digestion on chemical uptake. By reducing both the volume and chemical sorptive capacity of material remaining in the gastrointestinal lumen, uptake of dietary lipid and protein has the effect of increasing chemical fugacity in the material that remains. According to this *digestion hypothesis*, the resulting increase in chemical fugacity can result in chemical uptake across the gastrointestinal epithelium even when the lipid-normalized chemical concentration in the animal exceeds that of the diet, resulting in a biomagnification factor (BMF) >1 (Connolly and Pedersen, 1988; Gobas et al., 1993, 1999; see also *Bioaccumulation Assessment and Modeling, Bioconcentration, Bioaccumulation, and Biomagnification*).

Implementation of the digestion hypothesis within a PBTK modeling framework may be accomplished by assuming that a chemical equilibrium is established between fecal material contained within the terminal colon ($C_{FEC}$; μg/L) and venous blood draining the GIT ($C_{VGT}$; μg/L):

$$C_{FEC} = C_{VGT} P_{FB} \tag{2.70}$$

where $P_{FB}$ (unitless if feces and blood are assumed to have the same specific gravity) is a feces–blood equilibrium chemical partition coefficient (Nichols et al., 1998). This approach is like that proposed by Barber et al. (1991) to model PCB accumulation by lake trout, the principal difference being that an equilibrium is established with venous blood exiting the gut and not with the entire fish.

From mass-balance considerations, it follows that:

$$(Q_{FOOD}C_{FOOD}) - (Q_{FEC}C_{FEC}) = (Q_{GT}C_{VGT}) - (Q_{GT}C_{ART}) \tag{2.71}$$

where $Q_{FOOD}$ (L/h) is the feeding rate, $C_{FOOD}$ (μg/L) is the chemical concentration in the diet, $Q_{FEC}$ (L/h) is the fecal egestion rate, $Q_{GT}$ (L/h) is the blood flow rate to the lower intestine, and $C_{ART}$ (μg/L) is the chemical concentration in arterial blood. Substituting Equation 2.70 into Equation 2.71 and solving for $C_{VGT}$ yields:

$$C_{VGT} = (Q_{GT}C_{ART} + Q_{FOOD}C_{FOOD}) / (Q_{GT} + Q_{FEC}P_{FB}) \tag{2.72}$$

Venous blood draining the lower intestine is then assumed to mix with arterial blood flowing to the liver.

The primary advantage this simple gut description is that it is not necessary to fit the value of a first-order absorption rate constant. In a simplistic way, the model accounts for changes in chemical fugacity that accompany digestion (through its impact on $P_{FB}$) and the concentrating effect of a reduction in meal volume (by specifying both $Q_{FOOD}$ and $Q_{FEC}$). Moreover, because chemical can accumulate in blood flowing to the gut tissue compartment, the model provides for the possibility of decreased absorption due to a decline in the gradient for uptake. The disadvantage of this model is that it cannot account for kinetic limitations that may prevent an equilibrium from being established between blood and feces. In addition, because uptake of lipid and protein from the diet tends to be highly efficient, this description can be expected to result in $E_D$ values higher than most of the values shown in Figure 2.3A or 2.3B.

An alternative model for dietary uptake can be developed by incorporating a diffusion limitation on chemical flux between feces and tissues of the intestinal tract. For the feces:

$$dA_{FEC}/dt = (Q_{FOOD}C_{FOOD}) - (Q_{FEC}C_{FEC}) - k_{GT}(C_{FEC} - C_{GT}P_{FGT}) \quad (2.73)$$

where $A_{FEC}$ (μg/L) is the amount of chemical in feces, $k_{GT}$ (L/h) is a first-order clearance constant, $C_{GT}$ (μg/L) is the chemical concentration in gut tissues, and $P_{FGT}$ (unitless if feces and gut tissues are assumed to have the same specific gravity) is a feces–gut tissue equilibrium partition coefficient. For tissues of the intestinal tract:

$$dA_{GT}/dt = Q_{GT}(C_{ART} - C_{VGT}) + k_{GT}(C_{FEC} - C_{GT}P_{FGT}) \quad (2.74)$$

where $A_{GT}$ (μg/L) is the chemical concentration in gut tissues. The chemical concentration in venous blood draining the GIT is then determined from that of the gut tissues by assuming that chemical exchange between the blood and tissues is flow-limited. Using this approach, Nichols et al. (1998) simulated the uptake of TCDD by brook trout from a prepared diet. The same approach was used to simulate the effect of hepatic biotransformation on bioaccumulation of hydrophobic organic chemicals (Nichols et al. 2007). A disadvantage of this approach is that fitted values of $k_{GT}$ are difficult to interpret and could reflect a true diffusion limitation, slow chemical desorption from contents of the GIT, or some other kinetic constraint. It is difficult, therefore, to extrapolate fitted values of $k_{GT}$ to other chemicals and tested conditions.

## Hepatic and Renal Elimination

The foregoing descriptions of chemical exposure routes in fish are bidirectional with respect to chemical flux. Thus, blood and water flow limitations that define maximum rates of chemical uptake at the gills will also act to limit maximal rates of branchial elimination. Depending, however, on the way that a particular chemical is handled, a complete PBTK model may also have to incorporate other routes of elimination.

Hepatic and renal clearance may contribute substantially to whole-body chemical elimination, particularly if a substance undergoes biotransformation or is a substrate for active transporters that secrete chemicals into urine and bile. The kinetics of renal clearance by glomerular filtration are generally first order with respect to the unbound chemical concentration in plasma. In contrast, clearance pathways involving biotransformation and active secretion may exhibit saturable, nonlinear kinetics.

The first step in modeling elimination is to specify the organ(s) where it occurs and the kinetics (e.g., first-order, saturable) over the concentration range to which the tissue is exposed. Elimination may then be modeled by incorporating a clearance expression ($dA_{CL}/dt$; μg/h) into the mass-balance equation for the eliminating compartment:

$$dA_I/dt = Q_I(C_{ART} - C_{VI}) - dA_{CL}/dt \quad (2.75)$$

Referenced to the chemical concentration in venous blood draining the tissue, the kinetics of clearance under nonsaturating conditions can be written as:

$$dA_{CL}/dt = k_{FC}C_{VI}V_I \quad (2.76)$$

where $k_{FC}$ (1/h) is a first-order rate constant. Saturable pathways can be described using a Michaelis–Menten type of equation where the kinetic parameters $K_M$ (mg/L) and $V_{MAX}$ (mg/h) define a whole-organ clearance term ($CL_{ORGAN}$; L/h):

$$CL_{ORGAN} = V_{MAX}/(K_M + C_{VI}) \quad (2.77)$$

$$dA_{CL}/dt = V_{MAX}C_{VI}/(K_M + C_{VI}) = CL_{ORGAN}C_{VI} \quad (2.78)$$

Although often localized to the liver as a matter of convenience, biotransformation can be incorporated into any one or more of the individual tissue descriptions. In some cases, a modeler may be more interested in the kinetics of a metabolite than of the parent chemical from which it was derived. Under these circumstances, the disappearance of parent chemical (including both the rate of biotransformation and the site where it occurs) is equated to the production of the metabolite.

## Model Parameterization

Anatomical, physiological, and biochemical information required to develop PBTK models for fish is provided in numerous texts and hundreds of scientific papers (Mangold-Döring et al. (2021). Tissue volumes can be measured by dissecting a representative number of animals, provided the tissues exist as discrete entities. Methods for the estimation of adipose tissue content (a diffusely distributed tissue) were provided by Nichols et al. (1990). Other parameters, such blood flow rates to individual tissues (typically expressed as factions of $Q_C$), are more difficult to obtain. For these and other difficult-to-obtain parameters, it may be useful to construct a preliminary model and perform an analysis of model sensitivity to changes in model inputs. Based on this analysis, it may be possible to identify parameters to which the model is particularly sensitive. For example, simulations generated for hydrophobic organic chemicals are generally sensitive to the specified volume of the fat compartment (Mangold-Döring et al., 2021). The same analysis may lead to identification of parameters to which the model is relatively insensitive. For these parameters, estimates based on data from an unrelated fish species may be adequate. When performing this type of assessment, the modeled time scale can be very important. Chemical concentrations predicted at early time points in a simulated exposure may be sensitive to changes in $Q_W$, $Q_C$, and specified blood flow rates to tissues, while concentrations predicted a later time points may be relatively insensitive to changes in these parameters.

### *Allometric and Temperature Relationships*

Across mammals of varying size, the relationship between a biological parameter and body mass is commonly described by the equation $Y = a \times M^b$ where $Y$ is the parameter of interest, expressed on a whole-animal basis, $M$ is body mass, $a$ is a normalization coefficient (typically the value for a 1kg animal), and $b$ is the allometric scaling exponent. Physiological parameters linked to cellular metabolism (e.g., $Q_C$ and $VO_2$) tend to scale to a fractional exponent that ranges from 0.6 to 0.8 (Adolph, 1949; Schmidt-Nielsen, 1984). It is reasonable, therefore, to expect chemical clearance processes linked to these parameters to scale to body weight in a similar manner. For a substantial number of chemicals, this appears to be the case (Dedrick 1973; Espié et al., 2009; Mordenti, 1986; Rowland, 1985). The activities of enzymes responsible for hepatic biotransformation of chemical contaminants can vary widely among species. Thus, for chemicals that are cleared largely by hepatic biotransformation, a simple relationship between whole-body clearance and body weight is not expected. An exception could exist, however, for chemicals with high hepatic extraction ratios due to blood flow limitations on clearance.

It is difficult to evaluate the size dependence of physiological parameters in fish using data for different species due to the impact of temperature on metabolic and physiological processes. The strongest comparisons, therefore, have been made using different sized individuals of a single species held under identical conditions. Although highly variable among species, the average allometric exponent for $VO_2$ is about 0.8, or slightly higher than the average value for mammals (Schmidt-Nielsen, 1984). Within a single species, gill surface area also tends to scale to an exponent of about 0.8 (Hughes, 1984), as does the absorptive surface area of the intestinal mucosa (Buddington and Diamond, 1987). Data presented by Wood and Shelton (1980) suggest that $Q_C$ in rainbow trout scales to an exponent of about 0.9. Fitzsimmons et al. (2018) examined the size dependence of hepatic biotransformation in rainbow trout. Allometric exponents of −0.21 and −0.16 (nmol/min/g body weight basis) were determined for measured levels of UDP-glucuronosyltransferase (UGT) and glutathione-S-transferase (GST) activity. However, this apparent decrease in activity with increasing body weight was largely due to a size-dependent decrease in liver size as a fraction of total body weight. The intrinsic clearance of three polycyclic aromatic hydrocarbons, expressed on a mL/min/g body weight basis, did not differ significantly with body weight.

Barron et al. (1987a) evaluated the dependence of $Q_C$ on acclimation temperature in rainbow trout. The derived $Q_{10}$ value was 4.0 (i.e., a 10°C increase in temperature was associated with a fourfold increase in $Q_C$). The $Q_{10}$ values reported for $VO_2$ in fish and other poikilotherms generally average around 2.0 (Ott et al., 1980).

### *Chemical Partitioning*

Tissue–blood partition coefficients ($P_I$) can be estimated from measured chemical concentrations in animals that have been exposed to near steady state, provided that binding in blood and tissues does not saturate and that rapid biotransformation within a tissue (e.g., liver) does not reduce the concentration below that expected from simple partitioning. The disadvantage of this approach is that it requires prior exposure of animals to generate PBTK model simulations. A variety of *in vitro* methods have been developed to measure $P_I$ values, including gas phase vial equilibration (Gargas et al., 1989), ultrafiltration (Jepson et al., 1994), partitioning into an immiscible solvent phase (Murphy et al., 1995), equilibrium dialysis (Kousba and Sultatos, 2002), and solid-phase microextraction (Artola-Garicano et al., 2000; Tremblay et al., 2012). Several of these methods have been used to measure $P_I$ values for modeling efforts with fish (Hoffman et al., 1992; Law et al., 1991). The utility of any given method depends in part on characteristics of the test chemical including its volatility and hydrophobicity.

The availability of measured *in vitro* and *in vivo* partitioning information has led to the development of empirical models for the prediction of $P_I$. Most of these models employ a tissue composition-based approach. Early models for mammals emphasized the tissue and blood content of neutral lipid and phospholipid (DeJongh et al., 1997; Poulin and Krishnan, 1995). Partitioning was then predicted from a chemical's measured or estimated log $K_{OW}$ value. Over time, these models have been expanded to predict the partitioning of chemicals possessing a broader range of properties, including those that ionize at physiological pH and are substrates for membrane transporters (Peyret et al., 2010; Ruark et al., 2014; Schmitt, 2008). Empirical log $K_{OW}$-based models for the prediction of chemical binding in blood (Bertelsen et al., 1998; Fitzsimmons et al., 2001; Nichols et al., 2006) and tissues (Bertelsen et al., 1998) of fish have been provided by several authors. Other investigators have employed poly-parameter linear free energy relationships (pp-LFERs) in combination with blood and tissue composition data to predict $P_I$ values for fish (Endo et al., 2013; Larisch et al., 2017; Salmina et al., 2016; Zhao et al., 2018).

*Urinary and Biliary Excretion*

Radiolabeled microspheres have been used to estimate arterial blood flow to the kidneys of arctic grayling (*Thymallus arcticus*; Cameron, 1975), rainbow trout (Barron et al., 1987a), channel catfish (Schultz et al., 1999), northern pike (*Esox lucius*; Mangold-Döring et al., 2021), and white sturgeon (*Acipenser transmontanus*; Mangold-Döring et al., 2021). Renal portal blood flows in fish are poorly known and cannot be determined using the microsphere method. In most species, however, renal portal blood is supplied largely by the caudal vein, which drains both the skin and the trunk musculature (Satchell, 1992; Smith and Bell, 1975). An estimate of total blood flow to the kidney may therefore be determined by summing the arterial flow and the estimated flow in the caudal vein. In rainbow trout, this results in a total estimated flow equal to about 42% of $Q_C$ (Nichols et al., 1990). Reported urine flows in fish were summarized by Hickman and Trump (1969) and Hunn (1982). Urine flows in freshwater fish generally range from 1.0 to 10.0 mL/h/kg, while those in saltwater fish range from 0.1 to 1.0 mL/h/kg. Glomerular filtration rates in fish usually exceed urine flow by a factor of about 1.5 to 5.0 (Hickman and Trump, 1969). Measured bile flow rates in fish are generally quite low, ranging from 30 to 200 µL/h/kg (Boyer et al., 1976a, 1976b, 1976c; Gingerich et al., 1977; Sanz et al., 1993; Schmidt and Weber, 1973); nevertheless, secretion into bile represents an important route of elimination for some chemicals.

## PBTK Modeling in Practice

Confidence in a PBTK model for one fish species is gained if it performs well for several chemicals that possess different partitioning properties. Additional confidence is gained if the same model structure, parameterized for several species, still yields acceptable performance. For initial efforts of the type, it is helpful to start with chemicals that exhibit "simple" toxicokinetic behaviors including flow-limited distribution and an absence of biotransformation. These initial modeling efforts can be supported by a local sensitivity analysis (varying one parameter at a time) to gain insight into model behavior (Covington and Gearhart, 2020). Of special interest is the need to identify uncertain parameters to which the model is relatively sensitive.

If a model performs well for chemicals that behave simply, it can be used to simulate the kinetics of chemicals possessing more complex behavior. For example, it may be possible fit the value of a biotransformation rate constant (e.g., $k_{FC}$ in Equation 2.76) that "explains" measured chemical concentrations in blood. This parameter fitting exercise is generally performed using a statistical procedure (e.g., the nonlinear least squares method or maximization of a log-likelihood function) that generates the best fit of modeled simulations to measured data. Simultaneous fitting of multiple parameters (e.g., rate constants for hepatic metabolic clearance and renal secretion) may be possible if measured kinetic data are provided for several modeled quantities (e.g., chemical concentrations in plasma and urine); however, such methods must be used with caution due to the possible existence of more than one high-quality solution. Algorithms that perform this type of parameter fitting are provided as a part of most toxicokinetic software packages. These programs generally require initial estimates and plausible ranges for the parameters to be optimized. A detailed description of parameter fitting methods for PBTK model calibration was provided by Lin et al. (2020).

Bayesian methods for PBTK model calibration have also been described. This approach requires that a prior probability distribution be developed for each of the parameters to be optimized. These prior distributions are then updated by modeling to measured toxicokinetic data using a numerical method such as Markov Chain Monte Carlo (MCMC) analysis. This procedure yields a set of posterior probability distributions for the optimized model parameters that can be evaluated to assess the quality of each fitted term (typically characterized by a mean and standard deviation) as well as potential correlation among parameter estimates. MCMC analysis has been employed in several PBTK modeling efforts with fish (Li et al., 2011; Watanabe et al., 2009).

A second well-known use of Monte Carlo analysis involves the repeated simulation of a defined exposure scenario using model parameters drawn randomly from a set of statistical distributions. This method was used by Mangold-Döring et al. (2021) to demonstrate how differences in anatomy and physiology can lead to variation in BCFs for selected chemicals. The resulting variation in predicted BCFs was then compared that of empirical BCFs for the same chemicals to examine the extent of overlap.

## Utility of PBTK Models for Fish

The first PBTK models used to simulate the kinetics of environmental contaminants in fish were developed for rainbow trout (Law et al., 1991; Nichols et al., 1990). This initial focus on rainbow trout was due to the existence of substantial physiological information for this species, and because adult trout are large enough to employ investigative methods (e.g., chronic cannulation for the collection of blood and urine samples) needed to evaluate and calibrate the models. In the years following these initial efforts, PBTK models have been provided for at least 11 other species, including channel catfish (Nichols et al., 1993), fathead minnows (Lien and McKim, 1993), Japanese medaka (Lien and McKim, 1993), brook trout (Nichols et al., 1998), Chinook salmon (Law, 1999), lake trout (Lien et al., 2001), Crucian carp (*Carassius auratus*; Yang et al., 2012), Chinese sturgeon (*Acipenser sinensis*; Wan et al., 2013), zebrafish (Péry et al., 2014), European eel (*Anguilla anguilla*; Brinkmann et al., 2015), grass carp (*Ctenopharyngodon idella*; Xu et al., 2020), and white sturgeon (*Acipenser transmontanus*; Grimard et al., 2021).

Recognizing the inherent limitations of modeling one fish species at a time, Brinkmann et al. (2016) provided a multispecies modeling framework to facilitate cross-species extrapolation efforts. A similar framework given by Gretch et al. (2019) was used to examine the effects of physiological (e.g., growth) and environmental (e.g., temperature) factors relevant to environmental risk assessment efforts. More recently, Mangold-Döring et al. (2021) described a multi-species PBTK modeling approach that randomly samples from a database of anatomical and physiological information for 69 freshwater fish species to generate a population of "virtual species objects," each representing a plausible combination of parameter values. The model was then used to generate a distribution of BCF values for each of 34 organic chemicals. Mean BCFs predicted for 26 of 34 chemicals differed by less than a factor of 10 from measured values.

PBTK models are well suited to explore the impact of biotransformation on chemical uptake and accumulation. Several authors have extrapolated measured rates of *in vitro* biotransformation to the intact tissue (e.g., liver, intestinal mucosa, gill) and incorporated this information into the model to predict *in vivo* chemical accumulation (Law et al., 1991; Stadnicka-Michalak et al., 2018). Other investigators have used PBTK models to determine apparent *in vivo* rates of activity by a process of parameter fitting (Mittal and Ng, 2018; Wang et al., 2017a). Nichols et al. (2007) employed a PBTK model to explore the potential impact of hepatic and intestinal biotransformation on accumulation of organic chemicals representing a range of log $K_{OW}$ values.

The structure of a PBTK model provides for the possibility of relating chemical time-course predictions for a specific tissue to mode of action information for a particular chemical. For example, Nichols et al. (1998) developed a model that describes the maternal transfer of TCDD to developing ovaries of female brook trout (*Salvelinus fontinalis*) resulting from an aqueous exposure. Understanding this exposure-tissue residue linkage is critical because developmental effects on developing larvae represents the most sensitive endpoint for TCDD-induced toxicity to fish. PBTK models also can be used to better relate *in vitro* and *in vivo* effects information. Using a PBTK model for rainbow trout, Brinkmann et al. (2014) showed that *in vitro* EC50 data for induction of ethoxyresorufin-O-deethylase (EROD) activity and vitellogenin production were highly correlated with *in vivo* EC50 data from controlled aqueous exposures when these latter values were expressed as predicted chemical concentrations in liver tissue. *In vitro* EC50 values showed poor correlation to *in vivo* EC50 values expressed as aqueous chemical concentrations. A similar finding was reported by Stadnicka-Michalak et al. (2014), who used a PBTK model for trout to relate *in vitro* effects data from a gill cell line to external effect concentrations and predicted concentrations in gill tissue. The highest quality correlation was achieved by relating *in vitro* effects data to modeled chemical concentrations in the gill.

PBTK models for chemicals that possess endocrine disrupting activity have been developed by several authors (Breen et al., 2013; Li et al., 2011; Watanabe et al., 2009). These models are generally structured such that chemicals accumulating in a tissue interact with molecular targets present in the same tissue, thereby providing a complete toxicokinetic/toxicodynamic description. Additional models that link predicted chemical residues to known modes of toxic action include those given by Abbas and Hayton (1997) for paraoxon in rainbow trout (resulting in acetylcholinesterase inhibition) and by Tebby et al. (2019) for rainbow trout exposed to melamine and cyanuric acid (resulting in crystal formation and subsequent renal toxicity).

Several PBTK models for fish have been developed to describe the disposition of PFAS in fish. Models given by Ng and Hungerbühler (2013) and Khazaee and Ng (2018) highlighted the importance of specific binding and renal handling (active secretion and tubular reabsorption) as determinants of PFAS kinetics in fish. Utilizing an established PBTK model for trout, Mittal and Ng (2018) investigated the biotransformation of 8:2 fluorotelomer carboxylic acid (8:2 FTCA) to perfluorooctanoic acid (PFOA) and the subsequent elimination of PFOA to urine. Vidal et al. (2019) provided a PBTK model for perfluorooctane sulfonate (PFOS) in rainbow trout. Modeled results suggested that trout eliminate PFOS primarily by excretion into bile and subsequent loss in feces.

Law (1999) used a PBTK model for Chinook salmon to predict tissue residues and withdrawal times for fish exposed orally to oxytetracycline. A PBTK model for rainbow trout was used to predict withdrawal times for enrofloxacin after intravenous, oral, or bath administration (Yang et al., 2021). Additional models have been provided to predict withdrawal times for grass carp (*Ctenopharyngodon idella*) given an intravenous injection of doxycycline (Xu et al., 2020), and tilapia (*Oreochromis niloticus*) dosed orally with florfenicol (Lin and Chen, 2021). The concern in such cases relates to the length of time that fish must be held in clean water to assure that chemical concentrations in edible tissues are below a determined maximum residue limit. Because the withdrawal time is expressed in probabilistic terms, these models were run using a Monte Carlo simulation approach that accounts for parameter uncertainty and known or hypothesized interspecies variability.

## Semi-physiological Modeling Approaches

In a traditional PBTK model, rates of chemical delivery to and loss from each tissue compartment are related to the rate of tissue blood flow. Assuming that other simplifying assumptions are met (e.g., flow-limited chemical distribution), this approach can be used to predict kinetic behavior resulting from a defined exposure in the absence of any measured data. Alternatively, models with defined tissue compartments can be developed in the absence of detailed blood flow information, provided that sufficient kinetic data are available for the modeled tissues. Using this approach, intercompartmental rate constants are developed by a process of data fitting, typically using a stepwise procedure. Generally, these rate constants are expressed as clearance terms (e.g., L/day). In most cases, however, no assumptions are made regarding their mechanistic basis. Borrowing from Gretch et al. (2016), we refer to these descriptions as *semi-physiological models*.

A partial listing of semi-physiological models for fish includes those provided by Liao et al. (2005; for arsenic in tilapia), Franco-Uría et al. (2010; for cadmium in several species), Chen and Liao (2014; for copper in tilapia), Wang et al. (2017b; for methylmercury in black seabream, *Acanthopagrus schlegelii*), Le et al. (2018; for silver in the chub, *Squalius cephalus*), Zhang et al. (2019; for cadmium and lead in zebrafish), and Wang and Wang (2020; for inorganic and methylmercury in rabbitfish, *Siganus canaliculatus*). Reviews of this modeling approach have been given by Otero-Muras et al. (2010) and Wang and Tan (2019). Most models of this type have been developed for metals and metalloids, and many can be traced back to a model given by Thomann et al. (1997) for cadmium in rainbow trout. The special utility of this approach for metals and metalloids relates to the fact that these substances often exhibit complex kinetic behavior including restricted diffusion across cell membranes, binding to protein sulfhydryl groups in tissue, blood transport in association with specific binding proteins, and biotransformation (e.g., hepatic demethylation of methylmercury). Derived intercompartmental clearance terms for each substance, species, and dosing scenario may be viewed, therefore, as representing the net result of these processes.

Semi-physiological models can be linked to *in vivo* toxicity models as a means of assessing chemical risks to fish (Liao et al., 2005) and humans (Ling et al., 2005). A model developed for one substance and species may also have utility when making predictions for similar substances and species. Disadvantages include the requirement for substantial tissue-specific data to calibrate the models and the possible existence of multiple, equally good model solutions.

## Noncompartmental Analysis

Noncompartmental analysis can be used to estimate the $CL_{WB}$, $V_{SS}$, and persistence of a chemical in the body. Because it does not rely on curve-fitting techniques to fit model-based equations to experimental data, this approach avoids several statistical considerations such as the choice of weighting function and multiple minima in the sum of squares fitting criterion. Noncompartmental analysis was used by Barron et al. (1987b) to investigate the temperature dependence of di(2-ethylhexyl)phthalate kinetics in rainbow trout. Total body clearance values calculated in this manner were very close to those estimated using a three-compartment CV model. Alavi et al. (1993) used noncompartmental analysis to estimate kinetic parameters for sulfachlorpyridazine in channel catfish. Kleinow et al. (1994) employed noncompartmental analysis to compare the kinetics of oxolinic acid in rainbow trout and channel catfish. Fitzsimmons et al. (2009) used noncompartmental analysis to evaluate the kinetics of dichloroacetic acid (DCA) in rainbow trout pre-treated with the same substance. As in mammals, pretreatment with DCA increased the persistence of a subsequently injected dose, probably due to inhibition of enzymes responsible for DCA metabolism. The application of noncompartmental analysis to bolus dosing data is described below. Modifications to this approach are required to deal with more complex dosing regimens.

### Intravascular Bolus Administration

No explicit model structure is used in noncompartmental analysis; however, the chemical must be measured in and exit the body from the pool or compartment of its introduction. The method is based on the concept of mean residence time (MRT), where the MRT is the average time a molecule of the chemical is retained in the body. Following intravascular administration of a bolus dose, the rate of chemical elimination at any time is:

$$dX/dt = -CL_{WB}C_P \tag{2.79}$$

where $X$ represents the amount of chemical in the body, $CL_{WB}$ is the plasma clearance of the chemical and $C_P$ is the plasma concentration. From this relationship, the amount of chemical eliminated over a short time interval d$t$ is:

$$dX = -CL_{WB}C_P dt \tag{2.80}$$

The derivation of an equation that yields the MRT was provided by Kleinow et al. (2008). The resulting relationship is:

$$\mathrm{MRT} = \mathrm{AUMC}/\mathrm{AUC} \tag{2.81}$$

where AUC is the area under the plasma concentration–time profile and AUMC is the area under the moment curve, which is the integral of the product of plasma concentration and time.

In an intravascular bolus dosing scenario, the MRT can be used to calculate the $V_{SS}$ using the relationship:

$$V_{SS} = \mathrm{MRT}\ CL_{WB} \tag{2.82}$$

where

$$CL_{WB} = \mathrm{dose}/\mathrm{AUC} \tag{2.83}$$

Even when the plasma concentration–time profile is complex and irregular, it is generally possible to calculate the AUC and AUMC, provided the profile is completely characterized. Methods for estimating the AUC and AUMC are described by Kleinow et al. (2008).

## Bioaccumulation Assessment and Modeling

### Bioconcentration, Bioaccumulation, and Biomagnification

In this section, chemical bioconcentration, bioaccumulation, and biomagnification in fish are given special consideration because of their importance in the context of chemical regulations implemented world-wide. Various regulatory programs (Gobas et al., 2009; Moermond et al., 2012) and international treaties, like the Stockholm Convention on Persistent Organic Pollutants (UNEP, 2001), include bioaccumulation assessment as a part of the decision-making process for permitted chemical use in society. Bioaccumulation is the net result of chemical uptake and elimination by an organism, as determined by toxicokinetic processes. There are many bioaccumulation metrics (B-metrics) and regulatory criteria for categorizing chemicals with respect to their potential for bioaccumulation. Readers interested in the development, evolution, and use of these B-metrics and regulatory criteria are directed to several reviews (Gobas et al., 2009; Matthies et al., 2016; Moermond et al., 2012).

The term *bioconcentration* refers to chemical accumulation in an aquatic organism that results from an exposure to a chemical in water only (Veith et al., 1979). Bioconcentration is best measured in well-controlled laboratory experiments such as those described in the OECD 305 Test Guideline (OECD, 2012). The OECD 305 fish bioconcentration test is recommended for chemicals with log $K_{OW}$ <6, and the primary metric of regulatory interest is the steady-state bioconcentration factor ($BCF_{SS}$; L water/kg fish). In the literature, the abbreviation $BCF_{SS}$ is often simplified to BCF, with the implied assumption of steady-state conditions.

The BCF can be calculated as the ratio of the chemical concentration in a fish ($C_{FISH}$; μg/kg) to that in water ($C_W$; μg/L):

$$\text{BCF} = C_{\text{FISH}}/C_{\text{W}} \tag{2.84}$$

provided the concentration in the fish has achieved steady state (i.e., $dC_{\text{FISH}}/dt \approx 0$) and the concentration in water has remained stable throughout the exposure period. The BCF can also be determined kinetically as a ratio of the uptake rate constant from water ($k_1$; L of water cleared of chemical/kg fish/day) and the total elimination rate constant ($k_T$; 1/day):

$$\text{BCF}_{\text{K}} = k_1/k_T \tag{2.85}$$

When kinetic data are used to estimate a BCF, the organism does not have to achieve steady state; however, the chemical concentration in water must remain stable.

The lipid content of an organism can strongly influence measured BCF values. Reported whole-body lipid contents in fish (mass basis) range from a few percent to as high as 15%–20%; the median value for fish used in bioaccumulation testing efforts is approximately 5% (Arnot and Gobas, 2006; Arnot and Quinn, 2015). To account for this variability in lipid content, the BCF can be divided by a fish's measured lipid content to calculate a "lipid-normalized" value ($BCF_L$; L water/kg lipid). For example, if the measured BCF is 100 L water/kg fish and the tested fish has a whole-body lipid content of 5%, then the $BCF_L$ is 2,000 L water/kg lipid.

For very hydrophobic organic chemicals (log $K_{OW}$ >5), binding to DOC and POC requires consideration because this sorbed fraction is not available for uptake by fish (Barron, 1990; Arnot et al., 2010; Parkerton et al., 2008; also see *Basic Concepts, Chemical Hydrophobicity and the Octanol–Water Partition Coefficient ($K_{OW}$)*). The bioavailable fraction in water ($\phi$; unitless) is equal to the ratio of freely dissolved ($C_{WD}$; μg/L) and total ($C_W$; μg/L) chemical concentrations. A log $K_{OW}$-based algorithm for the prediction of $\phi$ was provided by Arnot and Gobas (2003):

$$\phi = 1/\left(1 + \chi_{\text{POC}}\alpha_{\text{POC}}K_{\text{OW}} + \chi_{\text{DOC}}\alpha_{\text{DOC}}K_{\text{OW}}\right) \tag{2.86}$$

where $\chi_{POC}$ (kg/L) and $\chi_{DOC}$ (kg/L) are the concentrations of POC and DOC in the water, respectively, and $\alpha_{POC}$ and $\alpha_{DOC}$ are unitless proportionality constants that describe the similarity in phase partitioning of POC and DOC in relation to that of octanol. Estimates of $\alpha_{POC}$ (0.35) and $\alpha_{DOC}$ (0.08) have been provided by Seth et al. (1999) and Burkhard (2000), respectively.

The importance of bioconcentration as a measure of chemical accumulation was recognized in the 1960s (Hamelink et al., 1971). In the 1970s, it became apparent that dietary uptake was also an important route of exposure, particularly for very hydrophobic (log $K_{OW}$ >5) chemicals (Bruggeman et al., 1981, 1984). The term *bioaccumulation* is used to refer to chemical accumulation resulting from all possible routes of exposure, as occurs in the natural environment. This usage is somewhat confusing since the same term may be employed in a broader sense to refer to any type of chemical accumulation (including bioconcentration); hence the expression *bioaccumulation assessment*. Nevertheless, this usage persists.

The bioaccumulation factor (BAF; L water/kg fish) is calculated as:

$$\text{BAF} = C_{\text{FISH}}/C_{\text{W}} \tag{2.87}$$

For field measurements, it is typically assumed that chemical concentrations in the fish and sampled exposure media (water, sediment, food) are not changing significantly over time; that is, steady state is approximated. In most cases, it is not feasible to statistically test this assumption due to the significant number of samples required; however, it is important to keep this assumption in mind when considering field B-metrics, like the BAF, particularly when sample sizes are small and chemical concentrations are near or below the limit of quantification. Bioaccumulation in the environment can also be referenced to a chemical concentration in sediment using a biota-sediment accumulation factor (BSAF; kg sediment/kg fish). The BSAF is calculated as:

$$\text{BSAF} = C_{\text{FISH}}/C_{\text{SEDIMENT}} \tag{2.88}$$

Fish lipid content and the nature of the sampled reference phase can have a large impact on calculated BAFs and

BSAFs. When examining this type of data, it is important to know whether the chemical concentration in a fish is expressed in terms of wet weight or lipid weight. For BAF calculations, a reported water concentration could refer to that in bulk water or a freely dissolved concentration. For BSAFs, measured concentrations in sediments may be referenced to sample wet weight, dry weight, or a measured organic carbon content. A critical review of several thousand BCFs and BAFs for fish and other aquatic organisms (Arnot and Gobas, 2006) highlights issues that require consideration when using this information for chemical assessments and model development. A critical review of measured BSAFs has also been provided (Burkhard et al., 2012a).

Another important B-metric is the biomagnification factor (BMF; kg diet/kg fish). A BMF can be calculated from field data as the ratio of the chemical concentration in a fish to that of its diet:

$$\mathrm{BMF} = C_{\mathrm{FISH}}/C_{\mathrm{DIET}} \tag{2.89}$$

Alternatively, a BMF can be determined from laboratory test data (OECD, 2012). Laboratory BMFs are typically calculated as the ratio of a dietary uptake rate constant ($k_D$; kg food cleared of chemical/kg fish/day) and $k_T$:

$$\mathrm{BMF_K} = k_D/k_T \tag{2.90}$$

It is important to note, however, that laboratory BMFs differ from field BMFs insofar as fish in the environment are exposed simultaneously to chemicals in water and in the diet.

A chemical in any system will seek to achieve a minimum free energy. When this condition is met, the chemical is at equilibrium and the chemical fugacity (activity or potential) will be equal in all components of the system. Field studies have shown, however, that the chemical fugacity in a high trophic level fish can exceed that in surrounding water by several orders of magnitude (Connolly and Pedersen, 1988; Kelly et al., 2007). Fish BMFs may be used to determine if chemical fugacity is increasing or decreasing between a consumer and its diet (Burkhard et al., 2012b; Connolly and Pedersen, 1988). For neutral organic chemicals, this assessment can be performed using BMFs expressed on a lipid-weight basis. However, when the organism lipid content is low (e.g., <~3%), other biological components like proteins and carbohydrates can contribute substantially to the estimation of chemical fugacity. In such cases, a "lipid-equivalent" normalization procedure that considers the contribution of lipid and nonlipid organic matter is recommended (deBruyn and Gobas, 2007; Kelly et al., 2007; Mackintosh et al. 2004). Another more explicit approach is to express the BMF as a fugacity ratio (Burkhard et al. 2012b). A chemical is said to have undergone biomagnification if the lipid normalized or fugacity-based BMF is >1.

Chemical biomagnification is thus an anomaly in nature and a primary concern when conducting bioaccumulation assessments. The process by which biomagnification occurs was described earlier in this chapter (see *Xenobiotic Absorption* and *Physiologically based Toxicokinetic Models*). Biotransformation and/or limitations on chemical uptake from the gastrointestinal tract can mitigate this process. A critical review of several hundred measured BMFs, $E_D$ estimates, and $k_T$ values for fish (Arnot and Quinn, 2015) highlights issues that require consideration when using measured BMFs for chemical assessment and model development.

The trophic magnification factor (TMF) is a B-metric that characterizes chemical accumulation across an entire food web. To determine a TMF, measured chemical concentrations in biota that comprise a food web are expressed as lipid-normalized values or fugacities. The resulting values are $\log_{10}$-transformed and regressed against the relative trophic positions of the sampled organisms. The TMF is then estimated from the slope ($x$) of the regression equation as $10^x$. Calculated in this manner, the TMF is essentially an "average" of the BMFs across all sampled trophic positions (Burkhard et al., 2013; Gobas et al., 2009). A chemical with a TMF >1 is generally considered bioaccumulative. The TMFs for some persistent organic chemicals have been reasonably well characterized (Walters et al., 2011); however, uncertainties remain with respect to TMF calculation and interpretation (Burkhard et al., 2013; Mackay et al., 2016). Furthermore, TMFs are very resource intensive due to the large sampling requirement. Figure 2.9 illustrates theoretical relationships between the BCF, BAF, BMF, and TMF. These relationships have also been described mathematically (Mackay et al., 2013).

## One-Compartment Bioaccumulation Model for Fish

The B-metrics used in regulatory decision-making are generally based on whole-body chemical concentrations and not the concentration present in a particular tissue (OECD, 2012). The rationale behind this approach is that organisms in the environment (e.g., a fish) typically consume whole organisms in their diet. Moreover, these B-metrics generally assume steady-state conditions implying, in turn, that the chemical exposure has taken place over a long period of time. In a long-term exposure, complexity associated differences in chemical distribution to different tissues largely disappear and an internal steady state becomes established. Under these conditions, the chemical kinetics in all tissues can be adequately represented by a one-compartment model (Nichols et al., 2007).

One-compartment models used in bioaccumulation assessments for fish have evolved over several decades beginning with the laboratory-based BCF models developed by Branson, Neely, and colleagues (Neely et al., 1974; Branson et al., 1975). These models were adapted by Bruggeman and colleagues (Bruggeman et al., 1981, 1984) to describe chemical bioaccumulation and biomagnification. One-compartment models for fish that occupy different trophic levels were then linked together to describe chemical bioaccumulation and biomagnification in aquatic

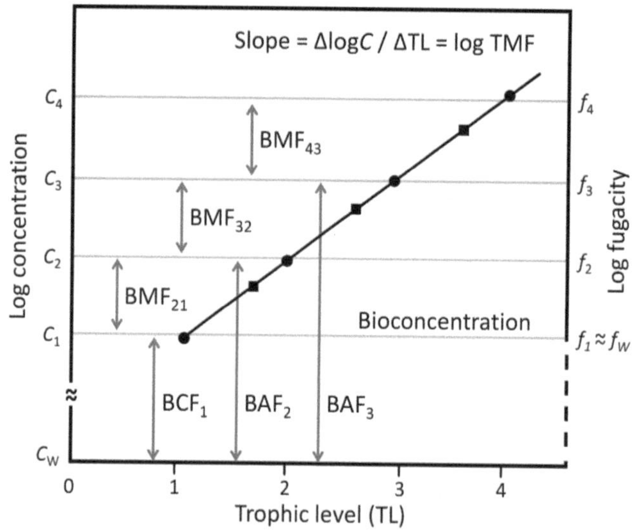


**FIGURE 2.9** Relationships between bioconcentration, bioaccumulation, biomagnification, and trophic magnification expressed as chemical concentrations and fugacities. The round symbols are for a linear food web, and the square symbols represent a nonlinear food web with multiple dietary items. BCF=bioconcentration factor, BAF=bioaccumulation factor, BMF=biomagnification factor, TMF=trophic magnification factor. Subscripts denote tropic levels (single number) or a difference between two trophic levels (two numbers; higher tropic level first). Adapted from Mackay et al. (2013).

food webs (Arnot and Gobas, 2004; Barber et al., 1991; Connolly and Pedersen, 1988; Gobas, 1993; Morrison et al., 1996; Thomann and Connolly, 1984; Thomann et al., 1992). Early models tended to rely on empirical, $K_{OW}$-based relationships to predict uptake and elimination rate constants (also see *Utility of Compartmental Models for Fish*). With time and a greater understanding of underlying processes, these models have become increasingly "physiological," with the result that uptake and elimination rate constants are predicted in part from knowledge of fish respiratory and gastrointestinal physiology. This mechanistic foundation confers greater flexibility to model fish of difference sizes with different attributes (e.g., dietary preferences) inhabiting a broader range of habitats (e.g., different temperatures). For a detailed description of models used to simulate chemical bioconcentration and bioaccumulation in fish, the reader is referred to the reviews provided by Barber (2003, 2008).

The following text describes a one-compartment bioaccumulation model given by Arnot and Gobas (2004). Since its introduction, this model has been amended to improve the prediction of gill ventilation volume (Arnot et al., 2008a). Nevertheless, core features remain largely unchanged and serve to illustrate current approaches to model development, parameterization, and use.

A general toxicokinetic mass-balance model for fish may be given as (Arnot and Gobas, 2004):

$$dM_F/dt = W_F\left(k_1C_{WD} + k_DC_D\right) - \left(k_2 + k_E + k_B\right)M_F \quad (2.91)$$

where $M_F$ (μg) is the mass of chemical in the fish, $W_F$ (kg) is fish weight, $k_1$ (L water/kg fish/day) is the gill uptake rate constant, $k_2$ (1/day) is the gill elimination rate constant, $k_D$ (kg food/kg fish/day) is the dietary uptake rate constant, $k_E$ (1/day) is the fecal egestion rate constant, $k_B$ (1/day) is the biotransformation rate constant, $C_{WD}$ (μg/L) is the freely dissolved (bioavailable) chemical concentration in the water, and $C_D$ (μg/kg food) is the total chemical concentration in the diet. The total elimination rate constant $k_T$ is the sum of the individual elimination rate constants:

$$k_T = k_2 + k_E + k_B \quad (2.92)$$

If $k_T$ is known, an elimination half-life ($t_{1/2}$; day) may be calculated as:

$$t_{1/2} = \ln 2/k_T = 0.693/k_T \quad (2.93)$$

The time to approach 95% of steady state is approximately $4k_T$. When a fish is near steady state, $dM_F/dt \approx 0$ and Equation 2.91 simplifies to:

$$W_F\left(k_1C_{WD} + k_DC_D\right) = \left(k_2 + k_E + k_B\right)M_F \quad (2.94)$$

Rearranging then gives:

$$M_F/W_F = C_F = \left(k_1C_{WD} + k_DC_D\right)/\left(k_2 + k_E + k_B\right) \quad (2.95)$$

where $C_F$ (μg/kg) is the chemical concentration in the fish.

Another process that should be considered when implementing steady-state calculations is the change in body mass, that is, fish growth. Fish growth is often considered to be a "pseudo-elimination" process because, although the chemical is not technically eliminated from the organism, the concentration in the organism is clearly a function of body mass. The effect of fish growth on modeled chemical concentrations is most pronounced for persistent and very hydrophobic chemicals because the other elimination rate constants tend to be very small. This issue is particularly relevant when examining laboratory test data because the fish are often well-fed and can grow rapidly. To address this issue, OECD testing guidance (OECD, 2012) requires that data be collected to estimate a growth rate constant ($k_G$; 1/day) when conducting BCF and BMF tests. If this growth rate is known, the appropriate steady-state calculation is:

$$C_F = (k_1 C_{WD} + k_D C_D)/(k_2 + k_E + k_B + k_G) \quad (2.96)$$

When exposure is from water only (as in a BCF test), Equation 2.96 simplifies to:

$$C_F = (k_1 C_{WD})/(k_2 + k_E + k_B + k_G) \quad (2.97)$$

When exposure is from the diet only (as in a dietary bioaccumulation test), the appropriate expression is:

$$C_F = (k_D C_D)/(k_2 + k_E + k_B + k_G) \quad (2.98)$$

Equations 2.96–2.98 can be rearranged to calculate the BCF, BMF, and BAF as a function of competing uptake and elimination rate constants. Each of these B-metrics effectively normalizes the chemical concentration in the fish to that in the exposure medium of interest. For example, the BCF based on the dissolved chemical concentration in water is $C_F/C_{WD}$. By rearranging, Equation 2.97 gives:

$$\text{BCF} = C_F/C_{WD} = k_1/(k_2 + k_E + k_B + k_G) = k_1/k_T \quad (2.99)$$

Using the same approach, the BMF may be calculated as:

$$\text{BMF} = C_F/C_D = k_D/k_T \quad (2.100)$$

and the BAF as:

$$\text{BAF}_{WD} = C_F/C_{WD} = (k_1 + k_D C_D/C_{WD})/k_T \quad (2.101)$$

For the BCF and BAF, equivalent expressions referenced to the total chemical concentration in water ($C_{WT}$; µg/L) may be written as:

$$\text{BCF} = k_1\,\phi/k_T \quad (2.102)$$

and,

$$\text{BAF}_{WT} = C_F/C_{WT} = (k_1\phi + k_D C_D/C_{WT})/k_T \quad (2.103)$$

Following the general approach outlined by Arnot and Gobas (2004), $k_1$ can be estimated as:

$$k_1 = E_W G_V/W_F \quad (2.104)$$

where $E_W$ (unitless) is the chemical transfer efficiency at the gill and $G_V$ (L water/day) is the ventilation rate. Direct measurements of $E_W$ in fish have been obtained for neutral organic chemicals across a range of $K_{OW}$ values (Figure 2.2). An empirical relationship that describes the observed $K_{OW}$ dependence of these measurements was provided by Gobas and Mackay (1987):

$$E_W = 1/(1.85 + (155/K_{OW})) \quad (2.105)$$

The mathematical form of this relationship is consistent with the lipid–water two-phase resistance model of membrane flux proposed Flynn and Yalkowsky (1972; see *Absorption, Distribution, and Excretion, Xenobiotic Absorption, The Gills*). To estimate $G_V$, Arnot and Gobas (2004) employed an allometric equation that was based on a review of oxygen consumption rate ($VO_2$) and $G_V$ data for fish and other aquatic species:

$$G_V = 1400 W_F^{0.65}/C_{OX} \quad (2.106)$$

Here, $C_{OX}$ (mg/L) is the dissolved $O_2$ concentration in water. This equation effectively estimates $G_V$ as the amount of water that must be respired by a fish to meet its biological $O_2$ demand.

In detail, this description of chemical uptake across the gills differs from the gill models proposed by Hayton and Barron (1990) and Erickson and McKim (1990a, 1990b), which were described earlier in this chapter (Equations 2.25 and 2.62–2.67). However, all three models share as a common feature a term that describes the rate of water flow across the gills, which establishes an upper limit on uptake. Additional terms then operate against this upper limit to yield the observed $K_{OW}$ dependence of uptake. The gill models provided by Hayton and Barron (1990) and Erickson and McKim (1990a, 1990b) have substantial utility from the perspective of understanding how different physiological (e.g., gill surface area, cardiac output, ventilation volume) and chemical (hydrophobicity, membrane diffusivity) attributes can impact chemical uptake. For screening-level bioaccumulation assessments, however, the relative simplicity of the description given by Arnot and Gobas (2004) is highly advantageous.

Once an estimate of $k_1$ has been obtained, $k_2$ may be estimated as:

$$k_2 = k_1/K_{FW} \quad (2.107)$$

where $K_{FW}$ is an equilibrium fish–water chemical partition coefficient (L water/kg fish). For neutral organic chemicals, the $K_{FW}$ may be approximated as:

$$K_{FW} = \nu_{LF} K_{OW} + \nu_{NF}\beta K_{OW} + \nu_{WF} \quad (2.108)$$

where $\nu_{LF}$, $\nu_{NF}$, and $\nu_{WF}$ (all unitless) are the fractional lipid, nonlipid organic matter (NLOM; assumed to be mostly protein), and water content of the fish, respectively, and $\beta$ (unitless) is a proportionality constant that relates chemical solubility in *n*-octanol to that in protein. In the model given by Arnot and Gobas (2004), $\beta$ was set to 0.035. Subsequent work has shown that $\beta$ is closer to 0.05 (deBruyn and Gobas, 2007).

The dietary uptake rate constant $k_D$ was calculated by Arnot and Gobas (2004) as:

$$k_D = E_D G_D / W_F \quad (2.109)$$

where $E_D$ (unitless) is the chemical transfer efficiency between contents of the GIT and the organism and $G_D$ (kg food/day) is the feeding rate. There are about 1,200 measurements of $E_D$ in fish for about 400 chemicals (e.g., Figure 2.3). The uncertainty in these measurements is relatively high, in part because of the variability in experimental methods (Arnot and Quinn, 2015; Barber, 2008). Nevertheless, several empirical relationships between $E_D$ and $K_{OW}$ have been developed (Barber, 2008). Mechanistic models that examine more explicitly the relationships between $E_D$ and chemical properties have also been provided (Arnot and Mackay, 2018). The bioaccumulation model given by Arnot and Gobas (2004) employs measured $E_D$ values when available. In the absence of a measured value, $E_D$ is estimated as:

$$E_D = 1/\left(2.0 + 3.0 \times 10^{-7} K_{OW}\right) \quad (2.110)$$

Equation 2.110 has the same mathematical form as Equation 2.105 and is consistent with the lipid–water two-phase resistance model of membrane flux.

The fecal egestion rate constant ($k_E$) was calculated by Arnot and Gobas (2004) as:

$$k_E = E_D G_E K_{GF} / W_F \quad (2.111)$$

where $G_E$ is the fecal egestion rate (kg feces/day) and $K_{GF}$ (kg/kg) is an equilibrium chemical partition coefficient that describes the sorptive capacity of the fish relative to that of the gut contents (feces) following digestion and absorption of dietary lipids, NLOM, and water. The $G_E$ is a function of $G_D$ and the digestibility of the diet according to:

$$G_E = G_D\left((1-\varepsilon_L)v_{LD} + (1-\varepsilon_N)v_{ND} + (1-\varepsilon_W)v_{WD}\right) \quad (2.112)$$

where $\varepsilon_L$, $\varepsilon_N$, $\varepsilon_W$ are the dietary absorption efficiencies of lipid, NLOM, and water, respectively, and $\nu_{LD}$, $\nu_{ND}$, and $\nu_{WD}$ are the fractions of lipid, NLOM, and water in the diet, respectively. The $K_{GF}$ was calculated as:

$$K_{GF} = \left(v_{LG}K_{ow} + v_{NG}\beta K_{OW} + v_{WG}\right) / \left(v_{LF}K_{OW} + v_{NF}\beta K_{OW} + v_{WF}\right) \quad (2.113)$$

where and $\nu_{LG}$, $\nu_{NG}$, and $\nu_{WG}$ are the fractions of lipid, NLOM, and water in the GIT after digestion (unitless). The fractions of lipid, NLOM, and water in the gut are estimated from:

$$v_{LG} = (1-\varepsilon_L)v_{LD} / \left((1-\varepsilon_L)v_{LD} + (1-\varepsilon_N)v_{ND} + (1-\varepsilon_W)v_{WD}\right) \quad (2.114)$$

$$v_{NG} = (1-\varepsilon_N)v_{ND} / \left((1-\varepsilon_L)v_{LD} + (1-\varepsilon_N)v_{ND} + (1-\varepsilon_W)v_{WD}\right) \quad (2.115)$$

$$v_{WG} = (1-\varepsilon_W)v_{WD} / \left((1-\varepsilon_L)v_{LD} + (1-\varepsilon_N)v_{ND} + (1-\varepsilon_W)v_{WD}\right) \quad (2.116)$$

When the model given by Arnot and Gobas (2004) is used to perform dynamic simulations of chemical accumulation over time, specific growth information is required to account for changes in body mass as well as changes in physiological processes that change with body size (e.g., $G_V$, as calculated by Equation 2.106). For steady-state solutions, a fixed growth rate constant ($k_G$) may be employed. The only input to Equation 2.91 not yet discussed is $k_B$. Biotransformation may substantially reduce the extent of chemical bioaccumulation in fish and is viewed as the principal source of uncertainty in many bioaccumulation assessments. *In vivo*, *in silico*, and *in vitro* methods for the estimation of $k_B$ are described below.

The Arnot and Gobas (2004) model was developed and evaluated for neutral organic chemicals. However, many contaminants ionize at environmental pH values (Armitage et al., 2017). To address this challenge, Armitage et al. (2013) modified the model given by Arnot and Gobas (2004) to simulate the bioconcentration of ionizable organic chemicals (IOCs). Mechanistic models for branchial uptake of IOCs must account for differences in the behavior of neutral and ionized chemical species. Complicating matters further, the fractional amount of neutral and ionized species in bulk water, at the gill surface, and within the organism may differ due to differences in pH. For highly dissociated IOCs, passive diffusion by a paracellular route (i.e., through gill epithelial tight junctions) may also be relevant. A sophisticated mechanistic model for branchial uptake of IOCs was provided by Erickson et al. (2006a, 2006b). Drawing on principles identified by Erickson et al. (2006a, 2006b), Armitage et al. (2013) amended the gill model given by Arnot and Gobas (2004) to predict the branchial uptake efficiency ($E_W$) of IOCs. The BCF model provided by Armitage et al. (2013) explicitly considers chemical partitioning to adipose lipid, membrane phospholipids, structural protein, and serum albumin. The separate consideration of membrane phospholipids is particularly relevant because cell membranes may serve as the dominant sorption phase for some IOCs (Escher et al., 2000; Timmer and Droge, 2017).

## A Weight of Evidence Approach for Bioaccumulation Assessment

Bioaccumulation assessments use a variety of methods, data, and categorization criteria. Uncertainty in the data used in these assessments exists whether the data are measured in laboratory experiments, determined from field measurements, or estimated with models. A weight of evidence (WoE) approach is therefore recommended in most regulatory programs to address uncertainty in decision-making. The WoE approach considers known lines of evidence (LoE) where a "weight" is assigned to each LoE based on the confidence (reliability) associated with the evidence (OECD, 2019). Evidence is combined and the overall strength of evidence (SoE) is determined to support or refute a hypothesis or question (e.g., is chemical "X" bioaccumulative?). The Bioaccumulation Assessment Tool (BAT) developed by Arnot and co-workers (Arnot et al., 2022) provides a formal WoE approach for bioaccumulation assessment that is consistent with OECD WoE principles (OECD, 2019). This tool guides an evaluator through the process of collecting, generating, and critically evaluating the reliability of various LoE required for assessing the bioaccumulation of neutral and ionizable organic chemicals in aquatic and terrestrial organisms. Included in the BAT are one-compartment toxicokinetic models for fish that derive from the Arnot and Gobas (2004) model as generally described above. The toxicokinetic models built into the BAT are also available online in EAS-E Suite (www.eas-e-suite.com).

Additional online tools can be used to address data gaps and provide information for chemical bioaccumulation assessments. For example, Gobas and co-workers developed the ADME-B(ioaccumulation) Calculator (Gobas et al., 2020), which is a toxicokinetic analysis tool to aid in the interpretation of data from dietary bioaccumulation tests with fish (OECD, 2012). Charles and co-workers developed the Modeling and StAtistical tools for ecotoxICology (MOSAIC) platform for quantifying key parameters following standard toxicokinetic and bioaccumulation experiments in aquatic organisms (Ratier et al., 2021). Krause and co-workers developed B-compass Fish and associated tools to estimate chemical bioconcentration factors in fish (Krause and Goss, 2020b). Like the BAT, B-compass Fish includes an option to incorporate data from *in vitro* biotransformation experiments.

## Estimating an Apparent Whole-Body Biotransformation Rate Constant ($k_B$) for Fish

Biotransformation was highlighted in the 1970s as a critical determinant of chemical bioaccumulation and toxicity in fish (Lech and Bend, 1980). Studies since have demonstrated the impact of biotransformation on chemical toxicokinetics and bioaccumulation, particularly for hydrophobic substances that possess high bioaccumulation potential due to their partitioning properties (Clark et al., 1990; de Wolf et al., 1993; Oliver and Niimi, 1985; Southworth et al., 1980). In a one-compartment bioaccumulation model for fish, biotransformation is generally represented as a first-order, whole-body biotransformation rate constant ($k_B$). Implied by this approach is an assumption that contaminant concentrations in fish that result from environmental exposures do not saturate enzymatic biotransformation pathways. Here, we describe model-based methods for estimating $k_B$ from *in vivo* and *in vitro* data, as well as Quantitative Structure Activity Relationships (QSARs) that predict $k_B$ from chemical structure and physicochemical properties.

### Estimating $k_B$ from *In Vivo* Data

Several authors have measured metabolic products retained by fish or eliminated in bile, urine, feces, and exposure water, and then used this information to estimate *in vivo* rates of biotransformation (Bradbury et al., 1986, 1993; Cravedi et al., 1999; McKim et al., 1986; Stehly and Hayton, 1989). In practice, however, these measurements are difficult to make due to low metabolite concentrations and incomplete extraction of samples. Alternatively, model-based approaches may be used to estimate an apparent *in vivo* $k_B$ from the impact that this activity has on the kinetics and/or accumulation of a parent chemical substance. One such approach involves an analysis of measured whole-body chemical residues during the depuration phase of an uptake/depuration study protocol (Arnot et al., 2008a; de Wolf et al., 1993; Fisk et al., 2000; Gert-Jan de Maagd et al., 1998; Konwick et al., 2006; Sijm et al., 1990; Tolls et al., 2000b). In this case, an apparent $k_B$ is calculated as the difference between a measured whole-body elimination rate constant ($k_T$) and an elimination rate constant ($k_X$) estimated for the same chemical, absent biotransformation:

$$k_B = k_T - k_X \tag{2.117}$$

An estimate of $k_X$ may be obtained by summing the elimination rate constants $k_2$, $k_E$, and $k_G$, calculated as described above. Alternatively, measured $k_X$ values for a set of recalcitrant chemicals can be regressed against log $K_{OW}$ to provide a predictive relationship (Lo et al., 2015; Saunders et al., 2020a). Ideally, this regression would be developed using data for chemicals incorporated into the same study. The accuracy of a $k_B$ value determined in this manner depends on both the reliability of the measured $k_T$ value and the validity of the approach used to estimate $k_X$.

A second approach, developed by Arnot et al. (2008a), involves the estimation of $k_B$ from a measured BCF value. In this case, the relevant equation is:

$$k_B = \left(k_1\,\phi/\text{BCF}\right) - k_X \tag{2.118}$$

where $k_X$ is defined as above. The first term on the right-hand side of this equation, $k_1\phi/\text{BCF}$, provides an alternative means of estimating $k_T$ (see Equation 2.102). In essence, Equation 2.117 estimates $k_B$ as the biotransformation rate constant required to "explain" an observed $k_T$ while Equation 2.118 estimates $k_B$ as the biotransformation rate constant required to "explain" a measured BCF. Arnot et al. (2008a) employed Equations 2.117 and 2.118 to evaluate $k_T$ and/or BCF data for a selected set of test chemicals. An uncertainty analysis, performed as part of this analysis, was used to propagate the uncertainty in measured and predicted data (e.g., $k_T$ and $k_X$, respectively) into estimated $k_B$ values to characterize the variance in these calculated terms and provide a basis for evaluating the domain of applicability of the method. These *in vivo* $k_B$ estimation procedures were then applied to a set of critically evaluated laboratory test data to develop a database of approximately 700 $k_B$ values (Arnot et al., 2008b). The results of this effort indicated that *in vivo* $k_B$ values in fish span approximately 6 orders of magnitude.

### Quantitative Structure Activity Relationships (QSARs) for the Prediction of $k_B$

The *in vivo* $k_B$ database developed by Arnot et al. (2008b) is notable for its size and the diversity of represented chemical structures. Using this information, several groups have developed QSARs that predict a chemical's biotransformation half-life (abbreviated $HL_B$, where $HL_B = \ln 2/k_B$) in fish. Initially, a multiple linear regression model was derived by the method of ordinary least squares using 59 descriptors: $K_{OW}$, molecular mass, and 57 molecular fragments (Arnot et al., 2009). The selected molecular fragments were based on those used by previous authors to predict chemical biodegradation by microorganisms (see Arnot et al., 2009 for citations). A second fragment-based $HL_B$-QSAR was developed by Brown et al. (2012). For this effort, however, the fragments were not selected by expert knowledge but rather by using an Iterative Fragment Selection (IFS) method. The IFS method selects the best fragments to fit the data using an automated process of fragment selection and model testing. A third $HL_B$-QSAR was developed using a combination of global descriptors and fragments (Papa et al., 2014). The advantage of using global descriptors is that they encode for holistic structural information considering the whole molecular structure and various intermolecular interactions, which cannot be captured using fragments. The statistical performance of these three $HL_B$-QSARs was shown to be similar; however, the total number of descriptors employed by Papa et al. (2014) was substantially smaller than that used by Arnot et al. (2009) or Brown et al. (2012). Papa et al. (2014) also showed that a consensus-based QSAR modeling approach provided improved predictions of $k_B$ when compared to using individual models separately. Mansouri et al. (2018) developed a fish $HL_B$-QSAR using a *k*-Nearest Neighbors (*k*-NN) algorithm. Molecular descriptors were developed using PaDEL, an open source software. The *k*-NN algorithm was then used to select an optimal model. All the QSARs described above were developed following OECD guidance for the validation of QSARs used in regulatory decision-making (OECD, 2007) and may be accessed through EAS-E Suite (www.eas-e-suite.com). Another fish $HL_B$-QSAR was developed by Kuo and Di Toro (2013) using Abraham solvation descriptors. Abraham descriptors are calculated directly from molecular structure and describe chemical partitioning as a linear combination of different intermolecular interactions. This approach assumes that internal chemical bioavailability substantially determines *in vivo* rates of chemical biotransformation.

### Estimating $k_B$ from *In Vitro* Data and *In Vitro*-to-*In Vivo* Extrapolation

As an alternative to QSARs for the prediction of $k_B$, an estimate of $k_B$ may be obtained by extrapolating a measured rate of *in vitro* intrinsic clearance ($CL_{IN\ VITRO,INT}$; µL/min/mg protein or $10^6$ cells) to the intact animal (Nichols et al., 2006; Cowan-Ellsberry et al., 2008). The methods used to perform this extrapolation are based on procedures developed by the pharmaceutical industry for preclinical screening of drug candidates (Rodrigues, 1997) and rely on the measurement of *in vitro* activity under first-order (nonsaturating) conditions (i.e., when the substrate concentration $C_S$ is $<< K_M$). Measurements of $CL_{IN\ VITRO,INT}$ can be made by quantifying the appearance of metabolic products, but this only provides information for one biotransformation pathway. By measuring the disappearance of chemical substrate, one can account for all pathways operating against the chemical in the selected *in vitro* test system (Iwatsubo et al., 1997; Obach et al., 1997). This substrate depletion approach is especially relevant when the study goal is to evaluate the potential for biotransformation to impact bioaccumulation of the parent chemical.

Generally, these measurements are made using an *in vitro* system derived from liver tissue (e.g., liver microsomes, S9 fraction, or isolated hepatocytes). Implied by this approach is an assumption that the liver is the primary site of biotransformation. If the assumption of first-order conditions is met, natural log-transformed substrate concentrations will tend to decline linearly over time. The slope of the line describe by these data yields a first-order elimination rate constant ($k_{EL}$; 1/min). Division of $k_{EL}$ by the *in vitro* concentration of protein or cells then provides an estimate of $CL_{IN\ VITRO,INT}$. Scaling factors (e.g., mg S9 protein/g liver) are used to extrapolate the $CL_{IN\ VITRO,INT}$ to an estimate of *in vivo* intrinsic clearance, expressed on a whole-organ basis ($CL_{IN\ VIVO,INT}$; L/h) or as the whole-organ clearance normalized to body weight (L/h/kg). This is a type of *in vitro*-to-*in vivo* extrapolation, commonly abbreviated IVIVE. The resulting estimate of $CL_{IN\ VIVO,INT}$ is then be used as an input to the physiological model of $CL_H$ given previously (Equation 2.16).

Because the *in vitro* systems used to make these measurements consist largely of water, early users of this method often assumed that $CL_{IN\ VIVO,INT}$ values obtained by IVIVE provide a good estimate of the unbound intrinsic clearance ($CL_{INT,U}$ in Equations 2.13, 2.15, 2.16 and 2.18). Subsequent work has shown that substantial binding can occur even in these dilute test systems and that prediction of $CL_H$ is improved by accounting for these binding effects (Obach et al., 1997). This is generally done by dividing the term $f_{U,B}$ in Equation 2.16 by a second term that describes *in vitro* chemical binding (e.g., $f_{U,MIC}$, $f_{U,S9}$ or $f_{U,HEP}$). An additional adjustment factor ($f_W = 0.99/0.84 = 1.18$) has been recommended to account for the difference in water content of blood (approx. 84%) and the *in vitro* measurement system (approx. 99%; Krause and Goss, 2018). For an S9 system, the amended version of Equation 2.16 becomes:

$$CL_H = Q_L f_U f_W CL_{IN\ VIVO,INT} / \left(Q_L + f_U f_W CL_{IN\ VIVO,INT}\right) \tag{2.119}$$

where $f_U = f_{U,B}/f_{U,S9}$.

An estimate of $k_B$ is obtained by dividing the derived $CL_H$ value by a chemical's apparent volume of distribution ($V$), referenced to the chemical concentration in blood

(Nichols et al., 2006). For a one-compartment model, $V$ may be estimated as the ratio of the fish–water partition coefficient ($P_{FW}$) and a blood–water partitioning value ($P_{BW}$). The $P_{FW}$ is the BCF (typically represented as $BCF_P$) that would be predicted from simple partitioning considerations, absent biotransformation and chemical egestion in feces. Empirical and composition-based equations that predict the $BCF_P$ and $P_{BW}$ from chemical log $K_{OW}$ have been provided by several authors (Fitzsimmons et al., 2001; Nichols et al., 2006; Han et al., 2007; Arnot and Gobas, 2004).

Using the forgoing methods, several research groups have extrapolated measured $CL_{IN\ VITRO,INT}$ values to the whole animal as a means of refining modeled BCFs predictions. BCFs predicted in this manner tend to be much closer to empirical values than are BCFs predicted in the absence of biotransformation (Cowan-Ellsberry et al., 2008; Han et al., 2007; Laue et al., 2014; Saunders et al., 2019). Additional modifications to these procedures have been suggested to account for biotransformation occurring in extrahepatic tissues such as the gills and intestinal epithelium (Saunders et al., 2020b; Krause and Goss, 2020a). Simplified approaches that ignore the rate limitation imposed by liver blood flow have also been provided (Trowell et al., 2018; Krause and Goss, 2020a).

The $CL_{IN\ VIVO,INT}$ calculated as above may also be used as an input to a fish PBTK model to develop a clearance expression for the liver (Nichols et al., 2006). For this application, the product term $f_W CL_{IN\ VIVO,INT}$ would be substituted for the product term $k_{FC}V_I$ in Equation 2.76. If chemical binding in blood and in the *in vitro* test system differed substantially, a correction factor like that described above would also be required. This extrapolation would not involve the consideration of liver blood flow, however, since this is already accounted for in the mass-balance description for liver tissue (Equation 2.75).

### Domain of Applicability of *In Vivo* and *In Vitro* Methods for $k_B$ Estimation

The approximate domains of applicability (DOAs) of *in vivo* and *in vitro* methods for the estimation of $k_B$ are shown in Figure 2.10 using gray shading and hatched lines, respectively. The methods used to estimate $k_B$ from *in vivo* data employ both measured data ($k_T$ or a BCF) and modeled kinetic rate constants ($k_X$, itself a function of $k_1$, $k_2$, and $k_E$). In general, there is greater certainty in the estimate of $k_B$ when $k_T$, determined directly or estimated from the BCF, is substantially larger than $k_X$ (Arnot et al., 2008a). Lower certainty in the $k_B$ estimate exists when $k_T$ values approach $k_X$. To better define the DOA of these methods, $k_B$ values were calculated as a function of $k_X$ to illustrate points the plot where $k_B$ is equal to 10 $k_X$ (upper curve), $k_X$ (middle curve), and 0.1 $k_X$ (lower). To perform this assessment, $k_X$ values were generated using the IVIVE/BCF model given by Nichols et al. (2013) assuming no growth. This model assumes a 10 g trout containing 5% whole-body lipid, living at 15°C. The lower margin of the gray shaded area corresponds to points where $k_B = 0.3\ k_X$.

Practical limitations on the use of *in vitro* metabolizing systems substantially determine their DOA with respect to $k_B$ estimation. When dealing with low rates of biotransformation, the principal challenge is to demonstrate a significant difference in "depletion" slopes for active samples and inactivated (negative) controls, incorporated to account for possible abiotic loss processes (e.g., adsorption). The lowest detectable rate of activity (assay sensitivity) is determined, therefore, by sample replication, the total number of sampling time points, and analytical variability. Increasing the total assay run time may increase assay sensitivity. However, all *in vitro* systems have a limited working lifetime. Current experience indicates that the lowest detectable depletion rate constant ($k_{EL}$) for a standard liver S9 assay (OECD, 2018) is about 0.15/h (Chen et al., 2016; Nichols et al., 2021). Greater sensitivity may be achieved by adding a protease inhibitor to the S9 system (Nichols et al., 2021) or by using an *in vitro* system (e.g., liver spheroids; Baron et al., 2017; Uchea et al., 2013) that has a longer working lifetime. Given typical levels of analytical variance, however, large increases in assay sensitivity are unlikely. For very high log $K_{OW}$ chemicals, chemical adsorption to the sides of the reaction vessel could limit the utility of these *in vitro* systems. Presently, however, this upper limit on substrate hydrophobicity is not known.

To define the DOA of the *in vitro* $k_B$ estimation method, $k_B$ values were predicted using $k_{EL}$ values that represent reasonable limits of assay performance (0.05/h–50/h) across log $K_{OW}$ values ranging from 0 to 7. These predictions were again obtained using the IVIVE/BCF model given by Nichols et al. (2013). An examination of Figure 2.10 indicates that different methods for $k_B$ estimation may be more-or-less well suited to a particular chemical. For example, *in vivo* methods may have greater utility for some very hydrophobic chemicals that present *in vitro* adsorption challenges. In contrast, *in vitro* methods may be better suited for use with some relatively hydrophilic chemicals (log $K_{OW}$ <3.5) because high rates of uptake and elimination across the gills make it difficult to estimate $k_B$ from *in vivo* data. Nevertheless, there is a large zone of overlap in the DOA of *in vitro* and *in vivo* methods for $k_B$ estimation, which creates an opportunity to compare $k_B$ values obtained using both procedures (Arnot et al., 2008a). In doing so, however, it is important to keep in mind that *in vitro* methods can only account for biotransformation occurring in the evaluated tissue(s), while $k_B$ estimates obtained from *in vivo* data reflect the biotransformation capacity of the entire fish.

## Reverse Toxicokinetics and Quantitative *In Vitro*-to-*In Vivo* Extrapolation

Increasingly, the field of toxicology is moving toward the use of *in vitro* test systems to evaluate the potential for chemical effects on exposed animals. To support this activity, toxicokinetic models are being used to estimate environmental exposures that could be expected to result in target tissue exposures equivalent to those associated measured

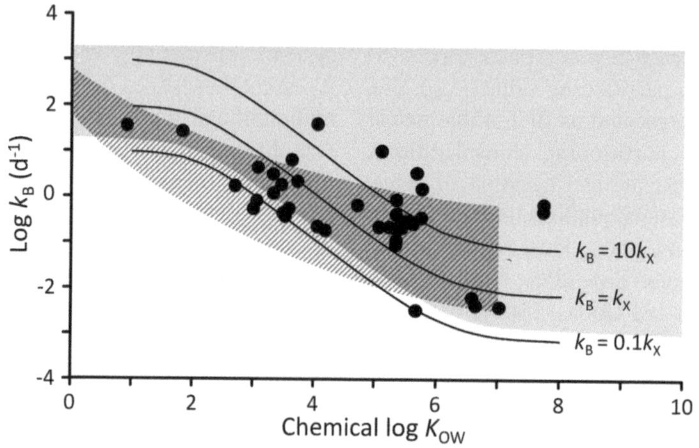


**FIGURE 2.10** Domain of applicability (DOA) of *in vivo* and *in vitro* methods for estimating whole-body biotransformation rates ($k_B$) in fish. The approximate DOA of *in vivo* methods is shown as the gray shaded area while that of *in vitro* methods is indicated by hatched lines. The methods used to estimate these DOAs are described in the text. Solid lines indicate points on the plot where the estimated $k_B$ is 10 times higher than, equal to, or one-tenth the sum ($k_X$) of the rate constants for branchial elimination ($k_2$) and fecal egestion ($k_E$). Individual points represent $k_B$ values estimated by modeling to *in vivo* BCFs and total elimination rate constants ($k_T$) for 37 test chemicals (Arnot et al., 2008a; Figure 4B in the cited work).

*in vitro* effects [e.g., the chemical concentration at 50% of maximum activity ($AC_{50}$) or the lowest effect concentration (LEC)]. In comparison with the traditional use of toxicokinetic models, which involves prediction of a target tissue dose associated with a defined whole-organism exposure (whole-organism exposure → *in vivo* tissue dose), this effort requires that the models be used "in reverse" (*in vitro* dose → *in vivo* tissue dose → whole-organism exposure), hence the expression *reverse toxicokinetics*. The overall process of relating *in vitro* effects information to a corresponding whole-organism exposure is commonly termed Quantitative *In Vitro*-to-*In Vivo* Extrapolation (QIVIVE; Yoon et al., 2012).

In mammals, QIVIVE is generally used to relate *in vitro* effects information to a continuous dietary dosing rate (e.g., mg/kg/day). These extrapolations can be performed using a multicompartment PBTK model (Louisse et al., 2017). However, screening-level assessments are often performed using simpler models that predict steady-state chemical concentrations in blood under the assumption of 100% oral bioavailability and one-compartment kinetic behavior (Rotroff et al., 2010; Wetmore et al., 2012, 2015). This approach relies on a clearance description given by Wilkinson and Schenker (1975):

$$C_{SS} = k_0 / \left[ \begin{array}{l} (\text{GFR}\, f_{U,B}) + \\ \left( Q_L f_U CL_{\text{IN VIVO,INT}} / \left( Q_L + f_U CL_{\text{IN VIVO,INT}} \right) \right) \end{array} \right] \quad (2.120)$$

where $C_{SS}$ is a steady-state chemical concentration in blood and $k_0$ is a zero-order dietary uptake rate (mg/kg/day). Here, renal clearance is calculated as the product of GFR and $f_{U,B}$, while hepatic clearance is estimated by extrapolating a measured rate of $CL_{\text{IN VITRO,INT}}$ to $CL_{\text{IN VIVO,INT}}$ (i.e., by IVIVE) followed by use of the well-stirred liver model (Equation 2.119; here ignoring the $f_W$ term). Equation 2.120 is used to calculate the $C_{SS}$ for an assumed $k_0$ of 1 mg/kg/day. An oral equivalent dose (mg/kg/day; i.e., an oral dose that can be associated with observed *in vitro* effects) is then calculated by reverse dosimetry as:

$$\text{oral equivalent dose} = AC_{50} \text{ or } LEC/C_{SS} \quad (2.121)$$

Finally, the oral equivalent dose calculated in this manner is compared to a known or hypothesized exposure to assess the potential for toxic effects. Implied by this procedure is an assumption that $AC_{50}$ or LEC values expressed as total chemical concentrations correspond to $C_{SS}$ values sufficient for bioactivity *in vivo* (Rotroff et al., 2010). This procedure may be adapted, however, to calculate an oral equivalent dose from unbound chemical concentrations *in vitro* and in blood. Binding within an *in vitro* system can be measured directly or predicted using an *in vitro* dosimetry model (Armitage et al., 2014; Fischer et al., 2017; Henneberger et al., 2020; Schreiber et al., 2008; Stadnicka-Michalak et al., 2014).

For a substantial number of chemicals, QIVIVE with fish presents a simpler challenge than it does with mammals. In a continuous aqueous exposure, the unbound chemical concentration in the animal ($C_{FISH,U}$) is close to that in the water ($C_{W,U}$), provided the exposure progresses to near steady state, biotransformation is negligible, and the rate of chemical elimination across the gills is substantially larger than rates of organism growth ($k_G$) and chemical egestion to feces ($k_E$). This last condition is generally satisfied for chemicals with log $K_{OW}$ values less than about 5. With these assumptions, the unbound chemical concentration that causes *in vitro* effects (e.g., an $AC_{50,U}$ value) can be related directly to $C_{FISH,U}$, and by extension $C_{W,U}$. Dividing $C_{W,U}$ by the free chemical fraction in water ($\phi$; Equation 2.86) then yields a corresponding total concentration ($C_{W,TOT}$). This extrapolation does not require a toxicokinetic model for the exposed organism. Instead, one need only account for nonspecific binding *in vitro* and in the exposure water.

Biotransformation complicates QIVIVE with fish by reducing the whole-body chemical concentration, resulting in a $C_{FISH,U} < C_{W,U}$. One may account for this activity, however, by using a 1-compartment bioconcentration model such as that given by Arnot and Gobas (2004; see earlier text for definitions of modeled terms). Under the conditions outlined above (steady-state aqueous exposure, slow or no growth, log $K_{OW}$ <5), the total chemical concentration in a fish ($C_{FISH}$) is approximated by the ratio $k_1C_{W,U}/k_2$, provided the chemical does not undergo biotransformation. If the same chemical undergoes biotransformation, this total concentration (here designated $C_{FISH,B}$) is approximated by the ratio $k_1C_{W,U}/(k_2+k_B)$. The ratio of these two calculated chemical concentrations (no biotransformation/biotransformation) is therefore:

$$C_{FISH}/C_{FISH,B} = \left(k_1C_{W,U}/k_2\right)/\left[k_1C_{W,U}/\left(k_2+k_B\right)\right] \qquad (2.122)$$

Assuming further that the unbound concentration in a fish varies in proportion to the total concentration, this ratio becomes:

$$C_{FISH,U}/C_{FISH,B,U} = \left(k_1C_{W,U}/k_2\right)/\left[k_1C_{W,U}/\left(k_2+k_B\right)\right] \qquad (2.123)$$

Absent biotransformation, $C_{W,U}$ is approximately equal to $C_{FISH,U}$. Substituting $C_{FISH,U}$ into the numerator of the right hand term in Equation 2.123 and rearranging yields:

$$C_{W,U} = C_{FISH,B,U}\left[\left(k_1/k_2\right)/\left(k_1/\left(k_2+k_B\right)\right)\right] \qquad (2.124)$$

To apply this relationship to QIVIVE, a measured $AC_{50,U}$ value would be set equal to $C_{FISH,B,U}$. The $C_{FISH,B,U}$ would then be multiplied by the ratio term $(k_1/k_2)/(k_1/(k_2+k_B))$ to calculate a corresponding $C_{W,U}$. For this assessment, $k_1$, $k_2$, and $k_B$ would be estimated using methods described above. Dividing $C_{W,U}$ by $\phi$ would then yield a corresponding total concentration in water ($C_{W,TOT}$).

Application of the QIVIVE approach to fish would require a more sophisticated modeling strategy if a steady state between the animal and surrounding water is unlikely to exist. The features of this strategy would depend on the modeled circumstances and could require dynamic simulation of chemical concentrations in the fish. However, the goal of the extrapolation would remain the same; namely, to link an *in vitro* effect concentration to a corresponding external exposure.

## ACKNOWLEDGMENTS

The authors gratefully acknowledge the contributions of Kevin Kleinow, James McKim, and William Hayton, who co-authored the original version of this chapter (Kleinow et al., 2008). We also thank Steven Lasee and Leslie Saunders for constructive reviews of this manuscript, and Keith Tarpley for his assistance in preparing the figures.

## DISCLAIMER

This chapter has been subjected to review by the Office of Research and Development, United States Environmental Protection Agency and approved for publication. Approval does not signify that the contents reflect the views of the Agency, nor does mention of any trade names or commercial products constitute endorsement or recommendation for use.

## NOTES

1. The processes of absorption, distribution, biotransformation, and excretion have been referred to by many authors using the abbreviation ADME, where the letter M stands for *metabolism*. However, this usage represents a potential source of confusion, since the term *metabolism* is widely used in other contexts (e.g., cellular metabolism associated with energy extraction). We have elected, therefore, to use the term *biotransformation* to refer to enzymatically mediated reactions that transform xenobiotic substrates to chemically distinct products.
2. For some xenobiotics (e.g., methylmercury), the extent of biomagnification is quantified by comparing whole-body chemical concentrations in fish and their prey. Biomagnification of hydrophobic organic chemicals is generally evaluated by comparing whole-body chemical concentrations that have been normalized to total lipid content.

## REFERENCES

Abbas, R., Hayton, W.L. (1996). Toxicokinetics of parathion and paraoxon in rainbow trout after intravascular administration and water exposure. *Toxicol. Appl. Pharmacol.* **136**, 194–199.

Abbas, R., Hayton, W.L. (1997). A physiologically based pharmacokinetic and pharmacodynamic model for paraoxon in rainbow trout. *Toxicol. Appl. Pharmacol.* **145**, 192–201.

Abraham, M.H., Poole, C.F., Poole, S.K. (1999). Classification of stationary phases and other materials by gas chromatography. *J. Chromatogr. A* **842**, 79–114.

Abraham, M.H., Ibrahim, A., Zissimos, A.M. (2004). Determination of sets of solute descriptors from chromatographic measurements. *J. Chromatogr. A* **1037**, 29–47.

Admire, B., Yalkowsky, S.H. (2013). Predicting the octanol solubility of organic compounds. *J. Pharm. Sci.* **102**, 2112–2119.

Adolph, E.F. (1949). Quantitative relations in the physiological constitutions of mammals. *Science* **109**, 579–585.

Alavi, F.K., Rolf, L.L., Clarke, C.R. (1993). The pharmacokinetics of sulfachlorpyridazine in channel catfish, *Ictalurus punctatus*. *J. Vet. Pharmacol. Therap.* **16**, 232–236.

Allen, J.L., Dawson, V.K., Hunn, J.B. (1979). Excretion of the lampricide Bayer 73 by rainbow trout. In *Aquatic Toxicology* (L.L. Marking and R. Kimerle, Eds.), pp. 52–61, Spec. Tech. Publ. 667, American Society for Testing and Materials, Philadelphia.

Andreeva, A.M. (2010). Structure of fish serum albumins. *J. Evol. Biochem. Physiol.* **46**, 135–144.

Anliker, R., Moser, P. (1987). The limits of bioaccumulation of organic pigments in fish - their relation to the partition-coefficient and the solubility in water and octanol. *Ecotoxicol. Environ. Saf.* **13**, 43–52.

Armitage, J.M., Arnot, J.A., Wania, F., Mackay, D. (2013). Development and evaluation of a mechanistic bioconcentration model for ionogenic organic chemicals in fish. *Environ. Toxicol. Chem.* **32**, 115–128.

Armitage, J.M., Wania, F., Arnot, J.A. (2014). Application of mass balance models and the chemical activity concept to facilitate the use of *in vitro* toxicity data for risk assessment. *Environ. Sci. Technol.* **48**, 9770–9779.

Armitage, J.M., Erickson, R.J., Luckenbach, T., Ng, C.A., Prosser, R.S., Arnot, J.A., Schirmer, K., Nichols, J.W. (2017). Assessing the bioaccumulation potential of ionizable organic compounds: Current knowledge and research priorities. *Environ. Toxicol. Chem.* **36**, 882–897.

Arnot, J.A., Gobas, F.A.P.C. (2003). A generic QSAR for assessing the bioconcentration potential of organic chemicals in aquatic food webs. *QSAR Comb. Sci.* **22**, 337–345.

Arnot, J.A., Gobas, F.A.P.C. (2004). A food web bioaccumulation model for organic chemicals in aquatic ecosystems. *Environ. Toxicol. Chem.* **23**, 2343–2355.

Arnot, J.A., Gobas, F.A.P.C. (2006). A review of bioconcentration factor (BCF) and bioaccumulation factor (BAF) assessments for organic chemicals in aquatic organisms. *Environ. Rev.* **14**, 257–297.

Arnot, J.A., Mackay, D., Bonnell, M. (2008a). Estimating metabolic biotransformation rates in fish from laboratory data. *Environ. Toxicol. Chem.* **27**, 341–351.

Arnot, J.A., Mackay, D., Parkerton, T.F., Bonnell, M. (2008b). A database of fish biotransformation rates for organic chemicals. *Environ. Toxicol. Chem.* **27**, 2263–2270.

Arnot, J.A., Meylan, W., Tunkel, J., Howard, P.H., Mackay, D., Bonnell, M., Boethling, R.S. (2009). A quantitative structure-activity relationship for predicting metabolic biotransformation rates for organic chemicals in fish. *Environ. Toxicol. Chem.* **28**, 1168–1177.

Arnot, J.A., Arnot, M.I., Mackay, D., Couillard, Y., MacDonald, D., Bonnell, M., Doyle, P. (2010). Molecular size cutoff criteria for screening bioaccumulation potential: Fact or fiction? *Integr. Environ. Assess. Manag.* **6**, 210–224.

Arnot, J.A., Quinn, C.L. (2015). Development and evaluation of a database of dietary bioaccumulation test data for organic chemicals in fish. *Environ. Sci. Technol.* **49**, 4783–4796.

Arnot, J.A., Mackay, D. (2018). The influence of chemical degradation during dietary exposures to fish on biomagnification factors and bioaccumulation factors. *Environ. Sci. Process. Impacts* **20**, 86–97.

Arnot, J.A., Toose, L., Armitage, J.M., Embry, M., Sangion, A., Hughes, L. (2022). A weight of evidence approach for bioaccumulation assessment. *Integr. Environ. Assess. Manag.* Accepted for publication and available at: *https://doi.org/10.1002/ieam.4583.* Accessed 04/28/2022.

Artola-Garicano, E., Vaes, W.H.J., Hermens, J.L.M. (2000). Validation of negligible depletion solid-phase microextraction as a tool to determine tissue/blood partition coefficients for semivolatile and nonvolatile organic chemicals. *Toxicol. Appl. Pharmacol.* **166**, 138–144.

Aslamkhan, A.G., Thompson, D.M., Perry, J.L., Bleasby, K., Wolff, N.A., Barros, S., Miller, D.S., Pritchard, J.B. (2006). The flounder organic anion transporter fOat has sequence, function, and substrate specificity similarity to both mammalian Oat1 and Oat3. *Am. J. Physiol., Regul. Integr. Comp. Physiol.* **291**, R1773–R1780.

Axelsson, M., Driedzic, W.R., Farrell, A.P., Nilsson, S. (1989). Regulation of cardiac output and gut blood flow in the sea raven, *Hemitripterus americanus. Fish Physiol. Biochem.* **6**, 315–326.

Axelsson, M., Fritsche, R. (1991). Effects of exercise, hypoxia and feeding on the gastrointestinal blood flow in the Atlantic cod (*Gadus morhua*). *J. Exp. Biol.* **158**, 181–198.

Axelsson, M., Thorarensen, H., Nilsson, S., Farrell, A.P. (2000). Gastrointestinal blood flow in the red Irish lord, *Hemilepidotus hemilepidotus*: Long-term effects of feeding and adrenergic control. *J. Comp. Physiol.* **170**, 145–152.

Barber, M.C., Suarez, L.A., Lassiter, R.R. (1988). Modeling bioconcentration of nonpolar organic pollutants by fish. *Environ. Toxicol. Chem.* **7**, 545–558.

Barber, M.C., Suarez, L.A., Lassiter, R.R. (1991). Modeling bioaccumulation of organic pollutants in fish with an application to PCBs in Lake Ontario salmonids. *Can. J. Fish. Aquat. Sci.* **48**, 318–337.

Barber, M.C. (2003). A review and comparison of models for predicting dynamic chemical bioconcentration in fish. *Environ. Toxicol. Chem.* **22**, 1963–1992.

Barber, M.C. (2008). Dietary uptake models used for modeling the bioaccumulation of organic contaminants in fish. *Environ. Toxicol. Chem.* **27**, 755–777.

Bard, S.M. (2000). Multixenobiotic resistance as a cellular defense mechanism in aquatic organisms. *Aquat. Toxicol.* **48**, 357–389.

Baron, M.G., Mintram, K.S., Owen, S.F., Hetheridge, M.J., Moody, A.J., Purcell, W.M., Jackson, S.K., Jha, A.N. (2017). Pharmaceutical metabolism in fish: Using a 3-D hepatic *in vitro* model to assess clearance. *PLoS One* **12(1)**, e0168837.

Barron, M.G., Tarr, B.D., Hayton, W.L. (1987a). Temperature dependence of cardiac output and regional blood flow in rainbow trout, *Salmo gairdneri* Richardson. *J. Fish Biol.* **31**, 735–744.

Barron, M.G., Tarr, B.D., Hayton, W.L. (1987b). Temperature-dependence of di-2-ethylhexyl phthalate (DEHP) pharmacokinetics in rainbow trout. *Toxicol. Appl. Pharmacol.* **88**, 305–312.

Barron, M.G., Schultz, I.R., Hayton, W.L. (1989). Presystemic branchial metabolism limits di-2-ethylhexyl phthalate accumulation in fish. *Toxicol. Appl. Pharmacol.* **98**, 49–57.

Barron, M.G. (1990). Bioconcentration. Will water-borne organic chemicals accumulate in aquatic animals? *Environ. Sci. Technol.* **24**, 1612–1618.

Barron, M.G., Stehly, G.R., Hayton, W.L. (1990). Pharmacokinetic modeling in aquatic animals. I. Models and concepts. *Aquat. Toxicol.* **18**, 61–86.

Bertelsen, S.L., Hoffman, A.D., Gallinat, C.A., Elonen, C.M., Nichols, J.W. (1998). Evaluation of log $K_{OW}$ and tissue lipid content as predictors of chemical partitioning to fish tissues. *Environ. Toxicol. Chem.* **17**, 1447–1455.

Beyenbach, K.W., Kirschner, L.B. (1976). The unreliability of mammalian glomerular markers in teleostean renal studies. *J. Exp. Biol.* **64**, 369–378.

Black, M.C., McCarthy, J.F. (1988). Dissolved organic macromolecules reduce the uptake of hydrophobic organic contaminants by the gills of rainbow trout (*Salmo gairdneri*). *Environ. Toxicol. Chem.* **7**, 593–600.

Black, M.C., Millsap, D.S., McCarthy, J.F. (1991). Effects of acute temperature change on respiration and toxicant uptake by rainbow trout (*Salmo gairdneri* Richardson). *Physiol. Zool.* **64**, 145–168.

Boethling, R.S., Mackay, D. (2000). *Handbook of Property Estimation Methods for Chemicals*. Lewis, Boca Raton.

Bourne, D.W.A. (2021). Pharmacokinetic software. Available at: pharmpk.com/soft.html. Accessed 04/28/2022.

Boyer, J.L., Schwartz, J., Smith, N. (1976a). Biliary secretion in elasmobranchs. I. Bile collection and composition. *Am. J. Physiol.* **230**, 970–973.

Boyer, J.L., Schwarz, J., Smith, N. (1976b). Biliary secretion in elasmobranchs. II. Hepatic uptake and biliary excretion of organic anions. *Am. J. Physiol.* **230**, 974–981.

Boyer, J.L., Schwarz, J., Smith, N. (1976c). Selective hepatic uptake and biliary excretion of$^{35}$S-sulfobromophthalein in marine elasmobranchs. *Gastroenterology* **70**, 254–256.

Bradbury, S.P., Coats, J.R., McKim, J.M. (1986). Toxicokinetics of fenvalerate in rainbow trout (*Salmo gairdneri*). *Environ. Toxicol. Chem.* **5**, 567–576.

Bradbury, S.P., Dady, J.M., Fitzsimmons, P.N., Voit, M.M., Hammermeister, D.E., Erickson, R.J. (1993). Toxicokinetics and metabolism of aniline and 4-chloroaniline in medaka (*Oryzias latipes*). *Toxicol. Appl. Pharmacol.* **118**, 205–214.

Branson, D.R., Blau, G.E., Alexander, H.C., Neely, W.B. (1975). Bioconcentration of 2,2',4,4'-tetrachlorobiphenyl in rainbow trout as measured by an accelerated test. *Trans. Am. Fish. Soc.* **104**, 785–792.

Breen, M., Villeneuve, D.L., Ankley, G.T., Bencic, D.C., Breen, M.S., Watanabe, K.H., Lloyd, A.L., Conolly, R.B. (2013). Developing predictive approaches to characterize adaptive responses of the reproductive endocrine axis to aromatase inhibition: II. Computational modeling. *Toxicol. Sci.* **133**, 234–247.

Brinkmann, M., Eichbaum, K., Buchinger, S., Reifferscheid, G., Bui, T., Schäffer, A., Hollert, H., Preuss, T.G. (2014). Understanding receptor-mediated effects in rainbow trout: *In vitro-in vivo* extrapolation using physiologically based toxicokinetic models. *Environ. Sci. Technol.* **48**, 3303–3309.

Brinkmann, M., Freese, M., Pohlmann, J.-D., Kammann, U., Preuss, T.G., Buchinger, S., Reifferscheid, G., Beiermeister, A., Hanel, R., Hollert, H. (2015). A physiologically based toxicokinetic (PBTK) model for moderately hydrophobic organic chemicals in the European eel (*Anguilla anguilla*). *Sci. Tot. Environ.* **536**, 279–287.

Brinkmann, M., Schlechtriem, C., Reininghaus, M., Eichbaum, K., Buchinger, S., Reifferscheid, G., Hollert, H., Preuss, T.G. (2016). Cross-species extrapolation of uptake and disposition of neutral organic chemicals in fish using a multispecies physiologically-based toxicokinetic model framework. *Environ. Sci. Technol.* **50**, 1914–1923.

Brown, T.N., Arnot, J.A., Wania, F. (2012). Iterative fragment selection: A group contribution approach to predicting fish biotransformation half-lives. *Environ. Sci. Technol.* **46**, 8253–8260.

Bruggeman, W.A., Martron, L.B.J.M., Kooiman, D., Hutzinger, O. (1981). Accumulation and elimination kinetics of di-, tri-, and tetra-chlorobiphenyls by goldfish after dietary and aqueous exposure. *Chemosphere* **10**, 811–832.

Bruggeman, W.A., Opperhuizen, A., Wijbenga, O., Hutzinger, A. (1984). Bioaccumulation of super-lipophilic chemicals in fish. *Environ. Toxicol. Chem.* **7**, 173–189.

Bucking, C., Anderson, W.G. (2020). The digestive system. In *The Physiology of Fishes* (S. Currie and D.H. Evans, Eds.), pp. 79–90, CRC Press, Boca Raton.

Buddington, R.K., Diamond, J.M. (1987). Pyloric ceca of fish: A 'new' absorptive organ. *Am. J. Physiol.* **252**, G65–G76.

Bungay, P.M., Dedrick, R.L., Guarino, A.M. (1976). Pharmacokinetic modeling of the dogfish shark (*Squalus acanthias*): Distribution and urinary and biliary excretion of phenol red and its glucuronide. *J. Pharmacokinet. Biopharmaceut.* **4**, 377–388.

Burkhard, L.P. (2000). Estimating dissolved organic carbon partition coefficients for nonionic organic chemicals. *Environ. Sci. Technol.* **34**, 4663–4668.

Burkhard, L.P., Arnot, J.A., Embry, M.R., Farley, K.J., Hoke, R.A., Kitano, M., Leslie, H.A., Lotufo, G.R., Parkerton, T.F., Sappington, K.G., Tomy, G.T., Woodburn, K.B. (2012a). Comparing laboratory- and field-measured biota-sediment accumulation factors. *Integr. Environ. Assess. Manag.* 8, 32–41.

Burkhard, L.P., Arnot, J.A., Embry, M.R., Farley, K.J., Hoke, R.A., Kitano, M., Leslie, H.A., Lotufo, G.R., Parkerton, T.F., Sappington, K.G., Tomy, G.T., Woodburn, K.B. (2012b). Comparing laboratory and field measured bioaccumulation endpoints. *Integr. Environ. Assess. Manag.* **8**, 17–31.

Burkhard, L.P., Borgå, K., Powell, D.E., Leonards, P., Muir, D.C.G., Parkerton, T.F., Woodburn, K.B. (2013). Improving the quality and scientific understanding of trophic magnification factors (TMFs). *Environ. Sci. Technol.* **47**, 1186–1187.

Cameron, J.N. (1975). Blood flow distribution as indicated by tracer microspheres in resting and hypoxic arctic grayling (*Thymallus arcticus*). *Comp. Biochem. Physiol.* **52A**, 441–444.

Cappelli, C.I., Benfenati, E., Cester, J. (2015). Evaluation of QSAR models for predicting the partition coefficient (log P) of chemicals under the REACH regulation. *Environ. Res.* **143**, 26–32.

Chen, W.-Y., Liao, C.-M. (2014). Interpreting copper bioaccumulation dynamics in tilapia using systems-level explorations of pulsed acute/chronic exposures. *Ecotoxicol.* **23**, 1124–1136.

Chen, Y., Hermens, J.L.M., Jonker, M.T.O., Arnot, J.A., Armitage, J.M., Brown, T., Nichols, J.W., Fay, K.A., Droge, S.T.J. (2016). Which molecular features affect the intrinsic hepatic clearance rate of ionizable organic chemicals in fish? *Environ. Sci. Technol.* **50**, 12722–12731.

Chiou, C.T. (1985). Partition coefficients of organic compounds in lipid-water systems and correlations with fish bioconcentration factors. *Environ. Sci. Technol.* **19**, 57–62.

Clark, K.E., Gobas, F.A.P.C., Mackay, D. (1990). Model of organic chemical uptake and clearance from fish from food and water. *Environ. Sci. Technol.* 24, 1203–1213.

Clewell, H.J., III, Clewell, R.A., Andersen, M.E. (2014). Physiologically based pharmacokinetic and toxicokinetic models. In *Hayes' Principles and Methods of Toxicology* (A.W. Hayes and C.L. Kruger, Eds.), pp. 247–293, Taylor & Francis, Boca Raton.

Collicutt, J.M., Eales, J.G. (1974). Excretion and enterohepatic cycling of $^{125}$I-L-thyroxine in channel catfish, *Ictalurus punctatus* Rafinesque. *Gen. Comp. Endocrinol.* **23**, 390–402.

Collier, T.K., Varanasi, U. (1991). Hepatic activities of xenobiotic metabolizing enzymes and biliary levels of xenobiotics in English sole (*Parophrys vetulus*) exposed to environmental contaminants. *Arch. Environ. Contam. Toxicol.* **20**, 462–473.

Connolly, J.P., Pedersen, C.J. (1988). A thermodynamic-based evaluation of organic chemical accumulation in aquatic organisms. *Environ. Sci. Technol.* **22**, 99–103.

Consoer, D.M., Hoffman, A.D., Fitzsimmons, P.N., Kosian, P.A., Nichols, J.W. (2014). Toxicokinetics of perfluorooctanoate (PFOA) in rainbow trout (*Oncorhynchus mykiss*). *Aquat. Toxicol.* **156**, 65–73.

Consoer, D.M., Hoffman, A.D., Fitzsimmons, P.N., Kosian, P.A., Nichols, J.W. (2016). Toxicokinetics of perfluorooctane sulfonate in rainbow trout (*Oncorhynchus mykiss*). *Environ. Toxicol. Chem.* **35**, 717–727.

Covington, T.R., Gearhart, J.M. (2020). Sensitivity and Monte Carlo analysis techniques and their use in uncertainty, variability, and population analysis. In *Physiologically Based Pharmacokinetic (PBPK) Modeling, Methods and Applications in Toxicology and Risk Assessment* (J.W. Fisher, J.M. Gearhart, Z. Lin, Eds.), pp. 211–242, Academic Press, London.

Cowan-Ellsberry, C.E., Dyer, S.D., Erhardt, S., Bernhard, M.J., Roe, A.L., Dowty, M.E., Weisbrod, A.V. (2008). Approach for extrapolating *in vitro* metabolism data to refine bioconcentration factor estimates. *Chemosphere* **70**, 1804–1817.

Cravedi, J.P., Lafuente, A., Baradat, M., Hillenweck, A., Perdu-Durand, E. (1999). Biotransformation of pentachlorophenol, aniline and biphenyl in isolated rainbow trout (*Oncorhynchus mykiss*) hepatocytes: Comparison with *in vivo* metabolism. *Xenobiotica* **29**, 499–509.

Cunningham, J.G. (1997). *Textbook of Veterinary Physiology*, W.B. Saunders, Philadelphia.

deBruyn, A.M.H., Gobas, F.A.P.C. (2007). The sorptive capacity of animal protein. *Environ. Toxicol. Chem.* **26**, 1803–1808.

Deeley, R.G., Westlake, C., Cole, S.P. (2006). Transmembrane transport of endo- and xenobiotics by mammalian ATP-binding cassette multidrug resistance proteins. *Physiol. Rev.* **86**, 849–899.

Dedrick, R.L. (1973). Animal scale-up. *J. Pharmacokinet. Biopharmaceut.* **1**, 435–461.

DeJongh, J., Verhaar, H.J.M., Hermens, J.L.M. (1997). A quantitative property-property relationship (QPPR) approach to estimate *in vitro* tissue-blood partition coefficients of organic chemicals in rats and humans. *Arch. Toxicol.* **72**, 17–25.

Derendorf, H., Schmidt, S. (2020). *Rowland and Tozer's Clinical Pharmacokinetics and Pharmacodynamics*, 5th ed, Wolters Kluwer, Philadelphia.

De Smet, H., Blust, R., Moens, L. (1998). Absence of albumin in the plasma of the common carp *Cyprinus carpio*: Binding of fatty acids to high density lipoprotein. *Fish Physiol. Biochem.* **19**, 71–81.

deWolf, W., Seinen, W., Hermens, J.L.M. (1993). Biotransformation and toxicokinetics of trichloroanilines in fish in relation to their hydrophobicity. *Arch. Environ. Contam. Toxicol.* **25**, 110–117.

Dimitrov, S.D., Dimitrova, N.C., Walker, J.D., Veith, G.D., Mekenyan, O.G. (2003). Bioconcentration potential predictions based on molecular attributes - An early warning approach for chemicals found in humans, birds, fish and wildlife. *QSAR Combinatorial Sci.* **22**, 58–68.

Doi, A.M., Lou, Z., Holmes, E., Li, C.L.J., Venugopal, C.S., James, M.O., Kleinow, K.M. (2000). Effect of micelle fatty acid composition and 3,4,3′,4′-tetrachlorobiphenyl (TCB) exposure on intestinal [$^{14}$C]-TCB bioavailability and biotransformation in channel catfish *in situ* preparations. *Toxicol. Sci.* **55**, 85–96.

Endo, S., Escher, B.I., Goss, K.-U. (2011). Capacities of membrane lipids to accumulate neutral organic chemicals. *Environ. Sci. Technol.* **45**, 5912–5921.

Endo, S., Goss, K.-U. (2011). Serum albumin binding of structurally diverse neutral organic compounds: Data and models. *Chem. Res. Toxicol.* **24**, 2293–2301.

Endo, S., Bauerfeind, J., Goss, K.-U. (2012). Partitioning of neutral organic compounds to structural proteins. *Environ. Sci. Technol.* **46**, 12697–12703.

Endo, S., Brown, T.N., Goss, K.-U. (2013). General model for estimating partition coefficients to organism and their tissues using the biological compositions and polyparameter linear free energy relationships. *Environ. Sci. Technol.* **47**, 6630–6639.

Endo, S., Goss, K.-U. (2014). Applications of polyparameter linear free energy relationships in environmental chemistry. *Environ. Sci. Technol.* **48**, 12477–12491.

Erickson, R.J., McKim, J.M. (1990a). A simple flow-limited model for exchange of organic chemicals at fish gills. *Environ. Toxicol. Chem.* **9**, 159–165.

Erickson, R.J., McKim, J.M. (1990b). A model for exchange of organic chemicals at fish gills: Flow and diffusion limitations. *Aquat. Toxicol.* **18**, 175–198.

Erickson, R.J., McKim, J.M., Lien, G.J., Hoffman, A.D., Batterman, S.L. (2006a). Uptake and elimination of ionizable organic chemicals at fish gills: I. Model formulation, parameterization, and behavior. *Environ. Toxicol. Chem.* **25**, 1512–1521.

Erickson, R.J., McKim, J.M., Lien, G.J., Hoffman, A.D., Batterman, S.L. (2006b). Uptake and elimination of ionizable organic chemicals at fish gills: II. Observed and predicted effects of pH, alkalinity, and chemical properties. *Environ. Toxicol. Chem.* **25**, 1522–1532.

Escher, B.I., Schwarzenbach, R.P., Westall, J.C. (2000). Evaluation of liposome-water partitioning of acids and bases. I. Development of a sorption model. *Environ. Sci. Technol.* **34**, 3954–3961.

Espié, P., Tytgat, D., Sargentini-Maier, M.-L., Poggesi, I., Watelet, J.-B. (2009). Physiologically based pharmacokinetics (PBPK). *Drug Metab. Rev.* **41**, 391–407.

Eurell, J.A., Haensly, W.E. (1982). The histology and ultrastructure of the liver of Atlantic croaker *Micropogon undulatus* L. *J. Fish Biol.* **21**, 113–125.

Evans, D.H. (1993). *The Physiology of Fishes*, CRC Press, Boca Raton.

Ferreira, M., Costa, J., Reis-Henriques, M.A. (2014). ABC transporters in fish species: A review. *Front. Physiol.* **5**, 266.

Fischer, F.C., Henneberger, L., König, M., Bittermann, K., Linden, L., Goss, K.-U., Escher, B.I. (2017). Modeling exposure in the Tox21 *in vitro* bioassays. *Chem. Res. Toxicol.* **30**, 1197–1208.

Fisher, J.W., Gearhart, J.M., Lin, Z. (2020). *Physiologically Based Pharmacokinetic (PBPK) Modeling, Methods and Applications in Toxicology and Risk Assessment*. Academic Press, London.

Fisk, A.T., Norstrom, R.J., Cymbalisty, C.D., Muir, D.C.G. (1998). Dietary accumulation and depuration of hydrophobic organochlorines: Bioaccumulation parameters and their relationship with the octanol/water partition coefficient. *Environ. Toxicol. Chem.* **17**, 951–961.

Fisk, A.T., Tomy, G.T., Cymbalisty, C.D., Muir, D.C.G. (2000). Dietary accumulation and quantitative structure-activity relationships for depuration and biotransformation of short ($C_{10}$), medium ($C_{14}$), and long ($C_{18}$) carbon-chain polychlorinated alkanes by juvenile rainbow trout (*Oncorhynchus mykiss*). *Environ. Toxicol. Chem.* **19**, 1508–1516.

Fitzsimmons, P.N., Fernandez, J.D., Hoffman, A.D., Butterworth, B.C., Nichols, J.W. (2001). Branchial elimination of superhydrophobic organic compounds by rainbow trout (*Oncorhynchus mykiss*). *Aquat. Toxicol.* **55**, 23–34.

Fitzsimmons, P.N., Hoffman, A.D., Lien, G.J., Hammermeister, D.E., Nichols, J.W. (2009). Kinetics and effects of dichloroacetic acid in rainbow trout. *Aquat. Toxicol.* **94**, 186–194.

Fitzsimmons, P.N., Hoffman, A.D., Fay, K.A., Nichols, J.W. (2018). Allometric scaling of hepatic biotransformation in rainbow trout. *Comp. Biochem. Physiol.* **214C**, 52–60.

Flynn, G.L., Yalkowsky, S.H. (1972). Correlation and prediction of mass transport across membranes. I. Influence of alkyl chain length on flux-determining properties of barrier and diffusant. *J. Pharm. Sci.* **61**, 838–852.

Franco-Uría, A., Otero-Muras, I., Balsa-Canto, E., Alonso, AA., Roca, E. (2010). Generic parameterization for a pharmacokinetic model to predict Cd concentrations in several tissues of different fish species. *Chemosphere* **79**, 377–386.

Freidig, A.P., Artola Garicano, E., Busser, F.J.M, Hermens, J.L.M. (1998). Estimating impact of humic acid on bioavailability and bioaccumulation of hydrophobic chemicals in guppies using kinetic solid-phase extraction. *Environ. Toxicol. Chem.* **17**, 998–1004.

Fricker, G., Wossner, R., Drewe, J., Fricker, R., Boyer, J.L. (1997). Enterohepatic circulation of scymnol sulfate in an elasmobranch, the little skate (*Raja erinacea*). *Am. J. Physiol.* **273**, G1023–G1030.

Gargas, M.L., Burgess, R.J., Voisard, D.E., Cason, G.H., Andersen, M.E. (1989). Partition coefficients of low molecular weight volatile chemicals in various liquids and tissues. *Toxicol. Appl. Pharmacol.* **98**, 87–99.

Geisler, A., Endo, S., Goss, K.-U. (2012). Partitioning of organic chemicals to storage lipids: Elucidating the dependence on fatty acid composition and temperature. *Environ. Sci. Technol.* **46**, 9519–9524.

Gerlowski, L.E., Jain, R.K. (1983). Physiologically based pharmacokinetic modeling: Principles and applications. *J. Pharmaceut. Sci.* **72**, 1103–1127.

Gert-Jan de Maagd, P., de Poorte, J., Opperhuizen, A., Sijm, D.T.H.M. (1998). No influence after various exposure times on the biotransformation rate constants of benzo(a)anthracene in fathead minnow (*Pimephales promelas*). *Aquat. Toxicol.* 40, 157–169.

Gibaldi, M., Perrier, D. (1982). *Pharmacokinetics*, 2nd ed., Macel Decker, New York.

Gingerich, W.H., Weber, L.J., Larson, R.E. (1977). Hepatic accumulation, metabolism, and biliary excretion of sulfobromophthalein by rainbow trout (*Salmo gairdneri*). *Comp. Biochem. Physiol.* **58C**, 113–120.

Gobas, F.A.P.C., Opperhuizen, A., Hutzinger, O. (1986). Bioconcentration of hydrophobic chemicals in fish: Relationship with membrane permeation. *Environ. Toxicol. Chem.* **5**, 637–646.

Gobas, F.A.P.C., Mackay, D. (1987). Dynamics of hydrophobic organic chemical bioconcentration in fish. *Environ. Toxicol. Chem.* **6**, 495–504.

Gobas, F.A.P.C, Muir, D.C.G., Mackay, D. (1988). Dynamics of dietary bioaccumulation and faecal elimination of hydrophobic organic chemicals in fish. *Chemosphere* **17**, 943–962.

Gobas, F.A.P.C. (1993). A model for predicting the bioaccumulation of hydrophobic organic chemicals in aquatic foodwebs: Application to Lake Ontario. *Ecol. Modell.* **69**, 1–17.

Gobas, F.A.P.C., Zhang, X., Wells, R. (1993). Gastrointestinal magnification: The mechanism of biomagnification and food chain accumulation of organic chemicals. *Environ. Sci. Technol.* **27**, 2855–2863.

Gobas, F.A.P.C., Wilcockson, J.B., Russell, R.W., Haffner, G.D. (1999). Mechanism of biomagnification in fish under laboratory and field conditions. *Environ. Sci. Technol.* **33**, 133–141.

Gobas, F.A.P.C., Morrison, H.A. (2000). Bioconcentration and biomagnification in the aquatic environment. In *Handbook of Property Estimation Methods for Chemicals: Environmental and Health Sciences* (R.S. Boethling and D. Mackay, Eds.), pp. 189–231, CRC Press, Boca Raton.

Gobas, F.A.P.C., de Wolf, W., Burkhard, L.P., Verbruggen, E., Plotzke, K. (2009). Revisiting bioaccumulation criteria for POPs and PBT assessments. *Integr. Environ. Assess. Manag.* **5**, 624–637.

Gobas, F.A.P.C., Lee, Y.-S., Lo, J.C., Parkerton, T.F., Letinski, D.J. (2020). A toxicokinetic framework and analysis tool for interpreting Organization for Economic Co-operation and Development guideline 305 dietary bioaccumulation tests. *Environ. Toxicol. Chem.* **39**, 171–188.

Goss, K.-U., Schwarzenbach, R.P. (2001). Linear free energy relationships used to evaluate equilibrium partitioning of organic chemicals. *Environ. Sci. Technol.* **35**, 1–9.

Gray, J.E., Doolittle, R.F. (1992). Characterization, primary structure, and evolution of lamprey plasma albumin. *Prot. Sci.* **1**, 289–302.

Gretch, A., Brochot, C., Dorne, J.-L., Quignot, N., Bois, F.Y., Beaudoin, R. (2016). Toxicokinetic models and related tools in environmental risk assessment of chemicals. *Sci. Tot. Environ.* **578**, 1–15.

Gretch, A., Tebby, C., Brochot, C., Bois, F.Y., Bado-Nilles, A., Dorne, J.-L., Quignot, N., Beaudouin, R. (2019). Generic physiologically-based toxicokinetic modelling for fish: Integration of environmental factors and species variability. *Sci. Tot. Environ.* **651**, 516–531.

Grimard, C., Mangold-Döring, A., Alharbi, H., Weber, L., Hogan, N., Jones, P.D., Giesy, J.P., Hecker, M., Brinkmann, M. (2021). Toxicokinetic models for bioconcentration of organic contaminants in two life stages of white sturgeon (*Acipenser transmontanus*). *Environ. Sci. Technol.* **55**, 11590–11600.

Grosell, M., McGeer, J.C., Wood, C.M. (2001). Plasma copper clearance and biliary copper excretion are stimulated in copper-acclimated trout. *Am. J. Physiol.* **280**, R796–R806.

Gruger, E.H., Karrick, N.L., Davidson, A.I., Hruby, T. (1975). Accumulation of 3,4,3′,4′-hexachlorobiphenyl and 2,4,5,2′,4′,5′- and 2,4,6,2′,4′,6′-hexachlorobiphenyl in juvenile Coho salmon. *Environ. Sci. Technol.* **9**, 121–127.

Guarino, A.M., Lech, J.J. (1986). Metabolism, disposition, and toxicity of drugs and other xenobiotics in aquatic species. *Vet. Hum. Toxicol.* **28** (Suppl. 1), 38–44.

Guiney, P.D., Melancon, Jr., M.J., Lech, J.J., Peterson, R.E. (1979). Effects of egg and sperm maturation and spawning on the distribution and elimination of a polychlorinated biphenyl in rainbow trout (*Salmo gairdneri*). *Toxicol. Appl. Pharmacol.* **47**, 261–272.

Guiney, P.D., Peterson, R.E. (1980). Distribution and elimination of a polychlorinated biphenyl after acute dietary exposure in yellow perch and rainbow trout. *Arch. Environ. Contam. Toxicol.* **9**, 667–674.

Hack, C.E., Efremenko, A.Y., Pendse, S.N, Ellison, C.A., Najjar, A., Hewitt, N., Schepky, A., Clewell, H.J., III. (2020). Physiologically based pharmacokinetic modeling software. In *Physiologically Based Pharmacokinetic Modeling, Methods and Applications in Toxicology and Risk Assessment* (J.W. Fischer, J.M. Gearhart, and Z. Lin, Eds.), pp. 81–126, Academic Press, London.

Hamelink, J.L., Waybrant, R.C., Ball, R.C. (1971). A proposal: Exchange equilibria control the degree chlorinated hydrocarbons are biologically magnified in lentic environments. *Trans. Am. Fish. Soc.* **100**, 207–214.

Hampton, J.A., Lantz, R.C., Hinton, D.E. (1989). Functional units in rainbow trout (*Salmo gairdneri*, Richardson) liver. III. Morphometric analysis of parenchyma, stroma, and component cell types. *Am. J. Anat.* **185**, 58–73.

Han, X., Nabb, D.L., Mingoia, R.T., Yang, C.-H. (2007). Determination of xenobiotic intrinsic clearance in freshly isolated hepatocytes from rainbow trout (*Oncorhynchus mykiss*) and rat and its application in bioaccumulation assessment. *Environ. Sci. Technol.* **41**, 3269–3276.

Hansch, C. (1972). Quantitative relationships between lipophilic character and drug metabolism. *Drug Metab. Rev.* **1**, 1–14.

Hansch, C., Leo, A., Hoekman, H. (1995). *Exploring QSAR: Volume 2: Hydrophobic, Electronic, and Steric Constants*. American Chemical Society, Washington.

Hayton, W.L. Barron, M.G. (1990). Rate-limiting barriers to xenobiotic uptake by the gill. *Environ. Toxicol. Chem.* **9**, 151–157.

He, M.-J., Luo, X.-J., Chen, M.-Y., Sun, Y.-X., Chen, S.-J., Mai, B.-X. (2012). Bioaccumulation of polybrominated diphenylethers and decabromodiphenyl ethane in fish from a river system in a highly industrialized area, South China. *Sci. Tot. Environ.* **419**, 109–115.

Hendriks, A.J., van der Linde, A., Cornelissen, G., Sijm, D.T. (2001). The power of size. 1. Rate constants and equilibrium ratios for accumulation of organic substances related to octanol-water partition ratio and species weight. *Environ. Toxicol. Chem.* **20**, 1399–420.

Henneberger, L., Mühlenbrink, M., Heinrich, J.J., Teixeira, A., Nicol, B., Escher, B.I. (2020). Experimental validation of mass balance models for *in vitro* cell-based bioassays. *Environ. Sci. Technol.* **54**, 1120–1127.

Henneberger, L., Klüver, N., Mühlenbrink, M., Escher, B. (2022). Trout and human plasma protein binding of selected pharmaceuticals informs the fish plasma model. *Environ. Toxicol. Chem.* **41**, 559–568.

Hickman, Jr., C.P., Trump, B.F. (1969). The kidney. In *Fish Physiology*, Vol. 1 (W.S. Hoar and D.J. Randall, Eds.), pp. 91–239, Academic Press, New York.

Hinton, D.E., Couch, J.A. (1998). Architectural pattern, tissue and cellular morphology in livers of fishes: Relationship to experimentally induced neoplastic responses. In *Fish Ecotoxicology* (T. Braunbeck, D.E. Hinton, and B. Streit, Eds.), pp. 141–164, Birkhauser Verlag, Basel.

Hodges, G., Eadsforth, C., Bossuyt, B., Bouvy, A., Enrici, M.-H., Geurts, M., Kotthoff, M., Michie, E., Miller, D., Müller, J., Oetter, G., Roberts, J., Schowanek, D., Sun, P., Venzmer, J. (2019). A comparison of log $K_{OW}$ (*n*-octanol-water partition coefficient) values for non-ionic, anionic, cationic and amphoteric surfactants determined using predictions and experimental methods. *Environ. Sci. Eur.* **31**, 1.

Hofer, R., Forstner, H., Rettenwander, R. (1982). Duration of gut passage and its dependence on temperature and food consumption in roach, *Rutilus rutilus* L.: Laboratory and field experiments. *J. Fish Biol.* **20**, 289–299.

Hoffman, A.D., Bertelsen, S.L., Gargas, M.L. (1992). An *in vitro* gas equilibration method for determination of chemical partition coefficients in fish. *Comp. Biochem. Physiol.* **101A**, 47–51.

Hughes, G.H. (1984). General anatomy of the gills. In *Fish Physiology*, Vol. 10, Part A (W.S. Hoar and D.J. Randall, Eds.), pp. 1–72, Academic Press, Orlando.

Hunn, J.B., Allen, J.L. (1974). Movement of drugs across the gills of fish. *Annu. Rev. Pharmacol.* **14**, 47–55.

Hunn, J.B., Allen, J.L. (1975). Residue dynamics of quinaldine and TFM in rainbow trout. *Gen. Pharmacol.* **6**, 15–18.

Hunn, J.B. (1982). Urine flow rate in freshwater salmonids: A review. *Prog. Fish-Culturist* **44**, 119–125.

Ingebrigtsen, K., Solbakken, J.E. (1985). Distribution and elimination of [$^{14}$C]-hexachlorobenzene after single oral exposure in cod (*Gadus morhua*) and flounder (*Platichthys flesus*). *J. Toxicol. Environ. Health* **16**, 197–205.

Ingebrigtsen, K., Solbakken, J.E., Norheim, G., Nafstad, I. (1988). Distribution and elimination of [$^{14}$C] octachlorostyrene in cod (*Gadus morhua*), rainbow trout (*Salmo gairdneri*), and blue mussel (*Mytilus edulis*). *J. Toxicol. Environ. Health* **25**, 361–372.

Iwatsubo, T., Hirota, N., Ooie, T., Suzuki, H., Shimada, N., Chiba, K., Ishizaki, T., Green, C.E., Tyson, C.A., Sugiyama, Y. (1997). Prediction of *in vivo* drug metabolism in the human liver from *in vitro* metabolism data. *Pharmacol. Ther.* **73**, 147–171.

James, M.O. (1987). Conjugation of organic pollutants in aquatic species. *Environ. Health Perspect.* **71**, 97–103.

Jepson, G.W., Hoover, D.K., Black, R.K., McCafferty, J.D., Mahle, D.A., Gearhart, J.M. (1994). A partition coefficient determination method for nonvolatile chemicals in biological tissues. *Fundam. Appl. Toxicol* **22**, 519–524.

Jobling, M. (1987). Influences of food particle size and dietary energy content on patterns of gastric evacuation in fish: Test of a physiological model of gastric emptying. *J. Fish Biol.* **30**, 299–314.

Johnsen, S., Kukkonen, J., Grande, M. (1989). Influence of natural aquatic humic substances on the bioavailability of benzo(a) pyrene to atlantic salmon. *Sci. Tot. Environ.* **81/82**, 691–702.

Kaka, J.S., Hayton, W.L. (1978). Temperature and surfactant dependence of accumulation of 4-aminoantipyrene and ethanol in fish. *J. Pharmaceut. Sci.* **67**, 1558–1563.

Kapoor, B.G., Smit, H., Verighira, I.A. (1975). The alimentary canal and digestion in teleosts. In *Advances in Marine Biology*, Vol. 13 (F.S. Russel and C.M. Yonge, Eds.), pp. 102–219, Academic Press, London.

Karara A.H., Hayton, W.L. (1984). Pharmacokinetic model for the uptake and disposition of di-2-ethylhexyl phthalate in sheepshead minnow *Cyprinodon variegatus*. *Aquat. Toxicol.* **5**, 181–195.

Kelly, B.C., Ikonomou, M.G., Blair, J.D., Morin, A.E., Gobas, F.A. (2007). Food web-specific biomagnification of persistent organic pollutants. *Science* **317**, 236–239.

Khazaee, M., Ng, C.A. (2018). Evaluating parameter availability for physiologically based pharmacokinetic (PBPK) modeling of perfluorooctanoic acid (PFOA) in zebrafish. *Environ. Sci. Process. Impacts* **20**, 105–119.

Kinter, W.B. (1966). Chlorophenol red influx and efflux: Microspectrophotometry of flounder kidney tubules. *Am. J. Physiol.* **211**, 1152–1164.

Kinter, W.B. (1975). Structure and function of the renal tubules isolated from fish kidney. *Fortschritte der Zoologie* **23**, 223–231.

Klaassen, C.D., Plaa, G.L. (1967). Determination of sulfobromophthalein storage and excretory rate in small animals. *J. Appl. Physiol.* **22**, 1151–1155.

Kleinow, K.M., Lech, J.J. (1988). A review of the pharmacokinetics and metabolism of sulfadimethoxine in the rainbow trout (*Salmo gairdneri*). *Vet. Hum. Toxicol.* **30**, 26–30.

Kleinow, K.M., Beilfuss, W.L., Jarboe, H.H., Droy, B.F., Lech, J.J. (1992). Pharmacokinetics, bioavailability, distribution, and metabolism of sulfadimethoxine in the rainbow trout (*Oncorhynchus mykiss*). *Can. J. Fish. Aquat. Sci.* **49**, 1070–1077.

Kleinow, K.M., Jarboe, H.H., Shoemaker, K.E., Greenless, K.J. (1994). Comparative pharmacokinetics and bioavailability of oxolinic acid in channel catfish (*Ictalurus punctatus*) and rainbow trout (*Oncorhynchus mykiss*). *Can. J. Fish. Aquat. Sci.* **51**, 1205–1211.

Kleinow, K.M., Smith, A.A., McElroy, A.E., Wiles, J.E. (1996). Role of the mucous surface coat, dietary bulk and mucosal cell turnover in the intestinal disposition of benzo(*a*)pyrene (BaP). *Mar. Environ. Res.* **42**, 65–73.

Kleinow, K.M., James, M.O. (2001). Response of the teleost gastrointestinal system to xenobiotics. In *Target Organ Toxicity in Marine and Freshwater Teleosts*, Vol. I (D. Schlenk and W.H. Benson, Eds.), pp. 269–362, Taylor & Francis, New York.

Kleinow, K.M., Nichols, J.W., Hayton, W.L., McKim, J.M., Barron, M.G. (2008). Toxicokinetics in fishes. In *The Toxicology of Fishes* (R.T. Di Giulio and D.E. Hinton, Eds.), pp. 55–152, CRC Press, Boca Raton.

Könemann, H., van Leeuwen, K. (1980). Toxicokinetics in fish: Accumulation and elimination of six chlorobenzenes by guppies. *Chemosphere* **9**, 3–19.

Konwick, B.J., Garrison, A.W., Black, M.C., Avants, J.K., Fish, A.T. (2006). Bioaccumulation, biotransformation, and metabolite formation of fipronil and chiral legacy pesticides in rainbow trout. *Environ. Sci. Technol.* **40**, 2930–2936.

Kousba, A., Sultatos, L.G. (2002). Continuous system modeling of equilibrium dialysis for determinations of tissue partitioning of parathion and paraoxon. *Toxicol. Lett.* **133**, 153–159.

Krause, S., Goss, K.-U. (2018). *In vitro-in vivo* extrapolation of hepatic metabolism for different scenarios. *Chem. Res. Toxicol.* **31**, 1195–1202.

Krause, S., Goss, K.-U. (2020a). Comparison of a simple and a complex model for BCF prediction using *in vitro* biotransformation data. *Chemosphere* **256**, 127048.

Krause, S., Goss, K.-U. (2020b). B-compass Fish v 1.2, Helmholtz Centre for Environmental Research-UFZ, Leipzig, Germany. Available at: https://www.ufz.de/index.php?en=47255. Accessed 04/28/2022.

Kuo, D.T.F., Di Toro, D.M. (2013). Biotransformation model of neutral and weakly polar organic compounds in fish incorporating internal partitioning. *Environ. Toxicol. Chem.* **32**, 1873–1881.

Landrum, P.F., Lee II, H., Lydy, M. (1992). Toxicokinetics in aquatic systems: Model comparisons and use in hazard assessment. *Environ. Toxicol. Chem.* **11**, 1709–1725.

Larisch, W., Brown, T.N., Goss, K.-U. (2017). A toxicokinetic model for fish including multiphase sorption features. *Environ. Toxicol. Chem.* **36**, 1538–1546.

Laue, H., Gfeller, H., Jenner, K.J., Nichols, J.W., Kern, S., Natsch, A. (2014). Predicting the bioconcentration of fragrance ingredients by rainbow trout using measured rates of *in vitro* intrinsic clearance. *Environ. Sci. Technol.* **48**, 9486–9495.

Law, F.C.P., Abenini, S., Kennedy, C.J. (1991). A biologically based toxicokinetic model for pyrene in rainbow trout. *Toxicol. Appl. Pharmacol.* **110**, 390–402.

Law, F.C.P. (1999). A physiologically based pharmacokinetic model for predicting the withdrawal period of oxytetracycline in cultured chinook salmon (*Oncorhynchus tshawytscha*). In *Xenobiotics in Fish* (D.J. Smith, W.H. Gingerich, and M.G. Beconi-Barker, Eds.), pp. 105–121, Kluwer Academic/Plenum Publishers, New York.

Layiwola, P.J., Linnecar, D.F., Knights, B. (1983). Hydrolysis of the biliary glucuronic acid conjugate of phenol by the intestinal mucus/flora of goldfish (*Carassius auratus*). *Xenobiotica* **13**, 27–29.

Le, T.T.Y., García, M.R., Nachev, M., Grabner, D., Balsa-Canto, E., Hendriks, A.J., Sures, B. (2018). Development of a PBPK model for silver accumulation in chub infected with Acanthocephalan parasites. *Environ. Sci. Technol.* **52**, 12514–12525.

Lech, J.J., Pepple, S.K., Statham, C.N. (1973). Fish bile analysis: A possible aid in monitoring water quality. *Toxicol. Appl. Pharmacol.* **25**, 430–434.

Lech, J.J., Bend, J.R. (1980). Relationship between biotransformation and the toxicity and fate of xenobiotic chemicals in fish. *Environ. Health Perspect.* **34**, 115–131.

Leo, A. (2000). Octanol/water partition coefficients. In *Handbook of Property Estimation Methods for Chemicals: Environmental and Health Sciences* (R.S. Boethling and D. Mackay, Eds.), pp. 89–114, CRC Press, Boca Raton.

Li, Z., Kroll, K.J., Jensen, K.M., Villeneuve, D.L., Ankley, G.T., Brian, J.V., Sepúlveda, M.S., Orlando, E.F., Lazorchak, J.M., Kostich, M., Armstrong, B., Denslow, N.D., Watanabe, K.H. (2011). A computational model of the hypothalamic - pituitary - gonadal axis in female fathead minnows (*Pimephales promelas*) exposed to 17α-ethynylestradiol and 17β-trenbolone. *BMC Sys. Biol.* **5**, 63.

Liao, C.-M., Liang, H.-M., Chen, B.-C., Singh, S., Tsai, J.-W., Chou, Y.-H., Lin, W.-T. (2005). Dynamical coupling of PBPK/PD and AUC-based toxicity models for arsenic in tilapia *Oreochromis mossambicus* from blackfoot disease area in Taiwan. *Environ. Pollut.* **135**, 221–233.

Lieb, J.A., Bills, D.D., Sinnhuber, R.O. (1974). Accumulation of dietary polychlorinated biphenyls (Arochlor 1254) by rainbow trout (*Salmo gairdneri*). *J. Agric. Food Chem.* **22**, 638–642.

Lien, G.J., McKim, J.M. (1993). Predicting branchial and cutaneous uptake of 2,2′,5,5′-tetrachlorobiphenyl in fathead minnow (*Pimephales promelas*) and Japanese medaka (*Oryzias latipes*): Rate-limiting factors. *Aquat. Toxicol.* **27**, 15–32.

Lien, G.J., Nichols, J.W., McKim, J.M., Gallinat, C.A. (1994). Modeling the accumulation of three waterborne chlorinated ethanes in fathead minnows (*Pimephales promelas*): A physiologically based approach. *Environ. Toxicol. Chem.* **13**, 1195–1205.

Lien, G.J., McKim, J.M., Hoffman, A.D., Jenson, C.T. (2001). A physiologically based toxicokinetic model for lake trout (*Salvelinus namaycush*). *Aquat. Toxicol.* **51**, 335–350.

Lin, Z., Cheng, Y.-H., Chou, W.-C., Li, M. (2020). Physiologically based pharmacokinetic model calibration, evaluation, and performance assessment. In *Physiologically Based Pharmacokinetic (PBPK) Modeling, Methods and Applications in Toxicology and Risk Assessment* (J.W. Fisher, J.M. Gearhart, Z. Lin, Eds.), pp. 243–279, Academic Press, London.

Lin, H.-C., Chen, W.-Y. (2021). Bayesian population-based pharmacokinetic model for robustness evaluation of withdrawal time in tilapia aquaculture administrated to florfenicol. *Ecotoxicol Environ. Saf.* **210**, 111867.

Ling, M.-P., Liao, C.-M., Tsai, J.-W., Chen, B.-C. (2005). A PBTK/TD modeling-based approach can assess arsenic bioaccumulation in farmed tilapia (*Oreochromis mossambicus*) and human health risks. *Integr. Environ. Assess. Manag.* **1**, 40–54.

Lloyd, R. (1961). Effect of dissolved oxygen concentrations on the toxicity of several poisons to rainbow trout (*Salmo gairdneri* Richardson). *J. Exp. Biol.* **38**, 447–455.

Lo, J.C., Campbell, D.A., Kennedy, C.J., Gobas, F.A.P.C. (2015). Somatic and gastrointestinal *in vivo* biotransformation rates of hydrophobic chemicals in fish. *Environ. Toxicol. Chem.* **34**, 2282–2294.

Louisse, J., Beekmann, K., Rietjens, I.M.C.M. (2017). Use of physiologically based kinetic modeling-based reverse dosimetry to predict *in vivo* toxicity from *in vitro* data. *Chem. Res. Toxicol.* **30**, 114–125.

Luckenbach, T., Fischer, S., Sturm, A. (2014). Current advances on ABC transporters in fish. *Comp. Biochem. Physiol.* **165C**, 28–52.

Mackay, D., Bobra, A., Shiu, W.Y., Yalkowsky, S.H. (1980). Relationships between aqueous solubility and octanol-water partition coefficients. *Chemosphere* **9**, 701–711.

Mackay, D. (1982). Correlation of bioconcentration factors. *Environ. Sci. Technol.* **16**, 274–278.

Mackay, D., Arnot, J.A., Gobas, F.A.P.C., Powell, D.E. (2013). Mathematical relationships between metrics of chemical bioaccumulation in fish. *Environ. Toxicol. Chem.* **32**, 1459–1466.

Mackay, D., Celsie, A.K.D., Arnot, J.A., Powell, D.E. (2016). Processes influencing chemical biomagnification and trophic magnification factors in aquatic ecosystems; implications for chemical hazard and risk assessment. *Chemosphere* **154**, 99–108.

Mackintosh, C.E., Maldonado, J., Hongwu, J., Hoover, N., Chong, A., Ikonomou, M., Gobas, F.A.P.C. (2004). Distribution of phthalate esters in a marine aquatic food web: Comparison to polychlorinated biphenyls. *Environ. Sci. Technol.* **38**, 2011–2020.

Mangold-Döring, A., Grimard, C., Green, D., Petersen, P., Nichols, J.W., Hogan, N., Weber, W., Hollert, H., Hecker, M., Brinkmann, M. (2021). A novel multispecies toxicokinetic modeling approach in support of chemical risk assessment. *Environ. Sci. Technol.* **55**, 9109–9118.

Mansouri, K., Grulke, C.M., Judson, R.S., Williams, A.J. (2018). OPERA models for predicting physicochemical properties and environmental fate endpoints. *J Cheminformatics* **10**, 10.

Matthies, M., Solomon, K., Vighi, M., Gilman, A., Tarazona, J.V. (2016). The origin and evolution of assessment criteria for persistent, bioaccumulative and toxic (PBT) chemicals and persistent organic pollutants (POPs). *Environ. Sci. Process. Impacts* **18**, 1114–1128.

McCarthy, J.F., Jimenez, B.D. (1985). Reduction in bioavailability to bluegills of polycyclic aromatic hydrocarbons bound to dissolved humic material. *Environ. Toxicol. Chem.* **4**, 511–521.

McDonald, D.G., McMahon, B.R. (1977). Respiratory development in Arctic char (*Salvelinus alpinus*) under conditions of normoxia and chronic hypoxia. *Can. J. Zool.* **55**, 1461–1467.

McDougal, J.N., Jepson, G.W., Clewell III, H.J., Andersen, M.E. (1990). Dermal absorption of organic chemical vapors in rats and humans. *Fundam. Appl. Toxicol.* **14**, 299–308.

McElman, J.F., Balon, E.K. (1980). Early ontogeny of white sucker, *Catostomus commersoni*, with steps of saltatory development. *Environ. Biol. Fishes* **5**, 191–224.

McKim, J.M., Goeden, H.M. (1982). A direct measure of the uptake efficiency of a xenobiotic chemical across the gills of brook trout (*Salvelinus fontinalis*) under normoxic and hypoxic conditions. *Comp. Biochem. Physiol.* **72C**, 65–74.

McKim, J.M., Heath, E.M. (1983). Dose determinations for waterborne 2,5,2′5′-[$^{14}$C] tetrachlorobiphenyl and related pharmacokinetics in two species of trout (*Salmo gairdneri* and *Salvelinus fontinalis*): A mass- balance approach. *Toxicol. Appl. Pharmacol.* **68**, 177–187.

McKim, J.M., Schmieder, P., Veith, G. (1985). Absorption dynamics of organic chemical transport across trout gills as related to octanol-water partition coefficient. *Toxicol. Appl. Pharmacol.* **77**, 1–10.

McKim, J.M., Schmieder, P.K., Erickson, R.J. (1986). Toxicokinetic modeling of [$^{14}$C] pentachlorophenol in rainbow trout (*Salmo gairdneri*). *Aquat. Toxicol.* **9**, 59–80.

McKim, J.M., Schmieder, P.K., Carlson, R.W., Hunt, E.P., Niemi, G.J. (1987a). Use of respiratory-cardiovascular responses of rainbow trout (*Salmo gairdneri*) in identifying acute toxicity syndromes in fish. Part 1. Pentachlorophenol, 2,4–dinitrophenol, tricaine, methanesulfonate, and 1-octanol. *Environ. Toxicol. Chem.* **6**, 295–312.

McKim, J.M., Schmieder, P.K., Niemi, G.J., Carlson, R.W., Henry, T.R. (1987b). Use of respiratory-cardiovascular responses of rainbow trout (*Salmo gairdneri*) in identifying acute toxicity syndromes in fish. Part 2. Malathion, carbaryl, acrolein, and benzaldehyde. *Environ. Toxicol. Chem.* **6**, 313–328.

McKim, J.M., Nichols, J.W., Lien, G.J., Hoffman, A.D., Gallinat, C.A., Stokes, G.N. (1996). Dermal absorption of three waterborne chloroethanes in rainbow trout (*Oncorhynchus mykiss*) and channel catfish (*Ictalurus punctatus*). *Fundam. Appl. Toxicol.* **31**, 218–228.

McKim, J.M., Kolanczyk, R.C., Lien, G.J., Hoffman, A.D. (1999). Dynamics of renal excretion of phenol and major metabolites in the rainbow trout (*Oncorhynchus mykiss*). *Aquat. Toxicol.* **45**, 265–277.

Metcalf, V.J., Brennan, S.O., George, P.M. (1999). The Antarctic toothfish (*Dissostichus mawsoni*) lacks plasma albumin and utilizes high density lipoprotein as its major palmitate binding protein. *Comp. Biochem. Physiol.* **124B**, 147–155.

Metcalf, V.J., George, P.M., Brennan, S.O. (2007). Lungfish albumin is more similar to tetrapod than to teleost albumins: Purification and characterization of albumin from the Australian lungish, *Neoceratodus forsteri. Comp. Biochem. Physiol.* **147B**, 428–437.

Meylan, W.M., Howard, P.H. (1995). Atom/fragment contribution method for estimating octanol-water partition coefficients. *J. Pharm. Sci.* **84**, 83–92.

Miller, D.S. (1987). Aquatic models for the study of renal transport function and pollutant toxicity. *Environ. Health Perspect.* **71**, 59–68.

Miller, M.A. (1993). Maternal transfer of organochlorine compounds in salmonines to their eggs. *Can. J. Fish. Aquat. Sci.* **50**, 1405–1413.

Miller, D.S., Pritchard, J.B. (1997). Dual pathways for organic anion secretion in renal proximal tubule. *J. Exp. Zool.* **279**, 462–470.

Mittal, V.K., Ng, C.A. (2018). Formation of PFAAs in fish through biotransformation: A PBPK approach. *Chemosphere* **202**, 218–227.

Moermond, C.T., Janssen, M.P., de Knecht, J.A., Montforts, M.H., Peijnenburg, W.J., Zweers, P.G., Sijm, D.T. (2012). PBT assessment using the revised annex XIII of REACH: A comparison with other regulatory frameworks. *Integr. Environ. Assess. Manag.* **8**, 359–371.

Mordenti, J. (1986). Man versus beast: Pharmacokinetic scaling in mammals. *J. Pharmaceut. Sci.* **75**, 1028–1040.

Morrison, H.A., Gobas, F.A.P.C., Lazar, R., Haffner, D.G. (1996). Development and verification of a bioaccumulation model for organic contaminants in benthic invertebrates. *Environ. Sci. Technol.* **30**, 3377–3384.

Murphy, P.G., Murphy, J.V. (1971). Correlations between respiration and direct uptake of DDT in the mosquito fish (*Gambusia affinis*). *Bull. Environ. Contam. Toxicol.* **6**, 581–588.

Murphy, J.E., Janszen, D.B., Gargas, M.L. (1995). An *in vitro* method for determination of tissue partition coefficients of non-volatile chemicals such as 2,3,7,8-tetrachlorodibenzo-*p*-dioxin and estradiol. *J. Appl. Toxicol.* **15**, 147–152.

Nagel, R., Urich, K. (1980). Kinetic studies on the elimination of different substituted phenols by goldfish (*Carassius auratus*). *Bull. Environ. Contam. Toxicol.* **24**, 374–378.

Neely, W.B., Branson, D.R., Blau, G.E. (1974). Partition coefficients to measure bioconcentration potential of organic chemicals in fish. *Environ. Sci. Technol.* **8**, 1113–1115.

Neely, W.B. (1979). Estimating rate constants for the uptake and clearance of chemicals by fish. *Environ. Sci. Technol.* **13**, 1506–1510.

Neumann, P., Holeton, G.F., Heisler, N. (1983). Cardiac output and regional blood flow in gills and muscles after exhaustive exercise in rainbow trout (*Salmo gairdneri*). *J. Exp. Biol.* **105**, 1–14.

Ng, C.A., Hungerbühler, K. (2013). Bioconcentration of perfluorinated alkyl acids: How important is specific binding? *Environ. Sci. Technol.* **47**, 7214–7223.

Nichols, J.W., McKim, J.M., Andersen, M.E., Gargas, M.L., Clewell III, H.J., Erickson, R.J. (1990). A physiologically based toxicokinetic model for the uptake and disposition of waterborne organic chemicals in fish. *Toxicol. Appl. Pharmacol.* **106**, 433–447.

Nichols, J.W., McKim, J.M., Lien, G.J., Hoffman, A.D., Bertelsen, S.L., Gallinat, C.A. (1993). Physiologically based toxicokinetic modeling of three waterborne chloroethanes in channel catfish, *Ictalurus punctatus*. *Aquat. Toxicol.* **27**, 83–112.

Nichols, J.W., McKim, J.M., Lien, G.J., Hoffman, A.D., Bertelsen, S.L., Elonen, C.M. (1996). A physiologically based toxicokinetic model for dermal absorption of organic chemicals by fish. *Fundam. Appl. Toxicol.* **31**, 229–242.

Nichols, J.W., Jensen, K.M., Tietge, J.E., Johnson, R.D. (1998). A physiologically based toxicokinetic model for maternal transfer of 2,3,7,8-tetrachlorodibenzo-*p*-dioxin in brook trout (*Salvelinus fontinalis*). *Environ. Toxicol. Chem.* **17**, 2422–2434.

Nichols, J.W., Fitzsimmons, P.N., Whiteman, F.W., Dawson, T.D., Babeu, L., Juenemann, J. (2004). A physiologically based toxicokinetic model for dietary uptake of hydrophobic organic compounds by fish. I. Feeding studies with 2,2′,5,5′-tetrachlorobiphenyl. *Toxicol. Sci.* **77**, 206–218.

Nichols, J.W., Schultz, I.R., Fitzsimmons, P.N. (2006). *In vitro-in vivo* extrapolation of hepatic extrapolation data for fish. I. A review of methods, and strategies for incorporating intrinsic clearance estimates into chemical kinetic models. *Aquat. Toxicol.* **78**, 74–90.

Nichols, J.W., Fitzsimmons, P.N., Burkhard, L.P. (2007). *In vitro-in vivo* extrapolation of quantitative hepatic biotransformation data for fish. II. Modeled effects on chemical bioaccumulation. *Environ. Toxicol. Chem.* **26**, 1304–1319.

Nichols, J.W., Huggett, D.B., Arnot, J.A., Fitzsimmons, P.N., Cowan-Ellsberry, C.E. (2013). Toward improved models for predicting bioconcentration of well metabolized compounds by rainbow trout using measured rates of *in vitro* intrinsic clearance. *Environ. Toxicol. Chem.* **32**, 1611–1622.

Nichols, J.W., Hoffman, A.D., Swintek, J.A., Droge, S.T.J., Fitzsimmons, P.N. (2021). Addition of phenylmethylsulfonyl fluoride increases the working lifetime of the trout liver S9 substrate depletion assay, resulting in improved detection of low intrinsic clearance rates. *Environ. Toxicol. Chem.* **40**, 148–161.

Nigam, S.K., Bush, K.T., Martovetsky, G., Ahn, S.Y., Lium H.C., Bhatnagar, R.E. (2015). The organic anion transporter (OAT) family: A systems biology perspective. *Physiol. Rev.* **95**, 83–123.

Niimi, A.J. (1983). Biological and toxicological effects of environmental contaminants in fish and their eggs. *Can. J. Fish. Aquat. Sci.* **40**, 306–312.

Niimi, A.J., Oliver, B.G. (1983). Biological half-lives of polychlorinated biphenyl (PCB) congeners in whole fish and muscle of rainbow trout (*Salmo gairdneri*). *Can. J. Fish. Aquat. Sci.* **40**, 1388–1394.

Nishimura, H., Imai, M. (1982). Control of renal function in freshwater and marine teleosts. *Fed. Proc.* **41**, 2355–2360.

Norstrom, R.J., McKinnon, A.E., DeFreitas, S.W. (1976). A bioenergetics-based model for pollutant accumulation by fish. Simulation of PCB and methyl mercury residue levels in Ottawa River yellow perch (*Perca flavescens*). *J. Fish. Res. Board Can.* **33**, 248–267.

Obach, R.S., Baxter, J.G., Liston, T.E., Silber, B.M., Jones, B.C., MacIntyre, F., Rance, D.J., Wastall, P. (1997). The prediction of human pharmacokinetic parameters from preclinical *in vitro* metabolism data. *J. Pharmacol. Exp. Ther.* **283**, 46–58.

Oikawa, S., Itazawa, Y. (1985). Gill and body surface areas of the carp in relation to body mass, with special reference to the metabolism-size relationship. *J. Exp. Biol.* **117**, 1–14.

Oliver, B.G., Niimi, A.J. (1985). Bioconcentration factors of some halogenated organics for rainbow trout: Limitations in their use for prediction of environmental residues. *Environ Sci. Technol.* **19**, 842–849.

Olson, K.R. (1996). Secondary circulation in fish: Anatomical organization and physiological significance. *J. Exp. Zool.* **275**, 172–185.

Olsson, P.-E., Kling, P., Hogstrand, C. (1998). Mechanisms of heavy metal accumulation and toxicity in fish. In *Metal Metabolism in Aquatic Environments* (W.J. Langston and M.J. Bebianno, Eds.), pp. 321–350, Chapman & Hall, London.

Opperhuizen, A., van der Velde, E.W., Gobas, F.A.P.C., Liem, D.A.K, van der Steen, J.M.D. (1985). Relationship between bioconcentration in fish and steric factors of hydrophobic chemicals. *Chemosphere* **14**, 1871–1896.

Opperhuizen, A, Schrap, S.M. (1987). Relationships between aqueous oxygen concentration and uptake and elimination rates during bioconcentration of hydrophobic chemicals in fish. *Environ. Toxicol. Chem.* **6**, 335–342.

OECD. (1995). *Test No. 107: Partition Coefficient (n-Octanol/Water): Shake Flask Method, OECD Guidelines for the Testing of Chemicals, Section 1*, OECD Publishing, Paris.

OECD. (2004). *Test No. 117: Partition Coefficient (n-Octanol/Water), HPLC Method,* OECD Guidelines for the Testing of Chemicals, Section 1, OECD Publishing, Paris.

OECD. (2006). *Test No. 123: Partition Coefficient (1-octanol/water): Slow-Stirring Method, OECD Guidelines for the Testing of Chemicals, Section 1*, OECD Publishing, Paris.

OECD. (2007). *Guidance Document on the Validation of (Quantitative) Structure-Activity Relationship [(Q)SAR] models*, OECD Series and Testing and Assessment, No. 69, OECD Publishing, Paris.

OECD. (2012). *Test No. 305: Bioaccumulation in Fish: Aqueous and Dietary Exposure, OECD Guidelines for the Testing of Chemicals, Section 3*, OECD Publishing, Paris, France.

OECD. (2018). *Test No. 319B: Determination of in vitro intrinsic clearance using rainbow trout liver S9 sub-cellular fractions (RT-S9), OECD Guidelines for the Testing of Chemicals, Section 3*, OECD Publishing, Paris.

OECD. (2019). *Guiding Principles and Key Elements for Establishing a Weight of Evidence for Chemical Assessment*, OECD Series on Testing and Assessment, No. 311, OECD Publishing Paris.

Otero-Muras, I., Franco-Uría, A., Alonso, A.A., Balso-Canto, E. (2010). Dynamic multi-compartmental modelling of metal bioaccumulation in fish: Identifiability implications. *Environ. Model. Softw.* **25**, 344–353.

Ott, M.E., Heisler, N., Ultsch, G.R. (1980). A re-evaluation of the relationship between temperature and the critical oxygen tension in freshwater fishes. *Comp. Biochem. Physiol.* **67A**, 337–340.

Papa, E., Van Der Wal, L., Arnot, J.A., and Gramatica, P. (2014). Metabolic biotransformation half-lives in fish: QSAR modeling and consensus analysis. *Sci. Tot. Environ.* ***470-471***, 1040–1046

Parkerton, T.F., Arnot, J.A., Weisbrod, A.V., Russom, C., Hoke, R.A., Woodburn, K., Traas, T., Bonnell, M., Burkhard, L.P., Lampi, M.A. (2008.) Guidance for evaluating *in vivo* fish bioaccumulation data. *Integr. Environ. Assess. Manag.* **4**, 139–155.

Perrier, H., Perrier, C., Peres, G., Gras, J. (1977). The perchloro-soluble proteins of the serum of the rainbow trout (*Salmo gairdneri* Richardson): Albumin like and hemoglobin binding fraction. *Comp. Biochem. Physiol.* **57B**, 325–327.

Persson, L. (1986). Patterns of food evacuation in fishes: A critical review. *Environ. Biol. Fishes* **16**, 51–58.

Péry, A.R.R., Devillers, J., Brochot, C., Mombelli, E., Palluel, O., Piccini, B., Brion, F., Beaudouin, R. (2014). A physiologically based toxicokinetic model for the zebrafish *Danio rerio*. *Environ. Sci. Technol.* **48**, 781–790.

Peyret, T., Poulin, P., Krishnan, K. (2010). A unified algorithm for predicting partition coefficients for PBPK modeling of drugs and environmental chemicals. *Toxicol. Appl. Pharm.* **249**, 197–207.

Pinsuwan, S., Li, A., Yalkowsky, S.H. (1995). Correlation of octanol/water solubility ratios and partition coefficients. *J. Chem. Eng. Data* **40**, 623–626.

Plakas, S.M., Stehly, G.R., Khoo, L. (1992a). Pharmacokinetics and excretion of phenol red in the channel catfish. *Xenobiotica* **22**, 551–557.

Plakas, S.M., Khoo, L., Barron, M.G. (1992b). 2,4-Dichlorophenoxyacetic acid disposition following oral administration in channel catfish. *J. Agric. Food Chem.* **40**, 1236–1239.

Popovic, M., Zaja, R., Fent, K., Smital, T. (2014). Interaction of environmental contaminants with zebrafish organic anion transporting polypeptide, Oatp1d1 (Slco1d1). *Toxicol. Appl. Pharmacol.* **280**, 149–158.

Poulin, P., Krishnan, K. (1995). An algorithm for predicting tissue:blood partition coefficients of organic chemicals from *n*-octanol:water partition coefficient data. *J. Toxicol. Environ. Health* **46**, 117–129.

Price, J.W. (1931). *Growth and gill development in the small-mouthed black bass*, Micropterus dolomieu *Lacapede*, Contribution No. 4 to the Franz Theodore Stone Laboratory, The Ohio State University Press, Columbus.

Pritchard, J.B. (1981). Renal handling of environmental chemicals. In *Toxicology of the Kidney* (J.B. Hook, Ed.), pp. 99–116, Raven Press, New York.

Pritchard, J.B. (2001). Renal handling of organic acids and bases. In *The Textbook of Nephrology*, 4th ed. (S.G. Massry and R.J. Glassock, Eds.), pp. 93–97, Lippincott Williams & Wilkins, Baltimore.

Pritchard, J.B., Renfro, J.L. (1984). Interactions of xenobiotics with teleost renal function. In *Aquatic Toxicology*, Vol. 2 (L.J. Weber, Ed.), pp. 51–106, Raven Press, New York, NY.

Pritchard, J.B., Miller, D.S. (1993). Mechanisms mediating renal secretion of organic anions and cations. *Physiol. Rev.* **73**, 765–796.

Qiao, P., Gobas, F.A.P.C., Farrell, A.P. (2000). Relative contributions of aqueous and dietary uptake of hydrophobic chemicals to the body burden in juvenile rainbow trout. *Arch. Environ. Contam. Toxicol.* **39**, 369–377.

Ramsey, J.C., Andersen, M.E. (1984). A physiologically based description of the inhalation pharmacokinetics of styrene in rats and humans. *Toxicol. Appl. Pharmacol.* **73**, 159–175.

Ran Y, Jain N, Yalkowsky SH. 2001. Prediction of aqueous solubility of organic compounds by the general solubility equation (GSE). *J. Chem. Inf. Comp. Sci.* **41**, 1208–1217.

Randall, D.J. (1982). The control of respiration and circulation in fish during exercise and hypoxia. *J. Exp. Biol.* **100**, 275–288.

Ratier, A., Lopes, C., Multari, G., Mazerolles, V., Carpentier, P., Charles, S. (2021). New perspectives on the calculation of bioaccumulation metrics for active substances in living organisms. *Integ. Environ. Assess. Manage.* **18**, 10–18.

Riggs, D.S. (1963). *The Mathematical Approach to Physiological Problems*, Williams & Wilkins, Baltimore.

Robertson, J.C., Bradley, T.M. (1992). Liver ultrastructure of juvenile Atlantic salmon (*Salmo salar*). *J. Morphol.* **211**, 41–54.

Rodgers, D.W., Beamish, F.W.H. (1981). Uptake of waterborne methylmercury by rainbow trout (*Salmo gairdneri*) in relation to oxygen consumption and methylmercury concentration. *Can. J. Fish. Aquat. Sci.* **38,** 1309–1315.

Rodrigues, A.D. (1997). Preclinical drug metabolism in the age of high-throughput screening: An industrial perspective. *Pharm. Res.* **14**, 1504–1510.

Roesijadi, G., Robinson, W.E. (1994). Metal uptake regulation in aquatic animals: Mechanisms of uptake, accumulation and release. In *Aquatic Toxicology: Molecular, Biochemical, and Cellular Perspectives* (D.C. Malins and G.K. Ostrander, Eds.), pp. 387–420, Lewis, Boca Raton.

Rombough, P.J., Moroz, B.M. (1990). The scaling and potential importance of cutaneous and branchial surfaces in respiratory gas exchange in young Chinook salmon (*Oncorhynchus tshawytscha*). *J. Exp. Biol.* **154**, 1–12.

Rombout, J.H., Lamers, C.H., Helfrich, M.H., Dekker, A., Taverne-Thiele, J.J. (1985). Uptake and transport of intact macromolecules in the intestinal epithelium of carp (*Cyprinus carpio* L.) and the possible immunological implications. *Cell Tissue Res.* **239**, 519–530.

Roth, M., Obaidat, A., Hagenbuch, B. (2012). OATPs, OATs, and OCTs: The organic anion and cation transporters of the SLCO and SLC22A gene superfamilies. *Br. J. Pharmacol.* **165**, 1260–1287.

Rotroff, D.M., Wetmore, B.A, Dix, D.J., Ferguson, S.S., Clewell, H.J., Houck, K.A., LeCluyse, E.L., Andersen, M.E., Judson, R.S., Smith, C.M., Sochaski, M.A., Kavlock, R.J., Boellmann, F., Martin, M.T., Reif, D.M., Wambaugh, J.F., Thomas, R.S. (2010). Incorporating human dosimetry and exposure into high-throughput *in vitro* toxicity screening. *Toxicol. Sci.* **117**, 348–358.

Rowland, M., Benet, L.Z., Graham, G.G. (1973). Clearance concepts in pharmacokinetics. *J. Pharmacokinet. Biopharm.* **1**, 123–136.

Rowland, M. (1985). Physiologic pharmacokinetic models and inter-animal species scaling. *Pharmacol. Therapeut.* **29**, 49–68.

Ruark, C.D., Hack, C.E., Robinson, P.J., Mahle, D.A., Gearhart, J.M. (2014). Predicting passive and active tissue:plasma partition coefficients: Interindividual and interspecies variability. *J. Pharm. Sci.* **103**, 2189–2198.

Ruckesbusch, Y., Phaneuf, L., *and* Dunlop, R. *(*1991*)*. Physiology of Small and Large Animals, Dekker, Philadelphia.

Saarikoski, J., Lindstrom, R., Tyynela, M., Viluksela, M. (1986). Factors affecting the absorption of phenolics and carboxylic acids in the guppy (*Poecilia reticulata*). *Ecotoxicol. Environ. Saf.* **11**, 158–173.

Sahin, A., Tencalla, F.G., Dietrich, D.R., Naegeli, H. (1996). Biliary excretion of biochemically active cyanobacteria (blue-green algae) hepatotoxins in fish. *Toxicology* **106**, 123–130.

Salmina, E.S., Wondrousch, D., Kühne, R., Potemkin, V.A., Schüürmann, G. (2016). Variation in predicted internal concentrations in relation to PBPK model complexity in rainbow trout. *Sci. Tot. Environ.* **550**, 586–597.

Samuelsen, O.B., Ervik, A., Wennevik, V. (1995). Absorption, tissue distribution, metabolism and excretion of ormetoprim and sulphadimethoxine in Atlantic salmon (*Salmo salar*) after intravenous and oral administration of Romet[30]. *Xenobiotica* **25**, 1169–1180.

Sangster, J. (1989). Octanol-water partition coefficients of simple organic compounds. *J. Phys. Chem. Ref. Data* **18**, 1111–1229.

Sanz, A., Cardenete, G., Hidalgo, F., Garcia-Gallego, M. (1993). Study of biliary secretion in trout (*Oncorhynchus mykiss*) by cannulation of bile duct. *Comp. Biochem. Physiol.* **104A**, 525–529.

Satchell, G.H. (1992). The venous system. In *Fish Physiology*, Vol. 12 (W.S. Hoar, D.J. Randall, and A.P. Farrell, Eds.), pp. 141–184, Academic Press, New York.

Saunders, L.J., Fontanay, S., Nichols, J.W., Gobas, F.A.P.C. (2019). Concentration dependence of *in vitro* biotransformation rates of hydrophobic organic sunscreen agents in rainbow trout S9 fractions: Implications for bioaccumulation assessment. *Environ. Toxicol. Chem.* **38**, 548–560.

Saunders, L.J., Hoffman, A.D., Nichols, J.W., Gobas, F.A.P.C. (2020a). Dietary bioaccumulation and biotransformation of hydrophobic organic sunscreen agents in rainbow trout. *Environ. Toxicol. Chem.* **39**, 574–586.

Saunders, L.J., Fitzsimmons, P.N., Nichols, J.W., Gobas, F.A.P.C. (2020b). *In vitro-in vivo* extrapolation of hepatic and gastrointestinal biotransformation rates of hydrophobic chemicals in rainbow trout. *Aquat. Toxicol.* **228**, 105629.

Schar, M., Maly, I.P., Sasse, D. (1985). Histochemical studies on metabolic zonation of the liver in the trout (*Salmo gairdneri*). *Histochemistry* **83**, 147–151.

Schlenk, D., Moore, C.T. (1993). Distribution and elimination of the herbicide propanil in the channel catfish (*Ictalurus punctatus*). *Xenobiotica* **23**, 1017–1024.

Schmidt, D.C., Weber, L.J. (1973). Metabolism and biliary excretion of sulfobromophthalein by rainbow trout (*Salmo gairdneri*). *J. Fish. Res. Board Can.* **30**, 1301–1308.

Schmidt-Nielsen, K. (1984). *Scaling: Why Is Animal Size So Important?*, Cambridge University Press, Cambridge.

Schmieder, P.K., Henry, T.R. (1988). Plasma binding of 1-butanol, phenol, nitrobenzene and pentachlorophenol in the rainbow trout and rat: A comparative study. *Comp. Biochem. Physiol.* **91C**, 413–418.

Schmieder, P.K., Weber, L.J. (1992). Blood and water flow limitations on gill uptake of organic chemicals in the rainbow trout (*Oncorhynchus mykiss*). *Aquat. Toxicol.* **24**, 103–122.

Schmitt, W. (2008). General approach for the calculation of tissue to plasma partition coefficients. *Toxicol. In Vitro* **22**, 457–467.

Schrap, S.M., Opperhuizen, A. (1990). Relationship between bioavailability and hydrophobicity: Reduction of the uptake of organic chemicals by fish due to the sorption on particles. *Environ. Toxicol. Chem.* **9**, 715–724.

Schreiber, R., Altenburger, R., Paschke, A., Küster, E. (2008). How to deal with lipophilic and volatile organic substances in microtiter plate assays. *Environ. Toxicol. Chem.* **27**, 1676–1682.

Schultz, I.R., Hayton, W.L. (1994). Body size and the toxicokinetics of trifluralin in rainbow trout. *Toxicol. Appl. Pharmacol.* **129**, 138–145.

Schultz, I.R., Hayton, W.L., Kemmenoe, B.H. (1995). Disposition and toxicokinetics of diquat in channel catfish. *Aquat. Toxicol.* **33**, 297–310.

Schultz, I.R., Hayton, W.L. (1997). Influence of body fat on trifluralin toxicokinetics in rainbow trout (*Oncorhynchus mykiss*). *Environ. Toxicol. Chem.* **16**, 997–1001.

Schultz, I.R., Barron, M.G., Newman, M.C., Vick, A.M. (1999). Blood flow distribution and tissue allometry in channel catfish (*Ictalurus punctatus*). *J. Fish Biol.* **54**, 1275–1286.

Schultz, I.R., Orner, G., Merdink, J.L., Skilman, A. (2001). Dose-response relationships and pharmacokinetics of vitellogenin in rainbow trout after intravascular administration of 17alpha-ethynylestradiol. *Aquat. Toxicol.* **51**, 305–318.

Schwarzenbach, R.P., Gschwend, P.M., Imboden, D.M. (2016). *Environmental Organic Chemistry*, 5th ed., Wiley, Hoboken.

Seth, R., Mackay, D., Muncke, J. (1999). Estimating the organic carbon partition coefficient and its variability for hydrophobic chemicals. *Environ. Sci. Technol.* **33**, 2390–2394.

Shrable, J.B., Tiemeier, O.W., Deyoe, L.W. (1969). Effects of temperature on the rate of digestion by channel catfish. *Prog. Fish-Culturist* **31**, 131–138.

Sijm, D, T, H, M, Yarechewskim, A.L., Muir, D.C.G., Webster, G.R.B., Seinen, W., Opperhuizen, A. (1990). Biotransformation and tissue distribution of 1,2,3,7-tetrachlorodibenzo-*p*-dioxin, 1,2,3,4,7-pentachlorodibenzo-*p*-dioxin and 2,3,4,7,8-pentachlorodibenzo-p-dioxin in rainbow trout. *Chemosphere* **21**, 845–866.

Smith, L.S. (1989). Digestive functions in teleost fishes. In *Fish Nutrition*, 2nd ed. (J.E. Halver, Ed.), pp. 331–421, Academic Press, San Diego.

Smith, L.S. and G.R. Bell. (1975). *A Practical Guide to the Anatomy and Physiology of Pacific Salmon*, Miscellaneous Special Publication 27, Department of the Environment, Fisheries and Marine Service, Ottawa.

Southworth, G.R., Keffer, C.C., Beauchamp, J.J. (1980). Potential and realized bioconcentration: A comparison of observed and predicted bioconcentration of azaarenes in the fathead minnow (*Pimephales promelas*). *Environ. Sci. Technol.* **14**, 1529–1531.

Stadnicka-Michalak, J., Tannenberger, K., Schirmer, K., Ashauer, R. (2014). Measured and modeled toxicokinetics in cultured fish cells and application to in vitro - in vivo toxicity extrapolation. *PLoS One* **9(3)**, e92303.

Stadnicka-Michalak, J., Weiss, F.T., Fischer, M., Tanneberger, K., Schirmer, K. (2018). Biotransformation of benzo[*a*]pyrene by three rainbow trout (*Oncorhynchus mykiss*) cell lines and extrapolation to derive a fish bioconcentration factor. *Environ. Sci. Technol.* **52**, 3091–3100.

Statham, C.N., Melancon, Jr., M.J., Lech, J.J. (1976). Bioconcentration of xenobiotics in trout bile: A proposed monitoring aid for some waterborne chemicals. *Science* **193**, 680–681.

Steffensen, J.F., Lomholt, J.P. (1992). The secondary vascular system. In *Fish Physiology*, Vol. 12 (W.S. Hoar, D.J. Randall, and A.P. Farrell, Eds.), pp. 185–213, Academic Press, San Diego.

Stehly G.R., Hayton W.L. (1989). Disposition of pentachlorophenol in rainbow trout (*Salmo gairdneri*): Effect of inhibition of metabolism. *Aquat. Toxicol.* **14**, 131–148.

Stehly, G.R., Plakas, S.M. (1992). Disposition of 1-naphthol in the channel catfish (*Ictalurus punctatus*). *Drug Metab. Dispos.* **20**, 70–73.

Stehly, G.R., Plakas, S.M. (1993). Pharmacokinetics, tissue distribution, and metabolism of nitrofurantoin in the channel catfish (*Ictalurus punctatus*). *Aquaculture* **113**, 1–10.

Stenzel, A., Goss, K.-U., Endo, S. (2013a). Determination of polyparameter linear free energy relationship (pp-LFER) substance descriptors for established and alternative flame retardants. *Environ. Sci. Technol.* **47**, 1399–1406.

Stenzel, A., Goss, K.-U., Endo, S. (2013b). Experimental determination of polyparameter linear free energy relationship (pp-LFER) substance descriptors for pesticides and other contaminants: New measurements and recommendations. *Environ. Sci. Technol.* **47**, 14204–14214.

Stroband, H.W., van deer Meer, H., Timmermans, L.P. (1979). Regional functional differentiation in the gut of the grasscarp, *Ctenopharyngodon idella* (Val.). *Histochemistry* **64**, 235–249.

Stroband, H.W., Kroon, A.G. (1981). The development of the stomach in *Clarias lazera* and the intestinal absorption of protein macromolecules. *Cell Tissue Res.* **215**, 397–415.

Tebby, C., Brochot, C., Dorne, J.-L., Beaudouin, R. (2019). Investigating the interaction between melamine and cyanuric acid using a physiologically-based toxicokinetic model in rainbow trout. *Toxicol. Appl. Pharmacol.* **370**, 184–195.

Teorell, T. (1937). Kinetics of distribution of substances administered to the body. *Archives Internationales de Pharmacodynamie et de Therapie* **57**, 205–225.

Thomann, R.V., Connolly, J.P. (1984). Model of PCB in the Lake Michigan lake trout food chain. *Environ. Sci. Technol.* **18**, 65–71.

Thomann, R.V., Connolly, J.P., Parkerton, T.F. (1992). An equilibrium model of organic chemical accumulation in aquatic food webs with sediment interaction. *Environ. Toxicol. Chem.* **11**, 615–629.

Thomann, R.V., Shkreli, F., Harrison, S. (1997). A pharmacokinetic model of cadmium in rainbow trout. *Environ. Toxicol. Chem.* **16**, 2268–2274.

Thomas, R.E., Rice, S.D. (1981). Metabolism and clearance of phenolic and mono-, di-, and polynuclear aromatic hydrocarbons by Dolly Varden char. In *Physiological Mechanisms of Marine Pollutant Toxicity* (W.B. Vernberg, A. Calabrese, F.P. Thurberg, and F.J. Vernberg, Eds.), pp. 425–448, Academic Press, New York.

Tietge, J.E., Johnson, R.D., Jensen, K.M., Cook, P.M., Elonen, G.E., Fernandez, J.D., Holcombe, G.W., Lothenbach, D.B., Nichols, J.W. (1998). Reproductive toxicity and disposition of 2,3,7,8-tetrachlorodibenzo-*p*-dioxin in adult brook trout (*Salvelinus fontinalis*) following a dietary exposure. *Environ. Toxicol. Chem.* **17**, 2395–2407.

Timmer, N., Droge, S.T.J. (2017). Sorption of cationic surfactants to artificial cell membranes: Comparing phospholipid bilayers with monolayer coatings and molecular simulations. *Environ. Sci. Technol.* **51**, 2890–2898.

Tolls, J., Haller, M., Seinen, W., Sijm, D.T.H.M. (2000a). LAS bioconcentration: Tissue distribution and effect of hardness-implications for processes. *Environ. Sci. Technol.* **34**, 304–310.

Tolls, J., Lehmann, M.P., Sijm, D.T.H.M. (2000b). Quantification of *in vivo* biotransformation of the anionic surfactant $C_{12}$-2-linear alkylbenzene sulfonate in fathead minnow. *Environ. Toxicol. Chem.* **19**, 2394–2400.

Tovell, P.W.A., Howes, D., Newsome, C.S. (1975). Absorption, metabolism and excretion by goldfish of the anionic detergent sodium lauryl sulfate. *Toxicology* **4**, 17–29.

Tremblay, R.T., Kim, D., Fisher, J.W. (2012). Determination of tissue to blood partition coefficients for nonvolatile herbicides, insecticides, and fungicides using negligible depletion solid-phase microextraction (nd-SPME) and ultrafiltration. *J. Toxicol. Environ. Health, Part A* **75**, 288–298.

Trowell, J.J., Gobas, F.A.P.C., Moore, M.M., Kennedy, C.J. (2018). Estimating the bioconcentration factors of hydrophobic organic compounds from biotransformation rates using rainbow trout hepatocytes. *Arch. Environ. Contam. Toxicol.* **75**, 295–305.

Uchea, C., Sarda, S., Schulz-Utermoehl, T., Owen, S., Chipman, K.J. (2013). *In vitro* models of xenobiotic metabolism in trout for use in environmental bioaccumulation studies. *Xenobiotica* **43**, 421–431.

Ulrich, N., Endo, S., Brown, T.N., Watanabe, N., Bronner, G., Abraham, M.H., Goss, K.-U. (2017). UFZ-LSER database v 3.2.1, Helmholtz Centre for Environmental Research-UFZ, Leipzig, Germany. Available at: https://www.ufz.de/lserd. Accessed 04/28/2022.

UNEP. 2001. *Final Act of the Conference of Plenipotentiaries on the Stockholm Convention on Persistent Organic Pollutants.* United Nations Environment Programme, Geneva.

Vaes, W.H.J., Ramos Urrestarazu, E., Verhaar, H.J.M., Hermens, J.L.M. (1998). Nonpolar versus polar narcosis: Is there a difference? *Environ. Toxicol. Chem.* **17**, 1380–1384.

Van Veld, P.A. (1990). Absorption and metabolism of dietary xenobiotics by the intestine of fish. *Rev. Aquat. Sci.* **2**, 185–203.

Varanasi, U., Uhler, M., Stranahan, S.I. (1978). Uptake and release of naphthalene and its metabolites in skin and epidermal mucus of salmonids. *Toxicol. Appl. Pharmacol.* **44**, 277–289.

Veith, G.D., DeFoe, D.L., Bergstedt, B.V. (1979). Measuring and estimating the bioconcentration factor of chemicals in fish. *J. Fish. Res. Bd. Can.* **36**, 1040–1048.

Vick, A.M., Hayton, W.L. (2001). Methyltestosterone pharmacokinetics and oral bioavailability in rainbow trout (*Oncorhynchus mykiss*). *Aquat. Toxicol.* **52**, 177–188.

Vidal, A., Babut, M., Garric, J., Beaudouin, R. (2019). Elucidating the fate of perfluorooctanoate sulfonate using a rainbow trout (*Oncorhynchus mykiss*) physiologically-based toxicokinetic model. *Sci. Tot. Environ.* **691**, 1297–1309.

Vodicnik, M.J., Peterson, R.E. (1985). The enhancing effect of spawning on elimination of a persistent polychlorinated biphenyl from female yellow perch. *Fundam. Appl. Toxicol.* **5**, 770–776.

Wagner, J.G. (1993). *Pharmacokinetics for the Pharmaceutical Scientist*, Technomic, Lancaster.

Walters, D.M., Mills, M.A., Cade, B.S., Burkhard, L.P. (2011). Trophic magnification of PCBs and its relationship to the octanol-water partition coefficient. *Environ. Sci. Technol.* **45**, 3917–3924.

Wan, Y., Zhang, K., Dong, Z., Hu, J. (2013). Distribution is a major factor affecting bioaccumulation of decabrominated diphenyl ether: Chinese sturgeon (*Acipenser sinensis*) as an example. *Environ. Sci. Technol.* **47**, 2279–2286.

Wang, G., Shi, H., Du, Z., Chen, H., Peng, J., Gao, S. (2017a). Bioaccumulation mechanism of organophosphate esters in adult zebrafish (*Danio rerio*). *Environ. Pollu.* **229**, 177–187.

Wang, X., Wang, W.-X. (2020). Determination of the low Hg accumulation in rabbitfish (*Siganus canaliculatus*) by various elimination pathways: Simulation by a physiologically based pharmacokinetic model. *Environ. Sci. Technol.* **54**, 7440–7449.

Wang, X., Wu, F., Wang, W.-X. (2017b). *In vivo* mercury demethylation in a marine fish (*Acanthopagrus schlegelii*). *Environ. Sci. Technol.* **51**, 6441–6451.

Wang, W.-X., Tan, Q.-G. (2019). Applications of dynamic models in predicting the bioaccumulation, transport and toxicity of trace metals in aquatic organisms. *Environ. Pollu.* **252**, 1561–1573.

Watanabe, K.H., Li, Z., Kroll, K.J., Villeneuve, D.L., Garcia-Reyero, N., Orlando, E.F., Sepúlveda, M.S., Collette, T.W., Ekman, D.R., Ankley, G.T., Denslow, N.D. (2009). A computational model of the hypothalamic-pituitary-gonadal axis in male fathead minnows exposed to 17α-ethinylestradiol and 17β-estradiol. *Toxicol. Sci.* **109**, 180–192.

Weisiger, R.A., Zacks, C.M, Smith, N.D., Boyer, J.L. (1984). Effect of albumin binding on extraction of sulfobromophthalein by perfused elasmobranch liver: Evidence for dissociation-limited uptake. *Hepatology* **4**, 492–501.

Wetmore, B.A., Wambaugh, J.F., Ferguson, S.S., Sochaski, M.A., Rotroff, D.M., Freeman, K., Clewell, H.J., Dix, D.J., Andersen, M.E., Houck, K.A., Allen, B., Judson, R.S., Singh, R., Kavlock, R.J., Richard, A.M. and Thomas, R.S. (2012). Integration of dosimetry, exposure, and high-throughput screening data in chemical toxicity assessment. *Toxicol Sci.* **125**, 157–174.

Wetmore, B.A., Wambaugh, J.F., Allen, B., Ferguson, S.S., Sochaski, M.A., Setzer, R.W., Houck, K.A., Strope, C.L., Cantwell, K., Judson, R.S., LeCluyse, E., Clewell, H.J., Thomas, R.S. and Andersen, M.E. (2015). Incorporating high-throughput exposure predictions with dosimetry-adjusted *in vitro* bioactivity to inform chemical toxicity testing. *Toxicol Sci.* **148**, 121–136.

Wilkinson, G.R., Schenker, S. (1975). Drug disposition and liver disease. *Drug Metab. Rev.* **4**, 139–175.

Wilkinson, G.R., Shand, D.G. (1975). A physiological approach to hepatic drug clearance. *Clin. Pharmacol. Ther.* **18**, 377–390.

Windell, J.T., Kitchell, J.F., Norris, D.O., Norris, J.S., Foltz, J.W. (1976). Temperature and rate of gastric evacuation by rainbow trout (*Salmo gairdneri*). *Trans. Am. Fish. Soc.* **6**, 712–717.

Wood, C.M., Shelton, G. (1980). Cardiovascular dynamics and adrenergic responses of rainbow trout *in vivo*. *J. Exp. Biol.* **87**, 247–270.

Wourms, J.P., Grove, B.D., Lombardi, J. (1988). The maternal-embryonic relationship in viviparous fishes. In *Fish Physiology*, Vol. 11 (W.S. Hoar and D.J. Randall, Eds.), pp. 1–134, Academic Press, New York.

Xu, N., Li, M., Chou, W.-C., Lin, Z. (2020). A physiologically based pharmacokinetic model of doxycycline for predicting tissue residues and withdrawal intervals in grass carp (*Ctenopharyngodon idella*). *Food Chem. Toxicol.* **137**, 111127.

Yang, F., Sun, N., Sun, Y.X., Shan, Q., Zhao, H.Y., Zeng, D.P., Zeng, Z.L. (2012). A physiologically based pharmacokinetic model for florfenicol in crucian carp and oral-to-intramuscular extrapolation. *J. Vet. Pharmacol. Therap.* **36**, 192–200.

Yang, F., Yang, F., Wang, D., Zhang, C.-S., Wang, H., Song, Z.-W., Shao, H.-T., Zhang, M., Yu, M.-L., Zheng, Y. (2021). Development and application of a water temperature related physiologically based pharmacokinetic model for enrofloxacin and its metabolite ciprofloxacin in rainbow trout. *Front. Vet. Sci.* **7**, 608348.

Yoon, M., Campbell, J.L., Andersen, M.E., Clewell, H.J. 2012. Quantitative *in vitro* to *in vivo* extrapolation of cell-based toxicity assay results. *Crit. Rev. Toxicol.* **42**, 633–652.

Zaharko, D.S., Dedrick, R.L., Oliverio, V.T. (1972). Prediction of the distribution of methotrexate in the sting rays *Dasyatidae sabina* and *sayi* by use of a model developed in mice. *Comp. Biochem. Physiol.* **42A**, 183–194.

Zhang, Y., Feng, J., Gao, Y., Liu, X., Qu, L., Zhu, L. (2019). Physiologically based toxicokinetic and toxicodynamic (PBTK-TD) modelling of Cd and Pb exposure in adult zebrafish *Danio rerio*: Accumulation and toxicity. *Environ. Pollu.* **249**, 959–968.

Zhao, S., Jones, K.C., Sweetman, A.J. (2018). Can poly-parameter linear-free energy relationships (pp-LFERs) improve modelling bioaccumulation in fish? *Chemosphere* **191**, 235–244.

# 3 Biotransformation in Fishes[1]

*Daniel Schlenk*
University of California Riverside

*Jed Goldstone*
Woods Hole Oceanographic Institution

*Margaret O. James*
University of Florida

*Peter van den Hurk*
Clemson University

**Abstract**

Chapter 3 in toxicology of fishes addresses the occurrence, regulation, and catalytic activities of biotransformation systems in fish and concludes with a focus on three model compounds and chemical classes and the contribution of biotransformation toward the toxicology of each compound. Biotransformation is a two-phase process catalyzed primarily through enzymatic reactions that often radically alter the chemistry of nonpolar lipophilic chemicals to polar water-soluble metabolites predominately leading to detoxification and elimination of the parent compounds. Unfortunately, the alteration of chemistry required for enhanced polarity often creates reactive intermediates through bioactivation, which can be more biologically hazardous than the initial parent compounds. This chapter provides a comprehensive update to what is known about phase I and phase II biotransformation enzymes in fish.

## INTRODUCTION

Biotransformation is a two-phase process catalyzed primarily through enzymatic reactions that often radically alter the chemistry of nonpolar lipophilic chemicals to polar water-soluble metabolites predominately leading to detoxification and elimination of the parent compounds. Unfortunately, the alteration of chemistry required for enhanced polarity often creates reactive intermediates through bioactivation, which can be more biologically hazardous than the initial parent compounds. The phase I process either adds or exposes polar atoms within a xeno- or endobiotic compound. Three general phase I reactions include oxidation, reduction, and hydrolysis (Table 3.1). When polarity has been enhanced through phase I reactions, phase II reactions generally attempt to further enhance polarity through conjugation of the phase I product with a bulky polar endogenous molecule. Alternatively, phase II reactions may protect against bioactivation by masking functional groups (i.e., amines) prone to reactive intermediate formation with groups that likely provide steric hindrance (i.e., methyl, acetyl) rather than augmented polarity (Table 3.1).

## PHASE I REACTIONS

### Oxidation

Various enzymes are involved in the oxidation of xeno- and endobiotic compounds. Dehydrogenases oxidize substrates transferring electrons to an electron-deficient acceptor that is typically an essential cofactor for catalysis (e.g., $NAD^+$). Oxygenases catalyze the incorporation of molecular oxygen into molecules. While water is the source of oxygen for oxidases, peroxidases derive oxygen from peroxide cofactors.

### Cytochrome P450 Family of Drug-Metabolizing Enzymes

#### Overview

The most dominant enzyme system responsible for oxidation processes in phase I biotransformation is the cytochrome P450 monooxygenases. The cytochrome P450s (CYPs) constitute a superfamily of heme-containing proteins that catalyze biological oxidation and reduction reactions. Klingenberg (1958) and Garfinkel (1958) first reported that hepatic microsomes contain a pigment that binds carbon monooxide with an unusual visible absorption maximum at 450 nm in its CO-reduced difference spectrum. Omura and Sato (1962) discovered that this pigment was a *b*-type cytochrome and called it cytochrome P450. The hepatic microsomal CYP system has broad substrate specificity and is responsible for oxidative metabolism of many structurally diverse endogenous and xenobiotic compounds. CYP enzymes are important for converting lipophilic foreign chemicals into more water-soluble products for excretion and, hence, detoxification. On the other hand, CYP enzymes catalyze the conversion of certain compounds such as polycyclic aromatic hydrocarbons (PAHs) and nitrosamines into more toxic intermediates. Constitutive CYP forms

[1] Adapted and updated from Chapter 4. Biotransformation in Fishes, Schlenk, Celander, Gallagher, George, James, Kullman, van den Hurk, and Willett. *The Toxicology of Fishes*, Eds. Di Giulio and Hinton



DOI: 10.1201/9781003160694-3

**TABLE 3.1**
**Phase I and Phase II Enzymatic Activities and Cofactors**

| Enzyme | Cofactors |
|---|---|
| **Phase I** | |
| *Oxidation* | |
| Cytochrome P450 | Oxygen, NADPH, cytochrome $b_5$ (optional) |
| Flavin-containing monooxygenase | Oxygen, NADPH |
| Monoamine oxidase | $H_2O$ |
| Aldehyde oxidase | $NAD^+$ |
| Alcohol dehydrogenase | |
| Aldehyde dehydrogenase | |
| Cyclooxygenase | Arachidonic acid, oxygen |
| Peroxidase (PGH synthetase, lipoxygenase) | Peroxide (lipid-OOH or $H_2O_2$) |
| *Reduction* | |
| NADPH oxidoreductaseAldo-keto reductase | NAD(P)H |
| Hydrolysis | |
| Carboxylesterase | $H_2O$ |
| Epoxide hydrolase | |
| **Phase II** | |
| UDP-glucuronosyl transferase | UDPGA |
| Sulfotransferase | PAPS |
| Amino acid conjugation | Amino acids (taurine, glycine, glutamine) |
| Glutathione *S*-transferase | Glutathione |
| Acetylation | Acetyl-coenzyme A |
| Methylation | *S*-Adenosylmethionine |

that appeared early in evolution are involved in biosynthesis (anabolism) of endogenous substances such as steroids, fatty acids, vitamins, bile acids, leukotrienes, thromboxanes, and prostaglandins (Ryan and Levin, 1990). Inducible CYP forms emerged later in evolution and are primarily involved in the breakdown (catabolism) of endobiotics as well as xenobiotics. The following section attempts to address what is known regarding specific regulation, and what has been discovered regarding substrate specificities, catalytic function, and, when available, physiological role in fish. Most of the characterization has occurred in zebrafish (Stegeman et al., 2010), and some families and isoforms have been better characterized than others. It is hoped that informational gaps are identified here that will stimulate further research in underrepresented CYP families and enhance our understanding of this important superfamily of enzymes.

The CYP superfamily is ancient, with the ancestral gene having existed more than 3.5 billion years ago (Nelson et al., 1993). Animals, plants, microorganisms, and even some viruses all contain CYP, and in mammals, they have been identified in every tissue. The emergence of new CYP genes results from a sequence of events, including speciation, gene and genome duplication, divergence, and drift as a function of mutation and fixation all withstanding evolutionary pressures (Nebert et al., 1989; Nelsen et al., 1996). CYPs are generally most prevalent in the liver in association with the endoplasmic reticulum or mitochondria, but tissue-specific expression may lead to enzyme activities that impact phenotypes, metabolism, and chemical disposition. A review of cytochrome P450 in fish was conducted in 2012 (Uno et al., 2012), and one specifically in zebrafish was performed in 2016 (Saad et al., 2016).

CYPs considered in the same gene family display more than 40% amino acid sequence identity, and those within a subfamily are more than 55% identical (Nelson et al., 1993). Nomenclature has been standardized so "*CYP*" indicates the gene, followed by an Arabic numeral for the gene family, a capital letter for the subfamily, and an Arabic numeral for the specific subfamily member; for example, CYP1A1 is responsible for the metabolic activation of benzo(*a*)pyrene (BaP) in most species, including mammals. CYPs in gene families CYP1 to CYP4 are typically found in the microsomal fraction and are generally responsible for oxidation or metabolism of steroids and xenobiotic metabolism. CYPs in families CYP5 to CYP51 have relatively narrow substrate specificity and function principally in physiological regulation.

The overall CYP-mediated reaction takes the form of

$$RH + O_2 + NADPH + H^+ \rightarrow ROH + H_2O + NADP^+$$

The reaction begins with the transfer of electrons from NAD(P)H to NADPH–cytochrome P450 reductase (POR) in the microsomal system (Figure 3.1). The active site of CYPs is generally a noncovalently bound heme iron in the form of protoporphyrin IX (Gonzalez and Gelboin, 1993; Henne et al., 2001). CYP-mediated reactions include aliphatic and aromatic carbon hydroxylation, epoxidation, heteroatom dealkylations, heteroatom oxygenation, substrate rearrangement, deamination, desulfuration, and dehalogenation (Figure 3.2).

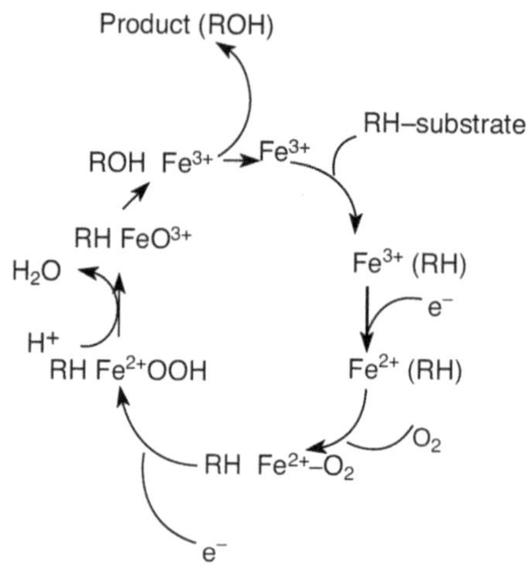


**FIGURE 3.1** CYP catalytic cycle.

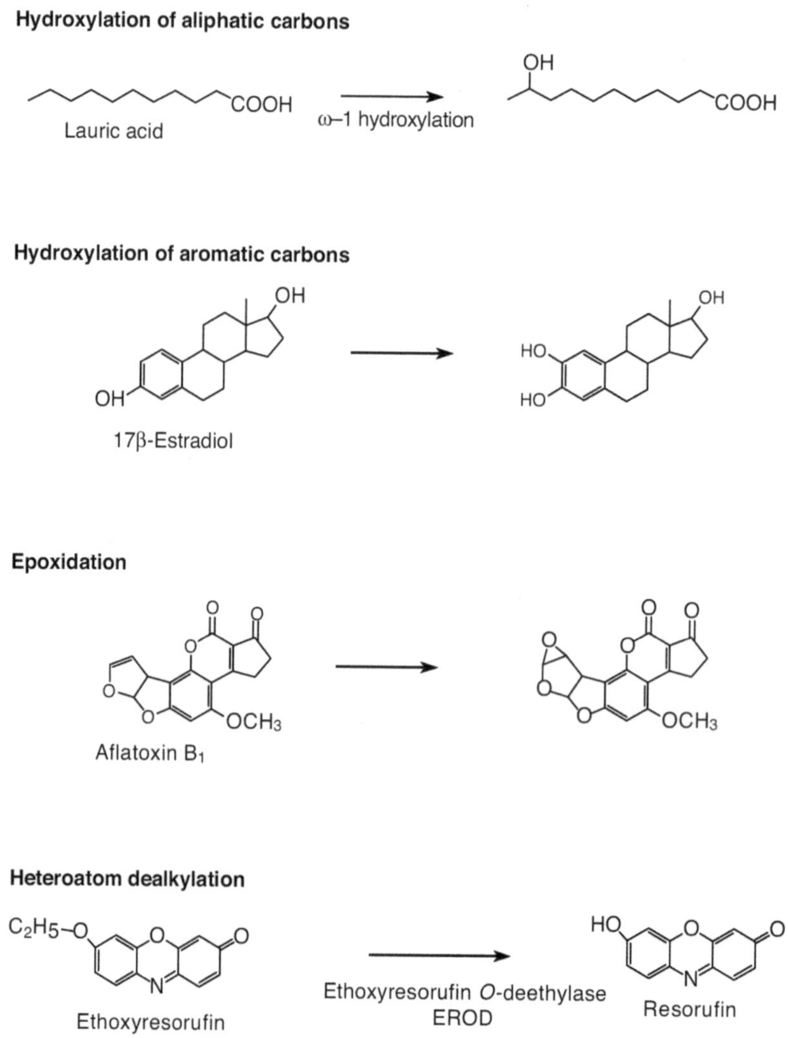


**FIGURE 3.2** Examples of reactions catalyzed by cytochrome P450.

*(Continued)*

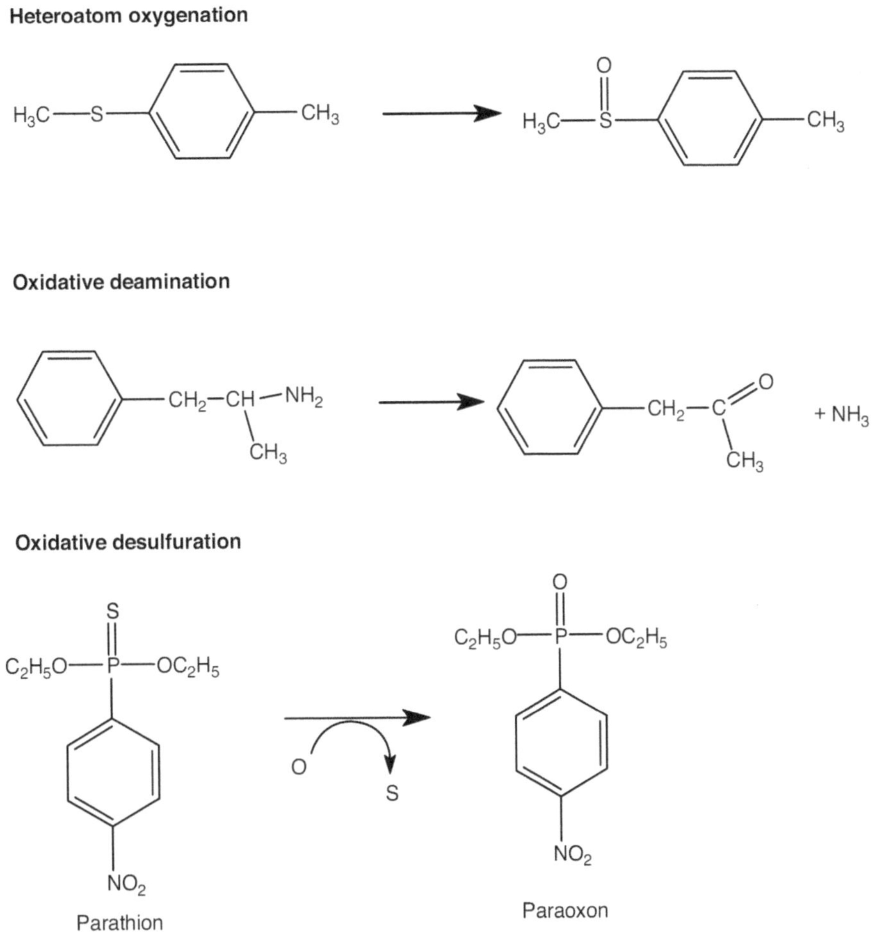


**FIGURE 3.2 (*Continued*)** Examples of reactions catalyzed by cytochrome P450.

The reaction cycle is initiated when the substrate binds to the oxidized ($Fe^{+3}$) CYP complex and facilitates an electron transfer from NADPH to the complex. Oxygen then binds to the reduced CYP complex with coordination to iron *trans* to thiolate. The second electron is contributed by either NADPH–cytochrome P450 reductase or NADH–cytochrome $b_5$ reductase. The next step involves cleavage of the oxygen–oxygen bond, uptake of two protons, and release of water. Oxygen is inserted into the substrate through the generation of hydroxyl and carbon free radicals. Finally, dissociation of ROH restores the P450 to the initial ferric state (Parkinson, 2001). In addition, the peroxide shunt can allow a peroxy compound to substitute for oxygen in substrate oxidation.

Expression of CYP genes is regulated by diverse mechanisms. Basal levels of individual CYP mRNAs and proteins are regulated via transcriptional and posttranscriptional processes, including mRNA and protein stabilization or degradation as well as epigenetically through microRNAs and histone modifications. Induction of CYP gene expression of isozymes in families one through three has been extensively investigated in vertebrates, including fish, and is described in the next sections of this chapter. Table 3.2 summarizes many inducers, inhibitors, and substrates of CYP1, CYP2, and CYP3, which can be used to measure the induction of specific CYP isozyme-dependent activities in fish research. As shown in the table, many inducers, such as PAHs, are substrates for the CYPs that they induce and therefore stimulate their own metabolism. However, this is not always the case as some inducers may not serve as substrates. CYP-mediated metabolism of some substrates can be highly complex and is dependent on both the species, life stage, and tissue or organ investigated. Figure 3.3 illustrates many of the oxidative metabolites of BaP that are catalyzed by CYP isoenzymes. The metabolite profiles are variable and dependent on species differences in the expression of CYP isoforms and their catalytic activities; for example, fish (such as the brown bullhead) treated with BaP preferentially metabolize the hydrocarbon to the more toxic 7,8- and 9,10-oxidation products (Willett et al., 2000). In contrast, mussels exposed to BaP preferentially (47% total metabolites) form 1–6, 3–6, and 6–12 BaP quinones (Michel et al., 1995).

## CYP1

*Regulation*—CYP1 gene expression is induced by structurally diverse aromatic hydrocarbons that bind to the aryl hydrocarbon receptor (AhR). The AhR is a basic helix–loop–helix

**TABLE 3.2**
**CYP Gene Families, Inducers, Inhibitors, and Substrates**

| P450 Gene | Substrates | Inducers | Inhibitors |
|---|---|---|---|
| CYP1A | BaP<br>Estradiol<br>Ethoxyresorufin<br>Dimethyl benzanthracene<br>Phenacetin<br>Chlorpyrifos | PAHs (BaP, BNF, 3-MC)<br>Retene<br>Dioxins/furans (TCDD)<br>PCBs (CB77, 126, 169) | 2-Aminoanthracene<br>Elipticine<br>PCB 77<br>Fluoranthene<br>Cadmium<br>Tributyltin<br>α-Naphthoflavone<br>Parathion<br>Ketoconazole<br>Miconazole<br>Clotrimazole<br>SKF525A<br>Fluoxetine<br>Erythromycin |
| CYP1B | Ethoxyresorufin<br>Methoxyresorufin<br>Benzyloxyresorufin<br>Pentoxyresorufin<br>17β-estradiol | BaP, TCDD | Fluoxetine<br>Erythromycin<br>α-Naphthoflavone |
| CYP1C | Ethoxyresorufin<br>Methoxyresorufin<br>Benzyloxyresorufin<br>Pentoxyresorufin<br>7-Methoxy-coumarin<br>7-Methoxy-4-methyl-coumarin<br>7-Methoxy-4-bromomethyl-coumarin<br>7-Methoxy-4-trifluoromethyl-coumarin<br>7-Ethoxy-4-trifluoromethyl-coumarin<br>BaP<br>Testosterone | BaP, TCDD | Fluoxetine<br>Erythromycin |
| CYP1D | Ethoxyresorufin<br>Methoxyresorufin<br>Benzyloxyresorufin<br>Pentoxyresorufin<br>BaP | | |
| CYP2AA1 | Testosterone | 16a-carbonitrile | |
| CYP2AA2 | | 1,4-Bis [2-(3,5-dichloropyridyloxy)] benzene<br>Phenobarbital<br>Pregnenolone | ? |
| CYP2AA12 | | Pregnenolone | |
| CYP2K1 | Lauric acid (ω–1)<br>17β-Estradiol<br>Benzphetamine<br>Progesterone (16α) | Diethyldithiocarbamate (activator)<br>BNF (decrease)<br>Testosterone (decrease mRNA)<br>Chlorpyrifos<br>Estrogens (decrease protein) | Ketoconazole<br>Miconazole<br>Clotrimazole<br>Cimetidine<br>Parathion<br>α-Naphthoflavone |
| CYP2K6 | | Pregnenolone | |
| CYP2M1 | Lauric acid (ω–6)<br>Progesterone | Estrogens (decrease protein) | ? |
| CYP2N1 | Arachidonic acid Benzphetamine Alkoxyresorufins | TPA/starvation (decrease mRNA) | ? |
| CYP2N2 | Arachidonic acid Benzphetamine Alkoxyresorufins | TPA/starvation (decrease mRNA) | ? |
| CYP2P1 | — | Fasting/refeed (increase mRNA) | ? |
| CYP2P3 | Arachidonic acid Benzphetamine | TPA/starvation (decrease mRNA) | ? |

*(Continued)*

**TABLE 3.2 (*Continued*)**
**CYP Gene Families, Inducers, Inhibitors, and Substrates**

| P450 Gene | Substrates | Inducers | Inhibitors |
|---|---|---|---|
| | Arachidonic acid Benzphetamine Alkoxyresorufins | TPA/starvation (decrease mRNA) | ? |
| CYP2X1 | Benzphetamine Aminopyrene | ? | ? |
| CYP3A27 | Testosterone (6 β, 16β–OH)<br>Progesterone<br>Estradiol<br>BaP<br>Nifedipine<br>Benzphetamine<br>Ethoxycoumarin | Ketoconazole (high)<br>Estrogens (decrease)<br>Chlorpyrifos | α-Naphthoflavone<br>Parathion<br>Ketoconazole (low)<br>Miconazole<br>Clotrimazole<br><br>SKF525A<br>Cimetidine<br>Elipticine<br>Piperonyl butoxide<br>Isosafrole<br>Gestoden<br>17α-Ethinylestradiol<br>1-Aminobenzotriazole<br>5,8,11,14-Eicosatetraynoic acid<br>Gemfibrozil<br>Ciprofloxacin<br>Fluoxetine |
| CYP3A38 | Testosterone (6 β, 16β–OH)<br>Benzyloxyresorufin<br>7-Benzyloxy-4-[trifluoromethyl]-coumarin | Estrogens (decrease) | Nonylphenol (>100 nM) |
| CYP3A40 | Testosterone (6 β)<br>Benzyloxyresorufin<br>7-Benzyloxy-4-[trifluoromethyl]-coumarin | — | — |
| CYP3A45 | Testosterone (6 β) | — | — |
| CYP3A65 | Testosterone<br>Microcystin | TCDD<br>Pregnenolone<br>Rifampicin<br>Dexamethasone | ketoconazole |
| CYP3C1 | | Pregnenolone<br>PCB126 | |

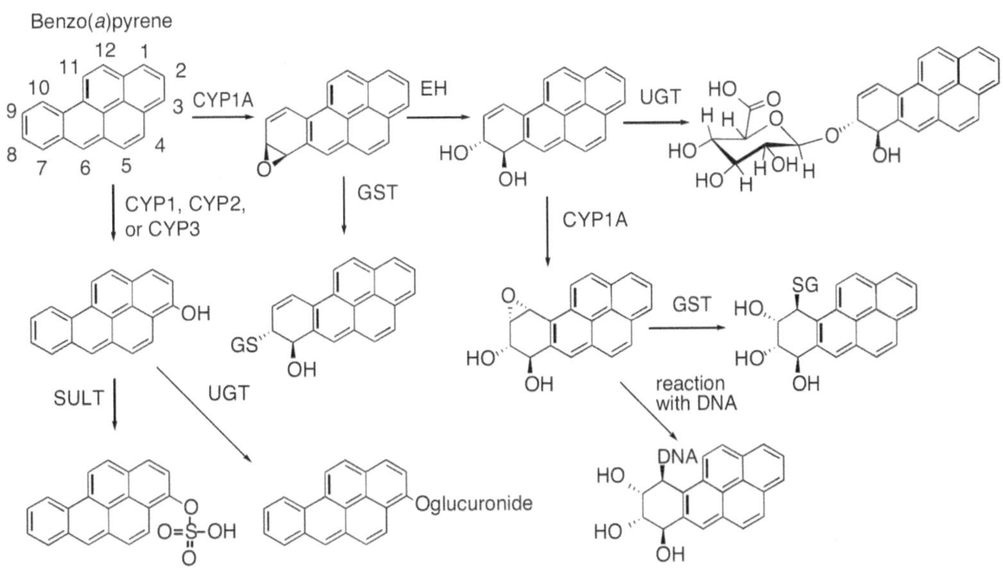


**FIGURE 3.3** Biotransformation pathway of benzo(*a*)pyrene (BaP).

DNA-binding protein that has been extensively characterized in veterbrates and zebrafish. In fish, at least three AhRs (AhR1a, AhR1b, AhR2) have been characterized through a number of expression and knockdown studies (reviewed in Shankar et al., 2020). Each form has unique regulatory roles in tissue expression, signaling, and development. In mammals, the AhR controls the transcription of the genes CYP1A1, CYP1A2, CYP1B1, as well as some phase II enzymes such as, some but not all, glutathione *S*-transferase, uridine diphosphate (UDP)-glucuronosyltransferase, and aldehyde-3-dehydrogenase enzymes (Shankar et al., 2020). Other CYPs in zebrafish, including CYP2AA12, CYP3A65, and CYP3C1, are also induced by AHR agonists (Kubota et al., 2015). Agonists for the AhR include both synthetic and naturally occurring compounds. The best characterized ligands include certain PAHs and planar halogenated aromatic hydrocarbons (HAHs), including polychlorinated dibenzo-*p*-dioxins (PCDDs) and -furans (PCDFs), as well as polychlorinated biphenyls (PCBs). Planar HAHs bind to the AhR in predictable structure–activity relationships. 2,3,7,8-Tetrachlorodibenzo-*p*-dioxin (TCDD) has the highest AhR-binding affinity among the HAHs and is also correlated with the high toxic potency of this congener (Safe, 1990). Structure–activity relationships are not as clear-cut for the PAHs because the toxic and genotoxic responses induced by these compounds are not all AhR-mediated; however, Billiard and coworkers (2002) found a positive relationship between PAH binding to the teleost AhR and PAH potency for CYP1A induction. Examples of naturally occurring dietary AhR ligands include indole-3-carbinol, indolo-(3,2-*b*)-carbazole, dibenzoylmethanes, curcumin, and carotinoids (Denison and Nagy, 2003; Jeuken et al., 2003).

A simplified mechanism of AhR-mediated induction of CYP1A involves a ligand entering the cell where it associates with the cytosolic AhR, which exists as a multiprotein complex with two molecules of heat shock protein 90 (Hsp90), aryl hydrocarbon receptor interacting protein (AIP), also known as the X-associated protein 2 (XAP2), and a 23-kDa cochaperone protein (p23) (Denison and Nagy, 2003). Binding of ligand causes a conformational change that facilitates nuclear translocation and association with one or more AhR nuclear translocator proteins (ARNTs). The AhR/ARNT heterodimer, in turn, associates with dioxin response elements (DREs; sometimes called AhR response elements, or AhREs) and with various coregulators (Carlson and Perdew, 2002) to initiate transcription and translation of AhR-responsive genes, including CYP1s. Induction of CYP1 protein can be determined at the transcriptional (mRNA) level using northern blot or quantitative reverse transcription polymerase chain reaction (qRT-PCR) and at the protein level by using western blot, enzyme-linked immunosorbent assays (ELISAs), or immunohistochemistry. Finally, CYP1-dependent catalytic activities such as ethoxyresorufin-*O*-deethylase (EROD) can also be used to measure induction.

*CYP1A*—Research on CYPs in fish is more limited than that reported for mammals. Like mammals, fish CYPs catalyze oxidation of many of the same endogenous substrates and environmental chemicals. The most studied CYP form in fish is CYP1A which was purified from rainbow trout liver in the early 1980s (Williams and Buhler 1983, 1984). The first fish CYP1A cDNA was cloned by Heilmann et al. (1988), and sequence analysis showed 57%–59% and 51%–53% homology to mammalian CYP1A1 and CYP1A2 genes, respectively. This cDNA hybridized to a 2.8-kb mRNA that was induced 10-fold by 3-methylcholanthrene (3-MC) in the rainbow trout. In fish, this form is generally recognized as CYP1A, and the terminology is supported because mammalian 1A1 and 1A2 are believed to have diverged 250 MYA by a gene-duplication event (Nebert and Gonzalez, 1987), whereas fish diverged from the mammalian line prior to that time. A hybrid CYP1A gene in fish is further suggested because some regions of the trout CYP1A sequence are identical to all mammalian CYP1A1 but not CYP1A2, while other regions are identical to all mammalian CYP1A2 but not CYP1A1 (Stegeman and Hahn, 1994). This homology pattern can be explained by a single CYP1A gene in fish that is ancestral to both mammalian CYP1A1 and CYP1A2. Berndtson and Chen (1993) reported a CYP1A2 gene cloned in rainbow trout; however, the sequence of this clone was only 4% different from the CYP1A1 clone isolated from the same species. Furthermore, the fish CYP1A2 was not orthologous to the mammalian 1A2 and was coordinately induced with CYP1A1 in trout treated with 3-MC (Berndtson and Chen, 1993). For historical reasons due to uncertainties in the original nomenclature, CYP1As within fish are not being provided a number following the subfamily.

Since the early reports on the effects of crude oil on brown trout (*Salmo trutta*) (Payne and Penrose, 1975), research correlating environmental pollution with the induction of CYP1A-mediated activities in fish has been used as a biomarker. The induction of hepatic CYP1A mRNA, immunoreactive protein, and EROD enzyme activities in fish has been extensively studied in both controlled laboratory and field experiments. Some of these studies are summarized in a review by Bucheli and Fent (1995), which found that 93% of the field studies (68/76) showed that CYP1A induction in fish was related to contaminant levels in the environment. It should also be noted that fish adapted to living in highly contaminated habitats (e.g., superfund sites with PAH or HAH contamination) are also refractory to CYP1A induction (Bello et al., 2001; Brammell et al., 2004; Meyer et al., 2002; Prince and Cooper, 1995). Prolonged exposure of rainbow trout to a PCB mixture resulted in unresponsiveness to 3-MC treatment and decreased CYP1A expression upon additional dosage of PCB (Celander and Forlin, 1995). The mechanisms of this inhibition are not entirely clear (Meyer et al., 2003; Powell et al., 2000). However, Stegeman and coworkers have suggested that increased reactive oxygen production by uncoupling of the CYP catalytic cycle results in the loss of CYP1A protein at high doses of halogenated AHR ligands (Schlezinger and Stegeman, 2001; Schlezinger et al., 2006).

*Inhibitors*—The type, time dependence, and the degree of inhibition caused by various CYP inhibitors are also species-dependent. Selective inhibitors of CYP isozymes have been used to characterize substrate specificity and modulation of toxicity and carcinogenicity of xenobiotics and in biochemical mechanistic studies of CYP enzymes. Both EROD activities and the mutagenicity of 3-MC are inhibited by ellipticine and its derivatives. Structural analysis of ellipticine derivatives reveled that methyl substitution in the 5 and 11 positions is essential for inhibitory responses (Lesca et al., 1980). Some environmental contaminants are also CYP1A inhibitors in fish; for example, 2-aminoanthracene (2-AA) caused a 67% inhibition of BNF-induced EROD activity in channel catfish (*Ictalurus punctatus*) (Watson et al., 1995). In both *in vitro* and *in vivo* experiments, 2-AA was a mechanism-based inhibitor of CYP1A. Similarly, 3,3′,4,4′-tetrachlorinated biphenyl (PCB 77) at high doses causes competitive inhibition of CYP1A enzyme activity and decreased induction of CYP1A protein (Gooch et al., 1989; White et al., 1997b). White and coworkers' (1997a) data suggested that, in scup, tetrachlorinated biphenyl (TCB) decreased CYP1A protein by enhancing protein degradation. TCB also initiated redox cycling through an uncoupling of CYP1A (Schlezinger and Stegeman, 2001). Finally, in killifish cotreated with BaP and the four-ring PAH fluoranthene (FL), both hepatic EROD activity and CYP1A immunoreactive protein levels were significantly inhibited (Willett et al., 2001). Although a covalent interaction between FL and CYP1A was not detected, the relative composition of DNA adducts changed in cotreated fish, suggesting that BaP metabolism is significantly affected when fish are coexposed to PAH mixtures. Antifungal imidazoles (clotrimazole and ketoconazole) inhibit CYP1A-mediated EROD activities in gizzard shad (*Dorosoma cepedianum*), rainbow trout, and Atlantic cod (Hasselberg et al., 2004; Hegelund et al., 2004; Levine et al., 1997). Acrylamide, an environmental contaminant, is also an inhibitor of CYP1A and suggests feedback regulation on CYP1A mRNA transcription (Haasch et al., 1992; Petersen and Lech, 1987).

*CYP1B*—Until the mid-1990s, the CYP1 gene family was believed to contain a single subfamily with the well-known members CYP1A1 and CYP1A2; however, in 1994, the cDNA for CYP1B1 was isolated from TCDD-induced human keratinocyte cells (Sutter et al., 1994). CYP1B1 is the only form of CYP1B identified in vertebrates and has been noted in a number of fish species (Goldstone et al., 2007). Heterologously expressed zebrafish CYP1B1 catalyzed a host of dealkylated resorufin analogs (Stegeman et al., 2015) as well as 2- and 4-hydroxy estradiol (Scornaienchi et al., 2010; Smith et al., 2012).

An elevated ratio of 4-/2-hydroxyestradiol formation in human mammary microsomes has been used as a risk factor for malignant breast cancer (Liehr and Ricci, 1996). Catechol estrogens, such as 4-hydroxyestradiol, are capable of undergoing metabolic redox cycling between hydroquinone and quinone forms, generating potentially mutagenic free radicals and oxidative stress. In channel catfish, the 4-/2-hydroxyestradiol ratio was statistically higher in microsomes from BaP-treated fish compared to controls (0.2 and 0.04, respectively) (Butala et al., 2004). The shift toward more 4-hydroxyestradiol production in exposed fish suggests induced production of the redox active estrogen metabolite. As with PAH metabolism, lower levels of CYP1B1 and less associated formation of 4-hydroxyestradiol may be indicative of lower genotoxicity. More studies are necessary to determine the physiological significance of CYP1B in fish. Fluoxetine and ethrythromycin were effective inhibitors of deakylation reactions catalyzed by expressed zebrafish CYP1B1 (Smith et al., 2012).

*CYP1C*—Two forms of CYP1C have been characterized from zebrafish (Stegeman et al., 2015; Scornaienchi et al., 2010; Smith et al., 2012). CYP1C1 and CYP1C2 were effective in dealkylating resorufins, coumarins, and 7-ethoxyfluorescein-ethylester (Stegeman et al., 2015). Each form also catalyzed the 6b-hydroxylation of testosterone and the hydroxylation of BaP (Stegeman et al., 2015). Tissue-specific expression and relative roles of CYP1A and CYP1C1 have been investigated in *Fundulus heteroclitus* and PAH-induced carcinogenesis (Wang et al., 2006; Wang et al., 2010). As with CYP1B, fluoxetine and erthryomycin were effective inhibitors of catalytic activities (Smith et al., 2012).

*CYP1D*—One form of CYP1D has been identified and expressed from zebrafish (Stegeman et al., 2015). While the hydroxylation of BaP was noted, dealkylation of coumarins or ethoxy-fluorescein-ethylester was not observed. Metabolism of testosterone was also reported, but the metabolite was not the 6-β hydroxide (Stegeman et al., 2015). CYP1D1 is not induced by AHR ligands (Goldstone et al., 2009) and is expressed at higher levels in male adult zebrafish than in females, as is the case also for CYP1A, CYP1B1, and CYP1C1, but not CYP1C2 (Kubota et al., 2019).

## CYP2

*Phylogeny*—CYP2 and CYP1 genes belong to the CYP2 clan. The CYP1 gene family is believed to have diverged from the CYP2 gene family more than 420 MYA (Nelson, 2003). CYP2 is the most diverse CYP gene family, with 13 known CYP2 subfamilies in fish: 2K, 2M, 2N, 2P, 2R, 2U, 2V, 2X, 2Y, 2Z, 2AA, 2AD, and 2AE (Kirischian et al., 2011) (Table 3.3). Two of these, CYP2R and CYP2U, also have mammalian representatives; thus, these subfamilies probably represent more conserved CYP2 genes that emerged over 420 million years of evolution. These earlier emerging CYP2 isozymes are most likely involved in the metabolism of endobiotics, including vitamin D (CYP2R1: Dong et al., 2021) and fatty acids and N-arachidonoylserotonin (Dhers et al., 2017).

Although mammalian and piscine CYP2 gene families have structurally diverged during vertebrate evolution, there are still some conserved structures as well as catalytic functions among certain CYP2 subfamilies; for example, fish CYP2N, CYP2P, CYP2V, and CYP2Z genes are

## TABLE 3.3
## CYP2 Gene Family in Fish

| Subfamily/Gene | Species | References |
|---|---|---|
| CYP2K | Rainbow trout (*O. mykiss*) | Buhler et al. (2000); Cok et al. (1998); Katchamart et al. (2002); Yang et al. (2000); |
| 2K1 | Killifish (*F. heteroclitus*) | |
| 2K2 | Rainbow trout | |
| 2K3-2K5 | Zebrafish (*D. rerio*) | |
| 2K6-2K8; 2K16-2K22 | Pufferfish (*F. rubripes*) | |
| 2K9-2K11 | Pufferfish | |
| 2K12-2K15 pseudogenes | | |
| CYP2M | Rainbow trout | Buhler et al. (2000); Cok et al. (1998); Katchamart et al. (2002); Yang et al. (1998); Zhu et al. (2006); De Anna et al. (2021) |
| 2M1 | | |
| CYP2N | Killifish | Oleksiak et al. (2000); Peterson and Bain (2004); Vrolijk et al. (1994) |
| 2N1 | Killifish | |
| 2N2 | Scup (*S. chrysops*) | |
| 2N3 | Butterfly fish (*Chaetodon* sp.) | DeBusk et al. (2008) |
| 2N4-2N8 | Pufferfish | |
| 2N9-2N11 | Zebrafish | |
| 2N13 | | |
| CYP2P | Killifish | Oleksiak et al. (2003) |
| 2P1-2P3 | Pufferfish | |
| 2P4 | Pufferfish | |
| 2P5 pseudogene | Zebrafish | |
| 2P6-2P10 | Largemouth bass (*M. salmoides*) | |
| 2P11 | Atlantic salmon (*S. salar*) | |
| 2P | | |
| CYP2R | Zebrafish | |
| 2R1 | Pufferfish | |
| 2R1 | Pufferfish | |
| 2R2-2R3 pseudogenes | | |
| CYP2U | Zebrafish | |
| 2U1 | Pufferfish | |
| 2U1 | | |
| CYP2V | Zebrafish | |
| CYP2V1 | | |
| CYP2X1 | Catfish (*I. punctatus*) | Schlenk et al. (2002) |
| CYP2X1 | Pufferfish | |
| 2X2-2X4 | Pufferfish | |
| 2.2X5 pseudogene | Zebrafish | |
| 2X6-2X11 | | |
| CYPY | Pufferfish | |
| 2Y1-2Y2 | Zebrafish | |
| 2Y3-2Y4 | | |
| CYPZ | Pufferfish | |
| 2ZI-2Z2 | | |
| CYP2AA | | |
| 2AA1-2AA8 | | |
| CYP2AD | Pufferfish | |
| 2AD1 (formerly 2N12) | Zebrafish | |
| 2AD2-2AD3; 2AD6 | Medaka (*O. latipes*) | |
| 2AD4 | Three-spined stickleback (*Gasterosteus aculeatus*) | |
| 2AD5 | | |
| CYP2AD12 | Marsh killifish (*Kryptolebas marmoratus*) | |
| CYP2AE | Zebrafish | |
| 2AE1 | | |

*Note*: Genes, gene fragments, and possible pseudogenes were obtained from the cytochrome P450 homepage: https://drnelson.uthsc.edu/.

related to mammalian CYP2D and CYP2J genes (Nelson, 2003). Phylogenetic and syntenic analyses show that the functionally characterized killifish CYP2P and CYP2N subfamilies (Oleksiak et al., 2003) are closely related to the vertebrate CYP2J subfamily and have likely arisen from a common ancestral gene by tandem duplication (Goldstone et al., 2010).

Members of the CYP2 gene family in fish are involved in the metabolism of endobiotics, such as arachidonic acid, lauric acid, and sex steroid hormones, as well as xenobiotics such as alkoxyresorufins, and benzphetamine (Buhler and Wang-Buhler, 1998; Oleksiak et al., 2003; Yang et al., 1998, 2000). The following paragraphs summarize highlights from functional studies of some of these CYP2 subfamilies.

*CYP2AA subfamily*—Studies in zebrafish have cloned and expressed two isoforms: CYP2AA1 and CYP2AA2 (Kubota et al., 2013). This subfamily does not show orthology to any human or mammalian CYP genes. CYP2AA1 mRNA was expressed in the intestine, while CYP2AA2 transcripts were higher in the kidney. As mentioned below, differences in regulation by classic CYP inhibitors were observed in the livers of zebrafish with CYP2AA1 inducible by PCN and CYP2AA2 increased by TCPOBOP and phenobarbital (Kubota et al., 2013). CYP2AA genes have also been examined in dietary studies in grass carp (Shi et al., 2018) and suggested to be hepatoxocity markers in zebrafish (Driessen et al., 2014).

*CYP2AD subfamily*—The CYP2AD family was original identified in zebrafish (Goldstone et al., 2010). *CYP2AD2* is one of the major CYP genes expressed in adult zebrafish liver (Kubota et al., 2019), and CYP2AD12 is induced by 4-octylphenol in mangrove killifish (Puthumana et al., 2017), yet nothing is known about the function of these enzymes. CYP2AD genes exhibit sexually dimorphic expression in both yellow catfish and zebrafish (Goldstone et al., 2010).

*CYP2B-Like Forms*—Fish liver microsomes have been shown to metabolize prototypical mammalian CYP2B substrates, including aldrin, benzphetamine, ethylmorphine, aminopyrine, and alkoxyresorufins (Buhler and Williams, 1989; Eisele et al., 1984; Elskus and Stegeman, 1989; Goksøyr et al., 1987; Haasch et al., 1994; Kleinow et al., 1990; Stegeman, 1981). The existence of piscine CYP2B-like enzymes further was supported by protein purification and immuno-cross-reactivity studies; however, there is no genomic support for a piscine CYP2B gene ortholog (Kirischian et al., 2011). This finding illustrates one major challenge in CYP research: the assignment of function to a particular enzyme.

CYP2B-like functional activities and antibody cross-reactivity have been demonstrated for a variety of species. Reconstitution of scup hepatic P450B, with 50% N-terminal sequence identity to rat CYP2Bs, demonstrated oxidation of testosterone at the 15α-position (Klotz et al., 1986) and cross-reactivity with antibodies against both scup P450B and rat CYP2B1 (Stegeman and Hahn, 1994). Reconstituted rainbow trout hepatic CYP isozymes catalyzed lauric acid hydroxylase activity and cross-reacted with rat CYP2B1 antibodies (Miranda et al., 1989, 1990). One of these enzymes was later assigned as CYP2M1 (Yang et al., 1998). Further studies are needed to assign these functions to the genes identified in genomic and transcriptomic studies.

The presence of hepatic CYP2B immunoreactive proteins and comparatively high activities toward mammalian prototypical CYP2B substrates (i.e., aminopyrene and pentoxyresorufin) were observed in some tropical fishes from the Bermuda Archipelago; however, great differences were found in aminopyrene-*N*-demethylase and pentoxyresorufin-*O*-depentylase activities among species, with the Bermuda chub (*Kyphosous sectatrix*) and sergeant major (*Abudefduf saxatilis*) displaying the highest activities. In addition, polyclonal antibody (pAb) against rat CYP2B1 cross-reacted most strongly with hepatic microsomal proteins in tomtate (*Haemulon aurolineatum*), pinfish (*Lagodan rhomboides*), Bermuda chub, and sergeant major (Stegeman et al., 1997). The reason for the observed differences in CYP2B-like expression among species is not known. Earlier, it was suggested that natural dietary compounds may be causing these differences, as higher CYP2B-like protein levels were observed in butterfly fish (*Chaetodon capistratus*) that consumed gorgonians (containing high levels of allelochemicals) compared to butterfly fish that avoided gorgonians (Vrolijk et al., 1994). CYP2N mRNA expression in *Chaetodon xanthurus* (CYP2N7) was significantly higher than in the facultative coralline-feeding butterfly fish *C. kleini, C. auriga* (CYP2N6), or *C. vagabundus*, as well as an obligate coralline feeding species (*C. punctofasciatus*) from the Great Barrier Reef in Australia (DeBusk, 2001). When each species, including *C. xanthurus*, was gavaged with gorgonian extracts from *Sinnularia maxima* for 3 days, with the exception of *C. punctofasciatus*, CYP2N mRNA expression was diminished (DeBusk, 2001). In the Bermuda species investigated, herbivorous fish had higher CYP levels (including CYP2B-like proteins) compared to carnivorous fish (Stegeman et al., 1997). It remains to be shown if natural dietary chemicals may act as inducers of CYP2B-like forms in fish or if other mechanisms are involved.

Phenobarbital (PB) and 1,4-*bis*(2-[3,5-dichloropyridyloxy])benzene (TCPOBOP) are powerful PB-type inducers of CYP2B genes in mammals (Poland et al., 1981). In mammals, induction of CYP2B by PB-type inducers proceeds through the activation of the constitutive androstane receptor (CAR) followed by nuclear translocation, dimerization with the retinoid X receptor (RXR), and binding to phenobarbital response elements (PBREMs) in the promoter region of the CYP2B genes (Honkakoski et al., 1997, 1998a, 1998b).

In fish, however, an apparent lack of response to PB-type inducers has been observed (Buhler and Williams, 1989; Eisele et al., 1984; Goksøyr et al., 1987; Haasch et al., 1994; Iwata et al., 2002; Stegeman, 1981). No induction of CYP protein levels, including scup P450B, or catalytic activities were seen in scup injected with TCPOBOB. In fact, TCPOBOB treatment had no effect on the translocation of the cytosolic CAR-immunoreactive protein in scup liver (Iwata et al., 2002). As mentioned above, studies with zebrafish CYP2AA2 have shown induction following TCPOBOB and phenobarbital (Kubota et al., 2013). These data point to functional differences, possibly in receptor activation or translocation, between fish and mammals.

Teleost fish do not possess a gene for CAR (Mathäs et al., 2012), and in fact, there are some teleost which lack PXR as well (Maglich et al., 2003; Eide et al., 2018). The lack of a piscine CAR receptor may be one explanation for the observed lack of PB-type as well as diminished CYP3A (see below) induction in fish.

*CYP2E-Like Forms*—Similar to CYP2B-like genes, while fish possess CYP2E1-like activities, and antibody cross-reactivity, fish do not have a direct ortholog of CYP2E1 (Kirischian et al., 2011). The possible existence of a CYP2E form in fish (*Poeciliopsis monacha-lucida*) was proposed based on hybridization with a rat CYP2E1 49-base oligonucleotide, antibodies to rat CYP2E1, as well as responsiveness to ethanol treatment. This CYP form was suggested to be involved in the CYP-mediated dealkylation of the fish carcinogen diethylnitrosamine (Kaplan et al., 1991). Furthermore, hepatic microsomal metabolism of the mammalian CYP2E substrate chlorzoxazone in winter flounder (*Pleuronectes americanus*) and in viviparous *Poeciliopsis monacha* and *Poeciliopsis viriosa* is indicative of the presence of CYP2E-like enzymes (Kaplan et al., 2001; Wall and Crivello, 1998). CYP2E1 is an important toxicological P450 in mammals, and the proper assignment of CYP2E1-like activity in fish is still an open question.

*The CYP2K Subfamily*—The CYP2K family is quite sizeable in fish (Kirischian et al., 2011, Goldstone et al., 2010). Rainbow trout CYP2K1 is one of the dominant CYP forms expressed in liver and trunk kidney, and it displays sexually dimorphic expression, with higher levels in sexually mature males compared to females (Buhler et al., 1994). In addition to liver and trunk kidney, CYP2K1 is also expressed, though at lower levels, in blood cells, upper intestine, head kidney, stomach, heart, gonads, and male muscle (Cok et al., 1998). Heterologous expression of CYP2K1 showed that CYP2K1 catalyzed hydroxylation of lauric acid primarily at the (ω-1) position but also to a minor extent at the (ω-2) position (Buhler et al., 1997). Treatment of rainbow trout with 17β-estradiol decreased CYP2K1 mRNA and protein levels as well as lauric acid hydroxylase activities. Treatment with testosterone resulted in slightly decreased CYP2K1 mRNA levels, whereas this treatment had no significant effect on CYP2K1 protein levels (Buhler et al., 2000). Thus, CYP2K1 is mainly expressed in the digestive tract and appears to be involved in the metabolism of endobiotics such as lauric acid as well as xenobiotics. In addition, CYP2K1 is also expressed in, for example, steroidogenic tissues; furthermore, hepatic CYP2K1 expression is affected by sex steroids, particularly 17β-estradiol (Buhler et al., 2000; Cok et al., 1998). This may have toxicological implications, as many natural fish populations are exposed to endocrine-disrupting chemicals (EDCs), including estrogenic compounds. In fact, treatment of rainbow trout with xenoestrogens (methoxychlor, diethylstilbestrol, 4-*tert*-octylphenol, and biochanin A) decreased hepatic acid hydroxylase activity and reduced CYP2K1 protein expression (Katchamart et al., 2002). CYP2K gene orthologs have also been cloned from killifish, pufferfish, and zebrafish (see Tables 3.2 and 3.3). CYP2Ks in zebrafish in particular appear to be important, as they are induced by known PXR agonists including the steroids pregnenolone and clobetasol (Faltermann et al., 2020; Salanga et al., 2020).

*The CYP2M Subfamily*—As mentioned earlier, CYP2M1 was originally isolated and cloned from rainbow trout liver (Miranda et al., 1989; Yang et al., 1998). This form was not identified in zebrafish (Saad et al., 2016). In trout, the highest expression of CYP2M1 was observed in the liver, but also the trunk kidney expresses CYP2M1. Interestingly, in the trunk kidney, a pronounced sexually dimorphic expression of CYP2M1 was seen, with juvenile females expressing 20-fold higher levels than juvenile males. Expression decreased in the trunk kidney in sexually mature animals (Yang et al., 1998). In addition to the liver and trunk kidney, CYP2M1 is also expressed, though at lower levels, in the head kidney, stomach, heart, gonads, and brain (Cok et al., 1998). Recombinant CYP2M1 catalyzed hydroxylation of lauric acid primarily at the (ω-6) position (Yang et al., 1998). Although CYP2M1 appears to be involved in the metabolism of fatty acids and fatty acid derivatives, the physiological function is unknown. Treatment of rainbow trout with 17β-estradiol resulted in decreased CYP2M1 protein and mRNA levels, whereas treatment with testosterone had no effect on CYP2M1 (Buhler et al., 2000). In addition, treatment with xenoestrogens (methoxychlor, diethylstilbestrol, 4-*tert*-octylphenol, and biochanin A) also reduced CYP2M1 protein levels as well as lauric acid hydroxylase activity (Katchamart et al., 2002). It should be noted, however, that the lauric acid hydroxylase activity assay used could not distinguish between CYP2K1, CYP2M1, and possible other CYP activities. Increased renal ω and ω-6 hydroxylation of lauric acid was observed in male channel catfish treated with ciprofibrate, whereas in male bluegill (*Lepomis macrochirus*), hepatic ω, ω-4, and ω-5 hydroxylation was induced by ciprofibrate (Haasch et al., 1998), implying CYP2M1-like activity. CYP2M transcripts have been identified in coho salmon olfactory rosette (Matsuo et al., 2008; Lavado et al., 2014), with the induction of CYP2M1 protein expression observed at increased salinities during seawater acclimation. Although sequence data are needed to verify the identity of the CYP enzymes induced, these data imply the induction of CYP2M-like activities by peroxisome proliferators.

*The CYP2N Subfamily*—CYP2N1 and CYP2N2 were originally cloned from killifish liver and heart cDNA libraries (Oleksiak et al., 2000). CYP2N1 mRNA expression was detected at high levels in liver and intestine (mid gut) and at low levels in heart and brain. CYP2N2 mRNA levels were highest in the heart and brain and present at lower levels in liver and intestine. Heterologous expression of CYP2N1 and CYP2N2 revealed transformation of arachidonic acid to epoxyeicosatrienoic acids. CYP2N1 preferentially metabolized arachidonic acid at the 8,9- and 11,12-olefins and to a lesser degree at the 14,15-olefin, and CYP2N2 preferentially metabolized arachidonic acid at the 8,9-olefin and to a lesser degree at the 11,12- and 14,15-olefins. In addition to being arachidonic acid epoxygenases and hydrolases, the CYP2Ns also metabolize xenobiotics. Thus, both CYP2N1 and CYP2N2 isoforms metabolize benzphetamine and also exhibit minimal alkoxyresorufin-*O*-dealkylase activities (Oleksiak et al., 2000). Intestinal CYP2N1 mRNA levels were decreased in starved animals and in animals treated with 12-*O*-tetradecanoylphorbol-13-acetate (TPA) or starved and treated with TPA-treated killifish. Intestinal CYP2N2 mRNA levels were diminished in animals treated with TPA, whereas starvation had no significant effect on this response; however, increased CYP2N2 mRNA expression was observed in the heart (but not in the brain) in killifish treated with TPA (Oleksiak et al., 2000). Hepatic CYP2N2 mRNA levels were increased in killifish exposed to anthracene in the lab and in killifish collected from a PAH-contaminated site (Peterson and Bain, 2004); thus, CYP2N genes are expressed in hepatic and extrahepatic organs and appear to be regulated by environmental agents. CYP2N gene orthologs also have been cloned from scup, butterfly fish (*Chaetodon xanthurus*), pufferfish, and zebrafish (see Table 3.3).

*The CYP2P Subfamily*—CYP2P1, CYP2P2, and CYP2P3 were originally cloned from a killifish genomic library (Oleksiak et al., 2003). CYP2P genes are predominantly expressed in the liver and intestine. Recombinant CYP2P3 catalyzed benzphetamine *N*-dealkylation and arachidonic acid oxidation. Arachidonic acid was oxidized to 14,15-, 11,12-, and 8,9-epoxyeicosatrienoic acids and 19-hydroxyeicosatetraenoic acid. The regiospecificity was similar to human CYP2J2 and rat CYP2J3 isozymes. Similar to that observed for CYP2N1 and CYP2N2, decreased levels of intestinal CYP2P2 and CYP2P3 mRNA were observed in killifish treated with TPA. Prolonged starvation or fasting (20 days) resulted in decreased levels of CYP2P2 and CYP2P3. Neither TPA nor fasting had any significant impact on CYP2P1 mRNA levels; however, fasting followed by refeeding resulted in increased levels of CYP2P1 mRNA but not CYP2P2 and CYP2P3 transcription levels. In rat, fasting resulted in reduced intestinal CYP2J3 and CYP2J4 protein levels; thus, intestinal expression of killifish CYP2P2 and CYP2P3 genes and arachidonic acid regiospecificity of the recombinant CYP2P3 isozyme show great similarities to mammalian CYP2J forms. Relatedness to mammalian CYP2J genes was further confirmed by phylogenetic analysis (Oleksiak et al., 2003, Kirischian et al., 2011, Goldstone et al., 2010). CYP2P gene orthologs also have been cloned from scup, butterfly fish, pufferfish, zebrafish, and largemouth bass (*Micropterus salmoides*), Atlantic salmon, and mangrove killifish (Table 3.3). In particular, CYP2P genes exhibit tandem duplication, resulting in large gene clusters in different species (Puthumana et al., 2017). Interestingly, CYP2P9 is among the major CYP genes expressed in untreated adult zebrafish liver (Kubota et al., 2019), while CYP2P18 and CYP2P19 in mangrove killifish *Kryptolebias marmoratus* are induced by bisphenol A, but not benzo[*a*]pyrene (Puthumana et al., 2017).

*The CYP2X Subfamily*—The presence of two CYP2-like isozymes in channel catfish was indicated based on immunoreactivity with antibodies against both rainbow trout CYP2K1 and CYP2M1. Furthermore, treatment with ethanol (Perkins and Schlenk, 1998) or the insecticide fenitrothion (Perkins, 1999) specifically decreased the expression of the lower (47-kDa) protein. Clofibrate treatment specifically increased the upper (51-kDa) protein. Furthermore, female catfish displayed higher levels of the 47-kDa protein compared to males (Perkins, 1999). A CYP2-immunoreactive protein, denoted CM-HA3, was next isolated from channel catfish liver. *N*-Terminal amino acid analyses followed by a BLAST search revealed sequence identity to both CYP2K1 and CYP2M1 (Perkins et al., 2000). By using degenerate PCR primers designed against this *N*-terminal followed by RACE, a CYP clone was isolated from catfish liver and designated as CYP2X1 (Schlenk et al., 2002); however, the derived amino acid sequence was different from the *N*-terminus of CM-HA3. CYP2X1 expressed in *Sf9* cells demonstrated benzphetamine demethylase activity, but testosterone, fenthion, and *p*-nitrophenol metabolism was not observed. CYP2X gene orthologs have also been cloned from pufferfish and zebrafish (see Table 3.3).

## CYP3

*Phylogeny*—Analyses of the complete sequences of several teleost genomes indicate that fish species contain a complement of CYP gene families similar to those found in mammals (Goldstone et al., 2010, Eide et al., 2021). To date, however, tissue distribution, mechanisms of gene regulation, and the catalytic function of many of these enzymes remain unknown. The CYP3 gene family is believed to have diverged between 800 and 110 MYA (Maurel, 1996), and four subfamilies have been identified, including CYP3A to CYP3D (McArthur et al., 2003; Yan and Cai, 2010). To date, more than 200 teleost CYP3A genes have been identified by sequence homologies. Additional subfamilies (CYP3B to CYP3D) have been discovered by data mining of various genome databases. The CYP3 identity of these genes was based on gene sequence homologies of less than 55% when compared to members of the CYP3A family. Current nomenclature for the CYP3 gene family, however, does not reflect orthologous relationships between organisms due to the presence of multiple CYP3A-like sequences in individual species; thus, CYP3A diversity is thought to include both orthology (diversification due to speciation) and paralogy (diversification due to gene duplication) (McArthur et al., 2003; Nelson et al., 1996).

Although teleost, diapsid, and mammalian CYP3A genes have undergone independent diversifications, there is clear evidence for conserved genomic loci linking the CYP3A and CYP3B-D regions between fish and tetrapods (Qiu et al., 2008; Yan and Cai, 2010).

Multiple CYP3A paralogs have been identified in several species, including medaka, rainbow trout, and killifish (Celander and Stegeman, 1997; Hegelund and Celander, 2003; Kullman and Hinton, 2001; Kullman et al., 2000; Lee and Buhler, 2003; Lee et al., 1998; Lemaire et al., 1996). Teleost CYP3A paralogs demonstrate high degrees of sequence similarity: 90% (CYP3A38 and CYP3A40), 94% (CYP3A27 and CYP3A45), and 98% (CYP3A30 and CYP3A56) for medaka, trout, and killifish, respectively. Each sequence conforms to the specific structural features associated with the cytochrome CYP gene superfamily and exhibits >40% sequence similarity to the CYP3A subfamily. The topologies of all CYP enzymes are similar, especially regarding structurally conserved regions such as the heme-binding domain, oxygen-binding region, and specific sites associated with redox interactions (Szklarz and Halpert, 1997). Differences in CYP catalytic activities are suggested to be determined predominantly by amino acid composition in six substrate-recognition sites (SRS1 to SRS6) (Gotoh, 1992), and although this idea has been challenged (Kirischian et al., 2011), this hypothesis has not been fully refuted.

Many crystal structures of the critically important human CYP3A4 enzyme exist. However, no crystal structures for fish CYP3A exist, but homology modeling has provided some guidance (Chng et al., 2012; Wu et al., 2021). Key amino acids associated with CYP3A substrate specificity, binding, and regio-specific catalysis have been identified in mammalian systems (Hsu et al., 2018; Wright et al., 2019; Sevrioukova and Poulos, 2017), vastly complicated by the catalytic complexity of CYP3A and the selection of conformational substates (Wang et al., 2021).

Multiple CYP3A-like teleost proteins have been observed using immunochemical detection in numerous other species. Although gene sequences for these species have not been cross-identified with the proteins, cross-reactivity with antibodies specific for either mammalian or teleost CYP3A proteins suggests that multiple CYP3A-like proteins are present in the liver and intestine of several teleosts (Celander et al., 1996ab).

*Function*—Functionally, CYP3A enzymes are among the most versatile forms of CYPs as they have unusually broad substrate specificities for both endogenous and exogenous substrates, including steroids, bile acids, eicosanoids, retinoids, xenobiotics such as pharmaceuticals, and procarcinogens (Gillam et al., 1993; Li et al., 1995; Smith et al., 1996; Waxman et al., 1998). CYP3A-like proteins were initially purified from several teleost species, including scup, rainbow trout, and Atlantic cod (Celander et al., 1989; Klotz et al., 1986; Miranda et al., 1989). Identification of these proteins as CYP3A-like was based predominantly on steroid hydroxylase activity and cross-reactivity with CYP3A-specific antibodies. In some instances, antibodies were additionally used as catalytic inhibitors. Purified cytochrome P450A from scup and LMC5 from rainbow trout exhibited specific steroid hydroxylase activity, similar to that observed with mammalian CYP3A enzymes. Each enzyme additionally demonstrated minimal benzo(*a*)pyrene hydroxylase and ethoxycoumarin *O*-deethylase activities, suggesting a functional difference from the previously identified and inducible CYP1A form. Further characterization of purified teleost enzymes was undertaken by comparative reciprocal western blot analysis. Cross-reactivity between teleost and mammalian antibodies further supported a close structural as well as functional similarity between teleost and human CYP3A enzymes (Celander et al., 1996b; Miranda et al., 1991). Functional characterization of recombinant CYP3A enzymes has been determined for CYP3A27, CYP3A45, CYP3A38, and CYP3A40 by heterologous expression in baculovirus systems. In reconstitution experiments, recombinant protein exhibited catalytic activities for the 6β-, 2β-, and 16β-hydroxylation of testosterone at 1.48, 0.043, and 0.034 nmol/min/nmol, respectively, as well as dehydrogenation of nifedipine at 50 pmol/min/nmol (Lee and Buhler, 2002). Although turnover rates were significantly higher with the recombinant enzyme, hydroxylase profiles were similar to that observed with purified CYP3A27 enzyme. Heterologous expression of the rainbow trout paralog CYP3A45 additionally exhibits testosterone hydroxylase activity with 6β-hydroxytestosterone as the major metabolite. Activity was higher than that observed with CYP3A27 and was significantly enhanced by the addition of cytochrome $b_5$ to the reconstitution assays (Lee and Buhler, 2003).

Recombinant medaka CYP3A38 and CYP3A40 catalyzed hydroxylation of testosterone, as well as the *O*-debenzylation of benzyloxyresorufin (BR) and 7-benzyloxy-4-(trifluoromethyl)-coumarin (BFC); however, efficiencies and specificities were significantly different between the two isoforms. Thus, $K_m$ and $V_{max}$ activities based on BFC *O*-debenzyloxylase were estimated to be 0.116 and 0.363 μM and 7.95 and 7.77 nmol/min/nmol P450 for CYP3A38 and CYP3A40, respectively. Medaka CYP3A38 preferentially catalyzed testosterone hydroxylation at the 6β- and 16β-positions, with minor hydroxylation at other positions within the steroid nucleus, whereas CYP3A40 catalysis was predominantly limited to the 6β-position (Kashiwada et al., 2005). Putative identification of CYP3A substrate-recognition sites indicated that 12 of the 49 amino acid differences between CYP3A38 and CYP3A40 occur in SRS1, SRS3, and SRS5, previously known to be associated with steroid hydroxylation (Kashiwada, et al., 2005; Kullman and Hinton, 2001). Functional analysis of teleost and mammalian CYP3A paralogs has demonstrated that gene-duplication events are tied to acquisition of new function and that convergent evolution of CYP3A function may be frequent among independent gene copies (McArthur et al., 2003).

The physiological role of CYP3A is yet to be fully determined. Given the robust steroid hydroxylase activity, there is speculation that CYP3A forms play an important part in steroid (hormones as well as bile acids) homeostasis.

However, zebrafish CYP3A65 exhibits low rates of 17beta-estradiol metabolism, especially in comparison to CYP1A (Scornaienchi et al., 2010), and thus CYP3As may be more important in androgen and progestogen metabolism. CYP3A genes are expressed in tissues that act as barriers to the environment (i.e., digestive and respiratory tracts), and together with the broad substrate specificities of CYP3A enzymes, these observations suggest that they evolved as biochemical defense to prevent bioaccumulation of both xenobiotics and toxic endobiotics. In a study comparing generalist and hard-coral feeding butterflyfish, generalists fed an allelochemical (5-episinulaptolide) from a soft-coral (*Sinularia* sp.) had higher survival and higher hepatic CYP3A protein concentrations that were induced following feeding (Maldonado et al., 2016a). In vitro incubations of 5-episinulaptolide with liver microsomes from a butterflyfish species that feeds on soft-corals (*Chaetodon unimaculatus*) led to the formation of a nontoxic epoxide metabolite that was predicted from a model of CYP3A4 with 5-episinulaptolide as a substrate (Maldonado et al., 2016a). These data indicate CYP3A may serve as a detoxifying mechanism allowing generalist and soft-coral feeding species of fish to consume chemically defended prey. Similar results in terms of CYP3A expression and toxicity were observed in fish treated with homarine, a hard-coral antifeeding compound (Maldonado et al., 2016b).

Another CYP3 family, the CYP3Bs were identified in medaka, expression of all forms was observed in all organs, particularly in the intestine and liver. The exception was CYP3B4 which was not observed in males (Shaya et al., 2014). Expression of CYP3B5 in olfactory organs was comparable to the liver. CYP3Cs have been examined in zebrafish (Corley-Smith et al., 2006; Shaya et al., 2014). Four CYP3C genes are present in the zebrafish genome (Goldstone et al., 2010), and CYP3C1 is prominent in zebrafish liver (Kubota et al., 2015; Shaya et al., 2019) Differences in the CYP3C function relative to CYP3A are not known.

*Regulation*—CYP3A are the major constitutive CYP forms expressed in the liver and intestine of most mammals and other species, including fish (Celander et al., 1996b, 1989; Hegelund and Celander, 2003; Hegelund et al., 2004; Husoy et al., 1994). Numerous studies have demonstrated, however, that constitutive expression of CYP3A genes between species, and paralogous CYP3A forms within species, is highly variable with age, gender, development, tissue localization, and between individuals. Variations in gene expression may be due to external environmental factors, including temperature, salinity, diet, or other environmental stressors; biological factors, such as circulating hormone levels; or tissue-specific factors associated with development or reproductive cycle. Depending on species, sexual dimorphic expression of CYP3A has been observed. Lee et al. (1998) reported higher levels of CYP3A expression in juvenile female rainbow trout intestine when compared to males. Gender differences were additionally observed in the kidney and liver but to a lesser degree than in the intestine. A similar trend was demonstrated in winter flounder (*Pleuronectes americanus*) hepatic microsomes and in Atlantic cod, where CYP3A expression was seven-fold higher in females than males (Gray et al., 1991; Hasselberg et al., 2004). In contrast, adult killifish males displayed up to 2.5-fold higher levels of hepatic and extrahepatic CYP3A30 and CYP3A56 mRNA and protein levels compared to females (Hegelund and Celander, 2003). This finding was in agreement with higher hepatic testosterone 6β-hydroxylase activities in male killifish (Gray et al., 1991; Stegeman and Woodin, 1984). Other species, including medaka and sexually mature rainbow trout, also exhibit a higher level of CYP3A expression in male liver microsomes compared to females (Celander et al., 1989, 1996b; Harlow and Halpert, 1998; Kullman and Hinton, 2001; Lee and Buhler, 2003; Li et al., 1995; Smith et al., 1996; Waxman et al., 1998). Variations in CYP3A expression in Atlantic salmon and turbot (*Scophthalmus maximus*) during reproductive cycles have been reported, implying a regulatory role of sex steroids on CYP3A expression (Arukwe and Goksøyr, 1997; Larsen et al., 1992). Steroid reproductive hormones and plasma growth hormones have been shown to influence the sexual dimorphic expression of CYP3A in rodents (Park et al., 1999; Sakuma et al., 2002; Wang and Strobel, 1997).

In teleosts, the effects of exposure to reproductive hormones on CYP3A expression differ depending on species. Numerous studies with rainbow trout, brook trout, medaka, and other fish species have demonstrated that exposure to 17β-estradiol results in a suppression of total microsomal CYP content. This correlated with decreased CYP3A mRNA expression, CYP protein levels, or steroid hydroxylase activities (Buhler et al., 2000; Celander et al., 1989; Pajor et al., 1990). Treatment of male Atlantic cod with 17β-estradiol resulted in an increase in CYP3A protein expression (Hasselberg et al., 2004). The mechanisms of hormonal regulation of CYP genes in teleosts have yet to be determined; however, it is possible that circulating hormone levels (steroids or growth hormone) are associated with sexual dimorphic differences in CYP expression in fish.

Numerous pharmaceuticals and xenobiotics have been demonstrated to alter CYP3A gene transcription in mammals via binding and transactivation of members of the nuclear receptor family NR1, including the pregnane X receptor (PXR), the constitutive androstane receptor (CAR), and the vitamin D receptor (VDR) (Pascussi et al., 2003; Plant and Gibson, 2003). As mentioned above, although Maglich et al. (2003) did not find CAR in the pufferfish genome, they identified a single PXR/CAR gene that was more PXR like, indicating CAR may have evolved after divergence of fish or may have been lost in some or all teleosts. To date, PXR and VDR have been identified in few teleost species. Prototypical mammalian NR1 (PXR) receptor agonists, including dexamethasone (DEX), rifampicin (RIF), and pregnenolone-16α-carbonitrile (PCN), are seemingly less effective at altering teleost CYP3A gene transcription (Celander et al., 1989, 1996a, 1996b).

In tilapia, PCN treatment resulted in two-fold induction of CYP3A proteins (Pathiratne and George, 1996). The ligand-binding region of PXR was isolated from zebrafish (Kliewer et al., 2002). This region was shown to be activated by some prototypical mammalian PXR agonists, including nifedipine, phenobarbital, clotrimazole, and some steroids, but PCN, DEX, and RIF did not activate the fish receptor (Moore et al., 2002). Induction of hepatic CYP3A expression *in vivo* by ketoconazole was observed in juvenile rainbow trout and Atlantic cod (Hasselberg et al., 2004; Hegelund et al., 2004). As mentioned above, response to 17β-estradiol resulted in either enhanced or diminished CYP3A expression, depending on the species examined (Buhler and Wang-Buhler, 1998; Hasselberg et al., 2004; Husoy et al., 1994; Pajor et al., 1990). Furthermore, xenoestrogens also affect CYP3A expression. Male Atlantic cod exposed to alkylphenols showed enhanced hepatic CYP3A expression, whereas treatment with ethinylestradiol and nonylphenol suppressed CYP3A expression in Atlantic salmon and Atlantic cod (Arukwe et al., 1997; Hasselberg et al., 2004). Slight increases in hepatic CYP3A expression have been demonstrated in rainbow trout exposed to cortisol, whereas cortisol treatment had no effect on CYP3A protein levels in Arctic char (*Salvelinus alpinus*) (Celander et al., 1989; Jorgensen et al., 2001).

Gene sequences for PXR and VDR have been identified in fish (Kliewer et al., 2002), and there is good evidence for functional data in transcriptional activation of CYP3 genes. Extensive in vitro binding studies for PXR and VDR from zebrafish, pufferfish, and medaka point to conservation in the DNA binding behaviors (Reschly et al., 2007). (For more on nuclear receptors in teleosts, see Chapter 3.) Transcriptionally, there are clear linkages between PXR and CYP3A induction: morpholino knockdown of PXR leads to the loss of CYP3A65 induction in zebrafish (Bainy et al., 2013, Kubota et al., 2015), although genomic ablation of PXR did not appear to have the same effect (Salanga et al., 2020). As mentioned above, the treatment of butterflyfish with the soft-coral-derived natural product 5-episinulaptolide induced CYP3A protein and catalytic activities in species of fish that were amenable to induction (Maldonado et al., 2016a). A similar response was observed in generalist and hard-coral species of butterflyfish treated with the antifeeding hard-coral metabolite, homarine (Maldonado et al., 2016b). Overall, the modulation of teleost CYP3A expression following xenobiotic exposure is highly variable and species-dependent. This variability may be additionally due to hidden allelic variation in population samples, as even laboratory strains of zebrafish exhibit more than 10-fold variation in PXR reporter gene activation (Lille-Langøy et al., 2019).

Studies examining the binding behavior of mammalian PXR have demonstrated transactivation and CYP3A induction by xenoestrogens, including bisphenol A, nonylphenol, DDT, and other organochlorine pesticides (Masuyama et al., 2000; Takeshita et al., 2001; You, 2004). The response of aquatic species to these and other PXR, CAR, and VDR ligands has been minimal; thus, the overall induction of CYP3A in teleosts appears to be considerably weaker than that reported in mammals. Compared to mammalian species, this suggests that teleost CYP3A expression may be governed by alternative transcriptional mechanisms.

Notably, AHR regulation of zebrafish CYP3A65 has been observed in a number of studies (Chang et al., 2013, Kubota et al., 2015). In several studies, discrepancies have been noted between changes in gene expression and changes in steroid hydroxylase activity. In a study by Hasselberg et al. (2005), exposure of juvenile Atlantic cod to ketoconazole resulted in slight increases in CYP3A expression but marked decreases in CYP3A-specific catalytic activity. Similar discrepancies have been noted in medaka, for which significant decreases in catalytic activity cannot be accounted for by corresponding reductions in CYP3A expression (Kashiwada et al., 2005). These discrepancies may arise due to interactive effects between the xenobiotic CYP3A inhibitors (e.g., antifungal azole P450 inhibitors) that may affect endobiotic CYP inducers (e.g., Wassmur et al., 2010). As noted in the section on CYP1A, these discrepancies between message level and catalytic activity necessitate the determination of expression at several levels prior to making conclusions about regulatory mechanisms.

As with function, much less is known about the expression of the CYP3C subfamily. However, it is clear that CYP3C genes are expressed and induced in sex-specific manners in adult zebrafish (Kubota et al., 2019, Shaya et al., 2019). In particular, exposure to AHR ligands led to dose-dependent increases of CYP3C genes in female, but not male, intestine, while estrogens exposure led to inducing of all CYP3Cs in male intestine and female liver. Interestingly, CYP3C3 and CYP3C4 were induced in the testis; CYP3C1 and CYP3C4 were slightly induced in the ovary, suggesting divergent functional roles for these tandemly duplicated genes (Shaya et al., 2019).

*Localization*—Localization of CYP3A expression is similar to that observed in mammalian species, with high levels of expression found in the liver and intestinal mucosa. Some differences are observed between specific CYP3A paralogs and may represent putative regulatory and functional differences that occurred during gene duplication events. Immunohistochemical and mRNA analysis of CYP3A27 has demonstrated strong responses in the intestinal ceca, proximal descending intestine, and liver, with minor expression occurring in the brain (Cok et al., 1998; Lee et al., 2001). The rainbow trout paralog CYP3A45 was predominantly expressed in the gastrointestinal tract, with weak expression occurring in the liver (Lee and Buhler, 2003). In killifish, CYP3A30 and CYP3A56 were coexpressed in the intestine and liver, consistent with that observed for rainbow trout CYP3A27. In both species, CYP3A expression is prominent in the digestive tract (intestinal mucosa and liver), suggesting a role for these enzymes in the first-pass metabolism of xenobiotics. Extrahepatic CYP3A expression is observed in the gill, kidney, brain, spleen, and ovary, suggesting a possible role for CYP3A enzymes in the fine-tuning of endogenous substrates at the site of synthesis or action as well as metabolic defense against xenobiotics (Hegelund et al, 2004).

*Development*—Biotransformation systems for embryonic and fetal tissues of vertebrates are not as well characterized as those for adults, although the reverse is true for zebrafish. Embryogenesis is a dynamic process that presents a continuously changing metabolic profile as enzymes are induced and repressed (Miller et al., 1996). In general, developing organisms lack many CYP forms present in adults; however, in some instances, they contain certain forms that are only expressed during development. These differences in total CYP content may explain developmental processes and altered sensitivities to toxic exposures during development. Teleost embryos are often sensitive to xenobiotics, and prolonged exposures result in numerous developmental malformations, early mortality, and delayed hatch (Cooke and Hinton, 1999; Villalobos et al., 1996). The nonspecific onset of many of these aberrations may reflect an inability of embryos to sufficiently metabolize and detoxify many of these xenobiotics, or the particular sensitivity of specific developmental pathways.

Developmental patterns of overall CYP expression have been studied in zebrafish (Goldstone et al., 2010), who found that there were waves of different CYP expression patterns during early development (to 48 hpf). Many of these CYPs are not considered directly of toxicological interest, as they are primarily involved in the synthesis and catabolism of various hormones. However, perturbations in these CYP levels may still lead to abnormal development. CYP1, CYP2, and CYP3 gene expression all exhibited different peaks during development, discussed below for CYP1 and CYP3, but not clearly dissected for CYP2s as of yet.

Expression patterns of the CYP1 and CYP3A genes have been studied in some detail throughout development. CYP3A65 expression was studied in detail by Buhler and collegues, who found that CYP3A65 was strongly inducible during early larval stages (post hatch), and eventually localized to the developing liver (Tseng et al., 2005). Other species display similar patterns: CYP3A27 is expressed early during rainbow trout development (Lee et al., 1998), while Kullman and Hinton (2001) demonstrated that CYP3A38 and CYP3A40 genes are differentially regulated during embryonic development in medaka. CYP3A40 is expressed early in embryonic development and continues throughout adult stages, while CYP3A38 is tightly suppressed during embryonic development and is only expressed posthatch. CYP3A40 mRNA expression was detected as early as stage 11, representing an early multicellular stage of development. Liver formation does not occur until stage 26, demonstrating that initial CYP3A40 expression occurs prior to organogenesis. CYP1 gene expression patterns were found to be similarly dynamic during zebrafish development (Jönsson et al., 2007). CYP1B1 was found to peak at 2 dpf, while the CYP1Cs peaked around hatching (3 dpf), and CYP1A after hatching (14–21 dpf). Induction potential for these genes differed at different times postfertilization.

*Mechanisms of Stimulation and Inhibition*—Crystallographic studies have demonstrated that CYP3A proteins have a large substrate-binding pocket in comparison to other members of the CYP superfamily (Wright et al., 2019). This large pocket is thought to enable CYP3A enzymes to bind multiple substrate molecules at any given time and results in unusual kinetic behaviors consistent with allosteric interaction. The atypical kinetic behavior of CYP3A enzymes results in either sigmoidal or convex rate–substrate concentration profiles indicative of positive or negative cooperativity (Houston and Kenworthy, 2000). With human CYP3A4, homotropic cooperativity (i.e., one substrate) has been observed with numerous chemicals and results in an initial lag in the rate–substrate concentration profile, thus generating sigmoidal profiles (Harlow and Halpert, 1997). Addition of secondary substrates such as α-naphthoflavone (ANF) can result in the modification of testosterone hydroxylation activity, suggesting heterotropic (two or more substrates) cooperative interaction (Harlow and Halpert, 1997, 1998). Structural studies indicate that there are two binding pockets in the CYP3A4 active site, leading to complex cooperativity in substrate or inhibitor binding (Fernando et al., 2006; Kapelyukh et al., 2008). Similar effects have been observed in fish. Activation of BFC *O*-debenzyloxylase activity by ANF was also observed in rainbow trout liver microsomes (Hegelund and Celander, 2003). In studies with CYP3A38 and CYP3A40, addition of higher (>25 μM) BFC concentrations resulted in a decrease of catalytic activity and a convex rate–substrate plot. This type of deviation from Michaelis–Menten kinetics is indicative of negative homotropic cooperativity and is due to an inability to maintain $V_{max}$ at higher substrate concentrations. In separate experiments, nonylphenol was used as a heterotropic effector. Results were biphasic, suggesting that nonylphenol acts as a concentration-dependent cooperative activator and inhibitor of BFC catalysis. The rate substrate plot demonstrates a sigmoidal curve at low concentrations (below 100 nM), indicative of heterotropic activation; however, at higher concentrations, activity levels are decreased, resulting in a convex rate–substrate plot that is indicative of heterotropic inhibition (Kullman et al., 2004). Rainbow trout and killifish exposed to ketoconazole, a potent antifungal agent, additionally exhibited significant decreases in CYP3A-mediated BFC *O*-debenzyloxylase activity, suggesting that this pharmaceutical is a potent heterotropic inhibitor of CYP3A activity (Hegelund et al., 2004). Alkylphenol additionally inhibited CYP3A activity in Atlantic cod liver microsomes with an $IC_{50}$ of 100 μM compared to ketoconazole, which had an $IC_{50}$ in a submicromolar range (Hasselberg et al., 2004). These studies demonstrate that teleost CYP3A enzymes exhibit unusual kinetic behaviors consistent with allosteric interaction and cannot be described by hyperbolic kinetic models. Homotropic cooperative inhibition of BFC at high concentrations suggests that the CYP3A protein in fish, as in humans, is capable of binding multiple substrate molecules, which may result in autoactivation or inhibition of catalysis. Interestingly, the addition of nonylphenol, alkylphenol, or ketokonazole results in heterotropic cooperative inhibition at environmentally relevant concentrations (Hasselberg et al., 2004; Hegelund et al., 2004). Given the putative role of CYP3A in maintaining the homeostatic balance for numerous endobiotics, enzymatic activation and inhibition by xenobiotic compounds may represent a (nongenomic) mechanism of altered metabolism and subsequent toxicity.

*Genomic technologies*

With the rise of genomics has come the application of new technologies applied to CYP function in fish toxicology. In addition to morpholino oligonucleotide (MO) knockdowns, genome editing technologies, in particular CRISPR-Cas, have been applied in fish. Most of these applications have been made in zebrafish, as it is an accepted biomedical and regulatory model species amenable to laboratory studies. However, there is an increased interest in applying these technologies to less commonly used species.

CYP functions have been studied using morpholinos, both the functional aspects and in particular regulatory studies. So, for example, CYP1A MO knockdown has been used to examine the role of CYP1A in heart deformation induced by TCDD (Carney et al., 2004) and cardiosulfa (Ko and Shin, 2012), while knockdown of CYP1B1 was used to demonstrate that CYP1B1 does not play a significant role in the embryotoxicity of selected PAHs (Timme-Laragy et al., 2008). MO knockdown of CYP1C1 and CYP1C2 demonstrated the roles that these proteins play in the reducing of blood flow to the developing zebrafish brain via the mesencephalic vein (Kubota et al., 2011). The roles of ligand-activated receptors have been probed by MO knockdown of AHRs, PXR, and VDRs, in a variety of fish species including zebrafish, killifish, and medaka (e.g., AHR: Garner et al., 2013, Brown et al., 2015, Massarsky et al., 2016); PXR: Kubota et al., 2015, VDR: Lin et al., 2012). MO knockdown of Cyp19ab1 was used to understand adverse outcomes associated with BaP-mediated aromatase inhibition (Alharthy et al., 2017).

Genome editing has also significantly contributed to the understanding of CYP function in toxicology, including the addition of fluorescent protein tags, human enzyme forms, or the induction of indels in a gene. For example, CYP1A spatiotemporal expression during development has been visualized using transgenic reporters not only in zebrafish (Shen et al., 2018) but also in marine medaka *Oryzias melastigma* (Zhou et al., 2020). Functional differences between zebrafish and human CYP3A have been examined in the zebrafish context by inserting a human CYP3A4 gene into a zebrafish line (Poon et al., 2017), although in a nonspecific manner potentially leading to significant overexpression of the human protein relative to zebrafish CYP3s. Finally, CRISPR-Cas technologies have been used to knockout a variety of fish CYPs, so far mostly in the study of endogenous functions, for example, CYP26B1 (Pogoda et al., 2018) and CYP27C1 (Enright et al., 2015). Knockout of the different AHRs in Atlantic killifish has contributed substantially to the understanding of the different roles of AHRs in toxicity (Aluru et al., 2015), despite the challenge of genetic manipulation over the 2-year generation time of these environmental marine models.

These technologies will enable the study of fish CYPs in toxicology primarily in those species amenable to genomic manipulation but contribute significantly to our overall understanding through phyologenetic relationships with nonmodel species.

## Flavin-Containing Monooxygenases

### Overview

Flavin-containing monooxygenases (FMOs) are a multigene family of enzymes involved in the monooxygenation of primarily soft-nucleophilic-heteroatom-containing compounds (see Table 3.4) and some inorganic compounds (Hines et al., 1994; Ziegler, 1988; 1993; Ziegler and Mitchell, 1972). Because FMOs are located in the smooth endoplasmic reticulum (microsomes) and require NADPH and oxygen as cofactors for catalysis, FMO-catalyzed reactions were thought at one time to merely be another mixed function oxidase reaction. FMOs were finally identified as unique enzymatic entities following purification of the enzyme by Ziegler and Mitchell (1972). Through an elegant series of experiments, Poulsen and Ziegler (1979) showed that FMOs have a distinct reaction mechanism that is not dependent on a reductase coenzyme (such as CYP), but they are directly reduced by NADPH. The reduction of the flavin by NADPH then allows binding of molecular oxygen, creating a hydroperoxyflavin that is very susceptible to nucleophilic attack by soft-nucleophilic-heteroatom-containing compounds and various inorganic species; thus, compounds with nondelocalized electronic features, such as tertiary amine- and sulfur-containing compounds, are excellent substrates for FMOs (Ziegler, 1988; 1993; Ziegler and Mitchell, 1972). Numerous substrates have been identified in mammalian systems, but few have been examined in fish (Schlenk, 1993, 1998).

One of the most surveyed, but more difficult to measure, FMO-catalyzed reactions in fish is that of trimethylamine oxidase (or more accurately TMA oxygenase) (for reviews, see Baker et al., 1963; Schlenk, 1998). Although FMO has yet to be purified to homogeneity in fish, numerous inhibition and correlative studies have indicated that TMA oxidase is catalyzed by FMO (Agustsson and Strom, 1981; Goldstein and Dewitt-Harley, 1973; Peters et al., 1995; Schlenk, 1994; Schlenk and Li-Schlenk, 1994; Schlenk et al., 1995). A direct relationship also exists between TMA content in fish tissues and enzyme expression (Larsen and Schlenk, 2001a; Raymond, 1998; Raymond and DeVries, 1998; Schlenk, 1998). Fish that possess high tissue (muscle, liver, blood) concentrations of TMA or TMA *N*-oxide have higher levels of expression and enzyme activity than fish that do not have either of these biomolecules; for example, TMA-lacking species, such as the channel catfish (*Ictalurus*

**TABLE 3.4**
**Putative Substrates for Flavin-Containing Monooxygenases in Fish**

| Substrate | Product |
|---|---|
| **Nitrogen-containing** | |
| Trimethylamine | Trimethylamine *N*-oxide |
| *N,N*-Dimethylaniline | *N,N*-Dimethylaniline *N*-oxide |
| 2-Aminofluorene | 2-Hydroxyaminofluorene[a] |
| **Sulfur-containing** | |
| Thiobencarb | Thiobencarb-*S*-oxide[a] |
| Thiourea | Thiourea-sulfonic acid[a] |
| Methimazole | Methimazole-sulfonic acid[a] |
| Aldicarb | Aldicarb sulfoxide[a] |
| **Selenium-containing** | |
| Dimethylselenide | Dimethylselenoxide[a] |
| Selenomethionine | Selenomethione-Se-oxide[a] |

[a] More toxic metabolite than parent compound.

*punctatus*), do not express FMO-like protein or enzymatic activity, whereas TMA-containing rainbow trout or various elasmobranchs possess relatively high levels of FMO activity and expression in various tissues (Schlenk and Buhler, 1991, 1993; Schlenk and Li-Schlenk, 1994; Schlenk et al., 1993). The relationship between TMA and FMO has several implications regarding the evolution and possible physiological functions of these enzymes, which are discussed further below.

Xenobiotics that have been shown to be substrates of the enzyme in fish include tertiary amines such as *N,N*-dimethylanaline (DMA) and TMA, thioether pesticides (e.g., aldicarb), thiocarbamates (e.g., eptam, thiobencarb), thiocarbamides (e.g., methimizole, thiourea), and thioamides (e.g., thiobenzamide) (Table 3.4); for a review, see Schlenk (1998). Currently, the most characterized diagnostic substrates for FMO activity in fish are thiourea and DMA (Schlenk, 1993, 1998). Enzymatic activities in fish appear to be sensitive to temperature and have a relatively high pH optimum of 8.0 to 9.6 in various fishes (Schlenk, 1993, 1998). Each of these assays is a simple spectrophotometric method; however, other more elaborate high-performance liquid chromatography (HPLC) methods utilizing enantioselective oxygenations have been identified in mammals to differentiate activities catalyzed by specific isoforms (Rettie et al., 1995). Recent studies have indicated unique stereochemical sulfoxidation reactions in trout, which have not been observed with any previous isoforms of mammalian FMOs (Schlenk et al., 2004). Typically, to validate FMO activity for an uncharacterized xenobiotic, coincubation of other FMO substrates or CYP inhibitors is necessary, because FMO and CYP share many substrates. It is imperative to note, however, that many putative inhibitors of mammalian enzymes are not as effective in fish and few have been well-characterized, so care should be taken when interpreting data regarding the use of enzyme inhibitors (especially, CYP and FMO) in studies with fish.

Currently, 12 FMO genes have been identified in mammals (classified as FMO1 to FMO12) and 8 genes identified in zebrafish (Kupsco and Schlenk, 2017). Seven genes were observed in rainbow trout, but with only one that has been cloned and characterized (Rodriguez-Fuentes et al., 2008). In Japanese medaka, five putative medaka fmos were characterized from the medaka genome through the National Center for Biotechnology Information (NCBI) database by protein motifs and alignments. They were identified as fmo4, fmo5A, fmo5B, fmo5C, and fmo5D (Kupsco and Schlenk, 2017).

## Regulation, Function, and Toxicological Relevance

Regulation of FMO expression appears to be extremely complex, with the enzyme often being expressed in a random manner (Baker et al., 1963). As mentioned earlier, FMO has been observed in all marine fish species, some euryhaline, and virtually no freshwater species. Strong correlations have been observed between FMO1-like proteins, enzymatic activity, and mRNA recognized by FMO1 cDNA in juvenile Atlantic flounder (*Platichthys flesus*) and turbot (*Scophthalmus maximus*) (Peters et al., 1995; Schlenk et al., 1996a, 1996b), but catalytic activity did not correspond with mRNA expression in sexually mature adult flounder. In fact, although FMO1-like mRNA was observed in all sexually mature animals, hepatic FMO activity was lacking in more than 40% of the animals, with males having more frequent expression. Although evidence suggests that various hormones may modulate the expression of FMO (El-Alfy and Schlenk, 2002; El-Alfy et al., 2002; Schlenk et al., 1997), no consistent induction of enzyme expression has been observed following xenobiotic treatment. In medaka, estradiol downregulated the hepatic FMO3-like protein but induced the gill FMO1-like form (El-Alfy and Schlenk, 2002). Testosterone downregulated both forms of FMO and activity in medaka. Arterial infusion of cortisol induced the expression of FMO1 in the rainbow trout gill and liver (El-Alfy et al., 2002).

As mentioned above, FMO activity and protein in several tissues, particularly the gill, appear to be directly correlated with serum osmolality. Several hypotheses have been put forth to explain this relationship. Four possibilities are as follows: (1) TMA *N*-oxide is produced as a cellular defense to prevent the enzyme inactivation by high cellular ion (Na, K) or urea that occurs in fish residing in subartic or subantarctic environments (Raymond, 1998; Raymond and DeVries, 1998); (2) TMA *N*-oxide is produced to counterbalance high tissue and serum levels of urea that are present regardless of temperature (e.g., in sharks) (Van Waarde, 1988; Yancey et al., 1982); (3) TMA *N*-oxide may be produced in muscle of deep-sea gadiform teleosts (which also produce urea) as an adaptive defense against high pressure (Gillett et al., 1997); and (4) TMA *N*-oxide may be formed by FMOs as a secondary organic osmolyte in response to shifts in salinity regimes (Schlenk, 1993). Each scenario is interrelated, with the common similarity being hyperosmolality. In some euryhaline fish, FMO activity and expression are directly related to the salinity regime in which the animal resides (Charest et al., 1988; Daikoku and Sakaguchi, 1990; Daikoku et al., 1988; El-Alfy and Schlenk, 2002; Lange and Fugelli, 1965; Schlenk and El-Alfy, 1998; Schlenk et al., 1996a, 1996b). FMO activity and protein expression are higher in the gills and kidneys than the liver in several species of euryhaline fish such as the Atlantic flounder (*Platichthys flesus*), Japanese medaka (*Oryzias latipes*), and rainbow trout (Larsen and Schlenk, 2001a; Schlenk, 1998; Schlenk and El-Alfy, 1998; Schlenk et al., 1995). In addition, FMO activity in trout and medaka is downregulated following the steroid treatment, which also downregulates osmoregulatory function (i.e., $Na^+/K^+$-ATPase) (McCormick, 1995; Schlenk et al., 1997). FMO activity and expression in rapid osmoconformers such as striped bass and tilapia do not respond to changes in salinity (Wang et al., 2001), which is also consistent with mechanisms of $Na^+/K^+$-ATPase regulation in these species. Testing scenarios 1 and 3, recent studies in rainbow trout have indicated that urea infusion or reductions in temperature induce FMO activity and increase muscular TMA *N*-oxide concentrations (Larsen and Schlenk, 2001b). Given the induction following cortisol treatment, several factors may be involved in FMO regulation in rainbow trout, including stress resulting from alterations in cellular redox potential due to urea or hyperosmotic conditions (El-Alfy et al., 2002; Larsen and Schlenk, 2001b). Promoter analyses upstream of a hepatic FMO gene isolated from rainbow trout indicated a glucocorticoid response element and treatment of primary trout hepatocytes with cortisol and sodium chloride-induced expression (Rodriguez-Fuentes et al., 2008). Thus, a relationship to osmoregulatory action and stress appears to be present with FMO.

The toxicological significance of FMO-catalyzed biotransformation reactions has not been extensively examined in fish. Recent studies with organoselenides and FMO from sharks have indicated that FMO may be involved in the oxidation and initiation of redox cycling in these species (Schlenk et al., 2003); however, most studies examining the toxicological roles of FMO have examined pesticides. The oxygenation of the pesticides thiobencarb and eptam in striped bass (*Morone saxatilis*) by hepatic FMO was shown to lead to the formation of a reactive-intermediate that covalently bound protein sulfhydral groups (Cashman et al., 1990; Perkins et al., 1999); however, protein binding was not observed *in vivo* by thiobencarb. *S*-Oxygenation of aldicarb to the sulfoxide by FMO significantly increased the inhibition of acetylcholinesterase in the rainbow trout (*Oncorhynchus mykiss*) (250-fold) and Japanese medaka (*Oryzias latipes*) (40-fold) (El-Alfy and Schlenk, 2002; Perkins et al., 1999). Elevated toxicity has been observed in FMO-containing fish which can activate aldicarb to the more potent sulfoxide compared to species that lack FMO and convert aldicarb to the less toxic sulfone or hydrolytic metabolites (Perkins and Schlenk, 2000; Schlenk, 1995).

Enhanced sulfoxidation may possibly explain the enhanced toxicity of aldicarb in higher salinity observed in medaka and trout, as FMO expression has been shown to be directly correlated to salinity in medaka (Larsen and Schlenk, 2001a; Schlenk and El-Alfy, 1998). Studies comparing the effects of salinity on aldicarb toxicity in trout and striped bass indicate that salinity significantly enhances the toxicity of aldicarb in trout but not in striped bass (Wang et al., 2001). In striped bass, aldicarb sulfoxide formation and FMO expression were unchanged by salinity, whereas salinity increased aldicarb sulfoxide formation, cholinesterase inhibition, and FMO expression in rainbow trout. Consequently, understanding factors that affect the expression patterns of FMO is important when considering species-specific sensitivities to xenobiotics and differential responses of organisms to environmental factors such as salinity and temperature regimes.

Analyses of promoting regions for medaka FMOs indicated binding sites for several transcription factors involved in cellular development and the unfolded protein response (UPR). Treatment of embryos (5–6 days postfertilization) with Tunicamycin, an inducer of endoplasmic reticulum stress significantly increased expression of *fmo4 and fmo5A*. In contrast, treatment of the embryos with dithiothreitol significantly reduced the expression of *fmo5A*, *fmo5B*, and *fmo5C* demonstrating potential functionality of sulfur redox metabolism (Kupsco and Schlenk, 2017).

## Monoamine Oxidases

Monoamine oxidases catalyze the oxidation and eventual elimination of alpha carbon groups from secondary amines. Monoamine oxidases have been characterized in several fish species, with most occurring in trout. Given the critical importance in catecholamine metabolism, most studies have focused on its endogenous role in the neurophysiology of fish. In contrast to terrestrial vertebrates, which have two forms of the enzyme (MAO A and MAO B), fish appear to only have a single form that is genetically distinct from terrestrial vertebrates. Although no specific studies have examined the role of MAO in xenobiotic biotransformation in fish, the effects of various organic and inorganic pollutants on enzyme activity have been examined (Senatori et al., 2003).

## Alcohol and Aldehyde Dehydrogenases

Alcohol dehydrogenase (ADH) catalyzes the oxidation of alcohols to aldehydes, which are subsequently converted to acids by aldehyde dehydrogenase (ALDH). $NAD^+$ is a cofactor for each enzyme. A class 3 ADH cDNA was first identified in sea bream, in which its expression was observed in all tissues as well as eggs and embryos. Expression decreased during early embryonic development but increased four-fold from day 1 to 21 after hatching, indicating that the maternal ADH mRNA is present in the eggs and embryos but diminishes as development occurs, allowing the larval tissue to express its own ADH (Funkenstein and Jakowiew, 1996). An additional ADH3 cDNA was also identified by RT-PCR in zebrafish (*Danio rerio*) (Dasmahapatra et al., 2001). Expression of the gene in zebrafish embryos appeared to correspond with temporal variations in zebrafish susceptibility to ethanol toxicity. In cod, an ADH enzyme was purified that displayed structural similarities to ADH3, but functionally, it was more like ADH1. Ethanol was an excellent substrate for the purified enzyme, and 4-methylpyrazole was a strong inhibitor ($K_i$=0.1 μM) (Danielsson et al., 1992). Allylic and acetylenic alcohols appear to be bioactivated through oxidation by trout liver ADH and may also act as inhibitors (Bradbury and Christensen, 1991). ALDH has been observed in all tissues of numerous fish species (Nagai et al., 1997). Similar to CYP1A, ALDH has been observed in mammals to be regulated by the Ah receptor. Studies in the dab (*Limanda limanda*), the sea bass (*Dicentrarchus labrax*), and the rainbow trout failed to observe increases in ALDH activities following 3-MC or BNF treatment (Lemaire et al., 1996); however, expression was significantly elevated in liver tumor tissues from adult rainbow trout treated with aflatoxin (Parker et al., 1993).

### Peroxidases

Lipoxygenases, cyclooxygenases, and peroxidases such as prostaglandin-H synthetase are typically involved in the oxidation of arachidonic acid to hydroxy and peroxyeicosatetraenoic acids, which are subsequently converted to prostaglandins, prostacylins, and thromboxanes. In mammals, each of these enzymes has been shown to activate various xenobiotics through cooxidation pathways requiring a hydroperoxide as a cofactor. Although the enzymes have been observed in fish, their contributions to xenobiotic biotransformation have not been examined. Examples of lipoxygenases observed in rainbow trout gills include the 15- and 12-lipoxygenase (German and Berger, 1990; Hsieh et al., 1988). PGH synthetase has also been observed in gills of freshwater fish (Christ and Van Dorp, 1972).

### Aldehyde Oxidase

Aldehyde oxidase is a molybdozyme that is similar to xanthine oxidase located in the cytosols of the liver in mammals and fish. Pyrroles, pyridines, pyrimidines, purines, and aromatic aldehydes derived from catecholamine metabolism tend to undergo oxidation reactions; however, under anaerobic conditions, 2-hydroxypyrimidine, $N^1$-methylnicotinamide, or butyraldehyde may act as electron donors in liver cytosols in fish, leading to substrate reduction under these conditions. Enzyme activity is inhibited by menadione, β-estradiol, and chlorpromazine. Recent studies in the goldfish (*Carassius auratus*) and the sea bream (*Pagrus major*) have demonstrated that aldehyde oxidase catalyzes the reduction of fenthion sulfoxide to the parent sulfide (Kitamura et al., 2003).

## REDUCTASES

One of the most common reductases involved in xenobiotic biotransformation is the cytochrome P450 reductase, which catalyzes single-electron reductions to substrates prone to accept single electrons (e.g., quinones, cyclic, or aromatic amines). Often, this pathway activates heterocyclic compounds to redox cycling intermediates. Limited studies have evaluated the role of this enzyme in biotransformation in fish.

### NADPH Oxido-reductase

NAD(P)H:[quinone acceptor] oxidoreductase (NQO1) plays a significant role in one- and two-electron reduction reactions, particularly in the liver. 4-Nitroquinoline 1-oxide and nitrofurantoin were both reduced by DT diaphorase to genotoxic metabolites in a brown bullhead fibroblast cell line (Hasspieler et al., 1997). A novel dicoumarol-sensitive oxidoreductase that catalyzes the reduction of phenanthrenequinone was purified from gastric cytosol in the channel catfish (*Ictalurus punctatus*) (Hasspieler and Di Giulio, 1994). Due to its likely AhR-mediated regulation, the enzyme has shown promise as a biomarker in fish. Whether or not bifunctional regulation through ARE occurs (as in mammals) is uncertain.

### Azo- and Nitroreductases

Catalyzing up to three sequential two-electron reductions, nitroreductases play a significant role in the biotransformation of primarily nitroaromatic compounds to corresponding amino aromatic compounds. When 2-nitrofluorene was incubated with liver microsomes or cytosol of sea bream (*Pagrus major*) in the presence of NADPH or 2-hydroxypyrimidine, 2-aminofluorene was formed (Ueda et al., 2002). Hepatic nitroreductases were also observed in catfish and catalyzed the formation of superoxide from nitrofurantoin, *p*-nitrobenzoic acid, and *m*-dinitrobenzene (Washburn and Di Giulio, 1988). Hepatic nitro- and azoreductases were observed in the marine teleosts barracuda (*Sphyraena barracuda*) and yellowtail snappers (*Ocyurus chrysurus*); however, only azoreductase activity was observed in elasmobranches, such as the lemon shark (*Negaprion brevirostris*) and stingray (*Dasyatis americana*) (Adamson et al., 1965). The lampricide 3-trifluoromethyl-4-nitrophenol was reduced by nitroreductases to 3-trifluoromethyl-4-aminophenol in rainbow trout (Lech and Costrini, 1972).

### Aldo-Keto Reductases

Aldo-keto reductases (AKR) are cytosolic proteins that use NADPH as a reducing cofactor. Substrates include carbonyls or aldehydes that are reduced to alcohols. One reaction that has toxicological implications is the conversion by AKR1 enzymes of PAH trans-dihydrodiols to ortho-quinones. Twelve AKR genes have been identified in zebrafish (Stegeman et al., 2010). While AKR1a1b plays an important role in gluconeogenesis in zebrafish (Li et al., 2020), AKR1a1a has been shown to detoxify the reactive carbonyl-aldehyde, acrolein, through knockout studies (Qi et al., 2021).

## HYDROLYSIS

### Epoxide Hydrolase

#### Overview

Xenobiotic epoxides and arene oxides are usually formed by cytochrome P450-dependent oxygenation of a double bond or an aromatic ring. Due to the strain of the three-membered oxirane ring, they readily react with cellular nucleophiles such as water, glutathione, or nucleophilic centers in DNA bases. The function of the epoxide hydrolase (EH) group of enzymes is to catalyze the addition of water to an epoxide or arene oxide. Epoxide hydrolase enzymes are considered part of a larger class of hydrolytic enzymes, including esterases, proteases, dehalogenases, and lipases (Beetham et al., 1995). Studies with mammalian enzymes have shown that two major epoxide hydrolase enzymes utilize xenobiotic epoxides as substrates. These are the cytosolic enzyme that utilizes *trans*-epoxides as substrates and the microsomal enzyme that prefers *cis*-epoxides and arene oxides. In both cases, the products are *trans*-dihydrodiols (Hammock and Hasagawa, 1983) (see Figure 3.4).

Microsomal epoxide hydrolase is of particular importance for arene oxides produced by the action of cytochrome P450 enzymes on polycyclic aromatic hydrocarbons. For most arene oxides, conversion to the dihydrodiol results in detoxification of the PAHs. In some cases, however, epoxide hydrolase plays a role in the formation of reactive diol epoxide metabolites; for example, conversion of benzo(*a*)pyrene-7,8-oxide to the 7,8-dihydrodiol is part of the pathway leading to the ultimate carcinogen, (+)-*anti*-benzo(*a*)pyrene-7,8-dihydrodiol-9,10-oxide.

Substrates commonly used for study of microsomal epoxide hydrolase activity are *cis*-stilbene oxide (shown in Figure 3.4) and styrene-7,8-oxide. The method most commonly used to measure epoxide hydrolase activity was to incubate the radiolabeled epoxide or arene oxide with microsomes at pH 9 and then measure the amount of product formed. The pH optimum of microsomal EH activity in most species that have been examined was 8.5–9.5 (Balk et al., 1980; James et al., 1979). For some substrates (e.g., styrene oxide and *cis*-stilbene oxide), unreacted substrate was separated from product by extraction (Gill et al., 1983; James et al., 2004) and for others, such as benzo(*a*)pyrene-4,5-oxide, by chromatography (Jerina and Dansette, 1977). Other methods have also been developed, such as gas chromatography, spectrophotometric, and fluorimetric methods (Dansette et al., 1979; Westkaemper and Hanzlik, 1981), although these tend to be of lower sensitivity than the radiochemical methods. In all methods, an important consideration is to keep the reaction pH 6 or higher at all steps, as epoxides undergo spontaneous hydrolysis at acidic pH.

In mammalian species, antibodies have been used to quantitate microsomal epoxide hydrolase enzyme in liver subcellular fractions (Bend et al., 1978). Such studies have not been conducted in fish. The cross reactivity of fish epoxide hydrolases with mammalian microsomal or cytosolic epoxide hydrolases is not known. To date, there is only one report of a cloned and expressed fish microsomal epoxide hydrolase, from zebrafish (Kim et al., 2005).

#### Enzyme Specificity, Regulation, and Inhibition

The first studies of epoxide hydrolase activity in fish were carried out with hepatic, branchial, gonadal, and intestinal microsomes from marine fish common to coastal Maine or Florida (Bend et al., 1978; Gill et al., 1982; James et al., 1974, 1976; Westkaemper and Hanzlik, 1981). These studies showed that although epoxides were readily hydrolyzed in the liver of most fish species, considerable variability exists between individuals as well as between species in the measured epoxide hydrolase activity. The gill, kidney, intestine, and gonads generally had lower epoxide hydrolase activity than the liver. Freshwater fish also had readily measured epoxide hydrolase activity in the liver and other

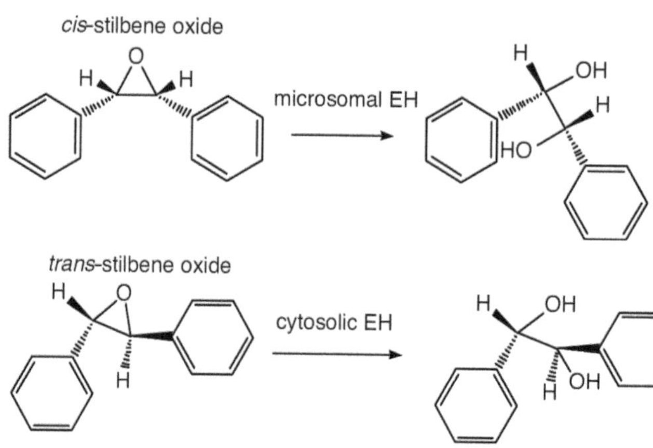


**FIGURE 3.4** Reactions catalyzed by epoxide hydrolases.

**TABLE 3.5**
**Microsomal Epoxide Hydrolase Activity with Styrene Oxide in Representative Fish Species**

| Species | Organ | Activity[a] | Refs. |
|---|---|---|---|
| Channel catfish (*Ictalurus punctatus*) | Liver | 2.62±1.0 | James et al. (1997); Willett et al. (2000) |
| | Intestine | 3.78±1.95 | |
| | Liver | 1.03±0.06 | |
| Brown bullhead (*Ameiurus nebulosus*) | Liver | 0.85±0.07 | Willett et al. (2000) |
| Trout (*Oncorhynchus mykiss*) | Liver | 1.2 | Perdudurand and Cravedi (1989) |
| Sturgeon (*Acipenser baeri*) | Liver | 2.77 | Perdudurand and Cravedi (1989) |
| | Kidney | 1.37 | |
| | Gill | 1.37 | |
| Sheepshead (*Archosargus probatocephalus*) | Liver | 5.6±2.4 | James et al. (1974) |
| | Kidney | 0.05 | |
| | Intestine | 0.74±0.33 | |
| | Gill | 0.16 | |
| | Ovary | 0.08 | |
| | Testis | 0.52 | |
| Southern flounder (*Paralichthyes lethostigma*) | Liver | 2.0±0.7 | James et al. (1974) |
| | Kidney | 1.33±0.11 | |
| Winter flounder (*Pseudopleuronectes americanus*) | Liver | 27.9±15.9 | James et al. (1974) |
| | Kidney | 8.25±6.11 | |
| Northern pike (*Esox lucius*) | Liver | 4 | Balk et al. (1980) |
| English sole (*Parophrys vetulus*) | Liver | 2.8±0.2 | Collier et al. (1986) |
| Scup (*Stenotomus chrysops*) | Liver | 2.6–9 | Stegeman and James (1985) |
| Dogfish shark (*Squalus acanthias*) | Liver | 6.3±2.1 | James et al. (1974) |
| | Kidney | 12.6±3.9 | |
| Atlantic stingray (*Dasyatis sabina*) | Liver | 6.2±1.8 | James et al. (1974) |
| | Kidney | 3.1±1.8 | |
| Large skate (*Raja ocellata*) | Liver | 1.8±0.2 | James et al. (1974) |
| | Kidney | 0.55 | |
| Small skate (*Raja erinacea*) | Liver | 0.46±0.4 | James et al. (1974) |
| | Kidney | 0.16 | |

[a] In units of $nmol \cdot min^{-1} \cdot mg\ protein^{-1}$.

organs (Parker et al., 1993; Perdudurand and Cravedi, 1989; Stott and Sinnhuber, 1978; Walker et al., 1978; Willett et al., 2000). Table 3.5 shows activities found in some representative marine and freshwater species in the liver and other organs.

The importance of epoxide hydrolase activity in the biotransformation of polycyclic aromatic hydrocarbons was highlighted in a study of the *in vitro* metabolism of benzo(*a*) pyrene in scup hepatic microsomes (Stegeman and James, 1985). The ratio of BaP-9,10-dihydrodiol to 9-hydroxy-BaP was found to correlate well with the epoxide hydrolase activity of hepatic microsomes. Individual fish with low epoxide hydrolase activity formed proportionally more of the phenolic rearrangement product, 9-hydroxy-BaP, than BaP-9,10-dihydrodiol; individuals with high epoxide hydrolase activity formed proportionally more BaP-9,10-dihydrodiol and less of the phenolic rearrangement product. This relationship was not found for the ratio of BaP-7,8-dihydrodiol to 7-hydroxy-BaP. It was thought that this was because of differences in the chemical reactivity of the arene oxides of BaP. BaP-9,10-oxide, unlike BaP-7,8-oxide, was not readily hydrolyzed in the absence of epoxide hydrolase, making the presence of the epoxide hydrolase enzyme critical to determining if BaP-9,10-oxide would be hydrolyzed to the dihydrodiol or rearrange to the 9-hydroxy product. The presence of compounds that modulate enzyme activity was found to influence the metabolites of BaP formed in hepatic microsomes from control or 3-methylcholanthrene-induced sheepshead (Little et al., 1981). Incubation of BaP with hepatic microsomes in the presence of naphthoimidazole, a substance that inhibits CYP1A but stimulates epoxide hydrolase, gave a higher ratio of BaP-9,10-dihydrodiol to 9-hydroxy-BaP than incubations in the absence of the modulating agent, presumably by routing the BaP-9,10-oxide formed by CYP1A to BaP-9,10-dihydrodiol rather than to 9-hydroxy-BaP (Little et al., 1981).

Modulation of epoxide hydrolase activity has been investigated following treatment with various xenobiotics. In mammalian species, several xenobiotics induce epoxide hydrolase activity, including phenobarbital, *trans*-stilbene oxide, and some aryl hydrocarbon receptor agonists, including PCBs (Bresnick et al., 1977; Gillette et al., 1987).

In fish, however, induction of epoxide hydrolase activity following administration of these agents has not been demonstrated (James and Little, 1981; James et al., 1997). Indeed, in stingrays treated with 3-methylcholanthrene at a dose that induced arylhydrocarbon hydroxylase (AHH) activity ten-fold, epoxide hydrolase activity was significantly lower in hepatic microsomes from treated fish (4.54±0.55 nmol/min/mg protein) relative to controls (5.81±0.76) (James and Bend, 1980). A similar trend was observed in 3-MC-treated sheepshead. Flatfish exposed to PAH- and PCB-contaminated Puget Sound sediments showed no increase in epoxide hydrolase activity (Collier and Varanasi, 1991; Collier et al., 1986). Likewise, channel catfish and brown bullhead treated with 10 mg/kg BaP showed no significant induction or species difference in liver microsomal *cis*-stilbene-oxide hydrolase activities (Willett et al., 2000). In the splake, treatment with the fish anesthetic tricaine methane sulfonate reduced epoxide hydrolase activity in liver and duodenum (Laitenen et al., 1981).

### Carboxylesterases

Hydrolytic biotransformation of xenobiotics by various forms of carboxylesterases in fish plays a significant role in the detoxification of various pesticides (Glickman et al., 1982; Straus and Chambers, 1995; Wallace and Dargan, 1987) and plasticizers (Barron et al., 1989) (Figure 3.5). Most of the studies examining this pathway in fish have focused on postmitochondrial or cytosolic enzymes (Salamastrakis and Haritos, 1988), and some studies have examined the microsomal activities (Soldano et al., 1992; Vittozzi et al., 2001). No carboxylesterase genes have been characterized in fish, but studies with purified proteins have been carried out (Boone and Chambers, 1997) and activities have been noted in zebrafish (Li et al., 2017). Differences in carboxylesterase activities among species have been hypothesized to be responsible for the acute toxicity of organophosphates (Keizer et al., 1991, 1993, 1995) and pyrethroids (Glickman et al., 1979, 1982). Carp were resistant to diazinon toxicity because of relatively high activity of hydrolyzing esterase activity, whereas trout was very sensitive to toxicity because of a lack of esterase activity and a sensitive acetylcholinesterase (Keizer et al., 1995). A similar relationship was observed in trout with permethrin (Glickman et al., 1979).

## PHASE II ENZYMES

### UDP-Glucuronosyltransferases

#### Overview

The UDP-glucuronosyltransferases (UGTs) represent a major group of phase II conjugating enzymes. Glucuronidation is principally a characteristic feature of vertebrates; invertebrates tend to prefer to utilize glycosylation. Glucuronidation is the major pathway for the conversion (and inactivation) of both endogenous and exogenous compounds to polar, water-soluble compounds that are then excreted in the bile (compounds>350 MW) or urine (compounds<300 MW). The UGTs are active in the metabolism of endogenous compounds such as steroid hormones, thyroid hormones, and waste products such as bilirubin (Dutton, 1990). In addition, an important role is played in the biotransformation of natural toxins and anthropogenic toxicants that are absorbed into the organism. Because of their importance in the breakdown of therapeutic drugs, UGTs are extensively researched in the medical and pharmaceutical fields (Rowland et al. 2013; Meech et al., 2019). This explains why so much is known about UGTs in mammalian systems, while research on UGTs in lower vertebrates and invertebrates is much more limited.

The glucuronosyltransferases are located in the endoplasmic reticulum (ER), with the active site facing inward into the lumen of the ER (Figure 3.6). The various isozymes have a common *C*-terminal, which anchors the enzyme in the membrane of the ER. This brings them in close proximity to phase I enzymes, such as CYP1A, which are also located on the ER; thus, phenolic phase I metabolites formed by CYPs can immediately be conjugated by the neighboring phenol-type UGT (UGT1A6). It has even been suggested that direct contact between CYP1A1 and UGT1A6 occurs and that these protein–protein interactions enhance the activity of the glucuronidation enzyme (Taura et al., 2004).

Glucuronidation involves the transfer of an activated sugar group to the substrate. This process increases the water solubility and the molecular weight of the substrates, thus facilitating transport and excretion. The preferred cofactor for most UGT isoforms is uridine diphosphoglucuronic acid (UDPGA). However, a number of other nucleotide–sugar cofactors have been reported for these enzymes; therefore, the name *UDP-glycosyltransferase* has been

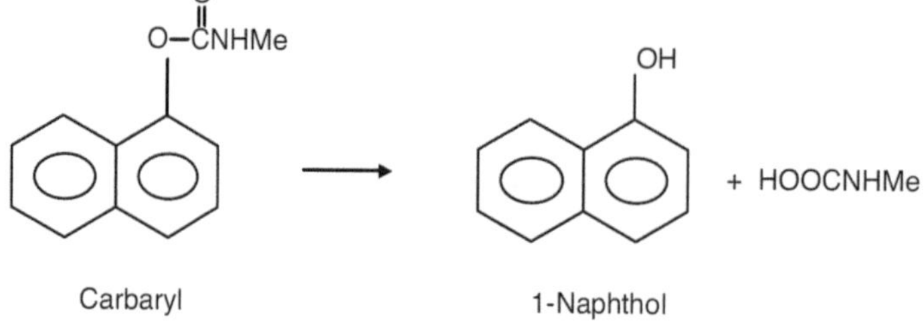


**FIGURE 3.5** Reactions catalyzed by carboxylesterases.

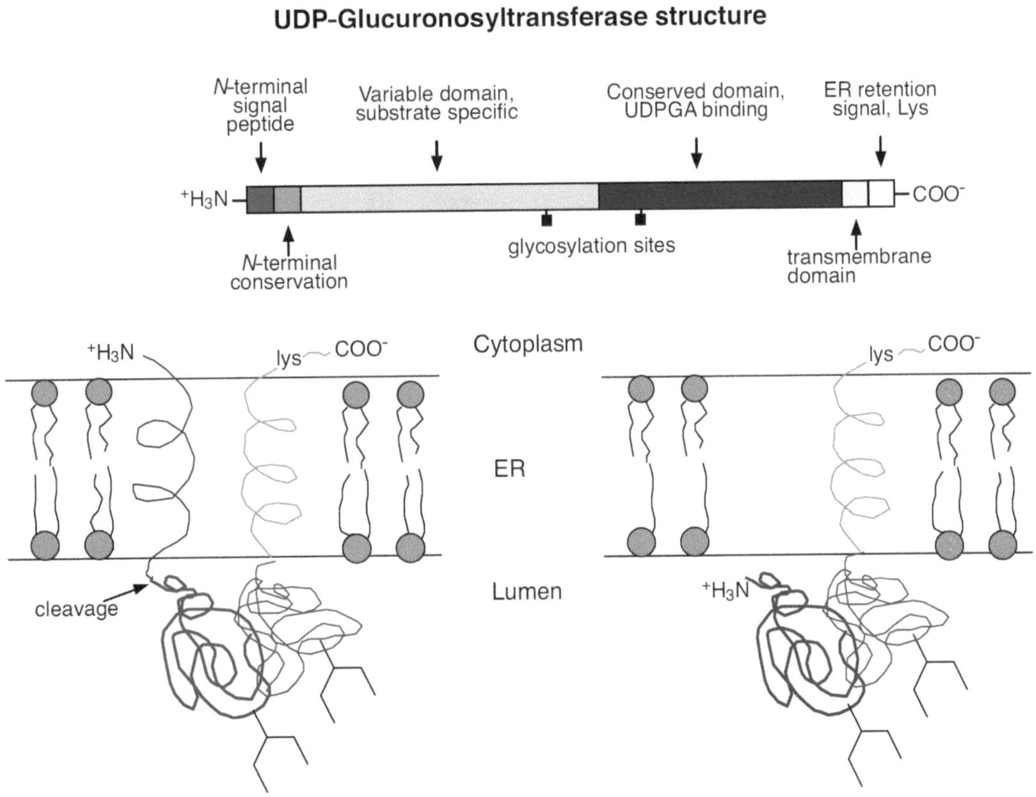


**FIGURE 3.6** Structure and organization of UDP–glucuronosyltransferases.

proposed as a better name for this superfamily (Mackenzie et al., 1997). UDPGA is synthesized in the cytosol from uridine-triphosphate and glucose-1-phosphate. Contrary to many phase I reactions, glucuronidation does not require additional energy supply in the form of ATP or NADPH. The synthesis of UDPGA in the cytosol introduces an extra threshold for the glucuronidation reaction rate, as the water-soluble cofactor has to be actively transported through the ER membrane to reach the active site on the UGT enzyme. Mechanisms surrounding the transport are still unknown.

## UGT Gene Structure

In fish species, about 40 UGT genes are found that are grouped in the UGT1, UGT2, and UGT5 families (Huang and Wu, 2010; Wang et al., 2014). As in mammalian species, the isozymes in the UGT1 and UGT2 families result through alternative splicing of a variable region onto a constant region. The UGT5 family is unique for fish and other lower vertebrates, and the genes consist of a single large exon (Huang and Wu, 2010). The proteins encoded by the fish UGT genes do not appear to be orthologous to any of the known mammalian UGT isozymes, which makes it difficult to predict substrate specificity. Functional analysis in zebrafish showed that UGT1A1, UGT1A7, and UGT1B1 had good activity towards phenolic substrates like 4-methylumbelliferone, 4-nitrophenol, and 1-naphthol (Wang et al., 2014). Members of the UGT5 family expressed good activity toward steroid hormones like estradiol and testosterone.

The gene encoding the major phenol-conjugating isoform of plaice has been cloned (UGT1B1) and found to share an almost identical five-exon structure with mammalian UGT1 family members (George and Leaver, 2002). When heterologously expressed, it displays specificity for planar phenols such as 1-naphthol and 4-nitrophenol as aglycones but no demonstrable activity with bilirubin or steroids (Leaver et al., 2007).

Expressed sequence tags (ESTs) for UGTs have subsequently been cloned from flounder, plaice, and zebrafish and reveal that at least 14 distinct UGTs are expressed (George and Taylor, 2002; Leaver et al., 2007). The zebrafish genes are divisible into three groups; the group with closest similarity to the characterized UGT1B1 gene comprises two alternatively spliced six-exon genes. Gene 6220 comprises six exon 1s, three of which are known to be expressed (Figure 3.7). The products arise by alternative splicing of the primary transcripts. Inspection of the nucleotide sequences of the intervening introns between the multiple exon 1s reveals the presence of peroxisomal proliferator response elements (PPREs) and xenobiotic response elements (XREs), which would support the plaice induction data (George and Leaver, 2002). A further three zebrafish genes comprise six exons and divide with the mammalian UGT2 family, thus inferring the closest homology. Interestingly, gene 4649 also appears to exhibit alternative splicing. Intronless UGT genes can also be identified in zebrafish (four putative genes, two reported as ESTs) and in

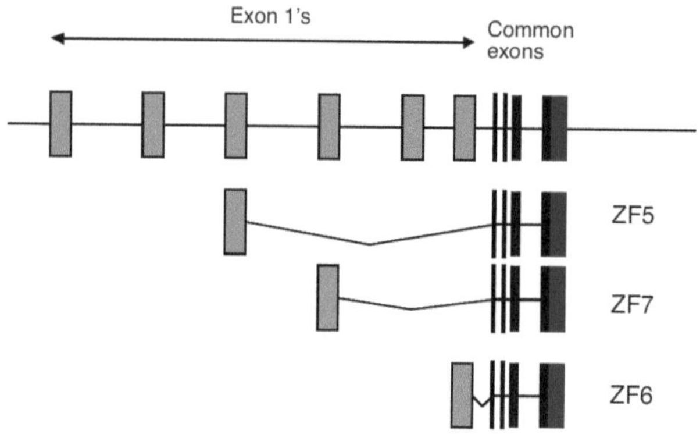


**FIGURE 3.7** Structure of the zebrafish UGT1B gene showing alternative splicing of primary transcripts.

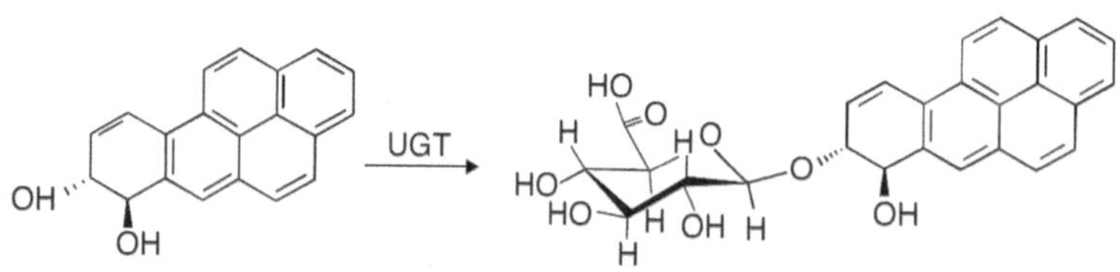


**FIGURE 3.8** Glucuronidation of benzo(*a*)pyrene-7,8-dihydrodiol catalyzed by UGT.

the pufferfish (two genes). These genes (family 3) have an unknown function and appear to be unique to fish (Huang and Wu, 2010).

### Reactions and Substrate Specificity

The spectrum of the acceptor substrates is very wide, and as with most enzymes that exhibit a broad specificity for structurally diverse compounds, multiple isoenzymes belonging to a number of multigene families are found (Mackenzie et al., 1997). The enzymes catalyze the transfer (conjugation) of glucuronic acid from the high-energy nucleotide UDP-glucuronic acid (UDPGA) to a wide variety of acceptor substrates (aglycones) to form β-glucuronides (Figure 3.8). The most common are the formation of *O*-glucuronides from alcohols, phenols, and carboxylic acids and *N*-glucuronides of carbamates, amides, and amines. *S*-Glucuronides of aryl mercaptans and thiocarbamates and *C*-glucuronides formed by conjugation of 1,3-dicarbonyls have also been identified in mammals but not so far in aquatic species. The endogenous roles of these enzymes are in the detoxification of toxic metabolites (e.g., bilirubin) formed from the degradation of heme, the secretion of bile acids (e.g., lithocholate), termination of hormone action (retinol, $T_3$, $T_4$, sex steroids), cessation of the action of olfactory stimulants, and in the case of fish, excretion of glucuronides of sex steroids as pheromones and spawning stimulants (Lambert and Resink, 1991). Many of the enzymes have a broad and overlapping substrate specificity, which gives them the ability to conjugate many xenobiotic compounds in addition to their preferred endobiotic substrates. This appears to be the case for those conjugating bilirubin and planar polyaromatic phenols, although several isoenzymes, particularly those metabolizing steroid hormones, appear to show a more restricted substrate specificity (Table 3.6).

Most investigations of fish have been *in vivo* studies concerned with the identification of the glucuronides of endobiotic and xenobiotic compounds. Glucuronide conjugates of bilirubin, 17β-estradiol, triiodothyronine ($T_3$), and thyroxine ($T_4$) have been identified in the bile of several fish species, whereas glucuronides of 3α-, 3β-, and 17β-hydroxy steroids and their metabolites have been identified in several tissues and body fluids, showing that the scope for endogenous compounds is the same as in mammals (George, 1994).

Quantitatively, glucuronidation is the most important pathway for detoxication and excretion of xenobiotic compounds in mammals, and studies to date indicate that this is also the case in fish. A recent listing of the wide and structurally diverse range of xenobiotic compounds whose glucuronides have been detected in bile, urine, and various tissues is given in Table 3.7. Compared with the large number of compounds shown to be glucuronidated *in vivo*, few studies have been reported on the scope of glucuronidation in isolated microsomal preparations from fish. The liver is the most active tissue (see later section). Planar phenols such as 4-nitrophenol, 1-naphthol, and 4-methylumbelliferone are readily conjugated, although the rate may vary by as much as an order of magnitude between species. One difficulty in intercomparison is the well-known latency observed in microsomes due to the lumenal orientation of the enzyme and inaccessibility of UDPGA. In fish, this latency, which varies in a tissue-specific manner, does not appear to be too great as in mammals. Maximal activity in microsomes is only obtained in the presence of an optimized amount of detergent; indeed, an excess of detergent has an inhibitory effect (Burchell and Coughtrie,

**TABLE 3.6**
**Compounds Glucuronidated in Fish**

| Chemical Class | Examples |
|---|---|
| Aromatic hydrocarbons | Benzene (metabolite phenol)<br>Naphthalene (metabolite 1-naphthol) |
| Aliphatic hydrocarbons | Hexachlorocyclohexane (lindane) |
| Polyaromatic hydrocarbons (phenol and diol metabolites) | Phenanthrene, pyrene, chrysene, benzo(*a*)pyrene, retene (7-isopropyl-1-methylphenanthrene) |
| *N*-Heteroaromatics | Quinoline, dimethylquinoline, carbazole |
| Aromatic amines | Aniline, 2,4-dichloroaniline, naphthylamine |
| Thioazole | 2-Amino-4-phenylthiazole (anesthetics, such as Piscaine™; the *N*-glucuronide) |
| *O*-Heteroaromatics (hydroxylated or dealkylated metabolites) | Dibenzofuran(s), tetrachlorodibenzofuran, 7-ethoxycoumarin |
| *S*-Heteroaromatics | Dibenzothiophene |
| Biphenyls (hydroxylated metabolites) | biphenyl, tetrachlorobiphenyl |
| Resin acids | Abietic, hehydroabietic, hydroabietic, isopimaric, pimaric acids (present in wood pulps) |
| Phenolics | Phenol 1-naphthol, 4-amino phenol, 1-chlorophenol, *penta*-chlorophenol (wood preservatives), chlorophenolics formed during paper bleaching, 4-nitrophenol, 3-trifluoromethylnitrophenol (lampreycide), phenolphthalein, phenolsulfonphthalein (dyes, slow) |
| Phenolic xenoestrogens | Bisphenol A, diethylstilbesterol, 4-nonylphenol, nonylphenol diethoxylate, *tert*-octylphenol (degradation products of alkylphenoxylate detergents) |
| Phytoestrogens | Coumesterol, genistein, biochinin A |
| Antibiotics | Chloramphenicol, oxolinic acid, dimethylquinoline, miloxacin |
| Insecticides | Organophosphates (e.g., fenitrothion malathion, chloropyriphos); carbamates (e.g., 1-naphthyl-*N*-methylcarbamate, Sevin™); pyrethroids (e.g., pyrethrin) |
| Fungicides | Imidazole (e.g., Prochloraz™), pentachlorophenol |
| Plasticizers | *mono*-Ethylhexylphthalate, di-2-ethylhexylphthalate |
| Miscellaneous industrial chemicals, drugs | Picric acid, picramic acids, morphine, valproic acid, pristane, digoxigenin monodigitoxide |
| Endobiotics | Bilirubin, bilirubin glucuronide; cholic acids, cholate, deoxycholate, lithocholate; retinoic acid; triodothyronine (T3), thyroxine (T4); 3α-hydroxysteroids (androsterone); 3-hydroxysteroids (estradiol, estrone); 17β-hydroxysteroids (testosterone, 17-methyltestosterone) |

**TABLE 3.7**
**Effects of β-Naphthoflavone (BNF) and 3-Methylcholanthrene (3-MC) on UGT Induction in Various Fish Species**

| Species | Inducer | Dose | Maximum induction | References |
|---|---|---|---|---|
| Rainbow trout (*Oncorhynchus mykiss*) | BNF | 100 mg/kg | 3× | Andersson et al. (1985) |
| | BNF | 50 mg/kg | 2× | Celander and Forlin (1995) |
| | BNF | 5 mg/kg | 1.7× | Lemaire et al. (1996) |
| Sea bass (*Dicentrarchus labrax*) | 3-MC | 20 mg/kg | 1.5× | Lemaire et al. (1996) |
| | BNF | 80 mg/kg | 1.8× | Novi et al. (1998) |
| | BNF | 0.3–0.9 μM aqueous | 3× | Gravato and Santos (2002) |
| Dab (*Limanda limanda*) | 3-MC | 20 mg/kg | No induction | Lemaire et al. (1996) |
| Plaice (*Pleuronectes platessa*) | 3-MC | 10 mg/kg | 1.4× | George and Young (1986) |
| Channel catfish (*Ictalurus punctatus*) | 3-MC | 50 mg/kg | 1.7× | Gaworecki et al. (2004) |
| Brown bullhead (*Ictalurus nebulosus*) | 3-MC | 20 mg/kg | 2× | Pangrekar and Sikka (1992) |
| Mummichog (*Fundulus heteroclitus*) | 3-MC | 50 mg/kg | 1.7–2× | Gaworecki et al. (2004) |

1989; Clarke et al., 1992b). Many published results are not comparable because the species differences are remarkable, most notable being the much lower capacity for conjugation of phenols in trout than plaice (George, 1994). This large variation in biotransformation rates was also observed between Cyprinid and Centrarchid species exposed to benzo[*a*]pyrene (van den Hurk et al., 2017).

The capability of fish to glucuronidate a variety of chemicals has been described in numerous publications; a recent review by Katagi (2020) gives an overview of pesticides and industrial chemicals that were found to be metabolized by phase 1 and phase 2 enzymes in fish species, including references to studies that measure conjugation by UGTs. An overview is given of the benefits and problems of measuring biotransformation in S9 fractions versus using microsomes, cytosol, hepatocytes, cell lines, or tissue slices. The glucuronidation of pharmaceuticals is a well-known biotransformation process in mammalian species, and because testing systems with fish are much easier to perform than with mammalian model species, a large interest has developed in recent years to use fish models for the evaluation of toxicity and biotransformation of pharmaceuticals, especially using zebrafish (Wang et al., 2014; Lee et al., 2021). While zebrafish appears to be an attractive model, caution needs to be observed when results are used for extrapolation to mammalian models, as was illustrated by Kantae et al. (2016) who found that in zebrafish larvae the nonsteroidal anti-inflammatory acetaminophen was five to six times more likely to be sulfonated than glucuronidated, while in most mammalian models, glucuronidation is responsible for the majority of acetaminophen biotransformation (Bertolini et al., 2006). Other recent overviews have been published on the biotransformation of steroids in fish (James, 2011), the antibiotic erythromycin in salmon (Rodrigues et al., 2019), and the upregulation of UGT in salmon exposed to perfluorooctane sulfonamide (Olufsen and Arukwe, 2015).

## Enzymology of Piscine UGTs

The UGTs are membrane-bound enzymes that are quite labile when isolated, requiring phospholipids to maintain activity; therefore, UGTs have proven to be notoriously difficult to purify and characterize. The only nonmammalian UGTs to be purified were from the plaice (Clarke et al., 1992c). At least six immunoreactive UGT peptides were visualized in plaice microsomes in western blots with mammalian antisera, indicating common structural epitopes (Clarke et al., 1992a). From purification studies, at least five UGTs have been identified in plaice liver (George, 1994), and the phenol-conjugating isoform that was purified to homogeneity displayed a very high activity toward planar phenols and no measurable activity toward bilirubin or steroids. The bilirubin-conjugating isoform appeared to exhibit both bilirubin and phenol-conjugating activity as found in mammals. Reviews of earlier studies on glucuronidation in fish can be found in Clarke et al. (1992a) and George (1994). The picture that emerges from these reviews is that UGT isoforms are found in a wide variety of fish species and that specific enzymatic activities are found for bilirubin, steroid and thyroid hormones, and phenolic compounds. As in mammalian systems, a diverse group of xenobiotic compounds has been identified as substrates for UGTs, among which are chlorinated phenols, aromatic hydrocarbon metabolites, phthalates, pesticides, and antibiotics. Comparison of UGT activities among species was found to be a problematic task primarily due to the nonoptimization of detergent concentrations in assays in most published studies. The enzymes also display maximal activity at 37°C (Clarke et al., 1992b); thus, corrections for temperature should be used for intercomparisons.

## Tissue Distribution

In fish species, UGT activity is usually high in the liver and intestine, but measurable activities have also been found in the gill, kidney, and muscle tissue (Clarke et al., 1992a; George et al., 1998; James et al., 1998; Singh et al., 1996). UGTs play an important role in gonadal tissues; in addition to the regulation of steroid hormones, UGTs play a role in the production of sex pheromones in fish (Lambert and Resink, 1991). Testosterone UGT activity is present in the liver, testis, and intestine (Clarke et al., 1992b), and glucuronidation of pregnenolone and androstenedione has been demonstrated *in vitro* with testicular preparations (Andersson, 1992). Glucuronidation of bilirubin is confined to liver in both plaice and salmon.

Based on the complete zebrafish UGT gene repertoire published by Huang and Wu (2010), real-time PCR primers were designed to map the expression of UGT isoforms in a wide variety of tissues in zebrafish (Wang et al., 2014). They found strong tissue-specific expression for a number of isoforms from each of the three UT families, with multiple isoforms expressed in the gills. Other tissues express UGT isoforms related to their specific functions, like biotransformation of steroids in the sex organs, and olfactory signaling compounds in the brain and gill. In a comparable study by Christen and Fent (2014), primers were designed for individual UGT5 isozymes, and for the UGT1a and UGT1b families, basal expression of these genes was measured in different zebrafish tissues. Most expression was found in the liver and intestine, with a second cluster in the brain and gonads.

## Regulation of UGTs

Coregulation of both CYPs and UGTs occurs in mammals, and prototypical inducers such as clofibrate, PAHs, phenobarbital, and pregnenolone-16α-carbonitrile (PCN) differentially induce the expression of both CYP and UGT isoforms (Meech et al., 2019). The degree of upregulation of UGT activity is generally some two- to three-fold. UGT1A1 (bilirubin conjugation) is induced by the hyperlipidemic agent (and peroxisomal proliferator) clofibrate (George). In common with a number of phase I and II genes, UGT1A6 (planar phenol conjugation) is induced by interaction of the Ah receptor with an XRE in its promoter region. This coregulation of CYP1A and UGT1A6 in mammals and

CYP1A and UGT1B1 in fish has been shown to facilitate detoxification of PAHs such as BaP. Mammalian CYP2B and CYP3A and UGT1A genes are also induced via a nuclear pregnane X receptor and a constitutive androstane receptor. Interestingly, these recognition motifs are also present in the zebrafish UGTs (George, unpublished data). Mammalian steroid UGTs are induced by PCN. Several reports indicate a modest induction of phenol UGT activity in fish from polluted environments and after experimental PAH exposure. To study the induction of AhR-activated enzymes, two PAHs are often used: 3-methylcholanthrene (3-MC) and β-naphthoflavone (BNF). Both compounds have been used as CYP1A and phenol UGT inducers in a number of fish species (Table 3.7). Maximal induction of CYP1A is usually found around three days after single treatment, but phenol UGT induction appears to be slower, with a maximum induction occurring around eight days after treatment. In general, induction of EROD activity can be up to 250-fold, but UGT activity is never induced more than three- to six-fold. Two species, cod (*Gadus morhua*) and gilthead sea bream (*Sparus aurata*), displayed little or no response to AhR ligands as phenol UGT inducers (Goksøyr et al., 1987; Pretti et al., 2001). It must be noted, however, that multiple isoforms in mammals, including the constitutive steroid isoforms, also conjugate 4-nitrophenol. Immunoblot and northern blot analyses of xenobiotic-treated plaice have shown that induction of UGTs appears to be tissue-specific (Clarke et al., 1992a). Treatment with the PAH (3-MC) increased phenol-conjugating activity and a 56-kDa immunoreactive peptide in the liver by approximately 1.7-fold. The bifunctional PAH-type inducer BNF caused an induction (approximately three- to fourfold) of UGT1B1 mRNA only in the intestine (Leaver et al., 2007). Although clofibrate did not appear to induce phenol- or bilirubin-conjugating activities (Clarke et al., 1992a), treatment with the potent peroxisome proliferator-activated receptor γ (PPARγ)-mediated peroxisomal proliferator perfluorooctanoate (PFOA) has not been found to induce UGT1B1 mRNA levels in tissues other than the kidney (Leaver et al., unpublished data). This pattern of induction by BNF and PFOA is identical to that observed for the glutathione *S*-transferase gene (GSTA), which contains peroxisomal proliferator response elements (PPREs) and antioxidant response elements (AREs) but not xenobiotic response elements (XREs) in the promoter region (Leaver et al., 1997). Exposure to PCBs generally induces UGT activity in fish species (Table 3.8). When tilapia (*Oreochromis niloticus*) were fed a diet amended with sewage sludge, tissue PCB concentrations were increased, together with an increase in UGT activity (Yang et al., 1993). Coplanar PCBs are AhR ligands; however, when coplanar PCBs are metabolized by CYP isozymes, the hydroxylated metabolites that are formed can be potent inhibitors of UGT and sulfotransferase (van den Hurk et al., 2002).

### Inhibition of UGTs

Of great environmental concern is inhibition of the steroid-type UGT by environmental pollutants. Inhibition of these enzymes could cause an accumulation of active hormone that may lead to disturbed gonadal cycles and even tumor formation in the hormone secreting organs. Some xenoestrogens were indeed demonstrated to affect UGTs in fish species; for example, nonylphenol inhibited steroid-conjugating UGT activity in Atlantic salmon (Arukwe et al., 1997), and nonylphenol diethoxylate inhibited UGT activity in rainbow trout hepatocytes (Sturm et al., 2001). Growth hormone caused an inhibition of UGT with testosterone as substrate (Cravedi et al., 1995), indicating that there may be interactions between different UGT isoforms through their respective substrates. In addition, liver UGTs are important in the metabolism and excretion of thyroid hormones in fish; thyroid glucuronides have been measured in the bile of a variety of fish species (George, 1994). In rainbow trout (*Oncorhynchus mykiss*), injection of thyroid hormones demonstrated that glucuronidation plays an important role in maintaining a homeostatic thyroid status (Finnson and Eales, 1999). Other environmental toxicants that inhibit UGTs are chlorinated hydrocarbons. Pentachlorophenol and other compounds in pulp mill effluents were shown to inhibit UGT activity in rainbow trout (Castren and Oikari, 1987). Hydroxylated PCBs form another group recognized as potent inhibitors of phenol-type UGT in channel catfish (van den Hurk et al., 2002). Inhibition of phenol-type UGT by PCB metabolites may hinder the detoxification of procarcinogens, as was demonstrated with benzo(*a*)pyrene-7,8-dihydrodiol in channel catfish (James et al., 1994).

**TABLE 3.8**
**Effects of PCBs on the Induction of UGTs in Various Fish Species**

| Species | Inducer | Dose (mg/kg) | Maximum Induction | References |
|---|---|---|---|---|
| Rainbow trout (*Oncorhynchus mykiss*) | PCB 77, 126 | 1–5 | 1.6× | Huuskonen et al. (1996) |
| | Clophen A50 | 100 | 2× | Forlin et al. (1996) |
| | 3,3′,4,4′-TCB | 0.3 | 1.75–2× | Blom and Forlin (1997) |
| Brook trout (*Salvelinus fontinalis*) | 3,3′,4,4′-TCB | 10 | 1.5× | Boyer et al. (2000) |
| Sand flathead (*Platycephalus bassensis*) | Arochlor® 1254 | 400 | 1.75× | Brumley et al. (1995) |

## Glutathione Transferases

### Overview

The glutathione transferases (GSTs; EC 2.5.1.18) are a supergene family of phase II enzymes that provide cellular protection against the toxic effects of a variety of endogenous and environmental chemicals. These dimeric enzymes are ubiquitously distributed and comprise approximately 2%–4% of total cytosolic proteins in the liver. The most important reaction catalyzed by all isoforms is the conjugation of the -SH group of the tripeptide glutathione (gamma-glutamyl-cysteinyl-glycine) with an electrophilic center that can be a C, N, or S atom; these are present in arene oxides, aliphatic and aromatic halides, and α,β-unsaturated carbonyls. Following the formation of the glutathione conjugate, the metabolite may undergo two separate amino acid cleavage reactions followed by *N*-acetylation to form mercapturic acid derivatives (Figure 3.9) (Tierbach et al., 2020). The substrate specificity of the GSTs is extremely broad. Notable toxic xenobiotic electrophilic compounds that are conjugated and of toxicological interest include carcinogens and their metabolites, such as benzo(*a*)pyrene, 7,12-dimethylbenzanthracene, 5-methylchrysene, and pesticides (e.g., alachlor, atrazine, DDT, lindane, methylparathion). A number of GSTs also catalyze biosynthetic reactions of the leukotrienes and prostaglandins, and others act as organic peroxidases and steroid isomerases. A major role in endogenous metabolism is the detoxification of products of oxidative stress arising from oxidation of lipids, nucleic acids, and proteins—for example, base propenals such as acrolein from DNA oxidation; cholesterol oxide, fatty acid hydroperoxides, and hydroxynonenals from lipid oxidation; and protein carbonyls from protein oxidation. Some isoforms can also exhibit a covalent binding rather than a catalytic role with some compounds. Reactive metabolites of carcinogens (e.g., PAHs) may bind covalently in a suicide reaction that prevents their reaction with DNA; however, many other neutral or lipophilic compounds that are not substrates, including steroid and thyroid hormones, bile acids, bilirubin, fatty acids, and heme, may bind noncovalently in a reversible manner. Functionally, the significance of this noncovalent binding is unknown, but considering the high concentration of the enzyme proteins in the cytosol, roles in intracellular transport, as a buffer for these compounds, and as an efflux system (via the ATP-dependent glutathione conjugate efflux pumps GS-X, MOAT, and Mrp), have been postulated (Tierbach et al., 2020).

### GST Gene Structure

Glutathione transferases are widely distributed in nature, and essentially all eukaryotic species contain multigene families, many of which contain further subfamilies of proteins. Both cytosolic and membrane-bound forms are present (Board et al., 1997, 2000; Hayes and Pulford, 1995). They are generally classified according to sequence homology and assigned to seven separate families of cytosolic enzymes (designated class alpha, mu, omega, pi, sigma, theta, and zeta) and to two membrane families—microsomal and mitochondrial (kappa). The native cytosolic enzymes are present as dimers of 24- to 26-kDa subunits, and a characteristic of the different families is that within

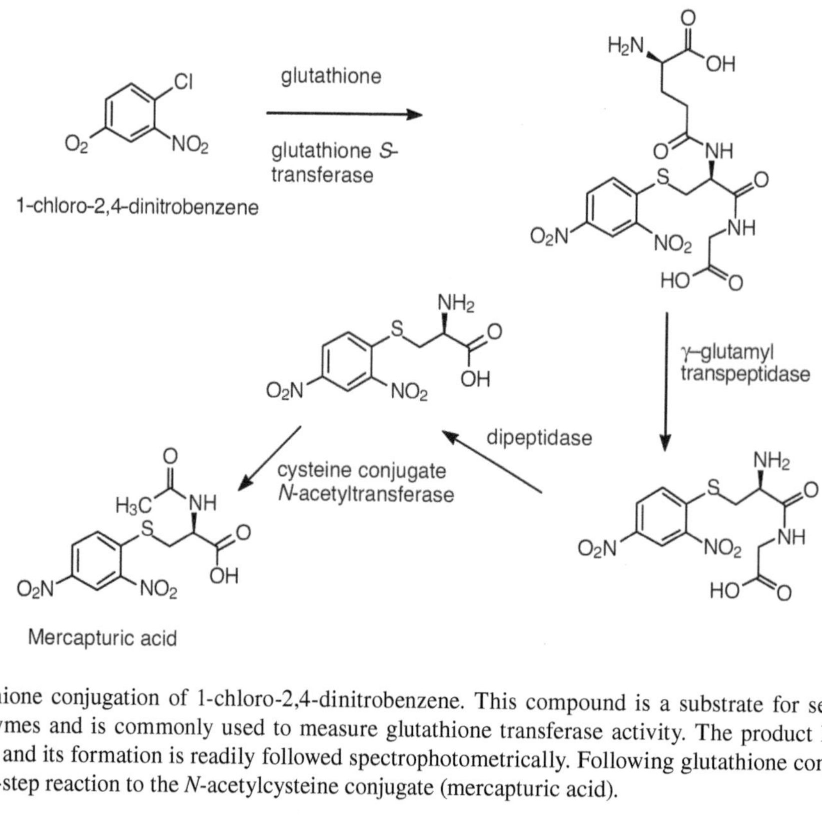


**FIGURE 3.9** Glutathione conjugation of 1-chloro-2,4-dinitrobenzene. This compound is a substrate for several distinct glutathione *S*-transferase enzymes and is commonly used to measure glutathione transferase activity. The product has a strong ultraviolet absorbance at 344 nm, and its formation is readily followed spectrophotometrically. Following glutathione conjugation, the conjugate is converted in a three-step reaction to the *N*-acetylcysteine conjugate (mercapturic acid).

each family, the proteins contain conserved amino acid residues that enable formation of both homo- and heterodimers of enzyme subunits. The microsomal enzymes are trimers of approximately 15-kDa subunits and are integral membrane proteins.

In lower vertebrates and invertebrates, few GSTs have been fully characterized, although on the basis of the broad spectrum of catalytic activities found, immunochemical comparisons, and nucleotide sequence homologies, the presence of multiple isoforms from a number of gene families in all phyla is a certainty. On the basis of immunochemical cross-reactivity with antisera exhibiting family specificity, proteins of the alpha, mu, pi, and theta-like families have been identified in many fish species (Table 3.9), and in three mollusk species, pi-class enzymes are the major isoforms. Recent work on the GSTs in zebrafish (*Danio rerio*) found 27 *gst* genes in this species, with seven classes of cytosolic GSTs (alpha, mu, pi, omega, theta, zeta, and rho), four mitochondrial genes (kappa), and eight microsomal membrane-bound isoforms (MAPEG) (Glisic et al., 2015).

The relative abundances of these isoforms differ between species (Trute et al., 2007; van den Hurk et al., 2017). GSTs are most abundant in the liver. The predominant isoform in the cyprinids, salmonids, and gadoids is a pi-class homolog, and the major isoform in flatfish, mullet, and bass is now designated as Rho-class (Konishi et al., 2005). From nucleotide and protein sequence data, genes for all nine families have been identified in fish to date. Table 3.10 shows for each species the number of GST isoforms assigned to each family.

From purification studies, pi-class GSTs have been identified as the major hepatic and intestinal GSTs in the livers of salmonids, brown bullhead, channel catfish, zebrafish, and

**TABLE 3.9**
**Immunochemical Analysis of GST Class Occurrence in Aquatic Animals**

| | Anti-Plaice-GST | | Anti-Rat GST Subunits | | | | |
|---|---|---|---|---|---|---|---|
| | | | Alpha Class | | | Mu Class | Pi Class |
| Species | A | B | A1 | A3 | A4 | M1 | P1 |
| Catfish | | | ++ | | | + | ++++ |
| Cod | ++++ | ± | + | NS | ± | ++ | +++ |
| Flounder | ++++ | + | – | – | ± | – | – |
| Mullet | | | | | | | |
| Plaice | ++++ | ++++ | +++ | – | ++++ | – | – |
| Turbot | ++++ | +++ | ± | ± | + | ± | – |
| Rainbow trout | + | ± | – | – | ± | – | ++++ |
| Sea bass | | | | | | | |
| Sea trout | + | ± | +++ | + | +++ | +++ | ++++ |
| Salmon | – | + | – | ++ | – | ++++ | – |
| Mussel | | | | | | | ++++ |
| Clam | | | | | | | ++++ |

*Source:* Adapted from George, S.G., in *Aquatic Toxicology: Molecular, Biochemical, and Cellular Perspectives*. Ostrander, G.K. and Malins, D.C., Eds., CRC Press, Boca Raton, FL, 1994, pp. 37–85.

**TABLE 3.10**
**GST Gene Family Homologs Identified by Cloning in Fish Species**

| | Gene Class | | | | | | | |
|---|---|---|---|---|---|---|---|---|
| Species | Alpha | Mu | Pi | Omega | Theta | Zeta | Rho | Microsomal |
| Zebrafish | 3 | 3 | 2 | 2 | 3 | 1 | 1 | 8 |
| Stickleback | 1 | 1 | - | 1 | 1 | 1 | 5 | 2 |
| Pufferfish | 1 | 2 | - | 1 | 2 | 1 | 3 | 2 |
| Medaka | 1 | 2 | - | 2 | 3 | 1 | 5 | 3 |
| Lancelet | 2 | 2 | - | 4 | 1 | 1 | 3 | 2 |
| Bass | 2 | - | - | - | - | - | 2 | - |
| Catfish | - | 1 | 1 | - | - | - | - | - |
| Flounder | 1 | 2 | - | - | 1 | - | 2 | 2 |

*Note:* Numbers represent quantity of homologs.

lamprey (Trute et al., 2007). These piscine pi-class enzymes display high activities with BaP epoxides and diol epoxides. The most highly characterized piscine GSTs are from the plaice, a pleuronectid flatfish, where purification and cloning studies identified a multigene cluster of enzymes originally designated as GSTA. These are most closely related to the mammalian, insect, and plant theta-class enzymes and have now been designated as a separate class, Rho. They appear to be unique to fish (Trute et al., 2007; Glisic et al., 2015). Their genetic organization appears to be the same in plaice, flounder, sea bream, and largemouth bass. The primary role of these theta-like enzymes is in oxidative defense. GST-A is most active toward hydroxynonenals, and GSTA1 (or GSTR2) exhibits a high peroxidase activity with organic peroxides. They are the major isoforms in most tissues of flatfish. Homologs have been identified by cDNA sequencing, purification, or immunochemical cross-reactivity in a number of other fish species, and quantitatively, they also represent the major isoforms in bass and mullet livers (Gallagher et al., 2000; Martinez-Lara et al., 1997). Recent work has identified and characterized GST isoforms in European eel *Anguilla anguilla* (Carletti et al., 2008), black rockfish *Sebastes schlegelii* (Jayasinghe et al., 2016), and zebrafish *Danio rerio* (Glisic et al., 2015), among numerous others that have cloned and characterized individual GST isoforms or isoform in a specific GST class.

## Reactions and Substrate Specificity

Before cloning studies and in the early days of purification, attempts were made to identify diagnostic substrates or reactions that may be diagnostic of the presence of certain isoenzymes in different tissues and species. It must be realized that great caution must be exercised when using catalytic activities (CDNB) to evaluate GST expression, as the activity toward any one substrate is a function of both the overlapping substrate specificities of the different isoforms and their different abundances within tissues; thus, such predictions may be unreliable. Consider, for example, the use of activity with ethacrynic acid (ETHA) which is diagnostic of pi-class GSTs but can also be conjugated by an alpha-class enzyme (GSTA4) in three fish species, including cod, plaice, and trout. The relative hepatic activity ratios are 40 for trout, 4 for cod, and 1 for plaice. Although this would correctly indicate the predominance of a pi-class enzyme in trout and not in plaice (despite the presence of a significant amount of a GSTA4 homolog), it does not reflect immunological data showing that the concentration of GST pi in cod liver is very little lower than that of trout. Thus, the usefulness of diagnostic substrates is limited, and this approach is less reliable than immunochemical investigations.

In lower vertebrates, such as elasmobranchs and teleosts, both the covalent and noncovalent binding activities of the GSTs are very much lower as compared to rodents (Foureman et al., 1987; George and Buchanan, 1990). This may be attributed to the lower abundance of a GSTA1 homolog or an evolutionary adaptation of the enzyme in terrestrial vertebrates, as it has been postulated that terrestrial plant phytoalexins are bound by this protein. Teleost species contain stores of polyunsaturated fatty acids that are readily oxidized by free-radical attack, and this may explain the high constitutive levels of isoforms that detoxify lipid peroxidation products such as the alkenals and hydroxynonenals (e.g., GSTA class and GSTA4 homologs) as they will be better protected against xenobiotic-induced oxidative damage. This is particularly relevant in fish such as the cod and plaice where the fat is stored in droplets within the hepatocytes and not in adipose tissue as in the salmonids.

The substrate specificities, primarily with prototypical and endogenous substrates, have been determined with a number of highly purified preparations or recombinant GSTs from several fish species. In common with mammalian GSTs, they show greatest activity with CDNB as the substrate. The alphaclass enzyme from sea bass conjugates the alkenal *trans*-non-2-enal (N2E) at a higher rate than the prototypical xenobiotic substrates, and of these, the highest rates were observed with ethacrynic acid (ETHA) and nitrobutyl chloride (NBC) (Angelucci et al., 2000). This is in agreement with an assignment of the sea bass enzyme as a GSTA4 homolog. The pi-class enzymes from salmon, trout, and catfish all exhibit relatively high rates of conjugation of ETHA, again following the pattern observed with a mammalian GST. The catfish enzyme has high activity with (±)-benzo(*a*)pyrene-4,5-oxide and *anti*benzo(*a*)pyrene-7,8-dihydrodiol-9,10-epoxide as substrates, showing that it is an effective detoxicant of the active carcinogenic metabolite of BaP (Gallagher et al., 1996).

In zebrafish, enzyme kinetics for five model substrates was established for nine cloned and purified GST isoforms (Glisic et al., 2015). The tested GST alpha and omega isoforms only showed activity toward CDNB, while mu, pi, theta, and rho class isoforms also had good activity with monochlorobimane (MCB). Ethacrynic acid appeared to be exclusively conjugated by pi class isoforms, while dichloroacetic acid was only used by a zeta isoform, and dehydroascorbate was uniquely metabolized by GST omega. GST zeta does not utilize CDNB as a substrate (Board and Anders, 2011).

## Fish GST and Oxidative Stress

As mentioned above, in addition to their protective activities toward electrophilic chemicals, certain GST isozymes can catalyze the reduction of cellular peroxides to their corresponding alcohols, as well as conjugate endogenous genotoxic unsaturated aldehydes formed during the peroxidation of membrane lipids (Alin et al., 1985; Hubatsch et al., 1998). Accordingly, the GST pathway in some species is an integral component of the cellular antioxidant defense system. Of the reactive intermediates produced during oxidative stress, 4-hydroxynonenal (4HNE) is a particularly reactive α,β-unsaturated aldehyde that is generated during lipid peroxidation as a result of the degradation of ω-6 polyunsaturated fatty acids (Esterbauer et al., 1991). 4HNE production is accelerated

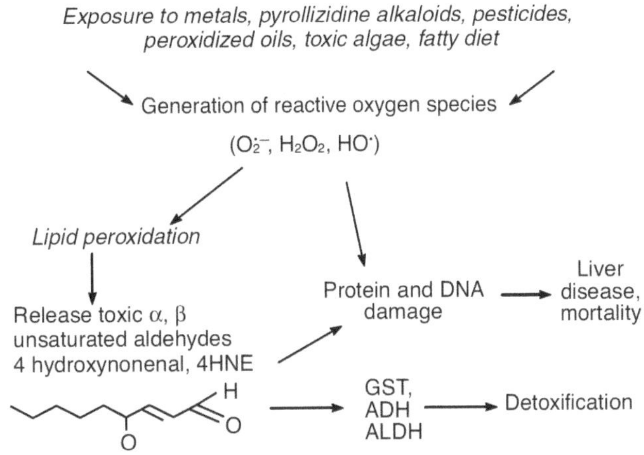


**FIGURE 3.10** Pathway for the generation and detoxification of 4-hydroxynonenal (4HNE). In mammals and certain fish species, GST constitutes a protective pathway against 4HNE injury.

during exposure to a variety of prooxidant environmental pollutants (Figure 3.10). Because of its high reactivity, 4HNE rapidly forms covalent adducts with biomolecules containing nucleophilic sites, such as sulfhydryl groups of glutathione, cysteine, lysine, and histidine residues of proteins, and nucleophilic sites of nucleic acids. In rodents and humans, the alpha-class GSTA4 subclass displays uniquely high catalytic activity toward 4HNE and other α,β-unsaturated aldehydes, suggesting that these enzymes may have distinctively evolved as a secondary line of defense against oxidative injury (Hubatsch et al., 1998). As discussed previously, studies with the marine fish plaice (*Pleuronectes platessa*) have revealed the presence of a GST enzyme (GSTR1) that is a relatively efficient catalyst for the conjugation of a series of unsaturated alkenals and hydroxyalkenals, including 4HNE, but displaying little or no activity toward model substrates for mammalian GST. The recombinant Rho class of enzymes in plaice displays higher rates of conjugation of the natural substrates *trans*-oct-2-enal (O2E), N2E, and 4-hydroxy-2,3-*trans*-nonenal (4HNE) than prototypical substrates (apart from CDNB) (Leaver and George, 1998; Martinez-Lara et al., 2002). They also exhibit a high glutathione dependent peroxidase activity with cumene hydroperoxide; substrate activity with phospholipid hydroperoxides has not been studied. Although both isoforms conjugate 4HNE at the same rate, GSTR1 displays a two- to ten-fold higher activity toward O2E and N2E and also shows low but measurable activity with the prototypical xenobiotic substrates, with ETHA being the highest. GSTA1 has a five-fold higher peroxidase activity (25 μmol/min/mg protein). The purified GSTA homolog of sea bass displays the same characteristic profile of a high activity with N2E and relatively high activity with ETHA compared with other substrates.

Largemouth bass (*Micropterus salmoides*) expresses a liver GST that shares extensive sequence identity to the aforementioned 4HNE-metabolizing GSTA isolated from plaice that detoxifies 4HNE. Interestingly, the bass GST exhibits a catalytic activity toward 4HNE that exceeds that of several mammalian and aquatic species (Doi et al., 2004; Pham et al., 2002). Bass GSTA is also similar to a GST form found in two other fish species (European flounder and fathead minnow); however, similar genes may not be present in other aquatic species. Bass GSTA exhibits little sequence identity (21% or less) to several mammalian GSTs, including the rat alpha-class GST form rGSTA4-4, which, as discussed, has a high catalytic efficiency toward conjugation of GSH with 4HNE. The presence of this interesting GST in several fish species suggests an important conservation of function that likely protects fish lipid membranes against the deleterious effects of oxidative injury.

### Tissue Distribution

With the exception of the mammalian theta- and zeta-class enzymes, all GSTs conjugate the prototypical substrate 1-chloro-2,4-dinitrobenzene (CDNB) with greatest activity (Figure 3.9). CDNB conjugating activity has been observed in all fish species examined to date (Table 3.11) and occurs in multiple tissues (George, 1994). Because CDNB is a general substrate and the rate of conjugation can vary between isoforms by up to two orders of magnitude, comparison of total CDNB conjugating activity in livers of different species, for example, is probably of little relevance in toxicological evaluation of environmentally relevant xenobiotics, as it will bear little relationship to the rate of metabolism of other compounds which may be isoform-specific. Functional comparisons of activity toward individual toxicants are perhaps more meaningful from a toxicological viewpoint. Whereas kidney is the most active toward styrene-7,8-oxide and BaP-4,5-oxide relative to other tissues in the little skate (*Raja erinacea*) (Bend et al., 1978; Foureman et al., 1987; Gill et al., 1982), this finding must be placed in the context of relevance to the animal as compounds of this

**TABLE 3.11**
**The Catalytic Properties of Glutathione *S*-Transferase in Cytosol and Glutathione Affinity-Column-Purified Fractions from Several Fish Species**

| Species | Preparation | References | Activity (nmol/min/mg protein) | | |
|---|---|---|---|---|---|
| | | | BPDE | EA | CDNB |
| Catfish *(Ictalurus punctatus)* | Liver cytosol | Gallagher et al. (1996) | 5.3 | 37 | 1325 |
| | Affinity purified fraction | | 128 | — | 24.000 |
| | Intestinal cytosol | Gadagbui and James (2000) | 8.3 | 65 | 1010 |
| | Affinity purified fraction | | 453 | 2880 | 107,000 |
| Brown bullhead *(Ameiurus nebulosus)* | Liver cytosol | Henson et al. (2001) | <5 | — | 1130 |
| | Affinity purified fraction | | 310 | — | 5,700 |
| Rainbow trout *(Oncorhynchus mykiss)* | Liver cytosol | Melgar Riol et al. (2001) | — | 20 | 630 |
| | Affinity purified fraction | | — | 2030 | 77,180 |
| Atlantic salmon *(Salmo salar)* | Liver cytosol | Novoa-Valinas et al. (2002) | — | 70 | 290 |
| | Affinity purified fraction | | — | 4350 | 16,600 |
| | Kidney cytosol | Novoa-Valinas et al. (2002) | — | 60 | 200 |
| | Affinity purified fraction | | — | 10,620 | 37,500 |
| Brown trout *(Salmo trutta)* | Liver cytosol | Novoa-Valinas et al. (2002) | — | 90 | 650 |
| | Affinity purified fraction | | — | 3790 | 27,690 |

*Note:* BPDE, (+)-*anti*-benzo(*a*)pyrene-7,8-dihydrodiol-9,10-oxide; EA, ethacrynic acid; CDNB, 1-chloro-2,4-dinitrobenzene.

type may not be present systemically. They will more likely be taken up in the diet or generated by metabolism in the intestine or liver and conjugated in these tissues. Of special interest is the identification and characterization of GSTs in olfactory tissue in coho salmon, which indicates that GSTs serve as detoxification pathways for xenobiotics as well as modulation of olfactory signals (Espinoza et al., 2013).

In zebrafish, GSTs were found in a variety of tissues with variable expression levels, with marked differences between males and females for a number of isoforms (Glisic et al., 2015; Tierbach et al., 2020). Very high expression was found for Pi class genes in male gills and ovaries, and similarly very high expression of Theta class and the fish-specific Rho class genes were observed in liver samples.

In general, most *in vitro* laboratory investigations of GST–CDNB conjugating activities in fish tissues use assay conditions and substrate concentrations similar to those proposed for mammals: 1 mM GSH and 1 mM CDNB) (Habig and Jakoby, 1981). Kinetic studies, however, have demonstrated that these substrate concentrations may not be at saturation with respect to initial rate kinetics for many isoforms because the $K_m$ values differ by up to two orders of magnitude. Because the enzymes contain both electrophile and nucleophile binding sites, GST detoxification rates are determined by the concentration of electrophilic substrate and by the concentrations of nucleophilic cosubstrate (GSH). Kinetic studies in largemouth bass and brown bullheads in the presence of variable electrophile concentrations suggest that *in vitro* saturation of hepatic GST–CDNB conjugation occurs at higher electrophile concentrations in brown bullheads than in largemouth bass (Gallagher et al., 2000). Such an observation is consistent with a higher capacity for brown bullheads to detoxify electrophilic GST substrates under conditions of high environmental exposure. These observations are important under the prototypical conditions of environmental chemical exposure when the amount of chemical reaching the liver is relatively low, and the rate of *in vivo* clearance is directly proportional to the amount of chemical concentrations *in vivo*. Also, as alluded to earlier, it is important to consider that the rate of GST–CDNB activity may not be reflective of rates of GST conjugation of environmentally relevant GST substrates, such as pesticides or epoxide carcinogens. For example, the rates of hepatic GST–CDNB conjugation by starry flounder and English sole are not correlated with the rates of hepatic GST conjugation of (+)-7β,8α-dihydroxy-9α,10α-oxy-7,8,9,10-tetrahydrobenzo(*a*)pyrene (BPDE) in those species. Specifically, starry flounder catalyzes the rate of CDNB conjugation at initial rates threefold higher than is observed in English sole, whereas GST–BPDE activities are threefold higher in English sole as compared to starry flounder (Gallagher et al., 1998). Thus, specific GST conjugation of environmental chemicals in fish must be quantitated under assay conditions where the conjugation of the chemical can be monitored directly.

Although glutathione conjugation is generally associated with detoxification, small halogenated alkenes, such as dibromoethane, are activated by glutathione conjugation. In mammals, theta-class GSTs are particularly efficient in the conversion of dibromoethane to the carcinogenic sulfonium ion metabolite, as shown in Figure 3.11 (Thier et al., 1996). It was shown that a small fish species, the medaka, was quite susceptible to the development of liver cancer following exposure to ethylenedibromide and that a form of GST in liver was increased in the exposed fish, suggesting the presence of a theta-like GST in the medaka (Hawkins et al., 1998). This has been confirmed with cloning studies. Other fish species, such as the pleuronectid flatfish, bass, and mullet, have been shown to have theta-like GST in their livers (Gallagher et al., 2000; Leaver et al., 1993; Martinez-Lara et al., 1997). It is tempting to speculate that this may account for the high tumor incidence observed in flatfish from some polluted environments.

## Regulation of GSTs

Many of the mammalian GST isoforms are inducible; however, GST regulation is complex, and expression is both developmentally regulated and tissue-specific (e.g., some are expressed in brain and testis only). Some GSTs are responsive to hormonal effects (growth hormone, thyroxine, insulin), and over 100 xenobiotic compounds have been shown to act as inducers (Hayes and Pulford, 1995). Moreover, in humans and the rat (but not in fish), GSTP1 is overexpressed in hepatocytes during the process of carcinogenesis. From structural and promoter analyses of GST genes, various functional regulatory elements have been identified. Promoter sequences targeted by prototypical GST antioxidant-inducing agents such as BHA not only include antioxidant response elements (AREs) but also electrophile response elements (EpREs) (Li and Jaiswal, 1993; Rushmore et al., 1991). The growing body of studies of GST induction in fish suggests that ARE-mediated GST gene induction may be conserved among many fish species. The GSTA1 gene (Figure 3.12) contains two peroxisomal proliferator response elements (PPREs) that coordinately upregulate expression of both GSTA and GSTA1 after exposure to peroxisomal proliferators (hyperlipidemic drugs, but potentially also environmental contaminants with binding ability such as synthetic lubricants, phthalate ester plasticizers, PCBs, certain pesticides, and alkylphenols). One of these PPREs is associated with an estrogen response element (ERE) that may produce cross-talk and thus explain the sex-dependent differences in expression that are observed. The GSTA gene contains four upstream antioxidant response elements that upregulate expression by compounds which *in vivo* include β-naphthoflavone (or probably a metabolite) and *trans*-stilbene oxide (Leaver et al., 1993). Two of these AREs are associated with CCAAT boxes, which may confer tissue specificity of response (Figure 3.12).

An important facet of the GST pathway from an environmental perspective is the potential for induction of GST isozymes on exposure to certain dietary and environmental chemicals. Typically, a modest induction (twofold or less) of overall GST–CDNB activity is observed in fish exposed to prototypical GST-inducing agents in the laboratory. Studies of GST–CDNB activities in fish may be complicated by variations in diet, water temperature, gender, and reproductive cycling (Swain and Melius, 1984). Furthermore, as discussed, GST–CDNB activity represents an integration of the activity of multiple isoforms, and treatment effects on GST isoforms may not always be distinguishable by the analysis of GST–CDNB activity. The importance in ascertaining

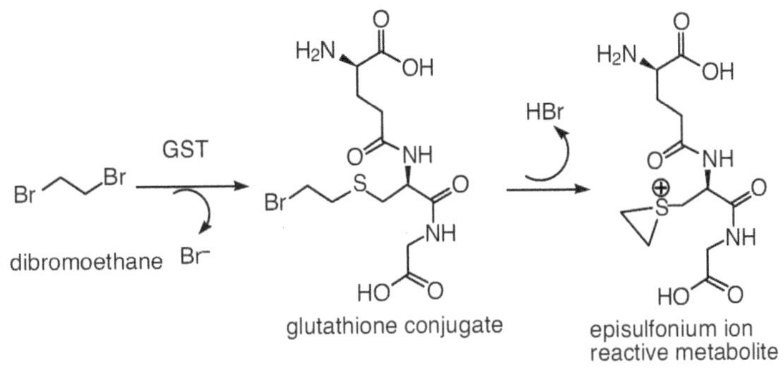


**FIGURE 3.11** Formation of a reactive metabolite by glutathione conjugation. Halogenated alkanes such as dibromoethane can undergo glutathione conjugation then eliminate the second halogen to give a reactive episulfonium ion. This electrophilic metabolite can react with cellular nucleophiles.

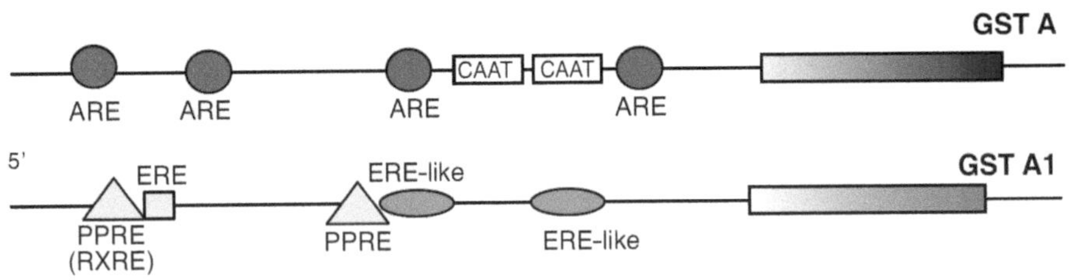


**FIGURE 3.12** Promoter regions of the plaice Rho-class genes.

effects of inducing agents on multiple fish GST isoforms is underscored by the fact that selective modulation of those GST isoforms with high specific activity toward environmental toxicants or their metabolites may be missed if GST–CDNB activity is the only endpoint. Studies of mRNA expression of GSTs in fish in response to chemical exposure provide another avenue of research. The expression profiles of seven GST genes in river pufferfish exposed to the heavy metal, cadmium, showed that mRNA expression of genes for GST-Mu, GST-Omega, and GST-Zeta was upregulated, but not GST-alpha or kappa (Kim et al., 2010). Ultimately, modulation of key GST isoforms that primarily contribute to the conjugation of environmental agents (or their metabolites) will be the critical determinants of chemical susceptibility.

Many laboratory studies of GST induction in fish have typically been conducted in conjunction with studies of CYP1A induction and have therefore involved AhR inducers, such as 3-methylcholanthrene (3-MC) and β-naphthoflavone (BNF) (George, 1994). Studies of GST induction in rodents indicate that antioxidant compounds such as ethoxyquin and butylated hydroxyanisole (BHA) are more effective inducers of GST expression than are AhR agonists (Buetler et al., 1995; Hayes and Pulford, 1995). This may not be the case in fish; for example, the level of induction in GST–CDNB activity in bullhead liver (1.6-fold) by ethoxyquin is consistent with previous studies of GST induction by other agents in other fish species. As an example, approximately two-fold inductions of GST–CDNB activity have been reported in trout exposed to BNF (Celander et al., 1993) and in plaice (*Pleuronectes platessa*) exposed to BHA or *trans*-stilbene oxide (Leaver et al., 1992). Equivocable results have been obtained in studies on the induction of fish GSTs by xenobiotics, partially due to their high constitutive levels and large interanimal variations in activities in wild populations, but primarily due to the use of substrates (e.g., HNE, ETHA) that are conjugated by several isoforms or the general substrate CDNB.

Attempts have been made to utilize GST induction in several invertebrates as biomarkers of xenobiotic exposure; indeed, the pi-class enzymes of Mollusca and Crustacea do show small (one- to two-fold) increases in CDNB conjugating activity to many compounds, especially pesticides (Table 3.12). Several studies have also reported increased fish GST activities in fish inhabiting polluted environments (Armknecht et al., 1998; Otto and Moon, 1996). Accordingly, induction of GST activity has been proposed as a biomarker of exposure to environmental pollutants under field conditions. Numerous studies report on the induction of GST in field collected fish (e.g., Mierzejewski et al., 2014; van den Hurk and Haney, 2017). And, a meta-analysis of biomarkers used as indicators for polycyclic aromatic hydrocarbons (PAH) effects concludes that the GST-CDNB assay is a robust biomarker (Santana et al., 2018). However, as mentioned above, careful characterization of the inducibility of individual fish GST isozymes under laboratory conditions and with respect to physiological status is necessary to interpret the field data of GST expression.

### Inhibition of GSTs

Very little literature exists regarding the inhibition of GST activity; however, compounds that bind covalently (e.g., carcinogens) or noncovalently (e.g., heme degradation products, hematin itself, and bilirubin) will inhibit the activity of isoforms to which they are bound. Some xenobiotics may also be inhibitors of GST. Although the binding constants for bilirubin and hematin for plaice liver cytosolic GSTs are lower than those for mammalian GSTA1, these endobiotics do bind and act as inhibitors with $I_{50}$ values of 320 and 10mM,

**TABLE 3.12**
**Induction of GSTs by Xenobiotics in Aquatic Organisms**

| | Inducers | Effect |
|---|---|---|
| PAHs | 3-MC | No effect on activity in minnows and killifish, variable pattern (±1.5-fold in flounder and plaice) |
| | BNF | No effect on activity, induction of GSTmRNA in plaice, no effect on activity in trout |
| | Petroleum/oil | Induced activity in scallop, mussel, and clam |
| PCBs | Tetrachlorobiphenyl | Induced activity in trout |
| | Arochlor® | Induced GSTA in flounder and plaice, activity in clams |
| | Clophen | Slight induction of activity in trout |
| Reactive epoxide | *trans*-Stilbene oxide | Induced GSTA in plaice and flounder |
| Antioxidants | BHA | Induced activity in plaice |
| | Ethoxyquin | Induced activity in salmon and possibly catfish |
| Pesticides | 2,4-D | Induced activity in carp and tilapia |
| | Azinphosmethyl | Induced activity in carp and tilapia |
| | Azoxystrobin | Induced activity in zebrafish |
| Peroxisomal proliferators (PPRE-γagonists) | Clofibrate, perflourooctanoic acid (PFOA) | Induced plaice GSTA and GSTA1 |

respectively (George and Buchanan, 1990). Binding constants of these compounds with purified GSTs from elasmobranchs were an order of magnitude lower than with plaice. Of particular note is the very high potency of organotin compounds (tributyltin, triphenyltin) for the inhibition of plaice cytosolic GST activity *in vitro* (George and Buchanan, 1990).

## SULFOTRANSFERASE

### Overview

The cytosolic sulfotransferase (SULT) family of enzymes catalyzes the transfer of the sulfonate group from 3′-phosphoadenosine-5′-phosphosulfate (PAPS) to hydroxyl (phenolic or alcoholic) and amine groups in a range of endogenous and exogenous substrates, giving rise to sulfate or sulfamate conjugates (Coughtrie, 2016; Gunal et al., 2019). Most of our knowledge of this pathway derives from studies in humans and laboratory animals such as rats; however, there are also several studies of the pathway in fish species as discussed below. The general sulfotransferase reaction is shown below, where R is the main part of the xenobiotic substrate and X=O or N.

$$RXH + PAPS \rightarrow RXSO_3H + PAP$$

The cosubstrate for sulfonation, PAPS, is synthesized from adenosine triphosphate (ATP) and inorganic sulfate in reactions catalyzed by PAPS synthase, PAPSS (Gunal et al., 2019). PAPSS is a bifunctional enzyme: one domain adds sulfate to ATP to form adenosine-5'-phosphosulfate (APS) and the other domain phosphorylates the APS to give PAPS (van den Boom et al., 2012). In humans, there are two forms of PAPSS, PAPSS1, and PAPSS2 (Gunal et al., 2019). This pathway has not been studied in fish, but a gene highly related to PAPSS2 has been identified in salmon and pike (Leong et al., 2010). There is no information on levels of PAPS in the fish liver, but in human liver, PAPS levels are in the low μM range (Cappiello et al., 1989). The low availability of PAPS is thought to be a factor in the relatively low effectiveness of sulfonation as a biotransformation pathway when organisms are exposed to a sudden increase of a potential substrate. Another consideration is that PAP, the desulfated reaction product of PAPS, is still able to bind to the active site of the enzyme and inhibit activity (Cook et al., 2012; James, 2021b).

Sulfate or sulfamate conjugates, once formed, are acidic molecules with $pK_a$ values in the range of 2–4, and they exist largely as anions at physiological pH. Like other anions, they are readily excreted in bile and urine with the assistance of transporter proteins (Deeley et al., 2006, Srimaroeng et al., 2008).

Xenobiotic compounds that are potential substrates for SULTs include phenol and phenol derivatives, phase I hydroxylated metabolites of polycyclic aromatic hydrocarbons, polychlorinated biphenyls and polybrominated diphenyl ethers, aryl amines and their hydroxylated metabolites, and other xenobiotics and their metabolites with hydroxyl or amino groups in the molecular structure. In addition, several physiologically important substrates such as steroids, bile acids, and neurotransmitters are substrates for sulfonation. Examples of xenobiotic and physiologically important substrates are shown in Figure 3.13.

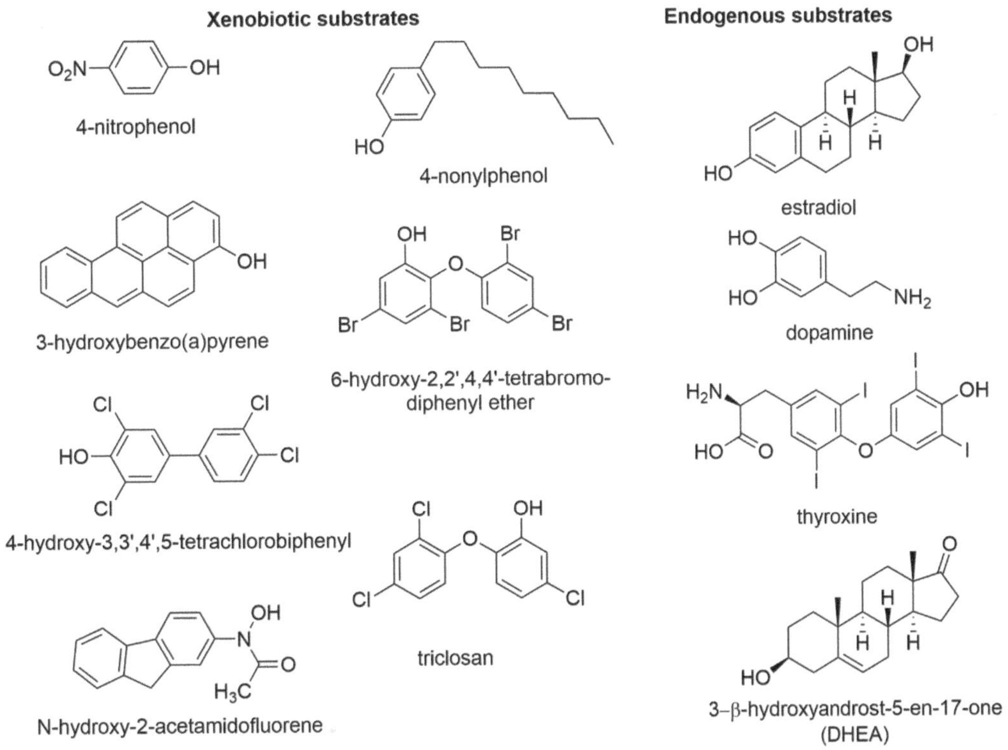


**FIGURE 3.13** Selected xenobiotic and endogenous substrates for PAPS–sulfotransferase.

In most cases, sulfonation is a detoxication pathway, but for some substrates, formation of the sulfate ester or the sulfamate leads to the activation of the chemical to a reactive metabolite. For a detoxication example, consider sulfonation of hydroxylated (phenolic) metabolites of PAHs (Figure 3.14). The conjugate is more water-soluble and more readily excreted than the parent compound. Similarly, sulfonate conjugates of ring hydroxylated metabolites of *N*-acetylaminofluorene are relatively stable, water-soluble metabolites that are readily excreted. On the other hand, sulfonation of benzylic hydroxy groups, formed from hydroxylation of the methyl group in methylated PAHs (e.g., 6-hydroxymethyl-benzo(*a*)pyrene, 7-hydroxymethyl-12-methyl-benz(*a,h*)anthracene), may be an activation pathway (Watabe et al., 1983) (Figure 3.14). This is because a sulfate group attached to an aliphatic carbon, especially a benzylic carbon, is a good leaving group and yields the carbocation that can alkylate DNA.

In addition to its importance in toxication and detoxication, sulfonation is an important pathway for transport and storage of several endogenous compounds. Estradiol, dehydroepiandrosterone (DHEA), and other hydroxylated steroids are readily sulfonated; they may be transported around the body as the sulfate conjugates and then hydrolyzed back to the active steroid at the target organ (Gilligan et al., 2017). The ability of the sulfonated steroid to cross cell membranes is sometimes superior to that of the unconjugated steroid. This is in part because the conjugates are substrates for several organic anion transporter proteins involved in the uptake as well as efflux of anions and in part because of more favorable solubility characteristics for diffusion conferred by the addition of a polar group to an otherwise poorly water-soluble molecule. The transport and hydrolysis of sulfonate conjugates of steroids have been described in mammals and proposed to be important in fish (James, 2011). Not only are steroid hormones readily sulfonated but also other physiologically important chemicals such as dopamine, related biogenic amines, and thyroid hormones. In the case of dopamine, the amino group and the ring hydroxyl groups are potential sites of sulfonation. In the shark, the sterol scymnol has been found as the sulfate ester, which appears to serve as a bile acid (Macrides et al., 1994). Rainbow trout liver contains SULT activity that is active with thyroid hormones, especially 3,3′,5-triiodothyronine ($T_3$) (Finnson and Eales, 1998).

Evidence for the sulfonation pathway in fish has been sought by examining bile, urine, and tank water of exposed fish for sulfate conjugates (Layiwola and Linnecar, 1981; Layiwola et al., 1983), as well as by studying enzyme activity *in vitro* with the substrate of interest as discussed below. Studies that demonstrate sulfonation have also been conducted in hepatocytes and with *in situ* isolated preparations.

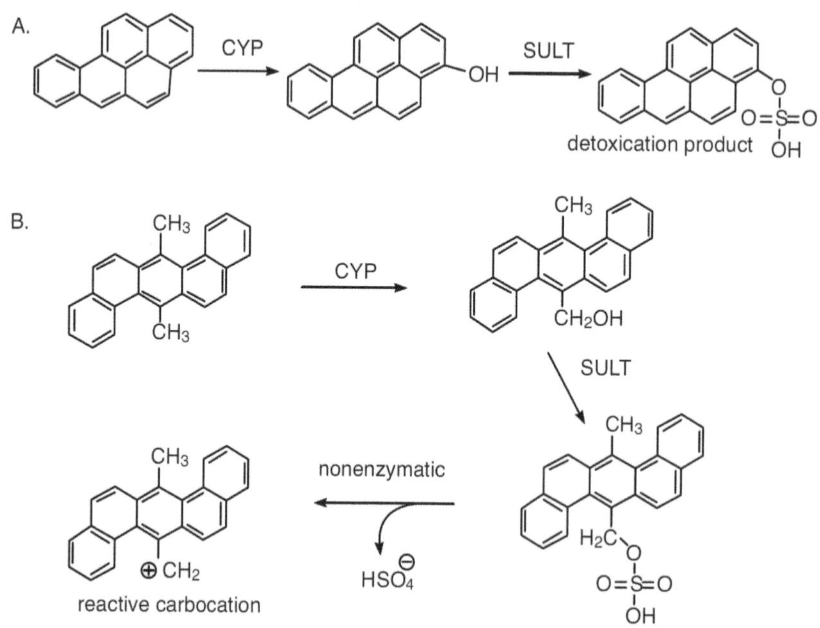


**FIGURE 3.14** Sulfonation as a toxication or detoxication pathway. This figure shows examples of polycyclic aromatic hydrocarbons that are metabolized to nontoxic (A) or reactive (B) sulfate esters. The example in (A) shows the oxygenation of benzo(*a*)pyrene to 3-hydroxybenzo(*a*)pyrene, followed by sulfonation of the phenolic hydroxyl group to the nontoxic sulfate ester. The example in (B) shows the oxygenation of one of the methyl groups in 7,12-dimethyl-benz(*a,h*)anthracene to give 7-hydroxymethyl-12-methylbenz(*a,h*) anthracene, followed by sulfonation of the benzylic (alcohol-like) hydroxyl group to the unstable sulfate ester. This ester can spontaneously eliminate the sulfate group, leaving the reactive carbocation shown. The carbocation will react with cellular nucleophiles, possibly including DNA bases.

## Gene Structure of SULT

Different organisms have been found to have differing numbers of SULT gene families. In humans, there is evidence for four SULT gene families, SULT1, SULT2, SULT4, and SULT6 (Suiko et al., 2017). SULT3 and SULT5 genes have been found in mice but are poorly characterized (Blanchard et al., 2004, Gamage, 2006). Most catalytically active SULT enzymes known to exist in the liver and other organs of humans are in the SULT1 and SULT2 families and have differing substrate selectivities (Coughtrie, 2016). A SULT4 protein has been found to be conserved in animals and expressed in the brain and other tissues, including zebrafish, where it was found to affect behavior (Crittenden et al., 2015); however, it does not appear to catalyze the sulfonation of any substrate and its function is still under investigation (Hossain et al., 2019).

Examination of the genomes of zebrafish, *Danio rerio*, mummichog, *Fundulus heteroclitus*, medaka, *Oryzias latipes*, marine medaka, *Oryzias melastigma*, and mangrove killifish, *Kryptolebias mormoratus*, has led to identification of six SULT gene families in fish. For most of these fish, the SULT genes and the enzymes they code for are not well studied. An exception is the zebrafish, where the research of C-M Liu and coworkers for a number of years has considerably advanced our knowledge of zebrafish SULTs as reviewed (Suiko et al., 2017; Kurogi et al., 2013).

## Zebrafish sulfotransferase activity

Liu and coworkers have cloned, expressed, and purified 20 distinct zebrafish SULTs and characterized several of them with respect to substrate selectivity and developmental expression (Suiko et al., 2017; Liu et al., 2008; Yasuda et al., 2008). Table 3.13 details some of the substrate selectivity characteristics. An earlier study by the Liu group demonstrated that a zebrafish SULT2 had high activity with dehydroepiandrosterone (DHEA), but no activity with phenolic substrates (Sugahara et al., 2003b).

From these results, it appears that the zebrafish SULT1 isoforms, which have good activity with phenolic substrates, are similar in properties to mammalian SULT1 isoforms. The SULT3 family showed activity with bulkier molecules with alcoholic hydroxyl groups, including DHEA and pregnenolone, while also showing activity with a few phenolic xenobiotics, such as bis-phenol A. Some mammalian SULT2 family members, particularly SULT2A1, have good activity with similar substrates (James and Ambadapadi, 2013).

Liu and coworkers have examined the developmental expression of several zebrafish SULTs from unfertilized

**TABLE 3.13**
**Substrate Selectivity of Several Zebrafish (Danio Rerio) Sulfotransferases**

| Substrate | SULT1ST7 | SULT1ST8 | SULT3ST1 | SULT3ST2 | SULT3ST3 |
|---|---|---|---|---|---|
| 17-β-Estradiol | N.D. | N.D. | 0.52 | 1.37 | 0.90 |
| Dehydroepiandrosterone | N.D. | N.D. | 0.13 | 8.04 | 9.37 |
| Pregnenolone | - | - | 0.04 | 2.67 | 11.24 |
| Diethylstilbestrol | N.D. | N.D. | 0.72 | 0.12 | 0.17 |
| Bisphenol A | N.D. | N.D. | 0.31 | N.D. | 0.19 |
| Ethynylestradiol | N.D. | N.D. | 0.15 | 0.85 | - |
| 3-Chloro-4-diphenylol | 1.85 | 3.65 | 0.04 | 0.05 | - |
| 3,3′,5,5′-Tetrachloro-4,4′-biphenyldiol | 1.82 | 2.81 | - | - | - |
| Acetaminophen | 1.09 | N.D. | - | - | - |
| Chlorogenic acid | 1.5 | 1.59 | - | - | - |
| Catechin | 9.89 | 0.65 | - | - | - |
| Daidzein | 0.99 | 0.11 | - | - | - |
| Epigallocatechin gallate | 2.17 | 1.32 | - | - | - |
| Quercetin | 1.5 | 0.78 | - | - | - |
| β-Naphthol | 1.38 | 0.17 | - | - | - |
| 4-Nitrophenol | N.D. | N.D. | - | - | - |
| *n*-Octylphenol | 0.20 | N.D. | - | - | - |
| *n*-Nonylphenol | N.D. | N.D. | - | - | - |
| 2-Butoxyethanol | - | - | 0.21 | - | - |

N.D. activity below detection limit, estimated at 0.01 nmol/min/mg protein; - not tested
Summarized from Liu et al. (2008), Yasuda et al. (2008), Xi et al. (2016).

eggs through the hours, days, and weeks postfertilization, by assessing relative mRNA expression relative to beta-actin mRNA expression (Table 3.14).

### SULTs in fish species other than zebrafish

SULT enzymes have been studied in several marine and freshwater fish. A SULT with activity toward the endogenous substrate scymnol was found in shark liver, but no xenobiotic substrates were studied. The shark SULT had a higher molecular weight (40,000) than mammalian SULT (Macrides et al., 1994). SULT proteins were isolated and purified from hepatic and intestinal cytosol of the channel catfish (Tong and James, 2000). Catalytically active fractions from the liver and intestine contained proteins of 41,000 MW that cross-reacted with an antibody to the human phenol-sulfating form of phenol sulfotransferase, SULT1A1 (Falany et al., 1990). Internal amino acid sequence from the liver enzyme indicated that it was in the SULT1 family (Tong and James, 2000). A SULT1 cDNA and promoter region was cloned from the liver of the chub, using conserved sequence of zebrafish SULT1 as primers, then expressed in *E. coli* and used to study substrate specificity (Assem et al., 2006). The chub SULT1 protein has activity with isoflavones and to a lesser extent with other phenolic compounds and dopamine, but not estrogens or thyroid hormones (Assem et al., 2006).

### Reactions and Substrate Specificity

The specificity of sulfonation by substrate has been studied in several fish species (Table 3.15) in additional to zebrafish, discussed above (Table 3.13). Sulfonation of *N*-hydroxy-*N*-acetylaminofluorene and 3-hydroxyacetylaminofluorene (procarcinogenic and noncarcinogenic metabolites,

**TABLE 3.14**
**Expression of mRNA at Several Times Postfertilization and in 3-Month Adults**

| Form | Egg | Hours Postfertilization | | | | | | | Weeks Postfertilization | | | | Adult | |
|---|---|---|---|---|---|---|---|---|---|---|---|---|---|---|
| | | 1 | 3 | 6 | 12 | 24 | 48 | 72 | 1 | 2 | 3 | 4 | Male | Female |
| SULT1ST7 | – | – | – | – | – | — | + | – | ++ | ++ | – | – | – | ++ |
| SULT1ST8 | – | – | – | – | – | – | – | – | ++ | + | + | – | – | – |
| SULT3ST1 | ++ | ++ | ++ | ++ | ++ | + | + | ++ | ++ | +++ | +++ | +++ | + | ++ |
| SULT3ST2 | + | – | – | +/– | – | – | + | ++ | ++ | ++ | +++ | ++ | + | +++ |

The number of + signs indicates the strength of the signal; – indicates no detection.
Data shown in this table are from publications Yasuda et al. (2008) and Liu et al. (2008).

**TABLE 3.15**
**Cytosolic Sulfotransferase Activities in Different Fish Species**

| Species | Preparation | Substrate | Activity (pmole/min/mg) | References |
|---|---|---|---|---|
| Guppy (*Poecilia reticulate*) | Liver cytosol | 4-Methylumbelliferone | 86 | James et al. (1994) |
| | 7-Hydroxy-2-acetylaminofluorene | | 121 | |
| | *N*-Hydroxy-2-acetylaminofluorene | | 8 | |
| Medaka (*Oryzias latipes*) | Liver cytosol | 4-Methylumbelliferone | 130 | James et al. (1994) |
| | 3-Hydroxy-2-acetylaminofluorene | | 25 | |
| Channel catfish (*Ictalurus punctatus*) | Liver cytosol | 9-Hydroxy-benzo(*a*)pyrene | 600 | Tong and James (2000) |
| | Intestinal cytosol | 9-Hydroxy-benzo(*a*)pyrene | 930 | Tong and James (2000) |
| | | Benzo(*a*)pyrene-7,8-dihydrodiol | 2 | van den Hurk and James (2000) |
| Mummichog (*Fundulus heteroclitus*) | Liver cytosol | 9-Hydroxy-benzo(*a*)pyrene | 8 | Gaworecki et al. (2004) |
| Lamprey (*Petromyzon marinus*) | Liver cytosol | Petromyzonol | 232 | Venkatachalam et al. (2004) |
| Rainbow trout (*Oncorhynchus mykiss*) | Liver cytosol | *N*-Hydroxy-2-acetylaminofluorene | 0.2 | Elmarakby et al. (1995) |
| | Liver cytosol | Thyroxine ($T_4$) | 7.6 | Finnson and Eales (1998) |
| | Liver cytosol | 13,5,3′-Triiodothyronine ($T_3$) | 14 | Finnson and Eales (1998) |

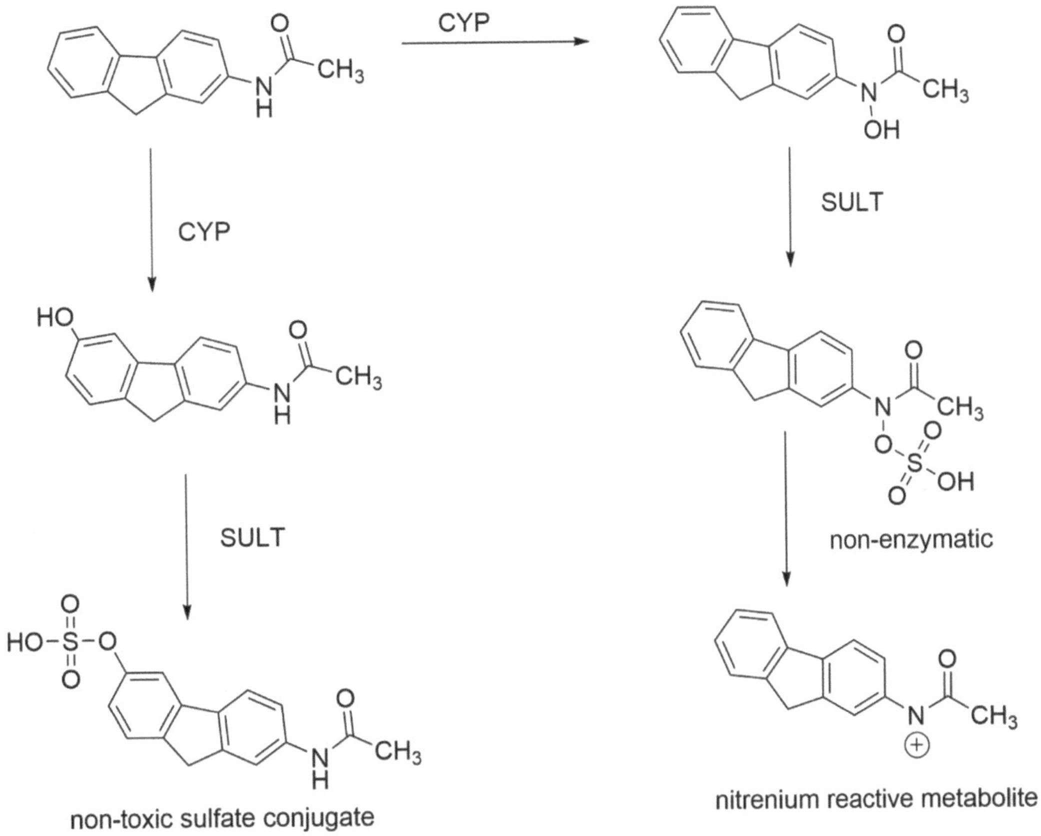


**FIGURE 3.15** Sulfonation as a pathway for toxication or detoxication pathway. The figure above illustrates two possible pathways of metabolism of the carcinogen *N*-acetamidofluorene. The *N*-hydroxy-acetamidofluorene forms an unstable sulfate conjugate that readily eliminated sulfate, resulting in a reactive, electrophilic nitrenium ion. The ring-hydroxylated *N*-acetaminofluorene forms a stable sulfate ester that is excreted.

respectively, of acetamidofluorene) has been studied in small fish used in carcinogenicity testing (see Figure 3.15).

As well as the acetamidofluorene metabolites, sulfonation of the model compound 4-methylumbelliferone was examined in these small fish. Sulfonation of *N*-hydroxy-*N*-acetamidofluorene was demonstrated in control and 2-acetamidofluorene (AAF)-treated guppy, indicating that the pathway to the reactive ultimate carcinogen was intact (James et al., 1994). SULT activity was higher in guppy than medaka, suggesting a possible biochemical basis for the higher sensitivity of guppy to the carcinogenic effects of AAF. Rainbow trout, a species that is relatively insensitive to the carcinogenicity of AAF, formed very little *N*-hydroxy-2-acetylaminofluorene (N-OH-AAF) or N-OH-AAF sulfate (Steward et al., 1995).

Sulfonation of several xenobiotic substrates has been studied in the channel catfish and in SULT enzymes isolated from the liver and intestine (Tong and James, 2000). These include 3,7- and 9-hydroxybenzo(*a*)pyrene, benzo(*a*)pyrene-7,8-dihydrodiol, 2-naphthol, acetaminophen, 6-hydroxymethylbenzo(*a*)pyrene, 7-hydroxymethyl-12-methylbenz(*a*,*h*)anthracene, and *p*-nitrophenol. The isolated intestinal and hepatic enzymes had nanomolar $K_m$ values with 3-, 7-, or 9-hydroxybenzo(*a*)pyrene as substrate and were also active with several other phenolic substrates. Higher activity was found with the phenolic substrates than the alcoholic substrates (Tong and James, 2000). In contrast to the benzo(*a*)pyrene phenols, the dihydrodiol procarcinogen BaP-7,8-dihydrodiol was a very poor substrate for sulfonation in the catfish intestine or liver (van den Hurk and James, 2000). Others have shown in mammalian systems that BaP-7,8-dihydrodiol inhibits SULT but is a poor substrate (Rao and Duffel, 1992). *In vivo* studies with the southern flounder, however, have shown that the sulfate conjugate of BaP-7,8-dihydrodiol was formed to some extent and was excreted more rapidly than the glucuronide (Pritchard and Bend, 1991).

### Tissue Distribution

Most studies of sulfonation have focused on activity in the liver, but in the catfish, the intestine was shown to be as active as the liver in catalyzing the sulfonation of several substrates, including hydroxylated benzo(*a*)pyrenes, dopamine, and 6-hydroxymethylbenzo(*a*)pyrene (James et al., 1997, Tong and James, 2000). Demonstration of activity in the intestine is important in the context of first-pass metabolism of dietary carcinogens, which are often present at low levels. When low concentrations of substrate are present in ingested material, biotransformation in the intestine becomes more important. The capacity of the fish intestine for metabolism is fairly low, because the liver is a larger organ, but if only small amounts of xenobiotic are present, as is the case in many environmental samples, the intestine, and other extrahepatic organs that are portals of entry, such as gills and skin, become important sites of first-pass metabolism.

### Regulation of SULT

Not enough studies of SULT in fish have been conducted to have a good understanding of their regulation by endogenous or exogenous factors. In mammalian species, SULTs are not induced by exposure to PAHs, and there are reports that AhR agonists such as PAHS repress SULT activity (Runge-Morris and Wilusz, 1994). The few studies that have been conducted in fish show no clear evidence of an effect of PAH-type inducing agents on SULT activity (James et al., 1997). In catfish exposed to β-naphthoflavone, there was a slight loss of activity with 7-hydroxybenzo(*a*)pyrene but no effect with the 3- and 9-hydroxy-BaP substrates. Largemouth bass from a site contaminated with PAH and lead showed similar SULT activities with 9-hydroxy-BaP to bass from clean lakes (Schreiber et al., 2006).

Investigations into the expression of phenol-type SULT in mummichog (*Fundulus heteroclitus*) have revealed that this species has a remarkably low expression of this enzyme (Gaworecki et al., 2004). The activity toward 3-hydroxybenzo(*a*)pyrene was very low compared to catfish and other fish species (Gaworecki et al., 2004). In addition, a polyclonal antibody that was generated against channel catfish phenol-type SULT did not recognize any SULT protein in mummichog samples. Despite the low expression of phenol-type SULT in mummichog from clean reference sites, animals from the Atlantic Wood site in the Elizabeth River in Virginia, a site heavily contaminated with PAHs and other contaminants, showed significantly increased SULT activity (Gaworecki et al., 2004). Treatment with single doses of 3-MC or BaP failed to induce the protein; thus, contrary to AhR-mediated CYP1A and UGT induction, SULT upregulation does not appear to be linked to AhR activation.

### Inhibition of SULT

Sulfotransferase is subject to inhibition from a variety of sources (James and Ambadapadi, 2013). *In vitro*, PAP derived from the cosubstrate or sulfonate donor molecule PAPS inhibits SULT activity. Many bulky substrates inhibit their own sulfonation; for example, in catfish, intestinal cytosol 3-hydroxybenzo(*a*)pyrene inhibited its own metabolism at concentrations above 0.5 μM (Tong and James, 2000). The nonsubstrate β-naphthoflavone inhibited SULT activity with 3-, 7-, and 9-hydroxybenzo(*a*)pyrene in channel catfish intestinal and hepatic cytosol, with an $IC_{50}$ of 48.0±3.0 μM (James et al., 1997). A natural flavonoid, quercetin, was a potent inhibitor of sulfonation of 3-, 7-, and 9-hydroxybenzo(*a*)pyrene in channel catfish intestinal cytosol, with an $IC_{50}$ of less than 1 μM (van den Hurk and James, 2000). Although pentachlorophenol was reported to inhibit phenol sulfonation in mammals, it was a poor inhibitor of the sulfonation of hydroxylated benzo(*a*)pyrenes in catfish and had an $IC_{50}$ of about 100 μM (van den Hurk and James, 2000). Studies showed that several hydroxylated PCBs inhibited estrogen SULT in mammalian systems with nanomolar $IC_{50}$ values (Kester et al., 2000). Studies of the effects of hydroxylated PCB on the sulfonation of hydroxylated benzo(*a*)pyrenes in catfish intestinal cytosol showed that several of these PCB metabolites inhibited SULT, with low-to-high micromolar $IC_{50}$ values (van den Hurk et al., 2002). Several metal cations, including mercury, cobalt, zinc, cadmium, copper, and lead, were shown to inhibit zebrafish SULT, although quite high concentrations (5 mM) were tested (Sugahara et al., 2003a).

Sulfonation of thyroid hormones was studied in liver cytosol from rainbow trout (*Oncorhynchus mykiss*) (Finnson and Eales, 1998). Based on optimum pH, thermal stability, and specific inhibitors, the investigators concluded that it is most likely that only one isozyme form is responsible for the sulfonation of thyroid hormones in trout, in contrast to the multiple isozyme forms found in mammals. No information is available yet on the inhibiting effects of hydroxylated PCBs on thyroid hormone sulfonation in fish, as has been reported for mammalian systems (Schuur et al., 1998).

### Activation of SULT

*In vitro*, SULT activity with some substrates can be increased by the addition of $Mg^{2+}$ and others by incorporation of BSA into the assay vial. This phenomenon was also observed with the isolated catfish SULT (Tong and James, 2000) and zebrafish SULT (Sugahara et al., 2003a). The *in vivo* significance of this activation is unclear (Konishi-Imamura et al., 1995).

## Acetylation

### Overview

This pathway utilizes acetyl-CoA to donate an acetyl group to an amino group in a xenobiotic. Xenobiotic substrates include arylamines, aliphatic amines, and hydrazines. The reaction pathway with 2-aminofluorene as the substrate is shown in Figure 3.16. The enzymology of *N*-acetylation has been extensively studied in mammals, but few studies of acetylation in fish have been performed, beyond

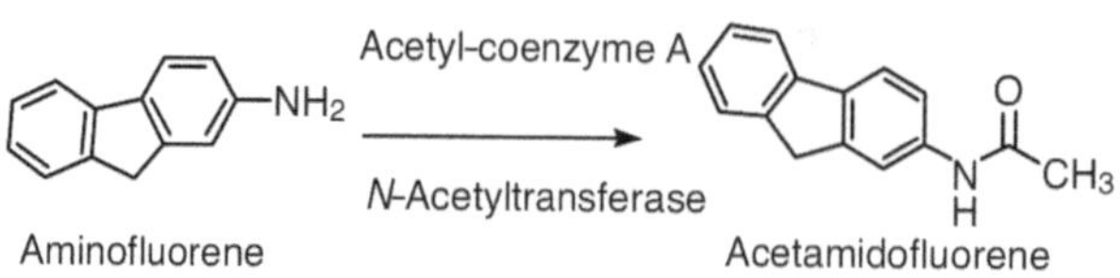


**FIGURE 3.16** Acetylation of arylamines. The arylamine aminofluorene forms acetamidofluorene in a reaction catalyzed by *N*-acetyltransferase. The cosubstrate is acetyl–coenzyme A.

demonstration of the pathway with model substrates (de Wolf et al., 1993; Gregus et al., 1983). In humans and lab animals, two *N*-acetyltransferases, NAT1 and NAT2, catalyze *N*-acetylation of xenobiotic substrates (Sim et al., 2008). Evolutionary biology studies have demonstrated the present of genes that align with NATs 1 and 2 in fish species (Sabbagh et al., 2013). Acetylation is also an important terminal step in the formation of mercapturic acid metabolites from glutathione conjugates (Figure 3.9). It is thought, however, that the enzyme that acetylates cysteine conjugates, NAT8, is different from the *N*-acetyltransferase that acetylates xenobiotics containing amino groups that are not part of an amino acid (Veiga-Da-Cunha et al., 2010).

### Enzyme Specificity, Regulation, and Inhibition

Acetylation has been shown to be a major pathway essential for zebrafish development (Ree et al., 2015) and neuroendocrine function in fish (Zilberman-Peled et al., 2007). It is also important for the metabolism of a number of xenobiotics containing amino groups. These include therapeutic drugs such as sulfadimethoxine (Droy et al., 1989; Kleinow et al., 1992) and pollutants such as aniline, chloroaniline, and quinolines (Birkholz et al., 1989, Bradbury et al., 1993). Acetylation of the procarcinogen 2-aminofluorene leaves the molecule open for further metabolism by *N*-hydroxylation and sulfonation to a reactive nitrenium ion, as described in the section on sulfonation. Another pathway of activation of acetamidofluorene is *N,O*-acetyl transfer, leading to the unstable *N,O*-acetyl metabolite. This is readily deacetylated to yield a nitrenium ion active metabolite. Thus, although acetylation of amines may improve the water solubility and ease of excretion, it may in some cases lead to the formation of more reactive metabolites. In mammalian species, the *N*-acetyltransferase enzymes are not inducible but are expressed constitutively. Although it is clear from molecular biology studies and functional studies that NAT enzymes are present in fish, the number of forms of *N*-acetyltransferase in fish and their regulation has not been well studied.

## Amino Acid Conjugation

### Overview

A relatively little studied pathway of phase II metabolism is the conjugation of carboxylic acids with amino acids. This pathway results in the formation of an amide bond between the carboxylate and the amino group of the amino acid (Figure 3.17). The resulting conjugate is an anion at physiological pH and is generally a better substrate for organic anion transporter proteins in kidney than the unconjugated xenobiotic; thus, amino acid conjugation facilitates the excretion of xenobiotic carboxylic acids (James and Pritchard, 1987). The amino acid utilized for this pathway in all fish species studied to date is taurine (James, 1987, Plakas and James, 1990, Burke et al., 1987, Cravedi and Tulliez, 1987, Acena et al., 2017; Fu et al., 2020; James, 1976). This differs from mammalian species, in which carboxylic acids can be metabolized to glycine, glutamine, or, less commonly, taurine conjugates (James, 2021). It has been suggested that glycine conjugates (the major metabolites of benzoic acids in mammals) are metabolites of *p*-aminobenzoate in fish (James, 1986), but no solid evidence supports this pathway in fish. The enzymes responsible for amino acid conjugation are found in the liver and kidney mitochondria (James, 2021). Two enzymes are involved: a carboligase that catalyzes the formation of an acyl-coenzyme A (CoA) intermediate in two steps (van der Sluis, 2018) and an aminoacyltransferase that catalyzes formation of the amide bond (Vessey, 1996; Kelley and Vessey, 2003). Figure 3.17 shows this pathway with benzoic acid as the substrate and benzoyltaurine as the product. Studies

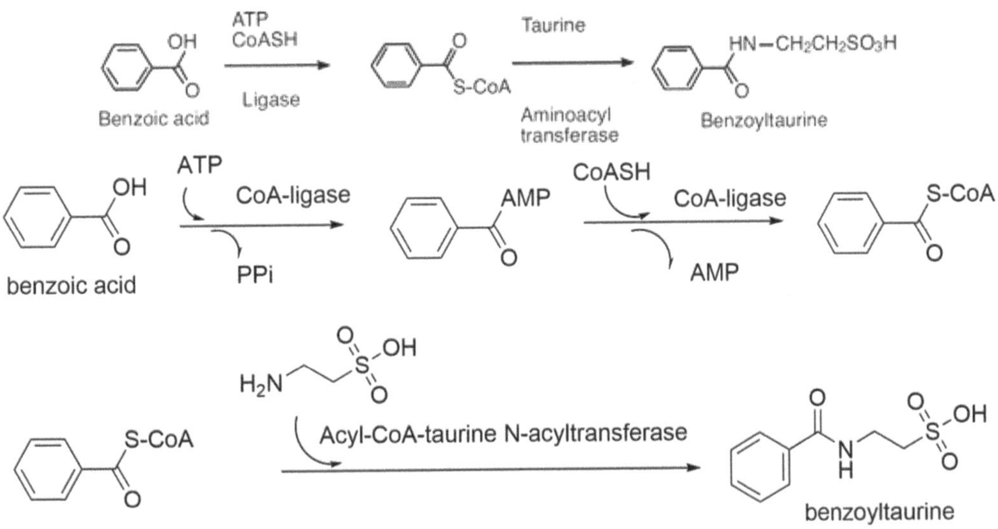


**FIGURE 3.17** The pathway of conjugation of carboxylic acids in fish, illustrated with benzoic acid. The benzoic acid forms a coenzyme A thioester, in a two-step reaction catalyzed by acyl–CoA ligase. The benzoyl–coenzyme A thioester then reacts with taurine, catalyzed by aminoacyl transferase, to give benzoyltaurine.

of the first step of this reaction—two steps leading to the formation of the acyl-coenzyme A—have not been carried out in fish. The second final step, catalyzed by acyl-CoA–taurine aminoacyltransferase, is located in the mitochondrial matrix in those fish that have been studied, namely, Southern flounder, *Paralichthys lethostigma*, Stingray, *Dasyatis sabina* and red drum, *Sciaenops ocellate*, and the kidney had higher activity than the liver (James and Pritchard, 1987; James, 1976). The activity has been studied with phenylacetyl-coenzyme A or benzoyl-coenzyme A as the substrate and taurine as the cosubstrate. Even when the incubations were supplemented with glycine, no glycine conjugates were found.

### Enzymes Specificity, Regulation, and Inhibition

Taurine conjugation of carboxylic acids is generally considered to be a detoxication pathway. Taurine conjugates are acidic and exist largely in the anionic form at physiological pH. It may be expected that they will therefore be readily excreted in urine. The taurine conjugate of benzoic acid was not, however, as good a substrate for the southern flounder renal organic acid transporter as hippuric acid (benzoylglycine), even though no hippurate was formed in flounder given benzoate (James and Pritchard, 1987). It was, nevertheless, excreted more readily than benzoic acid. Several xenobiotic substrates carboxylic acids are metabolized via this pathway in fish and include 2,4-dichlorophenoxyacetic acid (2,4-D), 2,4,5-trichlorophenoxyacetic acid (2,4,5-T), triclopyr, and *p*-aminobenzoic acid, a metabolite of the fish anesthetic tricaine methane sulfonates listed in Table 3.16. The taurine conjugates have been identified variously in whole animal, bile, urine, and tank water. In addition to xenobiotic substrates, there are some endogenous physiologically important substrates including bile acids, benzoic acid, and phenylacetic acid, the latter being metabolites of aromatic amino acids. Southern flounder have been shown to excrete taurine conjugates of benzoic acid (James and Pritchard, 1987), 2,4-D, and 2,4,5-T (James, unpublished data). Dogfish shark excreted 2,4,5-trichlorophenoxyacetyltaurine in urine following a dose of 2,4,5-T (Guarino et al., 1977). Winter flounder excreted phenylacetyltaurine and 2,4-D-taurine after doses of phenylacetic acid and 2,4-D, respectively (James and Bend, 1976; Pritchard and James, 1979). Channel catfish excreted benzoyltaurine in urine after administration of benzoic acid (Plakas and James, 1990; Fu et al., 2020; Brozinski et al., 2013; Miyamoto et al., 2015). The taurine conjugation pathway is not limited to fish: metabolism of the herbicide and fungicide, triclopyr, has not been examined reported in fish, but a crustacean, the crayfish, retained triclopyr as the taurine conjugate in hepatopancreas and slowly excreted the conjugate (Barron et al., 1991). A summary of xenobiotics shown to form taurine conjugates in fish is given in Table 3.16.

These examples demonstrate that the amino acid conjugation pathway is important for the elimination of carboxylic acids in fish; however, Nno studies have been performed on the regulation of this pathway in fish. Inhibitors and activators of the pathway are not known.

## TOXICOLOGICAL RELEVANCE

The previous sections have addressed the occurrence, regulation, and catalytic activities of biotransformation systems in fish. The following section focuses on three model compounds and chemical classes and the contribution of biotransformation toward the toxicology of each compound. Benzo(*a*)pyrene was selected as being representative of the large class of polynuclear aromatic hydrocarbons (PAHs). In addition, due to the significant potential for exposure to fish, the biotransformation of organophosphates and carbamates is discussed.

**TABLE 3.16**
**Taurine Conjugation of Xenobiotic Carboxylic Acids in Fish**

| Xenobiotic carboxylic acid | Fish species | Where identified | References |
|---|---|---|---|
| Benzoic acid | Southern flounder | Urine | James and Pritchard (1987) |
| | Rainbow trout | Tank water | Burke et al. (1987) |
| | Channel catfish | Urine | Plakas and James (1990) |
| Phenylacetic acid | Winter flounder | Urine | James and Bend (1976) |
| | Rainbow trout | Urine | Cravedi and Tulliez (1987) |
| 2,4-Dichloro-phenoxyacetic acid | Winter flounder | Urine | Pritchard and James (1979) |
| | Dogfish shark | Urine | James and Bend (1976) |
| 2,4,5-Trichloro-phenoxyacetic acid | Dogfish shark | Urine | Guarino et al. (1977) |
| Cyclohexylacetic acid | Rainbow trout | Urine | Cravedi and Tulliez (1987) |
| Ibuprofen | Rainbow trout | Bile | Brozinski et al. (2013) |
| | Sole | Bile | Acena et al. (2017) |
| Flumenequine | Channel catfish | Urine | Plakas et al. (1990) |
| Diclofenac | Rainbow trout | Hepatocytes | Fu et al. (2020) |
| Clofibrate | Zebrafish | Embryos | Brox et al. (2016) |
| Chrysanthemic acid | Bluegill sunfish | Whole body extract | Miyamoto et al. (2015) |

### Benzo(a)pyrene

The metabolic fate of PAHs in fish has been extensively studied because PAHs are common pollutants of the aquatic environment and are carcinogens in fish as well as higher vertebrates (for reviews in fish, see Buhler and Wang-Buhler, 1998; Buhler and Williams, 1989; Stegeman, 1981; Stegeman and Hahn, 1994). Benzo(*a*)pyrene (BaP), the model carcinogenic PAH, has received a great deal of attention and has served as a model to better understand PAH biotransformation. This compound is not in itself carcinogenic but becomes so upon biotransformation to reactive metabolites that form adducts to DNA bases and initiate the process of mutagenesis and carcinogenesis. BaP can also be metabolized to products that are not directly carcinogenic, such as quinones, phenols, and dihydrodiols (Figure 3.3). BaP possesses a structural "bay region" that impairs enzymatic detoxification of specific stereoisomer metabolites; for example, conversion of BaP to the (–)-7,8-dihydrodiol by consecutive reactions catalyzed by CYP1A and epoxide hydrolase (EH), respectively, may lead to an additional CYP1A oxygenation at the 9,10 position, producing the ultimate carcinogen, (+)-*anti*-BaP-7,8-dihydrodiol-9,10-epoxide (Jerina et al., 1984). These reactions are readily catalyzed by CYP1A and EH in the fish liver (Buhler and Wang-Buhler, 1998; Buhler and Williams, 1989; Stegeman, 1981; Stegeman and Hahn, 1994). This particular metabolite is recalcitrant to a second EH hydrolytic reaction due to steric hindrance resulting from the "bay region" ring system; thus, the reactive intermediate is free to interact with DNA leading to potential mutagenesis and carcinogenesis.

CYP1A is clearly the predominant enzyme responsible for each oxygenation of BaP, but other CYP isoforms are likely contributors. Studies with recombinant mammalian CYP isoforms indicate metabolic activation of several PAHs by CYP1B1 as well as CYP2C9, CYP2C19, and CYP3A4, albeit at slower rates than CYP1A and CYP1B enzymes (Shimada et al., 1999). CYP1A forms are readily inducible in fish by PAH and related planar polycyclic molecules, and this likely is a major factor in the sensitivity of fish to PAH-type carcinogens Scornaienchi et al. (2010). In addition, other phase I enzymes may catalyze oxidation reactions (see above). Inhibition of cytotoxicity with indomethacin within trout cell lines that lack EROD activity suggests that prostaglandin-H synthase may contribute to the activation of BaP (Schirmer et al., 2000). Clearly cooxidative pathways should be explored in more detail, particularly in fish that possess significant quantities of unsaturated fatty acids which are prone to oxidation.

Other metabolites with potential biological activity include quinones, which have been suggested to be derived from initial phenolic metabolites. BaP has been shown to be converted to several quinone metabolites, including the 1,6- and 3,6-quinones derived from the 1-hydroxy and 3-hydroxybenzo(*a*)pyrene precursors (Stegeman, 1981). Although numerous biotransformation studies have identified quinones as BaP metabolites (Stegeman and Hahn, 1994; Willett et al., 2000; Yuan et al., 1997), few have explored the potential effects of these metabolites on cell function. Lemaire et al. (1994) reported that BaP quinones were capable of creating hydroxyl radical in hepatic microsomes of flounder and perch. Oxidative damage in larval turbot was observed following BaP treatment with parallel formation of BaP quinone metabolites.

Although quinones and the (–)-7,8-dihydrodiol can lead to greater toxicity of BaP, it should be noted that the majority of the metabolites of BaP are nontoxic. Examples include phenols (at the 1 and 3 positions) which are primarily conjugated as glucuronides and potentially sulfate derivatives (Buhler and Williams, 1989; Stegeman and Hahn, 1994). Glutathione conjugates presumably of the various arene oxides (4,5, 7,8, and 9,10) have also been reported in fish (Gallagher et al., 1996). The 4,5 and 9,10 dihydrodiols of BaP also do not appear to undergo bioactivation.

Tremendous species differences exist in the bioactivation of BaP in fish. Liver microsomes of channel catfish, which appear less susceptible to PAH-induced liver cancer, produce lower levels of (+)-7,8-expoxide (based on lower concentrations of (–)-7,8-dihydrodiol) than cancer-prone brown bullhead catfish (Pangrekar and Sikka, 1992; Yuan et al., 1997). *In vivo* studies by Willett et al. (2000) confirmed the species differences in (–)-7,8-dihydrodiol formation. However, hepatic EROD activities did not correlate with 7,8-BaP dihydrodiol formation, indicating that another enzyme system may be involved in BaP oxygenation (Willett et al., 2000). Ploch et al. (1998) also reported that, although *in vitro* DNA binding by BaP was directly related to hepatic EROD activity in the two species, *in vivo* DNA adducts were not related to hepatic EROD. These results are consistent with studies in carp, which have significantly higher rates of phase I metabolite formation compared to brown bullhead but are more resistant to PAH-induced cancers (Sikka et al., 1990).

The effects of other pollutants and environmental conditions on other metabolic pathways of PAHs, or all xenobiotics, are largely unknown. Dieldrin significantly enhances the biliary elimination of BaP but does not alter the metabolite profile (Barnhill et al., 2003). Tributyltin significantly inhibits BaP biotransformation in Arctic char (Padros et al., 2003). Rainbow trout acclimated to hypersaline conditions had a higher level of phase I metabolites and a shift from dihydrodiol to phenolic metabolites with 3-hydroxybenzo(*a*)pyrene predominant (Seubert and Kennedy, 2000).

In addition to phase I enzymes, species-dependent differences in phase II pathways may also contribute to differences in BaP-induced toxicities between species. Although significant epidemiological and biochemical evidence exists in humans and rodents (GSTM1), few studies have demonstrated a correlation between adverse effects and species-specific phase II metabolism in fish; for example, hepatic EH and GST activities in channel catfish were higher than in brown bullhead (Gallagher et al., 1996; Willett et al., 2000).

When comparing overall biliary metabolite profiles in each species, significant differences were not observed in quantity or profile of conjugated BaP metabolites (Willett et al., 2000). Likewise, English sole from contaminated sites in Puget Sound appears to be more sensitive to liver lesions resulting from PAH exposure than starry flounder, but overall GST-catalyzed conjugation of BPDE is threefold higher in English sole (Gallagher et al., 1998). Although glucuronides represent the highest percentage of conjugated BaP metabolites in fish, few studies have identified specific metabolites or evaluated species differences in hepatic UGTs and whether this pathway contributes to species differences in BaP toxicity.

Although the liver is a major site of biotransformation of BaP, biotransformation to active metabolites in portals of entry, such as the intestine, gills, and skin, is also important. Studies in channel catfish intestine have shown that benzo(*a*)pyrene was metabolized to several metabolites, including BaP-7,8-dihydrodiol-9,10-epoxide, by intestinal microsomes and that metabolism was readily induced by exposure to the AhR agonist β-naphthoflavone (James et al., 1997; Kleinow et al., 1998). The major form of GST present in the intestine was a pi-class GST that had good activity with (+)-*anti*-BaP-7,8-diol-9,10-oxide (Gadagbui and James, 2000). Similar studies carried out with the channel catfish intestinal preparation in which (–)-BaP-7,8-dihydrodiol was infused showed that a major pathway for detoxification of the (–)-BaP-7,8-dihydrodiol was glucuronidation. In the catfish, (–)-BaP-7,8-dihydrodiol was more readily glucuronidated than sulfated, although the rate of glucuronidation of an environmentally relevant concentration of (–)-BaP-dihydrodiol was much lower than the rate of glucuronidation of a similar concentration of BaP phenols; the rate of conjugation of 2.5-μM (–)-BaP-7,8-dihydrodiol in catfish intestinal microsomes was $0.025 \pm 0.01$ nmol/min/mg protein, whereas that of 1-μM 7-hydroxybenzo(*a*)pyrene was $0.35 \pm 0.07$ nmol/min/mg protein (James et al., 1997; van den Hurk and James, 2000). In induced catfish, metabolism of (–)-BaP-7,8-dihydrodiol to the ultimate carcinogen (+)-*anti*-BaP-7,8-diol-9,10-oxide was increased, and the capacity of the intestine to detoxify the diol-epoxide by glutathione conjugation was apparently overwhelmed, as the induced fish had considerably higher DNA adducts than controls (James et al., 1988). It is likely that species differences in the rate of conjugation of BaP-7,8-dihydrodiol relative to the rate of activation by CYP to BaP-7,8-dihydrodiol-9,10-oxide are factors in the species differences in sensitivity to this carcinogen.

## Organophosphate Esters and Carbamates

Acetylcholinesterases are common targets for a host of xenobiotics, including various classes of pesticides. Biotransformation often plays a critical role in facilitating or impeding the binding of these compounds to cholinesterases. One class of pesticides whose toxicology is significantly affected by biotransformation is the organophosphate ester insecticides. In general, organophosphate esters possess a phosphorothionate group bound to at least two alkyl ethers (see Figure 3.18). The phosphorothionate is usually activated through oxidative desulfuration to an oxon, which irreversibly binds to the serine residues of the anionic site of cholinesterase (see Figure 3.19). The inability of cholinesterase to regenerate because of this covalent interaction results in irreversible binding and inactivation of the enzyme. The classical example is that of parathion, which is activated to paraoxon which inactivates acetylcholinesterase (Figure 3.20). This monooxygenation is classically catalyzed by cytochrome P450 isoforms, but depending on the electronegativity of adjacent functional groups, flavin-containing monooxygenases may also be involved in oxon formation (Levi and Hodgson, 1992). Alternatively, various hydrolytic processes (i.e., dearylation of parathion) may also occur which detoxify the compound, as each hydrolytic derivative does not inhibit cholinesterase. Hydrolysis may occur subsequent to or prior to desulfuration and is catalyzed by a host of carboxylesterases (Maxwell, 1992). An additional detoxification reaction involves the conjugation of certain compounds such as methyl parathion and methyl paraoxon with glutathione (Benke et al., 1974).

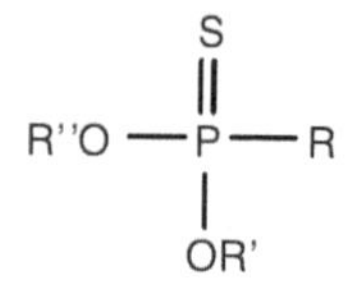


**FIGURE 3.18** Structure of organophosphate ester.

Although the toxicity of a host of organophosphates has been examined in fish, the biotransformation of these compounds and elucidation of specific inhibitory metabolites have only been examined in a very few cases. As these organophosphates demonstrate acute toxicity, it is likely that all are converted to oxonic metabolites, even though few have been characterized chromatographically. The most studied organophosphate ester in fish species is parathion. Several studies have reported paraoxon formation in a range of fish species including channel catfish (*Ictalurus punctatus*) (Straus et al., 2000), bluegill sunfish (*Lepomis macrochirus*), bullheads (*Ictalurus melas*), white flounder (*Pseudopleuronectes americanus*) (Hitchcock and Murphy, 1971), sculpin (*Leptacottus armatus*) (Murphy, 1966), rainbow trout (Abbas et al., 1996; Wallace and Dargan, 1987), and mosquitofish (*Gambusia affinis*) (Boone and Chambers, 1997). Studies in channel catfish have examined the relative differences between dearylation and desulfuration of parathion as well as chlorpyrifos and the role of specific CYP isoforms in these processes (Straus et al., 2000). Pretreatment of fish with Arochlor® 1254 failed to alter either reaction in chlorpyrifos or parathion, indicating CYP1A was not involved in dearylation or desulfuration of either compound. Diazinon is another insecticide in which *in vitro* and *in vivo* biotransformations have been

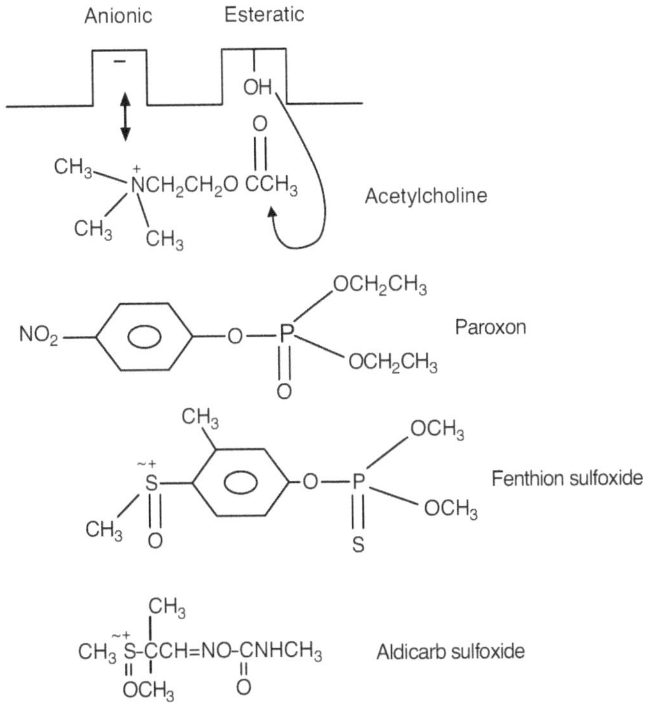


**FIGURE 3.19** Inhibition of acetylcholinesterase by aldicarb sulfoxide demonstrating partial positive-charge interactions with enzyme.

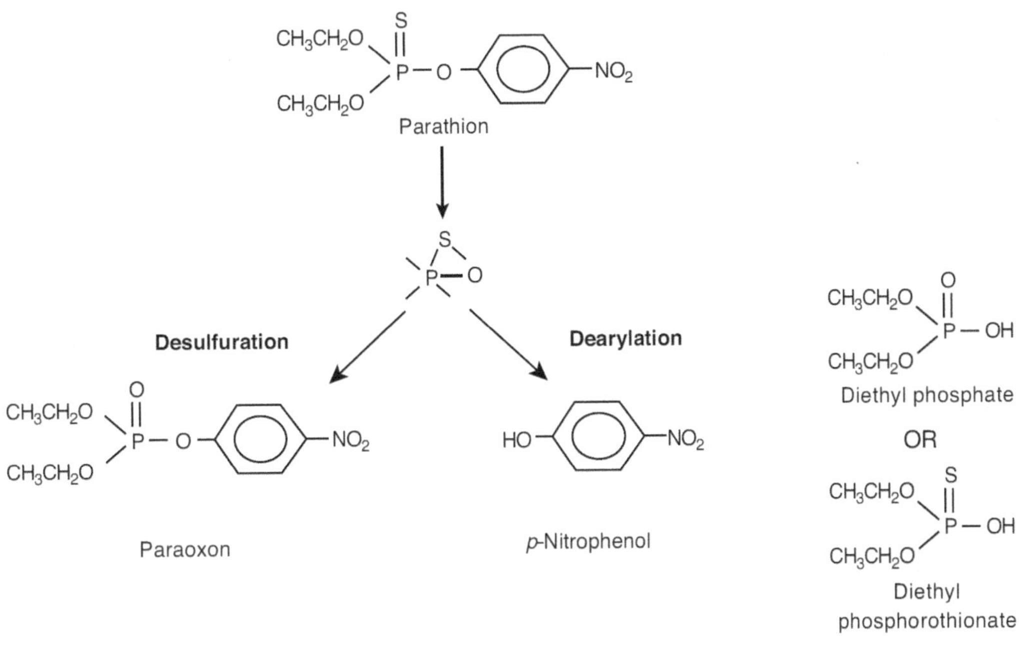


**FIGURE 3.20** Metabolic pathway of parathion in channel catfish. (From Straus et al., 2000. With permission.)

well documented in fish (Keizer et al., 1991, 1993). In fact, differences in oxidative desulfuration have been shown to be responsible for toxicity differences between fish species and developmental stages (Hamm et al., 1998).

Oxidative desulfuration is not the only transformation enhancing the inhibitory potency of cholinesterase-inhibiting xenobiotics. Sulfoxidation of xenobiotics anterior to the phosphorothionate moiety has also been shown to enhance the binding of the xenobiotic to the cationic site of cholinesterase (Figure 3.19). Recent studies in goldfish have shown that fenthion undergoes transformation to the sulfoxide and separately desulfuration to the oxon *in vivo* (Kitamura et al., 2000). *In vitro* studies using hepatopancreas microsomes from goldfish demonstrated conversion of fenthion to the sulfoxide, which was inhibited by SKF-525A and partially by alpha-naphthylthiourea, thus suggesting roles of CYP and FMO in the transformation (Kitamura et al., 1999). Studies with hepatopancreas cytosol indicated reduction of the sulfoxide to the sulfide through aldehyde oxidase (Kitamura et al., 1999). Fenthion sulfoxide was also observed in bioaccumulation studies using medaka treated with fenthion, but neither the sulfone nor oxon was observed (Tsuda et al., 1996).

Although aldicarb is a carbamate insecticide, oxygenation of the sulfur anterior to the carbamyl moiety significantly elevates the toxicity of this compound 40–150 times depending on the species (El-Alfy and Schlenk, 2002; El-Alfy et al., 2001; Perkins et al., 1999). *S*-Oxygenation of aldicarb has been observed in rainbow trout, Japanese medaka (*Oryzias latipes*), hybrid striped bass (*Morone saxatilis*×*chrysops*), channel catfish (*Ictalurus punctatus*), and rainbow trout (*Oncorhynchus mykiss*) (El-Alfy and Schlenk, 1998; Perkins and Schlenk, 2000; Perkins et al., 1999; Schlenk and Buhler, 1991; Wang et al., 2001). In each species except catfish, FMOs contributed to the *S*-oxygenation. As channel catfish and other predominantly freshwater fish do not have active FMOs (Schlenk et al., 1993), cytochrome P450 is exclusively responsible for this reaction (Perkins et al., 1999); however, no single CYP isoform appears to predominate in the sulfoxidation of aldicarb. Rather, it is likely that several isoforms participate in the reaction (Perkins et al., 1999). The reduced *S*-oxygenation of aldicarb in channel catfish relative to rainbow trout may explain its resistance against aldicarb toxicity compared to rainbow trout, which readily convert aldicarb to aldicarb sulfoxide (Perkins and Schlenk, 2000).

In summary, fish are capable of activating organophosphate esters to more potent cholinesterase inhibitors. Species differences in the bioactivation and esterase-mediated cleavage of organophosphates have been shown to greatly contribute to species differences in toxicity to these compounds. These data argue for more studies to better characterize specific enzymes responsible for these transformations to help identify sensitive populations of species that may be severely impacted by these compounds.

## CONCLUSIONS

Biotransformation can be a very important process in the disposition and mechanistic determinations of the mode of action of xenobiotics. Alterations in enzyme expression can dramatically affect the sensitivity of an organism to the toxic insult of a xenobiotic or the disposition of endogenous substrates. Alteration may occur as a result of genetics, diet, sex, environmental influences, or other xenobiotics. Understanding the latter is critical in risk evaluations of chemical mixtures. Very little is known regarding the substrate specificities or the regulation of biotransformation enzymes in fish. Through advances in genomic technologies and the use of fish models in human health research (e.g., zebrafish, medaka), numerous genotypic discoveries have recently occurred. It is likely that phenotypic functionality studies with heterologously expressed enzymes resulting from genomic examinations will help in better understanding biotransformation pathways in fish. In addition, more studies are needed with whole-animal systems to better characterize *in vivo* pathways of chemical biotransformation in species other than classic fish models. Such studies are imperative for physiologically based toxicokinetic (PBTK) models, which may help better estimate dose and aid regulators in reducing uncertainty between species, thus leading to more accurate evaluations of chemical risk.

## REFERENCES

Abbas, R. et al., Toxicokinetics of parathion and paraoxon in rainbow trout after intravascular administration and water exposure, *Toxicol. Appl. Pharmacol.*, 136, 194, 1996.

Acena, J. et al., Metabolite profiling of carbamazepine and ibuprofen in Solea senegalensis bile using high-resolution mass spectrometry, *Anal. Bioanal. Chem.*, 409, 5441–5450, 2017.

Adamson, R. H. et al., Comparative biochemistry of drug metabolism by azo and nitro reductase, *Proc. Natl. Acad. Sci. U.S.A.*, 54, 1386, 1965.

Agustsson, I. and Strom, A. R., Biosynthesis and turnover of trimethylamine oxide in the teleost cod *Gadus morhua*, *J. Biol. Chem.*, 256, 8045, 1981.

Alharthy, K. M. et al., Mechanistic evaluation of benzo(a)pyrene's developmental toxicities mediated by reduced CYP19a1b activity, *Toxicol. Sci.*, 155, 135–147, 2017.

Alin, P., Danielson, U. H., and Mannervik, B., 4-Hydroxyalk-2-enals are substrates for glutathione transferase, *FEBS Lett.*, 179, 267, 1985.

Andersson, T., Subcellular fractionation of rainbow trout gonads with emphasis on microsomal enzymes involved in steroid metabolism, *Cell Tissue Res.*, 268, 479, 1992.

Andersson, T., Pesonen, M., and Johansson, C., Differential induction of cytochrome P-450-dependent monooxygenase, epoxide hydrolase, glutathione transferase and UDP glucuronosyl transferase activities in the liver of the rainbow trout by beta-naphthoflavone or Clophen A50, *Biochem. Pharmacol.*, 34, 3309, 1985.

Aluru, N., et al., Targeted mutagenesis of aryl hydrocarbon receptor 2a and 2b genes in Atlantic killifish (Fundulus heteroclitus). *Aquat Toxicol.*, 158,192–201, 2015.

Angelucci, S. et al., Purification and characterization of glutathione transferases from the sea bass (*Dicentrarchus labrax*) liver, *Arch. Biochem. Biophys.*, 373, 435, 2000.

Armknecht, S. L., Kaattari, S. L., and Van Veld, P. A., An elevated glutathione *S*-transferase in creosoteresistant mummichog (*Fundulus heteroclitus*). *Aquat. Toxicol.*, 41, 1, 1998.

Arukwe, A., Forlin, L., and Goksøyr, A., Xenobiotic and steroid biotransformation enzymes in Atlantic salmon (*Salmo salar*) liver treated with an estrogenic compound, 4-nonylphenol. *Environ. Toxicol. Chem.*, 16, 2576, 1997.

Arukwe, A. and Goksøyr, A., Changes in three hepatic cytochrome P450 subfamilies during a reproductive cycle in turbot (*Scophthalmus maximus* L.), *J. Exp. Zool.*, 277, 313, 1997.

Assem, F. L., Kirk, C. J., and Chipman, J. K., Substrate characterisation of a recombinant sulfotransferase SULT1 and mRNA expression in chub (Leuciscus cephalus) tissues. *Biochem. Biophys. Res. Commun.*, 349, 900–905, 2006.

Baker, J. R., Streuempler, A., and Chaykin, S., A comparative study of trimethylamine-*N*-oxide biosynthesis, *Biochim. Biophys. Acta*, 71, 58, 1963.

Balk, L. et al., Initial characterization of drug-metabolizing systems in the liver of the northern pike, *Esox lucius*, *Drug Metab. Disp.*, 8, 98, 1980.

Barnhill, M. L. et al., Dieldrin stimulates biliary excretion of $^{14}$C-benzo(*a*)pyrene polar metabolites but does not change the biliary metabolite profile in rainbow trout (*Oncorhynchus mykiss*), *Toxicol. Sci.*, 75, 249, 2003.

Barron, M. G., Hansen, S. C., and Ball, T., Pharmacokinetics and metabolism of triclopyr in the crayfish (*Procambarus clarki*), *Drug Metab. Disp.*, 19, 163, 1991.

Barron, M. G., Schultz, I. R., and Hayton, W. L., Presystemic branchial metabolism limits di-2-ethylhexyl phthalate accumulation in fish, *Toxicol. Appl. Pharmacol.*, 98, 49, 1989.

Beetham, J. K. et al., Gene evolution of epoxide hydrolases and recommended nomenclature, *DNA Cell Biol.*, 14, 61, 1995.

Bello, S. M. et al., Acquired resistance to Ah receptor agonists in a population of Atlantic killifish (*Fundulus heteroclitus*) inhabiting a marine superfund site: *in vivo* and *in vitro* studies on the inducibility of xenobiotic metabolizing enzymes, *Toxicol. Sci.*, 60, 77, 2001.

Bend, J. R., James, M. O., and Dansette, P. M., *In vitro* metabolism of xenobiotics in some marine animals, *Ann. N.Y. Acad. Sci. U.S.A.*, 298, 505, 1978.

Benke, G. M. et al., Comparative toxicity, anticholinesterase action, and metabolism of methyl parathion and parathion in sunfish and mice, *Toxicol. Appl. Pharmacol.*, 28, 97, 1974.

Berndtson, A. K. and Chen, T. T., Two unique *CYP1* genes are expressed in response to 3-methylcholanthrene treatment in rainbow trout, *Arch. Biochem. Biophys.*, 310, 187, 1993.

Bertolini A, et al., Paracetamol: new vistas of an old drug. *CNS Drug Rev.* 12, 250–275, 2006.

Billiard, S. M. et al., Binding of polycyclic aromatic hydrocarbons (PAHs) to teleost aryl hydrocarbon receptors (ARHs), *Comp. Biochem. Physiol.*, 133, 55, 2002.

Birkholz, D. A., Coutts, R. T., and Hrudey, S. E., Uptake and biotransformation of 6,7-dimethylquinoline and 6,8-dimethylquinoline by rainbow trout (*Salmo gairdneri*), *Xenobiotica*, 19, 695, 1989.

Blanchard, R. L. et al., A proposed nomenclature system for the cytosolic sulfotransferase (SULT) superfamily, *Pharmacogenetics*, 14, 199, 2004.

Blom, S. and Forlin, L., Effects of PCB on xenobiotic biotransformation enzyme activities in the liver and 21-hydroxylation in the head kidney of juvenile rainbow trout, *Aquat. Toxicol.*, 39, 215, 1997.

Board, P. G. and Anders, M. W. Glutathione transferase zeta: discovery, polymorphic variants, catalysis, inactivation and properties of *GSTz1*$^{-/-}$ mice. *Drug Metabolism Reviews*, 43, 215, 2011.

Board, P. G. et al., Identification, characterization, and crystal structure of the omega class glutathione transferases, *J. Biol. Chem.*, 275, 24798, 2000.

Board, P. G. et al., Zeta, a novel class of glutathione transferases in a range of species from plants to humans, *Biochem. J.*, 328, 929, 1997.

Boone, J. S. and Chambers, J. E., Biochemical factors contributing to toxicity differences among chlorpyrifos, parathion, and methyl parathion in mosquitofish (*Gambusia affinis*), *Aquat. Toxicol.*, 39, 333, 1997.

Boyer, P.-M., Ndayibagira, A., and Spear, P. A., Dose-dependent stimulation of hepatic retinoic acid hydroxylation/oxidation and glucuronidation in brook trout, *Salvelinus fontinalis*, after exposure to 3,3′,4,4′-tetrachlorobiphenyl, *Environ. Toxicol. Chem.*, 19, 700, 2000.

Bradbury, S. P. and Christensen, G. M., Inhibition of alcohol dehydrogenase activity in acetylenic and allylic alcohols: concordance with *in vivo* electrophile reactivity in fish, *Environ. Toxicol. Chem.*, 10, 1155, 1991.

Bradbury, S. P. et al., Toxicokinetics and metabolism of aniline and 4-chloroaniline in medaka (*Oryzias latipes*), *Toxicol. Appl. Pharmacol.*, 118, 205, 1993.

Brammell, B. F. et al., Apparent lack of CYP1A response to high PCB body burdens in fish from a chronically contaminated PCB site, *Mar. Environ. Res.*, 58, 251, 2004.

Bresnick, E. H. et al., Effect of phenobarbital and 3-methylcholanthrene administration on epoxide hydrase levels in liver microsomes, *Biochem. Pharmacol.*, 26, 891, 1977.

Brown, D. R. et al., Zebrafish cardiotoxicity: the effects of CYP1A inhibition and AHR2 knockdown following exposure to weak aryl hydrocarbon receptor agonists. *Environ. Sci. Pollut. Res. Int.*, 22, 8329–8338, 2015.

Brox, S. et al., Metabolism of clofibric acid in zebrafish embryos (Danio rerio) as determined by liquid chromatography-high resolution-mass spectrometry. *Comp. Biochem. Physiol. C Toxicol. Pharmacol.*, 185–186, 20–28, 2016.

Brozinski, J. M. et al., Identification and dose dependency of ibuprofen biliary metabolites in rainbow trout. *Chemosphere*, 93, 1789–1795, 2013.

Brumley, C.-M. et al., Validation of biomarkers of marine pollution exposure in sand flathead using Aroclor 1254. *Aquat. Toxicol.*, 31, 249, 1995.

Bucheli, T. D. and Fent, K., Induction of cytochrome P450 as a biomarker for environmental contamination in aquatic ecosystems, *Crit. Rev. Environ. Sci. Technol.*, 25, 201, 1995.

Buetler, T. M. et al., Induction of phase I and phase II drug-metabolizing enzyme mRNA, protein, and activity by BHA, ethoxyquin, and oltipraz, *Toxicol. Appl. Pharmacol.*, 135, 45, 1995.

Buhler, D. R. and Wang-Buhler, J. L., Rainbow trout cytochrome P450s: purification, molecular aspects, metabolic activity, induction and role in environmental monitoring, *Comp. Biochem. Physiol.*, 121C, 107, 1998.

Buhler, D. R. and Williams, D. E., Enzymes involved in metabolism of PAH by fishes and other aquatic animals: oxidative enzymes (or phase I enzymes), in *Metabolism of Polycyclic Aromatic Hydrocarbons in the Aquatic Environment*, Varanasi, U., Ed., CRC Press, Boca Raton, FL, 1989, pp. 151–184.

Buhler, D. R. et al., Cloning and sequencing of the major rainbow trout constitutive cytochrome P450 (CYP2K1): identification of a new cytochrome P450 gene subfamily and its expression in mature rainbow trout liver and trunk kidney, *Arch. Biochem. Biophys.*, 312, 45, 1994.

Buhler, D. R. et al., Effects of 17β-estradiol and testosterone on hepatic mRNA/protein levels and catalytic activities of CYP2M1, CYP2K1, and CYP3A27 in rainbow trout (*Oncorhynchus mykiss*), *Toxicol. Appl. Pharmacol.*, 168, 91, 2000.

Buhler, J.-L. et al., Expression of a constitutive cytochrome P450 (CYP2K1) in livers of rainbow trout (*Oncorhynchus mykiss*) embryo and sac fry, *Aquat. Toxicol.*, 37, 237, 1997.

Burchell, B. and Coughtrie, M. W., UDP-glucuronosyltransferases, *Pharmacol. Therapeut.*, 43, 261, 1989.

Burke, A. B. et al., Formation of the taurine conjugate of benzoic acid in rainbow trout, Salmo gairdneri, *Drug Metab. Dispos.*, 15, 581–582, 1987.

Butala, H. et al., Microsomal estrogen metabolism in channel catfish, *Mar. Environ. Res.*, 58, 489, 2004.

Cappiello, M. et al., Distribution of 2-naphthol sulphotransferase and its endogenous substrate adenosine 3'-phosphate 5'-phosphosulphate in human tissues, *Eur. J. Clin. Pharmacol.*, 37, 317–320, 1989.
Carletti, E. et al., Glutathione transferases from Anguilla anguilla liver: identification, cloning and functional characterization, *Aquat. Toxicol.*, 90, 48–57, 2008.
Carlson, D. B. and Perdew, G. H., A dynamic role for the Ah receptor in cell signaling? Insights from a diverse group of Ah receptor interacting proteins, *J. Biochem. Mol. Toxicol.*, 16, 317, 2002.
Carney, S. A., Peterson, R. E., and Heideman, W., 2,3,7,8-Tetrachlorodibenzo-p-dioxin activation of the aryl hydrocarbon receptor/aryl hydrocarbon receptor nuclear translocator pathway causes developmental toxicity through a CYP1A-independent mechanism in zebrafish, *Mol. Pharmacol.*, 66, 512–521, 2004.
Cashman, J. R. et al., *S*-oxygenation of thiobencarb (Bolero) in hepatic preparations from striped bass (*Morone saxatilis*) and mammalian systems, *Chem. Res. Toxicol.*, 3, 433, 1990.
Castren, M. and Oikari, A., Changes of the liver UDP-glucuronosyltransferase activity in trout (*Salmo gairdneri* Rich.) acutely exposed to selected aquatic toxicants, *Comp. Biochem. Physiol.*, 86C, 357, 1987.
Celander, M. and Forlin, L., Decreased responsiveness of the hepatic cytochrome-P450 1A1 system in rainbow trout (*Oncorhynchus mykiss*) after prolonged exposure to PCB, *Aquat. Toxicol.*, 33, 141, 1995.
Celander, M., Hahn, M. E., and Stegeman, J. J., Cytochromes P450 (CYP) in the *Poeciliopsis lucida* hepatocellular carcinoma cell line (PLHC-1): dose- and time-dependent glucocorticoid potentiation of CYP1A induction without induction of CYP3A, *Arch. Biochem. Biophys.*, 329, 113, 1996a.
Celander, M., Ronis, M., and Forlin, L., Initial characterization of a constitutive cytochrome P450 isoenzyme in rainbow trout liver, *Mar. Environ. Res.*, 28, 9, 1989.
Celander, M. and Stegeman, J. J., Isolation of a cytochrome P450 3A cDNA sequence (CYP3A30) from the marine teleost *Fundulus heteroclitus* and phylogenetic analyses of CYP3A genes, *Biochem. Biophys. Res. Comm.*, 236, 306, 1997.
Celander, M. et al., Induction of cytochrome P450 1A1 and conjugating enzymes in rainbow trout (*Oncorhynchus mykiss*) liver: a time course study. *Comp. Biochem. Physiol.*, 106C, 343–349, 1993.
Celander, M. et al., Immunochemical relationships of cytochrome P450-like protein in teleost fish, *Fish Physiol. Biochem.*, 15, 323, 1996b.
Chang, C. T. et al., Regulation of zebrafish CYP3A65 transcription by AHR2, *Toxicol. Appl. Pharmacol.*, 270, 174–184, 2013.
Charest, R. P., Chenoweth, M., and Dunn, A., Metabolism of trimethylamines in kelp bass (*Paralabrax clathratus*) and marine and freshwater pink salmon (*Oncorhynchus gorbuscha*), *J. Comp. Physiol. B.*, 158, 609, 1988.
Chng, H. T., Ho, H. K., Yap, C. W., Lam, S. H., and Chan, E. C. An investigation of the bioactivation potential and metabolism profile of Zebrafish versus human, *J. Biomol. Screen.*, 17, 974–986, 2012.
Christ, E. J. and Van Dorp, D. A., Comparative aspects of prostaglandin biosynthesis in animal tissues, *Biochim. Biophys. Acta*, 270, 537, 1972.
Christen, V. and Fent, K., Tissue-, sex- and development-specific transcription profiles of eight UDP-glucuronosyltransferase genes in zebrafish (Danio rerio) and their regulation by activator of aryl hydrocarbon receptor, *Aquat Toxicol.*, 150, 93–102, 2014.
Clarke, D. J., Burchell, B., and George, S. G., Differential expression and induction of UDP-glucuronosyltransferase isoforms in hepatic and extrahepatic tissues of a fish: immunochemical and functional characterization, *Toxicol. Appl. Pharmacol.*, 115, 130, 1992b.
Clarke, D. J., Burchell, B., and George, S. G., Functional and immunological comparison of hepatic UDPglucuronosyltransferases in a piscine and a mammalian species, *Comp. Biochem. Physiol.*, 102B, 425, 1992a.
Clarke, D. J., George, S. G., and Burchell, B., Multiplicity of UDP-glucuronosyltransferases in fish: purification and characterization of a phenol UDP-glucuronosyltransferase from the liver of a marine teleost, *Pleuronectes platessa*, *Biochem. J.*, 284, 417, 1992c.
Cok, I. et al., Expression of CYP2M1, CYP2K1, and CYP3A27 in brain, blood, small intestine, and other tissues of rainbow trout, *Biochem. Biophys. Res. Comm.*, 244, 790, 1998.
Collier, T. K. and Varanasi, U., Hepatic activities of xenobiotic metabolizing enzymes and biliary levels of xenobiotics in English sole (*Parophrys vetulus*) exposed to environmental contaminants, *Arch. Environ. Contam. Toxicol.*, 20, 462, 1991.
Collier, T. K. et al., Xenobiotic metabolizing enzymes in spawning English sole (*Parophrys vetulus*) exposed to organic-solvent extracts of marine sediments from contaminated and reference areas, *Comp. Biochem. Physiol.*, 84C, 291, 1986.
Cook, I. et al., A nucleotide-gated molecular pore selects sulfotransferase substrates, *Biochemistry*, 51, 5674–5683, 2012.
Cooke, J. B. and Hinton, D. E., Promotion by 17 beta-estradiol and beta-hexachlorocyclohexane of hepatocellular tumors in medaka (*Oryzias latipes*), *Aquat. Toxicol.*, 45, 127, 1999.
Corley-Smith G.E., et al., CYP3C1, the first member of a new cytochrome P450 subfamily found in zebrafish (Danio rerio). *Biochem. Biophys. Res. Commun.*, 340, 1039–1046, 2006.
Coughtrie, M. W. H. 2016. Function and organization of the human cytosolic sulfotransferase (SULT) family, *Chem. Biol. Interact.*, 259, 2–7, 2016.
Coumoul, X., Diry, M., and Barouki, R., PXR-dependent induction of human CYP3A4 gene expression by organochlorine pesticides, *Biochem. Pharmacol.*, 64, 1513, 2002.
Cravedi, J. P. and Tulliez, J., Urinary metabolites of dodecylcyclohexane in Salmo gairdneri: evidence of aromatization and taurine conjugation in trout, *Xenobiotica*, 17, 1103–1111, 1987.
Cravedi, J.-P. et al., Influence of growth hormone on the hepatic mixed function oxidase and transferase systems of rainbow trout, *Fish Physiol. Biochem.*, 14, 259, 1995.
Crittenden, F. et al., Activity Suppression Behavior Phenotype in SULT4A1 Frameshift Mutant Zebrafish, *Drug Metab. Dispos.*, 43, 1037–1044, 2015.
Daikoku, T., Murata, M., and Sakaguchi, M., Effects of intraperitoneally injected and dietary trimethylamine on the biosynthesis of trimethylamine oxide in relation to seawater adaptation of the eel *Anguilla japonica* and the guppy *Poecilia reticulata*, *Comp. Biochem. Physiol.*, 89A, 261, 1988.
Daikoku, T. and Sakaguchi, M., Changes in level of trimethylamine and trimethylamine oxide during adaptation of young eel *Anguilla anguilla* to a seawater environment, *Nippon Suisan Gakkaishi*, 56, 1895, 1990.
Danielsson, O., Eklund, H., and Joernvall, H., The major piscine liver alcohol dehydrogenase has class-mixed properties in relation to mammalian alcohol dehydrogenases of classes I and III, *Biochemistry*, 31, 3751, 1992.

Dansette, P. M., DuBois, G. C., and Jerina, D. M., Continuous fluorometric assay of epoxide hydrase activity, *Anal. Biochem.*, 97, 340, 1979.

Dasmahapatra, A. K. et al., Developmental expression of alcohol dehydrogenase (ADH3) in zebrafish (*Danio rerio*), *Biochem. Biophys. Res. Comm.*, 286, 1082, 2001.

De Anna, J. S. et al., The insecticide chlorpyrifos modifies the expression of genes involved in the PXR and AhR pathways in the rainbow trout, Oncorhynchus mykiss. *Pest. Biochem. Physiol.*, 178, 104920, 2021.

de Wolf, W., Seinen, W., and Hermens, J. L., *N*-acetyltransferase activity in rainbow trout liver and *in vitro* biotransformation of chlorinated anilines and benzenes in fish, *Xenobiotica*, 23, 1045, 1993.

DeBusk, B., Biochemical and Molecular Responses of Aquatic Organisms to Diet-Derived Defensive Secondary Metabolites: Characterization of Critical Biotransformation Enzymes, Ph.D. dissertation, University of Mississippi, 2001.

DeBusk, B., Slattery, M., Ki, J.., Lee, J.S., Aparicio-Fabre, R., and Schlenk D. Species differences and effects of soft coral extracts from *Sinnularia maximus* on the expression of cytochrome P4501A and 2N in butterflyfishes (*Chaetodon* spp.). *Fish Physiol. Biochem.* 34, 483–492, 2008.

Deeley, R. G., Westlake, C., and Cole, S. P. Transmembrane transport of endo- and xenobiotics by mammalian ATP-binding cassette multidrug resistance proteins, *Physiol. Rev.*, 86, 849–899, 2006.

Denison, M. S. and Nagy, S. R., Activation of the aryl hydrocarbon receptor by structurally diverse exogenous and endogenous chemicals, *Ann. Rev. Pharmacol. Toxicol.*, 43, 309, 2003.

Dhers, L. et al., Cytochrome P450 2U1, a very peculiar member of the human P450s family. *Cell. Mol. Life Sci.*, 74, 1859–1869, 2017.

Doi, A. M. et al., Molecular cloning and characterization of a glutathione *S*-transferase from largemouth bass (*Micropterus salmoides*) liver that is involved in the detoxification of 4-hydroxynonenal, *Biochem. Pharmacol.*, 67, 2129, 2004.

Dong, A. N. et al., The CYP2R1 enzyme: structure, function, enzymatic properties and genetic polymorphism, *J. Pharm. Pharm. Sci.*, 24, 94–112, 2021.

Driessen, M. et al., Gene expression markers in the zebrafish embryo reflect a hepatotoxic response in animal models and humans, *Toxicol. Lett.*, 230, 48–56, 2014.

Droy, B. F. et al., Influence of ormetoprim on the bioavailability, distribution, and pharmacokinetics of sulfadimethoxine in rainbow trout (*Oncorhynchus mykiss*), *Comp. Biochem. Physiol.*, 94C, 303, 1989.

Dutton, G. J., *Glucuronidation of Drugs and Other Compounds*, CRC Press, Boca Raton, FL. 1990.

Eide, M. et al., Independent losses of a xenobiotic receptor across teleost evolution, *Sci. Rep.*, 8, 10404, 2018.

Eide, M., et al., The chemical defensome of five model teleost fish. *Sci. Rep.* 11, 10546, 2021.

Eisele, T. A. et al., The effects of route of exposure and combined exposure of mixed function oxidase inducers and suppressors on hepatic parameters in rainbow trout (*Salmo gairdneri*), *Aquat. Toxicol.*, 5, 211, 1984.

El-Alfy, A., Larsen, B. K., and Schlenk, D., Effect of cortisol and urea on flavin monooxygenase activity and expression in rainbow trout, *Oncorhynchus mykiss*, *Mar. Environ. Res.*, 54, 275, 2002.

El-Alfy, A. and Schlenk, D., Effect of 17β-estradiol and testosterone on the toxicity of aldicarb to Japanese medaka, *Oryzias latipes*, *Toxicol. Sci.*, 68, 381, 2002.

El-Alfy, A. and Schlenk, D., Potential mechanisms of the enhancement of aldicarb toxicity to Japanese medaka, *Oryzias latipes*, at high salinity, *Toxicol. Appl. Pharmacol.*, 152, 175, 1998.

El-Alfy, A. et al., Characterization of salinity enhanced toxicity of aldicarb to Japanese medaka: sexual and developmental differences, *Environ. Toxicol. Chem.*, 20, 2093, 2001.

Elmarakby, S.-A. et al., Comparative metabolism of 2-acetylaminofluorene by rainbow trout and rat liver subcellular fractions, *Aquat. Toxicol.*, 33, 1, 1995.

Elskus, A. A. and Stegeman, J. J., Further consideration of phenobarbital effects on cytochrome P-450 activity in the killifish, *Fundulus heteroclitus*, *Comp. Biochem. Physiol.*, 92C, 223, 1989.

Enright, J. M. et al., Cyp27c1 Red-Shifts the Spectral Sensitivity of Photoreceptors by Converting Vitamin A1 into A2, *Curr Biol.*, 25, 3048–3057, 2015.

Espinoza, H.M., Shireman, L. M., McClain, V., Atkins, W., Gallagher, E. P. Cloning, expression and analysis of the olfactory glutathione S-transferases in coho salmon, *Biochem. Pharmacol.*, 85(6), 839–848, 2013. doi: 10.1016/j.bcp.2012.11.018. Epub 2012 Dec 19. PMID: 23261526; PMCID: PMC3660137.

Esterbauer, H., Schaur, R. J., and Zollner, H., Chemistry and biochemistry of 4-hydroxynonenal, malonaldehyde and related aldehydes, *Free Radic. Biol. Med.*, 11, 81, 1991.

Falany, C. N. et al., Purification and characterization of human liver phenol-sulfating phenol sulfotransferase, *Arch. Biochem. Biophys.*, 278, 312, 1990.

Faltermann, S. et al., Effects of the glucocorticoid clobetasol propionate and its mixture with cortisol and different class steroids in adult female zebrafish. *Aquat Toxicol.* 218, 105372, 2020.

Fernando, H., Halpert, J. R., and Davydov, D. R., Resolution of multiple substrate binding sites in cytochrome P450 3A4: the stoichiometry of the enzyme-substrate complexes probed by FRET and Job's titration, *Biochemistry*, 45, 4199–4209, 2006.

Finnson, K. W. and Eales, J. G., Effect of T(3) treatment and food ration on hepatic deiodination and conjugation of thyroid hormones in rainbow trout, *Oncorhynchus mykiss*, *Gen. Comp. Endocrinol.*, 115, 379, 1999.

Finnson, K. W. and Eales, J. G., Sulfation of thyroid hormones by liver of rainbow trout, *Oncorhynchus mykiss*, *Comp. Biochem. Physiol.*, 120C, 415, 1998.

Forlin, L. et al., Effects on UDP glucuronosyl transferase, glutathione transferase, DT-diaphorase and glutathione reductase activities in rainbow trout liver after long-term exposure to PCB, *Mar. Environ. Res.*, 42, 213, 1996.

Foureman, G. L. et al., The stereoselectivity of four hepatic glutathione *S*-transferases purified from a marine elasmobranch (*Raja erinacea*) with several K-region polycyclic arene oxide substrates, *Biochim. Biophys. Acta.*, 914, 127, 1987.

Fu, Q. et al., Biotransformation changes bioaccumulation and toxicity of diclofenac in aquatic organisms. *Environ. Sci. Technol.*, 54, 4400–4408, 2020.

Funkenstein, B. and Jakowiew, S. B., Molecular cloning of fish alcohol dehydrogenase cDNA, *Gene*, 174, 159, 1996.

Gadagbui, B. K. and James, M. O., Activities of affinity-isolated glutathione *S*-transferase (GST) from channel catfish whole intestine, *Aquat. Toxicol.*, 49, 27, 2000.

Gallagher, E. P. et al., Channel catfish glutathione *S*-transferase isoenzyme activity toward (+)-*anti*-benzo(*a*)pyrene-trans-7,8-dihydrodiol-9m10-epoxide, *Aquat. Toxicol.*, 34, 135, 1996.

Gallagher, E. P. et al., *In vitro* kinetics of hepatic glutathione *S*-transferase conjugation in largemouth bass and brown bullheads, *Environ. Toxicol. Chem.*, 19, 319, 2000.
Gallagher, E. P. et al., Isolation and cloning of homologous glutathione *S*-transferase cDNAs from English sole and starry flounder liver, *Aquat. Toxicol.*, 44, 171, 1998.
Gamage, N., Human sulfotransferases and their role in chemical metabolism, *Toxicol. Sci.*, 90, 5–22, 2006.
Garfinkel, D., Studies on pig liver microsomes 1. Enzyme and pigment composition of different microsomal fractions, *Arch. Biochem. Biophys.*, 77, 493, 1958.
Garner, L. V., Brown, D. R., Di Giulio, R. T. Knockdown of AHR1A but not AHR1B exacerbates PAH and PCB-126 toxicity in zebrafish (Danio rerio) embryos, *Aquat. Toxicol.*, 142–143, 336–346, 2013.
Gaworecki, K. M., Rice, C. D., and van den Hurk, P., Induction of phenol-type sulfotransferase and glucuronosyltransferase in channel catfish and mummichog, *Mar. Environ. Res.*, 58, 525, 2004.
George, S. G., Enzymology and molecular biology of phase II xenobiotic-conjugating enzymes in fish, in *Aquatic Toxicology: Molecular, Biochemical and Cellular Perspectives*, Ostrander G. K. and Malins, D. C., Eds., CRC Press, Boca Raton, FL, 1994, pp. 37–85.
George, S. G. and Buchanan, G., Isolation, properties and induction of plaice liver cytosolic glutathione-Stransferases. *Fish Physiol. Biochem.*, 8, 437, 1990.
George, S. G. and Leaver, M. J., Allelic variations in the plaice UGT1B1 gene, *Mar. Environ. Res.*, 54, 259, 2002.
George, S. G., Leaver, M. J., and Wright, J., Structural studies of a UDP-glucuronosyltransferase gene from the plaice (*Pleuronectes platessa*), *Mar. Environ. Res.*, 46, 33, 1998.
George, S. G. and Taylor, A., Molecular evidence for multiple UDP-glucuronosyltransferase gene families in fish, *Mar. Environ. Res.*, 54, 253, 2002.
George, S. G. and Young, P., The time course of effects of cadmium and 3-methylcholanthrene on activities of enzymes of xenobiotic metabolism and metallothionein levels in the plaice, *Pleuronectes platessa*, *Comp. Biochem. Physiol.*, 83C, 37, 1986.
German, J. B. and Berger, R., Formation of 8,15-dihydroxy eicosatetraenoic acid via 15- and 12-lipoxygenases in fish, *Lipids*, 25, 849, 1990.
Gill, S. S., Ota, K., and Hammock, B. D., Radiometric assays for mammalian epoxide hydrolases and glutathione *S*-transferase, *Anal. Biochem.*, 131, 273, 1983.
Gill, S. S. et al., Rapid and sensitive enzyme-linked immunosorbent assay for the microsomal epoxide hydrolase, *Carcinogenesis*, 3, 1307, 1982.
Gillam, E. M. et al., Expression of modified human cytochrome P450 3A4 in *Escherichia coli* and purification and reconstitution of the enzyme, *Arch. Biochem. Biophys.*, 305, 123, 1993.
Gillett, M. B. et al., Elevated levels of trimethylamine oxide in muscles of deep-sea gadiform teleosts: a high pressure adaptation, *J. Exp. Zool.*, 279, 386, 1997.
Gillette, D. M. et al., Comparative toxicology of tetrachlorobiphenyls in mink and rats. I. Changes in hepatic enzyme activity and smooth endoplasmic reticulum volume, *Fundam. Appl. Toxicol.*, 8, 5, 1987.
Gilligan, L. C. et al., Estrone sulfate transport and steroid sulfatase activity in colorectal cancer: implications for hormone replacement therapy, *Front. Pharmacol.*, **8**, 103, 2017.
Glickman, A. H., Weitman, S. D., and Lech, J. J., Differential toxicity of *trans*-permethrin in rainbow trout and mice. I. Role of biotransformation, *Toxicol. Appl. Pharmacol.*, 66, 153, 1982.
Glickman, A. H. et al., *In vitro* metabolism of permethrin isomers by carp and rainbow trout liver microsomes, *J. Agric. Food Chem.*, 27, 1038, 1979.
Glisic, B. et al., Characterization of glutathione-S-transferases in zebrafish (Danio rerio). *Aquat. Toxicol.*, 158, 50–62, 2015.
Goksøyr, A. et al., Species characteristics of hepatic xenobiotic and steroid biotransformation systems of two teleost fish, Atlantic cod (*Gadus morhua*) and rainbow trout (*Salmo gairdneri*), *Toxicol. Appl. Pharmacol.*, 89, 347, 1987.
Goldstein, L. and Dewitt-Harley, S., Trimethylamine oxidase of nurse shark liver and its relation to mammalian mixed function amine oxidase, *Comp. Biochem. Physiol.*, 45B, 895, 1973.
Goldstone, J.V., et al., Cytochrome P450 1 genes in early deuterostomes (tunicates and sea urchins) and vertebrates (chicken and frog): origin and diversification of the CYP1 gene family. *Mol. Biol. Evol.*, 24, 2619–2631, 2007.
Goldstone, J. V. et al., Cytochrome P450 1D1: a novel CYP1A-related gene that is not transcriptionally activated by PCB126 or TCDD, *Arch. Biochem. Biophys.*, 482, 7–16, 2009.
Goldstone, J.V., et al., Identification and developmental expression of the full complement of Cytochrome P450 genes in Zebrafish. *BMC Genomics* 11, 643, 2010.
Gonzalez, F. J. and Gelboin, H. V., Role of human cytochrome P450s in risk assessment and susceptibility to environmentally based disease, *J. Toxicol. Environ. Health*, 40, 289, 1993.
Gooch, J. W. et al., Effects of *ortho*- and non-*ortho*-substituted polychlorinated biphenyl congeners on the hepatic monooxygenase system in the scup (*Stenotomus chrysops*), *Toxicol. Appl. Pharmacol.*, 98, 422, 1989.
Gotoh, O., Substrate recognition sites in cytochrome P450 family 2 (CYP2) proteins inferred from comparative analysis of amino acid and coding nucleotide sequences, *J. Biol. Chem.*, 267, 83, 1992.
Gravato, C. and Santos, M. A., Liver phase I and phase II enzymatic induction and genotoxic responses of beta-naphthoflavone water-exposed sea bass, *Ecotoxicol. Environ. Safety*, 52, 62, 2002.
Gray, E. S., Woodin, B. R., and Stegeman, J. J., Sex differences in hepatic monooxygenases in winter flounder (*Pseudopleuronectes americanus*) and scup (*Stenotomus chrysops*) and regulation of P450 forms by estradiol, *J. Exp. Zool.*, 259, 330, 1991.
Gregus, Z. et al., Hepatic phase I and phase II biotransformations in quail and trout: comparison to other species commonly used in toxicity testing, *Toxicol. Appl. Pharmacol.*, 67, 430, 1983.
Guarino, A. M., James, M. O., and Bend, J. R., Fate and distribution of the herbicides 2,4-dichlor-phenoxyacetic acid (2,4–D) and 2,4,5–trichlorophenoxyacetic acid (2,4,5–T) in the dogfish shark, *Xenobiotica*, 7, 623–631, 1977.
Gunal, S. et al, Sulfation pathways from red to green, *J. Biol. Chem.*, 294**,** 12293–12312, 2019.
Haasch, M. L., Henderson, M. C., and Buhler, D. R., Induction of lauric acid hydroxylase activity in catfish and bluegill by peroxisome proliferating agents, *Comp. Biochem. Physiol.*, 121C, 297, 1998.
Haasch, M. L. et al., Effect of acrylamide monomer on hepatic CYP1A1 monooxygenase induction in rainbow trout, *Comp. Biochem. Physiol.*, 102C, 281, 1992.
Haasch, M. L. et al., Use of 7-alkoxyphenoxazones, 7-alkoxycoumarins, and 7-alkoxyquinolines as fluorescent substrates for rainbow trout hepatic microsomes after treatment with various inducers, *Biochem. Pharmacol.*, 47, 893, 1994.

Habig, W. H. and Jakoby, W. B., Assays for differentiation of glutathione *S*-transferase, *Methods Enzymol.*, 77, 398, 1981.

Hamm, J. T., Wilson, B. W., and Hinton, D. E., Organophosphate-induced acetylcholinesterase inhibition and embryonic retinal cell necrosis *in vivo* in the teleost (*Oryzias latipes*), *Neurotoxicology*, 19, 853, 1998.

Hammock, B. D. and Hasagawa, L. S., Differential substrate selectivity of murine hepatic cytosolic and microsomal epoxide hydrolases, *Biochem. Pharmacol.*, 32, 1155, 1983.

Harlow, G. R. and Halpert, J. R., Alanine-scanning mutagenesis of a putative substrate recognition site in human cytochrome P450 3A4. Role of residues 210 and 211 in flavonoid activation and substrate specificity, *J. Biol. Chem.*, 272, 5396, 1997.

Harlow, G. R. and Halpert, J. R., Analysis of human cytochrome P450 3A4 cooperativity: construction and characterization of a site-directed mutant that displays hyperbolic steroid hydroxylation kinetics, *Proc. Natl. Acad. Sci. U.S.A.*, 95, 6636, 1998.

Hasselberg, L. et al., Effects of alkylphenols on CYP1A and CYP3A expression in first spawning Atlantic cod (*Gadus morhua*), *Aquat. Toxicol.*, 67, 303, 2004.

Hasselberg, L. et al., Interactions between xenoestrogens and ketoconazole on hepatic CYP1A and CYP3A in juvenile Atlantic cod (*Gadus morhua*), *Comp. Hepatol.*, 4, 2, 2005.

Hasspieler, B. M. and Di Giulio, R. T., Dicoumarol-sensitive NADPH: phenanthrenequinone oxidoreductase in channel catfish (*Ictalurus punctatus*), *Toxicol. Appl. Pharmacol.*, 125, 184, 1994.

Hasspieler, B. M., Haffner, G. D., and Adeli, K., Roles of DT diaphorase in the genotoxicity of nitroaromatic compounds in human and fish cell lines, *J. Toxicol. Environ. Health*, 52, 137, 1997.

Hawkins, W. E. et al., Carcinogenic effects of 1,2-dibromoethane (ethylene dibromide; EDB) in Japanese medaka (*Oryzias latipes*), *Mutat. Res.*, 399(2), 221, 1998.

Hayes, J. D. and Pulford, D. J., The glutathione *S*-transferase supergene family: regulation of GST and the contribution of the isoenzymes to cancer chemoprotection and drug resistance, *Crit. Rev. Biochem. Mol. Biol.*, 30, 445, 1995.

Hegelund, T. and Celander, M. C., Hepatic versus extrahepatic expression of CYP3A30 and CYP3A56 in adult killifish (*Fundulus heteroclitus*). *Aquat. Toxicol.*, 64, 277, 2003.

Hegelund, T. et al., Effects of the antifungal imidazole ketoconazole on CYP1A and CYP3A in rainbow trout and killifish, *Environ. Toxicol. Chem.*, 23, 1326, 2004.

Heilmann, L. J. et al., Trout P450IA1: cDNA and deduced protein sequence, expression in liver, and evolutionary significance, *DNA*, 7, 379, 1988.

Henne, K. R. et al., Covalent linkage of prosthetic heme to CYP4 family P450 enzymes, *Biochemistry*, 40, 12925, 2001.

Henson, K. L., Stauffer, G., and Gallagher, E. P., Induction of glutathione *S*-transferase activity and protein expression in brown bullhead (*Ameiurus nebulosus*) liver by ethoxyquin, *Toxicol. Sci.*, 62, 54, 2001.

Hines, R. N. et al., The mammalian flavin-containing monooxygenases: molecular characterization and regulation of expression, *Toxicol. Appl. Pharmacol.*, 125, 1, 1994.

Hitchcock, M. and Murphy, S. D., Activation of parathion and guthion by mammalian, avian, and piscine liver homogenates and cell fractions, *Toxicol. Appl. Pharmacol.*, 19, 37, 1971.

Honkakoski, P. and Negishi, M., Characterization of a phenobarbital-responsive enhancer module in mouse P450 CYP2B10 gene, *J. Biol. Chem.*, 272, 14943, 1997.

Honkakoski, P. et al., Activation by diverse xenochemicals of the 51-base-pair phenobarbital-responsive enhancer module in the CYP2B10 gene, *Mol. Pharmacol.*, 53, 597, 1998a.

Honkakoski, P. et al., The orphan nuclear receptor CAR-retinoid X receptor heterodimer activates the phenobarbital-responsive enhancer module of the CYP2B gene, *Mol. Cell. Biol.*, 18, 5652, 1998b.

Hossain, M. I. et al., SULT4A1 protects against oxidative-stress induced mitochondrial dysfunction in neuronal cells, *Drug Metab. Disp.*, 47, 949–953, 2019.

Houston, J. B. and Kenworthy, K. E., *In vitro-in vivo* scaling of CYP kinetic data not consistent with the classical Michaelis-Menten model, *Drug Metab. Disp.*, 28, 246, 2000.

Hsieh, R. J., German, J. B., and Kinsella, J. E., Lipoxygenase in fish tissue: some properties of the 12-lipoxygenase from trout gill, *J. Agric. Food Chem.*, 36, 680, 1988.

Hsu, M. H., Savas, U., and Johnson, E. F., The X-ray crystal structure of the human mono-oxygenase cytochrome P450 3A5-ritonavir complex reveals active site differences between P450s 3A4 and 3A5, *Mol. Pharmacol.*, 93, 14–24, 2018.

Huang, H. and Wu, Q., Cloning and comparative analyses of the zebrafish Ugt repertoire reveal its evolutionary diversity, *PLoS One*, 5, e9144, 2010.

Hubatsch, I., Ridderstrom, M., and Mannervik, B., Human glutathione transferase A4-4: an alpha class enzyme with high catalytic efficiency in the conjugation of 4-hydroxynonenal and other genotoxic products of lipid peroxidation, *Biochem. J.*, 330, 175, 1998.

Husoy, A. M. et al., Immunohistochemical localization of CYP1A and CYP3A-like isozymes in hepatic and extrahepatic tissues of Atlantic cod (*Gadus morhua* L.), a marine fish, *Toxicol. Appl. Pharmacol.*, 129, 294, 1994.

Huuskonen, S. et al., Effects of non-*ortho*-substituted polychlorinated biphenyls (congeners 77 and 126) on cytochrome P4501A and conjugation activities in rainbow trout (*Oncorhynchus mykiss*), *Comp. Biochem. Physiol.*, 113C, 205, 1996.

Iwata, H. et al., Species-specific responses of constitutively active receptor (CAR)-CYP2B coupling: lack of CYP2B inducer-responsive nuclear translocation of CAR in marine teleost, scup (*Stenotomus chrysops*), *Comp. Biochem. Physiol.*, 131C, 501, 2002.

James, M. O., Conjugation of organic pollutants in aquatic species, *Environ. Health Perspect.*, 71, 97–103, 1987.

James, M. O. Drug metabolism: phase II enzymes, in *Pharmacokinetics*, Riddick, D. S., Ed., Elsevier, Amsterdam, 2021a.

James, M. O. Enzyme kinetics of PAPS-sulfotransferase, *Methods Mol. Biol.*, 2342, 285–300, 2021b.

James, M. O., Steroid catabolism in marine and freshwater fish, *J. Steroid Biochem. Mol. Biol.*, 127, 167–175, 2011.

James, M. O., Taurine conjugation of carboxylic acids in some marine species, in *Conjugation Reactions in Drug Biotransformations*, Aitio, A., Ed., Elsevier, Amsterdam, pp. 121–129, 1976.

James, M. O., Xenobiotic conjugation in aquatic species, in *Xenobiotic Conjugation Chemistry*, Paulson, G. D., Caldwell, J., Hutson, D. H., and Menn, J. J., Eds., American Chemical Society, Washington, DC, 1986.

James, M. O. and Ambadapadi, S., Interactions of cytosolic sulfotransferases with xenobiotics. *Drug Metab. Rev.*, 45, 401–414, 2013.

James, M. O. and Bend, J. R., Polycyclic aromatic hydrocarbon induction of cytochrome P-450-dependent mixed-function oxidases in marine fish, *Toxicol. Appl. Pharmacol.*, 54, 117, 1980.

James, M. O. and Bend, J. R., Taurine conjugation of 2,4-dichlorophenoxyacetic acid and phenylacetic acid in two marine species, *Xenobiotica*, 6, 393–398, 1976.

James, M. O., Fouts, J. R., and Bend, J. R., Hepatic and extrahepatic metabolism, *in vitro*, of an epoxide (8-(14) C-styrene oxide) in the rabbit, *Biochem. Pharmacol.*, 25, 187, 1976.

James, M. O., Fouts, J. R., and Bend, J. R., *In vitro* epoxide metabolism in some marine species, *Bull. Mt. Desert Island Biol. Lab.*, 14, 41, 1974.

James, M. O., Hawkins, W. E., and Walker, W. W., Phase 1 and phase 2 biotransformation and carcinogenicity of 2-acetylaminofluorene in medaka and guppy, *Aquat. Toxicol.*, 28, 79, 1994.

James, M. O., Heard, C. S., Hawkins, W. E., Effect of 3-methylcholanthrene on mono-oxygenase, epoxide hydrolase, and glutathione *S*-transferase activities in small estuarine and freshwater fish, *Aquat. Toxicol.*, 12, 1, 1988.

James, M. O. and Little, P. J., Polyhalogenated biphenyls and phenobarbital: evaluation as inducers of drug metabolizing enzymes in the sheepshead, *Archosargus probatocephalus*, *Chem. Biol. Interact.*, 36, 229, 1981.

James, M. O. and Pritchard, J. B., *In vivo* and *in vitro* renal metabolism and excretion of benzoic acid by a marine teleost, the southern flounder, *Drug Metab. Disp.*, 15, 665, 1987.

James, M. O., Sikazwe, D., and Gadagbui, B., Isolation of a pi class glutathione *S*-transferase (GST) from catfish intestinal mucosa, *Mar. Environ. Res.*, 46, 57, 1998.

James, M. O. et al., Dietary modulation of phase 1 and phase 2 activities with benzo(*a*)pyrene and related compounds in the intestine but not the liver of the channel catfish, *Ictalurus punctatus*, *Drug Metab. Disp.*, 25, 346, 1997.

James, M. O. et al., Epoxide hydrase and glutathione *S*-transferase activities with selected alkene and adrene oxides in several marine species, *Chem. Biol. Interact.*, 25, 321, 1979.

James, M. O. et al., Increased toxicity of benzo(*a*)pyrene-7,8-dihydrodiol in the presence of polychlorobiphenylols, *Mar. Environ. Res.*, 58, 343, 2004.

Jayasinghe, J. D., Bathige, S. D., Nam, B. H., Noh, J. K., and Lee, J., Comprehensive characterization of three glutathione S-transferase family proteins from black rockfish (Sebastes schlegelii), *Comp. Biochem. Physiol. C Toxicol. Pharmacol.*, 189, 31–43, 2016.

Jerina, D. M. and Dansette, P. M., Hepatic microsomal epoxide hydrase: a sensitive radiometric assay for hydration of arene oxides of carcinogenic aromatic hydrocarbons. *Mol. Pharmacol.*, 13, 342, 1977.

Jerina, D. M. et al., Identification of the ultimate carcinogenic metabolites of the polycyclic aromatic hydrocarbons: bay-region (R, S)-diol-(S, R)-epoxides, in *Foreign Compound Metabolism*, Caldwell, J. and Paulson, G., Eds., London, Taylor & Francis, 1984, pp. 257–270.

Jeuken, A. et al., Activation of the Ah receptor by extracts of dietary herbal supplements, vegetables, and fruits, *J. Agric. Food Chem.*, 51, 5478, 2003.

Jönsson, M. E. et al., Role of AHR2 in the expression of novel cytochrome P450 1 family genes, cell cycle genes, and morphological defects in developing zebra fish exposed to 3,3',4,4',5–pentachlorobiphenyl or 2,3,7,8-tetrachlorodibenzo-p-dioxin, *Toxicol. Sci.*, 100, 180–193, 2007.

Jorgensen, E. H. et al., The effect of stress on toxicant-dependent cytochrome p450 enzyme responses in the Arctic charr (*Salvelinus alpinus*), *Environ. Toxicol. Chem.*, 20, 2523, 2001.

Kantae V, et al., Pharmacokinetic modeling of paracetamol uptake and clearance in Zebrafish Larvae: expanding the allometric scale in vertebrates with five orders of magnitude. *Zebrafish*, 13, 504–510, 2016.

Kapelyukh, Y. et al., Multiple substrate binding by cytochrome P450 3A4: estimation of the number of bound substrate molecules, *Drug Metab Dispos.*, 36, 2136–2144, 2008.

Kaplan, L. A., Fielding, E., and Crivello, J. F., The genetic regulation of liver microsomal CYP2E1 activity among strains of the viviparous fish *Poeciliopsis*, *Comp. Biochem. Physiol.*, 128C, 143, 2001.

Kaplan, L. A. et al., Nitrosodiethylamine metabolism in the viviparous fish *Poeciliopsis*: evidence for the existence of liver P450pj activity and expression, *Carcinogenesis*, 12, 647, 1991.

Kashiwada, S., Hinton, D. E., and Kullman, S. W., Functional characterization of medaka CYP3A38 and CYP3A40: kinetics and catalysis by expression in a recombinant baculovirus system, *Comp. Biochem. Physiol. C Toxicol. Pharmacol.*, 141, 338–348, 2005.

Katagi, T. *In vitro* metabolism of pesticides and industrial chemicals in fish, *J. Pestic. Sci.*, 45, 1–15, 2020.

Katchamart, S. et al., Effect of xenoestrogen exposure on the expression of cytochrome P450 isoforms in rainbow trout liver, *Environ. Toxicol. Chem.*, 21, 2445, 2002.

Keizer, J., D'Agostino, G., and Vittozzi, L., The importance of biotransformation in the toxicity of xenobiotics to fish. I. Toxicity and bioaccumulation of diazinon in guppy (*Poecilia reticulata*) and zebra fish (*Brachydanio rerio*), *Aquat. Toxicol.*, 21, 239, 1991.

Keizer, J. et al., Comparative diazinon toxicity in guppy and zebra fish: different role of oxidative metabolism, *Environ. Toxicol. Chem.*, 12, 1243, 1993.

Keizer, J. et al., Enzymological differences of AChE and diazinon hepatic metabolism: correlation of *in vitro* data with the selective toxicity of diazinon to fish species, *Sci. Total Environ.*, 171, 213, 1995.

Kelley, M. and Vessey, D. A., Techniques for measuring the activity of carboxylic acid:CoA ligase and acyl-CoA:amino acid N-acyltransferase: the amino acid conjugation pathway. *Curr. Protoc. Toxicol.*, Chapter 4, Unit4 11, 2003.

Kester, M. H. et al., 2000. Potent inhibition of estrogen sulfotransferase by hydroxylated PCB metabolites: a novel pathway explaining the estrogenic activity of PCBs, *Endocrinology*, 141, 1897–1900, 2000.

Kim, H. S. et al., Cloning and characterization of a fish microsomal epoxide hydrolase of Danio rerio and application to kinetic resolution of racemic styrene oxide, *J. Mol. Catal. B: Enzymatic*, 37, 30–35, 2005.

Kim, J.-H., Dahms, H.-U., Rhee, J.-S., Lee, Y.-M., Lee, J., Han, K.-N and Lee, J.-S. Expression profiles of seven glutathione S-transferase (GST) genes in cadmium-exposed river pufferfish (*Takifugu obscurus*), *Comp. Biochem. Physiol. Part C.*, 151, 99, 2010.

Kirischian, N. et al., Phylogenetic and functional analysis of the vertebrate cytochrome P450 2 family, *J. Mol. Evol.*, 72, 56–71, 2011.

Kitamura, S. et al., Interconversion between fenthion and fenthion sulfoxide in goldfish, *Carassius auratus*, *J. Health Sci.*, 45, 266, 1999.

Kitamura, S. et al., *In vitro* metabolism of fenthion and fenthion sulfoxide by liver preparations of sea bream, goldfish and rats, *Drug Metab. Disp.*, 31, 179, 2003.

Kitamura, S. et al., Whole-body metabolism of the organophosphorus pesticide, fenthion, in goldfish, *Carassius auratus*, *Comp. Biochem. Physiol.* 126C, 259, 2000.

Kleinow, K. M. et al., A comparison of hepatic P450 induction in rat and trout (*Oncorhynchus mykiss*): delineation of the site of resistance of fish to phenobarbital-type inducers, *Environ. Health Perspect.*, 71, 105, 1990.

Kleinow, K. M. et al., Bioavailability and biotransformation of benzo(*a*)pyrene in an isolated perfused *in situ* catfish intestinal preparation, *Environ. Health Perspect.*, 106, 155, 1998.

Kleinow, K. M. et al., Pharmacokinetics, bioavailability, distribution, and metabolism of sulfadimethoxine in the rainbow-trout (*Oncorhynchus mykiss*), *Can. J. Fish. Aquat. Sci.*, 49, 1070, 1992.

Kliewer, S. A., Goodwin, B., and Willson, T. M., The nuclear pregnane X receptor: a key regulator of xenobiotic metabolism, *Endocrine Rev.*, 23, 687, 2002.

Klingenberg, M., Pigments of rat liver microsomes, *Arch. Biochem. Biophys.*, 75, 376, 1958.

Klotz, A. V. et al., Cytochrome P-450 isozymes from the marine teleost *Stenotomus chrysops*: their roles in steroid hydroxylation and the influence of cytochrome $b_5$, Arch. Biochem. Biophys., 249, 326, 1986.

Ko, S. K. and Shin, I., Cardiosulfa induces heart deformation in zebrafish through the AhR-mediated, CYP1A-independent pathway, *ChemBioChem*, 13, 1483–1489, 2012.

Konishi, T., Kato, K., Araki, T., Shiraki, K., Takagi, M., and Tamura, Y., A new class of gluathione *S*-transferase from the hepatopancreas of the red sea bream, *Pagrus major*, *Biochem. J.*, 388, 299–307, 2005.

Konishi-Imamura, L. et al., Regulation of arylsulfate sulfotransferase from a human intestinal bacterium by nucleotides and magnesium ion, *J. Enzyme Inhib.*, 8, 233, 1995.

Kubota, A. et al., Role of pregnane X receptor and aryl hydrocarbon receptor in transcriptional regulation of pxr, CYP2, and CYP3 genes in developing zebrafish, *Toxicol. Sci.*, 143, 398–407, 2015.

Kubota, A. et al., Role of zebrafish cytochrome P450 CYP1C genes in the reduced mesencephalic vein blood flow caused by activation of AHR2, *Toxicol. Appl. Pharmacol.*, 253, 244, 2011.

Kubota, A. et al., Transcriptional profiling of cytochrome P450 genes in the liver of adult zebrafish, Danio rerio, *Toxicol. Sci.*, 44, 347–356, 2019.

Kubota, A. et al., The cytochrome P450 2AA gene cluster in zebrafish (Danio rerio): expression of CYP2AA1 and CYP2AA2 and response to phenobarbital-type inducers, *Toxicol. Appl. Pharmacol.*, 272, 172–179, 2013.

Kullman, S. W. and Hinton, D. E., Identification, characterization, and ontogeny of a second cytochrome P450 3A gene from the fresh water teleost medaka (*Oryzias latipes*), *Mol. Reprod. Develop.*, 58, 149, 2001.

Kullman, S. W., Hamm, J. T., and Hinton, D. E., Identification and characterization of a cDNA encoding cytochrome P450 3A from the fresh water teleost medaka (*Oryzias latipes*), *Arch. Biochem. Biophys.*, 380, 29, 2000.

Kullman, S. W., Kashiwada, S., and Hinton, D. E., Analysis of medaka cytochrome P450 3A homotropic and heterotropic cooperativity, *Mar. Environ. Res.*, 58, 469, 2004.

Kupsco, A., and Schlenk D., Developmental Expression and Regulation of Flavin-containing monooxygenase by the Unfolded Protein Response in Japanese Medaka (*Oryzias latipes*). *Comp. Biochem. Physiol.* Part C. 191, 7–13, 2017.

Kurogi, K. et al., The use of zebrafish as a model system for investigating the role of the SULTs in the metabolism of endogenous compounds and xenobiotics, *Drug Metab. Rev.*, 45, 431–440, 2013.

Laitenen, M. et al., Tricaine (MS-222) induced modification of the metabolism of foreign compounds in the liver and duodenal mucosa of the splake (*Salvelinus fontinalis*×*Salvelinus namaycush*), *Acta Pharmacol. Toxicol.*, 49, 92, 1981.

Lambert, J. G. D. and Resink, J. W., Steroid glucuronides as male pheromones in the reproduction of African catfish *Clarias gariepinus*: a brief review, *J. Steroid Biochem. Mol. Biol.*, 40, 549, 1991.

Lange, R. and Fugelli, K., The osmotic adjustment in the euryhaline teleosts, the flounder, *Pleuronectes flesus*, and the three-spined stickleback, *Gasterosteus aculeatus*, *Comp. Biochem. Physiol.*, 15, 283, 1965.

Larsen, B. K. and Schlenk, D., Effect of salinity on flavin-containing monooxygenase expression and activity in rainbow trout (*Oncorhynchus mykiss*), *J. Comp. Physiol. B*, 171, 421, 2001a.

Larsen, B. K. and Schlenk, D., Effect of urea and low temperature on the expression and activity of flavincontaining monooxygenase in the liver and gill of rainbow trout (*Oncorhynchus mykiss*), *Fish Physiol. Biochem.*, 25(1), 19–29, 2001b.

Larsen, H. E., Celander, M., and Goksøyr, A., The cytochrome-p450 system of Atlantic salmon (*Salmo salar*). 2. Variations in hepatic catalytic activities and isozyme patterns during an annual reproductive-cycle, *Fish Physiol. Biochem.*, 10, 291, 1992.

Lavado, R., Aparicio-Fabre, R., and Schlenk, D., Effects of salinity acclimation on the expression and activity of Phase I enzymes (CYP450 and FMOs) in coho salmon (Oncorhynchus kisutch), *Fish Physiol. Biochem.*, 40, 267–278, 2014.

Layiwola, P. J. and Linnecar, D. F., The biotransformation of [$^{14}$C] phenol in some freshwater fish, *Xenobiotica*, 11, 167–171, 1981.

Layiwola, P. J., Linnecar, D. F., and Knights, B., The biotransformation of three 14C-labelled phenolic compounds in twelve species of freshwater fish, *Xenobiotica*, 13, 107–113, 1983.

Leaver, M. J., Clarke, D. J., and George, S. G., Molecular studies of the phase II xenobiotic conjugative enzymes of marine *Pleuronectid* flatfish, *Aquatic. Toxicol.*, 22, 265, 1992.

Leaver, M. J. and George, S. G., A piscine glutathione *S*-transferase which efficiently conjugates the endproducts of lipid peroxidation, *Mar. Environ. Res.*, 46, 71, 1998.

Leaver, M. J., Scott, K., and George, S. G., Cloning and characterisation of the major hepatic glutathione *S*-transferase from a marine teleost flatfish, the plaice (*Pleuronectes platessa*) with structural similarities to plant, insect and mammalian theta class isoenzymes, *Biochem. J.*, 292, 189, 1993.

Leaver, M. J., Wright, J., and George, S. G., Structure and expression of a cluster of glutathione *S*-transferase genes from a marine fish, the plaice (*Pleuronectes platessa*), *Biochem. J.*, 321, 405, 1997.

Leaver, M. J., Wright, J., Hodgson, P., Boukouvala, E., and George, S. G., Piscine UDP-glucuronosyltransferase 1B, *Aquat. Toxicol.*, 84, 356–365, 2007.

Lech, J. J. and Costrini, N. V., *In vitro* and *in vivo* metabolism of 3-trifluoromethyl-4-nitrophenol (TFM) in rainbow trout, *Comp. Gen. Pharmacol.*, 3, 160, 1972.

Lee, H. C., Lin, C. Y., and Tsai, H. J. Zebrafish, an in vivo platform to screen drugs and proteins for biomedical use, *Pharmaceuticals (Basel)*, 14, 500, 2021.

Lee, S. J. and Buhler, D. R., Cloning, tissue distribution, and functional studies of a new cytochrome P450 3A subfamily member, CYP3A45, from rainbow trout (*Oncorhynchus mykiss*) intestinal ceca, *Arch. Biochem. Biophys.*, 412, 77, 2003.

Lee, S. J. and Buhler, D. R., Functional properties of a rainbow trout CYP3A27 expressed by recombinant baculovirus in insect cells, *Drug Metab. Disp.*, 30, 1406, 2002.

Lee, S. J. et al., Cloning, sequencing, and tissue expression of CYP3A27, a new member of the CYP3A subfamily from embryonic and adult rainbow trout livers, *Arch. Biochem. Biophys.*, 360, 53, 1998.

Lee, S. J. et al., Immunohistochemical localization and differential expression of cytochrome P450 3A27 in the gastrointestinal tract of rainbow trout, *Toxicol. Appl. Pharmacol.*, 177, 94, 2001.

Lemaire, P., Forlin, L., and Livingstone, D. R., Responses of hepatic biotransformation and antioxidant enzymes to CYP1A-inducers (3-methylcholanthrene, β-naphthoflavone) in sea bass (*Dicentrarchus labrax*), dab (*Limanda limanda*), and rainbow trout (*Oncorhynchus mykiss*), *Aquat. Toxicol.*, 36, 141, 1996.

Lemaire, P. et al., Stimulation of oxyradical production of hepatic microsomes of flounder (*Platichthys flesus*) and perch (*Perca fluviatilis*) by model and pollutant xenobiotics, *Arch. Environ. Cont. Toxicol.*, 26, 191, 1994.

Leong, J. S. et al, Salmo salar and Esox lucius full-length cDNA sequences reveal changes in evolutionary pressures on a post-tetraploidization genome, *BMC Genom.*, 11, 279, 2010.

Lesca, P. et al., Structure-activity relationships in the inhibitory effects of ellipticines on benzo(*a*)pyrene hydroxylase activity and 3-methylcholanthrene mutagenicity, *Biochem. Pharmacol.*, 29, 3231, 1980.

Levi, P. E. and Hodgson, E., Metabolism of organophosphorus compounds by the flavin-containing monooxygenase, in *Organophosphates: Chemistry, Fate and Effects*, Chambers, J. E. and Levi, P. E., Eds., Academic Press, San Diego, CA, 1992, pp. 141–154.

Levine, S. L., Czosnyka, H., and Oris, J. T., Effect of the fungicide clotrimazole on the bioconcentration of benzo[*a*]pyrene in gizzard shad (*Dorosoma cepedianum*): *in vivo* and *in vitro* inhibition of cytochrome P4501A activity, *Environ. Toxicol. Chem.*, 16, 306, 1997.

Li, A. P., Kaminski, D. L., and Rasmussen, A., Substrates of human hepatic cytochrome P450 3A4, *Toxicol. Appl. Pharmacol.*, 15, 1, 1995.

Li, D. et al., Imaging and detection of carboxylesterase in living cells and zebrafish pretreated with pesticides by a new near infrared fluorescence off-on probe, *J. Agric. Food Chem.*, 65, 4209–4215, 2017.

Li, X., et al., Regulation of Gluconeogenesis by Aldo-keto-reductase 1a1b in Zebrafish. *iScience.* 23,101763, 2020.

Li, Y. and Jaiswal, A. K., Regulation of human NAD(P)H:quinone oxidoreductase gene: role of AP1 binding site contained within human antioxidant response element, *J. Biol. Chem.*, 267, 15097, 1992 (erratum in *J. Biol. Chem.*, 268, 21454, 1993).

Liehr, J. G. and Ricci, M. J., 4-Hydroxylation of estrogens as a marker of human mammary tumors, *Proc. Natl. Acad. Sci. U.S.A.*, 93, 3294, 1996.

Lille-Langøy, R. et al., Sequence Variations in pxr (nr1i2) from zebrafish (Danio rerio) strains affect nuclear receptor function. *Toxicol Sci.* 168, 28–39, 2019.

Lin, C. H. et al., Action of vitamin D and the receptor, VDRa, in calcium handling in zebrafish (Danio rerio). *PLoS One* 7. e45650, 2012.

Little, P. J. et al., Imidazole derivatives as inhibitors of cytochrome P450-dependent oxidation and activators of epoxide hydrolase in hepatic microsomes from a marine fish, *Biochem. Pharmacol.*, 30, 2876, 1981.

Liu, T. A. et al., Identification and characterization of two novel cytosolic sulfotransferases, SULT1 ST7 and SULT1 ST8, from zebrafish. *Aquat. Toxicol.*, 89, 94–102, 2008.

Mackenzie, P. I. et al., The UDP glycosyltransferase gene superfamily: recommended nomenclature update based on evolutionary divergence, *Pharmacogenetics*, 7, 255, 1997.

Macrides, T. A. et al., Enzymic sulfation of bile salts: partial purification and characterization of an enzyme from the liver of the shark *Heterodontus portusjacksoni* that catalyses the sulfation of the shark bile steroid 5 beta-scymnol, *Comp. Biochem. Physiol.*, 107, 461, 1994.

Maglich, J. M. et al., The first completed genome from a teleost fish (*Fugu rubripes*) adds significant diversity to the nuclear receptor superfamily, *Nucl. Acids Res.*, 31, 4051, 2003.

Maldonado, A. et al., Biochemical mechanisms for geographical adaptations to novel toxin exposures in butterflyfish, *PLOS One*, 11, e0154208, 2016a.

Maldonado, A. et al., Hard coral (Porities lobata) extracts and homarine on cytochrome P450 expression in Hawaiian butterflyfishes with different feeding strategies, *Comp. Biochem. Physiol. Part C*, 179, 57–63, 2016b.

Martinez-Lara, E., Leaver, M., and George, S. G., Evidence from heterologous expression of glutathione *S*transferases A and A1 of the plaice (*Pleuronectes platessa*) that their endogenous role is in detoxification of lipid peroxidation products, *Mar. Environ. Res.*, 54, 263, 2002.

Martinez-Lara, E. et al., Purification and characterisation of multiple glutathione transferase isoenzymes from grey mullet liver, *Cell. Mol. Life Sci.*, 53, 759, 1997.

Massarsky, A. et al., AHR2 morpholino knockdown reduces the toxicity of total particulate matter to zebrafish embryos, *Toxicol. Appl. Pharmacol.*, 309, 63–76, 2016.

Masuyama, H. et al., Endocrine disrupting chemicals, phthalic acid and nonylphenol, activate pregnane X receptor-mediated transcription, *Mol. Endocrinol.*, 14, 421, 2000.

Mathäs, M. et al., Evolutionary history and functional characterization of the amphibian xenosensor CAR, *Mol. Endocrinol.*, 26, 14–26, 2012.

Matsuo, A.Y.O. et al., Characterization of Phase I biotransformation enzymes in coho salmon(Oncorhynchus kisutch) *Comp. Biochem.Physiol.*, Part C 147, 78–84, 2008.

Maurel, P., The CYP3 family, in *Cytochrome P450: Metabolic and Toxicological Aspects*, Ioannides, C., Ed., CRC Press, Boca Raton, FL, 1996.

Maxwell, D. M., Detoxication of organophosphorus compounds by carboxylesterase, in *Organophosphates: Chemistry, Fate, and Effects*, Chambers, J. E. and Levi, P. E., Eds., Academic Press, San Diego, CA, 1992, pp. 183–227.

McArthur, A. G. et al., Phylogenetic analysis of the cytochrome P450 3 (CYP3) gene family, *J. Mol. Evol.*, 57, 200, 2003.

McCormick, S. D., *Hormonal Control of Gill $Na^+$,$K^+$-ATPase and Chloride Cell Function*, Academic Press, San Diego, CA, 1995.

Meech, R., The UDP-glycosyltransferase (UGT) superfamily: new members, new functions and novel paradigms, *Physiol. Rev.*, 99, 1153–1222, 2019.

Melgar Riol, M. J. et al., Glutathione *S*-transferases from rainbow trout liver and freshly isolated hepatocytes: purification and characterization, *Comp. Biochem. Physiol.*, 128C, 227, 2001.

Meyer, J. N., Nacci, D., and Di Giulio, R. T., Cytochrome P4501A (CYP1A) in killifish (*Fundulus heteroclitus*): heritability of altered expression and relationship to survival in contaminated sediments, *Toxicol. Sci.*, 68, 69, 2002.

Meyer, J. N. et al., Expression and inducibility of aryl hydrocarbon receptor pathway genes in wild-caught killifish (*Fundulus heteroclitus*) with different contaminant-exposure histories, *Environ. Toxicol. Chem.*, 22, 2337, 2003.

Michel, X. R., Beasse, C., and Narbonne, J.-F., *In vivo* metabolism of benzo(*a*)pyrene in the mussel *Mytilus galloprovincialis*, *Arch. Environ. Contam. Toxicol.*, 28, 215, 1995.

Mierzejewski, J., Haney, D. C., and van den Hurk, P., Biomarker responses in sunfish species and largemouth bass from the Saluda River, South Carolina, *Ecotoxicol. Environ. Saf.*, 110, 8–15, 2014. doi: 10.1016/j.ecoenv.2014.08.003. Epub 2014 Aug 29. PMID: 25173848.

Miller, M. S. et al., Drug metabolic enzymes in developmental toxicology, *Fund. Appl. Toxicol.*, 34, 165, 1996.

Miranda, C. L. et al., Comparison of rainbow trout and mammalian cytochrome P450 enzymes: evidence for structural similarity between trout P450 LMC5 and human P450IIIA4, *Biochem. Biophys. Res. Comm.*, 176, 558, 1991.

Miranda, C. L. et al., Immunological characterization of constitutive isozymes of cytochrome P-450 from rainbow trout: evidence for homology with phenobarbital-induced rat P-450s, *Biochim. Biophys. Acta*, 1037, 155, 1990.

Miranda, C. L. et al., Purification and characterization of hepatic steroid hydroxylase from untreated rainbow trout, *Arch. Biochem. Biophys.*, 268, 227, 1989.

Miyamoto, M. et al., Bioconcentration and metabolism of trans-tetramethrin in fish, *J. Pest. Sci.*, 40, 130–137, 2015.

Moore, L. B. et al., Pregnane X receptor (PXR), constitutive androstane receptor (CAR), and benzoate X receptor (BXR) define three pharmacologically distinct classes of nuclear receptors, *Mol. Endocrinol.*, 16, 977, 2002.

Murphy, S. D., Liver metabolism and toxicity of thiophosphate insecticides in mammalian, avian, and piscine species, *Proc. Soc. Exp. Biol. Med.*, 123, 392, 1966.

Nagai, T. et al., Organ distribution of aldehyde dehydrogenase activity in several fish, *Fish. Res.*, 61, 1047, 1997.

Nelson, D. R., Comparison of P450s from human and *Fugu*: 420 million years of vertebrate evolution, *Arch. Biochem. Biophys.*, 409, 18, 2003.

Nelson, D. R. et al., The P450 superfamily: update on new sequences, gene mapping, accession numbers, and nomenclature, *Pharmacogenetics*, 6, 1, 1996.

Nelson, D. R. et al., The P450 superfamily: update on new sequences, gene mapping, accession numbers, early trivial names of enzymes, and nomenclature, *DNA Cell. Biol.*, 12, 1, 1993.

Nebert, D. W. and Gonzalez, F. J., P450 genes: structure, evolution, and regulation, *Ann. Rev. Biochem.*, 56, 945, 1987.

Nebert, D. W. et al., Evolution of the cytochrome P450 genes, *Xenobiotica*, 19, 1149, 1989.

Novi, S. et al., Biotransformation enzymes and their induction by beta-naphthoflavone in adult sea bass (*Dicentrarchus labrax*), *Aquat. Toxicol.*, 41, 63, 1998.

Novoa-Valinas, M. C. et al., Comparative study of the purification and characterization of the cytosolic glutathione *S*-transferases from two salmonid species: Atlantic salmon (*Salmo salar*) and brown trout (*Salmo trutta*), *Comp. Biochem. Physiol.*, 131C, 207, 2002.

Oleksiak, M. F. et al., Identification and regulation of a new vertebrate cytochrome P450 subfamily, the CYP2Ps, and functional characterization of CYP2P3, a conserved arachidonic acid epoxygenase/19-hydroxylase, *Arch. Biochem. Biophys.*, 411, 223, 2003.

Oleksiak, M. F. et al., Identification, functional characterization, and regulation of a new cytochrome P450 subfamily, the CYP2Ns, *J. Biol. Chem.*, 275, 2312, 2000.

Olufsen, M. and Arukwe, A., Endocrine, biotransformation, and oxidative stress responses in salmon hepatocytes exposed to chemically induced hypoxia and perfluorooctane sulfonamide (PFOSA), given singly or in combination. *Environ. Sci. Pollut. Res. Int.*, 22, 17350–17366, 2015.

Omura, T. and Sato, R., A new cytochrome in liver microsomes, *J. Biol. Chem.*, 237, 1375, 1962.

Otto, D. M. and Moon, T. W., Phase I and II enzymes and antioxidant responses in different tissues of brown bullheads from relatively polluted and non-polluted systems, *Arch. Environ. Contam. Toxicol.*, 31, 141, 1996.

Padros, J. et al., Metabolic interactions between low doses of benzo(*a*)pyrene and tributyltin in Arctic charr (*Salvelinus alpinus*): a long-term *in vivo* study, *Toxicol. Appl. Pharmacol.*, 192, 45, 2003.

Pajor, A. M. et al., Feminization of the hepatic microsomal cytochrome P-450 system in brook trout by estradiol, testosterone, and pituitary factors, *J. Exp. Zool.*, 253, 51, 1990.

Pangrekar, J. and Sikka, H. C., Xenobiotic metabolizing enzyme activity in the liver and kidney of the brown bullhead (*Ictalurus nebulosus*), *Mar. Environ. Res.*, 34, 287, 1992.

Park, S. H. et al., Distinctive roles of STAT5a and STAT5b in sexual dimorphism of hepatic P450 gene expression: impact of STAT5a gene disruption, *J. Biol. Chem.*, 274, 7421, 1999.

Parker, L. M. et al., Biochemical and histological properties of hepatic tumors of rainbow trout, *Carcinogenesis*, 14, 211, 1993.

Parkinson, A., Biotransformation of xenobiotics, in *Casarett and Doull's Toxicology: The Basic Science of Poisons*, Klaassen, C. D., Ed., McGraw-Hill, New York, 2001, pp. 133–224.

Pascussi, J. M. et al., The expression of CYP2B6, CYP2C9 and CYP3A4 genes: a tangle of networks of nuclear and steroid receptors, *Biochim. Biophys. Acta*, 1619, 243, 2003.

Pathiratne, A. and George, S. G., Comparison of xenobiotic metabolizing enzymes of tilapia with those of other fish species and interspecies relationships between gene families, *Mar. Environ. Res.*, 42, 293, 1996.

Payne, J. F. and Penrose, W. R., Induction of aryl hydrocarbon (benzo[*a*]pyrene) hydroxylase in fish by petroleum, *Bull. Environ. Contam. Toxicol.*, 14, 112, 1975.

Perdudurand, E. F. and Cravedi, J. P., Characterization of xenobiotic metabolizing enzymes in sturgeon (*Acipenser baeri*), *Comp. Biochem. Physiol.*, 93B, 921, 1989.

Perkins, E. J., Characterization of the Hepatic Cytochrome P450 Monooxygenase System and Its Role in the Toxicity of Aldicarb in the Channel Catfish (*Ictalurus punctatus*), Ph.D. dissertation, University of Mississippi, 1999.

Perkins, E. J., DeBusk, B. C., and Schlenk, D., Isolation of a novel hepatic CYP2-related protein from channel catfish, *Ictalurus punctatus*, *Fish Physiol. Biochem.*, 22, 199, 2000.

Perkins, E. J., El-Alfy, A., and Schlenk, D., *In vitro* sulfoxidation of aldicarb by hepatic microsomes of channel catfish *Ictalurus punctatus*, *Toxicol. Sci.*, 48, 67, 1999.

Perkins, E. J. and Schlenk, D., Immunochemical characterization of hepatic cytochrome P450 isozymes in the channel catfish: assessment of sexual, developmental and treatment-related effects, *Comp. Biochem. Physiol.*, 121C, 305, 1998.

Perkins, E. J. and Schlenk, D., *In vivo* metabolism, acetylcholinesterase inhibition, and toxicokinetics of aldicarb in channel catfish (*Ictalurus punctatus*), *Toxicol. Sci.*, 53, 308, 2000.

Peters, L. D. et al., Characterisation of hepatic flavin-containing monooxygenase from the marine teleost, turbot (*Scophthalmus maximus* L.), *Xenobiotica*, 25, 121, 1995.

Petersen, D. W. and Lech, J. J., Hepatic effects of acrylamide in rainbow trout, *Toxicol. Appl. Pharmacol.*, 89, 249, 1987.

Peterson, J. S. K. and Bain, L. J., Differential gene expression in anthracene-exposed mummichogs (*Fundulus heteroclitus*), *Aquat. Toxicol.*, 66, 345, 2004.

Pham, R. T., Gardner, J. L., and Gallagher, E. P., Conjugation of 4-hydroxynonenal by largemouth bass (*Micropterus salmoides*) glutathione *S*-transferases, *Mar. Environ. Res.*, 54, 291, 2002.

Plakas, S. M. and James, M. O., Bioavailability, metabolism, and renal excretion of benzoic-acid in the channel catfish (*Ictalurus punctatus*), *Drug Metab. Disp.*, 18, 552, 1990.

Plant, N. J. and Gibson, G. G., Evaluation of the toxicological relevance of CYP3A4 induction, *Curr. Opin. Drug Disc. Develop.*, 6, 50, 2003.

Ploch, S. A. et al., Comparative *in vitro* and *in vivo* benzo[*a*] pyrene-DNA adduct formation and its relationship to CYP1A activity in two species of ictalurid catfish. *Toxicol. Appl. Pharmacol.*, 149, 90, 1998.

Pogoda, H. M. et al., Direct activation of chordoblasts by retinoic acid is required for segmented centra mineralization during zebrafish spine development. *Development*. 145, dev159418, 2018.

Poland, A., Mak, I., and Glover, E., Species differences in responsiveness to 1,4-*bis*[2-(3,5-dichloropyridoxy)] benzene, a potent phenobarbital-like inducer of microsomal monooxygenase activity, *Mol. Pharmacol.*, 20, 442, 1981.

Poon, K. L. et al., Humanizing the zebrafish liver shifts drug metabolic profiles and improves pharmacokinetics of CYP3A4 substrates, *Arch Toxicol.*, 91, 1187–1197, 2017.

Poulsen, L. L. and Ziegler, D. M., The liver microsomal FAD-containing monooxygenase: Spectral characterization and kinetic studies, *J. Biol. Chem.*, 254, 6449, 1979.

Powell, W. H. et al., Developmental and tissue-specific expression of AHR1, AHR2, and ARNT2 in dioxinsensitive and -resistant populations of the marine fish, *Fundulus heteroclitus*, *Toxicol. Sci.*, 57, 229, 2000.

Pretti, C. et al., Effects of beta-naphthoflavone on the cytochrome P450 system and phase II enzymes in gilthead seabream (*Sparus aurata*), *Comp. Biochem. Physiol.*, 130C, 133, 2001.

Prince, R. and Cooper, K. R., Comparisons of the effects of 2,3,7,8-tetrachlorodibenzo-*p*-dioxin on chemically impacted and non-impacted subpopulations of *Fundulus heteroclitus*. II. Metabolic considerations, *Environ. Toxicol. Chem.*, 14, 589, 1995.

Pritchard, J. B. and James, M. O., Factors determining the rate of excretion of anionic xenobiotics, *Environ. Health Perspect.*, 33, 326, 1979.

Pritchard, J. B., and Bend, J.R., Relative Roles of Metabolism and Renal Excretory Mechanisms in Xenobiotic Elimination by Fish. *Environ. Health Perspect.*, 90, 85–92, 1991.

Puthumana, J. et al., Nine co-localized cytochrome P450 genes of the CYP2N, CYP2AD, and CYP2P gene families in the mangrove killifish Kryptolebias marmoratus genome: Identification and expression in response to B[α]P, BPA, OP, and NP, *Aquat Toxicol.*, 187, 132–140, 2017.

Qi, H., et al., Reduced acrolein detoxification in akr1a1a zebrafish mutants causes impaired insulin receptor signaling and microvascular alterations. *Adv Sci* (Weinh). 8, e2101281, 2021.

Qiu, H. et al., CYP3 phylogenomics: evidence for positive selection of CYP3A4 and CYP3A7. *Pharmacogenet. Genom.*, 18, 53–66, 2008.

Rao, S. I. and Duffel, M. W., Inhibition of rat hepatic aryl sulphotransferase IV by dihydrodiol derivatives of benzo[*a*] pyrene and naphthalene, *Xenobiotica*, 22, 247, 1992.

Raymond, J. A., Trimethylamine oxide and urea synthesis in rainbow smelt and some other northern fishes, *Physiol. Zool.*, 71, 515, 1998.

Raymond, J. A. and DeVries, A. L., Elevated concentrations and synthetic pathways of trimethylamine oxide and urea in some teleost fishes of McMurdo Sound, Antarctica, *Fish Physiol. Biochem.*, 18, 387, 1998.

Ree, R. et al., The N-terminal acetyltranferase Naa10 is essential for zebrafish development, *Biosci. Rep.*, 35, e00249, 2015.

Reschly E.J., et al., Functional evolution of the vitamin D and pregnane X receptors. *BMC Evol. Biol.*, 7, 222, 2007.

Rettie, A. E., Meier, G. P., and Sadeque, A. J. M., Prochiral sulfides as *in vitro* probes for multiple forms of the flavin-containing monooxygenase, *Chem. Biol. Interact.*, 96, 3, 1995.

Rodrigues, S., Antunes, S. C., Correia, A. T., and Nunes, B., Toxicity of erythromycin to Oncorhynchus mykiss at different biochemical levels: detoxification metabolism, energetic balance, and neurological impairment, *Environ. Sci. Pollut. Res. Int.*, 26, 227–239, 2019.

Rodrıguez-Fuentes, G., Aparicio-Fabre, R., Li, Q., and Schlenk D. Osmotic regulation of a novel flavin-containing monooxygenase in primary cultured cells from rainbow trout (*Oncorhynchus mykiss*). *Drug Metab. Disp.*, 36, 1212–1217, 2008.

Rowland, A., Miners, J. O., and Mackenzie, P. I., The UDP-glucuronosyltransferases: their role in drug metabolism and detoxification, *Int. J. Biochem. Cell. Biol.*, 45, 1121–1132, 2013.

Runge-Morris, M. and Wilusz, J., Suppression of hydroxysteroid sulfotransferase-a gene expression by 3-methylcholanthrene, *Toxicol. Appl. Pharmacol.*, 125, 133, 1994.

Rushmore, T. H., Morton, M. R., and Pickett, C. B., The antioxidant responsive element: activation by oxidative stress and identification of the DNA consensus sequence required for functional activity, *J. Biol. Chem.*, 266, 11632, 1991.

Ryan, D. E. and Levin, W., Purification and characterization of hepatic microsomal cytochrome P-450, *Pharmaceut. Therapeut.*, 45, 153, 1990.

Saad, M. et al., Xenobiotic metabolism in the zebrafish: a review of the spatiotemporal distribution, modulation and activity of Cytochrome P450 families 1 to 3, *J. Toxicol. Sci.*, 41, 1–11, 2016.

Sabbagh, A. et al., Rapid birth-and-death evolution of the xenobiotic metabolizing NAT gene family in vertebrates with evidence of adaptive selection, *BMC Evol. Biol.*, 13, 62, 2013.

Safe, S. H., Polychlorinated biphenyls (PCBs), dibenzo-*p*-dioxins (PCDDs), dibenzofurans (PCDFs) and related compounds: environmental and mechanistic considerations which support the development of toxic equivalency factors (TEFs), *CRC Crit. Rev. Toxicol.*, 21, 51, 1990.

Sakuma, T. et al., Regulation of the expression of two female-predominant CYP3A mRNAs (CYP3A41 and CYP3A44) in mouse liver by sex and growth hormones, *Arch. Biochem. Biophys.*, 404, 234, 2002.

Salamastrakis, S. S. and Haritos, A. A., Physiochemical characterization and tissue distribution of multiple molecular forms of fish (*Trachurus trachurus*) esterases, *Comp. Biochem. Physiol.*, 91B, 741, 1988.

Salanga, M. C. et al., CRISPR-Cas9-Mutated Pregnane X Receptor (pxr) Retains Pregnenolone-induced Expression of cyp3a65 in Zebrafish (Danio rerio) Larvae, *Toxicol. Sci.*, 174, 51–62, 2020.

Santana, M. S. et al., Biomarker responses in fish exposed to polycyclic aromatic hydrocarbons (PAHs): systematic review and meta-analysis, *Environ. Pollut.*, 242(Pt A), 449–461, 2018.

Schirmer, K. et al., Transitory metabolic disruption and cytotoxicity elicited by benzo(*a*)pyrene in two cell lines from rainbow trout liver, *J. Biochem. Mol. Toxicol.*, 14, 262, 2000.

Schlenk, D., A comparison of endogenous and exogenous substrates of the flavin-containing monooxygenases in aquatic organisms, *Aquat. Toxicol.*, 26, 157, 1993.

Schlenk, D., Occurrence of flavin-containing monooxygenases in non-mammalian eukaryotic organisms, *Comp. Biochem. Physiol.*, 121C, 185, 1998.

Schlenk, D., The effect of 2-methylisoborneol on cytochrome P450 expression in the channel catfish (*Ictalurus punctatus*), *Aquaculture* 120, 33, 1994.

Schlenk, D., Use of aquatic organisms as models to determine the *in vivo* contribution of flavin-containing monooxygenases to xenobiotic transformation, *Mol. Mar. Biol. Biotech.*, 4, 323, 1995.

Schlenk, D. and Buhler, D. R., Immunological characterization of flavin-containing monooxygenases from the liver of rainbow trout (*Oncorhynchus mykiss*): sexual- and age-dependent differences and the effect of trimethylamine on enzyme regulation, *Biochim. Biophys. Acta*, 1156, 103, 1993.

Schlenk, D. and Buhler, D. R., Role of flavin-containing monooxygenase in the *in vitro* biotransformation of aldicarb in rainbow trout (*Oncorhynchus mykiss*), *Xenobiotica*, 21, 1583, 1991.

Schlenk, D. and El-Alfy, A., Expression of branchial flavin-containing monooxygenase is directly correlated with salinity-induced aldicarb toxicity in the euryhaline fish (*Oryzias latipes*), *Mar. Environ. Res.*, 46, 103, 1998.

Schlenk, D., El-Alfy, A., and Buhler, D. R., Downregulation of hepatic flavin-containing monooxygenase activity by 17β-estradiol in rainbow trout (*Oncorhynchus mykiss*), *Comp. Biochem. Physiol.*, 118C, 199, 1997.

Schlenk, D. and Li-Schlenk, R., Characterization of liver flavin-containing monooxygenase of the smooth dogfish shark (*Squalus acanthias*) and partial purification of liver flavin-containing monooxygenase of the silky shark (*Carcharhinus falciformis*), *Comp. Biochem. Physiol.*, 109B, 655, 1994.

Schlenk, D., Peters, L. D., and Livingstone, D. R., Correlation of salinity with flavin-containing monooxygenase activity but not cytochrome P450 activity in the euryhaline fish (*Platichthys flesus*), *Biochem. Pharmacol.*, 52, 815, 1996a.

Schlenk, D., Peters, L. D., and Livingstone, D. R., Down-regulation of piscine flavin containing monooxygenase by decreased salinity in euryhaline flounder (*Platichthys flesus*), *Mar. Environ. Res.*, 42, 339, 1996b.

Schlenk, D., Yeung, C., and Rettie, A., Unique stereoselective sulfoxidation of thioethers indicates novel flavincontaining monooxygenase in liver of rainbow trout, *Mar. Environ. Res.*, 58, 499, 2004.

Schlenk, D., Zubcov, N., and Zubcov E., Effects of salinity on the uptake, biotransformation and toxicity of dietary seleno-L-methionine to rainbow trout, *Toxicol. Sci.*, 75, 309, 2003.

Schlenk, D. et al., Channel catfish liver monooxygenases: immunological characterization of constitutive cytochromes P450 and the absence of active flavin-containing monooxygenases, *Biochem. Pharmacol.*, 45, 217, 1993.

Schlenk, D. et al., Cloning and sequencing of cytochrome P450 2X1 from channel catfish (*Ictalurus punctatus*), *Mar. Environ. Res.*, 54, 391, 2002.

Schlenk, D. et al., Differential expression and activity of flavin-containing monooxygenases in euryhaline and stenohaline flatfishes, *Comp. Biochem. Physiol.*, 112C, 179, 1995.

Schlezinger, J. J. and Stegeman, J. J., Induction and suppression of cytochrome P450 1A by 3,3′,4,4′,5-pentachlorobiphenyl and its relationship to oxidative stress in the marine fish scup (*Stenotomus chrysops*), *Aquat. Toxicol.*, 52, 101, 2001.

Schlezinger, J. J. et al., Uncoupling of cytochrome P450 1A and stimulation of reactive oxygen species production by co-planar polychlorinated biphenyl congeners, *Aquat Toxicol.*, 77, 422–432, 2006.

Schreiber, E. A., Otter, R. R. and Van Den Hurk, P., A biomarker approach to measure biological effects of contaminant exposure in largemouth bass from Lake Conestee, South Carolina, U.S.A, *Environ. Toxicol. Chem.*, 25, 1926–1932, 2006.

Schuur, A. G. et al., *In vitro* inhibition of thyroid hormone sulfation by hydroxylated metabolites of halogenated aromatic hydrocarbons, *Chem. Res. Toxicol.*, 11, 1075, 1998.

Scornaienchi, M. L. et al., Cytochrome P450-mediated 17b-estradiol metabolism in zebrafish (Danio rerio), *J. Endocrinol.*, 206, 317–325, 2010.

Senatori, O. et al., Monoamine oxidase in teleosts, *Biogenic Amines*, 17, 199, 2003.

Seubert, J. M. and Kennedy, C. J., Benzo(*a*)pyrene toxicokinetics in rainbow trout (*Oncorhynchus mykiss*) acclimated to different salinities, *Arch. Environ. Cont. Toxicol.*, 38, 342, 2000.

Sevrioukova, I. F. and Poulos, T. L., Structural basis for regiospecific midazolam oxidation by human cytochrome P450 3A4. *Proc. Natl. Acad. Sci. U S A.*, 114, 486–491, 2017.

Shankar, P. et al., A review of the functional roles of the zebrafish aryl hydrocarbon receptors, *Toxicol. Sci.*, 178, 215–238, 2020.

Shaya, L., DeJong, C., and Wilson, J. Y., Expression patterns of cytochrome P450 3B and 3C genes in model fish species, *Comp. Biochem. Physiol.*, 166, 115–125, 2014.

Shaya, L., Jones, D. E., and Wilson, J. Y., CYP3C gene regulation by the aryl hydrocarbon and estrogen receptors in zebrafish, *Toxicol. Appl. Pharmacol.*, 362, 77–85, 2019.

Shen, C., et al., Generation of a Tg(cyp1a-12DRE:EGFP) transgenic zebrafish line as a rapid in vivo model for detecting dioxin-like compounds, *Aquat. Toxicol.*, 205, 174–181, 2018.

Shi, X. C. et al., Cytochrome P450 2AA molecular clone, expression pattern, and different regulation by fish oil and lard oil in diets of grass carp (Ctenopharyngodon idella). *Fish Physiol. Biochem.*, 44, 1019–1026, 2018.

Shimada, T. et al., Metabolism of benzo(*a*)pyrene to *trans*-7,8-dihydroxy-7,8-dihydrobenzo(*a*)pyrene by recombinant human cytochrome P450 1B1 and purified liver epoxide hydrolase, *Chem. Res. Toxicol.*, 12, 623, 1999.

Sikka, H. C. et al., Comparative metabolism of benzo(*a*)pyrene by liver microsomes from brown bullhead and carp. *Aquat. Toxicol.*, 16, 101, 1990.

Sim, E., Walters, K., and Boukouvala, S., Arylamine N-acetyltransferases: from structure to function, *Drug Metab. Rev.*, 40, 479–510, 2008.

Singh, Y. et al., Trout liver slices for metabolism and toxicity studies, *Drug Metab. Disp.*, 24, 7, 1996.

Smith, B. J. et al., Characterization of the effects of tebufelone on hepatic cytochromes P450 in the beagle dog, *Drug Metab. Disp.*, 24, 523, 1996.

Smith, E. M. et al., In vitro inhibition of cytochrome P450-mediated reactions by gemfibrozil, erythromycin, ciprofloxacin and fluoxetine in fish liver microsomes, *Aquatic Toxicol.*, 109, 259–266, 2012.

Soldano, S. et al., Xenobiotic metabolizing enzyme systems in test fish. IV. Comparative studies of liver microsomal and cytosolic hydrolases, *Comp. Biochem. Physiol.*, 101C, 117, 1992.

Srimaroeng, C., Perry, J. L., and Pritchard, J. B., Physiology, structure, and regulation of the cloned organic anion transporters, *Xenobiotica*, 38, 889–935, 2008.

Stegeman, J. J., Polynuclear aromatic hydrocarbons and their metabolism in the marine environment, in *Polycyclic Hydrocarbons and Cancer*, H. V. Gelboin, Ed., Academic Press, New York. Vol. 3, 1981, pp. 1–60.

Stegeman, J. J., Goldstone, J. V., and Hahn, M. E., Perspectives on zebrafish as a model in environmental toxicology, *Zebrafish Fish Physiol.*, 29, 367–439, 2010.

Stegeman, J. J. and Hahn, M. E., Biochemistry and molecular biology of monooxygenases: current perspectives on forms, functions, and regulation of cytochrome P450 in aquatic species, in *Aquatic Toxicology: Molecular, Biochemical, and Cellular Perspectives*, Malins, D. C. and Ostrander, G. K., Eds., Lewis Publishers, Boca Raton, FL, 1994, pp. 87–206.

Stegeman, J. J. and James, M. O., Individual variation in patterns of benzo(*a*)pyrene metabolism in the marine fish scup (*Stenotomus chrysops*), *Mar. Environ. Res.*, 17, 122, 1985.

Stegeman, J. J. and Woodin, B. R., Differential regulation of hepatic xenobiotic and steroid metabolism in marine teleost species, *Mar. Environ. Res.*, 14, 422, 1984.

Stegeman, J. J. et al., Cytochromes P450 (CYP) in tropical fishes: catalytic activities, expression of multiple CYP proteins and high levels of microsomal P450 in liver of fishes from Bermuda, *Comp. Biochem. Physiol.*, 116C, 61, 1997.

Stegeman, J. J. et al., Functional characterization of zebrafish cytochrome P450 1 family proteins expressed in yeast, *Biochim et Biophys Acta*, 1850, 2340–2352, 2015.

Steward, A. R. et al., Metabolism of 2-acetylaminofluorene by hepatocytes isolated from rainbow trout, *Toxicol. Appl. Pharmacol.*, 130, 188, 1995.

Stott, W. T. and Sinnhuber, R. O., Dietary protein levels and aflatoxin B metabolism in rainbow trout (*Salmo gairdneri*), *J. Environ. Pathol. Toxicol.*, 2(2), 379, 1978.

Straus, D. L. and Chambers, J. E., Inhibition of acetylcholinesterase and aliesterases of fingerling channel catfish by chlorpyrifos, parathion, and *S,S,S*-tributylphosphorotrithioate (DEF), *Aquat. Toxicol.*, 33, 311, 1995.

Straus, D. L., Schlenk, D., and Chambers, J. E., Hepatic microsomal desulfuration and dearylation of chlorpyrifos and parathion in fingerling channel catfish: lack of effect from Aroclor 1254, *Aquat. Toxicol.*, 50, 141, 2000.

Sturm, A. et al., Effects of prochloraz and nonylphenol diethoxylate on hepatic biotransformation enzymes in trout: a comparative *in vitro/in vivo*-assessment using cultured hepatocytes, *Aquat. Toxicol.*, 53, 229, 2001.

Sugahara, T. et al., Sulfation of hydroxychlorobiphenyls: molecular cloning, expression, and functional characterization of zebrafish SULT1 sulfotransferases, *Eur. J. Biochem.*, 270, 2404, 2003a.

Sugahara, T. et al., Sulphonation of dehydroepiandrosterone and neurosteroids: molecular cloning, expression, and functional characterization of a novel zebrafish SULT2 cytosolic sulphotransferase, *Biochem. J.*, 375, 785, 2003b.

Suiko, M. et al., Updated perspectives on the cytosolic sulfotransferases (SULTs) and SULT-mediated sulfation, *Biosci. Biotechnol. Biochem.*, 81, 63–72, 2017.

Sutter, T. R. et al., Complete cDNA sequence of a human dioxin-inducible mRNA identifies a new gene subfamily of cytochrome P450 that maps to chromosome 2, *J. Biol. Chem.*, 269, 13092, 1994.

Swain, L. and Melius, P., Characterization of benzo[*a*]pyrene metabolites formed by 3-methylcholanthreneinduced goldfish, black bullhead and brown bullhead, *Comp. Biochem. Physiol. C*, 79, 151, 1984.

Szklarz, G. D. and Halpert, J. R., Molecular modeling of cytochrome P450 3A4, *J. Comp. Aided Mol. Design*, 11, 265, 1997.

Takeshita, A. et al., Bisphenol-A, an environmental estrogen, activates the human orphan nuclear receptor, steroid and xenobiotic receptor-mediated transcription, *Eur. J. Endocrinol.*, 145, 513, 2001.

Taura, K. et al., Cytochrome P450 1A1 (CYP1A1) inhibitor alpha-naphthoflavone interferes with UDPglucuronosyltransferase (UGT) activity in intact but not in permeabilized hepatic microsomes from 3-methylcholanthrene-treated rats: possible involvement of UGT-P450 interactions, *Biol. Pharmaceut. Bull.*, 27, 56, 2004.

Thier, R. et al., Human glutathione *S*-transferase T1-1 enhances mutagenicity of 1,2-dibromoethane, dibromomethane and 1,2,3,4-diepoxybutane in *Salmonella typhimurium, Carcinogenesis*, 17, 163, 1996.

Tierbach, A. et al., Characterization of the mercapturic acid pathway, an important phase II biotransformation route, in a zebrafish embryo cell line, *Chem. Res. Toxicol.*, 33, 2863–2871, 2020.

Timme-Laragy, A. R. et al., CYP1B1 knockdown does not alter synergistic developmental toxicity of polycyclic aromatic hydrocarbons in zebrafish (Danio rerio), *Mar. Environ. Res.*, 66, 85, 2008.

Tong, Z. and James, M. O., Purification and characterization of hepatic and intestinal phenol sulfotransferase with high affinity for benzo[*a*]pyrene phenols from channel catfish, *Ictalurus punctatus*, *Arch. Biochem. Biophys.*, 376, 409, 2000.

Trute, M. et al., Characterization of hepatic glutathione S-transferases in coho salmon (Oncorhynchus kisutch), *Aquat. Toxicol.*, 81, 126–136, 2007.

Tseng, H.P., et al., Constitutive and xenobiotics-induced expression of a novel CYP3A gene from zebrafish larva. *Toxicol. Appl. Pharmacol.* 205, 247–258, 2005.

Tsuda, T. et al., Accumulation and excretion of fenthion, fenthion sulfoxide and fenthion sulfone by killifish (*Oryzias latipes*), *Comp. Biochem. Physiol.*, 113C, 45, 1996.

Ueda, O., Kitamura, S., and Ohta, S., Metabolism of 2-nitrofluorene, an environmental pollutant by liver preparations of sea bream (*Pagrus major*), *Xenobiotica*, 32, 667, 2002.

Uno, T., Ishizuka, M., and Itakura, T., Cytochrome P450 (CYP) in fish, *Environ. Toxicol. Pharmacol.*, 34, 1–13, 2012.

Van Den Boom, J. et al., 3′-Phosphoadenosine 5′-phosphosulfate (PAPS) synthases, naturally fragile enzymes specifically stabilized by nucleotide binding, *J. Biol. Chem.*, 287, 17645–17655, 2012.

van den Hurk, P., Gerzel, L. E., Calomiris, P., and Haney, D. C., Phylogenetic signals in detoxification pathways in Cyprinid and Centrarchid species in relation to sensitivity to environmental pollutants, *Aquat Toxicol.*, 188, 20–25, 2017.

van den Hurk, P. and Haney, D. C., Biochemical effects of pollutant exposure in fish from urban creeks in Greenville, SC (USA). *Environ Monit Assess.*, 189(5), 211, 2017. doi: 10.1007/s10661-017-5918-2. Epub 2017 Apr 8. PMID: 28391451.

van den Hurk, P. and James, M. O., Sulfation and glucuronidation of benzo[*a*]pyrene-7,8-dihydrodiol in intestinal mucosa of channel catfish (*Ictalurus punctatus*), *Mar. Environ. Res.*, 50, 11, 2000.

van den Hurk, P. et al., Hydroxylated polychlorinated biphenyls as inhibitors of the sulfation and glucuronidation of 3-hydroxybenzo(*a*)pyrene, *Environ. Health Perspect.*, 110, 343, 2002.

Van Der Sluis, R., Analyses of the genetic diversity and protein expression variation of the acyl: CoA medium-chain ligases, ACSM2A and ACSM2B, *Mol. Genet. Genom.*, 293, 1279–1292, 2018.

Van Waarde, A., Biochemistry of non-protein nitrogenous compounds in fish including the use of amino acids for anaerobic energy production, *Comp. Biochem. Physiol.*, 91B, 207, 1988.

Veiga-Da-Cunha, M. et al., Molecular identification of NAT8 as the enzyme that acetylates cysteine S-conjugates to mercapturic acids, *J. Biol. Chem.*, 285, 18888–18898, 2010.

Venkatachalam, K.-V. et al., Isolation, partial purification, and characterization of a novel petromyzonol sulfotransferase from *Petromyzon marinus* (lamprey) larval liver, *J. Lipid Res.*, 45, 486, 2004.

Vessey, D. A., Enzymes involved in the formation of amide bonds, in *Comprehensive Toxicology*, Vol. 3, Guengerich, F. P., Ed., Elsevier, Amsterdam, 1996.

Villalobos, A. S. et al., Dioxin-like properties of a trichloroethylene combustion-generated aerosol, *Environ. Health Perspect.*, 104, 734, 1996.

Vittozzi, L. et al., Mechanistic aspects of organophosphorothionate toxicity in fish and humans, *Environ. Int.*, 26, 125, 2001.

Vrolijk, N. H. et al., Toxicological and ecological implications of biotransformation enzymes in tropical teleost *Chaetodon capistratus*, *Mar. Biol.*, 119, 151, 1994.

Walker, C. H., Bentley, P., and Oesch, F., Phylogenetic distribution of epoxide hydratase in different vertebrate species, strains and tissues measured using three substrates, *Biochim. Biophys. Acta*, 539, 427, 1978.

Wall, K. L. and Crivello, J., Chlorzoxazone metabolism by winter flounder liver microsomes: evidence for existence of a CYP2E1-like isoform in teleosts, *Toxicol. Appl. Pharmacol.*, 151, 98, 1998.

Wallace, K. B. and Dargan, J. E., Intrinsic metabolic clearance of parathion and paraoxon by livers from fish and rodents, *Toxicol. Appl. Pharmacol.*, 90, 235, 1987.

Wang, H. and Strobel, H. W., Regulation of CYP3A9 gene expression by estrogen and catalytic studies using cytochrome P450 3A9 expressed in *Escherichia coli, Arch. Biochem. Biophys.*, 344, 365, 1997.

Wang, J., Grisle, S., and Schlenk, D., Effects of salinity on aldicarb toxicity to juvenile rainbow trout (*Oncorhynchus mykiss*) and striped bass (*Morone saxatilis×chrysops*), *Toxicol. Sci.*, 64, 200, 2001.

Wang, J. et al., Unraveling the structural basis of selective inhibition of human cytochrome P450 3A5, *J. Am. Chem. Soc.*, 143, 18467–18480, 2021.

Wang, L. et al., CYP1C1 messenger RNA expression is inducible by benzo(a)pyrene in fundulus heteroclitus embryos and adults, *Toxicol. Sci.*, 93, 331–340, 2006.

Wang, L. et al., Expression of CYP1C1 and CYP1A in fundulus heteroclitus during PAH-induced carcinogenesis. *Aquatic Toxicol.*, 99, 439–447, 2010.

Wang, Y., Huang, H., and Wu, Q., Characterization of the zebrafish Ugt repertoire reveals a new class of drug-metabolizing UDP glucuronosyltransferases, *Mol Pharmacol.*, 86, 62–75, 2014.

Washburn, P. C. and Di Giulio, R. T., Nitroaromatic stimulation of superoxide production in three species of freshwater fish, *Mar. Environ. Res.*, 24, 291, 1988.

Wassmur B., Grans J., Kling P., and Celander M.C., Interactions of pharmaceuticals and other xenobiotics on hepatic pregnane X receptor and cytochrome P450 3A signaling pathway in rainbow trout (*Oncorhynchus mykiss*). *Aquat. Toxicol.* 100, 91–100, 2010.

Watabe, T., Metabolic activation of 7,12-dimethylbenz[alpha] anthracene (DMBA) and 7-methylbenz[alpha]anthracene (7-MBA) by rat liver P-450 and sulfotransferase, *J. Toxicol. Sci.*, 8, 119, 1983.

Watson, D. E. et al., Aminoanthracene is a mechanism-based inactivator of CYP1A in channel catfish hepatic tissue, *Toxicol. Appl. Pharmacol.*, 135, 208, 1995.

Waxman, D. J. et al., Human liver microsomal steroid metabolism: identification of the major microsomal steroid hormone 6β-hydroxylase cytochrome P450 enzyme, *Arch. Biochem. Biophys.*, 263, 424, 1998.

Westkaemper, R. B. and Hanzlik, R. P., Mechanistic studies of epoxide hydrolase utilizing a continuous spectophotometric assay, *Arch. Biochem. Biophys.*, 208, 195, 1981.

White, R. D., Shea, D., and Stegeman, J. J., Metabolism of the aryl hydrocarbon receptor agonist 3,3′,4,4′-tetrachlorobiphenyl by the marine fish scup (*Stenotomus chrysops*) *in vivo* and *in vitro*, *Drug Metab. Disp.*, 25, 564, 1997b.

White, R. D. et al., Induction and post-transcriptional suppression of hepatic cytochrome P450 1A1 by 3,3′,4,4′-tetrachlorobiphenyl, *Biochem. Pharmacol.*, 53, 1029, 1997a.

Willett, K. L. et al., Comparative metabolism and excretion of benzo(*a*)pyrene in two species of Ictalurid catfish, *Toxicol. Sci.*, 58, 68, 2000.

Willett, K. L. et al., *In vivo* and *in vitro* inhibition of CYP1A-dependent activity in *Fundulus heteroclitus* by the polynuclear aromatic hydrocarbon (PAH) fluoranthene, *Toxicol. Appl. Pharmacol.*, 177, 264, 2001.

Williams, D. E. and Buhler, D. R., Benzo(*a*)pyrene-hydroxylase catalyzed by purified isozymes of cytochrome P450 from β-naphthoflavone-fed rainbow trout, *Biochem. Pharmacol.*, 33, 3743, 1984.

Williams, D. E. and Buhler, D. R., Purified form of cytochrome P-450 from rainbow trout with high activity toward conversion of aflatoxin $B_1$ to aflatoxin $B_1$-2,3-epoxide, *Cancer. Res.*, 43, 4752, 1983.

Wright, W. C., Chenge, J., and Chen, T., Structural perspectives of the CYP3A family and their small molecule modulators in drug metabolism, *Liver Res.*, 3, 132–142, 2019.

Wu, S., Li, H., and You, J., Identification of CYP isozymes involved in enantioselective metabolism of fipronil in fish liver: In vitro metabolic kinetics and molecular modeling, *Environ. Toxicol. Chem.*, 2021 Oct 29. doi: 10.1002/etc.5241.

Xi, Y., Seyoum, H., and Liu, M. C., Role of SULT-mediated sulfation in the biotransformation of 2-butoxyethanol and sorbitan monolaurate: A study using zebrafish SULTs, *Aquat. Toxicol.*, 177, 19–21, 2016.

Yan, J., and Cai, Z., Molecular evolution and functional divergence of the cytochrome P450 3 (CYP3) Family in Actinopterygii (ray-finned fish), *PLoS One*, 5, e14276, 2010.

Yancey, P. H. et al., Living with water stress: evolution of osmolyte systems, *Science*, 217, 1214, 1982.

Yang, M.-S. et al., Feasibility studies on the use of sewage sludge as supplementary feed for rearing tilapia: II. PcBs of the treated fish, and biochemical response in the fish liver, *Environ. Technol.*, 14, 1163, 1993.

Yang, Y.-H. et al., CYP2M1: cloning, sequencing, and expression of a new cytochrome P450 from rainbow trout liver with fatty acid (ω-6)-hydroxylation activity, *Arch. Biochem. Biophys.*, 352, 271, 1998.

Yang, Y.-H. et al., Heterologous expression of CYP2K1 and identification of the expressed protein (BVCYP2K1) as lauric acid (ω-1)-hydroxylase and aflatoxin $B_1$ *exo*-epoxidase, *Drug Metab. Disp.*, 28, 1279, 2000.

Yasuda, T. et al., Characterization and ontogenic study of novel steroid-sulfating SULT3 sulfotransferases from zebrafish. *Mol. Cell Endocrinol.*, 294, 29–36, 2008.

You, L., Steroid hormone biotransformation and xenobiotic induction of hepatic steroid metabolizing enzymes, *Chem. Biol. Interact.*, 147, 233, 2004.

Yuan, Z. X. et al., Comparative metabolism of benzo[*a*]pyrene by liver microsomes of channel catfish and brown bullhead, *Environ. Toxicol. Chem.*, 16, 835, 1997.

Zhou, Y. et al., Generation and application of a Tg(cyp1a:egfp) transgenic marine medaka (Oryzias melastigma) line as an in vivo assay to sensitively detect dioxin-like compounds in the environment. *J. Haz. Mat.*, 391,122192, 2020.

Zhu, S., King, C. S., and Haasch, L. M. 2006. Environmental induction of CYP1A-, CYP2M1-, and CYP2K1-like proteins in tropical fish species by produced formation water on the northwest shelf of Australia. *Mar. Environ. Res.*, 62, 322–326, 2006.

Ziegler, D. M., Flavin-containing monooxygenases: catalytic mechanisms and substrate specificities, *Drug Metab. Rev.*, 19, 1, 1988.

Ziegler, D. M., Recent studies on the structure and function of multisubstrate flavin-containing monooxygenases, *Ann. Rev. Pharmacol. Toxicol.*, 33, 179, 1993.

Ziegler, D. M. and Mitchell, C. H., Microsomal oxidase. IV. properties of a mixed function amine oxidase isolated from pig liver microsomes, *Arch. Biochem. Biophys.*, 150, 116, 1972.

Zilberman-Peled, B. et al., Transcriptional regulation of arylalkylamine-N-acetyltransferase-2 gene in the pineal gland of the gilthead seabream, *J. Neuroendocrinol.*, 19, 46–53, 2007.

# 4 Reactive Oxygen Species and Redox Stress

*Alicia R. Timme-Laragy*
University of Massachusetts Amherst

*Richard T. Di Giulio and Joel N. Meyer*
Duke University

**Abstract**

The purpose of Chapter 4 is to describe a particular set of phenomena that collectively comprise an important mechanism of chemical toxicity and cellular defense. These phenomena, referred to as *redox stress*, encompass oxidative and reductive stress and apply to a diverse array of chemicals, ultimately resulting in a diverse array of health outcomes. The study of redox stress broadly includes biological phenomena associated with the generation of reactive oxygen species (ROS), molecular systems designed to protect cells from ROS (i.e., *antioxidant defense systems*), and the active vs. passive deleterious impacts of ROS. As in many areas of mechanistic toxicology, understanding of redox stress in fishes has lagged behind understanding in mammals and, indeed, has depended to a large degree on studies in other biological systems for the development of methodologies and concepts; however, the study of redox stress in fishes is currently a vibrant area of investigation that is revealing a number of important similarities and differences with other organisms. Moreover, the multitude of chemicals entering freshwater and marine systems that can contribute to redox stress underlies the need for information on this phenomenon in fishes. In this chapter, we describe fundamental aspects of ROS chemistry and generation, antioxidant defense systems, cellular and organismal impacts, and specific mechanisms by which pollutants play roles in these processes. We also describe the current state of understanding of redox stress in fishes, technological advances, and potential future directions for the field.

## INTRODUCTION

Mechanisms underlying the toxicity of numerous environmental pollutants are discussed throughout this book. The purpose of this chapter is to describe a particular set of phenomena that collectively comprise an important mechanism of chemical toxicity and cellular defense. These phenomena, referred to as *redox stress* (Timme-Laragy et al. 2018), encompass oxidative and reductive stress and apply to a diverse array of chemicals, ultimately resulting in a diverse array of health outcomes. The study of redox stress broadly includes biological phenomena associated with the generation of reactive oxygen species (ROS), molecular systems designed to protect cells from ROS (i.e., *antioxidant defense systems*), and the active vs. passive deleterious impacts of ROS (Figure 4.1). Excellent detailed monographs on ROS are Halliwell and Gutteridge (2015), and Sies et al. (2017, 2022). As in many areas of mechanistic toxicology, understanding of redox stress in fishes has lagged behind understanding in mammals and, indeed, has depended to a large degree on studies in other biological systems for the development of methodologies and concepts; however, the study of redox stress in fishes is currently a vibrant area of investigation that is revealing a number of important similarities and differences with other organisms. Moreover, the multitude of chemicals entering freshwater and marine systems that can contribute to redox stress underlies the need for information on this phenomenon in fishes. In this chapter, we describe fundamental aspects of ROS chemistry and generation, antioxidant defense systems, cellular and organismal impacts, and specific mechanisms by which pollutants play roles in these processes. We also describe the current state of understanding of redox stress in fishes, technological advances, and potential future directions for the field.

## REACTIVE OXYGEN SPECIES AND FREE RADICALS

Oxygen (as dioxygen, $O_2$) currently accounts for about 21% of the Earth's atmosphere, and oxygen is the most abundant element in the planet's crust (at 54%); however, life first evolved under essentially anaerobic conditions, and $O_2$ is believed to have first appeared as a by-product of photosynthesis by blue-green algae (cyanobacteria) beginning about 2.5 billion years ago (Harman 1986). As atmospheric $O_2$ gradually rose over the ensuing millennia, several notable developments occurred:

1. Organisms remaining anaerobic either retreated to anaerobic microenvironments or perished due to the effects of $O_2$ such as oxidation of metabolic intermediates (such as thiols, iron, and pteridines), enzyme inhibition (such as occurs with nitrogenases), and effects of ROS generated during oxidations.
2. Aerobic organisms evolved that took advantage of the electron-accepting capacity of $O_2$. The energetic efficiency of $O_2$ as an electron acceptor over those employed by anaerobes (e.g., iron and sulfur) enhanced the evolution of more advanced life forms

DOI: 10.1201/9781003160694-4

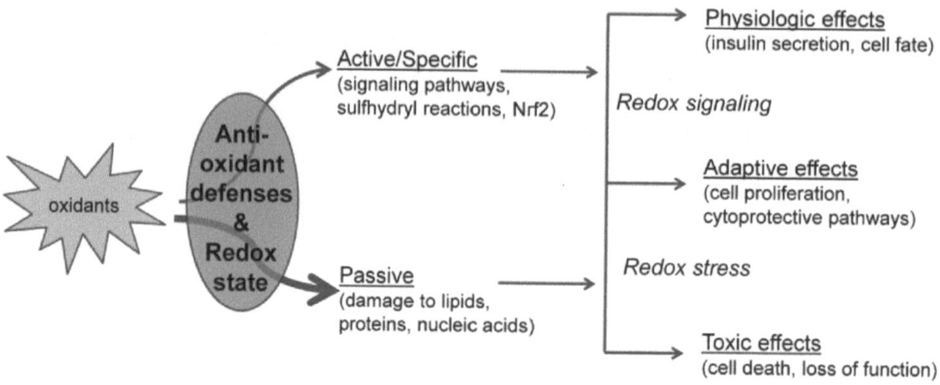


**FIGURE 4.1** Delineation of active and/or specific effects of redox signaling vs. the passive and toxic impacts. From Timme-Laragy et al. (2018).

(i.e., eukaryotic organisms). This switch required the concomitant development of mechanisms to protect these organisms from the toxic effects of $O_2$.

3. Ozone ($O_3$) levels in the atmosphere rose, giving organisms protection from intense solar ultraviolet radiation; this also enhanced eukaryotic evolution.

Although the focus of this chapter is on toxicities associated with oxygen species, the critical role played by $O_2$ in shaping and promoting advanced life forms as we know them is of fundamental importance.

The primary basis for the toxicity of oxygen lies in its propensity to undergo electron transfers that yield reactive intermediates, here termed *ROS*. On the other hand, the ability of $O_2$ to accept electrons underlies its utility to aerobic organisms; for example, a key function of $O_2$ is to serve as the terminal electron acceptor in mitochondrial electron transport, which drives the production of high-energy adenosine triphosphate (ATP). In this process, $O_2$ is reduced to $H_2O$; this is a four-electron reductive process that proceeds sequentially through the one-, two-, and three-electron products. These univalent reductions of $O_2$ to water are shown in Equations 4.1–4.4:

$$O_2 + e^- \rightarrow O_2^{\bullet-} \tag{4.1}$$

$$O_2^{\bullet-} + e^-\ 2H^+ \rightarrow H_2O_2 \tag{4.2}$$

$$H_2O_2 + e^-\ H^+ \rightarrow \bullet OH + H_2O \tag{4.3}$$

$$\bullet OH + e^-\ H^+ \rightarrow H_2O \tag{4.4}$$

$$\text{Sum: } O_2 + 4e^- \rightarrow H_2O \tag{4.5}$$

The one-, two-, and three-electron products shown in Equations 4.1, 4.2, and 4.3, respectively, are the *superoxide anion radical, hydrogen peroxide*, and the *hydroxyl radical*, respectively. The two radical species are free radicals; as defined by Halliwell and Gutteridge (2015), a free radical "is any species capable of independent existence that contains one or more unpaired electrons." The existence of unpaired electrons tends to make free radicals reactive, although the range of reactivity among free radicals is extensive. $O_2^{\bullet-}$, for example, is relatively weakly reactive, while ·OH is extremely reactive and is among the most potentially deleterious compounds among radicals and ROS encountered in cells. Thus, ·OH reacts indiscriminately with cellular constituents such as membrane lipids, proteins, and DNA. Due to its reactivity, it has an *in vivo* lifetime measured in nanoseconds (Pryor 1986). Interestingly, $O_2$ itself is by definition a free radical; it contains two unpaired electrons, each in a different $\pi$ orbital and having parallel spins. This last feature impedes the ability of $O_2$ to act as an oxidant via accepting electron pairs from typical molecules containing electron pairs with opposite spins. If it were not for this spin restriction, $O_2$ would qualify as an extremely potent ROS, and aerobic life undoubtedly would have a very different nature.

Although $H_2O_2$ is not a technically radical, it is an important ROS. It is a weak reducing and oxidizing agent, which lends to its use as a disinfectant at high concentrations. Its uncharged nature permits it to cross cell membranes and act as a signaling molecule; however, its ability to serve as a precursor to ·OH at physiologically meaningful levels underlies its significance to cellular oxidative stress. This ability of $H_2O_2$ to generate ·OH via reaction with $O_2^{\bullet-}$ was first postulated by Haber and Weiss (1934), who proposed the following reaction:

$$O_2^{\bullet-} + H_2O_2 \rightarrow \bullet OH + OH^- + O_2 \tag{4.6}$$

Reaction 4.6 is termed the *Haber–Weiss reaction*, and reaction rates in aqueous solution, although thermodynamically favorable, are very low (Haber and Weiss 1934); however, as noted by Weiss in 1935, this reaction can be effectively catalyzed by transition metals such as iron (Weiss 1935), as shown in Equations 4.7 and 4.8:

$$Fe^{3+} + O_2^{\bullet-} \rightarrow Fe^{2+} + O_2 \tag{4.7}$$

$$Fe^{2+} + H_2O_2 \rightarrow \bullet OH + OH^- + Fe^{3+} \tag{4.8}$$

$$\text{Net: } O_2^{\bullet-} + H_2O_2 \rightarrow \bullet OH + OH^- + O_2 \tag{4.9}$$

The net reaction, Equation 4.9, is the Haber–Weiss reaction. Equation 4.8 is referred to as the *Fenton reaction* (Walling 1975). Other transition metals present in cells, particularly copper, also can participate in Fenton-like reactions. The abilities of iron and copper to participate in cellular ·OH generation, in addition to their roles as essential nutrients, may underlie the existence of proteins (such as ferritin for iron and ceruloplasmin and metallothionein for copper) that tightly regulate their cellular transport and fate.

In addition to these three $O_2$ reduction products, other ROS and related chemicals can play important roles in oxidative stress. These include singlet oxygen, ozone, alkoxyl and peroxyl radicals, and reactive nitrogen species (RNS), such as nitric oxide and peroxynitrite (Halliwell and Gutteridge 2015). Additionally, carbon- and sulfur-centered free radicals occur and affect cellular function. Singlet oxygen (commonly denoted as $^1O_2$) is an excited state of $O_2$ at 22.4 kcal above ground state. In this form, the outer $\pi$ electrons are paired and in antiparallel spins; thus, $^1O_2$ is not a free radical, but the spin restriction of ground state $O_2$ is removed, making $^1O_2$ more reactive than $O_2$. A major source of $^1O_2$ in the environment is through excitation of ultraviolet-absorbing chemicals, including many polycyclic aromatic hydrocarbons (PAHs). This reaction has important ramifications in aquatic toxicology and is discussed later in this chapter.

Ozone ($O_3$) also is not a free radical but is a much more powerful oxidizing agent than $O_2$. Ozone is produced in the stratosphere by the photodissociation of $O_2$ into oxygen atoms, which then react with $O_2$ to produce $O_3$. Tropospheric (ground-level) $O_3$ is produced by sunlight-enhanced reactions between nitric oxides and volatile organic compounds; warmer temperatures and fossil-fuel combustion produce nitric oxides, which enhance this process. While stratospheric $O_3$ provides protection to organisms by reducing the solar ultraviolet radiation that reaches the Earth's surface, ground-level $O_3$ can be toxic, particularly to animal respiratory systems and plant photosynthesis.

Alkoxyl (RO·) and peroxyl (ROO·) radicals are potent oxidants that often arise in cells subsequent to ·OH attack of organic chemicals. Oftentimes, such attack initially produces carbon-based radicals, which under aerobic conditions react with $O_2$ to produce RO· and ROO·. This process underlies membrane lipid peroxidation, a hallmark cellular toxicity associated with ROS that is described below.

Although not the focus of this chapter, RNS also play important roles in cellular physiology and toxicity and interact with ROS. Nitric oxide (NO·) is synthesized in many organisms by *nitric oxide synthases* (NOSs); these enzymes catalyze the oxidation of arginine to produce NO· and citrulline (Muriel 2000). Three forms of NOS have been identified in mammals, referred to as type 1, constitutive, or neuronal NOS (cNOS, nNOS); type 2 or inducible NOS (iNOS); and type 3 or endothelial NOS (eNOS). These enzymes are also found in fish, although NOS3 appears to have been lost in the teleost radiation, and NOS2 is missing from some teleost fish, while has undergone a duplication event in other teleosts (*nos2a* and *nos2b*) (Cioni et al. 2019).

NO· has been studied predominantly from the standpoint of the normal physiological roles it plays. These include acting as mediators of phagocytosis by macrophages and neutrophils and functioning as a neurotransmitter to relax smooth muscle and thereby produce vasodilation (Neumann et al. 2001, Ignarro 2002, Toda and Okamura 2003). NO· is not particularly reactive with nonradicals but is highly reactive with other radicals. Toxicities associated with NO·, observed in some instances, are thought in large part to be due to its facile reaction with $O_2^{\bullet -}$. This reaction has at least two potentially deleterious consequences: (1) the loss of functional NO·, thereby interfering with vasodilation (and potentially leading to hypertension, for example), and (2) the production of peroxynitrite via the reaction:

$$O_2^{\bullet -} + NO\bullet \rightarrow ONOO^- \qquad (4.10)$$

Peroxynitrite is a very powerful oxidizing species that can elicit cytotoxicity through several mechanisms, including those described below for ROS as well as by nitration of DNA bases and aromatic amino acids (Beckman and Koppenol 1996).

## ENDOGENOUS CELLULAR SOURCES OF REACTIVE OXYGEN SPECIES

Before describing mechanisms by which chemical pollutants can generate ROS and redox stress in fishes, it is important to discuss mechanisms by which these species arise normally. As alluded to earlier, the cellular generation of ROS is an unavoidable cost of aerobic life—the price exacted for making use of the energetic efficiency of $O_2$ as an electron acceptor.

Perhaps quantitatively the most important sources of ROS during normal cellular metabolism are the electron transport chains of mitochondria, the endoplasmic reticulum, and in plants, chloroplasts. Photosynthesis within chloroplasts poses unique problems for plants in terms of oxidative damage, which can be exacerbated by many common air pollutants (Hippeli and Elstner 1996). In animals, mitochondrial respiration is usually the most important internal source of ROS. It is estimated that under normal cellular levels of $O_2$ consumption, about 0.1% of the $O_2$ utilized by mitochondria forms $O_2^{\bullet -}$ (Fridovich 2004). Depending on conditions, mitochondrial production of ROS can occur in at least 11 sites, depending on fuel source and other conditions, with complex I (with release to the matrix) and III (with release to both the matrix and intermembrane space) typically quantitatively most important (Wong et al. 2017). Most sites of ROS production are in the matrix and in the form of superoxide anion, which due to its charge and short half-life is not normally expected to exit the mitochondria (with the exception of during apoptosis). Conversion to $H_2O_2$ may permit exit from the matrix; in addition, some complex III production of ROS (when it occurs at site IIIQo), as well as ROS production by glycerol 3-phosphate dehydrogenase, results in release to the intermembrane space (Wong et al. 2017), which may facilitate signaling (Ježek et al. 2020). An excellent review of superoxide anion and hydrogen peroxide anion formation by mitochondria, their conversion to other ROS species and effects, and how to measure them and their molecular and organismal effects is provided by Brand (2020).

Endoplasmic reticula (*microsomes* in subcellular fractions) are another important source of ROS production. Oxidative metabolism of endogenous compounds and xenobiotics by the cytochrome P450 enzyme systems that function in this fraction are discussed in detail in Chapter 2. As described, P450s catalyze oxidations by cleaving $O_2$, with one oxygen atom ultimately added to the substrate and the other ultimately giving rise to $H_2O$. In this process, two electrons provided by NADPH–P450 reductase or cytochrome $b_5$ are sequentially added to drive catalysis. This cycle can be uncoupled, resulting in the diversion of electrons to give rise to $O_2^{\bullet-}$ and $H_2O_2$ (Goeptar et al. 1995). Some xenobiotics can greatly enhance this uncoupling, as described later in this chapter.

Several oxidative enzymes present in the cytosol, lysosomes, and peroxisomes (Di Meo et al. 2016) can also generate ROS during catalysis, including xanthine oxidase, tryptophan dioxygenase, diamine oxidase, guanyl cyclase, and glucose oxidase (Fridovich 1978, Halliwell and Gutteridge 2015, Sies et al. 2017). Additionally, a number of molecules with key roles in cellular function can auto-oxidize in the presence of $O_2$, yielding $O_2^{\bullet-}$; these include glyceraldehyde, $FMNH_2$, $FADH_2$, epinephrine and norepinephrine, dopamine, tetrahydropteridines, and thiols such as cysteine (Halliwell and Gutteridge 2015). These autoxidations are oftentimes substantially accelerated in the presence of transition metal ions such as iron and copper. The $O_2$-carrying proteins hemoglobin and myoglobin can also be sources of $O_2^{\bullet-}$ (Brantley et al. 1993, Balagopalakrishna et al. 1996). In these proteins, $O_2$ binding requires the reduced state of iron, Fe(II), in the heme group. Occasionally, oxygen will release as $O_2^{\bullet-}$ and concomitantly yield Fe(III) in the heme. These forms of hemoglobin and myoglobin are referred to as *methemoglobin* and *metmyoglobin*, respectively, and are unable to bind $O_2$. Approximately 3% of human hemoglobin is thought to undergo conversion to methemoglobin every day, suggesting a substantial basal flux of ROS in erythrocytes (Winterbourn 1985). As discussed later, many chemicals can significantly increase methemoglobin formation.

ROS are often cast as bad actors, but it is important to note that ROS play positive roles as well. These include signaling processes critical for stress response, metabolic regulation, cell division, and organismal development (Love et al. 2013, Reczek and Chandel 2015, Shadel and Horvath 2015). Another important example is the production of ROS by neutrophils and macrophages by many vertebrates wherein ROS are employed in the phagocytic activity of these cells (Babior 2000). Upon stimulation by pathogens, these cells increase $O_2$ consumption up to 20 times resting levels; this phenomenon is referred to as the *respiratory burst* (Pinegin et al. 2018). During the respiratory burst, NADPH provided by the pentose phosphate pathway is oxidized to NADP+ by NADPH oxidase, an enzyme complex associated with the plasma membrane (Henderson and Chappel 1996); the two electrons abstracted from each NADPH are transferred to $O_2$, yielding $O_2^{\bullet-}$ as indicated in Equation 4.11:

$$\text{NADPH} + 2O_2 \rightarrow \text{NADP}^+ + H^+ + 2O_2^{\bullet-} \quad (4.11)$$

Because $O_2^{\bullet-}$ is not highly reactive in aqueous solution, it is considered unlikely that $O_2^{\bullet-}$ itself is the ultimate phagocytic agent employed by neutrophils and macrophages; however, $O_2^{\bullet-}$ can readily react with itself and dismutate into hydrogen peroxide (Equation 4.12) or react with NO·, which is sometimes produced by these cells, to produce peroxynitrite (see Equation 4.10). Both products are bactericidal.

$$2O_2^{\bullet-} + 2H^+ \rightarrow H_2O_2 + O_2 \quad (4.12)$$

Additionally, $H_2O_2$ in the presence of iron within the bacterium may produce cytotoxic ·OH via the Fenton reaction (see Equation 4.8). The enzyme myeloperoxidase occurs in neutrophils but not macrophages (Hampton et al. 1998). This enzyme, first isolated from human pus, is a nonspecific heme-containing peroxidase that in the presence of $H_2O_2$ catalyzes the oxidation of chloride ion ($Cl^-$) to hypochlorous acid (HOCl), a potent bactericidal oxidant.

Although the generation of ROS by these phagocytic cells is generally beneficial to the host organism, chronic inflammation can produce damage in local tissues and can contribute to diseases associated with oxidative stress, including arthritis and some cancers. The activities of phagocytes in fishes, including ROS generation, are described in Chapter 6.

## ANTIOXIDANT DEFENSES

As described above, the generation of ROS is an inescapable part of aerobic life. To flourish, aerobic organisms have evolved mechanisms to regulate ROS, and healthy organisms generally can cope well with the normal flux of ROS associated with respiration, certain enzyme activities, autoxidations, phagocytosis, and so forth. These mechanisms include enzyme systems that act to quench ROS, low-molecular-weight compounds that directly scavenge ROS (either produced endogenously or obtained from the diet), and proteins that act to sequester prooxidants. A number of these mechanisms, particularly some antioxidant enzymes, have been highly conserved in the course of evolution. Among vertebrates, most mechanisms are very similar, with differences among phyla generally being more quantitative than qualitative in nature; thus, extensive research in this area with mammalian models has greatly informed work with fishes.

### Antioxidant Enzyme Systems

#### Superoxide Dismutases

A superoxide dismutase (SOD) was first isolated from bovine blood (McCord and Fridovich 1969); since that time, several forms of this enzyme have been identified and characterized (Fridovich 1995). These enzymes accelerate the dismutation of $O_2^{\bullet-}$ by accepting an electron from one $O_2^{\bullet-}$ and passing it to another, yielding the following net reaction:

$$O_2^{\bullet-} + O_2^{\bullet-} \rightarrow H_2O_2 + O_2 \quad (4.13)$$

The observations that practically all aerobic organisms contain SOD, that many organisms including vertebrates produce several distinct SOD proteins, and that these enzymes enhance a reaction that occurs rapidly and nonenzymatically support the theory that $O_2^{\bullet-}$ plays an important role in redox stress, despite its low reactivity vs. some other ROS (notably, ·OH).

Vertebrates contain three distinct forms of SOD: copper/zinc (CuZnSOD; SOD1), manganese (MnSOD; SOD2), and extracellular (ECSOD;SOD3). CuZnSOD occurs predominantly in the cytosol, with some present in lysosomes and the nucleus. Eukaryotic CuZnSODs have a molecular mass of about 32,000 and contain two protein subunits, each containing one Cu and one Zn ion. At physiological pH, the uncatalyzed dismutation of $O_2^{\bullet-}$ has a rate constant on the order of $5 \times 10^5$ $M^{-1}s^{-1}$; bovine erythrocyte CuZnSOD accelerates this reaction to about $1.6 \times 10^9$ $M^{-1}s^{-1}$, effectively reducing the half-life of cellular $O_2^{\bullet-}$ by several orders of magnitude. CuZnSOD is highly sensitive to inhibition by cyanide, a characteristic useful in distinguishing its activity from MnSOD in analytical assays. Diethyldithiocarbamate is also a potent inhibitor of CuZnSOD.

Manganese SOD in animals is highly associated with mitochondria; it also occurs in bacteria and plants. Amino acid sequences of MnSODs across these phyla are quite similar (and distinct from those of CuZnSODs), consistent with the endosymbiotic theory that proposes that mitochondria evolved from a symbiosis between an early CuZnSOD-containing eukaryote and a MnSOD-containing prokaryote (Fridovich 1995). Interestingly, the blue crab (*Callinectes sapidus*) and other marine crustaceans that rely on Cu-dependent hemocyanins for $O_2$ transport contain no CuZnSOD, but have MnSOD in both mitochondria and cytosol (Brouwer et al. 1997). Vertebrate MnSODs have a molecular mass of about 92,000, and they usually contain four protein subunits and 2 or 4 Mn ions per enzyme. MnSODs are insensitive to inhibition by both cyanide and diethyldithiocarbamate. Despite these differences between MnSOD and CuZnSOD, both catalyze essentially the same reaction (Equation 4.13).

Extracellular SOD is also a CuZnSOD but it has a far higher molecular weight than cytosolic CuZnSOD (135,000 vs. 32,000). ECSOD is a tetrameric glycoprotein in which each subunit contains one Cu and one Zn ion. It is particularly abundant in the extracellular space of blood vessels, bound to heparin sulfate, the glycosaminoglycan component of heparin sulfate proteoglycans that interact with various proteins on cell surfaces (Fukai et al. 2002). The large mass of this tetrameric protein and its affinity for cell surfaces prevent filtration by the kidneys (Fridovich 1998). In blood vessels, ECSOD appears to play a critical role in protecting NO· from $O_2^{\bullet-}$. NO· produced by endothelial cells stimulates smooth muscle relaxation; thus, diminished ECSOD expression or activity may play a role in cardiovascular diseases (Fukai et al. 2002).

Another SOD found in bacteria, algae, and higher plants (but not observed in animal tissues) contains iron; hence, it is termed FeSOD. Most FeSODs contain two subunits, and each enzyme molecule has one or two iron ions. The bacterium *Escherichia coli* also contains a hybrid dimeric SOD, with one subunit from MnSOD and the other from FeSOD; this hybridization is facilitated by the similar amino acid sequences shared by the two forms (Fridovich 1995).

## Catalases

Although SODs very effectively reduce cellular $O_2^{\bullet-}$ concentrations, the downside of this activity is that one of the two products produced is $H_2O_2$. Hydrogen peroxide can be damaging in its own right, and in the presence of transition metals such as copper and iron, it can also serve as the precursor to highly reactive ·OH. Most aerobic organisms and all vertebrates possess two enzyme systems that metabolize $H_2O_2$: the catalases and peroxidases (including glutathione peroxidases (GPXs) found in vertebrates, discussed below). Peroxidases generally can act on a variety of organic peroxides as well as $H_2O_2$, whereas catalases are largely restricted to $H_2O_2$.

Vertebrate catalases are large proteins with a molecular mass of 120,000 and consisting of four subunits, each containing a ferric heme group at its active site (Reid et al. 1981). The heme groups are deep within the subunits and accessed by narrow channels, which accounts for the substrate specificity of catalase. They catalyze a reaction conceptually similar to that catalyzed by SOD, a dismutation reaction (Equation 4.14). Aminotriazole is an effective catalase inhibitor (Darr and Fridovich 1986).

$$2H_2O_2 \rightarrow O_2 + 2H_2O \tag{4.14}$$

The major cellular location of vertebrate catalase is in peroxisomes. Peroxisomes are organelles that function in the β-oxidation of fatty acids, and $H_2O_2$ is produced as a by-product of this process; thus, catalase prevents damage to peroxisomes and also impedes the movement of $H_2O_2$ to other locations in the cell (due to its uncharged nature, $H_2O_2$ traverses organelle membranes more readily than other ROS). The role that catalases play in the metabolism of $H_2O_2$ produced outside of peroxisomes appears to vary among tissues. The next enzyme discussed, GPX, has a broader distribution within cells and likely plays a more important role in clearing hydrogen (and other) peroxides produced in some tissues, such as liver; however, under conditions elevating $H_2O_2$ production in some tissues including erythrocytes (that do not contain peroxisomes), lung, and eye, catalase activity may be particularly important (Halliwell and Gutteridge 2015).

## Glutathione Peroxidases and Transferases

GPXs provide another mechanism by which animals can detoxify $H_2O_2$; moreover, they can also reduce fatty acid peroxides (LOOH), an important manifestation of oxidative stress described later. Four selenium-dependent GPXs have been identified broadly in mammals: classical GPX (GPX1), gastrointestinal GPX (GPX2), plasma GPX (GPX3), and phospholipid hydroperoxide GPX (GPX4). All contain selenocysteine—an amino acid analogous to cysteine in which the sulfur atom has been replaced by selenium—at the active site. Most GPXs characterized in vertebrates contain four subunits, each containing one selenium residue; GPX4, however is monomeric. Three selenium-independent GPX enzymes have also been identified: GPX 5, GPX 7, and GPX8; GPX6 is unusual as it does contain a selenium in humans, but not in rats or mice (reviewed in Brigelius-Flohé and Maiorino (2013)).

Based on tissues examined and substrates used, it is likely that most selenium-dependent GPX activity measured in fish thus far corresponds to GPX1. In zebrafish, Gpx1a and Gpx1b are co-orthologs to the mammalian GPX1, and Gpx4a and Gpx4b are co-orthologs to the mammalian GPX4. Zebrafish also contain a Gpx 2, 3, 5, 7, and 8; for further identification and expression patterns during zebrafish embryonic development of glutathione (GSH)-related genes, the reader is referred to Timme-Laragy et al. (2013).

The reactions catalyzed by GPXs involve the reduction of a peroxide substrate to its corresponding alcohol, coupled with the oxidation of reduced GSH to glutathione disulfide (GSSG) (Chance et al. 1979). In the case of $H_2O_2$, the corresponding alcohol is water (Equation 4.15); with lipid peroxides, it is the corresponding lipid alcohol (LOH, Equation 4.16):

$$H_2O_2 + 2GSH \rightarrow GSSG + H_2O \quad (4.15)$$

$$LOOH \rightarrow 2GSH \rightarrow GSSG + LOH \quad (4.16)$$

An important distinction between catalase and GPX is the ability of only the latter to reduce lipid peroxides; however, most GPXs, including classical GPX, cannot metabolize lipid peroxides that are esterified to lipid molecules in membranes, for example. Such peroxides must first be released by lipase activity. An exception to this is GPX4, which can directly reduce lipid peroxides associated with membranes (and also free LOOH).

Some glutathione S-transferases (GSTs) exhibit peroxidase activity, which is specific for LOOH (they are unable to act on $H_2O_2$). This activity is sometimes referred to as *selenium-independent* or *nonselenium* GPX activity, as GSTs contain no selenium. Although it can account for an appreciable portion of total GPX activity measured in tissue preparations with standard substrates such as cumene hydroperoxide, its significance *in vivo* is unclear (Halliwell and Gutteridge 2015). Quantification of real GPX activity using $H_2O_2$ as substrate is sometimes preferred. Some GSTs play another role in antioxidant defense by conjugating breakdown products from lipid peroxidation such as 4-hydroxynonenal (Hayes et al. 2005).

## Glutathione: Synthesis and Maintenance

Glutathione (GSH) plays critical roles in the protection of cells from chemical insult. It also plays additional roles in homeostasis, cell fate decisions, metabolism, biosynthesis, transport, and cellular communication. GSH is the tripeptide γ-glutamyl-cysteinyl-glycine (Figure 4.1); concentrations vary widely among species and tissue but are generally in the low millimolar range (Timme-Laragy et al. 2013). GSH is synthesized through two reactions, the first catalyzed by glutamate cysteine ligase (GCL; previously termed γ-glutamylcysteine synthase, or GCS) and the second by GSH synthetase (Equation 4.17 and Equation 4.18, respectively):

$$\text{L-glutamate} + \text{L-cysteine} + \text{ATP} \rightarrow \text{L-}\gamma\text{-glutamylcysteine} + \text{ADP} + P_i \quad (4.17)$$

$$\text{L-}\gamma\text{-glutamy1} - \text{L-cysteine} + \text{L-glycine} + \text{ATP} \rightarrow \text{GSH} + \text{ADP} + P_i \quad (4.18)$$

Glutamate cysteine ligase is the rate-limiting enzyme of GSH synthesis; it is sensitive to inhibition by buthionine sulfoximine (BSO), which can be used experimentally to halt GSH synthesis (Meister 1995). GCL has been purified from a number of eukaryotes, and all are heterodimers composed of a heavy and a light subunit (molecular weights of about 73,000 and 31,000, respectively) (Griffith 1999, Griffith and Mulcahy 1999). The heavy subunit ($GCL_h$ or GCLC, where C stands for "catalytic") is responsible for catalytic activity; the light subunit ($GCL_l$ or GCLM, where M stands for "modifier") plays a regulatory function. In mammals, genes for the two are on separate chromosomes, and their transcription does not appear to be highly coordinated, although both have an antioxidant response element (ARE) in the promoter region (discussed below). Factors affecting GCL activity include availability of substrates (glutamate and cysteine) and feedback inhibition by GSH. GCLM modulates activity of the complete enzyme essentially by reducing $K_m$ values for substrate concentrations and increasing the $K_i$ for GSH.

As described above, GPX couples the reduction of $H_2O_2$ and LOOH to corresponding alcohols with the oxidation of GSH to GSSG. GSH can also directly scavenge ROS, including $O_2^{\bullet-}$, ·OH, RO·, ROO·, and $ONOO^-$ (Griffith 1999, Halliwell and Gutteridge 2015), with the thiol (–SH) group provided by cysteine as the active moiety that undergoes oxidation. In these cases as well, GSSG arises from GSH oxidation as a consequence. GSSG contains the backbones of two GSH molecules, covalently linked through a disulfide bond (Figure 4.1). Normal healthy cells not experiencing undue oxidative pressure typically exhibit GSH:GSSG ratios approaching or greater than 100:1, and declines in this ratio have been used as a marker of oxidative stress; the Nernst equation has also been used to calculate the glutathione redox potential (Kirlin et al. 1999, Timme-Laragy et al. 2013). This glutathione redox balance is determined by the net output of GSH synthesis, export of oxidized and conjugated glutathione, and activity of glutathione reductase (GR), which catalyzes the recycling of GSSG to 2 molecules of GSH through a reduction reaction:

$$GSSG + NADPH + H^+ \rightarrow 2GSH + NADP^+ \quad (4.19)$$

Glutathione reductase contains two subunits, each with fatty acid desaturase (FAD) at its active site (Thieme et al. 1981). Electrons provided by NADPH (produced by the pentose phosphate pathway) are passed through FAD *en route* to their reduction of GSSG. In addition to the direct effects of ROS described below, the energetic costs of antioxidant defense activity are significant, as illustrated by the energy required to produce, maintain, and utilize GSH for this role.

## Quinone Reductases

Quinone reductases do not directly act on ROS; however, by reducing quinones, they can diminish ROS that are generated by quinones during the process of redox cycling, described below. Because quinones comprise an important group of environmental contaminants (as well as natural products), the inclusion of quinone reductase here is warranted (although it is also grouped with phase II enzymes). Quinones, such as the natural product juglone (Figure 4.2), can exert toxicity via two mechanisms: reactions with –SH groups of GSH and proteins, leading to GSH depletion and enzyme inactivation, for example, and through generation of ROS via redox cycling (Bolton et al. 2000, Dinkova-Kostova and Talalay 2000).

In the process of redox cycling, quinones can gain an electron from a reduced source, such as NADPH–cytochrome P450 reductase, yielding a semiquinone radical (Figure 4.2). Under aerobic conditions, this radical can donate its unshared electron to $O_2$, producing $O_2^{\bullet-}$. Quinone reductases circumvent this process by catalyzing a two-electron reduction of the quinone to its corresponding hydroquinone. In addition to quenching ROS formation, the hydroquinone products thus formed are oftentimes excellent substrates for phase II enzymes, particularly sulfotransferases (ST) and UDP-glucuronosyltransferase (UGT); thus, quinone reductase activity also enhances elimination.

In mammals, two distinct quinone reductases have been identified; both are primarily cytosolic and both contain FAD (Foster et al. 2000). Earlier, only one had been recognized; it is formally termed NAD(P)H:quinone oxidoreductase, also referred to as DT diaphorase. Zhao et al. (1997), through comparisons with a cDNA clone isolated by Jaiswal (1994), demonstrated that a previously ignored protein with quinone reductase activity (Liao and Williams-Ashman 1961) had close sequence homology to DT diaphorase. This "rediscovered" protein structurally resembles a truncated version of DT diaphorase but is a separate gene product. These enzymes are now oftentimes referred to as *quinone reductase type 1* (QR1 or NQO1), which is DT diaphorase, and *quinone reductase type 2* (QR2 or NQO2). Major differences between QR1 and QR2 include functional electron donors (QR1 employs NADH and NADPH, but QR2 employs nonphosphorylated nicotinamides), inhibition (QR1 but not QR2 is very sensitive to dicoumarol), and regulation. QR1 is highly inducible through both xenobiotic response elements and AREs (described below), but QR2 is apparently not inducible.

## Redoxins

Redoxins are enzymes that regulate redox signaling sand largely fall into four groups: peroxiredoxins, sulfiredoxins, thioredoxins, and glutaredoxins. We provide a brief overview here, and direct the interested reader to other excellent reviews (Hanschmann et al. 2013, Lo Conte and Carroll 2013).

Peroxiredoxins (Prdxs) are oligomeric peroxide-sensing and -degrading enzymes present in high abundance (often nearly 1% of all cellular protein (Detienne et al., 2018)); in

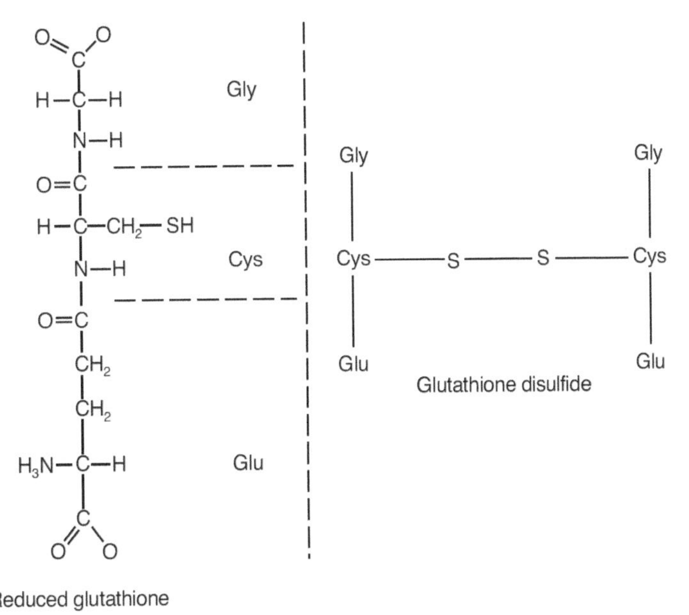


**FIGURE 4.2** Structures of reduced GSH and GSSG.

many cases, they may play larger roles than catalases and peroxidases in detoxifying $H_2O_2$ (Bolduc et al. 2021). Prdxs are evolutionarily ancient but well-conserved, appearing to have evolved with the rise in atmospheric oxygen and occurring in nearly all species from bacteria to mammals (Detienne et al. 2018). They are typically present as multiple genes (six in humans) encoding proteins expressed in most cellular compartments including the cytosol, nucleus, endoplasmic reticulum, mitochondria, and lysosomes. They are generally present in the soluble fraction, but some may be membrane-associated. They have a redox-sensitive cysteine that is oxidized by peroxides, including lipid peroxides, as well as peroxynitrite. They are reduced by thioredoxin, ascorbic acid, or glutathione (mediated by GST), depending on the specific Prdx. They serve important roles in both protecting against oxidative stress and regulating redox signaling by modulating localized, compartment-specific $H_2O_2$ concentrations and by oxidizing and thereby modulating the activity of signaling proteins such as kinases (Detienne et al. 2018, Bolduc et al. 2021).

Sulfiredoxins (Srxs) serve largely to reduce Prdx when in the hyperoxidized, sulfinic state to restore its activity, and can act as a switch between the antioxidant and chaperone functions of Prdx enzymes (Mishra et al. 2015). These enzymes also play a role in removing GSH from S-glutathionylated proteins (more on this in the "Protein Oxidations" section of this chapter).

Thioredoxins (Trxs) are small selenoproteins (~12 kDa) found in cytoplasm (Trx1) and mitochondria (Trx2). They have a common structural feature of four α-helices that surround a core comprising a 5-stranded β-sheet (Lu and Holmgren 2014). Thioredoxin reductase (TrxR) enzymes are homodimeric flavoproteins that reduce Trxs and other substrates; these enzymes—along with the glutathione–glutaredoxin system—provide a major component of redox homeostasis maintenance within cells.

Glutaredoxins (Grxs), sometimes referred to as thioltransferases, are highly conserved GSH-dependent enzymes that use GSH as a cofactor to donate electrons to maintain cellular redox homeostasis and reduce disulfide bridges. Grx1 is located in the cytoplasm, while Grx2 and Grx5 are in the mitochondria. Grxs serve essential physiological functions such as GSH-dependent catalysis, the enzymatic reversal of protein-*S*-glutathionylation, and even iron metabolism (Ouyang et al. 2018). Grxs have been utilized as redox biosensors by fusing them to fluorescent proteins; more on this can be found in this chapter in the "Transgenic Redox Biosensors" section. It is proposed that Grxs contain two GSH interaction sites—a glutathione-scaffold site that interacts with glutathionylated disulfide substrates, and a glutathione-activator site that interacts with GSH (Begas et al. 2017).

### Other Enzymes

The proteins discussed here are generally considered major players as antioxidant enzymes; however, they do not include all enzymes known to have some antioxidant function. For example, heme oxygenase (Kvam et al. 1999) and other enzymes may play important roles as well, at least under some circumstances.

## Low-Molecular-Weight Antioxidants

A number of biomolecules can directly scavenge ROS nonenzymatically. GSH, discussed above, is a very important component of this group, although it is also essential to GPX activity. Several other prominent antioxidants are vitamin-related and are obtained through the diet by most animals. These include ascorbic acid (vitamin C), tocopherols (vitamin E components), and carotenoids (vitamin A, or retinol, precursors), although this is not an all-inclusive list.

### Ascorbic Acid (Vitamin C)

Plants and many animals can synthesize ascorbic acid from glucose, although humans and other primates, bats, and passerine birds cannot (Nishikimi and Yagi 1996, Moreau and Dabrowski 2003). Among fishes, teleosts cannot perform this synthesis, but those retaining more ancestral characteristics such as lampreys, sharks, rays, lungfishes, sturgeons, paddlefishes, and bowfin can (Moreau and Dabrowski 1998). Those animals that cannot synthesize ascorbic acid lack the enzyme that catalyzes the terminal step in biosynthesis, gulonolactone oxidase. A DNA sequence resembling the gene for this enzyme in other animals has been identified in humans and guinea pigs, but it is extensively mutated and inactive (Nishikimi and Yagi 1996).

At physiological pH, ascorbic acid exists largely as ascorbate (Figure 4.3) and is highly water soluble. Ascorbate serves as a cofactor for a number of enzymes, including proline hydroxylase and lysine hydroxylase, that are involved in collagen synthesis, as well as for dopamine-β-hydroxylase, which catalyzes the conversion of dopamine into norepinephrine (Nishikimi and Yagi 1996). The hallmark human disease associated with vitamin C deficiency is scurvy, a common disease among sailors in earlier times (before the importance of fresh fruit was recognized); it is characterized by muscle weakness, blood vessel fragility, and bleeding from gums and other mucous membranes.

Ascorbate is a powerful reducing agent and has been shown with numerous *in vitro* studies to scavenge a number of ROS, including $O_2^{\bullet-}$, ·OH, ROO·, and HOCl (Halliwell 1996). Although unequivocally proving the significance of these reactions *in vivo* is difficult, the function of ascorbate as an antioxidant is widely accepted based on considerations of its *in vitro* reactivity, concentrations in tissues, experimental manipulations of its levels in animals and accompanying markers of oxidative stress, and related consequences in human deficiencies (Halliwell and Gutteridge 2015).

Upon reduction of an oxygen radical, ascorbate is oxidized to the ascorbyl radical (also known as semidehydroascorbate) (Bendich et al. 1986). This radical is relatively stable and unreactive, a feature central to its antioxidant function; that is, a much more reactive free radical reacts

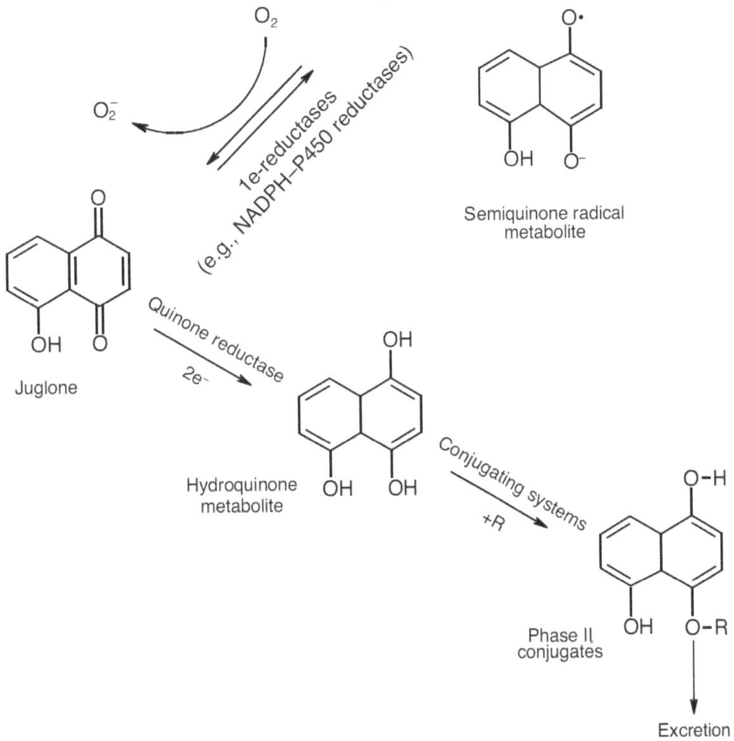


**FIGURE 4.3** Metabolism of the plant-derived quinone juglone. Upper right portion depicts redox cycling through the semiquinone radical intermediate, a one-electron reduction product. The lower pathway depicts the two-electron reduction pathway, which can be catalyzed by DT diaphorase (NAD(P)H:quinone oxidoreductase). This pathway yields phenolic compounds that are readily detoxified by phase II pathways (see Chapter 3).

with ascorbate, yielding a less reactive and hence more innocuous ascorbyl radical. The ascorbyl radical is readily oxidized again, yielding dehydroascorbate (DHA, the two-electron oxidation metabolite of ascorbate). DHA has little antioxidant activity and can break down to several products (notably oxalic acid and threonic acid) or be reduced back to ascorbate by an apparently nonspecific GSH-dependent reductase (Wells et al. 1990).

## Vitamin E

Vitamin E is a highly effective lipid-soluble scavenger of peroxyl radicals and hence a key protectant against membrane damage via lipid peroxidation (Niki and Matsuo 1993). It appears to be a dietary requirement of all animals. Vitamin E is actually a mixture of several tocopherols and tocotrienols; the major component acting in animal cells is α-tocopherol (Figure 4.3) (Diplock 1985). α-Tocopherol (in addition to other tocopherols and tocotrienols) is effective at protecting membranes from oxidative attack by lipid peroxyl radicals (LOO·) because it is more reactive with these radicals than are membrane lipids and proteins. Vitamin E components are also effective scavengers of $O_2^{\bullet -}$, ·OH, and $^1O_2$ and have been further shown to play a role in maintaining glutathione homeostasis (Zhang et al. 2021). As with ascorbate, α-tocopherol itself becomes a radical species upon reacting with an oxygen radical (Figure 4.3). α-Tocopherol can be regenerated from its radical form by mechanisms including reduction of the tocopherol radical by ascorbate; this reaction appears to represent an important cooperation between these vitamins to maintain active α-tocopherol within membranes (Traber 1994, Traber and Stevens 2011).

## Carotenoids

Carotenoids are a large group (over 600 described) of plant pigments, some of which are absorbed from the diet in many animals (Krinsky 1993). Carotenoid structure generally includes about 40 carbons, most of which occur in alternating single and double bonds, a feature that allows for electron delocalization and absorbance of light in the visible range (Britton 1995). Either or both ends of this carbon chain are often modified into cyclic rings, which may also contain oxygen groups; several carotenoids are illustrated in Figure 4.3. Carotenoids are very lipophilic and virtually insoluble in water. Many, including the most abundant carotenoid observed in humans, β-carotene (Figure 4.4), serve as precursors for vitamin A (retinol) and retinoic acid, which play important roles in cell growth, differentiation, and development. Some, such as astaxanthin (Figure 4.4), contribute to the red coloration of some fishes, including male sticklebacks, for which this coloration plays a role in sexual preference by female sticklebacks (Barber et al. 2000).

In terms of antioxidant function, the best established role for carotenoids is that of a scavenger of singlet oxygen in plant chloroplasts (Telfer et al. 1994); $^1O_2$ is generated at

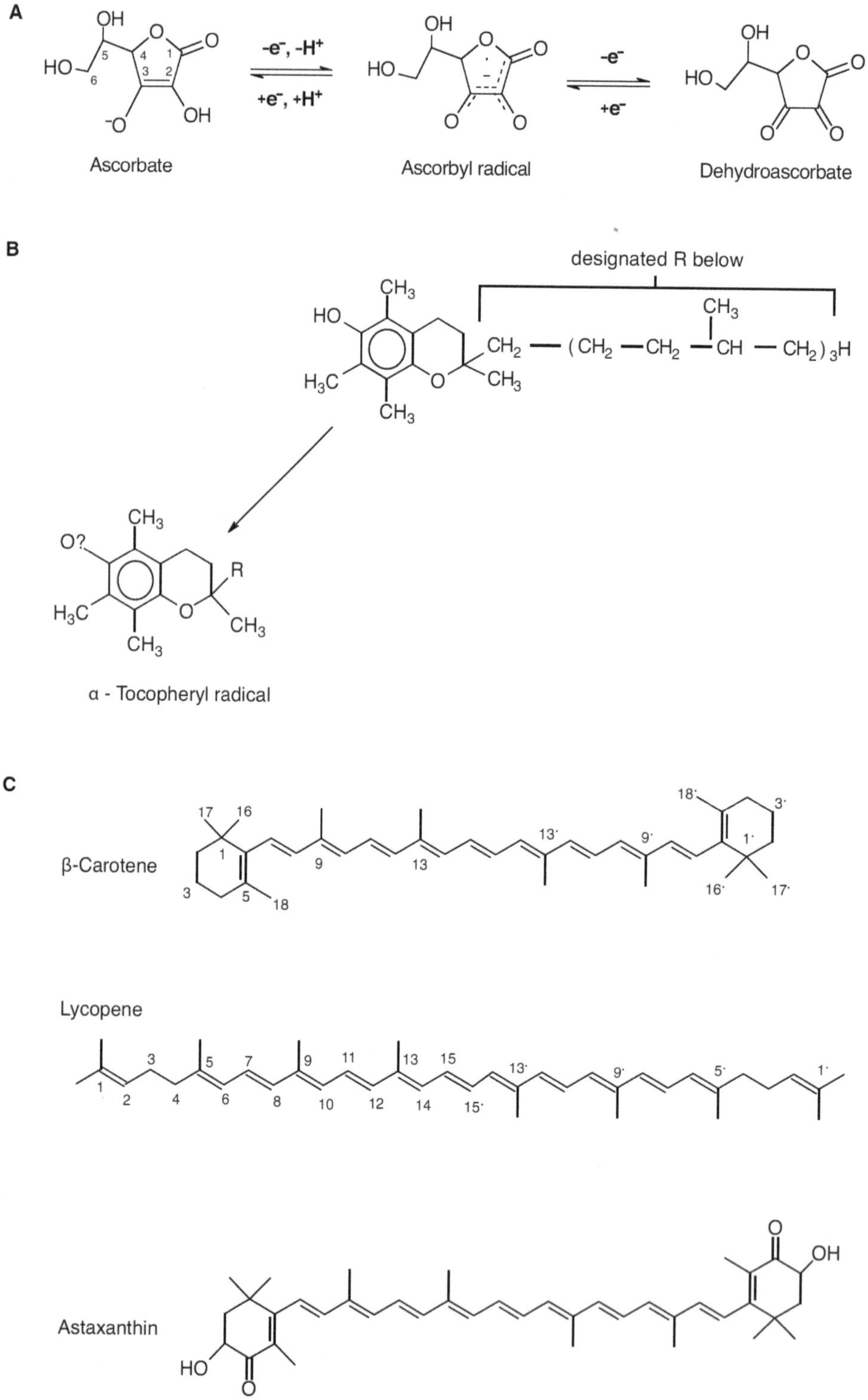


**FIGURE 4.4** Low-molecular-weight antioxidants. (A) Ascorbate (vitamin C) and its one-electron (ascorbyl radical) and two-electron (dehydroascorbate) oxidation products. (B) α-Tocopherol (vitamin E) and its radical product. (C) Representative carotenoids.

high rates during photosynthesis. The ability of carotenoids to scavenge a variety of oxygen radicals has been demonstrated *in vitro* (Liebler and McClure 1996). Although it is generally believed that they also serve a significant antioxidant function in animals *in vivo*, this has not been firmly established.

## Other Antioxidants

A number of endogenous compounds with various well-characterized functions other than acting as an antioxidant have been shown to be also capable of scavenging ROS in various *in vitro* systems. These include bilirubin, estradiol, lipoic acid, coenzyme Q (ubiquinol), and uric acid;

however, the significance of *in vitro* studies suggesting an antioxidant function to the *in vivo* situation is oftentimes difficult to gauge.

Another potentially important function of some compounds, other than direct ROS scavenging, is regulation of metals that catalyze ROS production (e.g., via Fenton-like chemistry). Iron and copper are the two essential transition metals that naturally occur in vertebrates at relatively high concentrations, and if they are free in cells, they can readily enhance ROS generation, lipid peroxidation, and other manifestations of oxidative stress. Consequently, cells have mechanisms to carefully regulate them, keeping them available for incorporation into appropriate metalloproteins, for example, while preventing them from participating in destructive oxidative reactions, such as Fenton reactions and autoxidations.

For iron, perhaps the most important regulatory protein is ferritin. Most intracellular iron is bound to ferritin, which forms a large shell-like structure comprised in mammals of 24 subunits each of about 20,000 molecular weight and distinguished into two types: H and L. Up to 4500 iron ions can be stored in the core of a protein. H subunits are involved in iron detoxification by virtue of ferroxidase activity that oxidizes $Fe^{2+}$ to $Fe^{3+}$. L subunits facilitate nucleation of $Fe^{3+}$ within the core of ferritin for long-term storage. Ferritins are a large family with six subfamilies in vertebrates distinguished by phylogeny and ferroxidative ability (Golan et al. 2021).

Ceruloplasmin serves an analogous function for copper (Harris 1995). Dietary copper entering the bloodstream largely binds to albumin; this complex is taken up by the liver, where the copper is incorporated into ceruloplasmin. Ceruloplasmin has a molecular weight of about 132,000 and tightly binds six copper ions. The copper–ceruloplasmin complex is secreted by the liver into the plasma and can contribute copper to cells requiring it following binding to cell surface receptors. Ceruloplasmin has been found in several fish species, including carp, plaice, mullet, tilapia, and European eel (Grosell et al. 1998).

Metallothioneins (MTs) are low-molecular-weight (about 6,500) proteins and very cysteine-rich (about 30% of amino acid residues), which underlies their high binding affinities for a number of metals, including copper, cadmium, zinc, silver, bismuth, and mercury. Two isoforms, MT-1 and MT-2, appear to occur in all animal tissues and are the most-studied MTs. They typically bind five to seven metal ions per molecule and are inducible by some metals (including those mentioned above), stress hormones (such as glucocorticoids), oxidants, and inflammation. Considerable debate currently exists regarding whether their primary role is regulation of essential metals, particularly copper and zinc, for use in metalloenzymes vs. protection against metal toxicity, particularly from cadmium. Additionally, several lines of experimental evidence support an antioxidant function for MTs, including inducibility by prooxidants such as hydrogen peroxide and paraquat, protection against such chemicals in MT-enriched vs. MT-depleted cells (e.g., via transfections and gene knockout models), and the ability of MT to scavenge ROS and spare GSH (Isani and Carpenè 2014).

## REACTIVE OXYGEN SPECIES AND GENE EXPRESSION

Antioxidant defenses and other physiological functions can be altered in response to exposure to oxidative stress in a variety of ways including at the levels of mRNA transcription, mRNA stabilization, protein activation, and more (Sies and Jones 2020). Here we provide examples of how ROS may result in changes at the transcriptional level. We note that fish have undergone genome duplication events over the course of evolution; particularly with regards to transcription factors, there are often duplicates retained that correspond to one mammalian gene, and these retained co-orthologs in some cases have undergone subfunctionalization and/or neofunctionalization (Postlethwait et al. 2004).

### Overview of ROS-Mediated Modulation of Gene Expression in Eukaryotes

The expression of many genes is affected by ROS; however, it is important to bear in mind that not all such changes necessarily represent adaptive, antioxidant responses. Some changes in gene expression simply result from oxidative damage; for example, low levels of damage may alter the activity of transcription factors (Ma 2010, Halliwell and Gutteridge 2015, Hopkins and Neumann 2019). Additionally, ROS are generated purposefully not only as physiological effector molecules (such as $O_2^{\bullet-}$ generated by phagocytes, discussed previously) but also as signaling molecules in pathways not necessarily related to redox stress *per se* (Palmer and Paulson 1997, Morel and Barouki 1999, Finkel 2000, Sauer et al. 2001, Holmstrom and Finkel 2014, Sies 2021). Thus, ROS may be generated by enzymes such as NADPH oxidase or in the mitochondria as part of normally functioning signaling pathways, and in some fish can be beneficial, as has been shown for H2O2 in fin regeneration (Niethammer et al. 2009). It may be useful to conceptualize the effect of the signaling elicited by ROS as varying qualitatively according to the amount and duration of the associated stress, with low magnitudes of deviation from redox homeostasis in some cases producing a cellular response of proliferation, intermediate levels often producing temporary growth arrest/senescence and an adaptive response, and higher magnitudes leading to apoptosis, and potentially necrosis (Davies 1999, Chandra et al. 2000, Martindale and Holbrook 2002, Halliwell and Gutteridge 2015, Timme-Laragy et al. 2018). Thus, ROS cannot be thought of as simply and exclusively damage-causing agents; rather, it is the production of inappropriate amounts of ROS or production in the wrong place or at the wrong time that is problematic, as well as the resulting deviations from the redox homeostatic set points. Furthermore, the ability of ROS-generating chemicals to cause damage is not limited to their ability to generate highly reactive toxic intermediates but is also related to their ability to alter the functioning of normal signaling pathways that utilize ROS as messenger molecules. Finally, signaling by reactive nitrogen, copper, iron, and other redox-active metal species, in addition to ROS, can also play important roles in modulating gene expression (Halliwell and Gutteridge 2015).

### Effects of ROS on Transcription Factors

The activity of many transcription factors is dependent upon the reduced or oxidized state of redox-responsive moieties such as thiols or Fe–S clusters; as a result, oxidative stress can alter the activity of those transcription factors by altering the redox status of the cell (Primiano et al. 1997, Arrigo 1999, Sen 2000, Schafer and Buettner 2001, Brigelius-Flohe and Flohe 2011). Transcription factors in eukaryotes reported to be affected in this fashion include NFE2 and related family members (Nrf1, Nrf2, Nrf3, NFE2), AP-1, Maf, Nrl, NF-IL6, Sp-1 family members, protein kinase C, glucocorticoid receptors, estrogen receptors, aryl hydrocarbon receptor, NF-κB, p53, and others (Primiano et al. 1997, Arrigo 1999, Dalton et al. 1999, Sen 2000, Crawford 2002, Michiels et al. 2002). ROS also increase the rate of transcription of many transcription factors, including members of the AP-1, NF-κB, and AP-2 families (Dalton et al. 1999). The upstream events leading to activation of transcription of these factors are complex and include alterations in phosphorylation of signaling molecules, release of arachidonic acid from cell membranes and subsequent metabolism of the arachidonic acid, and mobilization of $Ca^{2+}$ (for reviews, see Dalton et al. (1999), Sauer et al. (2001), Martindale and Holbrook (2002), Ma (2010)).

### Effects of ROS on mRNA Expression

The transcription factors mentioned above are involved in numerous signaling pathways, including pathways related to cell division and differentiation, immunological response, cytokine expression and inflammatory response, xenobiotic metabolism, and many others; thus, it is not surprising that toxicant-induced ROS modulate the expression of many genes. A small sampling of the many genes reported to be regulated (in at least some cases) in eukaryotes by ROS includes c-*fos* and c-*myc* (cellular protooncogenes), *gadd*45 and *gadd*153 (growth arrest and DNA damage genes), γ-glutamyl transpeptidase, GCL (both subunits), some GST isoforms, some UGT isoforms, epoxide hydrolase, heme oxygenase, MnSOD, and ECSOD, but not usually CuZnSOD, metallothionein, GPX1, interleukin 8, cytochrome IV, thioredoxin, and an aldo-keto reductase (Shull et al. 1991, Prestera et al. 1993, Brady et al. 1997, Burczynski et al. 1999, Dhakshinamoorthy et al. 2000, Crawford 2002). The expression of some of these (and other) genes is regulated by ROS via ARE, as discussed below. As indicated by this partial listing, although many of the genes identified as upregulated by ROS could clearly play important antioxidant roles, many others are not classical antioxidants.

The expression of many genes can also be downregulated by ROS; among these are many mitochondrial genes (Fujii and Taniguchi 1999, Morel and Barouki 1999, Crawford 2002) and a large number of nuclear-encoded genes involved in the immune response, cell replication, carbohydrate metabolism, hormonal responses, phase I xenobiotic metabolism, and other cellular activities (Morel and Barouki 1999, Nebert et al. 2000, Barouki and Morel 2001). As in the case of ROS-mediated gene induction, these results seem likely to reflect not only adaptive changes directed at decreasing oxidative damage but also alterations in pathways in which ROS normally play a role as signaling molecules or simple oxidative damage to cellular macromolecules that mediate gene expression in these pathways. The perturbation of gene expression in response to ROS can thus vary both in directional change, magnitude, as well as in duration of the response.

### Antioxidant Response Element (ARE)-Mediated Gene Regulation

The most well-studied molecular mechanism for specifically responding to ROS in eukaryotes is the modulation of gene expression via an enhancer element termed the ARE, or electrophile response element (EpRE) (Zhang 2006, Tebay et al. 2015, Yamamoto et al. 2018). ARE transcription factors include one or more CNC-bZIP proteins (e.g., Nrf1 and especially Nrf2), as part of a heterodimer with a small Maf proteins. Nrf2 is repressed in the cytosol by a Keap1 dimer that targets Nrf2 for ubiquitination and degradation by the 26S proteasome. Structural changes to Keap1 such as the oxidation of sulfhydryl groups lead to a break in the delivery of Nrf2 to the proteasome, permitting it to move to the nucleus, heterodimerize, and act as a transcription factor. Phosphorylation of Nrf2 also can regulate the transcriptional activity of Nrf2 (Huang et al. 2000, Kong et al. 2001), via an increase in Nrf2 stability (Nguyen et al. 2003). Electrophile-induced polyubiquitination of the Keap1 protein also plays a role in Nrf2 activation as well (Eggler et al. 2005, Hong et al. 2005). The NFE2-related proteins (Nrf1a, Nrf1b, Nrf2a, Nrf2b, Nrf3, NF-E2) and Keap1 (Keap1a, Keap1b) are present in zebrafish (Hahn et al. 2015, Fuse and Kobayashi 2017). Nrf2a is the co-ortholog to the mammalian Nrf2 and has been demonstrated in multiple studies to regulate the expression of genes via AREs in response to electrophilic inducers (Mukaigasa et al. 2012, Timme-Laragy et al. 2012, Sant et al. 2017, Mills et al. 2020).

### ROS-Mediated Modulation of Gene Expression: Summary

The alterations observed in eukaryotic gene regulation in response to ROS are highly complex. These alterations in gene expression may result from interaction with pathways that normally employ ROS as signaling molecules;

antioxidant responses, including those aimed at preventing the generation of ROS, increasing the scavenging of ROS, and repair of oxidative damage; and alterations resulting from oxidation of transcription factors or other important cellular macromolecules. The expression of a wide variety of genes is altered by exposure to ROS, including many genes coding for proteins that are not typically thought of as antioxidants, but rarely involves very large changes in expression. These characteristics may relate to the fact that ROS play a normal physiological role in eukaryotes, suggesting that repressing ROS to too low of a concentration might be problematic and to the fact that the transcription factors that mediate many of the apparently adaptive antioxidant responses are involved in other pathways as well, suggesting that large alterations in their activity might be disruptive to other important biological responses. Finally, although we have focused on ROS-mediated transcriptional regulation, ROS-mediated regulation of protein translation and degradation are also evolutionarily conserved. For example, upon oxidative damage, ribosomal translation is inhibited, and this response is protective against further oxidative stress (Silva et al. 2015). Furthermore, translation of antioxidant proteins may be spared from translational repression (Simões et al. 2021). The mechanisms by which oxidative stress can result in degradation of existing, damaged proteins via proteasomal, autophagic, and other pathways are also increasingly understood (Pajares et al. 2015, Manohar et al. 2019, Wang et al. 2022), and there is emerging evidence that proteins involved in regulating the response to oxidative stress can be stabilized under conditions of oxidative stress (McIntosh et al. 2018).

## DELETERIOUS CELLULAR EFFECTS OF REACTIVE OXYGEN SPECIES

Aerobic organisms have evolved a complex array of antioxidant defenses that effectively deal with typical fluxes of ROS encountered by aerobic cells. As these fluxes increase, due to both natural phenomena (such as increased aerobic metabolism during physical exertion, sharp rises in dissolved oxygen in aquatic environments, or pronounced inflammatory responses) as well as unnatural variables (such as anthropogenic chemicals), cellular antioxidant capacity can be overtaxed. As described above, through mechanisms including ROS-mediated changes in gene expression, cells can adapt to some extent by enhancing their antioxidant capacity. Upregulation of antioxidant defenses comprises an early response to elevated ROS fluxes; however, such fluxes can exceed even adapted antioxidant capacity, at which point detrimental cellular impacts may ensue.

A major difficulty in the study of redox stress and ROS is the variety of ways in which they can manifest; that is, it is far more difficult to determine if a particular chemical is acting, as a principal mechanism of toxic action, as a prooxidant causing oxidative stress vs. determining, for example, if it is an acetylcholinesterase inhibitor, an aryl hydrocarbon receptor agonist, or an estrogen mimic. A lack of effect as measured by any particular assay (or two or three) cannot be used to rule out the occurrence of redox stress. A similar caution applies to the measure of antioxidants as markers for exposure or adaptation to prooxidants, with the expectation that exposures below the threshold for toxicity will result in upregulations of antioxidants. Again, the complexity of mechanisms by which prooxidants may or may not augment various antioxidant system components presents major challenges to this approach. These are issues that have plagued the development of biomarkers for oxidative stress in fish, as described later.

Although redox stress is a complex and often difficult subject of inquiry, it remains an important underlying mechanism of cellular toxicity. Given the high reactivity of many ROS, such as ·OH, it is not surprising that they are often indiscriminant in terms of cellular targets. Principal cellular constituents, including lipids, proteins, and DNA, are subject to attack, as are cellular functions such as redox status and energetics and cellular signaling. Tools have been developed to study all of these phenomena. The following is not an exhaustive list of cellular effects but does address major ones that have received considerable research attention in a variety of organisms, including fishes.

### Lipid Peroxidation

Perhaps the most studied targets of ROS are polyunsaturated fatty acids (PUFAs), which are fatty acids containing two or more carbon–carbon double bonds. Of particular interest are those associated with the membranes of cells, including organelles such as mitochondria, lysosomes, and endoplasmic reticula. These membranes are complex and diverse structures, typically comprised of various phospholipids (such as phosphatidylcholine, or lecithin), cholesterol, lipoproteins, and proteins that serve various functions, particularly with respect to signal transduction and transport of materials across membranes.

In PUFAs, the hydrogen atoms on saturated carbons (allylic hydrogens) adjacent to carbons participating in double bonds are most prone to hydrogen abstraction by oxygen radicals such as ·OH, RO·, and ROO·. These allylic hydrogens form less stable bonds to carbon due to the adjacent carbon–carbon double bonds; an example within a hydrocarbon chain is denoted here:

$$-CH = CH - CH_2 - CH = CH -$$

Abstraction of an allylic hydrogen by an oxygen radical comprises the *initiation* step of lipid peroxidation, which is included in the summary schematic of lipid peroxidation (Figure 4.4); susceptibility of different PUFAs to lipid peroxidation increases with increasing number of unsaturated carbon double bonds. Initiation results in a lipid radical (R·), which then undergoes rearrangement to a conjugated diene radical; under typical aerobic conditions, this lipid radical will readily react with $O_2$, yielding a lipid peroxyl radical (ROO·). The peroxyl radical can react with another

PUFA, thereby abstracting hydrogen, becoming a lipid peroxide (LOOH), and generating another R·. This second R· can also react with $O_2$ to yield ROO·, and this process can be repeated many times, constituting a free-radical chain reaction termed *propagation* of lipid peroxidation; thus, initiation by one molecule of an oxygen radical can potentially result in the peroxidation of many PUFA molecules. Propagation is an important feature of many free-radical reactions whereby one radical can stimulate a cascade of potentially deleterious reactions in biological systems; this phenomenon is addressed again later.

A number of other important reactions are associated with the process of lipid peroxidation; for example, the peroxyl radical can react with other membrane lipids (e.g., cholesterol) or proteins, in addition to PUFA, thus altering these molecules while forming ROOH. Transition metals such as iron and copper, in addition to enhancing production of the powerful initiator ·OH through Fenton chemistry, can also react directly with ROOH to produce RO· and ROO·, which can initiate new radical chain reactions. In fact, it is thought that transition metals are required for lipid peroxidation to proceed at a significant rate (Sevanian and Ursini 2000). Lipid peroxidation can be terminated by the reaction of two lipid radicals to produce nonradical products. Additionally, lipid peroxidation can be slowed by the action of α-tocopherol, as described earlier. While stopping a particular radical chain reaction, donation of hydrogen to ROO· by α-tocopherol (yielding the relatively unreactive α-tocopherol radical) results in ROOH, which is still subject to metal-catalyzed radical generation. This is repaired by the action of GPXs, discussed earlier, that reduces ROOH to the corresponding alcohol (ROH), effectively preventing further lipid peroxidation. Major consequences of membrane lipid peroxidation include decreased membrane fluidity, increased permeability resulting in inappropriate leakiness to some molecules, and inhibition of membrane-bound enzymes (Richter 1987). Specific oxidized lipids such as 4-hydroxynonenal can also serve signaling purposes, for example activating stress-response signaling pathways (Zhang and Forman 2017) and mitochondrial uncoupling proteins (Cadenas 2018).

### Protein Oxidations

Many proteins are also susceptible to attack by ROS, which can have deleterious consequences. Important consequences of protein oxidations by ROS include enzyme inactivation, alteration of receptors and other proteins involved in signal transduction, and perturbed ion homeostasis (e.g., via effects on ATPases) (Halliwell and Gutteridge 2015). Direct effects of ROS on specific amino acids in proteins include oxidations of the reduced sulfur moiety of cysteine and methionine to produce disulfides and sulfoxides/sulfones, respectively (Dean et al. 1997), and production of carbonyl groups (aldehydes and ketones) on side chains of some amino acids, particularly of proline, arginine, lysine, and threonine (Dalle-Donne et al. 2003), and oxidations of the aromatic side chains of phenylalanine, tryptophan, tyrosine, and histidine (Stadtman and Levine 2003). Among these effects, the production of carbonyls is considered the most common effect, and increases in their production have been associated with a number of human diseases, including Alzheimer's disease, cystic fibrosis, diabetes, and amyotrophic lateral sclerosis; consequently, many assays have been developed to detect protein carbonyls (Dalle-Donne et al. 2003).

A number of laboratory studies of fishes have indicated protein carbonyls as markers of oxidative stress from exposures including pesticides (Parvez and Raisuddin 2005), selenium (Gobi et al. 2018), and silver nanoparticles (Valerio-García et al. 2017), to name a few. Their use in field studies is briefly described in "Biomarkers" section.

In addition to protein carbonyls, other redox modification of proteins (by ROS, RNS, and other post-translational modifications) occurs to modify function. Some of these "redox switches" occur as an endogenous regulatory mechanism, others comprise a toxic response leading to irreparable damage of the protein (Moldogazieva et al. 2018). The field of redox proteomics has rapidly expanded over the last decades, and has emerged as an important component of redox stress, with functional implications for diverse cellular activities such as regulation of protein kinase activity (Corcoran and Cotter 2013), mitochondrial fission and fusion (Wolf et al. 2020), and more. These redox modifications of proteins are firmly linked to cellular dysfunctions and various disease states (Butterfield and Dalle-Donne 2014).

Protein S-glutathionylation (Pr-SSG) occurs when a GSH molecule is transferred to a low-pKa Cys moiety on a given protein, and serves to protect protein integrity during periods of redox stress that would otherwise lead to Cys-oxidations. Increases in Pr-SSG are generally (but not always) in response to oxidative changes to the GSH redox potential; accumulation of GSSG promotes oxidation of Cys residues and increases S-glutathionylation, while a more reduced GSH redox potential can lead to reduction of the protein and release of GSH to form GSSG; this can also be performed enzymatically by glutaredoxin (Timme-Laragy et al. 2018). Direct oxidation of Cys on proteins leads to formation of sulfenic acid (Pr-SOH). Sulfenic acids are reversed via reduction by thioredoxin enzymes (Kalinina and Novichkova 2021), but they can also be further oxidized to sulfinic acids (Pr-$SO_2$H). Sulfinic acids can be remedied through enzymatic reduction (e.g., sulfiredoxins (Chang et al. 2004)), but can be further oxidized again to form a sulfonic acid (Pr-$SO_3$H). The sulfonic acid formation is irreversible and leads to loss of protein function (Lo Conte and Carroll 2013). More on the enzymes involved in these reactions is found earlier in the chapter under the "Redoxins" section.

### Oxidative DNA Damage

ROS can interact with DNA to produce oxidative DNA damage. RNS can also participate in these and related forms of DNA damage. Like oxidative damage to proteins and lipids, oxidative DNA damage can be repaired; however, if unrepaired, DNA damage can lead to mutations, which may be inherited. Mutations that occur in somatic cells are inherited only in that organism's life, but germ cell mutations may be inherited as genome sequence alterations in subsequent generations.

The species of ROS most associated with oxidative DNA damage is the hydroxyl radical (·OH), the most reactive of the ROS. However, most damage to the nuclear genome likely occurs after longer-lived, diffusible species such as $H_2O_2$ produced at one cellular localization (such as endoplasmic reticula or mitochondria) diffuse to another (such as the nucleus) and, via metal-catalyzed Fenton chemistry, yield DNA-damaging hydroxyl radicals some distance removed from ROS production. The inherent reactivity of ·OH precludes such direct mobility of this species, and charged species such as superoxide anion are limited in their ability to cross membranes. However, as discussed in more detail below, mitochondrial DNA (mtDNA) is located in close proximity to the electron transport chain, typically the major source of ROS production in the cell. Another oxidant with that has similar effects on DNA as ·OH is peroxynitrite ($ONO_2^-$), formed by reaction between nitric oxide (NO·) and $O_2^{\bullet -}$ (Koppenol et al. 1992). Other species, including singlet oxygen ($^1O_2$), NO·, and products of lipid peroxidation such as malondialdehyde and 4-hydroxynonenal, can also damage DNA (Friedberg and Friedberg 2006, Halliwell and Gutteridge 2015).

Direct oxidations of DNA bases, in large part via ·OH, comprise a major form of initial DNA damage by ROS and related oxidants (Halliwell and Gutteridge 2015) (see Figure 4.5). In addition to oxidative reactions, the interaction of $ONOO^-$ with DNA results in the nitration of guanine to form 8-nitroguanine (Figure 4.5). Dozens of forms of oxidative base damage have been identified (Bjelland and Seeberg 2003, Cadet et al. 2017). Measurements of oxidative DNA damage have been employed as markers for oxidative stress in tissues of animals, including fish; however, one must be careful when examining oxidative DNA data, as artifactual oxidations can occur during these analyses which has led to considerable debate concerning true background levels for various tissues and the most accurate approaches for analysis (Cadet et al. 1997, Marnett 2000). It is important to note that there is a significant amount of "background" DNA damage in most cells, representing a steady state of endogenous damage and repair processes, so control or reference populations should be chosen carefully. mtDNA is more susceptible than nuclear DNA to oxidation, and a subset of oxidative mtDNA damage persists longer than nDNA damage (Yakes and Van Houten 1997); for this reason, mtDNA damage has been suggested as a biomarker of exposure to oxidizing and other genotoxic agents (Meyer 2010). Changes in mtDNA copy number (the number of genomes per cell) have also been reported after oxidative damage (Furda et al. 2012); however, mtDNA copy number may either increase or decrease, complicating the use of this parameter as a biomarker (Meyer et al. 2017). Finally, it is also possible to measure DNA mutations and draw inferences about oxidative DNA damage. Although DNA damage and mutations result from many forms of damage (Di Giulio and Hinton 2008), not just oxidative damage, specific mutations have been identified that may serve as "signatures" of oxidative stress (De Bont and van Larebeke 2004).

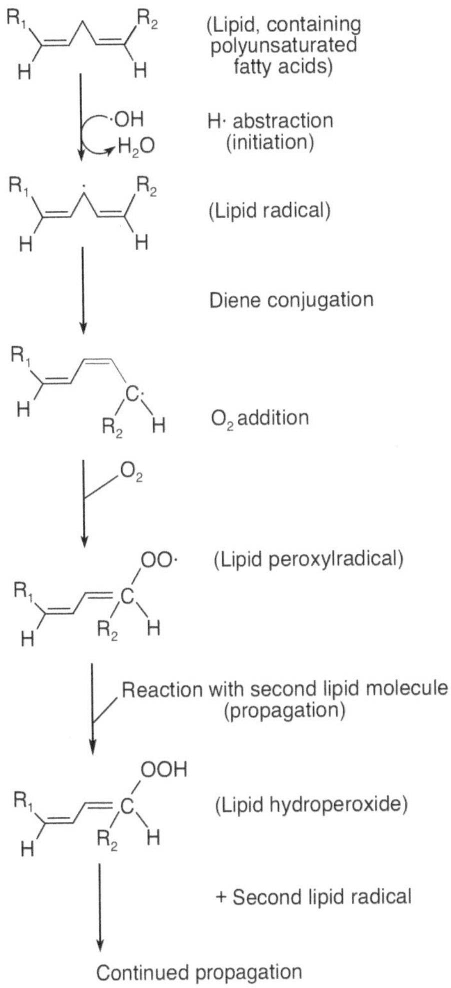


**FIGURE 4.5** The biochemical process of hydroxyl radical (·OH)-initiated lipid peroxidation.

Most oxidative DNA damage is relatively rapidly repaired (Cadet et al. 2017). Base excision repair is particularly important for the repair of oxidized DNA bases in both the nuclear and mitochondrial genomes; nucleotide excision repair may also play a role in repairing nuclear DNA damage (Van Houten et al. 2018). Although the DNA repair machinery is usually very effective in dealing with oxidative DNA damage, it is not completely effective, as evidenced by the presence of oxidized DNA in cells of every aerobic organism examined to date. Increases in the rates of damage above background, by ionizing radiation or ROS-generating chemicals, for example, can increase the levels of damage and associated probabilities of mutations. Also, this repair machinery itself can be perturbed by ROS and RNS; for example, NO· inhibits several DNA repair enzymes, which suggests a potential synergy for NO· to both enhance oxidative DNA damage (via peroxynitrite formation) and inhibit repair of that damage (Laval et al. 1997, Marnett 2000).

The great bulk of our knowledge concerning mechanisms of oxidative DNA damage and repair is derived from studies of simple models such as bacteria, yeast, and mammalian cell culture; however, it is clear that these phenomena are operative in fish as well, which is certainly not surprising given that fish are aerobic organisms. DNA repair pathways have not been carefully assessed in many fish species, but generally appear to be well-conserved, as illustrated in the well-characterized model *Danio rerio* (Pei and Strauss 2013, Canedo and Rocha 2021). Oxidative DNA damage has often been detected in fish. Nishimoto et al., for example, reported elevated concentrations of 8-oxo-dG in tissues of the English sole (*Parophrys vetulus*) injected with nitrofurantoin (Nishimoto et al. 1995). Likewise, Malin and colleagues (1990) detected the guanine-derived lesion 2,6-diamino-4-hydroxy-5-formamido-pyrimidine in neoplastic livers of feral English sole exposed to carcinogens. As in nonpiscine models, concentrations of 8-oxo-dG and other forms of oxidative DNA damage in tissues have also been adopted as a biomarker of oxidative stress and environmental pollution in many species of fish (Rodríguez-Ariza et al. 1999, Machella et al., 2005, de Lapuente et al. 2015).

### Redox Status and Energetics

The redox status of a cell (or cellular compartment, tissue, etc.) refers to the collective ratios of interconvertible oxidized and reduced forms of redox couples. Key redox couples underlying redox status include $NADH/NAD^+$, $NADPH/NADP^+$, and GSH/GSSG. Reduced forms of these couples are critical for a number of fundamental cellular functions including energy production (e.g., mitochondrial ATP production linked with electron release by NADH, resulting in oxidation to $NAD^+$), biosynthesis (many pathways employ NADPH), and protection from ROS (wherein GSH plays a central role as described previously); therefore, cells and tissues maintain overall reducing environments that favor the reduced forms of most redox couples, including those just mentioned.

ROS can drive redox status to a more oxidized state through a variety of direct and indirect mechanisms. GSH can be oxidized directly by ROS, as well as through GPX-catalyzed reductions of $H_2O_2$ and lipid peroxides (Equations 4.15 and 4.16). The reduction of GSSG resulting from GSH oxidations back to GSH is catalyzed by glutathione reductase, which requires the oxidation of NADPH. NADH and NADPH are also oxidized during the activity of quinone reductase (Equation 6.19). Both reduced pyridine nucleotides can be oxidized during the generation of ROS by redox-cycling chemicals, a phenomenon described later. There is also evidence that NADH and NADPH can act directly as ROS scavengers, resulting in their oxidation (Kirsch and De Groot 2001). This activity is suggested to be important in mitochondria, where their concentrations are similar to that of GSH (~5 mM). In all of these cases, however, oxidations of reduced intermediates impose an energetic cost.

In some cases, reductive stress may also occur within a specific organelle or cell type. For example, pancreatic β-cells maintain an oxidized redox environment to facilitate $H_2O_2$ as a signaling molecule linking glucose-stimulated ATP generation and insulin release, and $H_2O_2$ is also required for β-cell proliferation and neogenesis (Ahmed Alfar et al. 2017). Further, insulin biosynthesis requires oxidative folding of three disulfide bonds, which are subject to reductive misfolding (Weiss 2013). We and others have shown that β-cells have lower levels of antioxidant defenses than other islet cell types and are more vulnerable to redox stress (Lenzen et al. 1996, Pi et al. 2007, Miki et al. 2018, Rastogi et al. 2020). Because islets are highly vascularized to monitor blood glucose, there is efficient delivery of toxic products from the blood stream (Schaeffer et al. 2011). Based on these characteristics, β-cells are uniquely vulnerable to both oxidative and reductive stress (Timme-Laragy et al. 2018, Xiao and Loscalzo 2020).

## DELETERIOUS ORGANISMAL EFFECTS OF REACTIVE OXYGEN SPECIES

The preceding discussions have focused on molecular and cellular aspects of ROS and oxidative stress, and the complexity and varied nuances of these phenomena should be apparent; therefore, it is not surprising that a diverse array of diseases and pathologies have been associated with oxidative stress. It is beyond the scope of this chapter to discuss these in detail; however, a number of diseases (most described in humans), pathologies, and other effects are noted here to provide a sense of the breadth of health outcomes that are at least in part associated with oxidative stress.

Oxidative stress, as measured by cellular effects described above (e.g., oxidations of lipids, protein, DNA, perturbed redox status, and altered antioxidant capacity), has been observed in association with many disease states and tissue injuries, but it is important to bear in mind that such associations do not confer causality. Evidence of oxidative stress may be encountered in any tissue undergoing trauma due to infection, injury, or cell death (apoptosis or necrosis)—such tissues are inherently more prone to oxidative damage due, for example, to compromised energetics. Determining that oxidative stress is the cause of a disease or tissue damage rather than the result requires careful analysis. Moreover, many diseases and pathologies have complex etiologies, and attempts to pinpoint a single cause, such as ROS, are often erroneous.

### Health Relevance

Compelling evidence exists that oxidative stress is an important contributing factor for numerous human diseases and pathologies. These include neurodegenerative diseases (such as Alzheimer's disease, Parkinson's disease, and amyotrophic lateral sclerosis), atherosclerosis, arthritis, and cancer (Halliwell and Gutteridge 2015). For most of these, their relevance to fishes is lacking and/or unclear. However, some fishes, such as the zebrafish, may provide useful models for exploring underlying mechanisms (Bradford et al. 2017; Bailone et al. 2020). Cancer, however, which encompasses a diverse set of diseases with a complex etiology, has been studied extensively in fishes as well as humans. Considerable evidence supports a role for ROS and other free radicals in tumor initiation, promotion, and progression.

Aging is also a phenomenon of broad health relevance, to humans and other organisms. Considerable evidence supports an important role for oxidative stress in this phenomenon, although there are many other factors involved (Iakovou and Kourti 2022). Supporting evidence for the free-radical theory of aging includes the increased accumulation of oxidized proteins, lipids, and DNA with age; increased life span in organisms engineered to overexpress antioxidant enzymes (e.g., CuZnSOD and GR) in some studies; and increased life span associated with caloric restriction (which reduces mitochondrial production of ROS, a key component of this theory). Several fish species have been employed in aging research including zebrafish and species that exhibit rapid senescence such as Pacific salmon (*Oncorhynchus* spp.) (Morbey et al. 2005) and the African turquoise killifish (*Nothobranchius furzeri*) (Kim et al. 2016).

## MECHANISMS OF CHEMICAL-MEDIATED OXIDATIVE STRESS

As mentioned earlier, oxidative stress has been defined as "a disturbance in the prooxidant–antioxidant balance in favor of the former, leading to potential damage" (Sies and Cadenas 1985). Newer conceptualizations of changes to the cellular redox environment recognize the potential for altered redox homeostasis to include problematic reductive stress, describing alterations in either direction as "redox stress" (Timme-Laragy et al. 2018). In such scenarios, interference with production of prooxidant molecules with roles in normal physiology (e.g., signaling for development or immune response) may have deleterious effects. These perspectives also emphasize the dynamic nature of maintenance of redox state within a range necessary for normal physiological function, including adaptive responses to low-level increases in prooxidants, as "oxidative eustress" (Sies et al. 2022). In this section, we focus on pollutant-mediated deviations toward a supraphysiologically oxidized state (oxidative stress), because the great bulk of the fish literature is in this area. Two broad sets of mechanisms by which pollutants can produce oxidative stress can be delineated: mechanisms that enhance ROS production and mechanisms that reduce antioxidant capacity. Illustrative examples of both are described below.

### ENHANCED ROS PRODUCTION

Mechanisms by which chemicals can enhance ROS (and in some cases NOS) production include redox cycling, interactions with electron transport chains (notably in mitochondria and microsomes, as well as chloroplasts in plants), and photosensitization. Additionally, some chemicals are metabolized to carbon-based free radicals that can donate unpaired electrons to $O_2$, thereby generating $O_2^{\bullet-}$; carbon tetrachloride ($CCl_4$) is a notable example. Its toxicity is based on its conversion by cytochrome P450s, particularly CYP2E1, to a trichloromethyl radical $CCl_3^{\bullet}$ (Recknagel et al. 1989, Zangar et al. 2000).

#### Redox Cycling

Redox cycling is a mechanism by which a diverse array of chemicals including many environmental pollutants can generate intracellular ROS. Redox cycling chemicals include diphenols and quinones, nitroaromatics and azo compounds, aromatic hydroxylamines, bipyridyliums, and certain metal chelates, particularly of copper and iron (Di Giulio et al. 1989, Halliwell and Gutteridge 2015). These include large classes of compounds of broad industrial use, many pesticides, ubiquitous elements, and metabolic products of numerous pollutants. Examples of these redox-active chemicals are provided in Figure 4.6.

In the redox cycle, the parent compound accepts an electron from a reduced cofactor, such as NADH or NADPH; this reaction is typically catalyzed by a reductase such as xanthine oxidase or cytochrome P450 reductase (Kappus 1986). Cytochrome P450 reductase normally functions to transfer electrons from NADPH to cytochrome P450 via FAD and flavin-mononucleotide (FMN) contained in this reductase. However, it is also a key catalyst in xenobiotic redox cycling in which the electron donated by NADPH yields a radical of the xenobiotic. In the presence of $O_2$, the unpaired electron of the radical metabolite is donated to $O_2$, yielding $O_2^{\bullet-}$ and regenerating the parent compound; importantly, the parent compound can repeat this cycle until it is cleared or metabolized to an inactive product. In the course of each redox cycle, two potentially deleterious events occur—a high-energy reducing equivalent is expended (e.g., the oxidation of NADPH to $NAD^+$), and an oxygen radical is produced. Moreover, the proliferative nature of these potentially harmful outcomes associated with redox cycling (i.e., one molecule of xenobiotic causing the oxidation of many molecules of NADPH and the production of many molecules of $O_2^{\bullet-}$) is observed in other aspects of free radical biology, such as lipid peroxidation described earlier. Redox cycling of some chemicals also occurs in the mitochondria (Doroshow and Davies 1986). A generalized redox cycle that includes associations with cellular toxicities and antioxidant defenses is shown in Figure 4.7. It should be noted, however, that in some cases, redox cycling chemicals are toxic by mechanisms other than ROS production (Imlay and Fridovich 1992); thus, the *in vitro* capability of

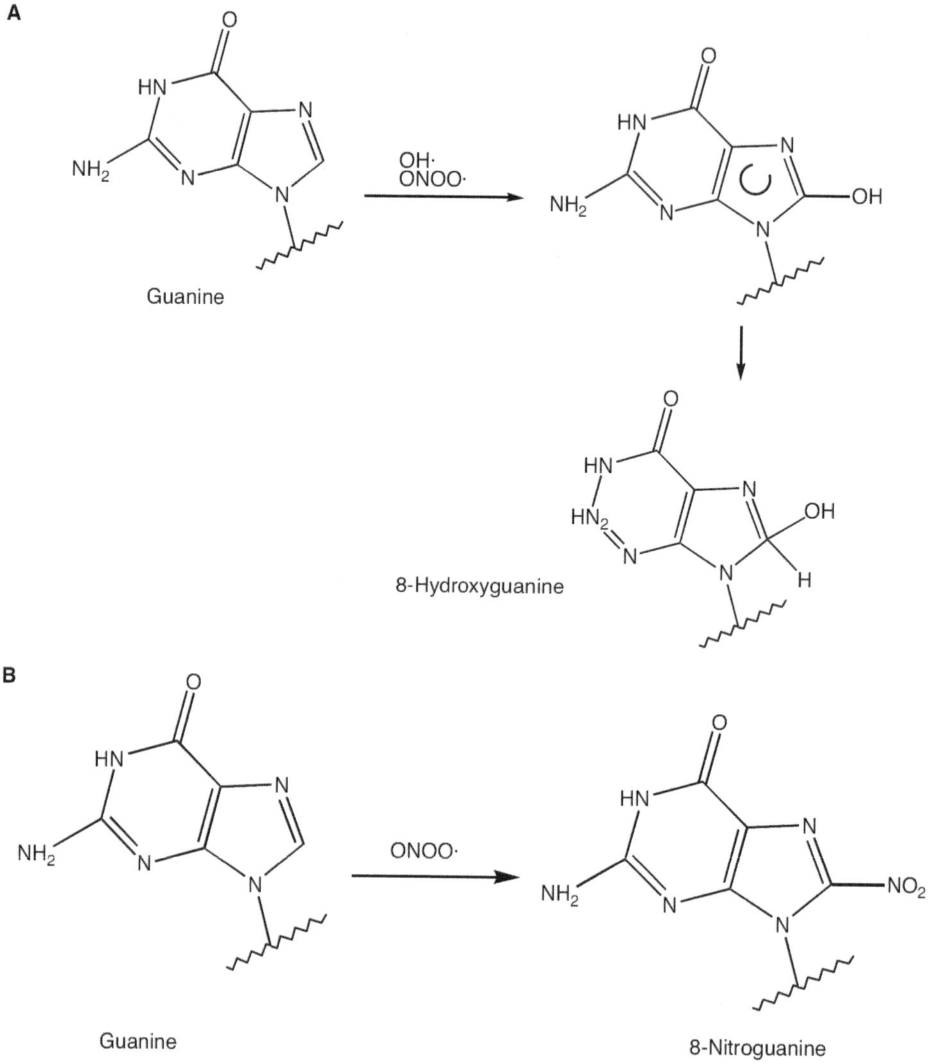


**FIGURE 4.6** (a) Hydroxyl radical (OH·)- and peroxynitrite ($ONOO^-$)-mediated oxidation of guanine. (b) Peroxynitrite ($ONOO^-$)-mediated nitration of guanine to form 8-nitroguanine.

a chemical to cause oxidative stress is not a definitive demonstration of an *in vivo* mechanism of toxicity.

## Uncoupling or Inhibition of Electron Transport

Another important mechanism by which chemicals can enhance ROS production is through interactions with electron transport chains, particularly those in mitochondria and endoplasmic reticula (microsomes) in animals. As discussed earlier, these systems are important sources of ROS during normal aerobic respiration due a degree of inherent uncoupling between NAD(P)H oxidation and substrate reduction. Chemicals can enhance ROS production by these chains in several ways. Planar halogenated aromatic hydrocarbons such as 2,3,7,8-tetrachlorodibenzo-*p*-dioxin (TCDD) and certain polychlorinated biphenyls (PCBs) that bind to the AhR and upregulate expression of genes for biotransformation enzymes including CYP1A (see Chapter 2) have been observed to enhance ROS generation in mitochondria, microsomes, and whole cells (Park et al. 1996, Nebert et al. 2000). In the case of TCDD-enhanced $H_2O_2$ production in mice mitochondria, this effect was found to be dependent on TCDD binding to the AhR but independent of CYP1A activity (Senft et al. 2002). Studies with 3,3′,4,4′-PCB and 3,3′,4,4′,5-PCB (PCB 77 and PCB 126, respectively, both established AhR ligands) in the marine fish scup (*Stenotomus chrysops*) demonstrated the ability of these AhR agonists to potently induce CYP1A and at higher doses to uncouple microsomal electron transfer, resulting in increased ROS production and inactivation of CYP1A (Schlezinger et al. 1999, Schlezinger and Stegeman 2001). In this case, the production of ROS was thought to be associated with interactions of the active site of the enzyme with these chlorinated chemicals that are recalcitrant to metabolism and thereby uncouple normal electron flow to the substrate (i.e., PCB), resulting in ROS (Figure 4.8).

Inhibition of electron transport in mitochondria can also lead to production of ROS; blockage of the flow of electrons leaves mitochondrial proteins in a highly reduced state, which can lead to the reduction of oxygen and ROS production. Chemicals known to produce ROS by this mechanism

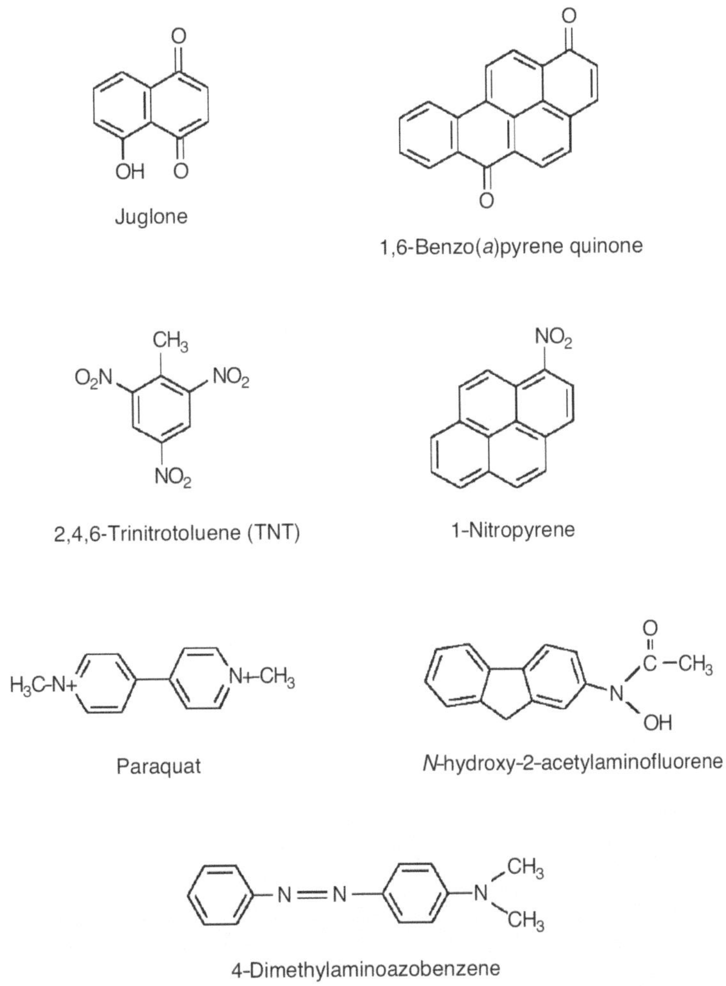


**FIGURE 4.7** Representative redox-cycling chemicals including a natural product (juglone), a metabolite of the procarcinogen benzo(*a*)pyrene (1,6-benzo(*a*)pyrene quinone), the explosive 2,4,6-trinitrotoluene (TNT), a common combustion by-product (1-nitropyrene), an herbicide (paraquat), a carcinogenic aromatic hydroxylamine (*N*-hydroxy-2-acetylaminofluorene), and an azo dye (4-dimethylaminobenzene).

include rotenone, an insecticide (Li et al. 2003) now used in the United States mostly as a nonselective piscicide; antimycin A, an antibiotic frequently used to study inhibition of mitochondrial electron transport (Chen et al. 2003); 1-methyl-4-phenylpyridinium (MPTP), a compound used to study Parkinson's disease (Smith and Bennett 1997); certain oxidatively modified PAHs (Tripuranthakam et al. 1999); and cadmium (Wang et al. 2004). Remarkably, inhibition of electron transport by hypoxia results in ROS production at Complex III, which activates hypoxia signaling (Chandel et al. 2000); upon reperfusion and reoxygenation, ROS production at Complex I is a major driver for oxidative damage (Brand et al. 2016, Chouchani et al. 2016). Hypoxia-reoxygenation has also been studied in a number of fish species (Podrabsky et al. 2012, Parente et al. 2013, Onukwufor et al. 2016), with particular mechanistic depth in shellfish (Sokolova 2018); fish species that are regularly exposed to hypoxic conditions may be particularly adapted to this stressor. Chronic over-reduction of the ETC due to high caloric intake and/or lack of exercise may also cause sufficient hyper-reduction of the ETC to generate ROS by this mechanism (Fisher-Wellman and Neufer 2012). Alternatively, xenobiotic-mediated uncoupling of proton pumping from oxidative phosphorylation in mitochondria can result in reduced production of ROS (Turrens 1997, Kowaltowski and Vercesi 1999), presumably due to the less-reduced state of the ETC proteins. Indeed, endogenous uncoupling proteins are upregulated by, and protective against, ROS (Brand and Esteves 2005, Cadenas 2018).

### Photosensitization

Ultraviolet (UV) radiation is typically divided into three general categories: UVC (200–280 nm), UVB (280–320 nm), and UVA (320–400 nm). UVC is not environmentally relevant, as it is effectively blocked by the Earth's atmosphere; however, UVB and UVA radiation to differing degrees do have the ability to penetrate the Earth's atmosphere and water columns to depths dependent on the wavelength of the radiation and the clarity of the water. Both are capable of generating ROS and free radicals either directly

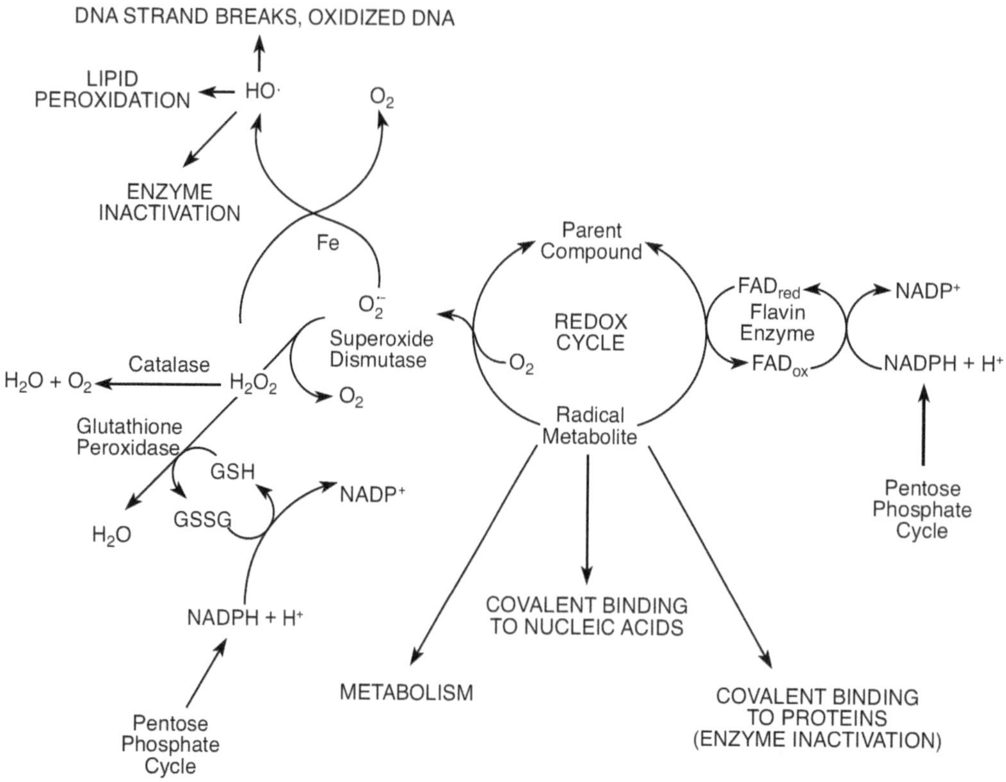


**FIGURE 4.8** An overview of ROS generation by redox cycling, key enzymatic antioxidant defenses, and cellular targets of ROS. Adapted from Kappus (1985).

or via excitation of photosensitizing chemicals, including both endogenously produced compounds (Young 1997) and many common pollutants of aquatic systems (Larson and Weber 1994). Many PAHs, for example, are orders of magnitude more acutely toxic to aquatic organisms, including fish, in the presence of ultraviolet radiation than in its absence (Bowling et al. 1983, Arfsten et al. 1996, Ankley et al. 1997, El-Alawi et al. 2002). More recently, the same has been found for some nanomaterials, such as titanium dioxide (Ma et al. 2013). Energy absorbed by the chemical bonds present in these compounds excites electrons to higher energy orbitals; this energy can lead to changes in the chemical structure of the PAH and in some cases reaction with other molecules, such as oxygen, in a process referred to as *photomodification* (Huang et al. 1997, McConkey et al. 1997). The extra energy can also cause the loss of an electron from the PAH, or the extra energy can be given off by the compound in a variety of ways (Larson and Weber 1994). As an example, photosensitization occurs when the energy gained by the PAH that absorbed the photon (the photosensitizer) is then passed on to another molecule that would not have been chemically able to absorb that energy directly from the photon, such as $O_2$ (El-Alawi et al. 2002). Both photosensitization and photomodification of PAHs can lead to the production of ROS; for example, when an electron is lost, it is often absorbed by oxygen, leading to the production both of superoxide anion and a radical compound. Similarly, when energy is lost from the excited compound, it can be transferred to oxygen, producing the reactive singlet oxygen, which is highly reactive and toxic (Briviba et al. 1997). Of course, high-energy radiation such as UV radiation can cause damage in the absence of photosensitizers; however, although UV radiation has been shown to cause oxidative stress, DNA damage, and other biological effects in fish (Fabacher et al. 1994, Charron et al. 2000, Lesser et al. 2001), the environmental relevance of this source of oxidative stress to fish is unclear (Williamson 1995).

### Diminished Antioxidant Defense

Although less studied than enhanced ROS production, interference with antioxidant defense system components represents another mechanism by which chemicals can exert oxidative stress in animals. Inhibition of antioxidant enzymes comprises an important component of this phenomenon; for example, quinones have been shown to inhibit SOD activities (Smith and Evans 1984). This effect may be mediated by $H_2O_2$ generated via redox cycling, as prolonged $H_2O_2$ exposure can inhibit CuZnSOD and FeSOD (Halliwell and Gutteridge 2015). Relatedly, extracts of diesel exhaust particles were observed to effectively inhibit CuZnSOD *in vitro*, and this effect was surmised to be associated with quinone components of the extracts (Kumagai et al. 1995). Dowla et al. examined the effects of several chemicals used in agriculture (acephate, cadmium, methamidophos, maleic hydrazide, and nicotine) on activities of human blood SOD, cholinesterase, and δ-aminolevulinic acid dehydrase (ALAD), and found

SOD to be the most sensitive to inhibition by all chemicals tested (Dowla et al. 1996). Plant polyphenols such as chalcones and tannic acids were shown to be potent GR inhibitors *in vitro*, with $IC_{50}$ values in the micromolar range (Zhang et al. 1997). The relevance of such *in vitro* studies to *in vivo* exposures remains speculative. Methylmercury, however, was shown to inhibit hepatic GPX activities and reduce GSH levels in rat pups; these effects were associated with increased lipid peroxidation and hepatotoxicity. These results support the purported prooxidant role of methylmercury, a metal complex that cannot directly generate ROS via redox cycling or Fenton chemistry, as is the case for metals such as Fe and Cu. Moreover, the reactive nature of the sulfhydryl group of GSH makes this critical antioxidant prone to depletion by a number of oxidants and electrophiles, including nonredox cycling metals such as mercury, mentioned above, and cadmium (Nemmiche 2017).

## STUDIES WITH FISHES

A large number of studies of oxidative stress in fish have been carried out in the last several decades, with different motivations. These motivations might be summarized as follows: (1) increasing the understanding of basic biochemical and molecular mechanisms related to oxidative stress in fish, (2) developing biomarkers related to oxidative stress fish, and (3) investigating the relationships between oxidative stress and the health of wild populations of fish and humans. Some of the results of the studies carried out are presented below with respect to these three concerns.

### Basic Biochemical and Molecular Mechanisms of Redox Stress in Fish

The study of the toxicity of prooxidant xenobiotics in fish serves both to inform risk assessment with regard to fish populations and to inform our understanding of the toxicity of such chemicals in other organisms, including humans. Mechanistic studies in fish can be further categorized as characterizing the ability of a xenobiotic to generate ROS in fish cells or living fish, as well as characterizing the antioxidant defenses present in different species of fish, either with or without prior xenobiotic exposure. It is often useful to employ comparative studies involving different species (piscine and nonpiscine) to better understand the impact of prooxidant chemicals in the environment.

#### Antioxidant Defenses in Fish

Many if not all of the basic antioxidant defenses (enzymatic and nonenzymatic) characterized in mammals are also present in fish. In addition, similar evidence supports the existence of many of the phase II enzymes that can be viewed as playing an antioxidant role or that are regulated by an ARE in mammals (discussed earlier in the section on the ARE and in Chapter 2). Much of the characterization of fish antioxidant defenses has occurred in biomedical fish models such as zebrafish (*Danio rerio*). For example, the glutathione antioxidant defense system has been extensively characterized in the zebrafish embryo including the GSH content, redox potential, and expression of genes involved in GSH synthesis, utilization, and recycling, across different stages of development (Timme-Laragy et al. 2013). The role of the transcription factors Nrf1 and Nrf2 in regulating and modulating the GSH system in zebrafish embryos has been identified, and the response to pro-oxidants and recovery of redox homeostasis at specific developmental stages have been elucidated (Sant et al. 2017). Zebrafish embryos have also been used to generate a spatiotemporal map of GSH utilization and content and the relation to Nrf2 activation (Rastogi et al. 2019, Rastogi et al. 2021), protein-*S*-glutathionylation using the biotinylated GSH molecule BioGEE (Rastogi et al. 2021), the catalase antioxidant defense system, and $H_2O_2$ generation (Gauron et al. 2016). Expression of GST proteins has been mapped throughout different life stages of zebrafish (Tierbach et al. 2018), and tissue-specific expression and enzyme kinetic activity measured (Glisic et al. 2015). Many of these methods are derived from mammalian systems, but are easily translated to fish species (Massarsky et al. 2017). As observed in mammals, tissue glutathione levels are often depleted after short-term oxidant exposures but elevated after long-term exposures. Furthermore, glutathione depletion sensitizes fish, like mammals, to the toxicity of prooxidant xenobiotics (Gallagher et al. 1992).

Components of antioxidant defense systems have been characterized in other fish species as well. In coho and chinook salmon, GST enzymes have been characterized from various tissues (Donham et al. 2005, Espinoza et al. 2013). The antioxidant capacity and enzyme activities of SOD, catalase, GPX, and glutathione reductase, glutathione-s-transferase, and GSH have been characterized in several freshwater species: tilapia (Oreochromis niloticus); carp (Cyprinus carpio); trout (Oncorhynchus mykiss); and catfish (Clarias gariepinus) (Atli et al. 2016, Hamed et al. 2016) as just a few examples.

Other important antioxidants, such as vitamin C, vitamin E, ubiquinone, and carotenoids, also play important roles (Parihar and Dubey 1995, Payne et al. 1998, Olsen et al. 1999, Bell et al. 2000, TOCHER et al. 2002); particularly, given studies in developmental redox biology and nutrition have established essential roles for these components (Head and Traber 2021). In addition, important differences clearly exist in terms of the ability of different fish species to synthesize vitamin C (Moreau-Regis and Dabrowski 1998). Although the majority of studies of nonenzymatic antioxidants in toxicology have historically focused on glutathione, it is increasingly clear that other compounds are also important. Studies of the total oxyradical scavenging capacity (Winston et al. 1998, Regoli and Winston 1999) of tissue homogenates from both aquatic and nonaquatic organisms have demonstrated that compounds other than the classical antioxidants such as glutathione, vitamin C, and vitamin E can contribute very significantly to the *in vitro* oxyradical scavenging capacity of those tissue preparations. As in mammals, it is likely that diet plays a very important role in determining the nonenzymatic antioxidant capacity of fish tissues (Nakano et al. 1999, Olsen et al. 1999, Mourente et al. 2000, Pascual et al. 2003).

## In Vivo Prooxidant Studies in Fish

As previously stated, the fact that a chemical produces ROS in an *in vitro* situation does not mean that the same chemical will cause oxidative stress *in vivo*. The additional complications introduced by kinetics of uptake, metabolism, and excretion in a living organism can be understood only by testing the effect of a chemical in an *in vivo* system. Table 4.1 is a representative list of stressors shown to exert oxidative stress in intact fish. It is important to bear in mind that not all chemicals in the chemical classes listed in column 1 are expected to be prooxidants. Specific chemicals that induced oxidative stress in the studies cited are listed in parentheses as examples. Generally speaking, the chemicals shown to cause oxidative stress in mammals have also done so in fish. Lushchak (2016) provides a thorough review of studies of concerning oxidative stress associated with exposures to environmental contaminants.

Many chemicals are more or less toxic depending on other factors. An example that has received considerable attention recently is phototoxicity, which is the greatly enhanced toxicity of specific chemicals (especially many PAHs) in the presence of ultraviolet radiation. UV radiation can excite an electron of such chemicals, and the resultant excited-state molecule is more reactive and often ultimately more toxic. Empirical studies with fish have strongly supported the hypothesis that the phototoxicity or photo-enhanced toxicity of PAHs resulting from the process of photosensitization is mediated by oxidative stress (Weinstein et al. 1997, Choi and Oris 2000). In addition, chemical structure alterations produced by UV irradiation can produce compounds such as quinones and phenols that are capable of redox cycling and interfering with electron flow in electron transport chains of mitochondria (Huang et al. 1997, Tripuranthakam et al. 1999), both processes that can lead to oxidative stress. Thus, phototoxicity represents a distinctive source of oxidative stress that may be of particular importance in aquatic systems with clear water columns. It is important to note, however, that most of the studies carried out thus far have been conducted under laboratory or constrained field conditions, so the ecological relevance of phototoxicity is controversial (McDonald and Chapman 2002).

Pro-oxidants have also been shown to potentiate toxicity of other compounds, such as that of TCDD. Cardiotoxicity in zebrafish was shown to be exacerbated by paraquat, rotenone, and $H_2O_2$, while the aspects of TCDD cardiotoxicity that are attributed to oxidative stress were inhibited by antioxidant treatments including activators of Nrf2 (Tanaka et al. 2021).

## Transgenic Redox Biosensors, Reporter Genes, Genetic Mutants

The use of transgenic fish models has greatly expanded over the last decade, primarily in laboratory zebrafish (*Danio rerio*). While comprehensive reviews on this topic

**TABLE 4.1**

**Representative List of Stressors (Chemical and Nonchemical) Shown to Cause Oxidative Stress,[a] Induce Production of ROS,[b] or Induce or Repress Antioxidant Defenses**

| Stressor (Common Examples) | References |
|---|---|
| Hydrogen peroxide | Winzer et al. (2000), Winzer et al. (2001) |
| Quinones (menadione, naphthoquinone, benzoquinone) | Akerman et al. (2003), Petřivalský et al. (1997), Stephensen et al. (2002), Schmieder et al. (2003) |
| Aromatic nitro compounds (nitrofurantoin) | Winzer et al. (2001) |
| Metals (cadmium, copper) | Radi and Matkovics (1988), Ariyoshi et al. (1990), Pedrajas et al. (1995), Carvan et al. (2001), Rau et al. (2004) |
| PAHs (β-naphthoflavone, benzo(*a*)pyrene, 3-methylcholanthrene) | Winzer et al. (2000), Carvan et al. (2001), Winzer et al. (2001), Stephensen et al. (2002), Hughes and Gallagher (2004) |
| Pesticides (paraquat, endosulfan, dieldrin, malathion, chlorothalonil) | Gallagher and Di Giulio (1992), Pedrajas et al. (1995), Pandey et al. (2001), Elia et al. (2002), Akerman et al. (2003), Dorval and Hontela (2003), Dorval et al. (2003), Luo et al. (2005) |
| Halogenated aromatic hydrocarbons (PCBs, dioxins) | Förlin et al. (1996), Schlezinger et al. (1999), Carvan et al. (2001), Schlezinger and Stegeman (2001) |
| Ozone | Ritola et al. (2002) |
| Hypoxia | Cooper et al. (2002) |
| Hyperoxia | Ritola et al. (2002) |
| Anoxia-reoxygenation | Lushchak et al. (2001), Ross et al. (2001) |
| Heat stress | Parihar and Dubey (1995), Parihar et al. (1996) |

[a] As indicated by markers such as lipid peroxidation.

[b] As indicated *in vitro* either by dyes whose fluorescent properties are altered by exposure to ROS or by reporter transgenes.

*Note*: Not all chemicals pertaining to the chemical classes listed are prooxidant; specific examples studied in the references cited are listed in parentheses.

are available elsewhere (e.g., (Breus and Dickmeis 2021)), this approach generally requires fluorescent microscopy, a tissue-specific approach, and biosensors tailored to change in response to specific redox signals, such as the glutathione redox potential (roGFP2) or $H_2O_2$ (HyPer; (Pak et al. 2020)), and ARE-gene regulation with *Tg(3EpRE:hsp70:mCherry)* (Panieri et al. 2017). Subcellular compartmental readouts are also possible, differentiating between redox responses in the cytosol, mitochondria, and nucleus (Panieri et al. 2017). In screening for toxicant-induced redox modulation, it is important to optimize the duration of exposure and the time post-exposure at which the assessment occurs to avoid false-negative findings.

Zebrafish have also been widely used in generating mutant strains, including in redox-sensitive genes. For example, there are mutant knockouts for Keap1a and Keap1b (Nguyen et al. 2020), and a line (fh318) with a point mutation in Nrf2a that disrupts ARE-binding (Mukaigasa et al. 2012). This area is rapidly expanding with the use of gene-editing technologies such as Crispr-CAS9 (Mills et al. 2020, Xu et al. 2020, Wang et al. 2021).

### Differences between Fish and Mammals

As we have seen, many of the basic mechanisms of ROS production and damage are shared between fish and mammals. Despite the overall similarities in antioxidant defenses, however, important differences exist between fish and mammalian species, some of which are summarized here. ROS production in microsomal fractions was higher in rainbow trout, as well as in two aquatic invertebrate species, with NADH than with NADPH as a source of reducing power, in contrast to the situation seen in rats (Jewell and Winston 1989, Kennish et al. 1989). Glutathione depletion under conditions of starvation was much slower in mullet (*Mugil cephalus*) (Thomas and Wofford 1984) and channel catfish (*Ictalurus punctatus*) (Gallagher et al. 1992) than in rodents. Perhaps because of the observed low (relative to rodents) rate of turnover of glutathione in catfish, administration of BSO, an inhibitor of GCL, did not completely deplete hepatic glutathione, although co-administration of BSO and diethyl maleate, a glutathione depletor, led to nearly complete depletion of glutathione (Gallagher et al. 1992). There is considerable variability in the reported catalytic activities of many of the antioxidant enzymes, both between fish and mammals and between different fish species (Di Giulio et al. 1989, Forlin et al. 1995, Winston 1991, Hasspieler et al. 1994a, 1994b, Gadagbui et al. 1996, Rana and Singh 1996, Ploch et al. 1999). Biochemical differences that are important in terms of sample preparation and treatment also exist; for example, mammalian CuZnSOD is a remarkably thermostable enzyme, but the CuZnSODs from two different fish species are much more labile at high temperatures (60°C and above) (Nakano et al. 1995, Capo et al. 1997). Vitamin E levels appear to be generally higher in fish cell membranes than in mammalian cell membranes, but it is unclear whether this difference is protective in the context of the typically higher levels of PUFA observed in fish cell membranes (Singh et al. 1992, Bell et al. 2000). Some studies have found that fish cells are more resistant than mammalian cells to oxidative stress *in vitro* (Singh et al. 1992, Venditti et al. 1999), while others have found the opposite (Rau et al. 2004).

## Biomarkers of Oxidative Stress

How is it possible to know if oxidative stress is a problem in a given field situation? A number of biomarkers have been tested (reviewed broadly in Sies 2017 and Murphy et al, 2022), including oxidized lipids, proteins, and nucleic acids, as well as induction of genes such as those regulated by AREs. More recently, integrative exposomics approaches have been implemented (Niedzwiecki 2019). However, despite a large number of laboratory exposures to pure compounds, environmentally relevant contaminant mixtures, and field studies, no single biomarker of oxidative stress has emerged that is as reliable and specific as other established biochemical biomarkers such as acetylcholinesterase activity for organophosphate insecticides, δ-ALAD activity for exposures to lead, or CYP1A expression/activity for aryl hydrocarbon receptor agonists (see Chapter 3). Various studies have identified GSSG:GSH ratios, levels of MT or lipid peroxidation, and activities of GR, microsomal GST, or microsomal GPX as sensitive indicators in the system being studied, but these markers have been completely unresponsive in other contexts. An extensive review of biomarker studies that included a specific review of biomarkers of oxidative stress in fishes (van der Oost et al. 2003) failed to identify any marker that was responsive in a high percentage of the studies reviewed, with lipid peroxidation being perhaps the most consistent marker. More recent reviews have identified and discussed possible reasons for discordant results between studies (Benedetti et al. 2015, Lushchak 2016).

Why have better markers of oxidative stress in wild fish not been identified? A variety of factors may be involved. First of all, as in mammals, large inductions (e.g., comparable to the inductions in CYP1A observed after exposure to certain xenobiotics, as discussed in Chapter 3) in antioxidant enzymes have not been observed in fish exposed to prooxidants. As a result, any confounding factors present have a strong chance of hiding an otherwise observable effect. Unfortunately, there are many such confounding factors. An environmental pollution mixture will rarely be expected to exert toxicity only by oxidative stress, so other forms of toxicity may be occurring. Additionally, sex and reproductive conditions as well as seasonal factors can affect many antioxidant parameters (Livingstone et al. 1995, McFarland et al. 1999, Winzer et al. 2001, Winzer et al. 2002a, 2002b, Meyer et al. 2003) but are not always taken into account. In that case, the variance associated with sex becomes "noise," potentially obscuring real differences. Similarly, temperature can greatly alter the metabolic capacity of poikilotherms and has been shown to affect antioxidant defenses in fish (Parihar and Dubey

1995, Parihar et al. 1996, Olsen et al. 1999, Heise et al. 2003). Diet alters the activity of many antioxidant enzymes (George et al. 2000, Mourente et al. 2000, Hidalgo et al. 2002, Mourente et al. 2002, Pascual et al. 2003), as well as altering the tissue concentrations of nonenzymatic antioxidants, as mentioned above. Dissolved oxygen (Lushchak et al. 2001, Ross et al. 2001, Cooper et al. 2002, Hermes-Lima and Zenteno-Savín 2002, Ritola et al. 2002) and salinity (Kolayli and Keha 1999, Martínez-Alvarez et al. 2002) have also been observed to affect antioxidant parameters. Seasonal effects have been observed (Ronisz et al. 1999, Bacanskas et al. 2004) and are likely to incorporate many other biological and environmental variables, such as temperature, reproductive status, developmental stage, and food sources. Additionally, time courses can be complicated; for example, although total glutathione levels can increase dramatically in response to prooxidant exposure, the initial response is often depletion, and degree of induction is very likely to be additionally affected by diet, as the availability of cysteine can be limiting for the production of GSH. Antioxidant enzymes have also been observed sometimes to be depressed at the level of activity or expression after exposure to prooxidants (Radi and Matkovics 1988, Pedrajas et al. 1995, Kim and Lee 1997, Zikic et al. 1997, Fujii and Taniguchi 1999, Stephensen et al. 2002). Hence, it is useful to conceptualize redox stress as an adverse deviation from homeostatic conditions (Timme-Laragy et al. 2018), rather than a sole directional change.

Physiological or genetic adaptation to pollution has been shown to lead to an altered response even in biomarkers that are usually fairly robust, such as CYP1A (Hahn 1998, Elskus et al. 1999, Roy et al. 2001, Meyer et al. 2002). Although the possibility of adaptation to oxidative stress in fish has been less studied, it may also be a significant factor in some cases (McFarland et al. 1999, Meyer et al. 2003, Bacanskas et al. 2004). There is strong evidence in other organisms for "mitohormetic" effects where mild mitochondrial stress, including oxidative damage, leads to upregulated antioxidant defenses, which may also imply resolution of markers of damage (Yun and Finkel 2014). Transcriptional changes may be persistent past the period of exposure via epigenetic mechanisms (Weinhouse 2021). Markers of damage, such as lipid peroxidation and DNA damage, may only reflect recent or constant damage, as both can be repaired. Another important concern is that most fish populations are highly outbred with a high degree of genetic variability (compared to typical laboratory mammalian models), which may contribute additional variability to the responses observed. Finally, as mentioned earlier, the various antioxidant isoenzymes potentially represented in fish are not well characterized, and it may be that potentially good biomarkers are being overlooked due to nonspecific substrates, antibodies, etc. An improved understanding of biological and environmental modulation of oxidative stress and antioxidant defenses in fish, along with careful experimental design and improved techniques, should enhance the utility of biomarkers for oxidative stress in the future; however, as discussed above, no single broadly applicable biomarker for oxidative stress is likely to emerge in the foreseeable future.

## Redox Stress and the Health of Wild Fish Populations

As described above, redox stress is known to play an important role in many disease processes and chemical-mediated toxicity in humans. Furthermore, it is clear that there are many similarities among humans, mammalian models, and fish in terms of production and cellular effects of oxidative stress, as well as in terms of antioxidant defenses; nonetheless, the importance of oxidative stress to fish populations in the wild is not yet fully understood. Biomarkers of effect of exposure to ROS such as lipid peroxidation (Di Giulio et al. 1993, Livingstone et al. 1993, Eufemia et al. 1997) and DNA damage (Di Giulio et al. 1993, Stein et al. 1993, Malins and Gunselman 1994, Sugg et al. 1996, Eufemia et al. 1997, Theodorakis et al. 1997, Payne et al. 1998, Rodríguez-Ariza et al. 1999, Theodorakis et al. 1999, Stephensen et al. 2000) indicate that fish inhabiting polluted environments exhibit oxidative stress (although not all of these DNA damage studies are necessarily strictly oxidative-stress related). Some common fish diseases, such as jaundice and methemoglobinemia (brown blood disease), have been associated with oxidative stress (Jensen 1996, Sakai et al. 1998).

Fish populations adapted to inhabit ecosystems contaminated with chemical mixtures containing prooxidants such as creosote (Meyer et al. 2003) and metals (Weis et al., 1999; (Xie and Klerks 2003) have been shown to be less fit than their counterparts from clean sites in several contexts. There are many published examples of fish populations inhabiting sites contaminated with prooxidant chemicals that exhibit high rates of DNA damage and cancer (Black 1983, Malins et al. 1984, Mix 1986, Vogelbein et al. 1990, Malins and Gunselman 1994). Although the chemicals present at such sites are not exclusively toxic via prooxidant mechanisms, a strong relationship exists between oxidative stress and cancer. Furthermore, the ability of channel catfish to avoid cancer in the same polluted environments that cause liver cancer in brown bullhead catfish has been associated with higher antioxidant defenses (Hasspieler et al. 1994a, 1994b, Ploch et al. 1999). Dietary exposure to hydrogen peroxide enhanced the development of oxidative DNA damage (8-OHdG) and hepatic tumors in rainbow trout exposed to the carcinogen *N*-methyl-*N′*-nitro-*N*-nitrosoguanidine in a laboratory study (Kelly et al. 1992). It is difficult to unequivocally link a given pollutant or form of toxicity with population-level measurements of health. Long-term laboratory exposures will likely be needed to better define the links between exposure to oxidative stress and ecologically relevant alterations in fish populations, and improved markers of oxidative stress will be necessary to analyze the responses of populations of wild fish to oxidative stress.

## FUTURE DIRECTIONS

A better understanding of redox-mediated signaling in fish, how it is regulated, and what biochemical characteristics modulate these mechanisms will greatly improve our ability to understand the impact of redox-active chemicals on fish. Similarly, application of the knowledge gained in studies of redox stress in mammalian models will inform improved generation of hypotheses and experimental design.

As discussed above, major advances have been made in the understanding of gene regulation in response to redox stress, transcription factor activation, redox potential, and the importance of distinguishing different sources and types of redox-mediated stress responses. These advances have been facilitated by improved methodology and transgenic redox biosensors. Still, there remains much to learn, particularly in the areas of gene transcription in response to xenobiotic exposures, sub-cellular compartment redox conditions, and post-translational redox regulation of protein activity and function. Future studies should also take into account (through experimental design and/or statistical analysis) as much as possible the effects of confounding biological and environmental factors such as sex, developmental stage, season, oxygen tension, and diet. Improved methodology for the measurement of DNA damage, lipid peroxidation, and other biomarkers of effect will permit more sensitive assays, and increased knowledge of fish-specific forms of antioxidant enzymes will provide more specific assays. Further, the use of transgenic redox biosensors and reporter genes may result in highly specific measurements of the generation on ROS *in vivo*, in the laboratory, or in the field. Assays that take advantage of fish-specific biology may also prove valuable; for example, the presence of nucleated blood cells in fish permits a nonlethal measurement of DNA damage.

At the same time as biochemical and molecular methodologies are improved and mechanistic understanding of redox stress in aquatic organisms is refined, it will be important to begin to more carefully explore relationships between redox stress and population-level effects. The associations between contaminant exposure and cancer epizootics and other fitness costs mentioned above suggest that redox stress is an ecologically important phenomenon; however, few attempts have been made to quantify the costs of the oxidative damage caused by environmental pollutants for individual fish, fish populations, and aquatic ecosystems. Although a mechanistic understanding of redox stress in fish is important from the standpoint of comparative toxicology and basic biology, a solid understanding of the ecological importance is dependent upon studies of higher level effects.

## REFERENCES


Ahmed Alfar, E., D. Kirova, J. Konantz, S. Birke, J. Mansfeld and N. Ninov (2017). "Distinct Levels of reactive oxygen species coordinate metabolic activity with beta-cell mass plasticity." *Sci Rep* **7**(1): 3994.

Akerman, G., P. Amcoff, U. Tjarnlund, K. Fogelberg, O. Torrissen and L. Balk (2003). "Paraquat and menadione exposure of rainbow trout (Oncorhynchus mykiss)--studies of effects on the pentose-phosphate shunt and thiamine levels in liver and kidney." *Chem Biol Interact* **142**(3): 269–283.

Ankley, G. T., R. J. Erickson, B. R. Sheedy, P. A. Kosian, V. R. Mattson and J. S. Cox (1997). "Evaluation of models for predicting the phototoxic potency of polycyclic aromatic hydrocarbons." *Aquat Toxicol* **37**(1): 37–50.

Arfsten, D. P., D. J. Schaeffer and D. C. Mulveny (1996). "The effects of near ultraviolet radiation on the toxic effects of polycyclic aromatic hydrocarbons in animals and plants: a review." *Ecotoxicol Environ Saf* **33**(1): 1–24.

Ariyoshi, T., S. Shiiba, H. Hasegawa and K. Arizono (1990). "Effects of the environmental pollutants on heme oxygenase activity and cytochrome P-450 content in fish." *Bull Environ Contam Toxicol* **44**(2): 189–196.

Arrigo, A. P. (1999). "Gene expression and the thiol redox state." *Free Radic Biol Med* **27**(9–10): 936–944.

Atli, G., E. G. Canli, A. Eroglu and M. Canli (2016). "Characterization of antioxidant system parameters in four freshwater fish species." *Ecotoxicol Environ Saf* **126**: 30–37.

Babior, B. M. (2000). "Phagocytes and oxidative stress." *Am J Med* **109**(1): 33–44.

Bacanskas, L. R., J. Whitaker and R. T. Di Giulio (2004). "Oxidative stress in two populations of killifish (Fundulus heteroclitus) with differing contaminant exposure histories." *Mar Environ Res* **58**(2–5): 597–601.

Balagopalakrishna, C., P. T. Manoharan, O. O. Abugo and J. M. Rifkind (1996). "Production of superoxide from hemoglobin-bound oxygen under hypoxic conditions." *Biochemistry* **35**(20): 6393–6398.

Barber, I., S. A. Arnott, V. A. Braithwaite, J. Andrew, W. Mullen and F. A. Huntingford (2000). "Carotenoid-based sexual coloration and body condition in nesting male sticklebacks." *J Fish Biol* **57**(3): 777–790.

Barouki, R. and Y. Morel (2001). "Repression of cytochrome P450 1A1 gene expression by oxidative stress: mechanisms and biological implications." *Biochem Pharmacol* **61**(5): 511–516.

Beckman, J. S. and W. H. Koppenol (1996). "Nitric oxide, superoxide, and peroxynitrite: the good, the bad, and ugly." *Am J Physiol* **271**(5 Pt 1): C1424–1437.

Begas, P., L. Liedgens, A. Moseler, A. J. Meyer and M. Deponte (2017). "Glutaredoxin catalysis requires two distinct glutathione interaction sites." *Nat Commun* **8**(1): 14835.

Bell, J. G., J. McEvoy, D. R. Tocher and J. R. Sargent (2000). "Depletion of α-tocopherol and astaxanthin in Atlantic salmon (Salmo salar) affects autoxidative defense and fatty acid metabolism." *J Nutr* **130**(7): 1800–1808.

Bendich, A., L. J. Machlin, O. Scandurra, G. W. Burton and D. D. M. Wayner (1986). "The antioxidant role of vitamin C." *Adv Free Rad Biol Med* **2**(2): 419–444.

Benedetti, M., M. E. Giuliani and F. Regoli (2015). "Oxidative metabolism of chemical pollutants in marine organisms: molecular and biochemical biomarkers in environmental toxicology." *Ann N Y Acad Sci* **1340**: 8–19.

Bjelland, S. and E. Seeberg (2003). "Mutagenicity, toxicity and repair of DNA base damage induced by oxidation." *Mutat Res* **531**(1–2): 37–80.

Black, J. J. (1983). "Field and laboratory studies of environmental carcinogenesis in Niagara river fish." *J Great Lakes Res* **9**(2): 326–334.

Bolduc, J., K. Koruza, T. Luo, J. Malo Pueyo, T. N. Vo, D. Ezeriņa and J. Messens (2021). "Peroxiredoxins wear many hats: factors that fashion their peroxide sensing personalities." *Redox Biol* **42**: 101959.

Bolton, J. L., M. A. Trush, T. M. Penning, G. Dryhurst and T. J. Monks (2000). "Role of quinones in toxicology." *Chem Res Toxicol* **13**(3): 135–160.

Bowling, J. W., G. J. Leversee, P. F. Landrum and J. P. Giesy (1983). "Acute mortality of anthracene-contaminated fish exposed to sunlight." *Aquat Toxicol* **3**(1): 79–90.

Brady, T. C., L. Y. Chang, B. J. Day and J. D. Crapo (1997). "Extracellular superoxide dismutase is upregulated with inducible nitric oxide synthase after NF-kappa B activation." *Am J Physiol* **273**(5): L1002–L1006.

Brand, M. D. (2020). "Riding the tiger - physiological and pathological effects of superoxide and hydrogen peroxide generated in the mitochondrial matrix." *Crit Rev Biochem Mol Biol* **55**(6): 592–661.

Brand, M. D., R. L. Goncalves, A. L. Orr, L. Vargas, A. A. Gerencser, M. Borch Jensen, Y. T. Wang, S. Melov, C. N. Turk, J. T. Matzen, V. J. Dardov, H. M. Petrassi, S. L. Meeusen, I. V. Perevoshchikova, H. Jasper, P. S. Brookes and E. K. Ainscow (2016). "Suppressors of superoxide-H(2) O(2) production at site I(Q) of mitochondrial complex I protect against stem cell hyperplasia and ischemia-reperfusion injury." *Cell Metab* **24**(4): 582–592.

Brand, M. D. and T. C. Esteves (2005). "Physiological functions of the mitochondrial uncoupling proteins UCP2 and UCP3." *Cell Metab* **2**(2): 85–93.

Brantley, R. E., Jr., S. J. Smerdon, A. J. Wilkinson, E. W. Singleton and J. S. Olson (1993). "The mechanism of autooxidation of myoglobin." *J Biol Chem* **268**(10): 6995–7010.

Breus, O. and T. Dickmeis (2021). "Genetically encoded thiol redox-sensors in the zebrafish model: lessons for embryonic development and regeneration." *Biol Chem* **402**(3): 363–378.

Brigelius-Flohe, R. and L. Flohe (2011). "Basic principles and emerging concepts in the redox control of transcription factors." *Antioxid Redox Signal* **15**(8): 2335–2381.

Brigelius-Flohé, R. and M. Maiorino (2013). "Glutathione peroxidases." *Biochim Biophys Acta* **1830**(5): 3289–3303.

Britton, G. (1995). "Structure and properties of carotenoids in relation to function." *FASEB J* **9**(15): 1551–1558.

Briviba, K., L. O. Klotz and H. Sies (1997). "Toxic and signaling effects of photochemically or chemically generated singlet oxygen in biological systems." *Biol Chem* **378**(11): 1259–1265.

Brouwer, M., T. H. Brouwer, W. Grater, J. J. Enghild and I. B. Thogersen (1997). "The paradigm that all oxygen-respiring eukaryotes have cytosolic CuZn-superoxide dismutase and that Mn-superoxide dismutase is localized to the mitochondria does not apply to a large group of marine arthropods." *Biochemistry* **36**(43): 13381–13388.

Burczynski, M. E., H. K. Lin and T. M. Penning (1999). "Isoform-specific induction of a human aldo-keto reductase by polycyclic aromatic hydrocarbons (PAHs), electrophiles, and oxidative stress: implications for the alternative pathway of PAH activation catalyzed by human dihydrodiol dehydrogenase." *Cancer Res* **59**(3): 607–614.

Butterfield, D. A. and I. Dalle-Donne (2014). "Redox proteomics: from protein modifications to cellular dysfunction and disease." *Mass Spectrom Rev* **33**(1): 1–6.

Cadenas, S. (2018). "Mitochondrial uncoupling, ROS generation and cardioprotection." *Biochim Biophys Acta Bioenerg* **1859**(9): 940–950.

Cadet, J., K. J. A. Davies, M. H. Medeiros, P. Di Mascio and J. R. Wagner (2017). "Formation and repair of oxidatively generated damage in cellular DNA." *Free Radic Biol Med* **107**: 13–34.

Cadet, J., T. Douki and J. L. Ravanat (1997). "Artifacts associated with the measurement of oxidized DNA bases." *Environ Health Perspect* **105**(10): 1034–1039.

Canedo, A. and T. L. Rocha (2021). "Zebrafish (Danio rerio) using as model for genotoxicity and DNA repair assessments: historical review, current status and trends." *Sci Total Environ* **762**: 144084.

Capo, C., M. E. Stroppolo, A. Galtieri, A. Lania, S. Costanzo, R. Petruzzelli, L. Calabrese, F. Polticelli and A. Desideri (1997). "Characterization of Cu,Zn superoxide dismutase from the bathophile fish, Lampanyctus crocodilus." *Comp Biochem Physiol B Biochem Mol Biol* **117**(3): 403–407.

Carvan, M. J., 3rd, D. M. Sonntag, C. B. Cmar, R. S. Cook, M. A. Curran and G. L. Miller (2001). "Oxidative stress in zebrafish cells: potential utility of transgenic zebrafish as a deployable sentinel for site hazard ranking." *Sci Total Environ* **274**(1–3): 183–196.

Chance, B., H. Sies and A. Boveris (1979). "Hydroperoxide metabolism in mammalian organs." *Physiol Rev* **59**(3): 527–605.

Chandel, N. S., D. S. McClintock, C. E. Feliciano, T. M. Wood, J. A. Melendez, A. M. Rodriguez and P. T. Schumacker (2000). "Reactive oxygen species generated at mitochondrial complex III stabilize hypoxia-inducible factor-1alpha during hypoxia: a mechanism of O2 sensing." *J Biol Chem* **275**(33): 25130–25138.

Chandra, J., A. Samali and S. Orrenius (2000). "Triggering and modulation of apoptosis by oxidative stress." *Free Radic Biol Med* **29**(3-4): 323–333.

Chang, T.-S., W. Jeong, H. A. Woo, S. M. Lee, S. Park and S. G. Rhee (2004). "Characterization of mammalian sulfiredoxin and its reactivation of hyperoxidized peroxiredoxin through reduction of cysteine sulfinic acid in the active site to cysteine." *J Biol Chem* **279**(49): 50994–51001.

Charron, R. A., J. C. Fenwick, D. R. Lean and T. W. Moon (2000). "Ultraviolet-B radiation effects on antioxidant status and survival in the zebrafish, Brachydanio rerio." *Photochem Photobiol* **72**(3): 327–333.

Chen, Q., E. J. Vazquez, S. Moghaddas, C. L. Hoppel and E. J. Lesnefsky (2003). "Production of reactive oxygen species by mitochondria: central role of complex III*." *J Biol Chem* **278**(38): 36027–36031.

Choi, J. and J. T. Oris (2000). "Evidence of oxidative stress in bluegill sunfish (Lepomis macrochirus) liver microsomes simultaneously exposed to solar ultraviolet radiation and anthracene." *Environ Toxicol Chem* **19**(7): 1795–1799.

Chouchani, E. T., V. R. Pell, A. M. James, L. M. Work, K. Saeb-Parsy, C. Frezza, T. Krieg and M. P. Murphy (2016). "A unifying mechanism for mitochondrial superoxide production during ischemia-reperfusion injury." *Cell Metab* **23**(2): 254–263.

Cioni, C., E. Angiulli and M. Toni (2019). "Nitric oxide and the neuroendocrine control of the osmotic stress response in teleosts." *Int J Mol Sci* **20**(3): 489.

Cooper, R. U., L. M. Clough, M. A. Farwell and T. L. West (2002). "Hypoxia-induced metabolic and antioxidant enzymatic activities in the estuarine fish Leiostomus xanthurus." *J Exp Marine Biol Ecol* **279**(1): 1–20.

Corcoran, A. and T. G. Cotter (2013). "Redox regulation of protein kinases." *FEBS J* **280**(9): 1944–1965.

Crawford, D. R. (2002). "Regulation of mammalian gene expression by reactive oxygen species." In *Reactive Oxygen Species in Biological Systems: An Interdisciplinary Approach*. D. L. Gilbert and C. A. Colton (Eds.). Boston, MA, Springer US: 155–171.

Dalle-Donne, I., R. Rossi, D. Giustarini, A. Milzani and R. Colombo (2003). "Protein carbonyl groups as biomarkers of oxidative stress." *Clin Chim Acta* **329**(1-2): 23–38.

Dalton, T. P., H. G. Shertzer and A. Puga (1999). "Regulation of gene expression by reactive oxygen." *Annu Rev Pharmacol Toxicol* **39**: 67–101.

Darr, D. and I. Fridovich (1986). "Irreversible inactivation of catalase by 3-amino-1,2,4-triazole." *Biochem Pharmacol* **35**(20): 3642.

Davies, K. J. (1999). "The broad spectrum of responses to oxidants in proliferating cells: a new paradigm for oxidative stress." *IUBMB Life* **48**(1): 41–47.

De Bont, R. and N. van Larebeke (2004). "Endogenous DNA damage in humans: a review of quantitative data." *Mutagenesis* **19**(3): 169–185.

de Lapuente, J., J. Lourenço, S. A. Mendo, M. Borràs, M. G. Martins, P. M. Costa and M. Pacheco (2015). "The Comet Assay and its applications in the field of ecotoxicology: a mature tool that continues to expand its perspectives." *Front Genet* **6**: 180.

Dean, R. T., S. Fu, R. Stocker and M. J. Davies (1997). "Biochemistry and pathology of radical-mediated protein oxidation." *Biochem J* **324**(Pt 1): 1–18.

Detienne, G., W. De Haes, L. Mergan, S. L. Edwards, L. Temmerman and S. Van Bael (2018). "Beyond ROS clearance: peroxiredoxins in stress signaling and aging." *Ageing Res Rev* **44**: 33–48.

Dhakshinamoorthy, S., D. J. Long, 2nd and A. K. Jaiswal (2000). "Antioxidant regulation of genes encoding enzymes that detoxify xenobiotics and carcinogens." *Curr Top Cell Regul* **36**: 201–216.

Di Giulio, R. T., C. Habig and E. P. Gallagher (1993). "Effects of black rock harbor sediments on indices of biotransformation, oxidative stress, and DNA integrity in channel catfish." *Aquat Toxicol* **26**(1): 1–22.

Di Giulio, R. T. and D. E. Hinton (2008). *The Toxicology of Fishes*. Boca Raton, FL, CRC Press.

Di Giulio, R. T., P. C. Washburn, R. J. Wenning, G. W. Winston and C. S. Jewell (1989). "Biochemical responses in aquatic animals: a review of determinants of oxidative stress." *Environ Toxicol Chem* **8**(12): 1103–1123.

Di Meo, S., T. T. Reed, P. Venditti and V. M. Victor (2016). "Role of ROS and RNS sources in physiological and pathological conditions." *Oxid Med Cell Longev* **2016**: 1245049–1245049.

Dinkova-Kostova, A. T. and P. Talalay (2000). "Persuasive evidence that quinone reductase type 1 (DT diaphorase) protects cells against the toxicity of electrophiles and reactive forms of oxygen." *Free Radic Biol Med* **29**(3-4): 231–240.

Diplock, A. T. (1985). *Fat-Soluble Vitamins*. London, Heinemann.

Donham, R. T., D. Morin, W. T. Jewell, M. W. Lamé, H. J. Segall and R. S. Tjeerdema (2005). "Characterization of cytosolic glutathione S-transferases in juvenile Chinook salmon (Oncorhynchus tshawytscha)." *Aquat Toxicol* **73**(3): 221–229.

Doroshow, J. H. and K. J. Davies (1986). "Redox cycling of anthracyclines by cardiac mitochondria. II. Formation of superoxide anion, hydrogen peroxide, and hydroxyl radical." *J Biol Chem* **261**(7): 3068–3074.

Dorval, J. and A. Hontela (2003). "Role of glutathione redox cycle and catalase in defense against oxidative stress induced by endosulfan in adrenocortical cells of rainbow trout (Oncorhynchus mykiss)." *Toxicol Appl Pharmacol* **192**(2): 191–200.

Dorval, J., V. S. Leblond and A. Hontela (2003). "Oxidative stress and loss of cortisol secretion in adrenocortical cells of rainbow trout (Oncorhynchus mykiss) exposed in vitro to endosulfan, an organochlorine pesticide." *Aquat Toxicol* **63**(3): 229–241.

Dowla, H. A., M. Panemangalore and M. E. Byers (1996). "Comparative inhibition of enzymes of human erythrocytes and plasma in vitro by agricultural chemicals." *Arch Environ Contam Toxicol* **31**(1): 107–114.

Eggler, A. L., G. Liu, J. M. Pezzuto, R. B. van Breemen and A. D. Mesecar (2005). "Modifying specific cysteines of the electrophile-sensing human Keap1 protein is insufficient to disrupt binding to the Nrf2 domain Neh2." *Proc Natl Acad Sci USA* **102**(29): 10070–10075.

El-Alawi, Y. S., X. D. Huang, D. G. Dixon and B. M. Greenberg (2002). "Quantitative structure-activity relationship for the photoinduced toxicity of polycyclic aromatic hydrocarbons to the luminescent bacteria Vibrio fischeri." *Environ Toxicol Chem* **21**(10): 2225–2232.

Elia, A. C., W. T. Waller and S. J. Norton (2002). "Biochemical responses of bluegill sunfish (Lepomis macrochirus, Rafinesque) to atrazine induced oxidative stress." *Bull Environ Contamin Toxicol* **68**(6): 809–816.

Elskus, A. A., E. Monosson, A. E. McElroy, J. J. Stegeman and D. S. Woltering (1999). "Altered CYP1A expression in Fundulus heteroclitus adults and larvae: a sign of pollutant resistance?" *Aquat Toxicol* **45**(2): 99–113.

Espinoza, H. M., L. M. Shireman, V. McClain, W. Atkins and E. P. Gallagher (2013). "Cloning, expression and analysis of the olfactory glutathione S-transferases in coho salmon." *Biochem Pharmacol* **85**(6): 839–848.

Eufemia, N. A., T. K. Collier, J. E. Stein, D. E. Watson and R. T. Di Giulio (1997). "Biochemical responses to sediment-associated contaminants in brown bullhead (Ameiurus nebulosus) from the Niagara River ecosystem." *Ecotoxicology* **6**(1): 13–34.

Fabacher, D. L., E. E. Little, S. B. Jones, E. C. DeFabo and L. J. Webber (1994). Ultraviolet-B radiation and the immune response of rainbow trout: Chapter 18. In *Modulators of Fish Immune Responses*. Joanna S. Stolen, Thema C. Fletcher (Ed.). Fair Haven, NJ, SOS Publications,.

Finkel, T. (2000). "Redox-dependent signal transduction." *FEBS Lett* **476**(1–2): 52–54.

Fisher-Wellman, K. H. and P. D. Neufer (2012). "Linking mitochondrial bioenergetics to insulin resistance via redox biology." *Trends Endocrinol Metab* **23**(3): 142–153.

Förlin, L., S. Blom, M. Celander and J. Sturve (1996). "Effects on UDP glucuronosyl transferase, glutathione transferase, DT-diaphorase and glutathione reductase activities in rainbow trout liver after long-term exposure to PCB." *Marine Environ Res* **42**(1): 213–216.

Foster, C. E., M. A. Bianchet, P. Talalay, M. Faig and L. M. Amzel (2000). "Structures of mammalian cytosolic quinone reductases." *Free Radic Biol Med* **29**(3–4): 241–245.

Fuse, Y. and M. Kobayashi (2017). "Conservation of the Keap1-Nrf2 system: an evolutionary journey through stressful space and time." *Molecules* **22**(3): 436.

Fridovich, I. (1978). "The biology of oxygen radicals." *Science* **201**(4359): 875–880.

Fridovich, I. (1995). "Superoxide radical and superoxide dismutases." *Annu Rev Biochem* **64**: 97–112.
Fridovich, I. (1998). "The trail to superoxide dismutase." *Protein Sci* **7**(12): 2688–2690.
Fridovich, I. (2004). "Mitochondria: are they the seat of senescence?" *Aging Cell* **3**(1): 13–16.
Friedberg, E. C. and E. C. Friedberg (2006). *DNA Repair and Mutagenesis*. Washington, DC, ASM Press.
Fujii, J. and N. Taniguchi (1999). "Down regulation of superoxide dismutases and glutathione peroxidase by reactive oxygen and nitrogen species." *Free Radic Res* **31**(4): 301–308.
Fukai, T., R. J. Folz, U. Landmesser and D. G. Harrison (2002). "Extracellular superoxide dismutase and cardiovascular disease." *Cardiovasc Res* **55**(2): 239–249.
Furda, A. M., A. M. Marrangoni, A. Lokshin and B. Van Houten (2012). "Oxidants and not alkylating agents induce rapid mtDNA loss and mitochondrial dysfunction." *DNA Repair (Amst)* **11**(8): 684–692.
Gadagbui, B. K.-M., M. Addy and A. Goksøyr (1996). "Species characteristics of hepatic biotransformation enzymes in two tropical freshwater teleosts, tilapia (Oreochromis niloticus) and mudfish (Clarias anguillaris)." *Comp Biochem Physiol Part C: Pharmacol, Toxicol Endocrin* **114**(3): 201–211.
Gallagher, E. P., B. M. Hasspieler and R. T. Di Giulio (1992). "Effects of buthionine sulfoximine and diethyl maleate on glutathione turnover in the channel catfish." *Biochem Pharmacol* **43**(10): 2209–2215.
Gallagher, E. P. and R. T. Di Giulio (1992). "A comparison of glutathione-dependent enzymes in liver, gills and posterior kidney of channel catfish (Ictalurus punctatus)." *Comp Biochem Physiol C Comp Pharmacol Toxicol* **102**(3): 543–547.
Gauron, C., F. Meda, E. Dupont, S. Albadri, N. Quenech'Du, E. Ipendey, M. Volovitch, F. Del Bene, A. Joliot, C. Rampon and S. Vriz (2016). "Hydrogen peroxide (H2O2) controls axon pathfinding during zebrafish development." *Dev Biol* **414**(2): 133–141.
George, S., C. Riley, J. McEvoy and J. Wright (2000). "Development of a fish in vitro cell culture model to investigate oxidative stress and its modulation by dietary vitamin E." *Mar Environ Res* **50**(1-5): 541–544.
Glisic, B., I. Mihaljevic, M. Popovic, R. Zaja, J. Loncar, K. Fent, R. Kovacevic and T. Smital (2015). "Characterization of glutathione-S-transferases in zebrafish (Danio rerio)." *Aquat Toxicol* **158**: 50–62.
Gobi, N., B. Vaseeharan, R. Rekha, S. Vijayakumar and C. Faggio (2018). "Bioaccumulation, cytotoxicity and oxidative stress of the acute exposure selenium in Oreochromis mossambicus." *Ecotoxicol Environ Saf* **162**: 147–159.
Goeptar, A. R., H. Scheerens and N. P. Vermeulen (1995). "Oxygen and xenobiotic reductase activities of cytochrome P450." *Crit Rev Toxicol* **25**(1): 25–65.
Golan, M. P., S. Piłsyk, A. Muszewska and A. Wawrzyniak (2021). "Ferritins in chordata: potential evolutionary trajectory marked by discrete selective pressures: history and reclassification of ferritins in chordates and geological events' influence on their evolution and radiation." *Bioessays* **43**(1): e2000207.
Griffith, O. W. (1999). "Biologic and pharmacologic regulation of mammalian glutathione synthesis." *Free Radic Biol Med* **27**(9–10): 922–935.
Griffith, O. W. and R. T. Mulcahy (1999). "The enzymes of glutathione synthesis: gamma-glutamylcysteine synthetase." *Adv Enzymol Relat Areas Mol Biol* **73**: 209–267, xii.
Grosell, M. H., C. Hogstrand and C. M. Wood (1998). "Renal Cu and Na excretion and hepatic Cu metabolism in both Cu acclimated and non-acclimated rainbow trout (Oncorhynchus mykiss)." *Aquat Toxicol* **40**(2): 275–291.
Haber, F. and J. Weiss (1934). "The catalytic decomposition of hydrogen peroxide by iron salts." *Proc R Soc Lond* **147**: 332–351.
Hahn, M. (1998). "Mechanisms of innate and acquired resistance to dioxin-like compounds." *Rev Toxicol* **2**(5): 395–443.
Hahn, M. E., A. R. Timme-Laragy, S. I. Karchner and J. J. Stegeman (2015). "Nrf2 and Nrf2-related proteins in development and developmental toxicity: insights from studies in zebrafish (Danio rerio)." *Free Radic. Biol. Med.* **88**: 275–289.
Halliwell, B. (1996). "Vitamin C: antioxidant or pro-oxidant in vivo?" *Free Radic Res* **25**(5): 439–454.
Halliwell, B. and J. M. C. Gutteridge (2015). *Free Radicals in Biology and Medicine*. Oxford, Oxford University Press.
Hamed, R. R., N. S. Saleh, A. Shokeer, R. A. Guneidy and S. S. Abdel-Ghany (2016). "Glutathione and its related enzymes in the gonad of Nile Tilapia (Oreochromis niloticus)." *Fish Physiol Biochem* **42**(1): 353–364.
Hampton, M. B., A. J. Kettle and C. C. Winterbourn (1998). "Inside the neutrophil phagosome: oxidants, myeloperoxidase, and bacterial killing." *Blood* **92**(9): 3007–3017.
Hanschmann, E.-M., J. R. Godoy, C. Berndt, C. Hudemann and C. H. Lillig (2013). "Thioredoxins, glutaredoxins, and peroxiredoxins--molecular mechanisms and health significance: from cofactors to antioxidants to redox signaling." *Antioxid Redox Signal* **19**(13): 1539–1605.
Harman, D. (1986). *Free Radicals, Ageing and Degenerative Disease*. New York, Alan R. Liss.
Harris, E. D. (1995). "The iron-copper connection: the link to ceruloplasmin grows stronger." *Nutr Rev* **53**(6): 170–173.
Hasspieler, B. M., J. V. Behar, D. B. Carlson, D. E. Watson and R. T. Di Giulio (1994a). "Susceptibility of channel catfish (Ictalurus punctatus) and brown bullhead (Ameiurus nebulosus) to oxidative stress: a comparative study." *Aquat Toxicol* **28**(1): 53–64.
Hasspieler, B. M., J. V. Behar and R. T. Di Giulio (1994b). "Glutathione-dependent defense in channel catfish (Ictalurus punctatus) and brown bullhead (Ameiurus nebulosus)." *Ecotoxicol Environ Saf* **28**(1): 82–90.
Hayes, J. D., J. U. Flanagan and I. R. Jowsey (2005). "Glutathione transferases." *Annu Rev Pharmacol Toxicol* **45**: 51–88.
Head, B. and M. G. Traber (2021). "Expanding role of vitamin E in protection against metabolic dysregulation: insights gained from model systems, especially the developing nervous system of zebrafish embryos." *Free Radic Biol Med* **176**: 80–91.
Heise, K., S. Puntarulo, H. O. Pörtner and D. Abele (2003). "Production of reactive oxygen species by isolated mitochondria of the Antarctic bivalve Laternula elliptica (King and Broderip) under heat stress." *Comp Biochem Physiol C Toxicol Pharmacol* **134**(1): 79–90.
Henderson, L. M. and J. B. Chappel (1996). "NADPH oxidase of neutrophils." *Biochim Biophys Acta* **1273**(2): 87–107.
Hermes-Lima, M. and T. Zenteno-Savín (2002). "Animal response to drastic changes in oxygen availability and physiological oxidative stress." *Comp Biochem Physiol C Toxicol Pharmacol* **133**(4): 537–556.
Hidalgo, M. C., A. Expósito, J. M. Palma and M. de la Higuera (2002). "Oxidative stress generated by dietary Zn-deficiency: studies in rainbow trout (Oncorhynchus mykiss)." *Int J Biochem Cell Biol* **34**(2): 183–193.

Hippeli, S. and E. F. Elstner (1996). "Mechanisms of oxygen activation during plant stress: biochemical effects of air pollutants." *J Plant Physiol* **148**(3): 249–257.

Hong, F., K. R. Sekhar, M. L. Freeman and D. C. Liebler (2005). "Specific patterns of electrophile adduction trigger Keap1 ubiquitination and Nrf2 activation." *J Biol Chem* **280**(36): 31768–31775.

Hopkins, B. L. and C. A. Neumann (2019). "Redoxins as gatekeepers of the transcriptional oxidative stress response." *Redox Biol* **21**: 101104.

Huang, H. C., T. Nguyen and C. B. Pickett (2000). "Regulation of the antioxidant response element by protein kinase C-mediated phosphorylation of NF-E2-related factor 2." *Proc Natl Acad Sci U S A* **97**(23): 12475–12480.

Huang, X.-D., B. J. McConkey, T. S. Babu and B. M. Greenberg (1997). "Mechanisms of photoinduced toxicity of photomodified anthracene to plants: inhibition of photosynthesis in the aquatic higher plant Lemna gibba (duckweed)." *Environ Toxicol Chem* **16**(8): 1707–1715.

Hughes, E. M. and E. P. Gallagher (2004). "Effects of beta-naphthoflavone on hepatic biotransformation and glutathione biosynthesis in largemouth bass (Micropterus salmoides)." *Mar Environ Res* **58**(2–5): 675–679.

Iakovou, E. and M. Kourti (2022). "A comprehensive overview of the complex role of oxidative stress in aging, the contributing environmental stressors and emerging antioxidant therapeutic interventions." *Front Aging Neurosci* **14**: 827900.

Ignarro, L. J. (2002). "Nitric oxide as a unique signaling molecule in the vascular system: a historical overview." *J Physiol Pharmacol* **53**(4 Pt 1): 503–514.

Imlay, J. and I. Fridovich (1992). "Exogenous quinones directly inhibit the respiratory NADH dehydrogenase in Escherichia coli." *Arch Biochem Biophys* **296**(1): 337–346.

Isani, G. and E. Carpenè (2014). "Metallothioneins, unconventional proteins from unconventional animals: a long journey from nematodes to mammals." *Biomolecules* **4**(2): 435–457.

Jaiswal, A. K. (1994). "Human NAD(P)H:quinone oxidoreductase2. Gene structure, activity, and tissue-specific expression." *J Biol Chem* **269**(20): 14502–14508.

Jensen, F. B. (1996). "Physiological effects of nitrite in teleosts and crustaceans." In *Toxicology of Aquatic Pollution: Physiological, Molecular and Cellular Approaches*. E. W. Taylor (Ed.). Cambridge, Cambridge University Press: 169–186.

Jewell, C. S. E. and G. W. Winston (1989). "Oxyradical production by hepatopancreas microsomes from the red swamp crayfish, Procambarus clarkii." *Aquat Toxicol* **14**(1): 27–46.

Ježek, P., B. Holendová and L. Plecitá-Hlavatá (2020). "Redox signaling from mitochondria: signal propagation and its targets." *Biomolecules* **10**(1): 93. doi: 10.3390/biom10010093. PMID: 31935965; PMCID: PMC7023504.

Kalinina, E. and M. Novichkova (2021). "Glutathione in protein redox modulation through s-glutathionylation and s-nitrosylation." *Molecules* **26**(2): 435. doi: 10.3390/molecules26020435. PMID: 33467703; PMCID: PMC7838997.

Kappus, H. (1985). "Lipid peroxidation : mechanisms, analysis, enzymology and biological relevance." In *Oxidative Stress*. H. Sies (Ed.). London, Academic Press: 273–310.

Kappus, H. (1986). "Overview of enzyme systems involved in bio-reduction of drugs and in redox cycling." *Biochem Pharmacol* **35**(1): 1–6.

Kelly, J. D., G. A. Orner, J. D. Hendricks and D. E. Williams (1992). "Dietary hydrogen peroxide enhances hepatocarcinogenesis in trout: correlation with 8-hydroxy-2′-deoxyguanosine levels in liver DNA." *Carcinogenesis* **13**(9): 1639–1642.

Kennish, J. M., M. L. Russell and S. J. Netzel (1989). "The role of iron chelates in the NAD(P)H-dependent oxidation of 2-keto-4-thiomethylbutyric acid (KMBA) by rainbow trout and pacific salmon microsomal fractions." *Marine Environ Res* **28**(1): 87–91.

Kim, K. B. and B. M. Lee (1997). "Oxidative stress to DNA, protein, and antioxidant enzymes (superoxide dismutase and catalase) in rats treated with benzo(a)pyrene." *Cancer Lett* **113**(1–2): 205–212.

Kim, Y., H. G. Nam and D. R. Valenzano (2016). "The short-lived African turquoise killifish: an emerging experimental model for ageing." *Dis Model Mech* **9**(2): 115–129.

Kirlin, W. G., J. Cai, S. A. Thompson, D. Diaz, T. J. Kavanagh and D. P. Jones (1999). "Glutathione redox potential in response to differentiation and enzyme inducers." *Free Radic Biol Med* **27**(11–12): 1208–1218.

Kirsch, M. and H. De Groot (2001). "NAD(P)H, a directly operating antioxidant?" *FASEB J* **15**(9): 1569–1574.

Kolayli, S. and E. Keha (1999). "A comparative study of antioxidant enzyme activities in freshwater and seawater-adapted rainbow trout." *J Biochem Mol Toxicol* **13**(6): 334–337.

Kong, A. N., E. Owuor, R. Yu, V. Hebbar, C. Chen, R. Hu and S. Mandlekar (2001). "Induction of xenobiotic enzymes by the MAP kinase pathway and the antioxidant or electrophile response element (ARE/EpRE)." *Drug Metab Rev* **33**(3-4): 255–271.

Koppenol, W. H., J. J. Moreno, W. A. Pryor, H. Ischiropoulos and J. S. Beckman (1992). "Peroxynitrite, a cloaked oxidant formed by nitric oxide and superoxide." *Chem Res Toxicol* **5**(6): 834–842.

Kowaltowski, A. J. and A. E. Vercesi (1999). "Mitochondrial damage induced by conditions of oxidative stress." *Free Radic Biol Med* **26**(3-4): 463–471.

Krinsky, N. I. (1993). "Actions of carotenoids in biological systems." *Annu Rev Nutr* **13**: 561–587.

Kumagai, Y., J. Taira and M. Sagai (1995). "Apparent inhibition of superoxide dismutase activity in vitro by diesel exhaust particles." *Free Radic Biol Med* **18**(2): 365–371.

Kvam, E., A. Noel, S. Basu-Modak and R. M. Tyrrell (1999). "Cyclooxygenase dependent release of heme from microsomal hemeproteins correlates with induction of heme oxygenase 1 transcription in human fibroblasts." *Free Radic Biol Med* **26**(5-6): 511–517.

Larson, R. A. and E. J. Weber (1994). *Reaction Mechanisms in Environmental Organic Chemistry*. Boca Raton, Fl, Lewis Publishers.

Laval, F., D. A. Wink and J. Laval (1997). "A discussion of mechanisms of NO genotoxicity: implication of inhibition of DNA repair proteins." *Rev Physiol Biochem Pharmacol* **131**: 175–191.

Lenzen, S., J. Drinkgern and M. Tiedge (1996). "Low antioxidant enzyme gene expression in pancreatic islets compared with various other mouse tissues." *Free Radic Biol Med* **20**(3): 463–466.

Lesser, M. P., J. H. Farrell and C. W. Walker (2001). "Oxidative stress, DNA damage and p53 expression in the larvae of Atlantic cod (Gadus morhua) exposed to ultraviolet (290-400 nm) radiation." *J Exp Biol* **204**(Pt 1): 157–164.

Li, N., K. Ragheb, G. Lawler, J. Sturgis, B. Rajwa, J. A. Melendez and J. P. Robinson (2003). "Mitochondrial complex I inhibitor rotenone induces apoptosis through enhancing mitochondrial reactive oxygen species production." *J Biol Chem* **278**(10): 8516–8525.

Liao, S. and H. G. Williams-Ashman (1961). "Enzymatic oxidation of some non-phosphorylated derivatives of dihydronicotinamide." *Biochem Biophys Res Commun* **4**: 208–213.

Liebler, D. C. and T. D. McClure (1996). "Antioxidant reactions of beta-carotene: identification of carotenoid-radical adducts." *Chem Res Toxicol* **9**(1): 8–11.

Livingstone, D. R., P. Lemaire, A. Matthews, L. Peters, D. Bucke and R. J. Law (1993). "Pro-oxidant, antioxidant and 7-ethoxyresorufin O-deethylase (EROD) activity responses in liver of Dab (Limanda limanda) exposed to sediment contaminated with hydrocarbons and other chemicals." *Marine Pollut Bull* **26**(11): 602–606.

Livingstone, D. R., P. Lemaire, A. Matthews, L. D. Peters, C. Porte, P. J. Fitzpatrick, L. Förlin, C. Nasci, V. Fossato, N. Wootton and P. Goldfarb (1995). "Assessment of the impact of organic pollutants on goby (Zosterisessor ophiocephalus) and mussel (Mytilus galloprovincialis) from the Venice Lagoon, Italy: biochemical studies." *Marine Environ Res* **39**(1): 235–240.

Lo Conte, M. and K. S. Carroll (2013). "The redox biochemistry of protein sulfenylation and sulfinylation." *J Biol Chem* **288**(37): 26480–26488.

Love, N. R., Y. Chen, S. Ishibashi, P. Kritsiligkou, R. Lea, Y. Koh, J. L. Gallop, K. Dorey and E. Amaya (2013). "Amputation-induced reactive oxygen species are required for successful Xenopus tadpole tail regeneration." *Nat Cell Biol* **15**(2): 222–228.

Lu, J. and A. Holmgren (2014). "The thioredoxin antioxidant system." *Free Radic Biol Med* **66**: 75–87.

Luo, Y., X. R. Wang, H. H. Shi, D. Q. Mao, Y. X. Sui and L. L. Ji (2005). "Electron paramagnetic resonance investigation of in vivo free radical formation and oxidative stress induced by 2,4-dichlorophenol in the freshwater fish Carassius auratus." *Environ Toxicol Chem* **24**(9): 2145–2153.

Lushchak, V. I. (2016). "Contaminant-induced oxidative stress in fish: a mechanistic approach." *Fish Physiol Biochem* **42**(2): 711–747.

Lushchak, V. I., L. P. Lushchak, A. A. Mota and M. Hermes-Lima (2001). "Oxidative stress and antioxidant defenses in goldfish Carassius auratus during anoxia and reoxygenation." *Am J Physiol Regul Integr Comp Physiol* **280**(1): R100–R107.

Ma, Q. (2010). "Transcriptional responses to oxidative stress: pathological and toxicological implications." *Pharmacol Ther* **125**(3): 376–393.

Machella, N., F. Regoli and R. M. Santella (2005). "Immunofluorescent detection of 8-oxo-dG and PAH bulky adducts in fish liver and mussel digestive gland." *Aquat Toxicol* **71**(4): 335–343.

Malins, D. C. and R. Haimanot (1990). "4,6-Diamino-5-formamidopyrimidine, 8-hydroxyguanine and 8-hydroxyadenine in DNA from neoplastic liver of English sole exposed to carcinogens." *Biochem Biophys Res Commun* **173**(2): 614–619.

Malins, D. C., B. B. McCain, D. W. Brown, S. L. Chan, M. S. Myers, J. T. Landahl, P. G. Prohaska, A. J. Friedman, L. D. Rhodes and et al. (1984). "Chemical pollutants in sediments and diseases of bottom-dwelling fish in Puget Sound, Washington." *Environ Sci Technol* **18**(9): 705–713.

Malins, D. C. and S. J. Gunselman (1994). "Fourier-transform infrared spectroscopy and gas chromatography-mass spectrometry reveal a remarkable degree of structural damage in the DNA of wild fish exposed to toxic chemicals." *Proc Natl Acad Sci* **91**(26): 13038–13041.

Manohar, S., S. Jacob, J. Wang, K. A. Wiechecki, H. W. L. Koh, V. Simões, H. Choi, C. Vogel and G. M. Silva (2019). "Polyubiquitin chains linked by lysine residue 48 (K48) selectively target oxidized proteins in vivo." *Antioxid Redox Signal* **31**(15): 1133–1149.

Marnett, L. J. (2000). "Oxyradicals and DNA damage." *Carcinogenesis* **21**(3): 361–370.

Martindale, J. L. and N. J. Holbrook (2002). "Cellular response to oxidative stress: signaling for suicide and survival." *J Cell Physiol* **192**(1): 1–15.

Martínez-Alvarez, R. M., M. C. Hidalgo, A. Domezain, A. E. Morales, M. García-Gallego and A. Sanz (2002). "Physiological changes of sturgeon Acipenser naccarii caused by increasing environmental salinity." *J Exp Biol* **205**(Pt 23): 3699–3706.

Massarsky, A., J. S. Kozal and R. T. Di Giulio (2017). "Glutathione and zebrafish: old assays to address a current issue." *Chemosphere* **168**: 707–715.

McConkey, B. J., C. L. Duxbury, D. G. Dixon and B. M. Greenberg (1997). "Toxicity of a pah photooxidation product to the bacteria photobacterium phosphoreum and the duckweed Lemna gibba: effects of phenanthrene and its primary photoproduct, phenanthrenequinone." *Environ Toxicol Chem* **16**(5): 892–899.

McCord, J. M. and I. Fridovich (1969). "Superoxide dismutase. An enzymic function for erythrocuprein (hemocuprein)." *J Biol Chem* **244**(22): 6049–6055.

McDonald, B. G. and P. M. Chapman (2002). "PAH phototoxicity--an ecologically irrelevant phenomenon?" *Mar Pollut Bull* **44**(12): 1321–1326.

McFarland, V. A., L. S. Inouye, C. H. Lutz, A. S. Jarvis, J. U. Clarke and D. D. McCant (1999). "Biomarkers of oxidative stress and genotoxicity in livers of field-collected brown bullhead, Ameiurus nebulosus." *Arch Environ Contam Toxicol* **37**(2): 236–241.

McIntosh, D. J., T. S. Walters, I. J. Arinze and J. Davis (2018). "Arkadia (RING finger protein 111) mediates sumoylation-dependent stabilization of Nrf2 through K48-linked ubiquitination." *Cell Physiol Biochem* **46**(1): 418–430.

Meister, A. (1995). "Mitochondrial changes associated with glutathione deficiency." *Biochim Biophys Acta* **1271**(1): 35–42.

Meyer, J. N. (2010). "QPCR: a tool for analysis of mitochondrial and nuclear DNA damage in ecotoxicology." *Ecotoxicology* **19**(4): 804–811.

Meyer, J. N., D. E. Nacci and R. T. Di Giulio (2002). "Cytochrome P4501A (CYP1A) in killifish (Fundulus heteroclitus): heritability of altered expression and relationship to survival in contaminated sediments." *Toxicol Sci* **68**(1): 69–81.

Meyer, J. N., J. D. Smith, G. W. Winston and R. T. Di Giulio (2003). "Antioxidant defenses in killifish (Fundulus heteroclitus) exposed to contaminated sediments and model prooxidants: short-term and heritable responses." *Aquat Toxicol* **65**(4): 377–395.

Meyer, J. N., T. C. Leuthner and A. L. Luz (2017). "Mitochondrial fusion, fission, and mitochondrial toxicity." *Toxicology* **391**: 42–53.

Michiels, C., E. Minet, D. Mottet and M. Raes (2002). "Regulation of gene expression by oxygen: NF-kappaB and HIF-1, two extremes." *Free Radic Biol Med* **33**(9): 1231–1242.

Miki, A., C. Ricordi, Y. Sakuma, T. Yamamoto, R. Misawa, A. Mita, R. D. Molano, N. D. Vaziri, A. Pileggi and H. Ichii (2018). "Divergent antioxidant capacity of human islet cell subsets: a potential cause of beta-cell vulnerability in diabetes and islet transplantation." *PLoS One* **13**(5): e0196570.

Mills, M. G., R. Ramsden, E. Y. Ma, J. Corrales, L. A. Kristofco, W. B. Steele, G. N. Saari, F. Melnikov, J. Kostal, T. J. Kavanagh, J. B. Zimmerman, A. M. Voutchkova-Kostal, B. W. Brooks, P. Coish, P. T. Anastas and E. Gallagher (2020). "CRISPR-generated Nrf2a loss- and gain-of-function mutants facilitate mechanistic analysis of chemical oxidative stress-mediated toxicity in zebrafish." *Chem Res Toxicol* **33**(2): 426–435.

Mishra, M., H. Jiang, L. Wu, H. A. Chawsheen and Q. Wei (2015). "The sulfiredoxin-peroxiredoxin (Srx-Prx) axis in cell signal transduction and cancer development." *Cancer Lett* **366**(2): 150–159.

Mix, M. C. (1986). "Cancerous diseases in aquatic animals and their association with environmental pollutants: a critical literature review." *Marine Environ Res* **20**(1): 1–141.

Moldogazieva, N. T., I. M. Mokhosoev, N. B. Feldman and S. V. Lutsenko (2018). "ROS and RNS signalling: adaptive redox switches through oxidative/nitrosative protein modifications." *Free Radic Res* **52**(5): 507–543.

Morbey, Y. E., C. E. Brassil and A. P. Hendry (2005). "Rapid senescence in pacific salmon." *Am Nat* **166**(5): 556–568.

Moreau, R. and K. Dabrowski (1998). "Fish acquired ascorbic acid synthesis prior to terrestrial vertebrate emergence." *Free Radic Biol Med* **25**(8): 989–990.

Moreau, R. and K. Dabrowski (2003). "Alpha-tocopherol downregulates gulonolactone oxidase activity in sturgeon." *Free Radic Biol Med* **34**(10): 1326–1332.

Morel, Y. and R. Barouki (1999). "Repression of gene expression by oxidative stress." *Biochem J* **342 Pt 3**(Pt 3): 481–496.

Mourente, G., E. Díaz-Salvago, D. R. Tocher and J. G. Bell (2000). "Effects of dietary polyunsaturated fatty acid/vitamin E (PUFA/tocopherol ratio on antioxidant defence mechanisms of juvenile gilthead sea bream (Sparus aurata L., Osteichthyes, Sparidae)." *Fish Physiol Biochem* **23**(4): 337–351.

Mourente, G., E. Díaz-Salvago, J. G. Bell and D. R. Tocher (2002). "Increased activities of hepatic antioxidant defence enzymes in juvenile gilthead sea bream (Sparus aurata L.) fed dietary oxidised oil: attenuation by dietary vitamin E." *Aquaculture* **214**(1): 343–361.

Mukaigasa, K., L. T. Nguyen, L. Li, H. Nakajima, M. Yamamoto and M. Kobayashi (2012). "Genetic evidence of an evolutionarily conserved role for Nrf2 in the protection against oxidative stress." *Mol Cell Biol* **32**(21): 4455–4461.

Muriel, P. (2000). "Regulation of nitric oxide synthesis in the liver." *J Appl Toxicol* **20**(3): 189–195.

Murphy, M. P., H. Bayir, V. Belousov, C. J. Chang, K. J. A. Davies, M. J. Davies, T. P. Dick, T. Finkel, H. J. Forman, Y. Janssen-Heininger, D. Gems, V. E. Kagan, B. Kalyanaraman, N. G. Larsson, G. L. Milne, T. Nyström, H. E. Poulsen, R. Radi, H. Van Remmen, P. T. Schumacker, P. J. Thornalley, S. Toyokuni, C. C. Winterbourn, H. Yin and B. Halliwell (2022). "Guidelines for measuring reactive oxygen species and oxidative damage in cells and in vivo." *Nat Metab* **4**(6): 651–662.

Nakano, T., M. Sato and M. Takeuchi (1995). "Unique molecular properties of superoxide dismutase from teleost fish skin." *FEBS Lett* **360**(2): 197–201.

Nakano, T., T. Kanmuri, M. Sato and M. Takeuchi (1999). "Effect of astaxanthin rich red yeast (Phaffia rhodozyma) on oxidative stress in rainbow trout." *Biochim Biophys Acta* **1426**(1): 119–125.

Nebert, D. W., A. L. Roe, M. Z. Dieter, W. A. Solis, Y. Yang and T. P. Dalton (2000). "Role of the aromatic hydrocarbon receptor and [Ah] gene battery in the oxidative stress response, cell cycle control, and apoptosis." *Biochem Pharmacol* **59**(1): 65–85.

Nemmiche, S. (2017). "Oxidative signaling response to cadmium exposure." *Toxicol Sci* **156**(1): 4–10.

Neumann, N. F., J. L. Stafford, D. Barreda, A. J. Ainsworth and M. Belosevic (2001). "Antimicrobial mechanisms of fish phagocytes and their role in host defense." *Dev Comp Immunol* **25**(8–9): 807–825.

Nguyen, V. T., L. Bian, J. Tamaoki, S. Otsubo, M. Muratani, A. Kawahara and M. Kobayashi (2020). "Generation and characterization of keap1a- and keap1b-knockout zebrafish." *Redox Biol* **36**: 101667.

Nguyen, T., P. J. Sherratt, H. C. Huang, C. S. Yang and C. B. Pickett (2003). "Increased protein stability as a mechanism that enhances Nrf2-mediated transcriptional activation of the antioxidant response element. Degradation of Nrf2 by the 26 S proteasome." *J Biol Chem* **278**(7): 4536–4541.

Niedzwiecki, M. M., D. I. Walker, R. Vermeulen, M. Chadeau-Hyam, D. P. Jones and G. W. Miller (2019). The exposome: molecules to populations." *Annu Rev Pharmacol Toxicol* **59**: 107–127.

Niethammer, P., C. Grabher, A. T. Look and T. J. Mitchison (2009). "A tissue-scale gradient of hydrogen peroxide mediates rapid wound detection in zebrafish." *Nature* **459**(7249): 996–999.

Niki, E. and M. Matsuo (1993). Rates and products of reactions of vitamin E with oxygen radicals In *Vitamin E in Health and Disease*. L. Packer and J. Fuchs (Eds.). New York, Dekker: 121–130.

Nishikimi, M. and K. Yagi (1996). "Biochemistry and molecular biology of ascorbic acid biosynthesis." *Subcell Biochem* **25**: 17–39.

Nishimoto, M., B.-T. Le Eberhart, H. R. Sanborn, C. Krone, U. Varanasi and J. E. Stein (1995). "Effects of a complex mixture of chemical contaminants on hepatic glutathione, L-cysteine and γ-glutamylcysteine synthetase in English sole (Pleuronectes vetulus)." *Environ Toxicol Chem* **14**(3): 461–469.

Olsen, R. E., E. Løvaas and Ø. Lie (1999). "The influence of temperature, dietary polyunsaturated fatty acids, α-tocopherol and spermine on fatty acid composition and indices of oxidative stress in juvenile Arctic char, Salvelinus alpinus (L.)." *Fish Physiol Biochem* **20**(1): 13–29.

Onukwufor, J. O., D. Stevens and C. Kamunde (2016). "Bioenergetic and volume regulatory effects of mitoKATP channel modulators protect against hypoxia-reoxygenation-induced mitochondrial dysfunction." *J Exp Biol* **219**(Pt 17): 2743–2751.

Ouyang, Y., Y. Peng, J. Li, A. Holmgren and J. Lu (2018). "Modulation of thiol-dependent redox system by metal ions via thioredoxin and glutaredoxin systems." *Metallomics* **10**(2): 218–228.

Pajares, M., N. Jiménez-Moreno, I. H. K. Dias, B. Debelec, M. Vucetic, K. E. Fladmark, H. Basaga, S. Ribaric, I. Milisav and A. Cuadrado (2015). "Redox control of protein degradation." *Redox Biol* **6**: 409–420.

Pak, V. V., D. Ezerina, O. G. Lyublinskaya, B. Pedre, P. A. Tyurin-Kuzmin, N. M. Mishina, M. Thauvin, D. Young, K. Wahni, S. A. Martinez Gache, A. D. Demidovich, Y. G. Ermakova, Y. D. Maslova, A. G. Shokhina, E. Eroglu, D. S. Bilan, I. Bogeski, T. Michel, S. Vriz, J. Messens and V. V. Belousov (2020). "Ultrasensitive genetically encoded indicator for hydrogen peroxide identifies roles for the oxidant in cell migration and mitochondrial function." *Cell Metab* **31**(3): 642–653, e646.

Palmer, H. J. and K. E. Paulson (1997). "Reactive oxygen species and antioxidants in signal transduction and gene expression." *Nutr Rev* **55**(10): 353–361.

Pandey, S., I. Ahmad, S. Parvez, B. Bin-Hafeez, R. Haque and S. Raisuddin (2001). "Effect of endosulfan on antioxidants of freshwater fish Channa punctatus Bloch: 1. Protection against lipid peroxidation in liver by copper preexposure." *Arch Environ Contam Toxicol* **41**(3): 345–352.

Panieri, E., C. Millia and M. M. Santoro (2017). "Real-time quantification of subcellular H(2)O(2) and glutathione redox potential in living cardiovascular tissues." *Free Radic Biol Med* **109**: 189–200.

Parente, V., S. Balasso, G. Pompilio, L. Verduci, G. I. Colombo, G. Milano, U. Guerrini, L. Squadroni, F. Cotelli, O. Pozzoli and M. C. Capogrossi (2013). "Hypoxia/reoxygenation cardiac injury and regeneration in zebrafish adult heart." *PLoS One* **8**(1): e53748.

Parihar, M. S. and A. K. Dubey (1995). "Lipid peroxidation and ascorbic acid status in respiratory organs of male and female freshwater catfish Heteropneustes fossilis exposed to temperature increase." *Comp Biochem Physiol C Pharmacol Toxicol Endocrinol* **112**(3): 309–313.

Parihar, M. S., A. K. Dubey, T. Javeri and P. Prakash (1996). "Changes in lipid peroxidation, superoxide dismutase activity, ascorbic acid and phospholipid content in liver of freshwater catfish Heteropneustes fossilis exposed to elevated temperature." *J Ther Biol* **21**(5): 323–330.

Park, J. Y., M. K. Shigenaga and B. N. Ames (1996). "Induction of cytochrome P4501A1 by 2,3,7,8-tetrachlorodibenzo-p-dioxin or indolo(3,2-b)carbazole is associated with oxidative DNA damage." *Proc Natl Acad Sci U S A* **93**(6): 2322–2327.

Parvez, S. and S. Raisuddin (2005). "Protein carbonyls: novel biomarkers of exposure to oxidative stress-inducing pesticides in freshwater fish Channa punctata (Bloch)." *Environ Toxicol Pharmacol* **20**(1): 112–117.

Pascual, P., J. R. Pedrajas, F. Toribio, J. López-Barea and J. Peinado (2003). "Effect of food deprivation on oxidative stress biomarkers in fish (Sparus aurata)." *Chem Biol Interact* **145**(2): 191–199.

Payne, J. F., D. C. Malins, S. Gunselman, A. Rahimtula and P. A. Yeats (1998). "DNA oxidative damage and vitamin A reduction in fish from a large lake system in Labrador, Newfoundland, contaminated with iron-ore mine tailings." *Marine Environ Res* **46**(1): 289–294.

Pedrajas, J. R., J. Peinado and J. López-Barea (1995). "Oxidative stress in fish exposed to model xenobiotics. Oxidatively modified forms of Cu,Zn-superoxide dismutase as potential biomarkers." *Chem Biol Interact* **98**(3): 267–282.

Pei, D. S. and P. R. Strauss (2013). "Zebrafish as a model system to study DNA damage and repair." *Mutat Res* **743–744**: 151–159.

Petřivalský, M., M. Machala, K. Nezveda, V. Piačka, Z. Svobodová and P. Drábek (1997). "Glutathione-dependent detoxifying enzymes in rainbow trout liver: search for specific biochemical markers of chemical stress." *Environ Toxicol Chem* **16**(7): 1417–1421.

Pi, J., Y. Bai, Q. Zhang, V. Wong, L. M. Floering, K. Daniel, J. M. Reece, J. T. Deeney, M. E. Andersen, B. E. Corkey and S. Collins (2007). "Reactive oxygen species as a signal in glucose-stimulated insulin secretion." *Diabetes* **56**(7): 1783–1791.

Pinegin, B., N. Vorobjeva, M. Pashenkov and B. Chernyak (2018). "The role of mitochondrial ROS in antibacterial immunity." *J Cell Physiol* **233**(5): 3745–3754.

Ploch, S. A., Y.-P. Lee, E. MacLean and R. T. Di Giulio (1999). "Oxidative stress in liver of brown bullhead and channel catfish following exposure to tert-butyl hydroperoxide." *Aquat Toxicol* **46**(3): 231–240.

Podrabsky, J. E., M. A. Menze and S. C. Hand (2012). "Long-term survival of anoxia despite rapid ATP decline in embryos of the annual killifish Austrofundulus limnaeus." *J Exp Zool A Ecol Genet Physiol* **317**(8): 524–532.

Postlethwait, J., A. Amores, W. Cresko, A. Singer and Y. L. Yan (2004). "Subfunction partitioning, the teleost radiation and the annotation of the human genome." *Trends Genet* **20**(10): 481–490.

Prestera, T., W. D. Holtzclaw, Y. Zhang and P. Talalay (1993). "Chemical and molecular regulation of enzymes that detoxify carcinogens." *Proc Natl Acad Sci U S A* **90**(7): 2965–2969.

Primiano, T., T. R. Sutter and T. W. Kensler (1997). "Redox regulation of genes that protect against carcinogens." *Comp Biochem Physiol B Biochem Mol Biol* **118**(3): 487–497.

Pryor, W. A. (1986). "Oxy-radicals and related species: their formation, lifetimes, and reactions." *Annu Rev Physiol* **48**: 657–667.

Radi, A. A. and B. Matkovics (1988). "Effects of metal ions on the antioxidant enzyme activities, protein contents and lipid peroxidation of carp tissues." *Comp Biochem Physiol C Comp Pharmacol Toxicol* **90**(1): 69–72.

Rana, S. V. S. and R. Singh (1996). "Specie's differences in glutathione-dependent enzymes in the liver and kidney of two fresh water fishes and their implications for cadmium toxicity." *Ichthyol Res* **43**(3): 223–229.

Rastogi, A., C. W. Clark, S. M. Conlin, S. E. Brown and A. R. Timme-Laragy (2019). "Mapping glutathione utilization in the developing zebrafish (Danio rerio) embryo." *Redox Biol* **26**: 101235.

Rastogi, A., E. G. Severance, H. M. Jacobs, S. M. Conlin, S. T. Islam and A. R. Timme-Laragy (2021). "Modulating glutathione thiol status alters pancreatic β-cell morphogenesis in the developing zebrafish (Danio rerio) embryo." *Redox Biol* **38**: 101788.

Rastogi, A., H. M. Jacobs, S. T. Islam, E. G. Severance, S. M. Conlin, M. E. Hahn and A. R. Timme-Laragy (2020). "Pancreatic Nrf2 expression and organ morphogenesis is altered by modulating glutathione in the developing zebrafish embryo." *The Toxicologist*, Supplement to Toxicological Sciences **174**(1): Abstract #1630.

Rau, M. A., J. Whitaker, J. H. Freedman and R. T. Di Giulio (2004). "Differential susceptibility of fish and rat liver cells to oxidative stress and cytotoxicity upon exposure to prooxidants." *Comp Biochem Physiol C Toxicol Pharmacol* **137**(4): 335–342.

Recknagel, R. O., E. A. Glende, J. A. Dolak and R. L. Waller (1989). "Mechanisms of carbon tetrachloride toxicity." *Pharmacol Ther* **43**(1): 139–154.

Reczek, C. R. and N. S. Chandel (2015). "ROS-dependent signal transduction." *Curr Opin Cell Biol* **33**: 8–13.

Regoli, F. and G. W. Winston (1999). "Quantification of total oxidant scavenging capacity of antioxidants for peroxynitrite, peroxyl radicals, and hydroxyl radicals." *Toxicol Appl Pharmacol* **156**(2): 96–105.

Reid, T. J., 3rd, M. R. Murthy, A. Sicignano, N. Tanaka, W. D. Musick and M. G. Rossmann (1981). "Structure and heme environment of beef liver catalase at 2.5 A resolution." *Proc Natl Acad Sci U S A* **78**(8): 4767–4771.

Richter, C. (1987). "Biophysical consequences of lipid peroxidation in membranes." *Chem Phys Lipid* **44**(2): 175–189.

Ritola, O., L. D. Peters, D. R. Livingstone and P. Lindström-Seppä (2002). "Effects of in vitro exposure to ozone and/or hyperoxia on superoxide dismutase, catalase, glutathione and lipid peroxidation in red blood cells and plasma of rainbow trout, Oncorhynchus mykiss (Walbaum)." *Aquacul Res* **33**(3): 165–175.

Rodríguez-Ariza, A., J. Alhama, F. M. Díaz-Méndez and J. López-Barea (1999). "Content of 8-oxodG in chromosomal DNA of Sparus aurata fish as biomarker of oxidative stress and environmental pollution." *Mutat Res* **438**(2): 97–107.

Ronisz, D., D. G. Larsson and L. Förlin (1999). "Seasonal variations in the activities of selected hepatic biotransformation and antioxidant enzymes in eelpout (Zoarces viviparus)." *Comp Biochem Physiol C Pharmacol Toxicol Endocrinol* **124**(3): 271–279.

Ross, S. W., D. A. Dalton, S. Kramer and B. L. Christensen (2001). "Physiological (antioxidant) responses of estuarine fishes to variability in dissolved oxygen." *Comp Biochem Physiol C Toxicol Pharmacol* **130**(3): 289–303.

Roy, N. K., S. Courtenay, Z. Yuan, M. Ikonomou and I. Wirgin (2001). "An evaluation of the etiology of reduced CYP1A1 messenger RNA expression in the Atlantic tomcod from the Hudson River, New York, USA, using reverse transcriptase polymerase chain reaction analysis." *Environ Toxicol Chem* **20**(5): 1022–1030.

Sakai, T., H. Murata, M. Endo, T. Shimomura, K. Yamauchi, T. Ito, T. Yamaguchi, H. Nakajima and M. Fukudome (1998). "Severe oxidative stress is thought to be a principal cause of jaundice of yellowtail Seriola quinqueradiata." *Aquaculture* **160**(3): 205–214.

Sant, K. E., J. M. Hansen, L. M. Williams, N. L. Tran, J. V. Goldstone, J. J. Stegeman, M. E. Hahn and A. Timme-Laragy (2017). "The role of Nrf1 and Nrf2 in the regulation of glutathione and redox dynamics in the developing zebrafish embryo." *Redox Biol* **13**: 207–218.

Sauer, H., M. Wartenberg and J. Hescheler (2001). "Reactive oxygen species as intracellular messengers during cell growth and differentiation." *Cell Physiol Biochem* **11**(4): 173–186.

Schaeffer, M., D. J. Hodson, C. Lafont and P. Mollard (2011). "Endocrine cells and blood vessels work in tandem to generate hormone pulses." *J Mol Endocrinol* **47**(2): R59. https://jme.bioscientifica.com/view/journals/jme/47/2/R59.xml

Schafer, F. Q. and G. R. Buettner (2001). "Redox environment of the cell as viewed through the redox state of the glutathione disulfide/glutathione couple." *Free Radic Biol Med* **30**(11): 1191–1212.

Schlezinger, J. J. and J. J. Stegeman (2001). "Induction and suppression of cytochrome P450 1A by 3,3',4,4',5-pentachlorobiphenyl and its relationship to oxidative stress in the marine fish scup (Stenotomus chrysops)." *Aquat Toxicol* **52**(2): 101–115.

Schlezinger, J. J., R. D. White and J. J. Stegeman (1999). "Oxidative inactivation of cytochrome P-450 1A (CYP1A) stimulated by 3,3',4,4'-tetrachlorobiphenyl: production of reactive oxygen by vertebrate CYP1As." *Mol Pharmacol* **56**(3): 588–597.

Schmieder, P. K., M. A. Tapper, R. C. Kolanczyk, D. E. Hammermeister, B. R. Sheedy and J. S. Denny (2003). "Discriminating redox cycling and arylation pathways of reactive chemical toxicity in trout hepatocytes." *Toxicol Sci* **72**(1): 66–76.

Sen, C. K. (2000). "Cellular thiols and redox-regulated signal transduction." *Curr Top Cell Regul* **36**: 1–30.

Senft, A. P., T. P. Dalton, D. W. Nebert, M. B. Genter, A. Puga, R. J. Hutchinson, J. K. Kerzee, S. Uno and H. G. Shertzer (2002). "Mitochondrial reactive oxygen production is dependent on the aromatic hydrocarbon receptor." *Free Radic Biol Med* **33**(9): 1268–1278.

Sevanian, A. and F. Ursini (2000). "Lipid peroxidation in membranes and low-density lipoproteins: similarities and differences." *Free Radic Biol Med* **29**(3–4): 306–311.

Shadel, G. S. and T. L. Horvath (2015). "Mitochondrial ROS signaling in organismal homeostasis." *Cell* **163**(3): 560–569.

Shull, S., N. H. Heintz, M. Periasamy, M. Manohar, Y. M. Janssen, J. P. Marsh and B. T. Mossman (1991). "Differential regulation of antioxidant enzymes in response to oxidants." *J Biol Chem* **266**(36): 24398–24403.

Sies, H. (2021). "Oxidative eustress: On constant alert for redox homeostasis." *Redox Biol* **41**: 101867.

Sies, H., C. Berndt and D. P. Jones (2017). "Oxidative stress." *Annu Rev Biochem* **86**: 715–748.

Sies, H. and E. Cadenas (1985). "Oxidative stress: damage to intact cells and organs." *Philos Trans R Soc Lond B Biol Sci* **311**(1152): 617–631.

Sies, H., V. V. Belousov, N. S. Chandel, M. J. Davies, D. P. Jones, G. E. Mann, M. P. Murphy, M. Yamamoto and C. Winterbourn (2022). "Defining roles of specific reactive oxygen species (ROS) in cell biology and physiology." *Nat Rev Mol Cell Biol.*

Sies, H. and D. P. Jones (2020). "Reactive oxygen species (ROS) as pleiotropic physiological signalling agents." *Nat Rev Mol Cell Biol* **21**(7): 363–383.

Silva, G. M., D. Finley and C. Vogel (2015). "K63 polyubiquitination is a new modulator of the oxidative stress response." *Nat Struct Mol Biol* **22**(2): 116–123.

Simões, V., L. Harley, B. K. Cizubu, Y. Zhou, J. Pajak, N. A. Snyder, J. Bouvette, M. J. Borgnia, G. Arya, A. Bartesaghi and G. M. Silva (2021). "Redox sensitive E2 Rad6 controls cellular response to oxidative stress via K63 ubiquitination of ribosomes." bioRxiv: 2021.2004.2022.439970.

Singh, Y., G. L. Hall and M. G. Miller (1992). "Species differences in membrane susceptibility to lipid peroxidation." *J Biochem Toxicol* **7**(2): 97–105.

Smith, M. T. and C. G. Evans (1984). "Inhibitory effect of superoxide-generating quinones on superoxide dismutase." *Biochem Pharmacol* **33**(19): 3109–3110.

Smith, T. S. and J. P. Bennett, Jr. (1997). "Mitochondrial toxins in models of neurodegenerative diseases. I: in vivo brain hydroxyl radical production during systemic MPTP treatment or following microdialysis infusion of methylpyridinium or azide ions." *Brain Res* **765**(2): 183–188.

Sokolova, I. (2018). "Mitochondrial adaptations to variable environments and their role in animals' stress tolerance." *Integr Comp Biol* **58**(3): 519–531.

Stadtman, E. R. and R. L. Levine (2003). "Free radical-mediated oxidation of free amino acids and amino acid residues in proteins." *Amino Acids* **25**(3–4): 207–218.

Stein, J. E., T. K. Collier, W. L. Reichert, E. Casillas, T. Hom and U. Varanasi (1993). "Bioindicators of contaminant exposure and sublethal effects in benthic fish from Puget Sound, WA, USA." *Marine Environ Res* **35**(1): 95–100.

Stephensen, E., J. Sturve and L. Förlin (2002). "Effects of redox cycling compounds on glutathione content and activity of glutathione-related enzymes in rainbow trout liver." *Comp Biochem Physiol C Toxicol Pharmacol* **133**(3): 435–442.

Stephensen, E., J. Svavarsson, J. Sturve, G. Ericson, M. Adolfsson-Erici and L. Förlin (2000). "Biochemical indicators of pollution exposure in shorthorn sculpin (Myoxocephalus scorpius), caught in four harbours on the southwest coast of Iceland." *Aquat Toxicol* **48**(4): 431–442.

Sugg, D. W., J. A. Brooks, C. H. Jagoe, M. H. Smith, R. K. Chesser, J. W. Bickham, M. D. Lomakin, C. E. Dallas and R. J. Baker (1996). "DNA damage and radiocesium in channel catfish from chernobyl." *Environ Toxicol Chem* **15**(7): 1057–1063.

Tanaka, K., H. Adachi, H. Akasaka, J. Tamaoki, Y. Fuse, M. Kobayashi, T. Kitazawa and H. Teraoka (2021). "Oxidative stress inducers potentiate 2,3,7,8-tetrachlorodibenzo-p-dioxin-mediated pre-cardiac edema in larval zebrafish." *J Vet Med Sci* **83**(7): 1050–1058.

Tebay, L. E., H. Robertson, S. T. Durant, S. R. Vitale, T. M. Penning, A. T. Dinkova-Kostova and J. D. Hayes (2015). "Mechanisms of activation of the transcription factor Nrf2 by redox stressors, nutrient cues, and energy status and the pathways through which it attenuates degenerative disease." *Free Radic Biol Med* **88**(Pt B): 108–146.

Telfer, A., S. Dhami, S. M. Bishop, D. Phillips and J. Barber (1994). "beta-Carotene quenches singlet oxygen formed by isolated photosystem II reaction centers." *Biochemistry* **33**(48): 14469–14474.

Theodorakis, C. W., B. G. Blaylock and L. R. Shugart (1997). "Genetic ecotoxicology I: DNA integrity and reproduction in mosquitofish exposed in situ to radionuclides." *Ecotoxicology* **6**(4): 205–218.

Theodorakis, C. W., T. Elbl and L. R. Shugart (1999). "Genetic ecotoxicology IV: survival and DNA strand breakage is dependent on genotype in radionuclide-exposed mosquitofish." *Aquat Toxicol* **45**(4): 279–291.

Thieme, R., E. F. Pai, R. H. Schirmer and G. E. Schulz (1981). "Three-dimensional structure of glutathione reductase at 2 A resolution." *J Mol Biol* **152**(4): 763–782.

Thomas, P. and H. W. Wofford (1984). "Effects of metals and organic compounds on hepatic glutathione, cysteine, and acid-soluble thiol levels in mullet (Mugil cephalus L.)." *Toxicol Appl Pharmacol* **76**(1): 172–182.

Tierbach, A., K. J. Groh, R. Schönenberger, K. Schirmer and M. J. Suter (2018). "Glutathione s-transferase protein expression in different life stages of zebrafish (Danio rerio)." *Toxicol Sci* **162**(2): 702–712.

Timme-Laragy, A. R., J. V. Goldstone, B. R. Imhoff, J. J. Stegeman, M. E. Hahn and J. M. Hansen (2013). "Glutathione redox dynamics and expression of glutathione-related genes in the developing embryo." *Free Radic Biol Med* **65**: 89–101.

Timme-Laragy, A. R., M. E. Hahn, J. M. Hansen, A. Rastogi and M. A. Roy (2018). "Redox stress and signaling during vertebrate embryonic development: regulation and responses." *Semin Cell Dev Biol* **80**: 17–28.

Timme-Laragy, A. R., S. I. Karchner, D. G. Franks, M. J. Jenny, R. C. Harbeitner, J. V. Goldstone, A. G. McArthur and M. E. Hahn (2012). "Nrf2b, novel zebrafish paralog of oxidant-responsive transcription factor NF-E2-related factor 2 (NRF2)." *J Biol Chem* **287**(7): 4609–4627.

Tocher, D. R., G. Mourente, A. Van der Eecken, J. O. Evjemo, E. Diaz, J. G. Bell, I. Geurden, P. Lavens and Y. Olsen (2002). "Effects of dietary vitamin E on antioxidant defence mechanisms of juvenile turbot (Scophthalmus maximus L.), halibut (Hippoglossus hippoglossus L.) and sea bream (Sparus aurata L.)." *Aquac Nutr* **8**(3): 195–207.

Toda, N. and T. Okamura (2003). "The pharmacology of nitric oxide in the peripheral nervous system of blood vessels." *Pharmacol Rev* **55**(2): 271–324.

Traber, M. G. (1994). "Determinants of plasma vitamin E concentrations." *Free Radic Biol Med* **16**(2): 229–239.

Traber, M. G. and J. F. Stevens (2011). "Vitamins C and E: beneficial effects from a mechanistic perspective." *Free Radic Biol Med* **51**(5): 1000–1013.

Tripuranthakam, S., C. L. Duxbury, T. S. Babu and B. M. Greenberg (1999). "Development of a mitochondrial respiratory electron transport bioindicator for assessment of aromatic hydrocarbon toxicity." (1364): 350–361. astm.org.

Turrens, J. F. (1997). "Superoxide production by the mitochondrial respiratory chain." *Biosci Rep* **17**(1): 3–8.

Valerio-García, R. C., A. L. Carbajal-Hernández, E. B. Martínez-Ruíz, V. H. Jarquín-Díaz, C. Haro-Pérez and F. Martínez-Jerónimo (2017). "Exposure to silver nanoparticles produces oxidative stress and affects macromolecular and metabolic biomarkers in the goodeid fish Chapalichthys pardalis." *Sci Total Environ* **583**: 308–318.

van der Oost, R., J. Beyer and N. P. Vermeulen (2003). "Fish bioaccumulation and biomarkers in environmental risk assessment: a review." *Environ Toxicol Pharmacol* **13**(2): 57–149.

Van Houten, B., G. A. Santa-Gonzalez and M. Camargo (2018). "DNA repair after oxidative stress: current challenges." *Curr Opin Toxicol* **7**: 9–16.

Venditti, P., C. M. Daniele, M. Balestrieri and S. Di Meo (1999). "Protection against oxidative stress in liver of four different vertebrates." *J Exp Zool* **284**(6): 610–616.

Vogelbein, W. K., J. W. Fournie, P. A. Van Veld and R. J. Huggett (1990). "Hepatic neoplasms in the mummichog Fundulus heteroclitus from a creosote-contaminated site." *Cancer Res* **50**(18): 5978–5986.

Walling, C. (1975). "Fenton's reagent revisited." *Acc Chem Res* **8**(4): 125–131.

Wang, X., A. Ma and J. Yang (2021). "Genetic parameter estimates for three antioxidant factors in cultured Takifugu rubripes." *Fish Shellfish Immunol* **119**: 645–650.

Wang, Y., J. Fang, S. S. Leonard and K. M. Rao (2004). "Cadmium inhibits the electron transfer chain and induces reactive oxygen species." *Free Radic Biol Med* **36**(11): 1434–1443.

Wang, Y. T., T. Y. Liu, C. H. Shen, S. Y. Lin, C. C. Hung, L. C. Hsu and G. C. Chen (2022). "K48/K63-linked polyubiquitination of ATG9A by TRAF6 E3 ligase regulates oxidative stress-induced autophagy." *Cell Rep* **38**(8): 110354.

Weinhouse, C. (2021). "The roles of inducible chromatin and transcriptional memory in cellular defense system responses to redox-active pollutants." *Free Radic Biol Med* **170**: 85–108.

Weinstein, J. E., J. T. Oris and D. H. Taylor (1997). "An ultrastructural examination of the mode of UV-induced toxic action of fluoranthene in the fathead minnow, Pimephales promelas." *Aquat Toxicol* **39**(1): 1–22.

Weiss, J. (1935). "Investigations on the radical HO2 in solution." *Trans Faraday Soc* **31**(0): 668–681.

Weiss, M. A. (2013). "Diabetes mellitus due to the toxic misfolding of proinsulin variants." *FEBS Lett* **587**(13): 1942–1950.

Wells, W. W., D. P. Xu, Y. F. Yang and P. A. Rocque (1990). "Mammalian thioltransferase (glutaredoxin) and protein disulfide isomerase have dehydroascorbate reductase activity." *J Biol Chem* **265**(26): 15361–15364.

Williamson, C. E. (1995). "What role does UV-B radiation play in freshwater ecosystems?" *Limnol Oceanogr* **40**(2): 386–392.

Winston, G. W. (1991). "Oxidants and antioxidants in aquatic animals." *Comp Biochem Physiol C Comp Pharmacol Toxicol* **100**(1-2): 173–176.

Winston, G. W., F. Regoli, A. J. Dugas, Jr., J. H. Fong and K. A. Blanchard (1998). "A rapid gas chromatographic assay for determining oxyradical scavenging capacity of antioxidants and biological fluids." *Free Radic Biol Med* **24**(3): 480–493.

Winterbourn, C. C. (1985). "Free-radical production and oxidative reactions of hemoglobin." *Environ Health Perspect* **64**: 321–330.

Winzer, K., C. J. Van Noorden and A. Köhler (2002a). "Glucose-6-phosphate dehydrogenase: the key to sex-related xenobiotic toxicity in hepatocytes of European flounder (Platichthys flesus L.)?" *Aquat Toxicol* **56**(4): 275–288.

Winzer, K., C. J. Van Noorden and A. Köhler (2002b). "Sex-specific biotransformation and detoxification after xenobiotic exposure of primary cultured hepatocytes of European flounder (Platichthys flesus L.)." *Aquat Toxicol* **59**(1–2): 17–33.

Winzer, K., G. W. Winston, W. Becker, C. J. Van Noorden and A. Köehler (2001). "Sex-related responses to oxidative stress in primary cultured hepatocytes of European flounder (Platichthys flesus L.)." *Aquat Toxicol* **52**(2): 143–155.

Winzer, K., W. Becker, C. J. Van Noorden and A. Köhler (2000). "Short-time induction of oxidative stress in hepatocytes of the European flounder (Platichthys flesus)." *Mar Environ Res* **50**(1-5): 495–501.

Wolf, C., V. López Del Amo, S. Arndt, D. Bueno, S. Tenzer, E. M. Hanschmann, C. Berndt and A. Methner (2020). "Redox modifications of proteins of the mitochondrial fusion and fission machinery." *Cells* **9**(4): 815. doi: 10.3390/cells9040815. PMID: 32230997; PMCID: PMC7226787.

Wong, H. S., P. A. Dighe, V. Mezera, P. A. Monternier and M. D. Brand (2017). "Production of superoxide and hydrogen peroxide from specific mitochondrial sites under different bioenergetic conditions." *J Biol Chem* **292**(41): 16804–16809.

Xiao, W. and J. Loscalzo (2020). "Metabolic responses to reductive stress." *Antioxid Redox Signal* **32**(18): 1330–1347.

Xie, L. and P. L. Klerks (2003). "Responses to selection for cadmium resistance in the least killifish, Heterandria formosa." *Environ Toxicol Chem* **22**(2): 313–320.

Xu, H., Y. Jiang, S. Li, L. Xie, Y. X. Tao and Y. Li (2020). "Zebrafish Oxr1a knockout reveals its role in regulating antioxidant defenses and aging." *Genes (Basel)* **11**(10): 1118. doi: 10.3390/genes11101118. PMID: 32987694; PMCID: PMC7598701.

Yakes, F. M. and B. Van Houten (1997). "Mitochondrial DNA damage is more extensive and persists longer than nuclear DNA damage in human cells following oxidative stress." *Proc Natl Acad Sci U S A* **94**(2): 514–519.

Yamamoto, M., T. W. Kensler and H. Motohashi (2018). "The KEAP1-NRF2 system: a thiol-based sensor-effector apparatus for maintaining redox homeostasis." *Physiol Rev* **98**(3): 1169–1203.

Young, A. R. (1997). "Chromophores in human skin." *Phys Med Biol* **42**(5): 789–802.

Yun, J. and T. Finkel (2014). "Mitohormesis." *Cell Metab* **19**(5): 757–766.

Zhang, D.D. (2006). "Mechanistic studies of the Nrf2-Keap1 signaling pathway." *Drug Metab Rev* **38**(4): 769–789.

Zangar, R. C., J. M. Benson, V. L. Burnett and D. L. Springer (2000). "Cytochrome P450 2E1 is the primary enzyme responsible for low-dose carbon tetrachloride metabolism in human liver microsomes." *Chem Biol Interact* **125**(3): 233–243.

Zhang, H. and H. J. Forman (2017). "4-Hydroxynonenal-mediated signaling and aging." *Free Radic Biol Med* **111**: 219–225.

Zhang, J., B. Head, S. W. Leonard, J. Choi, R. L. Tanguay and M. G. Traber (2021). "Vitamin E deficiency dysregulates thiols, amino acids and related molecules during zebrafish embryogenesis." *Redox Biol* **38**: 101784.

Zhang, K., E. B. Yang, W. Y. Tang, K. P. Wong and P. Mack (1997). "Inhibition of glutathione reductase by plant polyphenols." *Biochem Pharmacol* **54**(9): 1047–1053.

Zhao, Q., X. L. Yang, W. D. Holtzclaw and P. Talalay (1997). "Unexpected genetic and structural relationships of a long-forgotten flavoenzyme to NAD(P)H:quinone reductase (DT-diaphorase)." *Proc Natl Acad Sci U S A* **94**(5): 1669–1674.

Zikic, V., A. S. Stajn, B. I. Ognjanovic, S. Z. Pavlovic and Z. S. Saicic (1997). "Activities of superoxide dismutase and catalase in erythrocytes and transaminases in the plasma of carps (Cyprinus carpio L.) exposed to cadmium." *Physiol Res* **46**(5): 391–396.

# 5 Toxic Responses of the Fish Nervous System[1]

*Michael J. Carvan III*
University of Wisconsin-Milwaukee

*Steven P. Bradbury*
Iowa State University

**ABSTRACT**

Few of the approximately 86,000 chemicals on the Toxic Substances Control Act inventory have been tested for neurotoxicity to support risk assessments. Chapter 5 provides a framework to approach questions concerning the neurotoxic effects of chemicals in fish. First, a brief summary of the structural and functional attributes of the nervous system is provided, followed by an overview of neurotoxic mechanisms of action. The final section of the chapter summarizes mechanisms of action and manifestations of neurotoxic effects for several classes of compounds. In this latter section, examples highlight where structural and functional alterations to the nervous system at the subcellular to cellular level can be linked to physiological and behavioral effects. The integration of effects across levels of biological organization is essential for establishing the mechanistic basis underlying neurotoxicity, as well as for identifying and quantifying ecologically relevant neurotoxic effects in fish.

## INTRODUCTION

Few of the approximately 86,000 chemicals on the Toxic Substances Control Act inventory have been tested for neurotoxicity to support risk assessments (Krimsky, 2017; USEPA, 2022; Woodruff et al., 2023), even though many of them are likely to be neurotoxic. Neurotoxicity has been defined as the adverse effects of physical, biological, or chemical agents on the structure or function of the nervous system in developing or adult organisms (Philbert et al., 2000). From a human health risk assessment perspective, the potential for neurotoxic effects associated with synthetic chemicals has led to the development of valid, sensitive, and reproducible methods to identify neurotoxic chemicals, characterize neurological effects, and determine the mechanisms by which chemicals produce neurotoxicity. Similar efforts to develop methods for assessing neurotoxicity in fish may yield further insights into neurotoxic mechanisms (Carlson et al., 1998; Dasgupta et al., 2022; Fitzgerald, 2021; Hawkey et al., 2021; Khalili et al., 2022; Rice et al., 1997; Timme-Laragy et al., 2006; Weber et al., 1997) in addition to addressing ecological risk concerns.

This chapter builds on the previous edition (Bradbury et al., 2008) and aims to provide a framework to approach questions concerning the neurotoxic effects of chemicals in fish. First, a brief summary of the structural and functional attributes of the nervous system is provided, followed by an overview of neurotoxic mechanisms of action. The final section of the chapter summarizes mechanisms of action and manifestations of neurotoxic effects for several classes of compounds. In this latter section, an attempt is made to highlight examples where structural and functional alterations to the nervous system at the subcellular to cellular level can be linked to physiological and behavioral effects. The integration of effects across levels of biological organization is essential for establishing the mechanistic basis underlying neurotoxicity, as well as for identifying and quantifying ecologically relevant neurotoxic effects in fish.

## OVERVIEW OF FISH NERVOUS SYSTEM DEVELOPMENT, STRUCTURE, AND FUNCTION

### DEVELOPMENT OF FISH NERVOUS SYSTEM

The development of fish nervous system follows the same general design as the development of all other vertebrate nervous systems (Figure 5.1). Beginning with gastrulation, in a process known as neural induction, the ectoderm is specified into either surface epidermis or neuroectoderm. The surface epidermis ultimately forms skin, while neuroectoderm becomes the neural tissue. Evidence from several model organisms including the zebrafish (*Danio rerio*), a widely used developmental model, indicates that neural induction is mediated through the bone morphogenetic protein (BMP) signaling pathway. Surface epidermis is induced when ectoderm is exposed to BMP signaling. Extracellular antagonism of BMP ligands prevents BMP signaling, allowing the ectoderm to adopt a neuroectodermal fate (Blader and Strähle, 2000; Lewis and Eisen, 2003; Schmidt et al., 2013).

Following induction, the neuroectoderm undergoes several morphogenic movements to form the rudimentary nervous system (Lowery and Sive, 2004; Schmidt et al., 2013). Initially, the neuroectoderm forms a flat epithelial sheet called the neural plate (Figure 5.1A). As morphogenesis begins, the lateral edges of the neural plate rise to form the neural folds

[1] Adapted and updated from Chapter 9. Toxic Responses of the Fish Nervous System. Bradbury, Carlson, Henry, Padilla, Cowden. The Toxicology of Fishes, Eds. Di Giulio and Hinton.



DOI: 10.1201/9781003160694-5

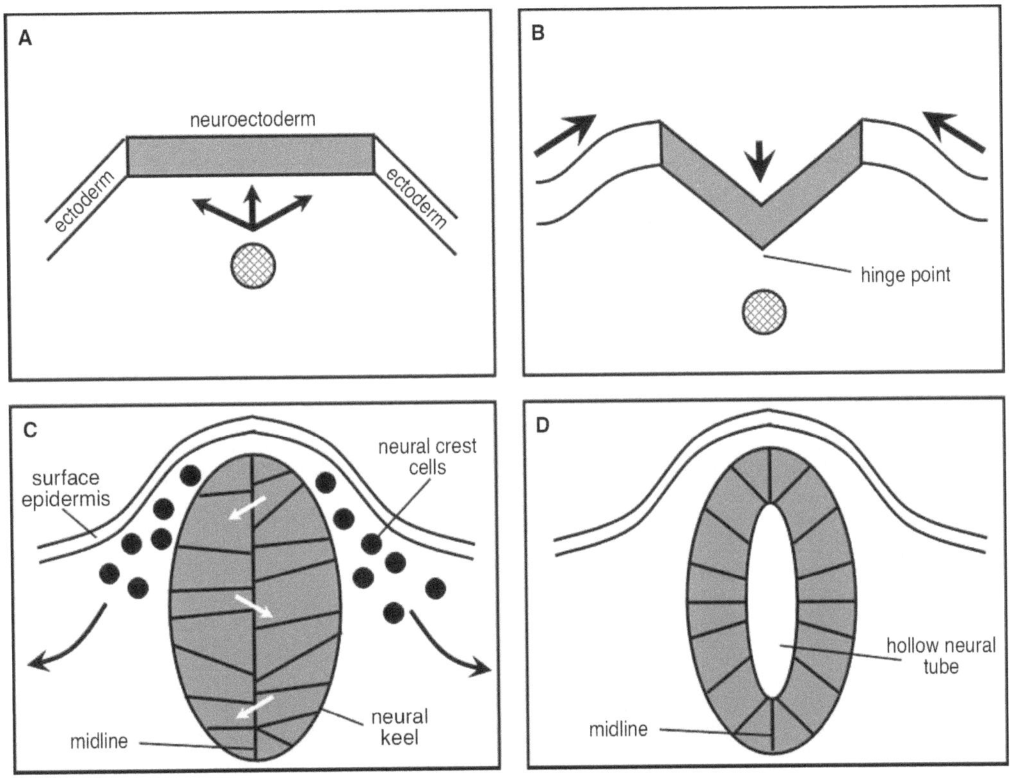


**FIGURE 5.1** Overview of neural tube development. (A) BMP signaling antagonism allows the neuroectoderm to form from ectoderm, forming the neural plate. Signals from the notochord begin patterning the neural plate. (B) Morphogenetic movements begin forming the neural keel. (C) The neural keel is a solid rod of neuroectoderm tissue. Neural crest cells have begun migrating from the dorsal regions, and surface epidermis covers the neural keel. Unlike most vertebrates, cells within the neural keel are capable of crossing the midline. (D) Apoptosis of cells within the neural keel forms the hollow neural tube.

(Figure 5.1B). Specialized neural crest cells are induced at the apex of each neural fold. These neural crest cells migrate from the neural folds and ultimately give rise to several structures, including neurons and glia of the peripheral nervous system (PNS). The neural folds are brought into apposition and ultimately join at the midline. Unlike other vertebrates, fish neural fold fusion produces a solid rod of cells called the neural keel (Figure 5.1C). Although the midline is distinct during neural keel formation, cells within the neural keel are capable of crossing the midline. Once covered by the surface epidermis, the neural keel subsequently undergoes programed cell death to become a hollow neural tube (Figure 5.1D). At the anterior end of the neural tube, bulging begins the subdivision of the brain into the prosencephalon (forebrain), mesencephalon (midbrain), and rhombencephalon (hindbrain) (Figure 5.2). The remainder of the neural tube becomes the spinal cord. Within the spinal cord, progenitor cells are patterned by secreted signaling molecules along the anterior–posterior and dorsal–ventral axes. Anterior–posterior patterning of the neuroectoderm begins during gastrulation in response to signals derived from the organizer (Blader and Strähle, 2000; H. Wang et al., 2019). The forebrain is thought to be the default state of neural tissue, with subsequent posteriorization involving retinoic acid, fibroblast growth factor, and Wnt signaling (Lewis and Eisen, 2003; Tuazon and Mullins, 2015). Along the dorsal–ventral axis, distinct neuronal cell types are located at stereotypic positions in the spinal cord, with sensory neurons specified dorsally and motorneurons specified ventrally (Blader and Strähle, 2000; Lewis and Eisen, 2003; Schmidt et al., 2013). With other vertebrates, sonic hedgehog protein secreted from the floor plate and notochord induces ventral motor neuron differentiation in a concentration-dependent manner (Blader and Strähle, 2000; Lewis and Eisen, 2003; Schmidt et al., 2013). In the dorsal neural tube, BMP signaling from the roof plate and adjacent non-neural ectoderm are hypothesized to induce sensory neuron fates in a concentration-dependent manner (Blader and Strähle, 2000; Lewis and Eisen, 2003; Schmidt et al., 2013).

Once neuronal fates have been specified in appropriate anterior–posterior and dorsal–ventral locations, the next step in nervous system development is axonogenesis when neurons extend axons to their innervation targets. Embryonic nervous system development has been particularly well characterized in zebrafish (Bernhardt, 1999; Bremer and Granato, 2016). Therefore, the summary of axonogenesis given here is necessarily restricted to zebrafish, although many facets of zebrafish axonogenesis are likely applicable to other teleosts. A tremendous amount of descriptive information is known about zebrafish axonogenesis, with axonal pathways and targets characterized for several neuron classes (Bernhardt, 1999; Hutson and Chien, 2002; Lewis and Eisen, 2003; Schmidt et al., 2013). The cellular and molecular basis for axon guidance is not as well understood. Several genetic screens in zebrafish have identified mutants with defective axonogenesis (Gulati-Leekha and Goldman, 2006; Hutson and Chien, 2002; Karlstrom et al.,

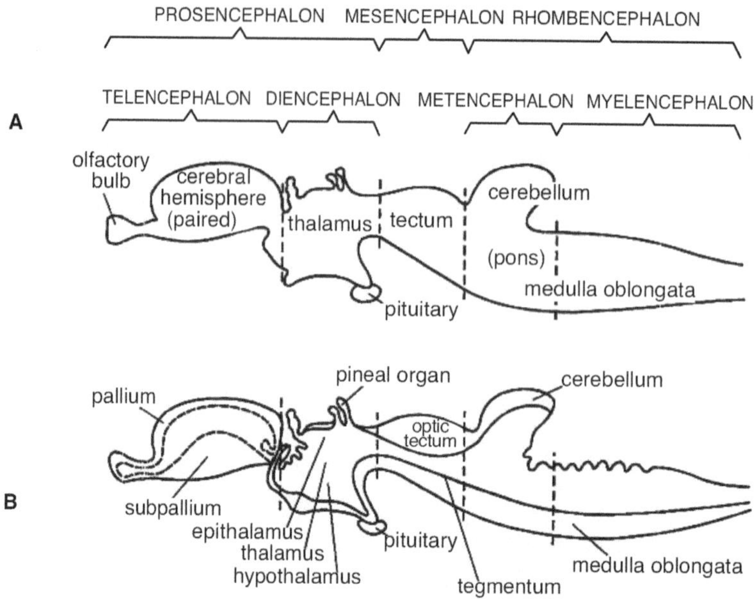


**FIGURE 5.2** Major anatomical features of the fish brain (B is median section of A). (Adapted from Bond 1996.)

1997) and several axonal guidance gene homologs have been identified in the zebrafish genome (Bernhardt, 1999; Chilton, 2006). Basic functional analysis of these gene homologs has revealed similarity to their mammalian counterparts (Fricke et al., 2001; Ingold et al., 2015; J. S. Lee et al., 2001; J. Zhang et al., 2004). More recent studies have explored mechanistic environmental guidance forces (chemoattraction, chemorepulsion, contact attraction, and contact repulsion) driven by the ligand-receptor process that mediates axon guidance and cell migration (Aberle, 2019; Bremer and Granato, 2016; Taku et al., 2016).

The final step in embryonic nervous system development occurs when an axon reaches its innervation target. Synapses are either chemical or electrical and are formed between neurons and their targets in a process known as synaptogenesis. Synaptogenesis is well-studied in aquatic vertebrates, including zebrafish. There are common molecular drivers regulating the formation of electrical and chemical synapses (Jabeen and Thirumalai, 2018). Junction formation requires (1) selecting the junction site; (2) adhering to the cellular membranes in close apposition; (3) anchoring to the cytoskeleton; and (4) coordinating protein recruitment between the two cells to form a functional junction (Martin et al., 2020). Each junction type has unique features to address these requirements, as well as a common foundation (Jabeen and Thirumalai, 2018; Martin et al., 2020). Much less is known about the formation of electrical synapses and zebrafish has been a critical tool for exploring and identifying the previously unknown components (Martin et al., 2020).

In adult fish, the central nervous system (CNS) includes the neuronal structures encased within the skull and spinal column. PNS is comprised of nerve ganglia lying outside the spinal column as well as nerve processes found elsewhere throughout the organism. As noted above, this subdivision of the nervous system occurs during embryogenesis through ectodermal morphogenesis and patterning. The coordination and control of physiological and behavioral processes reflect the integration of structural and functional attributes of the entire nervous system.

## Central Nervous System Anatomy

Although significant differences exist in anatomical features across fish species, the basic structural components of the developing CNS include prosencephalon (forebrain), mesencephalon (midbrain), rhombencephalon (hindbrain), and spinal cord (Figure 5.2).

### Prosencephalon (Forebrain)

Prosencephalon is comprised of telencephalon and diencephalon (Bond, 1996; Wullimann, 1998). Telencephalon possesses rostrally located olfactory bulbs, olfactory lobes, and, more caudally, cerebral hemispheres. Telencephalon is the site where the olfactory nerve (cranial nerve I) enters the olfactory bulbs. In actinopterygian fish, the cerebral hemispheres can be divided into dorsal and ventral components, which are homologous to pallium and subpallium, respectively, of other vertebrates. The majority of afferent neurons entering telencephalon are olfactory; however, there is some indication that non-olfactory ascending inputs exist, although their sites of integration have not been identified. It is now generally accepted that telencephalon contributes to the processing of visual inputs, in addition to olfactory information, and plays a role in feeding, defense, shoaling, socialization, and aggressive and reproductive behaviors (Bernstein, 1970; Bond, 1996; Davis and Kassel, 1983; Sharpe et al., 2010; Northcutt and Davis, 1983; Stednitz et al., 2018; Teles et al., 2016; Wullimann, 1998). Numerous studies have documented the presence of a variety of neurotransmitters in telencephalon, including acetylcholine, gamma-aminobutyrate (GABA), neuropeptides, catecholamines, excitatory amino acids (e.g., glutamate), and nitric oxide (summarized in reviews by Pedroni and Ampatzis (2019), Rico et al. (2011), Trudeau (2022), Vernier (2017)).

Diencephalon in actinopterygian fish has been subdivided into the preoptic area, hypothalamus, thalamus, epithalamus, and synencephalon (Bradford and Northcutt, 1983). Wullimann (1998), however, considers the preoptic area an intermediate zone between telencephalon and diencephalon in the teleost brain. Wullimann (1998) also considers synencephalon as part of the pretectum (see below), which is intermingled with diencephalic cell groups. Below the hypothalamus lies the pituitary gland, and the pineal organ arises from the epithalamus. The primary afferent neurons to the thalamus are the optic nerves (cranial nerve II), which enter the brain and cross anterior to the diencephalon. The hypothalamus receives a number of afferents from the telencephalon; the thalamus receives afferents from the cerebellum and serves as a relay center for the transfer of olfactory and striate body impulses to other parts of diencephalon and lower brain centers. In general, diencephalon appears to serve as a coordination center for incoming and outgoing signals associated with internal homeostasis and affect the endocrine system through the pituitary gland (Bernstein, 1970; Bond, 1996; Bradford and Northcutt, 1983; Wullimann, 1998). Acetylcholine, GABA, glutamate, nitric oxide, noradrenaline, dopamine, and 5-hydroxytryptamine have been identified as neurotransmitters in diencephalon, primarily in the hypothalamus (Brantley and Bass, 1988; Holmqvist and Ekstrom, 1997; Khan et al., 1996; Parent, 1983; Sas et al., 1990; Sloley et al., 1992).

### Mesencephalon (Midbrain)

Mesencephalon, comprised of the optic tectum and the ventral tegmentum, is the primary site for processing impulses received from the optic nerve. Extensive electrophysiological and evoked response studies have been undertaken to create maps of visual fields on the optic tectum in cyclostomes, plagiostomes, and actinopterygians. The optic nerve is decussated, with projections from the right eye projecting to the left tectal hemisphere and from the left eye projecting to the right tectal hemisphere. Ascending efferent fibers from the optic tectum have been traced to the thalamus and pretectum while descending efferent fibers proceed to the dorsolateral region of the ventral tegmentum. Some tectal outputs provide terminals to the reticular formation and cell groups in the medulla. Electrical stimulation and lesion studies support the conclusion that the optic tectum contains neuronal units that coordinate visual inputs with locomotor activity (Bernstein, 1970; Bond, 1996; Rodríguez et al., 2021; Vanegas, 1983; Wullimann, 1998). Reports suggest that acetylcholine, GABA, glutamate, nitric oxide, and glycine are predominant neurotransmitters within the optic tectum (Brantley and Bass, 1988; Rio et al., 1996; van Deusen and Meyer, 1990; Holmqvist and Ekstrom, 1997; Becker et al., 1991). Moderate catecholaminergic innervation has also been documented (Borgonovo et al., 2021; Sas et al., 1990).

### Rhombenchephalon (Hindbrain)

Rhombencephalon is comprised of metencephalon and myelencephalon. Metencephalon includes the cerebellum and pons, with the cerebellum being the major component. The structure of the cerebellum is highly variable in fish. In teleosts, it can be divided into three major sections that are arranged similarly to that of the cerebellar cortex in other vertebrates. These areas are termed valvula cerebelli, corpus cerebelli, and vestibulolateral lobe. Each area contains a molecular layer, Purkinje (ganglion) cell layer, and granule cell layer. Primary sensory fibers in hair cell sensory systems (see sensory organ systems discussion below) and neurons from the spinal cord are important ascending inputs to the cerebellum. Efferents from the cerebellum project to the ventral tegmentum, the thalamus, and the reticular formation. The cerebellum is associated with muscle tone, motor control, and lateral line sensory input (Bernstein, 1970; Bond, 1996; Finger, 1983; Wullimann, 1998) and is essential for spatial cognition and associative, sensorimotor learning and emotional conditioning (F. Rodríguez et al., 2005). The cerebellum has been reported to contain glutamate, aspartate, GABA, glycine, as well as catecholaminergic neurotransmitters (Ma, 1994; Sas et al., 1990). Myelencephalon, although containing portions of the cerebellum, is predominantly comprised of the medulla oblongata (Wullimann, 1998). The boundary between the medulla oblongata and spinal cord in fish is generally indistinct and can be most readily defined by the types of information transmitted with associated nerve columns. Information derived from the general sensory, cutaneous, vestibular, lateral line, and trigeminal (cranial nerve V) nerve fibers is carried in the somatic sensory column. Nerve fibers derived from chemoreceptors and nerves arising in the viscera are associated with the visceral sensory column. Sensory inputs associated with sight and olfaction do not input directly to the medulla. A visceral motor column in the medulla carries efferent fibers derived from the facial (cranial nerve VII), glossopharyngeal (cranial nerve IX), and vagus (cranial nerve X) nerves to the glands and musculature of the viscera. A somatic nerve column carries efferent nerve fibers to the ocular muscles and muscles of the pharyngeal complex. The medulla of the actinopterygians contains a pair of Mauthner cells, which are giant interneurons at the level of the vestibulocochlear nerve (cranial nerve VIII) that coordinate the startle response associated with sensory inputs (Hale et al., 2016). The dendrites of the Mauthner cells connect with fibers of the trigeminal nerve, facial nerve, glossopharyngeal nerve, and vagus nerve to the cerebellum and the optic tectum. Their axons pass the length of the spinal cord and coordinate the musculature in the tail and associated rapid swimming movements (Bernstein, 1970; Bond, 1996; Hale et al., 2016). Serotonin, catecholamines, glycine, GABA, and nitric oxide have been reported to be neurotransmitters in the medulla (Parent, 1983; Sas et al., 1990; Becker et al., 1991; Faber and Korn, 1988; Legendre and Korn, 1994; Triller et al., 1993; Schober et al., 1994).

### Spinal Cord

In fish, the spinal cord occupies the entire vertebral canal. The cord consists of a series of segments that form dorsal and ventral roots. The dorsal and ventral horns correspond to regions of sensory input (dorsal root) and motor output (ventral root). As is generally noted within vertebrates, these roots join to form the spinal nerves. Differentiation of the spinal cord increases with evolutionary progression from Cyclostomata (lampreys and hagfish) to Chondrichthyes (cartilaginous fish) to Osteichthyes (bony fish). In the latter fishes, the gray matter is clearly divided into dorsal and ventral horns. The dorsomedial gray matter innervates the trunk musculature and specialized areas, such as the pectoral fin. Many fibers ascend to the medulla oblongata and cerebellum. The descending fibers consist of many vestibulospinal and reticulospinal fibers and the giant Mauthner cells, which contain large-diameter, well-myelinated axons. The dendrites and cell bodies of Mauthner cells are found in the medulla oblongata. These cells have many collaterals that have extensive connections with interneurons and motor neurons in the spinal cord. The role of the Mauthner cells is to mediate sensory inputs through the startle response, as mentioned previously. As a final note, the spinal cords of adult and larval fish are unique from mammals in their capacity for anatomical and physiological regeneration (Cigliola et al., 2020; Tsata and Wehner, 2021).

## Peripheral Nervous System Anatomy and Function: The Autonomic Nervous System

The discussion of the CNS and associated components in the PNS has made reference to aspects of both the somatic and autonomic nervous systems, which are responsible for "voluntary" and "involuntary" actions. In mammals, the autonomic nervous system contributes to the regulation of organs and tissues such as the gut, heart, and blood vessels. In fish, the autonomic nervous system is similarly involved in the control of the gut and circulatory systems and control of the swim bladder and melanophores. Studies in teleosts suggest that functional attributes of the autonomic nervous system are similar to that noted in mammals; however, the presence of gills and a swim bladder in fish and a smaller number of nerves and neurotransmitters highlight some important differences (Donald, 1998; Nilsson and Holmgren, 1998; Olsson, 2009; Sandblom and Axelsson, 2011).

An anatomical differentiation of the fish autonomic nervous system into sympathetic and parasympathetic subsystems is difficult (reviewed in Donald, 1998; Nilsson and Holmgren, 1998; Olsson, 2009; Sandblom and Axelsson, 2011). Instead, a terminology that differentiates cranial nervous pathways from spinal nervous pathways has been proposed. In both the cranial and spinal pathways, a preganglionic neuron runs from the CNS to an autonomic ganglion. A postganglionic neuron then serves as the afferent to the appropriate tissue or organ. The spinal autonomic ganglia are typically located close to the spinal column in higher vertebrates and teleosts, while autonomic ganglia associated with cranial pathways are usually located near or within the effector organ. Within the fishes, there is great variability in the anatomy of the autonomic nervous system. In the cyclostomes, the autonomic nervous system is poorly developed, with no spinal ganglia readily apparent, and it is difficult to differentiate sensory and autonomic nerve fibers. In the elasmobranchs and teleosts, autonomic ganglia are observed along the spinal column. Cranial autonomic fibers have been associated with the oculomotor (cranial nerve III), facial (cranial nerve VII), glossopharyngeal (cranial nerve IX), and vagus (cranial nerve X) nerves in the elasmobranchs. The teleost autonomic system most closely resembles that of higher vertebrates. Of note are the spinal ganglia that enter the cranium and have connections with the trigeminal (cranial nerve V)/facial (cranial nerve VII), glossopharyngeal (cranial nerve IX), and vagus (cranial nerve X) nerves. It is thought that innervation of the head and gills is associated with spinal ganglia. In the posterior region of the abdomen, the sympathetic chains fuse, and the vesicular nerves leave the chain to innervate the gut and urinogenital organs. Cranial autonomic fibers have been reported in the oculomotor (cranial nerve III) and vagus nerves (cranial nerve X), both of which reach the gut and swim bladder.

The enteric nervous system (ENS) is a component of the autonomic nervous system and the largest part of PNS. ENS is developmentally derived from the neural crest and provides intrinsic innervation of the gut, regulating all essential gut functions (Ganz, 2018) and influencing gut microbiota function (Rolig et al., 2017). The neurons of the fish ENS are either scattered upon the longitudinal muscle layer of the gut or aggregated in small groups at the nodes of fiber connections (Ceccotti et al., 2018). ENS in fish varies somewhat by species but follows a similar overall arrangement as other vertebrate groups. The majority of nerve cell bodies are found in the myenteric plexus. These cells are sensory neurons, interneurons, or motor neurons that innervate muscles, blood vessels, and glands (Olsson, 2009).

The autonomic nervous system in fish contains numerous neuronal subtypes based on the existence of neurochemical markers (Ganz, 2018). These include neurotransmitters and neuropeptides such as acetylcholine, nitric oxide, serotonin, calretinin, calbindin, calcitonin gene-related peptide, tachykinins, substance P, neurokinin A, vasoactive intestinal peptide, pituitary adenylate cyclase–activating peptide, and bombesin (Bjenning and Holmgren, 1988; Ceccotti et al., 2018; Donald, 1998; Nilsson and Holmgren, 1998; Rolig et al., 2017; Shepherd and Eisen, 2011).

## Sensory Organ Systems

The sensory systems of fish can be divided into visual, auditory, mechanosensory, electrosensory, and chemoreception. In the assessment of neurotoxicity, these sensory organ systems are viewed as readouts of a functional nervous system. The eyes of fish generally have the same structures as those noted in the vertebrates, including the anterior chamber, iris, lens, and vitreous chamber that is lined by the retina (Nilsson, 2021). The retina is comprised of pigmented epithelium, which consists of photoreceptors (rod and cone cells), horizontal cells, bipolar cells, amacrine cells, and ganglion cells and nerve fibers that lead to the optic nerve. The optic nerve is comprised of four different types of fibers, most of which are afferent and project to the contralateral side of the brain (the optic tectum). Horizontal cells facilitate information flow between photoreceptors, and amacrine cells facilitate lateral flow of information among ganglion cells. Bipolar cells direct excitatory or inhibitory responses vertically to the ganglion cells, which subsequently send information to the CNS through the optic nerve. Fish have the most diverse range of eye functions among the vertebrates, reflecting the extreme variation of aquatic environments (Carleton et al., 2020; Corbo, 2021; Musilova et al., 2019). Intensity and spectral sensitivity are modulated by variations in the utilization of opsin gene(s) and changes in the chromophore prosthetic group (derivatives of vitamin $A_1$ or $A_2$) (Carleton et al., 2020; Corbo, 2021). Many fish have a single opsin gene, whereas, deep-sea fish have the highest number of visual opsins including silver spinyfin (*Diretmus argenteus*) with 40 opsins (Musilova et al., 2019). As reviewed by Hawryshyn (1998) and Lasater (1990), glutamate is the putative neurotransmitter in the photoreceptors that mediates synaptic communication with bipolar cells, while GABA has been well documented in horizontal and amacrine cells (reviewed in Hawryshyn, 1998; Lasater, 1990; Marc and Cameron, 2001). Glycine, dopamine, and acetylcholine have also been reported as neurotransmitters in the retina (Lasater, 1990). The potential role of neuropeptides has yet to be elucidated.

The auditory, mechanosensory, and electrosensory systems include the inner ear, the lateral line (comprised of neuromasts and canals), and the ampullary and tuberous organs of the electrosensory lateral line (Bond, 1996; Schellart and Wubbels, 1998). The ear functions contribute to balance and sound reception, while lateral line functions are related to sensing water displacement and pressure. The electrosensory lateral line is responsible for sensing electrical fields and signals. The inner ear of fish includes three otolith organs: saccule, lagena, and utricle organs. Sacculus and lagena are primarily responsible for sound reception. The sensory epithelium, termed the macula, is comprised of sensory hair cells that are the receptor cells. Auditory stimulus causes deflections of hair bundles, leading to depolarization of hair cells and a subsequent increase in the firing rate of afferent nerve fibers to the brain. The auditory nerve (vestibulocochlear nerve; cranial nerve VIII) innervates the macula. Those portions of the brain thought to integrate auditory inputs have been discussed previously. The mechanosensitive lateral line consists of neuromasts that are found over the entire body, either free on the skin or within pits, grooves, or canals. A neuromast consists of sensory hair cells with supporting cells, all of which are covered by cupula. Some neuromasts are open to water, via pores in the bone or scales. Water movement causes an impulse of fluid within the lateral line that in turn causes deformation of the cupulae. This deformation of hair cells results in a response in the nerve fiber innervating the neuromast. The number of innervating nerve fibers can range from a few to several hundred. The dorsal anterior lateral line nerve innervates the dorsal part of the head, the ventral lateral line nerve innervates the cheek and lower jaw, and the posterior lateral line nerve innervates the trunk and tail. The involvement of different portions of the brain in integrating responses to these inputs has been discussed previously. Of note are the projections of auditory and lateral line afferents to the Mauthner cells, which invoke the startle response.

Chemosensory systems in fishes are primarily divided into gustation and olfaction and are well developed in fish. Taste buds are comprised of 100–150 specialized epithelium cells, which include receptor cells, basal cells, and supporting cells. Taste buds are found primarily in the mouth, pharyngeal region, and gill arches; however, in some species, they may be found on barbels, fins, and other parts of the body surface. Taste buds in the oropharyngeal cavity are innervated by the glossopharyngeal (cranial nerve IX) and vagus (cranial nerve X) nerves, while the facial nerve (cranial nerve VII) innervates taste buds on external surfaces. Receptor cells within the taste buds respond to specific compounds, such as amino acids. Many taste receptors are thought to be ion channels or proteins associated with ion channels. The binding of chemicals to the appropriate receptor elicits depolarization of the receptor cell. Through chemical synapses, the potential of the innervating nerve fiber may be modulated prior to propagation to the CNS (Bond, 1996; Morais, 2017; Sorenson and Caprio, 1998). The olfactory organs are found in pits on the anterior portion of the head. The epithelium within the olfactory chambers is arranged in a series of lamellae that form the olfactory rosette. Both sensory and non-sensory cells are found in the epithelium. The non-sensory cells serve to protect the sensory cells and transport chemical stimuli. Ciliated and microvillous receptor cells have been noted in fish. These cells are innervated by the olfactory nerve, which provides input to the olfactory bulb. The receptor cells are responsive to amino acids, bile acids, gonadal steroids, and prostaglandins (Byrd and Caprio, 1982; Caprio and Byrd, 1984; Chang and Caprio, 1996; Morais, 2017; Ohno et al., 1984; Sato and Suzuki, 2001). A variety of alcohols, amines, carboxylic acids, nucleotides, and hydrocarbons can also elicit olfactory responses (Kolesnikov and Kosolapov, 1993; Rolen et al., 2003). Specific receptor proteins in sensitive cells are thought to be responsible for initial olfactory responses. These receptors initiate second-messenger-mediated signal transduction pathways that open ion channels and lead to the production of action potentials (Bond, 1996; Morais, 2017; Sorenson and Caprio, 1998). Taste receptors and signaling pathways have been found in the fish gastrointestinal tract but the ligands that activate these receptors and their physiological role remain to be discovered (Morais, 2017).

## BASIC MECHANISMS OF NEUROTOXICITY

Before considering some mechanisms of neurotoxicity in adult fish, it is important to remember that neurotoxic compounds are also likely to have developmental effects at lower concentrations. In particular, the potency and mode of action of neurotoxic compounds may be altered during the maturation of the nervous system. To underscore the importance of this concept, compounds with adverse effects on the immature nervous system are categorized as developmental neurotoxicants.

It is widely accepted that the developing nervous system is particularly susceptible to some neurotoxic compounds. This assertion is based largely on the fact that neurotoxicant exposures having little to no effect on the mature nervous system may produce significant and lasting effects on the developing nervous system. Several processes critical for the proper nervous system structure and function, including neuronal specification, migration, and synaptogenesis, all occur *in ovo*. Furthermore, several neurotransmitters have developmental roles that are distinct from their synaptic roles (e.g., acetylcholine) (Lauder and Schambra, 1999). Disruptions in any of these processes or neurotransmitters may have unique effects during development that are not seen in the mature nervous system and are potential mechanisms of developmental neurotoxicity.

In the mature nervous system, neurotoxic mechanisms of action can be associated with a variety of unique anatomical and physiological characteristics of the nervous system (Anthony et al., 1996; Philbert et al., 2000). Many aspects of the mammalian nervous system contribute to its vulnerability: maintenance of the blood–brain or blood–nerve barrier, high energy requirements, high lipid environment, complex spatial geometry, and the need for rapid transmittal of information between cells. Because the structural and functional aspects of neurons are highly conserved, the framework presented in Chapter 16 of *Cassarett & Doull's Toxicology*, "Toxic Responses of the Nervous System" (Moser et al., 2019), for understanding the mechanisms of neurotoxicity, as well as the manifestations of toxic effects, is relevant and briefly reviewed.

In terms of xenobiotic exposure, CNS and PNS are unique in that tissue and cellular attributes minimize the transport of potentially toxic compounds to the sites of action. The restricted interfaces between the blood and many components of the nervous system have been termed barriers: the blood–brain barrier encloses the CNS, and the blood–nerve barrier encloses the PNS. The basis for the blood–brain barrier is that the endothelial cells lining the vasculature of the brain have extremely tight intercellular junctions. As a consequence, to gain access, xenobiotics are required to diffuse through these cellular barriers. The lipid-rich myelin-surrounding axons, which are critical for insulating the nerve and ensuring rapid propagation of action potentials, may provide a barrier to hydrophilic xenobiotics. In general, the more hydrophilic the compound, the less likely it will cross one of these barriers; however, some hydrophilic compounds may gain access by piggybacking on specific transporters that are present in these barriers. Oligodendrocytes and Schwann cells found in CNS and PNS, respectively, are responsible for maintaining the myelin sheaths that exclude water and ions from the axons and for minimizing extracellular spaces. A wide variety of specialized proteins must be synthesized and complex lipid metabolism must be maintained to support the myelin environment. Neurons also need to maintain ion gradients to support neuronal transmission. These maintenance activities require a high aerobic metabolic rate. As a consequence, the nervous system requires high levels of oxygen and glucose to maintain homeostatic ion gradients and to restore ion gradients after depolarizations.

Finally, the unique spatial arrangement of the nervous system makes it vulnerable to toxic insults. The nervous system is comprised of relatively large cells as a result of evolutionary pressures to establish high-speed intracellular communication; consequently, axon volumes are typically much greater than those of the cell bodies and have a unique demand for macromolecule synthesis. The cell body of each neuron synthesizes everything that is needed by the rest of the cell (axons and dendrites); these products are then transported throughout the cell to replenish cellular constituents. In addition, spent cellular constituents are transported retrogradely from the processes to the cell body. The retrograde transport also contains biochemical information related to the status of the environment in the distal portion of the neuron. The spatial arrangement of the nervous system also requires maintenance of the transmission of electrical pulses along the length of axons and across the synapses between neurons. Maintenance of both intracellular and extracellular communication is, of course, essential for the proper function of the nervous system.

In conclusion, the nervous system has a variety of unique structural characteristics that meet the need for rapid cellular communication. These unique structural characteristics are primarily associated with maintaining the integrity of the cell body and axon and the ability of the nerves to support the propagation of action potentials and synaptic transmission. Xenobiotics capable of disrupting ion channels that are essential for maintaining and supporting proper ion balances or capable of disrupting the chemical transmission of potentials across synapses are capable of causing neurotoxic effects. As discussed previously, the maintenance of aerobic respiration, high rates of protein and lipid synthesis, and extensive transport of synthetic products from the cell body to the axon are critical for maintaining the structural and functional characteristics of the nervous system. Consequently, xenobiotics capable of disrupting neuron-specific synthetic and metabolic pathways, modifying the products of these reactions, or inhibiting or uncoupling aerobic metabolism can elicit adverse effects specific to the nervous system. It is interesting to note that some fish species have evolved specific metabolic strategies for anoxic conditions. Research suggests that anoxic-tolerant species may have decreased levels of excitatory neurotransmitters and increased levels of inhibitory transmitters in the brain that enable metabolic depression (Van Ginneken et al., 1996).

## MANIFESTATIONS OF NEUROTOXICITY IN FISH

Neurotoxic effects of chemicals are assessed by quantifying structural and functional responses at the subcellular to organismal levels of biological organization. Functional observations at the organismal and cellular levels can provide insights concerning potential sites and modes of action, while cellular and biochemical responses can provide insights into molecular mechanisms of action. Cellular and biochemical investigations can be used to identify neurotoxic potential, characterize the nature of neurological effects, and determine the mechanisms by which chemicals produce neurotoxic effects (Tilson, 1996). A significant challenge in developing investigative methods and associated bioassay techniques lies in linking neuromorphological, neurochemical, and neurophysiological alterations with functional (i.e., behavioral) observations (Legradi et al., 2018). Many chemically induced biochemical, physiological, or morphological perturbations have been reported in cellular and organismal systems, but consequent behavioral effects on organisms have not been established. Because behavioral responses are an integration of biochemical, physiological, and morphological processes, linking behavioral observations to these types of observations can provide the needed bridge between subcellular and cellular responses and ecological consequences (Fitzgerald, et al., 2021; Little, 1990).

The adverse outcome pathway (AOP) is a tool for organizing our understanding of (neuro)toxicological information across many levels of biological organization—from the molecular initiating event to responses of organisms and populations (Ankley et al., 2010; Mora-Zamorano et al., 2018a; Ramšak et al., 2022). An AOP focused on neurotoxicity necessitates linkages from exposure to behavior, and many AOPs have been proposed based on data from zebrafish (Mora-Zamorano et al., 2018b; Ramšak et al., 2022). Ecotoxicology and behavioral neurotoxicology are among the fields that will advance dramatically in the next few decades, thanks to technological innovations in the detection of biomolecules and continued rapid developments in computing. Biologically based systems are inherently non-linear and stochastic, making their mathematical, statistical, and computational modeling quite a formidable endeavor; however, integrated dynamic modeling of multiscale organizational and temporal changes will reveal important mechanistic information relevant to neurotoxicity (Carvan et al., 2008).

With the development of initial acute toxicity datasets for industrial organic chemicals, it was established that the potency of xenobiotics in fish was dependent on the hydrophobicity (or lipophilicity) of a chemical (Konemann, 1981; Veith et al., 1983). Subsequent experimental studies and modeling efforts have led to general acceptance that the relationships between hydrophobicity and lethality represent the minimum, or baseline, toxicity that a compound can elicit in the absence of a more specific mode of toxic action. With additional study, it became clear that there are subclasses that are more potent than would be predicted based on hydrophobicity or lipophilicity (Duarte et al., 2022) and would need further classification on the basis of either acute potency or physiological and behavioral characteristics of the neurotoxic response.

### STRUCTURAL MANIFESTATIONS OF NEUROTOXICITY IN FISH

As discussed previously, dynamic interactions between the neuronal cell body and the axon are critical for maintaining proper neuronal structure and function. Disruption of these interactions can result in a variety of pathologies. Neuronopathies result from toxicants capable of causing injury to the cell body followed by degeneration of the remaining cell processes. Neuronal loss is typically permanent and is manifested by global symptoms or dysfunctions consistent with the specific nervous tissue target. For example, 1-methyl-4-phenyl-1,2,3,6-tetrahydropyridine causes neuronopathies in mammals through its cytotoxic dihydropyridium ion metabolite; similar observations have been reported in fish (Matsui et al., 2014; Piña et al., 2022). Axonopathies are associated with toxicants whose primary site of action is the axon, causing degeneration of the axon along with surrounding myelin. Because the nerve cell body is not affected, recovery and regeneration of the axon are possible. The effects of neurotoxicants capable of eliciting axonopathies are sometimes described as causing chemical transections of the nerve (e.g., acrylamide, carbon disulfide, some diketones, and certain classes of organophosphorus esters). Likewise, compounds disrupting microtubles (such as colchicine) will perturb axonal transport and thereby cause pathology. Heavy metals have been shown to cause axonopathies in multiple species of fish (Green and Planchart, 2018). Myelinopathies are associated with compounds capable of causing intramyelinic edema, such as hexachlorophene (Kinoshita et al., 2000; Yoshikawa, 2001). Hexachlorophene-induced myelinopathies have also been reported in non-mammals (Reier et al., 1978). Pesticides were shown to cause myelinopathies in tilapia (Saleh et al., 2022), and several chemicals cause alteration in myelination pathways in zebrafish (Jin et al., 2021; Panlilio et al., 2020; F. Yang et al., 2021; L. Yang et al., 2022). These responses may be due to effects on the myelin itself or due to effects on the myelinating cells.

### PHYSIOLOGICAL MANIFESTATIONS OF NEUROTOXICITY IN FISH

#### Electrophysiology

Electrophysiological techniques measure electrical potentials and the transmission of impulses in the nervous system. These techniques can be applied to *in vitro* or *in vivo* preparations and used to assess responses of the brain, spinal cord, components of PNS, or sensory systems to determine whether or not a neurotoxic response can be elicited by a xenobiotic. Electrophysiological techniques can be used to determine if a toxicant acts pre- or postsynaptically on specific portions of a fiber or on specific ion channels.

Measurements can be made from the surface of the organism, extracellularly or intracellularly (Baker and Lowndes, 1986; Eisenbrandt et al., 1994; Fox et al., 1982; Tomasello and Wlodkowic, 2022). Bahr (1972, 1973), for example, reported an *in vivo* method to record the evoked electrical activity of the trunk lateral line in surgically prepared rainbow trout (*Oncorhynchus mykiss*) for periods up to 48 hours. Whole-cell patch recordings from zebrafish lateral line ganglia are also used to design adaptive systems for autonomous underwater robots (Haehnel-Taguchi et al., 2018). Evans and Hara (1985) described an *in vivo* rainbow trout model to quantify evoked electro-olfactograms, and, more recently, several studies have used similar techniques to investigate the effects of heavy metals on olfactory sensory neurons (Dew et al., 2016; Lazzari et al., 2019; Sommers et al., 2016). Kreft and coworkers (1985) described a technique to assess electroretinograms from photostimulated *in vitro* eye preparations, and more recently (Daniel N Weber et al., 2008) used patch clamp studies to describe the persistent effects of developmental methylmercury exposure on $I_A$ and $I_K$ currents in bipolar cells of the developmentally exposed adult. Schafer and coworkers (1995) described an *in vivo* anesthetized catfish (*Ameiurus nebulosus*) preparation to measure spontaneous afferent activity from electroreceptor organs. Several whole cell and membrane preparations have also been used to study ion channels and the role of specific neurotransmitters in channel function; for example, voltage clamp analysis has been used to study the role of glycine, GABA, and *N*-methyl-D-aspartate (NMDA) in isolated lamprey spinal cord neuron function and to identify calcium channel subtypes in lamprey sensory and motor neurons (Baev et al., 1992; El Manira and Bussieres, 1997; Niessen et al., 2016).

To date, most electrophysiological techniques applied to fish involve *in vitro* models or surgically invasive approaches using restrained or anesthetized animals. Although these methods yield data regarding cellular and molecular mechanisms of neurotoxic action, disruption of the nervous system or sensory organs can make it difficult to study the behavioral outcomes of perturbed neurological function. Two laboratories (Carlson et al., 1998; Featherstone et al., 1991, 1993) pioneered an approach relating electrophysiological responses to a defined behavioral endpoint. In these studies, the electrophysiological responses of Mauthner cells, motor neurons and interneurons, and white musculature to an invoked startle response were measured *in vivo* in unanesthetized, freely moving larval fish. These waveforms are easily triggered and recorded, yielding stereotypical, reproducible results. Techniques are being developed for noninvasive *in situ* electrophysiology using multielectrode arrays to enhance the understanding of the neurophysiological foundations responsible for the altered behavioral phenotypes (Haehnel-Taguchi et al., 2018). This technique was used by Tomasello and Sive (2020) to quantify brain and spinal cord activity changes in zebrafish in response to valproic acid exposure.

### Respiratory–Cardiovascular

The effects of toxicants on the CNS or PNS can elicit a wide variety of integrated respiratory–cardiovascular responses. In nearly all cases, toxic effects seen in multiple systems are considered to be coincident and a causative relationship is not explored. However, these responses can be evaluated to determine the extent to which they can be associated with specific mechanisms of neurotoxic action. Several elegant studies have demonstrated that specific suites of respiratory–cardiovascular responses of spinally transected rainbow trout were sufficient to differentiate the effects associated with industrial organic chemicals whose sites and mechanisms of action are distinct, including narcotics, acetylcholinesterase inhibitors, pyrethroid insecticides, cyclodiene insecticides, and strychnine (Bradbury et al., 1991a; Bradbury and Coats, 1989; Bradbury et al., 1991b; McKim et al., 1987). More specifically, distinct suites of effects on cough frequency, ventilation frequency, ventilation volume, oxygen consumption, and arterial blood oxygen, carbon dioxide, pH, hemoglobin, and hematocrit were associated with the different compound classes studied. Furthermore, these respiratory–cardiovascular responses were consistent with the neurodepressant, stimulant, or convulsant mechanisms of action of the xenobiotics studied.

### Behavioral

The behavior of a fish reflects the integrated output of the nervous system at the organismal level in response to stimuli perceived in the environment. The extent to which these chemically induced behavioral changes are ecologically relevant must be considered in terms of adverse effects on an organism's ability to survive and reproduce. Behavioral endpoints can be categorized into individual or interindividual responses (Rand, 1985). Individual responses include undirected and directed locomotion and feeding. Undirected locomotion refers to the movement of an animal that is not related to the intentional placement of a stimulus. The direction of undirected locomotion is considered random. Spontaneous activity can be affected by neurotoxicants and described with a variety of quantitative assays (e.g., embryo tail flexure, swimming in eleutheroembryos or fry) that have varying degrees of environmental relevance but have shown to be quite sensitive to the effects of neurotoxicants (Dasgupta et al., 2022; Fitzgerald, et al., 2021; Garcia et al., 2016; Klüver et al., 2015; K. Zhang et al., 2021).

Directed locomotion refers to movement in response to specific external stimuli and numerous behaviors have been used to detect neurological dysfunction induced by chemical exposure. A few of the more common behavioral endpoints are briefly described below. The touch response is a rapid escape, or startle, reflex evoked by a touch to the flank or head (Roberts et al., 2019). Visual behaviors include phototaxis (evoked by light/dark or color differences), optomotor response (visuomotor reflex characterized by the fish swimming in the same direction as a high-contrast visual stimulus), optokinetic response (visuomotor reflex characterized by the fish eye movements in the same direction as a high-contrast visual stimulus), and locomotion in response to a rapid light-dark transition (Fitzgerald, et al., 2021; LeFauve et al., 2021; Suzuki and Grillner, 2018). The ability of a fish to swim with or against a water current (negative and positive rheotaxis) is a commonly studied forced locomotive response (Coombs et al., 2020). Introducing fish into an unfamiliar open arena is often used to quantify anxious or exploratory behavior and is impacted by several neurotoxic chemicals (Lucon-Xiccato et al., 2022; Merola et al., 2021; Pompermaier et al., 2022).

The directed movement of fish due to avoidance or attractiveness to xenobiotics or natural chemical stimuli has been examined in detail. Chemosensory disruption can result in an overt preference for or avoidance of a xenobiotic. Avoidance of or preference for natural and xenobiotic chemical cues during or following xenobiotic exposure can also occur and be quantified by observing movements of an organism across chemical gradients. The locomotion of a fish in response to a xenobiotic is likely the result of interactions with the olfactory or gustatory receptors; however, avoidance can also be the result of irritation of mucous membranes. The ability of a xenobiotic to mask or counteract natural chemical signals used in migration, for example, may be the consequence of competition for receptor sites on sensory cell membranes. Alternatively, xenobiotics may be directly toxic to receptor cells (Hara et al., 1983). A wide variety of xenobiotics, including metals, petroleum constituents, detergents, and insecticides, have been observed to affect olfaction in fish across several biological scales from molecular biology to behavior (Calvo-Ochoa and Byrd-Jacobs, 2019; Tierney et al., 2010).

Of the interindividual responses, aggression, mating, territoriality, dominance, shoaling/schooling, and predator–prey interactions have been studied most extensively (Heath, 1995; Rand, 1985; Kalueff et al., 2013). Predator–prey interactions have been examined in studies in which either the predator or the prey is exposed to the xenobiotic, as well as in studies where both sets of organisms are exposed. Endpoints in these studies can include prey survival rates, predation rates, or prey handling times. Although effects on predator–prey interactions have been observed for a variety of insecticides, herbicides, and metals, the mechanistic basis for these effects is largely unknown. The need to understand fully the basis of xenobiotic effects on predator–prey interactions has been understood for decades (Sandheinrich and Atchison, 1990; Weis and Candelmo, 2012). In this regard, Carlson and coworkers (1998) reported an attempt to relate the effects of a wide variety of organic xenobiotics on prey survival in terms of electrophysiological responses of the Mauthner-cell-mediated startle response, which initiates an escape behavior in response to predation. Several studies have examined the effects of pollutants on predator–prey interactions in a laboratory setting (Pompermaier et al., 2020; Stoczynski and van den Hurk, 2020; Weis and Candelmo, 2012).

## EXAMPLES OF MAJOR CLASSES OF NEUROTOXICANTS

### Neuroactive Pharmaceuticals

Neuroactive pharmaceuticals and illicit drugs (illicit drugs are included here for completeness) are chemicals of emerging concern within the class of pharmaceuticals and personal care products (PPCPs). Toxic properties of PPCPs are those that occur at concentrations far in excess of the therapeutic window. Research reported during the early 1980s in both the United States and The Netherlands established that the majority of industrial organic chemicals (excluding pesticides and pharmaceutical agents) elicit acute toxic effects in fish through a narcosis mechanism (Konemann, 1981; Veith et al., 1983). Narcosis can be defined as a *reversible* state of stupor, drowsiness, or unconsciousness resulting from exposure to a xenobiotic. In the context of intact organism, the terms *narcosis* and *general anesthesia* were previously used interchangeably. The actual mechanism(s) of narcosis and anesthesia has been an active area of research for decades but remains unknown; however, there are a few strong theories that tend to coalesce around the disruption of transmembrane receptors (Jedlicka et al., 2021; Jiang-Xie et al., 2019; Pavel et al., 2020; Son, 2010). There are numerous pharmaceuticals and illicit drugs in wide use today that have been specifically designed to interact with transmembrane receptors or modulate their activity.

#### Mechanisms of Neuroactive Pharmaceutical Neurotoxicity

Both lipophilicity (Meyer–Overton rule) and thermodynamic activity (Ferguson's rule) (Ferguson, 1939; Jedlicka et al., 2021; Son, 2010) are related to the potency of neuroactive pharmaceuticals. Thus, varying aqueous concentrations of different compounds may be required to cause comparable effects, even though their thermodynamic activity at the site of action is proposed to be the same. There is, however, ample evidence of potency that is not consistent with thermodynamic theory (Duarte et al., 2022). These departures from thermodynamic consistency suggest that sometimes potency must be viewed as a more complex toxicological and physicochemical interaction between the xenobiotic and the site of action. As a result, investigators developing hypotheses to explain the molecular events of anesthesia/narcosis are increasingly acknowledging that their models must accommodate the likelihood of multiple sites and mechanisms of action.

Franks and Lieb (1990, 1994, and 2004) proposed that anesthesia is the result of direct interactions between xenobiotics and neuronal ion channels. This hypothesis evolved partly from x-ray diffraction experiments undertaken by these investigators showing insignificant changes in lipid membrane bilayers at physiologically relevant concentrations of narcotic xenobiotics (although see the study by Cantor (1997)). Subsequent studies with firefly luciferase established that inhibition of luciferase activity correlated with anesthetic potency and that the sites of action can have polar and nonpolar characteristics. These investigators also demonstrated that narcotics caused reversible inhibition of specific, spontaneously firing neurons in the giant snail. This inhibition was saturable and consistent with binding to a receptor site.

Over the last decade, these proteins have been identified as specific receptors in most cases, including $GABA_A$ receptor, opioid receptor, glycine receptor, glutamate receptor, nicotinic acetylcholine receptor, and TREK-1 (which is activated by disruption of the neuronal cell membrane perhaps lending some credence to the century-old lipid theory of Meyer and Overton) (Jedlicka et al., 2021; Jiang-Xie et al., 2019; Khan et al., 2014; Pavel et al., 2020; Son, 2010).

### Manifestations of Neuroactive Pharmaceutical Neurotoxicity in Fish

Our understanding of neurotoxicity has evolved to include numerous behavioral alterations and disruptions of neurobehavioral pathways with sublethal outcomes that are much more subtle than narcosis. The resulting behavioral changes could have profound impacts on wild fish population survival in consistently exposed environments.

Neuroactive pharmaceuticals have been found in aquatic systems for as long as these drugs have been manufactured. Quantitative assessments have identified neuroactive pharmaceuticals in and around sewage and livestock effluent for decades, and more recently in open waters as analytical techniques have advanced (Cerveny et al., 2021; Fent et al., 2006; Grabicová et al., 2020; Tanoue et al., 2020). These agents elicit biological responses at low levels of exposure due to high levels of therapeutic target evolutionary conservation between fishes and other vertebrates. There are hundreds of peer-reviewed publications investigating the effects of neuroactive pharmaceuticals in fish (summarized and analyzed in Duarte et al. (2022). The main findings are that (1) lipophilicity is not a good indicator of bioconcentration and neurotoxicity of neuroactive pharmaceuticals as a broad class and (2) principal component analysis reveals that more than 67% of bioconcentration factor variation can be explained by experimental temperature and exposure time (Duarte et al., 2022). This reveals a strong need for procedural standardization in xenobiotic exposure studies.

There are numerous published articles exploring the neurotoxic effects of a single pharmaceutical or illicit drug in a single species of fish. The drugs' mechanisms of action are generally conserved between fish and mammals. Based on the number of chemicals and the diversity of fish species, it is impossible to provide a succinct summary of the body of literature. However, some illustrative examples are helpful.

Benzodiazepines found in European surface waters have been evaluated singly and in mixtures for behavioral neurotoxicity in freshwater perch (*Perca fluviatilis*) (Cerveny et al., 2020). While none of the four test chemicals (oxazepam, bromazepam, temazepam, and clobazam) affected social behavior, bromazepam, temazepam, and clobazam increased boldness as indicated by a significantly reduced latency to leave a refuge and explore a novel and potentially dangerous environment ("danger zone"). A mixture of the four drugs increased boldness, increased time spent swimming in the danger zone, and increased swimming activity (Cerveny et al., 2020). In wild fish, these behavioral changes could reduce survival by increasing the chances of being preyed upon and needlessly increasing energy demand through excessive swimming.

A recent study (Henry et al., 2022) demonstrated the behavioral effects of low levels of anticonvulsants (gabapentin and carbamazepine), inflammatory drugs (ibuprofen), and antidepressants (fluoxetine and venlafaxine) commonly found in the environment on larval zebrafish (five days postfertilization). Behaviors measured were spontaneous swimming, two variations of the light/dark photomotor assay (Burgess and Granato, 2007; Jarema et al., 2022), one of which was a simulated predator assay, and an innovative thermotaxis assay. Gabapentin resulted in an increased distance traveled by larvae when presented with the simulated predator (in the dark) and in the more traditional light/dark assay during the light phases. Ibuprofen exposure resulted in an increased distance traveled during the dark phases of the light/dark assay and a reduced time spent in the higher temperatures of the thermotaxis chamber. The antidepressants tested are selective serotonin reuptake inhibitors (SSRIs) and the larvae showed an increase in mean-level behavioral responses to light with moderation of response to dark (Henry et al., 2022). This reduction in response magnitude and variation is similar to effects in humans and has also been documented in guppies (*Poecilia reticulata*) following a two-year multigenerational exposure (Polverino et al., 2021). The effects of SSRI antidepressants on interindividual behavior in the laboratory suggest that exposures in wild fish populations could interfere with their adaptive potential in the face of changing environments.

## Cholinesterase Inhibitors

Several neurotoxic compounds are known to target the cholinergic nervous system. By far the largest classes are the organophosphorus and carbamate compounds. Organophosphorus and carbamate compounds have been used extensively as insecticides, herbicides, and fungicides. Initial studies on the synthesis and activity of organophosphorus insecticides were undertaken in Germany during the 1930s. Agricultural use of insecticidal organophosphorus and carbamates increased in the 1960s as the use of organochlorine compounds declined. In the United States, the use of organophosphorus insecticides peaked in the mid-1970s and has declined with the increased use of pyrethroid and carbamate insecticides. Insecticidal carbamates were first introduced in 1954 and are still employed for agricultural and household applications (Malhotra et al., 2021).

Organophosphorus insecticides are organic phosphoric ester compounds with the general structure of $R_1P(=Y)(R_2)X$, where Y is an oxygen or thio group, X is a leaving group (e.g., halide, phenoxy, or other groups), and $R_1$ and $R_2$ are typically alkyl, alkoxy, aryl, or aryloxy substituents. Representative structures of organophosphorus insecticide classes are depicted in Figure 5.3. Carbamate insecticides are organic compounds that contain a RNC(=O)O moiety; examples of common *N*-methyl carbamate and *N*-methyl carbamoyl oximes are provided in Figure 5.4. While organophosphorus and carbamate insecticides are structurally diverse, their common neurotoxic mechanism is related to their ability to phosphorylate or carbamylate esterases.

### Mechanisms of Cholinesterase Inhibitor Neurotoxicity

The neurotoxic mechanism for organophosphorus and carbamate insecticides is the inhibition of acetylcholinesterase (AChE) in CNS and PNS (Figure 5.5) (reviewed in Ecobichon, 1996; Taylor, 2001). AChE is responsible for hydrolyzing the neurotransmitter acetylcholine; inhibition of esterase causes accumulation of acetylcholine in synapses and excessive stimulation of muscarinic and nicotinic receptors (Figure 5.5B). Increased levels of acetylcholine overstimulate nicotinic and muscarinic receptors in CNS, PNS, and neuromuscular junctions, causing a wide range of signs of poisoning. These signs may include changes in body temperature, heart rate, blood pressure, muscle twitching, or tremors. Death is usually due to the cessation of respiration due to anticholinesterase effects in both CNS and PNS. Although both organophosphates and carbamates target AChE, nicotine acts as a direct cholinergic agonist rather than by cholinesterase inhibition; nicotine directly binds nicotinic receptors to hyperstimulate cholinergic neurons.

Both organophosphosphorus and carbamate compounds inhibit AChE activity by acting as pseudosubstrates. Unlike the natural substrate, these compounds remain in the active site of the enzyme for much longer periods of time, thereby preventing the enzyme from hydrolyzing its natural substrate, acetylcholine (Aldridge and Reiner, 1969). Inhibition of AChE activity by organophosphorus insecticides is primarily a function of electrophilicity of the phosphorus atom, with increased electrophilicity (increased partial positive charge) associated with greater rates of bonding to the negatively charged oxygen in the serine hydroxyl group. Thus, for organophosphorus insecticides, the binding of the organophosphorus compound to AChE is reasonably fast. Hydrolysis of the organophosphorus esters, however, is very slow, leading to an accumulation of phosphorylated AChE. Phosphorylated AChE is, for all practical purposes, irreversibly inhibited due to the very low reversibility of the enzyme activity.

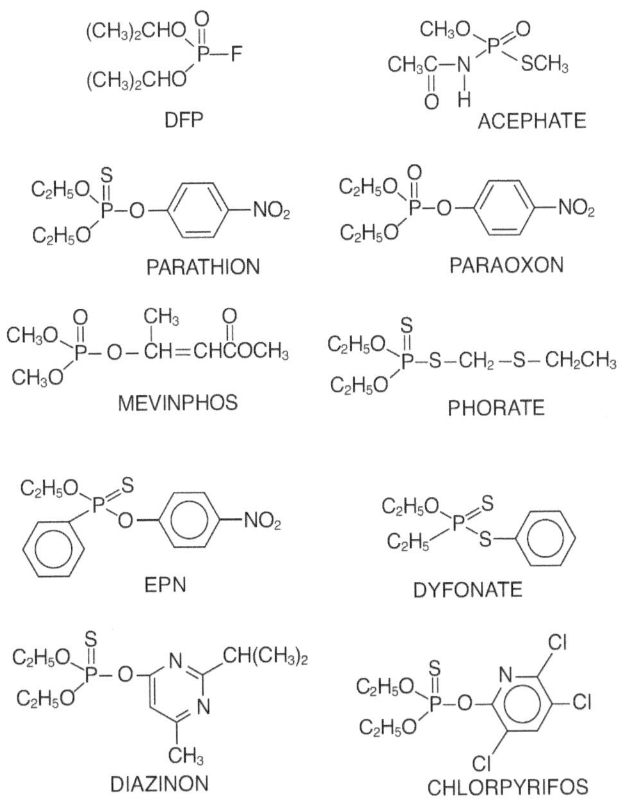


**FIGURE 5.3** Examples of organophosphorus insecticides.

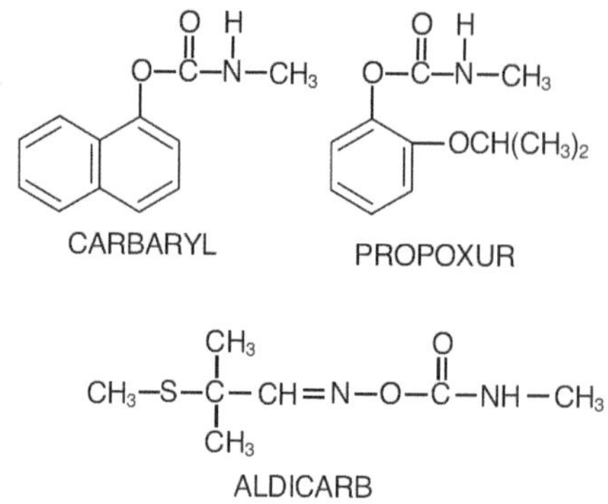


**FIGURE 5.4** Examples of *N*-methyl carbamate (carbaryl and propoxur) and *N*-methyl carbamoyl oxime (aldicarb) carbamate insecticides.

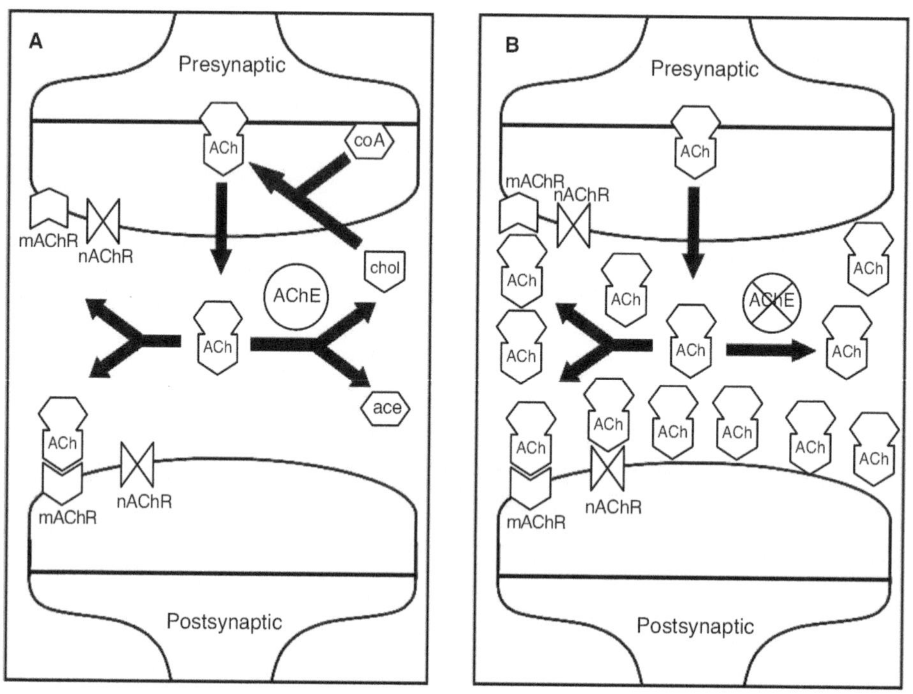


**FIGURE 5.5** Role of acetylcholinesterase at synaptic junctions. (A) Typical cholinergic synapse. Acetylcholine is secreted by the presynaptic neuron into the synapse, where it binds to presynaptic and postsynaptic muscarinic and nicotinic receptors. Acetylcholinesterase breaks down acetylcholine into cholinesterase and acetylcholine. Cholinesterase is transported to the presynaptic neuron, where acetyl coenzyme A converts it into acetylcholine. (B) Effect of acetylcholinesterase inhibitors on cholinergic synapses. Acetylcholine is secreted into the synapse but is not broken down by acetylcholinesterase, leading to an excess amount of acetylcholine in the synapse. ACh, acetylcholine; AChE, acetylcholinesterase; mAChR, muscarinic receptor; nAChR, nicotinic receptor; chol, cholinesterase; coA, acetyl coenzyme A; ace, acetate.

Recovery of active AChE activity is dependent on the synthesis of a new enzyme. In some cases, phosphorylated enzymes undergo an aging reaction when an alkoxy or aryloxy R group is dealkylated or dearylated and results in a negatively charged monoalkyl or aryl enzyme. In such instances, the hydrolysis step is no longer possible; thus, the enzyme is irreversibly altered. Kinetic studies on the inhibition of AChE activity, including aging and recovery, have been reported in a variety of fish species (Carr et al., 1995; Johnson and Wallace, 1987; Straus and Chambers, 1995; Wallace and Herzberg, 1988). Typically, recovery of AChE activity in fish is slower than that observed in mammals.

Inhibition of AChE activity by carbamate insecticides is a function of carbamylation of the enzyme. Kinetics of this inhibition are slightly different from that observed with organophosphorus insecticides. Hydrolysis of carbamates (i.e., decarbamylation), while significantly slower than that observed for acetylcholine, is much more rapid than that observed with most organophosphorus compounds; thus, although AChE inhibition elicited by an organophosphorus pesticide is, in effect, irreversible, carbamate inhibition of esterase activity may be reversed. Furthermore, aging of the carbamylated enzyme does not occur.

Interestingly, although anticholinesterase compounds all disrupt acetylcholine signaling, all anticholinesterase compounds do not produce identical responses to poisoning (reviewed in Pope, 1999; Moretto et al., 2011). Acetylcholine signaling disruption is the first step in the acute neurotoxic response but is not sufficient to explain all manifestations of toxicity. Several secondary, non-acetylcholinesterase targets have been recognized (Casida and Quisad, 2004; Pope, 1999) and are thought to play instrumental roles in the ultimate effects of cholinergic neurotoxicants. Furthermore, disruption of acetylcholine signaling alone may not fully explain the developmental neurotoxicity or other chronic effects of these compounds (Moretto et al., 2011). Careful consideration should be given to the non-cholinergic targets of these compounds when assessing toxicity, particularly in the developing embryo. As shown in Figure 5.5A, the biochemical mechanisms of acetylcholine production, secretion, and transmission are known in great detail. For the purposes of this discussion, only acetylcholine pathway components targeted by cholinergic agonists are addressed.

### Manifestations of Cholinesterase Inhibitor Neurotoxicity in Fish

Brief exposures to organophosphorus pesticides can produce long-lasting neurological effects because of the irreversibility of AChE inhibition; in fact, depression of AChE activity is considered a primary indicator of organophosphate pesticide exposure in fish. As a result, extensive research in fish neurotoxicity has sought to correlate acute toxicity in adult fish with the level of AChE inhibition (Heath, 1995; Murty and Ramani, 1992; Zinkl et al., 1991). Based on work in several different fish species and with several cholinergic poisons, it is generally accepted that 70%–80% AChE inhibition is lethal.

Acute exposures to high levels of organophosphorus and carbamate insecticides, in a variety of species including goldfish (*Carassius auratus*), fathead minnows (*Pimephales promelas*), medaka (*Oryzias latipes*), and rainbow trout, increased spontaneous locomotor activity with high incidences of convulsions, spasms, tetany, scoliosis, lordosis, and hemorrhage in the vertebral column, presumably due to damage resulting from spasms (Bradbury et al., 1991a, 1991b; Drummond and Russom, 1990; McKim et al., 1987; Rice et al., 1997; Saglio et al., 1996). Organophosphorus and carbamate insecticides have variable effects on cough rate. Chlorpyrifos (Bradbury et al., 1991a) and fenitrothion (Klaverkamp and Hobden, 1980) have been reported to increase cough response in rainbow trout, while malathion, carbaryl (McKim et al., 1987), and acephate (Klaverkamp and Hobden, 1980) did not. Increased cough rate, however, is not thought to be associated with AChE inhibition (Klaverkamp and Hobden, 1980). Spinally transected rainbow trout exposed to chlorpyrifos exhibited increased defecation and bile loss from the anal opening, consistent with the muscarinic effects of AChE inhibition (Bradbury et al., 1991a).

In spinally transected rainbow trout exposed to carbaryl, malathion (McKim et al., 1987), or chlorpyrifos (Bradbury et al., 1991a), decreased heart rate, decreased gill oxygen uptake efficiency, and increased ventilation volume were observed. Decreased heart rate has been attributed to the inhibition of the heart by the vagus nerve (cranial nerve X) through cholinergic synapses. Decreased oxygen uptake by the gills and a compensatory increase in ventilation volume have been proposed to be caused by continuous stimulation of neuromuscular junctions associated with sphincters at the base of the efferent filamental arteries to secondary lamellae of the gill. The resulting vasoconstriction is thought to reduce blood flow to the lamellae, effectively reducing respiratory surface area and oxygen uptake efficiency (McKim et al., 1987; Pavlov, 1994).

Despite the wealth of acute toxicity data, relatively little is known about the developmental neurotoxicity of cholinergic agonists in fish. Existing data demonstrate that exposure either *in ovo* or as juvenile fish has detrimental consequences on learning and motor neuron development. Chlorpyrifos exposure produced hypoactivity in zebrafish eleutheroembryo swimming behavior and touch-induced swimming bursts (Levin et al., 2004; D. Yang et al., 2011). In addition, developmental chlorpyrifos exposure in zebrafish embryos has long-term effects on learning. Adult zebrafish exposed to chlorpyrifos during development show reduced choice accuracy and spatial discrimination (Levin et al., 2003). Developmental exposure of zebrafish embryos to chlorpyrifos oxon metabolite also inhibits axonal growth of sensory neurons, primary motor neurons, and secondary motor neurons at concentrations that do not cause mortality, gross developmental defects, or aberrant somatic muscle differentiation (D. Yang et al., 2011).

Behavioral effects have been seen in juveniles of other fish species as well. At concentrations of carbaryl or chlorpyrifos up to ten times lower than 48-hour $LC_{50}$ values, larval medaka were more susceptible to predation, although a consistent dose–response relationship between carbaryl exposure and susceptibility to predation was not observed (Carlson et al., 1998). Developmental chlorpyrifos exposure was demonstrated to have long-term effects on forced swimming performance in medaka 20 and 40 days post exposure (Sastre et al., 2018). *In vivo* electrophysiological studies of sublethal chlorpyrifos and carbaryl effects on the Mauthner cell startle response in larval medaka (Carlson et al., 1998) demonstrated an effect on neuromuscular junctions, as evidenced by a dose-related increase in the ratio of startle response to stimuli. An increase in motor neuron to muscle delay with increased exposure concentration was also noted. Both responses are consistent with AChE inhibition in the neuromuscular junction.

Interestingly, evidence from zebrafish suggests that AChE inhibition may not be the sole mechanism of developmental neurotoxicity. Developmental exposure to nicotine causes morphological changes in zebrafish eleutheroembryos and impairs the swimming behavior and escape response (Svoboda et al., 2002; Mora-Zamorano et al., 2016). Using an *Islet1*-GFP transgenic zebrafish strain, nicotine was shown to delay the development of spinal neurons and cause disruptions in axonal pathfinding of secondary motor neurons by nicotinic receptor activation (Svoboda et al., 2002). The molecular mechanism of developmental neurotoxicity remains unidentified for the majority of cholinergic agonists.

## Pyrethroid Insecticides

Directed synthesis has produced insecticides derived from the natural pyrethrin esters of pyrethrum flowers (e.g., *Chrysanthemum cinerariifolium*) (Shafer et al., 2005). These synthetic, pyrethroid insecticides have greater stability in light and air than the natural pyrethrin esters yet maintain critical stereochemical characteristics required for alignment with target receptors (Soderlund et al., 2002). The synthetic pyrethroids are divided into two classes based on the presence or absence of a cyano group on the alpha carbon of the 3-phenoxybenzyl alcohol moiety (Figure 5.6). Type II pyrethroids all contain an α-cyano side group, while type I pyrethroids do not. The U.S. EPA has determined that ecological risk mitigation to reduce spray drift, reduce runoff, and promote soil conservation practices for the entire class of pyrethrins and pyrethroids is necessary to address potential risks for terrestrial/aquatic invertebrates and fish (USEPA, 2019). Because of their potent insecticidal activity, low mammalian and avian toxicity, and varying levels of environmental stability (Bradbury and Coats, 1989; Clark, 1995), synthetic pyrethroids represent nearly 30% of the U.S. dollar value of the world insecticide market (Bao et al., 2020).

### Mechanisms of Pyrethroid Neurotoxicity

In both mammals and insects, the primary mechanism of acute synthetic pyrethroid neurotoxicity is the disruption of voltage-sensitive sodium channels (VSSCs) (Clark, 1995; Narahashi, 1992; Shafer et al., 2005; Soderlund et al., 2002; Soderlund, 2012). Mammalian VSSCs consist of a single α subunit and two β subunits (Shafer et al., 2005). The α subunit forms the channel pore, and the β subunits modify channel properties and membrane location (Figure 5.7A). Pyrethroid insecticides bind the α subunit and disrupt sodium regulation (Figure 5.7B, C). Type I pyrethroid insecticides prolong VSSC opening, allowing more sodium to cross the membrane and leading to repetitive firing of action potentials. Conversely, type II pyrethroids delay VSSC inactivation, resulting in a depolarization-dependent block that prevents action potential generation (Shafer et al., 2005; Soderlund et al., 2002; Soderlund, 2012). Similar to rodents, type II pyrethroids are more toxic than type I pyrethroids to zebrafish embryos (Demicco et al., 2009). Although the primary mechanism of acute pyrethroid neurotoxicity is the disruption of VSSCs, evidence suggests that numerous secondary sites and mechanisms of action are possible (Soderlund et al., 2002; Soderlund, 2012); for example, deltamethrin and resmethrin are toxic to *Paramecium tetraurelia*, although this organism does not have VSSCs (Soderlund et al., 2002). In this case, the synthetic pyrethroids were acting through the calcium channels in the cilia membrane of the *P. tetraurelia* (Soderlund et al., 2002). In isolated presynaptic terminals, pairs of pyrethroids were found to be either antagonistic or additive in assays of calcium uptake but found to be additive, antagonistic, or synergistic in neurotransmitter release, suggesting a third mechanism of action that was later found to be associated with chloride channels (Soderlund, 2012). Synthetic pyrethroids have also been shown to affect ligand-operated channels such as the GABA receptor–ionophore complex, the nicotinic acetylcholine receptor, and the peripheral-type benzodiazepine receptor (Shafer et al., 2005; Soderlund et al., 2002; Soderlund, 2012). These effects may be limited to a subset of pyrethroids and sometimes associated with physiologically unrealistic mammalian pyrethroid exposures or nonspecific interactions. Both type I and type II pyrethroids cause repetitive firing in synapses, neuromuscular junctions, and the CNS. As a consequence, pyrethroid intoxication has been associated with releases of acetylcholine, GABA, dopamine, and norepinephrine. Type II compounds tend to elicit greater neurotransmitter releases than type I compounds because of the enhanced ability to depolarize sensory and presynaptic nerve endings (Clark, 1995).

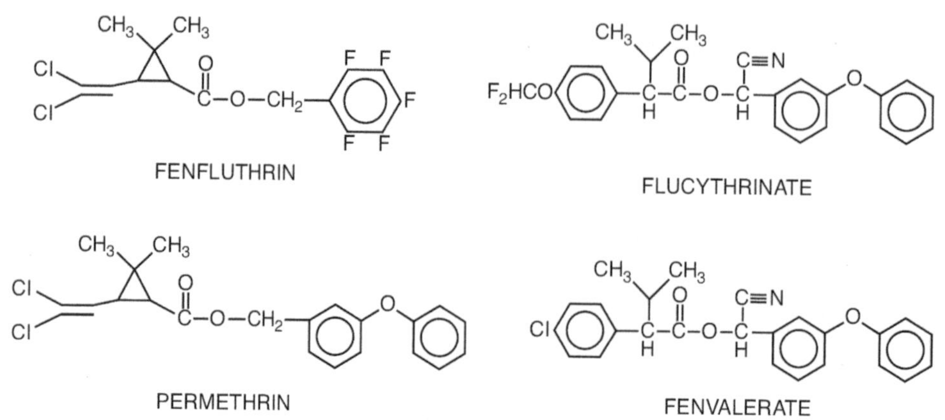


**FIGURE 5.6** Examples of type I (fenfluthrin and permethrin) and type II (fenvalerate and flucythrinate) pyrethroid insecticides.

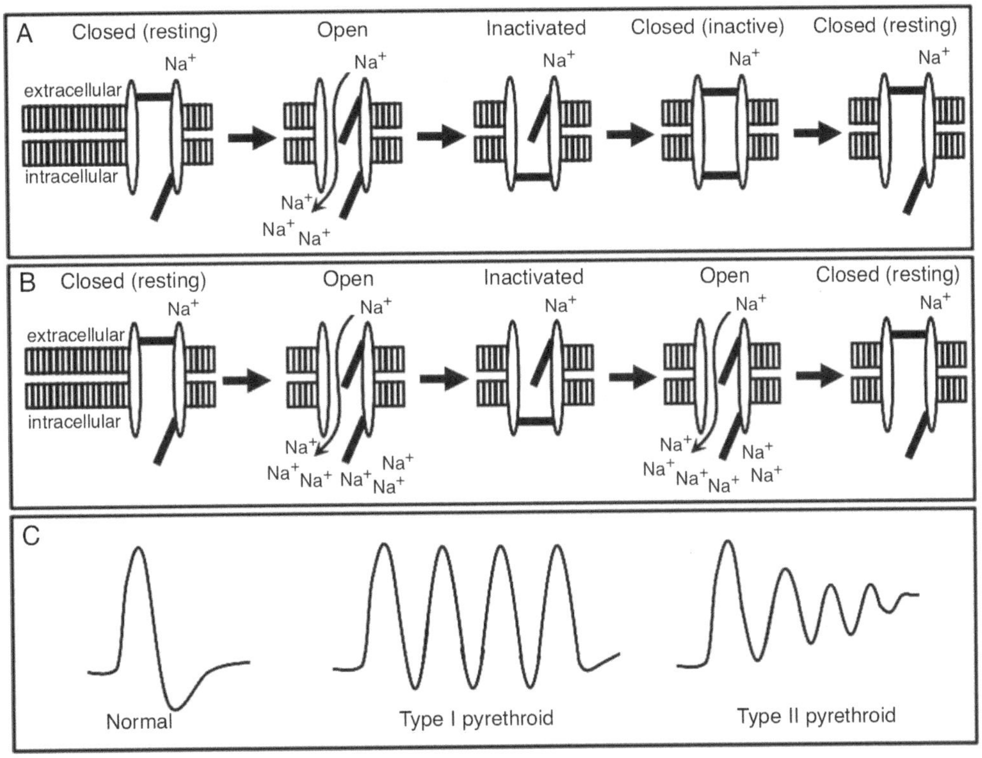


**FIGURE 5.7** Effects of pyrethroids on voltage-sensitive sodium channels. (A) Firing cycle of a voltage-sensitive sodium channel (VSSC). At resting potential, the channel is closed to the extracellular environment. In response to a stimulus, the channel opens to allow sodium ions ($Na^+$) to flow into the cell. When the stimulus has passed, the channel closes the intracellular and extracellular gates to prevent $Na^+$ influx, ultimately returning to the resting state. (B) Pyrethroids can affect the VSSCs in two ways. Type I pyrethroids prolong the opening of the sodium channel, leading to repetitive firing of action potentials. Type II pyrethroids prevent the closing of the channel to an inactive state. (C) Comparing VSSC action potentials of normal, type I pyrethroid exposures, and type II pyrethroid exposures. Whereas type I pyrethroids lead to repetitive action potential firings, type II pyrethroids show a gradual reduction in action potential amplitude, although never returning to the resting state.

In addition, perturbation of intraterminal calcium homeostasis, ATP-activated calcium sequestration, and responses of protein phosphorylation associated with calcium-dependent neurotransmitter release have also been suggested as playing a role in the pyrethroid mode of action. Some of these effects may further accentuate neurotransmitter release caused by repetitive firing (Clark, 1995).

Unfortunately, there is little consensus on which, if any, of the acute neurotoxicity mechanisms are applicable to pyrethroid developmental neurotoxicity (Shafer and Meyer, 2004). A subset of mammalian sodium channel isoforms have been identified with enhanced sensitivity to pyrethroids that may play an important role in developmental neurotoxic effects (Soderlund, 2012) Some pyrethroid insecticides show significant age differences in acute toxicity, with younger animals usually being more sensitive. Several studies also demonstrate persistent changes in motor activity, learning, and sexual activity following developmental pyrethroid exposure. None of these effects has been definitively associated with a putative neurotoxic mechanism.

## Manifestations of Pyrethroid Neurotoxicity in Fish

Although synthetic pyrethroids have minimal mammalian and avian toxicity, they are very toxic to fish (Bradbury and Coats, 1989). Because of their high lipophilicity, the synthetic pyrethroids are readily absorbed through the gills (Baser et al., 2003; Polat et al., 2002). Fish lack at least one enzyme that metabolizes pyrethroids, meaning that metabolic turnover is particularly slow (Baser et al., 2003; Tilak et al., 2003). As a result, fish are particularly susceptible to pyrethroids entering the aquatic ecosystem. An extensive amount of research has focused on determining acute toxic dosage in adult fish (Baser et al., 2003; Das and Mukherjee, 2003; David et al., 2004; Polat et al., 2002; Rebach, 1999; Saha and Kaviraj, 2003; Tandon et al., 2005; Tilak et al., 2003; Tripathi and Verma, 2004).

As reviewed by Bradbury and Coats (1989), acute pyrethroid intoxication in small aquarium fish typically causes loss of schooling behavior, followed by hyperactivity, erratic swimming, violent whole-body seizures, and loss of buoyancy. Consistent with the steep dose–response relationships typically observed for pyrethroids, Carlson and coworkers (1998) failed to note significant effects on medaka susceptibility to predation or electrophysiological responses associated with the Mauthner-cell-mediated startle response at sublethal levels. Spinally transected rainbow trout exhibited an elevated cough rate shortly after exposure (Bradbury et al., 1991a). Increased cough rates were typically associated with an elevated secretion of mucus. Further intoxication resulted in increased hyperexcitability followed by tremors that progressed to seizures anterior to the site of transection. During seizures, the opercula were flared and in a state of tetany. Prior to death, seizures subsided and fish became inactive. Similar responses have been noted in bluegill (*Lepomis macrochirus*) (Bradbury et al., 1987; Little et al., 1993). The stages of behavioral changes in fish are generally consistent with those observed in mammals (Bradbury and Coats, 1989); however, an insufficient number of compounds have been studied in fish to differentiate pyrethroid intoxication syndromes as has been done with mammals (Shafer et al., 2005; Soderlund et al., 2002). Because seizures are typically stimulus dependent in hypersensitized fish, it seems reasonable to assume that pyrethroid-induced coughs could trigger convulsions. The cough response itself could be a CNS-mediated component in the seizure syndrome, a side effect due to interactions with sensory receptors in the pharynx and gill arches, or direct irritation of gill tissue. Both fenvalerate and permethrin have been shown to cause gill damage consistent with irritation (Bradbury et al., 1987; Kumaraguru et al., 1982). As reviewed by Clark (1995), pyrethroids have also been reported to cause transient dermal tingling, itching, and burning in humans and irritation to the mucous lining of respiratory passages.

Disturbance of sodium, calcium, and chloride ion-gated channels by pyrethroids leads to neurotoxic effects and neurobehavioral changes. There are numerous publications that suggest these neurotoxic effects are also directly responsible for impacts on energy metabolism, homeostasis, teratogenesis, hematology, and other processes (Soderlund, 2012). In fish, the toxic effects are many and may include the production of reactive oxygen species, delayed development, cardiotoxicity, DNA damage, and endocrine disruption (Haverinen and Vornanen, 2016; Ullah et al., 2019). Changes in the activity of sodium, calcium, and chloride ion-gated channels will impact many cell and tissue types outside the nervous system and be completely independent of neurotoxicity. Understanding the significance of these secondary targets in chronic pyrethroid toxicity is urgently needed.

Unlike studies of acute toxicity, few studies have addressed the developmental neurotoxicity of synthetic pyrethroids in fish. Exposure to sublethal levels of permethrin *in ovo* caused a hatching delay in medaka. Medaka hatchlings demonstrated hyperactivity, uncoordinated movement, and an inability to respond to stimuli. These hatchlings also failed to inflate their swimming bladder and had spinal curvatures (González-Doncel et al., 2003). In zebrafish eleutheroembryos developmentally exposed to pyrethroids, a dose-dependent increase in mortality, craniofacial abnormalities and pericardial edema were observed (Demicco et al., 2009). At lower sublethal concentrations, zebrafish eleutheroembryos exhibited body axis curvature and spasms. Spasms were ameliorated with diazepam (an inhibitory neurotransmitter in CNS), while both axis curvature and spasms were ameliorated with MS-222 (a sodium channel antagonist) confirming the role of sodium channels in these two endpoints (Demicco et al., 2009). Unfortunately, other neurotoxic effects have yet to be correlated with a putative neurotoxicity mechanism, so it is unclear how pyrethroids may be inducing the spectrum of developmental neurotoxicity endpoints.

## Organochlorine Insecticides

Organochlorine insecticides are the largest category of insecticides and are used worldwide for public health (e.g., mosquito control) and agricultural production (Figure 5.8). Organochlorine insecticides are comprised of four distinct structural classes, which tend to be associated with unique mechanisms of action: (1) chlorinated ethane derivatives, (2) cyclodienes, (3) polychlorobornanes, and (4) lindane.

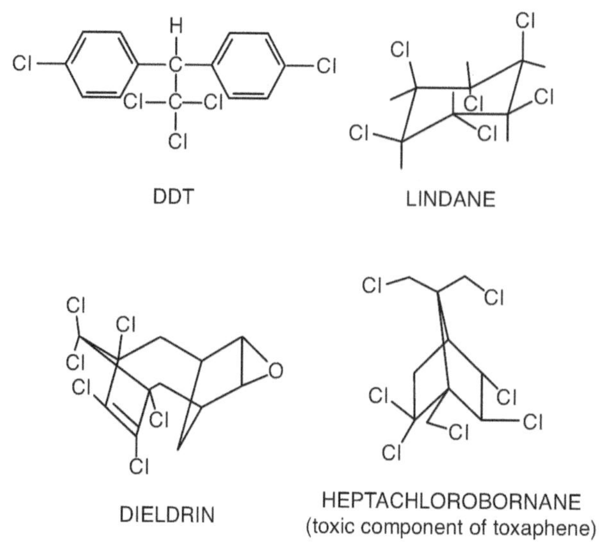


**FIGURE 5.8** Examples of organochlorine insecticides.

The use of organochlorines has declined dramatically because of insecticide resistance and environmental concerns. Due to their persistence in the environment and their continued use in some parts of the world, organochlorine residues in sediments, soils, and biota are still observed (Woolley, 1995). Due to their lipophilicity, organochlorines bioaccumulate in adipose tissue where they can remain for decades (Mrema et al., 2013).

### Mechanisms of Organochlorine Insecticide Neurotoxicity

As with synthetic pyrethroid insecticides, the neurotoxicity mechanism of chlorinated ethanes is the disruption of VSSCs. Studies with invertebrate and vertebrate preparations established that dichlorodiphenyltrichloroethane (DDT) prolongs the falling phase of the action potential, which typically produces repetitive firing. This repetitive neuron firing leads to hyperexcitability and tremor noted in intoxicated insects, mammals, birds, and fish. Studies by Narahashi (1994) demonstrated that DDT caused voltage-sensitive channels to remain in the open state longer than normal and close slowly resulting in an increased overall open time. As reviewed by Woolley (1995), it appears that type I pyrethroids and DDT likely affect the same target site on the sodium channel. Acute toxicity appears to be associated with blocking sodium channels and GABA receptors. Hyperexcitability is also associated with increased releases of neurotransmitters throughout the nervous system which likely reflect secondary responses but could represent additional neurotoxic mechanisms (Abreu-Villaça and Levin, 2017).

In mammals, cyclodienes and polychlorobornanes cause anorexia, salivation, vomiting, and convulsions. During convulsions, insect and amphibian preparations exhibit excessive uptake of calcium into synaptosomes. This synaptosomal calcium uptake is associated with an increased release of neurotransmitters. Subsequently, it was shown that cockroach strains resistant to cyclodiene insecticides, lindane, and toxaphene were also resistant to picrotoxin. Picrotoxin acts at one of the receptor sites in the GABA receptor–ionophore complex and is thought to block the channel. Normally, the GABA-gated chloride channel opens upon GABA binding, causing an increased flow of chloride, neuronal hypopolarization, and depression of excitability. Picrotoxin blockage of the GABA receptor eliminates normal chloride modulation of neuronal activity, causing hyperexcitability and, potentially, even convulsions. As these toxic effects are also induced by cyclodienes and polycholorbornanes, it was hypothesized that organochlorines may act by blocking GABA receptors. Confirmatory studies with insect and mammalian preparations established that several cyclodiene insecticides, toxaphene, and lindane act specifically at the picrotoxin site of the GABA receptor. In addition, binding at the picrotoxin site of the GABA receptor tends to correlate with toxic potency. Furthermore, these compounds inhibit GABA-stimulated flows of chloride across membrane vesicles (see reviews by Coats, 1990; Woolley, 1995).

Lindane is the gamma isomer of hexachlorocyclohexane; it is approximately 10 times more acutely toxic to the rat than the alpha isomer and roughly 100 times more potent than the beta and delta isomers. The gamma isomer is strongly excitatory in both insect and mammalian nervous systems, while the alpha, beta, and delta isomers are weakly active or have depressant effects. At acute doses, lindane increases neuronal activity and neurotransmitter release. In addition to convulsant effects, lindane also causes anorexia, diarrhea, and hyperthermia in the rat. As discussed above, lindane is thought to antagonize GABA receptors in CNS and the gastrointestinal tract. Consistent with this hypothesis, toxic effects of lindane can be attenuated or eliminated by pretreatment with GABA receptor agonists (Woolley, 1995).

Although disruption of VSSCs and GABA receptors is a well-characterized mechanism of acute organochlorine neurotoxicity, there is little evidence for the role of these mechanisms in developmental neurotoxicity. Lindane is known to inhibit mammalian embryonic development from the eight-cell stage up to the blastocyst stage (Scascitelli and Pacchierotti, 2003). Several potential mechanisms of developmental neurotoxicity have been proposed, including disruption of endocrine homeostasis and alterations in intercellular communication (Scascitelli and Pacchierotti, 2003; Abreu-Villaça and Levin, 2017).

### Manifestations of Organochlorine Insecticide Neurotoxicity in Fish

Organochlorine exposure affects behavior in many fish species. In unrestrained fish, cyclodiene exposure caused hyperactivity in response to stimuli, followed by recurrent tremors, rapid pectoral fin movement, and convulsions (Carlson et al., 1998). In spinally transected rainbow trout, endosulfan and endrin intoxication induced branchial tremors, increased cough rate, and increased pectoral fin movement, with eventual tetany and convulsions anterior to the site of transection (Bradbury et al., 1991a). Hyperactivity and increased cough frequency have also been reported with DDT intoxication in a number of fish species (Heath, 1995; Murty, 1986). DDT-induced hyperactivity is thought to contribute to decreased schooling behavior (Murty, 1986). At sublethal concentrations of endosulfan, medaka are less susceptible to predation, presumably due to hyperactivity; however, at endosulfan concentrations approximating the $LC_{20}$ level, medaka are more susceptible to predation than control fish (Carlson et al., 1998). Endosulfan exposure causes hyperactivity in zebrafish eleutheroembryos, which is ameliorated by co-exposure to vitamin E (alpha-tocopherol) (Dale et al., 2017) African catfish (*Clarias gariepinus*) exposed to endosulfan, dieldrin, or heptachlor for 96 hours showed behavioral changes including hyperactivity, jerky movement, abnormal swimming, disability of equilibrium, abnormalities in ventilatory function, mucus secretion as well as abnormalities in skin pigmentation (Islam et al., 2021).

Respiratory–cardiovascular responses of cyclodiene-intoxicated rainbow trout included increased cough rate and ventilation volume, with no change in oxygen uptake efficiency, resulting in an increase in oxygen consumption. In a consistent manner, arterial blood oxygen remained near control levels until near death, while arterial blood carbon dioxide and pH levels decreased only slightly. Overall, the seizure activity in the cyclodiene-exposed trout was not associated with a shift to anaerobic metabolism. These responses were similar to those elicited by uncouplers of oxidative phosphorylation and may suggest a secondary effect associated with the inhibition of ATPase activity (Bradbury et al., 1991a). Increased oxygen consumption has also been reported in a number of species following exposure to DDT and methoxychlor (Murty, 1986). *In vivo* electrophysiological studies of sublethal endosulfan exposures in medaka demonstrated increased motor neuron amplitude peaks and significantly increased stimulus-response ratio on the Mauthner cell startle response. The hyper-responsiveness of the Mauthner cell to stimuli is consistent with cyclodiene acting at the picrotoxin site in the GABA receptor–chloride complex. Furthermore, GABA is an important afferent inhibitory neurotransmitter to the Mauthner cell (Carlson et al., 1998).

## Ethanol

Ethanol is a well-known neurotoxicant in mammals. Developmental exposure to ethanol can result in neurodevelopmental, physical, psychological, and behavioral impairments known collectively as a fetal alcohol spectrum disorder (Riley et al., 2011). The developmental potency of alcohol consumption has led to intensive investigation into the mechanisms of ethanol neurotoxicity, with particular emphasis on developmental neurotoxicity.

### Mechanisms of Ethanol Neurotoxicity

Although many mechanisms have been postulated for the toxic effects of ethanol on the adult nervous system, ethanol is thought to produce neurotoxic effects mainly through interaction with the glutaminergic system, binding to the NMDA receptor and possibly interfering with the normal interaction of glycine with that receptor (Tsai and Coyle, 1998). Ethanol is a well-known developmental neurotoxicant in humans and laboratory animals. There is no single global mechanism that explains the effects of ethanol on fetal development and subsequent lifelong consequences (Chung et al., 2021).

Many hypotheses exist, including interaction with neurotrophins (Boschen and Klintsova, 2017), cell adhesion molecules (Licheri and Brigman, 2021), or specific receptors (Lotfullina and Khazipov, 2018); increased apoptosis and maladaptive plasticity (Granato and Dering, 2018); or increased oxidative stress (Brocardo et al., 2011). Several studies have delved into the mechanisms of the effects of ethanol on developing fish, especially zebrafish where impacts have been observed in many places during the embryo development from the stem cell gene expression at the blastula/gastrula stage, gastrulation cell movement, morphogenesis of CNS, and neuronal development (Alsakran and Kudoh, 2021). The earliest studies suggested that prechordal plate migration may be perturbed in ethanol-treated zebrafish embryos leading to craniofacial deformities (Blader and Strähle, 2000). Ethanol-induced apoptosis may also play a part in ethanol effects on zebrafish embryos through an evolutionarily conserved mechanism that requires the calcium-dependent activation of CaMKII (calcium/calmodulin-dependent protein kinase II) (Flentke et al., 2013). Numerous other mechanisms of ethanol developmental toxicity have been proposed and tested in zebrafish (as reviewed recently in Alsakran and Kudoh (2021)).

### Manifestations of Ethanol Neurotoxicity in Fish

Ethanol affects the function of the adult fish nervous system (Dlugos and Rabin, 2003; Gerlai et al., 2000), producing hyperactivity or hypoactivity (depending on dose), aggression, and changes in the startle response. Studies of ethanol-induced neurotoxicity in zebrafish have shown that ethanol is a teratogen that produces developmental delay (Reimers et al., 2004), craniofacial abnormalities (Bilotta et al., 2004; Blader and Strähle, 2000; Carvan et al., 2004), changes in eye development (Bilotta et al., 2002, 2004), cell death (Carvan et al., 2004; Loucks and Carvan, 2004), and behavioral changes (Carvan et al., 2004). Interestingly, the occurrence or severity of these effects varies with the particular strain of zebrafish exposed (Loucks and Carvan, 2004). Numerous behavioral deficits including learning and memory, anxiety, sociality, and shoaling have been described in zebrafish developmentally exposed to concentrations of ethanol low enough to not induce mortality or morphological abnormalities (Facciol and Gerlai, 2020).

## Metals

The neurotoxicological effects of metal exposure must be considered in the context of the chemical form of the metal, route and duration of exposure, and toxicokinetics. Effects observed in the field also require consideration of sediment and water quality characteristics associated with bioavailability (Dopp et al., 2004; Rüdel, 2003). Sources of neurotoxic metals in aquatic ecosystems include atmospheric deposition, point-source discharges, and non-point-source releases. Cadmium is primarily released through petroleum refining, fossil fuel combustion, and the use of cadmium in copper and nickel smelting (Wren et al., 1995). Mercury inputs are typically associated with atmospheric inputs from the combustion of fossil fuels and acid rain, point-source discharges from industry and gold mining operations, and non-point-source discharges from the agricultural use of organomercury seed treatments (Outridge et al., 2018). Organotin and copper compounds have been input directly into aquatic systems as biocides in the shipping, fishing, and aquaculture industries (Blunden and Chapman, 1986; Rüdel, 2003). Globally, lead inputs include wet and dry atmospheric deposition, sewage treatment effluents, industrial and mining discharges, and non-point releases from mine tailings and highways (Gidlow, 2004; Pain, 1995). Aluminum loading to aquatic ecosystems is primarily associated with acidification and resulting releases from rocks, soils, and sediments (Lukiw and McLachlan, 1995).

### Mechanisms of Metal Neurotoxicity

Generally, organismal-level effects elicited by metals are derived from multiple mechanisms. In addition to direct neurotoxic mechanisms, metals can damage respiratory surfaces and interfere with energy metabolism, iron homeostasis, osmoregulation, and endocrine function (Balali-Mood et al., 2021; Jaishankar et al., 2014; Schildroth et al., 2022). The extent to which metals are acting through neurotoxic mechanisms of action is a function of metal species, dose, and route of exposure. Comprehensive reviews of primary scientific literature on metal toxicology (Balali-Mood et al., 2021; Tchounwou et al., 2012) and neurotoxicology (Aschner and Costa, 2017) have been published, based primarily on mammalian model organisms. Below are brief overviews of neurotoxicity mechanisms for some selected metals whose neurotoxic mechanisms have been well studied.

A long environmental half-life and wide range of organ toxicity make cadmium one of the most toxic metals in the environment. Several mechanisms have been suggested for the acute neurotoxicity of cadmium (recently reviewed in Branca et al. [2018]). Cadmium interrupts calcium signaling in the cell by promoting calcium efflux from the endoplasmic reticulum and mitochondria (Branca et al., 2018). In the brain, cadmium inhibits magnesium and sodium/potassium ATPases and affects glutamate, acetylcholine, GABA, and dopamine neurotransmitter function. In addition, cadmium effects on phospholipid metabolism in synaptic membranes, axonal transport, and basal adenylate cyclase activity have been reported (Carmona et al., 2021). Cadmium-mediated disruption of cellular antioxidant defense mechanisms has also been proposed to increase neuronal damage through oxidative stress (Branca et al., 2018; Hastings, 1995).

Although all forms of mercury are neurotoxic to humans and animals, the most toxic forms are organic mercury compounds. Organic mercury compounds are more lipophilic than elemental mercury or inorganic mercury compounds and therefore bioaccumulate in animal tissues (Shafer, 2000). Of the organic mercury compounds, methylmercury has been particularly well studied. Accumulation of methylmercury in CNS and primary sensory tracts of mammals causes degeneration of visual cortices, dorsal root ganglia fibers, and the cerebellum. Numerous mechanisms have been proposed to explain the neurotoxicity of methylmercury in vertebrates, including disruption of the cell cycle and induction of apoptosis. Subcellularly, methylmercury is hypothesized to disrupt microtubule assembly and alter ion channel function. At the molecular level, cation homeostasis disruption and interference with synaptic transmission have been hypothesized to mediate methylmercury neurotoxicity. Additionally, methylmercury is highly reactive with sulfhydryl groups (Ajsuvakova et al., 2020). The binding of methylmercury to sulfhydryl groups in critical proteins or macromolecules has been hypothesized to disrupt protein, DNA, and RNA metabolism; perturb $Ca^{2+}$ homeostasis; induce oxidative damage; and alter protein phosphorylation. Many of these mechanisms are interrelated and likely interact to produce methylmercury neurotoxicity (Novo et al., 2021).

Lead is a potent neurotoxicant in humans and animals, associated with intellectual disability and behavioral disorders (Gundacker et al., 2021). A variety of mechanisms have been proposed for lead neurotoxicity, including vascular and neuroglia injury, oxidative stress, alterations in transcription, blockage of neurotransmitter release, and failure of neuronal migration (Gundacker et al., 2021; Ramírez Ortega et al., 2021). A common component of these neurotoxic mechanisms is the ability of lead to mimic biologically relevant metals and cations, particularly calcium (Rădulescu and Lundgren, 2019). Lead mimicry of calcium has two primary consequences leading to neurotoxicity. Initially, lead competes with calcium for binding sites on calcium-regulated proteins such as calmodulin, synaptotagmin, and cadherin, disrupting gene expression, proteomic synaptic machinery, and neuronal migration (Ramírez Ortega et al., 2021). Binding site competition displaces calcium ions, subsequently exposing cells to higher levels of free calcium and potentially leading to oxidative stress and impaired myelination. The second method by which lead mimicry of calcium induces neurotoxicity is the disruption of calcium transport. Disruption of chloride ion exchange causes hyperexcitation of neurons. Lead also blocks calcium ion flux across dopamine receptors, glutamate receptors, and voltage-sensitive calcium channels. This blockage affects neurotransmitter synthesis and release, ultimately leading to neuron hyperexcitability and damage (Carmona et al., 2021; Ramírez Ortega et al., 2021).

Most exposure to arsenic occurs via ingestion or inhalation of inorganic arsenic, which is enzymatically converted to mono- and di-methylated metabolites. These methylated arsenicals (specifically methylated AsIII) are substantially more cytotoxic and linked to both cancer and non-cancer health effects (Thomas, 2021). Chronic exposure to arsenic can result in the degradation of the axons of peripheral neurons (polyneuropathy) likely as a result of oxidative stress and mitochondrial dysfunction (Balali-Mood et al., 2021; Hu et al., 2020; Mochizuki, 2019). CNS effects can include psychosis, memory loss, and impaired intellectual function (Bjørklund et al., 2020; Rahaman et al., 2021). Arsenic exposure leads to overactivation of *N*-methyl-D-aspartate receptors in the hippocampus, which impairs synaptic plasticity leading to impacts on learning and memory (Liu and Wang, 2022). Much of our recent understanding of the effects of chronic arsenic exposure comes from multiple epidemiological studies in Bangladesh and other areas with high levels of arsenic in drinking water from wells (Ahmad et al., 2018).

Metals represent some of the most potent developmental neurotoxicants, particularly mercury and lead. Fetal methylmercury exposure in humans causes Minamata disease, characterized by cerebral palsy, intellectual disability, and seizures (Davidson et al., 2004). Autopsies of Minamata patients reveal cerebral and cortical brain lesions with peripheral neuropathy (Gochfeld, 2003). Furthermore, studies with model organisms suggest that low-level exposure to methylmercury may not manifest effects until much later in life (Davidson et al., 2004). The developing nervous system is also vulnerable to lead exposure. Children exposed to lead have poor coordination, behavioral problems, and reading disabilities (De Gennaro, 2002). Several aspects of the developing nervous system are sensitive to metal neurotoxicity. Increased absorption through an immature blood–brain barrier means that neurotoxicity may be induced at a much lower exposure level. Furthermore, sublethal metal exposure levels may alter key processes such as synapse formation, synapse refinement, and neurotransmitter release, causing effects that are manifested later in life (Marchetti, 2003).

## Manifestations of Metal Neurotoxicity in Fish

Numerous reviews have summarized the behavioral responses of fish to metal intoxication (most recently in Green and Planchart (2018), Hook et al.( 2014), Paschoalini and Bazzoli (2021)). In addition to direct neurotoxic mechanisms, alterations in avoidance or attraction responses, activity patterns, critical swimming speed, respiratory behavior, intraspecific social interactions, reproduction, feeding, and predator avoidance (Atchison et al., 1987) can be attributed to direct damage to respiratory surfaces and interference with energy metabolism, osmoregulation, and endocrine function (Heath, 1995; Weber and Spieler, 1994). Examples provided below attempt to link neurophysiological or behavioral responses to neuropathological or biochemical alterations.

### *Cadmium, Mercury, Lead, Arsenic*

Studies of chemoreception have quantified the extent to which metals attract or repel fish and the extent to which metals affect responses to endogenous chemical signals such as pheromones. Alterations in avoidance or attraction responses have been observed in response to a number of metals, including cadmium, copper, and mercury. In coho salmon (*Oncorhynchus kisutch*), cadmium induces apoptosis in the olfactory epithelium, decreases the expression of important olfactory and vomeronasal genes, and impairs alarm response behavior (Williams and Gallagher, 2013). In rainbow trout, cadmium-induced olfactory impairment affects social behavior and the formation of dominance hierarchies (Sloman et al., 2003). Cadmium is toxic to the lateral line neuromasts of sea bass (*Dicentrarchus labrax*) which inhibits response to lateral line stimulation (Faucher et al., 2006, 2008). Cadmium also affects auditory hair cell morphology and hearing sensitivity in fathead minnow (Low and Higgs, 2015). In fish, cadmium exposure has been correlated to changes in brain AChE activity, although these neurochemical changes have not been correlated with changes in swimming behavior in larval rainbow trout (Beauvais et al., 2001). Developmental cadmium exposure causes changes in the startle response of zebrafish to changes in light (LeFauve and Connaughton, 2017).

Although many metals elicit an avoidance response, mercuric chloride and methylmercury attract fish (Atchison et al., 1987; Heath, 1995). Exposure of mercuric chloride and methylmercury to the olfactory bulb and receptors of rainbow trout and Atlantic salmon (*Salmo salar*) depressed electrical responses (Baatrup et al., 1990; Hara et al., 1976; Sutterlin and Sutterlin, 1970). Methylmercury has also been found to preferentially accumulate in olfactory receptors and the olfactory nerve of Atlantic salmon following dietary exposure (Baatrup et al., 1990; Berntssen et al., 2003). Chronic dietary exposure reduced overall activity in the Atlantic salmon and caused preferential histopathological damage to the brain stem (Berntssen et al., 2003). Developmental mercury exposure in yellow perch (*Perca flavescens*) affects swimming behavior and foraging efficiency (Mora-Zamorano et al., 2017). In zebrafish, developmental mercury exposure causes changes in a startle response, locomotion, anxiety, and learning and memory with many of these phenotypes continuing into adulthood (Glazer and Brennan, 2021; Mora-Zamorano et al., 2016; Smith et al., 2010; Weber, 2006; Xu et al., 2012), with some effects persisting for three generations due to transgenerational epigenetic inheritance (Carvan et al., 2017).

Lead exposure caused a general increase in locomotor activity in mirror carp (*Cyprinus carpio*) (Rehman, 2003). The effects of lead on zebrafish were reviewed by Lee and Freeman (J. Lee and Freeman, 2014) in the context of early life exposure influencing adult neurodegenerative disease. Developmental exposure of zebrafish to lead also caused hyperactivity in eleutheroembryos and learning/memory deficits in adulthood (Chen et al., 2012). Developmentally exposed zebrafish also demonstrate a reduction in startle response that persists into adulthood, along with deficits in social behavior and aggression (Rice et al., 2011; Weber and Ghorai, 2013). In zebrafish and fathead minnows, lead exposure reduced feeding ability, as evidenced by feeding miscues and increased prey-handling times (Nyman, 1981; Weber et al., 1991). This reduction in feeding ability was attributed to psychomotor coordination, based on correlations between increased brain lead levels, increased serotonin and norepinephrine concentrations, and decreased feeding ability (Weber et al., 1991). Lead exposure induced similar increases in brain serotonin and decreases in GABA in walking catfish (*Clarias batrachus*) (Katti and Sathyanesan, 1986). Yet, although lead increased brain serotonin and norepinephrine levels, lead exposure did not increase dopamine levels in fathead minnows (Weber et al., 1991).

Arsenic exposure has been demonstrated to cause oxidative stress in several fish species. Studies investigating specific neurotoxicity or behavioral effects are more limited. Arsenic exposure affects swimming behavior in *Cnesterodon decemmaculatus* (González Núñez et al., 2022) and causes damage to the optic tectum in exposed zebrafish (Mondal et al., 2021). Developmental arsenic exposure in zebrafish leads to a reduction in spontaneous movement in larvae, changes in color preference, impairment in learning, and a decrease in exploratory behavior throughout the lifespan (Abu Bakar et al., 2022). Arsenic-induced changes in motor activity and anxiety-like behaviors were also found to persist into adulthood and continue for three generations (Valles et al., 2020).

The effects of a number of metals on prey susceptibility, schooling behavior, and aggression have been studied across a wide array of species (Heath, 1995; Weber and Spieler, 1994). Although many of these adverse effects are hypothesized to reflect alterations in sensory receptors or the structure and function of PNS and CNS, direct cause-and-effect relationships have not been established. Even less evidence is available for the developmental neurotoxicity of metals in fish.

## Neurotoxins

Through evolutionary pressure, plants and algae have developed a number of strategies to avoid consumption and maximize competitive advantages in accumulating resources. There are thousands of terrestrial plant species that are toxic to animals, with perhaps one-half of these species producing compounds targeting the nervous system. Several aquatic organisms produce neurotoxins that affect fish and are of increasing concern in many areas (Kazmi et al., 2022). The following discussion provides an overview of the neurotoxicity of several specific neurotoxins and illustrates how these compounds are used to assess fish brain function.

### Mechanisms of Neurotoxin Toxicity

Neurotoxins derived from aquatic organisms produce toxicity by a variety of mechanisms. Toxins derived from cyanobacteria, dinoflagellates, and coral act on membrane sodium channels directly or on membranes to create ion permeability. Toxins that activate sodium channels cause long-lasting sodium ion influx and correspondingly prolonged depolarization of excitable membranes, preventing complete inactivation. Failure of membrane repolarization causes periods of rapid, spontaneous impulse firing (Mackieh et al., 2021; Vilariño et al., 2018). Five neurotoxin binding sites have been identified on the functional α-subunit of voltage-gated ion channels (designated sites 1 through 5) and toxins are grouped according to their binding site affinities. Toxins that bind site 1 block the conductance of sodium ions across the channel, tetrodotoxins, saxitoxins, and μ-conotoxins. The subsequent inhibition of sodium conductance renders excitable membranes inactive and halts impulse propagation. Tetrodotoxin is produced by numerous species of marine microorganisms that have been isolated from marine and freshwater sediments or found to be associative or symbiotic with marine animals including some fish in the order Tetraodontiformes (Magarlamov et al., 2017; Pratheepa and Vasconcelos, 2013). Saxitoxins are responsible for paralytic shellfish poisonings and are primarily produced by dinoflagellates in the marine environment and by cyanobacteria in freshwater environments (Cusick and Sayler, 2013). μ-Conotoxins are produced by predatory *Conus* spp. snails and primarily target sodium channel subtypes found in skeletal muscle and the brain (Prashanth et al., 2017). Brevetoxins, derived from the marine dinoflagellates *Karenia brevis* and related species, and ciguatoxins, derived from *Gambierdiscus* spp. and *Fukuyoa* spp., are the most studied sodium channel activator toxins. The most commonly cited models suggest that brevetoxins and ciguatoxins both affect voltage-dependent sodium channels by binding to site 5 and lowering the membrane depolarization action potential. The resultant sodium influx causes plasma membrane depolarization and triggers spontaneous action potentials, and the presence of toxins slows inactivation (Stevens et al., 2011; Vilariño et al., 2018).

Some toxins create ion permeabilities without affecting ion channels. Palytoxin, derived from coral, irreversibly increases monovalent cation permeability, by converting the sodium/potassium ATPase pump to a passive channel (Fernández-Sánchez et al., 2021). Other aquatic neurotoxins have mechanisms that do not center on ion regulation; for example, cyanobacteria from numerous planktonic and benthic genera including *Anabaena* produce anatoxins that target nicotinic receptors at neuromuscular junctions. Anatoxin-a acts as an acetylcholine mimic, binding nicotinic acetylcholine receptors at neuromuscular junctions, and as an AChE inhibitor causing overstimulation of muscle cells in a manner similar to organophosphorus pesticides discussed previously (Colas et al., 2021; Jakubowska and Szeląg-Wasielewska, 2015).

Microcystins are a family of toxins prevalent in harmful algal blooms in freshwater systems and produced by cyanobacteria including *Microcystis*, *Anabaena*, *Oscillatoria*, and *Nostoc*. There are multiple mechanisms through which microcystins exert their neurotoxic effects (Hu et al., 2016). Microcystins have been shown to inhibit serine/threonine phosphatases which can increase acetylcholine synthesis and modulate the activity of ATP-sensitive potassium channels. Microcystin exposure also disrupts neuronal cytoskeleton in the brain, causes oxidative stress, increases calcineurin activity, and triggers neurons to undergo apoptosis (Hu et al., 2016).

### Manifestations of Neurotoxin Toxicity in Fish

In mammals, including humans, tetrodotoxins and saxitoxins block vasomotor nerve function, ultimately causing death by paralysis of respiratory muscles (Ritchie and Greene, 1985). Tetrodotoxin has been administered to numerous fish species to characterize neurophysiology; however, direct effects in fish have been rarely reported. It is known that sensitivity or resistance to tetrodotoxin is genetically based with zebrafish sodium channels ($Na_v1.4$) being sensitive and those of the puffer fish *Takifugu rubripes* and *Tetraodon nigroviridis* being resistant (Venkatesh et al., 2005). This resistance allows puffer fish to feed on organisms with tetrodotoxin in their tissues and for the toxin to accumulate in their own tissues. Accumulations of saxitoxins, the toxins responsible for paralytic shellfish poisoning, have been documented primarily because they are monitored within the context of human health protection. Toxic endpoints of saxitoxins in fish, however, are rarely assessed. In the laboratory, saxitoxin is highly toxic to cultured neurons from *Hoplias malabaricus* (Silva et al., 2014) and affects neurodevelopment in medaka and zebrafish eleutheroembryos (Lefebvre et al., 2004; Tian et al., 2014). Saxitoxin-induced paralysis is developmental stage dependent in zebrafish with effects lasting long after exposure (Lefebvre et al., 2004). Some predatory *Conus* spp. snails hunt fish with the assistance of conotoxins. The neurotoxic effects on fish from natural predator–prey interactions are rarely investigated. However, conotoxin-induced alteration of zebrafish behavior has been used to discover bioactive conotoxin peptides in *Conus* venom leading to subsequent drug development (Bosse et al., 2021; Dutt et al., 2019). Brevetoxins and ciguatoxins elicit similar effects in fish, consistent with their identical mechanisms of neurotoxicity summarized previously. Red tides caused by *Karenia brevis* are associated with massive fish kills due to brevetoxin exposure. In laboratory studies of zebrafish and mahi-mahi (*Coryphaena hippurus*) embryos, brevetoxin caused oxidative stress, alteration of neurotransmitter pathways, and altered energy metabolism. Interestingly, zebrafish were also more than twice as sensitive to brevetoxin as mahi-mahi (Annunziato et al., 2022). Exposure of ciguatera toxins to coney (*Epinephelus fulvus*), schoolmaster (*Lutjanus apodus*), mahogany snapper (*Lutjanus mahogoni*), largemouth bass (*Micropterus salmoides*), blueheads (*Thalassoma bifasciatum*) (Davin et al., 1986, 1988), and western mosquitofish (*Gambusia affinis*) (Lewis, 1992) caused skin color variations, rapid opercular movement, inactivity, loss of equilibrium, erratic swimming, jerky feeding movements, loss of orientation, and death. Mullet (*Mugil cephalus*) fed *Gambierdiscus polynesiensis* cells, which produce saxitoxin, demonstrated a loss of equilibrium, bursts of erratic swimming, convulsions, and twitching (Ledreux et al., 2014).

While palytoxins have been found in the tissues of fish and crabs (Fernández-Sánchez et al., 2021; Patocka et al., 2018), evidence of toxic effects in fish appears to be lacking in the scientific literature. In laboratory studies, anatoxin-a and anatoxin-a(s) elicit acetylcholine and AChE inhibition responses in electric rays (*Torpedo californica*) and in electric eels (*Electrophorus electricus*) (Hyde and Carmichael, 1991; Mahmood and Carmichael, 1987; Swanson et al., 1991). Medaka given high doses of anatoxin-a by gavage showed abnormal swimming, diminution of opercular movement, muscular rigidity, and paresis (Colas et al., 2020, 2021). Anatoxins appear to have little to no effect on zebrafish or carp development; however, cellular extracts were highly toxic to carp embryos (Oberemm et al., 1999; Osswald et al., 2009). Although anatoxin poisonings have been reported in livestock, domestic animals, and wild birds, anatoxin has little effect on fish in a natural setting (Carmichael, 1997; Colas et al., 2020, 2021).

Microcystins are often the predominant cyanotoxins associated with cyanobacterial blooms in natural surface freshwaters. These toxins are well-known (hepato)toxic agents in fish through inhibition of protein phosphatase (resulting in the collapse of the cytoskeleton and loss of cellular integrity) with indications of oxidative stress and disruption of ion regulation (Gélinas et al., 2012). Microcystin exposure disrupted brain AChE in juvenile rainbow trout and pearl cichlid (*Geophagus brasiliensis*) (Calado et al., 2019; Gélinas et al., 2012). Direct effects on brain ultrastructure and dysregulation of nerve myelination gene expression and neurotransmitter conduction protein expression pathways have been shown in adult zebrafish (M. Wang et al., 2010; L. Yang et al., 2022; Yu et al., 2021) along with an abnormality in swimming and social behaviors (Baganz et al., 2004; Wu et al., 2019).

## SUMMARY

The mechanisms and effects of neurotoxicants in fish have been studied with a wide variety of *in vitro* and *in vivo* methods. In addition to advancing the understanding of fundamental neurobiology, this research helps develop techniques to assess the adverse effects of neurotoxicants for ecological risk assessments. Of special note are investigations that link structural and functional alterations of the nervous system at the molecular or cellular level to physiological and behavioral effects. Research integrating information across levels of a biological organization will further the mechanistic understanding of neurotoxicants, allowing for extrapolation across species and chemicals and development of approaches to identifying ecologically relevant neurotoxic responses. One of the greatest challenges in developing methods for neurotoxicology is associating neuromorphological, neurochemical, and neurophysiological alterations with behavioral changes. The *in vitro* and *in vivo* fish systems described here are uniquely suited for characterizing chemical-induced neurotoxicity and understanding the biochemical, physiological, and morphological mechanisms underlying behavioral alterations.

## REFERENCES

Aberle, H. (2019). Axon guidance and collective cell migration by substrate-derived attractants. *Front. Mol. Neurosci.*, 12. https://doi.org/10.3389/fnmol.2019.00148

Abreu-Villaça, Y., and Levin, E. D. (2017). Developmental neurotoxicity of succeeding generations of insecticides. *Environ. Int.*, *99*, 55–77. https://doi.org/10.1016/j.envint.2016.11.019

Abu Bakar, N., Wan Ibrahim, W. N., Che Abdullah, C. A., Ramlan, N. F., Shaari, K., Shohaimi, S., Mediani, A., Nasruddin, N. S., Kim, C.-H., and Mohd Faudzi, S. M. (2022). Embryonic

Arsenic exposure triggers long-term behavioral impairment with metabolite alterations in Zebrafish. *Toxics*, *10*(9), 493. https://doi.org/10.3390/toxics10090493

Ahmad, S. A., Khan, M. H., and Haque, M. (2018). Arsenic contamination in groundwater in Bangladesh: implications and challenges for healthcare policy. *Risk Manage. Health. Policy*, *11*, 251–261. https://doi.org/10.2147/RMHP.S153188

Ajsuvakova, O. P., Tinkov, A. A., Aschner, M., Rocha, J. B. T., Michalke, B., Skalnaya, M. G., Skalny, A. V., Butnariu, M., Dadar, M., Sarac, I., Aaseth, J., and Bjørklund, G. (2020). Sulfhydryl groups as targets of mercury toxicity. *Coord. Chem. Rev.*, *417*, 213343. https://doi.org/10.1016/J.CCR.2020.213343

Aldridge, W. N., and Reiner, E. (1969). Acetylcholinesterase: two types of inhibition by an organophosphorus compound: one the formation of phosphorylated enzyme and the other analogous to inhibition by substrate. *Biochem. J.*, *115*, 147–162.

Alsakran, A., and Kudoh, T. (2021). Zebrafish as a model for fetal alcohol spectrum disorders. *Front. Pharmacol.*, 12. https://doi.org/10.3389/fphar.2021.721924

Ankley, G. T., Bennett, R. S., Erickson, R. J., Hoff, D. J., Hornung, M. W., Johnson, R. D., Mount, D. R., Nichols, J. W., Russom, C. L., Schmieder, P. K., Serrrano, J. a, Tietge, J. E., and Villeneuve, D. L. (2010). Adverse outcome pathways: a conceptual framework to support ecotoxicology research and risk assessment. *Environ. Toxicol. Chem./SETAC*, *29*(3), 730–741. https://doi.org/10.1002/etc.34

Annunziato, M., Eeza, M. N. H., Bashirova, N., Lawson, A., Matysik, J., Benetti, D., Grosell, M., Stieglitz, J. D., Alia, A., and Berry, J. P. (2022). An integrated systems-level model of the toxicity of brevetoxin based on high-resolution magic-angle spinning nuclear magnetic resonance (HRMAS NMR) metabolic profiling of zebrafish embryos. *Sci. Total Environ.*, *803*, 149858. https://doi.org/10.1016/j.scitotenv.2021.149858

Anthony, D. C., Montine, T. J., and Graham, D. G. (1996). Toxic responses of the nervous system. In *Casarett & Doull's Toxicology: The Basic Science of Poisons*, Klaassen, C. D., Amdur, M. O., and Doull, J., Eds., McGraw-Hill, New York, pp. 463–486.

Aschner, M., and Costa, L. G. (Eds.). (2017). *Neurotoxicity of Metals* (Vol. 18). Springer International Publishing. https://doi.org/10.1007/978-3-319-60189-2

Atchison, G. J., Henry, M. G., and Sandheinrich, M. B. (1987). Effects of metals on fish behavior: a review. *Environ. Biol. Fishes*, *18*, 11–25.

Baatrup, E., Doving, K. B., and Winberg, S. (1990). Differential effects of mercurial compounds on the electroolfactogram (EOG) of salmon (*Salmo salar L.*). *Ecotoxicol. Environ. Saf.*, *20*, 269–276.

Baev, K. V., Rusin, K. I., and Safronov, B. V. (1992). Primary receptor for inhibitory transmitters in lamprey spinal cord neurons. *Neuroscience*, *46*, 931–941.

Baganz, D., Staaks, G., Pflugmacher, S., and Steinberg, C. E. W. (2004). Comparative study of microcystin-LR-induced behavioral changes of two fish species, Danio rerio and Leucaspius delineatus. *Environ. Toxicol.*, *19*(6), 564–570. https://doi.org/10.1002/tox.20063

Bahr, T. G. (1972). Recording electrical activity from the lateral-line nerve in trout. *Prog. Fish-Cult.*, *34*, 59–61.

Bahr, T. G. (1973). Electrophysiological responses of trout to dissolved oxygen and cyanide. In *Bioassay Techniques and Environmental Chemistry*, Glass, G. E., Ed., Ann Arbor Scientific Publishers, Ann Arbor, MI, pp. 231–255.

Baker, T., and Lowndes, H. E. (1986). Electrophysiological correlates of sensorimotor system neurotoxicology. *Annu. Rev. Pharmacol. Toxicol.*, *26*, 517–545.

Balali-Mood, M., Naseri, K., Tahergorabi, Z., Khazdair, M. R., and Sadeghi, M. (2021). Toxic mechanisms of five heavy metals: mercury, lead, chromium, cadmium, and arsenic. *Front. Pharmacol.*, 12. https://doi.org/10.3389/fphar.2021.643972

Bao, W., Liu, B., Simonsen, D. W., and Lehmler, H.-J. (2020). Association between exposure to pyrethroid insecticides and risk of all-cause and cause-specific mortality in the general US adult population. *JAMA Intern. Med.*, *180*(3), 367. https://doi.org/10.1001/jamainternmed.2019.6019

Baser, S., Erkoç, F., Selvi, M., and Koçak, O. (2003). Investigation of acute toxicity of permethrin on guppies *Poecilia reticulata*. *Chemosphere*, *51*, 469–474.

Beauvais, S. L., Jones, S. B., Parris, J. T., Brewer, S. K., and Little, E. E. (2001). Cholinergic and behavioral neurotoxicity of carbaryl and cadmium to larval rainbow trout (*Oncorhynchus mykiss*). *Ecotoxicol. Environ. Saf.*, *49*, 84–90.

Becker, C. M., Seitanidou, T., and Triller, A. (1991). Conservation of antigenic epitopes of the inhibitory glycine receptor in rodent and goldfish CNS. *Brain Res. Mol. Brain Res.*, *11*, 327–333.

Bernhardt, R. R. (1999). Cellular and molecular bases of axonal pathfinding during embryogenesis of the fish central nervous system. *J. Neurobiol.*, *38*, 137–160.

Bernstein, J. J. (1970). *Anatomy and Physiology of the Central Nervous System*. Academic Press, New York.

Berntssen, M. H. G., Aatland, A., and Handy, R. D. (2003). Chronic dietary mercury exposure causes oxidative stress, brain lesions, and altered behavior in Atlantic salmon (*Salmo salar*) parr. *Aquat. Toxicol.*, *65*, 55–72.

Bilotta, J., Barnett, J. A., Hancock, L., and Saszik, S. (2004). Ethanol exposure alters zebrafish development: a novel model of fetal alcohol syndrome. *Neurotoxicol. Teratol.*, *26*, 737–743.

Bilotta, J., Saszik, S., Givin, C. M., Hardesty, H. R., and Sutherland, S. E. (2002). Effects of embryonic exposure to ethanol on zebrafish visual function. *Neurotoxicol. Teratol.*, *24*, 759–766.

Bjenning, C., and Holmgren, S. (1988). Neuropeptides in the fish gut: an immunohistochemical study of evolutionary patterns. *Histochemistry*, *88*, 155–163.

Bjørklund, G., Oliinyk, P., Lysiuk, R., Rahaman, M. S., Antonyak, H., Lozynska, I., Lenchyk, L., and Peana, M. (2020). Arsenic intoxication: general aspects and chelating agents. *Archiv. Toxicol.*, *94*(6), 1879–1897. https://doi.org/10.1007/s00204-020-02739-w

Blader, P., and Strähle, U. (2000). Zebrafish developmental genetics and central nervous system development. *Hum. Mol. Genet.*, *9*, 945–951.

Blunden, S. J., and Chapman, A. (1986). Organotin compounds in the environment. In *Organometallic Compounds in the Environment*, Craig, P. J., Ed., John Wiley & Sons, New York, pp. 111–127.

Bond, C. E. (1996). *Biology of Fishes*. Saunders College Publishing, Orlando, FL, pp. 241–258.

Borgonovo, J., Ahumada-Galleguillos, P., Oñate-Ponce, A., Allende-Castro, C., Henny, P., and Concha, M. L. (2021). Organization of the catecholaminergic system in the short-lived fish nothobranchius furzeri. *Front. Neuroanat.*, *15*. https://doi.org/10.3389/fnana.2021.728720

Boschen, K. E., & Klintsova, A. Y. (2017). Neurotrophins in the Brain: Interaction With Alcohol Exposure During Development. *Vitam. Horm.*, *104*, 197–242. https://doi.org/10.1016/bs.vh.2016.10.008.

Bosse, G. D., Urcino, C., Watkins, M., Flórez Salcedo, P., Kozel, S., Chase, K., Cabang, A., Espino, S. S., Safavi-Hemami, H., Raghuraman, S., Olivera, B. M., Peterson, R. T., and Gajewiak, J. (2021). Discovery of a potent conorfamide fromconus episcopatus using a novel Zebrafish larvae assay. *J. Nat. Prod.*, *84*(4), 1232–1243. https://doi.org/10.1021/acs.jnatprod.0c01297

Bradbury, S. P., Carlson, R. W., Henry, T. R., Padilla, S., and Cowden, J. (2008). Toxic responses of the fish nervous system. *Toxicol. Fish.* https://doi.org/10.1201/9780203647295

Bradbury, S. P., Carlson, R. W., Niemi, G. J., and Henry, T. R. (1991a). Use of respiratory-cardiovascular responses of rainbow trout (*Oncorhynchus mykiss*) in identifying acute toxicity syndromes in fish. Part 4. Central nervous system seizure agents. *Environ. Toxicol. Chem.*, *10*, 115–131.

Bradbury, S. P., and Coats, J. R. (1989). Comparative toxicology of the pyrethroid insecticides. *Rev. Environ. Contam. Toxicol.*, *108*, 133–177.

Bradbury, S. P., Henry, T. R., Niemi, G. J., Carlson, R. W., and Snarski, V. M. (1991b). Use of respiratory-cardiovascular responses of the rainbow trout (*Salmo gairdneri*) in identifying acute toxicity syndromes in fish. Part 3. Polar narcotics. *Environ. Toxicol. Chem.*, *8*, 247–261.

Bradbury, S. P., Symonik, D. M., Coats, J. R., and Atchison, G. J. (1987). Toxicity of fenvalerate and its constituent isomers to the fathead minnow, *Pimephales promelas*, and bluegill, *Lepomis macrochirus*. *Bull. Environ. Contam. Toxicol.*, *38*, 727–735.

Bradford, M. R., and Northcutt, R. G. (1983). Organization of the diencephalon and pretectum of the rayfinned fishes. In *Fish Neurobiology, Higher Brain Areas and Functions* (Vol. 2), Davis, R. E., and Northcutt, R. G, Ed., The University of Michigan Press, Ann Arbor, MI, pp. 117–164.

Branca, J. V., Morucci, G., and Pacini, A. (2018). Cadmium-induced neurotoxicity: still much ado. *Neural Regen. Res.*, *13*(11), 1879. https://doi.org/10.4103/1673-5374.239434

Brantley, R. K., and Bass, A. H. (1988). Cholinergic neurons in the brain of a teleost fish (*Porichthys notatus*) located with a monoclonal antibody to choline acetyltransferase. *J. Comp. Neurol.*, *275*, 87–105.

Bremer, J., and Granato, M. (2016). Myosin phosphatase fine-tunes zebrafish motoneuron position during axonogenesis. *PLoS Genet.*, *12*(11). https://doi.org/10.1371/journal.pgen.1006440

Brocardo, P. S., Gil-Mohapel, J., and Christie, B. R. (2011). The role of oxidative stress in fetal alcohol spectrum disorders. *Brain Res. Rev.*, *67*(1–2), 209–225. https://doi.org/10.1016/j.brainresrev.2011.02.001

Burgess, H. A., and Granato, M. (2007). Modulation of locomotor activity in larval zebrafish during light adaptation. *J. Exp. Biol.*, *210*(14), 2526–2539. https://doi.org/10.1242/jeb.003939

Byrd, Jr., R. P., and Caprio, J. (1982). Comparison of olfactory receptor (EOG) and bulbar (EEG) responses to amino acids in the catfish, *Ictalurus punctatus*. *Brain Res.*, *249*, 73–80.

Calado, S. L. de M., Vicentini, M., Santos, G. S., Pelanda, A., Santos, H., Coral, L. A., Magalhães, V. de F., Mela, M., Cestari, M. M., and Silva de Assis, H. C. (2019). Sublethal effects of microcystin-LR in the exposure and depuration time in a neotropical fish: Multibiomarker approach. *Ecotoxicol. Environ. Saf.*, *183*, 109527. https://doi.org/10.1016/J.ECOENV.2019.109527

Calvo-Ochoa, E., and Byrd-Jacobs, C. (2019). The olfactory system of zebrafish as a model for the study of neurotoxicity and injury: implications for neuroplasticity and disease. *Int. J. Mol. Sci.*, *20*(7), 1639. https://doi.org/10.3390/ijms20071639

Cantor, R. S. (1997). The lateral pressure profile in membranes: a physical mechanism of general anesthesia. *Biochemistry 36*, 2339–2344.

Caprio, J., and Byrd, Jr., R. P. (1984). Electrophysiological evidence for acidic, basic, and neutral amino acid olfactory receptor sites in the catfish. *J. Gen. Physiol.*, *84*, 403–422.

Carleton, K. L., Escobar-Camacho, D., Stieb, S. M., Cortesi, F., and Justin Marshall, N. (2020). Seeing the rainbow: Mechanisms underlying spectral sensitivity in teleost fishes. *J. Exp. Biol.*, *223*(8). https://doi.org/10.1242/jeb.193334

Carlson, R. W., Bradbury, S. P., Drummond, R. A., and Hammermeister, D. E. (1998). Neurological effects on startle response and escape from predation by medaka exposed to organic chemicals. *Aquat. Toxicol.*, *43*, 51–68.

Carmichael, W. W. (1997). The cyanotoxins. *Adv. Biochem. Res.*, *27*, 211–256.

Carmona, A., Roudeau, S., and Ortega, R. (2021). Molecular mechanisms of environmental metal neurotoxicity: A focus on the interactions of metals with synapse structure and function. *Toxics*, *9*(9). https://doi.org/10.3390/toxics9090198

Carr, R. L., Straus, D. L., and Chambers, J. E. (1995). Inhibition and aging of channel catfish brain acetylcholinesterase following exposure to two phosphorothionate insecticides and their active metabolites. *J. Toxicol. Environ. Health*, *45*, 325–336.

Carvan, M. J., Incardona, J. P., Rise, M. L., Iii, M. J. C., and John, P. (2008). Meeting the challenges of aquatic vertebrate ecotoxicology. *Bioscience*, *58*(11), 1015–1025.

Carvan, M. J., Kalluvila, T. A., Klingler, R. H., Larson, J. K., Pickens, M., Mora-Zamorano, F. X., Connaughton, V. P., Sadler-Riggleman, I., Beck, D., and Skinner, M. K. (2017). Mercury-induced epigenetic transgenerational inheritance of abnormal neurobehavior is correlated with sperm epimutations in zebrafish. *PLoS One*, *12*(5), e0176155. https://doi.org/10.1371/journal.pone.0176155

Carvan, 3rd, M. J., Loucks, E., Weber, D. N., and Williams, F. E. (2004). Ethanol effects on the developing zebrafish: neurobehavior and skeletal morphogenesis. *Neurotoxicol. Teratol.*, *26*, 757–768.

Casida, J. E., and Quisad, G. B. (2004). Organophosphate toxicology: safety aspects of nonacetylcholinesterase secondary targets. *Chem. Res. Toxicol.*, *17*, 983–998.

Ceccotti, C., Giaroni, C., Bistoletti, M., Viola, M., Crema, F., and Terova, G. (2018). Neurochemical characterization of myenteric neurons in the juvenile gilthead sea bream (Sparus aurata) intestine. *PLoS One*, *13*(8). https://doi.org/10.1371/journal.pone.0201760

Cerveny, D., Brodin, T., Cisar, P., McCallum, E. S., and Fick, J. (2020). Bioconcentration and behavioral effects of four benzodiazepines and their environmentally relevant mixture in wild fish. *Sci. Total Environ.*, *702*. https://doi.org/10.1016/j.scitotenv.2019.134780

Cerveny, D., Grabic, R., Grabicová, K., Randák, T., Larsson, D. G. J., Johnson, A. C., Jürgens, M. D., Tysklind, M., Lindberg, R. H., and Fick, J. (2021). Neuroactive drugs and other pharmaceuticals found in blood plasma of wild European fish. *Environ. Int.*, *146*, 106188. https://doi.org/10.1016/J.ENVINT.2020.106188

Chang, Q., and Caprio, J. (1996). Electrophysiological evidence for the broad distribution of specific odorant receptor molecules across the olfactory organ of the channel catfish. *Chem. Senses*, *21*, 519–527.

Chen, J., Chen, Y., Liu, W., Bai, C., Liu, X., Liu, K., Li, R., Zhu, J.-H., and Huang, C. (2012). Developmental lead acetate exposure induces embryonic toxicity and memory deficit in adult zebrafish. *Neurotoxicol. Teratol.*, *34*(6), 581–586. https://doi.org/10.1016/j.ntt.2012.09.001

Chilton, J. K. (2006). Molecular mechanisms of axon guidance. *Dev. Biol.*, *292*, 13–24.

Chung, D. D., Pinson, M. R., Bhenderu, L. S., Lai, M. S., Patel, R. A., and Miranda, R. C. (2021). Toxic and teratogenic effects of prenatal alcohol exposure on fetal development, adolescence and adulthood. *Int. J. Mol. Sci.*, *22*(16). https://doi.org/10.3390/ijms22168785

Cigliola, V., Becker, C. J., and Poss, K. D. (2020). Building bridges, not walls: Spinal cord regeneration in zebrafish. *Dis. Model. Mech.*, *13*(5). https://doi.org/10.1242/dmm.044131

Clark, J. M. (1995). Effects and mechanisms of pyrethrin and pyrethroid insecticides. In *Handbook of Neurotoxicology*, Chang, L. W., and Dyer, R. S., Eds., Marcel Dekker, New York, pp. 511–546.

Coats, J. R. (1990). Mechanisms of toxic action and structure-activity relationships for organochlorine and synthetic pyrethroid insecticides. *Environ. Health Perspect.*, *87*, 255–262.

Colas, S., Duval, C., and Marie, B. (2020). Toxicity, transfer and depuration of anatoxin-a (cyanobacterial neurotoxin) in medaka fish exposed by single-dose gavage. *Aquat. Toxicol.*, *222*. https://doi.org/10.1016/j.aquatox.2020.105422

Colas, S., Marie, B., Lance, E., Quiblier, C., Tricoire-Leignel, H., and Mattei, C. (2021). Anatoxin-a: Overview on a harmful cyanobacterial neurotoxin from the environmental scale to the molecular target. *Environ. Res.*, *193*, 110590. https://doi.org/10.1016/j.envres.2020.110590

Coombs, S., Bak-Coleman, J., and Montgomery, J. (2020). Rheotaxis revisited: A multi-behavioral and multisensory perspective on how fish orient to flow. *J. Exp. Biol.*, *223*(23). https://doi.org/10.1242/jeb.223008

Corbo, J. C. (2021). Vitamin A1/A2 chromophore exchange: Its role in spectral tuning and visual plasticity. *Dev. Biol.*, *475*(March), 145–155. https://doi.org/10.1016/j.ydbio.2021.03.002

Cusick, K. D., and Sayler, G. S. (2013). An overview on the marine neurotoxin, saxitoxin: Genetics, moleculartargets, methods of detection and ecological functions. *Mar. Drugs*, *11*(4), 991–1018. https://doi.org/10.3390/md11040991

Dale, K., Rasinger, J. D., Thorstensen, K. L., Penglase, S., and Ellingsen, S. (2017). Vitamin E reduces endosulfan-induced toxic effects on morphology and behavior in early development of zebrafish (Danio rerio). *Food Chem. Toxicol.*, *101*, 84–93. https://doi.org/10.1016/j.fct.2017.01.004

Das, B. K., and Mukherjee, S. C. (2003). Toxicity of cypermethrin in *Labeo rohita* fingerlings: biochemical, enzymatic and haematological consequences. *Comp. Biochem. Physiol. C.*, *134*, 109–121.

Dasgupta, S., Simonich, M. T., and Tanguay, R. L. (2022). Zebrafish behavioral assays in toxicology. *Methods Mol. Biol.*, *2474*, 109–122. https://doi.org/10.1007/978-1-0716-2213-1_11

David, M., Mushigeri, S. B., Shivakumar, R., and Philip, G. H. (2004). Response of *Cyprinus carpio* (Linn) to sublethal concentration of cypermethrin: alterations in protein metabolic profiles. *Chemosphere*, *56*, 347–352.

Davidson, P. W., Myers, G. J., and Weiss, B. (2004). Mercury exposure and child development outcomes. *Pediatrics*, *113*, 1023–1029.

Davin, Jr., W. T., Kohler, C. C., and Tindall, D. R. (1986). Effects of ciguatera toxins on the bluehead. *Trans. Am. Fish. Soc.*, *115*, 908–912.

Davin, Jr., W. T., Kohler, C. C., and Tindall, D. R. (1988). Ciguatera toxins adversely affect piscivorous fishes. *Trans. Am. Fish. Soc.*, *117*, 374–378.

Davis, R. E., and Kassel, J. (1983). Behavioral functions of the teleostean telencephalon. In *Fish Neurobiology, Higher Brain Areas and Functions* (Vol. 2), Davis, R. E., and Northcutt, R. G. Eds., The University of Michigan Press, Ann Arbor, MI, pp. 237–264.

De Gennaro, L. D. (2002). Lead and the developing nervous system. *Growth Dev. Aging*, *66*, 43–50.

Demicco, A., Cooper, K. R., Richardson, J. R., and White, L. A. (2009). Developmental neurotoxicity of pyrethroid insecticides in zebrafish embryos. *Toxicol. Sci.*, *113*(1), 177–186. https://doi.org/10.1093/toxsci/kfp258

Dew, W. A., Veldhoen, N., Carew, A. C., Helbing, C. C., and Pyle, G. G. (2016). Cadmium-induced olfactory dysfunction in rainbow trout: Effects of binary and quaternary metal mixtures. *Aquat. Toxicol.*, *172*, 86–94. https://doi.org/10.1016/j.aquatox.2015.12.018

Dlugos, C. A., and Rabin, R. A. (2003). Ethanol effects on three strains of zebrafish: model system for genetic investigations. *Pharmacol. Biochem. Behav.*, *74*, 471–480.

Donald, J. A. (1998). Autonomic nervous system. In *The Physiology of Fishes*, Evans, D. H., Ed., CRC Press, Boca Raton, FL, pp. 407–439.

Dopp, E., Hartmann, L. M., Florea, A.-M., Rettenmeier, A. W., and Hirner, A. V. (2004). Environmental distribution, analysis and toxicity of organometal(loid) compounds. *Crit. Rev. Toxicol.*, *34*, 301–333.

Drummond, R. A., and Russom, C. L. (1990). Behavioral toxicity syndromes: a promising tool for assessing toxicity mechanisms in juvenile fathead minnows. *Environ. Toxicol. Chem.*, *9*, 37–46.

Duarte, I. A., Fick, J., Cabral, H. N., and Fonseca, V. F. (2022). Bioconcentration of neuroactive pharmaceuticals in fish: Relation to lipophilicity, experimental design and toxicity in the aquatic environment. *Sci. Total Environ.*, *812*, 152543. https://doi.org/10.1016/J.SCITOTENV.2021.152543

Dutt, M., Giacomotto, J., Ragnarsson, L., Andersson, Å., Brust, A., Dekan, Z., Alewood, P. F., and Lewis, R. J. (2019). The α1-adrenoceptor inhibitor ρ-TIA facilitates net hunting in piscivorous Conus tulipa. *Sci. Reports*, *9*(1), 17841. https://doi.org/10.1038/s41598-019-54186-y

Ecobichon, D. J. (1996). Toxic effects of pesticides. In *Casarett & Doull's Toxicology: The Basic Science of Poisons*, Klaassen, C. D., Ed., McGraw-Hill, New York, pp. 643–689.

Eisenbrandt, D. L., Allen, S. L., Berry, P. H., Classen, W., Bury, D., Mellert, W., Millischer, R. J., Schuh, W., and Bontinck, W. J. (1994). Evaluation of the neurotoxic potential of chemicals in animals. *Food Chem. Toxicol.*, *32*, 655–669.

El Manira, A., and Bussieres, N. (1997). Calcium channel subtypes in lamprey sensory and motor neurons. *J. Neurophysiol.*, *78*, 1334–1340.

Evans, R. E., and Hara, T. J. (1985). The characteristics of the electro-olfactogram (EOG): its loss and recovery following olfactory nerve section in rainbow trout (*Salmo gairdneri*). *Brain Res.*, *330*, 65–75.

Faber, D. S., and Korn, H. (1988). Unitary conductance changes at teleost Mauthner cell glycinergic synapses: a voltage-clamp and pharmacologic analysis. *J. Neurophysiol.*, *60*, 1982–1999.

Facciol, A., and Gerlai, R. (2020). Zebrafish shoaling, its behavioral and neurobiological mechanisms, and its alteration by embryonic alcohol exposure: a review. *Front. Behav. Neurosci.*, *14*. https://doi.org/10.3389/fnbeh.2020.572175

Faucher, K., Fichet, D., Miramand, P., and Lagardère, J. P. (2006). Impact of acute cadmium exposure on the trunk lateral line neuromasts and consequences on the "C-start" response behaviour of the sea bass (Dicentrarchus labrax L.; Teleostei, Moronidae). *Aquat. Toxicol.*, *76*(3–4), 278–294. https://doi.org/10.1016/j.aquatox.2005.10.004

Faucher, K., Fichet, D., Miramand, P., and Lagardère, J.-P. (2008). Impact of chronic cadmium exposure at environmental dose on escape behaviour in sea bass (Dicentrarchus labrax L.; Teleostei, Moronidae). *Environ. Pollut.*, *151*(1), 148–157. https://doi.org/10.1016/j.envpol.2007.02.017

Featherstone, D., Drewes, C. D., Coats, J. R., and Bradbury, S. P. (1993). A non-invasive neurotoxicity assay using larval medaka. In *Environmental Toxicology and Risk Assessment*, (Vol. 2), Gorsuch, J. W., Dwyer, F. J., Ingersoll C. G., and LaPointe, T. W., Eds., ASTM STP 1216, American Society for Testing and Materials, Philadelphia, PA, pp. 275–288.

Featherstone, D., Drewes, C. D., Coats, J. R., Bradbury, S. P., Drewes, C. D., and Coats, J. R. (1991). Noninvasive detection of electrical events during the startle response in larval medaka. *J. Exp. Biol.*, *158*, 583–589.

Fent, K., Weston, A. A., and Caminada, D. (2006). Ecotoxicology of human pharmaceuticals. *Aquat. Toxicol.*, *76*(2), 122–159. https://doi.org/10.1016/J.AQUATOX.2005.09.009

Ferguson, J. (1939). The use of chemical potentials as indices of toxicity. *Proc. R. Soc. Lond. Ser. B Biol. Sci.*, *127*, 387–404.

Fernández-Sánchez, M.-T., Cabrera-García, D., Pérez-Gómez, A., and Novelli, A. (2021). Potential neurotoxins: Palytoxins. In A. Novelli, M.- T. Fernández-Sánchez, M. Aschner, and L. G. Costa (Eds.), *Marine Neurotoxins* (Vol. 6, pp. 223–273). Academic Press. https://doi.org/10.1016/bs.ant.2021.04.002

Finger, T. E. (1983). Organization of the teleost cerebellum. In *Fish Neurobiology, Brainstem and Sense Organs* (Vol. 1), Northcutt, R. G., and Davis, R. E., Eds., The University of Michigan Press, Ann Arbor, MI, pp. 261–284.

Fitzgerald, J. A., Könemann, S., Krümpelmann, L., Županič, A., and Vom Berg, C. (2021). Approaches to test the neurotoxicity of environmental contaminants in the zebrafish model: from behavior to molecular mechanisms. *Environ. Toxicol. Chem.*, *40*(4), 989–1006. https://doi.org/10.1002/etc.4951

Flentke, G. R., Klingler, R. H., Tanguay, R. L., Carvan III, M. J., and Smith, S. M. (2013). An evolutionarily conserved mechanism of calcium-dependent neurotoxicity in an early zebrafish model of FASD. *Alcohol. Clin. Exp. Res.*, *37*, 162A.

Fox, D. A., Lowndes, H. E., and Bierkamper, G. G. (1982). Electrophysiological techniques in neurotoxicity. In *Nervous System Toxicology*, Mitchell, C. L., Ed., Raven Press, New York, pp. 299–335.

Franks, N. P., and Lieb, W. R. (1990). Mechanisms of general anesthesia. *Environ. Health Perspect.*, *87*, 199–205.

Franks, N. P., and Lieb, W. R. (1994). Molecular and cellular mechanisms of general anaesthesia. *Nature*, *367*, 607–614.

Franks, N. P., and Lieb, W. R. (2004). Seeing the light. *Anesthesiology*, *101*, 235–237.

Fricke, C., Lee, J. S., Geiger-Rudolph, S., Bonhoeffer, F., and Chien, C. B. (2001). astray, a Zebrafish roundabout homolog required for retinal axon guidance. *Science*, *292*(5516), 507–510. https://doi.org/10.1126/science.1059496

Ganz, J. (2018). Gut feelings: studying enteric nervous system development, function, and disease in the zebrafish model system. *Develop. Dynam.*, *247*(2), 268–278. https://doi.org/10.1002/dvdy.24597

Garcia, G. R., Noyes, P. D., and Tanguay, R. L. (2016). Advancements in zebrafish applications for 21st century toxicology. *Pharmacol. Ther.*, *161*, 11–21. https://doi.org/10.1016/j.pharmthera.2016.03.009

Gélinas, M., Juneau, P., and Gagné, F. (2012). Early biochemical effects of Microcystis aeruginosa extracts on juvenile rainbow trout (Oncorhynchus mykiss). *Comp. Biochem. Physiol. Part B: Biochem. Mol. Biol.*, *161*(3), 261–267. https://doi.org/10.1016/j.cbpb.2011.12.002

Gerlai, R., Lahav, M., Guo, S., and Rosenthal, A. (2000). Drinks like a fish: zebra fish (*Danio rerio*) as a behavior genetic model to study alcohol effects. *Pharmacol. Biochem. Behav.*, *67*, 773–782.

Gidlow, D. A. (2004). Lead toxicity. *Occup. Med. (Lond.)*, *54*, 76–81.

Glazer, L., and Brennan, C. H. (2021). Developmental exposure to low concentrations of methylmercury causes increase in anxiety-related behaviour and locomotor impairments in Zebrafish. *Int. J. Mol. Sci.*, *22*(20), 10961. https://doi.org/10.3390/ijms222010961

Gochfeld, M. (2003). Cases of mercury exposure, bioavailability, and absorption. *Ecotoxicol. Environ. Saf.*, *56*, 174–179.

González-Doncel, M., de la Peña, E., Barrueco, C., and Hinton, D. E. (2003). Stage sensitivity of medaka (*Oryzias latipes*) eggs and embryos to permethrin. *Aquat. Toxicol.*, *62*, 255–268.

González Núñez, A. A., Ferro, J. P., Campos, L. B., Eissa, B. L., Mastrángelo, M. M., Ferrari, L., and Ossana, N. A. (2022). Evaluation of the Acute Effects of Arsenic on Adults of the Neotropical Native Fish *Cnesterodon decemmaculatus* Using a Set of Biomarkers. *Environ. Toxicol. Chem.*, *41*(5), 1246–1259. https://doi.org/10.1002/etc.5299

Grabicová, K., Grabic, R., Fedorova, G., Kolářová, J., Turek, J., Brooks, B. W., and Randák, T. (2020). Psychoactive pharmaceuticals in aquatic systems: A comparative assessment of environmental monitoring approaches for water and fish. *Environ. Pollut.,*, *261*. https://doi.org/10.1016/j.envpol.2020.114150

Granato, A., and Dering, B. (2018). Alcohol and the developing brain: why neurons die and how survivors change. *Int. J. Mol. Sci.*, *19*(10), 2992. https://doi.org/10.3390/ijms19102992

Green, A. J., and Planchart, A. (2018). The neurological toxicity of heavy metals: A fish perspective. *Comp. Biochem. Physiol. Part C: Toxicol. Pharmacol.*, *208*, 12–19. https://doi.org/10.1016/j.cbpc.2017.11.008

Gulati-Leekha, A., and Goldman, D. (2006). A reporter-assisted mutagenesis screen using α1-tubulin-GFP transgenic zebrafish uncovers missteps during neuronal development and axonogenesis. *Dev. Biol.*, *296*(1), 29–47. https://doi.org/10.1016/j.ydbio.2006.03.024

Gundacker, C., Forsthuber, M., Szigeti, T., Kakucs, R., Mustieles, V., Fernandez, M. F., Bengtsen, E., Vogel, U., Hougaard, K. S., and Saber, A. T. (2021). Lead (Pb) and neurodevelopment: A review on exposure and biomarkers of effect (BDNF, HDL) and susceptibility. *Int. J. Hygiene Environ. Health*, *238*. https://doi.org/10.1016/j.ijheh.2021.113855

Haehnel-Taguchi, M., Akanyeti, O., and Liao, J. C. (2018). Behavior, electrophysiology, and robotics experiments to study lateral line sensing in fishes. *Integr. Comp. Biol.*, *58*(5), 874–883. https://doi.org/10.1093/icb/icy066

Hale, M. E., Katz, H. R., Peek, M. Y., and Fremont, R. T. (2016). Neural circuits that drive startle behavior, with a focus on the Mauthner cells and spiral fiber neurons of fishes. *J. Neurogenet.*, *30*(2). https://doi.org/10.1080/01677063.2016.1182526

Hara, T. J., Brown, S. B., and Evans, R. E. (1983). Pollutants and chemoreception in aquatic organisms. In *Aquatic Toxicology*, Nriagu, J. O., Ed., John Wiley & Sons, New York, pp. 247–306.

Hara, T. J., Law, Y. M. C., and Macdonald, S. (1976). Effects of mercury and copper on the olfactory response in rainbow trout, *Salmo gairdneri*. *J. Fish. Res. Board Can.*, *33*, 1568–1573.

Hastings, L. (1995). Neurotoxicology of cadmium. In *Handbook of Neurotoxicology*, Chang, L. W., and Dyer, R. S., Eds., Marcel Dekker, New York, pp. 171–212.

Haverinen, J., and Vornanen, M. (2016). Deltamethrin is toxic to the fish (crucian carp, Carassius carassius) heart. *Pest. Biochem. Physiol.*, *129*, 36–42. https://doi.org/10.1016/j.pestbp.2015.10.014

Hawkey, A. B., Holloway, Z., and Levin, E. D. (2021). A behavioral test battery to assess larval and adult zebrafish after developmental neurotoxic exposure. *Neuromethods.*, *172*. https://doi.org/10.1007/978-1-0716-1637-6_16

Hawryshyn, C. W. (1998). Vision. In *The Physiology of Fishes*, Evans, D. H., Ed., CRC Press, Boca Raton, FL, pp. 345–374.

Heath, A. G. (1995). Behavioral and nervous system function. In *Water Pollution and Fish Physiology*, CRC Press, Boca Raton, FL, pp. 271–297.

Henry, J., Bai, Y., Kreuder, F., Saaristo, M., Kaslin, J., and Wlodkowic, D. (2022). Sensory-motor perturbations in Larval Zebrafish (Danio rerio) induced by exposure to low levels of neuroactive micropollutants during development. *Int. J. Mol. Sci.*, *23*(16). https://doi.org/10.3390/ijms23168990

Holmqvist, B., and Ekstrom, P. (1997). Subcellular localization of neuronal nitric oxide synthase in the brain of a teleost; an immunoelectron and confocal microscopical study. *Brain Res.*, *745*, 67–82.

Hook, S. E., Gallagher, E. P., and Batleyy, G. E. (2014). The role of biomarkers in the assessment of aquatic ecosystem health. *Integr. Environ. Assess. Manag.*, *10*, 327–341. https://doi.org/10.1002/ieam.1530

Hu, Y., Chen, J., Fan, H., Xie, P., and He, J. (2016). A review of neurotoxicity of microcystins. *Environ. Sci. Pollut. Res.*, *23*(8), 7211–7219. https://doi.org/10.1007/s11356-016-6073-y

Hu, Y., Li, J., Lou, B., Wu, R., Wang, G., Lu, C., Wang, H., Pi, J., and Xu, Y. (2020). The Role of Reactive Oxygen Species in Arsenic Toxicity. *Biomolecules*, *10*(2), 240. https://doi.org/10.3390/biom10020240

Hutson, L. D., and Chien, C. B. (2002). Wiring the zebrafish: axon guidance and synaptogenesis. *Curr. Opin. Neurobiol.*, *12*, 87–92.

Hyde, E. G., and Carmichael, W. W. (1991). Anatoxin-a(s), a naturally occurring organophosphate, is an irreversible active site-directed inhibitor of acetylcholinesterase (EC 3.1.1.7). *J. Biochem. Toxicol.*, *6*, 195–201.

Ingold, E., Berg-Maurer, C. M. V., Burckhardt, C. J., Lehnherr, A., Rieder, P., Keller, P. J., Stelzer, E. H., Greber, U. F., Neuhauss, S. C. F., and Gesemann, M. (2015). Proper migration and axon outgrowth of zebrafish cranial motoneuron subpopulations require the cell adhesion molecule MDGA2A. *Biology Open*, *4*(2), 146–154. https://doi.org/10.1242/bio.20148482

Islam, M. A., Amin, S. M. N., Brown, C. L., Juraimi, A. S., Uddin, M. K., and Arshad, A. (2021). Determination of median lethal concentration (LC50) for endosulfan, heptachlor and dieldrin pesticides to african catfish, clarias gariepinus and their impact on its behavioral patterns and histopathological responses. *Toxics*, *9*(12), 340. https://doi.org/10.3390/toxics9120340

Jabeen, S., and Thirumalai, V. (2018). The interplay between electrical and chemical synaptogenesis. *J. Neurophysiol.*, *120*, 1914–1922. https://doi.org/10.1152/jn.00398.2018.-Neurons

Jaishankar, M., Tseten, T., Anbalagan, N., Mathew, B. B., and Beeregowda, K. N. (2014). Toxicity, mechanism and health effects of some heavy metals. *Interdiscip. Toxicol.*, *7*(2), 60–72. https://doi.org/10.2478/intox-2014-0009

Jakubowska, N., and Szeląg-Wasielewska, E. (2015). Toxic Picoplanktonic Cyanobacteria-Review. *Mar. Drugs*, *13*(3), 1497–1518. https://doi.org/10.3390/md13031497

Jarema, K. A., Hunter, D. L., Hill, B. N., Olin, J. K., Britton, K. N., Waalkes, M. R., and Padilla, S. (2022). Developmental neurotoxicity and behavioral screening in Larval Zebrafish with a comparison to other published results. *Toxics*, *10*(5), 256. https://doi.org/10.3390/toxics10050256

Jedlicka, J., Groene, P., Linhart, J., Raith, E., Mustapha, D., and Conzen, P. (2021). Inhalational anaesthetics: an update on mechanisms of action and toxicity. *J. Exp. Neurol. Commen.*, *2*(2). https://www.scientificarchives.com/journal/journal-of-experimental-neurology

Jiang-Xie, L. F., Yin, L., Zhao, S., Prevosto, V., Han, B. X., Dzirasa, K., and Wang, F. (2019). A common neuroendocrine substrate for diverse general anesthetics and sleep. *Neuron*, *102*(5), 1053–1065.e4. https://doi.org/10.1016/j.neuron.2019.03.033

Jin, M., Dang, J., Paudel, Y. N., Wang, X., Wang, B., Wang, L., Li, P., Sun, C., and Liu, K. (2021). The possible hormetic effects of fluorene-9-bisphenol on regulating hypothalamic-pituitary-thyroid axis in zebrafish. *Sci. Total Environ.*, *776*. https://doi.org/10.1016/j.scitotenv.2021.145963

Johnson, J. A., and Wallace, K. B. (1987). Species-related differences in the inhibition of brain acetylcholinesterase by paraoxon and malaoxon. *Toxicol. Appl. Pharmacol.*, *88*, 234–241.

Kalueff, A. V, Gebhardt, M., Stewart, A. M., Cachat, J. M., Brimmer, M., Chawla, J. S., Craddock, C., Kyzar, E. J., Roth, A., Landsman, S., Gaikwad, S., Robinson, K., Baatrup, E., Tierney, K., Shamchuk, A., Norton, W., Miller, N., Nicolson, T., Braubach, O., ... Schneider, H. (2013). Towards a comprehensive catalog of zebrafish behavior 1.0 and beyond. *Zebrafish*, *10*(1), 70–86. https://doi.org/10.1089/zeb.2012.0861

Karlstrom, R. O., Trowe, T., and Bonhoeffer, F. (1997). Genetic analysis of axon guidance and mapping in the zebrafish. *Trends Neurosci.*, *20*(1), 3–8. https://doi.org/10.1016/S0166-2236(96)40005-4

Katti, S. R., and Sathyanesan, A. G. (1986). Lead nitrate induced changes in the brain constituents of the freshwater fish *Clarias batrachus* (L). *Neurotoxicology*, *7*, 47–51.

Kazmi, S. S. U. H., Yapa, N., Karunarathna, S. C., and Suwannarach, N. (2022). Perceived intensification in harmful algal blooms is a wave of cumulative threat to the aquatic ecosystems. *Biology*, *11*(6), 852. https://doi.org/10.3390/biology11060852

Khalili, A., van Wijngaarden, E., Youssef, K., Zoidl, G. R., and Rezai, P. (2022). Designing microfluidic devices for behavioral screening of multiple zebrafish larvae. *Biotechnol. J.*, *17*(1). https://doi.org/10.1002/biot.202100076

Khan, K. S., Hayes, I., and Buggy, D. J. (2014). Pharmacology of anaesthetic agents I: Intravenous anaesthetic agents. *Cont. Educ. Anaesth., Crit. Care Pain*, *14*(3), 100–105. https://doi.org/10.1093/bjaceaccp/mkt039

Khan, N. A., Troutand, D., Moulinoux, J.-P., and Deschaux, P. (1996). Characterization of serotonin receptors in fish brain: polyamines inhibit the binding process. *Neurosci. Res. Commun.*, *18*, 97–105.

Kinoshita, Y., Matsumura, H., Igisu, H., and Yokota, A. (2000). Hexachlorphene-induced brain edema in rat observed by proton magnetic resonance. *Brain Res.*, *873*, 127–130.

Klaverkamp, J. F., and Hobden, B. R. (1980). Brain acetylcholinesterase inhibition and hepatic activation of acephate and fenitrothion in rainbow trout (*Salmo gairdneri*). *Can. J. Fish. Aquat. Sci.*, *37*, 1450–1453.

Klüver, N., König, M., Ortmann, J., Massei, R., Paschke, A., Kühne, R., and Scholz, S. (2015). Fish Embryo Toxicity Test: Identification of Compounds with Weak Toxicity and Analysis of Behavioral Effects To Improve Prediction of Acute Toxicity for Neurotoxic Compounds. *Environ. Sci. Technol.*, *49*(11), 7002–7011. https://doi.org/10.1021/acs.est.5b01910

Kolesnikov, S. S., and Kosolapov, A. V. (1993). Cyclic nucleotide-activated channels in carp olfactory receptor cells. *Biochim. Biophys. Acta*, *1150*, 63–72.

Konemann, H. (1981). Quantitative structure-activity relationships in fish toxicity studies. Part 1. Relationship for 50 industrial pollutants. *Toxicology*, *19*, 209–221.

Kreft, W. D., Hoffert, J. R., and Fromm, P. O. (1985). Action of organophosphates on the electroretinogram of rainbow trout. *Exp. Biol.*, *44*, 19–27.

Krimsky, S. (2017). The unsteady state and inertia of chemical regulation under the US Toxic Substances Control Act. *PLoS Biol.*, *15*(12). https://doi.org/10.1371/journal.pbio.2002404

Kumaraguru, A. K., Beamish, F. W. H., and Ferguson, H. W. (1982). Direct and circulatory paths of permethrin (NRDC 143) causing histopathological changes in the gills of rainbow trout, *Salmo gairdneri* Richardson. *J. Fish. Biol.*, *20*, 87–91.

Lasater, E. M. (1990). Neurotransmitters and neuromodulators of the fish retina. In *The Visual System of Fish*, Douglas, R. H., and Djamgoz, M. B. A., Eds., Chapman & Hall, London, pp. 211–238.

Lauder, J. M., and Schambra, U. (1999). Morphogenic roles of acetylcholine. *Environ. Health Perspect.*, *107*, 65–69.

Lazzari, M., Bettini, S., Milani, L., Maurizii, M. G., and Franceschini, V. (2019). Differential nickel-induced responses of olfactory sensory neuron populations in zebrafish. *Aquat. Toxicol.*, *206*(June 2018), 14–23. https://doi.org/10.1016/j.aquatox.2018.10.011

Ledreux, A., Brand, H., Chinain, M., Bottein, M.-Y. D., and Ramsdell, J. S. (2014). Dynamics of ciguatoxins from Gambierdiscus polynesiensis in the benthic herbivore Mugil cephalus: Trophic transfer implications. *Harmful Algae*, *39*, 165–174. https://doi.org/10.1016/j.hal.2014.07.009

Lee, J., and Freeman, J. L. (2014). Zebrafish as a model for investigating developmental lead (Pb) neurotoxicity as a risk factor in adult neurodegenerative disease: A mini-review. *NeuroToxicology*, *43*, 57–64. https://doi.org/10.1016/j.neuro.2014.03.008

Lee, J. S., Ray, R., and Chien, C. Bin. (2001). Cloning and expression of three zebrafish roundabout homologs suggest roles in axon guidance and cell migration. *Dev. Dynam.*, *221*(2), 216–230. https://doi.org/10.1002/dvdy.1136

LeFauve, M. K., and Connaughton, V. P. (2017). Developmental exposure to heavy metals alters visually-guided behaviors in zebrafish. *Curr. Zoology*, *63*(2), 221–227. https://doi.org/10.1093/cz/zox017

LeFauve, M. K., Rowe, C. J., Crowley-Perry, M., Wiegand, J. L., Shapiro, A. G., and Connaughton, V. P. (2021). Using a variant of the optomotor response as a visual defect detection assay in zebrafish. *J. Biol. Methods*, *8*(1), e144. https://doi.org/10.14440/jbm.2021.341

Lefebvre, K. A., Trainer, V. L., and Scholz, N. L. (2004). Morphological abnormalities and sensorimotor deficits in larval fish exposed to dissolved saxitoxin. *Aquat. Toxicol.*, *66*(2), 159–170. https://doi.org/10.1016/j.aquatox.2003.08.006

Legendre, P., and Korn, H. (1994). Glycinergic inhibitory synaptic currents and related receptor channels in the zebrafish brain. *Eur. J. Neurosci.*, *6*, 1544–1557.

Legradi, J. B., Di Paolo, C., Kraak, M. H. S., van der Geest, H. G., Schymanski, E. L., Williams, A. J., Dingemans, M. M. L., Massei, R., Brack, W., Cousin, X., Begout, M. L., van der Oost, R., Carion, A., Suarez-Ulloa, V., Silvestre, F., Escher, B. I., Engwall, M., Nilén, G., Keiter, S. H., ... Hollert, H. (2018). An ecotoxicological view on neurotoxicity assessment. *Environ. Sci. Europe*, *30*(1), 1–34. https://doi.org/10.1186/s12302-018-0173-x

Levin, E. D., Chrysanthis, E., Yacisin, K., and Linney, E. (2003). Chlorpyrifos exposure of developing zebrafish: effects on survival and long-term effects on response latency and spatial discrimination. *Neurotoxicol. Teratol.*, *25*, 51–57.

Levin, E. D., Swain, H. A., Donerly, S., and Linney, E. (2004). Developmental chlorpyrifos effects on hatchling zebrafish swimming behavior. *Neurotoxicol. Teratol.*, *26*, 719–723.

Lewis, K. E., and Eisen, J. S. (2003). From cells to circuits: development of the zebrafish spinal cord. *Prog. Neurobiol.*, *69*, 419–449.

Lewis, R. J. (1992). Ciguatoxins are potent ichthyotoxins. *Toxicon*, *30*, 207–211.

Licheri, V., and Brigman, J. L. (2021). Altering cell-cell interaction in prenatal alcohol exposure models: insight on cell-adhesion molecules during brain development. *Front. Mol. Neurosci.*, *14*. https://doi.org/10.3389/fnmol.2021.753537

Little, E. E. (1990). Behavioral toxicology: stimulating challenges for a growing discipline. *Environ. Toxicol. Chem.*, *9*, 1–2.

Little, E. E., Dwyer, F. J., Fairchild, J. F., Delonay, A. J., and Zajicek, J. L. (1993). Survival of bluegill and their behavioral responses during continuous and pulsed exposures to esfenvalerate, a pyrethroid insecticide. *Environ. Toxicol. Chem.*, *12*, 871–878.

Liu, X., and Wang, J. (2022). N-Methyl-D-aspartate receptors mediate synaptic plasticity impairment of hippocampal neurons due to arsenic exposure. *Neuroscience*, *498*, 300–310. https://doi.org/10.1016/j.neuroscience.2022.07.017

Lotfullina, N., and Khazipov, R. (2018). Ethanol and the developing brain: inhibition of neuronal activity and neuroapoptosis. *The Neuroscientist*, *24*(2), 130–141. https://doi.org/10.1177/1073858417712667

Loucks, E., and Carvan, 3rd, M. J. (2004). Strain-dependent effects of developmental ethanol exposure in zebrafish. *Neurotoxicol. Teratol.*, *26*, 745–755.

Low, J., and Higgs, D. M. (2015). Sublethal effects of cadmium on auditory structure and function in fathead minnows (Pimephales promelas). *Fish Physiol. Biochem.*, *41*(2), 357–369. https://doi.org/10.1007/s10695-014-9988-6

Lowery, L. A., and Sive, H. (2004). Strategies of vertebrate neurulation and a re-evaluation of teleost neural tube formation. *Mech. Dev.*, *121*, 1189–1197.

Lucon-Xiccato, T., Loosli, F., Conti, F., Foulkes, N. S., and Bertolucci, C. (2022). Comparison of anxiety-like and social behaviour in medaka and zebrafish. *Sci. Reports*, *12*(1), 10926. https://doi.org/10.1038/s41598-022-14978-1

Lukiw, W. J., and McLachlan, D. R. (1995). Aluminum neurotoxicity. In *Handbook of Neurotoxicology*, Chang, L. W., and Dyer, R. S., Eds., Marcel-Dekker, New York, pp. 105–142.

Ma, P. M. (1994). Catecholaminergic systems in the zebrafish. II. Projection pathways and pattern of termination of the locus coeruleus. *J. Comp. Neurol.*, *344*, 256–269.

Mackieh, R., Abou-Nader, R., Wehbe, R., Mattei, C., Legros, C., Fajloun, Z., and Sabatier, J. M. (2021). Voltage-gated sodium channels: a prominent target of marine toxins. *Mar. Drugs*, *19*(10), 562. https://doi.org/10.3390/md19100562

Magarlamov, T., Melnikova, D., and Chernyshev, A. (2017). Tetrodotoxin-producing bacteria: detection, distribution and migration of the toxin in aquatic systems. *Toxins*, *9*(5), 166. https://doi.org/10.3390/toxins9050166

Mahmood, N. A., and Carmichael, W. W. (1987). Anatoxin-a(s), an anticholinesterase from the cyanobacterium *Anabaena flos-aquae* NRC-525-17. *Toxicon*, *25*, 1221–1227.

Malhotra, H., Kaur, S., & Phale, P. S. (2021). Conserved metabolic and evolutionary themes in microbial degradation of carbamate pesticides. *Front. Microbiol.*, *12*. Frontiers Media S.A. https://doi.org/10.3389/fmicb.2021.648868

Marc, R. E., and Cameron, D. (2001). A molecular phenotype atlas of the zebrafish retina. *J. Neurocytol.*, *30*, 593–654.

Marchetti, C. (2003). Molecular targets of lead in brain neurotoxicity. *Neurotox. Res.*, *5*, 221–236.

Martin, E. A., Lasseigne, A. M., and Miller, A. C. (2020). Understanding the molecular and cell biological mechanisms of electrical synapse formation. *Front. Neuroanat.*, *14*. https://doi.org/10.3389/fnana.2020.00012

Matsui, H., Uemura, N., Yamakado, H., Takeda, S., and Takahashi, R. (2014). Exploring the pathogenetic mechanisms underlying Parkinson's disease in medaka fish. *J. Parkinson's Dis.*, *4*(2), 301–310. https://doi.org/10.3233/JPD-130289

McKim, J. M., Schmieder, P. K., Niemi, G. J., Carlson, R. W., and Henry, T. R. (1987). Use of respiratory-cardiovascular responses of rainbow trout (*Salmo gairdneri*) in identifying acute toxicity syndromes in fish. Part 2. Malathion, carbaryl, acrolein, and benzaldehyde. *Environ. Toxicol. Chem.*, *6*, 313–328.

Merola, C., Lucon-Xiccato, T., Bertolucci, C., and Perugini, M. (2021). Behavioural effects of early-life exposure to parabens in zebrafish larvae. *J. Appl. Toxicol.*, *41*(11), 1852–1862. https://doi.org/10.1002/jat.4171

Mochizuki, H. (2019). Arsenic neurotoxicity in humans. *Int. J. Mol. Sci.*, *20*(14). https://doi.org/10.3390/ijms20143418

Mondal, P., Shaw, P., Dey Bhowmik, A., Bandyopadhyay, A., Sudarshan, M., Chakraborty, A., and Chattopadhyay, A. (2021). Combined effect of arsenic and fluoride at environmentally relevant concentrations in zebrafish (Danio rerio) brain: Alterations in stress marker and apoptotic gene expression. *Chemosphere*, *269*. https://doi.org/10.1016/j.chemosphere.2020.128678

Mora-Zamorano, F., Klingler, R., Murphy, C., Basu, N., Head, J., and Carvan III, M. J. (2016). Parental whole life cycle exposure to dietary methylmercury in zebrafish (Danio rerio) affects the behavior of offspring. *Environ. Sci. Technol.*, *In Press*. https://doi.org/10.1021/acs.est.6b00223

Mora-Zamorano, F. X., Klingler, R., Basu, N., Head, J., Murphy, C. A., Binkowski, F. P., Larson, J. K., and Carvan, M. J. (2017). Developmental methylmercury exposure affects swimming behavior and foraging efficiency of yellow perch (*Perca flavescens*) Larvae. *ACS Omega*, *2*(8), 4870–4877. https://doi.org/10.1021/acsomega.7b00227

Mora-Zamorano, F. X., Larson, J. K., and Carvan, M. J. (2018a). Neurobehavioral analysis methods for adverse outcome pathway (AOP) models and risk assessment. In *A Systems Biology Approach to Advancing Adverse Outcome Pathways for Risk Assessment*. https://doi.org/10.1007/978-3-319-66084-4_8

Mora-Zamorano, F. X., Larson, J. K., and Carvan, M. J. (2018b). Neurobehavioral analysis methods for adverse outcome pathway (AOP) models and risk assessment. In *A Systems Biology Approach to Advancing Adverse Outcome Pathways for Risk Assessment* (pp. 149–175). Springer International Publishing. https://doi.org/10.1007/978-3-319-66084-4_8

Mora-Zamorano, F. X., Svoboda, K. R., and Carvan, M. J. (2016). The nicotine-evoked locomotor response: a behavioral paradigm for toxicity screening in zebrafish (danio rerio) embryos and eleutheroembryos exposed to methylmercury. *PLoS One*, *11*(4), e0154570. https://doi.org/10.1371/journal.pone.0154570

Morais, S. (2017). The physiology of taste in fish: potential implications for feeding stimulation and gut chemical sensing. *Rev. Fisheries Sci. Aquacul.*, *25*(2), 133–149. https://doi.org/10.1080/23308249.2016.1249279

Moretto, A., Tiramani, M., and Colosio, C. (2011). Long-term neurotoxicological effects of anticholinesterases after either acute or chronic exposure. In *Anticholinesterase Pesticides* (pp. 97–108). John Wiley & Sons, Inc. https://doi.org/10.1002/9780470640500.ch8

Moser, V., Aschner, M., Richardson, J., Bowman, A., and Richardson, R. (2019). Toxic responses of the nervous system. In C. Klaassen (Ed.), *Casarett & Doull's Toxicology: The Basic Science of Poisons* (9th ed., pp. 839–875). McGraw Hill.

Mrema, E. J., Rubino, F. M., Brambilla, G., Moretto, A., Tsatsakis, A. M., and Colosio, C. (2013). Persistent organochlorinated pesticides and mechanisms of their toxicity. *Toxicology*, *307*, 74–88. https://doi.org/10.1016/j.tox.2012.11.015

Murty, A. S. (1986). *Toxicity of Pesticides to Fish*, Vol. *II*, CRC Press, Boca Raton, FL, pp. 55–100.

Murty, A. S., and Ramani, A. V. (1992). Toxicity of anticholinesterases to aquatic organisms. In *Clinical and Experimental Toxicology of Organophosphates and Carbamates*, Ballantyne, B., and Marrs, T. C., Eds., Butterworth-Heinemann, Oxford, pp. 305–317.

Musilova, Z., Cortesi, F., Matschiner, M., Davies, W. I. L., Patel, J. S., Stieb, S. M., De Busserolles, F., Malmstrøm, M., Tørresen, O. K., Brown, C. J., Mountford, J. K., Hanel, R., Stenkamp, D. L., Jakobsen, K. S., Carleton, K. L., Jentoft, S., Marshall, J., and Salzburger, W. (2019). Vision using multiple distinct rod opsins in deep-sea fishes. *Science*, *364*(6440), 588–592. https://doi.org/10.1126/science.aav4632

Narahashi, T. (1992). Nerve membrane $Na^+$ channels as targets of insecticides. *Trends Pharmacol. Sci.*, *13*, 236–241.

Narahashi, T. (1994). Role of ion channels in neurotoxicity. In *Principles of Neurotoxicology*, L. Chang, Ed., Marcel-Dekker, New York, pp. 609–655.

Niessen, K. V., Muschik, S., Langguth, F., Rappenglück, S., Seeger, T., Thiermann, H., and Worek, F. (2016). Functional analysis of Torpedo californica nicotinic acetylcholine receptors in multiple activation states by SSM-based electrophysiology. *Toxicol. Lett.*, *247*, 1–10. https://doi.org/10.1016/j.toxlet.2016.02.002

Nilsson, D. E. (2021). The diversity of eyes and vision. *Annu. Rev. Vision Sci.*, *7*(1), 19–41. https://doi.org/10.1146/annurev-vision-121820-074736

Nilsson, S., and Holmgren, S. (1998). Autonomic nerve function. In *The Physiology of Fishes*, Evans, D. H., Ed., CRC Press, Boca Raton, FL, pp. 279–313.

Northcutt, R. G., and Davis, R. E. (1983). Telencephalic organization in ray-finned fishes. In *Fish Neurobiology: Higher Brain Areas and Functions* (Vol. 2), Davis, R. E., and Northcutt, R. G., Eds., University of Michigan Press, Ann Arbor, MI, pp. 203–236.

Novo, J. P., Martins, B., Raposo, R. S., Pereira, F. C., Oriá, R. B., Malva, J. O., and Fontes-Ribeiro, C. (2021). Cellular and molecular mechanisms mediating methylmercury neurotoxicity and neuroinflammation. In *Int. J. Mol. Sci.*, *22*(6), 1–25. https://doi.org/10.3390/ijms22063101

Nyman, H.- G. (1981). Sublethal effects of lead (Pb) on size selective predation by fish-applications on the ecosystem level. *Int. Ver. Theor. Angew. Limnol. Verh.*, *21*, 1126–1130.

Oberemm, A., Becker, J., Codd, G. A., and Steinberg, C. (1999). Effects of cyanobacterial toxins and aqueous crude extracts of cyanobacteria on the development of fish and amphibians. *Environ. Toxicol.*, *14*(1), 77–88. https://doi.org/10.1002/(SICI)1522-7278(199902)14:*1<77::AID-TOX11>3.0.CO*;2–F

Ohno, T., Yoshii, K., and Kurihara, K. (1984). Multiple receptor types for amino acids in the carp olfactory cells revealed by quantitative cross-adaptation method. *Brain Res.*, *310*, 13–21.

Olsson, C. (2009). Autonomic innervation of the fish gut. *Acta Histochem.*, *111*(3), 185–195. https://doi.org/10.1016/j.acthis.2008.11.014

Osswald, J., Carvalho, A. P., Claro, J., and Vasconcelos, V. (2009). Effects of cyanobacterial extracts containing anatoxin-a and of pure anatoxin-a on early developmental stages of carp. *Ecotoxicol. Environ. Safety*, *72*(2), 473–478. https://doi.org/10.1016/j.ecoenv.2008.05.011

Outridge, P. M., Mason, R. P., Wang, F., Guerrero, S., and Heimbürger-Boavida, L. E. (2018). Updated Global and Oceanic Mercury Budgets for the United Nations Global Mercury Assessment 2018. *Environmental Science & Technology*, acs.est.8b01246. https://doi.org/10.1021/acs.est.8b01246

Pain, D. J. (1995). Lead in the environment. In *Handbook of Ecotoxicology*, Hoffmann, D. J., Rattner, B. A., Burton, Jr., G. A., and Cairns, Jr., J., Eds., CRC Press, Boca Raton, FL, pp. 356–391.

Panlilio, J. M., Aluru, N., and Hahn, M. E. (2020). Developmental neurotoxicity of the harmful algal bloom toxin domoic acid: cellular and molecular mechanisms underlying altered behavior in the zebrafish model. *Environ. Health Perspect.*, *128*(11), 1–19. https://doi.org/10.1289/EHP6652

Parent, A. (1983). The monoamine-containing neuronal systems in the teleostean brain. In *Fish Neurobiology: Higher Brain Areas and Functions*, Davis, R. E., and Northcutt, R. G., Eds., The University of Michigan Press, Ann Arbor, MI, pp. 285–316.

Paschoalini, A. L., and Bazzoli, N. (2021). Heavy metals affecting Neotropical freshwater fish: A review of the last 10 years of research. *Aquat. Toxicol.*, *237*, 105906. https://doi.org/10.1016/j.aquatox.2021.105906

Patocka, J., Nepovimova, E., Wu, Q., and Kuca, K. (2018). Palytoxin congeners. *Arch. Toxicol.*, *92*(1), 143–156. https://doi.org/10.1007/s00204-017-2105-8

Pavel, M. A., Petersen, E. N., Wang, H., Lerner, R. A., and Hansen, S. B. (2020). Studies on the mechanism of general anesthesia. *Proc. Natl. Acad. Sci.*, *117*(24), 13757–13766. https://doi.org/10.1073/pnas.2004259117

Pavlov, D. F. (1994). A simple pharmacological analysis of the cholinergic control of respiration in perch (*Perca fluviatilis* L.) exposed to DDVP, an organophosphorus insecticide. *Comp. Biochem. Physiol.*, *108C*, 107–111.

Pedroni, A., and Ampatzis, K. (2019). Large-scale analysis of the diversity and complexity of the adult spinal cord neurotransmitter typology. *IScience*, *19*. https://doi.org/10.1016/j.isci.2019.09.010

Philbert, M. A., Billingsley, M. L., and Reuhl, K. R. (2000). Mechanisms of injury in the central nervous system. *Toxicol. Pathol.*, *28*, 43–53.

Piña, B., Garcia-Reyero, N., Bashirzade, A. A. O., Cheresiz, S. V, Belova, A. S., Drobkov, A. V, Korotaeva, A. D., Azizi-Arani, S., Azimirad, A., Odle, E., Gild, E.-Y. V, Ardashov, O. V, Volcho, K. P., Bozhko, D. V, Myrov, V. O., Kolchanova, S. M., Polovian, A. I., Galumov, G. K., Salakhutdinov, N. F., ... Kalueff, A. V. (2022). *MPTP-Treated Zebrafish Recapitulate "Late-Stage" Parkinson's-like Cognitive Decline*. https://doi.org/10.3390/toxics10020069

Polat, H., Erkoç, F., Viran, R., and Koçak, O. (2002). Investigation of acute toxicity of beta-cypermethrin on guppies *Poecilia reticulata. Chemosphere*, *49*, 39–44.

Polverino, G., Martin, J. M., Bertram, M. G., Soman, V. R., Tan, H., Brand, J. A., Mason, R. T., and Wong, B. B. M. (2021). Psychoactive pollution suppresses individual differences in fish behaviour. *Proc. Royal Soc. B: Biol. Sci.*, *288*(1944). https://doi.org/10.1098/rspb.2020.2294

Pompermaier, A., Kirsten, K., Soares, S. M., Fortuna, M., Kalichak, F., Idalencio, R., Koakoski, G., Barreto, R. E., and Barcellos, L. J. G. (2020). Waterborne agrichemicals compromise the anti-predatory behavior of zebrafish. *Environ. Sci. Pollut. Res.*, *27*(31), 38559–38567. https://doi.org/10.1007/s11356-020-09862-2

Pompermaier, A., Varela, A. C. C., Mozzato, M. T., Soares, S. M., Fortuna, M., Alves, C., Tamagno, W. A., and Barcellos, L. J. G. (2022). Impaired initial development and behavior in zebrafish exposed to environmentally relevant concentrations of widely used pesticides. *Comp. Biochem. Physiol. Part C: Toxicol. Pharmacol.*, *257*, 109328. https://doi.org/10.1016/j.cbpc.2022.109328

Pope, C. N. (1999). Organophosphorus pesticides: do they all have the same mechanism of toxicity? *J. Toxicol. Environ. Health Part B2*, 161–181.

Prashanth, J. R., Dutertre, S., and Lewis, R. J. (2017). Pharmacology of predatory and defensive venom peptides in cone snails. *Mol. BioSystems*, *13*(12), 2453–2465. https://doi.org/10.1039/c7mb00511c

Pratheepa, V., and Vasconcelos, V. (2013). Microbial diversity associated with tetrodotoxin production in marine organisms. *Environ. Toxicol. Pharmacol.*, *36*(3), 1046–1054. https://doi.org/10.1016/j.etap.2013.08.013

Rădulescu, A., and Lundgren, S. (2019). A pharmacokinetic model of lead absorption and calcium competitive dynamics. *Sci. Reports*, *9*(1), 14225. https://doi.org/10.1038/s41598-019-50654-7

Rahaman, M. S., Rahman, M. M., Mise, N., Sikder, M. T., Ichihara, G., Uddin, M. K., Kurasaki, M., and Ichihara, S. (2021). Environmental arsenic exposure and its contribution to human diseases, toxicity mechanism and management. *Environ. Pollut.*, *289*. https://doi.org/10.1016/j.envpol.2021.117940

Ramírez Ortega, D., González Esquivel, D. F., Blanco Ayala, T., Pineda, B., Gómez Manzo, S., Marcial Quino, J., Carrillo Mora, P., and Pérez de la Cruz, V. (2021). Cognitive impairment induced by lead exposure during lifespan: mechanisms of lead neurotoxicity. *Toxics*, *9*(2), 23. https://doi.org/10.3390/toxics9020023

Ramšak, Ž., Modic, V., Li, R. A., vom Berg, C., and Zupanic, A. (2022). From causal networks to adverse outcome pathways: a developmental neurotoxicity case study. *Front. Toxicol.*, *4*. https://doi.org/10.3389/ftox.2022.815754

Rand, G. M. (1985). Behavior. In *Fundamentals of Aquatic Toxicology: Methods and Applications*, Rand G. M., and Petrocelli, S. R., Eds., Hemisphere Publishing, Washington, DC, pp. 221–307.

Rebach, S. (1999). Acute toxicity of permethrin/piperonyl butoxide on hybrid striped bass. *Bull. Environ. Contam. Toxicol.*, *62*, 448–454.

Rehman, S. (2003). Lead-exposed increases in movement, behavior, and brain lipid peroxidation in fish. *J. Environ. Sci. Health*, *A38*, 631–643.

Reier, P. J., Tabira, T., and Webster, H. (1978). Hexachlorphene-induced myelin lesions in the amphibian central nervous system: a freeze-fracture study. *J. Neurol. Sci.*, *35*, 257–274.

Reimers, M. J., Flockton, A. R., and Tanguay, R. L. (2004). Ethanol- and acetaldehyde-mediated developmental toxicity in zebrafish. *Neurotoxicol. Teratol.*, *26*, 769–781.

Rice, P. J., Drewes, C. D., Klubertanz, T. M., Coats, J. R., and Bradbury, S. P. (1997). Acute toxicity and behavioral effects of chlorpyrifos, phenol, strychnine, and 2,4-dinitrophenol to 30-day old Japanese medaka (*Oryzias latipes*). *Environ. Toxicol. Chem.*, *16*, 696–704.

Rice, C., Ghorai, J. K., Zalewski, K., and Weber, D. N. (2011). Developmental lead exposure causes startle response deficits in zebrafish. *Aquat. Toxicol.*, *105*(3–4), 600–608. https://doi.org/10.1016/j.aquatox.2011.08.014

Rico, E. P., Rosemberg, D. B., Seibt, K. J., Capiotti, K. M., Da Silva, R. S., and Bonan, C. D. (2011). Zebrafish neurotransmitter systems as potential pharmacological and toxicological targets. *Neurotoxicol. Teratol.*, *33*(6). https://doi.org/10.1016/j.ntt.2011.07.007

Riley, E. P., Infante, M. A., and Warren, K. R. (2011). Fetal Alcohol Spectrum Disorders: An Overview. *Neuropsychol. Rev.*, *21*(2), 73–80. https://doi.org/10.1007/s11065-011-9166-x

Rio, J. P., Vesselkin, N. P., Reperant, J., Kenigfest, N. B., Miceli, D., and Adanina, V. (1996). Retinal and non-retinal inputs upon retinopetal RMA neurons in the lamprey: a light and electron microscopic study combining HRP axonal tracing and GABA immunocytochemistry. *J. Chem. Neuroanat.*, *12*, 51–70.

Ritchie, J. M., and Greene, N. M. (1985). Local anesthetics. In *Goodman and Gilman's The Pharmacological Basis of Therapeutics*, Goodman, A. G., Goodman, L. S., Rall, T. W., and Murad, F., Eds., Macmillan, New York, p. 312.

Roberts, A. C., Chornak, J., Alzagatiti, J. B., Ly, D. T., Bill, B. R., Trinkeller, J., Pearce, K. C., Choe, R. C., Campbell, C. S., Wong, D., Deutsch, E., Hernandez, S., and Glanzman, D. L. (2019). Rapid habituation of a touch-induced escape response in Zebrafish (Danio rerio) Larvae. *PLoS One*, *14*(4), e0214374. https://doi.org/10.1371/journal.pone.0214374

Rodríguez, F., Durán, E., Gómez, A., Ocaña, F. M., Álvarez, E., Jiménez-Moya, F., Broglio, C., and Salas, C. (2005). Cognitive and emotional functions of the teleost fish cerebellum. *Brain Res. Bull.*, *66*(4–6), 365–370. https://doi.org/10.1016/j.brainresbull.2004.11.026

Rodríguez, F., Quintero, B., Amores, L., Madrid, D., Salas-Peña, C., and Salas, C. (2021). Spatial cognition in teleost fish: Strategies and mechanisms. *Animals*, *11*(8). https://doi.org/10.3390/ani11082271

Rolen, S. H., Sorensen, P. W., Mattson, D., and Caprio, J. (2003). Polyamines as olfactory stimuli in the goldfish *Carassius auratus*. *J. Exp. Biol.*, *206*, 1683–1696.

Rolig, A. S., Mittge, E. K., Ganz, J., Troll, J. V, Melancon, E., Wiles, T. J., Alligood, K., Stephens, W. Z., Eisen, J. S., and Guillemin, K. (2017). The enteric nervous system promotes intestinal health by constraining microbiota composition. *PLoS Biol.*, *15*(2). https://doi.org/10.1371/journal.pbio.2000689

Rüdel, H. (2003). Bioavailability of tin and tin compounds. *Ecotoxicol. Environ. Saf.*, *56*, 180–189.

Saglio, P., Trijasse, S., and Azam, D. (1996). Behavioral effects of waterborne carbofuran in goldfish. *Arch. Environ. Contam. Toxicol.*, *31*, 232–238.

Saha, S., and Kaviraj, K. A. (2003). Acute toxicity of synthetic pyrethroid cypermethrin in freshwater catfish *Heteropneustes fossilis* (Bloch). *Int. J. Toxicol.*, *22*, 325–328.

Saleh, S. M. M., Mohamed, I. A., Fathy, M., and Sayed, A. E. D. H. (2022). Neuro-hepatopathological changes in juvenile Oreochromis niloticus exposed to sublethal concentrations of commercial herbicides. *Environ. Toxicol. Pharmacol.*, *93*. https://doi.org/10.1016/j.etap.2022.103871

Sandblom, E., and Axelsson, M. (2011). Autonomic control of circulation in fish: A comparative view. *Auton. Neurosci.: Basic Clin.*, *165*(1), 127–139. https://doi.org/10.1016/j.autneu.2011.08.006

Sandheinrich, M. B., and Atchison, G. J. (1990). Sublethal toxicant effects on fish foraging behavior: empirical vs. mechanistic approaches. *Environ. Toxicol. Chem.*, *9*, 107–120.

Sas, E., Maler, L., and Tinner, B. (1990). Catecholaminergic systems in the brain of a gymnotiform teleost fish: an immunohistochemical study. *J. Comp. Neurol.*, *292*, 127–162.

Sastre, S., Fernández Torija, C., Atiénzar Pertusa, I., Beltrán, E. M., Pablos, M. V., and González-Doncel, M. (2018). Stage-dependent effects of chlorpyrifos on medaka (Oryzias latipes) swimming behavior using a miniaturized swim flume. *Aquat. Toxicol.*, *200*, 37–49. https://doi.org/10.1016/j.aquatox.2018.04.008

Sato, K., and Suzuki, N. (2001). Whole-cell response characteristics of ciliated and microvillous olfactory receptor neurons to amino acids, pheromone candidates and urine in rainbow trout. *Chem. Senses*, *26*, 1145–1156.

Scascitelli, M., and Pacchierotti, F. (2003). Effects of lindane on oocyte maturation and preimplantation embryonic development in the mouse. *Reprod. Toxicol.*, *17*, 299–303.

Schellart, N. A. M., and Wubbels, R. J. (1998). The auditory and mechanosensory lateral line system. In *The Physiology of Fishes*, Evans, D. H., Ed., CRC Press, Boca Raton, FL, pp. 283–312.

Schildroth, S., Kordas, K., Bauer, J. A., Wright, R. O., and Claus Henn, B. (2022). Environmental metal exposure, neurodevelopment, and the role of iron status: a review. *Curr. Environ. Health Rep.*, https://doi.org/10.1007/s40572-022-00378-0

Schmidt, R., Strähle, U., and Scholpp, S. (2013). Neurogenesis in zebrafish - from embryo to adult. *Neural Dev.*, *8*(1). https://doi.org/10.1186/1749-8104-8-3

Schafer, K., Braun, H. A., Peters, R. C., and Bretschneider, F. (1995). Periodic firing pattern in afferent discharges from electroreceptor organs of catfish. *Pflügers Arch.*, 429, 378–385.

Schober, A., Malz, C. R., Schober, W., and Meyer, D. L. (1994). NADPH-diaphorase in the central nervous system of the larval lamprey (*Lampetra planeri*). *J. Comp. Neurol.*, *345*, 94–104.

Shafer, T. J. (2000). Methylmercury effects on ion channels and electrical activity in neurons: future directions. *Cell. Mol. Biol.*, *46*, 855–864.

Shafer, T. J., and Meyer, D. A. (2004). Effects of pyrethroids on voltage-sensitive calcium channels: a critical evaluation of strengths, weaknesses, data needs, and relationship to assessment of cumulative neurotoxicity. *Toxicol. Appl. Pharmacol.*, *196*, 303–318.

Shafer, T. J., Meyer, D. A., and Crofton, K. M. (2005). Developmental neurotoxicity of pyrethroid insecticides: critical review and future research needs. *Environ. Health Perspect.*, *113*, 123–136.

Sharpe, R. L., Benskin, J. P., Laarman, A. H., MacLeod, S. L., Martin, J. W., Wong, C. S., and Goss, G. G. (2010). Perfluorooctane sulfonate toxicity, isomer-specific accumulation, and maternal transfer in zebrafish (Danio rerio) and rainbow trout (Oncorhynchus mykiss). *Environ. Toxicol. Chem.*, *29*(9), 1957–1966. https://doi.org/10.1002/etc.257

Shepherd, I., and Eisen, J. (2011). Development of the zebrafish enteric nervous system. *Methods Cell Biol.*, *101*. https://doi.org/10.1016/B978-0-12-387036-0.00006-2

Silva, C. A. da, Morais, E. C. P. de, Costa, M. D. M., Ribas, J. L. C., Guiloski, I. C., Ramsdorf, W. A., Zanata, S. M., Cestari, M. M., Oliveira Ribeiro, C. A., Magalhães, V. F., Trudeau, V. L., and Silva de Assis, H. C. (2014). Saxitoxins induce cytotoxicity, genotoxicity and oxidative stress in teleost neurons in vitro. *Toxicon*, *86*, 8–15. https://doi.org/10.1016/j.toxicon.2014.04.016

Sloley, B. D., Kah, O., Trudeau, V. L., Dulka, J. G., and Peter, R. E. (1992). Amino acid neurotransmitters and dopamine in brain and pituitary of the goldfish: involvement in the regulation of gonadotropin secretion. *J. Neurochem.*, *58*, 2254–2262.

Sloman, K. A., Scott, G. R., Diao, Z., Rouleau, C., Wood, C. M., and McDonald, D. G. (2003). Cadmium affects the social behaviour of rainbow trout, Oncorhynchus mykiss. *Aquat. Toxicol.*, *65*(2), 171–185. https://doi.org/10.1016/S0166-445X(03)00122-X

Smith, L. E., Carvan, M. J., Dellinger, J. A., Ghorai, J. K., White, D. B., Williams, F. E., and Weber, D. N. (2010). Developmental selenomethionine and methylmercury exposures affect zebrafish learning. *Neurotoxicol. Teratol.*, *32*(2), 246–255. https://doi.org/10.1016/j.ntt.2009.09.004

Soderlund, D. M. (2012). Molecular mechanisms of pyrethroid insecticide neurotoxicity: Recent advances. *Archiv. Toxicol.*, *86*(2), 165–181. https://doi.org/10.1007/s00204-011-0726-x

Soderlund, D. M., Clark, J. M., Sheets, L. P., Mullin, L. S., Piccirillo, V. J., Sargent, D., Stevens, J. T., and Weiner, M. L. (2002). Mechanisms of pyrethroid neurotoxicity: implications for cumulative risk assessment. *Toxicology*, *171*, 3–59.

Sommers, F., Mudrock, E., Labenia, J., and Baldwin, D. (2016). Effects of salinity on olfactory toxicity and behavioral responses of juvenile salmonids from copper. *Aquat. Toxicol.*, *175*, 260–268. https://doi.org/10.1016/j.aquatox.2016.04.001

Son, Y. (2010). Molecular mechanisms of general anesthesia. *Korean J. Anesthesiol.*, *59*(1), 3. https://doi.org/10.4097/kjae.2010.59.1.3

Sorenson, P. W., and Caprio, J. (1998). Chemoreception. In *The Physiology of Fishes*, Evans, D. H., Ed., CRC Press, Boca Raton, FL, pp. 375–405.

Stednitz, S. J., McDermott, E. M., Ncube, D., Tallafuss, A., Eisen, J. S., and Washbourne, P. (2018). Forebrain control of behaviorally driven social orienting in zebrafish. *Curr. Biol.*, *28*(15). https://doi.org/10.1016/j.cub.2018.06.016

Stevens, M., Peigneur, S., and Tytgat, J. (2011). Neurotoxins and their binding areas on voltage-gated sodium channels. *Front. Pharmacol.*, *2*. https://doi.org/10.3389/fphar.2011.00071

Stoczynski, L., and van den Hurk, P. (2020). Effects of selective serotonin reuptake inhibitor sertraline on hybrid striped bass predatory behavior and brain chemistry. *Aquat. Toxicol.*, *226*, 105564. https://doi.org/10.1016/j.aquatox.2020.105564

Straus, D. L., and Chambers, J. E. (1995). Inhibition of acetylcholinesterase and aliesterases of fingerling channel catfish by chlorpyrifos, parathion, and *S,S,S*-tributyl phosphorotrithionate (DEF). *Aquat. Toxicol.*, *33*, 311–324.

Sutterlin, A. M., and Sutterlin, N. (1970). Taste responses in Atlantic salmon (*Salmo salar*) parr. *J. Fish. Res. Board Can.*, *27*, 1927–1942.

Suzuki, D. G., and Grillner, S. (2018). The stepwise development of the lamprey visual system and its evolutionary implications. *Biol. Rev.*, *93*(3), 1461–1477. https://doi.org/10.1111/brv.12403

Svoboda, K. R., Vijayaraghavan, S., and Tanguay, R. L. (2002). Nicotinic receptors mediate changes in spinal motorneuron development and axonal pathfinding in embryonic zebrafish exposed to nicotine. *J. Neurosci.*, *22*, 10731–10741.

Swanson, K. L., Aronstam, R. S., Wonnacott, S., Rapoport, H., and Albuquerque, E. X. (1991). Nicotinic pharmacology of anatoxin analogs. I. Side chain structure-activity relationships at peripheral agonist and noncompetitive antagonist sites. *J. Pharmacol. Exp. Ther.*, *259*, 377–386.

Taku, A. A., Marcaccio, C. L., Ye, W., Krause, G. J., and Raper, J. A. (2016). Attractant and repellent cues cooperate in guiding a subset of olfactory sensory axons to a well-defined protoglomerular target. *Development (Cambridge)*, *143*(1), 123–132. https://doi.org/10.1242/dev.127985

Tandon, S. S., Srivastava, P. P., Mukherjee, S. C., and Saharan, N. (2005). Acute toxicity of synthetic pyrethroids to Indian major carp, *Catla catla* L. *Bull. Environ. Contam. Toxicol.*, *74*, 610–613.

Tanoue, R., Nozaki, K., Nomiyama, K., Kunisue, T., and Tanabe, S. (2020). Rapid analysis of 65 pharmaceuticals and 7 personal care products in plasma and whole-body tissue samples of fish using acidic extraction, zirconia-coated silica cleanup, and liquid chromatography-tandem mass spectrometry. *J. Chromatogr. A*, *1631*. https://doi.org/10.1016/j.chroma.2020.461586

Taylor, P. (2001). Anticholinesterase agents. In *Goodman and Gillman's The Pharmacological Basis of Therapeutics*, Hardman J. G., and Limbird, L. E., Eds., McGraw-Hill, New York, pp. 171–191.

Tchounwou, P. B., Yedjou, C. G., Patlolla, A. K., and Sutton, D. J. (2012). Heavy metal toxicity and the environment. In *Molecular, Clinical and Environmental Toxicology. Experientia Supplementum* (Vol. 101, pp. 133–164). A. Luch (Ed.), https://doi.org/10.1007/978-3-7643-8340-4_6

Teles, M. C., Cardoso, S. D., and Oliveira, R. F. (2016). Social plasticity relies on different neuroplasticity mechanisms across the brain social decision-making network in zebrafish. *Front. Behav. Neurosci.*, *10*(FEB). https://doi.org/10.3389/fnbeh.2016.00016

Thomas, D. J. (2021). Arsenic methylation - Lessons from three decades of research. *Toxicology*, *457*. https://doi.org/10.1016/j.tox.2021.152800

Tian, L., Cheng, J., Chen, X., Cheng, S. H., Mak, Y. L., Lam, P. K. S., Chan, L. L., and Wang, M. (2014). Early developmental toxicity of saxitoxin on medaka (Oryzias melastigma) embryos. *Toxicon*, *77*, 16–25. https://doi.org/10.1016/j.toxicon.2013.10.022

Tierney, K. B., Baldwin, D. H., Hara, T. J., Ross, P. S., Scholz, N. L., and Kennedy, C. J. (2010). Olfactory toxicity in fishes. In *Aquat. Toxicol.*, *96*(1), 2–26. https://doi.org/10.1016/j.aquatox.2009.09.019

Tilak, K. S., Veeraiah, K., and Vardhan, K. S. (2003). Toxicity and residue studies on fenvalerate to the freshwater fish *Channa punctatus* (Bloch). *Bull. Environ. Contam. Toxicol.*, *71*, 1207–1212.

Tilson, H. A. (1996). Evolution and current status of neurotoxicity risk assessment. *Drug Metab. Rev.*, *28*, 121–139.

Timme-Laragy, A. R., Levin, E. D., and Di Giulio, R. T. (2006). Developmental and behavioral effects of embryonic exposure to the polybrominated diphenylether mixture DE-71 in the killifish (*Fundulus heteroclitus*). *Chemosphere*, *62*, 1097–104.

Tomasello, D. L., and Sive, H. (2020). Noninvasive multielectrode array for brain and spinal cord local field potential recordings from live zebrafish larvae. *Zebrafish*, *17*(4), 271–277. https://doi.org/10.1089/zeb.2020.1874

Tomasello, D. L., and Wlodkowic, D. (2022). Noninvasive Electrophysiology: Emerging Prospects in Aquatic Neurotoxicity Testing. *Environ. Sci. Technol.*, *56*(8), 4788–4794. https://doi.org/10.1021/acs.est.1c08471

Triller, A., Sur, C., and Korn, H. (1993). Heterogeneous distribution of glycinergic and GABAergic afferents on an identified central neuron. *J. Comp. Neurol.*, *338*, 83–96.

Tripathi, G., and Verma, P. (2004). Fenvalerate-induced changes in catfish, *Clarias batrachus*: metabolic enzymes, RNA and protein. *Comp. Biochem. Physiol. C*, *138*, 75–79.

Trudeau, V. L. (2022). Neuroendocrine control of reproduction in teleost fish: concepts and controversies. *Annu. Rev. Animal Biosci.*, *10*. https://doi.org/10.1146/annurev-animal-020420-042015

Tsai, G., and Coyle, J. T. (1998). The role of glutamatergic neurotransmission in the pathophysiology of alcoholism. *Annu. Rev. Med.*, *49*, 173–184.

Tsata, V., and Wehner, D. (2021). Know how to regrow-axon regeneration in the Zebrafish spinal cord. *Cells*, *10*(6). https://doi.org/10.3390/cells10061404

Tuazon, F. B., and Mullins, M. C. (2015). Temporally coordinated signals progressively pattern the anteroposterior and dorsoventral body axes. *Semin. Cell Dev. Biol.*, *42*, 118–133. https://doi.org/10.1016/J.SEMCDB.2015.06.003

Ullah, S., Li, Z., Zuberi, A., Arifeen, M. Z. U., and Baig, M. M. F. A. (2019). Biomarkers of pyrethroid toxicity in fish. *Environ. Chem. Lett.*, *17*(2), 945–973. https://doi.org/10.1007/s10311-018-00852-y

USEPA. (2019). Docket Number *EPA-HQ-OPP-2008-0331, Pyrethroids and Pyrethrins: Ecological Risk Mitigation Proposal For 23 Chemicals*. Downloaded from Regulations.gov 25 September 2022.

Valles, S., Hernández-Sánchez, J., Dipp, V. R., Huerta-González, D., Olivares-Bañuelos, T. N., González-Fraga, J., and Bardullas, U. (2020). Exposure to low doses of inorganic arsenic induces transgenerational changes on behavioral and epigenetic markers in zebrafish (Danio rerio). *Toxicol. Appl. Pharmacol.*, *396*, 115002. https://doi.org/10.1016/j.taap.2020.115002

van Deusen, E. B., and Meyer, R. L. (1990). Pharmacologic evidence for NMDA, APB and kainate/quisqualate retinotectal transmission in the isolated whole tectum of goldfish. *Brain Res.*, *536*, 86–96.

Van Ginneken, V., Nieveen, M., Van Eersel, R., Van den Thillart, G., and Addink, A. (1996). Neurotransmitter levels and energy status in brain of fish species with and without the survival strategy of metabolic depression. *Comp. Biochem. Physiol. C*, *114A*, 189–196.

Vanegas, H. (1983). Organization and physiology of the teleostean optic tectum. In *Fish Neurobiology: Higher Brain Areas and Functions* (Vol. 2), Davis, R. E., and Northcutt, R. G., Eds., The University of Michigan Press, Ann Arbor, MI, pp. 43–90.

Veith, G. D., Call, D. J., and Brook, L. T. (1983). Structure-toxicity relationships for the fathead minnow, *Pimephales promelas*: narcotic industrial chemicals. *Can. J. Fish. Aquat. Sci.*, *40*, 743–748.

Venkatesh, B., Lu, S. Q., Dandona, N., See, S. L., Brenner, S., and Soong, T. W. (2005). Genetic basis of tetrodotoxin resistance in pufferfishes. *Curr. Biol.*, *15*(22), 2069–2072. https://doi.org/10.1016/j.cub.2005.10.068

Vernier, P. (2017). The brains of teleost fishes. In *Evolution of Nervous Systems: Second Edition*, *1–4*, 59–75. https://doi.org/10.1016/B978-0-12-804042-3.00004-X

Vilariño, N., Louzao, M., Abal, P., Cagide, E., Carrera, C., Vieytes, M., and Botana, L. (2018). Human poisoning from marine toxins: unknowns for optimal consumer protection. *Toxins*, *10*(8), 324. https://doi.org/10.3390/toxins10080324

Wallace, K. B., and Herzberg, U. (1988). Reactivation and aging of phosphorylated brain acetylcholinesterase from fish and rodents. *Toxicol. Appl. Pharmacol.*, *92*, 307–314.

Wang, H., Holland, P. W. H., and Takahashi, T. (2019). Gene profiling of head mesoderm in early zebrafish development: Insights into the evolution of cranial mesoderm. *EvoDevo*, *10*(1). https://doi.org/10.1186/s13227-019-0128-3

Wang, M., Wang, D., Lin, L., and Hong, H. (2010). Protein profiles in zebrafish (Danio rerio) brains exposed to chronic microcystin-LR. *Chemosphere*, *81*(6), 716–724. https://doi.org/10.1016/j.chemosphere.2010.07.061

Weber, D. N. (2006). Dose-dependent effects of developmental mercury exposure on C-start escape responses of larval zebrafish Danio rerio. *J. Fish Biol.*, *69*(1), 75–94. https://doi.org/10.1111/j.1095-8649.2006.01068.x

Weber, D. N., Connaughton, V. P., Dellinger, J. a, Klemer, D., Udvadia, A., Carvan III, M. J., and Carvan, M. J. (2008). Selenomethionine reduces visual deficits due to developmental methylmercury exposures. *Physiol. Behav.*, *93*(1), 250–260. https://doi.org/10.1016/j.physbeh.2007.08.023

Weber, D. N., Dingel, W. M., Panos, J. J., and Steinpreis, R. E. (1997). Alterations in neurobehavioral responses in fishes exposed to lead and lead-chelating agents. *Am. Zool.*, *37*, 354–362.

Weber, D. N., and Ghorai, J. K. (2013). Experimental design affects social behavior outcomes in adult zebrafish developmentally exposed to lead. *Zebrafish*, *10*(3), 294–302. https://doi.org/10.1089/zeb.2012.0780

Weber, D. N., Russo, A., Seale, D. B., and Spieler, R. E. (1991). Waterborne lead affects feeding abilities and neurotransmitter levels in juvenile fathead minnows (*Pimephales promelas*). *Aquat. Toxicol.*, *21*, 71–80.

Weber, D. N., and Spieler, R. E. (1994). Behavioral mechanisms of metal toxicity in fishes. In *Aquatic Toxicology: Molecular, Biochemical, and Cellular Perspectives*, Malins, D. C., and Ostrander, G. K., Eds., Lewis Publishers, Boca Raton, FL, pp. 421–467.

Weis, J. S., and Candelmo, A. (2012). Pollutants and fish predator/prey behavior: A review of laboratory and field approaches. *Curr. Zool.*, *58*(1), 9–20. https://doi.org/10.1093/czoolo/58.1.9

Williams, C. R., and Gallagher, E. P. (2013). Effects of cadmium on olfactory mediated behaviors and molecular biomarkers in coho salmon (Oncorhynchus kisutch). *Aquat. Toxicol.*, *140–141*, 295–302. https://doi.org/10.1016/j.aquatox.2013.06.010

Woodruff, T. J., Rayasam, S. D. G., Axelrad, D. A., Koman, P. D., Chartres, N., Bennett, D. H., Birnbaum, L. S., Brown, P., Carignan, C. C., Cooper, C., Cranor, C. F., Diamond, M. L., Franjevic, S., Gartner, E. C., Hattis, D., Hauser, R., Heiger-Bernays, W., Joglekar, R., Lam, J., … Zeise, L. (2023). A science-based agenda for health-protective chemical assessments and decisions: overview and consensus statement. *Environ. Health: Glob. Access Sci.*, *21*, 132. https://doi.org/10.1186/s12940-022-00930-3

Woolley, D. E. (1995). Organochlorine insecticides: neurotoxicity and mechanisms of action. In *Handbook of Neurotoxicology*, Chang, L. W. and Dyer, R. S., Eds., Marcel Dekker, New York, pp. 475–510.

Wren, C. D., Harris, S., and Harttrup, N. (1995). Ecotoxicology of mercury and cadmium. In *Handbook of Ecotoxicology*, Hoffmann, D. J., Rattner, B. A., Burton, Jr., G. A., and Cairns, Jr., J., Eds., CRC Press, Boca Raton, FL, pp. 392–423.

Wu, Q., Yan, W., Liu, C., Hung, T.-C., and Li, G. (2019). Co-exposure with titanium dioxide nanoparticles exacerbates MCLR-induced brain injury in zebrafish. *Sci. Total Environ.*, *693*, 133540. https://doi.org/10.1016/j.scitotenv.2019.07.346

Wullimann, M. F. (1998). The central nervous system. In *The Physiology of Fishes*, Evans, D. H., Ed., CRC Press, Boca Raton, FL, pp. 245–282.

Xu, X., C. Lamb, M. Smith, L. Schaefer, M.J. Carvan III, D. Weber. 2012. Developmental methylmercury exposure affects avoidance learning outcomes in adult zebrafish. *Journal of Toxicology and Environmental Health Sciences* 4:85-91. https://doi.org/10.5897/JTEHS12.004

Yang, D., Lauridsen, H., Buels, K., Chi, L. H., La Du, J., Bruun, D. A., Olson, J. R., Tanguay, R. L., and Lein, P. J. (2011). Chlorpyrifos-oxon disrupts zebrafish axonal growth and motor behavior. *Toxicol. Sci.*, *121*(1), 146–159. https://doi.org/10.1093/toxsci/kfr028

Yang, F., Yun, Y., Li, G., and Sang, N. (2021). Heavy metals in soil from gangue stacking areas increases children health risk and causes developmental neurotoxicity in zebrafish larvae. *Sci. Total Environ.*, *794*. https://doi.org/10.1016/j.scitotenv.2021.148629

Yang, L., Guo, H., Kuang, Y., Yang, H., Zhang, X., Tang, R., Li, D., and Li, L. (2022). Neurotoxicity induced by combined exposure of microcystin-LR and nitrite in male zebrafish (Danio rerio): Effects of oxidant-antioxidant system and neurotransmitter system. *Comp. Biochem. Physiol. Part C: Toxicol. Pharmacol.*, *253*, 109248. https://doi.org/10.1016/j.cbpc.2021.109248

Yoshikawa, H. (2001). Myelin-associated oligodendrocytic basic protein modulates the arrangement of radial growth of the axon and the radial component of myelin. *Med. Electron. Microsc.*, *34*, 160–164.

Yu, Y., Zhang, Q., Liu, G., Deng, Y., Kang, J., Zhang, F., Lu, T., Sun, L., and Qian, H. (2021). Proteomic analysis of zebrafish brain damage induced by Microcystis aeruginosa bloom. *Sci. Total Environ.*, *795*, 148865. https://doi.org/10.1016/J.SCITOTENV.2021.148865

Zhang, J., Lefebvre, J. L., Zhao, S., and Granato, M. (2004). Zebrafish unplugged reveals a role for muscle-specific kinase homologs in axonal pathway choice. *Nat. Neurosci.*, *7*(12), 1303–1309. https://doi.org/10.1038/nn1350

Zhang, K., Liang, J., Brun, N. R., Zhao, Y., and Werdich, A. A. (2021). Rapid Zebrafish behavioral profiling assay accelerates the identification of environmental neurodevelopmental toxicants. *Environ. Sci. Technol.*, *55*(3), 1919–1929. https://doi.org/10.1021/acs.est.0c06949

Zinkl, J., Lockhart, W., Kenny, S., and Ward, F. (1991). Effects of cholinesterase inhibiting insecticides on fish. In *Cholinesterase-Inhibiting Insecticides: Impacts on Wildlife and the Environment*, Mineau, P., Ed., Elsevier, Amsterdam, pp. 233–254.

# 6 The Immune System of Fish
## *A Target Organ of Toxicity*

*Charles D. Rice*
Clemson University

*Judy Zelikoff*
New York University School of Medicine

*Helmut Segner*
University of Bern

**Abstract**

The purpose of Chapter 6 is to provide a general overview of our current understanding of how xenobiotics target the immune system of teleostean fish and explore new frontiers toward understanding mechanisms of action and how these mechanisms may address the question of ecotoxicological consequences. There are over 20,000 different species of fishes, so it is impossible to draw simple or generalized conclusions about the immunobiology or immunotoxicology of all fish, because the differences between immune systems of major groups and between species within each group are substantial. Herein, we focused on those species of fish commonly used in research relevant to aquaculture, environmental health, and as models for human and veterinary biomedicine. Although seemingly primitive in comparison with mammals, the immune systems of fish are highly evolved to meet the demands of the environment in which they live. Surrounded by water of optimal temperature, fish exist in a relatively stable environment. Water, however, is teaming with microorganisms, some of which are pathogenic, while others may share a commensal or mutualistic relationship with the host, and they may form part of the microbiotic community (microbiome) on the epithelial surfaces of fish. The immune system must distinguish between these relationships. In addition, the water column contains dissolved inorganic and organic substances, some of which may affect immunocompetence. The importance of integument as an immune organ is maintained throughout the evolution of vertebrates, culminating in an extremely complex component of immunity and homeostasis in all modern vertebrates.

The purpose of this chapter is to provide a general overview of our current understanding of how xenobiotics target the immune system of teleostean fish and explore new frontiers toward understanding mechanisms of action and how these mechanisms may address the question of ecotoxicological consequences. There are over 20,000 different species of fish ranging from primitive agnathan cyclostomes (e.g., hagfish) to gnathan fish including chondrosteans (sturgeons, paddlefish), elasmobranchs (sharks, skates, and rays), and teleostean, or modern bony fish (cod, trout, catfish, tuna, killifish, zebrafish, blue gill, grouper, etc.). It is impossible to draw simple or generalized conclusions about the immunobiology or immunotoxicology of all fish, because the differences between immune systems of major groups and between species within each group are substantial. Herein, we focus on those species of fish commonly used in research relevant to aquaculture, environmental health, and as models for human and veterinary biomedicine.

Although seemingly primitive in comparison with mammals, the immune systems of fish are highly evolved to meet the demands of the environment in which they live. Surrounded by water of optimal temperature, fish exist in a relatively stable environment. Water, however, is teaming with microorganisms, some of which are pathogenic, while others may share a commensal or mutualistic relationship with the host, and they may form part of the macrobiotic community (microbiome) on the epithelial surfaces of fish. The immune system must distinguish between these relationships. In addition, the water column contains dissolved inorganic and organic substances, some of which may affect immunocompetence. Therefore, the mucosa of the alimentary canal, skin mucosa, and gill lamellae in contact with water collectively form perhaps the most important immune organ, the integument. The importance of the integument as an immune organ is maintained throughout the evolution of vertebrates, culminating in an extremely complex component of immunity and homeostasis in all modern vertebrates.

The immune system of fish must be able to replenish immuno-competent cells, and it must be in a sentinel location to process, present, and communicate information about the form and nature of foreign (non-self) materials, or altered homeostasis, to other cells of the immune system and other tissues. These cells of immune system of all gnathan vertebrates are either myeloid or lymphoid in origin and function. It is generally accepted that myeloid and lymphoid cells of fish are derived from a single pluripotent stem cell, as is the case with higher vertebrate systems. However, as discussed, we now appreciate that non-myeloid lymphoid cells such as epithelial cells, fibroblasts, endothelial cells, and other cells considered to be

DOI: 10.1201/9781003160694-6

stromal are also involved in immunophysiological homeostasis and especially along and within barrier tissues.

In higher vertebrates, myeloid and lymphoid cell types are generated in two separate compartments, the lymphoid organs (lymph nodes, thymus, and spleen) and the myeloid organ (the bone marrow). Fish do not have bone marrow or lymph nodes. Instead, the hematopoietic tissues of fish include spleen, thymus, anterior kidney (syn. pronephros/head kidney), posterior kidney (syn. mesonephros/trunk kidney/excretory kidney), and mucosa-associated lymphoid tissues (MALT) (Press and Evesen, 1999; Ferguson, 2006; Haugarvoll et al., 2008; Bjorgen and Koppang, 2021). Hematopoiesis is major role of the kidney, which, during development, differentiates into a hematopoietic anterior portion and a renal organ in the caudal region, which, in addition to the excretory function, contributes to blood cell formation. In some species of fish, such as the channel catfish, *Ictalurus punctatus*, these two areas are anatomically distinct. In others, there is a gradation of cell types along the renal tissue.

For the most part, early concepts in fish immunotoxicology dating back more than 50 years followed rapid advances made in rodent and human immunology and immunotoxicology. Logically, fish immunologists and immunotoxicologists initially focused on immune cell isolation and purification, fish cell culture, and cell-based functional assays, and ultimately to gene expression and gene regulation. Likewise, serological studies that examined the effects of xenobiotics on humoral immune responses were also modeled after mammalian systems. In large part, experimental designs using fish as aquatic models were intended to replicate as much as possible the various approaches used in mammalian immunotoxicology. As will be discussed throughout this chapter, much has been learned about fish immunology and immunotoxicology, but much remains unknown, and this is especially true when it comes to how we can use fish immunotoxicological data for environmental regulatory purposes. As we look into the future of fish immunotoxicology, the use of computational biology, toxicogenomics, and single-cell sequencing will be important to more appreciate the effects of xenobiotics on the immune system of fishes.

This chapter will also address how what we have learned over the past 50 years or so can be better appreciated considering a return to understanding and appreciating how the immune system is integrated with the whole organism and especially as we re-envision the role of physiological inflammation in homeostasis. As discussed, the immune system as a target organ for xenobiotics is difficult to understand in comparison to discrete organs such as the liver, eyes, gonads, and gills because the immune system is distributed with, and integrated across, the whole animal, and importantly, many cells of the immune system are trafficking throughout the animal. Under healthy conditions, this is not the case with the more succinctly located organs. In fact, the immune system shares many similarities with the neuro-endocrine system(s) and therefore may be considered the immune-neuro-endocrine systems in terms of whole-animal integration. To this point, the terms "meta-inflammation" and "metabolic inflammation" are adding a new dimension to our understanding of the natural role(s) of inflammation in whole organismal biology and will be discussed. Moreover, the ever-advancing field of barrier immunobiology informs us that a healthy mucosal barrier system also consists of symbiotic microflora (microbiome) and that one of several mechanisms of action associated with xenobiotic-induced immunotoxicity is a disruption of the healthy microbiome leading to dysbiosis. This new state of physiological dysfunction associated with dysbiosis results in a shift of barrier immunology toward a pro-inflammatory and immunotoxic environment that may have whole organismal effects. These new and exciting areas of immunobiology and immunotoxicology also may shed light on experimental observations made over the last half-century. A summary of the literature devoted to descriptive and historical studies in fish immunotoxicology has been reduced. Instead, the reader is referred to a partial listing of key reviews from the past several decades (Snieszko, 1974; Sindermann, 1979; Zeeman and Brindley, 1981; Sinderman *et al.*, 1982.; Porter *et al.*, 1984; Zeeman, 1986; Anderson, 1990; Weeks *et al.*, 1992; Dunier and Siwicki, 1993; Zelikoff, 1994; Rice, 2001; Rice and Arkoosh, 2002; Carlson and Zelikoff, 2008a; Segner *et al.*, 2011; Rehberger *et al.*, 2017b).

## IMMUNE CELLS

The immune system of vertebrates is physically and functionally integrated within the whole organism, consisting of cells organized among stromal cells, fibroblasts, and endothelial cells, and other support cells within tissues located throughout the organism that then comprise what is collectively referred to as the immune system. Another feature of the immune system is that many of these cells are constantly trafficking throughout the body, seeking damaged self, pathogens, or a lack of homeostasis. Being in the surface epithelia of gut, skin, and gills, as well as in the bloodstream, immune cells encounter a first-pass contact with toxicants during chemical uptake and distribution within the organism. In terms of fish immunology, many morphological and functional similarities exist between mammalian and teleostean leukocytes (Scapigliati *et al.*, 2018b). Lymphocytes, macrophages (MØs)/monocytes, granulocytes (neutrophilic, eosinophilic, and basophilic), and nonspecific cytotoxic cells (NCCs) have all been described in a variety of teleost models (Zhu and Su, 2022); however, species-specific differences in occurrence, morphology, and function are common. These differences among fishes make it difficult to generalize about fish leukocytes, overall. All these cell types are vitally important for inflammation and probably represent the most primitive group of immune-effector cells. Although excellent descriptions of cell types, including MØs (Ainsworth, 1992; Secombes and Fletcher, 1992; Suzuki and Iida, 1992; Havixbeck and Barreda, 2015; Hodgkinson *et al.*, 2015) granulocytes, NCCs (Evans and Jaso-Friedmann, 1992; Secombes *et al.*, 1996), T-cells and B-cells (Zwollo *et al.*, 2008; Scapigliati, 2013b; Yamaguchi *et al.*, 2019; Perdiguero *et al.*, 2021), can be found elsewhere, a brief review of immune cells and their functions follow.

It appears that jawed vertebrates (gnathostomes) are the most phylogenetically primitive organisms to possess lymphocytes. Both B- and T-lymphocytes have been described in teleosts primarily due to functional and structural similarities shared by their mammalian counterparts. The presence of antibody in bony fish led early investigators to believe that fish possessed antibody-producing B-lymphocytes. The expression of surface immunoglobulin (sIg) on fish lymphocytes has since been described in all fish species, including channel catfish (*Ictalurus punctatus*) (Lobb and Clem, 1982), carp (*Cyprinus carpio*) (Koumans-van Diepen *et al.*, 1994), sea bass (*Dicentrarchus labrax*) (Romestand, 1995). Teleost T-lymphocytes are generally recognized as surface immunoglobulin-negative lymphocytic cells, and T-lymphocyte-specific monoclonal antibodies have been produced for some species (Partula, 1999; Scapigliati *et al.*, 1999; Chaudhary *et al.*, 2012; Rebello *et al.*, 2016). The known functions of fish lymphocytes are described in detail later in this section.

## LYMPHOID ORGANS AND TISSUES

The lymphoid tissues and organs of teleost fish have previously been reviewed (Zapata, 1996; Press and Evensen, 1999; Soulliere and Dixon, 2017; Bjørgen and Koppang, 2021). Unlike mammals, fish lack hematopoietic bone marrow and identifiable lymph nodes, though recent work demonstrates the presence of a lymphatic system (Karpanen and Mäkinen, 2006; Yaniv *et al.*, 2006; Burggren *et al.*, 2017; Castranova *et al.*, 2021). Presumably, lymphatic circulation functions similarly to mammalian lymphatics through lipid absorption, interstitial tissue drainage, fluid balance, and a key component of immune functions. The primary site of hematopoiesis in teleosts is the pronephros or anterior portion of the kidney (commonly referred to as the head or cranial kidney), although our understanding of fish hematopoesis has improved greatly with the help of transgenic zebrafish lines (Davidson and Zon, 2004; He *et al.*, 2018; Lin *et al.*, 2020) and other common fish models of immunity (Zwollo *et al.*, 2005). Teleosts also possess a thymus and spleen (although splenic germinal centers are absent) along with mucosal lymphoid tissues. The morphology of fish lymphoid organs is highly species-specific, although some generalizations about structure and function can be made.

### Thymus

In most teleosts, the thymus is a paired organ located dorsolaterally above the opercular cavities (Chilmonczyk, 1992). The thymus has been found anywhere from just beneath the opercular epithelium, as in rainbow trout (*Oncorhynchus mykiss*) (Chilmonczyk, 1985), to a more internalized location, as described in angler fish (Fänge and Pulsford, 1985). Histologically, the thymus is poorly differentiated into cortical and medullary regions (Zapata, 1996; de Medeiros *et al.*, 2021), although in some species like seabass, *Dicentrachus labrax*, the cortico-medullar separation is clearly demarcated (Romano *et al.*, 1999a; Paiola *et al.*, 2017; Barraza *et al.*, 2021). In general, the thymus of lower vertebrates is less structured compared to higher vertebrates (da Costa Klosterhoff *et al.*, 2020). Figure 6.1 demonstrates the thymus from Japanese medaka (*Oryzias latipes*).

As in higher vertebrates, the fish thymus is believed to be the first lymphoid tissue to become populated by lymphocytes during development, and this seems to be true for most freshwater species, but not so for marine species. The ontogenesis of the thymus gland has been described for many fish species, including rainbow trout (*Oncorhynchus mykiss*) (Grace and Manning, 1980; Chilmonczyk, 1992), carp (*Cyprinus carpio*) (Van Loon *et al.*, 1982; Romano *et al.*, 1999a), zebrafish (*Danio rerio*) (Willett *et al.*, 1997; Willett *et al.*, 1999a), sea bass (*Dicentrarchus labrax*) (Abelli *et al.*, 1996), channel catfish (*Ictalurus punctatus*) (Petrie-Hanson and Ainsworth, 2001), tilapia (*Oreochromis niloticus*) (Fishelson, 1995), and the Atlantic killifish, (*Fundulus heteroclitus*) (Hunt and Rice, 2008). Involution of the teleost thymus has been described during the aging process (Ghoneum and Egami, 1982; Cooper *et al.*, 1983; Ellsaesser *et al.*, 1988; Fishelson, 1995; Kernen *et al.*, 2020) and following exposure to various environmental stimuli (Álvarez *et al.*, 1998; Kernen *et al.*, 2020). Besides being populated by both $sIg^-$ and $sIg^+$ lymphocytes (Scapigliati *et al.*, 1999), the thymus has been observed to possess epithelial cells of varying morphologies (Castillo *et al.*, 1990; Romano *et al.*, 1999b): MØs and monocytes (Castillo *et al.*, 1990); myeloid cells (Zapata, 1996);

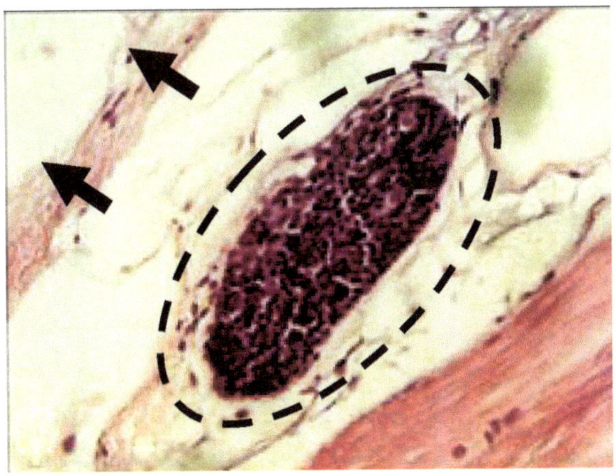

**FIGURE 6.1** Teleost thymus as represented by the Japanese medaka (*Oryzias latipes*). Note the close proximity of the organ (dashed circle) to the opercular epithelium (arrows). Hematoxylin and eosin-stained light micrograph (100× magnification).

Hassell's body-like structures (Romano *et al.*, 1999b); nurse cells (Flaño *et al.*, 1996; de Medeiros *et al.*, 2021); and neuroendocrine cells (Ottaviani *et al.*, 1995; Ottaviani *et al.*, 1997). Although the function of many of these thymic cell types remains to be determined, there is ample evidence that the teleost thymus serves the same role in T-lymphocyte maturation in fish as in mammals (Chilmonczyk, 1992; Nakanishi *et al.*, 2015; da Costa Klosterhoff *et al.*, 2020). Such evidence includes: (1) migration of radiolabeled lymphocytes through the thymus (Tatner, 1985), (2) Con A-induced proliferation of thymic lymphocytes (Ellsaesser *et al.*, 1988), (3) apoptosis of thymocytes (Abelli *et al.*, 1998), and (4) *rag* gene expression in the thymus (Willett *et al.*, 1997).

## Anterior Kidney

The kidney of teleost fish not only contains the excretory tissues also found in mammals (i.e., nephrons), but also houses the medullary and cortical adrenal homologs as well as stanniocalcin-secreting organs, the corpuscles of Stannius, and hematopoietic tissue (Figure 6.2). The fish kidney is often situated along the dorsal wall of the body cavity and can sometimes be divided into anterior (head or cranial) and posterior (trunk or caudal) sections. The anterior kidney is the primary hematopoietic organ in adult teleosts (Press and Evensen, 1999). The reticuloendothelial tissue of the head kidney forms an extensive system of sinusoids and dispersed populations of erythroid and lymphoid cells at various maturation stages and is sometimes considered as the analog of the human bone marrow. It appears from studies in cod (*Gadus morhua*) (Schroder et al., 1998), sea bass (*Dicentrarchus labrax*) (Breuil et al., 1997), rainbow trout (*Oncorhynchus mykiss*) (Castillo et al., 1993), and carp (*Cyprinus carpio*) (Koumans-van Diepen et al., 1994; Romano et al., 1997a, 1997b; Secombes et al., 1983) that the anterior kidney is the first lymphoid organ to possess sIg$^+$ cells during ontogenesis. These studies, along with those investigations examining the immune function of anterior kidney cells (Kaattari and Irwin, 1985; Zwollo *et al.*, 2005), suggest that this organ also serves as a secondary lymphoid organ.

The posterior kidney (*syn.* excretory kidney, trunk kidney), which corresponds to the mesonephros, is located retroperitoneally, between swim bladder and vertebral column. The organ is composed of glomeruli, renal tubuli, and hematopoietic tissue which forms a support matrix for the excretory structures. The hematopoietic part of the trunk kidney is often referred to as "interstitial tissue". It contains erythroid cells and myeloid cells as well as lymphocytes. The cells are randomly distributed, with no discernible specialized histological architecture.

## Spleen

The spleen may be regarded as the only bonafide secondary lymphoid organ shared among vertebrates (gnathans) (Neely and Flajnik, 2016). In most teleosts, the spleen is an encapsulated organ with abundant red pulp and poorly developed white pulp (Zapata, 1996; Press and Evensen, 1999) (Figure 6.3). Red pulp is populated by specialized macrophages and red blood cells, many of which are destined for recycling. White pulp contains lymphoid tissues where other specialized macrophages, T- and B-cells are located, and where antigen-specific immune responses (e.g., antibody production) take place. Unlike the thymus and anterior kidney, the teleost spleen seems to become organized late in the developmental process, depending on life history and whether freshwater or marine origins (Willett *et al.*, 1999b; Zapata *et al.*, 2006; Hunt and Rice, 2008). Although fish lack germinal centers, many immune cells can be found in both splenic red and white pulp (Zapata, 1982; Quesada *et al.*, 1990; Press *et al.*, 1994; Neely and Flajnik, 2016; Flajnik, 2018). The white pulp of teleosts contains poorly developed ellipsoids and may contain numerous MØ aggregations (Fänge and Nilsson, 1985; Press and Evensen, 1999; Sales *et al.*, 2017). These so-called MØ aggregates (MAs) or melanomacrophage centers have been observed in a variety of fish species; the occurrence and structure of MAs are species-, age-, environmental health-, and nutrition-dependent (Blazer *et al.*, 1987; Wolke, 1992; Camp *et al.*, 2000;

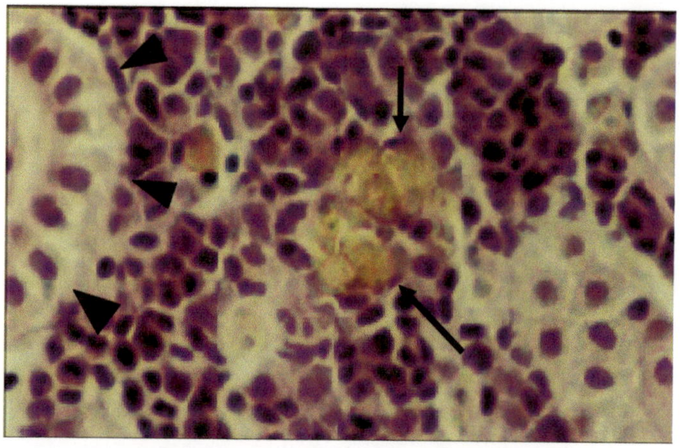

**FIGURE 6.2** Teleost anterior kidney as represented by the Japanese medaka (*Oryzias latipes*). Hematoxylin and eosin-stained light micrograph (100× magnification) demonstrating kidney hematopoietic tissue interspersed between renal tubules (arrowhead). Note the presence of small macrophage aggregates (arrow).

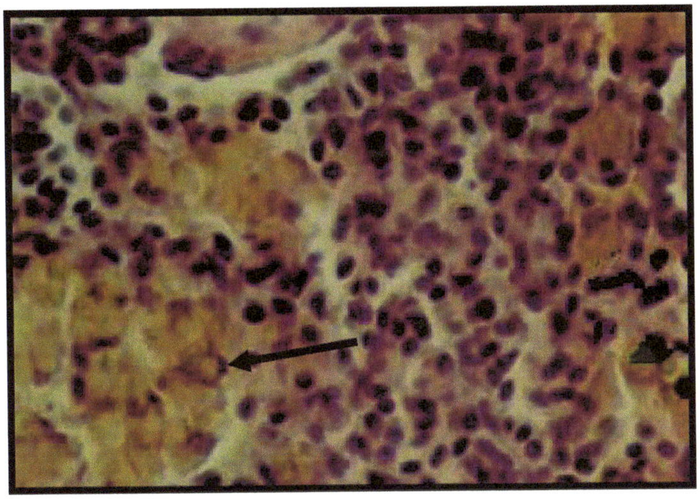

**FIGURE 6.3** Teleost spleen as represented by the Japanese medaka (*Oryzias latipes*). Hematoxylin and eosin-stained light micrograph (100× magnification) demonstrating hematopoietic splenocytes. Note the macrophage aggregates (arrow).

Fournie *et al.*, 2001; Agius and Roberts, 2003; Barst *et al.*, 2015; David and Kartheek, 2015; Steinel and Bolnick, 2017). Functions of macrophage aggregations are discussed in the section on innate immune functions.

### Mucosal Associated Lymphoid Tissues

Defining and defending the barrier between the outside and inside of an organism is one of the core aspects of immunity and homeostasis. Understanding the complexity of this barrier system has been one of the most intense areas of research in the field of immunology in recent times, leading to a new appreciation of mechanisms of barrier homeostasis. The classical view is that the barrier is protected first by an anatomical barrier, then a chemical barrier, and followed by a cellular defense. The chemical barriers consist of a plethora of constitutively expressed and inducible anti-microbial peptides (Valero *et al.*, 2020), and the cellular barrier consists of epithelial cells, MO, dendritic cells, natural killer cells, neutrophils, mast cell-like cells, and plasma cells (antibody-secreting cells) (Salinas, 2015). We now know that the barrier is covered by a layer of mucus products secreted by goblet cells, which themselves are under the control of epithelial signals and neuro-endocrine signals, as well as the healthy microbiome associated with the barrier. Perhaps one of the most intriguing findings of late is that the "true barrier" is both the anatomical barrier and the associated microbiome. Embedded within these mucosal areas are loose lymphoid-like tissues associated with the gut (GALT), skin (SALT), gill (GIALT), gill interbranchial (ILT), within the naso-pharynx (NALT), and most recently discovered, the cloacal bursa (CBur). Secretory/circulatory IgM is the primary immunoglobulin produced against specific immunogens/antigens, but within the last decade, a fish mucosal immunoglobulin (IgZ/IgT) has been described in several species, and will probably be discovered in most, if not all, teleosts (Fillatreau *et al.*, 2013). IgZ was described in zebrafish, while IgT was first described in rainbow trout, and most recently a consensus to name this class of immunoglobulin in all teleostean fishes as IgT (tau) has been proposed (Dornburg *et al.*, 2021). Functionally, IgT is similar in function to mucosal IgA of mammals (Mashoof and Criscitiello, 2016; Ji *et al.*, 2021). For a thorough examination of these subjects, the reader is referred to several recently published reviews (Salinas, 2015; Yu *et al.*, 2020; Bjørgen and Koppang, 2021). As discussed below in the section focused on mechanisms of action associated with xenobiotic-related immunotoxicology, it is the breach of barrier/mucosal/microbiome structure that imitates a homeostasis-seeking response to loss of structure, loss of function, and loss of regulation. In this light, it may be difficult to establish that a certain xenobiotic targets the immune "system" as a target organ above all other organ systems. Thus, one could argue that immunotoxicology is a result of inflammation associated with loss of structure, function, and regulation of the barrier system. This view may be valid when considering that absorption, distribution, metabolism, and elimination (ADMEK) is associated with the "epithelium" associated within various barriers near immune cells within an integrated organism.

## NONSPECIFIC IMMUNE FUNCTIONS

Nonspecific mechanisms of an immune response in teleosts include both soluble and cellular components. Soluble constituents are mainly evolutionarily conserved molecules that have broad cross-reactivity, recognizing molecular motifs shared by a wide variety of microorganisms. Nonspecific soluble molecules described in this chapter include complement, acute-phase proteins, and lysozyme. Although cytokines (e.g., interferons, interleukins, tumor necrosis factors, and transforming growth factors) are involved in the regulation of both innate and adaptive immune responses, they are included in this section for simplicity. Cellular components of the nonspecific immune response involved in maintaining internal homeostasis by removing dead or unwanted cells include MØs, granulocytes, and NCCs.

## Nonspecific Immune Responses: Soluble Components

### Complement

The complement system of fish has been reviewed extensively (Yano, 1996; Holland and Lambris, 2002; Boshra *et al.*, 2006; Najafpour *et al.*, 2020). Virucidal, bactericidal, parasiticidal, opsonic, and chemotactic activities of complement components have all been described in teleost species. Many teleost fish have been found to possess classical, alternative, and lytic complement components like those of mammals. Proteins involved in the lectin-activated pathway have also been described (Nakao *et al.*, 2000, 2006). Interestingly, the alternative complement pathways of several fish species are much more pronounced than those described for mammals, suggesting a greater importance for this pathway in fish immune defense (Yano, 1996; Sunyer *et al.*, 1998). Teleosts have been found to possess a more diverse array of complement components than higher vertebrates; for example, multiple C3 forms have been described for rainbow trout (*Oncorhynchus mykiss*), sea bream (*Sparus aurata*), and carp (*Cyprinus carpio*), and there is evidence that these isoforms are the products of separate genes (Sunyer *et al.*, 1996, 1997; Nakao *et al.*, 2000). Additionally, due to the presence of a molecule that shares characteristics of both C2 and factor B in bony fish, the emergence of separate molecules appears to have occurred following the divergence of teleost species from the tetrapod line (Boshra *et al.*, 2006).

### Cytokines (Interleukins)

The literature on fish cytokines has been reviewed extensively (Wang and Secombes, 2013; Bird and Tafalla, 2015; Secombes *et al.*, 2016; Zou and Secombes, 2016). From a classical mammalian perspective, cytokines and their cellular receptors are classified by function and regulation, which is dictated by protein structure. Cytokines are often redundant, pleiotropic, and even antagonistic depending on the expression patterns; locations/cell types/conditions; and, likewise, receptor expression patterns, locations/cell types/conditions. Interleukins (IL) are involved in the regulation of both innate and adaptive immunity (Secombes *et al.*, 2011). Historically, considerable effort has been directed toward the elucidation of interleukin genes in fish species (Secombes *et al.*, 1996, 2001; Secombes and Cunningham, 2004), and although amino acid sequence homologies between, for example, human or rodent IL-1-β and fish IL-1β may be low, the biological activities associated with innate immunity are high, reflecting higher conservation of receptors and their functions (Zou and Secombes, 2016).

In general, classifications of cytokines include those broadly associated with innate and/or adaptive immunity, while also being classified as the interleukin-1 family, hematopoeitin family, interferon family, tumor necrosis factor family, interleukin 17 family, and the chemokine family. Cytokines and their receptors may be expressed by/on cells typically associated with myeloid and lymphoid origin, but also fibroblasts, endothelial cells, and stromal cells. Of note, fish cytokines are involved in both innate and adaptive immunity (Secombes *et al.*, 2001; Wang and Secombes, 2013). Broadly, fish cytokine biology is similar in form and function to that of higher vertebrates.

One of the ways cytokines regulate immune functions is through the polarization of leukocytes and lymphocytes between functions that favor cell-mediated immune responses (Type 1), including degrees of acute inflammation, responses that favor humoral immunity, chronic inflammation, and wound healing (Type II). In terms of innate immunity, polarization of MØs, granuloctyes, and innate lymphoid-like cells toward Type 1 involve the pro-inflammatory type cytokines IFN-γ (interferon- γ), IL-1β, IL-12, and/or IL-18, to name a few. On the other hand, Type II responses are driven by IL-10, TGF-β (transforming growth factor- β), IL-4, and IL13 and others. The sources of these polarizing cytokines are key T-helper (Th) cells that are themselves polarized (e.g., Th1, Th2, Th-17, Th-9 etc.,) via expression of specific transcription factors in response to signals from stromal cells and antigen presenting cells (APCs). For more on the role of cytokines on innate immunity, the reader is referred to several key papers (Yang *et al.*, 2012; Ginhoux *et al.*, 2016; Qi *et al.*, 2016; Sitjà-Bobadilla *et al.*, 2016; Tafalla *et al.*, 2016; Levraud and Herbomel, 2019; Alesci *et al.*, 2020; Ferrero *et al.*, 2020; Rességuier *et al.*, 2020; Wentzel *et al.*, 2020a; Wentzel *et al.*, 2020b; Castranova *et al.*, 2021).

### Interferon Cytokines

In higher vertebrates, IFNs are typically classified as either type I, type II, or type III (Wack *et al.*, 2015; Lee and Ashkar, 2018). Type I IFNs are typified by IFN-α and IFN-β, but also IFN-ε, IFN-κ, and IFN-Ω, and collectively are antiviral. Type II is also known as IFN-γ, also referred to as immune IFN and strongly promotes cell-mediated immunity, while Type III is more recently described and may be involved with anti-fungal immunity. Evidence for the presence of interferon (IFN) in teleosts has been reviewed (Yano, 1996; Zou and Secombes, 2011). Teleost IFN activity was originally described in cell culture and by studies in rainbow trout (*Oncorhynchus mykiss*), demonstrating the induction of a protein with antiviral and MØ activating factor properties (Graham and Secombes, 1990). Recently, sequences of IFN-like genes have been identified in zebrafish (*Danio rerio*) (Altmann *et al.*, 2003) and catfish (*Ictalurus punctatus*) (Long *et al.*, 2004). Induction of antiviral fish Myxovius resistance (Mx) proteins by viral infection and polyinosinic:polycytidylic acid (poly I:C) incubation has also been demonstrated (Trobridge and Leong, 1995; Trobridge *et al.*, 1997; Lee *et al.*, 2000; Nygaard *et al.*, 2000). Due to the importance of understanding intracellular infection immunity in aquaculture species, advances in the field of fish IFNs are rapid (Zou and Secombes, 2016).

### Tumor Necrosis Factor (TNF) Cytokines

TNF family members and their receptors in fish stand out in discussion because they are involved in not only innate immunity and cell proliferation (Goetz *et al.*, 2004), but tissue remodeling (apoptosis) and lipolysis as well (Hong *et al.*, 2013; Liu *et al.*, 2015). Tumor necrosis factor sequences, like those of mammalian TNFα, have been described in several fish species (Hirono *et al.*, 2000; Bobe and Goetz, 2001; García-Castillo *et al.*, 2002; Zou *et al.*, 2003). Catfish TNFs were expressed in both lymphoid (spleen, head kidney, and thymus) and non-lymphoid (liver and gill) organs, as well as in peripheral blood leukocytes (Zou *et al.*, 2003), demonstrating a wide distribution.

### Transforming Growth Factor Cytokines

Members of the transforming growth factor (TGF) superfamily are involved in the regulation of numerous mammalian cellular responses including inflammation (Magor and Magor, 2001). Similarly, sequences homologous to mammalian TGF-β have been identified in several fish species (Daniels and Secombes, 1999; Harms et al., 2000; Laing et al., 1999; Magor and Magor, 2001; Zhan and Jimmy, 2000). In addition, mammalian TGF-$\beta_1$ was found to modulate rainbow trout MØ respiratory burst activity (Jang et al., 1994). One of the key functions of TGF-β, along with IL-10, in both fish and mammals is to suppress or regulate inflammatory and other immune responses, such as maintaining a T-regulatory cell population (Treg) (Nakanishi *et al.*, 2015; Sharma and Rudra, 2018).

### Acute-Phase Proteins, Anti-Microbial Peptides, and Other Soluble Factors

Lysozymes, low-molecular-weight antimicrobial factors that destroy Gram-positive bacterial cell walls, can also act as opsonins to enhance phagocytosis (Song *et al.*, 2021). Fish lysozyme has been found to be present in head kidney; skin; digestive tract; and gills; as well as intracellularly within neutrophils, monocytes, and eosinophilic granule cells (Yano, 1996; Marsh and Rice, 2010). One unique feature of fish lysozyme is that it appears to be effective against both Gram-positive and Gram-negative bacteria, although lysozyme-mediated Gram-negative bacterial lysis requires the presence of complement (Grinde, 1989). Other soluble compounds involved in teleost innate immunity—C-reactive proteins (CRPs) and other proteins of the acute phase responses, transferrins, lectins, $\alpha_2$-macroglobulin, and antimicrobial peptides—have been reported (Alexander and Ingram, 1992; Yano, 1996; Magor and Magor, 2001; Katzenback, 2015).

Serum CRP has been determined in various species of fish species, including rainbow trout (*Oncorhynchus mykiss*), channel catfish (*Ictalurus punctatus*), and plaice (*Pleuronectes platessa*). In addition, evidence exists for other compounds (present in serum and mucus of the digestive tract, gills, and skin) that may be involved in the acute-phase response. These include transferrin (Bayne and Gerwick, 2001), serum amyloid A and P (Jensen *et al.*, 1997; Lund and Olafsen, 1999), pre-cerebellin-like protein (Gerwick *et al.*, 2000), pleurocidin (Cole *et al.*, 1997; Douglas *et al.*, 2001), $\alpha_2$-macroglobulin (Mutsuro *et al.*, 2000), cathepsins (Aranishi and Mano, 2000), histone-like proteins (Noga *et al.*, 2001; Robinette and Noga, 2001), and lectins (Yano, 1996; Fujiki *et al.*, 2001; Inagawa *et al.*, 2001; Nakamura *et al.*, 2001). Beta defensins in fish have been recently described and shown to be remarkably conserved among vertebrates (Hao *et al.*, 2021).

## Nonspecific Immune Responses: Cellular Responses

### Monocytes/Macrophages

Macrophages, terminally differentiated mononuclear cells derived from circulating blood monocytes, have been well studied in mammals and a variety of fish species over the last several decades (Secombes and Fletcher, 1992; Enane *et al.*, 1993; Secombes, 1996; Forlenza *et al.*, 2011; Hodgkinson *et al.*, 2015). As in mammals, several distinct MØ subpopulations appear to exist in fish (Neumann *et al.*, 2000). Although investigations regarding the distribution of monocytes/MØs in adult fish and during development have been hampered due to a lack of readily available MØ-specific markers, studies have demonstrated MØs in carp (*Cyprinus carpio*) head kidney as early as 2 days after fertilization; by two weeks of age, MØs were observed in almost all lymphoid tissues (Romano *et al.*, 1997). More recently, the zebrafish model has demonstrated several types of macrophages exist, depending on the stage of development (Levraud and Herbomel, 2019). Phagocytosis, a critical function of monocytes/MØs and likely one of the most ancient immune mechanisms, has been the subject of intense investigation in many fish species (Secombes and Fletcher, 1992; Neumann *et al.*, 2001; Esteban *et al.*, 2015). Following uptake, engulfed pathogens are destroyed intracellularly via the production of reactive oxygen/nitrogen species and oxygen-independent mechanisms.

Several immunoregulatory molecules such as MØ-activating factor (MAF), TNF-α, TGF-β, β-glucans, growth hormone, and neurotransmitters have been implicated in the modulation of fish MØ phagocytosis (Secombes, 1994, 1996; Esteban *et al.*, 2015). Fish macrophages, along with B-cells and dendritic cells, are critically important as professional antigen processing and presenting cells (Lewis *et al.*, 2014; Kordon *et al.*, 2019) that connect innate immune functions with adaptive immunity, which is discussed under the section Adaptive Immunity.

Of interest is the occurrence in a variety of fish species of focal aggregations of pigmented MØ populations. MAs, sometimes referred to as centers, are found in a variety of ectothermic vertebrates and are commonly occurring among piscine species (Agius, 1980; Wolke, 1992). These so-called MØ aggregates, or melanomacrophage centers, have been observed in a variety of fish species; the occurrence and structure of MAs are species-, age-, environmental health-, and nutrition-dependent (Blazer *et al.*, 1987; Wolke, 1992; Camp *et al.*, 2000; Fournie *et al.*, 2001; Agius and Roberts, 2003; Barst *et al.*, 2015; David and Kartheek, 2015; Steinel and Bolnick, 2017). These aggregates, appearing anywhere from yellow to black in color, have been found to contain hemosiderin, lipofuscin (ceroid), or melanin pigments.

The gross anatomical and histological morphologies of MAs are highly species-dependent. Of note, these structures are often near rapidly proliferating lymphoid cells. Tissue distribution of MAs is species-dependent, but they are commonly observed in both lymphoid (i.e., spleen, kidney, and thymus) and non-lymphoid (i.e., liver, gonads, and heart) tissues.

The immunological relevance of MA in fish is highly controversial. Some believe that they serve as primordial germinal centers that function in antigen trapping and presentation, while others speculate that they serve as no more than "tissue dumpsters" that collect and harbor cellular byproducts and debris or as "recycling bins" involved in the destruction of red blood cells (Zapata, 1996; Steinel and Bolnick, 2017). To date, it is not clear if MA are part of the greater reticuloendothelial system (RES), also known as the monocyte/macrophage phagocytic system (MPS), a heterogenous population of phagocytes located systemically in fixed tissues and the circulatory system. These phagocytes are located within reticular connective tissues composed of type III collagen forming stromal connections for the phagocytes. In fish, these are found in the head kidney, the spleen, gills, brain as microglia, the peritoneal cavity, perhaps the liver, and probably part of the mucosal environment as well, and should not be confused with scavenger endothelial cells (Grace *et al.*, 1981; Seternes *et al.*, 2002; Nair *et al.*, 2003; Wu *et al.*, 2016; Seternes *et al.*, 2021).

## Granulocytes

Attempts to classify the granulocytic (heterophilic) cell types of teleosts according to mammalian nomenclature have been hampered by tremendous morphological variations in granulocyte subpopulations and the common occurrence of immature (blast) cells in fish (Ainsworth, 1992; Hine, 1992; Havixbeck and Barreda, 2015). Although some fish species (i.e., rainbow trout) do possess neutrophilic granulocytes with polymorphonuclear morphology akin to mammalian polymorphonuclear neutrophils (PMNs), the granulocytes of most fish species lack such a clear nuclear adaptation. Despite difficulties in classification and species-specific variations, neutrophilic, eosinophilic, and basophilic granulocytes have been described for many fish species (Ainsworth, 1992; Suzuki and Iida, 1992; Reite and Evensen, 1994; Reite, 1997; Reite and Evensen, 2006).

Typical functions of mammalian PMN during inflammation such as phagocytosis, respiratory burst activity, and chemotaxis, have also been observed for fish granulocytes. As in mammals, the PMN of many teleost species are the first immune cell type to emigrate to the site of inflammation (Havixbeck and Barreda, 2015). In some cases, this has been followed by a subsequent decrease in neutrophil number, indicating that certain hematopoietic tissues may serve in neutrophil storage (Suzuki and Iida, 1992).

Like their mammalian counterparts, evidence exists in fish that eosinophilic granulocytes can also be phagocytic and destroy ingested pathogens (Bodammer, 1986; Sharp *et al.*, 1989; Reite and Evensen, 2006). In addition, eosinophilic granulocytes of tilapia (*Oreochromis niloticus*) (Matsuyama and Iida, 1999), rainbow trout (*Oncorhynchus mykiss*) (Ellis, 1985; Powell *et al.*, 1991, 1993), and Atlantic salmon (*Salmo salar*) (Reite and Evensen, 1994) have been found to degranulate upon inoculation with certain noxious or infectious agents. Degranulation in tilapia was correlated with increased neutrophil migration to the injection site, suggesting that fish eosinophils may be analogous to mammalian mast cells (Matsuyama and Iida, 1999). Clearly, the conservation of neutrophil-like granulocytes throughout vertebrates indicate that form and functions of these cell types have been important to immunity and homeostasis from the earliest vertebrate ancestors (Fingerhut *et al.*, 2020).

## Nonspecific Cytotoxic Cells (NCC)

Many teleost species possess immune cells involved in spontaneous, nonspecific cytotoxicity against transformed mammalian cells, parasites, viral/bacterially infected and tumorous cells (Evans and Jaso-Friedmann, 1992; Secombes, 1996; Shen et al., 2002). A cell surface marker (isolated from catfish) found to be important in teleost NCC activity appears to be conserved between mammalian species and fish (Jaso-Friedmann et al., 1997, 2001). Nonspecific cytotoxic cells are usually found in the head kidney, spleen, and blood (Evans and Jaso-Friedmann, 1992); circulating NCCs are morphologically and functionally distinct from those of the pronephros and spleen (Hogan et al., 1996; Shen et al., 2002). Different from cytotoxic T cells, NCC do not depend on specific antigen presentation for recognizing infected cells. While the NK cells are part of the innate arm of the immune system, they belong to the lymphoid lineage.

## Innate Recognition of Pathogen- and Danger-Associated Molecular Pattern Receptors (PAMPs, DAMPs, PRRs)

Cells of the innate immune response, namely macrophages, neutrophils, natural killer cells (nonspecific cytotoxic cells), tissue resident dendritic cells and mast cells, and even epithelial cells, associated with barrier tissues, recognize damaged tissues and/or pathogens via a fixed-action-pattern-like response via innate receptors. The subject has been reviewed extensively (Mogensen, 2009; Lavelle *et al.*, 2010). Sequestered proteins within damaged and dying cells (DAMPs) released during the loss of structure may lead to a process known as pyroptosis or even necrosis. Examples of DAMPs include adenosine triphosphate (ATP), protein crystals, and hyaluronic acid. Many PAMPs recognize aspects of external pathogen patterns (LPS (lipopolysaccharide), flagellins, polysaccharides, carbohydrates) as well as molecular patterns within the pathogen (double-stranded deoxyribonucleic acid (DNA), single and doubled stranded viral messenger ribonucleic acid (mRNA). Regardless of the molecular patterns, the result is almost always the same: release of type I interferons and/or release of antimicrobial peptides; inflammatory cytokines (IL-1, IL-6, TNF-α, IL-8, IL-12, granulocyte-macrophage colony stimulating factors (GM-CSFs), others) and chemokines; and inducible enzymes, such as inducible nitric oxide synthase ( iNOS) and cyclo-oxygenase type 2 (COX-2). Both local and systemic effects follow the release of these factors, including the induction and release of the aforementioned acute phase proteins by the liver. These innate responses to pathogens are largely under the control of interferon releasing factor (IRF3) and/or NFκβ and AP-1, transcription factors activated by binding of ligand binding of the PPR. Toll-like receptors (TLRs) are typically located on the plasma membrane or endosomal membrane, while nucleotide-binding domain and leucine-rich repeat containing receptors (NLRs) and retinoic acid-inducible gene-1-like receptors (RLRs) are cytoplasmic. Others, like DNA-dependent activator of IFN regulatory factors (DAI), absent in melanoma-2 (AIM2), and RNA-activated protein kinase (PKR) are also cytoplasmic.

Research into the form and function of PAMPs and DAMPs is intensive and ongoing, and comparative immunologists now realize that many of the functions of TLRs (as an example) ascribed to mammalian TLRs evolved after fish and tetrapods split roughly 400 million years ago. As an example, in mammals TLR4 recognition of LPS leading to NFκβ activation through the myeloid differentiation 88 (aMyD88) pathway is well characterized as a proinflammatory sequelae. However, it seems that fish do not recognize LPS in a similar fashion and that LPS treatment actually suppresses NFκβ signaling (Sepulcre *et al.*, 2009). These same investigators suggest that in fish, this mechanism would be adaptive in response to the high concentrations of LPS in water surrounding the organism and note that the pro-inflammatory properties of LPS only appear later in evolution. For further reading on the role of PPRs in innate immunity of fish, the reader is referred to the following reviews and research. (Sepulcre *et al.*, 2009; Hansen *et al.*, 2011; Palti, 2011; Rajendran *et al.*, 2012; Aoki *et al.*, 2013; Phelps *et al.*, 2020; Sahoo, 2020; Hao *et al.*, 2021; Shan *et al.*, 2021; Shivam *et al.*, 2021).

## ADAPTIVE IMMUNE RESPONSES

Adaptive (acquired) immunity in mammalian species involves: (1) recognition of foreign, non-self entities by antigen-specific lymphocytes, (2) proliferation and differentiation of antigen-specific lymphocyte clones, and (3) memory and enhanced response upon subsequent exposure to antigen (due to greater numbers of antigen-specific lymphocytes). The nature of the adaptive immune response depends upon the activation of either T-lymphocytes (i.e., cell-mediated immunity) or B-lymphocytes (i.e., humoral immunity). T-cells, through the TCR (T-cell receptor), are specific to small internal and hydrophobic regions of larger proteins, while B-cells, through the BCR (B-cell receptor), are specific to stretches of hydrophilic regions. Teleost fish possess the genetic machinery necessary to mount both humoral- and cell-mediated immune responses (Kaattari and Piganelli, 1996; Manning and Nakanishi, 1996; Secombes and Belmonte, 2016; Smith *et al.*, 2019). This includes genes encoding recombination activating genes (RAGs) (Willett *et al.*, 1997), major histocompatibility complex (MHC) class I and II (Grimholt and Lie, 1998; Stet *et al.*, 1998; Grimholt and Lukacs, 2020), T-cell receptors and B-cell receptors, and their coreceptors (Wilson *et al.*, 1998; Haire *et al.*, 2000; Laing and Hansen, 2011; Quiniou *et al.*, 2011; Scapigliati *et al.*, 2018b; Wu *et al.*, 2020). Immunoglobulin heavy and light chains involve typical vertebrate gene rearrangement processes (Pilstrom and Bengtén, 1996; Edholm *et al.*, 2011; Fillatreau *et al.*, 2013; Rego *et al.*, 2020), cytokines (Zou and Secombes, 2016), and various immune coreceptors (Hansen and Strassburger, 2000; Yoder *et al.*, 2001; Laing and Hansen, 2011). In addition, the functional equivalents of APCs are present in fish (Vallejo *et al.*, 1991; Vallejo *et al.*, 1992; Zhu *et al.*, 2014; Kato *et al.*, 2018). Fish species have also been shown to exhibit antigen-specific antibody production (Kaattari and Piganelli, 1996), cytotoxic T-lymphocyte (CTL) activity (Verlhac *et al.*, 1990; Yoshida *et al.*, 1995; Stuge *et al.*, 2000; Taylor *et al.*, 2020), mixed-lymphocyte responses (Miller *et al.*, 1986; Scapigliati, 2013a; Scapigliati *et al.*, 2018a), delayed hypersensitivity (Manning and Nakanishi, 1996), graft-vs.-host reactions (Nakanishi, 1994), and xenograft/allograft rejection (Manning and Nakanishi, 1996). Clearly, fish have many of the characteristics of adaptive cellular and humoral immunity seen in higher vertebrates, though the structures, functions, and mechanisms of regulation may be different (Boehm *et al.*, 2012; Castro and Tafalla, 2015; Magadan *et al.*, 2015; Wilson, 2017). These evolutionary differences from the canonical mammalian model create ample opportunities for immunological and immunotoxicological research.

## EMERGING VIEWS OF INFLAMMATION IN IMMUNOTOXICOLOGY

The word "inflammation" is at the forefront of many chronic diseases affecting humans, including cancers, diabetes, obesity, metabolic syndrome, cardiovascular diseases, arthritis, and others. Historical use of the word inflammation was restricted to what we understood about acute innate responses to localized tissue damage and pathogen entry across anatomical barriers, and we understood chronic inflammation in terms of severe long-term disease outcomes. Understanding inflammation in fish is difficult when applying the canonical symptoms of these inflammatory responses in mammals such as swelling, redness, heat, pain, and perhaps coagulation as well. However, the inflammatory process includes recruitment and activation of white blood cells at the site of tissue injury and pathogen entry, including expression of inflammatory-mediating transcription factors (e.g., NF-κβ, activating protein-1 (AP-1), phosphorylation protein 38 (p38), IRFs, others), expression of classical inflammatory cytokines (IL-1β, TNF-α, IL-6, IL-12, GM-CSF, others), vascularization, and wound healing across diverse vertebrate phyla, including fish (Suzuki and Iida, 1992; Koppang *et al.*, 2007; Campos-Sánchez and Esteban, 2021; Xie *et al.*, 2021).

In recent years, the biological processes involved in inflammation have been revealed in surprising circumstances such as normal embryonic development; neurodevelopment; tissue regeneration; and in stressful conditions, such as starvation, sleep deprivation, shivering thermogenesis, and other cases, we may collectively refer to as physiological inflammation (metabolic inflammation) (Medzhitov, 2021). Physiological or metabolic inflammation is neither acute nor chronic, but rather meta-inflammatory, occurs in the absence of the above-described symptoms of acute or chronic inflammation, and seems to be associated with UPR (unfolded protein response) stress in sentinel cells of the immune system (Antonelli and Kushner, 2017). Of note, while circulation CRP (C-reactive) levels are indicative of acute and/or chronic inflammation, CRP levels in metabolic inflammation may be only slightly evaluated.

Mechanistically, the events associated with metabolic inflammation are associated with reprograming (polarizing) cells of the immune system toward new steady state responses following external and internal triggers. These triggers involve the loss of structure (e.g., altered gut epithelial structure and commensal microbiome), followed by loss of function (e.g., malabsorption of nutrients, altered peristalsis), and ultimately the loss of regulation (e.g., new steady state of lowered function and/or metabolism and dysbiosis of normal microflora) (Medzhitov, 2021). Moreover, a role of the UPR in driving meta-inflammation is now appreciated (Chaudhari *et al.*, 2014).

To understand and treat various diseases associated with inflammation, there has been much attention on anti-inflammatory small molecules that target key events during inflammatory processes. Several promising drug candidates are derived from natural products or repurposed drugs approved for other diseases. Some of these candidate drugs are active through pathways like known strong immunotoxic compounds, such as polyaromatic hydrocarbons (PAHs), 2,3,7,8-tetrachlorodibenzo-p-dioxin (TCDD), polychlorinated biphenyls (PCBs), pefluorochemicals (PFCs), and synthetic phenolic antioxidants to name a few (discussed elsewhere). Thus, the toxic effects of many environmental contaminants may be associated with meta- and chronic inflammation, and at the forefront of these toxic responses may be the reticuloendothelial-system (RES), also known as the mononuclear phagocytic system (MPS), involving macrophages—key players in inflammation. With environmental species such as fish and wildlife, the environment contains not only contaminants of interest, but also the local microbial consortium, and these interactions may lead to a state of inflammatory toxicity (Silva *et al.*, 2021).

## IMMUNOMODULATION BY CHEMICALS

Considerable evidence demonstrates the adverse effect of immune-compromise, whether stemming from inherited primary immunodeficiency, or acquired secondary immunodeficiency, such as chemical-induced immunotoxicity or disrupted neuro-endocrine-immune interactions. The possible adverse outcome from these immune dysfunction(s) is increased susceptibility to pathogens and possibly death or persistent levels of inflammation. From this point, one can intuit the effects of reduced survival in terms of local population size, and the resulting negative impact on community structure and function—not to mention the possible spread of pathogens throughout the ecological system. In addition, unlike other physiological systems, parenchymal cells of the immune system continually traffic from primary lymphoid organs throughout the organism in search of pathogens concentrated within the secondary lymphoid organs. Thus, immunocompetent cells may have a different exposure history to contaminants compared to localized cells of other physiological systems. To date, however, definite molecular initiating events have not been identified that could predict an adverse outcome pathway (AOP), and this is in large part because whole organism immunity has redundant mechanisms to maintain homeostasis. The future of AOPs in fish immunotoxicology will be discussed in subsequent sections.

Environmental immunotoxicology using fish models is rooted in technological approaches modeled after those first used in mammalian immunology and pharmacology/toxicology (Rice, 2001; Carlson and Zelikoff, 2008a; Segner *et al.*, 2011). For cell-based assays, most approaches use cells isolated from peripheral blood, the anterior kidney, the spleen, or peritoneum because these are readily available to the investigator. The thymus is much more difficult to examine because of its anatomical location behind the upper operculum epithelial and separate from the open cavity exposed for isolating other lymphoid tissues and cells. Furthermore, the thymus is anatomically and functionally most pronounced in early life stages of fish and changes dramatically on a seasonal basis (Kernen *et al.*, 2020). However, the thymus is crucial to T-cell development and maturation and therefore indispensable for adaptive immunity. Not only are developing thymocytes nurtured and educated in the thymus, but there is also evidence of, at least in teleost fishes, a role for B-cells, phagocytes, natural killer-like cells, and several types of critical stroma.

From a historical perspective, the literature on teleostean fish immunotoxicology in terms of common fish models and techniques to evaluate immune function has been extensively compiled and reviewed (Rehberger *et al.*, 2017b). Despite the large number of teleostean fish species, the most common species used in fish immunotoxicology are *Oncorhynchus mykiss* (20%), *Cyprinus carpio* (11%), *Sparus aurata* (11%), *Danio rerio* (9%), *Oncorhynchus tschawytscha* (7%), *Tilapia* (6%), *Dicentrarchus labrax* (3%), *Carassius auratus* (2%), *Oryzias latipes* (2%), *Ictalurus punctatus* (2%), and other species with less representation in literature comprising 24% of studies. Most of these species are of aquaculture interest, with *Oryzias latipes* and *Danio rerio* representing model species for both aquaculture and biomedical interests. Thus, much of our fish immunotoxicological knowledge is based on species that are accessible and easily cultured and not necessarily based on relative sensitivity to xenobiotics, or that are commonly used in environmental risk assessment (e.g., fathead minnow, *Pimephales promelas*) (Doering *et al.*, 2013; Thornton Hampton *et al.*, 2020). As global attention to the links between oceans and human health (OHH) increases, other species of fish should be examined as well (Landrigan *et al.*, 2020). Moreover, Rehberger et al.'s review indicates that many of the techniques and immunological endpoints historically used to assess immunity comprised of tier I-like approaches used in mammalian immunotoxicology, but also included pathogen challenges and induced innated immune functions. One of the key conclusions of the review is that toxicogenomics, single cell RNAseq, and computational biology are needed to fully understand mechanisms of action. After this review, another study from the group demonstrated that some common assays in immunotoxicology are not very predictive or at least not very discriminatory in determining whether certain classes of compounds were selectively immunotoxic (Rehberger *et al.*, 2020, 2021).

Fish immunotoxicology has been stimulated by field studies which reported disturbed immune functions and elevated prevalence of infectious and non-infectious diseases in fish populations from contaminated habitats (Vethaak and ap Rheinallt, 1992; Luebke *et al.*, 1997; Arkoosh *et al.*, 1998b, 2001b; Jacobson *et al.*, 2003; Iwanowicz *et al.*, 2012; Whitehead, 2013; Martínez-Gómez and Vethaak, 2019a; Gao *et al.*, 2020). These studies as well as others provide strong evidence that chronic, low-level contamination of aquatic environments can compromise fish immunity. Also, effluents from wastewater treatment works were found to result in immunotoxic effects in fish (Hoeger *et al.*, 2004; Liney *et al.*, 2006; Hébert *et al.*, 2008). Importantly, chemical effects on the fish immune system may cumulate with the impacts of other environmental stressors targeting the immune system, for instance, increasing water temperature or stress caused by habitat degradation (Petitjean *et al.*, 2020). Thus, immunotoxicity is a real and relevant ecological issue, as impaired immunity can lead to increased disease prevalence in wild fish stocks—a mechanism that is likely to contribute to the fact that fishes are the vertebrate class that currently experiences the highest rates of disease emergence. Pollution can not only increase the susceptibility of fishes to infectious and non-infectious diseases, but it can also act as a co-stressor enhancing the health impact of existing diseases (Martinez-Gomez and Vethaak, 2019). Finally, chemical impacts on fish immunity can also interfere via resource-trade-offs with other life-history traits such as growth and reproduction (Segner et al., 2017), thereby modulating the Darwinian fitness of organisms.

Despite the gaps in our knowledge of the mechanisms of action of various aquatic immunotoxic xenobiotics, there are classical examples of classes that have been extensively studied in fish that can perhaps be viewed from a different perspective than appreciated at the time of original publication. As discussed in the section of basic fish immunology, we now appreciate that anatomical, chemical, and cellular components of barrier immune responses (including symbiotic microbiome) are part of an intimate physical and functional relationship with epithelial cells of various types, including the hepatic sinusoid epithelium, gill epithelium, intestinal epithelium, skin and nasal epithelium, and others. Cells of the immune system collectively referred to as "sentinel cells" are distributed throughout the barrier, either in direct contact with epithelial cells or nearby in connective tissues, and include macrophages, dendritic cells, mast cells/eosinophilic cells, intra-epithelial innate-like lymphocytes, and $\gamma/\delta$ T-cells; each of which can be exposed to xenobiotics and/or their metabolites. As a key example, the presence of liver macrophages, or Kupffer cells, in fish has been a subject of debate for many years and primarily because in mammals, they reside in the lumen of the sinusoids, adherent to the endothelial lining by loose reticular collagen type III fibers, and comprise the majority (80–90%) of fixed tissue macrophages of the body (Phillips, 1989). Most of these fixed Kupffer cells in mammals are in the periportal area of the liver and as such are exposed to xenobiotics (or pathogen components) before other sentinel cells of the body. This has not been described in fish, and, therefore, it is generally assumed that fish do not have true Kupffer cells, but rather scavenger endothelial cells instead, as previously reviewed (Seternes *et al.*, 2021). This notion has been challenged in recent years by the observation of Kupffer-like cells in perfused livers of salmonids, suggesting that fish livers may have a role in immunity as an immune organ (Möller *et al.*, 2014) and also suggesting that isolated fixed sentinel immune cells in general should be examined as a part of routine immunotoxicology assessments. Kupffer cells, whether found in all teleosts, are one of the several types of macrophages comprising the previously discussed RES in the section on monocytes/macrophages.

As described in the introduction section, concerns about the effect of known and unknown environmental contaminants on immune functions of fish have been raised for nearly 50 years, culminating in volumes of publications during the late 1970s, the 1980s, and beyond to the late 1990s and early 2000s when it became clear that many of the contaminants under study by immunotoxicologists also affected reproduction in fish. Thus, the field and era of endocrine disrupting compounds (EDCs) began, and much of the attention from environmental toxicologists focused on immunotoxicology quickly shifted to EDCs by early 2000s (Swedenborg *et al.*, 2009; Marty *et al.*, 2011; Hampl *et al.*, 2016). It was about this time in the history of fish immunology that neuroendocrine-immune (NEI) interactions, from an environmental stress perspective, were demonstratable and important, and the field continues (Arnold and Rice, 2000; Harris and Bird, 2000; Rice and Arkoosh, 2002; Guo and Dixon, 2021), though not to the level of attention on EDCs. Moreover, studies focused on the immunotoxicity of metals more or less gave way to understanding the effects of metals, metal oxides, and carbon-based particles associated with nanoparticles (Jovanović and Palić, 2012). Interest in particle-associated toxicity followed as microplastic and nanomaterial pollution (and absorbed contaminants) came to the forefront in environmental toxicology, and consequently limited attention focused exclusively on the immunotoxic effects of particles in fish.

## METALS

Contamination of aquatic habitats with heavy metals from various industrial and mineral mining sources has been a problem for many years. The current interests in mineral mining, energy development and use, and dredging will undoubtedly result in further pollution of aquatic environments by such metals as arsenic (As), cadmium (Cd), lead (Pb), mercury (Hg), and zinc (Zn). The effects of metal pollution are measurable on both ecological and economic scales. Ecosystem impacts include contamination of sediments and the water column, accumulation of pollutants in biota over a wide area, and apparent increases in pollutant-related anomalies in the residing species.

Municipal wastes, industrial discharges, surface runoff, damage and weathering of vessel-protective paints, ocean dumping, and aerial inputs account for most ocean metal pollution. Although some of the routes of entry for metals into the oceans have been slowed or stopped in recent years, too little regulation has been implemented to date. The growing environmental pollution by potentially toxic metals gives rise to problems in the aquatic environment. Aquatic organisms' close contact with these metal pollutants and the reactivity of these pollutants results in accumulation within the organism that is dangerous not only for their own survival but also for humans. As a part of metal homeostasis required for normal cellular functions, metals usually bind to one or more metallothioneins (MT) that subsequently bind to metal response elements (MRE) via MRE-binding transcription factor (MTF-1) (Giedroc *et al.*, 2001). While essential nutrient metals influence the response of immune cells to pathogens and have important functions in orchestrating inflammation (Monteith and Skaar, 2021), excessive exposure to metals may overwhelm the ability to regulate metal homeostasis, leading to oxidative stress. Oxidative stress also leads to activation of the MRE, thereby exasperating the condition. In recent years, it has become clear that metals induce oxidative stress, weaken anti-oxidative defense mechanisms, inactivate enzymes, and interfere with DNA damage repair (Balali-Mood *et al.*, 2021).

The literature up to about 20 years ago on the immunotoxicity of metals in fish has been reviewed extensively (Carlson and Zelikoff, 2008a). The literature on metal immunotoxicity in fishes since then has been scant and mostly because of the above-mentioned switch by investigators to newer-emerging contaminants of concern. Frequently observed immune effects of metal exposure in fish include alterations of phagocytosis activity, respiratory burst activity, antibody production, and lymphoproliferation (see Carlson and Zelikoff, 2008a). For mammals, it has been shown that metals can activate the innate immune system through direct binding to pathogen recognition receptors, activation of the inflammasome, or the induction of cellular death and release of alarmins (McKee and Fontenot, 2016), and these mechanisms are likely to be active in fish as well. Beyond, there is a link between metal exposure, oxidative stress, and immune functions (Jin *et al.*, 2015). More recent studies using genomic approaches validate these findings by demonstrating general immunosuppression in common carp following exposure to cadmium, and close to 2,000 genes both upregulated and down regulated, including inflammatory genes, oxidative stress genes, and genes associated with metabolism (Jin *et al.*, 2015; Chen *et al.*, 2019). As with cadmium, exposure to other metals leads to typical oxidative stress and suppression of immune functions in zebrafish (Jin *et al.*, 2015) Clearly more studies are necessary to better understand the immunotoxicity of metals in fish systems at the genomic level and especially using single cell sequencing to determine immune cell type effects.

## PESTICIDES

Pesticides represent a diverse group of poisonous chemicals that are primarily used to control insects, weeds, or fungus pests. Many of the more persistent pesticides, especially the chlorinated hydrocarbons, are common pollutants of aquatic ecosystems. Pesticides enter the aquatic environment and become rapidly distributed through intentional application, aerial drift, runoff from agricultural fields following application, or accidental release. Fish residing in contaminated areas frequently encounter and accumulate sublethal levels of pesticides from the surrounding water and food, and they may then enter the food chain. In addition to the carcinogenic potential of some pesticides and their toxicity for the nervous and reproductive systems, pesticides represent a potential immunotoxic threat to all directly exposed organisms. Yet, despite this and their potential presence in the aquatic environment, relatively few studies have examined the toxic effects of pesticides on the immune response of fish.

### Organophosphate Insecticides

Organophosphorus insecticides, discovered in Germany just before World War II, have become the most widely used insecticides and have replaced organochlorines as they eliminate problems associated with environmental persistence, bioconcentration, and biomagnification. Inhibition of acetylcholinesterase (AChE) is the basis for organophosphate (OP) insecticide toxicity (Albers *et al.*, 1999), but mammalian literature also provides strong evidence for their immunosuppressive effects (Beaman *et al.*, 1999). It remains less clear as to what extent nonmammalian species (aquatic organisms) may be impacted by these chemicals. Most early studies in fish examined the immunomodulating effects of trichlorfon and malathion (Mishra and Srivastava, 1983; Areechon and Plumb, 1990; Cossarini-Dunier *et al.*, 1990; Plumb and Areechon, 1990; Siwicki *et al.*, 1990; Dutta *et al.*, 1992). Although somewhat limited in nature, these studies provided the groundwork for demonstrating the immunosuppressive effects of OP insecticides in fish. Malathion continues to be examined along with some less well-studied OP insecticides such as chlorpyrifos, edifenphos, and others though many now have been banned due to their extreme toxicity (Yang *et al.*, 2021).

*Malathion*—Diethyl(dimethoxyphosphinothioylthio) succinate (malathion) is a wide-spectrum OP insecticide used worldwide to control fruit- and vegetable-sucking and chewing insects, disease vectors such as mosquitoes and flies, and ectoparasites of various animal species. The most well-studied effects of malathion are its ability to inhibit AChE activity (likely through the binding of malathion to the AchE receptor) (Shao-Nan and De-Fang, 1996) and disrupt nerve impulse transmission (Richmonds and Dutta, 1992), but malathion has also been shown to modify mammalian host-defense mechanisms (Corsini *et al.*, 2008). Although studies are limited, malathion appears to affect the immune system of fish in a manner like that seen in mammals (Mishra and Srivastava, 1983; Areechon and Plumb, 1990; Plumb and Areechon, 1990; Dutta *et al.*, 1992; Beaman *et al.*, 1999). These collective studies demonstrated that exposure of Japanese medaka (*Oryzias latipes*) for 7 or 14 days to sublethal concentrations of malathion (0.2 or 0.8 ppm) dose-dependently depressed T cell-dependent antibody forming colony (AFC) numbers. Changes in humoral immunity occurred in the absence of effects on hematocrit/leukocrit or T-lymphocyte proliferative responses. Malathion has also been shown to reduce antibody titers against *Edwardsiella ictaluri* in channel catfish (*Ictalurus punctatus*). Studies also demonstrated that a 14- or 21-day exposure of medaka to either 0.1 or 0.3 ppm malathion significantly reduced host resistance against *Yersinia ruckeri* infection. Although studies indicate that malathion-induced toxicity in fish may not be mediated by the same detoxification enzymes important for bringing about effects in mammals (Pathiratne and George, 1998), it appears that mammalian immune assays can be used successfully in a teleost model for predicting malathion-induced immunosuppression.

*Chlorpyrifos*—Chlorpyrifos, developed in 1962 to control a wide variety of agricultural- and urban-associated insects, has wide application as a termiticide (Albers *et al.*, 1999). The health effects of chlorpyrifos on exposed humans are highly controversial, although it appears that relatively high doses are needed before any effects are observed. In addition to neurological outcomes, a few clinical and epidemiological studies have suggested that exposure to chlorpyrifos can alter hematologic and immunologic parameters in exposed individuals (Broughton, 1990; Fiedler *et al.*, 1992; Thrasher *et al.*, 1993). Chlorpyrifos has been found to contaminate aquatic environments (usually via runoff from treated sites), but very few studies have examined the effects of this OP insecticide on the immune response of fish. Across the spectrum of several studies, Chlorpyrifos alters inflammatory cytokine gene expression, increases reactive oxygen species formation, suppresses phagocytosis by macrophages, alters lysozyme activity, and alters IgM levels in circulation (Eder *et al.*, 2009; Li *et al.*, 2013; Raibeemol and Chitra, 2020; Zahran *et al.*, 2020).

## Pyrethroids

Pyrethroids are pharmacologically active synthetic compounds structurally based on the naturally occurring substance pyrethrum, which is derived from chrysanthemum flower extracts. The recently introduced synthetic pyrethroids (i.e., *S*-biallethrin, permethrin, cysmethrin, biosresmethrin, cypermethrin, deltamethrin, and fenvalerate) have attracted wide use by farmers primarily because of their potent insecticidal nature and apparent lack of human toxicity because synthetic pyrethroids are neither fully metabolized nor quickly detoxified; they have the potential to create a serious residue accumulation problem (Yang *et al.*, 2021). Moreover, it has been suggested that covalent binding of the reactive pyrethroids to liver proteins can create new haptens and potent antigens, resulting in hypersensitivity reactions (Hoellinger *et al.*, 1987; Diel *et al.*, 1998).

Although the new generation of pyrethroids appears to pose a limited threat to humans, they are for the most part highly toxic to fish (Bradbury *et al.*, 1985; Agnihothrudu, 1988). Exposure of *Cyprinus carpio* eggs to increasing concentrations of cypermethrin, deltamethrin, or fenvalerate resulted in reduced hatching ability and viable hatch, as well as inactive or abnormal larvae (i.e., increased vertebral column flexure, enlarged yolk and pericardial sacs, and stunted tail). Most laboratory studies examining the effects of synthetic pyrethroids on fish have focused on a single member of the newest classes of type I pyrethroids, permethrin. Since its synthesis in 1973, permethrin has gained in popularity due to its photostability, low mammalian and avian toxicity, and relatively short environmental half-life (i.e., a half-life of 28 days in soil and 14 days in seawater exposed to sunlight) (Rebach, 1999). Unfortunately, even though permethrin appears to degrade rapidly in water, it tends to bioconcentrate in fish (Wei et al., 1995). In fact, the major contraindication to permethrin use is its relatively high toxicity to nontarget organisms, including macroinvertebrates (ranging from 0.02 to 0.73 ppb in species tested) and estuarine fishes (ranging from 2.2 to 12 ppb in species tested).

Several studies have shown that exposure of fish to permethrin also modulates immune function (Sopinska and Guz, 1998). Exposure of carp (*Cyprinus carpio*), for example, for 96 hours to the synthetic pyrethroid ambush 25EC (at a concentration of 0.03 or 1.1 ppb) induced leukopenia, neutrocytosis, and reduced the percentage of kidney cells active in phagocytosis. Given that shorter and longer exposure times failed to produce similar effects, however, the results of these studies are questionable. The immunomodulating effects of permethrin at concentrations ranging from 0.01 to 2.0 ppb have been studies in Japanese medaka (*Oryzias latipes*) (Zelikoff, 1998). Although no effects were observed on body or immune organ weight, kidney and spleen cell viability, or hematocrit and leukocrit levels, permethrin concentrations ≥0.5 ppb tended to reduce plasma Ig levels and kidney and spleen cellularity after 7 days (compared to solvent-exposed controls); thymic cellularity was reduced by about 30% in fish exposed to the highest permethrin concentration. Alternatively, exposure to permethrin for 48 hours tended to increase the proliferative response of splenic lymphocytes. Probably, the most dramatic effect of permethrin exposure in this study was on host survival following bacterial challenge. Exposure of medaka to either 0.05 or 0.01 ppb permethrin for 2 or 7 days, respectively, increased host mortality by about 30% in response to *Yersinia ruckeri* infection (compared to unexposed, bacterially challenged fish). Results in medaka resemble those produced in BALB/c mice exposed to permethrin by oral gavage (Blaylock *et al.*, 1995). In the latter case, exposure to 1.0%, 0.1%, or 0.01% permethrin for 10 days had no effect on body and spleen weight or mitogen-stimulated lymphocyte proliferation. Alternatively, those responses in mice requiring specific antigen recognition or effector function (i.e., MLR (mixed lymphocyte reaction), CTL, or NK (natural killer) activity) were depressed by permethrin exposure. More recent studies collectively demonstrate that pyrethroids induce macrophage ROS, decrease phagocytosis, suppress IgM levels, and increase pro-inflammatory gene expression (Nayak *et al.*, 2004; Jin *et al.*, 2013; Guardiola *et al.*, 2014; Park *et al.*, 2020).

### Organochlorine Insecticides

The organochlorines (OCs), popular for their potent insecticidal activity and cost efficiency, consist of three major families of compounds: (1) those related to DDT, including dicofol and methoxychlor; (2) cyclodienes, such as aldrin and dieldrin; and (3) hexachlorocyclohexanes, such as lindane. To varying degrees, OCs are all characterized by low water solubility and high chemical stability that contribute to their persistence in the environment. The highly lipophilic nature of these compounds makes them subject to bioaccumulation such that organisms of sequentially higher trophic levels concentrate the pesticides from surrounding soil and water or through the consumption of organisms in lower trophic levels. In mammals, the most observed non-immune effects of OC pesticides relate to hepatotoxicity and neurotoxicity, with some difference in the actual symptoms produced by a given agent or isomer of that agent.

*Lindane*—Five isomers of 1,2,3,4,5,6-hexachlorocyclohexane (HCH)—alpha (α), beta (β), gamma (γ), delta (δ), and epsilon (ε)—were originally contained in a commercial preparation of HCH known as benzene hexachloride. This compound has not been marketed since 1978, but OC γ-HCH is commercially available as the insecticide lindane. Lindane is used as an insecticide treatment for crop protection and in human and veterinary medicine against ectoparasites. Excessive lindane can be persistent in food chains and readily accumulated by most animal species; fish can absorb lindane directly from the water or by ingestion of contaminated food, and it can bioaccumulate in fish at ratios of 500- to 1200-fold (Betoulle *et al.*, 2000). A review of organochlorine insecticide immunotoxicity in fish, including DDT metabolites p,p'-DDE, HCH, endosulfan, dieldrein (Yang *et al.*, 2021), demonstrates that these chemicals affect immunity in similar fashion as other insecticides. The primary effects are increased macrophage ROS, decreased phagocytosis, suppressed IgM levels, and increases in pro-inflammatory gene expression.

## Halogenated Aromatic Hydrocarbons

Halogenated aromatic hydrocarbons (HAHs) are low-molecular-weight compounds that are classified into several major families: polychlorinated dibenzodioxins (PCDDs), polychlorinated dibenzofurans (PCDFs), polychlorinated biphenyls (PCBs), polybrominated biphenyls (PBBs), polychlorinated terphenyls (PCTs), polychlorinated naphthalenes (PCNs), and polychlorinated diphenyl ethers (PDEs). In recent years, halogenated aromatic compounds have engendered increasing concern about their impact as environmental pollutants. The physical and chemical stability of the higher chlorinated compounds along with their lipophilicity contribute to their nearly ubiquitous occurrence and to efficient bioaccumulation via the aquatic and terrestrial food chains. Polychlorinated biphenyls have appeared as frequent contaminants of soil and water, and chlorophenols have been detected in surface and drinking water. The extremely toxic 2,3,7,8-tetrachlorodibenzo-*p*-dioxin (TCDD, or dioxin) has contaminated large areas of both water and soil through industrial accidents, improper waste disposal, and wide-scale application of herbicides containing small quantities of the chemical as a contaminant, in addition to being a trace byproduct of combustion. Few data exist concerning the effects of PCTs, PCNs, and PDEs on the immune response in fish, but the mammalian literature is replete with studies demonstrating the immunotoxicity of PCDDs, PCDFs, and PCBs.

### 2,3,7,8-Tetrachlorodibenzo-*p*-Dioxin

2,3,7,8-Tetrachlorodibenzo-*p*-dioxin (2,3,7,8-TCDD) is the most biologically potent of the HAHs. Exposure to TCDD (and structurally related HAHs) produces several toxic effects in laboratory mammals, including a generalized wasting syndrome, lymphoid involution (especially of the thymus), pancytopenia, immunosuppression, hepatomegaly and hepatoxicity chloracne, hyperkeratosis, gastric lesions, urinary tract hyperplasia, edema, tumor promotion, teratogenicity, and decreased spermatogenesis (Birnbaum, 1995; Baan *et al.*, 2009; White and Birnbaum, 2009). In addition to the potency differences associated with the structure-activity relationship, HAH-induced toxicity is characterized by wide variations in potency among different species of laboratory animals, for example, the $LD_{50}$ (lethal dose at 50% concentration) for TCDD varies approximately

5,000-fold between the most sensitive (guinea pigs, 0.6–2 μg/kg) and the least sensitive (hamsters, 5 mg/kg) species. Although the spectrum of toxicity can vary markedly among species, certain toxic effects have been observed as correlates to exposure in almost all species tested. The most notable of these effects are the generalized slow-wasting syndrome, thymic atrophy, immunosuppression, and hepatomegaly. In fact, the observation that lymphoid involution is among the earliest and most sensitive manifestations of TCDD (and related HAH) exposure and that it is almost universally observed among animal species provided the impetus to characterize HAH immunotoxicity.

Surprisingly, few studies have been performed examining the effects of dioxin on the immune responses of fish, despite the common occurrence of TCDDs in the aquatic environment; potential accumulation in the food chain; and effects on fish growth, survival, and reproduction (Giesy and Kannan, 2002); and demonstrated immunotoxicity in multiple mammalian species (Kerkvliet and Burleson, 1994; Dearstyne and Kerkvliet, 2002; Kerkvliet, 2002; North *et al.*, 2008; Kerkvliet, 2012). Most of those studies, however, have demonstrated TCDD-induced lymphoid organ hypocellularity (Grinwis *et al.*, 2000) and, in some cases, thymic involution (Spitsbergen *et al.*, 1988a). In a study employing two different strains of juvenile rainbow trout (Shasta and Wytheville), Spitsbergen et al. demonstrated that i.p. injection of fish with 1 μg TCDD per kg produced numerous lymphomyeloid lesions, including thymic involution, splenic lymphoid depletion, and hypocellularity of kidney hematopoietic tissues. In association with decreased hematopoiesis, Shasta strain yearling trout also demonstrated peripheral leukopenia and thrombocytopenia. It was concluded that lymphomyeloid and epithelial tissues were the primary targets for TCDD-induced pathologic lesions in rainbow trout (*Oncorhynchus mykiss*) and that lesion prevalence and severity were influenced by trout strain and hatchery source. In another study performed by the same laboratory, yearling trout were injected i.p. with up to 10 μg TCDD per kg, and the effects on plaque forming cell (PFC) numbers, MØ-mediated phagocytosis, and mitogen-stimulated blastogenesis of thymic and splenic lymphocytes were evaluated (Spitsbergen *et al.*, 1986). At the highest TCDD dose, fish exhibited a variety of morphologic and histologic alterations but few changes in immune function (except for reduced pokeweed mitogen-induced splenocyte proliferation). It was concluded that although trout appeared to be among the more sensitive species in terms of the lethal effects, they were relatively resistant to the immunosuppressive potential of TCDD.

Even though host resistance against viral infection is an extremely sensitive endpoint for TCDD-induced immunotoxicity in mammals (TCDD-induced effects occurring at nanogram concentrations) (Nohara *et al.*, 2002), trout fingerlings fed a diet of ≤1 μg/kg TCDD and subsequently challenged with infectious hematopoietic necrosis virus had no change in disease resistance compared to control fish (Spitsbergen *et al.*, 1988a). The general failure of TCDD to suppress rainbow trout (*Oncorhynchus mykiss*) immune function at doses below those causing clinical toxicity parallels findings with other pharmacologic immunomodulators (Sakai *et al.*, 2000). Studies performed with teleost species other than trout also demonstrate the sensitivity of lymphomyeloid tissue to dioxin exposure, as well as the relative resistance of other fish species to TCDD-induced immunosuppression, for example, i.p. exposure of yellow perch (*Perca flavescens*) to 5 μg TCDD per kg produced a variety of histologic pathologies, including decreased splenic cellularity. Thymic atrophy and decreased head kidney hematopoiesis (among other histologic and morphologic lesions) were also observed in this study, but only at TCDD doses producing 95% mortality (25 and 125 μg TCDD/kg). In another study, oral exposure of European flounder (*Platichthys flesus*) to dioxin (at a 50-fold higher dose than that needed to produce acute toxicity in trout—500 μg/kg) increased mitotic activity and hepatosomatic index, but only slightly reduced thymus size (Grinwis *et al.*, 2000). These studies demonstrate that fish (like their mammalian counterparts) vary dramatically in their overall sensitivity to dioxin.

### Polychlorinated Biphenyls (PCB)

There are 209 PCB congeners divided into three classes based on positioning of the chlorine molecules—namely, coplanar, mono-ortho coplanar, and noncoplanar. The number of chlorine molecules determine the solubility and therefore persistence of each congener in water. As the chlorine content of the congener increases, the congener becomes less water-soluble (Swain, 1991). Coplanar congeners appear to have the greatest potential for toxicity, as well as the greatest affinity for the aryl hydrocarbon receptor (AHR). Fish, like their mammalian counterparts, also appear sensitive to the immunotoxic effects of PCBs. Despite some inconsistencies between studies, exposure to PCBs frequently produces tissue atrophy (Grinwis et al., 2001), suppressed immune function, and reduced host resistance in exposed fish (Thuvander *et al.*, 1993; Arkoosh *et al.*, 1994; Rice and Schlenk, 1995; P. Regala, 2001; Burton *et al.*, 2002; Duffy *et al.*, 2002). Field investigations of PCB-exposed feral fish populations have demonstrated effects like those seen in controlled laboratory studies (Anderson et al., 1997, 2003; Arkoosh et al., 1998; Barron et al., 2000). In general, coplanar PCBs are the most immunotoxic in both fish and mammals.

## Polycyclic Aromatic Hydrocarbons

Polycyclic aromatic hydrocarbons (PAHs), well known for their carcinogenic and mutagenic actions, readily enter the environment via incomplete combustion of organic matter. Because individuals can be exposed to significant amounts in the atmosphere, soil, waterways and oceans, and the food chain, PAHs represent a serious health concern for humans. The primary sources of PAHs in the atmosphere are transportation emissions, coal- and oil-fired power stations, domestic heating, cigarette smoke, refuse burning, forest fires, volcanic activity, and industrial processes (i.e., petroleum products and tar).

In addition to their carcinogenic and mutagenic activities, many PAHs are also potent immunosuppressants. Perhaps the most toxic and best-studied PAHs are benzo-a-pyrene (BaP), 7,12-dimethylbenz(*a*)anthracene (DMBA), and 3-methylcholanthrene (3-MC). Both BaP and 3-MC are found environmentally, whereas certain methylated PAHs, such as DMBA, have been synthesized as model compounds. Benzo(*a*)pyrene and DMBA have been consistently shown in mammalian systems to suppress a variety of immune functional activities, as well as host resistance against microbial pathogens and tumors (White *et al.*, 1994). Indeed, the hallmark immunotoxic effect of most carcinogenic PAHs is suppressed antibody forming cell responses. Several mechanisms of PAH immunotoxicity have been hypothesized, including membrane perturbation, altered interleukin production, and disruption of $[Ca^{2+}]_i$ mobilization (White *et al.*, 1994; Reynaud *et al.*, 2003); however, the mechanism receiving most attention concerns AHR binding, subsequent activation of the AHR gene complex, and the cytochrome P450 1 (CYP1) family of P450 monooxygenases (Silkworth *et al.*, 1984; Willett *et al.*, 2001). For a review of many past examples of PAH-related immunotoxicity studies in fish, the reader is referred the following select sources (Reynaud and Deschaux, 2006; Carlson and Zelikoff, 2008b).

## NANOPARTICLES

Nanoparticles are intensively used for a broad array of applications and technologies. The increasing use of these materials has led to a growing release into aquatic environments, and, consequently, exposure of aquatic organisms including fish continues to increase (Scown *et al.*, 2010). Bioaccumulation of nanoparticles by fish has been demonstrated in a number of studies (Handy *et al.*, 2008). Fish can uptake nanoparticles through the barrier epithelia of skin and gills as well as through the epithelia of the gut. As these epithelia have barrier immune function, they represent an initial site of an interaction between nanomaterials and the immune system (Løvmo *et al.*, 2017; Torrealba *et al.*, 2019). Also, when the nanoparticles, after epithelial uptake, are distributed with the fish body via the blood stream, nanoparticles are in direct contact with these materials. A key determinant for the interaction of absorbed nanoparticles with and their internalization by immune cells is the formation of a coating around the particles, also referred to as "corona". The corona can be formed from plasma proteins, complement factors, lectins, but also from bacterial lipopolysaccharide. The corona composition strongly influences the nanoparticle immune effects (Rastgar *et al.*, 2022). For instance, if the particles are associated with lipopolysaccharides, they can activate TLR of phagocytes. Of note, the fate and activity of internalized nanoparticles inside the immune cell are largely influenced by the corona (Yue *et al.*, 2017). Interestingly, female fish with their much higher levels of vitellogenin and other egg yolk precursors appear to have different corona composition, and, consequently different nanoparticle effects, than male fish (Hayashi *et al.*, 2017). This aspect of sex-specific immune effects of environmental contaminants in fish has been largely neglected in research to date, although evidence is increasing that sex plays a critical role particularly for immune responses (Rehberger *et al.*, 2017b; Ye *et al.*, 2018). Concerning the effects of nanoparticles on fish immunity, in the innate immune system, particularly respiratory burst activity, phagocytotic activity, and cytokine production appear to be responsive (Torrealba *et al.*, 2019; Rastgar *et al.*, 2022). With respect to the adaptive immunity, there is initial evidence that nanomaterials may interfere with the development and function of lymphocytes (Rastgar *et al.*, 2022). Overall, while the current database clearly demonstrates that the fish immune system is a target of nanoparticles, our current understating is too fragmentary to identify which immune components and functions are particularly at risk.

## IMPORTANCE OF NUCLEAR RECEPTORS IN UNDERSTANDING CHEMICAL BIOLOGICAL INTERACTIONS

Over recent years, knowledge accumulated that fish immune cells express a variety of nuclear receptors (NRs) which serve as endogenous and environmental sensors. NRs are ligand dependent transcription factors responsible, among other functions, for induction of phase I, II, and III xenobiotic and endogenous substrate-metabolizing enzymes and other proteins affecting growth, differentiation, and cellular metabolism. Within the larger field of aquatic toxicology, the AHR is one of the most iconic NRs. Upon cytoplasmic binding to a variety of planar compounds, the AHR translocates to the nucleus, dimerizes with ARNT, and binds DNA at the XRE (xenobiotic response element) to then imitate the expression of CYP1A, glutathione-S-transferases (GST), uridine 5'-diphospho-glucuronosyltransferases (UGT), nuclear factor erythroid 2-related factor (Nrf2, multi-drug transporter proteins, and others including AHR-repressor that negatively controls the expression/function of AHR (Hahn *et al.*, 2017). As will be discussed in the following section on AHR-binding xenobiotics and immunity, some of the most potent immunotoxic xenobiotics are active through this NR.

There are 48 NRs in humans, 49 in mice and rats, and reportedly 73 in zebrafish, 67 in medaka, 74 in tilapia, and 66 in stickleback (Zhao *et al.*, 2015). Some of the most common classes of aquatic xenobiotics have been shown to bind to specific NRs, with classical down-stream effects on tissues, organs, and systems (Giulivo *et al.*, 2016; Maqbool *et al.*, 2016; Manibusan and Touart, 2017). Limited examples as demonstrated through Indio Biosciences, Inc. (Heuvel, 2018) include: (1) organochlorine pesticides (e.g., DDT) with estrogenic, antiestrogenic, or antiandrogenic activity, that bind to the NRs estrogen receptor-alpha (Erα) and androgen receptor (AR); (2) dioxins and PAHs (e.g., BAP, planar PCBs, PCDFs, PCDDs) that bind primarily with the AHR, but also the NR peroxisome proliferator-activated receptor gamma (PPAR-γ); (3) Organotins (e.g., tributyltin (TBT), triphenyltin oxide (TPTO)) that bind the NRs retinoic acid X receptor (RXR) and PPARγ; (4) polyfluoroalkyl compounds (e.g., PFCs) that bind the NRs PPARs, thyroxine receptor (TR), and constitutive androstane receptor (CAR); (5) Brominated flame retardants (e.g., polybrominated diphenyl ethers (PBDE), hexabromocyclododecane (HBCD), tetrabromobisphenol A (TBBPA-A)) that bind the NRs PXR, ERs, and TR); (6) Alkylphenols (e.g., 4-*n*-nonylphenol, 4-*n*-octylphenol) that bind the NRs ERs, AR, CAR; (7) Bisphenol A (BPA) that bind the NRs ERs, AR, TR, and glucocorticoid receptor (GR); and (8) Phthalates (e.g., di(2-ethylhexyl)phthalate (DEHP) that bind the NRs PPARs, CAR/PXR, and GR. Most of the effects of these xenobiotics are associated with reproductive effects and metabolic dysfunction through altered steroid or hormone metabolism (CYPs), but these same classes have been shown to impair immune functions when examined experimentally (Milla *et al.*, 2011; Rehberger *et al.*, 2020, 2021). The observation that common xenobiotic classes affect both reproduction and immune function may speak to the integration of the greater neuroendocrine-immune system. Beyond, it needs to be kept in mind that several environmental chemicals can interact with multiple receptors (Corsini *et al.*, 2014; MacKay and Abizaid, 2018) and that contaminant-receptor interaction can affect multiple immunological and physiological targets (Segner *et al.*, 2013; Chaves-Pozo *et al.*, 2018).

## NUCLEAR RECEPTORS AND IMMUNOTOXICITY IN FISH

Fish immune cells possess an ecosystem of receptors which are cross-linked through diverse signaling pathways. Examples of nuclear receptors being active in fish immune cells include the GR (Stolte *et al.*, 2008; Verburg-van Kemenade *et al.*, 2017), opioid receptors (Verburg-van Kemenade *et al.*, 2009), thyroid receptors (Quesada-García *et al.*, 2014), retinoid-related orphan receptors ROR (Du *et al.*, 2012), or PPARs (Qiu *et al.*, 2019; Wu *et al.*, 2022). The physiological function of these receptors is to orchestrate immune cell differentiation and functioning as well as to integrate the immune system with other physiological systems of the organism (Segner *et al.*, 2017; Verburg-van Kemenade *et al.*, 2017; Villani *et al.*, 2018). However, these receptors can also bind environmental chemicals and, thereby, could mediate toxic effects of the contaminants on the fish immune system. An example is provided by the ERs which are expressed in diverse immune cell populations of fish (Shelley *et al.*, 2013; Iwanowicz *et al.*, 2014; Massart *et al.*, 2014; Paiola *et al.*, 2017; Casanova-Nakayama *et al.*, 2018; Chaves-Pozo *et al.*, 2018), and these receptors are involved in mediating the modulating effects of estrogen-active compounds on fish immune functions (Milla *et al.*, 2011; Yang *et al.*, 2015; Chaves-Pozo *et al.*, 2018; Maciuszek *et al.*, 2020; Moreira *et al.*, 2021; Kernen *et al.*, 2022). Also, chemicals and nanoparticles can bind to the pattern recognition receptors of the immune cells and thereby disturb immune functioning (Zhao *et al.*, 2020; Rastgar *et al.*, 2022).

Here, we focus on the possible role of the AHR in mediating immunotoxic effects of chemicals in fish. The AHR is a member of the bHLH (basic helix-loop-helix)-PAS (*Per -Arnt-Sim*) family of signal sensors. Historically, the AHR was detected as a hepatic protein that binds chlorinated aromatic hydrocarbons, and that induces the activity of enzymes which can metabolize aromatic compounds. Historically, research on the vertebrate AHR has focused on its role as xenobiotic sensor, its function in regulating xenobiotic metabolizing enzymes, and its role in toxicity (Budinsky *et al.*, 2014). However, the view of the AHR as a primarily xenobiotic sensor is rapidly changing. More recently, it became evident that the AHR has multiple functions beyond toxicology, including cellular differentiation, coordination of the cell stress response, interaction with the intestinal microbiome, longevity, and that it plays a prominent role in the immune system. For mammals, it has been shown that the AHR is a critical transcription factor for the normal development and function of the mammalian immune system (Stockinger *et al.*, 2014; Esser *et al.*, 2018; Rothhammer and Quintana, 2019). The receptor is expressed in many cells of the hematopoietic lineage, both innate and adaptive arms of immunity, and it has a prominent role in regulating differentiation and activity of certain T cell populations. The immunological role of the AHR is particularly relevant in barrier organs like the gut where the organism is confronted with natural AHR ligands contained in the diet and with microbial molecules produced by the gut microbiome (Tian *et al.*, 2015). Activation of the AHR by environmental signals has an important role in maintaining immune homeostasis, preventing inflammatory damage, and maintaining barrier integrity (Metidji *et al.*, 2018). Given the role of the AHR in the immune system, it is not surprising that xenobiotic AHR ligands were found to act as immunotoxicants in mammals (Hall *et al.*, 2018; Suzuki *et al.*, 2020).

Growing evidence demonstrates that the immunological functions of the AHR are not restricted to mammals but exist in fish as well (Segner *et al.*, 2021). Evidence for this comes from the fact that the AHR is present in immune cells of teleost fish and that its cellular expression is regulated under exposure to xenobiotic AHR ligands (Holen and Olsvik, 2014; Song *et al.*, 2020). The AHR in piscine immune cells appears to be functional, as indicated by the upregulation of the xenobiotic metabolizing enzyme, cytochrome P4501A, in immune cells of fish exposed to AHR ligands (Carlson *et al.*, 2004; Nakayama *et al.*, 2008; Phalen *et al.*, 2017; Song *et al.*, 2020). Interestingly, exposure of rainbow trout liver cells and head kidney immune cells to equal concentrations of the AHR ligand, benzo(a)pyrene), results in a differential regulation of gene transcripts, while in the hepatocytes, pathways related to functions such as xenobiotic metabolism or stress response are prominently regulated; pathways being responsive in the head kidney cells relate more to immune functions (Figure 6.4). Fish exposed to AHR-binding xenobiotics showed, among others, alterations of phagocytotic cells of respiratory burst activity, of cytolytic activity, as well as changes of immune cell proliferation and numbers (Segner *et al.*, 2021). Also, immune pathologies like thymus atrophy and altered disease susceptibility were observed (Spitsbergen *et al.*, 1988b; Maule *et al.*, 2005; Bravo *et al.*, 2011; Song *et al.*, 2020). Finally, supporting evidence for immunotoxic effects of xenobiotic AHR ligands comes from field observations on immunosuppressed fish living in habitats contaminated with PAHs or PCBs (Arkoosh *et al.*, 2001a; Bayha *et al.*, 2017; Martínez-Gómez and Vethaak, 2019b).

Overall, xenobiotics, by agonizing or antagonizing AHR signaling in immune cells, interfere with the physiological role of the AHR in the fish immune system and thereby cause immune dysfunction and disease. Such knowledge provides a basis for a mode of action (MOA) or AOP framework for receptor-mediated fish immunotoxicity, comparable to MOA/AOP concepts for AHR-mediated tumorigenicity (Budinsky *et al.*, 2014) or AHR-mediated fish early life mortality (Doering *et al.*, 2019). Importantly, the approach discussed here for the case of the AHR is applicable to the many other endogenous receptors being active in fish immune cells and would improve the risk assessment and prediction of immunotoxic effects of chemicals on fish.

## FUTURE DIRECTIONS

The findings discussed in the present chapter show that a broad variety of environmental chemicals can impact the immune system of fishes. This applies to chemicals that have been the focus of ecotoxicology from its beginnings, such as metals, persistent organic pollutants, or pesticides, but it applies also to emerging contaminants like nanomaterials, endocrine disruptors, personal care products, and pharmaceuticals, many of them being prioritized as global high-risk chemicals (Yang *et al.*, 2022). Contaminant-induced immune disturbances are also a frequent observation in fish populations living in contaminated habitats (Arkoosh *et al.*, 1998a; Leaver *et al.*, 2010; Iwanowicz *et al.*, 2012; Whitehead, 2013; Gao *et al.*, 2020). Likewise, effluents of wastewater treatment plants have been reported to cause immunotoxic effects in fish, partly at concentrations lower

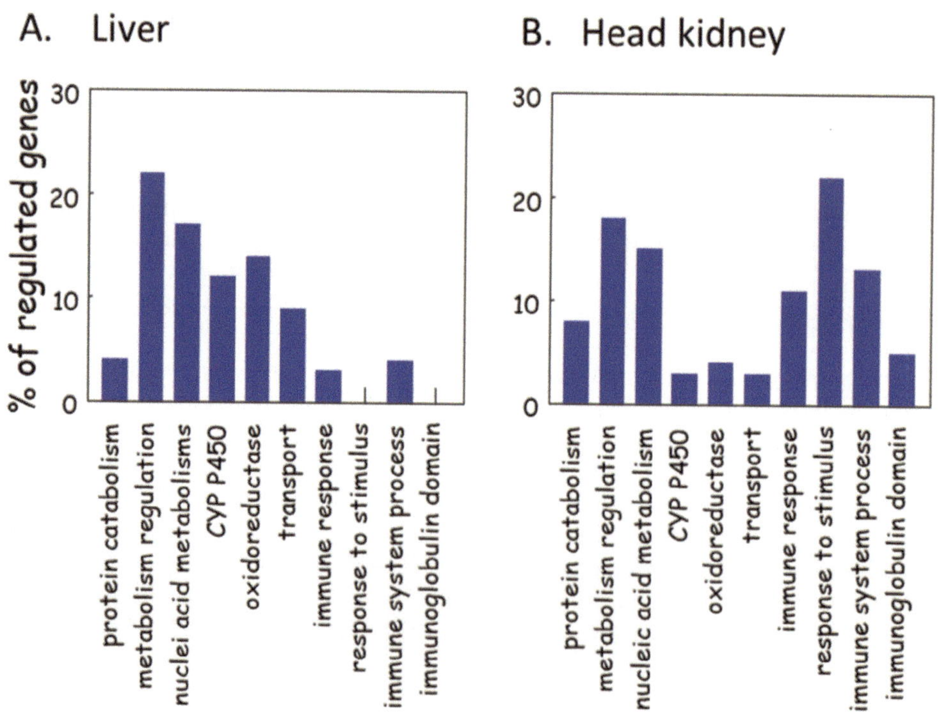


**FIGURE 6.4** Cell type-specific AhR-mediated regulation of gene expression in liver cells and in immune cells isolated from the head kidney. The cells were obtained from the same individual rainbow trout. After isolation, they were incubated for a 24h-recovery and attachment, and were then exposed for 48h to non-cytotoxic, equipotent concentrations of the AhR ligand, benzo(a)pyrene (BaP). Equipotency was assessed as those BaP concentrations that yielded the identical fold induction of the AhR-regulated xenobiotic metabolism gene, cytochrome P4501A, in liver cells and immune cells. After exposure, the RNA was extracted and the transcriptome was analyzed using two color microarray. The figure shows gene clusters that were up- or downregulated more than 2-fold in BaP-exposed immune and liver cells compared to control cells, based on the DAVID functional annotation clustering tool. The number of regulated genes that were assigned to the respective terms varied between 4 and 276. While in the liver cells, the BaP-dependent activation of the AhR predominantly regulates gene clusters involved in general and xenobiotic metabolism, the same molecular event in head kidney immune cells results in a prominent regulation of immune-related gene clusters. (Anja Maria Möller and Helmut Segner, unpublished data.)

than those required to induce recognizable changes in the structure and function of the reproductive endocrine system (Hoeger *et al.*, 2004; Liney *et al.*, 2006; Hébert *et al.*, 2008). The manifestation of immunotoxic effects is not restricted to fish but represents a widespread problem in contaminant-exposed wildlife (Acevedo-Whitehouse and Duffus, 2009; McMahon *et al.*, 2011; Mason *et al.*, 2013; Desforges *et al.*, 2016; De Guise *et al.*, 2021). Given the undisputed importance of immunity for phenotypic health and Darwinian fitness of organisms (Lochmiller, 1996; Graham *et al.*, 2011; Jackson, 2015; Palmer, 2018), immunotoxicity provides an opportunity to connect ecotoxicology to eco-immunology, disease ecology, and the actual discussions on the causes of the growing emergence of fish and wildlife diseases (Riley *et al.*, 2008; Martin *et al.*, 2010; Okamura and Feist, 2011; Martínez-Gómez and Vethaak, 2019a).

Given the relevance of immunotoxicology, the question is how to detect the potential of chemicals to interfere with the immune system of fish. Here, a first expectation of future research in fish immunotoxicology is to focus on exposure concentrations that are well below the threshold concentrations for general toxicity. Immune responses observed at test concentrations that are high enough to induce apical effects such as lethality (in vivo) or cytotoxicity (in vitro) do not necessarily represent an immunotoxic mode of action by itself but may represent general toxicity (Rehberger *et al.*, 2017a). Unfortunately, much of the published literature on fish immunotoxicity has used comparatively high exposure concentrations which makes it difficult to judge if the described effects are indeed due to an immunotoxic mode of action.

The next challenge for future research in fish immunotoxicology is the identification of endpoints and assays that are suitable: first, to screen whether a test agent has the potential to interact with the immune system, and second, to evaluate if this initial interaction propagates to a phenotype with compromised immunocompetence. The selection of appropriate endpoints and assays for the initial screening step as well as for the adverse outcome step is a complex task, for several reasons. The diffuse organization and the integrated functioning of the immune system complicates immunotoxicity assessment as it raises questions on the targets within the system which are impacted by the chemical. A further complicating fact in immunotoxicity assessment is that the overall capacity of the immune system to maintain homeostasis and health as well as to defend the organism against external stressors depends not so much on an individual immune element but on the balance between individual components and the functioning at the systemic level. By means of networked self-control, the immune system can buffer to some extent stressor-induced disturbances, and this makes it difficult to define thresholds of adversity or to extrapolate from the toxicant-induced modulation of a specific immune component to an overall impairment of immune functionality and capacity (Hartung and Corsini, 2013). In addition, toxic impacts often may not be visible in the resting immune system, but only when the immune system is activated, e.g., after a pathogen challenge (Köllner *et al.*, 2002; Hartung and Corsini, 2013). Finally, immune responses may not be directly triggered by a toxicant, but as indirect response to other forms of toxicity. For instance, an inflammatory reaction may be a secondary response to a toxicant-induced tissue damage (Redfern *et al.*, 2021). Also, toxicant effects on the microbiome can have consequences for immune system function (Redfern *et al.*, 2021). Another critical, although not immunotoxicity-specific, aspect to be considered is the argument recently raised (Johnson *et al.*, 2020) that for practical and ethical reasons, it is not realistic to add more and more tests to the existing ecotoxicity test procedures. Instead, we must aim for intelligent approaches that provide information without the need for additional resources or animal usage. All these factors given, the development of a screening and testing strategy to assess immunotoxic hazards will be a major challenge to future research in fish immunotoxicology.

Pragmatically, the most frequently used parameters to study immunotoxic effects in fish include phagocytotic activity; respiratory burst activity; and transcript levels of immune-related genes, particularly cytokines (Carlson and Zelikoff, 2008a; Segner *et al.*, 2011; Rehberger *et al.*, 2017a). While these parameters focus on innate immunity, assays such as in vitro lymphoproliferation assays address impacts on adaptive immunity. These various immune parameters can be measured either in vivo, ex vivo (fish exposure takes place in vivo, but the effect endpoint is measured in vitro), or in vitro (Zelikoff, 2000; Bols *et al.*, 2001; Segner *et al.*, 2011; Segner *et al.*, 2021).

For a screening step, in vitro assays would be most preferable. However, there exists no knowledge in which in vitro immune assays or in which combination of in vitro assays would be representative to detect immunotoxic potentials of chemicals to fish. For human immunotoxicology, a tiered in vitro testing strategy has been suggested (Corsini and Roggen, 2017), but we are far from that in fish immunotoxicology. Beyond, as recently shown for the in vitro phagocytes assay with rainbow trout leukocytes, the existing assays are often not validated with respect to their specificity, i.e., if they produce false positive/negative results (Rehberger *et al.*, 2021). Therefore, in a recent review on piscine in vitro immune assays, it was concluded that these assays are not yet ready for immunotoxicity screening (Segner *et al.*, 2021).

Alternative to or in combination with in vitro immunotoxicity assays, the incorporation of immune-specific endpoints into the zebra fish embryo test might be a valuable tool to screen for immunotoxic effects in fish. The fish embryo test is increasingly used as an alternative test for the ecotoxicological hazard assessment (Embry *et al.*, 2010). Since the innate immune system develops already during the embryonic period of zebrafish (Lam *et al.*, 2004), the embryo test should be able to detect toxic impacts on this arm of the immune system. Since there exist several hematopoietic-specific transgenic zebrafish lines, this model provides ample opportunities to study chemical effects on hematopoesis. In fact, the zebrafish embryo test has been extensively applied in drug development (Cassar *et al.*, 2020), but its potential of toxicological screening in environmental risk assessment has been little used, although this appears to change over time (Pang *et al.*, 2020; Namit *et al.*, 2022). Future research clearly should more give emphasis on the potential of the fish embryo test for immunotoxicity screening.

Another, yet unexplored option to provide an initial alert on potential immunotoxic actions of chemicals to fish would be reliance on trigger-based approaches. This is used, for instance, in human toxicological risk assessment, which includes immunotoxicity tests not used routinely but on a case-by-case basis (Boverhof *et al.*, 2014). The execution of immune tests is triggered upon the observation of altered immune-related parameters—for instance, altered white blood counts, immune organ weights, or histopathological changes—in chronic toxicity tests. Similarly, chronic fish toxicity tests like prolonged tests designed for endocrine disruptors could include certain immune parameters, as it was recently exemplified for zebrafish (Rehberger *et al.*, 2020; Kernen *et al.*, 2022). Such an approach would also enable the direct comparison of effect thresholds for different modes of toxic action such as reproductive toxicity, neurotoxicity, and immunotoxicity. A further possibility for a trigger approach could be based on "biological read-across" (Rand-Weaver *et al.*, 2013). This approach builds on the fact that many pathways through which chemicals interact with the human immune system are evolutionary conserved and do exist in fish as well. In this way, fish immunotoxicology could take advantage from the usually broader database in human toxicology and identify chemicals that may be immunotoxic not only in humans but also in fish. Finally, mechanistic information could serve as a trigger. If we know, for instance, that a chemical can activate a specific receptor pathway and if we know while this receptor plays a physiological role in the fish immune system, this may alert to possible immunotoxic activities of this compound.

These discussions bring us to another critical knowledge gap in fish immunotoxicology, which is the insufficient understanding of immunotoxic mechanisms vs. descriptive studies. The challenge in fish immunotoxicology studies is not so much to find a parameter that responds to the chemical exposure but to understand what the response means with respect to immune functioning and organism health. Therefore, the plea for more mechanistic understanding is not simply driven by academic curiosity but such knowledge is crucial, on the one hand, to be able to systematically group chemicals that share the same mechanistic pathway into one category rather than evaluating them on a compound-by-compound basis, categorize observations rather than collecting unconnected singular findings, and, on the other hand, to be able to predict adverse consequences of molecular and cellular responses (Hahn, 2011; Segner *et al.*, 2011). Mechanistic knowledge also provides an opportunity to better understand the connection between immune and stress responses. Toxic agents can activate stress-responsive molecules like heat shock proteins or signaling cascades like JNK (c-Jun-N-terminal kinases) or MAPK (mitogen-activated protein kinases), and these are often intimately connected with receptors involved in immune pathway regulation like NFkB (Pockley, 2003; Simmons *et al.*, 2009). In this context, recent observations (Rehberger *et al.*, 2021) found that non-immunotoxic baseline toxicants like butanol and ethylene glycol caused immune responses in isolated rainbow trout leukocytes, which may be explained by the nonspecific interaction of these chemicals with the cell membrane fluidity and, as a result, the activation of membrane-based stress signaling pathways.

Fostering the mechanistic understanding of immunotoxic effects in fish means also to give enhanced attention to the physiological interaction and communication between the various immune players. An example of the case is the immunity within barrier tissues like skin, gills, and gut. To date, fish immunotoxicology studies focused mainly on internal immune organs like head kidney and spleen, or the blood. Barrier tissues, however, are the sites where the uptake of environmental chemicals takes place, and at the same time, they are important sites of the fish's immune defense. Thus, it is here where the first contact between foreign chemicals and the immune system happens. Determining the effects of xenobiotics on the various sentinel cells within the barrier tissues should be a future priority for aquatic immunotoxicologists. As front-line cells of the immune system with roles in structure, function, and regulation of the barrier milieu, they are directly exposed to xenobiotics and/or their metabolites. At the barrier tissues, another player is located which may have relevance for the immune response to toxicants, that is the microbiome. While for mammals, the importance of the microbiome for barrier immunity is well established, research on the role of the microbiome on fish barrier tissues, their immunity, and the toxicant effects is still very much in its beginnings, but most likely this will be an active area in future research.

While the bad news is that our current mechanistic understanding of immunotoxic effects is still limited, the good news is that there is a steadily growing knowledge base on the functioning of the fish immune system and a steadily growing repertoire of immunological methods and tools. Thus, we are no longer restrained to measuring rather crude parameters, but fish immunotoxicologists have advanced tools available such as, e.g., single cell sequencing to analyze cell-type specific responses (Perdiguero *et al.*, 2021).

Another research need in fish immunotoxicology is to better understand the mechanistic linkages between molecular and cellular immune responses and an altered immunocompetence at the organism level. The majority of published immunotoxicity studies with fish focused on the measurement of molecular and cellular responses, but provide no answer whether these changes propagate into an adverse immune phenotype, or what is the ecologically relevant outcome. A molecular or cellular immune alteration is per se not an adverse effect, but these alterations may well be compensated and balanced within the complex network of the immune system. Therefore, alterations of molecular and cellular immune parameters are not necessarily predictive for a dysfunctioning at the immune system level, but the adversity needs to be demonstrated (Selgrade, 2007; Rehberger *et al.*, 2017b). In fish immunotoxicology, to date, the ultimate prove to demonstrate the adversity of a chemical impact on immune parameters is to perform a host resistance test (Köllner *et al.*, 2002). In this test, control and toxicant-treated fishes are exposed to infectious pathogens and if the chemical-treated groups show a higher pathogen susceptibility than the control fishes, this is taken as demonstration of the immunotoxicity of the chemical. The test has several drawbacks, for instance, it needs appropriate experimental facilities and the permit to work with infectious agents, and it is an animal experiment that causes severe suffering to the fish. Thus, it would be advantageous to have alternative approaches for predicting if a sublethal change in an immune parameter is predictive of an adverse immune phenotype of the organism. Here, the AOP framework may provide an option. Based on mechanistic information, an AOP established causal relationships between "molecular initiating events" (MIE), i.e., the initial chemico-biological interaction, and adverse outcomes at levels that bear ecological relevance, e.g., organism survival or population recruitment (Ankley *et al.*, 2010). In fish toxicology, AOPs have been established, for instance, for the linkages between xenobiotic activation of estrogen receptors and impaired reproductive performance (https://aopwiki.org). However, since immune phenotypes are cascading from a complex network of responses and are dependent on the relative balance of the various pathways and processes, establishing an AOP on immunotoxic effects represents a challenge. Recently, this was exemplified in the process of inflammation (Villeneuve *et al.*, 2018) and in the case of AHR-mediated immunotoxicity (Segner *et al.*, 2021). Nevertheless, the successful establishing of AOPs involving immune responses, e.g., the AOP for skin sensitization (Maxwell *et al.*, 2014), indicate that such challenges can be overcome provided there exists sufficient mechanistic knowledge. In any case, the AOP framework may be instrumental to design immunotoxicological studies and to select the immune endpoints to be measured among the myriads of possible immune parameters and to formulate hypotheses on the immune functions that might be affected by the molecular actions of the toxicants. In this context, it needs to be emphasized how much progress has been achieved over recent years in methods to study fish immunology as well as in knowledge of the functioning of the fish immune system. Thus, we are no longer restrained to measuring rather crude parameters, but fish immunotoxicologists have advanced tools available such as, e.g., transgenic zebrafish lines (Xu *et al.*, 2018) or single cell sequencing to analyze cell-type specific responses (Perdiguero *et al.*, 2021).

Finally, a further task for future fish immunotoxicology research is to expand our studies from the current focus on a few model species like zebrafish or rainbow trout to a broader range of fish species. While molecular targets of immunotoxicants may be conserved between species, pronounced species differences may exist in immune functioning. Species extrapolation is a principal challenge in ecotoxicology, but it may be particularly relevant in immunotoxicology. With the current knowledge, it is difficult to assess how similar or diverse immunotoxic responses are between fish species. More comparative studies that consider commonalities and differences in immunotoxic reactions between fish species are much needed.

## CONCLUSIONS

The immune system is exquisitely sensitive for assessing the toxic effects of chemicals and physical stresses of environmental concern. Adverse effects on the immune system are, for the most part, observed before more routine markers of chemical stress and toxicity, such as enzyme induction, tissue cellularity, and hematological changes. The sophistication and complexity of the immune system enable it to (potentially) be the most sensitive and therefore most prominent body function to detect harmful effects from environmental stressors. The sensitivity of this system seems to reside in the complex interactions that must occur for the mobilization of immune defenses. Disruption of any of these processes could offset the balance necessary for immunoregulation in the host and thus produce a cascade of detrimental secondary events, including compromised host resistance against infectious diseases and cancer.

Because of the sensitivity of the immune response to environmental toxicants and its importance for maintaining host resistance against disease, chemical-induced immune dysfunction can be predictive of the toxicological hazards and risks associated with pollutant exposure. The previous establishment of highly sensitive immune assays to enumerate these alterations has culminated in a well-characterized battery of endpoints that can possibly be used to predict biological impact and adverse health outcomes in exposed populations. Such sensitivities also have major applications in efficacy-testing programs including those following remediation. Information on host immunocompetence generated from such programs could aid in management decisions regarding the effectiveness of any remedial activities and the rates of recovery of affected sites. It could be assumed that changes in directions indicating decreased exposure and effects of affected sites precede an improvement in the ecological health of the environment; thus, assays that measure immune dysfunction can serve as rapid indicators of the direction of change in the toxic exposure and effects at a particular monitoring site. It is important to recognize that, although alterations in host immunocompetence and immune function are measured in a single individual, these effects often result in decreased resistance to infectious agents and neoplasia, and thus affect the population at large. Conclusions from field and laboratory studies suggest that chemical contaminants weaken hosts' immune defenses, making them more susceptible to epizootic infections.

## REFERENCES


Abelli, L., Baldassini, M.R., Meschini, R., Mastrolia, L., 1998. Apoptosis of thymocytes in developing sea bass dicentrarchus labrax (L.). *Fish & Shellfish Immunology* **8**, 13–24.

Abelli, L., Picchietti, S., Romano, N., Mastrolia, L., Scapigliati, G., 1996. Immunocytochemical detection of thymocyte antigenic determinants in developing lymphoid organs of sea bassDicentrarchus labrax (L.). *Fish & Shellfish Immunology* **6**, 493–505.

Acevedo-Whitehouse, K., Duffus, A.L.J., 2009. Effects of environmental change on wildlife health. *Philosophical Transactions of the Royal Society of London. Series B, Biological Sciences* **364**, 3429–3438.

Agius, C., 1980. Phylogenetic development of melano–macrophage centres in fish. *Journal of Zoology* **191**, 11–31.

Agius, C., Roberts, R.J., 2003. Melano-macrophage centres and their role in fish pathology. *Journal of Fish Diseases* **26**, 499–509.

Agnihothrudu, V., 1988. Pyrethroids: their future and toxicity, Advances in toxicology and environmental health. Proc. of the VI Annual Conf. of the society of toxicology, Guwahati, p. 69.

Ainsworth, A.J., 1992. Fish granulocytes: morphology, distribution, and function. *Annual Review of Fish Diseases* **2**, 123–148.

Albers, J.W., Cole, P., Greenberg, R.S., Mandel, J.S., Monson, R.R., Ross, J.H., Snodgrass, W.R., Spurgeon, A., Van Gemert, M., 1999. Analysis of chlorpyrifos exposure and human health: expert panel report. *Journal of Toxicology and Environmental Health Part B: Critical Reviews* **2**, 301–324.

Alesci, A., Lauriano, E.R., Aragona, M., Capillo, G., Pergolizzi, S., 2020. Marking vertebrates langerhans cells, from fish to mammals. *Acta Histochemica* **122**, 151622.

Alexander, J.B., Ingram, G.A., 1992. Noncellular nonspecific defence mechanisms of fish. *Annual Review of Fish Diseases* **2**, 249–279.

Altmann, S.M., Mellon, M.T., Distel, D.L., Kim, C.H., 2003. Molecular and functional analysis of an interferon gene from the zebrafish, Danio rerio. *Journal of Virology* **77**, 1992–2002.

Álvarez, F., Razquin, B.E., Villena, A.J., Zapata, A.n.G., 1998. Seasonal changes in the lymphoid organs of wild brown trout, Salmo trutta L: a morphometrical study. *Veterinary Immunology and Immunopathology* **64**, 267–278.

Anderson, D.P., 1990. Immunological indicators: effects of environmental stress on immune protection and disease outbreaks. *American Fisheries Society Symposium* **8**, 38.

Anderson, D.P. 1997. Environmental Factors in Fish Health: Immunological Aspects. In George, I., Teruyuki, N. (Eds.), *Fish Physiology*. Academic Press, New York, pp. 289–310.

Ankley, G.T., Bennett, R.S., Erickson, R.J., Hoff, D.J., Hornung, M.W., Johnson, R.D., Mount, D.R., Nichols, J.W., Russom, C.L., Schmieder, P.K., Serrrano, J.A., Tietge, J.E., Villeneuve, D.L., 2010. Adverse outcome pathways: a conceptual framework to support ecotoxicology research and risk assessment. *Environmental Toxicology and Chemistry* **29**, 730–741.

Antonelli, M., Kushner, I., 2017. It's time to redefine inflammation. *FASEB Journal* **31**, 1787–1791.

Aoki, T., Hikima, J., Hwang, S.D., Jung, T.S., 2013. Innate immunity of finfish: primordial conservation and function of viral RNA sensors in teleosts. *Fish Shellfish Immunol* **35**, 1689–1702.

Aranishi, F., Mano, N., 2000. Antibacterial cathepsins in different types of ambicoloured Japanese flounder skin. *Fish & Shellfish Immunology* **10**, 87–89.

Areechon, N., Plumb, J., 1990. Sublethal effects of malathion on channel catfish, Ictalurus punctatus. *Bulletin of Environmental Contamination and Toxicology* **44**, 435–442.

Arkoosh, M.R., Casillas, E., Clemons, E., Kagley, A.N., Olson, R., Reno, P., Stein, J.E., 1998a. Effect of pollution on fish diseases: potential impacts on salmonid populations. *Journal of Aquatic Animal Health* **10**, 182–190.

Arkoosh, M.R., Casillas, E., Huffman, P., Clemons, E., Evered, J., Stein, J.E., Varanasi, U., 1998b. Increased susceptibility of juvenile chinook salmon from a contaminated estuary to Vibrio anguillarum. *Transactions of the American Fisheries Society* **127**, 360–374.

Arkoosh, M.R., Clemons, E., Huffman, P., Kagley, A.N., Casillas, E., Adams, N., Sanborn, H.R., Collier, T.K., Stein, J.E., 2001a. Increased susceptibility of juvenile chinook salmon to vibriosis after exposure to chlorinated and aromatic compounds found in contaminated urban estuaries. *Journal of Aquatic Animal Health* **13**, 257–268.

Arkoosh, M.R., Clemons, E., Huffman, P., Kagley, A.N., Casillas, E., Adams, N.G., Sanborn, H.R., Collier, T., Stein, J.E., 2001b. Increased susceptibility of Juvenile Chinook Salmon to vibriosis after exposure to chlorinated and aromatic compounds found in contaminated Urban Estuaries. *Journal of Aquatic Animal Health* **13**, 257–268.

Arkoosh, M.R., Clemons, E., Myers, M., Casillas, E., 1994. Suppression of B-cell mediated immunity in Juvenile Chinook Salmon (Oncorhynchus Tshawytscha) after exposure to either a polycyclic aromatic hydrocan or to polychlorinated biphenyls. *Immunopharmacology and Immunotoxicology* **16**, 293–314.

Arnold, R., Rice, C., 2000. Channel catfish, Ictalurus punctatus, leukocytes secrete immunoreactive adrenal corticotropin hormone (ACTH). *Fish Physiology and Biochemistry* **22**, 303–310.

Baan, R., Grosse, Y., Straif, K., Secretan, B., El Ghissassi, F., Bouvard, V., Benbrahim-Tallaa, L., Guha, N., Freeman, C., Galichet, L., 2009. A review of human carcinogens-part F: chemical agents and related occupations. *The Lancet Oncology* **10**, 1143–1144.

Balali-Mood, M., Naseri, K., Tahergorabi, Z., Khazdair, M.R., Sadeghi, M., 2021. Toxic mechanisms of five heavy metals: mercury, lead, chromium, cadmium, and arsenic. *Frontiers in Pharmacology* **12**, 1–19.

Barraza, F., Montero, R., Wong-Benito, V., Valenzuela, H., Godoy-Guzmán, C., Guzmán, F., Köllner, B., Wang, T., Secombes, C.J., Maisey, K., Imarai, M., 2021. Revisiting the teleost thymus: current knowledge and future perspectives. *Biology* **10**, 8.

Barron, M.G., Anderson, M.J., Cacela, D., Lipton, J., Teh, S.J., Hinton, D.E., Zelikoff, J.T., Dikkeboom, A.L., Tillitt, D.E., Holey, M., 2000. PCBs, liver lesions, and biomarker responses in adult walleye (Stizostedium vitreum vitreum) collected from Green Bay, Wisconsin. *Journal of Great Lakes Research* **26**, 250–271.

Barst, B.D., Bridges, K., Korbas, M., Roberts, A.P., Van Kirk, K., McNeel, K., Drevnick, P.E., 2015. The role of melano-macrophage aggregates in the storage of mercury and other metals: An example from yelloweye rockfish (Sebastes ruberrimus). *Environmental Toxicology and Chemistry* **34**, 1918–1925.

Bayha, K.M., Ortell, N., Ryan, C.N., Griffitt, K.J., Krasnec, M., Sena, J., Ramaraj, T., Takeshita, R., Mayer, G.D., Schilkey, F., Griffitt, R.J., 2017. Crude oil impairs immune function and increases susceptibility to pathogenic bacteria in southern flounder. *PLoS One* **12**, e0176559.

Bayne, C.J., Gerwick, L., 2001. The acute phase response and innate immunity of fish. *Developmental & Comparative Immunology* **25**, 725–743.

Beaman, J., Finch, R., Gardner, H., Hoffmann, F., Rosencrance, A., Zelikoff, J., 1999. Mammalian immunoassays for predicting the toxicity of malathion in a laboratory fish model. *Journal of Toxicology and Environmental Health, Part A* **56**, 523–542.

Betoulle, S., Duchiron, C., Deschaux, P., 2000. Lindane differently modulates intracellular calcium levels in two populations of rainbow trout (Oncorhynchus mykiss) immune cells: head kidney phagocytes and peripheral blood leucocytes. *Toxicology* **145**, 203–215.

Bird, S., Tafalla, C., 2015. Teleost chemokines and their receptors. *Biology* **4**, 756–784.

Birnbaum, L.S., 1995. Developmental effects of dioxins and related endocrine disrupting chemicals. *Toxicology Letters* **82**, 743–750.

Bjørgen, H., Koppang, E.O., 2021. Anatomy of teleost fish immune structures and organs. *Immunogenetics* **73**, 53–63.

Blaylock, B., Abdel-Nasser, M., McCarty, S., Knesel, J., Tolson, K., Ferguson, P., Mehendale, H., 1995. Suppression of cellular immune responses in BALB/c mice following oral exposure to permethrin. *Bulletin of Environmental Contamination and Toxicology* **54**, 768–774.

Blazer, V.S., Wolke, R.E., Brown, J., Powell, C.A., 1987. Piscine macrophage aggregate parameters as health monitors: effect of age, sex, relative weight, season and site quality in largemouth bass (Micropterus salmoides). *Aquatic Toxicology* **10**, 199–215.

Bobe, J., Goetz, F.W., 2001. Molecular cloning and expression of a TNF receptor and two TNF ligands in the fish ovary. *Comparative Biochemistry and Physiology Part B: Biochemistry and Molecular Biology* **129**, 475–481.

Bodammer, J., 1986. Ultrastructural observations on peritoneal exudate cells from the striped bass. *Veterinary Immunology and Immunopathology* **12**, 127–140.

Boehm, T., Hess, I., Swann, J.B., 2012. Evolution of lymphoid tissues. *Trends in Immunology* **33**, 315–321.

Bols, N.C., Brubacher, J.L., Ganassin, R.C., Lee, L.E., 2001. Ecotoxicology and innate immunity in fish. *Developmental Comparative Immunology* **25**, 853–873.

Boshra, H., Li, J., Sunyer, J.O., 2006. Recent advances on the complement system of teleost fish. *Fish Shellfish Immunol* **20**, 239–262.

Boverhof, D.R., Ladics, G., Luebke, B., Botham, J., Corsini, E., Evans, E., Germolec, D., Holsapple, M., Loveless, S.E., Lu, H., van der Laan, J.W., White, K.L., Jr., Yang, Y., 2014. Approaches and considerations for the assessment of immunotoxicity for environmental chemicals: a workshop summary. *Regulatory Toxicolology and Pharmacology* **68**, 96–107.

Bradbury, S.P., Coats, J.R., McKim, J.M., 1985. Differential toxicity and uptake of two fenvalerate formulations in fathead minnows (Pimephales promelas). *Environmental Toxicology and Chemistry: An International Journal* **4**, 533–541.

Breuil, G., Vassiloglou, B., Pepin, J., Romestand, B., 1997. Ontogeny of IgM-bearing cells and changes in the immunoglobulin M-like protein level (IgM) during larval stages in sea bass (Dicentrarchus labrax). *Fish & Shellfish Immunology* **7**, 29–43.

Broughton, A., 1990. Chronic health effects and immunological alternations associated with exposures to pesticide. *Comments on Toxicology* **4**, 59–71.

Bravo, C.F., Curtis, L.R., Myers, M.S., Meador, J.P., Johnson, L.L., Buzitis, J., Collier, T.K., Morrow, J.D., Laetz, C.A., Loge, F.J., Arkoosh, M.R., 2011. Biomarker responses and disease susceptibility in juvenile rainbow trout Oncorhynchus mykiss fed a high molecular weight PAH mixture. *Environmental Toxicology and Chemistry* **30**, 704–714.

Budinsky, R.A., Schrenk, D., Simon, T., Van den Berg, M., Reichard, J.F., Silkworth, J.B., Aylward, L.L., Brix, A., Gasiewicz, T., Kaminski, N., Perdew, G., Starr, T.B., Walker, N.J., Rowlands, J.C., 2014. Mode of action and dose-response framework analysis for receptor-mediated toxicity: The aryl hydrocarbon receptor as a case study. *Critical Reviews in Toxicology* **44**, 83–119.

Burggren, W.W., Dubansky, B., Bautista, N.M., 2017. *Cardiovascular Development in Embryonic and Larval Fishes, Fish Physiology* **36**, 107–184. Elsevier, Amsterdam, pp. 107–184.

Burton, J.E., Dorociak, I.R., Schwedler, T.E., Rice, C.D., 2002. Circulating lysozyme and hepatic CYP1A activities during a chronic dietary exposure to tributyltin (TBT) and 3,3',4,4',5–pentachlorobiphenyl (PCB-126) mixtures in channel catfish, Ictalurus punctatus. *Journal of Toxicology and Environmental Health, Part A* **65**, 589–602.

Camp, K.L., Wolters, W.R., Rice, C.D., 2000. Survivability and immune responses after challenge with Edwardsiella ictaluriin susceptible and resistant families of channel catfish, Ictalurus punctatus. *Fish & Shellfish Immunology* **10**, 475–487.

Campos-Sánchez, J.C., Esteban, M.Á., 2021. Review of inflammation in fish and value of the zebrafish model. *Journal of Fish Diseases* **44**, 123–139.

Carlson, E., Zelikoff, J.T., 2008a. The immune system of fish: a target organ of toxicity. *The Toxicology of Fishes*, 489–529.

Carlson, E.A., Li, Y., Zelikoff, J.T., 2004. Benzo[a]pyrene-induced immunotoxicity in Japanese medaka (Oryzias latipes): relationship between lymphoid CYP1A activity and humoral immune suppression. *Toxicology and Applied Pharmacology* **201**, 40–52.

Carlson, R., Zelikoff, J.T., 2008b. The immune system of fish: a target organ of toxicology. In Di Giulio, R.T., Hinton, D.E., (Eds.), *The Toxicolog of Fishes*. CRC Press, Boca Raton, pp. 489–530.

Casanova-Nakayama, A., Wernicke von Siebenthal, E., Kropf, C., Oldenberg, E., Segner, H., 2018. Immune-specific expression and estrogenic regulation of the four estrogen receptor isoforms in female rainbow trout (Oncorhynchus mykiss). *International Journal of Molecular Sciences* **19**, 932.

Cassar, S., Adatto, I., Freeman, J.L., Gamse, J.T., Iturria, I., Lawrence, C., Muriana, A., Peterson, R.T., Van Cruchten, S., Zon, L.I., 2020. Use of zebrafish in drug discovery toxicology. *Chemical Research in Toxicology* **33**, 95–118.

Castillo, A., Razquin, B., Lopez-Fierro, P., Alvarez, F., Zapata, A., Villena, A., 1990. Enzyme-and immuno-histochemical study of the thymic stroma in the rainbow trout, Salmo gairdneri, Richardson. *Thymus* **15**, 153–166.

Castillo, A., Sánchez, C., Dominguez, J., Kaattari, S.L., Villena, A.J., 1993. Ontogeny of IgM and IgM-bearing cells in rainbow trout. *Developmental & Comparative Immunology* **17**, 419–424.

Castranova, D., Samasa, B., Venero Galanternik, M., Jung, H.M., Pham, V.N., Weinstein, B.M., 2021. Live imaging of intracranial lymphatics in the zebrafish. *Circulation Research* **128**, 42–58.

Castro, R., Tafalla, C., 2015. Mucosal health in aquaculture. In Beck, P., Peatman, E., (Eds.), *Overview of Fish Immunity*. Elsevier, Amsterdam, pp. 3–54.

Chaudhari, N., Talwar, P., Parimisetty, A., Lefebvre d'Hellencourt, C., Ravanan, P., 2014. A molecular web: endoplasmic reticulum stress, inflammation, and oxidative stress. *Frontiers in Cellular Neuroscience* **8**.

Chaudhary, D.K., Sood, N., Pradhan, P.K., Agarwal, N.K., Rathore, G., 2012. Production and characterization of a monoclonal antibody against putative T lymphocytes of Catla catla. In vitro cellular & developmental biology. *Animal* **48**, 483–492.

Chaves-Pozo, E., García-Ayala, A., Cabas, I., 2018. Effects of sex steroids on fish leukocytes. *Biology* **7**, 9.

Chen, J., Xu, Y., Han, Q., Yao, Y., Xing, H., Teng, X., 2019. Immunosuppression, oxidative stress, and glycometabolism disorder caused by cadmium in common carp (Cyprinus carpio L.): Application of transcriptome analysis in risk assessment of environmental contaminant cadmium. *Journal of Hazardous Materials* **366**, 386–394.

Chilmonczyk, S., 1985. *Evolution of the Thymus in Rainbow Trout, Fish Immunology*. Academic Press, London, pp. 285–292.

Chilmonczyk, S., 1992. The thymus in fish: development and possible function in the immune response. *Annual Review of Fish Diseases* **2**, 181–200.

Cole, A.M., Weis, P., Diamond, G., 1997. Isolation and characterization of pleurocidin, an antimicrobial peptide in the skin secretions of winter flounder. *Journal of Biological Chemistry* **272**, 12008–12013.

Cooper, E., Zapata, A., Barrutia, M.G., Ramirez, J., 1983. Aging changes in lymphopoietic and myelopoietic organs of the annual cyprinodont fish, Nothobranchius guentheri. *Experimental Gerontology* **18**, 29–38.

Corsini, E., Liesivuori, J., Vergieva, T., Loveren, H., Colosio, C., 2008. Effects of pesticide exposure on the human immune system. *Human & Experimental Toxicology* **27**, 671–680.

Corsini, E., Luebke, R.W., Germolec, D.R., DeWitt, J.C., 2014. Perfluorinated compounds: emerging POPs with potential immunotoxicity. *Toxicology Letters* **230**, 263–270.

Corsini, E., Roggen, E., 2017. Overview of in vitro assessment of immunotoxicity. *Current Opinion in Toxicology* **5**, 13–18.

Cossarini-Dunier, M., Demael, A., Siwicki, A.K., 1990. In vivo effect of the organophosphorus insecticide trichlorphon on the immune response of carp (Cyprinus carpio): I. Effect of contamination on antibody production in relation to residue level in organs. *Ecotoxicology and Environmental Safety* **19**, 93–98.

da Costa Klosterhoff, M., de Medeiros, A.F.F., Guerreiro, A., Pedrosa, V.F., Romano, L.A., 2020. Comparative histology of the human and teleost fish thymus. *Brazilian Journal of Development* **6**, 68460–68481.

Daniels, G.D., Secombes, C.J., 1999. Genomic organisation of rainbow trout, Oncorhynchus mykiss TGF-β. *Developmental & Comparative Immunology* **23**, 139–147.

David, M., Kartheek, R.M., 2015. Histopathological alterations in spleen of freshwater fish Cyprinus carpio exposed to sublethal concentration of sodium cyanide. *Open Veterinary Journal* **5**, 1–5.

Davidson, A.J., Zon, L.I., 2004. The 'definitive' (and 'primitive') guide to zebrafish hematopoiesis. *Oncogene* **23**, 7233–7246.

De Guise, S., Levin, M., Jasperse, L., Herrman, J., Wells, R.S., Rowles, T., Schwacke, L., 2021. Long-term immunological alterations in bottlenose dolphin a decade after the deepwater horizon oil spill in the Northern Gulf of Mexico: potential for multigenerational effects. *Environmental Toxicology and Chemistry* **40**, 1308–1321.

de Medeiros, A.F.F., Pedrosa, V.F., Romano, L.A., 2021. Ultrastructural study of the thymus of Mugil liza Valenciennes, 1836 juveniles. *Fish and Shellfish Immunology Reports* **2**, 100005.

Dearstyne, E., Kerkvliet, N., 2002. Mechanism of 2, 3, 7, 8-tetrachlorodibenzo-p-dioxin (TCDD)-induced decrease in anti-CD3-activated CD4+ T cells: the roles of apoptosis, Fas, and TNF. *Toxicology* **170**, 139–151.

Desforges, J.-P.W., Sonne, C., Levin, M., Siebert, U., De Guise, S., Dietz, R., 2016. Immunotoxic effects of environmental pollutants in marine mammals. *Environment International* **86**, 126–139.

Diel, F., Detscher, M., Schock, B., Ennis, M., 1998. In vitro effects of the pyrethroid S-bioallethrin on lymphocytes and basophils from atopic and nonatopic subjects. *Allergy* **53**, 1052–1059.

Doering, J., Giesy, J., Wiseman, S., Hecker, M., 2013. Predicting the sensitivity of fishes to dioxin-like compounds: possible role of the aryl hydrocarbon receptor (AhR) ligand binding domain. *Environmental Science and Pollution Research* **20**, 1219–1224.

Doering, J., Hecker, M., Villeneuve, D., Zhang, X., 2019. Adverse Outcome Pathway on aryl hydrocarbon receptor activation leading to early life stage mortality, via increased. OECD Series on Adverse Outcome Pathways 12, OECD Publishing. COX-2.

Dornburg, A., Ota, T., Criscitiello, M.F., Salinas, I., Sunyer, J.O., Magadán, S., Boudinot, P., Xu, Z., Flajnik, M.F., Singer, A., Gambón-Deza, F., Hansen, J.D., Yoder, J.A., 2021. From IgZ to IgT: a call for a common nomenclature for immunoglobulin heavy chain genes of ray-finned fish. *Zebrafish* **18**, 343–345.

Douglas, S., Gallant, J., Gong, Z., Hew, C., 2001. Cloning and developmental expression of a family of pleurocidin-like antimicrobial peptides from winter flounder, Pleuronectes americanus (Walbaum). *Developmental & Comparative Immunology* **25**, 137–147.

Du, L., Yang, X., Yang, L., Wang, X., Zhang, A., Zhou, H., 2012. Molecular evidence for the involvement of RORα and RORγ in immune response in teleost. *Fish Shellfish Immunology* **33**, 418–426.

Duffy, J., Carlson, E., Li, Y., Prophete, C., Zelikoff, J., 2002. Impact of polychlorinated biphenyls (PCBs) on the immune function of fish: age as a variable in determining adverse outcome. *Marine Environmental Research* **54**, 559–563.

Dunier, M., Siwicki, A.K., 1993. Effects of pesticides and other organic pollutants in the aquatic environment on immunity of fish: a review. *Fish & Shellfish Immunology* **3**, 423–438.

Dutta, H., Dogra, J., Singh, N., Roy, P., Nasar, S., Adhikari, S., Munshi, J., Richmonds, C., 1992. Malathion induced changes in the serum proteins and hematological parameters of an Indian catfish Heteropneustes fossilis (Bloch). *Bulletin of Environmental Contamination and Toxicology* **49**, 91–97.

Eder, K.J., Leutenegger, C.M., Köhler, H.-R., Werner, I., 2009. Effects of neurotoxic insecticides on heat-shock proteins and cytokine transcription in Chinook salmon (Oncorhynchus tshawytscha). *Ecotoxicology and Environmental Safety* **72**, 182–190.

Edholm, E.-S., Bengten, E., Wilson, M., 2011. Insights into the function of IgD. *Developmental & Comparative Immunology* **35**, 1309–1316.

Ellis, A., 1985. Eosinophilic granular cells (EGC) and histamine responses to Aeromonassalmonicida toxins in rainbow trout. *Developmental & Comparative Immunology* **9**, 251–260.

Ellsaesser, C.F., Bly, J.E., Clem, L., 1988. Phylogeny of lymphocyte heterogeneity: the thymus of the channel catfish. *Developmental & Comparative Immunology* **12**, 787–799.

Embry, M.R., Belanger, S.E., Braunbeck, T.A., Galay-Burgos, M., Halder, M., Hinton, D.E., Léonard, M.A., Lillicrap, A., Norberg-King, T., Whale, G., 2010. The fish embryo toxicity test as an animal alternative method in hazard and risk assessment and scientific research. *Aquatic Toxicology* **97**, 79–87.

Enane, N., Frenkel, K., O'connor, J., Squibb, K., Zelikoff, J., 1993. Biological markers of macrophage activation: applications for fish phagocytes. *Immunology* **80**, 68.

Esser, C., Haarmann-Stemmann, T., Hochrath, K., Schikowski, T., Krutmann, J., 2018. AHR and the issue of immunotoxicity. *Current Opinion in Toxicology* **10**, 91–97.

Esteban, M.Á., Cuesta, A., Chaves-Pozo, E., Meseguer, J., 2015. Phagocytosis in teleosts. Implications of the new cells involved. *Biology* **4**, 907–922.

Evans, D.L., Jaso-Friedmann, L., 1992. Nonspecific cytotoxic cells as effectors of immunity in fish. *Annual Review of Fish Diseases* **2**, 109–121.

Fänge, R., Nilsson, S., 1985. The fish spleen: structure and function. *Experientia* **41**, 152–158.

Fänge, R., Pulsford, A., 1985. The thymus of the angler fish, Lophius piscatorius (pisces: teleostei) a light and electron microscopic study. *Fish Immunology*, 293–311.

Ferguson, H.W., 1976. The relationship between ellipsoids and melano-macrophage centres in the spleen of turbot (Scophthalmus maximus). *Journal of Comparative Pathology* **86**, 377–380.

Ferrero, G., Gomez, E., Lyer, S., Rovira, M., Miserocchi, M., Langenau, D.M., Bertrand, J.Y., Wittamer, V., 2020. The macrophage-expressed gene (mpeg) 1 identifies a subpopulation of B cells in the adult zebrafish. *Journal of Leukocyte Biology* **107**, 431–443.

Fiedler, N., Maccia, C., Kipen, H., 1992. Evaluation of chemically sensitive patients. *Journal of Occupational Medicine*, 529–538.

Fillatreau, S., Six, A., Magadan, S., Castro, R., Sunyer, J.O., Boudinot, P., 2013. The astonishing diversity of Ig classes and B cell repertoires in teleost fish. *Frontiers in Immunology* **4**, 28. doi: 10.3389/fimmu.2013.00028.

Fingerhut, L., Dolz, G., de Buhr, N., 2020. What is the evolutionary fingerprint in neutrophil granulocytes? *International Journal of Molecular Sciences* **21**, 4523.

Fishelson, L., 1995. Cytological and morphological ontogenesis and involution of the thymus in cichlid fishes (Cichlidae, Teleostei). *Journal of Morphology* **223**, 175–190.

Flajnik, M.F., 2018. A cold-blooded view of adaptive immunity. *Nature Reviews Immunology* **18**, 438–453.

Flaño, E., Alvarez, F., López-Fierro, P., Razquin, B.E., Villena, A.J., Zapata, A.G., 1996. In vitro and in situ characterization of fish thymic nurse cells. *Developmental Immunology* **5**, 17–24.

Forlenza, M., Fink, I.R., Raes, G., Wiegertjes, G.F., 2011. Heterogeneity of macrophage activation in fish. *Developmental & Comparative Immunology* **35**, 1246–1255.

Fournie, J., Summers, J., Courtney, L., Engle, V., Blazer, V., 2001. Utility of splenic macrophage aggregates as an indicator of fish exposure to degraded environments. *Journal of Aquatic Animal Health* **13**, 105–116.

Fujiki, K., Bayne, C.J., Shin, D.-H., Nakao, M., Yano, T., 2001. Molecular cloning of carp (Cyprinus carpio) C-type lectin and pentraxin by use of suppression subtractive hybridisation. *Fish & Shellfish Immunology* **11**, 275–279.

Gao, Y., Xu, H., Li, L., Niu, C., 2020. Immune defense parameters of wild fish as sensitive biomarkers for ecological risk assessment in shallow sea ecosystems: A case study with wild mullet (Liza haematocheila) in Liaodong Bay. *Ecotoxicology and Environmental Safety* **194**, 110337.

García-Castillo, J., Pelegrín, P., Mulero, V., Meseguer, J., 2002. Molecular cloning and expression analysis of tumor necrosis factor α from a marine fish reveal its constitutive expression and ubiquitous nature. *Immunogenetics* **54**, 200–207.

Gerwick, L., Reynolds, W.S., Bayne, C.J., 2000. A precerebellin-like protein is part of the acute phase response in rainbow trout, Oncorhynchus mykiss. *Developmental & Comparative Immunology* **24**, 597–607.

Ghoneum, M.M.H., Egami, N., 1982. Age related changes in morphology of the thymus of the fish, Oryzias latipes. *Experimental Gerontology* **17**, 33–40.

Giedroc, D.P., Chen, X., Apuy, J.L., 2001. Metal response element (MRE)-binding transcription factor-1 (MTF-1): structure, function, and regulation. *Antioxidants & Redox Signaling* **3**, 577–596.

Giesy, J.P., Kannan, K., 2002. Peer reviewed: perfluorochemical surfactants in the environment. *Environmental Science & Technology* **36**, 146A–152A.

Ginhoux, F., Schultze, J.L., Murray, P.J., Ochando, J., Biswas, S.K., 2016. New insights into the multidimensional concept of macrophage ontogeny, activation and function. *Nature Immunology* **17**, 34–40.

Giulivo, M., López de Alda, M., Capri, E., Barceló, D., 2016. Human exposure to endocrine disrupting compounds: Their role in reproductive systems, metabolic syndrome and breast cancer. A review. *Environmental Research* **151**, 251–264.

Goetz, F.W., Planas, J.V., MacKenzie, S., 2004. Tumor necrosis factors. *Developmental & Comparative Immunology* **28**, 487–497.

Grace, M.F., Botham, J.W., Manning, M.J., 1981. Ontogeny of lymphoid organ function in fish. In Solomon, J.B., (Ed.), *Aspects of Developmental and Comparative Immunology*. Pergamon Press, Oxford, pp. 467–468.

Grace, M.F., Manning, M.J., 1980. Histogenesis of the lymphoid organs in rainbow trout, Salmo gairdneri Rich. 1836. *Developmental & Comparative Immunology* **4**, 255–264.

Graham, A.L., Shuker, D.M., Pollitt, L.C., Auld, S.K.J.R., Wilson, A.J., Little, T.J., 2011. Fitness consequences of immune responses: strengthening the empirical framework for ecoimmunology. *Functional Ecology* **25**, 5–17.

Graham, S., Secombes, C., 1990. Do fish lymphocytes secrete interferon-γ? *Journal of Fish Biology* **36**, 563–573.

Grimholt, U., Lukacs, M., 2020. Fate of MHCII in salmonids following 4WGD. *Immunogenetics* **73**: 79–91.

Grimholt, Y., Lie, O., 1998. The major histocompatibility complex in fish. *Revue Scientifique Et Technique* **17**, 121–127.

Grinde, B., 1989. Lysozyme from rainbow trout, Salmo gairdneri Richardson, as an antibacterial agent against fish pathogens. *Journal of Fish Diseases* **12**, 95–104.

Grinwis, G., Besselink, H., Van den Brandhof, E., Bulder, A., Engelsma, M., Kuiper, R., Wester, P., Vaal, M., Vethaak, A., Vos, J., 2000. Toxicity of TCDD in European flounder (Platichthys flesus) with emphasis on histopathology and cytochrome P450 1A induction in several organ systems. *Aquatic Toxicology* **50**, 387–401.

Grinwis, G., Van den Brandhof, E., Engelsma, M., Kuiper, R., Vaal, M., Vethaak, A., Wester, P., Vos, J., 2001. Toxicity of PCB-126 in European flounder (Platichthys flesus) with emphasis on histopathology and cytochrome P4501A induction in several organ systems. *Archives of Toxicology* **75**, 80–87.

Guardiola, F., Gónzalez-Párraga, P., Meseguer, J., Cuesta, A., Esteban, M., 2014. Modulatory effects of deltamethrin-exposure on the immune status, metabolism and oxidative stress in gilthead seabream (Sparus aurata L.). *Fish & Shellfish Immunology* **36**, 120–129.

Guo, H., Dixon, B., 2021. Understanding acute stress-mediated immunity in teleost fish. *Fish and Shellfish Immunology Reports* **2**, 100010.

Hahn, M.E., 2011. Mechanistic research in aquatic toxicology: perspectives and future directions. *Aquatic Toxicology* **105**, 67–71.

Hahn, M.E., Karchner, S.I., Merson, R.R., 2017. Diversity as opportunity: Insights from 600 million years of AHR evolution. *Current Opinion in Toxicology* **2**, 58–71.

Haugarvoll, E., Bjerkås, I., Nowak, B.F., Hordvik, I., Koppang, E.O., 2008. Identification and characterization of a novel intraepithelial lymphoid tissue in the gills of Atlantic salmon. *Journal of Anatomy* **213**, 202–209.

Haire, R.N., Rast, J.P., Litman, R.T., Litman, G.W., 2000. Characterization of three isotypes of immunoglobulin light chains and T-cell antigen receptor α in zebrafish. *Immunogenetics* **51**, 915–923.

Hall, A.J., McConnell, B.J., Schwacke, L.H., Ylitalo, G.M., Williams, R., Rowles, T.K., 2018. Predicting the effects of polychlorinated biphenyls on cetacean populations through impacts on immunity and calf survival. *Environmental Pollution* **233**, 407–418.

Hampl, R., Kubátová, J., Stárka, L., 2016. Steroids and endocrine disruptors-History, recent state of art and open questions. *The Journal of Steroid Biochemistry and Molecular Biology* **155**, 217–223.

Handy, R.D., Henry, T.B., Scown, T.M., Johnston, B.D., Tyler, C.R., 2008. Manufactured nanoparticles: their uptake and effects on fish--a mechanistic analysis. *Ecotoxicology* **17**, 396–409.

Hansen, J.D., Strassburger, P., 2000. Description of an ectothermic TCR coreceptor, CD8α, in rainbow trout. *The Journal of Immunology* **164**, 3132–3139.

Hansen, J.D., Vojtech, L.N., Laing, K.J., 2011. Sensing disease and danger: a survey of vertebrate PRRs and their origins. *Developmental & Comparative Immunology* **35**, 886–897.

Hao, X., Chi, H., Tang, X., Xing, J., Sheng, X., Zhan, W., 2021. The functions of β-defensin in flounder (Paralichthys olivaceus): antibiosis, chemotaxis and modulation of phagocytosis. *Biology* **10**, 1247.

Harms, C.A, et al., 2000. Cloning and sequencing hybrid striped bass (Morone saxatilis x M. chrysops) transforming growth factor-β (TGF-β), and development of a reverse transcription quantitative competitive polymerase chain reaction (RT-qcPCR) assay to measure TGF-β mRNA of teleost fish. *Fish & Shellfish Immunology* **10**(1), 61–85.

Harris, J., Bird, D.J., 2000. Modulation of the fish immune system by hormones. *Veterinary Immunology and Immunopathology* **77**, 163–176.

Hartung, T., Corsini, E., 2013. Immunotoxicology: challenges in the 21st century and in vitro opportunities. *ALTEX - Alternatives to Animal Experimentation* **30**, 411–426.

Havixbeck, J.J., Barreda, D.R., 2015. Neutrophil development, migration, and function in teleost fish. *Biology* **4**, 715–734.

Hayashi, Y., Miclaus, T., Murugadoss, S., Takamiya, M., Scavenius, C., Kjaer-Sorensen, K., Enghild, J.J., Strähle, U., Oxvig, C., Weiss, C., Sutherland, D.S., 2017. Female versus male biological identities of nanoparticles determine the interaction with immune cells in fish. *Environmental Science: Nano* **4**, 895–906.

He, S., Chen, J., Jiang, Y., Wu, Y., Zhu, L., Jin, W., Zhao, C., Yu, T., Wang, T., Wu, S., Lin, X., Qu, J.Y., Wen, Z., Zhang, W., Xu, J., 2018. Adult zebrafish Langerhans cells arise from hematopoietic stem/progenitor cells. *Elife* **7**, e36131.

Hébert, N., Gagné, F., Cejka, P., Cyr, D., Marcogliese, D.J., Blaise, C., Pellerin, J., Fournier, M., 2008. The effects of a primary-treated municipal effluent on the immune system of rainbow trout (Oncorhynchus mykiss): exposure duration and contribution of suspended particles. *Comparative Biochemical and Physiology C: Toxicology and Pharmacology* **148**, 258–264.

Heuvel, J.V., 2018. Nuclear receptors & endocrine/metabolic disruption. In Biosciences, I., (Ed.). nrresource.org, pp 1–16.

Hine, P.M., 1992. The granulocytes of fish. *Fish & Shellfish Immunology* **2**, 79–98.

Hirono, I., Nam, B.-H., Kurobe, T., Aoki, T., 2000. Molecular cloning, characterization, and expression of TNF cDNA and gene from Japanese flounder Paralychthys olivaceus. *The Journal of Immunology* **165**, 4423–4427.

Hodgkinson, J.W., Grayfer, L., Belosevic, M., 2015. Biology of bony fish macrophages. *Biology* **4**, 881–906.

Hoeger, B., van den Heuvel, M.R., Hitzfeld, B.C., Dietrich, D.R., 2004. Effects of treated sewage effluent on immune function in rainbow trout (Oncorhynchus mykiss). *Aquatic Toxicology* **70**, 345–355.

Hoellinger, H., Lecorsier, A., Sonnier, M., Leger, C., Do-Cao-Thang, Nguyen-Hoang-Nam, 1987. Cytotoxicity, cytogenotoxicity and allergenicity tests on certain pyrethroids. *Drug and Chemical Toxicology* **10**, 291–310.

Hogan, R.J., Stuge, T.B., Clem, L.W., Miller, N.W., Chinchar, V.G., 1996. Anti-viral cytotoxic cells in the channel catfish (Ictalurus punctatus). *Developmental & Comparative Immunology* **20**, 115–127.

Holen, E., Olsvik, P.A., 2014. Aryl hydrocarbon receptor protein and Cyp1A1 gene induction by LPS and phenanthrene in Atlantic cod (Gadus morhua) head kidney cells. *Fish Shellfish Immunol* **40**, 384–391.

Holland, M.C., Lambris, J.D., 2002. The complement system in teleosts. *Fish Shellfish Immunol* **12**, 399–420.

Hong, S., Li, R., Xu, Q., Secombes, C.J., Wang, T., 2013. Two types of TNF-α exist in teleost fish: phylogeny, expression, and bioactivity analysis of type-II TNF-α3 in rainbow trout Oncorhynchus mykiss. *Journal of Immunology* **191**, 5959–5972.

Hunt, L.R., Rice, C.D., 2008. Lymphoid tissue ontogeny in the mummichog, Fundulus heteroclitus. *Anatomical Record (Hoboken)* **291**, 1236–1245.

Inagawa, H., Kuroda, A., Nishizawa, T., Honda, T., Ototake, M., Yokomizo, Y., Nakanishi, T., Soma, G.-I., 2001. Cloning and characterisation of tandem-repeat type galectin in rainbow trout (Oncorhynchus mykiss). *Fish & Shellfish Immunology* **11**, 217–231.

Iwanowicz, L.R., Blazer, V.S., Hitt, N.P., McCormick, S.D., DeVault, D.S., Ottinger, C.A., 2012. Histologic, immunologic and endocrine biomarkers indicate contaminant effects in fishes of the Ashtabula River. *Ecotoxicology* **21**, 165–182.

Iwanowicz, L.R., Stafford, J.L., Patiño, R., Bengten, E., Miller, N.W., Blazer, V.S., 2014. Channel catfish (Ictalurus punctatus) leukocytes express estrogen receptor isoforms ERα and ERβ2 and are functionally modulated by estrogens. *Fish & Shellfish Immunology* **40**, 109–119.

Jackson, J.A., 2015. Immunology in wild nonmodel rodents: an ecological context for studies of health and disease. *Parasite Immunology* **37**, 220–232.

Jacobson, K.C., Arkoosh, M.R., Kagley, A.N., Clemons, E.R., Collier, T.K., Casillas, E., 2003. Cumulative effects of natural and anthropogenic stress on immune function and disease resistance in Juvenile Chinook Salmon. *Journal of Aquatic Animal Health* **15**, 1–12.

Jang, S., Hardie, L., Secombes, C., 1994. Effects of transforming growth factor β1 on rainbow trout Oncorhynchus mykiss macrophage respiratory burst activity. *Developmental & Comparative Immunology* **18**, 315–323.

Jaso-Friedmann, L., Leary III, J.H., Evans, D.L., 1997. NCCRP-1: a novel receptor protein sequenced from teleost nonspecific cytotoxic cells. *Molecular Immunology* **34**, 955–965.

Jaso-Friedmann, L., Leary III, J.H., Evans, D.L., 2001. The non-specific cytotoxic cell receptor (NCCRP-1): molecular organization and signaling properties. *Developmental & Comparative Immunology* **25**, 701–711.

Jensen, L.E., Hiney, M.P., Shields, D.C., Uhlar, C.M., Lindsay, A.J., Whitehead, A.S., 1997. Acute phase proteins in salmonids: evolutionary analyses and acute phase response. *The Journal of Immunology* **158**, 384–392.

Ji, J.-f., Hu, C.-b., Zhang, N., Huang, X., Shao, T., Fan, D.-d., Lin, A.-f., Xiang, L.-x., Shao, J.-z., 2021. New insights into IgZ as a maternal transfer Ig contributing to the early defense of fish against pathogen infection. *The Journal of Immunology* **206**, 2001.

Jin, Y., Liu, Z., Liu, F., Ye, Y., Peng, T., Fu, Z., 2015. Embryonic exposure to cadmium (II) and chromium (VI) induce behavioral alterations, oxidative stress and immunotoxicity in zebrafish (Danio rerio). *Neurotoxicology and Teratology* **48**, 9–17.

Jin, Y., Pan, X., Cao, L., Ma, B., Fu, Z., 2013. Embryonic exposure to cis-bifenthrin enantioselectively induces the transcription of genes related to oxidative stress, apoptosis and immunotoxicity in zebrafish (Danio rerio). *Fish & Shellfish Immunology* **34**, 717–723.

Johnson, A.C., Jin, X., Nakada, N., Sumpter, J.P., 2020. Learning from the past and considering the future of chemicals in the environment. *Science* **367**, 384–387.

Jovanović, B., Palić, D., 2012. Immunotoxicology of non-functionalized engineered nanoparticles in aquatic organisms with special emphasis on fish--review of current knowledge, gap identification, and call for further research. *Aquatic Toxicology* **118–119**, 141–151.

Kaattari, S.L., Irwin, M.J., 1985. Salmonid spleen and anterior kidney harbor populations of lymphocytes with different B cell repertoires. *Developmental & Comparative Immunology* **9**, 433–444.

Kaattari, S.L., Piganelli, J., 1996. The specific immune system: humoral defense. The fish immune system: organism, pathogen, and environment Academic Press, London, 207–254.

Karpanen, T., Mäkinen, T., 2006. Regulation of lymphangiogenesis-from cell fate determination to vessel remodeling. *Experimental Cell Research* **312**, 575–583.

Kato, G., Miyazawa, H., Nakayama, Y., Ikari, Y., Kondo, H., Yamaguchi, T., Sano, M., Fischer, U., 2018. A novel antigen-sampling cell in the teleost gill epithelium with the potential for direct antigen presentation in mucosal tissue. *Frontiers in Immunology* **9**, 2116. doi: 10.3389/fimmu.2018.02116.

Katzenback, B.A., 2015. Antimicrobial peptides as mediators of innate immunity in teleosts. *Biology (Basel)* **4**, 607–639.

Kerkvliet, N., 2002. T lymphocytes are direct, aryl hydrocarbon receptor (AhR)-dependent targets of 2,3,7,8-tetrachlorodibenzo-p-dioxin (TCDD): AhR expression in both CD4+ and CD8+ T cells is necessary for full suppression of a cytotoxic T lymphocyte response by TCDD. *Toxicology and Applied Pharmacology* **185**, 146–152.

Kerkvliet, N., Burleson, G., 1994. Immunotoxicity of TCDD and related halogenated aromatic hydrocarbons. *Immunotoxicology and Immunopharmacology* **2**, 97–121.

Kerkvliet, N.I., 2012. TCDD: an environmental immunotoxicant reveals a novel pathway of immunoregulation--a 30-year odyssey. *Toxicology Pathology* **40**, 138–142.

Kernen, L., Phan, A., Bo, J., Herzog, E.L., Huynh, J., Segner, H., Baumann, L., 2022. Estrogens as immunotoxicants: 17α-ethinylestradiol exposure retards thymus development in zebrafish (Danio rerio). *Aquatic Toxicology* **242**, 106025.

Kernen, L., Rieder, J., Duus, A., Holbech, H., Segner, H., Bailey, C., 2020. Thymus development in the zebrafish (Danio rerio) from an ecoimmunology perspective. *Journal of Experimental Zoology Part A: Ecological and Integrative Physiology* **333**, 805–819.

Köllner, B., Wasserrab, B., Kotterba, G., Fischer, U., 2002. Evaluation of immune functions of rainbow trout (Oncorhynchus mykiss)--how can environmental influences be detected? *Toxicology Letters* **131**, 83–95.

Koppang, E.O., Fischer, U., Satoh, M., Jirillo, E., 2007. Inflammation in fish as seen from a morphological point of view with special reference to the vascular compartment. *Current Pharmaceutical Design* **13**, 3649–3655.

Kordon, A., Karsi, A., Pinchuk, L., 2019. Innate immune responses in fish: antigen presenting cells and professional phagocytes. *Turkish Journal of Fisheries and Aquatic Sciences* **18**, 1123–1139.

Koumans-van Diepen, J.E., van de Lisdonk, M.H., Taverne-Thiele, A.J., Verburg-van Kemenade, B.L., Rombout, J.H., 1994. Characterisation of immunoglobulin-binding leucocytes in carp (Cyprinus carpio L.). *Developmental & Comparative Immunology* **18**, 45–56.

Laing, K.J., Hansen, J.D., 2011. Fish T cells: recent advances through genomics. *Developmental & Comparative Immunology* **35**, 1282–1295.

Lam, S.H., Chua, H.L., Gong, Z., Lam, T.J., Sin, Y.M., 2004. Development and maturation of the immune system in zebrafish, Danio rerio: a gene expression profiling, in situ hybridization and immunological study. *Developmental & Comparative Immunology* **28**, 9–28.

Landrigan, P.J., Stegeman, J.J., Fleming, L.E., Allemand, D., Anderson, D.M., Backer, L.C., Brucker-Davis, F., Chevalier, N., Corra, L., Czerucka, D., Bottein, M.D., Demeneix, B., Depledge, M., Deheyn, D.D., Dorman, C.J., Fénichel, P., Fisher, S., Gaill, F., Galgani, F., Gaze, W.H., Giuliano, L., Grandjean, P., Hahn, M.E., Hamdoun, A., Hess, P., Judson, B., Laborde, A., McGlade, J., Mu, J., Mustapha, A., Neira, M., Noble, R.T., Pedrotti, M.L., Reddy, C., Rocklöv, J., Scharler, U.M., Shanmugam, H., Taghian, G., van de Water, J., Vezzulli, L., Weihe, P., Zeka, A., Raps, H., Rampal, P., 2020. Human health and ocean pollution. *Annals of Global Health* **86**, 151.

Lavelle, E.C., Murphy, C., O'Neill, L., Creagh, E.M., 2010. The role of TLRs, NLRs, and RLRs in mucosal innate immunity and homeostasis. *Mucosal Immunology* **3**, 17–28.

Leaver, M.J., Diab, A., Boukouvala, E., Williams, T.D., Chipman, J.K., Moffat, C.F., Robinson, C.D., George, S.G., 2010. Hepatic gene expression in flounder chronically exposed to multiply polluted estuarine sediment: absence of classical exposure 'biomarker' signals and induction of inflammatory, innate immune and apoptotic pathways. *Aquatic Toxicology* **96**, 234–245.

Lee, A.J., Ashkar, A.A., 2018. The dual nature of Type I and Type II interferons. *Frontiers in Immunology* **9**, 2061. doi: 10.3389/fimmu.2018.02061.

Lee, J.-Y., Hirono, I., Aoki, T., 2000. Cloning and analysis of expression of Mx cDNA in Japanese flounder, Paralichthys olivaceus. *Developmental & Comparative Immunology* **24**, 407–415.

Levraud, J.P., Herbomel, P., 2019. Resident macrophage lookalikes of unexpected origin. *Developmental Cell* **49**, 501–502.

Lewis, K.L., Del Cid, N., Traver, D., 2014. Perspectives on antigen presenting cells in zebrafish. *Developmental and Comparative Immunology* **46**, 63–73.

Li, X., Liu, L., Zhang, Y., Fang, Q., Li, Y., Li, Y., 2013. Toxic effects of chlorpyrifos on lysozyme activities, the contents of complement C3 and IgM, and IgM and complement C3 expressions in common carp (Cyprinus carpio L.). *Chemosphere* **93**, 428–433.

Lin, X., Zhou, Q., Lin, G., Zhao, C., Wen, Z., 2020. Endoderm-derived myeloid-like metaphocytes in zebrafish gill mediate soluble antigen-induced immunity. *Cell Reports* **33**, 108227.

Liney, K.E., Hagger, J.A., Tyler, C.R., Depledge, M.H., Galloway, T.S., Jobling, S., 2006. Health effects in fish of long-term exposure to effluents from wastewater treatment works. *Environmental Health Perspectives* **114**, 81–89.

Liu, D., Mai, K., Ai, Q., 2015. Tumor necrosis factor alpha is a potent regulator in fish adipose tissue. *Aquaculture* **436**, 65–71.

Lobb, C.J., Clem, L.W., 1982. Fish lymphocytes differ in the expression of surface immunoglobulin. *Developmental & Comparative Immunology* **6**, 473–479.

Lochmiller, R.L., 1996. Immunocompetence and animal population regulation. *Oikos* **76**, 594–602.

Long, S., Wilson, M., Bengten, E., Bryan, L., Clem, L., Miller, N., Chinchar, V., 2004. Identification of a cDNA encoding channel catfish interferon. *Developmental & Comparative Immunology* **28**, 97–111.

Løvmo, S.D., Speth, M.T., Repnik, U., Koppang, E.O., Griffiths, G.W., Hildahl, J.P., 2017. Translocation of nanoparticles and Mycobacterium marinum across the intestinal epithelium in zebrafish and the role of the mucosal immune system. *Developmental & Comparative Immunology* **67**, 508–518.

Luebke, R.W., Hodson, P.V., Faisal, M., Ross, P.S., Grasman, K.A., Zelikoff, J., 1997. Aquatic pollution-induced immunotoxicity in wildlife species. *Fundamental and Applied Toxicology* **37**, 1–15.

Lund, V., Olafsen, J.A., 1999. Changes in serum concentration of a serum amyloid P-like pentraxin in Atlantic salmon, Salmo salar L., during infection and inflammation. *Developmental & Comparative Immunology* **23**, 61–70.

Maciuszek, M., Pijanowski, L., Pekala-Safinska, A., Kemenade, B., Chadzinska, M., 2020. 17β-Estradiol affects the innate immune response in common carp. *Fish Physiology and Biochemistry* **46**, 1775–1794.

MacKay, H., Abizaid, A., 2018. A plurality of molecular targets: the receptor ecosystem for bisphenol-A (BPA). *Hormones and Behavior* **101**, 59–67.

Magadan, S., Sunyer, O.J., Boudinot, P., 2015. Unique features of fish immune repertoires: particularities of adaptive immunity within the largest group of vertebrates. In Hsu, E., Du Pasquier, L., (Eds.), *Pathogen-Host Interactions: Antigenic Variation vs. Somatic adaptations. Results and Problems in Cell Differentiaton.* Springer, Cham, pp. 235–264.

Magor, B.G., Magor, K.E., 2001. Evolution of effectors and receptors of innate immunity. *Developmental & Comparative Immunology* **25**, 651–682.

Manibusan, M.K., Touart, L.W., 2017. A comprehensive review of regulatory test methods for endocrine adverse health effects. *Critical Reviews in Toxicology* **47**, 433–481.

Manning, M., Nakanishi, T., 1996. *The Fish Immune System.* by G. Iwama and T. Nakanishi, Academic Press, New York, p. 159.

Maqbool, F., Mostafalou, S., Bahadar, H., Abdollahi, M., 2016. Review of endocrine disorders associated with environmental toxicants and possible involved mechanisms. *Life Science* **145**, 265–273.

Marsh, M.B., Rice, C.D., 2010. Development, characterization, and technical applications of a fish lysozyme-specific monoclonal antibody (mAb M24-2). *Comparative Immunology, Microbiology and Infectious Diseases* **33**, e15–e23.

Martin, L.B., Hopkins, W.A., Mydlarz, L.D., Rohr, J.R., 2010. The effects of anthropogenic global changes on immune functions and disease resistance. *Annals of New York Academy of Sciences* **1195**, 129–148.

Martínez-Gómez, C., Vethaak, A.D., 2019a. Understanding the impact of chemicals on marine fish populations: the need for an integrative approach involving population and disease ecology. *Current Opinion in Environmental Science & Health* **11**, 71–77.

Martínez-Gómez, C., Vethaak, A.D., 2019b. Understanding the impact of chemicals on marine fish populations: the need for an integrative approach involving population and disease ecology. *Current Opinion in Environmental Science & Health* **11**, 1–7.

Marty, M.S., Carney, E.W., Rowlands, J.C., 2011. Endocrine disruption: historical perspectives and its impact on the future of toxicology testing. *Toxicological Sciences* **120**, S93–S108.

Mashoof, S., Criscitiello, M.F., 2016. Fish immunoglobulins. *Biology* **5**, 45.

Mason, R., Tennekes, H., Sánchez-Bayo, F., Jepsen, P., 2013. Immune suppression by neonicotinoid insecticides at the root of global wildlife declines. *Journal of Environmental Immunology and Toxicology* **1**, 3–12.

Massart, S., Milla, S., Kestemont, P., 2014. Expression of gene, protein and immunohistochemical localization of the estrogen receptor isoform ERα1 in male rainbow trout lymphoid organs; indication of the role of estrogens in the regulation of immune mechanisms. *Comparative Biochemistry and Physiology. Part B, Biochemistry & Molecular Biology* **174**, 53–61.

Matsuyama, T., Iida, T., 1999. Degranulation of eosinophilic granular cells with possible involvement in neutrophil migration to site of inflammation in tilapia. *Developmental & Comparative Immunology* **23**, 451–457.

Maule, A.G., Jorgensen, E.H., Vijayan, M.M., Killie, J.E.A., 2005. Aroclor 1254 exposure reduces disease resistance and innate immune responses in fasted arctic charr. *Environmental Toxicology and Chemistry* **24**, 117–124.

Maxwell, G., MacKay, C., Cubberley, R., Davies, M., Gellatly, N., Glavin, S., Gouin, T., Jacquoilleot, S., Moore, C., Pendlington, R., 2014. Applying the skin sensitisation adverse outcome pathway (AOP) to quantitative risk assessment. *Toxicology In Vitro* **28**, 8–12.

McKee, A.S., Fontenot, A.P., 2016. Interplay of innate and adaptive immunity in metal-induced hypersensitivity. *Current Opinion in Immunology* **42**, 25–30.

McMahon, T.A., Halstead, N.T., Johnson, S., Raffel, T.R., Romansic, J.M., Crumrine, P.W., Boughton, R.K., Martin, L.B., Rohr, J.R., 2011. The fungicide chlorothalonil is nonlinearly associated with corticosterone levels, immunity, and mortality in amphibians. *Environmental Health Perspectives* **119**, 1098–1103.

Medzhitov, R., 2021. The spectrum of inflammatory responses. *Science* **374**, 1070–1075.

Metidji, A., Omenetti, S., Crotta, S., Li, Y., Nye, E., Ross, E., Li, V., Maradana, M.R., Schiering, C., Stockinger, B., 2018. The environmental sensor AHR protects from inflammatory damage by maintaining intestinal stem cell homeostasis and barrier integrity. *Immunity* **49**, 353-362.e355.

Milla, S., Depiereux, S., Kestemont, P., 2011. The effects of estrogenic and androgenic endocrine disruptors on the immune system of fish: a review. *Ecotoxicology* **20**, 305–319.

Miller, N., Deuter, A., Clem, L., 1986. Phylogeny of lymphocyte heterogeneity: the cellular requirements for the mixed leucocyte reaction with channel catfish. *Immunology* **59**, 123.

Mishra, J., Srivastava, A.K., 1983. Malathion induced hematological and biochemical changes in the Indian catfish Heteropneustes fossilis. *Environmental Research* **30**, 393–398.

Mogensen, T.H., 2009. Pathogen recognition and inflammatory signaling in innate immune defenses. *Clinical Microbiology Reviews* **22**, 240–273.

Möller, A.-M., Korytář, T., Köllner, B., Schmidt-Posthaus, H., Segner, H., 2014. The teleostean liver as an immunological organ: Intrahepatic immune cells (IHICs) in healthy and benzo[a]pyrene challenged rainbow trout (Oncorhynchus mykiss). *Developmental & Comparative Immunology* **46**, 518–529.

Monteith, A.J., Skaar, E.P., 2021. The impact of metal availability on immune function during infection. *Trends in Endocrinology & Metabolism* **32**, 916–928.

Moreira, C., Paiola, M., Duflot, A., Varó, I., Sitjà-Bobadilla, A., Knigge, T., Pinto, P., Monsinjon, T., 2021. The influence of 17β-oestradiol on lymphopoiesis and immune system ontogenesis in juvenile sea bass, Dicentrarchus labrax. *Developmental & Comparative Immunology* **118**, 104011.

Mutsuro, J., Nakao, M., Fujiki, K., Yano, T., 2000. Multiple forms of α 2-macroglobulin from a bony fish, the common carp (Cyprinus carpio): striking sequence diversity in functional sites. *Immunogenetics* **51**, 847–855.

Nair, M., George, K.C., Sanil, N.K., K. S, S., 2003. Reticuloendothelial system of defense in Oreochromis mossambicus (Peters). *Indian Journal of Fisheries* **50**, 139–147.

Najafpour, B., Cardoso, J.C.R., Canário, A.V.M., Power, D.M., 2020. Specific evolution and gene family expansion of complement 3 and regulatory factor H in fish. *Frontiers in Immunology* **11**.

Nakamura, O., Watanabe, T., Kamiya, H., Muramoto, K., 2001. Galectin containing cells in the skin and mucosal tissues in Japanese conger eel, Conger myriaster: an immunohistochemical study. *Developmental & Comparative Immunology* **25**, 431–437.

Nakanishi, T., 1994. The graft-versus-host reaction in teleost. *Developmental & Comparative Immunology* **18**, 16–18.

Nakanishi, T., Shibasaki, Y., Matsuura, Y., 2015. T cells in fish. *Biology* **4**, 640–663.

Nakao, M., Mutsuro, J., Obo, R., Fujiki, K., Nonaka, M., Yano, T., 2000. Molecular cloning and protein analysis of divergent forms of the complement component C3 from a bony fish, the common carp (Cyprinus carpio): presence of variants lacking the catalytic histidine. *European Journal of Immunology* **30**, 858–866.

Nakao, M., Kajiya, T., Sato, Y., Somamoto, T., Kato-Unoki, Y., Matsushita, M., Nakata, M., Fujita, T., Yano, T., 2006. Lectin pathway of bony fish complement: identification of two homologs of the mannose-binding lectin associated with MASP2 in the common carp (Cyprinus carpio). *Journal of Immunology* **177**, 5471–5479.

Nakayama, A., Riesen, I., Kollner, B., Eppler, E., Segner, H., 2008. Surface marker-defined head kidney granulocytes and B lymphocytes of rainbow trout express benzo[a]pyrene-inducible cytochrome P4501A protein. *Toxicological Sciences* **103**, 86–96.

Namit, A., Dowell, W., Matiasek, S., Webster, J., Stachura, D.L., 2022. Pentachlorophenol has significant adverse effects on hematopoietic and immune system development in zebrafish (Danio rerio). *PLoS One* **17**, e0265618.

Nayak, A., Das, B., Kohli, M., Mukherjee, S., 2004. The immunosuppressive effect of α-permethrin on Indian major carp, rohu (Labeo rohita Ham.). *Fish & Shellfish Immunology* **16**, 41–50.

Neely, H.R., Flajnik, M.F., 2016. Emergence and evolution of secondary lymphoid organs. *Annual Review of Cell and Developmental Biology* **32**, 693–711.

Neumann, N.F., Barreda, D.R., Belosevic, M., 2000. Generation and functional analysis of distinct macrophage sub-populations from goldfish (Carassius auratus L.) kidney leukocyte cultures. *Fish & Shellfish Immunology* **10**, 1–20.

Neumann, N.F., Stafford, J.L., Barreda, D., Ainsworth, A.J., Belosevic, M., 2001. Antimicrobial mechanisms of fish phagocytes and their role in host defense. *Developmental & Comparative Immunology* **25**, 807–825.

Noga, E., Fan, Z., Silphaduang, U., 2001. Histone-like proteins from fish are lethal to the parasitic dinoflagellate Amyloodinium ocellatum. *Parasitology* **123**, 57–65.

Nohara, K., Izumi, H., Tamura, S.-i., Nagata, R., Tohyama, C., 2002. Effect of low-dose 2, 3, 7, 8-tetrachlorodibenzo-p-dioxin (TCDD) on influenza A virus-induced mortality in mice. *Toxicology* **170**, 131–138.

North, C.M., Kim, B.-S., Snyder, N., Crawford, R.B., Holsapple, M.P., Kaminski, N.E., 2008. TCDD-mediated suppression of the in vitro anti-sheep erythrocyte IgM antibody forming cell response is reversed by interferon-gamma. *Toxicological Sciences* **107**, 85–92.

Nygaard, R., Husgard, S., Sommer, A.-I., Leong, J.-A.C., Robertsen, B., 2000. Induction of Mx protein by interferon and double-stranded RNA in salmonid cells. *Fish & Shellfish Immunology* **10**, 435–450.

Okamura, B., Feist, S.W., 2011. Emerging diseases in freshwater systems. *Freshwater Biology* **56**, 627–637.

Ottaviani, E., Franchini, A., Franceschi, C., 1995. Evidence for the presence of immunoreactive POMC-derived peptides and cytokines in the thymus of the goldfish (Carassius c. auratus). *The Histochemical Journal* **27**, 597–601.

Ottaviani, E., Franchini, A., Franceschi, C., 1997. Evolution of neuroendocrine thymus: studies on POMC-derived peptides, cytokines and apoptosis in lower and higher vertebrates. *Journal of Neuroimmunology* **72**, 67–74.

P. Regala, C.D.R.T.E.S.R., 2001. The effects of tributyltin (TBT) and 3,3',4,4',5-pentachlorobiphenyl (PCB-126) mixtures on antibody responses and phagocyte oxidative burst activity in channel catfish, ictalurus punctatus. *Archives of Environmental Contamination and Toxicology* **40**, 386–391.

Paiola, M., Knigge, T., Picchietti, S., Duflot, A., Guerra, L., Pinto, P.I.S., Scapigliati, G., Monsinjon, T., 2017. Oestrogen receptor distribution related to functional thymus anatomy of the European sea bass, Dicentrarchus labrax. *Developmental & Comparative Immunology* **77**, 106–120.

Palmer, C.V., 2018. Immunity and the coral crisis. *Communications Biology* **1**, 91.

Palti, Y., 2011. Toll-like receptors in bony fish: from genomics to function. *Developmental & Comparative Immunology* **35**, 1263–1272.

Pang, S., Gao, Y., Li, A., Yao, X., Qu, G., Hu, L., Liang, Y., Song, M., Jiang, G., 2020. Tetrabromobisphenol A perturbs erythropoiesis and impairs blood circulation in zebrafish embryos. *Environmental Science & Technology* **54**, 12998–13007.

Park, S., Lee, J.-Y., Park, H., Song, G., Lim, W., 2020. Bifenthrin induces developmental immunotoxicity and vascular malformation during zebrafish embryogenesis. *Comparative Biochemistry and Physiology Part C: Toxicology & Pharmacology* **228**, 108671.

Partula, S., 1999. Surface markers of fish T-cells. *Fish & Shellfish Immunology* **9**, 241–257.

Pathiratne, A., George, S.G., 1998. Toxicity of malathion to Nile tilapia, Oreochromis niloticus and modulation by other environmental contaminants. *Aquatic Toxicology* **43**, 261–271.

Perdiguero, P., Morel, E., Tafalla, C., 2021. Diversity of rainbow trout blood B cells revealed by single cell RNA sequencing. *Biology* **10**, 511.

Petitjean, Q., Jean, S., Côte, J., Larcher, T., Angelier, F., Ribout, C., Perrault, A., Laffaille, P., Jacquin, L., 2020. Direct and

indirect effects of multiple environmental stressors on fish health in human-altered rivers. *Science Total Environmental* **742**, 140657.
Petrie-Hanson, L., Ainsworth, A.J., 2001. Ontogeny of channel catfish lymphoid organs. *Veterinary Immunology and Immunopathology* **81**, 113–127.
Phalen, L.J., Köllner, B., Hogan, N.S., van den Heuvel, M.R., 2017. Transcriptional response in rainbow trout (Oncorhynchus mykiss) B cells and thrombocytes following in vivo exposure to benzo[a]pyrene. *Environonmental Toxicology Pharmacol* **53**, 212–218.
Phelps, D.W., Fletcher, A.A., Rodriguez-Nunez, I., Balik-Meisner, M.R., Tokarz, D.A., Reif, D.M., Germolec, D.R., Yoder, J.A., 2020. In vivo assessment of respiratory burst inhibition by xenobiotic exposure using larval zebrafish. *Journal of Immunotoxicology* **17**, 94–104.
Phillips, N.C., 1989. Kupffer cells and liver metastasis. Optimization and limitation of activation of tumoricidal activity. *Cancer Metastasis Reviews* **8**, 231–252.
Pilstrom, L., Bengtén, E., 1996. Immunoglobulin in fish-genes, expression and structure. *Fish & Shellfish Immunology* **6**, 243–262.
Plumb, J.A., Areechon, N., 1990. Effect of malathion on humoral immune response of channel catfish. *Developmental & Comparative Immunology* **14**, 355–358.
Pockley, A.G., 2003. Heat shock proteins as regulators of the immune response. *Lancet* **362**, 469–476.
Porter, W.P., Hinsdill, R., Fairbrother, A., Olson, L.J., Jaeger, J., Yuill, T., Bisgaard, S., Hunter, W.G., Nolan, K., 1984. Toxicant-disease-environment interactions associated with suppression of immune system, growth, and reproduction. *Science* **224**, 1014–1017.
Powell, M.D., Briand, H.A., Wright, G.M., Burka, J.F., 1993. Rainbow trout (Oncorhynchus mykiss Walbaum) intestinal eosinophilic granule cell (EGC) response to Aeromonas salmonicida and Vibrio anguillarum extracellular products. *Fish & Shellfish Immunology* **3**, 279–289.
Powell, M.D., Wright, G.M., Burka, J.F., 1991. Degranulation of eosinophilic granule cells induced by capsaicin and substance P in the intestine of the rainbow trout (Oncorhynchus mykiss Walbaum). *Cell and Tissue Research* **266**, 469–474.
Press, C.M., Dannevig, B., Landsverk, T., 1994. Immune and enzyme histochemical phenotypes of lymphoid and nonlymphoid cells within the spleen and head kidney of Atlantic salmon (Salmo salar L.). *Fish & Shellfish Immunology* **4**, 79–93.
Press, C.M., Evensen, Ø., 1999. The morphology of the immune system in teleost fishes. *Fish & Shellfish Immunology* **9**, 309–318.
Qi, P., Xie, C., Guo, B., Wu, C., 2016. Dissecting the role of transforming growth factor-β1 in topmouth culter immunobiological activity: a fundamental functional analysis. *Scientific Reports* **6**, 27179.
Qiu, W., Yang, M., Liu, J., Xu, H., Luo, S., Wong, M., Zheng, C., 2019. Bisphenol S-induced chronic inflammatory stress in liver via peroxisome proliferator-activated receptor γ using fish in vivo and in vitro models. *Environmental Pollution* **246**, 963–971.
Quesada, J., Viilena, M., Agulleiro, B., 1990. Structure of the spleen of the sea bass (Dicentrarchus labrax): a light and electron microscopic study. *Journal of Morphology* **206**, 273–281.
Quesada-García, A., Valdehita, A., Kropf, C., Casanova-Nakayama, A., Segner, H., Navas, J.M., 2014. Thyroid signaling in immune organs and cells of the teleost fish rainbow trout (Oncorhynchus mykiss). *Fish Shellfish Immunol* **38**, 166–174.
Quiniou, S.M.A., Sahoo, M., Edholm, E.-S., Bengten, E., Wilson, M., 2011. Channel catfish CD8α and CD8β co-receptors: Characterization, expression and polymorphism. *Fish & Shellfish Immunology* **30**, 894–901.
Raibeemol, K., Chitra, K., 2020. Induction of immunological, hormonal and histological alterations after sublethal exposure of chlorpyrifos in the freshwater fish, Pseudetroplus maculatus (Bloch, 1795). *Fish & Shellfish Immunology* **102**, 1–12.
Rajendran, K.V., Zhang, J., Liu, S., Kucuktas, H., Wang, X., Liu, H., Sha, Z., Terhune, J., Peatman, E., Liu, Z., 2012. Pathogen recognition receptors in channel catfish: I. Identification, phylogeny and expression of NOD-like receptors. *Developmental & Comparitive Immunol* **37**, 77–86.
Rand-Weaver, M., Margiotta-Casaluci, L., Patel, A., Panter, G.H., Owen, S.F., Sumpter, J.P., 2013. The read-across hypothesis and environmental risk assessment of pharmaceuticals. *Environmental Science & Technology* **47**, 11384–11395.
Rastgar, S., Alijani Ardeshir, R., Segner, H., Tyler, C.R., J.G.M. Peijnenburg, W., Wang, Y., Salati, A.P., Movahedinia, A., 2022. Immunotoxic effects of metal-based nanoparticles in fish and bivalves. *Nanotoxicology*, **16**, 88–113.
Rebach, S., 1999. Acute toxicity of permethrin/piperonyl butoxide on hybrid striped bass. *Bulletin of Environmental Contamination and Toxicology* **62**, 448–454.
Rebello, S.C., Rathore, G., Punia, P., Sood, N., 2016. Development and characterization of a monoclonal antibody against the putative T cells of Labeo rohita. *Cytotechnology* **68**, 469–480.
Redfern, L.K., Jayasundara, N., Singleton, D.R., Di Giulio, R.T., Carlson, J., Sumner, S.J., Gunsch, C.K., 2021. The role of gut microbial community and metabolomic shifts in adaptive resistance of Atlantic killifish (Fundulus heteroclitus) to polycyclic aromatic hydrocarbons. *Science of Total Environment* **776**, 145955.
Rego, K., Bengtén, E., Wilson, M., Hansen, J.D., Bromage, E.S., 2020. Characterization of immunoglobulin light chain utilization and variable family diversity in rainbow trout. *Developmental & Comparative Immunology* **104**, 103566.
Rehberger, K., Escher, B.I., Scheidegger, A., Werner, I., Segner, H., 2021. Evaluation of an in vitro assay to screen for the immunotoxic potential of chemicals to fish. *Scientific Reports* **11**, 3167.
Rehberger, K., Werner, I., Hitzfeld, B., Segner, H., Baumann, L., 2017a. 20 Years of fish immunotoxicology - what we know and where we are. *Critical Reviews in Toxicology* **47**, 509–535.
Rehberger, K., Werner, I., Hitzfeld, B., Segner, H., Baumann, L., 2017b. 20 Years of fish immunotoxicology - what we know and where we are. *Critical Reviews in Toxicology* **47**, 516–542.
Rehberger, K., Wernicke von Siebenthal, E., Bailey, C., Bregy, P., Fasel, M., Herzog, E.L., Neumann, S., Schmidt-Posthaus, H., Segner, H., 2020. Long-term exposure to low 17α-ethinylestradiol (EE2) concentrations disrupts both the reproductive and the immune system of juvenile rainbow trout, Oncorhynchus mykiss. *Environment International* **142**, 105836.
Reite, O.B., 1997. Mast cells/eosinophilic granule cells of salmonids: staining properties and responses to noxious agents. *Fish & Shellfish Immunology* **7**, 567–584.
Reite, O., Evensen, Ø., 1994. Mast cells in the swimbladder of Atlantic salmon Salmo salar: histochemistry and responses to compound 48/80 and formalin-inactivated Aeromonas salmonicida. *Diseases of Aquatic Organisms* **20**, 95–100.

Reite, O.B., Evensen, O., 2006. Inflammatory cells of teleostean fish: a review focusing on mast cells/eosinophilic granule cells and rodlet cells. *Fish Shellfish Immunol* **20**, 192–208.

Sales, C., Silva, R., Amaral, M., Domingos, F., Ribeiro, R., Thomé, R., Santos, H., 2017. Comparative histology in the liver and spleen of three species of freshwater teleost. *Neotropical Ichthyology* **15**.

Salinas, I., 2015. The mucosal immune system of teleost fish. *Biology (Basel)* **4**, 525–539.

Scapigliati, G., 2013a. Functional aspects of fish lymphocytes. *Developmental & Comparative Immunology* **41**, 200–208.

Scapigliati, G., 2013b. Functional aspects of fish lymphocytes. *Developmental & Comparative Immunology* **41**, 200–208.

Scapigliati, G., Fausto, A.M., Picchietti, S., 2018a. Fish lymphocytes: an evolutionary equivalent of mammalian innate-like lymphocytes? *Frontiers in Immunology* **9**, 971–971.

Scapigliati, G., Fausto, A.M., Picchietti, S., 2018b. Fish lymphocytes: an evolutionary equivalent of mammalian innate-like lymphocytes? *Frontiers in Immunology* **9**.

Scapigliati, G., Romano, N., Abelli, L., 1999. Monoclonal antibodies in fish immunology: identification, ontogeny and activity of T-and B-lymphocytes. *Aquaculture* **172**, 3–28.

Scown, T.M., van Aerle, R., Tyler, C.R., 2010. Review: do engineered nanoparticles pose a significant threat to the aquatic environment? *Critical Reviews in Toxicology* **40**, 653–670.

Secombes, C., 1996. The nonspecific immune system: cellular defenses. *The Fish Immune System: Organism, Pathogen and Environment* **15**, 63–103.

Secombes, C., Cunningham, C., 2004. Cytokines: an evolutionary perspective *Dev Comp Immunol* **3**, 373-374.

Secombes, C., Fletcher, T., 1992. The role of phagocytes in the protective mechanisms of fish. *Annual Review of Fish Diseases* **2**, 53–71.

Secombes, C., Hardie, L., Daniels, G., 1996. Cytokines in fish: an update. *Fish & Shellfish Immunology* **6**, 291–304.

Secombes, C., Wang, T., Hong, S., Peddie, S., Crampe, M., Laing, K., Cunningham, C., Zou, J., 2001. Cytokines and innate immunity of fish. *Developmental & Comparative Immunology* **25**, 713–723.

Secombes, C.J., 1994. Enhancement of fish phagocyte activity. *Fish & Shellfish Immunology* **4**, 421–436.

Secombes, C.J., Belmonte, R., 2016. Overview of the fish adaptive immune system. In Adams, A., (Ed.), *Fish Vaccines*. Springer Basel, Basel, pp. 35–52.

Secombes, C.J., Wang, T., Bird, S., 2011. The interleukins of fish. *Developmental & Comparative Immunology* **35**, 1336–1345.

Secombes, C.J., Wang, T., Bird, S., 2016. Chapter 5- vertebrate cytokines and their evolution. In Malagoli, D., (Ed.), *The Evolution of the Immune System*. Academic Press Cambridge, Massachussetts, pp. 87–150.

Secombes, C., Zou, J., Laing, K., Daniels, G., Cunningham, C., 1999. Cytokine genes in fish. *Aquaculture* (Amsterdam) **172**, 93–102.

Segner, H., Bailey, C., Tafalla, C., Bo, J., 2021. Immunotoxicity of xenobiotics in fish: a role for the Aryl Hydrocarbon Receptor (AhR)? *International Journal of Molecular Sciences* **22**, 9460.

Segner, H., Casanova-Nakayama, A., Kase, R., Tyler, C.R., 2013. Impact of environmental estrogens on Yfish considering the diversity of estrogen signaling. *General Comparative Endocrinology* **191**, 190–201.

Segner, H., Verburg-van Kemenade, B.M.L., Chadzinska, M., 2017. The immunomodulatory role of the hypothalamus-pituitary-gonad axis: proximate mechanism for reproduction-immune trade offs? *Developmental & Comparative Immunology* **66**, 43–60.

Segner, H., Wenger, M., Möller, A.M., Köllner, B., Casanova-Nakayama, A., 2011. Immunotoxic effects of environmental toxicants in fish - how to assess them? *Environmental Science Pollution Research International* **19**, 2465–2476.

Selgrade, M.K., 2007. Immunotoxicity-the risk is real. *Toxicological Sciences* **100**, 328–332.
Sepulcre, M.P., Alcaraz-Pérez, F., López-Muñoz, A., Roca, F.J., Meseguer, J., Cayuela, M.L., Mulero, V., 2009. Evolution of lipopolysaccharide (LPS) recognition and signaling: fish TLR4 does not recognize LPS and negatively regulates NF-κB activation. *The Journal of Immunology* **182**, 1836–1845.
Seternes, T., Bøgwald, J., Dalmo, R.A., 2021. Scavenger endothelial cells of fish, a review. *Journal of Fish Diseases* **44**, 1385–1397.
Seternes, T., Sørensen, K., Smedsrød, B., 2002. Scavenger endothelial cells of vertebrates: a nonperipheral leukocyte system for high-capacity elimination of waste macromolecules. *Proceedings of the National Academy of Sciences* **99**, 7594.
Shan, S., Liu, R., Feng, H., Meng, F., Aizaz, M., Yang, G., 2021. Identification and functional characterization of a fish-specific tlr19 in common carp (Cyprinus carpio L.) that recruits TRIF as an adaptor and induces ifn expression during the immune response. *Veterinary Research* **52**, 88.
Shao-Nan, L., De-Fang, F., 1996. Correlation between biochemical parameters and susceptibility of freshwater fish to malathion. *Journal of Toxicology and Environmental Health* **48**, 413–418.
Sharma, A., Rudra, D., 2018. Emerging functions of regulatory T cells in tissue homeostasis. *Frontiers in Immunology* **9**. https://doi.org/10.3389/fimmu.2018.00883
Sharp, G., Pike, A., Secombes, C., 1989. The immune response of wild rainbow trout, Salmo gairdneri Richardson, to naturally acquired plerocercoid infections of Diphyllobothrium dendriticum (Nitzsch, 1824) and D. ditremum (Creplin, 1825). *Journal of Fish Biology* **35**, 781–794.
Shelley, L.K., Osachoff, H.L., van Aggelen, G.C., Ross, P.S., Kennedy, C.J., 2013. Alteration of immune function endpoints and differential expression of estrogen receptor isoforms in leukocytes from 17β-estradiol exposed rainbow trout (Oncorhynchus mykiss). *General and Comparative Endocrinology* **180**, 24–32.
Shen, L., Stuge, T.B., Zhou, H., Khayat, M., Barker, K.S., Quiniou, S.M., Wilson, M., Bengtén, E., Chinchar, V.G., Clem, L.W., 2002. Channel catfish cytotoxic cells: a mini-review. *Developmental & Comparative Immunology* **26**, 141–149.
Shivam, S., El-Matbouli, M., Kumar, G., 2021. Kinetics of parasite-specific antibody and B-cell-associated gene expression in brown trout, salmo trutta during proliferative kidney disease. *Biology* **10**, 1244.
Silkworth, J.B., Antrim, L., Kaminsky, L.S., 1984. Correlations between polychlorinated biphenyl immunotoxicity, the aromatic hydrocarbon locus, and liver microsomal enzyme induction in C57BL6 and DBA2 mice. *Toxicology and Applied Pharmacology* **75**, 156–165.
Silva, R.A., Di Giulio, R.T., Rice, C.D., 2021. The in vitro proinflammatory properties of water accommodated sediment extracts from a creosote-contaminated US environmental protection agency superfund site. *Environmental Toxicology and Chemistry* **40**, 1576–1585.
Simmons, S.O., Fan, C.Y., Ramabhadran, R., 2009. Cellular stress response pathway system as a sentinel ensemble in toxicological screening. *Toxicology Science* **111**, 202–225.
Sindermann, C.J., 1979. Pollution-associated diseases and abnormalities of fish and shellfish: a review. *Fishery Bulletin* **76**, 717–749.
Sinderman, C.J., Esser, S.C., Gould, E., McCain, B.B., McHugh, J.L., Morgan, R.P., Murchelano, R.A., Sherwood, M.J., Sptizer, P.R., 1982. Effects of pollutants on fishes., Proceedings of the Symposium of Ecological Stress. Estuarine Res. Fed., Columbia, pp. 23–28.
Sitjà-Bobadilla, A., Estensoro, I., Pérez-Sánchez, J., 2016. Immunity to gastrointestinal microparasites of fish. *Developmental & Comparative Immunology* **64**, 187–201.
Siwicki, A.K., Cossarini-Dunier, M., Studnicka, M., Demael, A., 1990. In vivo effect of the organophosphorus insecticide trichlorphon on immune response of carp (Cyprinus carpio): II. Effect of high doses of trichlorphon on nonspecific immune response. *Ecotoxicology and Environmental Safety* **19**, 99–105.
Smith, N.C., Rise, M.L., Christian, S.L., 2019. A comparison of the innate and adaptive immune systems in cartilaginous fish, ray-finned fish, and lobe-finned fish. *Frontiers in Immunology* **10**, 2292.
Snieszko, S.F., 1974. The effects of environmental stress on outbreaks of infectious diseases of fishes. *Journal of Fish Biology* **6**, 197–208.
Song, J.-Y., Casanova-Nakayama, A., Möller, A.-M., Kitamura, S.-I., Nakayama, K., Segner, H., 2020. Aryl hydrocarbon receptor signaling is functional in immune cells of rainbow trout (Oncorhynchus mykiss). *International Journal of Molecular Sciences* **21**, 6323.
Song, Q., Xiao, Y., Xiao, Z., Liu, T., Li, J., Li, P., Han, F., 2021. Lysozymes in fish. *Journal of Agricultural and Food Chemistry* **69**:15039–15051.
Sopinska, A., Guz, L., 1998. Effects of synthetic pyrethroids on carp (Russian). (1998). *Medcyna Wet* **54**, 126–128.
Soulliere, C., Dixon, B., 2017. *Immune System Organs of Bony Fishes, Reference Module in Life Sciences.* Elsevier, Amsderdam. https://doi.org/10.3389/fimmu.2018.00883
Spitsbergen, J., Schat, K., Kleeman, J., Peterson, R., 1986. Interactions of 2, 3, 7, 8-tetrachlorodibenzo-p-dioxin (TCDD) with immune responses of rainbow trout. *Veterinary Immunology and Immunopathology* **12**, 263–280.
Spitsbergen, J., Schat, K., Kleeman, J., Peterson, R., 1988a. Effects of 2, 3, 7, 8-tetrachlorodibenzo-p-dioxin (TCDD) or Aroclor 1254 on the resistance of rainbow trout, Salmo gairdneri Richardson, to infectious haematopoietic necrosis virus. *Journal of Fish Diseases* **11**, 73–83.
Spitsbergen, J.M., Kleeman, J.M., Peterson, R.E., 1988b. 2,3,7,8-Tetrachlorodibenzo-p-dioxin toxicity in yellow perch (Perca flavescens). *Journal of Toxicology and Environmental Health* **23**, 359–383.
Steinel, N.C., Bolnick, D.I., 2017. Melanomacrophage centers as a histological indicator of immune function in fish and other poikilotherms. *Frontiers in Immunology* **8**. https://doi.org/10.3389/fimmu.2017.00827
Stet, R.J., Kruiswijk, C.R., Saeij, J.P., Wiegertjes, G.F., 1998. Major histocompatibility genes in cyprinid fishes: theory and practice. *Immunological Reviews* **166**, 301–316.
Stockinger, B., Di Meglio, P., Gialitakis, M., Duarte, J.H., 2014. The aryl hydrocarbon receptor: multitasking in the immune system. *Annual Review of Immunology* **32**, 403–432.
Stolte, E.H., Nabuurs, S.B., Bury, N.R., Sturm, A., Flik, G., Savelkoul, H.F., Verburg-van Kemenade, B.L., 2008. Stress and innate immunity in carp: corticosteroid receptors and proinflammatory cytokines. *Molecular Immunology* **46**, 70–79.
Stuge, T.B., Wilson, M.R., Zhou, H., Barker, K.S., Bengtén, E., Chinchar, G., Miller, N.W., Clem, L.W., 2000. Development and analysis of various clonal alloantigen-dependent cytotoxic cell lines from channel catfish. *The Journal of Immunology* **164**, 2971–2977.
Sunyer, J.O., Tort, L., Lambris, J.D., 1997. Structural C3 diversity in fish: characterization of five forms of C3 in the diploid fish Sparus aurata. *The Journal of Immunology* **158**, 2813–2821.
Sunyer, J.O., Zarkadis, I.K., Lambris, J.D., 1998. Complement diversity: a mechanism for generating immune diversity? *Immunology Today* **19**, 519–523.

Sunyer, J.O., Zarkadis, I.K., Sahu, A., Lambris, J.D., 1996. Multiple forms of complement C3 in trout that differ in binding to complement activators. *Proceedings of the National Academy of Sciences* **93**, 8546–8551.

Suzuki, T., Hidaka, T., Kumagai, Y., Yamamoto, M., 2020. Environmental pollutants and the immune response. *Nature Immunology* **21**, 1486–1495.

Suzuki, Y., Iida, T., 1992. Fish granulocytes in the process of inflammation. *Annual Review of Fish Diseases* **2**, 149–160.

Swain, W.R., 1991. Effects of organochlorine chemicals on the reproductive outcome of humans who consumed contaminated Great Lakes fish: an epidemiologic consideration. *Journal of Toxicology and Environmental Health, Part A Current Issues* **33**, 587–639.

Swedenborg, E., Rüegg, J., Mäkelä, S., Pongratz, I., 2009. Endocrine disruptive chemicals: mechanisms of action and involvement in metabolic disorders. *Journal of Molecular Endocrinol* **43**, 1–10.

Tafalla, C., Leal, E., Yamaguchi, T., Fischer, U., 2016. T cell immunity in the teleost digestive tract. *Developmental& Compartitve Immunolology* **64**, 167–177.

Tatner, M.F., 1985. The migration of labelled thymocytes to the peripheral lymphoid organs in the rainbow trout, Salmogairdneri richardson. *Developmental & Comparative Immunology* **9**, 85–91.

Taylor, E.B., Chinchar, V.G., Quiniou, S.M.A., Wilson, M., Bengtén, E., 2020. Cloning and characterization of antiviral cytotoxic T lymphocytes in channel catfish, Ictalurus punctatus. *Virology* **540**, 184–194.

Thornton Hampton, L.M., Martyniuk, C.J., Venables, B.J., Sellin Jeffries, M.K., 2020. Advancing the fathead minnow (Pimephales promelas) as a model for immunotoxicity testing: Characterization of the renal transcriptome following Yersinia ruckeri infection. *Fish Shellfish Immunol* **103**, 472–480.

Thrasher, J.D., Madison, R., Broughton, A., 1993. Immunologic abnormalities in humans exposed to chlorpyrifos: preliminary observations. *Archives of Environmental Health: An International Journal* **48**, 89–93.

Thuvander, A., Wiss, E., Norrgren, L., 1993. Sublethal exposure of rainbow trout (Oncorhynchus mykiss) to Clophen A50: effects on cellular immunity. *Fish & Shellfish Immunology* **3**, 107–117.

Tian, J., Feng, Y., Fu, H., Xie, H.Q., Jiang, J.X., Zhao, B., 2015. The aryl hydrocarbon receptor: a key bridging molecule of external and internal chemical signals. *Environ Sci Technol* **49**, 9518–9531.

Torrealba, D., More-Bayona, J.A., Wakaruk, J., Barreda, D.R., 2019. Innate immunity provides biomarkers of health for teleosts exposed to nanoparticles. *Frontiers in Immunology* **9**. https://doi.org/10.3389/fimmu.2018.03074

Trobridge, G., Chiou, P., Kim, C., Leong, J., 1997. Induction of the Mx protein of rainbow trout Oncorhynchus mykiss in vitro and in vivo with poly I: C dsRNA and infectious hematopoietic necrosis virus. *Diseases of Aquatic Organisms* **30**, 91–98.

Trobridge, G.D., Leong, J.-A.C., 1995. Characterization of a rainbow trout Mx gene. *Journal of Interferon & Cytokine Research* **15**, 691–702.

Valero, Y., Saraiva-Fraga, M., Costas, B., Guardiola, F.A., 2020. Antimicrobial peptides from fish: beyond the fight against pathogens. *Reviews in Aquaculture* **12**, 224–253.

Vallejo, A.N., Miller, N.W., Clem, L.W., 1991. Phylogeny of immune recognition: processing and presentation of structurally defined proteins in channel catfish immune responses. *Developmental immunology* **1**, 137–148.

Vallejo, A.N., Miller, N.W., Clem, L.W., 1992. Antigen processing and presentation in teleost immune responses. *Annual Review of Fish Diseases* **2**, 73–89.

Van Loon, J., Secombes, C., Egberts, E., Van Muiswinkel, W., 1982. Ontogeny of the immune system in fish-role of the thymus. *In Vivo Immunology*, 335–341.

Verburg-van Kemenade, B.M., Savelkoul, H.F., Chadzinska, M., 2009. Function of the opioid system during inflammation in carp. *Annals of the New York Academy of Sciences* **1163**, 528–532.

Verburg-van Kemenade, B.M.L., Cohen, N., Chadzinska, M., 2017. Neuroendocrine-immune interaction: Evolutionarily conserved mechanisms that maintain allostasis in an ever-changing environment. *Developmental & Comparative Immunology* **66**, 2–23.

Verlhac, V., Sage, M., Deschaux, P., 1990. Cytotoxicity of carp (Cyprinus carpio) leucocytes induced against TNP-modified autologous spleen cells and influence of acclimatization temperature. *Developmental & Comparative Immunology* **14**, 475–480.

Vethaak, A.D., ap Rheinallt, T., 1992. Fish disease as a monitor for marine pollution: the case of the North Sea. *Reviews in Fish Biology and Fisheries* **2**, 1–32.

Villani, A.C., Sarkizova, S., Hacohen, N., 2018. Systems immunology: learning the rules of the immune system. *Annual Review of Immunology* **36**, 813–842.

Villeneuve, D.L., Landesmann, B., Allavena, P., Ashley, N., Bal-Price, A., Corsini, E., Halappanavar, S., Hussell, T., Laskin, D., Lawrence, T., Nikolic-Paterson, D., Pallardy, M., Paini, A., Pieters, R., Roth, R., Tschudi-Monnet, F., 2018. Representing the process of inflammation as key events in adverse outcome pathways. *Toxicology Science* **163**, 346–352.

Wack, A., Terczyńska-Dyla, E., Hartmann, R., 2015. Guarding the frontiers: the biology of type III interferons. *Nature Immunology* **16**, 802–809.

Wang, T., Secombes, C.J., 2013. The cytokine networks of adaptive immunity in fish. *Fish Shellfish Immunology* **35**, 1703–1718.

Weeks, B.A., Anderson, D.P., DuFour, A.P., Ann, F., Goven, A.J., Lahvls, G.P., Gabriele, P., 1992. *Immunological Biomarkers to Assess Environmental Stress, Biomarkers*. Lewis Publishers, Baca Raton FL USA, pp. 211–234.

Wentzel, A.S., Janssen, J.J.E., de Boer, V.C.J., van Veen, W.G., Forlenza, M., Wiegertjes, G.F., 2020a. Fish macrophages show distinct metabolic signatures upon polarization. *Frontiers in Immunology* **11**. https://doi.org/10.3389/fimmu.2020.00152

Wentzel, A.S., Petit, J., van Veen, W.G., Fink, I.R., Scheer, M.H., Piazzon, M.C., Forlenza, M., Spaink, H.P., Wiegertjes, G.F., 2020b. Transcriptome sequencing supports a conservation of macrophage polarization in fish. *Scientific Reports* **10**, 13470.

White, K., Kawabata, T., Ladics, G., 1994. Mechanisms of polycyclic aromatic hydrocarbon immunotoxicity. *Immunotoxicology and Immunopharmacology* **2**, 123–149.

White, S.S., Birnbaum, L.S., 2009. An overview of the effects of dioxins and dioxin-like compounds on vertebrates, as documented in human and ecological epidemiology. *Journal of Environmental Science and Health, Part C*, **7**, 197–211.

Whitehead, A., 2013. Interactions between Oil-Spill pollutants and natural stressors can compound ecotoxicological effects. *Integrative and Comparative Biology* **53**, 635–647.

Willett, C.E., Cortes, A., Zuasti, A., Zapata, A.G., 1999a. Early hematopoiesis and developing lymphoid organs in the zebrafish. *Developmental Dynamics* **214**, 323–336.

Willett, C.E., Cortes, A., Zuasti, A., Zapata, A.G., 1999b. Early hematopoiesis and developing lymphoid organs in the zebrafish. *Developmental Dynamics: An Official Publication of the American Association of Anatomists* **214**, 323–336.
Willett, C.E., Zapata, A.G., Hopkins, N., Steiner, L.A., 1997. Expression of zebrafishrag genes during early development identifies the thymus. *Developmental Biology* **182**, 331–341.
Willett, K.L., Wassenberg, D., Lienesch, L., Reichert, W., Di Giulio, R.T., 2001. In vivo and in vitro inhibition of CYP1A-dependent activity in Fundulus heteroclitus by the polynuclear aromatic hydrocarbon fluoranthene. *Toxicology and Applied Pharmacology* **177**, 264–271.
Wilson, A.B., 2017. MHC and adaptive immunity in teleost fishes. *Immunogenetics* **69**, 521–528.
Wilson, M., Zhou, H., Bengten, E., Clem, L., Stuge, T., Warr, G., Miller, N.W., 1998. T-cell receptors in channel catfish: structure and expression of TCR α and β genes. *Molecular Immunology* **35**, 545–557.
Wolke, R., 1992. Piscine macrophage aggregates: a review. *Annual Review of Fish Diseases* **2**, 91–108.
Wu, L., Qin, Z., Liu, H., Lin, L., Ye, J., Li, J., 2020. Recent advances on phagocytic B cells in teleost fish. *Frontiers in Immunology* **11**. https://doi.org/10.3389/fimmu.2020.00824
Wu, N., Song, Y.-L., Wang, B., Zhang, X.-Y., Zhang, X.-J., Wang, Y.-L., Cheng, Y.-Y., Chen, D.-D., Xia, X.-Q., Lu, Y.-S., Zhang, Y.-A., 2016. Fish gut-liver immunity during homeostasis or inflammation revealed by integrative transcriptome and proteome studies. *Scientific Reports* **6**, 36048.
Wu, X.Y., Nie, L., Lu, X.J., Fei, C.J., Chen, J., 2022. Molecular characterization, expression and functional analysis of large yellow croaker (Larimichthys crocea) peroxisome proliferator-activated receptor gamma. *Fish Shellfish Immunology* **123**, 50–60.
Xie, Y., Meijer, A.H., Schaaf, M.J.M., 2021. Modeling inflammation in zebrafish for the development of anti-inflammatory drugs. *Frontiers in Cell and Developmental Biology* **8**. https://doi.org/10.3389/fcell.2020.620984
Xu, H., Zhang, X., Li, H., Li, C., Huo, X.-J., Hou, L.-P., Gong, Z., 2018. Immune response induced by major environmental pollutants through altering neutrophils in zebrafish larvae. *Aquatic Toxicology* **201**, 99–108.
Yamaguchi, T., Takizawa, F., Furihata, M., Soto-Lampe, V., Dijkstra, J.M., Fischer, U., 2019. Teleost cytotoxic T cells. *Fish & Shellfish Immunology* **95**, 422–439.
Yang, C., Lim, W., Song, G., 2021. Immunotoxicological effects of insecticides in exposed fishes. *Comparative Biochemistry and Physiology Part C: Toxicology & Pharmacology* **247**, 109064.
Yang, M., Qiu, W., Chen, B., Chen, J., Liu, S., Wu, M., Wang, K.J., 2015. The in vitro immune modulatory effect of bisphenol A on fish macrophages via estrogen receptor α and nuclear factor-κB signaling. *Environmental Science Technology* **49**, 1888–1895.
Yang, M., Wang, X., Chen, D., Wang, Y., Zhang, A., Zhou, H., 2012. TGF-β1 exerts opposing effects on grass carp leukocytes: implication in teleost immunity, receptor signaling and potential self-regulatory mechanisms. *PLoS One* **7**, e35011.
Yang, Y., Zhang, X., Jiang, J., Han, J., Li, W., Li, X., Yee Leung, K.M., Snyder, S.A., Alvarez, P.J.J., 2022. Which micropollutants in water environments deserve more attention globally? *Environmental Science & Technology* **56**, 13–29.
Yaniv, K., Isogai, S., Castranova, D., Dye, L., Hitomi, J., Weinstein, B.M., 2006. Live imaging of lymphatic development in the zebrafish. *Nature Medicine* **12**, 711–716.
Yano T (1996): The nonspecific immune system: Humoral defence. In: Iwama G, Nakanishi T (eds.): The Fish Immune System: Organism, Pathogen and Environment. Academic Press, San Diego. 105–157.
Ye, R.R., Peterson, D.R., Kitamura, S.-I., Segner, H., Seemann, F., Au, D.W.T., 2018. Sex-specific immunomodulatory action of the environmentalestrogen 17α-ethynylestradiol alongside with reproductive impairment in fish. *Aquatic Toxicology (Amsterdam, Netherlands)* **203**, 95–106.
Yoder, J.A., Mueller, M.G., Wei, S., Corliss, B.C., Prather, D.M., Willis, T., Litman, R.T., Djeu, J.Y., Litman, G.W., 2001. Immune-type receptor genes in zebrafish share genetic and functional properties with genes encoded by the mammalian leukocyte receptor cluster. *Proceedings of the National Academy of Sciences* **98**, 6771–6776.
Yoshida, S.H., Stuge, T.B., Miller, N.W., Clem, L.W., 1995. Phylogeny of lymphocyte heterogeneity: cytotoxic activity of channel catfish peripheral blood leukocytes directed against allogeneic targets. *Developmental & Comparative Immunology* **19**, 71–77.
Yu, Y., Wang, Q., Huang, Z., Ding, L., Xu, Z., 2020. Immunoglobulins, mucosal immunity and vaccination in teleost fish. *Frontiers in Immunology* **11**. https://doi.org/10.3389/fimmu.2020.567941
Yue, Y., Li, X., Sigg, L., Suter, M.J.F., Pillai, S., Behra, R., Schirmer, K., 2017. Interaction of silver nanoparticles with algae and fish cells: a side by side comparison. *Journal of Nanobiotechnology* **15**, 16.
Zahran, E., Elbahnaswy, S., Risha, E., El-Matbouli, M., 2020. Antioxidative and immunoprotective potential of Chlorella vulgaris dietary supplementation against chlorpyrifos-induced toxicity in Nile tilapia. *Fish Physiology and Biochemistry* **46**, 1549–1560.
Zapata, A., 1982. Lymphoid organs of teleost fish. III. Splenic lymphoid tissue of Rutilus rutilus and Gobio gobio. *Developmental & Comparative Immunology* **6**, 87–94.
Zapata, A., Diez, B., Cejalvo, T., Gutiérrez-de Frías, C., Cortés, A., 2006. Ontogeny of the immune system of fish. *Fish & Shellfish Immunology* **20**, 126–136.
Zapata, A.G., 1996. Cells and tissues of the immune system of fish. The fish immune system. Organism, pathogen, and environment Academic Press.Inc. S.Diego, 255–287.
Zeeman, M., 1986. Modulation of the immune response in fish. *Veterinary immunology and immunopathology* **12**, 235–241.
Zeeman, M.G., Brindley, W.A., 1981. Effects of of toxic agents upon fish immune systems: a review. In Sharma, P.P., (Ed.), *Immunologic Considerations in Toxicology*. CRC Press, Boca Raton, pp. 1–60.
Zelikoff, J., 1994. Fish immunotoxicology. *Immunotoxicology and Immunopharmacology* **2**, 71–95.
Zelikoff, J.T., 1998. Biomarkers of immunotoxicity in fish and other non-mammalian sentinel species: predictive value for mammals? *Toxicology* **129**, 63–71.
Zelikoff, J.T., 2000. *Fish Immune Response as Biomarkers*. New York Univ Medical Center, Tuxedo.
Zhan, Y., Jimmy, K., 2000. Molecular isolation and characterisation of carp transforming growth factor β1 from activated leucocytes. *Fish & Shellfish Immunology* **10**, 309–318.
Zhao, C.S., Fang, D.A., Xu, D.P., 2020. Toll-like receptors (TLRs) respond to tributyltin chloride (TBT-Cl) exposure in the river pufferfish (Takifugu obscurus): evidences for its toxic injury function. *Fish Shellfish Immunology* **99**, 526–534.
Zhao, Y., Zhang, K., Giesy, J.P., Hu, J., 2015. Families of nuclear receptors in vertebrate models: characteristic and comparative toxicological perspective. *Scientific Reports* **5**, 8554–8554.

Zhu, L.Y., Lin, A.F., Shao, T., Nie, L., Dong, W.R., Xiang, L.X., Shao, J.Z., 2014. B cells in teleost fish act as pivotal initiating APCs in priming adaptive immunity: an evolutionary perspective on the origin of the B-1 cell subset and B7 molecules. *Journal of Immunology* **192**, 2699–2714.

Zhu, W., Su, J., 2022. Immune functions of phagocytic blood cells in teleost. *Reviews in Aquaculture* **14**, 630–646.

Zou, J., Secombes, C., 2016. The function of fish cytokines. *Biology* **5**, 23.

Zou, J., Secombes, C.J., 2011. Teleost fish interferons and their role in immunity. *Developmental & Comparative Immunology* **35**, 1376–1387.

Zou, J., Secombes, C.J., Long, S., Miller, N., Clem, L., Chinchar, V., 2003. Molecular identification and expression analysis of tumor necrosis factor in channel catfish (Ictalurus punctatus). *Developmental & Comparative Immunology* **27**, 845–858.

Zwollo, P., Cole, S., Bromage, E., Kaattari, S., 2005. B cell heterogeneity in the teleost kidney: evidence for a maturation gradient from anterior to posterior kidney. *Journal of Immunology* **174**, 6608–6616.

Zwollo, P., Haines, A., Rosato, P., Gumulak-Smith, J., 2008. Molecular and cellular analysis of B-cell populations in the rainbow trout using Pax5 and immunoglobulin markers. *Developmental and Comparative Immunology* **32**, 1482–1496.

# 7 Toxicity Resistance

## *Physiological Acclimations and Evolutionary Adaptations to Chemical Stress*

*Diane E. Nacci and Bryan W. Clark*
US Environmental Protection Agency

*Andrew Whitehead*
University of California Davis

**Abstract**

Exposure to environmental pollutants or xenobiotics is often toxic to fish, reducing their ability to develop, grow, reproduce, or survive. While variation among fish species in their sensitivities to pollutant classes sometimes reflects inherent differences in biological targets, it may also reflect variation in the regulation or functionality of defensive mechanisms that mitigate the effects of pollutant exposures. This chapter focuses on general mechanisms and specific examples by which fish protect against, tolerate, or resist the effects of toxic pollutants. For example, transient exposures may induce physiological acclimations that reduce sensitivity to similar exposures within one generation, or may be passed to one (or more) generations by genomic regulatory mechanisms that affect the accessibility of genes to induction, such as methylation. While genetic adaptation or evolved tolerance refers to the reduced sensitivity displayed by some fish populations following multi-generational toxic exposures and reflects inherited (genetic) changes in the regulation or functionality of defense mechanisms or biological targets. However, physiological acclimations are often energetically costly and the evolution of tolerance to toxic pollutants is rarely sufficiently rapid to rescue most fish species. Thus, the study of toxicity defense helps us to characterize and predict more holistically the ecological impacts of pollutants.

## KEY POINTS

- Fish species vary in their inherent sensitivity to classes of environmental chemicals (pollutants or xenobiotics); within species, pollution exposure history can result in reduced sensitivity, referred to as tolerance or resistance.
- Xenobiotic exposures can trigger a variety of defensive responses that affect tolerance by minimizing toxic concentrations that reach the biological target (detoxifying) and/or altering the functionality of the target (desensitizing).
- Physiological acclimation is a conditional (plastic) response to xenobiotic exposure that reduces sensitivity in individuals by transitory alterations in the activity of biochemical detoxification; these biologically costly processes are often effective defenses for short-duration exposures.
- Genetic adaptation is a fixed reduction in the sensitivity of populations resulting from the selection of individuals with changes in protein-coding or gene regulatory sequences that result in enhanced detoxification and/or relatively insensitive biological targets; these responses to chronic, multigenerational exposures are potentially effective but are sometimes accompanied by trade-offs.
- Real-world examples suggest that toxicity defense is often more complex than a single biochemical reaction, single-gene sequence change, or simple acclimation/adaptation dichotomy; whole-genome approaches are revealing more holistically the breadth of protective strategies that permit some fish to survive in a rapidly changing world.

## INTRODUCTION

Plants and animals, including fishes, have evolved in an environment that has included intermittent and continuous exposures to toxic materials (Ames et al. 1990a, 1990b, Menzie et al. 1992, Gribble 1994, Shaw 1999). Many basic mechanisms for processing and responding to exogenous chemicals are shared across the tree of life and are therefore ancient and highly evolutionarily conserved (Roesijadi 1992, Borst et al. 1993, Doehmer et al. 1993, Doige and Ames 1993, Hayes and Pulford 1995, Sheehan et al. 2001). Such defense responses to toxic chemical exposures have evolved over geological time equivalent to thousands of generations and have contributed to the observed variation in sensitivities to chemical classes among fish species or interspecific variation. Rarely among fish species, genetically based defense responses have evolved over contemporary time equivalent to only a few generations and have contributed to observed variation among populations within a single fish species or intraspecific variation. In this chapter, we consider resistance mechanisms that are inducible or fixed features of fish species and consider how rapidly



DOI: 10.1201/9781003160694-7

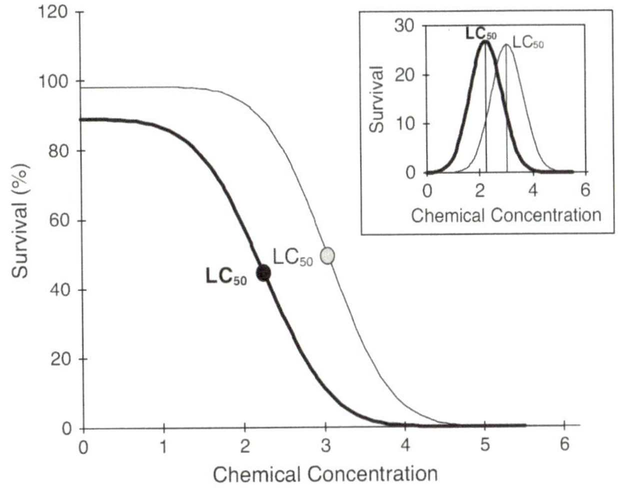


**FIGURE 7.1** Relative tolerance shown as cumulative and noncumulative (inset) distributions indicating population responses to chemical exposures. Lethal response curves and 50% lethal effect concentrations ($LC_{50}$) are shown for a relatively sensitive group (bold curve) and a more tolerant (light curve) group. Adapted from Van Veld and Nacci (2008).

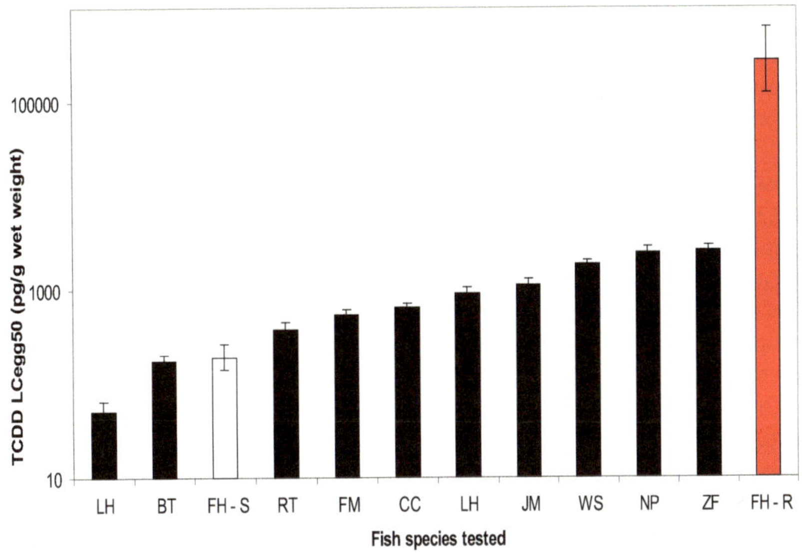


**FIGURE 7.2** Inter- and intraspecific variation in response to TCDD ($LC_{50}$ and 95% confidence intervals) in early life stages of fish species: lake trout (LT), brook trout (BT), *Fundulus heteroclitus* (FH, white bar), rainbow trout (RT), fathead minnow (FM), channel catfish (CC), lake herring (LH), Japanese medaka (JM), white sucker (WS), northern pike (NP), zebrafish (ZF), and *F. heteroclitus*-resistant population (FH-R, red bar). Adapted from Van Veld and Nacci (2008).

evolved resistance in contemporary environments often involves adaptive modification of these same mechanisms.

## Defining Tolerance

Operationally, relative sensitivity to toxicants is determined by conducting standardized bioassays and expressing results as the chemical concentration producing an adverse biological effect in some proportion of the tested representative group of fish species or population. The $LC_{50}$, or median lethal concentration, is the chemical concentration killing 50% of the tested group (Figure 7.1). $LC_{50}$ values provide a convenient way of comparing or ranking the sensitivities of tested groups. Lethal concentrations are high (or shifted right) in tolerant relative to more sensitive groups; however, all individuals within tested groups do not respond identically, and even within sensitive groups, some individuals are more tolerant or resistant to toxic effects than others (Figure 7.1 inset).

Different species of fish can vary dramatically in their inherent sensitivity to the toxic effects of specific chemicals. For example, the $LC_{50}$ values for the early life stages of ten species of freshwater fish from exposure to the potent organic contaminant 2,3,7,8-tetrachlorodibenzo-*p*-dioxin (TCDD, referred to informally as 'dioxin') vary by over 100-fold (Figure 7.2). This means that dioxin concentrations that kill a large proportion (50%) of exposed individuals from a sensitive fish species would kill very few but perhaps some individuals from a more tolerant fish species. Considered in this way, differences in sensitivities between fish species or populations reflect changes in the proportion of sensitive individuals within the group.

Within a single fish species, the sensitivities of different populations can be affected by prior chemical exposures. To illustrate one well-studied example, early life stage dioxin $LC_{50}$ values have been added to Figure 7.2 for two populations of the nonmigratory estuarine fish *Fundulus heteroclitus* (Atlantic killifish, referred to as killifish or mummichog). Within this context, a killifish population residing in an uncontaminated or reference site is relatively sensitive to dioxin in comparison with other fish species tested and ~1,000 times more sensitive than another killifish population resident in an extremely contaminated site. This killifish intraspecific sensitivity exceeds by more than ten-fold the interspecies variation in dioxin sensitivity among fishes; the mechanism by which this has occurred is explored in greater detail later in this chapter.

There are a variety of strategies for toxicity resistance that benefit short-term and long-term xenobiotic exposures (Table 7.1) (Chambers and Yarbrough 1979, Fernández-Delgado 1989, Winston 1991, Winston and Di Giulio 1991, Kurelec 1992, Roesijadi 1992, Doehmer et al. 1993, Fernandez-Salguero et al. 1995, Taylor and Feyereisen 1996, Chevillon et al. 1997, Eaton and Bammler 1999). Consistent with the duration of xenobiotic exposure and the need for protection from toxic effects, defense strategies also vary in their duration of action, i.e., within the exposed generation or across one (trans)- or more (multi)generations (Weis and Weis 1989, Hahn 1998b, Wirgin and Waldman 1998, Elskus 2001, Wirgin and Waldman 2004, Whitehead et al. 2017) (Table 7.1). The term physiological acclimation refers to the induction of a variety of defense mechanisms in an exposed individual that results in rapid but temporary (short-term) toxicity reduction. This is an example of phenotypic plasticity, a general term that refers to the capacity of a single genotype to produce different phenotypes conditionally (Hua et al. 2015b). Genetic adaptation describes a strategy for long-term or multigenerational toxicity resistance, where natural selection causes changes in allele frequencies within and between populations, leading to heritable evolved tolerance. Genetic adaptation or evolved tolerance occurs in response to the selection of tolerant individuals, where the genetic basis for toxicity resistance is passed on to progeny and retained through successive generations, even when progeny are not exposed to chemicals (Endler 1986). Thus, these genetic changes provide protection to future generations from subsequent or continuing xenobiotic exposures. In addition, toxicity resistance can persist beyond initial exposure (intra-generationally) or in one generation of progeny (trans-generationally) via persistent changes in the regulation of gene expression. For example, epigenetic modifications, such as microRNAs and methylation of promotor regions, regulate the expression of genes at the transcriptional and posttranscriptional levels. The roles and mechanisms of epigenetic modifications in fishes are the subject of the current study (see Chapter 10, 'Epigenetics'), although their roles in chemical tolerance among fishes are not yet well understood (Arzuaga et al. 2004, Timme-Laragy et al. 2005, Aluru et al. 2011, Hua et al. 2015a). Finally, nongenetic but inherited biological or chemical factors may contribute to the appearance of tolerance in offspring. In particular, the maternal transfer of messenger RNA (mRNA) and/or complex mixtures of xenobiotics in eggs may mitigate the response of progeny to additional xenobiotic exposures, which may resemble inherited tolerance but does not persist beyond one generation.

Together, these strategies for toxicity defense can provide protection of varying duration and associated ecological benefit; their implementation can determine whether xenobiotic exposure results in population persistence, decline, or failure, referred to as local extirpation (Figure 7.3). Physiological acclimation or intragenerational toxicity resistance may occur in direct and often immediate response to xenobiotic exposures, conferring transitory toxicity resistance and its costs. Under these conditions, healthy individuals can mount a robust transient toxicity resistance response. However, this response often comes at costs, e.g., reflecting the production of defense system components, which may not be sustainable even under optimal conditions. Thus, populations subject to chronic multigenerational pollutant exposures may require different defense strategies to persist, yet this evolutionary rescue may be successful under a very limited set of conditions. In particular, when some proportion

**TABLE 7.1**
**Processes and Conditions Contributing to Toxicity Resistance**

| Type of Resistance | Description |
|---|---|
| Resistance or tolerance | Relative ability of individuals or populations to function and survive during toxicant exposures that affect the fitness of susceptible (sensitive) individuals, populations, or species |
| Physiological acclimation | Toxicity resistance based on short-term physiological or biochemical responses to toxicant exposure such as upregulation of detoxification processes |
| Genetic adaptation (evolved tolerance) | Genetically heritable alterations that may affect the functionality or expression of target sites, toxicant receptors, and proteins involved in detoxification, toxicant binding, and efflux, resulting in toxicity resistance over successive generations |
| Epigenetic alterations | Toxicity resistance resulting from modifications in DNA such as hyper- or hypo-methylation that do not involve changes in nucleotide sequence but alter gene expression, resulting in reduced sensitivity in later life stages or a subsequent unexposed generation |
| Nongenetic, inherited tolerance | Resistance resulting from maternal transfer of toxicant or mRNA from parent to offspring; resistance does not likely last beyond one generation |

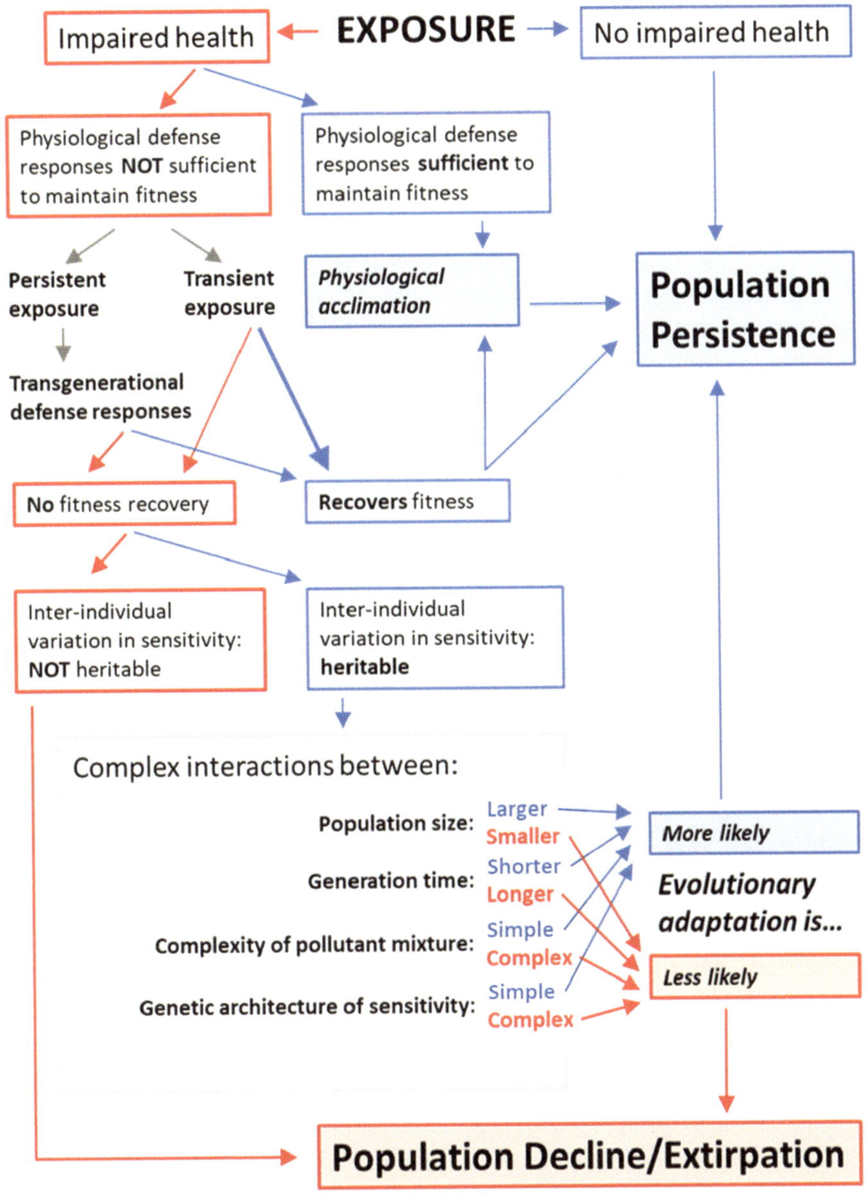


FIGURE 7.3 Exposure outcomes: The nature of exposures and the features of species interact to govern how acclimation and/or adaptation may unfold over physiological and ecological evolutionary time to enable or disable population abundance and persistence in polluted environments.

of interindividual variation in sensitivity is heritable or genetically encoded, long-term pollutant exposure may act as an agent of natural selection, resulting in the adaptive evolution of tolerant populations that are dominated by more tolerant individuals.

### Genetic Adaptation or Evolved Tolerance

While genetic adaptation may seem like the ultimate defense strategy to overcome toxic exposures (Figure 7.3), this strategy is only possible for a highly restricted number of fish species and toxic pollutants. Species characteristics related to life history, e.g., diet, behavior, reproduction, migration, and demography, e.g., population size and genetic diversity, contribute both to the potential for exposure and the ability to tolerate exposure. For example, exposure to toxic chemicals may be transient due to the properties of the xenobiotic, such as non-persistence or lability, or the species, such as the ability to migrate away from the source. In these cases, fish populations may persist through chemical-specific toxicity resistance acclimation strategies. Migratory fish populations whose habitats become lethally polluted with persistent toxic chemicals may emigrate or may be sustained locally through immigration from nearby uncontaminated sites, where possible. Rarely, populations of certain nonmigratory fish species may persist through the evolution of tolerance, i.e., the selective survival and reproductive fitness of tolerant individuals. For most fish species, evolutionary rescue from extreme pollution is unlikely. This is because the pace of chemical environmental change is very swift and the fitness challenge can be very large. As such, heritable genetic variation for resistance (e.g., Brazzola et al. 2014) must preexist in the population, because environmental change is too swift for a population to 'wait' for relevant genetic variation to arrive by mutation or migration. The more genetic variation that stands in a population, the more likely it is that some fraction of that variation may affect fitness when the environment changes. Since larger populations can harbor more genetic diversity (Leffler et al. 2012), species with very large population sizes and short generation times are more likely to evolve quickly enough to persist in chemically polluted habitats (e.g., see review by Whitehead et al. (2017)). This explains why there are many examples of insects, weeds, bacteria, and viruses evolving resistance to toxic chemicals, but relatively few examples of vertebrate animals evolving such resistance. Generally speaking, insects, weeds, bacteria, and viruses have much larger genetic gas tanks (and much shorter generation times) than vertebrate animals including fish (Leffler et al. 2012). An exception that supports the rule is killifish: They are vertebrate animals with populations that have rapidly and repeatedly evolved resistance to pollution. Importantly, compared to other vertebrates, killifish populations have among the highest levels of genetic diversity ever measured, such that the depth of their genetic gas tanks is on par with that typically observed in insects (Reid et al. 2017). Clearly, many species most vulnerable to human-mediated stressors do not always share these characteristics and face local extirpation or extinction in this rapidly changing world.

Numerous factors related to the type and amount of contaminant, exposure route and duration, the complexity of the contaminant mixture, and features of the exposed species can contribute to the likelihood, mode, and effectiveness of toxicity resistance in fish populations (Whitehead et al. 2017). Persistent contaminants can produce long-term exposures, sometimes generations after discharges have ceased; even non-persistent chemicals can produce persistent exposures when discharges are continuous. The toxic potency of the specific class of chemicals and the frequency, duration, and magnitude of exposure also contribute to the potential for and mechanisms of defense. Importantly, different categories of toxic chemicals affect fish by different mechanisms and require different mechanisms to mitigate toxicity. In particular, it is hypothesized that the likelihood of rapid, contemporary evolved tolerance is greater for xenobiotic classes whose toxicity may be altered by mutations in a small number of controlling loci and where the ancillary adverse outcomes may be limited because mutations affect the function of specialized vs. more generalized pathways (Nacci et al. 2016, Whitehead et al. 2017). A real-world example that tests some of these hypotheses is the demonstration of evolved tolerance to highly toxic dioxin-like compounds (DLCs) in the Atlantic killifish and their sister species, Gulf killifish, *Fundulus grandis*, described later in this chapter.

## TOXICITY DEFENSE MECHANISMS

Mechanisms of toxicity resistance have been studied in diverse organisms because of issues relating to drug resistance in the medical field, e.g., cancer, infectious diseases, and issues related to insecticide resistance in the fields of agriculture, forestry, and public health. Examples from these fields are informative to understanding toxicity resistance in fishes because many mechanisms underlying resistance are conserved across phyla (Doehmer et al. 1993, Taylor and Feyereisen 1996, Shaw 1999, Daborn et al. 2002, Gottesman 2002, Bunger et al. 2003). While specific mechanisms vary across species and xenobiotic classes, processes that minimize interactions between toxic chemicals and their biological target can reduce toxic effect. Thus, toxicity resistance strategies can be grouped as those that reduce the effective concentration of toxic chemical at biological target (detoxification) and those that result from alterations in the inherent sensitivity of biological target (desensitization). The processes, regulation, and outcome of some of the major mechanisms of detoxification in fishes are described briefly here. Also included here is an introduction to the evolutionary strategy of evolved tolerance by which fish populations with genetically distinct biological targets that are desensitized to specific contaminants can evolve via continuous directional selection (Table 7.2).

### Detoxification

Detoxification is an important strategy of toxicity resistance that involves the activation of defense processes, such as sequestration, biotransformation, elimination, and antioxidant cycling, which reduce the amount of toxic chemical reaching the biological target. Toxic concentrations can be reduced by specific detoxification processes, but a more holistic perspective takes into account the integration of multiple protective processes, i.e., as a 'defensome', which may provide adequate defense against chemical exposures. This section provides brief descriptions and examples of some major biochemical mechanisms by which xenobiotics are detoxified or neutralized; for more detailed discussions, also see other chapters in this volume, including Chapter 2, 'Biotransformation in Fishes', and Chapter 3, 'Reactive Oxygen Species and Oxidative Stress'.

#### Sequestration

Sequestration is a general toxicity defense strategy to reduce the accessibility of the toxic compound biochemically by binding with an agent that blocks further interactions and/or physically by intracellular partitioning. Sequestration is considered essential to metal regulation and detoxification, and in diverse microorganisms, plants, and animals is largely accomplished by metal-binding proteins referred to as metallothioneins (MTs) (Leber and Miya 1976, Shaw 1999). The presence of MTs across all domains of life largely reflects the ubiquitous, ancient challenge of managing exposures to metals that can serve both essential and toxic functions.

MTs play an integral role in the homeostasis of essential metals and the sequestration and defense against metal toxicity in fishes (Roesijadi 1992, Pavic et al. 2019). MTs have a high affinity or binding strength for a variety of metals (Castaldo et al. 2020), which supports their role in essential metal homeostasis, i.e., as donors for transcription factors, including steroid hormone receptors and zinc finger transcription factors (Pavic

**TABLE 7.2**
**General Mechanisms Addressed in Studies of Toxicity Resistance**

| Mechanism | Description |
|---|---|
| Sequestration | Proteins and other molecules in the cell can reduce the availability of toxicants to target sites and receptors through toxicant sequestration or binding. Metallothioneins (MTs) are a family of metal-binding proteins involved in reducing metal toxicity. Upregulation of MTs is often observed in metal-tolerant fish |
| Biotransformation | The fate and effect of many toxicants are governed by their susceptibility to enzymatic biotransformation. Biotransformation by phase I enzymes (e.g., cytochromes P450) and phase II enzymes (e.g., glutathione *S*-transferases) generally yields products that are less toxic and easier to eliminate. Examples of activation (i.e., increased toxicity) of relatively harmless compounds exist, however. Differences in the expression of these enzymes have a large effect on cell concentrations of toxicants and therefore on toxicity. |
| Elimination | Proteins such as P-glycoprotein (Pgp) residing on the plasma membrane of cells are involved in ATP-driven efflux of diverse drugs and toxicants from the cell. One form of Pgp, the multidrug-resistant protein (Mdr), is a major obstacle to the treatment of cancer because of its efficiency in pumping cancer drugs out of the cell. P-glycoprotein expression has been reported in a number of aquatic organisms, including fish. |
| Antioxidation | The antioxidant system accomplishes feedback-driven oxidation–reduction cycling to minimize damaging reactive oxygen species (ROS), which are produced directly or indirectly by toxic xenobiotic exposures. |
| Altered target or receptor | Upon entering a cell, a toxicant must interact with a cellular target or receptor to exert effects; for example, binding of dioxin-like compounds (DLCs) with the aryl hydrocarbon receptor (AHR) mediates the toxic effects of DLCs. Alterations (e.g., polymorphisms) in the AHR could alter toxicity via altered strength of binding or other mechanisms. |

et al. 2019). Relative to toxicity defense, metals are made inaccessible to biological targets not only by association with MTs but also by physical isolation into 'metabolically detoxified cellular fractions' of some metals (Goto and Wallace 2010, Webster et al. 2013, Fan et al. 2015). In particular, MTs are largely responsible for the cellular sequestration of zinc, copper, cadmium, and to a lesser extent iron, lead, and nickel in fishes (Wood 2012). Metal binding by MTs also mitigates metal redox potential, contributing to protection from fish metal toxicity through their role in antioxidant cycling (Tang et al. 2016, Pavic et al. 2019, Castaldo et al. 2020).

MT synthesis can be induced by exposure to many metals, including cadmium, copper, zinc, mercury, and nickel (Roesijadi 1992, 1994, 1996). Classically, the induction of MTs is triggered by metal binding to metal transcription factor 1 (MTF1), which initiates transcription via interaction with the metal response element (MRE) in the MT promotor region (Zhu et al. 2018, Pavic et al. 2019). However, oxidative stress and the glucocorticoid-driven pathway can also directly induce MT expression through respective antioxidant and glucocorticoid response elements (ARE and GRE) in the promotor region of MT genes (Pavic et al. 2019). Thus, increased expression of MTs has multiple benefits relative to metal toxicity resistance, and many studies indicate an association between increased metal tolerance and MT expression in fish and other aquatic organisms (Bryan 1976, Dixon and Sprague 1981a, 1981b, 1981c, McCarter and Roch 1983, Stegeman et al. 1992). More recently, Uren Webster and colleagues (2013) observed that increased MT synthesis has been shown extensively to occur in response to many metals, in both short-term laboratory exposures and chronically exposed wild fish, and it is the most consistent mechanism of metal tolerance in fish.

## Biotransformation

The fate and effects of many drugs and toxicants are governed by their susceptibility to enzymatic biotransformation (Stegeman et al. 1992, Di Giulio et al. 1995, Stegeman et al. 2010, Lushchak et al. 2018) (also see Chapter 3, 'Biotransformation in Fishes'). Biotransformation is often characterized as a sequence of events involving phase I, II, and III reactions. Phase I enzymes primarily catalyze oxidation, reduction, and hydrolysis reactions, often introducing a polar group into hydrophobic molecules. In fish, phase I reactions are catalyzed primarily by cytochrome P450 (CYP)-mediated monooxygenase (MO) systems; certain classes of xenobiotics can act as inducers and sometimes substrates for CYPs (Stegeman and Hahn 1994, Di Giulio et al. 1995). Biotransformation usually results in products that are less toxic and easier to eliminate than parent compounds; however, numerous examples exist in which relatively harmless compounds are transformed by specific reactions to yield cytotoxic and genotoxic products.

Compounds with polar side groups or compounds that acquire polar groups during phase I (i.e., CYP-mediated) biotransformation are susceptible to phase II reactions (Stegeman et al. 1992, Di Giulio et al. 1995, Stegeman et al. 2010) (see Chapter 2). Phase II conjugating enzymes add functional groups and form more readily excreted metabolites, e.g., using the polar group as a handle to link the compound to various polar endogenous compounds (sugar derivatives, amino acids, peptides, and sulfate). These reactions serve to reduce toxicity and to enhance water solubility and elimination. Among the phase II enzyme systems, the glutathione S-transferases (GSTs) are a particularly versatile, multifunctional family of detoxification enzymes (Mannervik et al. 1985, Hayes and Pulford 1995, Eaton and Bammler 1999, Salinas and Wong 1999). Reactions involving conjugation of epoxide-containing compounds to glutathione reduce the likelihood of cytotoxicity and genotoxicity resulting from the electrophilic attack of epoxide groups on DNA and other cellular macromolecules (Ito et al. 1988, Eriksson and Andersson 1992). Numerous products of oxidative stress, resulting from phase I reactions and lipid peroxidation, are GST substrates as well (Hayes and Strange 1995). Furthermore, most GSTs are soluble proteins and can sometimes participate in the transport of lipophilic toxicants to phase I enzymes, thereby enhancing phase I reactions. In addition, GSTs play a sacrificial role by covalently binding to activated phase I products, reducing the likelihood that these products will bind to and damage other cellular macromolecules. Phase III components include efflux transporters that actively eliminate parent compounds and biotransformation products (Lushchak 2011a, Lushchak et al. 2018).

This system of biotransformation is largely regulated by the aryl hydrocarbon receptor (AHR) signal transduction pathway (but also see later discussion regarding crosstalk with other regulatory pathways). As an initiating event, the AHR transduces the environmental signal of xenobiotic exposure into the feedback-regulated activation of phase I and II biotransformation enzymes. The outcome of these events is considered a major route by which lipophilic xenobiotics are detoxified (Vanden Heuvel and Lucier 1993, Birnbaum 1994, Kerkvliet 1995, Mimura et al. 1999, Thurmond et al. 1999, Hahn 2001, Toomey et al. 2001, Karchner et al. 2002, Bunger et al. 2003). A more holistic view of the outcome of AHR activation, including interactions with other defense pathways, must take into account many genes or pathways that are regulated by AHR unrelated to biotransformation, such as in hypoxia, developmental, and immune signaling, and activation during development can cause developmental defects (Chan et al. 1999, McMillan and Bradfield 2007, Furness and Whelan 2009, Matsumura 2009, Stevens et al. 2009, Denison et al. 2011). Thus, AHR activation via xenobiotic ligand binding can produce adverse effects through inappropriate activation of these alternate pathways; however, the focus of this discussion is limited to the role of AHR in xenobiotic transformation.

The current conceptual model for the activation of the AHR detoxification pathway retains focus on the receptor. AHR resides in the cytoplasm stably complexed with the molecular chaperone heat shock protein 90 (Hsp90) and other proteins, including AHR-interacting protein (AIP/XAP2/Ara9, referred hereafter as AIP). Upon ligand binding, the AHR/Hsp90 complex translocates to the nucleus and Hsp90 is exchanged for the AHR nuclear translocator (ARNT). The AHR/ARNT heterodimer regulates the transcription of downstream genes via xenobiotic response elements (XREs) (also known as aryl hydrocarbon response elements) in their promoter regions. A large battery of genes is XRE-regulated, including CYP1s and many other phase I (CYPs) and phase II (GSTs) biotransformation enzymes; however, the identities and functions of these genes are still being discovered (Mimura et al. 1999, Nebert et al. 2000, Hahn 2001, Karchner et al. 2002, King-Heiden et al. 2012, Shankar et al. 2020, Silvestre 2020). It is now recognized that AHR also may induce antioxidant responses, for example, through the presence of XRE(s) in genes including nuclear factor erythroid 2 p45-related factor 2 (Nrf2), a transcription factor that regulates the expression of several antioxidant genes (Figure 7.4). Conversely, some AHR genes contain antioxidant response elements (AREs) in their regulatory regions (King-Heiden et al. 2012, Goodale et al. 2015, Dietrich 2016, Shankar et al. 2020, Silvestre 2020). This crosstalk not only provides an important mechanism by which multiple toxic effects are generated by xenobiotic exposures but also provides support for the concept of the defensome (Stegeman et al. 2010, Eide et al. 2021), an integrated network of toxicity pathways that is discussed later in this section.

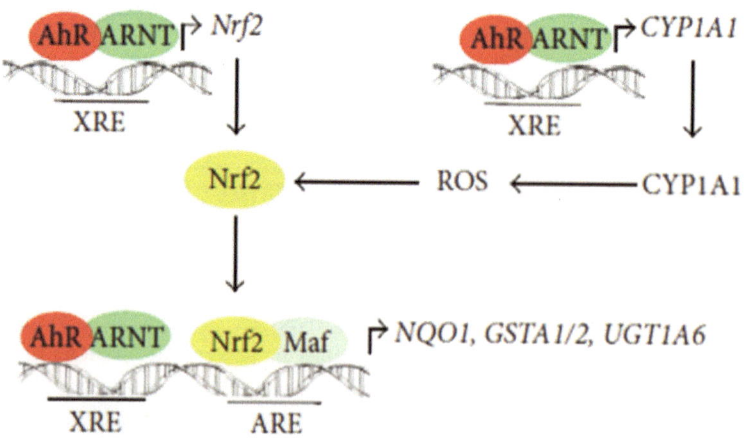


**FIGURE 7.4** Coordinated defense pathways: The proposed model for coordinate induction of nuclear factor erythroid 2 p45-related factor 2 (Nrf2)-dependent genes by the aryl hydrocarbon receptor (AhR). Activation of the AhR leads to dimerization with the aryl hydrocarbon receptor nuclear translocator (ARNT) and transcriptional activation of both Nrf2 and cytochrome P4501A1 (CYP1A1). CYP1A1 increases intracellular reactive oxygen species (ROS), thereby stabilizing Nrf2 protein. Nrf2, in association with musculoaponeurotic fibrosarcoma (Maf), binds to antioxidant response elements (AREs), and the AhR/ARNT complex binds to xenobiotic response elements (XREs) in the promoter regions of NADPH/quinone oxidoreductase 1 (NQO1), glutathione-S-transferases A 1, 2 (GSTA1/2), or uridine 5'-diphospho (UDP)-glucuronosyltransferase 1A6, (UGT1A6). Adapted from Dietrich (2016), open access.

### Elimination: *p*-Glycoproteins

The ATP-binding cassette (ABC) gene family is comprised of ATP-dependent transporters that mediate the cellular efflux of a wide range of organic chemicals and certain metals in aquatic organisms (Luckenbach et al. 2014). Early investigators speculated that the activity of Pgps may contribute to pollution tolerance in populations of aquatic species (Kurelec 1992). In support of this hypothesis, Pgp-like proteins were detected in excretory tissues of several pollution- or toxin-tolerant aquatic invertebrates (Kurelec 1992, Toomey and Epel 1993, Cornwall et al. 1995, Galgani et al. 1996). Further evidence is associated with elevated Pgp expression and activity in aquatic invertebrates exposed to toxicants (Minier et al. 1993, Kurelec et al. 1995). Additionally, altered Pgp expression has also been studied in fish, focusing on populations inhabiting contaminated environments (Cooper et al. 1996, 1999a, 1999b, Vogelbein et al. 1999, Bard 2000, Bard et al. 2002a, 2002b, Grans et al. 2015). More recently, transporters within this family including P-glycoprotein (Pgp or ABCB) and multidrug resistance-associated proteins (mrps) have been found to be highly expressed in the embryonic, larval, and adult stages of many aquatic organisms; non-specificity with respect to substrate supports their role in defense against xenobiotics including some metals and organic pollutants (Lerebours et al. 2016, Yin et al. 2016, Hu et al. 2019). The role of ABC transporters can provide both the first line of defense in the efflux of xenobiotics and the last line of defense in promoting the efflux of toxic metabolites (Epel et al. 2008, Luckenbach et al. 2014). In zebrafish embryos, the expressions and activities of ABC transporters (pgp, mrp1, and mrp2) are enhanced by metal exposures, which are mediated mainly via Pxr, upregulating pgp, and, to a lesser extent, Nrf2, affecting mrp1 and mrp2 (Hu et al. 2019). However, the nuclear receptors AHR1B and peroxisome proliferator-activated receptor-B (PPAR-B) have also been implicated in the regulation of mrps (Tian et al. 2019).

Both the upregulation and inhibition by xenobiotics of transporters within this grouping, also referred to as small molecule transporters (SMTs), which also include the solute carrier family (SLCs), have been reported (Nicklisch and Hamdoun 2020). In fact, intense study of xenobiotics producing inhibitory effects has been prompted by human medical interests in 'chemosensitizers' (or more generally transporter-interfering chemicals, TICs) that minimize the activities of SMTs and thereby increase the efficacy of cancer therapeutic drugs (Nicklisch and Hamdoun 2020). TICs in fish include xenobiotics such as organochlorine pesticides, brominated flame retardants (brominated diphenyl ethers or BDEs), and polychlorinated biphenyls (PCBs) (Nicklisch et al. 2021).

### Antioxidant Cycling

A special case of biotransformation involves the biological reduction in toxic reactive oxygen species (ROS) produced by xenobiotics or their metabolism. Through a variety of mechanisms, exposure to several classes of environmental contaminants results in elevated concentrations of ROS, which include hydroxyl radicals, superoxide anion, hydrogen peroxide, and nitric oxide (Di Giulio et al. 1995, Hassoun and Stohs 1996, Kelly et al. 1998) (see also Chapter 4, 'Reactive Oxygen Species and Oxidative Stress'). In general, some xenobiotics can increase ROS production as a by-product of biotransformation reactions, via alterations in the electron chain components, or the inhibition of function or synthesis of antioxidant components (Lushchak et al. 2018). For example, inefficient use of oxygen and electron transfer during phase I, such as CYP-mediated biotransformation reactions, results in the formation of ROS. Quinone metabolites produced during CYP-mediated biotransformation of polycyclic aromatic hydrocarbons (PAHs) and other chemicals also contribute to the formation of ROS. Autooxidation (redox cycling) is posited as a central mechanism of toxicity for some pesticides (Lushchak et al. 2018). A variety of ROS-mediated toxicities, including lipid peroxidation of cell membranes and oxidative damage to proteins and DNA, can result when chemical exposures overwhelm antioxidant defense mechanisms.

Cells contain multiple antioxidant defense mechanisms to protect themselves from the potentially harmful effects of these unstable compounds. These defenses include a variety of enzymes (e.g., superoxide dismutase, catalase, glutathione peroxidase, and glutathione reductase) and small molecules, such as ascorbic acid and glutathione, involved in either ROS scavenging or transformation of ROS into harmless products. Together, these interacting components are referred to as the antioxidant system. The activation of the antioxidant system in response to many classes of xenobiotics has been reported in aquatic species including fishes (Silvestre 2020), and alterations in most components of the antioxidant system have been reported as correlates of toxicity resistance in fishes (Lushchak 2011b, Lushchak et al. 2018).

Briefly, antioxidant homeostasis involves the reduction of toxic ROS (via superoxide dismutase, SOD) to hydrogen peroxide, which can be further reduced to water, while concurrently oxidizing glutathione (GSH) to glutathione disulfide (GSSG) via glutathione peroxidase (GPx); GSSG may be reduced via glutathione reductase (GR) to GSH while oxidizing the reduced form of nicotinamide adenine dinucleotide phosphate (NADPH) to NADP+ (Lushchak et al. 2018). The antioxidant system is activated via the Nrf2/Keap1 pathway, composed of the transcription factor, nuclear factor E2-related factor 2 (Nrf2), and its cytosolic regulatory partner, Kelch-like ECH-associated protein (Keap1). Keap1, a redox-sensitive adapter, binds and inactivates Nrf2, except under ROS conditions, when Keap1 is oxidized and releases Nrf2, resulting in nuclear translocation. Thus, the Nrf2/Keap1 pathway regulates the expression of antioxidant response element (ARE)-driven genes including antioxidant system genes, GR and GPx, but also genes important in other toxicity resistance pathways, including some biotransformation enzymes (e.g., glutathione transferase) and multidrug resistance transporters.

## Integrated Defense Network

As described above, specific biochemical mechanisms that are largely evolutionarily conserved across fish species act as major contributors to toxicity resistance for certain categories of xenobiotics (Eide et al. 2021). Yet, a more complex understanding of the holistic biology of toxicity resistance is evolving. The chemical defensome (Stegeman et al. 2010) is a system of interacting toxicity resistance pathways regulated through mechanisms that include biochemical feedback and genomic crosstalk. Thus, this system includes not only the effectors of toxicity resistance, such as biotransformation enzymes, transporters, antioxidants, and metal-responsive genes, but also genomic regulatory elements such as receptors and transcription factors (Figure 7.5). Eide et al. (2021) systematically evaluated elements of the defensome to explain interspecific variation in the mechanisms and degrees of toxicity resistance to specific categories of xenobiotics in five fish species. For example, exposure to the PAH benzo(a)pyrene (BAP) can not only result in activation of the AHR pathway but also can result in increased ROS via products and processes of biotransformation; zebrafish embryos and larvae respond to exposure to BAP with different expression patterns for genes from multiple defense pathways.

An important mechanism by which xenobiotics induce defense responses is initiated via interaction with receptors, which (in the nucleus) can interact with specific genetic response elements that reduce toxicity by upregulating proteins that sequester or biotransform toxic parent compounds. Crosstalk between multiple defense pathways can also be affected by the presence of regulatory sequences of multiple, specific genetic response elements. Together, this defense system provides the potential for redundant and integrated responses to the multiple toxic effects induced by single classes of xenobiotics and the complex mixtures of realistic environmental exposures. One example involves cross-regulation between the antioxidant and AHR pathways, including genes that contain both regulatory sequences ARE and XRE, and response elements associated with both xenobiotic and antioxidant response pathways (Tonelli et al. 2018, Shankar et al. 2020, Silvestre 2020). Dietrich (2016) proposed that response integration could be mediated via coordinated induction of Nrf-dependent genes by AHR and Nrf2 (Figure 7.4).

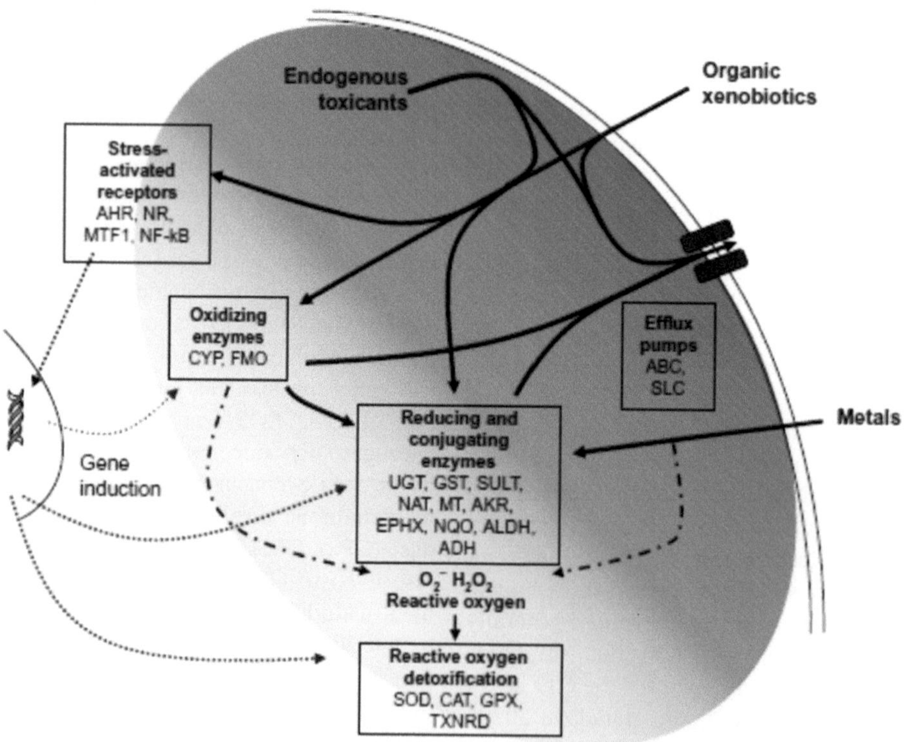


**FIGURE 7.5** Chemical defensome: This generalized scheme indicates the complex integration of pathways involved in oxidation, reduction, conjugation, and excretion of toxic compounds; both exogenous and endogenous toxicants are metabolized and/or excreted by members of the gene families detailed here. (nuclear factor kappa B (NF-kB), cytochrome P450 (CYP), flavoprotein monooxygenase (FMO), UDP-glucuronosyl transferase (UGT), glutathione-*S*-transferase (GST), sulfotransferase (SULT), *N*-acetyl transferase (NAT), metallothionein (MT), aldo-keto reductase (AKR), epoxide hydrolase (EPHX), NAD(P)H-quinone oxidoreductase (NQO), aldehyde dehydrogenase (ALDH), superoxide dismutase (SOD), catalase (CAT), glutathione peroxidase (GPX), ATP-binding cassette (ABC), solute carrier (SLC). Adapted with permission from Stegeman et al. (2010).

### Altered Biological Targets

The inherent sensitivity of a fish species to any class of xenobiotics is regulated by defense processes that alter the toxic concentration reaching the biological target, as well as the inherent sensitivity of the biological target(s). Genetic variants of biological targets may differ in their sensitivity to specific classes of xenobiotics; low-sensitivity variants may be beneficial under conditions of constant exposure. For example, one successful strategy for dichlorodiphenyltrichloroethane (DDT) resistance in insects involves modification of ion channels that represent toxicity targets for this pesticide, where adaptive mutations affect the ability of DDT to bind and/or inhibit the ion channel (ffrench-Constant et al. 2004, ffrench-Constant 2013). Similarly, evolved resistance to pyrethroid pesticide runoff in an aquatic crustacean is associated with mutations in voltage-gated sodium channels (VGSCs) that are normally inhibited (with toxic consequences) by these pesticides, and resistance to organophosphate (OP) pesticides is associated with mutations in acetylcholinesterase proteins that are normally inhibited by OPs (Weston et al. 2015, Fung et al. 2021). (Also see other contributors to resistance, e.g., Fung et al. (2021)). Rarely, regulatory changes in the expression of target (instead of changes in target functionality) have been associated with genetically evolved desensitization (Zhang et al. 2015). In addition to alterations in the apical biological target, desensitization may be achieved by variation in other molecular components that interact with the xenobiotic. In particular, tolerance might also be achieved by modification of a cellular receptor complex that normally recognizes and interacts with the chemical, triggering a signal transduction cascade that leads to a toxic response. In this case, modification of the receptor can result in reduced toxicant–receptor interaction, producing a concomitant reduction in toxicity. Alterations in one or more of the signal transduction components that become activated by toxicant–receptor interaction could also contribute to tolerance. The most well-studied example of evolved chemical tolerance in fish concerns the AHR and activation of its signal transduction pathway; further discussion on this example is included under Case Studies. It is important to note that in most cases described here the target site sensitivity is a genetically fixed feature of populations and species and is generally not an inducible defense, although Zhang et al. (2015) provide a rare example of regulatory changes in the expression of the biological target associated with insect pesticide tolerance.

## TOXICITY RESISTANCE TO THREE MAJOR POLLUTANT CLASSES

Toxicity resistance in fish has been explored in controlled laboratory studies and in field populations using a variety of approaches, experimental designs, and methodologies. As reported here, the study of toxicity resistance in fishes has expanded over recent decades from isolated observations of a few survivors in toxic agricultural ditches to sophisticated genome-wide mechanistic cross-species explorations. However, the scientific approaches that provide the strongest support for causality, especially with regard to genetic adaptation, have remained constant and may be useful recalling before reviewing these case studies.

Toxicity resistance is typically viewed by physiological toxicologists through the lens of defenses that are inducible by individuals following toxicant exposures, such as upregulation of detoxification. These inducible defenses may be transient or may enable longer-term acclimation that protects from subsequent exposures (acclimation) and may even extend to the protection of offspring. Comparative toxicologists typically search for mechanistic explanations for species variation in sensitivity to toxicants, which may include evolved variation in inducible defenses, but also evolved variation in the molecular targets of toxicants, which affects their sensitivity to perturbation; these are features of species that have evolved by natural selection and genetic drift over geological time. Eco-evolutionary toxicologists draw from the knowledge base established by physiological and comparative toxicologists and consider the mechanisms through which species may rapidly evolve resistance to anthropogenic chemicals in contemporary environments (evolutionary adaptation). Evolved resistance may be achieved through mutations that improve detoxification or that alter the biochemical targets of toxicants and thereby provide fitness advantages in habitats polluted with anthropogenic toxicants. It is important to note that distinguishing evolved resistance from exposure-induced acclimation is difficult, and sufficient evidence for concluding one over the other is often missing from many reports of elevated tolerance of animals captured from polluted sites compared with those from clean sites. To definitively distinguish acclimation from adaptation requires common-garden experiments where relative sensitivity to exposures between populations is compared, and tested animals should be at least one and ideally two generations removed from the field (Garland and Adolph 1991). Such studies may then be followed by physiological, biochemical, transcriptional, or genetic contrasts to offer insight into mechanisms that support evolved divergence in sensitivity.

Organized by chemical class, the following discussions describe toxicity resistance evaluated in laboratory- and field-based studies. Beginning with historical observations and continuing through to molecular, biological, and evolutionary mechanisms shown or proposed to contribute to toxicity resistance in fish populations, major xenobiotic classes represented are metals, pesticides, and polycyclic or polyhalogenated aromatic hydrocarbons (PHAHs), focusing on those with a shared, well-studied toxic mechanism.

### Metal Toxicity Resistance

Environmental exposures to high concentrations of metals, even essential metals that are key components of structural components and proteins, such as zinc and copper, can cause toxicity to fish through several mechanisms (Wood 2012, Tang et al. 2016, Castaldo et al. 2020). At the cellular level, transported metals can adversely affect ion (sodium, potassium, and calcium) homeostasis (Tang et al. 2016, Pavic et al. 2019, Castaldo et al. 2020). Thus, metal transporters such as copper transporter, divalent metal transporter, and copper transporting ATPases can limit metal uptake and toxic sequelae (Tang et al. 2016). Metal toxicity is also associated with oxidative stress, which can damage DNA, proteins, and membrane lipid homeostasis (Tang et al. 2016, Pavic et al. 2019, Castaldo et al. 2020). Some metals can induce oxidative stress by impairing electron transport chain processes or by acting as catalysts in ROS-producing reactions (e.g., Haber–Weiss and Fenton). In addition, metals may interact directly with sulfhydryl groups to deplete antioxidant defense components such as GSH (Benedetti et al. 2015).

Early evidence for toxicity resistance to metals in fish included laboratory exposures conducted to explain fish populations surviving chronic metal exposures. For example, pretreatment of rainbow trout eggs with zinc (Sinley et al. 1974) or with cadmium (Beattie and Pascoe 1978) increased the survivability of hatching and larval fish to subsequent metal exposures. Similar results were reported by Spehar and colleagues (1978) in studies with flagfish (*Jordanella floridae*). In another study, preexposure of larval rainbow trout to sublethal doses of cadmium produced fish with resistance to higher doses of cadmium (Pascoe and Beattie 1979). Further evidence of metal toxicity resistance began to appear in the literature shortly thereafter; for example, toxicity resistance to copper, arsenic, or cyanide was investigated in rainbow trout (Dixon and Sprague 1981a, 1981b). Preexposure to these toxicants resulted in enhanced survivability during subsequent, elevated exposures. Acclimation to copper was lost within 3 weeks of maintenance in clean water, indicating physiological acclimation. Numerous other accounts followed, describing acclimation responses following preexposure to copper, cadmium, mercury, zinc, aluminum, and selenium (Buckley et al. 1982, Duncan and Klaverkamp 1983, Orr et al. 1986), as well as mixtures of metals (Roch and McCarter 1984).

Dixon and Sprague (1981c) provided some of the first mechanistic evidence for the involvement of metal-binding proteins in physiological acclimation to metals. In that study, the authors reported increased synthesis of a low-molecular-weight soluble protein in the liver of copper-tolerant fish. Other investigators subsequently addressed the specific involvement of MTs and concluded that these proteins play important or dominant roles in reducing metal toxicity (Roch and McCarter 1984). In one study, McCarter and Roch (1983) demonstrated a strong relationship between $LC_{50}$ values for copper in Coho salmon and hepatic MT levels.

Benson and Birge (1985) provided some of the most convincing evidence for the role of MTs in physiological acclimation to metals. Fathead minnows (*Pimephales promelas*) collected from a metal-contaminated fly ash pond were much more tolerant of cadmium and copper than hatchery-raised fathead minnows. The maintenance of the metal-resistant individuals in clean water resulted in a significant decline in tolerance within a week. Conversely, tolerance in hatchery-raised individuals increased following exposure to sublethal concentrations of cadmium and declined within a week following their return to clean water. Throughout these experiments, a strong relationship was demonstrated between changes in metal tolerance and expression in the gills of MT or some other metal-binding protein.

In field investigations, Weis and Weis (1977b, 1977a) studied the effects of methylmercury (meHg) and inorganic mercury on embryonic development in killifish. In early studies, they used eggs from the relatively clean estuarine environment of Montauk and Southampton on Long Island, New York. They reported that both forms of mercury produced a variety of teratogenic effects, with meHg being more toxic. Later, these authors reported that significant variation in embryonic meHg tolerance occurred between batches of eggs produced by different females (Weis et al. 1982). Since heritable variation in tolerance is necessary for tolerance to evolve, these data provided some support that Hg tolerance was 'evolvable' in killifish. Furthermore, embryos hatched from the eggs of females from a contaminated estuary (Piles Creek, a tributary of the Arthur Kill in Linden, New Jersey) were much more tolerant of meHg than were embryos from a relatively unpolluted habitat (Weis et al. 1981). These authors presented several pieces of evidence, suggesting that meHg resistance in Piles Creek fish might have an inherited component (Weis et al. 1982).

Adeyemi and Klerks (2013) conducted laboratory metal preexposures to investigate mechanisms of acclimation in the least killifish (*Heterandria formosa*). They not only concluded that MT induction was associated with observed toxicity resistance in copper-acclimated fish but also noted that other factors that play a role in altered copper accumulation dynamics may play a role in the resistance. For example, Grosell et al. (2001) reported enhanced biliary copper excretion in copper-acclimated rainbow trout. Other fish studies (e.g., as reviewed in Webster et al. (2013)) have shown that metal exposure-induced increases in the expression of MTs and other genes encoding metal-specific transporting proteins and ATPases are associated with enhanced metal metabolism, storage, and excretion, supporting their protective role in metal toxicity.

MTs have been implicated in many studies as key components in acute metal toxicity resistance, but other defense pathways have also been implicated. For example, Zhu et al. (2018) reported that Cd exposure to zebrafish resulted in upregulation of mRNA and protein levels of MTs, speculating that the exchange of toxic metals for co-factor metal ions may trigger the upregulation of MTs through the MTF1/MT pathway. They also noted that transcriptional upregulation of Cu and Zn transporters might also play roles in defense against metal toxicity. In addition to MT and metal transporters, metal-induced upregulation of components of the antioxidant system, i.e., glutathione and dependent enzymes such as GR, GPx, and GST, has been reported in many laboratory studies of metal exposure in diverse fish species (Srikanth et al. 2013). Other studies have associated tolerance following short-term, laboratory exposures with altered expression of genes encoding metal-specific transporting proteins and ATPases (as reviewed in Webster et al. (2013)). Together, these findings suggest that fish toxicity resistance strategies following short-term exposures include those that limit the uptake of metals from the environment, slow onset of toxicity, and increase metal metabolism, storage, and excretion.

Furthermore, several studies reported enhanced protective responses by metal preexposures via mechanisms that include but are not limited to MT. Relative to intermittent or closely spaced exposures, MTs synthesized in response to a metal exposure can be maintained for several weeks (Castaldo et al. 2020). Several studies support MT induction as one mechanism by which prior exposures may enhance resistance to later exposures and/or alternate stressors (McGeer et al. 2007, Adeyemi and Klerks 2013). Zheng et al. (2016) reported protective acclimation responses in large yellow croaker (*Pseudosciaena crocea*) from zinc preexposures in the laboratory; their results suggested that Nrf2 played an important role in regulating the coordinated upregulation of components of the antioxidant system, such as Cu/Zn-SOD, CAT, and GPx and GR proteins. Furthermore, Wang et al. (2019) reported that Cd toxicity resistance in zebrafish was significantly enhanced by Zn preexposure that they termed 'priming hormesis' through multiple stress response pathways, as demonstrated by the upregulation of transcription factors (hsf1, hsf2, and mtf-1, and nrf2) and their target genes (sod1, cat, hsp70, and mt2) and the downregulation of several metal transport genes. Taken together, studies of metal-exposed fish suggest multiple interacting protective responses that may sequester metals from targets and protect against oxidative stress. This more complex response to metal exposures that reflects coordinately regulated feedback-driven response systems has been referred to as the defensome (Figure 7.5).

The defensome concept is not meant to suggest a fixed response network; rather, the importance of strategies that mitigate metal toxicity has been shown to vary by fish species and life stages. Tang et al. (2016) noted that in Persian sturgeon (*Acipenser persicus*) the induction of MT expression occurred in a concentration-dependent manner, predominantly at later life stages. In addition, there was an inverse relationship between MT expression and oxidative stress, measured as the concentration of lipid peroxidase (LPO) across all three life stages exposed to copper, supporting the importance of MT, both in sequestering Cu to reduce oxidative stress and in scavenging ROS to protect sturgeon against oxidative damage. Other mechanisms likely contribute to differences in metal sensitivities among life stages; for example, in adult fish, the upregulation of metal transporters might be an important compensatory mechanism that slows the uptake of metals (Webster et al. 2013). Shekh et al. (2020) investigated the basis for variation in metal sensitivity among fish species and life stages. They concluded that while early (*versus* later) life stages are more metal-sensitive, white sturgeon (*Acipenser transmontanus*) are more susceptible to copper damage (via disruption of ion homeostasis) than rainbow trout (*Oncorhynchus mykiss*). From their measurements of indicators, they infer that sturgeon increasingly rely on MT to neutralize Cu-induced oxidative stress and/or Cu loading, and trout predominantly use the antioxidant, GSH, to combat Cu-induced oxidative stress as they mature.

Laboratory studies have shown that the progeny of metal-exposed maternal fish sometimes demonstrate increased metal tolerance, suggestive of trans- or multigenerational toxicity resistance. For example, tilapia (*Oreochromis mossambicus*) hatchlings from Cd-treated mothers were more tolerant to Cd (Lin et al. 2000) as were the progeny of fathead minnow (*Pimephales promelas*) exposed to copper (Peake et al. 2004). Mechanistically, this tolerance was hypothesized to be caused by maternal transfer of MT mRNA or MT protein (Lin et al. 2000, Peake et al. 2004), although maternal deposition of metals per se could have primed a defensive response in progeny. It is worth noting that trans-generational differences may reflect genetic and/or epigenetic changes, such as alterations of the regulation, function, or copy number of defense genes such as MTs. Individual variation in metal sensitivity to manganese, independent of prior metal exposure, was reportedly transmitted to progeny in European whitefish (*Coregonus lavaretus*) (Arola et al. 2017). Studies like these build evidence toward understanding defense mechanisms; i.e., when variation in metal sensitivity is heritable, it may serve as the genetic foundation for adapted fish population that arise in the wild from the selection of most tolerant progeny.

Some general mechanisms of tolerance have been evaluated in fish populations subject to multigenerational exposures to metals. For example, selection for cadmium resistance was evaluated following multiple generations of laboratory exposures of the least killifish (Xie and Klerks 2003). Inherited resistance was observed: MT-like protein levels were elevated in cadmium-tolerant lines and hypothesized to play an important role in the observed tolerance and fitness costs (Xie and Klerks 2003). More often, multigeneration tolerance has been studied in wild populations inhabiting metal-contaminated environments, often with expectations that enhanced metal protective systems play an important role in population persistence. For example, Goto and Wallace (2010) examined killifish living in metal-polluted sites, focusing on the occurrence and inferred protective value of intracellular metal partitioning by MTs and other proteins. Varying responses to metals were also investigated in wild three-spined stickleback (*Gasterosteus aculeatus*) from two rivers with different histories of metal (and estrogenic) contamination (Webster et al. 2017). Laboratory exposures of progeny raised under clean conditions showed no differences in estrogen responsiveness but altered metal responses. Progeny from both populations responded to laboratory exposure to copper by upregulation responses in genes for both general stress, such as heat shock proteins and other molecular chaperones, and metal-specific genes, such as MTs, and downregulated genes for translation and energy storage functions. However, no mechanisms were proposed for the modestly elevated baseline and lower metal-induced response in the expression of MT in the historically metal-exposed population.

Other studies of wild fish populations residing in metal-contaminated waters have followed up observational studies with genomic studies designed to provide mechanistic insight into the role of altered metal defense in population persistence. For example, Pierron et al. (2009) described variation in the expression of MTs and other metal-responsive proteins in yellow perch (*Perca flavescens*) populations that were proportional to the level of industrial or mining activities and associated contamination in residence lakes. A genome-wide examination of protein sequences showed evidence for selection in these perch populations resident to metal contamination, but, surprisingly, metal defense genes, such as MT, were not among those under selection (Belanger-Deschenes et al. 2013). However, this study did not examine variation in regulatory sequences (promoters and enhancers), which may have contributed to adaptive variation in MT expression. Those authors inferred that the genes found under selection were consistent with adaptation to life in resource-poor environments, highlighting the costs of toxicity resistance (as discussed later in this chapter).

In another example, Webster et al. (2013) examined wild brown trout (*Salmo trutta*) populations using whole-transcriptome profiling. Here, MTb (MT isoform b) was among the most strongly upregulated genes, suggesting MT sequestration as an important mechanism of metal tolerance in this wild population, while the upregulation of the antioxidant system seemed to play a relatively minor role. In a subsequent study, Paris and Usher (2019) used a yeast (*Saccharomyces cerevisiae*)-based synthetic genetic array system to explore the role of MTs in conferring metal tolerance in wild brown trout. In particular, they used DNA from trout resident to a relatively clean (Camel) and a highly metal-polluted (Hayle) river to amplify MTa and MTb and then real-time PCR to determine that the genes were being expressed alongside a metal-tolerant phenotypic screen. They showed that the increased expression of MTa from both populations resulted in an increase in tolerance to low concentrations of metals. However, increased expression of the MTb Hayle, only, was associated with increased tolerance to high concentrations of some metals (Figure 7.6). Furthermore, there were pronounced differences in genetic

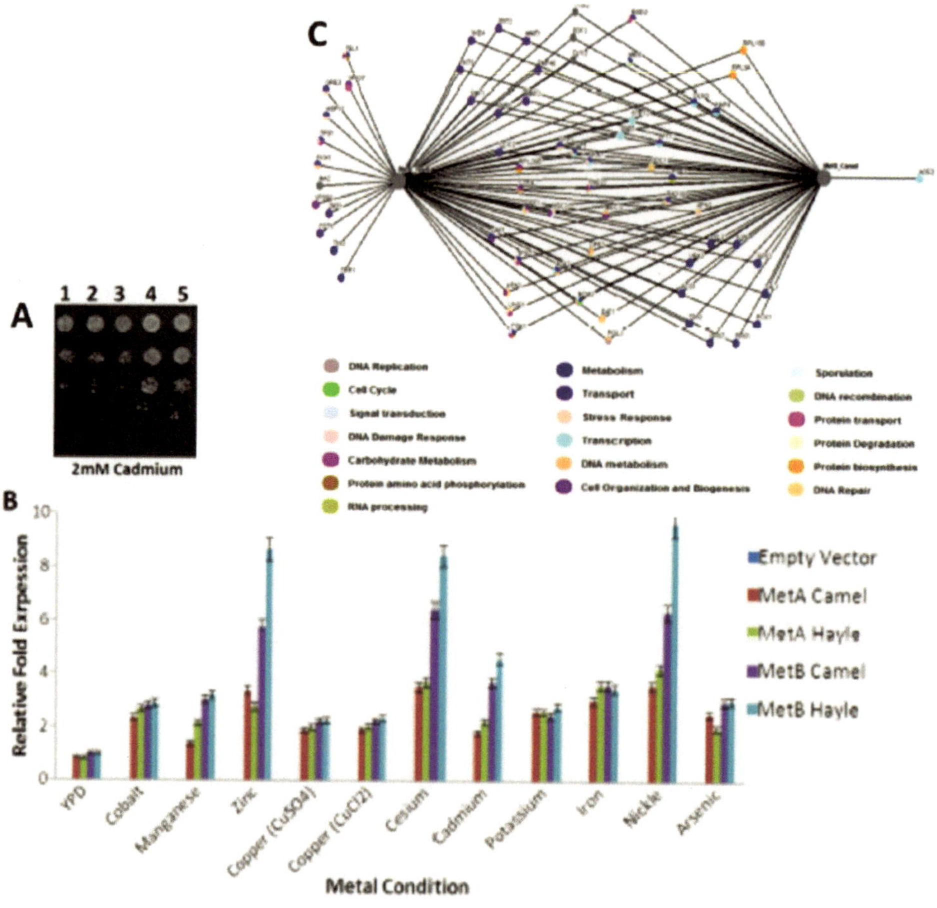


**FIGURE 7.6** Metallothionein (Met) gene isoform induction and interactions. (A) Metal-tolerant phenotypic screen (survival = white) for empty vector (1), MetA Camel (2), MetA_ Hayle (3), MetB_Camel (4), and MetB_Hayle 5). (B) Levels of expression of MetA and MetB from the rivers Camel and Hayle in a yeast-based expression system. (C) Genetic interaction network of MetB from Camel and Hayle trout; genes are represented by nodes that are color-coded according to their Saccharomyces Genome Database (SGD) cellular roles (www.yeastgenome.org). Interactions are represented by edges. Adapted from Paris and Usher (2019), open access.

interaction maps for MTb from Camel and Hayle (Figure 7.6): MTb Camel had only one unique interaction, while MTb Hayle had many interactions unique from those of MTb Camel. These differences in interactions suggested functional, regulatory, or copy number changes in the MT isoform associated with an evolved metal tolerance to metal in historically metal-exposed trout (MTb Hayle). These findings are consistent with many studies demonstrating the importance of MTs in protective responses to metal exposures in fishes, but more specifically link MTs to complex, adaptive metal tolerance in wild trout.

### Pesticide Toxicity Resistance

Organochlorine (OC) pesticides are a class of insecticides for which toxicity resistance in target (insects) and nontarget species was first reported. DDT, the first highly effective, modern OC pesticide, came into widespread use shortly after World War II (Hites and Eisenreich 1987, Ware 1991). Early results indicated exceptional performance in diverse agriculture, forestry, and human health-related applications. However, DDT accumulated in food chains and was distributed globally via atmospheric transport and other processes. In 1972, when DDT was banned in the United States, other organochlorine pesticides, structurally similar to DDT, were either in production or were produced shortly thereafter. These include chlordane, aldrin, dieldrin, heptachlor, toxaphene, and chlordecone (Kepone), all of which were subsequently banned in the United States because of problems such as those associated with DDT. Currently, in 2021, many of these pesticides are still being used worldwide.

DDT and other OC pesticides share properties of poor water solubility, high lipid solubility, slow rates of decomposition, and high bioaccumulation. Although they exhibit a broad range of well-documented toxicities, they are best known as potent neurotoxicants that are designed to selectively perturb specific molecular receptors within the central nervous system, although those target receptors can vary between chemical classes. Organochlorine pesticides may affect the function of sodium and calcium channels and transporters and interfere with γ-aminobutyric acid (GABA) neurotransmission by blocking specific GABA receptors (Casida 2009, Casida and Durkin 2013). OC pesticides have been reported to act as environmental estrogens that influence reproductive function via interference with numerous pathways including the production, release, transport, metabolism, action, or elimination of hormones (Lundholm 1997, Rosselli et al. 2000, Chedrese and Feyles 2001, Lushchak et al. 2018).

Other classes of pesticides, including organophosphates, carbamates, and pyrethroids, were also designed as neurotoxicants in target species, yet their mechanisms of toxicity are often conserved in nontarget species including fishes (e.g., Bharati and Saha (2021)). For example, organophosphates such as malathion and chlorpyrifos and carbamate insecticides are extremely neurotoxic as potent cholinesterase inhibitors that bind with the active site of acetylcholinesterase and prevent its ability to function in the normal catabolism of neurotransmitters. Pyrethroids such as permethrin and cypermethrin are neurotoxic via their ability to alter sodium channels and cause repetitive nervous discharges or membrane depolarization (Katagi 2010, Lushchak et al. 2018, Lu et al. 2021). In addition to the intended targeted action of pesticides, exposures also often produce toxicity through more generalized mechanisms. Oxidative stress, generated by a variety of mechanisms, may be an important contributor to the toxic outcomes of pesticide exposures for nontarget species (Lushchak 2011b, 2016, Wang et al. 2016, Lushchak et al. 2018). For example, pesticides may increase ROS production by direct interference with the electron transport chain, as an outcome of pesticide autooxidation (redox cycling), or as by-products of biotransformation. Pesticides may also disrupt ROS homeostasis by inhibiting the biosynthesis or the function of antioxidants and associated enzymes of the antioxidant system. For example, carbamates have been reported to limit the responsiveness of the antioxidant system by inhibiting glutathione reductase, which results in increased levels of GSSG, and inactivating the nuclear transcription factor nuclear factor kappa B (NF-κB) (Lushchak et al. 2018). Thus, pesticide-mediated toxicity may be maintained by the production of ROS and a compromised antioxidant system that cannot restore homeostasis.

Like pesticide toxicity, there are strong economic and human health incentives to understand the biochemical and evolutionary mechanisms of insecticide resistance in insects (e.g., Muhammad et al. (2021)). The biochemical and genetic mechanisms that permit adaptive or genetically evolved tolerance in wild insect populations have been shown to vary widely mechanistically and are often capable of developing over a relatively low number of generations. Common biochemical mechanisms of toxicity resistance include both increased metabolic detoxification and decreased sensitivity of biological targets (ffrench-Constant et al. 2004, Bass and Field 2011, Liu 2015, Hawkins et al. 2019). Enhanced biotransformation in insects is often accomplished by overexpression of detoxification or mitigation genes, which can evolve via alterations in regulatory or coding loci and/or amplification (two or more copies) of structural loci (Liu 2015, Bharati and Saha 2021). Some of the biochemical mechanisms of pesticide defenses may be conserved across target and nontarget species and are therefore relevant to our understanding of pesticide toxicity resistance in fishes.

Several processes that mitigate pesticide toxicity are induced by pesticide exposures. For example, pesticide exposures may induce biotransformation enzymes that detoxify toxic parent compounds, replacing them with less toxic metabolites and/or increasing the elimination of parent and metabolite compounds. For example, organochlorine pesticides may be reductively dehalogenated, and organophosphates and pyrethroids may be hydrolyzed by esterases (Lushchak et al. 2018). The three main enzyme families that have been implicated in pesticide toxicity resistance are esterases, GSTs, and CYP MOs (P450s) (Bass and Field 2011, Lu et al. 2021). In addition, pesticide-mediated ROS production can induce antioxidant defenses in exposed fish (Lushchak 2016). In particular, Lu et al. (2021) noted that enhanced induction and constitutive overexpression of P450 genes, especially in the CYP6 family (found in insects and not fishes, but possibly related to CYP families 3 and 5 in mammals and fish (Goldstone et al. 2010, Uno et al. 2012, Pan et al. 2016)), are frequently found in insecticide-resistant insect strains. Bass and Field (2011) noted that resistance to pesticides with ester bonds (organophosphates, carbamates, and pyrethroids) is often related to enhanced production of esterases, most commonly via gene amplification. Toxicity resistance to these same pesticides has also been reported to occur via increased production of GSTs, which may act in antioxidation and in phase II metabolism (Lu et al. 2021). The evolution of increased expression of detoxification genes may be achieved through mutations in the proximal promoter (e.g., Daborn et al. (2002)), by gene duplication (e.g., Zimmer et al. (2018)), or by loss-of-function mutations in repressors of transcription (e.g., Grant and Hammock (1992)). Upregulation of the efflux transporters, Pgps, has also been implicated in toxicity resistance among insects to pesticides, including pyrethroids such as cypermethrin and fenvalerate, the organophosphate methyl parathion, and the carbamate thiodicarb (reviewed in Clark and Di Giulio (2012)).

Additionally, alterations in target sites contribute to resistance in several arthropod species. Bass and Field (2011) noted that several highly conserved mutations have been described that give varying degrees of insensitivity, e.g., acetylcholinesterase for organophosphates or carbamates, the VGSC for pyrethroids, and the GABA receptor for cyclodienes. Furthermore, there are predominating mechanistic patterns among pesticide classes, such as mutations in the target site encoding genes (fungicides), polygenic metabolic resistance (herbicides), and insecticide resistance reflecting alterations in the target site or major metabolic resistance genes (Hawkins et al. 2019). Furthermore, Liu (2015) summarized that pesticide toxicity resistance in insects is not only conferred via multiple resistance mechanisms but also mediated through the interaction of regulatory and structural elements of complementary xenobiotic defense systems.

A large body of evidence has accumulated on mechanisms of pesticide tolerance in nontarget aquatic species other than fishes. For example, a series of studies was conducted for about a decade to characterize the underlying mechanism(s) of pesticide tolerance in wild populations of the aquatic invertebrate, *Hyalella azteca*. An initial population identified as thriving in an agricultural creek with high sediment levels of the organophosphate pesticide, chlorpyrifos, was investigated for acquired genetic resistance or evolved tolerance to the pesticide. Pyrethroid resistance was primarily attributed to the heritable mutation at a single locus of the VGSC and was associated with up to 550-fold increased pesticide tolerance in resistant populations compared with susceptible strains (Weston et al. 2013). Furthermore, Major et al. (2018) surveyed additional populations of *H. azteca* and identified a common mutation in all populations resident at highly pesticide-contaminated sites; a specific amino acid substitution was observed that has been shown to confer organophosphate and carbamate resistance in mosquitoes and has been associated with resistance in other insects. Furthermore, multiple copies of the resistance gene were found in the most tolerant individuals (Major et al. 2018). However, reduced tolerance (but not complete sensitivity) in laboratory-reared offspring suggested alternate mechanisms for tolerance other than the inherited genetic changes already identified. In particular, Fung et al. (2021) found that enhanced enzyme-mediated detoxification through CYP MOs and general esterases contributed to tolerance in the parental generation, but was not inherited by the F1 generation. These findings suggest that both acclimations and adaptations contribute to the pesticide tolerance observed in some populations of *H. azteca* (Fung et al. 2021).

Amphibian pesticide tolerance has also been often studied extensively. For example, Hua et al. (2015b) reviewed examples of toxicity resistance in amphibians, such as leopard frogs (*Lithobates sphenocephalus*) tolerant to carbaryl and wood frogs (*Lithobates sylvaticus*) tolerant to insecticides such as chlorpyrifos, carbaryl, and malathion. Interestingly, only a few amphibian studies have focused directly on the biochemical mechanisms of pesticide tolerance (Hua et al. 2013). However, some mechanistic information has been inferred from studies of cross-tolerance, within groups of pesticides with similar mechanisms of toxicity (Hua et al. 2013, 2015b). Also, several amphibian studies have examined the role of generalized stress responses in toxicity resistance. For example, Jones et al. (2017) observed enhanced toxicity resistance to carbaryl when introduced with predator cues and hypothesized the mechanism as a generalized stress-induced enhanced metabolism. Similarly, Billet and Hoverman (2020) observed general stress responses associated with delayed lethality to carbaryl in wood frogs (*Rana sylvatica*). They speculated on multiple biochemical mechanisms of glucocorticoid-mediated pesticide toxicity resistance such as upregulation of the activity of cytochromes P450s or increased production of AChE via the glucocorticoid regulatory response element in the AChE gene. These studies suggest that multiple biochemical pathways may contribute to pesticide toxicity resistance in amphibians and perhaps other nontargeted species, including fishes, as well.

One of the earliest and most influential reports of pesticide toxicity resistance in fishes occurred in response to DDT exposures. Resistance to DDT was initially viewed as a curious agricultural problem restricted to insects (Brown 1958); however, DDT resistance in populations of mosquitofish was reported shortly thereafter (Ferguson 1963, Vinson et al. 1963). Subsequent accounts described populations of mosquitofish that were resistant to other commonly used organochlorine, organophosphate, and pyrethroid insecticides (Boyd and Ferguson 1964, Ferguson and Bingham 1966, Culley and Ferguson 1969, Finley et al. 1970, Fabacher and Chambers 1973). The development of or cross-resistance among pesticides was commonly reported (Culley and Ferguson 1969, Ferguson 1969, Fabacher and Chambers 1973, Andreasen 1985). Other fish species, including green sunfish (*Lepomis cyanellus*), bluegill sunfish (*Lepomis macrochirus*), golden shiner (*Notemigonus crysoleucas*), and yellow bullhead (*Ictalurus natalis*) inhabiting polluted habitats, exhibited chemical resistance, as well (Ferguson et al. 1964, Ferguson and Bingham 1966). In laboratory studies, guppies (*Poecilia reticulata*) exposed to sublethal concentrations of pentachlorophenol developed resistance similar to that observed in natural populations collected from field sites (Norup 1972). More recent research on fish has focused on the toxicological effects of pesticide exposures, and studies on wild fish population demonstrating toxicity resistance have been limited (Klerks et al. 2011). For example, Nusbaumer et al. (2021) found sensitivities to a pesticide varied among populations of brown trout (*Salmo trutta*), but they were unrelated to historical pesticide exposure.

A variety of mechanisms of resistance to organochlorine pesticides in fish have been proposed in response to historical observations of pesticide toxicity resistance in fish. Fabacher and Chambers (1973) reported that mosquitofish resistant to the insecticide endrin had higher lipid content than sensitive populations and speculated that sequestration of toxicants in lipid could reduce toxicity and contribute to resistance. Pyrethroid-resistant fish exhibited higher mixed-function oxidase (CYP-dependent biotransformation) activity (Fabacher and Chambers 1973). In that study, laboratory-reared offspring of resistant fish exhibited some degree of tolerance. Although CYP activity was not measured in offspring, the authors speculated that elevated CYP might be an inherited trait and a contributor to the apparent genetic adaptation. They also suggested that desensitization of nervous tissue plays a role in resistance, although no supporting data were presented. In another study, Andreasen (1985) reported that the laboratory-reared F1 progeny of toxaphene-resistant fish maintained a pronounced resistance. The author suggested that resistance was due to the natural selection of a genetic trait present in the fish population and that elevated mixed-function oxidase (CYP) may contribute to resistance; however, resistance in generations beyond F1 was not measured, leaving open the possibility of the involvement of epigenetic mechanisms, maternal inheritance of genetic material, or maternal transfer of the toxicant to progeny. Weis and Weis (1989) reviewed much of the early literature on toxicity resistance and concluded that activation of the CYP system during toxicant exposure may contribute to physiological acclimation in some cases. Bonner and Yarbrough (1989) provided some evidence that organochlorine-resistant fish may possess an effective central nervous system barrier, leading to reduced target tissue exposure.

More recently, Clark and Di Giulio (2012) examined pesticide toxicity resistance and the role of adaptive variation in the regulation of the CYP system in Atlantic killifish inhabiting a highly PAH-contaminated site, Elizabeth River (ER) (VA, USA). They hypothesized that the well-studied constitutively downregulated CYP system in ER killifish, critical to their evolved resistance to organic xenobiotics (see later section), should produce predictable changes in their responsiveness to pesticides. They predicted that toxicity in ER killifish should be reduced relative to reference site killifish upon exposure to pesticides for which CYP-mediated metabolites are more toxic, represented by the organophosphate chlorpyrifos; conversely, ER killifish toxicity should be relatively increased for pesticides for which CYP-mediated metabolites are less toxic, represented by the pyrethroid permethrin. However, their results were mixed relative to predicted responses, suggesting that evolved tolerance in ER killifish might reflect multiple interacting adaptive changes. Furthermore, they cite other studies of ER killifish showing upregulation of other enzymes and proteins, including phase II enzymes (such as GST) and proteins of the antioxidant system, and hypothesize their roles in pesticide toxicity resistance. Furthermore, they hypothesized that elevated levels of the efflux transporter Pgp in ER fish may play a role in demonstrated toxicity resistance.

A transient, within-generational response to pesticide toxicity resistance, termed 'detoxification pathway priming', was reported in Gulf killifish (Oziolor et al. 2017). In particular, low-level carbaryl exposure in embryonic fish elicited a stronger-induced response to a lethal concentration of carbaryl days later. Mechanistically, this toxicity resistance was largely attributed to a short-term enhancement of biotransformation of carbaryl via the AHR pathway-mediated upregulation of CYP1A. However, the contributions of other systems, associated with or independent of AHR pathway activation, were not ruled out. In fact, lessons learned from pesticide toxicity resistance in insects suggest that complex and interactive defense responses are likely in fish and other nontarget species.

## PAH/PHAH Toxicity Resistance

PAHs, produced during the combustion of fossil fuels in industry, internal combustion engines, and home heating, are the most abundant toxicants in the marine environment (Manahan 1999). They are also important components of petroleum and can enter the marine environment through marine transportation activities, accidental spills, improper disposal of used crankcase oil, urban runoff, and numerous other routes, including creosote use and spills (O'Connor and Huggett 1988, Huggett et al. 1992, Huntley et al. 1995). However, some of the most persistent environmental contaminants are members of a large class of compounds referred to as PHAHs (Rappe 1991, Fletcher and McKay 1993, Eisler and Belisle 1996). In the 1960s, during extensive investigations of chlorinated pesticides in fish and wildlife, chemical analysis often revealed the presence of similar, although unidentified, compounds in environmental samples. These compounds, later identified as PCBs, are among the most abundant and widespread contaminants in our environment. PCBs are mixtures of synthetic industrial compounds that were produced for use as plasticizers, dielectric fluids in capacitors and transformers, flame retardants, and many other purposes. Approximately 102 PCB congeners, or chemical forms, exist in the environment. Congeners are identified by the position and number of chlorine atoms and by the spatial orientation (coplanar vs. noncoplanar) of the phenyl rings. Most toxicological research focuses on coplanar congeners whose toxicological potency exceeds that of noncoplanar congeners although coplanar congeners are a very small proportion by weight of PCB mixtures. The mode of toxicity of coplanar PCBs allows them to be grouped within a broad, mechanistic-based class of environmental contaminants referred to as DLCs. DLCs are named for their toxicological similarity to dioxin (TCDD), the most potent member of the DLC family (Safe 1986). Further discussions will focus on the highly toxic DLC family, which includes coplanar PCBs, polychlorinated dibenzodioxins (PCDDs), and polychlorinated dibenzofurans (PCDFs), and whose mechanisms of toxicity and toxicity resistance have been extensively studied.

### DLCs: AHR-Mediated Toxicity Resistance

DLCs are extremely toxic to vertebrates including fishes, with adverse effects noted for the most important biological processes, i.e., growth, development, immunity, and reproduction, which are reviewed in detail elsewhere (Peterson et al. 1993, King-Heiden et al. 2012, Shankar et al. 2020). In adult vertebrates, including fish, DLC effects are characterized by 'wasting syndrome', reduced feeding and extreme weight loss, and reduced reproduction, associated with altered steroid hormone homeostasis, delayed maturation, and reductions in female reproductive quantity and quality. Although broad variation in sensitivity among fish species has been noted (see Figure 7.2 and associated discussion), fish are most sensitive to the toxic effects of DLCs during development (King-Heiden et al. 2012). Typically, embryos develop normally through primary organogenesis and then, as development proceeds, exhibit increasing malformations, particularly to the cardiovascular system and cartilaginous structures in the craniofacial region, observed as yolk sac edema, hemorrhage, and craniofacial malformations, resulting in stunted growth and post-hatch mortality (Spitsbergen et al. 1991, Walker and Peterson 1991, Peterson et al. 1993, Carney et al. 2006, Tillitt et al. 2008, King Heiden et al. 2009).

In vertebrates, most of the toxic effects of DLCs and some of the toxic effects of PAHs are believed to be mediated by their binding to the AHR and the activation of the AHR signal transduction pathway (Birnbaum 1994, Kerkvliet 1995, Thurmond et al. 1999, Bunger et al. 2003, King-Heiden et al. 2012, Shankar et al. 2020). There is a close relationship between DLC–AHR strength of binding and degree of toxicity of individual DLCs; for example, the most toxic DLC known, TCDD, binds with a higher affinity than other DLCs. PCB 126 (3,3′,4,4′,5-pentachlorobiphenyl) and PCB 77 (3,3′,4,4′-tetrachlorobiphenyl) are among the most potent AHR agonists and most toxic PCB congeners. Invertebrates lacking an AHR that binds DLCs (Hahn 1998a) or strains of mice that carry a defective AHR (Fernandez-Salguero et al. 1995, Fernandez-Salguero et al. 1996) are relatively insensitive to DLCs. Two forms of AHR (AHR1 and AHR2) have been identified in some fish species, including Atlantic killifish for which DLC-adapted populations have been identified (Hahn et al. 1997, Karchner et al. 1999). In their review, Shankar et al. (2020) noted that additional duplicates of AHR1 and AHR2 have resulted from a fish-specific whole-genome duplication; retention of these gene duplicates in modern fishes may represent specialization of the multiple functions of their common ancestor or the evolution of new functions. In many fish species, AHR2 is considered the functional receptor for xenobiotic interactions; however, specific functionalities may differ by ligand, species, and toxic endpoint, e.g., Clark et al. (2010) and Goodale et al. (2015). Recently, Zhang et al. (2019) investigated the genetic basis for fish species variation in sensitivity to AHR-mediated toxicity by conducting a comprehensive, systematic study of insensitivity in the freshwater climbing perch (*Anabas testudineus*). Their analysis of elements of the AHR pathway revealed that (unlike in some other fish species) AHR2a was not functional in response to TCDD, while AHR1a, AHR1b, and AHR2b were functional as were ARNT (partner of AHR) and AHRR (repressor of AHR). Furthermore, they localized the TCDD insensitivity to the AHR2a ligand-binding domain and they speculated about the functional extent of this alteration, such as binding to other xenobiotic ligands.

While functionalities associated with specific ligands or agonists may differ upon AHR binding, it is generally useful to consider how the subsequent signal transduction process can contribute to toxicity resistance. Of the genes whose regulatory elements contain XREs, the biotransformation enzymes (described earlier in this chapter; for more details, see Chapter 2) can play a central role in both resistance and toxicity. One of the most well-documented toxicity resistance responses to organic xenobiotics is the AHR-mediated elevation or induction of specific CYP forms and concomitant increases in rates of CYP-mediated catalytic activity (Bucheli and Fent 1995). CYP1A is the major CYP form induced by exposure to specific PAHs and DLCs in fish (Stegeman and Hahn 1994). CYP1A levels are usually estimated via immunodetection or by measurement of ethoxyresorufin-O-deethylase (EROD), a CYP1-dependent activity (Loerracher and Braunbeck 2021). CYP1A induction is the most sensitive and best-characterized response to both DLC and PAH exposure and is often used as a biomarker of exposure to both classes of compounds (Stegeman and Hahn 1994, Whyte et al. 2000). However, early reports began to reveal suppression of CYP1A in some fish populations subject to chronic pollution. For example, perch (*Perca fluviatilis*) collected from Lake Johnson, a Swedish lake heavily contaminated with

PCBs, exhibited low levels of CYP1A and EROD compared with those collected from clean sites (Celander and Förlin 1995, Celander et al. 1996). Unlike fish from reference sites, the treatment of Lake Johnson perch with PCB 77 or β-naphthoflavone (BNF; a PAH and AHR inducer) did not result in detectable elevation of CYP1A (Förlin and Celander 1995). Wirgin et al. (1992) reported that CYP1A in PCB-contaminated tomcod was refractory to induction by PCB 77 but not BNF. These investigators suggested that prior exposure to PCBs but not PAHs affects the subsequent response of CYP1A to DLCs and that differential responsiveness to PAHs and DLCs might also indicate the existence of multiple molecular pathways for the induction of CYP1A. In the laboratory, exposure to high doses of DLCs and PAHs was reported to suppress CYP1A activity in multiple fish species (Willett et al. 1998), perhaps through toxicant binding to active CYP1A sites or other mechanisms (Stegeman and Hahn 1994). Prolonged (15-week) exposure of rainbow trout to PCBs yielded fish that were refractory to subsequent CYP1A induction by PCBs and the PAH 3-methylcholanthrene (Celander and Förlin 1995, Celander et al. 1996). Suppression of CYP1A reported in these and other studies (Gooch et al. 1989, Monosson and Stegeman 1991, White et al. 1997) was suggested as a form of physiological acclimation to PCBs and other compounds (Hahn 1998b). CYP1A was thought to play an integral role in AHR-mediated toxicity and toxicity resistance. CYP1A is now known not to be causal to developmental TCDD toxicity, while the role of CYPs in PAH-mediated developmental toxicity is still not well understood (Goodale et al. 2015, Shankar et al. 2020). However, despite the observed suppression of CYP in PAH-tolerant fish populations, inhibition of CYP, either chemically or by gene knockdown, actually dramatically increased the DLC-like developmental toxicity of some PAHs (Wassenberg and Di Giulio 2004, Billiard et al. 2006, Matson et al. 2008).

Canonically, biotransformation results in a decrease in signal (AHR ligand) that provides the feedback that ends the transduction cycle, thereby contributing to toxicity resistance through detoxification. Effective toxicity resistance ensues when ligands are good substrates for biotransformation enzymes, biotransformation products are less toxic and more easily eliminated than their parents, and oxidation byproducts can be managed by the antioxidant system. Many AHR agonists do not fit this profile; for example, PAHs and DLCs may be strong agonists, but DLCs are generally poor substrates for biotransformation, instead inducing redox cycling without termination. There is increasing awareness that AHR-mediated responses can contribute effectively to toxicity resistance in certain scenarios, i.e., in response to certain AHR agonists at limited exposure concentrations and durations. More generally, defense strategies that are effective for short-term sublethal exposures may not be effective for highly toxic chronic exposures.

The following examples of genetically adapted tolerance or evolved toxicity resistance in fish populations subject to intense multigenerational exposure to potent AHR agonists have been revealed through a series of collaborative studies conducted over more than two decades that proceed from field observations to molecular mechanisms underlying tolerance.

### Evolved Tolerance to DLCs

AHR binding, as the control point for pathway activation, is the foundation of variation in response to DLC exposures among populations of Atlantic tomcod (*Microgadus tomcod*). In particular, tomcod inhabiting the heavily polluted Hudson River are less sensitive to toxic xenobiotics than populations inhabiting less-contaminated sites (Wirgin and Waldman 1998). Like killifish, tomcod reside in or close to natal estuaries throughout their life cycle and are found in large ($10^6$) populations (Yuan et al. 2006a). Tomcod residing in the Hudson are highly contaminated with DLCs and PAHs and (in some years) exhibit a high prevalence of tumors (Cormier and Racine 1990). Compared to tomcod from a reference population, those collected from the Hudson River and depurated in the laboratory exhibit reduced induction of CYP1A1 by DLCs; however, induction by the PAH benzo[a]pyrene was not relatively reduced (Courtenay et al. 1999). In one study, responsiveness was regained in progeny from these fish (Roy et al. 2001). However, later evidence indicated that Hudson River progeny ($F_1$ and $F_2$) were resistant to waterborne PCBs at concentrations that are environmentally relevant to the field site and toxic to reference fish (Chambers et al. 2003). In addition, early developmental gene expression differed in Hudson River tomcod vs. reference populations when similarly exposed to DLCs (Carlson et al. 2003). In particular, CYP1A and AHRR were not induced in $F_1$ and $F_2$ Hudson River tomcod embryos exposed to DLCs (Carlson et al. 2003). Together, these findings suggest that the Atlantic tomcod of the Hudson River estuary has developed resistance to DLCs (but not to PAHs), as demonstrated by refraction to CYP1A inducibility (Yuan et al. 2006b). This resistance is at least partly heritable (Roy et al. 2001, Yuan et al. 2006a) and likely reflective of resistance to the early life stage effects of DLCs. Furthermore, these tolerant populations occupy a large spatial region, much of which is highly contaminated (200 miles of the Hudson is a superfund site (Yuan et al. 2006a)).

Heritability of evolved tolerance was confirmed in Hudson River tomcod and was correlated with mutations in AHR2 (AHR2-1) that were found almost exclusively in tolerant Hudson River fish but not in nearby sensitive tomcod populations. The variant AHR2-1 protein had a low binding affinity for TCDD, consistent with a large deletion that resulted in amino acid substitutions at critical residues within the AHR ligand-binding domain. Interestingly, this region of AHR2 is the site of interaction with AIP, a known AHR pathway partner that stabilizes AHR and enhances its ability to bind ligand and activate transcription (Wirgin et al. 2011). Wirgin et al. (2011) identified a single locus that correlated with tolerance in Hudson River tomcod, although this does not rule out the potential contribution to tolerance of other loci within or outside of the AHR pathway. Brown et al. (2017b) noted that the large variation between tolerant and sensitive tomcod populations in differentially expressed genes may reflect both direct and indirect effects of tolerance-associated alterations in the AHR pathway. The effects of chronic multigenerational exposure to environmental toxicants on the genome of Hudson River tomcod are much greater than previously expected.

### Killifish Evolved DLC Tolerance

The Atlantic killifish became an early model for the study of evolutionary divergence, e.g., reviewed in Mitton (1997), and stress tolerance, e.g., Nordlie (2006 and Burnett et al. (2007). The killifish has been the subject of extensive investigations on adaptation to environmental variables such as temperature along large geographic and geological time scales (Powers and Place 1978, Place and Powers 1979, Powers et al. 1986). It has now become one of the most well-studied models of contemporary evolution (Whitehead et al. 2017, Crawford et al. 2020). In particular, much of our early understanding of evolved tolerance to DLCs comes from studies of killifish inhabiting the most heavily contaminated estuaries along the Atlantic Coast of the USA (reviewed in Hahn (1998b), Elskus (2001), Nacci et al. 2002a), Hahn et al. (2004)). Initially, killifish from Newark Bay, which is polluted with a complex mixture of chemicals, were shown to tolerate the highly toxic contaminant mercury (e.g., Weis and Weis (1977a), Weis et al. (1981), Weis and Weis (1989)). This foundational research suggesting toxicity resistance to local contaminants led to experiments that revealed differences in the response to TCDD by the progeny of Newark Bay versus reference site killifish (Prince and Cooper 1995a). When exposed to TCDD, fish from a reference site exhibited a higher prevalence of embryonic deformities, lesions, and death. These results could not be accounted for by differences in TCDD uptake between groups (Prince and Cooper 1995b). The authors also demonstrated that adult fish from Newark Bay were unresponsive to EROD induction following TCDD exposure. Downregulation of EROD, a CYP1A-dependent activity, led the investigators to believe that an altered AHR pathway may be involved in resistance to TCDD. In subsequent studies, Elskus et al. (1999) reported no induction of CYP1A or EROD in wild-caught adult or F1 larval Newark fish following PCB exposures that caused induction in reference fish. Together with the findings of Prince and Cooper (1995a), these results suggested that Newark Bay fish were resistant to the sublethal effects of TCDD and that an altered AHR pathway may be involved in resistance to DLCs.

These intriguing findings initiated the development of a uniquely comprehensive story of evolved DLC tolerance in the Atlantic killifish. This story, itself evolving over decades, was revealed by multi-institutional collaborations, notably, the Fundulus Genome Consortium, which supported the development of the well-annotated and critically important genome (https://www.ncbi.nlm.nih.gov/genome/annotation_euk/Fundulus_heteroclitus/100/). Two killifish populations from highly contaminated estuaries were the subject of parallel and intensive research efforts that advanced our understanding of evolved tolerance in the wild: those resident to the ER, Atlantic Wood (AW) Industries, Portsmouth, Virginia (VA, USA), and New Bedford Harbor (NBH), Massachusetts (MA, USA). However, as similarities between these populations began to emerge, the study of evolved tolerance in killifish expanded to include comparisons between these and additional independent populations, which are described as this section continues.

### Evolved Tolerance in ER Killifish

The ER in Virginia is a heavily industrialized subestuary of the lower Chesapeake Bay (Huggett et al. 1992), located in the vicinity of Norfolk and Portsmouth. Since the early 1900s, numerous wood treatment facilities along the ER have used creosote as a biocide to protect telephone poles, pilings, and railroad ties from microbial decay. Although few records exist before the 1970s, at least two massive creosote spills occurred in the Southern Branch of the ER between 1963 and 1967. ER sediments exhibit an exponential increase in sedimentary PAHs from the river mouth to approximately 20 km upstream (O'Connor and Huggett 1988). In addition, ER sediments contain other contaminants commonly found in creosote, including nitrogen-, oxygen-, and sulfur-containing heterocyclic compounds, phenols, and pentachlorophenol. PCBs and metals are also present but are considered minor contaminants relative to the aromatic compounds associated with creosote.

Di Giulio and Clark (2015) provide comprehensive details concerning the history of ER contamination and its impact on resident killifish; the following description focuses on research progress in understanding evolved tolerance in ER killifish. Initially, while investigating pathologies in ER finfish, Vogelbein et al. (1990) discovered a population of killifish inhabiting a lagoon adjacent to an abandoned wood treatment facility (AW Industries). Sediment PAH concentrations (2,200 mg PAH per kg dry weight sediment) indicated that AW was one of the most heavily contaminated sites known worldwide with respect to PAHs and creosote. Adult killifish at the site exhibited an extremely high prevalence of hepatocellular carcinoma and other lesions (Vogelbein et al. 1990, Fournie and Vogelbein 1994, Vogelbein and Fournie 1994, Vogelbein et al. 1999). These lesions were not observed in killifish collected from a much less-contaminated site directly across the river (Scuffletown Creek), suggesting a chemical etiology to these lesions and little migration between sites. Although AW fish develop hepatocellular carcinoma and other chemically induced cancers, they are resistant to acute toxicity caused by exposure to AW sediment, PAHs, and PCB-126 (Williams 1994, Van Veld and Westbrook 1995, Meyer and Di Giulio 2002, Meyer et al. 2002, Ownby et al. 2002, Meyer and Di Giulio 2003). The ability of AW killifish to tolerate exposure to contaminated sediments was demonstrated to be heritable to some extent in laboratory-reared offspring (Meyer and Di Giulio 2002, Meyer et al. 2002, Ownby et al. 2002, Meyer and Di Giulio 2003, Meyer et al. 2003b, Matson et al. 2008, Clark et al. 2010). Hybrid larvae (i.e., AW×reference site) demonstrated sensitivity to contaminated sediments intermediate between those of the parental AW and reference populations; furthermore, the sensitivity of hybrids was similar regardless of which parent was from AW, suggesting a limited contribution from non-inherited maternal effects (Meyer and Di Giulio 2003). Tolerance declined in the $F_2$ laboratory progeny relative to that of $F_1$, suggesting that both physiological acclimation and genetic adaptation contribute to resistance (Meyer and Di Giulio 2002), although subsequent studies showed that tolerance to PAH- and PCB-induced teratogenesis in ER fish is strongly inherited in F2 fish (Nacci et al. 2010, Clark et al. 2014). More recently, Brown and coauthors demonstrated that F1 AW fish are resistant to heart damage, swimming performance deficits, and some behavioral alterations observed in juvenile reference fish caused by exposure to sub-teratogenic levels of contaminated sediment extracts during development (Brown et al. 2016, 2017a).

Several studies indicate that AW embryos and larvae are resistant to the teratogenic effects of contaminated sediments and individual contaminants. Embryos produced in the laboratory from reference site adults and exposed to AW sediment developed severe cardiac malformations and exhibited 100% mortality, while embryos from AW parents exhibited only 26% mortality and no cardiac malformations under these conditions (Williams 1994). Ownby et al. (2002) reported no difference between the first- and second-generation AW killifish in their ability to resist the teratogenic effects of creosote-contaminated sediment, suggesting a genetic component to resistance. AW embryos are also highly resistant to the teratogenic effects of PCB 126 (Meyer and Di Giulio 2002, Nacci et al. 2010, Clark et al. 2013). Because PCBs are not present at high concentrations at the AW site, resistance to this DLC suggests that some similarities may exist in mechanisms of resistance to PAHs from creosote and DLCs.

The mechanistic basis for toxicity resistance in the AW killifish has been addressed by several investigators. In a series of studies, Van Veld and coworkers investigated biological toxicity-resistant AW killifish. These fish exhibit altered expression of proteins involved in biotransformation and efflux; for example, immunoblot analysis indicated reduced expression of CYP1A (Van Veld et al. 1992). GST activity was elevated approximately fivefold in the liver of AW killifish compared with that of reference site fish (Van Veld et al. 1991, Armknecht et al. 1998). Monoclonal antibodies against killifish GST indicated a sixfold elevation of GST in the liver of AW fish relative to that measured in reference fish. In addition to alterations in phase I and II enzymes, Pgp transporters were elevated in the liver of AW fish relative to reference fish (Cooper et al. 1996, 1999a, 1999b). Immunohistochemical staining of killifish liver demonstrated this antigen on the canalicular surface of hepatocytes in normal liver, suggesting a transport role for this protein. A complementary DNA (cDNA) fragment was amplified in the liver (Cooper et al. 1999a) that was most similar to the mammalian sister gene of P-glycoprotein (Childs et al. 1995). Another cDNA fragment was amplified from the killifish liver and intestine that was a homolog of the mammalian *mdr*, the mammalian multidrug-resistant transporter (Cooper et al. 1999a). These results indicate that a xenobiotic transporter is likely elevated in the livers of a pollution-tolerant vertebrate population, consistent with the proposed role of these proteins in xenobiotic resistance.

Exposure to AW sediments caused increases in many of the antioxidant defenses tested in laboratory-raised killifish from both the AW and a reference site population (Meyer et al. 2003a). In that study, AW killifish were more resistant to the toxicity of a model prooxidant than were reference site fish, and they exhibited elevated levels of several antioxidant defenses in the absence of any chemical exposure. Bacanskas et al. (2004) also noted seasonal and sex-specific responses to prooxidants in xenobiotic-tolerant and xenobiotic-sensitive killifish from Virginia residence sites after collection and following laboratory holding. These results suggest that elevated antioxidant defenses might contribute to the resistant phenotype.

Furthermore, Van Veld and Westbrook (1995) reported low-level expression and insensitivity of CYP1A in AW killifish to inducing agents, suggesting that alterations in the AHR-mediated pathways contribute to chemical resistance. Lack of CYP1A inducibility persisted in laboratory-reared $F_1$ embryos and larvae (Meyer et al. 2002); however, inducibility was regained as AW $F_1$ matured into adults. Furthermore, inducibility was observed in all life stages of later generations raised in the laboratory (Meyer and Di Giulio 2002, Meyer et al. 2002). This pattern of reversible heritability of the CYP1A phenotype was not explained by maternal effects (Meyer et al. 2002), altered transcription of known CYP1A transcription factors (AHH and AHRR) (Meyer et al. 2003b), or hypermethylation of specific cytosines in the CYP1A promoter region (Timme-Laragy et al. 2005). Furthermore, Clark et al. (2014) used qPCR to show that CYP 1A, 1B1, and 1C1 were recalcitrant to upregulation by inducers in AW F1 killifish, although some recovery of expression in F2 was observed. Differential molecular techniques following AW sediment exposures to AW and local reference killifish revealed sex- and/or population-specific responses, including (potentially) novel genes and those not thought previously to be associated with AHR responsiveness (Meyer et al. 2005).

Additional studies provided further clarification on the association between recalcitrant CYP1A and the heritability of evolved tolerance in the ER killifish. For example, AW F1 and F2 embryos were highly resistant to cardiac teratogenesis; however, only partial inhibition of induction of CYP1A was observed in AW F2 embryos (Clark et al. 2014). Their results suggest that evolved tolerance is conferred by multiple factors, not all of which appear to be fully genetically heritable. Consistent with this conclusion, F1 embryos from killifish from throughout the ER, i.e., including populations resident to less-contaminated sites than AW, were highly resistant to teratogenesis although they demonstrated variation in induction of CYP1 activity (Clark et al. 2013).

To provide further mechanistic information concerning evolved tolerance in ER killifish, Clark et al. (2010) used a gene knockdown approach to conclude that the cardiac teratogenesis characteristic of DLC and some PAH exposures is primarily mediated by AHR2 (Figure 7.7). In particular, they demonstrated rescue from teratogenicity by morpholino gene knockdown of AHR2 but not AHR1. Four AHR paralogs were eventually described in killifish, and the relative roles of AHR1a, AHR1b, AHR2a, and AHR2b in mediating toxicity have yet to be determined (Reitzel et al. 2014).

The molecular basis for evolved tolerance in ER fish was advanced by a restriction site-associated DNA sequencing (RADseq) analysis of AW killifish (Osterberg et al. 2018). Here, RADseq, a reduced-representation genome approach (but also see Lowry et al. (2017) where RADseq provides the ability to detect signatures of selection, was used to identify genetic variation in the AW versus local reference site killifish in loci generally consistent with our expectations of AHR-mediated toxicity resistance. In particular, Osterberg et al. (2018) identified unique areas in AW killifish associated with cardiac structure and function (key cardiac and muscle genes), a target for PAH/DLC developmental toxicity. More broadly, they identified unique loci associated with AHR pathway components, including AIP and ARNT, and regulated elements, GST (phase II) and cytochrome b5 (CYB5, which minimizes incomplete oxidation of xenobiotics). These findings agree well with those of others (Nacci et al. 2016, Reid et al. 2016) who found AIP, ARNT1c, and GSTT1 as candidate genes in significant regions. However, they contrast with other genomic analyses of AW killifish that identified unique SNPs in the region of AHRs (Proestou et al. 2014) and, more specifically, a large deletion in the tandem paralogs AHR2a and AHR1a (Reid et al. 2016). In contrast to other studies (Williams and Oleksiak 2008, 2011, Proestou et al. 2014, Reid et al. 2016), CYP1a genes were not identified as unique in the study of AW killifish by Osterberg et al. (2018). Further mechanistic insight into AW-evolved tolerance was provided in a more comprehensive whole-genome sequencing approach (Reid et al. 2016) discussed later ('Convergent evolution').

## Evolved Tolerance in NBH Killifish

Over the same period of time that the mechanisms by which AW killifish resisted the toxicity of local contamination, the toxicity and tolerance of DLCs were being investigated in killifish resident to NBH, designated a superfund site because of extensive and heavy PCB contamination (Weaver 1984, Pruell et al. 1990, Lake et al. 1995, Nelson et al. 1996). Sediment PCB levels in the most contaminated areas of NBH have been measured as high as 190,000 mg PCB per kg sediment or 19% by weight (Weaver 1984). Killifish inhabiting

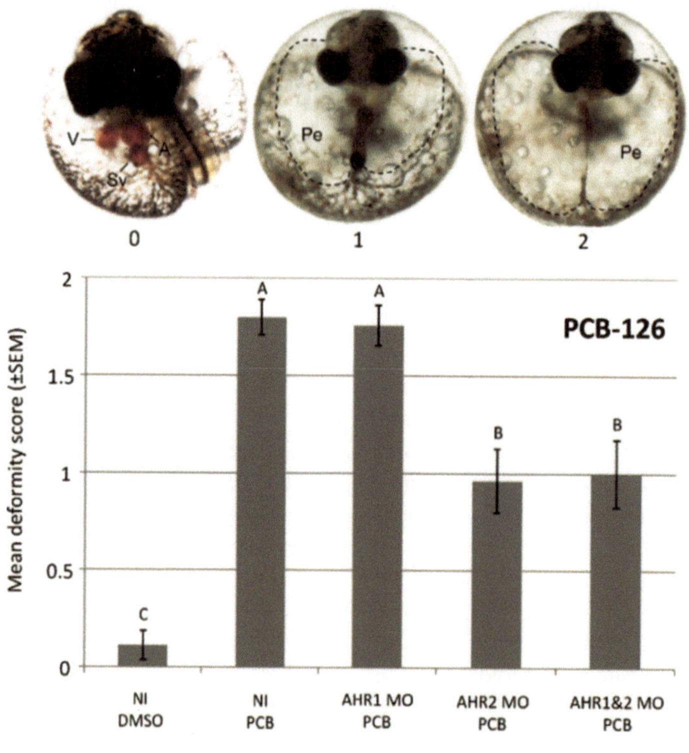


**FIGURE 7.7** Developmental tolerance upon exposure to 1 μg/L PCB-126 in AHR2 morpholino (AHR2MO) in comparison with non-injected killifish, manifested as reduced mean deformity score, where a score of O corresponds to a normal heart, 1 to a moderately deformed heart, and 2 to a severely deformed heart. Clearly visible in the normal heart are the ventricle (V), atrium (A), and sinus venosus (Sv). Pericardial edema (Pe, margin marked by a dashed line) surrounds elongated hearts in embryos scored as 1 and 2. Adapted with permission from Clark et al. (2010).

sites throughout NBH are heavily exposed to PCBs, including some of the most toxic (dioxin-like) congeners (Lake et al. 1995, Black et al. 1998a). NBH killifish carried levels of specific PCB congeners that were lethal to reference fish when the latter were exposed via injection or diet (Black et al. 1998a, Gutjahr-Gobell et al. 1999). As a result, Black et al. (1998b) suggested that NBH fish may also exhibit abnormally high mortality in the field; however, a series of studies confirmed that, despite extreme tissue PCB concentrations and modeled predictions of contaminant risks (Munns et al. 1997), NBH killifish are reproductively prolific (Nacci et al. 2002a) and have high condition indices and nearly normal levels of stored vitamin A (Nacci et al. 2001), which is considered a sensitive indicator of DLC toxicity in many vertebrate species (Fletcher et al. 2001). Complementary studies suggest that demographic compensation (e.g., increased reproductive effort) and migration (i.e., from less-contaminated populations) do not play important roles in supporting a persistent population in NBH (Nacci et al. 2002a, 2002c, 2007); therefore, tolerance to PCBs was proposed to explain NBH killifish population persistence.

In subsequent studies with NBH killifish, Nacci et al. (1999) reported that the acute toxicity of PCB 126 (one of the most toxic DLCs) was approximately 100-fold higher in the progeny of reference fish than in the progeny of NBH fish. The significance of the difference in sensitivity between populations within a single fish species is revealed when compared to the sensitivity variation among fish species for TCDD (Figure 7.2). When $LC_{50}$ values are transformed into toxic equivalency values that reflect sensitivity to TCDD (using values from (Nacci et al. 2002b, Van den Berg et al. 2006)), the intraspecies range for this single fish species (killifish) (>1,000 fold range) exceeds the interspecies sensitivity range among tested freshwater fish (<100 fold range) (Figure 7.2).

Tolerance to DLCs persisted in $F_1$ and $F_2$ generations of NBH killifish and therefore appears to involve genetic adaptation (Nacci et al. 1999, 2002). Similarly, transcriptomic responses to PCB-126 exposure during development were very different between NBH embryos and those from a nearby reference site, but were identical between NBH embryos that were 1 and 2 generations removed from the field (Whitehead et al. 2012). Furthermore, concentrations of PCB 126 producing early life stage lethality in hybrid embryos (i.e., NBH×reference killifish) were intermediate between the parental populations, and sensitivities to PCB 126 were similar in hybrids whether of maternal or paternal origins from NBH, suggesting minimal contributions from non-inherited maternal effects (Nacci et al. 1999). Laboratory exposures were transformed into estimates of tissue concentrations and compared to concentrations measured in eggs from NBH killifish to estimate the effects of field exposures, which were estimated from chemical measurements of sediment cores (Nacci et al. 2002a). These results confirmed that killifish embryos from a reference population, but not native NBH killifish, would be killed by concentrations of PCB 126 accumulated from NBH sediments (Nacci et al. 2002a). In particular, PCBs in NBH sediment have been at toxic levels since the 1940s (shortly after the discharge of PCBs into the harbor began) and continue to be toxic, even though discharges of PCBs ceased in the 1970s (Figure 7.8). Taken together, these studies are consistent with the conclusion that PCB contamination has resulted in a genetically adapted NBH killifish population.

The results of several studies indicate that DLC resistance in the NBH killifish involves alterations in the responsiveness of the AHR pathway; for example, reduced responsiveness to DLCs was investigated using a nondestructive *in ovo* EROD assay (Nacci et al. 1998, Nacci et al. 2005). Like Newark Bay killifish, NBH embryos were relatively insensitive to EROD induction by DLCs or PAHs (Nacci et al. 1999), although there was significant variation in EROD responsiveness among individual embryos within each population. Transcriptomic experiments in

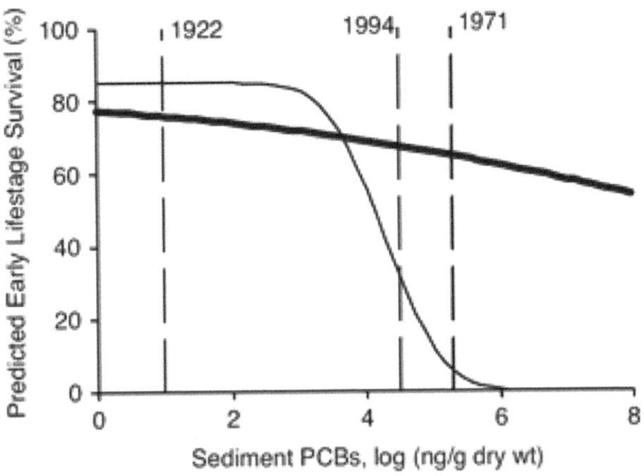


**FIGURE 7.8** Rapid evolution consistent with a survival advantage for temporal increases in sediment PCB126 for a resident tolerant population of *Fundulus heteroclitus* (NBH, heavy line) in comparison with a reference site (light line), based on laboratory studies of toxicity to PCB 126 and bioaccumulation of sediment PCBs (Nacci et al. 2002a). Estimated sediment PCB concentrations based on sediment cores from NBH are shown for dates before 1922, at the end of PCB deposition (1971), and in recent times (1994). Adapted from Van Veld and Nacci (2008).

embryos showed that many canonical targets for AHR activation (CYP1A1, CYP1C1, CYP1B, UDPGT (UDP-glucuronosyltransferase), GST, CYB5) were upregulated upon exposure to PCB-126 in reference site fish, but not in NBH fish until much higher (2–3 orders of magnitude) doses (Whitehead et al. 2012). Consistent with studies in embryos, adult fish from NBH were also less responsive to AHR agonists in comparison with reference site killifish. Nacci et al. (2002d) showed that exposure of adult killifish to the PAH benzo(*a*)pyrene (BaP) resulted in relatively fewer mutagenic BaP–DNA adducts in NBH killifish, consistent with reduced CYP1A-mediated activation of BaP required for DNA adduct formation.

Sensitivity to DLCs was also evaluated in cultured hepatocytes from adult NBH fish and reference killifish (Bello 1999, Bello et al. 2001). CYP1A was inducible in NBH hepatocytes but required a 14-fold higher concentration of TCDD than that required to induce CYP1A in reference hepatocytes. This indicates that the AHR pathway in NBH parental fish is functional but less sensitive or responsive to AHR agonists than in reference fish. Like parental fish, progeny from NBH were generally refractory to CYP1A induction, suggesting that some mechanisms of tolerance are inherited (Bello 1999, Bello et al. 2001). Other evidence suggested that short-term or noninheritable mechanisms may contribute to the observed differences between the NBH and reference fish (Bello 1999); for example, a PAH was a relatively effective inducer of CYP1A, and 2,3,7,8-tetrachlorodibenzofuran (TCDF) was an effective inducer of the AHR in $F_1$ progeny raised in the laboratory from NBH parents (Bello 1999). In addition, although freshly caught NBH fish had a lower hepatic content of AHR than reference fish, there were no differences in AHR levels between fish from different sites after a 90-day depuration period in the laboratory.

A series of studies conducted by Hahn and colleagues have addressed variations in the structure, regulation, and function of the AHR signal transduction pathway in NBH and reference populations. These studies have focused on the receptor (AHR1 and AHR2), partner (ARNT), and repressor (AHRR) pathway components, as well as the enzymes (e.g., CYP1A) regulated by the pathway. No differences were observed between local reference and NBH killifish in ARNT expression or AHR2 expression, but an altered pattern of tissue-specific expression of AHR1 was reported in NBH killifish compared with reference fish (Karchner et al. 1999, Powell et al. 2000). This pattern of expression was not inherited, although resistance in NBH killifish is inherited. Subsequent studies indicate genetic differences between AHR1 in NBH and reference killifish. AHR1 in NBH vs. reference killifish differed genetically but not functionally with respect to binding capacity and affinities for dioxin, suggesting that these differences cannot account for differences in dioxin sensitivity between populations (Hahn et al. 2004).

In addition to AHR1, there appear to be differences in the regulation of AHRR in NBH and reference fish. AHRR negatively regulates the transcriptional activity of the AHR (Mimura et al. 1999, Karchner et al. 2002). DLCs appear to induce AHRR in reference fish but not in NBH fish (Karchner et al. 2002), suggesting that upregulation of AHRR is not involved in resistance. More recently developed molecular tools were used to explore mechanisms by which killifish populations have adapted to DLCs; for example, the promoter region of CYP1A was cloned and characterized functionally as containing xenobiotic or dioxin response elements (XREs) and glucocorticoid response elements (GREs) (Powell et al. 2004). Sensitive, specific polyclonal antibodies were developed for AHR1, AHR2, and AHR repressor protein for killifish (Merson et al. 2006). Merson et al. (2006) described the application of these antibodies to demonstrate the occurrence of AHR pathway proteins in the heart, brain, ovary, and other tissues of killifish. Further evidence suggests that AHR2 plays an important role in DLC toxicity in zebrafish (Prasch et al. 2003) and killifish (Clark et al. 2010). Taken together, the body of work by Hahn and colleagues has provided a great deal of information on components of the AHR pathway in killifish and other species, including important clues to the molecular basis for inherited tolerance or genetic adaptation to DLCs.

Nacci et al. (2016) investigated the genetic basis for DLC tolerance in NBH killifish using a quantitative trait locus (QTL) approach that links genetic variants to phenotypes. In this classic breeding strategy, killifish from sensitive and tolerant populations are interbred, and then, their F1 progeny are interbred to produce families of F2 progeny. F2 progeny are then exposed to a single concentration of PCB126 that is toxic to the sensitive but not the tolerant originating populations. The exposed progeny were characterized as tolerant or sensitive by phenotype and genotyped using a dense panel of genetic markers. Correlations between specific genetic markers and phenotypes were interpreted using the Fundulus genome and shared synteny with other fish species. In total, QTLs explained 69% of the observed phenotypic variation. Consistent with mechanistic knowledge of DLC toxicity in fish and other vertebrates, the AHR2 region accounted for 17% of trait variation. Another strongly supported QTL occurred close to the gene coding for the AHR-interacting protein, AIP, which interacted significantly with another QTL near the gene for the AHR complex component, the chaperone, hsp90ab1. Consistently, functional variation near the AIP-AHR binding region was associated with Hudson River tomcod DLC tolerance (Wirgin et al. 2011). Furthermore, several QTLs for DLC tolerance in NBH killifish were similar to those identified for zebrafish (Waits and Nebert 2011). Taken together, these results suggest that DLC tolerance is a complex trait that may be affected by multiple genes in the AHR and other pathways (Miller et al. 2023).

## Convergent Evolution

While intensive studies of unique tolerant killifish populations have advanced our understanding of evolved tolerance greatly, the potential relevance and impact of findings have been greatly increased by more holistic examinations of multiple, independent tolerant killifish populations. Several studies compared genomic similarities among killifish from (up to) four highly contaminated sites (Proestou et al. 2014, Reitzel et al. 2014, Reid et al. 2016), but the basis for grouping populations together was provided in studies detailing phenotypic similarities among tolerant killifish in contrast to killifish resident to less-contaminated sites. For example, Nacci et al. (2010) synthesized the results of identically conducted developmental assays of PCB126 responses using killifish embryos raised in the laboratory from 24 wild populations. These populations originated from sites that ranged in sediment PCBs from 1 to 67,000 ng/g dry weight, representing a range of east coast killifish habitats from pristine to superfund levels. The sites included the residence sites for four killifish populations shown in various prior studies to demonstrate some form of toxicity resistance to local contaminants, previously uncharacterized sites and sites used as local reference sites (Figure 7.9). Consistent with expectations, embryos from these four killifish populations were highly tolerant of PCB126 exposures, with LC20 values (an inverse measure of sensitivity) up to 8,000 times greater than the most sensitive killifish populations tested. In addition, CYP1A recalcitrant phenotypes, as measured by *in ovo* EROD (Nacci et al. 1998), were also shared among tolerant killifish populations. However, their responses were not identical: LC20 values ranked ER as the most resistant population, followed by NBH, with Newark and Bridgeport ranking third with similar levels of resistance. Importantly, the level of resistance in resident fish was correlated with levels of DLC pollution of their sediments (Nacci et al. 2010), suggesting that natural selection has operated to derive resistance that scales directly with the selection pressure (contaminant concentrations). Across tolerant populations, inheritance was demonstrated in both lethal and CYP1A responsiveness in F1 and F2 progeny, although there was apparent erosion in tolerance in at least one of these populations (Table 7.3). Subsequent studies have used these phenotypic comparisons to evaluate in greater detail the molecular underpinnings of these parallel demonstrations of evolved tolerance, but perhaps there are broader implications from the development of models correlating sediment contamination with observed tolerance to PCB126 (Figure 7.9 inset). These results suggest killifish populations that reside in representative locations, i.e., those whose sediments are below levels correlated with ecological degradation (Long et al. 1995), may also be responding adaptively to the evolutionary pressures from exposure to contaminants that act through the AHR pathway.

Two genetic surveys of tolerant and sensitive killifish populations were designed to reveal selection and specifically targeted AHR pathway components. Reitzel et al. (2014) conducted genetic analyses of single-nucleotide polymorphic (SNP) markers comparing NBH and Newark, NJ, tolerant with nearby sensitive killifish populations. Their results confirmed genetic differentiation among sites, selection at specific nucleotides in AHR1 and AHR2, and specific AHR2 alterations associated with the PCB-resistant phenotype in the NBH population. However, these changes in allele frequencies were not fixed as such changes were, for example, in DLC-adapted Hudson River tomcod. Proestou et al. (2014) used a broader array (42 genes) of AHR-targeted SNPs to compare four tolerant killifish populations and their nearby reference populations. Pairwise comparisons of nearby tolerant and sensitive populations revealed high differentiation among sensitive and tolerant populations at loci including AHR1 and AHR2, and CYPs (CYP1A and 3A30); shifts in minor allele frequencies were observed at AHR2 and CYP1A loci across most sensitive or tolerant pairs, but only AHR2 exhibited shifts in the same direction across all pairs. Population-specific variation in CYP1A may reflect differential fitness benefits under site-specific environmental conditions; alternately, variation in

**TABLE 7.3**
**Tolerance in Four Killifish Populations**

| Site Code | Site Name | Sensitivity to PCB126,nh/L<br>Survival LC20 (F1) | Survival LC20 (F2) | EROD EC50 (F2) |
|---|---|---|---|---|
| 7 | Bar Harbor | 23 | NT | <2[b] |
| 19 | Newark | 13,836 | 12,491 | 2,591 |
| 22 | Bridgeport | 13,900 | 939 | <2,000 |
| 23 | New Bedford | 31,290 | 47,863 | 5,914 |
| 18 | Elizabeth River | >200,000[a] | >200,000 | >200,000 |

*NT:* not tested.
[a] No effect at the highest tested concentration, 200,000 ng/L.
[b] F1 embryos tested; effect at the lowest tested concentration, 2 ng/L.
*Note:* Responses of laboratory-reared (F1 and F2 generations) killifish (*F. heteroclitus*) early life stages following exposure to the dioxin-like compound (DLC), congener PCB126, showing survival and a biomarker of aryl hydrocarbon receptor (AHR) activation, ethoxyresorufin-O-deethylase, EROD, for a DLC-sensitive population (Bar Harbor, Maine) and three DLC-tolerant populations (Newark, New Jersey, Bridgeport, Connecticut, and New Bedford, Massachusetts). Adapted with permission from Nacci et al. (2010).

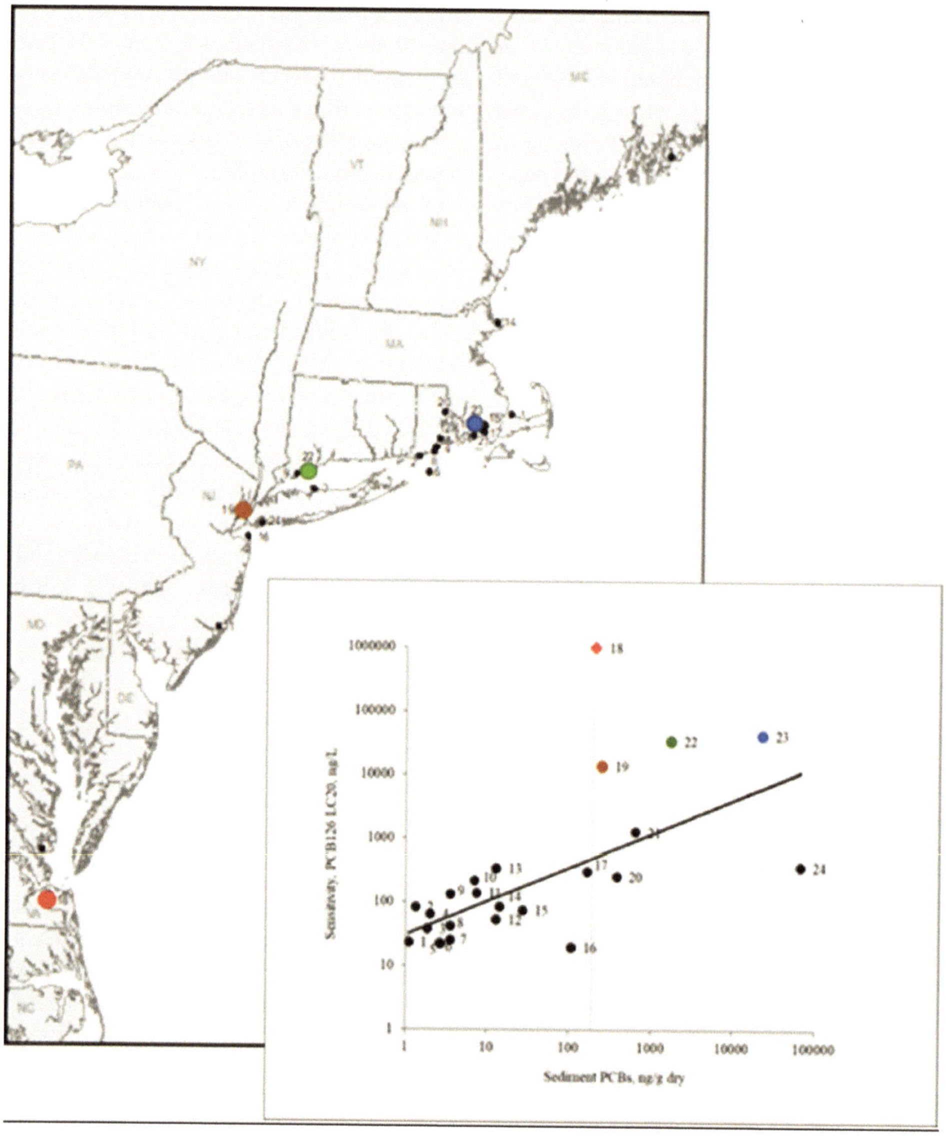


**FIGURE 7.9** Adaptive variation in killifish populations along the east coast of the United States (map) that have been measured for habitat contamination (sediment PCBs as proxy) and their early life stage sensitivity (LC20) to the model DLC, PCB126 (inset). Extreme adaptive tolerance has been studied in populations from highly DLC-contaminated sites, NBH, Massachusetts (#23, blue), Bridgeport, Connecticut (#22, green), and Newark, New Jersey (#19, orange). The ER (including the AW site) in Norfolk, Virginia (#18, red), is contaminated with PAHs and other contaminants associated with creosote. Adapted with permission from Nacci et al. (2010).

CYP1A may have no functional outcome (and therefore not a target of selection) because of the recalcitrant phenotype. Variation in CYP3A, which is responsive to non-DLC-like PCBs (Chang et al. 2013, Grans et al. 2015) that predominate PCB-contaminated sites, may also be related to the minimal responsiveness to non-DLC PCB congeners observed in NBH (and perhaps other tolerant) killifish populations relative to sensitive killifish from reference sites (Celander et al. 2021).

The mechanistic underpinnings of the convergent evolution of DLC tolerance in killifish were further explored through a particularly comprehensive approach comparing the four identified DLC-tolerant killifish populations using transcriptomic responses to DLC exposures of laboratory-reared killifish and whole-genome sequencing of field-collected killifish from these same populations (Reid et al. 2016). Consistent with various studies characterizing gene expression and phenotypic observations of DLC-exposed killifish, this study revealed the expected PCB exposure-induced gene expression in sensitive populations (including canonical transcriptional targets of AHR), but most of this PCB-induced transcriptional response was absent in fish from resistant populations (Figure 7.9). Prior studies not only showed a similar pattern of genome-wide gene expression in response to PCB-126 exposure but also showed that at extremely high PCB-126 exposures the transcriptional response of resistant populations became very similar to that of sensitive populations exposed to much lower doses (Whitehead et al. 2010, Whitehead et al. 2012). These studies were able to then conclude that resistance is associated with AHR desensitization, but not a full disabling of AHR responsiveness, and that toxicity that emerged at extremely high doses in resistant populations was achieved through similar mechanisms of action as toxicity induced at much lower doses in sensitive populations, i.e., a shift in sensitivity (as shown in Figure 7.1).

Using a population genomics approach, Reid et al. (2016) reported that hundreds of genomic regions showed signatures of strong recent natural selection in each of the resistant populations, and many of these signatures were unique to a single population, suggesting that recent evolutionary adaptation to urban estuaries is highly polygenic. However, the strongest signatures of natural selection in each resistant population, which also tended to be shared among populations, included key components of the AHR signaling pathway including AHR2a, AHR1a, AIP, and CYP1A. The strongest signal of selection across all populations was associated with AIP, consistent with evidence from both the NBH QTL (Nacci et al. 2016). Consistent with the important role for AIP in adaptive tolerance to DLCs in fish species, the AIP-AHR binding site was associated with tolerance in the Hudson River tomcod (Wirgin et al. 2011) studies. AHR (2a and 1a) paralogs, also identified as outliers in all tolerant populations, included deletions spanning the last several exons of AHR2a and the first several exons of AHR1a in three tolerant populations. Although three different deletions in AHR sequences were present in several resistant populations, only the deletions observed in the ER population were swept to high frequency. Other population-specific outliers were apparent, particularly in the ER killifish, which have the most different genetic backgrounds from the other populations studied and reside in the most toxicologically different habitat among the resistant populations. For ER killifish, unique signatures of selection included additional AHR pathway genes (ARNT1c and HSP90) and other biotransformation genes, some of which may be AHR-regulated, i.e., CYP1C1/1C2 and GSTT1. Alterations in these genes may reflect PAH defense strategies that are uniquely beneficial to the ER environment, which is the only site with extremely high concentrations of low-molecular-weight PAHs. Unlike the PCB/DLCs that predominate the toxic profiles at sites where other tolerant killifish reside, some PAH toxicity may be mediated via AHR while some toxic effects are not (Incardona et al. 2005, Clark et al. 2010), i.e., through disruption of voltage-gated potassium channels, as demonstrated in isolated cardiomyocytes (Brette et al. 2014). ER is the only killifish population that showed strong signatures of recent natural selection in two voltage-gated potassium channel genes (Reid et al. 2016).

Clearly, the AHR pathway has been a strong target for natural selection in all populations, suggesting that adaptive modification of AHR signaling has solved many of the adaptive challenges posed by life in urbanized estuaries. However, other genetic regions were also identified as under selection in each population. While such changes may reflect responses to site-specific stressors, they may also reflect the polygenic nature of pollution adaptation where different variants affect similar phenotypic outcomes in different individuals and populations in response to parallel selection pressures. These population-specific adaptations may also be indicative of population-specific adaptive compensation for the initial fitness costs of adaptation to pollution. Overall, Reid et al. (2016) reinforced increasingly compelling evidence of the AHR pathway as a shared target of selection (Figure 7.10). This study also showed that parallel selection pressures by PCB and PAH pollution in urban estuaries can lead to parallel phenotypic outcomes observed as similar levels of toxicity resistance and parallel functional responses to chemical exposures characterized as AHR pathway desensitization. However, that parallelism breaks down at the genic level; although some genes are held in common across populations, there are many population-specific genes involved in adaptation. This disparity is heightened at the level of individual mutations, where there are no shared adaptive mutations between ER and other populations. More broadly, Reid et al. (2016) concluded that the identified molecular changes are consistent with selection on high-standing genetic variation that characterizes *Fundulus* (e.g., Mitton (1997), Reid et al. (2017)) and have permitted this rapid and repeated adaptation to contemporary and highly toxic xenobiotics.

The phylogenetic tree at the top was estimated from genome-wide bi-allelic SNP frequencies and shows the expected biogeographic pattern of relatedness among populations: Geographically nearby populations cluster together, and northern populations are distinct from southern (Reid et al. 2016). The top right panel illustrates the key proteins associated with the AHR signaling pathway. The middle panels show that key genes in the AHR signaling pathway (AIP, AHR, and CYP1A) repeatedly show strong signatures of recent natural selection in all resistant populations. For each sensitive-resistant population pair, signatures of selection include fixation index, Fst (a measure of genetic differentiation between populations, where high Fst indicates recent strong selection between populations) and delta pi (a measure of change in nucleotide diversity between sensitive and resistant populations, where negative delta pi indicates a signature of selection in the resistant population). The bottom row of panels shows that different molecular variants underlie the resistant genotype in different populations. Though each population has converged on similar levels of pollution resistance, and some genomic regions show patterns of parallel evolution (middle panels), the parallelism of resistance starts to break down at the level of molecular variants. Though the northern population tends to have similar molecular variants associated with signatures of selection, the southern population tends to have different molecular variants associated with shared signatures of selection (adapted with permission from Reid et al. (2016) and Whitehead et al. (2017)).

While the evolution of tolerance in the Atlantic killifish has been extensively explored, more recent research has started to examine the nature of recently evolved pollution

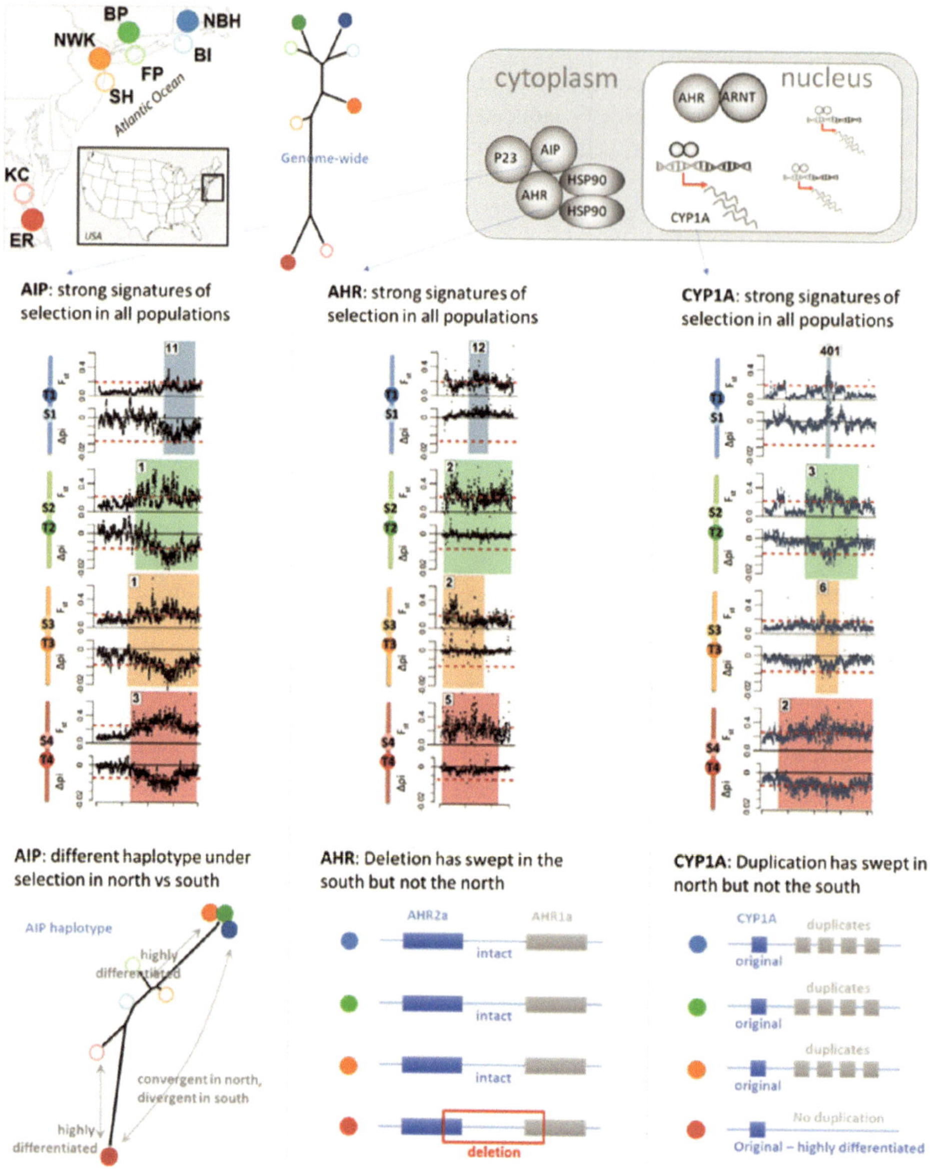


**FIGURE 7.10** Common targets of selection in pollution-tolerant killifish populations from highly contaminated sites (map). New Bedford, Massachusetts (T1, blue), Bridgeport, Connecticut (T2, green), and Newark, New Jersey (T3, orange), and the Elizabeth River, Norfolk, Virginia (T4, red)

resistance in their sister species Gulf killifish (*Fundulus grandis*), particularly in populations that span a gradient of pollution through the Houston (Texas, TX, USA) Shipping Channel (HSC). These studies have not only revealed similar molecular mechanisms of evolved resistance compared with Atlantic killifish but also uncovered an interesting and unexpected evolutionary mechanism permitting their rapid adaptation to toxic xenobiotics. Oziolor et al. (2014) reported that Gulf killifish embryos from the heavily polluted HSC were resistant to the model DLC, PCB126, and a PAH mixture, compared with a reference population, with reciprocal crosses between reference and contaminated populations exhibiting an intermediate level of toxicity resistance. These results resembled initial reports of genetically based and biparentally inherited evolved tolerance in NBH Atlantic killifish (Nacci et al. 1999). Further mechanistic similarities with the tolerance observed in Atlantic killifish emerged, including recalcitrant EROD responsiveness, which supported explanations of causal AHR pathway variation. Another study (Oziolor et al. 2019) provided further evidence of mechanistic parallelism between DLC tolerance in Atlantic and Gulf killifish including that tolerance in the HSC region scales with pollution and negatively correlates the inducibility of AHR signaling (e.g., Clark et al. (2013). AHR pathway targets of selection in tolerant populations of sister species were also similar, including AHR1a and AHR2a, ARNT, and AIP. Surprisingly, further species comparisons showed that adaptive variants at the AHR locus and at the ARNT locus, which are on different chromosomes,

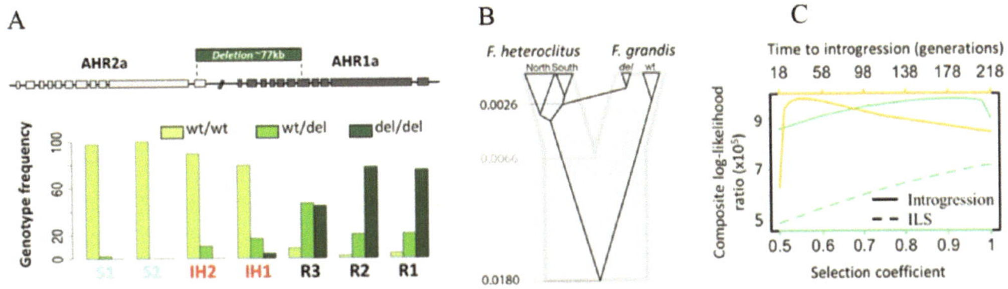


**FIGURE 7.11** Adaptive introgression of Gulf killifish (*F. grandis*) populations. (A) A deletion in *F. grandis* (~77 kb spanning AHR2a and AHR1a on chromosome (1) has swept to high frequency in resistant populations (bar plot, wt=wild type, del=deletion), where S1 and S2 are sensitive populations from relatively clean habitats at the mouth of the Houston Ship Channel (HSC; TX, USA), IH1 and IH2 are intermediately resistant populations from moderately contaminated sites in the middle of the HSC, and R1, R2, and R3 are highly resistant populations from severely polluted sites in the upper reaches of the HSC. (B) Estimated divergence time between species *F. grandis* and *F. heteroclitus* in mutational units is 0.0066 mutations/site. The mean time to coalescence in the genomic region surrounding the AHR deletion between *F. grandis* deletion alleles and southern *F. heteroclitus* is 0.0026 and that between intact *F. grandis* alleles and all of *F. heteroclitus* is 0.018. Since the coalescence times between species cannot postdate population divergence without gene flow, this is evidence that the deletion haplotype introgressed from southern *F. heteroclitus* populations. (C) Many simulations of different evolutionary scenarios indicated that the best fitting introgression model is much more likely than incomplete lineage sorting (ILS) as the source of the adaptive haplotype and reveals that the deletion haplotype offered a massive selective advantage in the most polluted HSC sites (selection coefficient of 0.8) and introgressed very recently into polluted HSC *F. grandis* populations (~34 generations before the time of sampling). Adapted with permission from Oziolor et al. (2019).

had become integrated into HSC Gulf killifish populations through a single recent (~1970s) hybridization event with Atlantic killifish (Figure 7.11). Analysis of these variants in Gulf killifish populations showed that they offered a massive selective advantage in the polluted HSC and are a rare example of evolutionary rescue being enabled by introgression between two species.

Taken together, the story of evolved tolerance in the Atlantic and Gulf killifish has provided an opportunity for an extraordinary degree of collaborative research, maturing from phenotypic observations to biochemical and genomic mechanisms, which has provided a uniquely holistic perspective on the physiological, evolutionary, and ecological impacts of persistent toxic xenobiotic pollution.

## COSTS OF TOXICITY RESISTANCE

Ecological and evolutionary theories suggest that adaptation to one form of environmental stress may increase susceptibility to other stressors (Futuyma 1997). This theory is based on the assumption that adaptation to a new environment involves alterations in a previous, optimal phenotype that has been shaped by the various selection pressures of the ancestral environment (Coustau et al. 2000). Alterations present in the new phenotype should be deleterious to survival in the ancestral environment. In general, adaptation to a stressful environment may include selection for altered life-history traits. For example, Belanger-Deschenes et al. (2013) found that in perch populations inhabiting a metal-contaminated environment the genes under selection were associated with a 'live fast die young' life-history strategy. Evidence for such changes in life-history traits was not found in NBH killifish adapted to DLCs (Nacci et al. 2002a); however, adaptations of this nature may be common to poor-quality environments, where the costs of defense proteins are greater than their benefits.

Within the context of adaptation to chemical contaminants, traits that are adaptive for specific chemical contaminants may present disadvantages under alternate conditions. These disadvantages are often referred to as *fitness costs* or *fitness trade-offs*. There are several possible mechanisms by which the development of toxicity resistance could present disadvantages to adapted individuals (Bergelson and Purrington 1996, Taylor and Feyereisen 1996, Shaw 1999, Shirley and Sibly 1999, Coustau et al. 2000, Strauss et al. 2002). These include (1) adaptations that involve alterations in cell targets (e.g., ion channels) or receptors that could disrupt their normal function in the cell and alter higher level functions in the organism; (2) resistance involving overproduction (induction) of proteins involved in detoxification or sequestration that requires energy and may redirect resources away from growth, reproduction, or response to other challenges; (3) genes involved in toxicity resistance that might be linked to other cellular processes, i.e., through pleiotropy become altered to a suboptimal state during adaptation to contaminants; and (4) contaminant-induced selection for tolerance that may alter population genetic structure and reduce genetic diversity in fish populations. The costs associated with alterations in genetic structure would be dependent on the nature of the alterations themselves.

Despite abundant theory, not a great deal of experimental evidence exists that links toxicity resistance with reduced fitness in vertebrates. Most progress in this area has been made in studies of insecticide resistance in insects where pleiotropic effects of insecticide resistance on putative fitness components have been identified (Raymond et al. 2001). For example, specific alleles involved in pesticide resistance in mosquitoes have been associated with longer development times and shorter wing spans (Bourguet et al. 2004). In addition, single-point mutations conferring target site mutation resistance to DDT in aphids and houseflies were associated with reduced overwintering survival and decreased response to alarm pheromones (Foster et al. 2003). Fitness costs have been inferred in pesticide-resistant populations of *H. azteca*, aquatic invertebrates, whose pesticide tolerance is primarily associated with a heritable mutation at the primary pyrethroid target site, the VGSC (Weston et al. 2013, Major et al. 2018), but also other non-inherited mechanisms (Fung et al. 2021). Laboratory studies showed that, in comparison with wild-type (pesticide-sensitive) *H. azteca*, individuals from resistant populations have lower upper thermal and salinity tolerances and greater sensitivity to other pollutants, e.g., DDT and copper (Heim et al. 2018). Fulton et al. (2021) confirmed that pyrethroid-resistant *H. azteca* exhibited reduced thermal tolerance and lower survival and lipid levels under a high salinity regime; however, resistant *H. azteca* produced significantly more offspring than sensitive individuals regardless of salinity treatment. Thus, modest, conditional differences in fitness were identified in resistant vs. sensitive in *H. azteca*, and despite speculation about reduced genetic diversity and its potential impacts, resistant populations of *H. azteca* are persistent and widespread.

### Tolerance and Individual Fitness Components

Experiments designed to study fitness trade-offs in fish have involved capturing fish from clean and contaminated sites and comparing their responses to various stressors under clean laboratory conditions. While it is hypothesized that fish adapted to contaminated environments are less fit in uncontaminated environments, experimental evidence is mixed. Similar observations in insects and plants have been called *between-environment trade-offs*, referring to responses observed after the transfer of the study population from the environment to which they have adapted to a less stressful environment, presumably resembling their ancestral conditions (Shirley and Sibly 1999).

The documentation of fitness costs in pollution-tolerant killifish populations is mixed. Nacci and coworkers have demonstrated that NBH killifish have high fecundity in clean environments (Nacci et al. 2002a) and respond normally to pathogen challenges (Nacci et al. 2009). There is some evidence that lipid homeostasis is altered in NBH killifish, with both field- and laboratory-raised individuals maintaining greater amounts of abdominal fat and liver lipids than fish from reference populations. However, while plausible, it is not yet clear whether these differences have consequences for the adapted population. Between-environment trade-offs have been more frequently observed in studies with the creosote-adapted ER killifish. Poor condition (microbial infection, poor growth, emaciation, and increased mortality) relative to reference fish is routinely observed in AW killifish maintained in clean estuarine water (reviewed in Di Giulio and Clark (2015)). Meyer and Di Giulio (2003) reported similar observations, noting the poor survival of AW $F_1$ offspring compared with the survival of reference $F_1$ offspring in artificial seawater. In addition, AW offspring were more susceptible to acute, short-term exposure to ultra-low hypoxia and to toxicity associated with photoactivation of PAHs (Meyer and Di Giulio 2003). The authors concluded that, although the AW killifish are adapted to their contaminated environment, they are maladapted to life in clean environments. A series of more recent studies have identified potential bioenergetic costs associated with adaptation to polluted habitats throughout the ER. Juvenile F1 AW killifish had reduced swimming performance (measured by Ucrit) relative to reference fish (Brown et al. 2017a). Jayasundara et al. (2017) examined bioenergetic differences in wild-caught adult killifish from three pollutant-tolerant subpopulations from the ER estuary in comparison with two reference populations. Basal oxygen consumption was elevated in the hearts (ex vivo) of AW fish compared with reference fish. Additionally, individuals from all three pollution-tolerant ER subpopulations exhibited reduced swimming performance (Ucrit), elevated routine metabolic rates, and reduced aerobic scope compared with reference fish. Similarly, basal oxygen consumption was elevated compared with a reference population in F1 embryos from the Republic Creosoting site (another highly contaminant-resistant subpopulation in the ER estuary); these Republic F1 individuals also exhibited lower maximal metabolic rates and aerobic scope than fish from a reference population (Lindberg et al. 2017). Additionally, RADseq analyses comparing the AW population to two reference populations identified multiple candidate genes involved with metabolic processes and mitochondrial function associated with regions that appear to be under selection in the AW population (Jayasundara et al. 2017). Collectively, these studies show consistent evidence of differences in multiple measures of bioenergetic capability for killifish subpopulations from throughout the ER. However, the relevance of these and previously mentioned observations to actual field conditions is uncertain. In the field, the AW population appears to thrive and does not appear to be experiencing a great deal of mortality due to phototoxicity, microbial infection, or low oxygen, although the ER regularly experiences significant periods of low dissolved oxygen (Zhang et al. 2018). Furthermore, it is difficult for most studies comparing the fitness of pollution-tolerant and reference populations to distinguish between specific costs of pollution adaptation versus local adaptation to other ecological factors that may be co-incident with the pollution.

Several studies over the past several years have addressed between-environment trade-offs in metal-tolerant fish populations. Laboratory studies indicate that killifish collected from the methylmercury-contaminated environment of Piles Creek in New Jersey appear to be less capable of capturing natural prey than fish from less-contaminated sites (Weis and Khan 1991, Smith and Weis 1997). These studies suggest a possible relationship between reduced ability to capture prey with reduced levels of serotonin in the brain (Weis et al. 2001) and altered thyroid function (Zhou et al. 1999a, Zhou et al. 1999b). However, rather than trade-offs to adaptations, poor performance in fish collected from highly contaminated habitats could be related to the toxic effects of residual contamination (even in laboratory-depurated fish) or other adverse effects related to the poor water and dietary quality common in highly stressed environments. Trade-offs can be more easily interpreted when comparing the progeny of field-collected fish. For example, laboratory-raised F1 Piles Creek killifish also appear to reach maturity earlier than reference fish but have decreased growth rates and shorter life spans (Toppin et al. 1987). Xie and Klerks (2003) also reported that laboratory-reared populations of cadmium-resistant least killifish (*Heterandria formosa*) were less tolerant of heat stress than reference populations.

Further studies are needed to determine whether the mechanisms responsible for the poor performance of fish in between-environment trade-off experiments are related to the fitness costs of toxicity resistance or to other factors, including direct contaminant-mediated effects. Toxic chemicals can exert multiple effects on fish, including adapted individuals, through numerous pathways; for example, abundant literature documents the negative impacts of chemical contaminants on the immune system in fish, including species collected from the ER (Rice 2001) (see Chapter 5). Frederick et al. (2007) reported that AW killifish exhibit very low titers of antibodies against several common microbial pathogens (e.g., *Vibrio* spp.) compared with the antibody responses observed in reference fish. This observation may help to explain the susceptibility of these fish to microbial infection and altered performance in clean laboratory conditions (Meyer and Di Giulio 2003). In this case, it is unclear whether the altered immune function in AW fish and increased susceptibility to infection are a contaminant-mediated effect or a cost associated with toxicity resistance. Future studies with fish are needed to identify specific genetic loci associated with both resistance and fitness.

## Selection and Genetic Diversity

Some evidence exists that exposure to environmental contaminants can alter genetic diversity and population structure in marine organisms (Nevo et al. 1984, Nadig et al. 1998, Theodorakis 2003, Puritz and Toonen 2011). The primary mechanisms of genetic change could be through limiting migration, population declines, or natural selection. Pollution could effectively fragment a previously contiguous habitat by creating areas of unsuitable habitat, which could result in the erosion of genetic diversity in remaining patches due to enhanced genetic drift and reduced genetic exchange across the region (e.g., Puritz and Toonen (2011)). Pollution could also impair fitness, leading to population decline, and severe population declines or population bottlenecks can cause an erosion of genetic diversity genome-wide as a result of elevated genetic drift. Whether population decline causes reduced genetic diversity depends on the severity of the decline and the amount of genetic variation present in the population before the decline. *F. heteroclitus* populations harbor extremely high levels of genetic variation (Reid et al. 2017), such that genetic erosion in populations that occupy polluted sites is very subtle (Reid et al. 2016). In contrast, *F. grandis* harbors lower levels of genetic variation compared with *F. heteroclitus*, and the erosion of genetic diversity in populations occupying polluted sites is more apparent (Oziolor et al. 2019). Because genetic adaptation involves selection for tolerant genotypes, adaptation could result in the loss of sensitive genotypes and reduced genetic diversity. In contrast to genetic consequences of fragmentation and population decline, where diversity is impacted across the genome, natural selection reduces genetic variation only at loci that affect fitness. Therefore, the genetic signatures of natural selection are needles in the proverbial haystack. Regardless of the cause, evolutionary theory suggests that populations exhibiting decreased genetic diversity may be less suited to function in variable environments because of altered physiological efficiency, viability, fecundity, mating success, or resistance to disease (Mitton 1997) and may be limited from adapting to further environmental change.

Early studies assessing the genetic impacts of contaminants on fish populations used available techniques such as allozymes and microsatellite markers, which were extremely limited in their ability to interrogate much of the genome. Numerous field studies using these historical techniques associated environmental contamination with altered population structure and genetic diversity in fish (Gillespie and Guttman 1989, Kopp et al. 1992, Guttman 1994, Murdoch and Hebert 1994). However, even then, it was recognized that factors other than contamination could affect fish population genetics (Diamond et al. 1991, Gillespie and Guttman, 1989, Kopp et al. 1992, Mulvey and Diamond, 1991). Thus, field studies were often supported by laboratory studies providing limited evidence consistent with contaminant-associated selection, i.e., differential survival by genotype (Chagnon and Guttman 1989, Gillespie and Guttman 1989, Newman et al. 1989, Diamond et al. 1991, Sullivan and Lydy 1999). Overall, literature from this period suggests that the effect of environmental xenobiotics on genetic diversity is mixed (van Straalen and Timmermans 2002), with some researchers suggesting that more rigorous experimental designs are required to establish relationships between genetic alterations and environmental contamination (Belfiore and Anderson 2001, Theodorakis 2003) and distinguish genetic impacts caused by contemporary environmental change (e.g., pollution) from historical natural processes that structured populations (biogeographic barriers to migration, isolation by distance, etc.) (e.g., Whitehead et al. (2003)).

By the early 2000s, strong evidence was building for genetic adaptation to pollutants in populations of killifish residing in the NBH and ER superfund sites, supporting high expectations for evidence of strong selection, i.e., reduced genetic diversity, despite limitations imposed by genetic technologies available then. For example, killifish from the most heavily contaminated site, AW, were genetically distinct from fish collected from other ER sites. The data suggested that a locally stable population existed at the AW site; however, no evidence for decreased genetic diversity in this population was observed. Furthermore, no relationship was observed between sediment PAH concentrations and genetic diversity in fish collected from any of the study sites (Mulvey et al. 2002, 2003). Similar results were reported in studies with NBH killifish (McMillan et al. 2006, Roark, 2003). Differences in the genetic structure of the populations under investigation were related to the geographic distances separating them. No relationships were observed among genetic structure, sediment contamination at the collection sites, and PCB tolerance in the populations inhabiting those sites. Genetic diversity did not differ significantly between sites, and the NBH fish did not exhibit a reduction in genetic diversity relative to those collected from relatively clean sites. In fact, large historical effective population sizes and no evidence of recent population bottlenecks were found for 15 killifish populations resident to sites along the Atlantic coast of the United States, including NBH (Adams et al. 2006).

More recently, genome-scale tools that interrogate many loci genome-wide have provided nuanced insight into the genetic consequences of pollution-induced habitat fragmentation, population decline, and natural selection. Some of the best evidence comes from killifish populations that occupy polluted sites and have evolved resistance. They show evidence of subtle erosion of genetic diversity genome-wide consistent with a recent population bottleneck, and genetic diversity is nearly completely purged at a few specific genomic locations that provide large fitness advantages in polluted sites (Reid et al. 2016, Oziolor et al. 2019). Guided by what we know from pollution genetics studies published so far, and from the broader population genetics literature, we can expect that as populations decline following pollution, a decline in genetic diversity may follow, where the severity of erosion of genetic diversity will be scaled by the severity of the demographic bottleneck and will be larger for populations with smaller starting effective population sizes (lower levels of standing genetic variation). If natural selection is influential, then we can expect a reduction in genetic diversity at loci that confer an adaptive advantage (which likely represent a very small fraction of the overall genome) and in the regions of the genome physically linked to such loci (linked selection) (Nielsen 2005). The size of the flanking region affected by linked selection should scale with the size of the fitness advantage afforded by the embedded causative variant. That is, for loci that confer a very large selective advantage, linked selection will be stronger, where the adaptive haplotype is large, and therefore, the region of reduced genetic diversity flanking the adaptive locus will be large. With time, recombination will recover genetic diversity in flanking DNA, though genetic variation will remain reduced surrounding the adaptive locus. Loss of genetic variation genome-wide (from genetic erosion) and at adaptive loci and the DNA surrounding them (from linked selection) may have negative consequences for populations. This is because genetic variation is the raw material for evolutionary change. Reductions in genetic diversity may affect loci that underpin traits that may be important for fitness in future environments, such that these losses may affect the future evolvability of populations. This may be especially important in the Anthropocene where the pace and severity of environmental change are large.

## Evolutionary and Ecological Impacts of Tolerance

Aquatic environments modified by intense human industrial and agricultural activities are typically characterized by physical disruption (e.g., channelization), loss of habitat, excessive nutrient and sediment enrichment, and elevated levels of diverse contaminants. Because any of these factors can contribute to alterations in community structure and function, it is difficult to evaluate the relative contribution of any single stressor to changes observed at degraded sites; however, demonstration and characterization of toxicity resistance may allow identification of relationships between (1) specific biochemical and molecular responses (biomarkers) of toxicity and resistance, (2) survival of tolerant species, and (3) loss of sensitive species. Thus, toxicity resistance to specific contaminants in specific species inhabiting these sites may be one of the most informative observations regarding the contribution of contaminant exposure to altered community structure.

Chesser and Sugg (1996) developed an ecosystem dynamics model to evaluate the impacts of selective agents on populations and community dynamics. They considered the roles that species redundancy, adaptation, and immigration play in the ability of an ecosystem to cope with toxic insults. Toxic exposures that threaten sensitive genotypes and species can change both the structure (number of species and population sizes) and the function (energy flow) of an ecosystem. The loss of sensitive species and genotypes results in simplified ecosystems that are less predictable and demonstrate lowered resiliency and lowered ability to rebound from subsequent perturbations.

Other disturbances at the ecosystem scale and beyond may be associated with the occurrence of chemically adapted fish. Fish inhabiting contaminated sites can become vehicles for the transfer of toxic chemicals through the food web. This may be particularly important with persistent, lipophilic contaminants (e.g., organochlorine pesticides and DLCs) because of their tendency to accumulate in fish. Although trophic transfer is not restricted to adapted populations of fish, adapted individuals may be capable of accumulating and transferring higher levels of toxicants than non-adapted individuals. The transfer of chemicals from adapted individuals to non-adapted individuals can be lethal. In one study, for example, insecticide-resistant sunfish tolerated a diet of insecticide-resistant mosquitofish containing concentrations of endrin that killed control sunfish fed the same diet (Ferguson et al. 1966). In another study, a single-resistant mosquitofish eliminated enough endrin into 10 liters of water to kill all control mosquitofish maintained in that water (Culley and Ferguson 1969). Trophic transfer can result in the bioaccumulation of contaminants in top predators, fish-eating birds, and other wildlife that use the ecosystem as a feeding ground (Johanif et al. 2021).

Concern has also arisen over the acceleration in the evolution of drug-resistant disease organisms and pesticide-resistant insects (Palumbi 2001). Documented cases of toxicity resistance in fish appear to primarily involve populations of nonmigratory fish species limited to specific highly contaminated sites; however, considerable movement and breeding between local populations of fish are expected. Increases in the frequencies of resistance genes could therefore increase the frequency of deleterious alleles in the gene pool of non-adapted populations. In addition, it is likely that fish inhabiting moderately contaminated sites or migratory species that pass through contaminated areas seasonally may experience contaminant-mediated selection pressures that go unnoticed. In insects, selection for tolerant individuals appears to be associated with numerous costs related to survival and life-history strategies. The long-term consequences of continuous chemical exposure in populations of fish are unknown but could include similar costs.

## SUMMARY AND CONCLUSIONS

Most fish species cannot avoid exposure to complex mixtures of xenobiotics; individual fitness and population persistence are dependent on a battery of defense strategies and mechanisms. Physiological acclimation and genetic adaptation are general strategies of toxicity resistance that can be beneficial, although often biologically costly, under certain conditions related to species and xenobiotic exposure attributes. General mechanisms by which tolerance is produced often reflect alterations in pathways by which toxicity is exerted; for example, increased tolerance may be the result of altered regulation of proteins involved in toxicant sequestration (binding), toxicant biotransformation (degradation), or efflux (elimination). Toxicity resistance can also result from alterations to biological targets including cellular receptors, which can reduce the effectiveness of toxicant–receptor interactions and subsequent signal transduction. Much of our understanding of chemical resistance mechanisms in diverse organisms comes from the fields of medicine (drug and antibiotic resistance) and agriculture (pesticide resistance in insects). The highly conserved nature of the pathways involved will allow us to continue to learn from these systems. However, our understanding of toxicity resistance in fishes in a rapidly changing world is maturing beyond isolated mechanisms to cross-regulated integrated defense networks. Evolved tolerance is an unlikely outcome of chronic pollution exposure in fish, but a few exceptional well-studied cases reveal fish populations that are highly desensitized to specific classes of xenobiotics relative to other populations of the same species, and these genetic adaptations may also come at a cost. For example, pollution-adapted fish, derived from populations with histories of multigenerational toxic exposures, when maintained in clean laboratory environments sometimes have a reduced ability to respond to environmental stressors such as low oxygen, microbial exposure, or to perform vital functions such as prey capture (e.g., capture prey) when compared with responses in observed individuals. Furthermore, pollution-induced population decline or natural selection can erode genetic diversity that could limit the fuel for adaptation to future environmental change. However, these costs and consequences have not been observed in all adapted fish populations; recent studies comparing populations of the same fish species uniquely adapted to different local exposures are providing opportunities to correlate specific relationships (e.g., pleiotropic relationships) between resistance and performance in field situations. Future studies will focus on toxicity resistance strategies for fish species and populations in the moderately contaminated but rapidly changing environments that are now typical of the Anthropocene. Additional studies will move beyond a focus on highly toxic xenobiotics that exist as legacies of industrial contamination and take into consideration contaminants of immediate and emerging concerns (e.g., personal-care products, pharmaceuticals, and wastewater organic contaminants) that likely exert subtle effects during the long-term, low-level exposures that are now common. Studies on adaptation in Atlantic and Gulf killifish populations chronically exposed to persistent chemical contaminants and the genomic tools that have been developed to advance their study are providing a unique opportunity to understand more holistically the impacts of xenobiotic exposures on our environment.

## ACKNOWLEDGMENTS AND DISCLAIMER

The authors are indebted to Dr. Peter Van Veld who coauthored the original chapter with DN (Toxicology of Fishes, 1st edition) upon which this chapter was based. The research described in this chapter has been funded in part by the U.S. Environmental Protection Agency (EPA) (DEN, BWC). It has been subjected to review by the Office of Research and Development and approved for publication. Approval does not signify that the contents reflect the views of the agency, nor does mention of trade names or commercial products constitute endorsement or recommendation for use.

## REFERENCES

Adams, S. M., Lindmeier, J. B., and Duvernell, D. D. 2006. Microsattelite analysis of the phylogeography and secondary contact hypotheses for the killifish, *Fundulus heteroclitus*. *Molecular Ecology* **15**:1109–1123.

Adeyemi, J. A., and P. L. Klerks. 2013. Occurrence of copper acclimation in the least killifish *Heterandria formosa*, and associated biochemical and physiological mechanisms. *Aquatic Toxicology* **130**:51–57.

Aluru, N., S. I. Karchner, and M. E. Hahn. 2011. Role of DNA methylation of AHR1 and AHR2 promoters in differential sensitivity to PCBs in Atlantic Killifish, Fundulus heteroclitus. *Aquatic Toxicology* **101**:288–294.

Ames, B. N., M. Profet, and L. S. Gold. 1990a. Dietary pesticides (99.99% all natural). *Proceedings of the National Academy of Sciences* **87**:7777–7781.

Ames, B. N., M. Profet, and L. S. Gold. 1990b. Nature's chemicals and synthetic chemicals: comparative toxicology. *Proceedings of the National Academy of Sciences* **87**:7782–7786.

Andreasen, J. K. 1985. Insecticide resistance in mosquitofish of the lower Rio Grande Valley of Texas - an ecological hazard. *Archives of Environmental Contamination and Toxicology* **14**:573–577.

Armknecht, S. L., S. L. Kaattari, and P. A. Van Veld. 1998. An elevated glutathione S-transferase in creosote-resistant mummichog (*Fundulus heteroclitus*). *Aquatic Toxicology* **41**:1–16.

Arola, H. E., Karjalainen, J., EEva-Riikka, V., Vaisanen, A., Kukkonen, J. V. K., Karjalainen, A. K. 2017. Tolerance of whitefish (Coregonus lavaretus) early life stages to manganese sulfate is affected by the parents. *Environmental Toxicology and Chemistry* **36** (5):1343–1353.

Arzuaga, X., W. Calcano, and A. Elskus. 2004. The DNA demethylating agent 5-azacytidine does not restore CYP1A induction in PCB resistant Newark Bay killifish (*Fundulus heteroclitus*). *Marine Environmental Research* **58**:517–520.

Bacanskas, L. R., J. Whitaker, and R. T. Di Giulio. 2004. Oxidative stress in two populations of killifish (*Fundulus heteroclitus*) with differing contaminant exposure histories. *Marine Environmental Research* **58**:597–601.

Bard, S. M. 2000. Multixenobiotic resistance as a cellular defense mechanism in aquatic organisms. *Aquatic Toxicology* **48**:357–389.

Bard, S. M., S. M. Bello, M. E. Hahn, and J. J. Stegeman. 2002a. Expression of P-glycoprotein in killifish (*Fundulus heteroclitus*) exposed to environmental xenobiotics. *Aquatic Toxicology* **59**:237–251.

Bard, S. M., B. R. Woodin, and J. J. Stegeman. 2002b. Expression of P-glycoprotein and cytochrome P450 1A in intertidal fish (*Anoplarchus purpurescens*) exposed to environmental contaminants. *Aquatic Toxicology* **60**:17–32.

Bass, C., and L. M. Field. 2011. Gene amplification and insecticide resistance. *Pest Management Science* **67**:886–890.

Beattie, J. H., and D. Pascoe. 1978. Cadmium uptake by rainbow trout, Salmo gairdneri eggs and alevins. *Journal of Fish Biology* **13**:631–637.

Belanger-Deschenes, S., P. Couture, P. G. C. Campbell, and L. Bernatchez. 2013. Evolutionary change driven by metal exposure as revealed by coding SNP genome scan in wild yellow perch (*Perca flavescens*). *Ecotoxicology* **22**:938–957.

Belfiore, N. M., and S. L. Anderson. 2001. Effects of contaminants on genetic patterns in aquatic organisms: a review. *Mutation Research-Reviews in Mutation Research* **489**:97–122.

Bello, S. M. 1999. *Characterization of Resistance to Halogenated Aromatic Hydrocarbons in a Population of Fundulus Heteroclitus from a Marine Superfund Site*. Woods Hole Oceanographic Institute, Woods Hole, MA.

Bello, S. M., D. G. Franks, J. J. Stegeman, and M. E. Hahn. 2001. Acquired resistance to Ah receptor agonists in a population of Atlantic killifish (Fundulus heteroclitus) inhabiting a marine Superfund site: In vivo and in vitro studies on the inducibility of xenobiotic metabolizing enzymes. *Toxicological Sciences* **60**:77–91.

Benedetti, M., M. E. Giuliani, and F. Regoli. 2015. Oxidative metabolism of chemical pollutants in marine organisms: molecular and biochemical biomarkers in environmental toxicology. *Annals N.Y. Academy Science*, **1340**:8–19. https://doi.org/10.1111/nyas.12698

Benson, W. H., and W. J. Birge. 1985. Heavy metal tolerance and metallothionein in fathead minnows - results from field and laboratory investigations. *Environmental Toxicology and Chemistry* **4**:209–217.

Bergelson, J., and C. B. Purrington. 1996. Surveying patterns in the cost of resistance in plants. *The American Naturalist* **148**:536–558.

Bharati, M., and D. Saha. 2021. Insecticide resistance status and biochemical mechanisms involved in Aedes mosquitoes: a scoping review. *Asian Pacific Journal of Tropical Medicine* **14**:52–63.

Billet, L. S., and J. T. Hoverman. 2020. Pesticide tolerance induced by a generalized stress response in wood frogs (*Rana sylvatica*). *Ecotoxicology* **29**:1476–1485.

Billiard, S. M., A. R. Timme-Laragy, D. M. Wassenberg, C. Cockman, and R. T. Di Giulio. 2006. The role of the aryl hydrocarbon receptor pathway in mediating synergistic developmental toxicity of polycyclic aromatic hydrocarbons to zebrafish. *Toxicological Sciences* **92**:526–536.

Birnbaum, L. S. 1994. The mechanism of dioxin toxicity: relationship to risk assessment. *Environ Health Perspect* **102**:157–167.

Black, D. E., R. Gutjahr-Gobell, R. J. Pruell, B. Bergen, and A. E. McElroy. 1998a. Effects of a mixture of non-ortho- and mono-ortho-polychlorinated biphenyls on reproduction in *Fundulus heteroclitus* (Linnaeus). *Environmental Toxicology and Chemistry* **17**:1396–1404.

Black, D. E., R. Gutjahr-Gobell, R. J. Pruell, B. Bergen, L. Mills, and A. E. McElroy. 1998b. Reproduction and polychlorinated biphenyls in *Fundulus heteroclitus* (Linnaeus) from New Bedford Harbor, Massachusetts, USA. *Environmental Toxicology and Chemistry* **17**:1405–1414.

Bonner, J. C., and J. D. Yarbrough. 1989. Role of the brain t-butylbicyclophosphorothionate receptor in vertebrate resistance to endrin, 1, 1, 1-trichloro-2, 2-bis (p-chlorophenyl) ethane and cypermethrin. *Journal of Pharmacology and Experimental Therapeutics* **249**:149–154.

Borst, P., A. Schinkel, J. Smit, E. Wagenaar, L. Van Deemter, A. Smith, E. Eijdems, F. Baas, and G. Zaman. 1993. Classical and novel forms of multidrug resistance and the physiological functions of P-glycoproteins in mammals. *Pharmacology & therapeutics* **60**:289–299.

Bourguet, D., T. Guillemaud, C. Chevillon, and M. Raymond. 2004. Fitness costs of insecticide resistance in natural breeding sites of the mosquito *Culex pipiens*. *Evolution* **58**:128–135.

Boyd, C. E., and D. E. Ferguson. 1964. Susceptibility and resistance of mosquito fish to several insecticides. *Journal of Economic Entomology* **57**:430–431.

Brazzola, G., N. Chevre, and C. Wedekind. 2014. Additive genetic variation for tolerance to estrogen pollution in natural populations of Alpine whitefish (Coregonus sp., Salmonidae). *Evolutionary Applications* **7**:1084–1093.

Brette, F., B. Machado, C. Cros, J. P. Incardona, N. L. Scholz, and B. A. Block. 2014. Crude oil impairs cardiac excitation-contraction coupling in fish. *Science* **343**:772–776.

Brown, A. W. A. 1958. *Insecticide Resistance in Arthropods*. Monograph Series 38, World Health Organization, Geneva, CH. 240 p.

Brown, D. R., J. M. Bailey, A. N. Oliveri, E. D. Levin, and R. T. Di Giulio. 2016. Developmental exposure to a complex PAH mixture causes persistent behavioral effects in naive Fundulus heteroclitus (killifish) but not in a population of PAH-adapted killifish. *Neurotoxicology and Teratology* **53**:55–63.

Brown, D. R., J. Thompson, M. Chernick, D. E. Hinton, and R. T. Di Giulio. 2017a. Later life swimming performance and persistent heart damage following subteratogenic PAH mixture exposure in the Atlantic killifish (Fundulus heteroclitus). *Environmental Toxicology and Chemistry* **36**:3246–3253.

Brown, S. M., A. Heguy, P. Zappile, H. Chen, A. Goradia, Y. Wang, Y. Hao, N. K. Roy, K. Vitale, R. C. Chambers, and I. Wirgin. 2017b. A dramatic difference in global gene expression between TCDD-treated Atlantic tomcod larvae from the resistant Hudson river and a nearby sensitive population. *Genome Biology and Evolution* **9**:2251–2264.

Bryan, G. 1976. Some aspects of heavy metal tolerance in aquatic organisms. *Effects of Pollutants on Aquatic Organisms* **2**:7–34.

Bucheli, T. D., and K. Fent. 1995. Induction of cytochrome P450 as a biomarker for environmental contamination in aquatic ecosystems. *Critical Reviews in Environmental Science and Technology* **25**:201–268.

Buckley, J., M. Roch, J. McCarter, C. Rendall, and A. Matheson. 1982. Chronic exposure of Coho salmon to sublethal concentrations of copper. 1. Effect of growth, on accumulation and distribution of copper, and on copper tolerance. *Comparative Biochemistry & Physiology*, **72C**:15–19.

Bunger, M. K., S. M. Moran, E. Glover, T. L. Thomae, G. P. Lahvis, B. C. Lin, and C. A. Bradfield. 2003. Resistance to 2,3,7,8-tetrachlorodibenzo-p-dioxin toxicity and abnormal liver development in mice carrying a mutation in the nuclear localization sequence of the aryl hydrocarbon receptor. *Journal of Biological Chemistry* **278**:17767–17774.

Burnett, K. G., L. J. Bain, W. S. Baldwin, G. V. Callard, S. Cohen, R. T. Di Giulio, D. H. Evans, M. Gomez-Chiarri, M. E. Hahn, C. A. Hoover, S. I. Karchner, F. Katoh, D. L. MacLatchy, W. S. Marshall, J. N. Meyer, D. E. Nacci, M. F. Oleksiak, B. B. Rees, T. D. Singer, J. J. Stegeman, D. W. Towle, P. A. Van Veld, W. K. Vogelbein, A. Whitehead, R. N. Winn, and D. L. Crawford. 2007. *Fundulus* as the premier teleost model in environmental biology: Opportunities for new insights using genomics. *Comparative Biochemistry and Physiology D-Genomics & Proteomics* **2**:257–286.

Carlson, E. A., Roy, N. K. Marzinke, M., and Wirgin, I.. 2003. Identification of novel genes associated with natural resistance to PCB-induced early life stage toxicity in the Atlantic tomcod populations. *Proceedings of the 24th Annual Meeting of the Society of Environmental Toxicology and Chemistry*, November 9-13, Austin, TX.

Carney, S. A., J. Chen, C. G. Burns, K. M. Xiong, R. E. Peterson, and W. Heideman. 2006. Aryl hydrocarbon receptor activation produces heart-specific transcriptional and toxic responses in developing zebrafish. *Molecular Pharmacology* **70**:549–561.

Casida, J. E. 2009. Pest toxicology: the primary mechanisms of pesticide action. *Chemical Research in Toxicology* **22**:609–619.

Casida, J. E., and K. A. Durkin. 2013. Neuroactive insecticides: targets, selectivity, resistance, and secondary effects. *Annual Review of Entomology* **58**:99–117.

Castaldo, G., G. Flipkens, M. Pillet, R. M. Town, L. Bervoets, R. Blust, and G. D. Boeck. 2020. Antagonistic bioaccumulation of waterborne Cu(II) and Cd(II) in common carp (*Cyprinus carpio*) and effects on ion-homeostasis and defensive mechanisms. *Aquatic Toxicology* **226**, 105561.

Celander, M., and L. Förlin. 1995. Decreased responsiveness of the hepatic cytochrome P450 1A1 system in rainbow trout (*Oncorhynchus mykiss*) after prolonged exposure to PCB. *Aquatic Toxicology* **33**:141–153.

Celander, M., J. J. Stegeman, and L. Förlin. 1996. CYP1A1-, CYP2B-and CYP3A-like proteins in rainbow trout (*Oncorhynchus mykiss*) liver: CYP1A1-specific down-regulation after prolonged exposure to PCB. *Marine Environmental Research* **42**:283–286.

Celander, M. C., J. V. Goldstone, N. R. Brun, B. Clark, S. Jayaraman, D. Nacci, and J. J. Stegeman. 2021. Resistance to Cyp3a induction by polychlorinated biphenyls, including non-dioxin-like PCB153, in gills of killifish (*Fundulus heteroclitus*) from New Bedford Harbor. *Environmental Toxicology and Pharmacology* **83**:103580.

Chagnon, N. L., and S. I. Guttman. 1989. Differential survivorship of allozyme genotypes in mosquitofish populations exposed to copper or cadmium. *Environmental Toxicology and Chemistry: An International Journal* **8**:319–326.

Chambers, J. E., and J. D. Yarbrough. 1979. A seasonal study of microsomal mixed-function oxidase components in insecticide-resistant and susceptible mosquitofish, Gambusia affinis. *Toxicology and Applied Pharmacology* **48**:497–507.

Chambers, R., D. Witing, D. Cerino, and I. Wirgin. 2003. PCB-induced early life-stage toxicity in Atlantic tomcod from the Hudson River and the Miramichi River. Pages 9-13 *in Proceedings of the 24th Annual Meeting of the Society of Environmental Toxicology and Chemistry*, November 9-13, Austin, TX.

Chan, W. K., G. Yao, Y. Z. Gu, and C. A. Bradfield. 1999. Cross-talk between the aryl hydrocarbon receptor and hypoxia inducible factor signaling pathways. *Demonstration of Competition and Compensation. Journal of Biological Chemistry* **274**:12115–12123.

Chang, C. T., H. Y. Chung, H. T. Su, H. P. Tseng, W. S. Tzou, and C. H. Hu. 2013. Regulation of zebrafish CYP3A65 transcription by AHR2. *Toxicology and Applied Pharmacology* **270**:174–184.

Chedrese, P. J., and F. Feyles. 2001. The diverse mechanism of action of dichlorodiphenyldichloroethylene (DDE) and methoxychlor in ovarian cells in vitro. *Reproductive Toxicology* **15**:693–698.

Chesser, R. K., and D. W. Sugg. 1996. Toxicant as selective agents in population and community dynamics. Pages 293–317 in M. C. Newman and C. H. Jagoe, editors. *Ecotoxicology: A Hierarchical Treatment*. Taylor and Francis Publishers, Inc., Washington, DC.

Chevillon, C., D. Bourguet, F. Rousset, N. Pasteur, and M. Raymond. 1997. Pleiotropy of adaptive changes in populations: comparisons among insecticide resistance genes in *Culex pipiens*. *Genetics Research* **70**:195–204.

Childs, S., R. L. Yeh, E. Georges, and V. Ling. 1995. Identification of a sister gene to P-glycoprotein. *Cancer Research* **55**:2029–2034.

Clark, B. W., A. J. Bone, and R. T. Di Giulio. 2014. Resistance to teratogenesis by F1 and F2 embryos of PAH-adapted *Fundulus heteroclitus* is strongly inherited despite reduced recalcitrance of the AHR pathway. *Environmental Science and Pollution Research* **21**:13898–13908.

Clark, B. W., E. M. Cooper, H. M. Stapleton, and R. T. Di Giulio. 2013. Compound- and mixture-specific differences in resistance to polycyclic aromatic hydrocarbons and PCB-126 among *Fundulus heteroclitus* Subpopulations throughout the Elizabeth River Estuary (Virginia, USA). *Environmental Science & Technology* **47**:10556–10566.

Clark, B. W., and R. T. Di Giulio. 2012. *Fundulus heteroclitus* adapted to PAHs are cross-resistant to multiple insecticides. *Ecotoxicology* **21**:465–474.

Clark, B. W., C. W. Matson, D. Jung, and R. T. Di Giulio. 2010. AHR2 mediates cardiac teratogenesis of polycyclic aromatic hydrocarbons and PCB-126 in Atlantic killifish (*Fundulus heteroclitus*). *Aquatic Toxicology* **99**:232–240.

Cooper, P., K. Reece, W. Vogelbein, and P. Van Veld. 1999a. P-glycoprotein related sequences from mummichog liver and intestine. Pages 25–29 *in* Proceedings of the Tenth International Symposium on Pollutant Responses in Marine Organisms, April.

Cooper, P. S., W. Vogelbein, and P. A. Van Veld. 1996. Immunohistochemical and immunoblot detection of P-glycoprotein in normal and neoplastic fish liver. *Techniques in Aquatic Toxicology* **1**:307–325.

Cooper, P. S., W. K. Vogelbein, and P. A. Van Veld. 1999b. Altered expression of the xenobiotic transporter P-glycoprotein in liver and liver tumours of mummichog (*Fundulus heteroclitus*) from a creosote-contaminated environment. *Biomarkers* **4**:48–58.

Cormier, S. M., and R. N. Racine. 1990. Histopathology of Atlantic tomcod: a possible monitor of xenobiotics in northeast tidal rivers and estuaries. Pages 59–71 *in* J. F. McCarthy and L. R. Shugart, editors. *Biomarkers of Environmental Contamination*. CRC Press, Boca Raton, FL.

Cornwall, R., B. H. Toomey, S. Bard, C. Bacon, W. M. Jarman, and D. Epel. 1995. Characterization of multixenobiotic/multidrug transport in the gills of the mussel Mytilus californianus and identification of environmental substrates. *Aquatic Toxicology* **31**:277–296.

Courtenay, S. C., C. M. Grunwald, G. L. Kreamer, W. L. Fairchild, J. T. Arsenault, M. Ikonomou, and Wirgin, II. 1999. A comparison of the dose and time response of CYP1A1 mRNA induction in chemically treated Atlantic tomcod from two populations. *Aquatic Toxicology* **47**:43–69.

Coustau, C., C. Chevillon, and R. ffrench-Constant. 2000. Resistance to xenobiotics and parasites: can we count the cost? *Trends in Ecology & Evolution* **15**:378–383.

Crawford, D. L., P. M. Schulte, A. Whitehead, and M. F. Oleksiak. 2020. Evolutionary physiology and genomics in the highly adaptable killifish (fundulus heteroclitus). *Comprehensive Physiology* **10**:637–671.

Culley, D. D., and D. E. Ferguson. 1969. Patterns of insecticide resistance in mosquitofish, *Gambusia affinis. Journal of the Fisheries Research Board of Canada* **26**:2395.

Daborn, P., J. Yen, M. Bogwitz, G. Le Goff, E. Feil, S. Jeffers, N. Tijet, T. Perry, D. Heckel, and P. Batterham. 2002. A single P450 allele associated with insecticide resistance in Drosophila. *Science* **297**:2253–2256.

Denison, M. S., A. A. Soshilov, G. He, D. E. DeGroot, and B. Zhao. 2011. Exactly the same but different: promiscuity and diversity in the molecular mechanisms of action of the aryl hydrocarbon (dioxin) receptor. *Toxicological Sciences* **124**:1–22.

Di Giulio, R., W. Benson, B. Sanders, and P. A. Van Veld. 1995. Xenobiotic metabolism and biochemical effects. Pages 523–561 *in* G. Rand, editor. *Fundamentals of Aquatic Toxicology: Effects, Environmental Fate and Risk Assessment*. CRC Press, Washington, DC.

Di Giulio, R. T., and B. W. Clark. 2015. The Elizabeth river story: a case study in evolutionary toxicology. *Journal of Toxicology and Environmental Health. Part B, Critical Reviews* **18**:259–298.

Diamond, S., M. Newman, M. Mulvey, and S. Guttman. 1991. Allozyme genotype and time-to-death of mosquitofish, *Gambusia holbrooki*, during acute inorganic mercury exposure: a comparison of populations. *Aquatic Toxicology* **21**:119–134.

Dietrich, C. 2016. Antioxidant functions of the aryl hydrocarbon receptor. *Stem Cells International* **2016**. Article ID 7943495, 10 pages http://dx.doi.org/10.1155/2016/7943495.

Dixon, D., and J. Sprague. 1981a. Acclimation-induced changes in toxicity of arsenic and cyanide to rainbow trout, *Salmo gairdneri* Richardson. *Journal of Fish Biology* **18**:579–589.

Dixon, D., and J. Sprague. 1981b. Acclimation to copper by rainbow trout (*Salmo gairdneri*)-a modifying factor in toxicity. *Canadian Journal of Fisheries and Aquatic Sciences* **38**:880–888.

Dixon, D., and J. Sprague. 1981c. Copper bioaccumulation and hepatoprotein synthesis during acclimation to copper by juvenile rainbow trout. *Aquatic Toxicology* **1**:69–81.

Doehmer, J., A. R. Goeptar, and N. P. Vermeulen. 1993. Cytochromes P450 and drug resistance. *Cytotechnology* **12**:357–366.

Doige, C. A., and G. F.-L. Ames. 1993. ATP-dependent transport systems in bacteria and humans: relevance to cystic fibrosis and multidrug resistance. *Annual Review of Microbiology* **47**:291–320.

Duncan, D., and J. Klaverkamp. 1983. Tolerance and resistance to cadmium in white suckers (*Catostomus commersoni*) previously exposed to cadmium, mercury, zinc, or selenium. *Canadian Journal of Fisheries and Aquatic Sciences* **40**:128–138.

Eaton, D. L., and T. K. Bammler. 1999. Concise review of the glutathione S-transferases and their significance to toxicology. *Toxicological Sciences: An Official Journal of the Society of Toxicology* **49**:156–164.

Eide, M., X. Zhang, O. A. Karlsen, J. V. Goldstone, J. Stegeman, I. Jonassen, and A. Goksoyr. 2021. The chemical defensome of five model teleost fish. *Scientific Reports* **11**:10546.

Eisler, R., and A. A. Belisle. 1996. *Planar PCB Hazards to Fish, Wildlife, and Invertebrates: A Synoptic Review*. US Department of the Interior, National Biological Service, Washington, DC.

Elskus, A. A. 2001. Toxicant resistance in wildlife: fish populations. Pages 273–276 in L.W. Robertson and L.G. Hansen, editors. *PCBs: Recent Advances in Environmental Toxicology and Health Effects*. University Press of Kentucky, Lexington, KY.

Elskus, A. A., E. Monosson, A. E. McElroy, J. J. Stegeman, and D. S. Woltering. 1999. Altered CYP1A expression in *Fundulus heteroclitus* adults and larvae: a sign of pollutant resistance? *Aquatic Toxicology* **45**:99–113.

Endler, J. A. 1986. *Natural Selection in the Wild*. Princeton University Press, Princeton, NJ. 336 pp.

Epel, D., T. Luckenbach, C. N. Stevenson, L. A. Macmanus-Spencer, A. Hamdoun, and T. Smital. 2008. Efflux transporters: Newly appreciated roles in protection against pollutants. *Environmental Science & Technology* **42**:3914–3920.

Eriksson, L. C., and G. N. Andersson. 1992. Membrane biochemistry and chemical hepatocarcinogenesis. *Biochemistry and Molecular Biology* **27**:1–55.

Fabacher, D. L., and H. Chambers. 1973. Apparent resistance to pyrethroids in organochlorine-resistant mosquitofish. *in* Proceedings of the Annual Conference of the Southeast Associated Game Fisheries Commission. pp. 461–464. Knoxville, Tennessee, USA, October 22-25.

Fan, W., Z. Xu, and W.-X. Wang. 2015. Contrasting metal detoxification in polychaetes, bivalves and fish from a contaminated bay. *Aquatic Toxicology* **159**:62–68.

Ferguson, D. 1963. In less than 20 years: Mississippi Delta wildlife developing resistance to pesticides. *Agricultural Chemicals* **18**:32–34.

Ferguson, D. E. 1969. Characteristics and significance of resistance to insecticides in fishes. *Proceedings of the Reservoir Fishery Resources Symposium*. April 5-7, Athens, GA, USA.

Ferguson, D. E., and C. R. Bingham. 1966. Endrin resistance in the yellow bullhead, *Ictalurus natalis. Transactions of the American Fisheries Society* **95**:325–326.

Ferguson, D. E., D. D. Culley, W. D. Cotton, and R. P. Dodds. 1964. Resistance to chlorinated hydrocarbon insecticides in three species of freshwater fish. *Bioscience* **14**:43–44.

Ferguson, D. E., J. L. Ludke, and G. G. Murphy. 1966. Dynamics of endrin uptake and release by resistant and susceptible strains of mosquitofish. *Transactions of the American Fisheries Society* **95**:335–344.

Fernández-Delgado, C. 1989. Life-history patterns of the salt-marsh killifish *Fundulus heteroclitus* (L.) introduced in the estuary of the Guadalquivir river (South West Spain). *Estuarine, Coastal and Shelf Science* **29**:573–582.

Fernandez-Salguero, P., T. Pineau, D. M. Hilbert, T. McPhail, S. S. T. Lee, S. Kimura, D. W. Nebert, S. Rudikoff, J. M. Ward, and F. J. Gonzalez. 1995. Immune system impairment and hepatic fibrosis in mice lacking the dioxin-binding AH receptor. *Science* **268**:722–726.

Fernandez-Salguero, P. M., D. M. Hilbert, S. Rudikoff, J. M. Ward, and F. J. Gonzalez. 1996. Aryl-hydrocarbon receptor-deficient mice are resistant to 2,3,7,8-tetrachlorodibenzo-*p*-dioxin-induced toxicity. *Toxicology and Applied Pharmacology* **140**:173–179.

ffrench-Constant, R. H. 2013. The molecular genetics of insecticide resistance. *Genetics* **194**:807.

ffrench-Constant, R. H., P. J. Daborn, and G. Le Goff. 2004. The genetics and genomics of insecticide resistance. *Trends in Genetics* **20**:163–170.

Finley, M. T., D. E. Ferguson, and J. L. Ludke. 1970. Possible selective mechanisms in the development of insecticide-resistant fish. *Pesticides Monitoring Journal* **3**:212–218.

Fletcher, C., and W. McKay. 1993. Polychlorinated dibenzo-p-dioxins (PCDDs) and dibenzofurans (PCDFs) in the aquatic environment-a literature review. *Chemosphere* **26**:1041–1069.

Fletcher, N., A. Hanberg, and H. Håkansson. 2001. Hepatic vitamin A depletion is a sensitive marker of 2,3,7,8-tetrachlorodibenzo-p-dioxin (TCDD) exposure in four rodent species. *Toxicological Sciences* **62**:166–175.

Förlin, L., and M. Celander. 1995. Studies of the inducibility of P450 1A in perch from the PCB-contaminated lake Järnsjön in Sweden. *Marine Environmental Research* **39**:85–88.

Foster, S., S. Young, M. Williamson, I. Duce, I. Denholm, and G. Devine. 2003. Analogous pleiotropic effects of insecticide resistance genotypes in peach-potato aphids and houseflies. *Heredity* **91**:98–106.

Fournie, J. W., and W. K. Vogelbein. 1994. Exocrine pancreatic neoplasms in the mummichog (*Fundulus heteroclitus*) from a creosote-contaminated site. *Toxicologic Pathology* **22**:237–247.

Frederick, L. A., P. A. Van Veld, and C. D. Rice. 2007. Bioindicators of immune function in creosote-adapted estuarine killifish, *Fundulus heteroclitus*. *Journal of Toxicology and Environmental Health-Part A-Current Issues* **70**:1433–1442.

Fulton, C. A., K. E. H. Hartz, N. W. Fuller, L. N. Kent, S. E. Anzalone, T. M. Miller, R. E. Connon, H. P. Poynton, and M. J. Lydy. 2021. Fitness costs of pesticide resistance in Hyalella azteca under future climate change scenarios. *Science of the Total Environment* **753**:141945.

Fung, C. Y., K. Y. Zhu, K. Major, H. C. Poynton, K. E. H. Hartz, G. Wellborn, and M. J. Lydy. 2021. The contribution of detoxification pathways to pyrethroid resistance in *Hyalella azteca*. *Environmental Pollution* **284**:117158.

Furness, S. G., and F. Whelan. 2009. The pleiotropy of dioxin toxicity--xenobiotic misappropriation of the aryl hydrocarbon receptor's alternative physiological roles. *Pharmacol Ther* **124**:336–353.

Futuyma, D. 1997. *Evolutionary Biology*, 3rd edition. Sinauer Associates, Inc., Sunderland, MA, 763 pp.

Galgani, F., R. Cornwall, B. H. Toomey, and D. D. Epel. 1996. Interaction of environmental xenobiotics with a multixenobiotic defense mechanism in the bay mussel *Mytilus galloprovincialis* from the coast of California. *Environmental Toxicology and Chemistry: An International Journal* **15**:325–331.

Garland, T. J., and S. Adolph. 1991. Physiological differentiation of vertebrate populations. *Annual Review of Ecology and Systematics* **22**:193–228.

Gillespie, R. B., and S. I. Guttman. 1989. Effects of contaminants on the frequencies of allozymes in populations of the central stoneroller. *Environmental Toxicology and Chemistry: An International Journal* **8**:309–317.

Goldstone, J. V., A. G. McArthur, A. Kubota, J. Zanette, T. Parente, M. E. Jönsson, D. R. Nelson, and J. J. Stegeman. 2010. Identification and developmental expression of the full complement of Cytochrome P450 genes in Zebrafish. *BMC Genomics* **11**:643.

Gooch, J. W., A. A. Elskus, P. J. Kloepper-Sams, M. E. Hahn, and J. J. Stegeman. 1989. Effects of *ortho* and non-*ortho* substituted polychlorinated biphenyl congeners on the hepatic monooxygenase system in scup (*Stenotomus chrysops*). *Toxicology and Applied Pharmacology* **98**:422–433.

Goodale, B. C., J. La Du, S. C. Tilton, C. M. Sullivan, W. H. Bisson, K. M. Waters, and R. L. Tanguay. 2015. Ligand-specific transcriptional mechanisms underlie aryl hydrocarbon receptor-mediated developmental toxicity of oxygenated PAHs. *Toxicological Sciences* **147**:397–411.

Goto, D., and W. G. Wallace. 2010. Metal intracellular partitioning as a detoxification mechanism for mummichogs (*Fundulus heteroclitus*) living in metal-polluted salt marshes. *Marine Environmental Research* **69**:163–171.

Gottesman, M. M. 2002. Mechanisms of cancer drug resistance. *Annual Review of Medicine* **53**:615–627.

Grans, J., B. Wassmur, M. Fernandez-Santoscoy, J. Zanette, B. R. Woodin, S. I. Karchner, D. E. Nacci, D. Champlin, S. Jayaraman, M. E. Hahn, J. J. Stegeman, and M. C. Celander. 2015. Regulation of pregnane-X-receptor, CYP3A and P-glycoprotein genes in the PCB-resistant killifish (*Fundulus heteroclitus*) population from New Bedford Harbor. *Aquatic Toxicology* **159**:198–207.

Grant, D. F., and B. D. Hammock. 1992. Genetic and molecular evidence for a trans-acting regulatory locus controlling glutathione S-transferase-2 expression in Aedes aegypti. *Molecular and General Genetics* **234**:169–176.

Gribble, G. W. 1994. The natural production of chlorinated compounds. *Environmental Science & Technology* **28**:310A–319A.

Grosell, M., J. C. McGeer, and C. M. Wood. 2001. Plasma copper clearance and biliary copper excretion are stimulated in copper-acclimated trout. *American Journal of Physiology. Regulatory, Integrative and Comparative Physiology* **280**:R796–R806.

Gutjahr-Gobell, R. E., D. E. Black, L. J. Mills, R. J. Pruell, B. K. Taplin, and S. Jayaraman. 1999. Feeding the mummichog (*Fundulus heteroclitus*) a diet spiked with non-ortho- and mono-ortho-substituted polychlorinated biphenyls: Accumulation and effects. *Environmental Toxicology and Chemistry* **18**:699–707.

Guttman, S. I. 1994. Population genetic structure and ecotoxicology. *Environmental Health Perspectives* **102**:97–100.

Hahn, M. E. 1998a. The aryl hydrocarbon receptor: A comparative perspective. *Comparative Biochemistry and Physiology C-Toxicology & Pharmacology* **121**:23–53.

Hahn, M. E. 1998b. Mechanisms of innate and acquired resistance to dioxin-like compounds. *Reviews in Toxicology* **2**:395–443.

Hahn, M. E. 2001. Dioxin toxicology and the aryl hydrocarbon receptor: Insights from fish and other non-traditional models. *Marine Biotechnology* **3**:S224–S238.

Hahn, M. E., S. I. Karchner, D. G. Franks, and R. R. Merson. 2004. Aryl hydrocarbon receptor polymorphisms and dioxin resistance in Atlantic killifish (Fundulus heteroclitus). *Pharmacogenetics* **14**:131–143.

Hahn, M. E., S. I. Karchner, M. A. Shapiro, and S. A. Perera. 1997. Molecular evolution of two vertebrate aryl hydrocarbon (dioxin) receptors (AHR1 and AHR2) and the PAS family. *Proceedings of the National Academy of Sciences of the United States of America* **94**:13743–13748.

Hassoun, E., and S. Stohs. 1996. TCDD, endrin and lindane induced oxidative stress in fetal and placental tissues of C57BL/6J and DBA/2J mice. *Comparative Biochemistry and Physiology Part C: Pharmacology, Toxicology and Endocrinology* **115**:11–18.

Hawkins, N. J., C. Bass, A. Dixon, and P. Neve. 2019. The evolutionary origins of pesticide resistance. *Biological Reviews* **94**:135–155.

Hayes, J. D., and D. J. Pulford. 1995. The glutathione S-transferase supergene family: regulation of GST and the contribution of the Isoenzymes to cancer chemoprotection and drug resistance part I. *Critical Reviews in Biochemistry and Molecular Biology* **30**:445–520.

Hayes, J. D., and R. C. Strange. 1995. Invited commentary potential contribution of the glutathione S-transferase supergene family to resistance to oxidative stress. *Free Radical Research* **22**:193–207.

Heim, J. R., D. P. Weston, K. Major, H. Poynton, K. E. H. Hartz, and M. J. Lydy. 2018. Are there fitness costs of adaptive pyrethroid resistance in the amphipod, *Hyalella azteca*? *Environmental Pollution* **235**:39–46.

Hites, R. A., and S. J. Eisenreich. 1987. Advances in Chemistry Series, 216. Source and fates of aquatic pollutants. *Symposium Held at the 190th Meeting of the American Chemical Society*, Chicago, Illinois, USA, September 8-13, 1985. Xiii+ 558p. American Chemical Society, Washington, DC, USA.

Hu, J., J. Tian, F. Zhang, H. Wang, and J. Yin. 2019. Pxr- and Nrf2-mediated induction of ABC transporters by heavy metal ions in zebrafish embryos. *Environmental Pollution* **255**:113329.

Hua, J., R. Cothran, A. Stoler, and R. Relyea. 2013. Cross-tolerance in amphibians: Wood frog mortality when exposed to three insecticides with a common mode of action. *Environmental Toxicology and Chemistry* **32**:932–936.

Hua, J., D. K. Jones, B. M. Mattes, R. D. Cothran, R. A. Relyea, and J. T. Hoverman. 2015a. The contribution of phenotypic plasticity to the evolution of insecticide tolerance in amphibian populations. *Evolutionary Applications* **8**:586–596.

Hua, J., D. K. Jones, B. M. Mattes, R. D. Cothran, R. A. Relyea, and J. T. Hoverman. 2015b. Evolved pesticide tolerance in amphibians: predicting mechanisms based on pesticide novelty and mode of action. *Environmental Pollution* **206**:56–63.

Huggett, R. J., P. A. van Veld, C. L. Smith, W. J. Hargis, Jr., W. K. Vogelbein, and B. A. Weeks. 1992. The effects of contaminated sediments in the Elizabeth River. Pages 403–430 *in* G. A. Burton, Jr., editor. *Sediment Toxicity Assessment*. Lewis Publishers, Boca Raton, FL.

Huntley, S., N. Bonnevie, and R. Wenning. 1995. Polycyclic aromatic hydrocarbon and petroleum hydrocarbon contamination in sediment from the Newark Bay Estuary, New Jersey. *Archives of Environmental Contamination and Toxicology* **28**:93–107.

Incardona, J. P., M. G. Carls, H. Teraoka, C. A. Sloan, T. K. Collier, and N. L. Scholz. 2005. Aryl hydrocarbon receptor-independent toxicity of weathered crude oil during fish development. *Environmental Health Perspectives* **113**:1755–1762.

Ito, N., H. Tsuda, M. Tatematsu, T. Inoue, Y. Tagawa, T. Aoki, S. Uwagawa, M. Kagawa, T. Ogiso, and T. Masui. 1988. Enhancing effect of various hepatocarcinogens on induction of preneoplastic glutathione S-transferase placental form positive foci in rats-an approach for a new medium-term bioassay system. *Carcinogenesis* **9**:387–394.

Jayasundara, N., P. W. Fernando, J. S. Osterberg, K. M. Cammen, T. F. Schultz, and R. T. Di Giulio. 2017. Cost of tolerance: physiological consequences of evolved resistance to inhabit a polluted environment in teleost fish fundulus heteroclitus. *Environmental Science & Technology* **51**:8763–8772.

Johanif, N., K. E. Huff Hartz, A. E. Figueroa, D. P. Weston, D. Lee, M. J. Lydy, R. E. Connon, and H. C. Poynton. 2021. Bioaccumulation potential of chlorpyrifos in resistant Hyalella azteca: Implications for evolutionary toxicology. *Environmental Pollution (Barking, Essex: 1987)* **289**:117900–117900.

Jones, D. K., W. D. Hintz, M. S. Schuler, E. K. Yates, B. M. Mattes, and R. A. Relyea. 2017. Inducible tolerance to agrochemicals was paved by evolutionary responses to predators. *Environmental Science & Technology* **51**:13913–13919.

Karchner, S. I., D. G. Franks, W. H. Powell, and M. E. Hahn. 2002. Regulatory interactions among three members of the vertebrate aryl hydrocarbon receptor family: AHR repressor, AHR1, and AHR2. *Journal of Biological Chemistry* **277**:6949–6959.

Karchner, S. I., W. H. Powell, and M. E. Hahn. 1999. Identification and functional characterization of two highly divergent aryl hydrocarbon receptors (AHR1 and AHR2) in the teleost *Fundulus heteroclitus* - Evidence for a novel subfamily of ligand-binding basic helix loop helix-Per-ARNT-Sim (bHLH-PAS) factors. *Journal of Biological Chemistry* **274**:33814–33824.

Katagi, T. 2010. Bioconcentration, bioaccumulation, and metabolism of pesticides in aquatic organisms. *Reviews of Environmental Contamination and Toxicology* **204**:1–132.

Kelly, K., C. M. Havrilla, T. C. Brady, K. H. Abramo, and E. D. Levin. 1998. Oxidative stress in toxicology: established mammalian and emerging piscine model systems. *Environmental Health Perspectives* **106**:375–384.

Kerkvliet, N. I. 1995. Immunological effects of chlorinated dibenzo-p-dioxins. *Environmental Health Perspectives* **103**:47–53.

King-Heiden, T. C., V. Mehta, K. M. Xiong, K. A. Lanham, D. S. Antkiewicz, A. Ganser, W. Heideman, and R. E. Peterson. 2012. Reproductive and developmental toxicity of dioxin in fish. *Molecular and Cellular Endocrinology* **354**:121–138.

King Heiden, T. C., J. Spitsbergen, W. Heideman, and R. E. Peterson. 2009. Persistent adverse effects on health and reproduction caused by exposure of zebrafish to 2,3,7,8-tetrachlorodibenzo-p-dioxin during early development and gonad differentiation. *Toxicological Sciences* **109**:75–87.

Klerks, P. L., L. Xie, and J. S. Levinton. 2011. Quantitative genetics approaches to study evolutionary processes in ecotoxicology; a perspective from research on the evolution of resistance. *Ecotoxicology* **20**:513–523.

Kopp, R. L., S. I. Guttman, and T. E. Wissing. 1992. Genetic indicators of environmental stress in central mudminnow (*Umbra limi*) populations exposed to acid deposition in the Adirondack Mountains. *Environmental Toxicology and Chemistry: An International Journal* **11**:665–676.

Kurelec, B. 1992. The multixenobiotic resistance mechanism in aquatic organisms. *CRC Critical Reviews in Toxicology* **22**:23–43.

Kurelec, B., D. Lucić, B. Pivčević, and S. Krča. 1995. Induction and reversion of multixenobiotic resistance in the marine snail Monodonta turbinata. *Marine Biology* **123**:305–312.

Lake, J., R. McKinney, C. Lake, F. Osterman, and J. Heltshe. 1995. Comparisons of patterns of polychlorinated biphenyl congeners in water, sediment, and indigenous organisms from New Bedford Harbor, Massachusetts. *Archives of Environmental Contamination and Toxicology* **29**:207–220.

Leber, A. P., and T. S. Miya. 1976. A mechanism for cadmium-and zinc-induced tolerance to cadmium toxicity: involvement of metallothionein. *Toxicology and Applied Pharmacology* **37**:403–414.

Leffler, E. M., K. Bullaughey, D. R. Matute, W. K. Meyer, L. Segurel, A. Venkat, P. Andolfatto, and M. Przeworski. 2012. Revisiting an old riddle: what determines genetic diversity levels within species? *Plos Biology* **10**:e1001388.

Lerebours, A., T. Van Vinh, and J.-P. Bourdineaud. 2016. Danio rerio ABC transporter genes abcb3 and abcb7 play a protecting role against metal contamination. *Journal of Applied Toxicology* **36**:1551–1557.

Lin, H. C., S. C. Hsu, and P. P. Hwang. 2000. Maternal transfer of cadmium tolerance in larval Oreochromis mossambicus. *Journal of Fish Biology* **57**:239–249.

Lindberg, C. D., N. Jayasundara, J. S. Kozal, T. C. Leuthner, and R. T. Di Giulio. 2017. Resistance to polycyclic aromatic hydrocarbon toxicity and associated bioenergetic consequences in a population of Fundulus heteroclitus. *Ecotoxicology* **26**:435–448.

Liu, N. 2015. Insecticide resistance in mosquitoes: impact, mechanisms, and research directions. *Annual Review of Entomology* **60**:537–559.

Loerracher, A.-K., and T. Braunbeck. 2021. Cytochrome P450-dependent biotransformation capacities in embryonic, juvenile and adult stages of zebrafish (Danio rerio)-a state-of-the-art review. *Archives of Toxicology* **95**:2299–2334.

Long, E. R., D. D. Macdonald, S. L. Smith, and F. D. Calder. 1995. Incidence of adverse biological effects within ranges of chemical concentrations in marine and estuarine sediments. *Environmental Management* **19**:81–97.

Lowry, D. B., S. Hoban, J. L. Kelley, K. E. Lotterhos, L. K. Reed, M. F. Antolin, and A. Storfer. 2017. Breaking RAD: an evaluation of the utility of restriction site-associated DNA sequencing for genome scans of adaptation. *Molecular Ecology Resources* **17**:142–152.

Lu, K., Y. Song, and R. Zeng. 2021. The role of cytochrome P450-mediated detoxification in insect adaptation to xenobiotics. *Current Opinion in Insect Science* **43**:103–107.

Luckenbach, T., S. Fischer, and A. Sturm. 2014. Current advances on ABC drug transporters in fish. *Comparative Biochemistry and Physiology C-Toxicology & Pharmacology* **165**:28–52.

Lundholm, C. 1997. DDE-induced eggshell thinning in birds: effects of p, p′-DDE on the calcium and prostaglandin metabolism of the eggshell gland. *Comparative Biochemistry and Physiology Part C: Pharmacology, Toxicology and Endocrinology* **118**:113–128.

Lushchak, V. I. 2011a. Adaptive response to oxidative stress: Bacteria, fungi, plants and animals. *Comparative Biochemistry and Physiology C-Toxicology & Pharmacology* **153**:175–190.

Lushchak, V. I. 2011b. Environmentally induced oxidative stress in aquatic animals. *Aquatic Toxicology* **101**:13–30.

Lushchak, V. I. 2016. Contaminant-induced oxidative stress in fish: a mechanistic approach. *Fish Physiology and Biochemistry* **42**:711–747.

Lushchak, V. I., T. M. Matviishyn, V. V. Husak, J. M. Storey, and K. B. Storey. 2018. Pesticide toxicity: a mechanistic approach. *Excli Journal* **17**:1101–1136.

Major, K. M., D. P. Weston, M. J. Lydy, G. A. Wellborn, and H. C. Poynton. 2018. Unintentional exposure to terrestrial pesticides drives widespread and predictable evolution of resistance in freshwater crustaceans. *Evolutionary Applications* **11**:748–761.

Manahan, S. E. 1999. *Environmental Chemistry*. Lewis Publishers, Boca Raton, FL. 7th ed.

Mannervik, B., P. Alin, C. Guthenberg, H. Jensson, M. K. Tahir, M. Warholm, and H. Jörnvall. 1985. Identification of three classes of cytosolic glutathione transferase common to several mammalian species: correlation between structural data and enzymatic properties. *Proceedings of the National Academy of Sciences* **82**:7202–7206.

Matson, C. W., B. W. Clark, M. J. Jenny, C. R. Fleming, M. E. Hahn, and R. T. Di Giulio. 2008. Development of the morpholino gene knockdown technique in Fundulus heteroclitus: A tool for studying molecular mechanisms in an established environmental model. *Aquatic Toxicology* **87**:289–295.

Matsumura, F. 2009. The significance of the nongenomic pathway in mediating inflammatory signaling of the dioxin-activated Ah receptor to cause toxic effects. *Biochemical Pharmacology* **77**:608–626.

McCarter, J., and M. Roch. 1983. Hepatic metallothionein and resistance to copper in juvenile coho salmon. *Comparative Biochemistry and Physiology Part C: Comparative Pharmacology* **74**:133–137.

McGeer, J. C., S. Nadella, D. H. Alsop, L. Hollis, L. N. Taylor, D. G. McDonald, and C. A. Wood. 2007. Influence of acclimation and cross-acclimation of metals on acute Cd toxicity and Cd uptake and distribution in rainbow trout (*Oncorhynchus mykiss*). *Aquatic Toxicology* **84**:190–197.

McMillan, A. M., Bagley, M. J., Jackson, S. A., and Nacci, D. E. 2006. Genetic diversity and structure of an estuarine fish (Fundulus heteroclitus) indigenous to sites associated with a highly contaminated urban harbor. *Ecotoxicology* **15**:539–548.

McMillan, B. J., and C. A. Bradfield. 2007. The aryl hydrocarbon receptor sans xenobiotics: endogenous function in genetic model systems. *Molecular Pharmacology* **72**:487–498.

Menzie, C. A., B. B. Potocki, and J. Santodonato. 1992. Exposure to carcinogenic PAHs in the environment. *Environmental Science & Technology* **26**:1278–1284.

Merson, R. R., D. G. Franks, S. I. Karchner, and M. E. Hahn. 2006. Development and characterization of polyclonal antibodies against the aryl hydrocarbon receptor protein family (AHR1, AHR2, and AHR repressor) of Atlantic killifish *Fundulus heteroclitus*. *Comparative Biochemistry and Physiology C-Toxicology & Pharmacology* **142**:85–94.

Meyer, J., and R. Di Giulio. 2002. Patterns of heritability of decreased EROD activity and resistance to PCB 126-induced teratogenesis in laboratory-reared offspring of killifish (*Fundulus heteroclitus*) from a creosote-contaminated site in the Elizabeth River, VA, USA. *Marine Environmental Research* **54**:621–626.

Meyer, J. N., and R. T. Di Giulio. 2003. Heritable adaptation and fitness costs in killifish (*Fundulus beteroclitus*) inhabiting a polluted estuary. *Ecological Applications* **13**:490–503.

Meyer, J. N., D. E. Nacci, and R. T. Di Giulio. 2002. Cytochrome P4501A (CYP1A) in killifish (*Fundulus heteroclitus*): Heritability of altered expression and relationship to survival in contaminated sediments. *Toxicological Sciences* **68**:69–81.

Meyer, J. N., J. D. Smith, G. W. Winston, and R. T. Di Giulio. 2003a. Antioxidant defenses in killifish (*Fundulus heteroclitus*) exposed to contaminated sediments and model prooxidants: short-term and heritable responses. *Aquatic Toxicology* **65**:377–395.

Meyer, J. N., D. C. Volz, J. H. Freedman, and R. T. Di Giulio. 2005. Differential display of hepatic mRNA from killifish (*Fundulus heteroclitus*) inhabiting a Superfund estuary. *Aquatic Toxicology* **73**:327–341.

Meyer, J. N., D. M. Wassenberg, S. I. Karchner, M. E. Hahn, and R. T. Di Giulio. 2003b. Expression and inducibility of aryl hydrocarbon receptor pathway genes in wild-caught killifish (*Fundulus heteroclitus*) with different contaminant-exposure histories. *Environmental Toxicology and Chemistry* **22**:2337–2343.

Miller, J. R., B. W. Clark, N. M. Reid, S. I. Karchner, J. L. Roach, M. E. Hahn, D. Nacci, and A. Whitehead. 2023. Independently evolved pollution resistance in four killifish populations is largely explained by few variants of large effect. *BioRxiv*. doi: 10.1101/2023.04.07.536079.

Mimura, J., M. Ema, K. Sogawa, and Y. Fujii-Kuriyama. 1999. Identification of a novel mechanism of regulation of Ah (dioxin) receptor function. *Genes & Development* **13**:20–25.

Minier, C., F. Akcha, and F. Galgani. 1993. P-glycoprotein expression in *Crassostrea gigas* and *Mytilus edulis* in polluted seawater. *Comparative Biochemistry and Physiology. B, Comparative Biochemistry* **106**:1029–1036.

Mitton, J. B. 1997. *Selection in Natural Populations*. Oxford University Press, Oxford.

Monosson, E., and J. J. Stegeman. 1991. Cytochrome P450E (P450IA) induction and inhibition in winter flounder by 3, 3′, 4, 4′-tetrachlorobiphenyl: Comparison of response in fish from Georges Bank and Narragansett bay. *Environmental Toxicology and Chemistry: An International Journal* **10**:765–774.

Muhammad, A., Ibrahim, S.S., Muktar, M.M., Irving, H., Abajue, M.C., Edith, N.M.A., Da'u. S.S., Paine, M.J.I., Wondji, C.S. 2021. High pyrethroid/DDT resistance in major malaria vector *Anopheles coluzzii* from Niger-Delta of Nigeria is probably driven by metabolic resistance mechanisms. *PLoS ONE* **16**(3):e0247944. https://doi.org/10.1371/journal.pone.0247944.

Mulvey, M., Diamond, S. A. 1991. Genetic factors and tolerance acquisition in populations exposed to metals and metalloids. Pages 301–321 in M.C. Newman and A.W. McIntosh, editors. *Metal Ecotoxicology: Concepts and Applications*. Lewis Publishers, Inc, Chelsea, MI.

Mulvey, M., M. C. Newman, W. Vogelbein, and M. A. Unger. 2002. Genetic structure of *Fundulus heteroclitus* from PAH-contaminated and neighboring sites in the Elizabeth and York Rivers. *Aquatic Toxicology* **61**:195–209.

Mulvey, M., M. C. Newman, W. K. Vogelbein, M. A. Unger, and D. R. Ownby. 2003. Genetic structure and mtDNA diversity of *Fundulus heteroclitus* populations from polycyclic aromatic hydrocarbon-contaminated sites. *Environmental Toxicology and Chemistry* **22**:671–677.

Munns, W. R., D. E. Black, T. R. Gleason, K. Salomon, D. Bengtson, and R. Gutjahr-Gobell. 1997. Evaluation of the effects of dioxin and PCBs on *Fundulus heteroclitus* populations using a modeling approach. *Environmental Toxicology and Chemistry* **16**:1074–1081.

Murdoch, M. H., and P. D. Hebert. 1994. Mitochondrial DNA diversity of brown bullhead from contaminated and relatively pristine sites in the Great Lakes. *Environmental Toxicology and Chemistry: An International Journal* **13**:1281–1289.

Nacci, D., D. Champlin, and S. Jayaraman. 2010. Adaptation of the estuarine fish *Fundulus heteroclitus* (Atlantic Killifish) to polychlorinated biphenyls (PCBs). *Estuaries and Coasts* **33**:853–864.

Nacci, D., L. Coiro, D. Champlin, S. Jayaraman, R. McKinney, T. R. Gleason, W. R. Munns, J. L. Specker, and K. R. Cooper. 1999. Adaptations of wild populations of the estuarine fish *Fundulus heteroclitus* to persistent environmental contaminants. *Marine Biology* **134**:9–17.

Nacci, D., L. Coiro, A. Kuhn, D. Champlin, W. Munns, J. Specker, and K. Cooper. 1998. Nondestructive indicator of ethoxyresorufin-*O*-deethylase activity in embryonic fish. *Environmental Toxicology and Chemistry* **17**:2481–2486.

Nacci, D., L. Coiro, D. Wassenber, and R. T. Di Giulio. 2005. A non-destructive technique to measure cytochrome P4501A enzyme activity in living embryos of the estuarine fish *Fundulus heteroclitus*. Pages 209–225 in G.K. Ostrander, editor. *Techniques in Aquatic Toxicology*, Volume 2. Taylor and Francis, Boca Raton, FL.

Nacci, D., T. Gleason, R. Gutjahr-Gobell, M. Huber, and W. Munns. 2002a. Effects of chronic stress on wildlife populations: a population modeling approach and case study. Pages 247–272 *in* M. Newman, editor. *Coastal and Estuarine Risk Assessment: Risk on the Edge*. CRC Press/Lewis Publishers, Boca Raton, FL.

Nacci, D., M. Huber, D. Champlin, S. Jayaraman, S. Cohen, E. Gauger, A. Fong, and M. Gomez-Chiarri. 2009. Evolution of tolerance to PCBs and susceptibility to a bacterial pathogen (*Vibrio harveyi*) in Atlantic killifish (*Fundulus heteroclitus*) from New Bedford (MA, USA) *harbor*. *Environmental Pollution* **157**:857–864.

Nacci, D., S. Jayaraman, and J. Specker. 2001. Stored retinoids in populations of the estuarine fish *Fundulus heteroclitus* indigenous to PCB-contaminated and reference sites. *Archives of Environmental Contamination and Toxicology* **40**:511–518.

Nacci, D., D. Proestou, D. Champlin, J. Martinson, and E. R. Waits. 2016. Genetic basis for rapidly evolved tolerance in the wild: adaptation to toxic pollutants by an estuarine fish species. *Molecular Ecology* **25**:5467–5482.

Nacci, D., S. Walters, T. Gleason, and J. Munns, W. R.. 2007. Using a spatial modeling approach to explore ecological factors relevant to the persistence of estuarine fish (*Fundulus heteroclitus*) in a PCB-contaminated estuary. Pages 184–199 *in* R. Akcakaya, J. D. Stark, and T.S. Bridges, editor. *Population-Level Ecotoxicological Risk Assessment: Case Studies*. Oxford University Press, New York, NY.

Nacci, D. E., D. Champlin, L. Coiro, R. McKinney, and S. Jayaraman. 2002b. Predicting the occurrence of genetic adaptation to dioxinlike compounds in populations of the estuarine fish *Fundulus heteroclitus*. *Environmental Toxicology and Chemistry* **21**:1525–1532.

Nacci, D. E., T. R. Gleason, and W. R. Munns. 2002c. Evolutionary and ecological effects of multi-generational exposures to anthropogenic stressors. *Human and Ecological Risk Assessment* **8**:91–97.

Nacci, D. E., M. Kohan, M. Pelletier, and E. George. 2002d. Effects of benzo[a]pyrene exposure on a fish population resistant to the toxic effects of dioxin-like compounds. *Aquatic Toxicology* **57**:203–215.

Nadig, S. G., K. Lee, and S. Adams. 1998. Evaluating alterations of genetic diversity in sunfish populations exposed to contaminants using RAPD assay. *Aquatic Toxicology* **43**:163–178.

Nebert, D. W., A. L. Roe, M. Z. Dieter, W. A. Solis, Y. Yang, and T. P. Dalton. 2000. Role of the aromatic hydrocarbon receptor and [Ah] gene battery in the oxidative stress response, cell cycle control, and apoptosis. *Biochemical Pharmacology* **59**:65–85.

Nelson, W., B. Bergen, S. Benyi, G. Morrison, R. Voyer, C. Strobel, S. Rego, G. Thursby, and C. Pesch. 1996. New Bedford harbor long-term monitoring assessment report: baseline sampling. US Environmental Protection Agency, National Health and Environmental Effects Research Laboratory, Atlantic Ecology Division, Narragansett, RI. EPA/600/R-96/097.

Nevo, E., R. Ben-Shlomo, and B. Lavie. 1984. Mercury selection of allozymes in marine organisms: prediction and verification in nature. *Proceedings of the National Academy of Sciences* **81**:1258–1259.

Newman, M. C., S. A. Diamond, M. Mulvey, and P. Dixon. 1989. Allozyme genotype and time to death of mosquitofish, *Gambusia affinis* (Baird and Girard) during acute toxicant exposure: a comparison of arsenate and inorganic mercury. *Aquatic Toxicology* **15**:141–156.

Nicklisch, S. C. T., and A. Hamdoun. 2020. Disruption of small molecule transporter systems by Transporter-Interfering Chemicals (TICs). *FEBS Letters* **594**:4158–4185.

Nicklisch, S. C. T., A. K. Pouv, S. D. Rees, A. P. McGrath, G. Chang, and A. Hamdoun. 2021. Transporter-interfering chemicals inhibit P-glycoprotein of yellowfin tuna (Thunnus albacares). *Comparative Biochemistry and Physiology C-Toxicology & Pharmacology* **248**.

Nielsen, R. 2005. Molecular signatures of natural selection. *Annual Review of Genetics* **39**:197–218.

Nordlie, F. G. 2006. Physicochemical environments and tolerances of cyprinodontoid fishes found in estuaries and salt marshes of eastern North America. *Reviews in Fish Biology and Fisheries* **16**:51–106.

Norup, B. 1972. Toxicity of chemicals in paper factory effluents. *Water Research*, **6**:1585–1588.

Nusbaumer, D., L. Marques da Cunha, and C. Wedekind. 2021. Testing for population differences in evolutionary responses to pesticide pollution in brown trout (*Salmo trutta*). *Evolutionary Applications* **14**:462–475.

O'Connor, J. M., and R. J. Huggett. 1988. Aquatic pollution problems, North Atlantic coast, including Chesapeake Bay. *Aquatic Toxicology* **11**:163–190.

Orr, P., R. Bradley, J. Sprague, and N. Hutchinson. 1986. Acclimation-induced change in toxicity of aluminum to rainbow trout (*Salmo gairdneri*). *Canadian Journal of Fisheries and Aquatic Sciences* **43**:243–246.

Osterberg, J. S., K. M. Cammen, T. F. Schultz, B. W. Clark, and R. T. Di Giulio. 2018. Genome-wide scan reveals signatures of selection related to pollution adaptation in non-model estuarine Atlantic killifish (*Fundulus heteroclitus*). *Aquatic Toxicology* **200**:73–82.

Ownby, D. R., M. C. Newman, M. Mulvey, W. K. Vogelbein, M. A. Unger, and L. F. Arzayus. 2002. Fish (*Fundulus heteroclitus*) populations with different exposure histories differ in tolerance of creosote-contaminated sediments. *Environmental Toxicology and Chemistry* **21**:1897–1902.

Oziolor, E. M., E. Bigorgne, L. Aguilar, S. Usenko, and C. W. Matson. 2014. Evolved resistance to PCB- and PAH-induced cardiac teratogenesis, and reduced CYP1A activity in Gulf killifish (*Fundulus grandis*) populations from the Houston Ship Channel, Texas. *Aquatic Toxicology* **150**:210–219.

Oziolor, E. M., W. Howard, R. Lavado, and C. W. Matson. 2017. Induced pesticide tolerance results from detoxification pathway priming. *Environmental Pollution* **224**:615–621.

Oziolor, E. M., N. M. Reid, S. Yair, K. M. Lee, S. G. VerPloeg, P. C. Bruns, J. R. Shaw, A. Whitehead, and C. W. Matson. 2019. Adaptive introgression enables evolutionary rescue from extreme environmental pollution. *Science* **364**:455.

Palumbi, S. R. 2001. Evolution - Humans as the world's greatest evolutionary force. *Science* **293**:1786–1790.

Pan, S.-T., D. Xue, Z.-L. Li, Z.-W. Zhou, Z.-X. He, Y. Yang, T. Yang, J.-X. Qiu, and S.-F. Zhou. 2016. Computational identification of the paralogs and orthologs of human cytochrome P450 superfamily and the implication in drug discovery. *International Journal of Molecular Sciences* **17**:1020.

Paris, J. R., and J. Usher. 2019. Functional genomic characterization of metallothioneins in brown trout (Salmo trutta L.). using synthetic genetic analysis. *Scientific Reports* **9**:11827.

Pascoe, D., and J. Beattie. 1979. Resistance to cadmium by pretreated rainbow trout alevins. *Journal of Fish Biology* **14**:303–308.

Pavic, M., P. Turcic, and M. Ljubojevic. 2019. Forgotten partners and function regulators of inducible metallothioneins. *Arhiv Za Higijenu Rada I Toksikologiju-Archives of Industrial Hygiene and Toxicology* **70**:256–264.

Peake, E. B., J. C. Locke, L. L. Tierney, and A. S. Kolok. 2004. Copper tolerance in fathead minnows: II. Maternal transfer. *Environmental Toxicology and Chemistry* **23**:208–211.

Peterson, R. E., H. M. Theobald, and G. L. Kimmel. 1993. Developmental and reproductive toxicity of dioxins and related compounds - cross-species comparisons. *CRC Critical Reviews in Toxicology* **23**:283–335.

Pierron, F., V. Bourret, J. St-Cyr, P. G. C. Campbell, L. Bernatchez, and P. Couture. 2009. Transcriptional responses to environmental metal exposure in wild yellow perch (Perca flavescens) collected in lakes with differing environmental metal concentrations (Cd, Cu, Ni). *Ecotoxicology* **18**:620–631.

Place, A. R., and D. A. Powers. 1979. Genetic variation and relative catalytic efficiencies of lactate dehydrogenase-B allozymes of *Fundulus heteroclitus*. *Proceedings of the National Academy of Sciences of the United States of America* **76**:2354–2358.

Powell, W. H., R. Bright, S. M. Bello, and M. E. Hahn. 2000. Developmental and tissue-specific expression of AHR1, AHR2, and ARNT2 in dioxin-sensitive and -resistant populations of the marine fish *Fundulus heteroclitus*. *Toxicological Sciences* **57**:229–239.

Powell, W. H., H. G. Morrison, E. J. Weil, S. I. Karchner, M. L. Sogin, J. J. Stegeman, and M. E. Hahn. 2004. Cloning and analysis of the CYP1A promoter from the Atlantic killifish (*Fundulus heteroclitus*). *Marine Environmental Research* **58**:119–124.

Powers, D. A., and A. R. Place. 1978. Biochemical genetics of Fundulus heteroclitus (L.). I. Temporal and spatial variation in gene frequencies of Ldh-B, Mdh-A, Gpi-B, and Pgm-A. *Biochemical Genetics* **16**:593–607.

Powers, D. A., I. Ropson, D. C. Brown, R. Van Beneden, R. Cashon, L. I. Gonzalez-Villaseñor, and J. A. Dimichele. 1986. Genetic variation in *Fundulus heteroclitus*: geographic distribution. *American Zoologist* **26**:131–144.

Prasch, A. L., H. Teraoka, S. A. Carney, W. Dong, T. Hiraga, J. J. Stegeman, W. Heideman, and R. E. Peterson. 2003. Aryl hydrocarbon receptor 2 mediates 2,3,7,8-tetrachlorodibenzo-p-dioxin developmental toxicity in zebrafish. *Toxicological Sciences* **76**:138–150.

Prince, R., and K. R. Cooper. 1995a. Comparisons of the effects of 2,3,7,8-tetrachlorodibenzo-p-dioxin on chemically impacted and nonimpacted subpopulations of *Fundulus heteroclitus*.1. *TCDD toxicity. Environmental Toxicology and Chemistry* **14**:579–587.

Prince, R., and K. R. Cooper. 1995b. Comparisons of the effects of 2,3,7,8-tetrachlorodibenzo-*p*-dioxin on chemically impacted and nonimpacted subpopulations of *Fundulus heteroclitus*.2. Metabolic considerations. *Environmental Toxicology and Chemistry* **14**:589–595.

Proestou, D. A., P. Flight, D. Champlin, and D. Nacci. 2014. Targeted approach to identify genetic loci associated with evolved dioxin tolerance in Atlantic Killifish (*Fundulus heteroclitus*). *BMC Evolutionary Biology* **14**:1–17.

Pruell, R. J., C. B. Norwood, R. D. Bowen, W. S. Boothman, P. F. Rogerson, M. Hackett, and B. C. Butterworth. 1990. Geochemical Study of Sediment Contamination in New Bedford Harbor, Massachusetts. *Marine Environmental Research* **29**:77–101.

Puritz, J. B., and R. J. Toonen. 2011. Coastal pollution limits pelagic larval dispersal. *Nature Communications* **2**:226.

Rappe, C. 1991. Sources of and human exposure to PCDDs and PCDFs. *Banbury Report*.

Raymond, M., C. Berticat, M. Weill, N. Pasteur, and C. Chevillon. 2001. Insecticide resistance in the mosquito Culex pipiens: what have we learned about adaptation? Pages 287–296 *in* A.P. Hendry and M.T. Kinnison, Editors. *Microevolution Rate, Pattern, Process*. Springer, New York, NY.

Reid, N. M., C. E. Jackson, D. Gilbert, P. Minx, M. J. Montague, T. H. Hampton, L. W. Helfrich, B. L. King, D. E. Nacci, N. Aluru, S. I. Karchner, J. K. Colbourne, M. E. Hahn, J. R. Shaw, M. F. Oleksiak, D. L. Crawford, W. C. Warren, and A. Whitehead. 2017. The landscape of extreme genomic variation in the highly adaptable atlantic killifish. *Genome Biology and Evolution* **9**:659–676.

Reid, N. M., D. A. Proestou, B. W. Clark, W. C. Warren, J. K. Colbourne, J. R. Shaw, S. I. Karchner, M. E. Hahn, D. Nacci, M. F. Oleksiak, D. L. Crawford, and A. Whitehead. 2016. The genomic landscape of rapid repeated evolutionary adaptation to toxic pollution in wild fish. *Science* **354**:1305–1308.

Reitzel, A. M., S. I. Karchner, D. G. Franks, B. R. Evans, D. Nacci, D. Champlin, V. M. Vieira, and M. E. Hahn. 2014. Genetic variation at aryl hydrocarbon receptor (AHR) loci in populations of Atlantic killifish (*Fundulus heteroclitus*) inhabiting polluted and reference habitats. *BMC Evolutionary Biology* **14**:1–18.

Rice, C. D. 2001. Fish immunotoxicology. *Target Organ Toxicity in Marina and Freshwater Teleosts* **2**:97–138.

Roark, S. 2003. Life history influences the genetic structure of populations of *Fundulus heteroclitus, Menidia menidia, and Geukensia demikssa* in a PCB-contaminated estuary, New Bedford Harbor, MA, USA. Dissertation. Miami University, Miami, OH, USA.

Roch, M., and J. McCarter. 1984. Hepatic metallothionein production and resistance to heavy metals by rainbow trout (Salmo gairdneri)--I. Exposed to an artificial mixture of zinc, copper and cadmium. *Comparative Biochemistry and Physiology. C, Comparative Pharmacology and Toxicology* **77**:71–75.

Roesijadi, G. 1992. Metallothioneins in metal regulation and toxicity in aquatic animals. *Aquatic Toxicology* **22**:81–113.

Roesijadi, G. 1994. Metal regulation in aquatic animals: mechanism of uptake, accumulation and release. Pages 387–420. *Molecular Mechanisms in Aquatic Toxicolog*. Springer, New York, NY.

Roesijadi, G. 1996. Metallothionein and its role in toxic metal regulation. *Comparative Biochemistry and Physiology Part C: Pharmacology, Toxicology and Endocrinology* **113**:117–123.

Rosselli, M., K. Reinhart, B. Imthurn, P. J. Keller, and R. K. Dubey. 2000. Cellular and biochemical mechanisms by which environmental oestrogens influence reproductive function. *Human Reproduction Update* **6**:332–350.

Roy, N. K., S. Courtenay, Z. P. Yuan, M. Ikonomou, and I. Wirgin. 2001. An evaluation of the etiology of reduced CYP1A1 messenger RNA expression in the Atlantic tomcod from the Hudson River, New York, USA, using reverse transcriptase polymerase chain reaction analysis. *Environmental Toxicology and Chemistry* **20**:1022–1030.

Safe, S. H. 1986. Comparative toxicology and mechanism of action of polychlorinated dibenzo-p-dioxins and dibenzofurans. *Annual Review of Pharmacology and Toxicology* **26**:371–399.

Salinas, A. E., and M. G. Wong. 1999. Glutathione S-transferases-a review. *Current Medicinal Chemistry* **6**:279–310.

Shankar, P., S. Dasgupta, M. E. Hahn, and R. L. Tanguay. 2020. A review of the functional roles of the zebrafish aryl hydrocarbon receptors. *Toxicological Sciences* **178**:215–238.

Shaw, A. J. 1999. The evolution of heavy metal tolerance in plants: adaptations, limits, and costs. *Genetics and Ecotoxicology*:9–30.

Sheehan, D., G. Meade, V. M. Foley, and C. A. Dowd. 2001. Structure, function and evolution of glutathione transferases: implications for classification of non-mammalian members of an ancient enzyme superfamily. *Biochemical Journal* **360**:1–16.

Shekh, K., A. J. Alcaraz, S. Niyogi, and M. Hecker. 2020. Comparative analyses of oxidative stress response and metallothionein induction in white sturgeon and rainbow trout during acute waterborne copper exposure. *Comparative Biochemistry and Physiology C-Toxicology & Pharmacology* **231**:108723.

Shirley, M. D., and R. M. Sibly. 1999. Genetic basis of a between-environment trade-off involving resistance to cadmium in *Drosophila melanogaster. Evolution* **53**:826–836.

Silvestre, F. 2020. Signaling pathways of oxidative stress in aquatic organisms exposed to xenobiotics. *Journal of Experimental Zoology Part A-Ecological and Integrative Physiology* **333**:436–448.

Sinley, J. R., J. P. Goettl, and P. H. Davies. 1974. The effects of zinc on rainbow trout (*Salmo gairdneri*) in hard and soft water. *Bulletin of Environmental Contamination and Toxicology* **12**:193–201.

Smith, G. M., and J. S. Weis. 1997. Predator-prey relationships in mummichogs (Fundulus heteroclitus (L)): effects of living in a polluted environment. *Journal of Experimental Marine Biology and Ecology* **209**:75–87.

Spehar, R. L., E. N. Leonard, and D. L. Defoe. 1978. Chronic effects of cadmium and zinc mixtures on flagfish (Jordanella floridae). *Transactions of the American Fisheries Society* **107**:354–360.

Spitsbergen, J. M., M. K. Walker, J. R. Olson, and R. E. Peterson. 1991. Pathological alterations in early life stages of lake trout, *Salvelinus namaycush*, exposed to 2,3,7,8-tetrachlorodibenzo-para-dioxin as fertilized eggs. *Aquatic Toxicology* **19**:41–71.

Srikanth, K., E. Pereira, A. C. Duarte, and I. Ahmad. 2013. Glutathione and its dependent enzymes' modulatory responses to toxic metals and metalloids in fish-a review. *Environmental Science and Pollution Research* **20**:2133–2149.

Stegeman, J., M. Bouwer, R. Di Giulio, L. Forlin, B. Fowler, B. Sanders, and P. Van Veld. 1992. Molecular responses to environmental contamination: proteins and enzymes as indicators of contaminant exposure and effects. Pages 235–335 *in* R. Huggett, editor. *Biomarkers: Biochemical, Physiological and Histological Markers of Anthropogenic Stress*. Lewis Publishers, Boca Raton, FL.

Stegeman, J. J., J. V. Goldstone, and M. E. Hahn. 2010. 10-Perspectives on zebrafish as a model in environmental toxicology. Pages 367–439 *in* S. F. Perry, M. Ekker, A. P. Farrell, and C. J. Brauner, editors. *Fish Physiology*. Academic Press, Cambridge, MA.

Stegeman, J. J., and M. E. Hahn. 1994. Biochemistry and molecular biology of monoxygenases: current perspectives on forms, functions, and regulation of cytochrome P450 in aquatic species. Pages 87–206 *in* D. C. Malins and G. K. Ostrander, editors. *Aquatic Toxicology: Molecular, Biochemical, and Cellular Perspectives*. CRC Press, Boca Raton, FL.

Stevens, E. A., J. D. Mezrich, and C. A. Bradfield. 2009. The aryl hydrocarbon receptor: a perspective on potential roles in the immune system. *Immunology* **127**:299–311.

Strauss, S. Y., J. A. Rudgers, J. A. Lau, and R. E. Irwin. 2002. Direct and ecological costs of resistance to herbivory. *Trends in Ecology & Evolution* **17**:278–285.

Sullivan, K. B., and M. J. Lydy. 1999. Differences in survival functions of mosquitofish (*Gambusia affinis*) and sand shiner (*Notropis ludibundus*) genotypes exposed to pesticides. *Environmental Toxicology and Chemistry: An International Journal* **18**:906–911.

Tang, S., J. A. Doering, J. Sun, S. C. Beitel, K. Shekh, S. Patterson, S. Crawford, J. P. Giesy, S. B. Wiseman, and M. Hecker. 2016. Linking oxidative stress and magnitude of compensatory responses with life-stage specific differences in sensitivity of white sturgeon (*Acipenser transmontanus*) to copper or cadmium. *Environmental Science & Technology* **50**:9717–9726.

Taylor, M., and R. Feyereisen. 1996. Molecular biology and evolution of resistance to toxicants. *Molecular Biology and Evolution* **13**:719–734.

Theodorakis, C. W. 2003. Establishing causality between population genetic alterations and environmental contamination in aquatic organisms. *Human and Ecological Risk Assessment* **9**:37–58.

Thurmond, T., A. Silverstone, R. Baggs, F. Quimby, J. Staples, and T. Gasiewicz. 1999. A chimeric aryl hydrocarbon receptor knockout mouse model indicates that aryl hydrocarbon receptor activation in hematopoietic cells contributes to the hepatic lesions induced by 2, 3, 7, 8-tetrachlorodibenzo-p-dioxin. *Toxicology and Applied Pharmacology* **158**:33–40.

Tian, J., J. Hu, G. Liu, H. Yin, M. Chen, P. Miao, P. Bai, and J. Yin. 2019. Altered gene expression of ABC transporters, nuclear receptors and oxidative stress signaling in zebrafish embryos exposed to CdTe quantum dots. *Environmental Pollution* **244**:588–599.

Tillitt, D. E., P. M. Cook, J. P. Giesy, W. Heideman, and R. E. Peterson. 2008. Reproductive impairment of Great Lakes lake trout by dioxin-like chemicals. Pages 819–876 *in* R.T. Di Giulio and D. E. Hinton, editors, *The Toxicology of Fishes*. Taylor and Francis, Washington, DC.

Timme-Laragy, A. R., J. N. Meyer, R. A. Waterland, and R. T. Di Giulio. 2005. Analysis of CpG methylation in the killifish CYP1A promoter. *Comparative Biochemistry and Physiology C-Toxicology & Pharmacology* **141**:406–411.

Tonelli, C., I. I. C. Chio, and D. A. Tuveson. 2018. Transcriptional regulation by Nrf2. *Antioxidants & Redox Signaling* **29**:1727–1745.

Toomey, B. H., and D. Epel. 1993. Multixenobiotic resistance in *Urechis caupo* embryos - Protection from environmental toxins. *Biological Bulletin* **185**:355–364.

Toomey, B. H., S. Bello, M. E. Hahn, S. Cantrell, P. Wright, D. E. Tillitt, and R. T. Di Giulio. 2001. 2,3,7,8-Tetrachlorodibenzo-p-dioxin induces apoptotic cell death and cytochrome P4501A expression in developing *Fundulus heteroclitus* embryos. *Aquatic Toxicology* **53**:127–138.

Toppin, S. V., M. Heber, J. S. Weis, and P. Weis. 1987. Changes in reproductive biology and life history in *Fundulus heteroclitus* in a polluted environment. Pages 171–184 *in* W. Vernbcrg, A. Calabrese, F. Thurberg, and FJ Vernberg, editors. *Pollution Physiology of Estuarine Organisms*. University of South Carolina Press, Columbia, SC.

Uno, T., M. Ishizuka, and T. Itakura. 2012. Cytochrome P450 (CYP) in fish. *Environmental Toxicology Pharmacology* **34**:1–13.

Van den Berg, M., L. S. Birnbaum, M. Denison, M. De Vito, W. Farland, M. Feeley, H. Fiedler, H. Hakansson, A. Hanberg, L. Haws, M. Rose, S. Safe, D. Schrenk, C. Tohyama, A. Tritscher, J. Tuomisto, M. Tysklind, N. Walker, and R. E. Peterson. 2006. The 2005 World Health Organization reevaluation of human and mammalian toxic equivalency factors for dioxins and dioxin-like compounds. *Toxicological Sciences* **93**:223–241.

van Straalen, N. M., and M. J. T. N. Timmermans. 2002. Genetic variation in toxicant-stressed populations: an evaluation of the "genetic erosion" hypothesis. *Human and Ecological Risk Assessment: An International Journal* **8**:983–1002.

Van Veld, P. A., U. C. Ko, W. K. Vogelbein, and D. J. Westbrook. 1991. Glutathione-s-transferase in intestine, liver and hepatic lesions of mummichog (*Fundulus heteroclitus*) from a creosote-contaminated environment. *Fish Physiology and Biochemistry* **9**:369–376.

Van Veld, P. A., and D. Nacci. 2008. Toxicity resistance. Pages 597–641 *in* R. T. Di Giulio and D. E. Hinton, editors. *The Toxicology of Fishes*. CRC Press, Boca Raton, FL.

Van Veld, P. A., W. K. Vogelbein, R. Smolowitz, B. R. Woodin, and J. J. Stegeman. 1992. Cytochrome P4501a1 in hepatic lesions of a teleost fish (*Fundulus heteroclitus*) collected from a polycyclic aromatic hydrocarbon-contaminated site. *Carcinogenesis* **13**:505–507.

Van Veld, P. A., and D. J. Westbrook. 1995. Evidence for depression of cytochrome P4501A in a population of chemically resistant mummichog (*Fundulus heteroclitus*). *Environmental Sciences* **3**:221–234.

Vanden Heuvel, J., and G. Lucier. 1993. Environmental toxicology of polychlorinated dibenzo-p-dioxins and polychlorinated dibenzofurans. *Environmental Health Perspectives* **100**:189–200.

Vinson, S. B., C. E. Boyd, and D. E. Ferguson. 1963. Resistance to DDT in mosquitofish, *Gambusia affinis*. *Science* **139**:217.

Vogelbein, W. K., and J. W. Fournie. 1994. Ultrastructure of normal and neoplastic exocrine pancreas in the mummichog, *Fundulus heteroclitus*. *Toxicologic Pathology* **22**:248–260.

Vogelbein, W. K., J. W. Fournie, P. S. Cooper, and P. A. Van Veld. 1999. Hepatoblastomas in the mummichog, *Fundulus heteroclitus* (L.), from a creosote-contaminated environment: a histologic, ultrastructural and immunohistochemical study. *Journal of Fish Diseases* **22**:419–431.

Vogelbein, W. K., J. W. Fournie, P. A. Vanveld, and R. J. Huggett. 1990. Hepatic neoplasms in the mummichog *Fundulus heteroclitus* from a creosote-contaminated site. *Cancer Research* **50**:5978–5986.

Waits, E. R., and D. W. Nebert. 2011. Genetic architecture of susceptibility to PCB126-induced developmental cardiotoxicity in zebrafish. *Toxicological Sciences* **122**:466–475.

Walker, M. K., and R. E. Peterson. 1991. Potencies of polychlorinated dibenzo-para-dioxin, dibenzofuran, and biphenyl congeners, relative to 2,3,7,8-tetrachlorodibenzo-para-dioxin, for producing early life stage mortality in rainbow trout (*Oncorhynchus mykiss*). *Aquatic Toxicology* **21**:219–238.

Wang, C.-C., L.-F. Si, S.-N. Guo, and J.-L. Zheng. 2019. Negative effects of acute cadmium on stress defense, immunity, and metal homeostasis in liver of zebrafish: the protective role of environmental zinc dpre-exposure. *Chemosphere* **222**:91–97.

Wang, S.-B., Y.-J. Wen, W.-L. Ren, Y.-L. Ni, J. Zhang, J.-Y. Feng, and Y.-M. Zhang. 2016. Mapping small-effect and linked quantitative trait loci for complex traits in backcross or DH populations via a multi-locus GWAS methodology. *Scientific Reports* **6**:29951.

Ware, G. W. 1991. *Fundamentals of Pesticides: A Self Instruction Guide*. Thomson Publications, Fresno, CA.

Wassenberg, D. M., and R. T. Di Giulio. 2004. Teratogenesis in Fundulus heteroclitus embryos exposed to a creosote-contaminated sediment extract and CYP1A inhibitors. *Marine Environmental Research* **58**:163–168.

Weaver, G. 1984. PCB Contamination in and around New-Bedford, Mass. *Environmental Science & Technology* **18**:A22–A27.

Webster, T. M. U., N. Bury, R. van Aerle, and E. M. Santos. 2013. Global transcriptome profiling reveals molecular mechanisms of metal tolerance in a chronically exposed wild population of brown trout. *Environmental Science & Technology* **47**:8869–8877.

Webster, T. M. U., T. D. Williams, L. Katsiadaki, A. Lange, C. Lewis, J. A. Shears, C. R. Tyler, and E. M. Santos. 2017. Hepatic transcriptional responses to copper in the three-spined stickleback are affected by their pollution exposure history. *Aquatic Toxicology* **184**:26–36.

Weis, J., and P. Weis. 1977a. Effects of heavy metals on development of the killifish, *Fundulus heteroclitus*. *Journal of Fish Biology* **11**:49–54.

Weis, J., and P. Weis. 1977b. Methylmercury teratogenesis in killifish, *Fundulus heteroclitus*. *Teratology* **16**:317–325.

Weis, J. S., and A. A. Khan. 1991. Reduction in prey capture ability and condition of mummichogs from a polluted habitat. *Transactions of the American Fisheries Society* **120**:127–129.

Weis, J. S., G. Smith, T. Zhou, C. Santiago-Bass, and P. Weis. 2001. Effects of contaminants on behavior: biochemical mechanisms and ecological consequences: killifish from a contaminated site are slow to capture prey and escape predators; altered neurotransmitters and thyroid may be responsible for this behavior, which may produce population changes in the fish and their major prey, the grass shrimp. *Bioscience* **51**:209–217.

Weis, J. S., and P. Weis. 1989. Tolerance and stress in a polluted environment. *Bioscience* **39**:89–95.

Weis, J. S., P. Weis, and M. Heber. 1982. Variation in response to methylmercury by killifish (*Fundulus heteroclitus*) embryos. Pages 109–119 *in* J. G. Pearson, R. Foster, and W. E. Bishop, editors. Aquatic Toxicology and Hazard Assessment: Fifth Conference, ASTM STP 766. American Society for Testing and Materials, Philadelphia, PA.

Weis, J. S., P. Weis, M. Heber, and S. Vaidya. 1981. Methylmercury tolerance of killifish (*Fundulus heteroclitus*) embryos from a polluted vs non-polluted environment. *Marine Biology* **65**:283–287.

Weston, D., H. Poynton, M. Lydy, and G. Wellborn. 2015. Adaptation, not acclimation, is the likely mechanism for reduced sensitivity of some wild *Hyalella* populations to pyrethroid insecticides. *Environmental Toxicology and Chemistry* **34**:2188–2190.

Weston, D. P., H. C. Poynton, G. A. Wellborn, M. J. Lydy, B. J. Blalock, M. S. Sepulveda, and J. K. Colbourne. 2013. Multiple origins of pyrethroid insecticide resistance across the species complex of a nontarget aquatic crustacean, *Hyalella azteca*. *Proceedings of the National Academy of Sciences of the United States of America* **110**:16532–16537.

White, R. D., D. Shea, A. R. Coloww, and J. J. Stegeman. 1997. Induction and post-transcriptional suppression of hepatic cytochrome P450 1A1 by 3, 3′, 4, 4′-tetrachlorobiphenyl. *Biochemical Pharmacology* **53**:1029–1040.

Whitehead, A., S. L. Anderson, K. M. Kuivila, J. L. Roach, and B. May. 2003. Genetic variation among interconnected populations of Catostomus occidentalis: implications for distinguishing impacts of contaminants from biogeographical structuring. *Molecular Ecology* **12**:2817–2833.

Whitehead, A., B. W. Clark, N. M. Reid, M. E. Hahn, and D. Nacci. 2017. When evolution is the solution to pollution: key principles, and lessons from rapid repeated adaptation of killifish (*Fundulus heteroclitus*) populations. *Evolutionary Applications* **10**:762–783.

Whitehead, A., W. Pilcher, D. Champlin, and D. Nacci. 2012. Common mechanism underlies repeated evolution of extreme pollution tolerance. *Proceedings of the Royal Society B-Biological Sciences* **279**:427–433.

Whitehead, A., D. A. Triant, D. Champlin, and D. Nacci. 2010. Comparative transcriptomics implicates mechanisms of evolved pollution tolerance in a killifish population. *Molecular Ecology* **19**:5186–5203.

Whyte, J., R. Jung, C. Schmitt, and D. Tillitt. 2000. Ethoxyresorufin-O-deethylase (EROD) activity in fish as a biomarker of chemical exposure. *CRC Critical Reviews in Toxicology* **30**:347–570.

Willett, K. L., K. Randerath, G.-D. Zhou, and S. H. Safe. 1998. Inhibition of CYP1A1-dependent activity by the polynuclear aromatic hydrocarbon (PAH) fluoranthene. *Biochemical Pharmacology* **55**:831–839.

Williams, C. A. 1994. *Toxicity Resistance in Mummichog (Fundulus heteroclitus) from a Chemically Contaminated Environment*. The College of William and Mary. Masters Thesis. 121 pp.

Williams, L. M., and M. F. Oleksiak. 2008. Signatures of selection in natural populations adapted to chronic pollution. *BMC Evolutionary Biology* **8**.

Williams, L. M., and M. F. Oleksiak. 2011. Ecologically and evolutionarily important SNPs identified in natural populations. *Molecular Biology and Evolution* **28**:1817–1826.

Winston, G. W. 1991. Oxidants and antioxidants in aquatic animals. *Comparative Biochemistry and Physiology. C, Comparative Pharmacology and Toxicology* **100**:173–176.

Winston, G. W., and R. T. Di Giulio. 1991. Prooxidant and antioxidant mechanisms in aquatic organisms. *Aquatic Toxicology* **19**:137–161.

Wirgin, II, G. L. Kreamer, C. Grunwald, K. Squibb, S. J. Garte, and S. Courtenay. 1992. Effects of prior exposure history on cytochrome-P4501A messenger RNA induction by PCB congener 77 in Atlantic tomcod. *Marine Environmental Research* **34**:103–108.

Wirgin, I., N. K. Roy, M. Loftus, R. C. Chambers, D. G. Franks, and M. E. Hahn. 2011. Mechanistic basis of resistance to PCBs in Atlantic Tomcod from the Hudson River. *Science* **331**:1322–1325.

Wirgin, I., and J. R. Waldman. 1998. Altered gene expression and genetic damage in North American fish populations. *Mutation Research-Fundamental and Molecular Mechanisms of Mutagenesis* **399**:193–219.

Wirgin, I., and J. R. Waldman. 2004. Resistance to contaminants in North American fish populations. *Mutation Research-Fundamental and Molecular Mechanisms of Mutagenesis* **552**:73–100. ISSN 0027-5107, https://doi.org/10.1016/j.mrfmmm.2004.06.005.

Wood, C. M. 2012. An introduction to metals in fish physiology and toxicology: basic principles. Pages 1–51 *in* C. M. Wood, A. P. Farrell, and C. J. Brauner, editors. *Homeostasis and Toxicology of Essential Metals* (Vol. 1). Academic Press, London.

Xie, L. T., and P. L. Klerks. 2003. Responses to selection for cadmium resistance in the least killifish, *Heterandria formosa. Environmental Toxicology and Chemistry* **22**: 313–320.

Yin, H., P. Bai, P. Miao, M. Chen, J. Hu, X. Deng, and J. Yin. 2016. Functional expressions of adenosine triphosphate-binding cassette transporters during the development of zebrafish embryos and their effects on the detoxification of cadmium chloride and beta-naphthoflavone. *Journal of Applied Toxicology* **36**:925–935.

Yuan, Z. P., S. Courtenay, R. C. Chambers, and I. Wirgin. 2006a. Evidence of spatially extensive resistance to PCBs in an anadromous fish of the Hudson River. *Environ Health Perspect* **114**:77–84.

Yuan, Z. P., S. Courtenay, and I. Wirgin. 2006b. Comparison of hepatic and extra hepatic induction of cytochrome P4501A by graded doses of aryl hydrocarbon receptor agonists in Atlantic tomcod from two populations. *Aquatic Toxicology* **76**:306–320.

Zhang, Q., P. J. Tango, R. R. Murphy, M. K. Forsyth, R. Tian, J. Keisman, and E. M. Trentacoste. 2018. Chesapeake Bay dissolved oxygen criterion attainment deficit: three decades of temporal and spatial patterns. *Frontiers in Marine Science* **5**:422.

Zhang, W., H. Q. Xie, Y. Li, T. Jin, J. Li, L. Xu, Z. Zhou, S. Zhang, D. Ma, M. E. Hahn, and B. Zhao. 2019. Transcriptomic analysis of Anabas testudineus and its defensive mechanisms in response to persistent organic pollutants exposure. *Science of the Total Environment* **669**:621–630.

Zhang, Y., X. Wang, B. Yang, Y. Hu, L. Huang, C. Bass, and Z. Liu. 2015. Reduction in mRNA and protein expression of a nicotinic acetylcholine receptor α8 subunit is associated with resistance to imidacloprid in the brown planthopper, Nilaparvata lugens. *Journal of Neurochemistry* **135**:686–694.

Zheng, J.-L., Q.-L. Zhu, C.-W. Wu, A.-Y. Zhu, B. Shen, and L. Zeng. 2016. Zinc acclimation mitigated high zinc induced oxidative stress by enhancing antioxidant defenses in large yellow croaker *Pseudosciaena crocea. Aquatic Toxicology* **172**:21–29.

Zhou, T., H. John-Alder, P. Weis, and J. Weis. 1999a. Thyroidal status of mummichogs from a polluted vs reference environment. *Environmental Toxicology and Chemistry* **18**:2817–2823.

Zhou, T., D. J. Rademacher, R. E. Steinpreis, and J. S. Weis. 1999b. Neurotransmitter levels in two populations of larval Fundulus heteroclitus after methylmercury exposure. *Comparative Biochemistry and Physiology Part C: Pharmacology, Toxicology and Endocrinology* **124**:287–294.

Zhu, Q.-L., W.-Y. Li, and J.-L. Zheng. 2018. Life-cycle exposure to cadmium induced compensatory responses towards oxidative stress in the liver of female zebrafish. *Chemosphere* **210**:949–957.

Zimmer, C. T., W. T. Garrood, K. S. Singh, E. Randall, B. Lueke, O. Gutbrod, S. Matthiesen, M. Kohler, R. Nauen, T. G. E. Davies, and C. Bass. 2018. Neofunctionalization of duplicated P450 genes drives the evolution of insecticide resistance in the brown planthopper. *Current Biology* **28**:268–274.e265.

# 8 Endocrine Disruption in Teleosts
## *Mechanisms of Endocrine Toxicity and Experimental Approaches*

*Megan E. Solan and Ramon Lavado*
Baylor University

**Abstract**

The scope of this chapter includes a brief introduction to the endocrine system of fishes, mechanisms of endocrine toxicity in teleosts, and experimental approaches for characterizing the effects of endocrine disruption. The mechanisms of toxicity explored in this chapter will focus on steroidogenesis, hormone metabolism, effects on the hypothalamic-pituitary-thyroid axis, and physiological modifications to reproductive systems by endocrine-disrupting chemicals. An overview of evidence of endocrine disruption in wild fish populations will be explored. Finally, *in vivo* and *in vitro* approaches for characterizing endocrine-disrupting effects in laboratories will be discussed. The chapter will conclude with suggestions for new paths forward for experimental approaches to facilitate the development of policies and practices that are much needed to adequately protect the integrity of our valuable aquatic ecosystems.

## INTRODUCTION

The endocrine system is essential to the physiological processes necessary for survival, growth, and reproduction in animals. This system facilitates communication across the various levels of biological organization within the animal from cells to tissues and organs. The nervous system and endocrine system are both involved in the regulation of physiological processes. These systems are heavily integrated; however, the primary means through which these systems facilitate communication differ. The nervous system relies on nerves and neurotransmitters whereas the endocrine system relies on specific chemical messengers and hormones, which are released into the circulatory system by endocrine glands.

The glands in the endocrine system are responsible for the synthesis, storage, and release of hormones. Hormones act at specific target tissues, where they preferentially bind to receptor molecules in target cells. The specific receptors will react to the hormone binding by initiating an alteration to cell function, which ultimately results in a physiological response within the organism. Different types of endocrine glands in fishes include the hypothalamus, pancreas, pituitary glands, thyroid gland, adrenal gland, and pineal glands. It is important to note that the nervous system and endocrine system are not separate and distinct; rather, they are heavily intertwined and thus sometimes referred to as the neuroendocrine system. For example, regulation of thyroid hormone levels occurs via the hypothalamus–pituitary–thyroid (HPT) axis and the hypothalamic–pituitary–gonadal (HPG) axis controls the development of reproductive tissues. Considering the significant role of these functions in homeostatic maintenance in a changing environment, understanding disruptions to the endocrine system is paramount to the sustainability of fish populations.

The assumption that the vertebrate endocrine system changes its functions under the influence of the compounds introduced anthropogenically is confirmed by extensive evidence collected over more than 30 years (Porte et al., 2006). These compounds have been named "endocrine-disrupting chemicals" (EDCs). The Environmental Protection Agency (EPA) refers to them as compounds of exogenous nature that interfere in the processing and activities of endogenous hormones that take part in the homeostasis, maintenance, and development control (Kavlock et al., 1996). EDCs include heavy metals, pharmaceuticals, nanoparticles, and industrial chemicals including phthalates, organophosphate pesticides, organochlorines, and bisphenol A.

In recent decades, researchers have gained insight into the mechanism of action of EDCs. A consensus statement from the scientific community described in La Merrill et al. (2020) identifies ten key characteristic features of EDCs:

1. Activating hormone receptors or interacting with them.
2. Producing an antagonistic effect on the receptors.
3. Changing the expression of hormone receptors.
4. Modifying hormone transport across cell membranes.
5. Modifying hormone synthesis.
6. Modifying transduction of signals in hormone-responsive cells.
7. Altering hormone level circulation or distributing hormones.
8. Altering clearance or metabolism of hormones.
9. Altering the fate or existence of the cells that produce hormones or are responsive to them.
10. Introducing epigenetic changes in cells that produce hormones or are responsive to them.

A rising interest in EDCs during the last decades can be attributed to the growing evidence for impacts on humans


DOI: 10.1201/9781003160694-8

and wildlife (Colborn et al., 1993). While sexual differentiation and determination are genetically established, these processes can be impacted by numerous environmental factors (Munday et al., 2006; Rajendiran et al., 2021). This feature, especially in piscine species, increases their vulnerability to environmental EDCs, which pollute aquatic habitats (Celino-Brady et al., 2021).

According to research on endocrine toxicity in fish, the mechanisms of action for most EDCs are primarily related to thyroid, gonadal, and adrenocortical functions. The hypothalamus secretes selected hormones, which, together with the pituitary gland, regulate the physiology of all endocrine glands; hence, the focus of this chapter is not on the chemical interrelation of the endocrine axes that monitor the endocrine functions but on the evidence collected from *in vivo* and *in vitro* research studies. Still, the sensitivity to chemicals that disrupt the endocrine system during such early developmental periods as gonadal sex differentiation and phenotypic sex determination is extremely high (Lau et al., 2016; Li et al., 2013; Nishimura & Tanaka, 2014). Therefore, a major focus of this chapter is describing the major mechanisms of reproductive endocrine function that get altered by chemicals, as there is already extensive evidence of the process of endocrine disruption in vertebrates, widely analyzed in the literature. The intent of this chapter is not to give an exhaustive account of endocrine physiology in fishes and detail every study on effects in the literature; instead, it will focus on providing a broad overview of current information on the effects of environmental exposure on endocrine functions.

## MECHANISMS OF ENDOCRINE TOXICITY IN TELEOSTS

### Effects of EDCs on Steroidogenesis and Hormone Metabolism

In all vertebrates, and teleosts in particular, steroid hormones play a critical role in the regulation of various physiological processes, including stress response, gonadal and embryonic development, and gametogenesis (Guiguen et al., 2010; Nagahama et al., 1995; Rajakumar & Senthilkumaran, 2020; Tokarz et al., 2015). Steroidogenesis takes place in gonadal and non-gonadal tissues. The latter include the tissues of the liver, kidney, adipose, and intestine (Swart et al., 2013). Steroidogenesis in teleosts occurs in a particular type of gonad cell, where the process of processing cholesterol into pregnenolone occurs. Afterward, it is transformed into specific types of steroids, particularly androgens, estrogens, and progestins, while the location for conversion to mineralocorticoids and corticosteroids takes place in peripheral tissues (Blázquez & Somoza, 2010; Hoffman et al., 2011; Leet et al., 2011). A schematic representation of steroidogenesis in teleosts, showing the major steroidogenic pathways for the synthesis of steroids is shown in Figure 8.1. Regulation of steroidogenesis is crucial to sex-specific steroid biosynthesis and can be influenced by many factors such as the preference of a co-factor, the limitation of the substrate, the expression of steroidogenic enzyme genes, or biochemical allosteric mechanisms (Tenugu et al., 2021).

Gonadal steroidogenesis is mostly regulated by the hypothalamo-hypophysial axis, although the sex-specific and maturity-stage steroidogenesis also depends on the

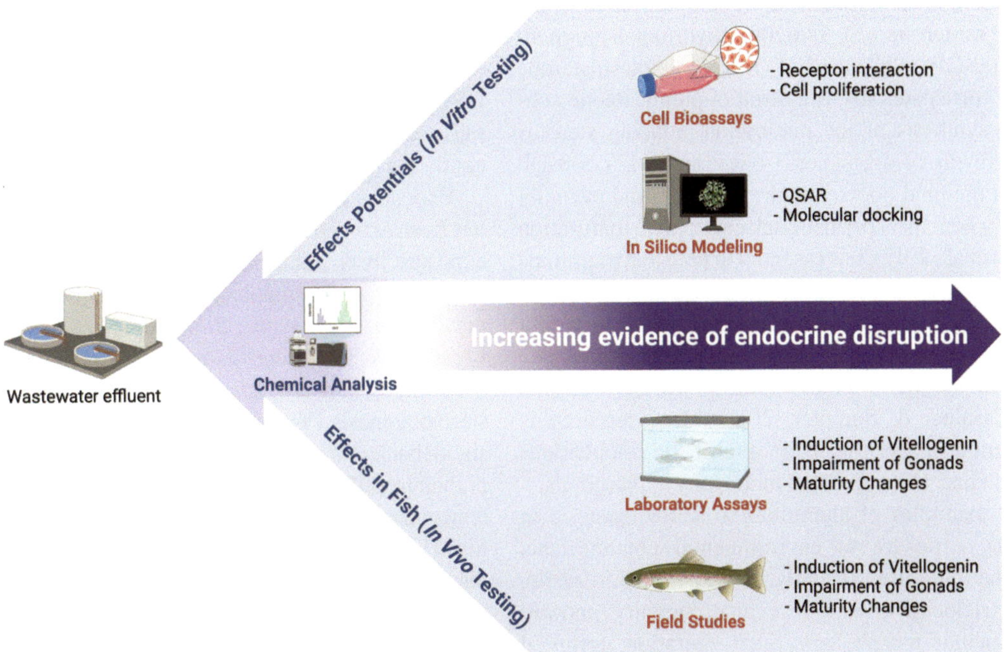


**FIGURE 8.1** Schematic representation of steroidogenesis in teleosts showing the major steroidogenic enzymes catalyzing the reaction for the synthesis of steroids. Blue boxes indicate major sex steroids and red boxes indicate maturation-inducing hormones (*cyp*, cytochrome P450; OH-, hydroxy-; *hsd*, hydroxysteroid dehydrogenase). Modified from Tenugu et al. (2021).

microenvironment and the steroidogenic cells, including paracrine or autocrine regulation (Ge, 2005; Scott et al., 2010). Gonadal development and sex determination in mammals are determined by genetics, and they demonstrate either restricted or no gonadal plasticity at all (Tanaka & Nishinakamura, 2014). This is in contrast to teleosts, which do display gonadal plasticity and often retain the ability to modify the gonadal sex either in adulthood or at other development stages. Influenced by environmental factors or exposure to hormone treatments, some species may even undergo either an entire transformation of the ovary into the testis or vice versa regardless of the chromosomal or genetic sex (García-Alonso et al., 2009; Soyano et al., 2022; Tenugu et al., 2021). Additionally, the development of the gonads without a specific testis or ovary differentiation is determined by the levels of hormones in the course of the critical stage of gonadal differentiation (Devlin & Nagahama, 2002).

Gonadal differentiation is primarily regulated by the expression of steroidogenic enzymes. Therefore, there is a direct effect of the impact of sex steroids on the role of particular transcription factors in regulating the expression of steroidogenic enzyme genes. The initial stage of steroidogenesis is a rate-limiting step as it implies transporting cholesterol with rate-limiting functions into the mitochondria by StAR (Steroidogenic Acute Regulatory protein) with further conversion into pregnenolone, which plays a key role as the first precursor in the cascade of steroidogenesis (Stocco, 2000). However, there is some uncertainty in the role of additional rate-limiting stages of steroidogenesis.

Cholesterol conversion to pregnenolone involves CYP11A1 (Figure 8.1) and it gives a start to the steroidogenesis process as it is after StAR (Tokarz et al., 2015). A few steroidogenic enzymes act sequentially to obtain such active steroids as testosterone (T), 17α,20β-dihydroxy-4-pregnen-3-one (17α,20β-DP), 17β-estradiol (E2), 11-ketotestosterone (11-KT), and corticosteroids as a result of pregnenolone conversion. The synthesis of the maturation-inducing steroids 17α,20β,21-trihydroxy-4-pregnen-3-one and 17α,20β-DP takes place at the final oocyte maturation stage and partially at the stage when the sperm of teleosts reach maturation (Sreenivasulu et al., 2012). It is essential to note that mammals also have 11-KT (potent androgen in fish) and 11-hydroxytestosterone (11-OHT) (Rege et al., 2019). However, a potent androgen in mammals, DHT (5α-dihydrotestosterone), has also been identified in the fathead minnow plasma (Margiotta-Casaluci & Sumpter, 2011). More research is needed to gain a deeper understanding of the implications of DHT in fish and 11-oxygenated androgens in mammals.

There are examples of alterations to steroidogenesis in fish following exposure to environmental contaminants throughout the literature. Recently, zebrafish (*Danio rerio*), exposed to triclocarban and inorganic mercury showed modified testicular morphology. Such alteration stemmed from alterations in the transcription of a few steroidogenic genes, namely *cyp19a*, *3β-hsd*, *cyp17*, and *17β-hsd* (Wang et al., 2016). In *Gobiocypris rarus*, the exposure to bisphenol A (BPA) modulated the mRNA expression of *StAR*, *cyp11a1*, *3β-hsd*, *cyp17a1*, and *cyp19a1a*. It is worth emphasizing that the increase in the expression of several mRNAs encoding testicular steroidogenic enzymes indicated an increased production of androgens in this study. However, a minor downregulation of *cyp17a1* mRNA expression could balance such overproduction (Liu et al., 2014). The correlation analysis of the mRNA expression levels showed that the BPA-mediated effects on testicular steroidogenesis entail sex steroid hormone receptor signaling, gonadotropin/gonadotropin receptor pathway, and transcription factors. The latter include the nuclear receptor subfamily 5, group A (Nr5a), and the forkhead box protein L2 (Foxl2) (Forner-Piquer et al., 2019). Exposure to ethynylestradiol (EE2) altered the expression of the steroidogenic genes *StAR*, *cyp11a1*, *3β-HSD*, *cyp17a1*, and *cyp19a1a*, similar to the BPA exposure. The researchers anticipated that such suppression might have happened due to negative feedback of EE2 on the pituitary follicle-stimulating hormone (FSH). Indeed, a direct EE2 effect on the testis may have also produced the described inhibition (Liu et al., 2012).

In another relevant study, the endocrine-disrupting chemical di(2-ethylhexyl)phthalate (DEHP) was able to alter the normal hormonal profile in goldfish. A 30-day exposure to DEHP reduced the levels of 11-KT and the luteinizing hormone (LH), indicating that that contaminant disrupts the testicular hormone production and reduces sperm quality. However, DEHP did not affect the gonadotropin-releasing hormone (GnRH) and related regulatory activities. Furthermore, it did not stimulate any estrogenic responses and did not act through sex hormone receptors (Golshan et al., 2015). In another study, focused on goldfish exposed to three nominal vinclozolin (VZ) concentrations (100, 400, and 800 μg/L), disruption of steroidogenesis and decreased sperm quality were observed in a dose-dependent manner. Another research evidenced the potential effects of VZ by the direct inhibition of 11-KT biosynthesis (Hatef et al., 2012). Furthermore, exposure to low levels of BPA (0.6, 4.5, and 11 μg/L) considerably reduced T and 11-KT, reducing significantly the sperm motility and velocity in the same species.

Exposure to diethylstilbestrol and flutamide in zebrafish has been associated with spermatogenesis decline. Flutamide exposure was associated with several outcomes, namely a decline in plasmatic 11-KT, deregulation of *aldh1a2*, *cyp26a1*, *nanos1*, *sycp3*, *dmc1*, *bax*, and *bcl2* genes encoding meiosis- and apoptosis-regulating signals and *dmrt1*, *sf1*, *cyp11b2*, *cyp17a1*, *ar*, *fshr*, and *lhr* genes controlling gonadal steroidogenesis (Yin et al., 2017). In the male three-spined stickleback (*Gasterosteus aculeatus*), the exposure to dibutyl phthalate (DBP) substantially increased plasma testosterone concentrations and reduced the levels of glue protein, spiggin. Hence, it can be stated that DBP acts as an endocrine-disrupting chemical in sexually developed three-spined sticklebacks (Aoki et al., 2011; Nendza et al., 2016).

In newborn guppies, *Poecilia reticulata*, a 3-month exposure to a sub-lethal dose of nonylphenol (NP) led to an abnormal transcription of hepatic vitellogenin associated with a marked decline in the gonadosomatic index (GSI). Males did not approach females for reproduction, which indicated impaired courtship behaviors. In fact, normal courtship behavior was delayed for 3 months, implying that NP has enough estrogenic potency to temporarily undermine reproductive functions (Cardinali et al., 2004).

## Effects of EDCs on the Hypothalamic–Pituitary–Gonadal Axis

Hormones and other signaling molecules secreted by the hypothalamic-pituitary-gonadal (HPG) axis regulate the proper development of reproductive tissues and align annual reproductive cycles in teleost fish, leading to the fertilization of mature gametes. In fish, the fundamental characteristics of the HPG axis resemble those of other vertebrates. The peripheral and internal sensory systems detect environmental signals, like changes in photoperiod, water temperature and its chemical composition, and physiological changes in nutritional state. The information passes through neural pathways to the hypothalamus and related areas of the brain. The hypothalamus integrates internal and external information about the organism's state and induces the secretion of multiple neurotransmitters and neuropeptides that alter the activities of the gonadotropin-releasing hormone (GnRH) (Figure 8.2) (Xu et al., 2011). GnRH is a decapeptide and is the primary neurohormone for reproductive system regulation. The GnRH neurons stimulate the gonadotropin-producing cells and control the synthesis and secretion of gonadotropins (Muñoz-Cueto et al., 2020). GnRH is discharged from nerve terminals into the gonadotropic area and attaches to specific receptors on the plasma membrane, activating intracellular signaling pathways. Monoamine neurotransmitters, i.e., dopamine and serotonin, affect the pituitary gland, controlling the secretion of GnRH and gonadotropins (Murugananthkumar & Sudhakumari, 2022).

The monoaminergic system and neurosteroids use positive and negative feedback to regulate the production of sex hormones. Changes in CYP19A1B, which controls the aromatization of androgens, lead to increased availability of estrogens (Böhne et al., 2013; Mouriec et al., 2009), and influence GnRH neurons in fish (Terasawa & Kenealy, 2012). One of the most important systems in mammals, the cholinergic system has been identified in teleosts (Toledo-Ibarra et al., 2013), but its role in gonadal function remains unknown.

Endocrine-disrupting chemicals affect the HPG axis by targeting genes related to monoaminergic systems, the gonadotropin-releasing hormone-gonadotropin (GnRH-GTH) axis, and steroidogenesis. Consequently, they modify the feedback regulation connected to the gonadal physiology. There is a growing body of evidence of adverse effects, including selective inhibition of genes, on the HPG axis by organophosphate and organochlorine EDCs. Organochlorine and organophosphate families of pesticides directly affect the HPG axis, targeting its primary role in regulating reproductive activity (Senthilkumaran, 2015). Organochlorine pesticides include DDT, dichlorodiphenylethylene (DDE), dieldrin, malathion, and endosulfan.

Several studies on fish revealed the toxic effects of endosulfan on testis- and ovary-determining pathways (Chakrabarty et al., 2012). Males exposed to endosulfan exhibited a reduction of spermatogenesis and testes development in the juvenile Asian catfish, *Clarias batrachus* (Rajakumar et al., 2012). Endosulfan exposure led to advanced ovarian maturation

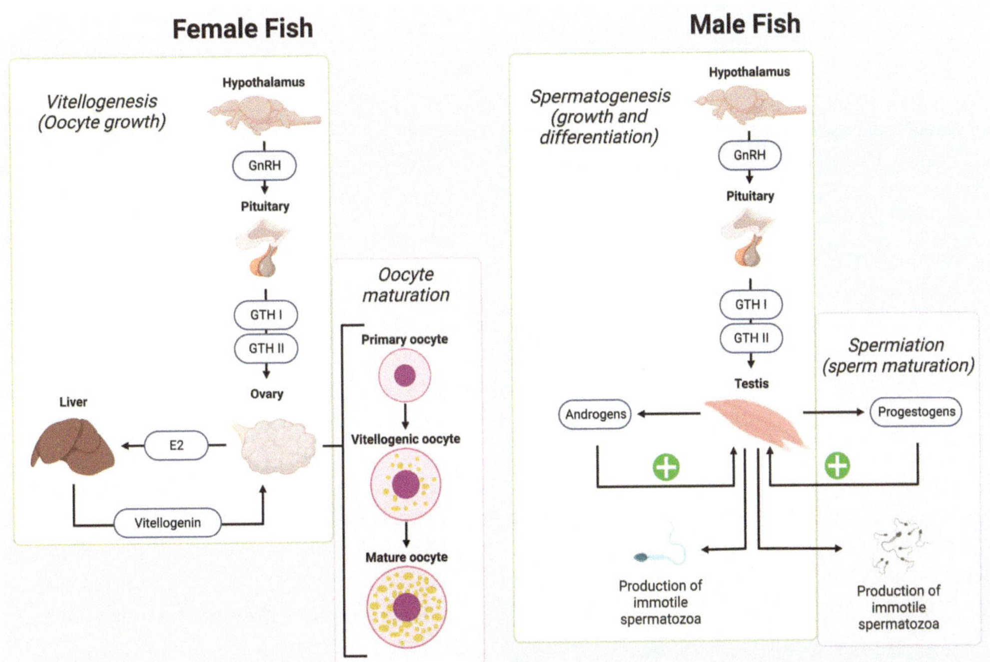


**FIGURE 8.2** Schematic representation of the hormonal control of the hypothalamus-pituitary-gonad axis of female fish and morphological changes associated with vitellogenesis and final oocyte maturation and male fish and the major events in the testes during spermatogenesis and spermiation. GnRH: Gonadotropin-releasing hormone. GTH I: Gonadotropin I. GTH II: Gonadotropin II. E2: 17β-Estradiol. Created with a license from Biorender.com.

and modification of ovary-specific genes in females. In all cases, serotonin, a monoamine neurotransmitter for GnRH-GTH release was altered by endosulfan exposure. Thus, any change to serotonin metabolism/synthesis is capable of affecting gonadal differentiation and development in fish (Fontaine et al., 2020; Kobayashi et al., 2013).

Research focused on malathion demonstrated its varied effects on lipid metabolism and growth. In teleosts, malathion decreases hormone levels targeting the reproductive system (Lal et al., 2013; Patil & David, 2010). Dose-dependent alterations of malathion exposure on *cfgnrh*, *tph,* and tyrosine hydroxylase were observed in male and female brains of the air-breathing catfish, *C. gariepinus.* Prathibha et al. (2014) also found that malathion blocked the reproductive development of *C. gariepinus.* Considering serotonin and catecholamines regulate GnRH-GTH release, malathion exposure may modulate the expression of genes and factors related to neurosteroidogenesis (Senthilkumaran, 2015).

Other pesticides have been reported to have endocrine-disrupting effects. For example, the commonly used organotin acaricide, azocyclotin, modified the transcription of several genes in the HPG axis, causing the inhibition of oocyte maturation and decelerated spermatogenesis in zebrafish (Ma et al., 2016). Recent research showed that exposure to the herbicide glyphosate in Japanese medaka modified neuroendocrine-associated genes (*kiss1*, *kiss2*) in the female brain in addition to reproductive, developmental, and epigenetic modifications from exposure (Smith et al., 2019).

Like some pesticides, selected pharmaceuticals in the aquatic environment and sewage treatment plants mimic sex steroids, which can affect non-target organisms, including fish. Recent studies in hybrid striped bass and fathead minnows demonstrated significant changes in brain serotonin levels after exposure to antidepressants (Bisesi et al., 2014). Similarly, sex hormone analogs such as EE2 and methyltestosterone (MT) modified the monoaminergic-GnRH system in air-breathing catfish (Swapna & Senthilkumaran, 2009). In female zebrafish, dydrogesterone (a synthetic progesterone) substantially affected the transcription of GnRH and steroidogenesis genes (Shi et al., 2019). In addition, estrogen and progestin in the environment have been observed to affect brain aromatase, modulating the GnRH-GTH release (Cano-Nicolau et al., 2016). Likewise, the exposure of zebrafish embryos to the industrial chemical, bisphenol A (BPA), modulated several brain neuroendocrine-associated genes, namely *kiss1, kiss1r, gnrh3, lhβ, fshβ,* and *er1* (Qiu et al., 2016).

Heavy metals and metallic compounds can adversely affect the HPG axis, altering sex steroid levels and expressions of several brain-specific genes while disrupting the serotonin-GnRH pathways. Comparable to sex hormones, heavy metals (Zn, Cu, Fe, Hg, Mn, As, Cd, Cr, and Pb) are fundamental to regulating biological functions but can be toxic to living organisms, including fishes. The prolonged exposure and the peculiarities of the organism's development are core factors defining the toxicity of the detailed elements (Deepa et al., 2019; Denier et al., 2009). Recent studies have documented a disruptive effect of heavy metals such as Cd, Cu, Fe, Mn, Pb, Hg, Zn, and methyl mercury (MeHg) on the HPG axis (Crump & Trudeau, 2009; Gárriz et al., 2019; Gárriz & Miranda, 2020). It has been found that the metallic compound $BaCl_2$ adversely affected brain-related genes in zebrafish (Kwon et al., 2016). Another report on the Atlantic croaker, *Micropogonias undulatus,* showed that $PbCl_2$ and Aroclor 1254 changed the stimulatory serotonin-GnRH pathway, which might result in reproductive impairment (Prasad et al., 2015).

It has been demonstrated that there is a multitude of compounds with endocrine-disrupting abilities that impact the HPG axis by altering GnRH and gene expression. While brain aromatase, *cyp19a1b* has been identified as a potential biomarker of EDC-induced abnormalities specifically at the HPG axis, other biomarker genes have yet to be identified and warrant further study. It is also worth noting that GTH receptors in ovarian follicles have been identified in fish (Oba et al., 1999). However, how EDCs affect these receptors is not clear. Hence, focusing on them would provide an insight into the additional impacts of EDCs on the HPG axis.

## Effects of EDC Exposure on Fish Ovarian Physiology

Modifications to fish sex physiology that result in the feminization of males and masculinization of females are a major threat to the reproductive success of fish populations. Effects of EDCs in fishes are often reported with respect to oocyte growth and maturation, modulation of hormone functioning, and subsequent impacts on fertility. A plausible mechanism for these effects may be the ability of xenobiotics to behave like weak estrogens.

Exposure of *Gobious rarus*, a minnow species, to low doses of BPA, was linked to modified ovarian structure in female fish following 14 and 35-day exposures (Zhang et al., 2014). In particular, the exposure led to a moderately increased ratio of premature oocytes and the occurrence of extra atretic follicles. In female zebrafish, *Danio rerio*, regular exposure to 25 ng/L EE2 and 0.02, 0.2, 2, 20, and 40 μg/L of DEHP damaged ovulation and embryogenesis. Upon exposure, fish were crossed, and the number of obtained embryos reached about 1% of those produced by the control group, indicating an apparent fertility reduction (Carnevali et al., 2010). Also, in *Gobious rarus*, the herbicide 2,4-dichloro-6-nitrophenol (DCNP) resulted in an increase in the number of damaged follicles and deterioration of vitellogenic oocytes. (Chen et al., 2016). Histological examinations have indicated rudimentary or deformed ovaries with advanced oocyte development following exposure of catfish to EE2 (50 ng/L) (Sridevi et al., 2015). Suppressive effects on oocyte development may delay spawning time and reduce fecundity.

While delayed ovarian development and damaged gametocytes are commonly reported vitellogenic effects in fish have also been observed. Diuron, a urea-substituted herbicide with reported antiandrogenic activity, and three of its metabolites were assessed in tilapia. It was found that contaminants increased the gonadosomatic index, whereas the ovary demonstrated an increase in vitellogenic oocytes and a decrease in germinative cells. Thus, diuron metabolites stimulate estrogenic activity and advance species' ovarian development (Margiotta-Casaluci et al., 2013). In marine medaka, *Oryzias latipes*, exposure to 8.5 μg/L of 3,3′-diindolylmethane (DIM) increased vitellogenin storage but decreased cathepsin enzyme activities, indicating an obstruction in vitellogenin uptake in developing female gametocytes. The decline emerged concurrently with a decrease in levels of eggshell proteins, contributing to the suppression of primary oocyte development and decreased fertility rates (Lahnsteiner et al., 2005). Developmental issues in offspring following parental exposure have also been reported (Wang et al., 2015).

### Effects of EDC Exposure on Fish Testis Physiology

Exposure of marine medaka, *Oryzias melastima*, to low quantities (0 and 8.5 μg/L) of 3,3′-diindolylmethane (DIM), a bioactive metabolite, entailed the overproduction of vitellogenin and eggshell proteins in the testis, emphasizing its estrogenic activity. In zebrafish, the consequences of regular exposure to DEHP (0.2 and 20 μg/L) were analyzed, underlining DEHP's capacity to undermine reproduction causing a mitotic arrest during spermatogenesis and diminishing embryo generation (Corradetti et al., 2013). Likewise, exposure of zebrafish to a variety of environmentally relevant concentrations of inorganic mercury caused intense histological impairments in the testis, namely a decrease in the size and quality of sperm. Additionally, the exposure to diethylstilbestrol, flutamide, and their combination undermined spermatogenesis, reducing sperm concentration (Yin et al., 2017).

The potential retardation of spermatogenesis was hypothesized in male carp, *Cyprinus carpio*, which was exposed to 2 mg/L perfluorooctanoic acid (PFOA) for almost 2 months. In that species, spermatogonia and spermatocytes served as the dominant germ cells (Giari et al., 2016). Additionally, an enlargement of interstitial tissue and multiplication of somatic cells occurred in the testis. Similarly, 2-week exposure of carp to different BPA concentrations, ranging from 1 to 1000 μg/L, induced intense alterations in the lobular structures and spermatogenic cysts (Mandich et al., 2007). Notably, in BPA-exposed goldfish, gonadosomatic index, hepatosomatic index (HSI), and E2 levels were not impacted. Such an observation implies that BPA was able to alter testicular spermatogenesis, affecting sperm maturation. Alterations to the physiology of testes can greatly diminish the chances for successful mating – an outcome that can have devastating impacts on populations.

## EVIDENCE OF ENDOCRINE DISRUPTION IN WILD FISH POPULATIONS

Fishes are among the most broadly studied aquatic vertebrates, serving as a sensible model for investigating the toxicity of EDCs on living organisms. Laboratory experiments are fundamental to understanding the effects of particular contaminants on fish physiology. However, in natural ecosystems, fish are exposed to multiple xenobiotics, chemicals that are extrinsic to the normal metabolism of that organism. EDC effects vary from gametogenesis suppression and development of intersex gonads to changes in the gonadosomatic index and decreased fecundity. This section will highlight studies that have provided evidence for endocrine disruption in wild fishes.

Jobling et al. (1998) was a pioneering study of intersex in fish caused by EDC exposure. It linked the increased prevalence of gonadal intersex and modifications in secondary sex features in male roach (*Rutilus rutilus*) from the UK rivers due to wastewater treatment plants. Subsequent studies demonstrated that contamination of the river ecosystem delayed spermatogenesis in male fish, whereas the oogenesis in female fish was less affected (Jobling et al., 2002). Findings of sexual disruptions in fish populations raise concerns over impacts on the sustainability of populations.

During the last two decades, an increasing number of studies have reported evidence of endocrine disruption in wild fish in UK rivers (Liney et al., 2005; van Aerle et al., 2001), in wild bream (*Abramis abrama*) from the Elbe River (Hecker et al., 2002), in perch (*Perca fluviatilis*) and roach (*Rutilus rutilus*) from a remote Swedish lake (Noaksson et al., 2003), in barb (*Barbus plebejus*) from the Po River (Viganò et al., 2001) and carp (*Cyprinus carpio*) from the Ebro river in Spain (Lavado et al., 2004), among others. One notable study utilized low concentrations of EE2 releases into an experimental lake in Canada over two field seasons. In addition to abrupt decreases in population size, the data illustrated that male fathead minnows (*Pimephales promelas*) presented intersex changes and modifications in testicular structure (Kidd et al., 2007). There was a correlation between exposure to estrogenic chemicals and the development of gonadal intersex in male fish. The effects of estrogens on vitellogenin induction and intersex development in wild fish populations were consistent with their laboratory findings.

Recent studies reported an elevated prevalence of male intersex in two species of darters (*Etheostoma* sp.) collected downstream of water treatment plants. VTG induction was one of the main responses of male darters to municipal wastewater. Unlike the roach, darters spend their entire life span within a limited territory. Thus, they serve as valid indicators of chemical contamination (Hicks et al., 2017). It is worth emphasizing that renovations in sewage treatment plants have reduced the prevalence of intersex in the darters and advancements in the wastewater treatment systems led to sufficient EDC filtration (Barber et al., 2012).

These initial field studies led to many studies on wild fish populations from various habitats where the effects of pollutants were investigated. These studies analyzed reproductive outcomes in response to acute and chronic exposures to chemical pollutants. Invariably, those studies confirmed that pulp and paper mill effluents cause irreversible alterations to reproductive outcomes (Hewitt et al., 2008). Though the controlled laboratory studies verified the effects observed in field studies, they did not identify specific chemical compounds responsible for the reproductive degradation. Prior research revealed bacterial transformation products of phytosterols in the pulp and paper mill effluents (Jenkins et al., 2004) but these compounds were not detected in adequate concentrations to clarify their potential effects on the reproductive system.

Recently, bioassays that measure chemical toxicity have been standardized, including those focused on substance effects on the reproductive system (Robitaille et al., 2022). These bioassays study a few species out of nearly 30,000 existing teleosts, most of which are small-bodied fish with short lifespans and reproductive periods. However, the bioassays have been deemed valuable considering the findings of the bioassays -corroborated observations of EDC effects in wild fish populations. Evidence suggests that the effluents comprise numerous ligands that bind to sex hormones and xenobiotic receptors (Hewitt et al., 2008). Thus, EDCs impact physiological processes by interacting with multiple pathways.

Notably, there are no compulsory EDC tests in many countries that identify EDC activity in sewage wastewater treatment plants. Attempts to mitigate the EDC impact on aquatic wildlife have been deemed insufficient in the USA (Robitaille et al., 2022). Developing a robust body of evidence for these effects enables the development of risk management approaches that are protective of aquatic ecosystems.

## EXPERIMENTAL APPROACHES FOR CHARACTERIZING THE EFFECTS OF ENDOCRINE DISRUPTORS COMPOUNDS (EDCS) IN FISH: NEW PATHS FORWARD

There has been substantial progress in the analysis of the effects of endocrine-disrupting chemicals in fish. As shown in Figure 8.3, approaches to EDC analysis are varied, comprising *in vivo* assessments, *in vitro* assays, and the application of *in silico* predictive models. The in vivo assessments evaluate the impact of EDCs during early life and adulthood, along with long-term and transgenerational consequences. In addition, *in vitro* assays describe the effects of chemicals on hormone receptors, whereas in silico models focus on EDCs-sensitive pathways.

### *In Vivo* Studies

While laboratory tests are able to analyze the correlation between chemically contaminated environments and dysfunction in fish, such aspects as the diversity of fish species, their habitat, and the wide spectrum of EDCs found in natural ecosystems call for methodologically diverse research methods. Each approach has a particular scope of application. The *in vivo* assessment employs different experimental protocols, including the exposure to chemicals observed in the natural environment, the addition of specific substances to the diet, or the delivery of chemical compounds through intraperitoneal injection. Notably, the duration of exposure is case-specific. The concentration of chemicals is either environmentally identical or capable

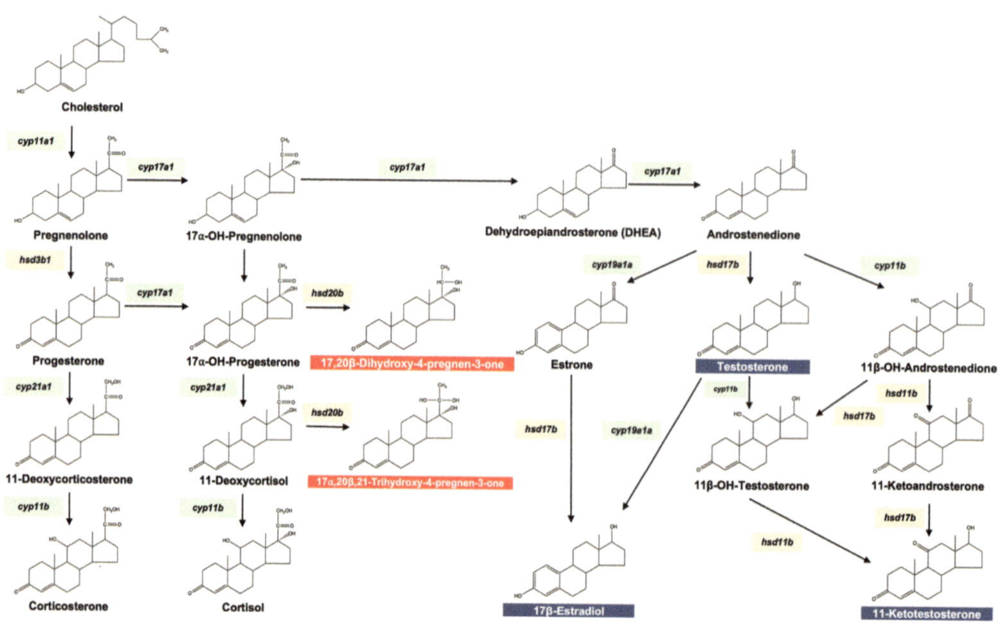


**FIGURE 8.3** Model of the study design for detection of effects of EDCs in the environment. This figure gives an overview of the experimental approaches to get evidence of endocrine disruption in fish species. Modified from Henneberg et al. (2014).

of stimulating a physiological response. In this approach, research models vary considerably and can be classified into those focused on large and small species. Salmonids, tilapia, and bass exemplify the first group, whereas the latter comprises zebrafish, medaka, and guppies. Research on larger fish is economically significant and provides sufficient tissue samples for analyses. However, considering the longer life spans and the need for larger rearing areas, trials on larger fish species are higher in maintenance costs. By comparison, small fish have a shorter life span, quicker maturation, higher reproductive rates, and lower maintenance costs. The enumerated factors suggest that smaller fish are more applicable for research purposes. Undeniably, the advances in genetics have enabled the analysis of different species at various life stages as models for investigating the effects of EDCs, regardless of the size of the fish and any experimental limitations.

In fish, levels of plasma sex hormones, steroidogenic enzymes, vitellogenin (VTG), and GSI comprise the criteria for analyzing the reactions of the reproductive system to direct or mediated EDC exposure. As mentioned above, VTG facilitates the identification of estrogens exposure. Under normal conditions, VTG is found exclusively in females. In males, VTG emerges as a response to environmental estrogens (Davis et al., 2009). It has been found that steroid hormone receptors control steroid and exogenous hormones (Gross & Yee, 2002; Park et al., 2007). The activation of estrogen receptors (ERs) elicits the synthesis of VTG (Bowman et al., 2002; Flouriot et al., 1996; Jalabert, 2005). Thus, VTG and ERs have been employed in both *in vivo* and *in vitro* experimental studies (Jobling & Sumpter, 1993; Jones et al., 2000; Leet et al., 2011; Tyler et al., 1996). The growth hormone/insulin-like growth factor-I (GH/IGF-I) axis is the fundamental endocrine system controlling the development of vertebrates, including teleosts (Reindl & Sheridan, 2012). Upon attachment to the growth hormone receptor (GHR), GH boosts the IGF production. Notably, GH and IGFs stimulate growth in target tissues (Duan, 1998; Fan et al., 2009). IGF binding proteins (IGFBPs) manage the actions of IGFs by defining their accessibility and functions (Duan & Xu, 2005). Alongside such morphological indicators as size and weight, factors involved in the GH/IGF system can be used as molecular markers for the estimation of the effects of EDCs on growth in fish.

### Adult Exposure

For several decades, studies on EDCs have focused on defining the correlation between chemical pollutants and morphological abnormalities in fish. Multiple empirical studies have described the effects of EDCs on sub-mature and fully mature fish. For instance, a mature fish exposed to EDCs may exhibit remarkably different outcomes compared to fish exposed to EDCs at an early-life stage or a developmentally critical phase. It is also worth noting that such chemicals affect not only an individual but also its offspring. The respectively larger body mass of sub-adult and adult individuals simplifies the aggregation and subsequent analysis of biological markers. It is also worth noting that adult fish with fully differentiated tissues and a highly developed endocrine system can mitigate the influence of exogenous chemicals, limiting the number of observable EDC outcomes (Palaniappan et al., 2008; Sun et al., 2009).

Some studies have exposed adult fish to EDCs for prolonged periods, while others have focused on brief or acute exposure. Certainly, definitions of regular and acute exposure differ depending on the life cycle of the given organism. In this chapter, periods of acute exposure to EDCs are equal to 1 week or less. In contrast, sub-regular/regular exposure defines a period ranging from a week to several months/years, regardless of the life stage. In both models, EDC exposure occurs in the aquatic environment or through diet/intraperitoneal injection. Physiological responses of fish may deviate from case to case due to bioaccumulation and bioavailability of chemical contaminants. Bioaccumulation may vary since different elimination mechanisms serve as its co-determinants. For example, direct equilibrium exchange of the pollutants at the gill epithelium might alleviate the accumulation of an aquatic contaminant by a non-dietary route (Streit, 1998). Consequently, dietary exposure might be different from aqueous uptake in quantitative terms.

### Early-Life Exposure

The analysis of the exposure to EDCs during the early-life phase enables the identification of abnormalities across the fish's lifespan. It clarifies their sensitivity to particular chemicals at an embryonic stage. In early-life trials, the long-term effects of EDCs have been observed immediately after exposure or upon depuration. Moreover, approaches employing larval fish decrease the experimental footprint and associated expenditure, as they require smaller rearing containers. Some studies used Petri dishes to conduct exposure experiments (Bhandari et al., 2015; Johns et al., 2011). It is important to note that the size of the fish at the initial stages of its development hinders the process of obtaining essential samples. Sex identification at the early stage of development can pose another challenge. Though some studies use the whole animal to compensate for the limited quantity of tissue (Duffy et al., 2014; Johns et al., 2011; Wang et al., 2020), this approach is not feasible in the given case. Notably, *in situ* hybridization can be performed to specify the expression patterns in different organs. The determination of the sex of the fish can be made at the later stages of development. However, high mortality rates in the early-life phase represent another burden to this experimental strategy.

### Transgenerational Studies

Several studies in fish have confirmed the adverse effects of EDCs on parental and offspring generations. Recent research findings suggest that EDCs influence not only the exposed individual but also its progeny, verifying the phenomenon of transgenerational inheritance (Crews et al.,

2007; Horan et al., 2017). It is worth noting that transgenerational trials feature the EDC exposure focused on the parental or F0 generation. Such experiments provided data on direct and indirect outcomes of EDC exposure. Moreover, transgenerational effects are not confined to epigenetic modifications such as DNA methylation and the changes in the expression of non-coding RNAs (Singh & Li, 2012) but also comprise of the impact on the direct transfer of oocyte reserves.

### *In Vitro* Studies

*In vivo* experiments consider all physiological processes and interchanges appearing in the animal after exposure to EDCs. In parallel, *in vitro* methods define the direct effects of chemicals on a particular tissue or cell type. Given that *in vitro* studies are more controlled and have fewer confounders, they are applicable to analyze the EDC mechanisms under near-ideal conditions. Furthermore, the manipulation of cellular pathways, namely the release of inhibitors/ activators, facilitates the identification of molecular target/s of EDCs within the cell, which is a critical measure in EDC categorization. Together with methods measuring morphological modifications, disrupted cell development, and hormone synthesis, such techniques can highlight the effects of different EDCs on cells or organs.

Contrary to *in vivo* approaches, *in vitro* experiments are more time-dependent, as most cell incubations have limited viability. In *in vitro* systems, the exposure time differs based on the viability of cells and the time needed to affect biomarkers or evoke specific physiological responses. There have been several cell lines derived from fish that are frequently used to study molecular endpoints of exposure including PHLC1, RTH-149, RTG-2, ZFL, RTgill-W1, Hepa-E1, and RTL-W1. Scholz et al. (2013) provide an overview of *in vitro* assays for endocrine-active substances in fish. Such testing screens EDCs and effluent samples and serves as a guide for higher testing fidelity.

*In vitro* screenings feature cell proliferation assays, yeast-based testing, and gene assays employing fish and human cell lines (Andersen et al., 1999; Lascombe et al., 2000; Murk et al., 2002; Routledge & Sumpter, 1996; Zacharewski, 1997). In human cells, E-screenings based on the proliferation of MCF-7 human breast cancer cells have been employed to investigate the estrogenicity of the aquatic environment, sediment, and effluents (Henneberg et al., 2014). However, primary cells directly isolated from fish are considered the gold standard for *in vitro* testing and have demonstrated great value in understanding EDC effects. *In vitro* assays using primary hepatocytes helped evaluate the influence of EDCs on the GH/IGF system in fish. In rainbow trout juveniles, for instance, hepatocytes exposed to EE2 for 2 days demonstrated a dose-specific increase in the regulation of *er* and *vtg* and a decrease in the normal regulation of *eghr1* and *igfbp1* (Hultman et al., 2015). Bream (*Abramis brama*) and carp hepatocytes exposed to EE2 for 4 days exhibited a dose-proportional induction of VTG (Rankouhi et al., 2004). Exposure of testicular cells from Japanese eel (*Anguilla japonica*) to NP with or without 11-KT for 2 or 4 weeks resulted in spermatogonial regeneration and caused hypertrophy in somatic cells while reducing the number of reproductive cells (Miura et al., 2005). Employing the same *in vitro* system, testicular fragments treated with or without hCG for 1–2 weeks evoked a concentration-specific suppression of hCG-induced reproductive cell multiplication and caused cell death and substantial DNA damage (Celino et al., 2009).

This part of the chapter is further divided into subsections, covering binding assays, in silico approaches, structural alerts, and molecular docking approaches. *In vitro* studies are reviewed to analyze the immediate effects of EDCs on hormone synthesis, ER/*er*, VTG/*vtg*, and GH/IGF-specific aspects and to clarify the endocrine-disrupting potency of complex environmental chemicals. Notably, a synthesis of such *in vitro* assays broadens the perspective on the effect of EDCs on the maturation and reproductive system in fish.

#### Binding Assays

The use of receptors linked to reporter genes in transformed cellular systems to detect biologically active xenobiotics has been proposed as a strategy to define chemicals by their functional properties. This approach facilitates the detection of not only receptor-ligand binding but also response element occupancy and gene activation.

An *in vitro* test employing a recombinant yeast system exposed for 48 h to NP demonstrated significant ER binding and induction of hepatocyte VTG (Petit et al., 1997). Another *in vitro* screening examined the chemical nature of the estrogenic environment utilizing the cell line of a rainbow trout gonad that expressed ERα. Findings identified that EE2, NP, and wastewater treatment effluents had marked estrogenicity (Ackermann et al., 2002). Rutishauser et al.'s research reviewed other *in vitro* assays for evaluating estrogenicity in effluents (Rutishauser et al., 2004). Indeed, Miyagawa et al. (2014) designed *in vitro* ERα gene screening frameworks based on HEK293 cells as hosts for ERα of different fish species, ranging from medaka, common carp, and goldfish guppy to stickleback, bluegill, and zebrafish, to investigate the vulnerability of certain species to estrogenic chemicals.

In zebrafish, *in vitro* screenings were based on human cancer cell transfections with an estrogen-inducible expression system in yeast cells. Such trials defined specific concentrations of chemicals to be used in *in vivo* assays. Both *in vitro* experiments considered EE2 and NP to be highly estrogenic. They compared to the *in vivo* assays, where the discussed steroidal estrogens induced plasma VTG and decreased the GSI in female zebrafish (Van den Belt et al., 2004). In a separate trial, *in vitro* assays were utilized to investigate the endocrine activity of aquatic environments, sediment, and effluents. The adverse effects on the reproductive system indicated by *in vitro* trials were further confirmed by *in vivo* experiments on juvenile brown trout (Henneberg et al., 2014).

Unlike most *in vivo* methods, the *in vitro* assays are rapid, cost-efficient, and can be conducted under controlled conditions, decreasing the chance of contamination by chemicals other than test substances. Some trials include mammalian cell lines in *in vitro* assessments to estimate dose-specific reactions to chemical pollutants. It is worth emphasizing that most of these *in vitro* trials have been developed to identify reproduction-specific genes.

### *In Silico* Approaches

Recently, computational approaches in drug development have been proposed for the EDCs field. Such *in silico* methodologies identify EDCs without any histological material, serving as an efficient screening strategy that alleviates the cost and time limitations of *in vivo* and *in vitro* frameworks. Factors such as molecular weight, topology, and electronic properties have served as bioactivity markers. *In silico* approaches include quantitative structure-property relationship (QSAR), structural alerts/ pharmacophore modeling, and molecular docking (Cavasotto & Abagyan, 2004; Lewis et al., 2015; Lyne, 2002). QSAR projection frameworks have been designed to predict the activities or properties of the molecule (Stahura & Bajorath, 2005). This method aligns a set of known ligands without the receptor network to detect characteristics indicative of bioactivity. In parallel, molecular docking focuses on the ligand conformation and its position to analyze the potency of binding affinity. This method serves as a model for predicting the relations between a small molecule and a protein focusing on the molecular activities in the binding site of target proteins (McConkey et al., 2002). *In silico* methods are relatively new but their utility has recently been used for describing the effects of EDCs in fish (Coady et al., 2017; Schneider et al., 2019; Vuorinen et al., 2013).

### QSAR/Structural Alerts

EDCs in fish were detected employing information from structural alerts and confirmed by *in vitro* and *in vivo* toxicological trials (Nendza et al., 2016). Structural alerts describe feasible estrogenic and androgenic endocrine activities based on *in vitro* studies and indicate potential endocrine disruptors in fish. Notably, the developed model has been applied to a database of substances to identify specific compounds with adverse estrogenic or androgenic outcomes. The highlighted methods enabled a comprehensive assessment of potential estrogenic and androgenic EDCs. As an example, in largemouth bass, transcriptome profiling and computational analysis outlined the impact of chemical substances on the ovary in detail. As the initial step, a dynamic model of ovarian development in unexposed fish was designed. The results of the transcriptome profiling on fish from contaminated ecosystems were analyzed to identify genes impacted at distinct phases of ovarian maturation. Then, a database that provides data on chemical relations with genes and proteins was utilized to identify chemicals that might have led to specific molecular responses (Basili et al., 2018). Notably, another study that used the QSAR method to predict EDC toxicity in fish suggested that estrogen receptor binding affinity depends on the correlation between lipophilicity and LC50 (Au - Bohlen et al., 2019). This study provided a key insight into understanding the molecular dynamics of EDCs and health outcomes in fish.

### Molecular Docking

Molecular docking is suitable for identifying EDCs binding to hormone receptors and proteins. The method has been employed to develop a more comprehensive understanding of the molecular interactions of environmental xenoestrogens (Sheikh et al., 2017). In killifish, for example, molecular docking indicated that BPA established interrelation with residues in *foxl2* (Cervantes-Camacho et al., 2020). Chen et al. (2019) assessed the estrogenic potency of NP and BPA using molecular docking analysis alongside a reduced life expectancy model in mature zebrafish. The study indicated an interchange potential of both NP and BPA to ERα (Tohyama et al., 2015). Another research employing a 3D-based computational approach demonstrated that the ligand-binding parts of medaka ERs have almost identical architecture. Notably, such computer-aided models might be particularly efficient in predicting endocrine-disrupting mechanisms.

## CONCLUSION

The purpose of this chapter was to describe the mechanisms and effects of endocrine toxicity in fishes, highlight evidence for disruptions in wild populations, and provide an overview of experimental approaches toward characterizing the effects. As demonstrated by this chapter, EDC exposure can result in physiological disruptions to systems essential for reproduction, and consequently have impacts on mating success. Fish tend to be especially susceptible to EDCs, which may be attributed to the unique features of their endocrine system physiology. This susceptibly warrants deep concern considering many EDCs are accumulative in aquatic environments and an increased body burden will result in irreversible impacts on ecosystems. There has been significant progress in the field of ecotoxicology to identify EDCs, elucidate mechanisms, and describe potential outcomes. However, there is still much uncertainty in the literature that may be hindering the creation of policies that can be protective of our aquatic ecosystems. To date, many countries still lack compulsory tests to identify EDC activity in wastewater treatment effluent – a major source of EDCs. Further developing a robust body of evidence for these effects enables the development of risk management approaches that are protective of aquatic ecosystems. As research into endocrine disruption in fishes moves forward, and we develop tools that enable a deeper understanding of immediate and long-term impacts of EDC exposure, we will gain insight into what strategies are necessary to effectively protect wild fish populations.

## REFERENCES


Ackermann, G. E., Brombacher, E., & Fent, K. (2002). Development of a fish reporter gene system for the assessment of estrogenic compounds and sewage treatment plant effluents. *Environmental Toxicology and Chemistry*, *21*(9), 1864–1875. https://doi.org/10.1002/etc.5620210914

Andersen, H. R., Andersson, A. M., Arnold, S. F., Autrup, H., Barfoed, M., Beresford, N. A., Bjerregaard, P., Christiansen, L. B., Gissel, B., Hummel, R., Jørgensen, E. B., Korsgaard, B., Le Guevel, R., Leffers, H., McLachlan, J., Møller, A., Nielsen, J. B., Olea, N., Oles-Karasko, A.,... et al. (1999). Comparison of short-term estrogenicity tests for identification of hormone-disrupting chemicals. *Environmental Health Perspectives*, *107*(Suppl 1), 89–108. https://doi.org/10.1289/ehp.99107s189

Aoki, K. A., Harris, C. A., Katsiadaki, I., & Sumpter, J. P. (2011). Evidence suggesting that di-n-butyl phthalate has antiandrogenic effects in fish. *Environmental Toxicology and Chemistry*, *30*(6), 1338–1345. https://doi.org/10.1002/etc.502

Au - Bohlen, M.-L., Au - Jeon, H. P., Au - Kim, Y. J., & Au - Sung, B. (2019). *In silico* modeling method for computational aquatic toxicology of endocrine disruptors: a software-based approach using QSAR toolbox. *Journal of Visualized Experiments 150*, e60054. https://doi.org/10.3791/60054

Barber, L. B., Vajda, A. M., Douville, C., Norris, D. O., & Writer, J. H. (2012). Fish endocrine disruption responses to a major wastewater treatment facility upgrade. *Environmental Science & Technology*, *46*(4), 2121–2131. https://doi.org/10.1021/es202880e

Basili, D., Zhang, J. L., Herbert, J., Kroll, K., Denslow, N. D., Martyniuk, C. J., Falciani, F., & Antczak, P. (2018). In silico computational transcriptomics reveals novel endocrine disruptors in largemouth bass (*Micropterus salmoides*). *Environmental Science and Technology*, *52*(13), 7553–7565. https://doi.org/10.1021/acs.est.8b02805

Bhandari, R. K., vom Saal, F. S., & Tillitt, D. E. (2015). Transgenerational effects from early developmental exposures to bisphenol A or 17α-ethinylestradiol in medaka, *Oryzias latipes*. *Scientific Reports*, *5*, 9303. https://doi.org/10.1038/srep09303

Bisesi, J. H., Jr., Bridges, W., & Klaine, S. J. (2014). Reprint of: effects of the antidepressant venlafaxine on fish brain serotonin and predation behavior. *Aquatic Toxicology*, *151*, 88–96. https://doi.org/10.1016/j.aquatox.2014.02.015

Blázquez, M., & Somoza, G. M. (2010). Fish with thermolabile sex determination (TSD) as models to study brain sex differentiation. *General and Comparative Endocrinology*, *166*(3), 470–477. https://doi.org/10.1016/j.ygcen.2009.10.004

Böhne, A., Heule, C., Boileau, N., & Salzburger, W. (2013). Expression and sequence evolution of aromatase cyp19a1 and other sexual development genes in East African cichlid fishes. *Molecular Biology and Evolution*, *30*(10), 2268–2285. https://doi.org/10.1093/molbev/mst124

Bowman, C. J., Kroll, K. J., Gross, T. G., & Denslow, N. D. (2002). Estradiol-induced gene expression in largemouth bass (*Micropterus salmoides*). *Molecular and Cellular Endocrinology*, *196*(1-2), 67–77. https://doi.org/10.1016/s0303-7207(02)00224-1

Cano-Nicolau, J., Vaillant, C., Pellegrini, E., Charlier, T. D., Kah, O., & Coumailleau, P. (2016). Estrogenic effects of several BPA analogs in the developing zebrafish brain. *Frontiers in Neuroscience*, *10*, 112. https://doi.org/10.3389/fnins.2016.00112

Cardinali, M., Maradonna, F., Olivotto, I., Bortoluzzi, G., Mosconi, G., Polzonetti-Magni, A. M., & Carnevali, O. (2004). Temporary impairment of reproduction in freshwater teleost exposed to nonylphenol. *Reproductive Toxicology*, *18*(4), 597–604. https://doi.org/10.1016/j.reprotox.2004.03.001

Carnevali, O., Tosti, L., Speciale, C., Peng, C., Zhu, Y., & Maradonna, F. (2010). DEHP impairs zebrafish reproduction by affecting critical factors in oogenesis. *PloS One*, *5*(4), e10201. https://doi.org/10.1371/journal.pone.0010201

Cavasotto, C. N., & Abagyan, R. A. (2004). Protein flexibility in ligand docking and virtual screening to protein kinases. *Journal of Molecular Biology*, *337*(1), 209–225. https://doi.org/10.1016/j.jmb.2004.01.003

Celino, F. T., Yamaguchi, S., Miura, C., & Miura, T. (2009). Arsenic inhibits *in vitro* spermatogenesis and induces germ cell apoptosis in Japanese eel (*Anguilla japonica*). *Reproduction*, *138*(2), 279–287. https://doi.org/10.1530/rep-09-0167

Celino-Brady, F. T., Lerner, D. T., & Seale, A. P. (2021). Experimental approaches for characterizing the endocrine-disrupting effects of environmental chemicals in fish [Review]. *Frontiers in Endocrinology*, *11*, 619361. https://doi.org/10.3389/fendo.2020.619361

Cervantes-Camacho, I., Guerrero-Estévez, S. M., López, M. F., Alarcón-Hernández, E., & López-López, E. (2020). Effects of bisphenol A on Foxl2 gene expression and DNA damage in adult viviparous fish *Goodea atripinnis*. *Journal of Toxicology and Environmental Health, Part A*, *83*(3), 95–112. https://doi.org/10.1080/15287394.2020.1730282

Chakrabarty, S., Rajakumar, A., Raghuveer, K., Sridevi, P., Mohanachary, A., Prathibha, Y., Bashyam, L., Dutta-Gupta, A., & Senthilkumaran, B. (2012). Endosulfan and flutamide, alone and in combination, target ovarian growth in juvenile catfish, Clarias batrachus. *Comparative Biochemistry and Physiology. Part C: Pharmacology, Toxicology and Endocrinology*, *155*(3), 491–497. https://doi.org/10.1016/j.cbpc.2011.12.007

Chen, H., Zhao, L., & Yu, Q. J. (2019). Determination and reduced life expectancy model and molecular docking analyses of estrogenic potentials of 17β-estradiol, bisphenol A and nonylphenol on expression of vitellogenin gene (*vtg1*) in zebrafish. *Chemosphere*, *221*, 727–734. https://doi.org/10.1016/j.chemosphere.2019.01.093

Chen, R., Liu, C., Yuan, L., Zha, J., & Wang, Z. (2016). 2, 4-Dichloro-6-nitrophenol, a photonitration product of 2, 4-dichlorophenol, caused anti-androgenic potency in Chinese rare minnows (*Gobiocypris rarus*). *Environmental Pollution*, *216*, 591–598. https://doi.org/10.1016/j.envpol.2016.06.016

Coady, K. K., Biever, R. C., Denslow, N. D., Gross, M., Guiney, P. D., Holbech, H., Karouna-Renier, N. K., Katsiadaki, I., Krueger, H., Levine, S. L., Maack, G., Williams, M., Wolf, J. C., & Ankley, G. T. (2017). Current limitations and recommendations to improve testing for the environmental assessment of endocrine active substances. *Integrated Environmental Assessment and Management*, *13*(2), 302–316. https://doi.org/10.1002/ieam.1862

Colborn, T., vom Saal, F. S., & Soto, A. M. (1993). Developmental effects of endocrine-disrupting chemicals in wildlife and humans. *Environmental Health Perspectives*, *101*(5), 378–384. https://doi.org/10.1289/ehp.93101378

Corradetti, B., Stronati, A., Tosti, L., Manicardi, G., Carnevali, O., & Bizzaro, D. (2013). Bis-(2-ethylexhyl) phthalate impairs spermatogenesis in zebrafish (*Danio rerio*). *Reproductive Biology*, *13*(3), 195–202. https://doi.org/10.1016/j.repbio.2013.07.003

Crews, D., Gore, A. C., Hsu, T. S., Dangleben, N. L., Spinetta, M., Schallert, T., Anway, M. D., & Skinner, M. K. (2007). Transgenerational epigenetic imprints on mate preference. *Proceedings of the National Academy of Sciences*, *104*(14), 5942–5946. https://doi.org/10.1073/pnas.0610410104

Crump, K. L., & Trudeau, V. L. (2009). Mercury-induced reproductive impairment in fish. *Environmental Toxicology and Chemistry*, *28*(5), 895–907. https://doi.org/10.1897/08-151.1

Davis, L. K., Fox, B. K., Lim, C., Hiramatsu, N., Sullivan, C. V., Hirano, T., & Grau, E. G. (2009). Induction of vitellogenin production in male tilapia (*Oreochromis mossambicus*) by commercial fish diets. *Comparative Biochemistry and Physiology. Part A: Comparative Physiology*, *154*(2), 249–254. https://doi.org/10.1016/j.cbpa.2009.06.009

Deepa, S., Murugananthkumar, R., Raj Gupta, Y., Gowda, K. S. M., & Senthilkumaran, B. (2019). Effects of zinc oxide nanoparticles and zinc sulfate on the testis of common carp, *Cyprinus carpio*. *Nanotoxicology*, *13*(2), 240–257. https://doi.org/10.1080/17435390.2018.1541259

Denier, X., Hill, E. M., Rotchell, J., & Minier, C. (2009). Estrogenic activity of cadmium, copper and zinc in the yeast estrogen screen. *Toxicology in Vitro*, *23*(4), 569–573. https://doi.org/10.1016/j.tiv.2009.01.006

Devlin, R. H., & Nagahama, Y. (2002). Sex determination and sex differentiation in fish: an overview of genetic, physiological, and environmental influences. *Aquaculture (Amsterdam, Netherlands)*, *208*(3), 191–364. https://doi.org/10.1016/S0044-8486(02)00057-1

Duan, C. (1998). Nutritional and developmental regulation of insulin-like growth factors in fish. *Journal of Nutrition*, *128*(2 Suppl), 306s–314s. https://doi.org/10.1093/jn/128.2.306S

Duan, C., & Xu, Q. (2005). Roles of insulin-like growth factor (IGF) binding proteins in regulating IGF actions. *General and Comparative Endocrinology*, *142*(1–2), 44–52. https://doi.org/10.1016/j.ygcen.2004.12.022

Duffy, T. A., Iwanowicz, L. R., & McCormick, S. D. (2014). Comparative responses to endocrine disrupting compounds in early life stages of Atlantic salmon, *Salmo salar*. *Aquatic Toxicology*, *152*, 1–10. https://doi.org/10.1016/j.aquatox.2014.03.015

Fan, Y., Menon, R. K., Cohen, P., Hwang, D., Clemens, T., DiGirolamo, D. J., Kopchick, J. J., Le Roith, D., Trucco, M., & Sperling, M. A. (2009). Liver-specific deletion of the growth hormone receptor reveals essential role of growth hormone signaling in hepatic lipid metabolism. *Journal of Biological Chemistry*, *284*(30), 19937–19944. https://doi.org/10.1074/jbc.M109.014308

Flouriot, G., Pakdel, F., & Valotaire, Y. (1996). Transcriptional and post-transcriptional regulation of rainbow trout estrogen receptor and vitellogenin gene expression. *Molecular and Cellular Endocrinology*, *124*(1–2), 173–183. https://doi.org/10.1016/s0303-7207(96)03960-3

Fontaine, R., Royan, M. R., von Krogh, K., Weltzien, F.-A., & Baker, D. M. (2020). Direct and indirect effects of sex steroids on gonadotrope cell plasticity in the teleost fish pituitary [Review]. *Frontiers in Endocrinology*, *11*, 605068. https://doi.org/10.3389/fendo.2020.605068

Forner-Piquer, I., Fakriadis, I., Mylonas, C. C., Piscitelli, F., Di Marzo, V., Maradonna, F., Calduch-Giner, J., Pérez-Sánchez, J., & Carnevali, O. (2019). Effects of dietary bisphenol a on the reproductive function of gilthead sea bream (*Sparus aurata*) testes. *International Journal of Molecular Sciences*, *20*(20), 5003. https://doi.org/10.3390/ijms20205003

García-Alonso, J., Ruiz-Navarro, A., Chaves-Pozo, E., Torralva, M., & García-Ayala, A. (2009). Gonad plasticity and gametogenesis in the endangered Spanish toothcarp *Aphanius iberus* (Teleostei: Cyprinodontidae). *Tissue and Cell*, *41*(3), 206–213. https://doi.org/10.1016/j.tice.2008.11.001

Gárriz, Á., Del Fresno, P. S., Carriquiriborde, P., & Miranda, L. A. (2019). Effects of heavy metals identified in Chascomús shallow lake on the endocrine-reproductive axis of pejerrey fish (*Odontesthes bonariensis*). *General and Comparative Endocrinology*, *273*, 152–162. https://doi.org/10.1016/j.ygcen.2018.06.013

Gárriz, Á., & Miranda, L. A. (2020). Effects of metals on sperm quality, fertilization and hatching rates, and embryo and larval survival of pejerrey fish (Odontesthes bonariensis). *Ecotoxicology*, *29*(7), 1072–1082. https://doi.org/10.1007/s10646-020-02245-w

Ge, W. (2005). Intrafollicular paracrine communication in the zebrafish ovary: the state of the art of an emerging model for the study of vertebrate folliculogenesis. *Molecular and Cellular Endocrinology*, *237*(1–2), 1–10. https://doi.org/10.1016/j.mce.2005.03.012

Giari, L., Vincenzi, F., Badini, S., Guerranti, C., Dezfuli, B. S., Fano, E. A., & Castaldelli, G. (2016). Common carp *Cyprinus carpio* responses to sub-chronic exposure to perfluorooctanoic acid. *Environmental Science and Pollution Research*, *23*(15), 15321–15330. https://doi.org/10.1007/s11356-016-6706-1

Golshan, M., Hatef, A., Socha, M., Milla, S., Butts, I. A., Carnevali, O., Rodina, M., Sokołowska-Mikołajczyk, M., Fontaine, P., Linhart, O., & Alavi, S. M. (2015). Di-(2-ethylhexyl)-phthalate disrupts pituitary and testicular hormonal functions to reduce sperm quality in mature goldfish. *Aquatic Toxicology*, *163*, 16–26. https://doi.org/10.1016/j.aquatox.2015.03.017

Gross, J. M., & Yee, D. (2002). How does the estrogen receptor work? *Breast Cancer Research*, *4*(2), 62–64. https://doi.org/10.1186/bcr424

Guiguen, Y., Fostier, A., Piferrer, F., & Chang, C. F. (2010). Ovarian aromatase and estrogens: a pivotal role for gonadal sex differentiation and sex change in fish. *General and Comparative Endocrinology*, *165*(3), 352–366. https://doi.org/10.1016/j.ygcen.2009.03.002

Hatef, A., Zare, A., Alavi, S. M., Habibi, H. R., & Linhart, O. (2012). Modulations in androgen and estrogen mediating genes and testicular response in male goldfish exposed to bisphenol A. *Environmental Toxicology and Chemistry*, *31*(9), 2069–2077. https://doi.org/10.1002/etc.1919

Hecker, M., Tyler, C. R., Hoffmann, M., Maddix, S., & Karbe, L. (2002). Plasma biomarkers in fish provide evidence for endocrine modulation in the Elbe river, Germany. *Environmental Science & Technology*, *36*(11), 2311–2321. https://doi.org/10.1021/es010186h

Henneberg, A., Bender, K., Blaha, L., Giebner, S., Kuch, B., Köhler, H. R., Maier, D., Oehlmann, J., Richter, D., Scheurer, M., Schulte-Oehlmann, U., Sieratowicz, A., Ziebart, S., & Triebskorn, R. (2014). Are *in vitro* methods for the detection of endocrine potentials in the aquatic environment predictive for *in vivo* effects? Outcomes of the Projects SchussenAktiv and SchussenAktivplus in the Lake Constance Area, Germany. *PloS One*, *9*(6), e98307. https://doi.org/10.1371/journal.pone.0098307

Hewitt, L. M., Kovacs, T. G., Dubé, M. G., MacLatchy, D. L., Martel, P. H., McMaster, M. E., Paice, M. G., Parrott, J. L., van den Heuvel, M. R., & van der Kraak, G. J. (2008).

Altered reproduction in fish exposed to pulp and paper mill effluents: roles of individual compounds and mill operating conditions. *Environmental Toxicology and Chemistry*, *27*(3), 682–697. https://doi.org/10.1897/07-195.1

Hicks, K. A., Fuzzen, M. L. M., McCann, E. K., Arlos, M. J., Bragg, L. M., Kleywegt, S., Tetreault, G. R., McMaster, M. E., & Servos, M. R. (2017). Reduction of intersex in a wild fish population in response to major municipal wastewater treatment plant upgrades. *Environmental Science & Technology*, *51*(3), 1811–1819. https://doi.org/10.1021/acs.est.6b05370

Hoffman, K., Webster, T. F., Bartell, S. M., Weisskopf, M. G., Fletcher, T., & Vieira, V. M. (2011). Private drinking water wells as a source of exposure to perfluorooctanoic acid (PFOA) in communities surrounding a fluoropolymer production facility. *Environmental Health Perspectives*, *119*(1), 92–97. https://doi.org/10.1289/ehp.1002503

Horan, T. S., Marre, A., Hassold, T., Lawson, C., & Hunt, P. A. (2017). Germline and reproductive tract effects intensify in male mice with successive generations of estrogenic exposure. *PLoS Genetics*, *13*(7), e1006885. https://doi.org/10.1371/journal.pgen.1006885

Hultman, M. T., Song, Y., & Tollefsen, K. E. (2015). 17α-Ethinylestradiol (EE2) effect on global gene expression in primary rainbow trout (Oncorhynchus mykiss) hepatocytes. *Aquatic Toxicology*, *169*, 90–104. https://doi.org/10.1016/j.aquatox.2015.10.004

Jalabert, B. (2005). Particularities of reproduction and oogenesis in teleost fish compared to mammals. *Reproduction Nutrition Development*, *45*(3), 261–279. https://doi.org/10.1051/rnd:2005019

Jenkins, R. L., Wilson, E. M., Angus, R. A., Howell, W. M., Kirk, M., Moore, R., Nance, M., & Brown, A. (2004). Production of androgens by microbial transformation of progesterone *in vitro*: a model for androgen production in rivers receiving paper mill effluent. *Environmental Health Perspectives*, *112*(15), 1508–1511. https://doi.org/10.1289/ehp.7161

Jobling, S., Coey, S., Whitmore, J. G., Kime, D. E., Van Look, K. J. W., McAllister, B. G., Beresford, N., Henshaw, A. C., Brighty, G., Tyler, C. R., & Sumpter, J. P. (2002). Wild intersex roach (*Rutilus rutilus*) have reduced fertility. *Biology of Reproduction*, *67*(2), 515–524. https://doi.org/10.1095/biolreprod67.2.515

Jobling, S., Nolan, M., Tyler, C. R., Brighty, G., & Sumpter, J. P. (1998). Widespread sexual disruption in wild fish. *Environmental Science and Technology*, *32*(17), 2498–2506. https://doi.org/10.1021/es9710870

Jobling, S., & Sumpter, J. P. (1993). Detergent components in sewage effluent are weakly estrogenic to fish - An *in vitro* study using rainbow trout (*Oncorhynchus mykiss*) hepatocytes. *Aquatic Toxicology*, *27*(3–4), 361–372. <Go to ISI>://A1993MT91000010

Johns, S. M., Denslow, N. D., Kane, M. D., Watanabe, K. H., Orlando, E. F., & Sepúlveda, M. S. (2011). Effects of estrogens and antiestrogens on gene expression of fathead minnow (Pimephales promelas) early life stages. *Environmental Toxicology*, *26*(2), 195–206. https://doi.org/10.1002/tox.20545

Jones, P. D., De Coen, W. M., Tremblay, L., & Giesy, J. P. (2000). Vitellogenin as a biomarker for environmental estrogens. *Water Science and Technology*, *42*(7–8), 1–14. https://doi.org/10.2166/wst.2000.0546

Kavlock, R. J., Daston, G. P., DeRosa, C., Fenner-Crisp, P., Gray, L. E., Kaattari, S., Lucier, G., Luster, M., Mac, M. J., Maczka, C., Miller, R., Moore, J., Rolland, R., Scott, G., Sheehan, D. M., Sinks, T., & Tilson, H. A. (1996). Research needs for the risk assessment of health and environmental effects of endocrine disruptors: a report of the U.S. EPA-sponsored workshop. *Environmental Health Perspectives*, *104*(Suppl 4), 715–740. https://doi.org/10.1289/ehp.96104s4715

Kidd, K. A., Blanchfield, P. J., Mills, K. H., Palace, V. P., Evans, R. E., Lazorchak, J. M., & Flick, R. W. (2007). Collapse of a fish population after exposure to a synthetic estrogen. *Proceedings of the National Academy of Sciences*, *104*(21), 8897–8901. https://doi.org/10.1073/pnas.0609568104|ISSN 0027-8424

Kobayashi, Y., Nagahama, Y., & Nakamura, M. (2013). Diversity and plasticity of sex determination and differentiation in fishes. *Sexual Development*, *7*(1–3), 115–125. https://doi.org/10.1159/000342009

Kwon, B., Ha, N., Jung, J., Kim, P. G., Kho, Y., Choi, K., & Ji, K. (2016). Effects of barium chloride exposure on hormones and genes of the hypothalamic-pituitary-gonad axis, and reproduction of zebrafish (*Danio rerio*). *Bulletin of Environmental Contamination and Toxicology*, *96*(3), 341–346. https://doi.org/10.1007/s00128-016-1731-9

La Merrill, M. A., Vandenberg, L. N., Smith, M. T., Goodson, W., Browne, P., Patisaul, H. B., Guyton, K. Z., Kortenkamp, A., Cogliano, V. J., Woodruff, T. J., Rieswijk, L., Sone, H., Korach, K. S., Gore, A. C., Zeise, L., & Zoeller, R. T. (2020). Consensus on the key characteristics of endocrine-disrupting chemicals as a basis for hazard identification. *Nature Reviews in Endocrinology*, *16*(1), 45–57. https://doi.org/10.1038/s41574-019-0273-8

Lahnsteiner, F., Berger, B., Kletzl, M., & Weismann, T. (2005). Effect of bisphenol A on maturation and quality of semen and eggs in the brown trout, *Salmo trutta f. fario*. *Aquatic Toxicology*, *75*(3), 213–224. https://doi.org/10.1016/j.aquatox.2005.08.004

Lal, B., Sarang, M. K., & Kumar, P. (2013). Malathion exposure induces the endocrine disruption and growth retardation in the catfish, Clarias batrachus (Linn.). *General and Comparative Endocrinology*, *181*, 139–145. https://doi.org/10.1016/j.ygcen.2012.11.004

Lascombe, I., Beffa, D., Rüegg, U., Tarradellas, J., & Wahli, W. (2000). Estrogenic activity assessment of environmental chemicals using *in vitro* assays: identification of two new estrogenic compounds. *Environmental Health Perspectives*, *108*(7), 621–629. https://doi.org/10.1289/ehp.00108621

Lau, E. S., Zhang, Z., Qin, M., & Ge, W. (2016). Knockout of zebrafish ovarian aromatase gene (cyp19a1a) by TALEN and CRISPR/cas9 leads to all-male offspring due to failed ovarian differentiation. *Scientific Reports*, *6*, 37357. https://doi.org/10.1038/srep37357

Lavado, R., Thibaut, R., Raldua, D., Martin, R., & Porte, C. (2004). First evidence of endocrine disruption in feral carp from the Ebro River. *Toxicology and Applied Pharmacology*, *196*(2), 247–257. https://doi.org/10.1016/j.taap.2003.12.012

Leet, J. K., Gall, H. E., & Sepúlveda, M. S. (2011). A review of studies on androgen and estrogen exposure in fish early life stages: effects on gene and hormonal control of sexual differentiation. *Journal of Applied Toxicology*, *31*(5), 379–398. https://doi.org/10.1002/jat.1682

Lewis, S. N., Garcia, Z., Hontecillas, R., Bassaganya-Riera, J., & Bevan, D. R. (2015). Pharmacophore modeling improves virtual screening for novel peroxisome proliferator-activated receptor-gamma ligands. *Journal of Computer-Aided Molecular Design*, *29*(5), 421–439. https://doi.org/10.1007/s10822-015-9831-x

Li, M. H., Yang, H. H., Li, M. R., Sun, Y. L., Jiang, X. L., Xie, Q. P., Wang, T. R., Shi, H. J., Sun, L. N., Zhou, L. Y., & Wang, D. S. (2013). Antagonistic roles of Dmrt1 and Foxl2 in sex differentiation via estrogen production in tilapia as demonstrated by TALENs. *Endocrinology*, *154*(12), 4814–4825. https://doi.org/10.1210/en.2013-1451

Liney, K. E., Jobling, S., Shears, J. A., Simpson, P., & Tyler, C. R. (2005). Assessing the sensitivity of different life stages for sexual disruption in roach (*Rutilus rutilus*) exposed to effluents from wastewater treatment works. *Environmental Health Perspectives*, *113*(10), 1299–1307. https://doi.org/10.1289/ehp.7921

Liu, S., Qin, F., Wang, H., Wu, T., Zhang, Y., Zheng, Y., Li, M., & Wang, Z. (2012). Effects of 17α-ethinylestradiol and bisphenol A on steroidogenic messenger ribonucleic acid levels in the rare minnow gonads. *Aquatic Toxicology*, *122–123*, 19–27. https://doi.org/10.1016/j.aquatox.2012.05.010

Liu, Y., Chen, S., Liu, S., Zhang, Y., Yuan, C., & Wang, Z. (2014). DNA methylation in the 5' flanking region of cytochrome P450 17 in adult rare minnow Gobiocypris rarus - tissue difference and effects of 17α-ethinylestradiol and 17α-methyltestoterone exposures. *Comparative Biochemistry and Physiology. Part C: Pharmacology, Toxicology and Endocrinology*, *162*, 16–22. https://doi.org/10.1016/j.cbpc.2014.03.001

Lyne, P. D. (2002). Structure-based virtual screening: an overview. *Drug Discovery Today*, *7*(20), 1047–1055. https://doi.org/10.1016/s1359-6446(02)02483-2

Ma, Y. N., Cao, C. Y., Wang, Q. W., Gui, W. J., & Zhu, G. N. (2016). Effects of azocyclotin on gene transcription and steroid metabolome of hypothalamic-pituitary-gonad axis, and their consequences on reproduction in zebrafish (*Danio rerio*). *Aquatic Toxicology*, *179*, 55–64. https://doi.org/10.1016/j.aquatox.2016.08.006

Mandich, A., Bottero, S., Benfenati, E., Cevasco, A., Erratico, C., Maggioni, S., Massari, A., Pedemonte, F., & Viganò, L. (2007). *In vivo* exposure of carp to graded concentrations of bisphenol A. *General and Comparative Endocrinology*, *153*(1–3), 15–24. https://doi.org/10.1016/j.ygcen.2007.01.004

Margiotta-Casaluci, L., Hannah, R. E., & Sumpter, J. P. (2013). Mode of action of human pharmaceuticals in fish: the effects of the 5-alpha-reductase inhibitor, dutasteride, on reproduction as a case study. *Aquatic Toxicology*, *128–129*, 113–123. https://doi.org/10.1016/j.aquatox.2012.12.003

Margiotta-Casaluci, L., & Sumpter, J. P. (2011). 5α-Dihydrotestosterone is a potent androgen in the fathead minnow (*Pimephales promelas*). *General and Comparative Endocrinology*, *171*(3), 309–318. https://doi.org/10.1016/j.ygcen.2011.02.012

McConkey, B. J., Sobolev, V., & Edelman, M. (2002). The performance of current methods in ligand-protein docking. *Current Science*, *83*(7), 845–856. https://www.jstor.org/stable/24107087

Miura, C., Takahashi, N., Michino, F., & Miura, T. (2005). The effect of para-nonylphenol on Japanese eel (*Anguilla japonica*) spermatogenesis in vitro. *Aquatic Toxicology*, *71*(2), 133–141. https://doi.org/10.1016/j.aquatox.2004.10.015

Miyagawa, S., Lange, A., Hirakawa, I., Tohyama, S., Ogino, Y., Mizutani, T., Kagami, Y., Kusano, T., Ihara, M., Tanaka, H., Tatarazako, N., Ohta, Y., Katsu, Y., Tyler, C. R., & Iguchi, T. (2014). Differing species responsiveness of estrogenic contaminants in fish is conferred by the ligand binding domain of the estrogen receptor. *Environmental Science and Technology*, *48*(9), 5254–5263. https://doi.org/10.1021/es5002659

Mouriec, K., Gueguen, M.-M., Manuel, C., Percevault, F., Thieulant, M.-L., Pakdel, F., & Kah, O. (2009). Androgens upregulate cyp19a1b (aromatase b) gene expression in the brain of zebrafish (*Danio rerio*) through estrogen receptors. *Biology of Reproduction*, *80*(5), 889–896. https://doi.org/10.1095/biolreprod.108.073643

Munday, P. L., Buston, P. M., & Warner, R. R. (2006). Diversity and flexibility of sex-change strategies in animals. *Trends in Ecology and Evolution*, *21*(2), 89–95. https://doi.org/10.1016/j.tree.2005.10.020

Muñoz-Cueto, J. A., Zmora, N., Paullada-Salmerón, J. A., Marvel, M., Mañanos, E., & Zohar, Y. (2020). The gonadotropin-releasing hormones: lessons from fish. *General and Comparative Endocrinology*, *291*, 113422. https://doi.org/10.1016/j.ygcen.2020.113422

Murk, A. J., Legler, J., van Lipzig, M. M., Meerman, J. H., Belfroid, A. C., Spenkelink, A., van der Burg, B., Rijs, G. B., & Vethaak, D. (2002). Detection of estrogenic potency in wastewater and surface water with three in vitro bioassays. *Environmental Toxicology and Chemistry*, *21*(1), 16–23.

Murugananthkumar, R., & Sudhakumari, C.-C. (2022). Understanding the impact of stress on teleostean reproduction. *Aquaculture and Fisheries*, *7*(5), 553–561. https://doi.org/10.1016/j.aaf.2022.05.001

Nagahama, Y., Yoshikuni, M., Yamashita, M., Tokumoto, T., & Katsu, Y. (1995). Regulation of oocyte growth and maturation in fish. *Current Topics in Developmental Biology*, *30*, 103–145. https://doi.org/10.1016/s0070-2153(08)60565-7

Nendza, M., Wenzel, A., Müller, M., Lewin, G., Simetska, N., Stock, F., & Arning, J. (2016). Screening for potential endocrine disruptors in fish: evidence from structural alerts and in vitro and in vivo toxicological assays. *Environmental Sciences Europe*, *28*(1), 26. https://doi.org/10.1186/s12302-016-0094-5

Nishimura, T., & Tanaka, M. (2014). Gonadal development in fish. *Sexual Development*, *8*(5), 252–261. https://doi.org/10.1159/000364924

Noaksson, E., Linderoth, M., Bosveld, A. T. C., Norrgren, L., Zebühr, Y., & Balk, L. (2003). Endocrine disruption in brook trout (*Salvelinus fontinalis*) exposed to leachate from a public refuse dump. *Science of the Total Environment*, *305*(1), 87–103. https://doi.org/10.1016/S0048-9697(02)00503-X

Oba, Y., Hirai, T., Yoshiura, Y., Yoshikuni, M., Kawauchi, H., & Nagahama, Y. (1999). Cloning, functional characterization, and expression of a gonadotropin receptor cDNA in the ovary and testis of amago salmon (*Oncorhynchus rhodurus*). *Biochemical and Biophysical Research Communications*, *263*(2), 584–590. https://doi.org/10.1006/bbrc.1999.1346

Palaniappan, P. L., Sabhanayakam, S., Krishnakumar, N., & Vadivelu, M. (2008). Morphological changes due to Lead exposure and the influence of DMSA on the gill tissues of the freshwater fish, *Catla catla*. *Food Chemical Toxicology*, *46*(7), 2440–2444. https://doi.org/10.1016/j.fct.2008.03.028

Park, C. B., Takemura, A., Aluru, N., Park, Y. J., Kim, B. H., Lee, C. H., Lee, Y. D., Moon, T. W., & Vijayan, M. M. (2007). Tissue-specific suppression of estrogen, androgen and glucocorticoid receptor gene expression in feral vitellogenic male Mozambique tilapia. *Chemosphere*, *69*(1), 32–40. https://doi.org/10.1016/j.chemosphere.2007.04.072

Patil, V. K., & David, M. (2010). Behavioral and morphological endpoints: as an early response to sublethal malathion intoxication in the freshwater fish, *Labeo rohita*. *Drug and Chemical Toxicology*, *33*(2), 160–165. https://doi.org/10.3109/01480540903196816

Petit, F., Le Goff, P., Cravédi, J. P., Valotaire, Y., & Pakdel, F. (1997). Two complementary bioassays for screening the estrogenic potency of xenobiotics: recombinant yeast for trout estrogen receptor and trout hepatocyte cultures. *Journal of Molecular Endocrinology*, *19*(3), 321–335. https://doi.org/10.1677/jme.0.0190321

Porte, C., Janer, G., Lorusso, L. C., Ortiz-Zarragoitia, M., Cajaraville, M. P., Fossi, M. C., & Canesi, L. (2006). Endocrine disruptors in marine organisms: approaches and perspectives. *Comparative Biochemistry and Physiology. Part C: Pharmacology, Toxicology and Endocrinology*, *143*(3), 303–315. https://doi.org/10.1016/j.cbpc.2006.03.004

Prasad, P., Ogawa, S., & Parhar, I. S. (2015). Role of serotonin in fish reproduction [Review]. *Frontiers in Neuroscience*, *9*, 195. https://doi.org/10.3389/fnins.2015.00195

Prathibha, Y., Murugananthkumar, R., Rajakumar, A., Laldinsangi, C., Sudhakumari, C. C., Mamta, S. K., Dutta-Gupta, A., & Senthilkumaran, B. (2014). Gene expression analysis in gonads and brain of catfish *Clarias batrachus* after the exposure of malathion. *Ecotoxicology and Environmental Safety*, *102*, 210–219. https://doi.org/10.1016/j.ecoenv.2013.12.029

Qiu, W., Zhao, Y., Yang, M., Farajzadeh, M., Pan, C., & Wayne, N. L. (2016). Actions of bisphenol A and bisphenol s on the reproductive neuroendocrine system during early development in zebrafish. *Endocrinology*, *157*(2), 636–647. https://doi.org/10.1210/en.2015-1785

Rajakumar, A., & Senthilkumaran, B. (2020). Steroidogenesis and its regulation in teleost-a review. *Fish Physioly and Biochemistry*, *46*(3), 803–818. https://doi.org/10.1007/s10695-019-00752-0

Rajakumar, A., Singh, R., Chakrabarty, S., Murugananthkumar, R., Laldinsangi, C., Prathibha, Y., Sudhakumari, C. C., Dutta-Gupta, A., & Senthilkumaran, B. (2012). Endosulfan and flutamide impair testicular development in the juvenile Asian catfish, *Clarias batrachus*. *Aquatic Toxicology*, *110–111*, 123–132. https://doi.org/10.1016/j.aquatox.2011.12.018

Rajendiran, P., Jaafar, F., Kar, S., Sudhakumari, C., Senthilkumaran, B., & Parhar, I. S. (2021). Sex determination and differentiation in teleost: roles of genetics, environment, and brain. *Biology*, *10*(10), 973. https://doi.org/10.3390/biology10100973

Rankouhi, T. R., Sanderson, J. T., van Holsteijn, I., van Leeuwen, C., Vethaak, A. D., & van den Berg, M. (2004). Effects of natural and synthetic estrogens and various environmental contaminants on vitellogenesis in fish primary hepatocytes: comparison of bream (*Abramis brama*) and carp (*Cyprinus carpio*). *Toxicological Sciences*, *81*(1), 90–102. https://doi.org/10.1093/toxsci/kfh176lISSN 1096-6080

Rege, J., Garber, S., Conley, A. J., Elsey, R. M., Turcu, A. F., Auchus, R. J., & Rainey, W. E. (2019). Circulating 11-oxygenated androgens across species. *Journal of Steroid Biochemistry and Molecular Biology*, *190*, 242–249. https://doi.org/10.1016/j.jsbmb.2019.04.005

Reindl, K. M., & Sheridan, M. A. (2012). Peripheral regulation of the growth hormone-insulin-like growth factor system in fish and other vertebrates. *Comparative Biochemistry and Physiology. Part A: Comparative Physiology*, *163*(3–4), 231–245. https://doi.org/10.1016/j.cbpa.2012.08.003

Robitaille, J., Denslow, N. D., Escher, B. I., Kurita-Oyamada, H. G., Marlatt, V., Martyniuk, C. J., Navarro-Martín, L., Prosser, R., Sanderson, T., Yargeau, V., & Langlois, V. S. (2022). Towards regulation of Endocrine Disrupting chemicals (EDCs) in water resources using bioassays - a guide to developing a testing strategy. *Environmental Research*, *205*, 112483. https://doi.org/10.1016/j.envres.2021.112483

Routledge, E. J., & Sumpter, J. P. (1996). Estrogenic activity of surfactants and some of their degradation products assessed using a recombinant yeast screen. *Environmental Toxicology and Chemistry*, *15*(3), 241–248. https://doi.org/10.1002/etc.5620150303

Rutishauser, B. V., Pesonen, M., Escher, B. I., Ackermann, G. E., Aerni, H. R., Suter, M. J., & Eggen, R. I. (2004). Comparative analysis of estrogenic activity in sewage treatment plant effluents involving three *in vitro* assays and chemical analysis of steroids. *Environmental Toxicology and Chemistry*, *23*(4), 857–864. https://doi.org/10.1897/03-286

Schneider, M., Pons, J. L., Labesse, G., & Bourguet, W. (2019). *In silico* predictions of endocrine disruptors properties. *Endocrinology*, *160*(11), 2709–2716. https://doi.org/10.1210/en.2019-00382

Scholz, S., Renner, P., Belanger, S. E., Busquet, F., Davi, R., Demeneix, B. A., Denny, J. S., Léonard, M., McMaster, M. E., Villeneuve, D. L., & Embry, M. R. (2013). Alternatives to *in vivo* tests to detect endocrine disrupting chemicals (EDCs) in fish and amphibians-screening for estrogen, androgen and thyroid hormone disruption. *Critical Reviews in Toxicology*, *43*(1), 45–72. https://doi.org/10.3109/10408444.2012.737762

Scott, A. P., Sumpter, J. P., & Stacey, N. (2010). The role of the maturation-inducing steroid, 17,20beta-dihydroxypregn-4-en-3-one, in male fishes: a review. *Journal of Fish Biology*, *76*(1), 183–224. https://doi.org/10.1111/j.1095-8649.2009.02483.x

Senthilkumaran, B. (2015). Pesticide- and sex steroid analogue-induced endocrine disruption differentially targets hypothalamo-hypophyseal-gonadal system during gametogenesis in teleosts - a review. *General and Comparative Endocrinology*, *219*, 136–142. https://doi.org/10.1016/j.ygcen.2015.01.010

Sheikh, I. A., Tayubi, I. A., Ahmad, E., Ganaie, M. A., Bajouh, O. S., AlBasri, S. F., Abdulkarim, I. M. J., & Beg, M. A. (2017). Computational insights into the molecular interactions of environmental xenoestrogens 4-tert-octylphenol, 4-nonylphenol, bisphenol A (BPA), and BPA metabolite, 4-methyl-2, 4-bis (4-hydroxyphenyl) pent-1-ene (MBP) with human sex hormone-binding globulin. *Ecotoxicology and Environmental Safety*, *135*, 284–291. https://doi.org/10.1016/j.ecoenv.2016.10.005

Shi, W. J., Hu, L. X., Huang, G. Y., Liu, Y. S., Zhang, J. N., Xie, L., & Ying, G. G. (2019). Dydrogesterone affects the transcription of genes in GnRH and steroidogenesis pathways and increases the frequency of atretic follicles in zebrafish (*Danio rerio*). *Chemosphere*, *216*, 725–732. https://doi.org/10.1016/j.chemosphere.2018.10.202

Singh, S., & Li, S. S. (2012). Epigenetic effects of environmental chemicals bisphenol A and phthalates. *International Journal of Molecular Sciences*, *13*(8), 10143–10153. https://doi.org/10.3390/ijms130810143

Smith, C. M., Vera, M. K. M., & Bhandari, R. K. (2019). Developmental and epigenetic effects of Roundup and glyphosate exposure on Japanese medaka (*Oryzias latipes*). *Aquatic Toxicology*, *210*, 215–226. https://doi.org/10.1016/j.aquatox.2019.03.005

Soyano, K., Amagai, T., Yamaguchi, T., Mushirobira, Y., Xu, W. G., Pham, N. T., & Murata, R. (2022). Endocrine regulation of maturation and sex change in groupers. *Cells*, *11*(5), 825. https://doi.org/10.3390/cells11050825

Sreenivasulu, G., Senthilkumaran, B., Sridevi, P., Rajakumar, A., & Rasheeda, M. K. (2012). Expression and immunolocalization of 20β-hydroxysteroid dehydrogenase during testicular cycle and after hCG induction, *in vivo* in the catfish, *Clarias gariepinus*. *General and Comparative Endocrinology*, *175*(1), 48–54. https://doi.org/10.1016/j.ygcen.2011.09.002

Sridevi, P., Chaitanya, R. K., Prathibha, Y., Balakrishna, S. L., Dutta-Gupta, A., & Senthilkumaran, B. (2015). Early exposure of 17α-ethynylestradiol and diethylstilbestrol induces morphological changes and alters ovarian steroidogenic pathway enzyme gene expression in catfish, *Clarias gariepinus*. *Environmental Toxicology*, *30*(4), 439–451. https://doi.org/10.1002/tox.21920

Stahura, F. L., & Bajorath, J. (2005). New methodologies for ligand-based virtual screening. *Current Pharmaceutical Design*, *11*(9), 1189–1202. https://doi.org/10.2174/1381612053507549

Stocco, D. M. (2000). The role of the StAR protein in steroidogenesis: challenges for the future. *Journal of Endocrinology*, *164*(3), 247–253. https://doi.org/10.1677/joe.0.1640247

Streit, B. (1998). Bioaccumulation of contaminants in fish. *EXS*, *86*, 353–387. https://doi.org/10.1007/978-3-0348-8853-0_12

Sun, P. L., Hawkins, W. E., Overstreet, R. M., & Brown-Peterson, N. J. (2009). Morphological deformities as biomarkers in fish from contaminated rivers in Taiwan. *International Journal of Environmental Research and Public Health*, *6*(8), 2307–2331. https://doi.org/10.3390/ijerph6082307

Swapna, I., & Senthilkumaran, B. (2009). Influence of ethynylestradiol and methyltestosterone on the hypothalamo-hypophyseal-gonadal axis of adult air-breathing catfish, *Clarias gariepinus*. *Aquatic Toxicology*, *95*(3), 222–229. https://doi.org/10.1016/j.aquatox.2009.09.008

Swart, A. C., Schloms, L., Storbeck, K. H., Bloem, L. M., Toit, T., Quanson, J. L., Rainey, W. E., & Swart, P. (2013). 11β-hydroxyandrostenedione, the product of androstenedione metabolism in the adrenal, is metabolized in LNCaP cells by 5α-reductase yielding 11β-hydroxy-5α-androstanedione. *Journal of Steroid Biochemistry and Molecular Biology*, *138*, 132–142. https://doi.org/10.1016/j.jsbmb.2013.04.010

Tanaka, S. S., & Nishinakamura, R. (2014). Regulation of male sex determination: genital ridge formation and *Sry* activation in mice. *Cellular and Molecular Life Sciences*, *71*(24), 4781–4802. https://doi.org/10.1007/s00018-014-1703-3

Tenugu, S., Pranoty, A., Mamta, S.-K., & Senthilkumaran, B. (2021). Development and organisation of gonadal steroidogenesis in bony fishes - A review. *Aquaculture and Fisheries*, *6*(3), 223–246. https://doi.org/10.1016/j.aaf.2020.09.004

Terasawa, E., & Kenealy, B. P. (2012). Neuroestrogen, rapid action of estradiol, and GnRH neurons. *Frontiers in Neuroendocrinology*, *33*(4), 364–375. https://doi.org/10.1016/j.yfrne.2012.08.001

Tohyama, S., Miyagawa, S., Lange, A., Ogino, Y., Mizutani, T., Tatarazako, N., Katsu, Y., Ihara, M., Tanaka, H., Ishibashi, H., Kobayashi, T., Tyler, C. R., & Iguchi, T. (2015). Understanding the molecular basis for differences in responses of fish estrogen receptor subtypes to environmental estrogens. *Environmental Science and Technology*, *49*(12), 7439–7447. https://doi.org/10.1021/acs.est.5b00704

Tokarz, J., Möller, G., Hrabě de Angelis, M., & Adamski, J. (2015). Steroids in teleost fishes: A functional point of view. *Steroids*, *103*, 123–144. https://doi.org/10.1016/j.steroids.2015.06.011

Toledo-Ibarra, G. A., Rojas-Mayorquín, A. E., & Girón-Pérez, M. I. (2013). Influence of the cholinergic system on the immune response of teleost fishes: potential model in biomedical research. *Clinical and Developmental Immunology*, *2013*, 536534. https://doi.org/10.1155/2013/536534

Tyler, C. R., van der Eerden, B., Jobling, S., Panter, G., & Sumpter, J. P. (1996). Measurement of vitellogenin, a biomarker for exposure to oestrogenic chemicals, in a wide variety of cyprinid fish. *Journal of Comparative Physiology B: Biochemical, Systemic, and Environmental Physiology*, *166*(7), 418–426. https://doi.org/10.1007/BF02337886

van Aerle, R., Nolan, T. M., Jobling, S., Christiansen, L. B., Sumpter, J. P., & Tyler, C. R. (2001). Sexual disruption in a second species of wild cyprinid fish (the gudgeon, *Gobio gobio*) in United Kingdom freshwaters. *Environmental Toxicology and Chemistry*, *20*(12), 2841–2847.

Van den Belt, K., Berckmans, P., Vangenechten, C., Verheyen, R., & Witters, H. (2004). Comparative study on the *in vitro/in vivo* estrogenic potencies of 17beta-estradiol, estrone, 17alpha-ethynylestradiol and nonylphenol. *Aquatic Toxicology*, *66*(2), 183–195. https://doi.org/10.1016/j.aquatox.2003.09.004

Viganò, L., Arillo, A., Bottero, S., Massari, A., & Mandich, A. (2001). First observation of intersex cyprinids in the Po River (Italy). *Science of the Total Environment*, *269*(1), 189–194. https://doi.org/10.1016/S0048-9697(00)00821-4

Vuorinen, A., Odermatt, A., & Schuster, D. (2013). In silico methods in the discovery of endocrine disrupting chemicals. *Journal of Steroid Biochemistry and Molecular Biology*, *137*, 18–26. https://doi.org/10.1016/j.jsbmb.2013.04.009

Wang, L., Yan, R., Yang, Q., Li, H., Zhang, J., Shimoda, Y., Kato, K., Yamanaka, K., & An, Y. (2020). Role of GH/IGF axis in arsenite-induced developmental toxicity in zebrafish embryos. *Ecotoxicology and Environmental Safety*, *201*, 110820. https://doi.org/10.1016/j.ecoenv.2020.110820

Wang, P., Du, Z., Gao, S., Zhang, X., & Giesy, J. P. (2016). Impairment of reproduction of adult zebrafish (Danio rerio) by binary mixtures of environmentally relevant concentrations of triclocarban and inorganic mercury. *Ecotoxicology and Environmental Safety*, *134*, 124–132. https://doi.org/10.1016/j.ecoenv.2016.08.026

Wang, Q., Lam, J. C., Han, J., Wang, X., Guo, Y., Lam, P. K., & Zhou, B. (2015). Developmental exposure to the organophosphorus flame retardant tris(1,3-dichloro-2-propyl) phosphate: estrogenic activity, endocrine disruption and reproductive effects on zebrafish. *Aquatic Toxicology*, *160*, 163–171. https://doi.org/10.1016/j.aquatox.2015.01.014

Xu, J., Huang, W., Zhong, C., Luo, D., Li, S., Zhu, Z., & Hu, W. (2011). Defining global gene expression changes of the hypothalamic-pituitary-gonadal axis in female sGnRH-antisense transgenic common carp (*Cyprinus carpio*). *PloS One*, *6*(6), e21057. https://doi.org/10.1371/journal.pone.0021057

Yin, P., Li, Y. W., Chen, Q. L., & Liu, Z. H. (2017). Diethylstilbestrol, flutamide and their combination impaired the spermatogenesis of male adult zebrafish through disrupting HPG axis, meiosis and apoptosis. *Aquatic Toxicology*, *185*, 129–137. https://doi.org/10.1016/j.aquatox.2017.02.013

Zacharewski, T. (1997). *In vitro* bioassays for assessing estrogenic substances. *Environmental Science & Technology*, *31*(3), 613–623. https://doi.org/10.1021/es960530o

Zhang, Y., Gao, J., Xu, P., Yuan, C., Qin, F., Liu, S., Zheng, Y., Yang, Y., & Wang, Z. (2014). Low-dose bisphenol A disrupts gonad development and steroidogenic genes expression in adult female rare minnow *Gobiocypris rarus*. *Chemosphere*, *112*, 435–442. https://doi.org/10.1016/j.chemosphere.2014.04.089

# 9 Multi- and Transgenerational Health Effects of Exposure to Toxicants in Fish

*Tracie R. Baker and Danielle Meyer*
College of Public Health and Health Professions
University of Florida

**Abstract**

A steady climb in toxicant-induced adverse health outcomes is not surprising considering the rising environmental presence of emerging contaminants, concomitant with the persistent environmental and body burden of heavy metals and legacy organic pollutants. Many of these toxicants are classified as endocrine-disrupting chemicals capable of interfering with endogenous hormonal function, as well as reproduction, neurodevelopment, and metabolic function. Exposure to environmental toxicants, whether legacy or emergent, can induce disease outcomes that persist to indirectly exposed and unexposed generations. Fish species are frequently used to model the mechanisms underlying heritable toxicant-induced disease, due to factors including inexpensive maintenance, rapid generation time, and ease of exposure. This chapter reviews and presents the challenges of this research in a fish model, and health outcomes in subsequent generations following parental exposure to dioxins, PCBs, PAHs, pharmaceuticals and personal care products, heavy metals, BPA, hormones, steroids, pesticides, herbicides, PFAS, radiation, and other toxicants. Despite the growing evidence of multigenerational health effects of many contaminants, it is clear that further mechanistic and epigenetic analysis is needed to interpret existing phenotypic data and understand implications for human and fish health.

## INTRODUCTION

In 2021, the World Health Organization (WHO) reported a staggering global health cost of toxicant exposure for 2019, estimating >2 million deaths (3.6% of total 2019 deaths) and >53 million DALYs (disability-adjusted life years) preventable by the management or mitigation of environmental toxicants (WHO, 2021). Despite years of progress in pollution control, the burden on public health has only increased over the last decade: 2019 estimates increased by 29% in deaths and 19% in DALYs from 2016, which had already seen a rise of 19% in deaths and 4% in DALYs above 2012 estimates (WHO, 2016, 2018, and 2021). Coupled with the rising cost of human life is the financial impact of toxicant-induced adverse health outcomes; Preker et al. (2017) estimated that 2013 healthcare costs related to air, water, and soil pollution ranged from US$240 to 630 billion, about 3–9% of global expenditures.

A steady climb in toxicant-induced adverse health outcomes is not surprising considering the rising environmental presence of emerging contaminants, e.g., biologically active pharmaceuticals and personal care products (PPCPs), concomitant with the persistent environmental and body burden of heavy metals and legacy organic pollutants, e.g., dioxins, polychlorinated biphenyls (PCBs), and pesticides (Phillips et al., 2021; Ebele et al, 2017; Vallero, 2005; Ross and Birnbaum, 2003). Many of the above toxicants are classified as endocrine-disrupting chemicals (EDCs) capable of interfering with endogenous hormonal function, with reproduction, neurodevelopment, and metabolic function as just a few of the resulting downstream targets (Welshons et al., 2003; Bhandari et al., 2015). Per the Developmental Origins of Health and Disease (DOHaD) paradigm, early-life exposure to toxicants or other stressors can disrupt organ systems when most susceptible to insult: during sensitive windows of development. Thus, a relatively brief, low-level EDC exposure can have disproportionate effects on the trajectory of development, resulting in lifelong or adult-onset disease (Barker and Osmond, 1986; Needleman et al., 1990).

One notable example is the aftermath of *in utero* exposure to the synthetic estrogen diethylstilbestrol (DES), prescribed to millions of women from 1938 to 1971 to prevent pregnancy complications. While some effects were reported in the mothers, DES had the most potent effects on developing offspring: the daughters of women who used DES were found to have ~40 times the risk of developing clear cell adenocarcinoma as adults as well as increased risk for multiple reproductive and fertility-related adverse outcomes (Herbst et al., 1971; Hoover et al., 2011). Later research suggests the indirectly exposed DES grandchildren were not immune to these effects, revealing adverse reproductive and neurodevelopmental health outcomes that included increased risk of ADHD (Attention-Deficit/Hyperactivity Disorder), hypospadias, and delay of menstrual regularity (Kioumourtzoglou et al., 2018; Titus-Ernstoff et al., 2006).

Alarming findings like these were a major driving force behind the growth in DOHaD and fetal programming studies, which revealed epigenetic regulation as one of the mechanisms underlying persistent and heritable effects (Dolinoy et al., 2007; Rogers et al., 2018). Epigenetic regulation can control gene expression networks without directly altering DNA sequence, most frequently via nucleosome positioning and covalent modifications to histones and DNA, and is capable of persisting over time and across generations (Lacal and Ventura, 2018). Thus, toxicant exposures of past generations



DOI: 10.1201/9781003160694-9

can interact with personal exposure history to intensify the burden of disease, compelling investigation into the characteristics of these persistent health outcomes, the exposure circumstances under which they are generationally transferred, and the relevant mechanisms behind such transfer.

Human studies of such effects are the gold standard for relevance, but inherent limitations restrict the pool of available clinical knowledge, including long generation times, latency from exposure to disease, confounds of genetics and co-exposures, and lack of specific information on dose or exposure window (Hernández and Tsatsakis, 2017). While multiple animal models, including rodent species (Bruner-Tran et al, 2014; Manikkam et al., 2013), *C. elegans* (Zhang et al., 2020), and a variety of aquatic organisms (Baker et al., 2014; De Liguoro et al., 2019) have been used to complement available clinical data by addressing each of these limitations, this review focuses on the use of fish species as a sensitive, rapid, and economical means to evaluate multi- and transgenerational toxicity of environmental toxicants, including persistent organic pollutant and heavy metal exposure (Bhandari, 2016).

## Fish as Model Organisms for Toxicant Exposure

A variety of teleost fish have been introduced as models for developmental and toxicological studies, including: cyprinids [zebrafish (*Danio rerio*), fathead minnow (*Pimephales promelas*)], salmonids [rainbow trout (*Oncorhynchus mykiss*), Atlantic salmon (*Salmo salar*)], poeciliids [guppy (*Poecilia reticulata*), western mosquitofish (*Gambusia affinis*)], cyprinodontids [mummichog (*Fundulus heteroclitus*, sheepshead minnow (*Cyprinodon variegatus*)], adrianichthyids such as medaka (*Oryzias latipes*; *O. melastigma*), and atherinopsids such as inland silversides (*Menidia beryllina*). Zebrafish and medaka are the most well-established model fish species to date, with fully sequenced genomes, an assortment of transgenic reporter lines available for screening toxicant-induced effects, and valuable developmental characteristics for large-scale, long-term exposure studies, including small size, inexpensive maintenance, short generation time (~3 months), high fecundity, and rapid, transparent early development (Bhandari, 2016; Lin et al., 2016). While zebrafish is currently the central model utilized in multi- and transgenerational aquatic toxicology studies, medaka is increasing in popularity due to certain advantageous traits: a smaller genome (800 Mb vs 1,700 Mb in zebrafish), readily available inbred lines, and an XY system of sex determination, as opposed to the environmentally-guided polygenic sex determination of domesticated zebrafish, who lost the WZ:ZZ system of their wild counterparts (Matsui et al., 2012; Wilson et al., 2014). Another model of note is the fathead minnow, distinct for their use in Tier 1 EPA 21-day reproductive toxicity testing. These fish are convenient for their sexual dimorphism (shared by medaka) and slightly larger size (50–75 mm) compared to zebrafish or medaka, allowing for easier blood sampling. However, their generation time of ≥4 months is slower than that of zebrafish or medaka (Ankley and Johnson, 2004). One species commonly studied in the context of population dynamics is the mummichog: unique due to a remarkable ability to adapt to a variety of environments highly contaminated with toxicants including dioxins, mercury, PCBs, and PAHs (polycyclic aromatic hydrocarbons; Weis, 2002). Mummichog (often referred to as killifish) typically remain close to their spawning location, and as such are excellent models of the effects of regional contamination and population adaptation.

While controlled and well-characterized studies performed on laboratory-bred or cultured model fish species are critical to understanding the mechanisms of generational toxicology, some degree of relevance to real-world exposure events is unavoidably lost. These factors include the quantity and route of naturally occurring toxicant exposure, variability in genetics and degree of contaminant exposure, changing population dynamics, and resilience of the population over time and generations (Galic et al., 2012; Van den Brink, 2013). Studying such interactions with limited interference by researchers provides a more accurate picture of the true environmental burden of such exposures and the subsequent fate of fish populations. Aside from direct effects on the organism of interest, toxicant exposures within the natural habitat can affect multiple organisms and trophic levels across the ecosystem, resulting in potentially altered predator-prey dynamics, resource availability, abiotic factors, and myriad other interactions that impact fish population status (Olff et al., 2009). The complex, emergent nature of these interactions within ecosystems and the ways through which they are disrupted by toxicant exposure across generations remain to be fully characterized.

## The Challenge of Defining Multi- and Transgenerational Effects

One challenge inherent to the relatively novel study of generational toxicology is establishing consistent, universal definitions for commonly used generational terms, such as "inter-", "trans-", and "multigenerational". For example, the word "transgenerational" has been used in some instances to indicate that effects were reported in more than one generation of fish, or that toxicants were passed down from parent to offspring generations. In other contexts, it describes the transmission of effects to an unexposed generation via epigenetic mechanisms and/or the persistence of effects across both exposed and unexposed generations. In yet another variation, it refers to multiple generations continuously exposed to a toxicant. Similar issues arise with the term "multigenerational": in some cases, it is defined as multiple generations of continuous exposure, while others use the term to refer to any study with scope beyond the exposed F0 generation. For our purposes, a "multigenerational" study is defined as between one and two full generational cycles separating the F0 window of exposure from the stage at which the F1 is evaluated for health effects; for example, a study in which F0 fish were exposed as juveniles would need to age F1 offspring to juvenilehood to qualify, while examining F1 egg survival or hatch rate would not. As outlined in Figure 9.1, "transgenerational", then, is defined as two or more full generational cycles between exposure and assessment.

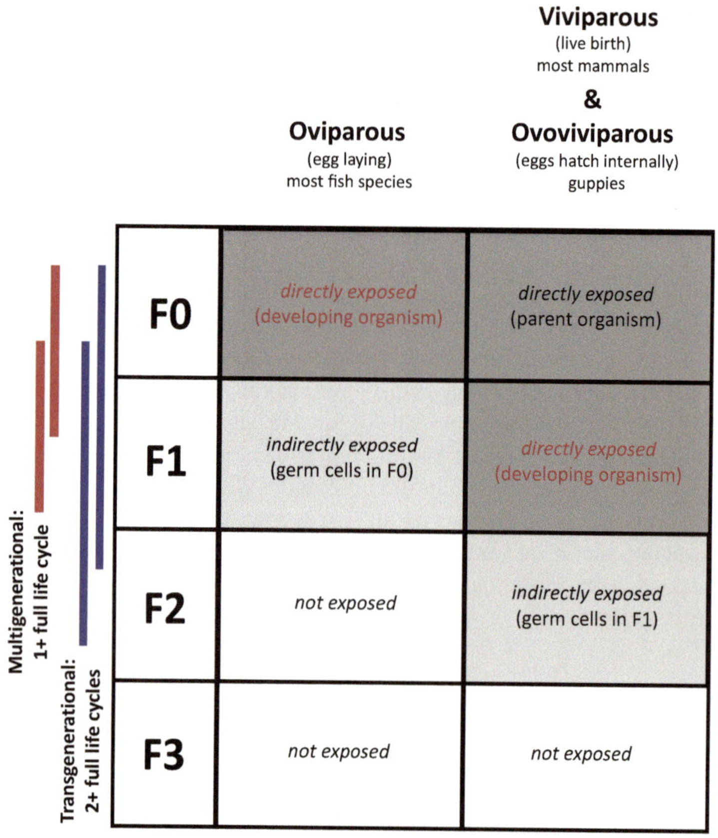


**FIGURE 9.1** Comparison of F0–F3 exposure paradigm between species that can be externally exposed to toxicants during development (oviparous) and species that must be exposed within the parent to target early development (viviparous, ovoviviparous). For each category, the organism targeted by exposure is in red text. Gray shading highlights the degree of exposure across successive generations. In descendants with no exposure, heritable epigenetic effects can be distinguished from effects of direct or indirect exposure. To the left, colored bars illustrate definitions used here for multi- and transgenerational study design: there must be 1+ or 2+ full life cycles, respectively, between exposure and assessment.

Another challenge involves the systematic evaluation and comparison of such studies. Factors such as specific critical windows targeted, exposure routes, and contaminant persistence in the organism and its environment are crucial to evaluating generational outcomes, but the degree to which they are reported and/or critically examined varies. Potential tools to address this issue include checklists of suggested minimum information for study publication, as well as databases that categorize and compile studies based on relevant factors – both methods that have been successfully utilized in related branches of toxicology, including developmental and epigenetic research (Fostel et al, 2007; Wang et al., 2018). The National Institute of Environmental Health Sciences (NIEHS)-funded Toxicant Exposures and Responses by Genomic and Epigenomic Regulators of Transcription (TaRGET) II consortium is an ongoing database of toxicant-induced epigenomic changes across a variety of tissues, cell types, and experimental conditions in the mouse model; a similarly structured database would be a useful method of systematically evaluating multi- and transgenerational effects of toxicants across fish species. Implementing a specific nomenclature and incorporating methods for systematic analysis will allow for the clear communication and interpretation of experimental findings and, more broadly, help to synthesize outcomes across studies, evaluate the strength of current evidence, and determine which areas in the field deserve further research.

## Continuous Exposure Studies

Continuous exposure, i.e., the consecutive exposure of multiple generations to a contaminant, is garnering increased interest within the field of aquatic generational toxicology. This type of study provides greater environmental relevance: exposures in nature are rarely isolated to one generation, due to continuous contamination events and environmental persistence. Generations of descendants will also be exposed to contaminants and mixtures their ancestors did not experience, as compounds are newly developed and/or phased out of use. Will contaminant exposure across multiple generations intensify the health burden for descendants, or can subsequent generations adapt to the stressor – and through what mechanisms? How will adaptation (or lack thereof) to previous exposures alter their ability to respond to novel exposures, and how do the structural and functional properties of the toxicants play into that response? One ongoing study addressing these questions involves certain Atlantic killifish populations that have developed an increased tolerance not only to the high levels of PCBs and other PAHs contaminating their ancestral habitat but also to other structurally similar compounds (Nacci et al., 2010; Arzuaga and Elskus, 2010); This resistance, however, developed at a cost of decreased basal metabolic fitness and increased sensitivity to other contaminants (Lindberg et al., 2017). Exploring the outcomes and mechanisms of continuous exposure across different fish species and contaminant mixtures, whether in wild fish or laboratory models, will be crucial for understanding the fate of exposed populations.

A critical disadvantage of the continuous exposure model, however, is the inability to target specific exposure windows. For example, exposure to a compound across two full generations (F0 and F1) obscures which effects in the subsequent F2 generation are specifically caused by exposure of developing germ cells, which are inherited from the original generation, and which are due to interactions between the two. Isolating precise critical windows and generations allows us to narrow down mechanisms and routes of effect, which is complementary to more environmentally relevant work with continuous and interactive exposures or challenge assays. Thus, for specificity's sake, further discussion of multi- and transgenerational studies will be limited to those with clearly defined exposure windows that do not extend beyond a single generation. Figure 9.2 summarizes the outcomes and characteristics of these studies.

## MULTI- AND TRANSGENERATIONAL STUDIES

### Dioxins, PCBs, & PAHs

Dioxins, dioxin-like PCBs, and other PAHs, classified both as EDCs and persistent organic pollutants (POPs), are some of the most notorious "bad actors" of the last century. These generally lipophilic, degradation-resistant compounds have persisted in environmental compartments for decades despite near worldwide limits on production, resulting in environmental presence and bioaccumulation across aquatic and terrestrial food webs. Thus, high trophic level consumers such as large fish, seals, birds, mink, and human populations have exhibited toxicity across reproductive, endocrine, and immune systems (Ross and Birnbaum, 2003; Vallero, 2005).

Well-characterized as a multigenerational reproductive toxicant across a variety of model organisms, TCDD, or 2,3,7,8-tetrachlorodibenzo-*p*-dioxin, was first demonstrated by King Heiden et al. (2009) to have effects extending beyond the exposed generation in fish: repeated acute exposures throughout F0 juvenile development increased mortality, induced developmental and craniofacial abnormalities, and impaired fertility in F0 zebrafish and their F1 offspring (exposed as developing germ cells). Further studies expanded the scope to the unexposed F2 generation, revealing that impaired fertility could persist transgenerationally through the male germline; these phenotypic outcomes were accompanied by altered reproductive, metabolic, and epigenetic transcriptomic networks (Baker et al., 2014, 2016; Meyer et al., 2018; Akemann et al., 2019, Akemann et al., 2020). Minimal evidence exists to date for TCDD-induced multi- or transgenerational changes in DNA methylation on a global scale; however, site-specific methylation changes were particularly distinct in the F1 generation, suggesting gamete development as a sensitive window for TCDD-induced effects on the methylome (Akemann et al, 2020). While multi- and transgenerational reproductive outcomes have been phenotypically well-characterized for TCDD in zebrafish, the underlying mechanisms and epigenetic regulators require further profiling, and effects on other fish species and physiological systems targeted by TCDD, including metabolic, immune, and cardiovascular systems, deserve exploration.

Some evidence exists for the multi- and transgenerational effects of exposure to PCB mixtures on metabolism, reproduction, and behavior in fish species. In mummichog, the main effects of adult F0 exposure presented in adult F1 (increased weight, minor effects on fertility) and larval F2 (decreased survival) specifically, suggesting that this PCB mixture may distinctly target developing germ cells (Matta et al., 2001). In zebrafish, F0 larvae were exposed to a PCB/PBDE (polybrominated diphenyl ethers) mixture through adulthood, resulting in some degree of transcriptomic or behavioral toxicity across five generations of fish (F0–F4). While F1 larvae were hyperactive, F2–F4 generations exhibited persistent hypoactivity, suggesting first: a heritable epigenetic modification underlying PCB/PBDE-induced larval hypoactivity, and second: that the effects of indirect germ cell exposure may have a distinct, potentially opposing mechanism leading to F1-specific hyperactivity (Alfonso et al., 2019; Blanc et al., 2021). Intriguingly, larval *dnmt*, or DNA methyltransferases, responsible for developmental and methylome maintenance, showed alternating patterns of activity across F2–F4 generations. Such shifts may be a sign of repeated generational efforts to compensate for dysregulation, as the pendulum swings past homeostasis for each generation. Changes in *dnmt* activity may induce epigenomic effects downstream; thus interrogating epigenomic profiles across affected generations may provide more insight into the behavioral and transcriptomic effects of PCB/PBDE exposure.

Exposure to the PAH benzo[*a*]pyrene (B[*a*]P) is known to influence multi- and transgenerational skeletal, behavioral, and metabolic outcomes in zebrafish and medaka models. In medaka, F0 adult exposure resulted in heritable skeletal dysregulation in F1–F3 offspring, including dysregulated bone metabolism genes, an increase in compressed vertebrae, and decreased bone thickness (Seemann et al., 2015; Seemann et al., 2017). In zebrafish, F0 B[*a*]P exposure in adulthood increased mortality and incidence of developmental abnormalities in the F1 generation primarily, with some lingering developmental effects in the F2 (Corrales et al., 2014). While these more severe outcomes had all but disappeared by the F3 and F4 generations, larval exposure to B[*a*]P resulted in metabolic and behavioral dysregulation persisting across F0–F2 generations; exposure-lineage larvae exhibited hyperactivity, decreased heart rate, and decreased OCR (oxygen consumption rate), while adults had increases in BMI (body mass index), OCR, and anxiety-like behaviors (Knecht et al., 2017). This persistent, life stage-dependent shift in metabolic profile likely has epigenetic underpinnings, a target for future work.

Other PAH exposures have also induced multi- and transgenerational developmental and behavioral effects in fish models.

(A)

**Targets of Multi- and Transgenerational Toxicant Exposure in Fish Models**

| Toxicant Class | Systems/Organs Affected | Sources Reviewed |
|---|---|---|
| Dioxins, PCBs, PAHs | Mortality and development; Craniofacial and skeletal system; Epigenome (methylome); Reproduction; Larval behavior; Metabolism | [2-4, 7, 18, 26, 29, 30, 38, 39, 42, 51, 52, 68] |
| PPCPs | Mortality and development; Larval and adult behavior; Reproduction; Cortisol response; Steroidogenesis | [ 6, 17, 21-23, 27, 36, 58, 63, 65-67] |
| Heavy Metals | Neurodevelopment; Reproduction; Epigenome (methylome) | [10, 14, 38, 40, 42, 64, 73, 75, 76] |
| BPA | Mortality and development; Cardiovascular; Osmoregulation; Reproduction; Metabolic; Epigenome (methylome) | [1, 5, 35, 46, 47, 59, 62, 71, 78] |
| Hormones, Steroids | Mortality and development; Behavior; Osmoregulation; Reproduction; Metabolic; Epigenome (methylome) | [5, 8, 19, 33, 37, 49, 50, 59, 69-71, 74] |
| Pesticides, Herbicides | Neurodevelopment; Reproduction; Epigenome (methylome) | [6, 16, 19, 32, 37, 48, 72] |
| PFASs | Mortality and development; Epigenome (methylome); Reproduction; Thyroid; Gastrointestinal | [11-13, 53, 54] |
| Radiation, Mutagens | Mortality and development; Inflammation; Reproduction; Epigenome (methylome, histone PTMs) | [20, 24, 25, 28, 34, 55, 56] |
| Other toxicants | Mortality and development; Oxidative stress/inflammation; Epigenome (methylome); Reproduction; Larval behavior; Metabolism/thyroid activity | [9, 15, 31, 41, 43-45, 57, 60, 61, 77] |

(B)

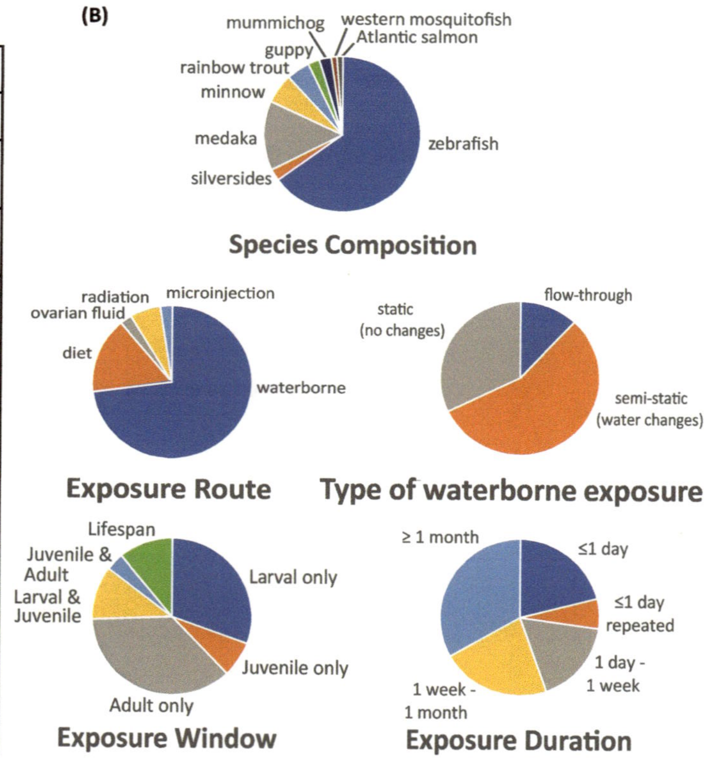


**FIGURE 9.2** Summary of the reviewed literature indicating outcomes and organ systems targeted by multi- and transgenerational toxicant exposure in fish models (**A**). Pie charts break down multi- and transgenerational studies according to species composition, exposure window, duration, and route (**B**).

The exposure of adult sheepshead minnows to a mixture of PAHs designed to mimic a crude oil spill (HEWAF, or high-energy water accommodated fraction) dysregulated F1 larval development, decreased somatic measurements in F1 adult females, and impaired prey capture ability in F1 and F2 larval offspring. Such heritable effects on prey capture have concerning implications for population maintenance in crude oil-contaminated areas (Jasperse et al., 2019). In zebrafish, F0 larval exposures to a variety of PAH mixtures extracted from river sediments or oil fractions caused dose- and mixture-dependent changes in F1 larval and juvenile behavior and impaired aromatase expression in response to EE2 (ethinylestradiol) challenge; however, F2 larval fish exhibited minimal behavioral effects of ancestral exposure (Vignet et al., 2015). Table 9.1 summarizes the multi- and transgenerational health outcomes of exposure to dioxin, PCBs, and PAHs reported above.

## PPCPs

Pharmaceuticals and personal care products are comprised of a broad variety of chemicals and products with industrial, agribusiness, personal health, medical, and cosmetic purposes. Such products are heavily used by an aging, expanding population, and are ubiquitously detected in environmental waterbodies due to a combination of factors, including improper disposal procedures and the inability of wastewater systems to address these relatively novel contaminants of emerging concern (CECs). While they are generally present in the environment at low levels, the lack of knowledge about the long-term, low-dose effects of these contaminants, coupled with their potential additive or synergistic effects, raises concerns about the health outcomes of all utilizing these waterbodies –including the fish species constantly submersed in this "contaminant soup" (Richmond et al., 2017). Table 9.2 summarizes the known multi- and transgenerational health effects of exposure to various PPCPs across model fish species.

Some of the most prevalent environmental PPCPs are prescription drugs, including carbamazepine, risperidone, and the SSRI (selective serotonin reuptake inhibitor) antidepressants venlafaxine and fluoxetine. There is evidence that SSRI exposure can result in multi- and transgenerational changes to stress response, and behavioral and steroidogenic

**TABLE 9.1**
**Summary of the Reviewed Literature Indicating Outcomes of Multi- and Transgenerational Exposure to Dioxins, PCBs, and PAHs in Fish Models**

| Dioxin (TCDD) | PCBs | PAHs |
|---|---|---|
| Increased mortality (F0, F1) | Increased weight (F1) | Decreased survival (F1) |
| Developmental and craniofacial abnormalities (F0, F1) | Impaired fertility (F1) | Increased developmental abnormalities (F1, F2) |
| | | Impaired prey capture; larvae (F1, F2) |
| Impaired fertility (F0–F2) | Decreased survival (F2) | Altered metabolic parameters; larvae and adults (F0–F2) |
| Altered reproductive, metabolic and epigenetic transcriptome; adult testes (F0–F2) | Altered metabolic, neurologic, and epigenetic transcriptome; larvae & adult brains (F0–F4) | Behavioral changes; larvae, juveniles, & adults (F0–F2) |
| Site-specific changes in methylome; adult testes (F0–F2) | Altered photomotor activity; larvae (F0–F4) | Altered skeletal phenotype and transcriptome; larvae and adults (F1–F3) |

**TABLE 9.2**
**Summary of the Reviewed Literature Indicating Outcomes of Multi- and Transgenerational Exposure across Various PPCPs in Fish Models**

| Venlafaxine & Fluoxetine | Carbamazepine | Risperidone |
|---|---|---|
| Decreased exploratory and locomotor behaviors; adults (F0–F2) | Reduced sperm motility and decreased whole body 11- KT (F0, F1) | Decreased survival (F1) |
| Altered timing of cortisol response; larvae (F0–F2) | Decreased fecundity (F0–F3) | Decreased activity; larvae (F1) |
| Changes to stress response, behavioral, and steroidogenic transcriptome; larvae and adults (F0–F3) | Altered male spawning behaviors and sperm morphology (F0–F4) | Increased exploratory activity; adults (F1) |
| Blunted cortisol response; adults (F0–F4) | | **Triclosan** |
| **Gemfibrozil** | **Clofibric acid** | Decreased size; larvae (F1) |
| Changes in 11-KT levels (F0–F2) | Increased abnormalities (F0, F1) | Decreased survival (F1) |
| Decreased fecundity (F0, F1, F3) | Altered weights and condition factors; adults (F0, F1) | **Coumarin 47** |
| Altered reproductive behaviors (F0, F1, F4) | Minor effects on lipid transcriptome; adults (F0, F1) | Increased length and weight; adults (F2, F3) |
| Altered sperm morphology (F0–F4) | | Increased activity; larvae (F2, F3) |

transcriptomic pathways in zebrafish, extending through the F3 generation (Vera-Chang et al., 2018, 2019a; Luu et al., 2021). Ancestral larval fluoxetine and venlafaxine exposures are respectively capable of blunting the cortisol response through the F4 generation (Vera-Chang et al., 2018, 2019a, 2019b) and altering its timing through the F2 (Thompson and Vijayan, 2021), while fluoxetine exposure can decrease exploratory and locomotor behaviors through the F2 generation (Vera-Chang et al., 2018). Challenging the F4 generation of fluoxetine-lineage fish with venlafaxine exacerbates the blunting of the cortisol response, suggesting that ancestral fluoxetine exposure sensitizes fish to other SSRIs, rather than inducing an adaptive response over generations (Vera-Chang et al., 2019b). Exposure to the anticonvulsant and mood stabilizer carbamazepine targets a spectrum of reproductive endpoints through multiple generations of descendants, including altered male spawning behaviors and sperm morphology (F0–F4), decreased fecundity (F0–F3), reduced sperm motility, and decreases in whole-body 11-KT (ketotestosterone) levels (F0–F1; Galus et al., 2014; Fraz et al., 2019a), while F0 larval exposure to the antipsychotic drug risperidone primarily targets the F1, reducing survival and activity at larval stages while increasing exploratory activity later in life (Kalichak et al., 2019). Together, these findings suggest that fish populations with such legacy exposures may be slow to return to baseline across the first several generations, facing challenges in such crucial survival-oriented behaviors as reproduction, exploring and interacting with their environment, and responding appropriately to natural stressors. The epigenetic mechanisms behind this inheritance will be of great interest in further studies.

Also frequently detected in the environment are blood lipid-lowering agents and their metabolites, including gemfibrozil and clofibric acid. As with carbamazepine, F0 exposure to gemfibrozil alters reproductive endpoints through the F4 generation in zebrafish, with specific targets including sperm morphology (F0–F4), fecundity (F0, F1, and F3 males), reproductive behaviors (F0, F1, and F4 males), and 11-KT levels (F0–F2; Galus et al., 2014; Fraz et al., 2019b). Some evidence supports the multigenerational effects of exposure to clofibric acid, a metabolite of PPARa (peroxisome proliferator-activated receptor alpha) agonist clofibrate; specifically: increased abnormalities, altered weights and condition factors, and minor effects on lipid metabolism transcriptome (Coimbra et al., 2015). Other PPCPs with evidence of multigenerational outcomes in fish models include triclosan and coumarin 47, two compounds primarily found in personal care products. Exposures to triclosan and coumarin 47 have respectively resulted in multi- and transgenerational effects in zebrafish; the antimicrobial agent triclosan decreases larval size and causes high F1 mortality (Stenzel et al., 2019), while coumarin 47, a compound used in dyes, increases fish length, weight, and activity in F2 and F3 generations (Blanc et al., 2020).

Prescription drugs and compounds from personal care products have demonstrated behavioral and reproductive effects that last even to the third or fourth generation after the initial toxicant exposure (F3–F4), with concerning implications for the generations of humans and wildlife that continue to utilize and live in PPCP-contaminated waterbodies. The contaminants discussed here are only a small fraction of known PPCPs and are nearly all modeled in zebrafish, highlighting the need to incorporate a greater variety of contaminants and model fish species into multi- and transgenerational study designs.

## Heavy Metals: Mercury, Lead, & Arsenic

Certain heavy metals, including mercury, lead, and arsenic, are of particular toxicological concern due to their relatively widespread environmental presence and ability to cause harmful physiological outcomes even at low levels. Some fish species readily accumulate mercury, leading to health effects not only within the population but in consumer species –including humans. In fact, most mercury exposures in humans occur through seafood consumption, as well as dental amalgam leaching, accidental inhalation, or dermal contact (Jannetto, 2011). Mercury primarily targets the nervous system with no known threshold of exposure; however, evidence suggests the cardiovascular system may also be disrupted (Bernhoft, 2012). Most multi- and transgenerational heavy metal research in fish species to date has been performed on mercury. The most straightforward support for inherited effects of mercury exposure in fish models is for neurodevelopmental outcomes; zebrafish have presented with visual deficits, hyperactivity, and lack of avoidance learning through the F2 generation (Xu et al., 2016a; Carvan et al., 2017). Other generational outcomes are somewhat mixed: one study reported no effects on reproductive, survival, somatic measurement, or gene expression endpoints for F1 and F2 zebrafish (Chen et al., 2021), while others indicate reproductive and somatic dysregulation in the F1 of mummichog and female zebrafish (Matta et al., 2001; Xie et al., 2020). While some of this discrepancy may be attributable to species differences and window of exposure in the case of mummichog, the rest deserves further investigation. Although there is minimal support for global changes in DNA methylation (Olsvik et al., 2014), there is evidence for site-specific changes in the methylome; F2 zebrafish sperm showed differentially methylated regions linked to metabolism and neuroactivity while F1 and F2 embryos displayed site-specific changes in DNA methylation (Olsvik et al., 2014; Carvan et al., 2017).

Although lead is not a POP, it is bioaccumulative and persistent, frequently released into the environment through anthropogenic activity, aged urban infrastructure, contaminated food and water, and remnants of leaded paint and gasoline. The neurodevelopmental effects of lead exposure are most devastating, with no theorized threshold of exposure; however, multiple other systems can exhibit toxicity, including reproductive, cardiovascular, kidney, and respiratory (White et al., 2007; Dietrich et al., 2001; Wani et al., 2015). F0 lead exposure has demonstrable effects on avoidance learning and hyperactivity in F2 adult zebrafish, coupled with sex-specific dysregulation of synaptic, neurogenesis, endocrine, and epigenetic genes (Xu et al., 2016b; Meyer et al., 2020). Further characterizing these transgenerational neurological deficits would be a useful next step, along with determining the extent to which endocrine and epigenetic mechanisms are disrupted.

Lastly, inorganic arsenic is also a highly potent neurotoxicant, a reproductive and cardiovascular toxicant, and a carcinogen, primarily released into the environment through the use of pesticides and insecticides, as well as cigarette smoke (Shalat et al., 1996; ATSDR, 2007; Jannetto, 2011; Zhang et al., 2020). Arsenic exposure also induced transgenerational outcomes in zebrafish, dysregulating site-specific methylation in F2 females and altering F2 patterns of activity across a variety of developmental windows (Valles et al., 2020). While much work remains in characterizing the transgenerational effects of heavy metal exposures in fish species, neurodevelopmental networks are evidently a highly sensitive target, resulting in persistent and heritable disruption.

## BPA

Human exposure to the plasticizer bisphenol A (BPA), used in the production of items including polycarbonate plastics, the epoxy resin linings of metal cans and jars, and thermal paper, is widespread, primarily through oral, dermal, and inhalation routes (Konieczna et al., 2015). Despite the short half-life of BPA, such ubiquity results in its pseudo-persistence in environmental compartments (Flint et al., 2012). BPA is a known endocrine-disrupting compound that can interfere with estrogenic signaling in humans and animal models, leading to reproductive and metabolic consequences including cancers, obesity, cardiovascular disease, and diabetes (Michałowicz, 2014; Konieczna et al., 2015). Considerable evidence is accumulating to support multi- and transgenerational outcomes of BPA exposure across a variety of animal models, including fish species (Xin et al., 2015; Zhou, 2018).

In zebrafish, evidence is strongest for multi- and transgenerational reproductive effects of BPA exposure, with developmental, cardiovascular, and metabolic pathways also affected. Reproductive outcomes include impaired fertility (F0–F2), disruption of ovarian transcriptome and methylome (F0–F2), increased gonadal abnormalities (F1), decreased ovarian gonadosomatic index (F2), abnormal sperm morphology (F2), and decreased sperm count (F2; Lombó et al., 2015; Akhter et al., 2018; Santangeli et al., 2019). Altered site-specific embryonic methylation and transcription (F0–F3) and increased embryonic malformations (F1, F2) were the primary developmental outcomes; other findings included F1 cardiac abnormalities and altered gene expression, as well as metabolic outcomes of insulin receptor downregulation (F0–F2) and decreased BMI (F2; Lombó et al., 2015; Akhter et al., 2018; Santangeli et al., 2019).

Similarly, evidence for transgenerational effects in medaka mostly implicates the reproductive system, in addition to developmental and osmoregulatory outcomes; fertility was reduced in F2–F3 generations and HPG (hypothalamus-pituitary-gonadal) axis genes were differentially expressed and methylated in the brain, pituitary, and testes of F2 males (Bhandari et al., 2015; Thayil et al., 2020). Other heritable outcomes include decreased embryonic survival through the F4 generation (Bhandari et al., 2015) and dysregulated expression of osmoregulatory genes in the gills of F3 females (Wang et al., 2019). Reproductive effects of F0 BPA exposure in minnow, however, were not inherited by the F2 generation: only F1 BPA-lineage females had immature ovaries with dysregulated steroidogenic networks and altered DNA and histone methylation pathways (Zhu et al., 2021). Interestingly, most generation-spanning effects were metabolic in the rainbow trout model; F1 and F2 generations displayed irregular growth rates, altered lipid and steroidogenesis content, dysregulated renal and hepatic gene expression, and altered stress response pathways (Sadoul et al., 2017; Thomas et al., 2018).

As summarized in Table 9.3, the transgenerational effects of BPA exposure on fish reproductive and metabolic systems are evidently both pervasive and persistent, indicating the risk of lasting health consequences in environmental populations. DNA and histone methylation appear to be promising mechanisms of inheritance and should be further targeted for study, alongside other potential routes of epigenetic modification.

## Hormones & Steroids

The synthetic hormone 17a-ethinylestradiol (17a-EE2) is one of the most potent estrogens, utilized widely as a component of human contraceptives and in aquaculture and livestock farming (Aris et al., 2014). Due to this pattern of use, it is widespread throughout wastewater intake and often inefficiently removed from effluent, resulting in downstream waterbody contamination (Ternes et al., 1999). In fish, exposure to 17a-EE2 and other artificial estrogens can disrupt gonadal development, impair fertility, feminize males, and even collapse population structures entirely, making this category of EDCs one of the most disruptive to aquatic life (Kidd et al., 2007; Schwindt et al., 2014).

Such effects can also persist across generations: exposure to 17a-EE2 induced multi- and transgenerational infertility in inland silversides (F1) and medaka (F1, F2) with accompanying transcriptomic and methylomic effects on HPG and EDC-linked pathways (Bhandari et al., 2015; Thayil et al., 2020; DeCourten et al., 2020; Major et al., 2020). In addition, 17a-EE2 exposure resulted in multi- and transgenerational increases in larval mortality through the F1 generation in rainbow trout (Brown et al., 2009), the F2 generation in inland silversides (DeCourten et al., 2020), and the F3-F4 generations in medaka (Bhandari et al., 2015). Other organ systems were also targeted by 17a-EE2 exposure: medaka showed transgenerational (F3) sex-specific

**TABLE 9.3**
**Summary of the Reviewed Literature Indicating Outcomes of Multi- and Transgenerational Exposure to BPA, Hormones, or Steroids in Fish Models**

| BPA | | 17-EE2 |
|---|---|---|
| Abnormal cardiac development and transcriptome; larvae (F1) | Increased developmental malformations (F1, F2) | Decreased fertility (F1, F2) |
| Increased gonadal abnormalities (F1) | Disruption of ovarian transcriptome and methylome (F0-F2) | Altered HPG and EDC-linked transcriptome and methylome; larvae and adults (F1, F2) |
| Decreased ovarian gonadosomatic index (F2) | Insulin receptor downregulation; sperm and embryos (F0-F2) | Altered skeletal, cardiovascular, neurological, epigenetic, and metabolic methylome; larvae (F1, F2) |
| Abnormal sperm morphology & reduced sperm count (F2) | Impaired fertility (F0-F3) | Increased anxiety behaviors; adults (F1, F2) |
| Decreased BMI; adults (F2) | Altered transcription and site-specific methylation; embryos (F0-F3) | Altered osmoregulation transcriptome in gills; adults (F3) |
| Dysregulation and differential methylation of HPG axis genes; adults (F2) | Decreased embryonic survival (F0-F4) | Increased larval mortality (F1–F4) |
| Altered osmoregulatory gene expression in gills; adults (F3) | **Zearalenone** | **Trenbolone & Levonorgestrel** |
| Dysregulated lipid and steroidogenesis content; adults (F1, F2) | Impaired reproduction (F0) | Increased developmental abnormalities (F1; through F2 for Levo.) |
| Altered renal and hepatic transcriptomes; adults (F1, F2) | Minor changes in somatic measurements; adults (F1) | Decreased fecundity (F1) |
| **Testosterone** | **Letrozole** | Dysregulated HPG axis transcriptome; larvae (F1) |
| Dysregulates glucose homeostasis; adults (F1) | Skews sex ratio towards males (F1) | Altered epigenetic, metabolic, endocrine, and system morphology methylome; larvae (F2) |
| Dysregulates ovarian Methylome; adults (F1) | | |

effects on osmoregulation transcriptomics (Wang et al., 2019), while in inland silversides, genes involved in skeletal, cardiovascular, neurological, epigenetic, and metabolic function were differentially methylated across F1 and F2 generations (Major et al., 2020). Behaviorally, zebrafish and guppy both presented with increased F1 anxiety outcomes, which extended to the F2 in guppy alone (Volkova et al., 2015a, 2015b).

Other hormonally active drugs, including testosterone, the aromatase inhibitor/antiestrogen letrozole, the androgenic steroid trenbolone, and the progestogen levonorgestrel, also persist in wastewater effluent and subsequently in aquatic waterbodies. Testosterone can dysregulate F1 glucose homeostasis and methylome in zebrafish (Xu et al., 2014), and letrozole is known to skew F1 sex ratios towards males in medaka (Liao et al., 2014). Both trenbolone and levonorgestrel induced developmental abnormalities, decreased fecundity, and dysregulated HPG gene expression in F1 inland silversides (DeCourten et al., 2020; Major et al., 2020); for levonorgestrel, developmental effects persisted into the F2 generation (DeCourten et al., 2020). There is also evidence that both trenbolone and levonorgestrel can induce transgenerational changes to the F2 methylome linked to epigenetic, metabolic, endocrine, and system morphology genes, a small subset of which were consistent across generations (Major et al., 2020). Finally, the estrogenic mycotoxin zearalenone induced reproductive effects in F0 but only mild effects on somatic measurements in F1 zebrafish, suggesting no or minimal inheritance of reproductive effects for this xenoestrogen (Schwartz et al., 2013). As summarized in Table 9.3, hormones and steroids can induce extensive multi- and transgenerational effects on a range of organ systems across multiple fish species, with disturbing implications for contaminated ecosystems.

## Pesticides & Herbicides

Pesticides and herbicides are beneficial for maintaining agricultural yield and preventing human disease, but their prolific use has resulted in heavy contamination across environmental compartments and unintended health consequences for off-target organisms, including humans and aquatic species (Kumar et al., 2013; Syafrudin et al., 2021). Synthetic pyrethroids, such as permethrin and bifenthrin, are commonly used insecticides due to their comparatively limited toxicity in humans (Ramchandra et al., 2019). However, aquatic species are highly susceptible, with negative health outcomes including disruption of neurological, hepatic, respiratory, reproductive, cardiovascular, and endocrine systems (Ullah et al., 2019). This increased susceptibility is also evident across generations: permethrin exposure in zebrafish altered larval activity through the F3 generation (Blanc et al., 2020), while bifenthrin exposure in inland silversides altered F1 fecundity and dysregulated the transcriptome and methylome through the F2, disrupting the methylation patterns of genes involved in skeletal, neurological, cardiovascular, endocrine, epigenetic, and metabolic pathways (DeCourten et al., 2020; Major et al., 2020). Chlorpyrifos-oxon is a metabolite of the organophosphate chlorpyrifos, another frequently used insecticide that is often detected in environmental waterbodies (Syafrudin et al., 2021). Although fish species are generally considered less susceptible to chlorpyrifos (Barron and Woodburn, 1995), effects on activity, global methylation, and acetylcholinesterase concentration were found to persist through the F2 generation in exposed zebrafish (Schmitt et al., 2020).

The herbicide atrazine is another heavily used agent frequently detected in the environment (Graymore et al., 2001). The presence and degree of risk to aquatic species from atrazine exposure remains controversial, although there is support for primarily sublethal outcomes, including changes in activity, immune function, gonadal morphology, and sex hormone concentrations (Rohr and McCoy, 2010). In fish species, there is some evidence of comparable multi- and transgenerational effects of atrazine exposure, including altered size, decreased activity, and serotonergic dysregulation in F1 zebrafish (Wirbisky et al., 2016; Lamb et al., 2020) as well as impaired fertility, decreased sperm count and motility, and dysregulated HPG genes expression in F2 medaka (Cleary et al., 2019). Research in fish species to determine the extent of these effects across different physiological systems, as well as potential epigenetic mechanisms underlying their inheritance, will help to clarify the degree of risk to aquatic populations.

## PFASs

Per- and polyfluoroalkyl substances (PFASs), organofluorine compounds used in non-stick, stain- and water-repellant products, cosmetics, and firefighting foam, are highly persistent and bioaccumulative pollutants found nigh-ubiquitously in environmental compartments, humans, and wildlife (Fromme et al., 2009; Pelch et al., 2019; Meegoda et al., 2020). PFAS exposure is linked to a growing array of negative health outcomes across species, including metabolic, reproductive, thyroid, immune, and neurological dysfunction (Pelch et al., 2019; ATSDR, 2021; Blake and Fenton, 2020).

In fish species, there is evidence that certain PFASs have persistent generational effects on the thyroid, reproductive, and digestive systems. For instance, F-53B (6:2 chlorinated polyfluorinated ether sulfonate), a relatively novel alternative to PFOS (perfluorooctane sulfonate), can decrease survival, increase malformations, and alter HPT (hypothalamus-pituitary-thyroid) axis functioning through the F2 generation, while impairing fertility and altering HPG gene expression through the F1 (Shi et al., 2018; 2019). Similarly, marine medaka exposed to PFBS (perfluorobutane sulfonic acid), another PFOS alternative, display transgenerational effects on HPT axis function and multigenerational increases in mortality and developmental abnormalities (Chen et al., 2018a; 2019), alongside multigenerational changes in intestinal biomarkers and microbial community (Chen et al., 2018b). Other multigenerational changes in marine medaka induced by PFBS exposure include global hypomethylation; investigation of specific epimutations inherited across generations would be a useful future direction of study (Chen et al., 2019). Transgenerational effects on the HPT axis appear reasonably consistent across the two species and compounds studied, which prompts exploration across a variety of species and compounds to determine whether there is a conserved mechanism of inheritance for thyroid dysregulation.

## Radiation & Mutagens

Ionizing radiation (IR), or high-energy radiation that can remove electrons from the materials through which it passes, is naturally present at low levels throughout the environment. However, its use in anthropogenic industrial and medical activities, such as medical imaging or the generation of nuclear power, can result in exposures at levels of concern for humans and wildlife (Adam-Guillermin et al., 2018). IR is capable of directly reacting with and damaging DNA and other genomic material, as well as damaging them indirectly through oxidative processes (Singh et al., 2021). Mutations to the genetic sequence, if heritable, can result in health outcomes persisting across generations.

Multiple studies have assessed the multi- and transgenerational effects of IR exposure across a variety of fish models. Initial work in medaka determined that irradiation of the male germ line at later developmental stages resulted in the inheritance of a lethal recessive mutation through at least the F4 generation (Ishikawa and Hyodo-Taguchi, 1997), as well as delayed somatic mutations in the F1 generation alone (Shimada and Shima, 2004). Intriguingly, embryonic and larval irradiation of rainbow trout was able to induce a bystander effect, in which an irradiated organism, tissue, or group of cells can initiate certain effects of radiation exposure in a naïve counterpart via signaling proteins. These outcomes were detected through the F2 generation, albeit dependent on the exposure window, tissue, and sex of the irradiated parent (Smith et al., 2016). In zebrafish, irradiation of F0 adults resulted in increased inflammation, eye deformities, genomic instability, and associated transcriptomic and site-specific methylomic changes in F1 offspring, the latter of which persisted through the F3 generation (Kamstra et al., 2018). Embryonic and juvenile irradiation similarly resulted in F1 outcomes of genomic instability and altered transcriptomics, in addition to adverse reproductive outcomes, genome-wide effects on DNA methylation, and conserved histone post-translational modifications (PTMs; Lindeman et al., 2019). While effects of early life irradiation were considerably more severe than outcomes of adult irradiation, likely due to targeting critical windows of germ line development, they were not transgenerationally persistent.

Exposure to the mutagen/clastogen MMS (methyl methanesulfonate) resulted in decreased survival, impaired fertility, and skewed sex ratio in the F1 generation, with increased developmental abnormalities persisting in the F2 in zebrafish (Fassbender and Braunbeck, 2013). The potent mutagen ENU (*N*-ethyl-*N*-nitrosourea) did not appear to affect guppy to the same extent; the only F1 endpoints of exposure were delayed maturity in males and fewer offspring in females (Herdegen and Radwan, 2015). Factors such as species sensitivity, species reproductive strategy, and length of exposure may contribute to the difference in mutagen effect.

## Other Toxicants

### Nanomaterials

Nanomaterials (NMs), commonly defined as smaller than 100 nanometers in diameter in at least one dimension, are comprised of a variety of different materials, including metals, plastics, and carbon (Jeevanandam et al., 2018). Due to unique properties (conductive, magnetic, etc.) emergent from their size, NMs are intentionally engineered for biomedical, industrial, and commercial uses, but can also be produced by the breakdown of larger materials (Sharifi et al., 2012). NMs are near-ubiquitous across environmental compartments, yet their potential health effects on humans and aquatic species are minimally understood; some evidence suggests potential for neuro-, hepato-, cardio-, and gut toxicity through means of oxidative stress and inflammation pathways, and respiratory effects if inhaled (Sharifi et al., 2012; Jeevanandam et al., 2018).

In zebrafish, F0 exposure to multiwalled carbon nanotubes reduced survival in the following generation (Cheng et al., 2009), and zinc oxide nanoparticle exposure resulted in transgenerational (through the F2) impacts on survival and larval development. Intriguingly, F0 larval co-exposure with dissolved organic matter (DOM), a substance with antioxidant properties, mitigated zinc oxide-induced mortality in the F2 generation only, and developmental malformations in both F1 and F2 generations (Kteeba et al., 2018). The heritable mechanism of DOM mitigation and the way through which it interacts with zinc oxide nanoparticle toxicity across generations deserve further study.

Another mechanism of NM toxicity can be as a vehicle for other adsorbed toxicants, allowing their transport into and bioaccumulation in environmental compartments and organ systems. One co-exposure of zebrafish to polystyrene nanoplastics (PS-NPs) and ethylhexyl salicylate (EHS), a UV absorbant compound typically found in sunscreens, determined that PS-NPs resulted in minimal multigenerational toxicity independently. However, they intensified EHS-induced outcomes of decreased survival, increased developmental abnormalities, and upregulated antioxidant and thyroid activity in the F1 generation (Zhou et al., 2021).

### Cannabinoids

Tetrahydrocannabinol (THC) and cannabidiol (CBD) are the two major cannabinoids, compounds derived from the *Cannabis* plant. THC is inhaled or consumed for psychogenic properties, while the non-intoxicating CBD is frequently taken for anxiety and pain relief (Kelly and Nappe, 2021). However, only one CBD-based drug has been approved to date by the FDA, for the purpose of treating childhood epilepsy (Pandelides et al., 2020a). While the use of THC and CBD is widespread, resulting in frequent detection in waterbodies, there has been limited work on its potential generational outcomes in humans and much less in fish (How and Gamal El-Din, 2021). Exposure of F0 zebrafish to CBD resulted in F1 hypoactivity and minor sex-specific changes in liver inflammation genes (Carty et al., 2019; Pandelides et al., 2020a), while THC exposure caused minor hyperactivity, decreased anxiety-like behaviors, and decreased survival of aged fish (Carty et al., 2019; Pandelides et al., 2020b). Of most concern is the complete inability of THC-lineage F1 to reproduce, which has troubling implications for fish populations in contaminated waterbodies (Pandelides et al., 2020b).

### Selenium

Selenium, a ubiquitous trace element commonly released into the environment through anthropogenic activities, is essential for maintaining enzyme function and homeostasis in humans and fish species. At excess levels, selenium is considered a highly toxic aquatic contaminant, linked to developmental, neurological, and endocrine dysregulation in humans and fish (Janz et al., 2010; Vinceti et al., 2018). Most exposure is dietary; after release into waterbodies, selenium can undergo biotransformation to selenomethionine, capable of bioaccumulation and both trophic and maternal transfer (Janz et al., 2010). While such characteristics indicate the potential for multi- and transgenerational effects, there are few studies evaluating these risks in aquatic species.

In F1 offspring of exposed F0 zebrafish, there is currently evidence for developmental mortality and abnormalities, impaired metabolic and aerobic function, and reward learning deficits, likely linked to a hyperfunctioning dopaminergic system and upregulation of antioxidant activity in the brain (Thomas and Janz, 2014a, 2014b; Naderi et al., 2018). While F2 descendants did not show increased mortality or abnormalities, exploring metabolic and neurobehavioral endpoints beyond the F1 will help to determine whether they persist transgenerationally. Species sensitivity distribution graphing suggests zebrafish as the most sensitive fish species evaluated to date; therefore, the inclusion of other fish models should give a broader, more representative picture of selenium multigenerational toxicity and the potential fate of fish populations (Thomas and Janz, 2014b).

### 4TPP

4-*Tert*-pentylphenol (4TPP) is an alkylphenol used in the production of phenolic resins and as a germicide in certain cleaning supplies. Although not highly persistent or bioaccumulative, 4TPP has weak estrogenic activity and is considered toxic to aquatic organisms (Crane et al., 2008). There is some evidence in fathead minnow of multigenerational (F1) endocrine disruption due to 4TPP exposure, including decreased prominence of male secondary sexual characteristics, reduced gonad weight, and increased vitellogenin in males (Panter et al., 2010).

### Wastewater Effluent

Wastewater treatment plants (WWTPs) are generally designed to remove suspended solids, organic matter, certain contaminants, and bacteria from effluent, reducing the pollutant load released back into waterbodies (Bothfeld, 2021). However, some PPCPs and micropollutants are not efficiently removed, due to their relative novelty as contaminants of concern, while other toxicants are added to the mixture through the process of disinfection, such as nitrosamines and trihalomethanes (Watson et al., 2012; Kim and Farnazo, 2017). Most studies examining the multigenerational effects of wastewater effluent exposure utilize continuous exposure, which more closely mimics environmental conditions at the cost of specificity. One study, however, reported that the F1 generation of F0 fathead minnow exposed to effluent from a municipal WWTP in Greenville, SC, had altered male secondary sexual characteristics and dysregulated reproduction, suggesting that this effluent mixture had multigenerational endocrine disrupting activity (Sowers et al., 2009). Further characterization of effluent mixtures and examination of their potential multi- or transgenerational effects could reveal ecologically relevant outcomes not detected in traditional single-toxicant exposures.

## CONCLUSIONS

The increase in novel environmental contaminants, persistence of legacy contaminants, and influence of past generational exposures entail a concerning burden of toxicant exposure for both humans and aquatic animals that shows no signs of abating. In response, the NIEHS (National Institute of Environmental Health Sciences) has prioritized mechanistic studies to understand persistent multi- and transgenerational toxicant-induced disease (Haugen et al., 2015). While fish models, with relatively quick generation times and transparent embryos, are a useful tool in examining these outcomes; few studies have been performed altogether, and most in a zebrafish or medaka model to date. Most evidence has been collected for toxicants including BPA, 17a-EE2, and SSRIs, while a wide array of PPCPs, PFASs, nanomaterials, and mixtures deserve further study. This calls for an assortment of studies systematically targeting a variety of fish species, critical windows, and organ systems in order to better characterize these effects. Existing studies are promising, with intriguing patterns of effect; however, follow-up mechanistic and epigenetic work is needed to best understand study implications for fish and human health.

## LITERATURE CITED

Adam-Guillermin C, Hertal-Aas T, Oughton D, Blanchard L, Alonzo F, Armant O, Horemans N (2018). Radiosensitivity and transgenerational effects in non-human species. *Ann ICRP*. 47(3–4):327–341.

Akemann C, Meyer DN, Gurdziel K, Baker TR (2019). Developmental dioxin exposure alters the methylome of adult male zebrafish gonads. *Front Genet*. 9:719.

Akemann C, Meyer DN, Gurdziel K, Baker TR (2020). TCDD-induced multi- and transgenerational changes in the methylome of male zebrafish gonads. *Environ Epigenet*. 6(1): dvaa010.

Akhter A, Rahaman M, Suzuki R, Murono Y, Tokumoto T (2018). Next-generation and further transgenerational effects of bisphenol A on zebrafish reproductive tissues. *Heliyon*. 4:e00788.

Alfonso S, Blanc M, Joassard L, Keiter SH, Munschy C, Loizeau V, Bégout ML, Cousin X (2019). Examining multi- and transgenerational behavioral and molecular alterations resulting from parental exposure to an environmental PCB and PBDE mixture. *Aquat Toxicol*. 208:29–38.

Ankley GT and Johnson RD (2004). Small fish models for identifying and assessing the effects of endocrine-disrupting chemicals. *ILAR J*. 45(4):469–483.

Aris AZ, Shamsuddin AS, Praveena SM (2014). Occurrence of 17α-ethynylestradiol (EE2) in the environment and effect on exposed biota: a review. *Environ Int*. 69:104–119.

Arzuaga X, Elskus AA (2010). Polluted-site killifish (Fundulus heteroclitus) embryos are resistant to organic pollutant-mediated induction of CYP1A activity, reactive oxygen species, and heart deformities. *Environ Toxicol Chem*. 29(3):676–682.

[ATSDR] Agency for Toxic Substances and Disease Registry (2007). Toxicological profile for arsenic. (Draft for Public Comment). Atlanta GA: U.S. Department of Health and Human Services, Public Health Service.

[ATSDR] Agency for Toxic Substances and Disease Registry (2021). Toxicological profile for perfluoroalkyls. (Draft for Public Comment). Atlanta GA: U.S. Department of Health and Human Services. Public Health Service.

Baker TR, King-Heiden TC, Peterson RE, Heideman W (2014). Dioxin induction of transgenerational inheritance of disease in zebrafish. *Mol Cell Endocrinol*. 398:36–41.

Baker BB, Yee JS, Meyer DN, Yang D, Baker TR (2016). Histological and transcriptomic changes in male zebrafish testes due to early life exposure to low level 2,3,7,8-tetrachlorodibenzo-*p*-dioxin. *Zebrafish*. 13:413–423.

Barker DJ, Osmond C (1986). Infant mortality, childhood nutrition, and ischaemic heart disease in England and Wales. *Lancet*. 1:1077–1081.

Barron MG, Woodburn KB (1995). Ecotoxicology of chlorpyrifos. *Rev Environ Contam Toxicol*. 144:1–93.

Bernhoft RA (2012). Mercury toxicity and treatment: a review of the literature. *J Environ Public Health*. 2012:460508.

Bhandari RK, vom Saal FS, Tillitt DE (2015). Transgenerational effects from early developmental exposures to bisphenol A or 17α-ethinylestradiol in medaka, Oryzias latipes. *Sci Rep*. 5:9303.

Bhandari RK (2016). Medaka as a model for studying environmentally induced epigenetic transgenerational inheritance of phenotypes. *Environ Epigenet*. (2)1:1–9.

Blake BE, Fenton SE (2020). Early life exposure to per- and polyfluoroalkyl substances (PFAS) and latent health outcomes: a review including the placenta as a target tissue and possible driver of peri- and postnatal effects. *Toxicology*. 443:152565.

Blanc M, Cormier B, Hyötyläinen T, Krauss M, Scherbak N, Cousin X, Keiter SH (2020). Multi- and transgenerational effects following early-life exposure of zebrafish to permethrin and coumarin 47: impact on growth, fertility, behavior and lipid metabolism. *Ecotoxicol Environ Saf*. 205:111348.

Blanc M, Alfonso S, Bégout ML, Barrachina C, Hyötyläinen T, Keiter SH, Cousin X (2021). An environmentally relevant mixture of polychlorinated biphenyls (PCBs) and polybrominated diphenylethers (PBDEs) disrupts mitochondrial function, lipid metabolism and neurotransmission in the brain of exposed zebrafish and their unexposed F2 offspring. *Sci Total Environ.* 754:142097.

Bothfeld F (2021). Contaminants of Emerging Concern and Wastewater Treatment. For the Water Quality Program, Washington State Department of Ecology. Publication 20-10-006.

Brown KH, Schultz IR, Nagler JJ (2009). Lack of a heritable reproductive defect in the offspring of male rainbow trout exposed to the environmental estrogen 17alpha-ethynylestradiol. *Aquat Toxicol.* 2009;91(1):71–74.

Bruner-Tran KL, Ding T, Yeoman KB, Archibong A, Arosh JA, Osteen KG (2014). Developmental exposure of mice to dioxin promotes transgenerational testicular inflammation and an increased risk of preterm birth in unexposed mating partners. *PLoS One* 9:e105084.

Carty DR, Miller ZS, Thornton C, Pandelides Z, Kutchma ML, Willett KL (2019). Multigenerational consequences of early-life cannabinoid exposure in zebrafish. *Toxicol Appl Pharmacol.* 364:133–143.

Carvan MJ III, Kalluvila TA, Klingler RH, Larson JK, Pickens M, Mora-Zamorano FX, Connaughton VP, Sadler-Riggleman I, Beck D, Skinner MK (2017). Mercury-induced epigenetic transgenerational inheritance of abnormal neurobehavior is correlated with sperm epimutations in zebrafish. *PLoS One* 12(5):e0176155.

Chen L, Hu C, Tsui MMP, Wan T, Peterson DR, Shi Q, Lam PKS, Au DWT, Lam JCW, Zhou B (2018a). Multigenerational disruption of the thyroid endocrine system in marine medaka after a life-cycle exposure to perfluorobutanesulfonate. *Environ Sci Technol.* 52(7):4432–4439.

Chen L, Lam JCW, Hu C, Tsui MMP, Wang Q, Giesy JP, Lam PKS (2018b). Perfluorobutanesulfonate exposure causes durable and transgenerational dysbiosis of gut microbiota in marine medaka. *Environ Sci Technol Lett.* 5(12):731–738.

Chen L, Lam JCW, Hu C, Tsui MMP, Lam PKS, Zhou B (2019). Perfluorobutanesulfonate exposure skews sex ratio in fish and transgenerationally impairs reproduction. *Environ Sci Technol.* 53(14):8389–8397.

Chen Q, An J, Xie D, Gong S, Lian X, Liu Z, Shen Y, Li Y (2021). Suppression and recovery of reproductive behavior induced by early life exposure to mercury in zebrafish. *Comp Biochem Physiol C Toxicol Pharmacol.* 239:108876.

Cheng J, Chan CM, Veca LM, Poon WL, Chan PK, Qu L, Sun YP, Cheng SH (2009). Acute and long-term effects after single loading of functionalized multi-walled carbon nanotubes into zebrafish (Danio rerio). *Toxicol Appl Pharmacol.* 235(2): 216–225.

Cleary JA, Tillitt DE, Vom Saal FS, Nicks DK, Claunch RA, Bhandari RK (2019). Atrazine induced transgenerational reproductive effects in medaka (Oryzias latipes). *Environ Pollut.* 251:639–650.

Coimbra AM, Peixoto MJ, Coelho I, Lacerda R, Carvalho AP, Gesto M, Lyssimachou A, Lima D, Soares J, André A, Capitão A, Castro LF, Santos MM (2015). Chronic effects of clofibric acid in zebrafish (Danio rerio): a multigenerational study. *Aquat Toxicol.* 160:76–86.

Corrales J, Thornton C, White M, Willett KL (2014). Multigenerational effects of benzo[a]pyrene exposure on survival and developmental deformities in zebrafish larvae. *Aquat Toxicol.* 148:16–26.

Crane M, Fisk P, Maycock D, Watts C, Wildey R (contractors), Jordinson H (EA), Ridgway P ( HSE) (2008). Environmental risk evaluation report: 4-tert-pentylphenol (CAS no. 80-46-6). Published by: Environmental Agency, Almondsbury, Bristol.

De Liguoro M, Maraj S, Merlanti R (2019). Transgenerational toxicity of flumequine over four generations of Daphnia magna. *Ecotoxicol Environ Saf.* 169:814–821.

DeCourten BM, Forbes JP, Roark HK, Burns NP, Major KM, White JW, Li J, Mehinto AC, Connon RE, Brander SM (2020). Multigenerational and transgenerational effects of environmentally relevant concentrations of endocrine disruptors in an estuarine fish model. *Environ Sci Technol.* 54(21):13849–13860.

Dietrich KN, Douglas RM, Succop PA, Berger OG, Bornschein RL (2001). Early exposure to lead and juvenile delinquency. *Neurotoxicol Teratol.* 23(6):511–518.

Dolinoy DC, Huang D, Jirtle RL (2007). Maternal nutrient supplementation counteracts bisphenol A-induced DNA hypomethylation in early development. *Proc Natl Acad Sci USA.* 104:13056–13061.

Ebele AJ, Abou-Elwafa Abdallah M, Harrad S (2017). Pharmaceuticals and personal care products (PPCPs) in the freshwater aquatic environment. *Emerg Contam.* 3(1):1–16.

Fassbender C, Braunbeck T (2013). Reproductive and genotoxic effects in zebrafish after chronic exposure to methyl methanesulfonate in a multigeneration study. *Ecotoxicology.* 22(5):825–837.

Flint S, Markle T, Thompson S, Wallace E (2012). Bisphenol A exposure, effects, and policy: a wildlife perspective. *J Environ Manage.* 104:19–34.

Fostel JM, Burgoon L, Zwickl C, Lord P, Corton JC, Bushel PR, Cunningham M, Fan L, Edwards SW, Hester S, Stevens J, Tong W, Waters M, Yang CH, Tennant R (2007). Toward a checklist for exchange and interpretation of data from a toxicology study, *Toxicol Sci.* 99(1):26–34.

Fraz S, Lee AH, Pollard S, Srinivasan K, Vermani A, David E, Wilson JY (2019a). Paternal exposure to carbamazepine impacts zebrafish offspring reproduction over multiple generations. *Environ Sci Technol.* 53(21):12734–12743.

Fraz S, Lee AH, Pollard S, Srinivasan K, Vermani A, Wilson JY (2019b). Parental gemfibrozil exposure impacts zebrafish $F_1$ offspring, but not subsequent generations. *Aquat Toxicol.* 212:194–204.

Fromme H, Tittlemier SA, Völkel W, Wilhelm M, Twardella D (2009). Perfluorinated compounds--exposure assessment for the general population in Western countries. *Int J Hyg Environ Health.* 212(3):239–270.

Galic N, Baveco H, Hengeveld GM, Thorbek P, Bruns E, Van den Brink PJ (2012). Simulating population recovery of an aquatic isopod: effects of timing of stress and landscape structure. *Environ Poll.* 163:91–99.

Galus M, Rangarajan S, Lai A, Shaya L, Balshine S, Wilson JY (2014). Effects of chronic, parental pharmaceutical exposure on zebrafish (Danio rerio) offspring. *Aquat Toxicol.* 151:124–134.

Graymore M, Stagnitti F, Allinson G (2001). Impacts of atrazine in aquatic ecosystems. *Environ Int.* 26(7–8):483–495.

Haugen A, Schug T, Collman G, Heindel J (2015). Evolution of DOHaD: the impact of environmental health sciences. *J Dev Origin Health Dis.* 6(2):55–64.

Herbst AL, Ulfelder H, Poskanzer DC (1971). Adenocarcinoma of the vagina. Association of maternal stilbestrol therapy with tumor appearance in young women. *N Engl J Med.* 284(15):878–881.

Herdegen M, Radwan J (2015). Effect of induced mutations on sexually selected traits in the guppy, Poecilia reticulata. *Anim Behav.* 110:105–111.

Hernández AF, Tsatsakis AM (2017). Human exposure to chemical mixtures: challenges for the integration of toxicology with epidemiology data in risk assessment. *Food Chem Toxicol.* 103:188–193.

Hoover RN, Hyer M, Pfeiffer RM, Adam E, Bond B, Cheville AL, Colton T, Hartge P, Hatch EE, Herbst AL, Karlan BY, Kaufman R, Noller KL, Palmer JR, Robboy SJ, Saal RC, Strohsnitter W, Titus-Ernstoff L, Troisi R (2011). Adverse health outcomes in women exposed in utero to diethylstilbestrol. *N Engl J Med.* 365(14):1304–1314.

How ZT, Gamal El-Din M (2021). A critical review on the detection, occurrence, fate, toxicity, and removal of cannabinoids in the water system and the environment. *Environ Pollut.* 268(Pt A):115642.

Ishikawa Y, Hyodo-Taguchi Y (1997). Heritable malformations in the progeny of the male medaka (Oryzias latipes) irradiated with X-rays. *Mutat Res.* 389(2–3):149–155.

Jannetto PJ (2011). Heavy Metals (Lead, Arsenic, and Mercury). [Powerpoint Transcript]. Clinical Chemistry Trainee Council: Pearls of Laboratory Medicine

Janz DM, DeForest DK, Brooks ML, Chapman PM, Gilron G, Hoff D, Hopkins WD, McIntyre DO, Mebane CA, Palace VP, Skorupa JP, Wayland M, (2010). Selenium toxicity to aquatic organisms. In PM Chapman, WJ Adams, M Brooks, CG Delos, SN Luoma, WA Maher, HM Ohlendorf, TS Presser, and P Shaw (Eds.): *Ecological Assessment of Selenium in the Aquatic Environment.* Boca Raton, FL, CRCPress, pp. 141–231.

Jasperse L, Levin M, Rogers K, Perkins C, Bosker T, Griffitt RJ, Sepúlveda MS, De Guise S (2019). Transgenerational effects of polycyclic aromatic hydrocarbon exposure on sheepshead minnows (Cyprinodon variegatus). *Environ Toxicol Chem.* 38(3):638–649.

Jeevanandam J, Barhoum A, Chan YS, Dufresne A, Danquah MK (2018). Review on nanoparticles and nanostructured materials: history, sources, toxicity and regulations. *Beilstein J Nanotechnol.* 9:1050–1074.

Kalichak F, de Alcantara Barcellos HH, Idalencio R, Koakoski G, Soares SM, Pompermaier A, Rossini M, Barcellos LJG (2019). Persistent and transgenerational effects of risperidone in zebrafish. *Environ Sci Pollut Res Int.* 26(25):26293–26303.

Kamstra JH, Hurem S, Martin LM, Lindeman LC, Legler J, Oughton D, Salbu B, Brede DA, Lyche JL, Aleström P (2018). Ionizing radiation induces transgenerational effects of DNA methylation in zebrafish. *Sci Rep.* 8(1):15373.

Kelly BF, Nappe TM (2021). Cannabinoid toxicity. In: *StatPearls* [Internet]. Treasure Island, FL: StatPearls Publishing.

Kidd KA, Blanchfield PJ, Mills KH, Palace VP, Evans RE, Lazorchak JM, Flick RW (2007). Collapse of a fish population after exposure to a synthetic estrogen. *Proc Natl Acad Sci USA*, 104(21):8897–8901.

Kim Y, Farnazo DM (2017). Toxicity characteristics of sewage treatment effluents and potential contribution of micropollutant residuals. *J Ecology Environ* 41:39.

King Heiden TC, Spitsbergen J, Heideman W, Peterson RE (2009). Persistent adverse effects on health and reproduction caused by exposure of zebrafish to 2,3,7,8-tetrachlorodibenzo-p-dioxin during early development and gonad differentiation. *Toxicol Sci.* 109:75–87.

Kioumourtzoglou M, Coull BA, O'Reilly ÉJ, Ascherio A, Weisskopf MG (2018). Association of exposure to diethylstilbestrol during pregnancy with multigenerational neurodevelopmental deficits. *JAMA Pediatr.* 172(7):670–677.

Knecht AL, Truong L, Marvel SW, Reif DM, Garcia A, Lu C, Simonich MT, Teeguarden JG, Tanguay RL (2017). Transgenerational inheritance of neurobehavioral and physiological deficits from developmental exposure to benzo[a] pyrene in zebrafish. *Toxicol Appl Pharmacol.* 329:148–157.

Konieczna A, Rutkowska A, Rachoń D (2015). Health risk of exposure to Bisphenol A (BPA). *Rocz Panstw Zakl Hig.* 66(1):5–11.

Kteeba SM, El-Ghobashy AE, El-Adawi HI, El-Rayis OA, Sreevidya VS, Guo L, Svoboda KR (2018). Exposure to ZnO nanoparticles alters neuronal and vascular development in zebrafish: Acute and transgenerational effects mitigated with dissolved organic matter. *Environ Pollut.* 242(Pt A):433–448.

Kumar S, Sharma AK, Rawat S, Jain D, Ghosh S (2013). Use of pesticides in agriculture and livestock animals and its impact on environment of India. *Asian J Environ Sci.* 8:51–57.

Lacal I, Ventura R (2018). Epigenetic inheritance: concepts, mechanisms and perspectives. *Front Mol Neurosci.* 11:292.

Lamb SD, Chia JHZ, Johnson SL (2020). Paternal exposure to a common herbicide alters the behavior and serotonergic system of zebrafish offspring. *PLoS One* 15(4):e0228357.

Liao PH, Chu SH, Tu TY, Wang XH, Lin AYC, Chen PJ (2014). Persistent endocrine disruption effects in medaka fish with early life-stage exposure to a triazole-containing aromatase inhibitor (letrozole). *J Hazard Mater.* 277:141–149.

Lin CY, Chiang CY, Tsai HJ (2016). Zebrafish and Medaka: new model organisms for modern biomedical research. *J Biomed Sci.* 23:19

Lindberg CD, Jayasundara N, Kozal JS, Leuthner TC, Di Giulio RT (2017). Resistance to polycyclic aromatic hydrocarbon toxicity and associated bioenergetic consequences in a population of Fundulus heteroclitus. *Ecotoxicology.* 26(3):435–448.

Lindeman LC, Kamstra JH, Ballangby J, Hurem S, Martín LM, Brede DA, Teien HC, Oughton DH, Salbu B, Lyche JL, Aleström P (2019). Gamma radiation induces locus specific changes to histone modification enrichment in zebrafish and Atlantic salmon. *PLoS One.* 14(2):e0212123.

Lombó M, Fernández-Díez C, González-Rojo S, Navarro C, Robles V, Paz Herráez M (2015). Transgenerational inheritance of heart disorders caused by paternal bisphenol A exposure. *Environ Pollut.* 206:667–678.

Luu I, Ikert H, Craig PM (2021). Chronic exposure to anthropogenic and climate related stressors alters transcriptional responses in the liver of zebrafish (Danio rerio) across multiple generations. *Comp Biochem Physiol C Toxicol Pharmacol.* 240:108918.

Major KM, DeCourten BM, Li J, Britton M, Settles ML, Mehinto AC, Connon RE, Brander SM (2020). Early life exposure to environmentally relevant levels of endocrine disruptors drive multigenerational and transgenerational epigenetic changes in a fish model. *Front Marine Sci.* 7(471):1–17.

Manikkam M, Tracey R, Guerrero-Bosagna C, Skinner MK (2013). Plastics derived endocrine disruptors (BPA, DEHP and DBP) induce epigenetic transgenerational inheritance of obesity, reproductive disease and sperm epimutations. *PLoS One.* 8:1–18.

Matsui H, Gavinio R, Takahashi R (2012). Medaka fish Parkinson's disease model. *Exp Neurobiol.* 21(3):94–100.

Matta MB, Linse J, Cairncross C, Francendese L, Kocan RM (2001). Reproductive and transgenerational effects of methylmercury or Aroclor 1268 on Fundulus heteroclitus. *Environ Toxicol Chem.* 20(2):327–335.

Meegoda JN, Kewalramani JA, Li B, Marsh RW (2020). A review of the applications, environmental release, and remediation technologies of per- and polyfluoroalkyl substances. *Int J Environ Res Public Health.* 17(21):8117.

Meyer DN, Baker BB, Baker TR (2018). Ancestral TCDD exposure induces multigenerational histologic and transcriptomic alterations in gonads of male zebrafish. *Toxicol Sci.* 164:603–612.

Meyer DN, Crofts EJ, Akemann C, Gurdziel K, Farr R, Baker BB, Weber D, Baker TR (2020). Developmental exposure to $Pb^{2+}$ induces transgenerational changes to zebrafish brain transcriptome. *Chemosphere*. 244:125527.

Michałowicz J (2014). Bisphenol A--sources, toxicity and biotransformation. *Environ Toxicol Pharmacol.* 37(2):738–758.

Nacci DE, Champlin D, Jayaraman S (2010). Adaptation of the estuarine fish fundulus heteroclitus (Atlantic Killifish) to Polychlorinated Biphenyls (PCBs). *Estuaries Coast.* 33:853–864.

Naderi M, Ferrari MCO, Chivers DP, Niyogi S (2018). Maternal exposure to dietary selenium causes dopaminergic hyperfunction and cognitive impairment in zebrafish offspring. *Environ Sci Technol.* 52(22):13574–13583.

Needleman HL, Schell A, Bellinger D, Leviton A, Allred EN (1990). The long-term effects of exposure to low doses of lead in childhood. An 11-year follow-up report. *N Engl J Med.* 322:83–88.

Olff H, Alonso D, Berg MP, Eriksson BK, Loreau M, Piersma T, Rooney N (2009). Parallel ecological networks in ecosystems. *Philos Trans R Soc Lond B Biol Sci.* 364(1524):1755–1779.

Olsvik PA, Williams TD, Tung HS, Mirbahai L, Sanden M, Skjaerven KH, Ellingsen S (2014). Impacts of TCDD and MeHg on DNA methylation in zebrafish (Danio rerio) across two generations. *Comp Biochem Physiol Part C Toxicol Pharmacol.* 165:17–27.

Pandelides Z, Thornton C, Faruque AS, Whitehead AP, Willett KL, Ashpole NM (2020a). Developmental exposure to cannabidiol (CBD) alters longevity and health span of zebrafish (Danio rerio). *Geroscience.* 42(2):785–800.

Pandelides Z, Thornton C, Lovitt KG, Faruque AS, Whitehead AP, Willett KL, Ashpole NM (2020b). Developmental exposure to $\Delta^9$-tetrahydrocannabinol (THC) causes biphasic effects on longevity, inflammation, and reproduction in aged zebrafish (Danio rerio). *Geroscience.* 42(3):923–936.

Panter GH, Hutchinson TH, Hurd KS, Bamforth J, Stanley RD, Wheeler JR, Tyler CR (2010). Effects of a weak oestrogenic active chemical (4-tert-pentylphenol) on pair-breeding and F1 development in the fathead minnow (Pimephales promelas). *Aquat Toxicol.* 97(4):314–323.

Pelch KE, Reade A, Wolffe TAM, Kwiatkowski CF (2019). PFAS health effects database: Protocol for a systematic evidence map. *Environ Int.* 130:104851

Phillips J, Haimbaugh A, Baker TR (2021). Occurrence of pharmaceuticals and emerging contaminants in Detroit waterbodies and health effects on fish living in these waters. Manuscript in submission.

Preker AS, Adeyi OO, Lapetra MG, Simon D-C, Keuffel E (2017). Health care expenditures associated with pollution: exploratory methods and findings. *Ann Glob Health.* 82(5):711–721.

Ramchandra AM, Chacko B, Victor PJ (2019). Pyrethroid poisoning. *Indian J Crit Care Med*, 23(Suppl 4):S267–S271.

Richmond EK, Grace MR, Kelly JJ, Reisinger AJ, Rosi EJ, Walters DM (2017). Pharmaceuticals and personal care products (PPCPs) are ecological disrupting compounds (EcoDC). *Elementa: Sci Anthrop.* 5(66):1–8.

Rogers JM, Lau C, Ellis-Hutchings RG (2018). 5.08- Epigenetics and the developmental origins of health and disease, In CA McQueen (Eds): *Comprehensive Toxicology (Third Edition)*, Elsevier, 118–136.

Rohr JR, McCoy KA (2010). A qualitative meta-analysis reveals consistent effects of atrazine on freshwater fish and amphibians. *Environ Health Perspect.* 118(1):20–32.

Ross PS, Birnbaum LS (2003). Integrated human and ecological risk assessment: a case study of Persistent Organic Pollutants (POPs) in humans and wildlife. *Human Ecol Risk Assess: An Int J.* 9(1):303–324.

Sadoul B, Birceanu O, Aluru N, Thomas JK, Vijayan MM (2017). Bisphenol A in eggs causes development-specific liver molecular reprogramming in two generations of rainbow trout. *Sci Rep.* 7(1):14131.

Santangeli S, Consales C, Pacchierotti F, Habibi HR, Carnevali O (2019). Transgenerational effects of BPA on female reproduction. *Sci Total Environ.* 685:1294–1305.

Schmitt C, Peterson E, Willis A, Kumar N, McManus M, Subbiah S, Crago J (2020). Transgenerational effects of developmental exposure to chlorpyrifos-oxon in zebrafish (DANIO RERIO). *Toxicol Appl Pharmacol.* 408:115275.

Schwartz P, Bucheli TD, Wettstein FE, Burkhardt-Holm P (2013). Life-cycle exposure to the estrogenic mycotoxin zearalenone affects zebrafish (Danio rerio) development and reproduction. *Environ Toxicol.* 28(5):276–289.

Schwindt AR, Winkelman DL, Keteles K, Murphy M, Vajda AM, Frid C (2014). An environmental oestrogen disrupts fish population dynamics through direct and transgenerational effects on survival and fecundity. *J Appl Ecol.* 51(3):582–591.

Seemann F, Peterson DR, Witten PE, Guo BS, Shanthanagouda AH, Ye RR, Zhang G, Au DWT (2015). Insight into the transgenerational effect of benzo[a]pyrene on bone formation in a teleost fish (Oryzias latipes). *Comp Biochem Physiol C Toxicol Pharmacol.* 178:60–67.

Seemann F, Jeong CB, Zhang G, Wan MT, Guo B, Peterson DR, Lee JS, Au DW (2017). Ancestral benzo[a]pyrene exposure affects bone integrity in F3 adult fish (Oryzias latipes). *Aquat Toxicol.* 183:127–134.

Shalat SL, Walker DB, Finnell RH (1996). Role of arsenic as a reproductive toxin with particular attention to neural tube defects. *J Toxicol Env Health.* 48(3):253–272.

Sharifi S, Behzadi S, Laurent S, Forrest ML, Stroeve P, Mahmoudi M (2012). Toxicity of nanomaterials. *Chem Soc Rev.* 41(6):2323–2343.

Shi G, Guo H, Sheng N, Cui Q, Pan Y, Wang J, Guo Y, Dai J (2018). Two-generational reproductive toxicity assessment of 6:2 chlorinated polyfluorinated ether sulfonate (F-53B, a novel alternative to perfluorooctane sulfonate) in zebrafish. *Environ Pollut.* 243(Pt B):1517–1527.

Shi G, Wang J, Guo H, Sheng N, Cui Q, Pan Y, Guo Y, Sun Y, Dai J (2019). Parental exposure to 6:2 chlorinated polyfluorinated ether sulfonate (F-53B) induced transgenerational thyroid hormone disruption in zebrafish. *Sci Total Environ.* 665:855–863.

Shimada A, Shima A (2004). Transgenerational genomic instability as revealed by a somatic mutation assay using the medaka fish. *Mutat Res.* 552(1–2):119–124.

Singh VK, Swartz HM, Seed TM (2021). Radiation medical countermeasures and use of epr biodosimetry to facilitate effectiveness of applied clinical procedures. *Appl Magn Reson.* 53:289–303.

Smith R, Seymour CB, Moccia RD, Mothersill C (2016). Irradiation of rainbow trout at early life stages results in trans-generational effects including the induction of a bystander effect in non-irradiated fish. *Environ Res.* 145:26–38.

Sowers AD, Gaworecki KM, Mills MA, Roberts AP, Klaine SJ (2009). Developmental effects of a municipal wastewater effluent on two generations of the fathead minnow, Pimephales promelas. *Aquat Toxicol.* 95(3):173–181.

Stenzel A, Wirt H, Patten A, Theodore B, King-Heiden T (2019). Larval exposure to environmentally relevant concentrations of triclosan impairs metamorphosis and reproductive fitness in zebrafish. *Reprod Toxicol*. 87:79–86.

Syafrudin M, Kristanti RA, Yuniarto A, Hadibarata T, Rhee J, Al-Onazi WA, Algarni TS, Almarri AH, Al-Mohaimeed AM (2021). Pesticides in drinking water-a review. *Int J Environ Res Publ Health*. 18(2):468.

Ternes TA, Stumpf M, Mueller J, Haberer K, Wilken RD, Servos M (1999). Behavior and occurrence of estrogens in municipal sewage treatment plants--I. Investigations in Germany, Canada and Brazil. *Sci Total Environ*. 225(1–2):81–90.

Thayil AJ, Wang X, Bhandari P, Vom Saal FS, Tillitt DE, Bhandari RK (2020). Bisphenol A and 17α-ethinylestradiol-induced transgenerational gene expression differences in the brain-pituitary-testis axis of medaka, Oryzias latipes†. *Biol Reprod*. 103(6):1324–1335.

Thomas JK, Janz DM (2014a). In ovo exposure to selenomethionine via maternal transfer increases developmental toxicities and impairs swim performance in F1 generation zebrafish (Danio rerio). *Aquat Toxicol*. 152:20–29.

Thomas JK, Janz DM (2014b). Developmental and persistent toxicities of maternally deposited selenomethionine in zebrafish (Danio rerio). *Environ Sci Technol*. 49(16):10182–10189.

Thomas JK, Birceanu O, Sadoul B, Vijayan MM (2018). Bisphenol A in eggs impairs the long-term stress performance of rainbow trout in two generations. *Environ Sci Technol*. 52(14):7951–7961.

Thompson WA, Vijayan MM (2021). Zygotic exposure to venlafaxine disrupts cortisol stress axis activity in multiple generations of zebrafish. *Environ Pollut*. 274:116535.

Titus-Ernstoff L, Troisi R, Hatch EE, Wise LA, Palmer J, Hyer M, Kaufman R, Adam E, Strohsnitter W, Noller K, Herbst AL, Gibson-Chambers J, Hartge P, Hoover RN (2006). Menstrual and reproductive characteristics of women whose mothers were exposed in utero to diethylstilbestrol (DES). *Int J Epidemiol*. 35(4):862–868.

Ullah S, Li Z, Zuberi A, Zain Ul Arifeen M, Faran Ashraf Baig MM (2019). Biomarkers of pyrethroid toxicity in fish. *Environ Chem Lett*. 17:945–973.

Vallero DA (2005). Persistent, bioaccumulative, and toxic pollutants. In: AccessScience. McGraw-Hill Education, pp. 1–8.

Valles S, Hernández-Sánchez J, Dipp VR, Huerta-González D, Olivares-Bañuelos TN, González-Fraga J, Bardullas U (2020). Exposure to low doses of inorganic arsenic induces transgenerational changes on behavioral and epigenetic markers in zebrafish (Danio rerio). *Toxicol Appl Pharmacol*. 396:115002.

Van den Brink PJ (2013). Assessing aquatic population and community-level risks of pesticides. *Environ Toxicol Chem*. 32:972–973.

Vera-Chang MN, St-Jacques AD, Gagné R, Martyniuk CJ, Yauk CL, Moon TW, Trudeau VL (2018). Transgenerational hypocortisolism and behavioral disruption are induced by the antidepressant fluoxetine in male zebrafish *Danio rerio*. *Proc Natl Acad Sci USA*. 115(52):E12435–E12442.

Vera-Chang MN, Moon TW, Trudeau VL (2019a). Ancestral fluoxetine exposure sensitizes zebrafish to venlafaxine-induced reductions in cortisol and spawning. *Endocrinology*. 160(9):2137–2142.

Vera-Chang MN, Moon TW, Trudeau VL. (2019b). Cortisol disruption and transgenerational alteration in the expression of stress-related genes in zebrafish larvae following fluoxetine exposure. *Toxicol Appl Pharmacol*. 382:114742.

Vignet C, Joassard L, Lyphout L, Guionnet T, Goubeau M, Le Menach K, Brion F, Kah O, Chung BC, Budzinski H, Bégout ML, Cousin X (2015). Exposures of zebrafish through diet to three environmentally relevant mixtures of PAHs produce behavioral disruptions in unexposed F1 and F2 descendant. *Environ Sci Pollut Res Int*. 22(21):16371–16383.

Vinceti M, Filippini T, Wise LA (2018). Environmental selenium and human health: an update. *Curr Environ Health Rep*. 5(4):464–485.

Volkova K, Reyhanian Caspillo N, Porseryd T, Hallgren S, Dinnétz P, Porsch-Hällström I (2015a). Developmental exposure of zebrafish (Danio rerio) to 17α-ethinylestradiol affects non-reproductive behavior and fertility as adults and increases anxiety in unexposed progeny. *Horm Behav*. 73:30–38.

Volkova K, Reyhanian Caspillo N, Porseryd T, Hallgren S, Dinnétz P, Olsen H, Porsch-Hällström I (2015b). Transgenerational effects of 17α-ethinyl estradiol on anxiety behavior in the guppy, Poecilia reticulata. *Gen Comp Endocrinol*. 223:66–72.

Wang T, Pehrsson EC, Purushotham D, Li D, Zhuo X, Zhang B, Lawson HA, Province MA, Krapp C, Lan Y, Coarfa C, Katz TA, Tang WY, Wang Z, Biswal S, Rajagopalan S, Colacino JA, Tsai ZT, Sartor MA, Neier K, Dolinoy DC, Pinto J, Hamanaka RB, Mutlu GM, Patisaul HB, Aylor DL, Crawford GE, Wiltshire T, Chadwick LH, Duncan CG, Garton AE, McAllister KA; TaRGET II Consortium, Bartolomei MS, Walker CL, Tyson FL (2018). The NIEHS TaRGET II Consortium and environmental epigenomics. *Nat Biotechnol*. 36(3):225–227.

Wang X, Hill D, Tillitt DE, Bhandari RK (2019). Bisphenol A and 17α-ethinylestradiol-induced transgenerational differences in expression of osmoregulatory genes in the gill of medaka (Oryzias latipes). *Aquat Toxicol*. 211:227–234.

Wani AL, Ara A, Usmani JA (2015). Lead toxicity: a review. *Interdiscip Toxicol*. 8(2):55–64.

Watson K, Shaw G, Leusch FDL, Knight NL (2012). Chlorine disinfection by-products in wastewater effluent: bioassay-based assessment of toxicological impact. *Water Res*. 46(18):6069–6083.

Weis J (2002). Tolerance to environmental contaminants in the mummichog Fundulus heteroclitus. *Human Ecol Risk Assess*. 8(5):933–953.

Welshons WV, Thayer KA, Judy BM, Taylor JA, Curran EM, vom Saal FS (2003). Large effects from small exposures. I. Mechanisms for endocrine-disrupting chemicals with estrogenic activity. *Environ Health Perspect*. 111:994–1006.

White LD, Cory-Slechta DA, Gilbert ME, Tiffany-Castiglioni E, Zawia NH, Virgolini M, Rossi-George A, Lasley SM, Qian YC, Basha MR (2007). New and evolving concepts in the neurotoxicology of lead. *Toxicol Appl Pharmacol*. 225(1):1–27.

[WHO] World Health Organization Chemical Safety and Health Unit. (2016, May 23). The public health impact of chemicals: knowns and unknowns. WHO (Ed.), pp. 1–13. Retrieved from: https://www.who.int/publications/i/item/WHO-FWC-PHE-EPE-16.01-eng.

[WHO] World Health Organization Global Chemicals and Health Network. (2018, January 1). The public health impact of chemicals: knowns and unknowns: data addendum for 2016. WHO (Ed.), pp. 1–2. Retrieved from: https://www.who.int/publications/i/item/WHO-CED-PHE-EPE-18.09.

[WHO] World Health Organization Chemical Safety and Health Unit, Environment, Climate Change and Health. (2021, July 6). The public health impact of chemicals: knowns and unknowns: data addendum for 2019. WHO (Ed.), pp. 1–4. Retrieved from: https://www.who.int/publications/i/item/WHO-HEP-ECH-EHD-21.01.

Wilson CA, High SK, McCluskey BM, Amores A, Yan YL, Titus TA, Anderson JL, Batzel P, Carvan MJ 3rd, Schartl M, Postlethwait JH (2014). Wild sex in zebrafish: loss of the natural sex determinant in domesticated strains. *Genetics.* 198(3):1291–1308.

Wirbisky SE, Weber GJ, Sepúlveda MS, Lin TL, Jannasch AS, Freeman JL (2016). An embryonic atrazine exposure results in reproductive dysfunction in adult zebrafish and morphological alterations in their offspring. *Sci Rep.* 6:21337.

Xie D, Chen Q, Gong S, An J, Li Y, Lian X, Liu Z, Shen Y, Giesy JP (2020). Exposure of zebrafish to environmentally relevant concentrations of mercury during early life stages impairs subsequent reproduction in adults but can be recovered in offspring. *Aquat Toxicol.* 229:105655.

Xin F, Susiarjo M, Bartolomei MS (2015). Multigenerational and transgenerational effects of endocrine disrupting chemicals: A role for altered epigenetic regulation? *Semin Cell Dev Biol.* 43:66–75.

Xu N, Chua AK, Jiang H, Liu NA, Goodarzi MO (2014). Early embryonic androgen exposure induces transgenerational epigenetic and metabolic changes. *Mol Endocrinol.* 28(8):1329–1336.

Xu X, Weber D, Martin A, Lone D (2016a). Trans-generational transmission of neurobehavioral impairments produced by developmental methylmercury exposure in zebrafish (Danio rerio). *Neurotoxicol Teratol.* 53:19–23.

Xu X, Weber D, Burge R, VanAmberg K (2016b). Neurobehavioral impairments produced by developmental lead exposure persisted for generations in zebrafish (Danio rerio). *Neurotoxicology.* 52:176–185.

Zhang X, Zhong HQ, Chu ZW, Zuo X, Wang L, Ren XL, Ma H, Du RY, Ju JJ, Ye XL, Huang CP, Zhu JH, Wu HM (2020). Arsenic induces transgenerational behavior disorders in Caenorhabditis elegans and its underlying mechanisms. *Chemosphere.* 252:126510.

Zhou D (2018). Ecotoxicity of bisphenol S to Caenorhabditis elegans by prolonged exposure in comparison with bisphenol A. *Environmental Toxicology and Chemistry.* 37(10):2560–2565.

Zhou R, Lu G, Yan Z, Jiang R, Sun Y, Zhang P (2021). Interactive transgenerational effects of polystyrene nanoplastics and ethylhexyl salicylate on zebrafish. *Environ Sci: Nano.* 8:146–159.

Zhu L, Liu Y, Xue X, Yuan C, Wang Z (2021). BPA's transgenerational disturbance to transcription of ovarian steroidogenic genes in rare minnow Gobiocypris rarus via DNA and histone methylation. *Sci Total Environ.* 762:143055.

## FIGURE CITATION LIST

1. Akhter A, Rahaman M, Suzuki R, Murono Y, Tokumoto T (2018). Next-generation and further transgenerational effects of bisphenol A on zebrafish reproductive tissues. *Heliyon.* 4:e00788.
2. Akemann C, Meyer DN, Gurdziel K, Baker TR (2020). TCDD-induced multi- and transgenerational changes in the methylome of male zebrafish gonads. *Environ Epigenet.* 6(1): dvaa010.
3. Alfonso S, Blanc M, Joassard L, Keiter SH, Munschy C, Loizeau V, Bégout ML, Cousin X (2019). Examining multi- and transgenerational behavioral and molecular alterations resulting from parental exposure to an environmental PCB and PBDE mixture. *Aquat Toxicol.* 208:29–38.
4. Baker TR, King-Heiden TC, Peterson RE, Heideman W (2014). Dioxin induction of transgenerational inheritance of disease in zebrafish. *Mol Cell Endocrinol.* 398:36–41.
5. Bhandari RK, vom Saal FS, Tillitt DE (2015). Transgenerational effects from early developmental exposures to bisphenol A or 17α-ethinylestradiol in medaka, Oryzias latipes. *Sci Rep.* 5:9303.
6. Blanc M, Cormier B, Hyötyläinen T, Krauss M, Scherbak N, Cousin X, Keiter SH (2020). Multi- and transgenerational effects following early-life exposure of zebrafish to permethrin and coumarin 47: impact on growth, fertility, behavior and lipid metabolism. *Ecotoxicol Environ Saf.* 205:111348.
7. Blanc M, Alfonso S, Bégout ML, Barrachina C, Hyötyläinen T, Keiter SH, Cousin X (2021). An environmentally relevant mixture of polychlorinated biphenyls (PCBs) and polybrominated diphenylethers (PBDEs) disrupts mitochondrial function, lipid metabolism and neurotransmission in the brain of exposed zebrafish and their unexposed F2 offspring. *Sci Total Environ.* 754:142097.
8. Brown KH, Schultz IR, Nagler JJ (2009). Lack of a heritable reproductive defect in the offspring of male rainbow trout exposed to the environmental estrogen 17alpha-ethynylestradiol. *Aquat Toxicol.* 91(1):71–74.
9. Carty DR, Miller ZS, Thornton C, Pandelides Z, Kutchma ML, Willett KL (2019). Multigenerational consequences of early-life cannabinoid exposure in zebrafish. *Toxicol Appl Pharmacol.* 364:133–143.
10. Carvan MJ III, Kalluvila TA, Klingler RH, Larson JK, Pickens M, Mora-Zamorano FX, Connaughton VP, Sadler-Riggleman I, Beck D, Skinner MK (2017). Mercury-induced epigenetic transgenerational inheritance of abnormal neurobehavior is correlated with sperm epimutations in zebrafish. *PLoS One* 12(5):e0176155.
11. Chen L, Hu C, Tsui MMP, Wan T, Peterson DR, Shi Q, Lam PKS, Au DWT, Lam JCW, Zhou B (2018a). Multigenerational disruption of the thyroid endocrine system in marine medaka after a life-cycle exposure to perfluorobutanesulfonate. *Environ Sci Technol.* 52(7):4432–4439.
12. Chen L, Lam JCW, Hu C, Tsui MMP, Wang Q, Giesy JP, Lam PKS (2018b). Perfluorobutanesulfonate exposure causes durable and transgenerational dysbiosis of gut microbiota in marine medaka. *Environ Sci Technol Lett.* 5(12):731–738.
13. Chen L, Lam JCW, Hu C, Tsui MMP, Lam PKS, Zhou B (2019). Perfluorobutanesulfonate exposure skews sex ratio in fish and transgenerationally impairs reproduction. *Environ Sci Technol.* 53(14):8389–8397.
14. Chen Q, An J, Xie D, Gong S, Lian X, Liu Z, Shen Y, Li Y (2021). Suppression and recovery of reproductive behavior induced by early life exposure to mercury in zebrafish. *Comp Biochem Physiol C Toxicol Pharmacol.* 239:108876.
15. Cheng J, Chan CM, Veca LM, Poon WL, Chan PK, Qu L, Sun YP, Cheng SH (2009). Acute and long-term effects after single loading of functionalized multi-walled carbon nanotubes into zebrafish (Danio rerio). *Toxicol Appl Pharmacol.* 235(2):216–225.
16. Cleary JA, Tillitt DE, Vom Saal FS, Nicks DK, Claunch RA, Bhandari RK (2019). Atrazine induced transgenerational reproductive effects in medaka (Oryzias latipes). *Environ Pollut.* 251:639–650.
17. Coimbra AM, Peixoto MJ, Coelho I, Lacerda R, Carvalho AP, Gesto M, Lyssimachou A, Lima D, Soares J, André A, Capitão A, Castro LF, Santos MM (2015). Chronic effects of clofibric acid in zebrafish (Danio rerio): a multigenerational study. *Aquat Toxicol.* 160:76–86.

18. Corrales J, Thornton C, White M, Willett KL (2014). Multigenerational effects of benzo[a]pyrene exposure on survival and developmental deformities in zebrafish larvae. *Aquat Toxicol.* 148:16–26.
19. DeCourten BM, Forbes JP, Roark HK, Burns NP, Major KM, White JW, Li J, Mehinto AC, Connon RE, Brander SM (2020). Multigenerational and transgenerational effects of environmentally relevant concentrations of endocrine disruptors in an estuarine fish model. *Environ Sci Technol.* 54(21):13849–13860.
20. Fassbender C, Braunbeck T (2013). Reproductive and genotoxic effects in zebrafish after chronic exposure to methyl methanesulfonate in a multigeneration study. *Ecotoxicology.* 22(5):825–837.
21. Fraz S, Lee AH, Pollard S, Srinivasan K, Vermani A, David E, Wilson JY (2019a). Paternal exposure to carbamazepine impacts zebrafish offspring reproduction over multiple generations. *Environ Sci Technol.* 53(21):12734–12743.
22. Fraz S, Lee AH, Pollard S, Srinivasan K, Vermani A, Wilson JY (2019b). Parental gemfibrozil exposure impacts zebrafish $F_1$ offspring, but not subsequent generations. *Aquat Toxicol.* 212:194–204.
23. Galus M, Rangarajan S, Lai A, Shaya L, Balshine S, Wilson JY (2014). Effects of chronic, parental pharmaceutical exposure on zebrafish (Danio rerio) offspring. *Aquat Toxicol.* 151:124–134.
24. Herdegen M, Radwan J (2015). Effect of induced mutations on sexually selected traits in the guppy, Poecilia reticulata. *Animal Behav.* 110:105–111.
25. Ishikawa Y, Hyodo-Taguchi Y (1997). Heritable malformations in the progeny of the male medaka (Oryzias latipes) irradiated with X-rays. *Mutat Res.* 389(2–3):149–155.
26. Jasperse L, Levin M, Rogers K, Perkins C, Bosker T, Griffitt RJ, Sepúlveda MS, De Guise S (2019). Transgenerational effects of polycyclic aromatic hydrocarbon exposure on sheepshead minnows (Cyprinodon variegatus). *Environ Toxicol Chem.* 38(3):638–649.
27. Kalichak F, de Alcantara Barcellos HH, Idalencio R, Koakoski G, Soares SM, Pompermaier A, Rossini M, Barcellos LJG (2019). Persistent and transgenerational effects of risperidone in zebrafish. *Environ Sci Pollut Res Int.* 26(25):26293–26303.
28. Kamstra JH, Hurem S, Martin LM, Lindeman LC, Legler J, Oughton D, Salbu B, Brede DA, Lyche JL, Aleström P (2018). Ionizing radiation induces transgenerational effects of DNA methylation in zebrafish. *Sci Rep.* 8(1):15373.
29. King Heiden TC, Spitsbergen J, Heideman W, Peterson RE (2009). Persistent adverse effects on health and reproduction caused by exposure of zebrafish to 2,3,7,8-tetrachlorodibenzo-p-dioxin during early development and gonad differentiation. *Toxicol Sci.* 109:75–87.
30. Knecht AL, Truong L, Marvel SW, Reif DM, Garcia A, Lu C, Simonich MT, Teeguarden JG, Tanguay RL (2017). Transgenerational inheritance of neurobehavioral and physiological deficits from developmental exposure to benzo[a]pyrene in zebrafish. *Toxicol Appl Pharmacol.* 329:148–157.
31. Kteeba SM, El-Ghobashy AE, El-Adawi HI, El-Rayis OA, Sreevidya VS, Guo L, Svoboda KR (2018). Exposure to ZnO nanoparticles alters neuronal and vascular development in zebrafish: acute and transgenerational effects mitigated with dissolved organic matter. *Environ Pollut.* 242(Pt A):433–448.
32. Lamb SD, Chia JHZ, Johnson SL (2020). Paternal exposure to a common herbicide alters the behavior and serotonergic system of zebrafish offspring. *PLoS One* 15(4):e0228357.
33. Liao PH, Chu SH, Tu TY, Wang XH, Lin AYC, Chen PJ (2014). Persistent endocrine disruption effects in medaka fish with early life-stage exposure to a triazole-containing aromatase inhibitor (letrozole). *J Hazard Mater.* 277:141–149.
34. Lindeman LC, Kamstra JH, Ballangby J, Hurem S, Martín LM, Brede DA, Teien HC, Oughton DH, Salbu B, Lyche JL, Aleström P (2019). Gamma radiation induces locus specific changes to histone modification enrichment in zebrafish and Atlantic salmon. *PLoS One.* 14(2):e0212123.
35. Lombó M, Fernández-Díez C, González-Rojo S, Navarro C, Robles V, Paz Herráez M (2015). Transgenerational inheritance of heart disorders caused by paternal bisphenol A exposure. *Environ Pollut.* 206:667–678.
36. Luu I, Ikert H, Craig PM (2021). Chronic exposure to anthropogenic and climate related stressors alters transcriptional responses in the liver of zebrafish (Danio rerio) across multiple generations. *Comp Biochem Physiol C Toxicol Pharmacol.* 240:108918.
37. Major KM, DeCourten BM, Li J, Britton M, Settles ML, Mehinto AC, Connon RE, Brander SM (2020). Early life exposure to environmentally relevant levels of endocrine disruptors drive multigenerational and transgenerational epigenetic changes in a fish model. *Front Marine Sci.* 7(471):1–17.
38. Matta MB, Linse J, Cairncross C, Francendese L, Kocan RM (2001). Reproductive and transgenerational effects of methylmercury or Aroclor 1268 on Fundulus heteroclitus. *Environ Toxicol Chem.* 20(2):327–335.
39. Meyer DN, Baker BB, Baker TR (2018). Ancestral TCDD exposure induces multigenerational histologic and transcriptomic alterations in gonads of male zebrafish. *Toxicol Sci.* 164:603–612.
40. Meyer DN, Crofts EJ, Akemann C, Gurdziel K, Farr R, Baker BB, Weber D, Baker TR (2020). Developmental exposure to $Pb^{2+}$ induces transgenerational changes to zebrafish brain transcriptome. *Chemosphere.* 244:125527.
41. Naderi M, Ferrari MCO, Chivers DP, Niyogi S (2018). Maternal exposure to dietary selenium causes dopaminergic hyperfunction and cognitive impairment in zebrafish offspring. *Environ Sci Technol.* 52(22):13574–13583.
42. Olsvik PA, Williams TD, Tung HS, Mirbahai L, Sanden M, Skjaerven KH, Ellingsen S (2014). Impacts of TCDD and MeHg on DNA methylation in zebrafish (Danio rerio) across two generations. *Comp Biochem Physiol Part C Toxicol Pharmacol.* 165:17–27.
43. Pandelides Z, Thornton C, Faruque AS, Whitehead AP, Willett KL, Ashpole NM (2020a). Developmental exposure to cannabidiol (CBD) alters longevity and health span of zebrafish (Danio rerio). *Geroscience.* 42(2):785–800.
44. Pandelides Z, Thornton C, Lovitt KG, Faruque AS, Whitehead AP, Willett KL, Ashpole NM (2020b). Developmental exposure to $\Delta^9$-tetrahydrocannabinol (THC) causes biphasic effects on longevity, inflammation, and reproduction in aged zebrafish (Danio rerio). *Geroscience.* 42(3):923–936.
45. Panter GH, Hutchinson TH, Hurd KS, Bamforth J, Stanley RD, Wheeler JR, Tyler CR (2010). Effects of a weak oestrogenic active chemical (4-tert-pentylphenol) on pair-breeding and F1 development in the fathead minnow (Pimephales promelas). *Aquat Toxicol.* 97(4):314–323.
46. Sadoul B, Birceanu O, Aluru N, Thomas JK, Vijayan MM (2017). Bisphenol A in eggs causes development-specific liver molecular reprogramming in two generations of rainbow trout. *Sci Rep.* 7(1):14131.

47. Santangeli S, Consales C, Pacchierotti F, Habibi HR, Carnevali O (2019). Transgenerational effects of BPA on female reproduction. *Sci Total Environ.* 685:1294–1305.
48. Schmitt C, Peterson E, Willis A, Kumar N, McManus M, Subbiah S, Crago J (2020). Transgenerational effects of developmental exposure to chlorpyrifos-oxon in zebrafish (DANIO RERIO). *Toxicol Appl Pharmacol.* 408:115275.
49. Schwartz P, Bucheli TD, Wettstein FE, Burkhardt-Holm P (2013). Life-cycle exposure to the estrogenic mycotoxin zearalenone affects zebrafish (Danio rerio) development and reproduction. *Environ Toxicol*. 28(5):276–289.
50. Schwindt AR, Winkelman DL, Keteles K, Murphy M, Vajda AM, Frid C (2014). An environmental oestrogen disrupts fish population dynamics through direct and transgenerational effects on survival and fecundity. *J Appl Ecol.* 51(3):582–591.
51. Seemann F, Peterson DR, Witten PE, Guo BS, Shanthanagouda AH, Ye RR, Zhang G, Au DWT (2015). Insight into the transgenerational effect of benzo[a]pyrene on bone formation in a teleost fish (Oryzias latipes). *Comp Biochem Physiol C Toxicol Pharmacol.* 178:60–67.
52. Seemann F, Jeong CB, Zhang G, Wan MT, Guo B, Peterson DR, Lee JS, Au DW (2017). Ancestral benzo[a]pyrene exposure affects bone integrity in F3 adult fish (Oryzias latipes). *Aquat Toxicol.* 183:127–134.
53. Shi G, Guo H, Sheng N, Cui Q, Pan Y, Wang J, Guo Y, Dai J (2018). Two-generational reproductive toxicity assessment of 6:2 chlorinated polyfluorinated ether sulfonate (F-53B, a novel alternative to perfluorooctane sulfonate) in zebrafish. *Environ Pollut.* 243(Pt B):1517–1527.
54. Shi G, Wang J, Guo H, Sheng N, Cui Q, Pan Y, Guo Y, Sun Y, Dai J (2019). Parental exposure to 6:2 chlorinated polyfluorinated ether sulfonate (F-53B) induced transgenerational thyroid hormone disruption in zebrafish. *Sci Total Environ.* 665:855–863.
55. Shimada A, Shima A (2004). Transgenerational genomic instability as revealed by a somatic mutation assay using the medaka fish. *Mutat Res.* 552(1–2):119–124.
56. Smith R, Seymour CB, Moccia RD, Mothersill C (2016). Irradiation of rainbow trout at early life stages results in trans-generational effects including the induction of a bystander effect in non-irradiated fish. *Environ Res.* 145:26–38.
57. Sowers AD, Gaworecki KM, Mills MA, Roberts AP, Klaine SJ (2009). Developmental effects of a municipal wastewater effluent on two generations of the fathead minnow, Pimephales promelas. *Aquat Toxicol.* 95(3):173–181.
58. Stenzel A, Wirt H, Patten A, Theodore B, King-Heiden T (2019). Larval exposure to environmentally relevant concentrations of triclosan impairs metamorphosis and reproductive fitness in zebrafish. *Reprod Toxicol.* 87: 79–86.
59. Thayil AJ, Wang X, Bhandari P, Vom Saal FS, Tillitt DE, Bhandari RK (2020). Bisphenol A and 17α-ethinylestradiol-induced transgenerational gene expression differences in the brain-pituitary-testis axis of medaka, Oryzias latipes†. *Biol Reprod.* 103(6):1324–1335.
60. Thomas JK, Janz DM (2014a). In ovo exposure to selenomethionine via maternal transfer increases developmental toxicities and impairs swim performance in F1 generation zebrafish (Danio rerio). *Aquat Toxicol.* 152:20–29.
61. Thomas JK, Janz DM (2014b). Developmental and persistent toxicities of maternally deposited selenomethionine in zebrafish (Danio rerio). *Environ Sci Technol.* 49(16):10182–10189.
62. Thomas JK, Birceanu O, Sadoul B, Vijayan MM (2018). Bisphenol A in eggs impairs the long-term stress performance of rainbow trout in two generations. *Environ Sci Technol.* 52(14):7951–7961.
63. Thompson WA, Vijayan MM (2021). Zygotic exposure to venlafaxine disrupts cortisol stress axis activity in multiple generations of zebrafish. *Environ Pollut.* 274:116535.
64. Valles S, Hernández-Sánchez J, Dipp VR, Huerta-González D, Olivares-Bañuelos TN, González-Fraga J, Bardullas U (2020). Exposure to low doses of inorganic arsenic induces transgenerational changes on behavioral and epigenetic markers in zebrafish (Danio rerio). *Toxicol Appl Pharmacol.* 396:115002.
65. Vera-Chang MN, St-Jacques AD, Gagné R, Martyniuk CJ, Yauk CL, Moon TW, Trudeau VL (2018). Transgenerational hypocortisolism and behavioral disruption are induced by the antidepressant fluoxetine in male zebrafish *Danio rerio. Proc Natl Acad Sci USA.* 115(52):E12435–E12442.
66. Vera-Chang MN, Moon TW, Trudeau VL (2019a). Ancestral fluoxetine exposure sensitizes zebrafish to venlafaxine-induced reductions in cortisol and spawning. *Endocrinology.* 160(9): 2137–2142.
67. Vera-Chang MN, Moon TW, Trudeau VL. (2019b). Cortisol disruption and transgenerational alteration in the expression of stress-related genes in zebrafish larvae following fluoxetine exposure. *Toxicol Appl Pharmacol.* 382:114742.
68. Vignet C, Joassard L, Lyphout L, Guionnet T, Goubeau M, Le Menach K, Brion F, Kah O, Chung BC, Budzinski H, Bégout ML, Cousin X (2015). Exposures of zebrafish through diet to three environmentally relevant mixtures of PAHs produce behavioral disruptions in unexposed F1 and F2 descendant. *Environ Sci Pollut Res Int.* 22(21):16371–16383.
69. Volkova K, Reyhanian Caspillo N, Porseryd T, Hallgren S, Dinnétz P, Porsch-Hällström I (2015a). Developmental exposure of zebrafish (Danio rerio) to 17α-ethinylestradiol affects non-reproductive behavior and fertility as adults and increases anxiety in unexposed progeny. *Horm Behav.* 73:30–38.
70. Volkova K, Reyhanian Caspillo N, Porseryd T, Hallgren S, Dinnétz P, Olsen H, Porsch-Hällström I (2015b). Transgenerational effects of 17α-ethinyl estradiol on anxiety behavior in the guppy, Poecilia reticulata. *General Comp Endocrinol.* 223:66–72.
71. Wang X, Hill D, Tillitt DE, Bhandari RK (2019). Bisphenol A and 17α-ethinylestradiol-induced transgenerational differences in expression of osmoregulatory genes in the gill of medaka (Oryzias latipes). *Aquat Toxicol.* 211:227–234.
72. Wirbisky SE, Weber GJ, Sepúlveda MS, Lin TL, Jannasch AS, Freeman JL (2016). An embryonic atrazine exposure results in reproductive dysfunction in adult zebrafish and morphological alterations in their offspring. *Sci Rep.* 6:21337.
73. Xie D, Chen Q, Gong S, An J, Li Y, Lian X, Liu Z, Shen Y, Giesy JP (2020). Exposure of zebrafish to environmentally relevant concentrations of mercury during early life stages impairs subsequent reproduction in adults but can be recovered in offspring. *Aquat Toxicol.* 229:105655.
74. Xu N, Chua AK, Jiang H, Liu NA, Goodarzi MO (2014). Early embryonic androgen exposure induces transgenerational epigenetic and metabolic changes. *Mol Endocrinol.* 28(8):1329–1336.

75. Xu X, Weber D, Martin A, Lone D (2016a). Transgenerational transmission of neurobehavioral impairments produced by developmental methylmercury exposure in zebrafish (Danio rerio). *Neurotoxicol Teratol.* 53:19–23.
76. Xu X, Weber D, Burge R, VanAmberg K (2016b). Neurobehavioral impairments produced by developmental lead exposure persisted for generations in zebrafish (Danio rerio). *Neurotoxicology.* 52:176–185.
77. Zhou R, Lu G, Yan Z, Jiang R, Sun Y, Zhang P (2021). Interactive transgenerational effects of polystyrene nanoplastics and ethylhexyl salicylate on zebrafish. *Environ Sci: Nano.* 8:146–159.
78. Zhu L, Liu Y, Xue X, Yuan C, Wang Z (2021). BPA's transgenerational disturbance to transcription of ovarian steroidogenic genes in rare minnow Gobiocypris rarus via DNA and histone methylation. *Sci Total Environ.* 762:143055.

# 10 Epigenetics

*Neelakanteswar Aluru*
Woods Hole Oceanographic Institution

*Kristine L. Willett*
University of Mississippi

**Abstract**

The term epigenetics was originally used to describe developmental processes. However, in the past couple of decades, the definition has shifted to focus on ways in which environmental factors can induce heritable phenotypic traits without altering the DNA sequence. The most widely studied epigenetic mechanisms include DNA methylation, histone modifications, and noncoding RNAs. Studies on epigenetic mechanisms have expanded from understanding the fundamental roles of these processes on physiology and development to assessing how environmental stressors can modulate epigenetic regulation of gene expression and phenotypes. Across taxa, stressor-mediated epigenetic changes have been associated with adverse outcomes at the cellular, organismal, and population levels. Fish are being used both as human and ecological health models to better understand the potential toxicological implications of contaminant exposure. This chapter will provide an overview of epigenetic mechanisms, summarize the existing fish studies that have investigated stressor-mediated epigenetic changes in fish, consider how epigenetic change can be incorporated into adverse outcome paradigms and risk assessment, and suggest research needs to link epigenetic changes causally to adverse toxicological outcomes.

## INTRODUCTION

Over the past half-century, the field of aquatic toxicology has played a number of important roles in understanding the effects of environmental chemical exposure on aquatic life. Some of the important developments include acute and chronic toxicity studies documenting susceptibility of individual species to toxicants, identification of biomarkers of exposure and effect, and the discovery of endocrine-mimicking properties of toxicants. These and many other developments have played a pivotal role in assessing environmental chemical exposure and acted as early warning indicators of threat to wildlife. However, in the last couple of decades, mechanistic toxicology has made significant contributions to our understanding of toxicant effects. This was possible due to rapid technological advances in large-scale (high throughput) quantification of biochemical and molecular changes possible in any species of interest. Understanding *how* environmental chemical exposure induced adverse phenotypes initially focused on quantification of gene expression patterns. These studies identified the molecular initiating events of individual chemicals, dysregulated physiological pathways associated with exposure, and causally linked gene expression changes to adverse phenotypes.

Gene expression is an intricate process and is regulated by transcriptional, post-transcriptional, and translational changes. In addition, epigenetic mechanisms (e.g., DNA methylation and histone modifications) also play an important role in regulating gene expression (Figure 10.1). The term "epigenetics" was originally coined by Conrad Waddington to describe the interpretation of the genotype during development (epigenesis) to give the phenotype (Deans and Maggert 2015). However, in the past few decades, the idea behind the term epigenetics has changed considerably. In its new, narrower, and simplified sense, epigenetics has been redefined as the set of mechanisms involved in regulating gene expression (Bird and Wolffe 1999). Specifically, this term is being used to refer to those

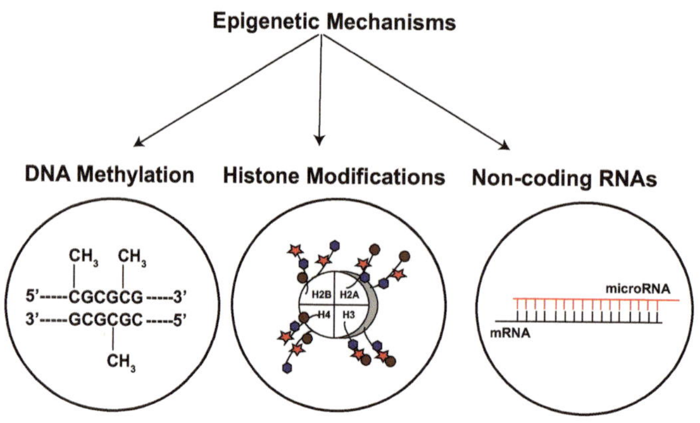


**FIGURE 10.1** Representation of some of the commonly studied epigenetic mechanisms.



DOI: 10.1201/9781003160694-10

mechanisms that cause phenotypic variation by altering the expression of genes rather than the underlying DNA sequence. Interest in understanding the effects of toxicants on epigenetic regulation of gene expression has grown rapidly in the last few years because several studies have demonstrated that exposure to toxicants during sensitive windows of development could lead to persistent changes in gene expression. In some instances, toxicants are known to induce multi-generational and transgenerational phenotypes. These observations together with the role of epigenetic mechanisms in stable alterations to gene expression have led to an intense focus on understanding the role of epigenetics in toxicant-induced phenotypes.

Epigenetic mechanisms can integrate genomic and environmental inputs to generate the instructions for producing a particular phenotype. In toxicology, epigenetic mechanisms are investigated to demonstrate how changes in gene expression patterns mediated by toxicant exposures can cause adverse phenotypes. This chapter will provide an overview of epigenetic mechanisms, summarize the existing fish studies that have investigated stressor-mediated epigenetic changes in fish, consider how epigenetic change can be incorporated into adverse outcome paradigms and risk assessment, and suggest research needs to be able to definitively link epigenetic change causally to adverse toxicological outcomes.

## EPIGENETIC MECHANISMS

### DNA Methylation

DNA methylation is one of the most well-studied epigenetic modifications in aquatic organisms (Figure 10.2). In vertebrates and most invertebrates (with some exceptions), DNA can be covalently modified by methylation of the cytosine base in the dinucleotide sequence 5'-CpG-3' (Bird 1980). CpG is an abbreviation for cytosine and guanine separated by a phosphate, which links the two nucleotides together in DNA. Cytosine methylation occurs predominantly in CpG dinucleotides. In vertebrates, DNA methylation patterns are established during embryonic development by *de novo* DNA methylating enzymes called DNMT3 (DNA methyltransferase 3). The established DNA methylation patterns are maintained during cell division by DNMT1, a maintenance DNA methylase (Bird 1993). Orthologs of both maintenance and *de novo* DNMTs have been characterized in a number fish species. Due to genome duplication in the fish lineage, there are two or more orthologs of many mammalian genes. In contrast to two *de novo* DNMTs (DNMT3A and DNMT3B) in mammals, fish possess multiple homologs. For example, in zebrafish there are two DNMT3A (*dnmt3aa* and *dnmt3ab*) and four DNMT3B (*dnmt3ba*, *3bb1*, *3bb2*, and *3bb3*) genes (Goll and Halpern 2011). In addition to establishing and maintaining DNA methylation during germline development and early embryogenesis, DNA is demethylated by both active and passive demethylation processes. Active DNA demethylation refers to an enzymatic process resulting in the removal of the methyl group from 5mC. It involves Ten-eleven translocation (TET) family of dioxygenases that biochemically convert 5-methylcytosine (5mC) to 5-hydroxymethylcytosine (5hmC). In successive steps, TET enzymes further hydroxylate 5hmC to generate 5-formylcytosine (5fC) and 5-carboxylcytosine (5caC). The oxidized 5mC derivatives cannot serve as substrates for DNMT1; they can be lost by replication-dependent passive demethylation (Wu and Zhang 2010).

DNA across the vertebrate genome (somatic tissues) is methylated at 65–80% of all CpG sites (Bird 1980). Recent advances in sequencing technology have allowed us to measure DNA methylation at single CpG site resolution in any target organism or tissue (Harrison and Parle-McDermott 2011). These studies indicate that highly methylated sequences include satellite DNAs, repetitive elements including transposons, non-repetitive intergenic DNA, and exons. Most sequences are methylated according to their frequency of CpG dinucleotides, except CpG islands (CGIs). CGIs are GC-rich sequences of ~1 kb in length that are unmethylated (Bird 1980). Multiple mechanisms exist in protecting CGIs from *de novo* methylation by DNMTs. For instance, the CXXC domain contains DNA-binding proteins such as Cfp and Kdm2a that bind specifically to unmethylated CpGs in CGIs and protect them from *de novo* methylation and create a transcriptionally competent chromatin environment (Blackledge et al. 2010). Until recently, CGIs were generally thought to be present mainly in the promoters and 5' regions of genes. Recent nucleotide-level resolution methylome sequencing studies revealed the presence of CGIs in the intergenic regions (Orphan CGIs)

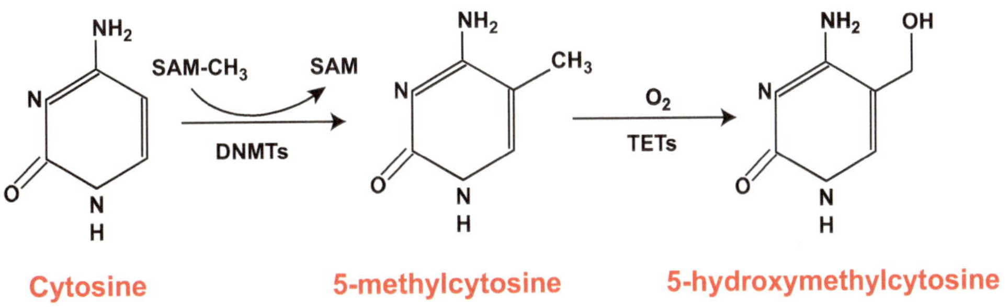


**FIGURE 10.2** DNA methylation reactions. DNA methylation is controlled by the functional interplay between two families of enzymes - DNA methyltransferases (DNMTs) and TET methylcytosine dioxygenases, which control DNA methylation and demethylation, respectively. DNMTs transfer a methyl group from the universal methyl donor, SAM, to the 5′-position of cytosine residues (5mC) in DNA, the bulk of which is present in the CpG sequence context, whereas TET proteins are iron (II) and 2-oxoglutarate-dependent dioxygenases that oxidize the methyl group of 5mC to 5-hydroxymethylcytosine (5hmC), 5-formylcytosine (5fC), and 5-carboxylcytosine (5caC) in DNA.

(Bell and Vertino 2017). Also, other differentially methylated regions termed shores (up to 2 kb away from a CGI) or shelves (located 2–4 kb of a CGI) were identified (Irizarry et al. 2009). The relevance of these CGIs on transcriptional regulation remains to be characterized.

One of the widely established roles for DNA methylation is in the regulation of gene expression. DNA methylation of gene promoter regions is associated with transcriptional repression. The presence of CpG sites in the gene promoter regions interferes with the binding of the transcription factors that activate transcription from a specific gene. In addition, recruitment of methyl-CpG binding proteins and other repressor complexes to the methylated DNA could also cause methylation-dependent repression. Several proteins that can bind to methyl CpGs have been identified (Bird and Wolffe 1999). They include methyl-CpG binding protein complex (MeCP1 and MeCP2) and methyl-CpG binding domain (MBD) proteins (MBD1, MBD2, MBD3, and MBD4). Each of these proteins associate with different corepressor complexes and alter chromatin structure. For example, MeCP1 is a multiprotein complex that includes the Nucleosome Remodeling Demethylase (NuRD) corepressor complex and MBD2 (Wade et al. 1999). NuRD comprises histone deacetylases (HDAC) and a large chromatin remodeling protein. NuRD is recruited to DNA by several DNA-binding proteins. Given that DNA and histones are integral components of chromatin, it is not surprising that DNA methylation interacts with histone modifications (Cedar and Bergman 2009).

## Histone Modifications

DNA within the nucleus is compacted into chromatin containing the histone proteins. The macromolecules that regulate chromatin fall into distinct classes of molecules. These include (a) ATP-dependent remodeling proteins that mobilize the histones within the chromatin, (b) histone chaperones that insert and remove generic or variant histones into chromatin, (c) post-transcriptional modification enzymes that add and remove chemical groups to the DNA or histone components of chromatin, and (d) noncoding RNA molecules that bind and modulate chromatin regulatory proteins (Altaf et al. 2007; Ho and Crabtree 2010; Vicent et al. 2010). All these macromolecules work in a highly coordinated fashion to regulate distinct chromatin functions.

The inevitable inaccessibility of DNA in chromatin may appear to be a problem, but evolution has turned it into an asset by inventing enzymes that are able to remodel nucleosomes. This allows access to DNA to be regulated locally as necessary. Nucleosome remodeling involves changing histone-DNA interactions as a means of disrupting, assembling, or moving nucleosomes. Nucleosome remodeling enzymes may liberate segments of DNA by complete or partial disassembly of nucleosomes, may alter the composition of nucleosomes with respect to histone variants, and may also affect the folding of nucleosomes. The nucleosome remodeling requires biochemical coupling to ATP hydrolysis (Becker and Workman 2013). Like all biochemical reactions, they are reversible, and the outcomes of remodeling are largely determined by the specification of remodeling enzymes and the involvement of cofactors. Nucleosomal remodeling ATPases are involved in the regulation of developmental gene expression patterns, the fast transcriptional response to environmental signals, and surveillance of the genome for DNA damage. Defects in nucleosomal remodeling enzymes during development can cause morphological defects (Ho and Crabtree 2010). Histone-binding proteins also act as chaperones and are involved in diverse functions including histone

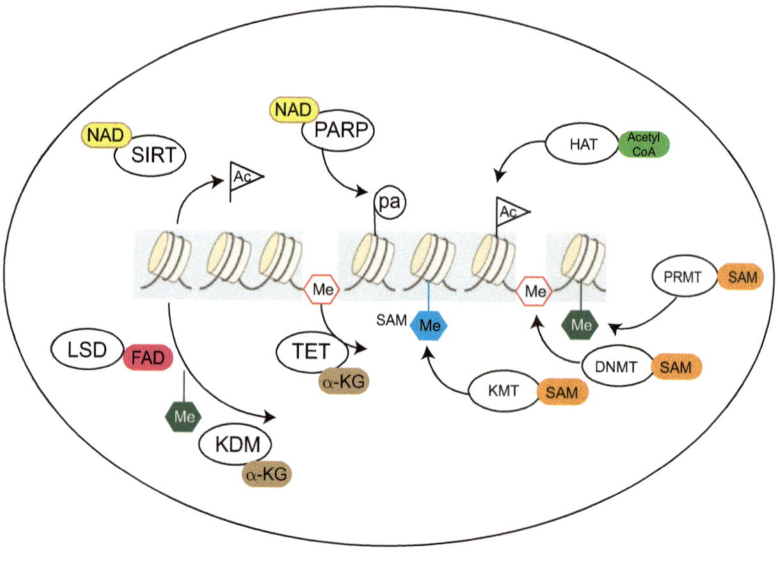


**FIGURE 10.3** Histone modifying enzymes. Covalent histone modifications are transduced by histone-modifying enzymes (writers) and removed by antagonizing activities (erasers). They are classified into groups according to the type of enzymatic action (e.g., methylation, acetylation, phosphorylation). Proteins with a specific affinity for a histone tail modification are termed "readers" (not shown in the figure). DNMT, DNA methyltransferase; KMT, lysine methyltransferase; HAT, histone acetyltransferase; KDM, lysine demethylase; TET, ten-eleven translocation; PARP, poly(ADP-ribose) polymerase; PRMT, protein arginine methyltransferases; LSD, lysine-specific histone demethylase; SIRT, sirtuin; Me, methylation, Ac, acetylation; pa, poly(ADP-ribose); SAM, S-adenosyl methionine; α-KG, alpha-ketoglutarate; FAD, flavin adenine dinucleotide; NAD, nicotinamide adenine dinucleotide.

storage, transport, deposition, and eviction. These functions have been mainly demonstrated during DNA replication.

The post-translational modification enzymes include proteins that add chemical groups as well as those that remove them. The enzymes that mediate histone modification (i.e., writers) include acetyltransferases, methyltransferases, kinases, and ubiquitinases (Figure 10.3). For example, histone acetyltransferases (HATs), also referred to as lysine acetyltransferases (KATs), are a group of enzymes that acetylate the side-chain amino group of lysine residues on histones (Bannister and Kouzarides 2011; Carrozza et al. 2003). These enzymes contribute to several different transcription-mediated biological processes including cell-cycle progression and hormone signaling. The histone lysine acetylation is highly reversible, and the acetylated lysine residues are removed by a group of enzymes known as histone deacetylases (HDACs), also known as erasers (Cress and Seto 2000). There are several different HDACs, but most of them use metal or $NAD^+$-dependent hydrolysis of acetylated residues. In addition to deacetylases, there are phosphatases, demethylases, and deubiquitinases that also act as erasers (Bannister and Kouzarides 2011). In addition to writers and erasers, there are proteins (i.e., readers) that can recognize specific histone modifications (Yap and Zhou 2010). For example, the acetylation marks on lysine residues are selectively read by bromodomain-containing proteins. These domains are highly conserved among chromatin associated proteins, including HATs. All these proteins play significant roles in normal physiology and in disease states (Bannister and Kouzarides 2011). An increasing number of studies have pointed to roles for other ligands that behave as enhancers or competitors to post-transcriptional modifications (Glatt et al. 2011). Many proteins that recognize histone modifications are prone to covalent modifications, including methylation, acetylation, phosphorylation, and sumoylation (Zhang et al. 2015). Such covalent modifications may influence the ability of the protein to recognize histone tails and serve to recruit additional factors that have further histone modifying or recognition functions.

Numerous studies in the past decade have identified long noncoding RNAs (lncRNA) to play a major role in gene transcription. Evidence indicates that they are crucial for the recruitment and proper function of chromatin remodeling complexes (Khalil et al. 2009; Rinn et al. 2007) and that noncoding RNA interactions are mediated in part by the same modules that recognize histone modifications (Yap et al. 2010). One of the well-studied examples of lncRNA-based epigenetic regulation is mammalian dosage compensation. A long intergenic noncoding RNA (lincRNA) termed XIST (X inactive-specific transcript) is expressed from one female X chromosome, resulting in the recruitment of polycomb group complexes to this chromosome with concomitant transcriptional silencing across a majority of the X chromosome (Bousard et al. 2019). LncRNAs are emerging as critical layers of epigenetic regulation in which different lncRNAs are associated with distinctive epigenetic states. They share a common mechanism where they physically associate with chromatin-modifying and -remodeling complexes and guide them to specific genomic loci that are crucial for proper cellular functions (Gil and Ulitsky 2020; Kurth and Mochizuki 2009).

## Noncoding RNAs

The importance of non-protein coding RNAs (ncRNAs) in the regulation of various biological processes in eukaryotes is well documented (Mattick 2003) (Figure 10.4). The number of known ncRNAs continues to grow, and the most widely known regulatory ncRNAs include microRNAs (miRNAs), small interfering RNAs (siRNAs), piwi-associated RNAs (piRNAs), and long noncoding RNAs (lncRNAs). MiRNAs are the most widely studied group in developmental toxicology, and a variety of environmental chemicals are known to affect their expression (Schraml et al. 2017; Yokoi and Nakajima 2011). Most of these studies have been conducted in zebrafish and have determined the expression patterns after acute developmental exposure to toxicants; very few studies have characterized the role of individual miRNAs in toxicant-induced phenotypic changes later in life. Similarly,

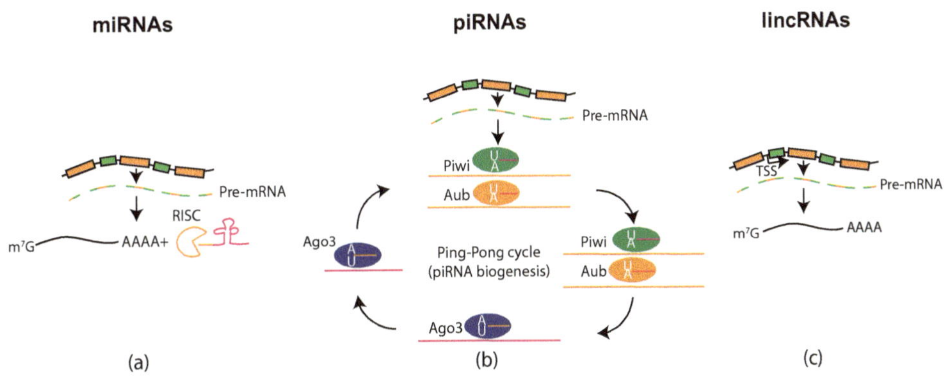


**FIGURE 10.4** Biogenesis of non-coding RNAs. (A) microRNAs transcribed from pre-microRNAs are bound to Argonaute (AGO) proteins to form the RNA-induced silencing complex (RISC) complex and inhibit mRNA expression. (B) piRNAs recognize their targets and recruit piwi proteins. It results in the cleavage of the primary piRNA transcript, producing the secondary piRNA in an amplification mechanism called as ping-pong mechanism. (C) Long intergenic noncoding RNAs are transcribed by polymerase II in a pre-RNA immature form. Once mature, they are poly-adenylated.

in a few other fish species, some specific miRNAs are differentially expressed upon toxicant exposure in experimental settings. However, the promise of miRNAs as biomarkers of exposure is yet to be realized. As epigenetic changes are considered to mediate long-term, multi-, and transgenerational effects, it is also essential to determine persistent changes that are transmitted to subsequent generations. Recently, a few studies in mammalian models have demonstrated that sperm miRNAs are involved in the transgenerational transmission of stress-induced phenotypes (Brieno-Enriquez et al. 2015; Rodgers et al. 2015).

Another group of ncRNAs that are expressed in germ cells and could potentially play a role in transgenerational effects are piRNAs. They originate from the intergenic repetitive elements in the genome and associate with the PIWI subfamily members of the argonaute family of proteins. Their role is to repress transposable elements in the germline and maintain genomic integrity (Ernst et al. 2017). The biogenesis of piRNAs in various model systems, including zebrafish, is well known, but there are no studies on the effects of toxicants on piRNA expression. Recent studies in *D. virilis* and *C. elegans* have demonstrated piRNA-mediated stable long-term gene silencing as a potential mechanism of transgenerational effects (Ashe et al. 2012; Grentzinger et al. 2012). However, the hypothesis that ncRNAs act as a mediator of developmental origins of health and diseases (DOHaD) and multigenerational effects needs further experimentation.

## EPIGENETICS AND ENERGY METABOLISM

One of the potential mechanisms by which toxicants can alter epigenetic patterns is by altering energy metabolism (Figure 10.5). Recent studies have identified that nutrients and their metabolites exert an important influence on the epigenome, because they serve as substrates and/or coenzymes for epigenetic-modifying enzymes. Given the fact that many environmental chemicals can affect energy metabolism, it is not surprising to see the effects of exposure on the epigenome.

DNA and histone methylation are derived from the metabolism of the essential amino acid methionine (Met) (Figure 10.6), which is almost exclusively obtained from the diet. Met is converted into the methyl-donor metabolite SAM (S-adenosyl-L-methionine), which is then used as a substrate for DNA and histone methyltransferases, producing S-adenosylhomocysteine (SAH), which competitively inhibits DNA and histone methyltransferases (HMTs). Disruptions in Met metabolism and one-carbon metabolism, such as changes to threonine or methionine intake or the activation or inhibition of metabolic enzymes in these pathways, have been shown in a variety of cellular systems to affect intracellular concentrations of SAM and SAH, thereby changing levels of both DNA and histone methylation (Caudill et al. 2001; Serefidou et al. 2019). Nutrients, such as glucose, fatty acids, amino acids, and vitamins, are utilized by cellular metabolic pathways to produce metabolites that are used as substrates or activity modulators of chromatin and DNA modifying enzymes.

Metabolism can also influence the activity of enzymes responsible for DNA and histone demethylation (Lu and Thompson 2012). They include the TET family of DNA demethylases (TETs) and histone demethylases, including JmjC-domain-containing demethylases (JHDMs) and amine oxidases (LSDs). TETs and JHDMs belong to the class of α-ketoglutarate (αKG)-dependent dioxygenases that use αKG and oxygen as substrates (Matuleviciute et al. 2021; Shmakova et al. 2014). Catalysis by TETs and JHDMs is activated by ascorbate and also requires ferrous iron, Fe(II), as a cofactor, whereas lysine-specific histone demethylases (LSD) rely on flavin adenine dinucleotide (FAD) (Macheroux et al. 2011).

Metabolic regulation of DNA and chromatin methylation is achieved by the regulation of redox balance and

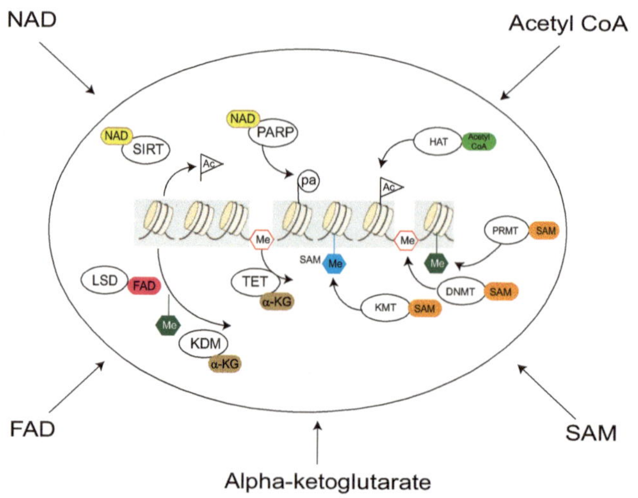


**FIGURE 10.5** Role of energy metabolism in epigenetic regulation of gene expression. The majority of the substrates involved in epigenetic modifications come from energy metabolism.

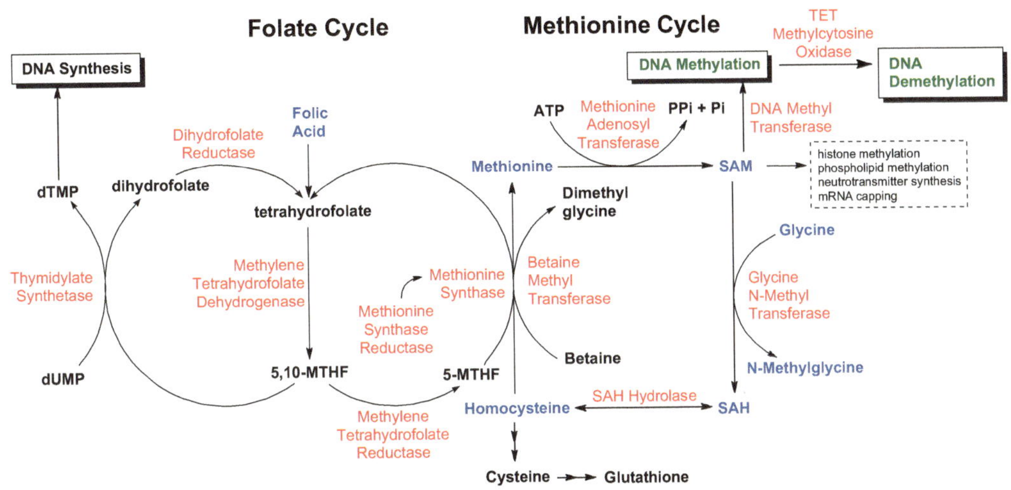


**FIGURE 10.6** Role of folate and methionine cycles in epigenetics. The physiological and epigenetic roles of the folate/methionine cycle include DNA synthesis (purines and thymidylate) and as a methyl donor in DNA and protein methylation.

hypoxia (Cyr and Domann 2011). In *C. elegans*, reactive oxygen species (ROS) generation during early development modulates transcription of the genes required for long-term survival by inhibiting the Histone H3 lysine K4 (H3K4) methyltransferases and reducing global levels of H3K4 trimethylation (H3K4me3) (Bazopoulou et al. 2019). Whether ROS generation as a result of metabolic reprogramming involves the oxidation and depletion of SAM remains to be seen, but several potential mechanisms exist, including the redox-responsiveness of the vitamin-B12-utilizing enzyme methionine synthase. In addition, hypoxia inhibits TETs as well as JHDMs such as KDM5A, KDM6A, and KDM6B (KDM-lysine demethylase) by limiting oxygen, resulting in global DNA and histone hypermethylation (Hancock et al. 2015; Thienpont et al. 2016).

In addition to DNA and histone methylation, acetylation of histones is dependent on acetyl-CoA (acetyl coenzyme A, (Etchegaray and Mostoslavsky 2016)). Acetyl CoA participates in many important biochemical reactions in protein, carbohydrate, and lipid metabolism. Histone acetylation involves transfer of an acetyl group derived from acetyl-CoA to the ε-amino group of a histone lysine, catalyzed by histone acetyltransferases (HATs). Acetyl-CoA is derived primarily from carbon units provided by extracellular glucose, which feeds into mitochondrial metabolism in order to generate citrate. Citrate is then exported to and lysed within the cytosol by the enzyme ATP-citrate lyase (ACLY) to generate acetyl-CoA. The availability of glucose and glycolytic activity subsequently influences global levels of histone acetylation through the generation of acetyl-CoA. The short-chain fatty acid (SCFA) acetate is an additional source of acetyl-CoA for histone acetylation, through the function of the enzyme acetyl-CoA synthetase 2 (ACSS2). Acetate is produced *de novo* from pyruvate, providing another pathway from glucose to histone acetylation. Both acetate and acetyl-CoA can also be generated from ethanol metabolism in the liver, which supports histone acetylation in the brain and other organs, or by the oxidative catabolism of lipids, which may contribute up to 90% of the acetylation of certain histone lysines (Etchegaray and Mostoslavsky 2016).

Like histone methylation, histone acetylation is actively turned over, in this case, by a class of enzymes called histone deacetylases (HDACs). HDACs are inhibited by metabolites, including butyrate and β-hydroxybutyrate (β-OHB), that are produced during fatty acid oxidation or ketogenesis, or are derived from the commensal microbiota (Alenghat and Artis 2014). Sirtuins, another class of HDACs, utilize NAD+ (nicotinamide adenine dinucleotide) to deacetylate histones and are sensitive to intracellular NAD+ levels (Jing and Lin 2015). Overall, the interdependency between metabolites and DNA and chromatin modifying enzymes plays an important role in maintaining homeostasis and mediating responses to changing environmental conditions and hence cannot be ignored.

## KNOWN EFFECTS OF EPIGENETIC MECHANISMS ON PHENOTYPIC CHANGE

Across the diversity of life forms, some epigenetic changes have been mechanistically linked to causing phenotypic change. For example, in insects, social/caste phenotypes are regulated by a diversity of epigenetic mechanisms including DNA methylation, histone acetylation, and miRNAs (reviewed in Villagra and Frias-Lasserre (2020)). A temperature-dependent histone demethylase regulates expression of their sex-determining gene in reptiles (Ge et al. 2018). In the field of toxicology, the Agouti mouse model has been used to assess epigenetic inheritance of exposure to phytoestrogens and bisphenol A (BPA) (Dolinoy 2008). An intra-cisternal A retrotransposon in the agouti protein controls expression of yellow fur based on methylation status, obesity, and diabetic phenotypes (Morgan et al. 1999). More recently, this model was used in a longitudinal aging study, and the persistence of the epigenetic variability at the retrotransposons was found not to be predictive of maternal dietary exposure to BPA or methyl donors (Bertozzi et al. 2021). Yet, when Sprague-Dawley rats were exposed to a low dose of BPA as neonates and then fed a "Western-style diet" as adults, the male but not female developmentally exposed adults developed dyslipidemia and metabolic disease (Trevino et al. 2020). The developmental susceptibility to later life dietary challenge was mechanistically linked to histone epigenetic reprogramming of the early growth response 1 transcriptome and metabolic dysfunction.

In human cohorts, certain epigenetic changes are recognized as biomarkers of adverse outcomes. For example, the maternal plasma histone modification H3K27me3 was associated with offspring neural tube defects and arsenic exposure, particularly if the mothers were folate deficient (Tauheed et al. 2017). Similarly, site-specific hypomethylation of the aryl hydrocarbon receptor repressor (AHRR) cg05575921 and coagulation factor II (thrombin) receptor-like 3 (F2RL3) cg03636183 are associated with smoking (Alhamdow et al. 2018; Bosse et al. 2012; Lee et al. 2017), $PM_{2.5}$ exposure (Tantoh et al. 2020), and lung cancer (Zhang et al. 2016). Differential miRNA expression has also been associated with both diseases or environmental toxicant exposure (e.g., see review by Addo and Fry 2020 for placental microRNAs (Addo et al. 2020)), but confirmed mechanistic linkages between exposure-mediated epigenetic change and birth outcomes and/or later life health effects are still needed (Bommarito et al. 2017). Similar studies linking epigenetic mechanisms to morphological or disease phenotype are lacking in the field of fish toxicology. With increased focus on understanding the role of epigenetic mechanisms in toxicant-induced alterations in phenotypes, it is possible to link epigenetic changes to organismal phenotypes. In the following section, we will summarize the status of the field of fish epigenetic toxicology.

## TOXICANT-INDUCED EFFECTS ON EPIGENETIC MACHINERY IN FISH

Several studies using zebrafish as a model system have demonstrated that toxicants can interfere with DNA methylation by affecting the expression of DNMT genes and altering gene-specific methylation and/or global DNA methylation. For example, developmental exposures to several toxicants alter DNA methyltransferase (DNMT) expression patterns in zebrafish (Table 10.1 and references therein). However, the impact of altered DNMT expression and DNA methylation patterns on any developmental or physiological

**TABLE 10.1**
**List of Zebrafish Studies Reporting Toxicant Effects on DNA Methylation. This includes the expression of DNMTs, gene-specific or genome wide DNA methylation.**

| Toxicant | Life Stage | DNMTs | Gene-Specific or Genome-Wide Methylation | References |
|---|---|---|---|---|
| Atenolol and venlafaxine | Adults | √ | Global DNA methylation | Lin et al. (2022) |
| Atrazine | Embryos | √ | Global DNA methylation | Wirbisky-Hershberger et al. (2017) |
| BaP | Adults | - | Gene-specific methylation | Gao et al. (2018) |
| | Embryos | √ | Gene-specific and global DNA methylation | Fang et al. (2013) |
| | | √ | Gene-specific methylation | Corrales et al. (2014) |
| Bis(2-ethylhexyl)-2,3,4,5-tetrabromophthalate | Embryos | - | Gene-specific methylation | Guo et al. (2020) |
| Bisphenol AF | Adults | √ | - | Zhang et al. (2022) |
| Bisphenol-A (BPA), diethylstilbestrol, 17α-ethynylestradiol, nickel, cadmium, tributyltin, arsenite, perfluorooctanoic acid, valproic acid, flusilazole | Embryos | - | Gene-specific and global methylation | Bouwmeester et al. (2016) |
| Bisphenols, perfluorinated chemicals, methoxychlor, permethrin, vinclozolin, and coumarin 47 | ZFL cell line | Histone demethylase kdm5ba expression | Global DNA methylation | Blanc et al. (2019) |
| BPA | Adults | - | Gene-specific methylation | Zhao et al. (2017) |
| | | - | Gene-specific methylation | Santangeli et al. (2019) |
| | | √ | Global DNA methylation | Laing et al. (2016) |
| | Embryos | √ | RRBS | Olsvik et al. (2019) |
| | | √ | Global DNA methylation | Liu et al. (2016) |
| BPA | Embryos | √ | Gene-specific methylation | Gyimah et al. (2021) |
| BPS | | | | |
| Cadmium | Embryos | √ | Ca2+, MAPK-JUK, p38 MAP kinase, PI3K-AKT, and VEGF signaling | Bian and Gao (2021) |
| | | √ | Gene expression | Park et al. (2020) |
| | Life Cycle | - | Gene-specific methylation | Pierron et al. (2021) |

*(Continued)*

**TABLE 10.1 (*Continued*)**
**List of Zebrafish Studies Reporting Toxicant Effects on DNA Methylation. This includes the expression of DNMTs, gene-specific or genome wide DNA methylation.**

| Toxicant | Life Stage | DNMTs | Gene-Specific or Genome-Wide Methylation | References |
|---|---|---|---|---|
| Carbon nanoparticles | Adults | - | Genome-wide methylation | Zhou et al. (2019) |
| 5-chloro-2-methyl-4-isothiazoline-3-one/2methyl-4-isothiazolin-3-one (CMIT/MIT) | Embryos | √ | Global DNA methylation | Chatterjee et al. (2021) |
| Copper | Adults | - | Genome-wide methylation | Tai et al. (2022) |
| | Embryos | - | Gene-specific methylation | Zhang et al. (2020) |
| | | - | Genome-wide methylation | Jin et al. (2021) |
| Copper and temp | Embryos | √ | - | Dorts et al. (2016) |
| Di-(2-ethylhexyl) phthalate | Adults | - | Gene-specific methylation | Ma et al. (2018) |
| 17α-ethinylestradiol | Adults | - | Gene-specific methylation | Stromqvist et al. (2010) |
| Fipronil | Embryos | - | Genome-wide DNA methylation | Qian et al. (2017, 2019) |
| Flutolanil | Adults | - | Global DNA methylation | Teng et al. (2020) |
| Lead | Embryos | √ | - | Sanchez et al. (2017) |
| Lead and temperature stress | Embryos | √ | Gene-specific DNA methylation | Park et al., (2020) |
| Metal mixtures | Adults | - | Global and gene-specific methylation | Hu et al. (2021b) |
| Metformin | Adults | √ | - | Lin et al. (2021) |
| Mono(2-ethylhexyl) phthalate | Embryos | - | Global, genome-wide, and locus-specific DNA methylation analyses | Kamstra et al. (2017) |
| PCB126 | Adults | - | Genome-wide methylation | Aluru et al. (2018), Aluru and Engelhardt (2022) |
| Permethrin | Embryos | - | Genome wide methylation | Blanc et al. (2021) |
| Phenanthrene | Embryos | - | Gene-specific methylation | Chen et al. (2021) |
| Probiotic and perfluorobutanesulfonate | Larva | - | Metabolites | Hu et al. (2022) |
| Raw and ozonated oil sands process-affected water | Life Cycle | - | Global DNA methylation | Philibert et al. (2019) |
| Silver nanoparticles | Embryos | | Gene-specific methylation | Xu et al. (2018) |
| TCDD | Juvenile | √ | Genome-wide methylation | Akemann et al. (2019, 2020) |
| | Embryos | √ | Gene-specific methylation | Aluru et al. (2015) |
| TCDD, MeHg | Adult | | Genomewide and gene-specific methylation | Olsvik et al. (2014) |
| Tris(1,3-dichloro-2-propyl)phosphate | Embryos | - | Gene-specific methylation | Kupsco et al. (2017) |
| | | √ | Gene-specific methylation | Li et al. (2020) |
| | | - | Global and gene-specific methylation | Avila-Barnard et al. (2022) |
| Uranium | Adults | - | Global DNA methylation | Gombeau et al. (2016, 2017) |

endpoints has not been clearly demonstrated. Nevertheless, these studies suggest that toxicant exposure could affect DNA methylation processes and potentially have long-term consequences. Several studies have investigated the role of DNA methylation on latent effects of developmental exposure to toxicants by conducting multi- and transgenerational studies. For instance, developmental exposure to 2,3,7,8-tetrachlorodibenzo-p-dioxin (TCDD) in zebrafish resulted in altered sex ratios and increased skeletal defects in F1 and F2 generations (Baker et al. 2014). This study also identified a sex-specific transgenerational phenotype resulting in decreased fecundity and fertilization rate. These phenotypes were observed only when TCDD lineage F2 males were crossed with unexposed females, suggesting that the male germline was transgenerationally affected by exposure. Investigations into the role of DNA methylation in this transgenerational phenotype suggest that multiple epigenetic pathways including DNA methylation could play a role in toxicant-induced transgenerational phenotypes. This is based on the observation that DNA methylation patterns of genes associated with histone modifiers were altered in males from the TCDD lineage.

In contrast to the studies in zebrafish where multiple epigenetic mechanisms have been studied for multiple drugs and contaminants, in other fish species (e.g., goldfish, carp, medaka, killifish, salmon) relatively fewer studies have investigated the impacts of toxicants on epigenetic change within each species (Table 10.2 and references therein). Most of these studies have assessed the impact of exposure on DNMT expression, global methylation, and

**TABLE 10.2**
**List of Studies in Non-Model Fish Species Where Toxicant-Exposure Induced Changes in DNA Methylation Changes Were Investigated**

| Toxicant | Species | Epigenetic Modifiers[a] | DNA Methylation[b] | References |
|---|---|---|---|---|
| Arsenic | Goldfish (*Carassius auratus*) | - | √ | Bagnyukova et al. (2007) |
| Atrazine | Common carp (*Cyprinus carpio*) | √ | - | Wang et al. (2014) |
| | Medaka (*Oryzias latipes*) | √ | √ | Cleary et al. (2019) |
| Benzo[a]pyrene | Atlantic killifish (*Fundulus heteroclitus*) | √ | - | Fang et al. (2010a, 2010b) |
| Bifenthrin<br>Trenbolone and Levonorgestrel | Inland silversides (*Menidia beryllina*) | √ | - | DeCourten et al. (2020) |
| Bisphenol A | Atlantic Salmon (*Salmo salar*) | √ | - | Olsvik et al. (2017) |
| | Grass carp (*Ctenopharyngodon exica*) | - | √ | Fan et al. (2020) |
| | Medaka (*Oryzias latipes*) | √ | √ | Thayil et al. (2020) |
| | Rare minnow (*Gobiocypris rarus*) | √ | √ | Liu et al. (2014, 2020), Yuan et al. (2016, 2017, 2019), Zhang et al. (2017, 2018, 2021); Zhu et al. (2020, 2021a, 2021b) |
| Bitumen | Yellow perch (*Perca flavescens*) | - | √ | McDonnell et al. (2019) |
| Cadmium | European eel (*Anguilla exicana*) | - | √ | Pierron et al. (2014) |
| | Tilapia (*Oreochromis niloticus*) | √ | √ | Hu et al. (2021a) |
| Chlorpyrifos | Common carp (*Cyprinus carpio*) | √ | - | Wang et al., (2014) |
| Complex mixtures | Tilapia (*Oreochromis niloticus*) | - | √ | Fuzinatto et al. (2015) |
| Copper | Grass carp (*Ctenopharyngodon exica*) | - | √ | Xu et al. (2020) |
| 2,4-Dichlorophenol | Goldfish (*Carassius auratus*) | √ | - | Zhang et al. (2014) |
| Di-2-ethylhexyl phthalate | Fathead minnow (*Pimephales promelas*) | √ | - | Wood et al. (2016) |
| Dicamba | Rainbow trout (*Oncorhynchus mykiss*) | - | √ | Miller et al. (2020) |
| Ethanol | Japanese rice fish (*Oryzias latipes*) | √ | - | Dasmahapatra and Khan (2015, 2016); Hu et al. (2009) |
| 17α-Ethynylestradiol | Fathead minnows (*Pimephales promelas*) | - | √ | Fetke et al. (2021) |
| | Inland silversides (*Menidia beryllina*) | √ | - | DeCourten et al. (2020) |
| | Medaka (*Oryzias latipes*) | √ | √ | Thayil et al. (2020) |
| | Medaka (*Oryzias latipes*) | - | √ | Contractor et al. (2004) |

(*Continued*)

**TABLE 10.2 (*Continued*)**
**List of Studies in Non-Model Fish Species Where Toxicant-Exposure Induced Changes in DNA Methylation Changes Were Investigated**

| Toxicant | Species | Epigenetic Modifiers[a] | DNA Methylation[b] | References |
|---|---|---|---|---|
| Genistein | Atlantic Salmon (*Salmo salar*) | √ | - | Olsvik et al. (2017) |
| Glyphosate | Rainbow trout (*Oncorhynchus mykiss*) | - | √ | Le Du-Carree et al. (2021) |
| Hexabromocyclododecane and 17-beta estradiol | Three-spine stickleback (*Gasterosteus aculeatus*) | √ | - | Aniagu et al. (2008) |
| Hydrocarbons and metalloids | Shoal flounder (*Syacium gunteri*) | - | √ | Quintanilla-Mena et al. (2020) |
| Hydrogen sulfide-rich springs | *Guppy* (*Poecilia exicana*) | - | √ | Kelley et al. (2021) |
| Hypoxia | Rainbow trout (*Oncorhynchus mykiss*) | - | √ | Leger et al. (2021) |
| Industrial solid waste | Tilapia (*Oreochromis niloticus*) | - | √ | Flohr et al. (2012) |
| Mine tailings (metal mixture) | Atlantic cod (*Gadus morhua*) | √ | - | Reinardy et al. (2019) |
| p,p'-DDE | Atlantic Salmon (*Salmo salar*) | - | √ | Olsvik and Softeland (2018) |
| PAHs | Atlantic killifish (*Fundulus heteroclitus*) | √ | √ | Aluru et al. (2011), Bugel et al. (2014), Glazer et al. (2018) |
| Tributyltin, Triphenyltin | *False kelpfish* (*Sebastiscus marmoratus*) | - | √ | Wang et al. (2009) |
| Triclosan | Medaka (*Oryzias latipes*) | √ | - | Song et al. (2020) |

[a] Epigenetic modifiers include enzymes involved in DNA methylation and demethylation.
[b] DNA methylation represents studies where either global 5mC is measured using ELISA or HPLC or gene-specific or genome-wide sequencing approaches are used.

gene-specific methylation. The reason for the lack of studies in non-model species is mainly due to unavailability of genomic information. For instance, for investigating gene-specific DNA methylation, one needs sequence information for the proximal promoter of a gene to identify the CpG dense regions and to design PCR (polymerase chain reaction) primers needed for quantification of methylation status. This process is not trivial for species without sequenced genomes. Hence, it is important to first sequence the promoter region for the gene of interest and confirm the number of CpG sites before quantifying the changes in DNA methylation in response to stressors. One example where this approach was used was in quantification of DNA methylation patterns in the vitellogenin (vtg) promoter. Xenoestrogen exposure induces vtg mRNA and protein expression in a wide variety of fish species. Several studies have tested the hypothesis that xenoestrogens regulate vtg gene expression by altering DNA methylation levels (Gao et al. 2018; Guo et al. 2021; Stromqvist et al. 2010). Thus, the vtg promoter region was sequenced and DNA methylation patterns assessed in response to toxicant exposure in many non-model systems. In recent years, with the availability of sequenced genomes for many species, it has become more feasible for quantifying DNA methylation by using gene-specific approaches or whole genome bisulfite sequencing methods. It still remains to be determined how DNA methylation changes in response to toxicant exposure alter physiology and reproduction in fish species inhabiting a variety of environments.

## EPIGENETICS, ADVERSE OUTCOME PATHWAYS, AND RISK ASSESSMENT

The Adverse Outcome Pathway (AOP) conceptual framework (see Chapter 12, Adverse Outcome Pathways in this book) was proposed in 2010 for use in environmental risk assessment of chemical contaminants as a way to link direct molecular initiating events (MIE), through a path of key events (KE), and key event relationships (KERs) to an ultimate adverse outcome (AO) (Ankley et al. 2010). The scientific community-sourced AOP knowledge base (AOP-KB; aopwiki.org) provides a platform to track AOP development as well as query connectivity across KEs via network analysis (Pollesch et al. 2019). Furthermore, high-content omics datasets have helped support the AOP framework by validating KE relevance across species, more precisely defining MIEs, and informing biomarker selection for use in ecological risk assessments (Brockmeier et al. 2017). Examples of the utility of the AOP approach in human health risk assessment include predicting chemical carcinogenicity (Kang et al. 2018), phthalates (Clewell et al. 2020) and radiation (Chauhan et al. 2021) toxicity, and acetylcholine esterase inhibition (Hodges et al. 2018). That said, incorporating epigenetic alterations into AOP pathways does present certain challenges that aquatic animal studies are helping inform. In order to establish an AOP incorporating epigenetic alterations, there must be understanding of the appropriate developmental stage and tissue subtypes to assess, the dose-dependence and the causal relationship between epigenetic change and both short- and long-term adverse outcomes, and the combinatorial effects of epigenetic changes (Angrish et al. 2018). Two examples using aquatic animal models to formulate AOPs including epigenetic events, DOHaD, and transgenerational effects are highlighted below.

### Zebrafish Embryonically Exposed to Cadmium

Ruiter and colleagues (2016) used an AOP framework to predict key events underlying the persistent adult effects from an embryonic exposure to cadmium. In their study, zebrafish embryos were exposed from 0–72 hpf to 1–32 μM cadmium chloride ($CdCl_2$) and DNA promoter methylation in three target genes, and DNA oxidation was measured at the end of the exposure. Following maturation in clean conditions for 10 weeks, the adverse outcomes of altered neurobehavior (increased baseline movement, novel tank diving) and reduced oxidative stress defense were assessed. The proposed AOP included $Cd^{2+}$-thiol binding, depleted antioxidants, glutathione, and S-adenosylmethionine, while oxidative damage to DNA accumulated collectively causing alterations in the methylome that contributed to altered neurobehavior by altering brain morphology and/or biochemistry. While the proposed key event relationships, particularly the specific and biologically plausible targets of altered DNA methylation, remain to be confirmed, the study highlights how the AOP approach can be useful in understanding contaminant effects on the DOHaD of environmentally relevant contaminants.

### Daphnia Magna and 5-Azacytidine-Mediated DNMT-Inhibition Transgenerational AOP

The water flea (*D. magna*) has many advantages as research model, including their rapid reproductive cycles via parthenogenesis and ecological relevance, that have led to their standardized use in toxicity testing. Song and colleagues (Song et al. 2021) exposed *D. magna* for 7 days to 5–80 μM of the established DNA-methyltransferase inhibitor 5-azacytidine (5AC) to identify novel AOP networks for transgenerational effects. Parental (F0) exposure to 5AC resulted in decreased reproduction in the F0–F3 generations. Using Bayesian network modeling and bioinformatics, the authors formulated AOP networks with a molecular initiating event of reduced DNMT activity and KEs of decreased global methylation and increased caspase expression leading to somatic cell and oocyte apoptosis, decreased oogenesis, and ultimately F0 and F3 adverse outcomes of decreased fecundity (APO wiki #336). While this AOP was developed using an established model DNMT inhibitor, the AOP can now be used to mechanistically understand other environmental chemicals that share common epigenetic pathways.

## RESEARCH CHALLENGES AND FUTURE DIRECTIONS

There is convincing evidence suggesting that chemical stressors alter epigenetic mechanisms in both model and non-model fish species. However, the results cannot be generalizable to other fish species because of several methodological limitations. For instance, the majority of the studies investigating DNA methylation have quantified either global DNA methylation levels or locus-specific DNA methylation patterns using PCR-based quantification methods (e.g., bisulfite conversion PCR, methylation-specific PCR). Even though these are well-established methods, they only provide either global methylated cytosine content or locus-specific DNA methylation profiles. The results from these studies cannot be generalizable because of innate differences in noncoding regions of the genome (introns, regulatory elements) and DNA methylation levels between fish species. One alternative that has been widely used in model organism studies is genome-wide sequencing approaches that evaluate methylation status at nucleotide resolution. They include whole genome bisulfite sequencing (WGBS), which evaluates every cytosine nucleotide in a genome, and reduced representation bisulfite sequencing (RRBS), a cost-effective alternative to WGBS that interrogates only a representative fraction of the genome. These tools are readily available for profiling DNA methylation in species with sequenced genomes. Application of these tools in toxicology research will not only help in identifying the genome-wide changes in DNA methylation in response to chemical exposure, but it will also help in generating new hypotheses for testing. However, these methods are still expensive for large-scale adoption and limited to organisms with a well-annotated genome.

Another significant limitation of epigenetic research is lack of bioinformatic tools to link differential methylation or chromatin modifications to gene expression. Even in more studied species such as zebrafish, connecting epigenetic changes to gene expression is based on tools that were originally designed for mammalian model systems. Adopting these bioinformatic tools for other fish species is not trivial and needs advanced computational biology expertise. The future for non-model fish species research is still bright as genome sequencing and computational biology and bioinformatics tools are becoming common.

In addition to the methodological challenges, there is very little focus beyond understanding the impacts of environmental chemical exposure on epigenetic mechanisms. In order to understand the effects of exposure to multiple stressors under natural conditions, it is imperative to understand the role of genetic and epigenetic variation on short- and long-term effects of exposures. Several recent studies using model systems have made observations that genetic variation may facilitate rapid adaptation via alterations to epigenetic patterns [e.g., plants (Bender and Fink 1995; Johannes et al. 2009; Reinders et al. 2009; Verhoeven et al. 2010), *C. elegans*; (Greer et al. 2011), and mammals (Daxinger and Whitelaw 2012)]. This genetic variation could occur by many mechanisms including local chromatin states altering the expression of a gene in the absence of a genetic polymorphism; modification of the phenotypic effects of mutations by key epigenetic regulator; or the action of protein-based heritable elements. For instance, variation in genes that are actively expressed in response to a chemical exposure might be more relevant to determining chemical susceptibility. In order to understand the interaction of genetic and epigenetic variation in eliciting a phenotype, the availability of comprehensive measurements of chromatin state and gene expression level on a tissue-specific basis is necessary.

Another mode of pervasive epigenetic modification is through changes in the expression or activity of a key regulatory protein. One of the best-studied examples of this phenomenon is the protein chaperone, heat shock protein 90 kDa (Hsp90) (Wong and Houry 2006). Hsp90 is involved in the proper folding of a wide array of proteins and is intimately connected to diverse cellular processes. Thus, perturbations in Hsp90 activity have the potential to alter the phenotypic effects of many polymorphisms. The same is likely true for other cellular proteins, although Hsp90 is unusually compelling because its activity is altered by environmental conditions, and its large client repertoire is biased toward key signaling factors for growth and development. The finding that Hsp90 is a capacitor for the (adaptive) loss of eyes in cavefish highlights the role of epigenetic processes in evolutionary processes (Rohner et al. 2013). Identifying key regulatory proteins that play critical role in toxicant or stressor responses is the first step in elucidating genetic and epigenetic mechanisms of regulation.

## CONCLUSIONS

Over the past few decades, considerable progress has been made in understanding the effects of environmental chemicals on epigenetic mechanisms in aquatic organisms. Increasing accessibility of genomic resources to non-model organisms has made it possible to investigate the effects of exposure on DNA methylomes and other epigenetic mechanisms directly in species of interest. However, the field of epigenetic toxicology is still in its infancy. One of the major challenges in the field is determining the causative role of epigenetic changes in toxicant-induced phenotypes. In addition, it is necessary to identify and validate epigenetic biomarkers of exposure and effect in order to incorporate epigenetic end-points in toxicological risk assessment, which requires thorough experimental studies to demonstrate reproducibility of the responses. One of the hypothesized roles of epigenetic mechanisms is in the regulation of persistent gene expression and phenotypic changes. Several studies have identified persistent phenotypes (including multi- and transgenerational) in response to developmental and pre-conceptional exposures to contaminants. However, causal mechanistic links between exposure-induced phenotypes and epigenetic changes are still not well understood. Innovations in genomic and sequencing technologies will help in addressing some of these knowledge gaps.

## ACKNOWLEDGMENTS

The authors would like to acknowledge Ms. Cammi Thornton for her help in researching and formatting the tables as well as thorough proofreading of our drafts. Additionally, we appreciate the helpful suggestions of our peer reviewers Drs. Tracie Baker and Danielle Meyer.

## REFERENCES

Addo, K. A., Palakodety, N., Hartwell, H. J., Tingare, A., and Fry, R. C. 2020. Placental microRNAs: responders to environmental chemicals and mediators of pathophysiology of the human placenta. *Toxicol Rep* 7, 1046–1056.

Akemann, C., Meyer, D. N., Gurdziel, K. and Baker T. R. 2019. Developmental dioxin exposure alters the methylome of adult male zebrafish gonads. *Front Genet* 9, 719.

Akemann, C., Meyer, D. N., Gurdziel, K., and Baker, T. R. 2020. TCDD-induced multi- and transgenerational changes in the methylome of male zebrafish gonads. *Environ Epigenet* 6(1), dvaa010.

Alenghat, T., and Artis, D. 2014. Epigenomic regulation of host-microbiota interactions. *Trends Immunol* 35(11), 518–525.

Alhamdow, A., Lindh, C., Hagberg, J., Graff, P., Westberg, H., Krais, A. M., Albin, M., Gustavsson, P., Tinnerberg, H., and Broberg, K. 2018. DNA methylation of the cancer-related genes F2RL3 and AHRR is associated with occupational exposure to polycyclic aromatic hydrocarbons. *Carcinogenesis* 39(7), 869–878.

Altaf, M., Saksouk, N., and Cote, J. 2007. Histone modifications in response to DNA damage. *Mutat Res* 618(1-2), 81–90.

Aluru, N., Engelhardt, J. 2022. Exposure to 3,3′,4,4′,5–pentachlorobiphenyl (PCB126) causes widespread DNA hypomethylation in adult zebrafish testis. *Toxicol Sci* 188(1), 75–87.

Aluru, N., Karchner, S. I., and Hahn, M. E. 2011. Role of DNA methylation of AHR1 and AHR2 promoters in differential sensitivity to PCBs in Atlantic Killifish, Fundulus heteroclitus. *Aquat Toxicol* 101(1), 288–294.

Aluru, N., Karchner, S. I., Krick, K. S., Zhu, W., and Liu, J. 2018. Role of DNA methylation in altered gene expression patterns in adult zebrafish (Danio rerio) *exposed to* 3, 3', 4, 4', 5–pentachlorobiphenyl (PCB 126). *Environ Epigenet* 4(1), dvy005.

Aluru, N., Kuo, E., Helfrich, L. W., Karchner, S. I., Linney, E. A., Pais, J. E., and Franks, D. G. 2015. Developmental exposure to 2,3,7,8-tetrachlorodibenzo-p-dioxin alters DNA methyltransferase (dnmt) expression in zebrafish (Danio rerio). *Toxicol Appl Pharmacol* 284(2), 142–151.

Angrish, M. M., Allard, P., McCullough, S. D., Druwe, I. L., Chadwick, L. H., Hines, E., and Chorley, B. N. (2018). Epigenetic applications in adverse outcome pathways and environmental risk evaluation. *Environ Health Perspect* 126(4), 045001-1-045001-045012. https://doi.org/10.1289/EHP2322.

Aniagu, S. O., Williams, T. D., Allen, Y., Katsiadaki, I., and Chipman, J. K. 2008. Global genomic methylation levels in the liver and gonads of the three-spine stickleback (Gasterosteus aculeatus) after exposure to hexabromocyclododecane and 17-beta oestradiol. *Environ Int* 34(3), 310–317.

Ankley, G. T., Bennett, R. S., Erickson, R. J., Hoff, D. J., Hornung, M. W., Johnson, R. D., Mount, D. R., Nichols, J. W., Russom, C. L., Schmieder, P. K., Serrrano, J. A., Tietge, J. E., and Villeneuve, D. L. (2010). Adverse outcome pathways: a conceptual framework to support ecotoxicology research and risk assessment. *Environ Toxicol Chem* 29(3), 730–741. https://doi.org/10.1002/etc.34

Ashe, A., Sapetschnig, A., Weick, E. M., Mitchell, J., Bagijn, M. P., Cording, A. C., Doebley, A. L., Goldstein, L. D., Lehrbach, N. J., Le Pen, J., Pintacuda, G., Sakaguchi, A., Sarkies, P., Ahmed, S., and Miska, E. A. 2012. piRNAs can trigger a multigenerational epigenetic memory in the germline of C. elegans. *Cell* 150(1), 88–99.

Avila-Barnard, S., Dasgupta, S., Cheng, V., Reddam, A., Wiegand, J. L., and Volz, D. C. 2022. Tris(1,3-dichloro-2-propyl) phosphate disrupts the trajectory of cytosine methylation within developing zebrafish embryos. *Environ Res* 211, 113078.

Bagnyukova, T. V., Luzhna, L. I., Pogribny, I. P., and Lushchak, V. I. 2007. Oxidative stress and antioxidant defenses in goldfish liver in response to short-term exposure to arsenite. *Environ Mol Mutagen* 48(8), 658–665.

Baker, T. R., King-Heiden, T. C., Peterson, R. E., and Heideman, W. 2014. Dioxin induction of transgenerational inheritance of disease in zebrafish. *Mol Cell Endocrinol* 398(1-2), 36–41.

Bannister, A. J., and Kouzarides, T. 2011. Regulation of chromatin by histone modifications. *Cell Res* 21(3), 381–395.

Bazopoulou, D., Knoefler, D., Zheng, Y., Ulrich, K., Oleson, B. J., Xie, L., Kim, M., Kaufmann, A., Lee, Y. T., Dou, Y., Chen, Y., Quan, S., and Jakob, U. 2019. Developmental ROS individualizes organismal stress resistance and lifespan. *Nature* 576(7786), 301–305.

Becker, P. B., and Workman, J. L. 2013. Nucleosome remodeling and epigenetics. *Cold Spring Harb Perspect Biol* 5(9), a017905.

Bell, J. S. K., and Vertino, P. M. 2017. Orphan CpG islands define a novel class of highly active enhancers. *Epigenetics* 12(6), 449–464.

Bender J, Fink GR. Epigenetic control of an endogenous gene family is revealed by a novel blue fluorescent mutant of Arabidopsis. *Cell* 1995 Dec 1;83(5), 725–734.

Bertozzi, T. M., Becker, J. L., Blake, G. E. T., Bansal, A., Nguyen, D. K., Fernandez-Twinn, D. S., Ozanne, S. E., Bartolomei, M. S., Simmons, R. A., Watson, E. D., and Ferguson-Smith, A. C. 2021. Variably methylated retrotransposons are refractory to a range of environmental perturbations. *Nat Genet* 53(8), 1233–1242.

Bian, X., and Gao, Y. 2021. DNA methylation and gene expression alterations in zebrafish embryos exposed to cadmium. *Environ Sci Pollut Res Int* 28(23), 30101–30110.

Bird, A. P. 1980. DNA methylation and the frequency of CpG in animal DNA. *Nucleic Acids Res* 8(7), 1499–1504.

Bird, A. P. 1993. Functions for DNA methylation in vertebrates. *Cold Spring Harb Symp Quant Biol* 58, 281–285.

Bird, A. P., and Wolffe, A. P. 1999. Methylation-induced repression--belts, braces, and chromatin. *Cell* 99(5), 451–454.

Blackledge, N. P., Zhou, J. C., Tolstorukov, M. Y., Farcas, A. M., Park, P. J., and Klose, R. J. 2010. CpG islands recruit a histone H3 lysine 36 demethylase. *Mol Cell* 38(2), 179–190.

Blanc, M., Antczak, P., Cousin, X., Grunau, C., Scherbak, N., Rüegg, J., and Keiter, S. H. 2021. The insecticide permethrin induces transgenerational behavioral changes linked to transcriptomic and epigenetic alterations in zebrafish (Danio rerio). *Sci Total Environ* 779, 146404.

Blanc, M., Rüegg, J., Scherbak, N., and Keiter, S. H. 2019. Environmental chemicals differentially affect epigenetic-related mechanisms in the zebrafish liver (ZF-L) cell line and in zebrafish embryos. *Aquat Toxicol* 215, 105272.

Bommarito, P. A., Martin, E., and Fry, R. C. 2017. Effects of prenatal exposure to endocrine disruptors and toxic metals on the fetal epigenome. *Epigenomics* 9(3), 333–350.

Bosse, Y., Postma, D. S., Sin, D. D., Lamontagne, M., Couture, C., Gaudreault, N., Joubert, P., Wong, V., Elliott, M., van den Berge, M., Brandsma, C. A., Tribouley, C., Malkov, V., Tsou, J. A., Opiteck, G. J., Hogg, J. C., Sandford, A. J., Timens, W., Pare, P. D., and Laviolette, M. 2012. Molecular signature of smoking in human lung tissues. *Cancer Res* 72(15), 3753–3763.

Bousard, A., Raposo, A. C., Zylicz, J. J., Picard, C., Pires, V. B., Qi, Y., Gil, C., Syx, L., Chang, H. Y., Heard, E., and da Rocha, S. T. 2019. The role of Xist-mediated Polycomb recruitment in the initiation of X-chromosome inactivation. *EMBO Rep* 20(10), e48019.

Bouwmeester, M. C., Ruiter, S., Lommelaars, T., Sippel, J., Hodemaekers, H. M., van den Brandhof, E. J., Pennings, J. L., Kamstra, J. H., Jelinek, J., Issa, J. P., Legler, J., and van der Ven, L. T. 2016. Zebrafish embryos as a screen for DNA methylation modifications after compound exposure. *Toxicol Appl Pharmacol* 291, 84–96.

Brieno-Enriquez, M. A., Garcia-Lopez, J., Cardenas, D. B., Guibert, S., Cleroux, E., Ded, L., Hourcade Jde, D., Peknicova, J., Weber, M., and Del Mazo, J. 2015. Exposure to endocrine disruptor induces transgenerational epigenetic deregulation of microRNAs in primordial germ cells. *PLoS One* 10(4), e0124296.

Brockmeier, E. K., Hodges, G., Hutchinson, T. H., Butler, E., Hecker, M., Tollefsen, K. E., Garcia-Reyero, N., Kille, P., Becker, D., Chipman, K., Colbourne, J., Collette, T. W., Cossins, A., Cronin, M., Graystock, P., Gutsell, S., Knapen, D., Katsiadaki, I., Lange, A., ... Falciani, F. (2017). The role of omics in the application of adverse outcome pathways for chemical risk assessment. *Toxicol Sci* 158(2), 252–262. https://doi.org/10.1093/toxsci/kfx097

Bugel, S. M., Bonventre, J. A., White, L. A., Tanguay, R. L., and Cooper, K. R. 2014. Chronic exposure of killifish to a highly polluted environment desensitizes estrogen-responsive reproductive and biomarker genes. *Aquat Toxicol* 152, 222–231.

Carrozza, M. J., Utley, R. T., Workman, J. L., and Cote, J. 2003. The diverse functions of histone acetyltransferase complexes. *Trends Genet* 19(6), 321–329.

Caudill, M. A., Wang, J. C., Melnyk, S., Pogribny, I. P., Jernigan, S., Collins, M. D., Santos-Guzman, J., Swendseid, M. E., Cogger, E. A., and James, S. J. 2001. Intracellular S-adenosylhomocysteine concentrations predict global DNA hypomethylation in tissues of methyl-deficient cystathionine beta-synthase heterozygous mice. *J Nutr* 131(11), 2811–2818.

Cedar, H., and Bergman, Y. 2009. Linking DNA methylation and histone modification: patterns and paradigms. *Nat Rev Genet* 10(5), 295–304.

Chatterjee, N., Lee, H., Kim, J., Kim, D., Lee, S., and Choi, J. 2021. Critical window of exposure of CMIT/MIT with respect to developmental effects on zebrafish embryos: Multi-level endpoint and proteomics analysis. *Environ Pollut* 268(Pt A), 115784.

Chauhan, V., Villeneuve, D., and Cool, D. (2021). Collaborative efforts are needed among the scientific community to advance the adverse outcome pathway concept in areas of radiation risk assessment. *Int J Rad Biol* 97(6), 815–823. https://doi.org/10.1080/09553002.2020.1857456.

Chen, Y., Zhang, Y., Yu, Z., Guan, Y., Chen, R., and Wang, C. 2021. Early-life phenanthrene exposure inhibits reproductive ability in adult zebrafish and the mechanism of action. *Chemosphere* 272, 129635.

Cleary, J. A., Tillitt, D. E., Vom Saal, F. S., Nicks, D. K., Claunch, R. A., and Bhandari, R. K. 2019. Atrazine induced transgenerational reproductive effects in medaka (Oryzias latipes). *Environ Pollut* 251, 639–650.

Clewell, R. A., Leonard, J. A., Nicolas, C. I., Campbell, J. L., Yoon, M., Efremenko, A. Y., McMullen, P. D., Andersen, M. E., Clewell, H. J., Phillips, K. A., and Tan, Y.-M. (2020). Application of a combined aggregate exposure pathway and adverse outcome pathway (AEP-AOP) approach to inform a cumulative risk assessment: a case study with phthalates. *Toxicol Vitro*, 66(March), 104855. https://doi.org/10.1016/j.tiv.2020.104855.

Contractor, R. G., Foran, C. M., Li, S., and Willett, K. L. 2004. Evidence of gender-and tissue-specific promoter methylation and the potential for ethinylestradiol-induced changes in Japanese medaka (Oryzias latipes) estrogen receptor and aromatase genes. *J Toxicol Environ Health A* 67(1), 1–22.

Corrales, J., Fang, X., Thornton, C., Mei, W., Barbazuk, W. B., Duke, M., Scheffler, B. E., and Willett, K. L. 2014. Effects on specific promoter DNA methylation in zebrafish embryos and larvae following benzo[a]pyrene exposure. *Comp Biochem Physiol C Toxicol Pharmacol* 163, 37–46.

Cress, W. D., and Seto, E. 2000. Histone deacetylases, transcriptional control, and cancer. *J Cell Physiol* 184(1), 1–16.

Cyr, A. R., and Domann, F. E. 2011. The redox basis of epigenetic modifications: from mechanisms to functional consequences. *Antioxid Redox Signal* 15(2), 551–589.

Dasmahapatra, A. K., and Khan, I. A. 2015. DNA methyltransferase expressions in Japanese rice fish (Oryzias latipes) embryogenesis is developmentally regulated and modulated by ethanol and 5-azacytidine. *Comp Biochem Physiol C Toxicol Pharmacol* 176–177, 1–9.

Dasmahapatra, A. K., and Khan, I. A. 2016. Modulation of DNA methylation machineries in Japanese rice fish (Oryzias latipes) embryogenesis by ethanol and 5-azacytidine. *Comp Biochem Physiol C Toxicol Pharmacol* 179, 174–183.

Daxinger L, Whitelaw E. Understanding transgenerational epigenetic inheritance via the gametes in mammals. *Nat Rev Genet.* 2012 Jan 31;13(3), 153–162.

Deans, C., and Maggert, K. A. 2015. What do you mean, "epigenetic"? *Genetics* 199(4), 887–896.

DeCourten, B. M., Forbes, J. P., Roark, H. K., Burns, N. P., Major, K. M., White, J. W., Li, J., Mehinto, A. C., Connon, R. E., and Brander, S. M. 2020. Multigenerational and transgenerational effects of environmentally relevant concentrations of endocrine disruptors in an estuarine fish model. *Environ Sci Technol* 54(21), 13849–13860.

Dolinoy, D. C. 2008. The agouti mouse model: an epigenetic biosensor for nutritional and environmental alterations on the fetal epigenome. *Nutr Rev* 66(Suppl 1), S7–S11.

Dorts, J., Falisse, E., Schoofs, E., Flamion, E., Kestemont, P., and Silvestre, F. 2016. DNA methyltransferases and stress-related genes expression in zebrafish larvae after exposure to heat and copper during reprogramming of DNA methylation. *Sci Rep* 6, 34254.

Ernst, C., Odom, D. T., and Kutter, C. 2017. The emergence of piRNAs against transposon invasion to preserve mammalian genome integrity. *Nat Commun* 8(1), 1411.

Etchegaray, J. P., and Mostoslavsky, R. 2016. Interplay between metabolism and epigenetics: a nuclear adaptation to environmental changes. *Mol Cell* 62(5), 695–711.

Fan, X., Hou, T., Jia, J., Tang, K., Wei, X., and Wang, Z. 2020. Discrepant dose responses of bisphenol A on oxidative stress and DNA methylation in grass carp ovary cells. *Chemosphere* 248, 126110.

Fang, X., Dong, W., Thornton, C., Scheffler, B., and Willett, K. L. 2010a. Benzo(a)pyrene induced glycine N-methyltransferase messenger RNA expression in Fundulus heteroclitus embryos. *Mar Environ Res* 69(Suppl), S74–S76.

Fang, X., Dong, W., Thornton, C., and Willett, K. L. 2010b. Benzo[a]pyrene effects on glycine N-methyltransferase mRNA expression and enzyme activity in Fundulus heteroclitus embryos. *Aquat Toxicol* 98(2), 130–138.

Fang, X., Thornton, C., Scheffler, B. E., and Willett, K. L. 2013. Benzo[a]pyrene decreases global and gene specific DNA methylation during zebrafish development. *Environ Toxicol Pharmacol* 36(1), 40–50.

Fetke, J. K., Martinson, J. W., Flick, R. W., Huang, W., Bencic, D. C., See, M. J., Pilgrim, E. M., Debry, R. W., and Biales, A. D. 2021. DNA methylation and expression of estrogen receptor alpha in fathead minnows exposed to 17alpha-ethynylestradiol. *Aquat Toxicol* 233, 105788.

Flohr, L., Fuzinatto, C. F., Melegari, S. P., and Matias, W. G. 2012. Effects of exposure to soluble fraction of industrial solid waste on lipid peroxidation and DNA methylation in erythrocytes of Oreochromis niloticus, as assessed by quantification of MDA and m(5)dC rates. *Ecotoxicol Environ Saf* 76(2), 63–70.

Fuzinatto, C. F., Flohr, L., Melegari, S. P., and Matias, W. G. 2015. Oxidative stress and hypermethylation induced by exposure of Oreochromis niloticus to complex environmental mixtures of river water from Cubatao do Sul, Brazil. *Ecotoxicol Environ Saf* 114, 190–197.

Gao, D., Lin, J., Ou, K., Chen, Y., Li, H., Dai, Q., Yu, Z., Zuo, Z., and Wang, C. 2018. Embryonic exposure to benzo(a)pyrene inhibits reproductive capability in adult female zebrafish and correlation with DNA methylation. *Environ Pollut* 240, 403–411.

Ge, C., Ye, J., Weber, C., Sun, W., Zhang, H., Zhou, Y., Cai, C., Qian, G., and Capel, B. 2018. The histone demethylase KDM6B regulates temperature-dependent sex determination in a turtle species. *Science* 360(6389), 645–648.

Gil, N., and Ulitsky, I. 2020. Regulation of gene expression by cis-acting long non-coding RNAs. *Nat Rev Genet* 21(2), 102–117.

Glatt, S., Alfieri, C., and Muller, C. W. 2011. Recognizing and remodeling the nucleosome. *Curr Opin Struct Biol* 21(3), 335–341.

Glazer, L., Kido Soule, M. C., Longnecker, K., Kujawinski, E. B., and Aluru, N. 2018. Hepatic metabolite profiling of polychlorinated biphenyl (PCB)-resistant and sensitive populations of Atlantic killifish (Fundulus heteroclitus). *Aquat Toxicol* 205, 114–122.

Goll, M. G., and Halpern, M. E. 2011. DNA methylation in zebrafish. *Prog Mol Biol Transl Sci* 101, 193–218.

Gombeau, K., Bourdineaud, J. P., Ravanat, J. L., Armant, O., Camilleri, V., Cavalie, I., Floriani, M., and Adam-Guillermin, C. 2017. Epigenetic, histopathological and transcriptomic effects following exposure to depleted uranium in adult zebrafish and their progeny. *Aquat Toxicol* 184, 14–25.

Gombeau, K., Pereira, S., Ravanat, J. L., Camilleri, V., Cavalie, I., Bourdineaud, J. P., and Adam-Guillermin, C. 2016. Depleted uranium induces sex- and tissue-specific methylation patterns in adult zebrafish. *J Environ Radioact* 154, 25–33.

Greer EL, Maures TJ, Ucar D, Hauswirth AG, Mancini E, Lim JP, Benayoun BA, Shi Y, Brunet A. Transgenerational epigenetic inheritance of longevity in Caenorhabditis elegans. *Nature* 2011 Oct 19;479(7373), 365–371.

Grentzinger, T., Armenise, C., Brun, C., Mugat, B., Serrano, V., Pelisson, A., and Chambeyron, S. 2012. piRNA-mediated transgenerational inheritance of an acquired trait. *Genome Res* 22(10), 1877–1888.

Guo, C. Y., Tseng, P. W., Hwang, J. S., Wu, G. C., and Chang, C. F. 2021. Potential role of DNA methylation of cyp19a1a promoter during sex change in protogynous orange-spotted grouper, Epinephelus coioides. *Gen Comp Endocrinol* 311, 113840.

Guo, W., Han, J., Wu, S., Shi, X., Wang, Q., and Zhou, B. 2020. Bis(2-ethylhexyl)-2,3,4,5-tetrabromophthalate affects lipid metabolism in zebrafish larvae via DNA methylation modification. *Environ Sci Technol* 54(1), 355–363.

Gyimah, E., Dong, X., Xu, H., Zhang, Z., and Mensah, J. K. 2021. Embryonic exposure to low concentrations of bisphenol A and S altered genes related to pancreatic β-cell development and DNA methyltransferase in zebrafish. *Arch Environ Contam Toxicol* 80(2), 450–460.

Hancock, R. L., Dunne, K., Walport, L. J., Flashman, E., and Kawamura, A. 2015. Epigenetic regulation by histone demethylases in hypoxia. *Epigenomics* 7(5), 791–811.

Harrison, A., and Parle-McDermott, A. 2011. DNA methylation: a timeline of methods and applications. *Front Genet* 2, 74.

Ho, L., and Crabtree, G. R. 2010. Chromatin remodelling during development. *Nature* 463(7280), 474–484.

Hodges, G., Gutsell, S., Taylor, N., Brockmeier, E., Butler, E., Rendal, C., and Colbourne, J. (2018). Invertebrate model species in AOP development. In N. Garcia-Reyero and C. A. Murphy (Eds.), *A Systems Biology Approach to Advancing Adverse Outcome Pathways for Risk Assessment* (pp. 75–106). Springer International Publishing. https://doi.org/10.1007/978-3-319-66084-4_5.

Hu, C., Huang, Z., Sun, B., Liu, M., Tang, L., and Chen, L. 2022. Metabolomic profiles in zebrafish larvae following probiotic and perfluorobutanesulfonate coexposure. *Environ Res* 204(Pt D), 112380.

Hu, F., Yin, L., Dong, F., Zheng, M., Zhao, Y., Fu, S., Zhang, W., and Chen, X. 2021a. Effects of long-term cadmium exposure on growth, antioxidant defense and DNA methylation in juvenile Nile tilapia (Oreochromis niloticus). *Aquat Toxicol* 241, 106014.

Hu, J., Liu, J., Li, J., Lv, X., Yu, L., Wu, K., and Yang, Y. 2021b. Metal contamination, bioaccumulation, ROS generation, and epigenotoxicity influences on zebrafish exposed to river water polluted by mining activities. *J Hazard Mater* 405, 124150.

Hu, Y., Willett, K. L., Khan, I. A., Scheffler, B. E., and Dasmahapatra, A. K. 2009. Ethanol disrupts chondrification of the neurocranial cartilages in medaka embryos without affecting aldehyde dehydrogenase 1A2 (Aldh1A2) promoter methylation. *Comp Biochem Physiol C Toxicol Pharmacol* 150(4), 495–502.

Irizarry, R. A., Ladd-Acosta, C., Wen, B., Wu, Z., Montano, C., Onyango, P., Cui, H., Gabo, K., Rongione, M., Webster, M., Ji, H., Potash, J., Sabunciyan, S., and Feinberg, A. P. 2009. The human colon cancer methylome shows similar hypo- and hypermethylation at conserved tissue-specific CpG island shores. *Nat Genet* 41(2), 178–186.

Jin, X., Liu, W., Miao, J., Tai, Z., Li, L., Guan, P., and Liu, J. X. 2021. Copper ions impair zebrafish skeletal myofibrillogenesis via epigenetic regulation. *Faseb J* 35(7), e21686.

Jing, H., and Lin, H. 2015. Sirtuins in epigenetic regulation. *Chem Rev* 115(6), 2350–2375.

Johannes F, Porcher E, Teixeira FK, Saliba-Colombani V, Simon M, Agier N, Bulski A, Albuisson J, Heredia F, Audigier P, Bouchez D, Dillmann C, Guerche P, Hospital F, Colot V. Assessing the impact of transgenerational epigenetic variation on complex traits. *PLoS Genet* 2009 Jun;5(6), e1000530.

Kamstra, J. H., Sales, L. B., Alestrom, P., and Legler, J. 2017. Differential DNA methylation at conserved non-genic elements and evidence for transgenerational inheritance following developmental exposure to mono(2-ethylhexyl) phthalate and 5-azacytidine in zebrafish. *Epigenet Chromatin* 10, 20.

Kang, D. S., Yang, J. H., Kim, H. S., Koo, B. K., Lee, C. M., Ahn, Y.-S., Jung, J.-H., and Seo, Y. R. (2018). Application of the adverse outcome pathway framework to risk assessment for predicting carcinogenicity of chemicals. *J Cancer Prev* 23(3), 126–133. https://doi.org/10.15430/jcp.2018.23.3.126.

Kelley, J. L., Tobler, M., Beck, D., Sadler-Riggleman, I., Quackenbush, C. R., Arias Rodriguez, L., and Skinner, M. K. 2021. Epigenetic inheritance of DNA methylation changes in fish living in hydrogen sulfide-rich springs. *Proc Natl Acad Sci U S A* 118(26), e2014929118.

Khalil, A. M., Guttman, M., Huarte, M., Garber, M., Raj, A., Rivea Morales, D., Thomas, K., Presser, A., Bernstein, B. E., van Oudenaarden, A., Regev, A., Lander, E. S., and Rinn, J. L. 2009. Many human large intergenic noncoding RNAs associate with chromatin-modifying complexes and affect gene expression. *Proc Natl Acad Sci U S A* 106(28), 11667–11672.

Kupsco, A., Dasgupta, S., Nguyen, C., and Volz, D. C. 2017. Dynamic alterations in DNA methylation precede tris(1,3-dichloro-2-propyl)phosphate-induced delays in zebrafish epiboly. *Environ Sci Technol Lett* 4(9), 367–373.

Kurth, H. M., and Mochizuki, K. 2009. Non-coding RNA: a bridge between small RNA and DNA. *RNA Biol* 6(2), 138–140.

Laing, L. V., Viana, J., Dempster, E. L., Trznadel, M., Trunkfield, L. A., Uren Webster, T. M., van Aerle, R., Paull, G. C., Wilson, R. J., Mill, J., and Santos, E. M. 2016. Bisphenol A causes reproductive toxicity, decreases dnmt1 transcription, and reduces global DNA methylation in breeding zebrafish (Danio rerio). *Epigenetics* 11(7), 526–538.

Le Du-Carree, J., Boukhari, R., Cachot, J., Cabon, J., Louboutin, L., Morin, T., and Danion, M. 2021. Generational effects of a chronic exposure to a low environmentally relevant concentration of glyphosate on rainbow trout, Oncorhynchus mykiss. *Sci Total Environ* 801, 149462.

Lee, D. H., Hwang, S. H., Lim, M. K., Oh, J. K., Song, D. Y., Yun, E. H., and Park, E. Y. 2017. Performance of urine cotinine and hypomethylation of AHRR and F2RL3 as biomarkers for smoking exposure in a population-based cohort. *PLoS One* 12(4), e0176783.

Leger, J. A. D., Athanasio, C. G., Zhera, A., Chauhan, M. F., and Simmons, D. B. D. 2021. Hypoxic responses in Oncorhynchus mykiss involve angiogenesis, lipid, and lactate metabolism, which may be triggered by the cortisol stress response and epigenetic methylation. *Comp Biochem Physiol Part D Genomics Proteomics* 39, 100860.

Li, R., Guo, W., Lei, L., Zhang, L., Liu, Y., Han, J., Chen, L., and Zhou, B. 2020. Early-life exposure to the organophosphorus flame-retardant tris (1,3-dichloro-2-propyl) phosphate induces delayed neurotoxicity associated with DNA methylation in adult zebrafish. *Environ Int* 134, 105293.

Lin, W., Huang, Z., Ping, S., Zhang, S., Wen, X., He, Y., and Ren, Y. 2022. Toxicological effects of atenolol and venlafaxine on zebrafish tissues: bioaccumulation, DNA hypomethylation, and molecular mechanism. *Environ Pollut* 299, 118898.

Lin, W., Yan, Y., Ping, S., Li, P., Li, D., Hu, J., Liu, W., Wen, X., and Ren, Y. 2021. Metformin-induced epigenetic toxicity in zebrafish: experimental and molecular dynamics simulation studies. *Environ Sci Technol* 55(3), 1672–1681.

Liu, Y., Wang, L., Zhu, L., Ran, B., and Wang, Z. 2020. Bisphenol A disturbs transcription of steroidogenic genes in ovary of rare minnow Gobiocypris rarus via the abnormal DNA and histone methylation. *Chemosphere* 240, 124935.

Liu, Y., Yuan, C., Chen, S., Zheng, Y., Zhang, Y., Gao, J., and Wang, Z. 2014. Global and cyp19a1a gene specific DNA methylation in gonads of adult rare minnow Gobiocypris rarus under bisphenol A exposure. *Aquat Toxicol* 156, 10–16.

Liu, Y., Zhang, Y., Tao, S., Guan, Y., Zhang, T., and Wang, Z. 2016. Global DNA methylation in gonads of adult zebrafish Danio rerio under bisphenol A exposure. *Ecotoxicol Environ Saf* 130, 124–132.

Lu, C., and Thompson, C. B. 2012. Metabolic regulation of epigenetics. *Cell Metab* 16(1), 9–17.

Ma, Y. B., Jia, P. P., Junaid, M., Yang, L., Lu, C. J., and Pei, D. S. 2018. Reproductive effects linked to DNA methylation in male zebrafish chronically exposed to environmentally relevant concentrations of di-(2-ethylhexyl) phthalate. *Environ Pollut* 237, 1050–1061.

Macheroux, P., Kappes, B., and Ealick, S. E. 2011. Flavogenomics--a genomic and structural view of flavin-dependent proteins. *FEBS J* 278(15), 2625–2634.

Mattick, J. S. 2003. Challenging the dogma: the hidden layer of non-protein-coding RNAs in complex organisms. *Bioessays* 25(10), 930–939.

Matuleviciute, R., Cunha, P. P., Johnson, R. S., and Foskolou, I. P. 2021. Oxygen regulation of TET enzymes. *FEBS J* 288(24), 7143–7161.

McDonnell, D., Madison, B. N., Baillon, L., Wallace, S. J., Brown, S. R., Hodson, P. V., and Langlois, V. S. 2019. Comparative toxicity of two diluted bitumens to developing yellow perch (Perca flavescens). *Sci Total Environ* 655, 977–985.

Miller, J. G. P., Jamwal, A., Ilnytskyy, Y., Hontela, A., and Wiseman, S. B. 2020. Dicamba elevates concentrations of S-adenosyl methionine but does not induce oxidative stress or alter DNA methylation in rainbow trout (Oncorhynchus mykiss) hepatocytes. *Comp Biochem Physiol Part D Genom Proteom* 36, 100744.

Morgan, H. D., Sutherland, H. G., Martin, D. I., and Whitelaw, E. 1999. Epigenetic inheritance at the agouti locus in the mouse. *Nat Genet* 23(3), 314–318.

Olsvik, P. A., Skjaerven, K. H., and Softeland, L. 2017. Metabolic signatures of bisphenol A and genistein in Atlantic salmon liver cells. *Chemosphere* 189, 730–743.

Olsvik, P. A., and Softeland, L. 2018. Metabolic effects of p,p'-DDE on Atlantic salmon hepatocytes. *J Appl Toxicol* 38(4), 489–503.

Olsvik, P. A., Whatmore, P., Penglase, S. J., Skjaerven, K. H., Angles d'Auriac, M., and Ellingsen, S. 2019. Associations between behavioral effects of bisphenol A and DNA methylation in zebrafish embryos. *Front Genet* 10, 184.

Olsvik, P. A., Williams, T. D., Tung, H. S., Mirbahai, L., Sanden, M., Skjaerven, K. H., and Ellingsen, S. 2014. Impacts of TCDD and MeHg on DNA methylation in zebrafish (Danio rerio) across two generations. *Comp Biochem Physiol C Toxicol Pharmacol* 165, 17–27.

Park, K., Han, E. J., Ahn, G., and Kwak, I. S. 2020. Effects of thermal stress-induced lead (Pb) toxicity on apoptotic cell death, inflammatory response, oxidative defense, and DNA methylation in zebrafish (Danio rerio) embryos. *Aquat Toxicol* 224, 105479.

Philibert, D. A., Lyons, D. D., Qin, R., Huang, R., El-Din, M. G., and Tierney, K. B. 2019. Persistent and transgenerational effects of raw and ozonated oil sands process-affected water exposure on a model vertebrate, the zebrafish. *Sci Total Environ* 693, 133611.

Pierron, F., Baillon, L., Sow, M., Gotreau, S., and Gonzalez, P. 2014. Effect of low-dose cadmium exposure on DNA methylation in the endangered European eel. *Environ Sci Technol* 48(1), 797–803.

Pierron, F., Lorioux, S., Heroin, D., Daffe, G., Etcheverria, B., Cachot, J., Morin, B., Dufour, S., and Gonzalez, P. 2021. Transgenerational epigenetic sex determination: Environment experienced by female fish affects offspring sex ratio. *Environ Pollut* 277, 116864.

Pollesch, N. L., Villeneuve, D. L., and O'Brien, J. M. (2019). Extracting and benchmarking emerging adverse outcome pathway knowledge. *Toxicol Sci* 168(2), 349–364. https://doi.org/10.1093/toxsci/kfz006

Qian, Y., Ji, C., Yue, S., and Zhao, M. 2019. Exposure of low-dose fipronil enantioselectively induced anxiety-like behavior associated with DNA methylation changes in embryonic and larval zebrafish. *Environ Pollut* 249, 362–371.

Qian, Y., Wang, C., Wang, J., Zhang, X., Zhou, Z., Zhao, M., and Lu, C. 2017. Fipronil-induced enantioselective developmental toxicity to zebrafish embryo-larvae involves changes in DNA methylation. *Sci Rep* 7(1), 2284.

Quintanilla-Mena, M., Gold-Bouchot, G., Zapata-Perez, O., Rubio-Pina, J., Quiroz-Moreno, A., Vidal-Martinez, V. M., Aguirre-Macedo, M. L., and Puch-Hau, C. 2020. Biological responses of shoal flounder (Syacium gunteri) to toxic environmental pollutants from the southern Gulf of Mexico. *Environ Pollut* 258, 113669.

Reinardy, H. C., Pedersen, K. B., Nahrgang, J., and Frantzen, M. 2019. Effects of mine tailings exposure on early life stages of atlantic cod. *Environ Toxicol Chem* 38(7), 1446–1454.

Reinders J, Wulff BB, Mirouze M, Marí-Ordóñez A, Dapp M, Rozhon W, Bucher E, Theiler G, Paszkowski J. Compromised stability of DNA methylation and transposon immobilization in mosaic Arabidopsis epigenomes. *Genes Dev* 2009 Apr 15;23(8), 939–950.

Rinn, J. L., Kertesz, M., Wang, J. K., Squazzo, S. L., Xu, X., Brugmann, S. A., Goodnough, L. H., Helms, J. A., Farnham, P. J., Segal, E., and Chang, H. Y. 2007. Functional demarcation of active and silent chromatin domains in human HOX loci by noncoding RNAs. *Cell* 129(7), 1311–1323.

Rodgers, A. B., Morgan, C. P., Leu, N. A., and Bale, T. L. 2015. Transgenerational epigenetic programming via sperm microRNA recapitulates effects of paternal stress. *Proc Natl Acad Sci U S A* 112(44), 13699–13704.

Rohner, N., Jarosz, D. F., Kowalko, J. E., Yoshizawa, M., Jeffery, W. R., Borowsky, R. L., Lindquist, S., and Tabin, C. J. 2013. Cryptic variation in morphological evolution: HSP90 as a capacitor for loss of eyes in cavefish. *Science* 342(6164), 1372–1375.

Ruiter, S., Sippel, J., Bouwmeester, M., Lommelaars, T., Beekhof, P., Hodemaekers, H., Bakker, F., van den Brandhof, E.-J., Pennings, J., and van der Ven, L. (2016). Programmed effects in neurobehavior and antioxidative physiology in zebrafish embryonically exposed to cadmium: observations and hypothesized adverse outcome pathway framework. *Int J Mol Sci* 17(11), 1830. https://doi.org/10.3390/ijms17111830.

Sanchez, O. F., Lee, J., Yu King Hing, N., Kim, S. E., Freeman, J. L., and Yuan, C. 2017. Lead (Pb) exposure reduces global DNA methylation level by non-competitive inhibition and alteration of dnmt expression. *Metallomics* 9(2), 149–160.

Santangeli, S., Consales, C., Pacchierotti, F., Habibi, H. R., and Carnevali, O. 2019. Transgenerational effects of BPA on female reproduction. *Sci Total Environ* 685, 1294–1305.

Schraml, E., Hackl, M., and Grillari, J. 2017. MicroRNAs and toxicology: a love marriage. *Toxicol Rep* 4, 634–636.

Serefidou, M., Venkatasubramani, A. V., and Imhof, A. 2019. The impact of one carbon metabolism on histone methylation. *Front Genet* 10, 764.

Shmakova, A., Batie, M., Druker, J., and Rocha, S. 2014. Chromatin and oxygen sensing in the context of JmjC histone demethylases. *Biochem J* 462(3), 385–395.

Song, X., Wang, X., and Bhandari, R. K. 2020. Developmental abnormalities and epigenetic alterations in medaka (Oryzias latipes) embryos induced by triclosan exposure. *Chemosphere* 261, 127613.

Song, Y., Kamstra, J. H., Cao, Y., Asselman, J., Anglès d'Auriac, M., and Friberg, N. (2021). High-throughput analyses and Bayesian network modeling highlight novel epigenetic adverse outcome pathway networks of DNA methyltransferase inhibitor mediated transgenerational effects. *J Hazard Mater* 408(October 2020). https://doi.org/10.1016/j.jhazmat.2020.124490.

Stromqvist, M., Tooke, N., and Brunstrom, B. 2010. DNA methylation levels in the 5' flanking region of the vitellogenin I gene in liver and brain of adult zebrafish (Danio rerio)--sex and tissue differences and effects of 17alpha-ethinylestradiol exposure. *Aquat Toxicol* 98(3), 275–281.

Tai, Z., Guan, P., Zhang, T., Liu, W., Li, L., Wu, Y., Li, G., and Liu, J. X. 2022. Effects of parental environmental copper stress on offspring development: DNA methylation modification and responses of differentially methylated region-related genes in transcriptional expression. *J Hazard Mater* 424(Pt C), 127600.

Tantoh, D. M., Wu, M. C., Chuang, C. C., Chen, P. H., Tyan, Y. S., Nfor, O. N., Lu, W. Y., and Liaw, Y. P. 2020. AHRR cg05575921 methylation in relation to smoking and PM2.5 exposure among Taiwanese men and women. *Clin Epigenet* 12(1), 117.

Tauheed, J., Sanchez-Guerra, M., Lee, J. J., Paul, L., Ibne Hasan, M. O. S., Quamruzzaman, Q., Selhub, J., Wright, R. O., Christiani, D. C., Coull, B. A., Baccarelli, A. A., and Mazumdar, M. 2017. Associations between post translational histone modifications, myelomeningocele risk, environmental arsenic exposure, and folate deficiency among participants in a case control study in Bangladesh. *Epigenetics* 12(6), 484–491.

Teng, M., Wang, C., Song, M., Chen, X., Zhang, J., and Wang, C. 2020. Chronic exposure of zebrafish (Danio rerio) to flutolanil leads to endocrine disruption and reproductive disorders. *Environ Res* 184, 109310.

Thayil, A. J., Wang, X., Bhandari, P., Vom Saal, F. S., Tillitt, D. E., and Bhandari, R. K. 2020. Bisphenol A and 17alpha-ethinylestradiol-induced transgenerational gene expression differences in the brain-pituitary-testis axis of medaka, Oryzias latipesdagger. *Biol Reprod* 103(6), 1324–1335.

Thienpont, B., Steinbacher, J., Zhao, H., D'Anna, F., Kuchnio, A., Ploumakis, A., Ghesquiere, B., Van Dyck, L., Boeckx, B., Schoonjans, L., Hermans, E., Amant, F., Kristensen, V. N., Peng Koh, K., Mazzone, M., Coleman, M., Carell, T., Carmeliet, P., and Lambrechts, D. 2016. Tumour hypoxia causes DNA hypermethylation by reducing TET activity. *Nature* 537(7618), 63–68.

Trevino, L. S., Dong, J., Kaushal, A., Katz, T. A., Jangid, R. K., Robertson, M. J., Grimm, S. L., Ambati, C. S. R., Putluri, V., Cox, A. R., Kim, K. H., May, T. D., Gallo, M. R., Moore, D. D., Hartig, S. M., Foulds, C. E., Putluri, N., Coarfa, C., and Walker, C. L. 2020. Epigenome environment interactions accelerate epigenomic aging and unlock metabolically restricted epigenetic reprogramming in adulthood. *Nat Commun* 11(1), 2316.

Verhoeven KJ, Jansen JJ, van Dijk PJ, Biere A. Stress-induced DNA methylation changes and their heritability in asexual dandelions. *New Phytol* 2010 Mar;185(4), 1108–1118.

Vicent, G. P., Nacht, A. S., Zaurin, R., Ballare, C., Clausell, J., and Beato, M. 2010. Minireview: role of kinases and chromatin remodeling in progesterone signaling to chromatin. *Mol Endocrinol* 24(11), 2088–2098.

Villagra, C., and Frias-Lasserre, D. 2020. Epigenetic molecular mechanisms in insects. *Neotrop Entomol* 49(5), 615–642.

Wade, P. A., Gegonne, A., Jones, P. L., Ballestar, E., Aubry, F., and Wolffe, A. P. 1999. Mi-2 complex couples DNA methylation to chromatin remodelling and histone deacetylation. *Nat Genet* 23(1), 62–66.

Wang, C., Zhang, Z., Yao, H., Zhao, F., Wang, L., Wang, X., Xing, H., and Xu, S. 2014. Effects of atrazine and chlorpyrifos on DNA methylation in the liver, kidney and gill of the common carp (Cyprinus carpio L.). *Ecotoxicol Environ Saf* 108, 142–151.

Wang, Y., Wang, C., Zhang, J., Chen, Y., and Zuo, Z. 2009. DNA hypomethylation induced by tributyltin, triphenyltin, and a mixture of these in Sebastiscus marmoratus liver. *Aquat Toxicol* 95(2), 93–98.

Wirbisky-Hershberger, S. E., Sanchez, O. F., Horzmann, K. A., Thanki, D., Yuan, C., and Freeman, J. L. 2017. Atrazine exposure decreases the activity of DNMTs, global DNA methylation levels, and dnmt expression. *Food Chem Toxicol* 109(Pt 1), 727–734.

Wong, K. S., and Houry, W. A. 2006. Hsp90 at the crossroads of genetics and epigenetics. *Cell Res* 16(9), 742–749.

Wood, R. K., Crowley, E., and Martyniuk, C. J. 2016. Developmental profiles and expression of the DNA methyltransferase genes in the fathead minnow (Pimephales promelas) following exposure to di-2-ethylhexyl phthalate. *Fish Physiol Biochem* 42(1), 7–18.

Wu, S. C., and Zhang, Y. 2010. Active DNA demethylation: many roads lead to Rome. *Nat Rev Mol Cell Biol* 11(9), 607–620.

Xu, Q. H., Guan, P., Zhang, T., Lu, C., Li, G., and Liu, J. X. 2018. Silver nanoparticles impair zebrafish skeletal and cardiac myofibrillogenesis and sarcomere formation. *Aquat Toxicol* 200, 102–113.

Xu, Y. H., Xu, Y. C., Hogstrand, C., Zhao, T., Wu, L. X., Zhuo, M. Q., and Luo, Z. 2020. Waterborne copper exposure up-regulated lipid deposition through the methylation of GRP78 and PGC1alpha of grass carp Ctenopharyngodon idella. *Ecotoxicol Environ Saf* 205, 111089.

Yap, K. L., Li, S., Munoz-Cabello, A. M., Raguz, S., Zeng, L., Mujtaba, S., Gil, J., Walsh, M. J., and Zhou, M. M. 2010. Molecular interplay of the noncoding RNA ANRIL and methylated histone H3 lysine 27 by polycomb CBX7 in transcriptional silencing of INK4a. *Mol Cell* 38(5), 662–674.

Yap, K. L., and Zhou, M. M. 2010. Keeping it in the family: diverse histone recognition by conserved structural folds. *Crit Rev Biochem Mol Biol* 45(6), 488–505.

Yokoi, T., and Nakajima, M. 2011. Toxicological implications of modulation of gene expression by microRNAs. *Toxicol Sci* 123(1), 1–14.

Yuan, C., Wang, L., Zhu, L., Ran, B., Xue, X., and Wang, Z. 2019. N-acetylcysteine alleviated bisphenol A-induced testicular DNA hypermethylation of rare minnow (Gobiocypris rarus) by increasing cysteine contents. *Ecotoxicol Environ Saf* 173, 243–250.

Yuan, C., Zhang, Y., Liu, Y., Wang, S., and Wang, Z. 2017. DNA demethylation mediated by down-regulated TETs in the testes of rare minnow Gobiocypris rarus under bisphenol A exposure. *Chemosphere* 171, 355–361.

Yuan, C., Zhang, Y., Liu, Y., Zhang, T., and Wang, Z. 2016. Enhanced GSH synthesis by Bisphenol A exposure promoted DNA methylation process in the testes of adult rare minnow Gobiocypris rarus. *Aquat Toxicol* 178, 99–105.

Zhang, T., Cooper, S., and Brockdorff, N. 2015. The interplay of histone modifications - writers that read. *EMBO Rep* 16(11), 1467–1481.

Zhang, T., Guan, P., Liu, W., Zhao, G., Fang, Y., Fu, H., Gui, J. F., Li, G., and Liu, J. X. 2020. Copper stress induces zebrafish central neural system myelin defects via WNT/NOTCH-hoxb5b signaling and pou3f1/fam168a/fam168b DNA methylation. *Biochim Biophys Acta Gene Regul Mech* 1863(10), 194612.

Zhang, T., Guan, Y., Wang, S., Wang, L., Cheng, M., Yuan, C., Liu, Y., and Wang, Z. 2018. Bisphenol A induced abnormal DNA methylation of ovarian steroidogenic genes in rare minnow Gobiocypris rarus. *Gen Comp Endocrinol* 269, 156–165.

Zhang, T., Liu, Y., Chen, H., Gao, J., Zhang, Y., Yuan, C., and Wang, Z. 2017. The DNA methylation status alteration of two steroidogenic genes in gonads of rare minnow after bisphenol A exposure. *Comp Biochem Physiol C Toxicol Pharmacol* 198, 9–18.

Zhang, X., Li, H., Qiu, Q., Qi, Y., Huang, D., and Zhang, Y. 2014. 2,4–Dichlorophenol induces global DNA hypermethylation through the increase of S-adenosylmethionine and the upregulation of DNMTs mRNA in the liver of goldfish Carassius auratus. *Comp Biochem Physiol C Toxicol Pharmacol* 160, 54–59.

Zhang, Y., Elgizouli, M., Schottker, B., Holleczek, B., Nieters, A., and Brenner, H. 2016. Smoking-associated DNA methylation markers predict lung cancer incidence. *Clin Epigenetics* 8, 127.

Zhang, Y., Li, T., Pan, C., Khan, I. A., Chen, Z., Yue, Y., and Yang, M. 2022. Intergenerational toxic effects of parental exposure to bisphenol AF on offspring and epigenetic modulations in zebrafish. *Sci Total Environ* 823, 153714.

Zhang, Y., Zhang, M., Zhu, Z., Yang, H., Wei, W., and Li, B. 2021. Bisphenol A regulates apolipoprotein A1 expression through estrogen receptors and DNA methlylation and leads to cholesterol disorder in rare minnow testis. *Aquat Toxicol* 241, 105999.

Zhao, F., Wei, P., Wang, J., Yu, M., Zhang, X., Tian, H., Wang, W., and Ru, S. 2017. Estrogenic effects associated with bisphenol a exposure in male zebrafish (Danio rerio) is associated with changes of endogenous 17β-estradiol and gene specific DNA methylation levels. *Gen Comp Endocrinol* 252, 27–35.

Zhou, W., Tian, D., He, J., Yan, X., Zhao, J., Yuan, X., and Peng, S. 2019. Prolonged exposure to carbon nanoparticles induced methylome remodeling and gene expression in zebrafish heart. *J Appl Toxicol* 39(2), 322–332.

Zhu, L., Liu, Y., Xue, X., Yuan, C., and Wang, Z. 2021a. BPA's transgenerational disturbance to transcription of ovarian steroidogenic genes in rare minnow Gobiocypris rarus via DNA and histone methylation. *Sci Total Environ* 762, 143055.

Zhu, L., Wang, L., Fan, X., Dong, C., Wang, G., and Wang, Z. 2021b. Chronic exposure to Bisphenol A resulted in alterations of reproductive functions via immune defense, oxidative damage and disruption DNA/histone methylation in male rare minnow Gobiocypris rarus. *Aquat Toxicol* 236, 105849.

Zhu, L., Yuan, C., Wang, M., Liu, Y., Wang, Z., and Seif, M. M. 2020. Bisphenol A-associated alterations in DNA and histone methylation affects semen quality in rare minnow Gobiocypris rarus. *Aquat Toxicol* 226, 105580.

# 11 Proteomics and Metabolomics in Fish

*Christopher J. Martyniuk and Nancy D. Denslow*
University of Florida

**Abstract**

Omics technologies have vastly improved our understanding of chemical toxicity in fishes. These approaches include proteomics and metabolomics/lipidomics which aim to measure the totality of proteins or metabolites in a tissue. Proteomics and metabolomics have shaped knowledge of how fishes respond to their chemical environment. Proteomics in fish toxicology has adapted both label (e.g., tandem mass tags, isobaric tags for relative, and absolute quantitation) and label-free methods (e.g., spectral counting and multiple reaction monitoring) to detect proteins responsive to chemical stressors in both the laboratory and in the natural environment. Both targeted and untargeted metabolomics have been conducted to study toxicant exposure in fishes, and the field has seen significant advances in lipidomics. In this chapter, we focus on these two unique research disciplines that utilize analytical chemistry and mass spectrometry. We introduce proteomics and metabolomics, discussing the analytical and quantitative methods used to capture molecular information in fishes in response to toxicants. We discuss recent case studies and innovative approaches that underscore their strengths and weaknesses. Lastly, we present studies using multiomics approaches in fish that integrate proteomic and metabolomic data to yield a more comprehensive framework for risk assessment. The advantage of a multiomics approach is improved understanding of the molecular events that underscore physiological adaptation and compensation to chemical stressors. Recommendations for future studies utilizing proteomics and metabolomics are described. Such mechanistic data are essential for advancing adverse outcome pathways in fish ecotoxicology.

## OMICS IN FISH TOXICOLOGY: AN OVERVIEW

The molecular revolution propelled environmental toxicology into a new era of discovery and investigation. At the turn of the century, omics technologies were at the forefront of discussions related to biological effects in wildlife (e.g., biomarkers of effect), environmental risk assessment, and environmental monitoring of wild populations. Although there was great excitement about these emerging technologies at the time, there was also skepticism about the reliability, repeatability, and relevance to higher-level endpoints. Concerns were also noted about the technical knowledge required to conduct such in depth and elaborate analyses. As with any new science, there is a period of uncertainty with many questions to be answered. Such was true for omics, but over time, there has been greater acceptance in diverse scientific disciplines. This was facilitated by advances in computational biology for data interpretation, improved experimental designs, and more refined quality control of the data, beginning with sample preparation and ending with statistical testing methods for large data sets. Despite a new level of acceptance and interest, the discipline continues to refine and hone methodology.

Freshwater and marine fishes are exposed to a diverse array of chemicals in the environment. Sources of chemical pollution include pesticide runoff from agricultural regions, nonmetabolized pharmaceuticals from sewage effluent, stormwater events collecting fertilizers and industrial products, chemical deposition from factory exhaust, industrial activities on the oceans like oil transport and exploration, and plastic debris and e-waste among other sources. Aquatic pollution is no longer isolated to regions with heavy industrial activity, but rather all water bodies are potential sinks for chemical waste. To combat such a diverse array of potential toxicants, researchers must be equipped with flexible and innovative approaches to detect biological responses in fishes that are exposed to chemical pollution. Several tools have been utilized in environmental monitoring assessments of fish populations by academics, industry, and governments, for example, whole body and morphometric indicators of chemical exposure (i.e., gonadosomatic index) and physiological endpoints such as steroid hormones, metabolism, and oxidative damage response. The science of omics provides novel insights and unique perspectives regarding the mechanism of action of aquatic pollutants. No longer are these techniques restricted to model species; in a relatively short period of time, there has been a universal effort toward incorporating omics data into environmental assessments of fish populations. A search of the NCBI database (May 2022) using the search term ("transcriptomics OR *proteomics* OR *metabolomics OR lipidomics*"+fishes+toxic*) reveals that over the past two decades, transcriptomics remains the most widely used method for assessing chemical stressors in fish (Figure 11.1) to elucidate molecular responses underlying complex exposures in a wide variety of fish species. These studies report on transcripts that are increased or decreased in relative abundance in exposed individuals compared to unexposed fish in the laboratory or fish collected from a reference site. Over the past decade, there has been an increase in the number of studies utilizing proteomics and metabolomics in fish toxicology; more recently, lipidomics is increasing in popularity as an approach to study the impact of chemical exposures on energy utilization and fat stores. Proteomic and metabolomic profiles uncovered by omics can present new clues into how a chemical induces toxicity. In essence, these approaches reveal novel mechanisms of action of chemicals in fishes. Sensitive pathways related to hormone, immune, metabolic, and cell-signaling pathways can then be identified and investigated more


DOI: 10.1201/9781003160694-11

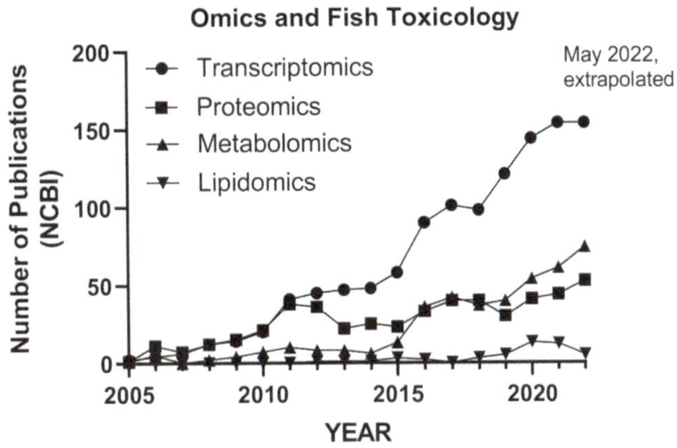


**FIGURE 11.1** Number of publications in NCBI for omics technologies, toxicology, and fishes over the past two decades. The year 2022 is extrapolated, from the end of May 26th, 2022, when the search was conducted, to December 31, 2022.

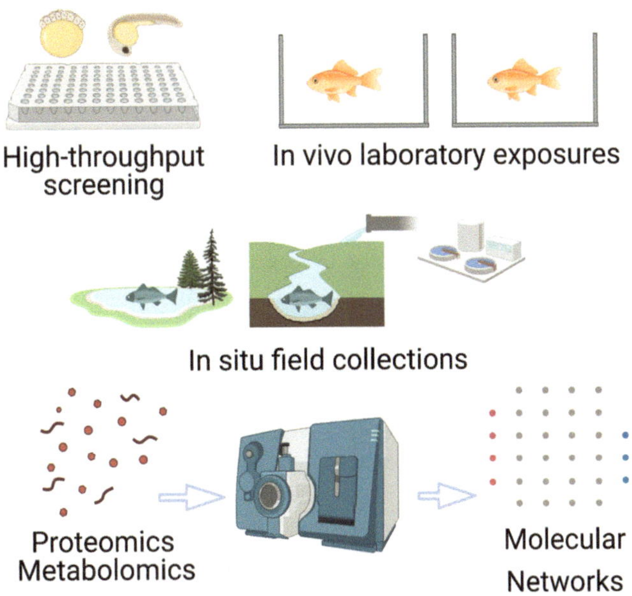


**FIGURE 11.2** Different proteomics and metabolomics experiments in fish toxicology. High-throughput screening assays have been developed for embryos and larvae to screen chemicals. Samples from such studies can be examined for protein and metabolite composition. Similarly, laboratory exposures with contaminated water and field exposures with fish sampled from polluted lakes can yield tissue to be analyzed using mass spectrometry.

intensely to achieve a greater understanding of risks associated with chemical exposures.

In this chapter, we focus on two related but distinct omics disciplines, each of which utilize principles of analytical chemistry and mass spectrometry. Proteomics and metabolomics/lipidomics have vastly improved our understanding of how fishes respond to their chemical environment. These studies include both laboratory- and field-based experiments to elucidate new pathways of toxicity (Figure 11.2). High-throughput toxicity testing using embryos and larvae of small-bodied fish species such as zebrafish (*Danio rerio*), fathead minnow (*Pimephales promelas*), and Japanese medaka (*Oryzias latipes*) in addition to other fishes has been used to reveal toxicity pathways for a wide array of environmental contaminants. These plate-based assays that include 100s of embryos have also been highly beneficial as a tool for pharmaceutical testing and elucidating the effects of illicit drugs in vertebrates. In parallel, laboratory-based investigations expose diverse fishes to chemicals using dietary and water-borne exposure paradigms. Omics technologies have also been used in field experiments to describe the impact of polluted environments on resident fish populations and to compare aquatic systems with different levels of anthropogenic input (e.g., industrial, commercial, and sewage treatment effluent) (Figure 11.2). Such experiments have the overarching objective of identifying proteins and metabolites that can be used as biomarkers that inform on acute and chronic toxicity, and ultimately, predict long-term effects on populations in the wild. In the following sections, we briefly introduce the disciplines of proteomics and metabolomics, discussing the methodology used to capture information about each. We provide case studies of their use, underscoring their strengths and weaknesses, and pointing out challenges that are unique to fishes. It is anticipated that proteomics and metabolomics will continue to contribute greatly to studies in fish toxicology.

## PROTEOMICS

Proteomics is the study of the totality of the protein complement within the cell. The basic building blocks of proteins are amino acids, and quantitative proteomics relies on analytical detection of amino acids to quantify relative protein abundance. Proteomics also includes studies characterizing posttranslational modifications on proteins. These include acetylation and methylation, ubiquitination, glycosylation, phosphorylation, and sumoylation. Posttranslational modifications of proteins have high relevance to the physiology of the organism, as they activate or inhibit enzyme activity (e.g., phosphorylation) or function as molecular signals to transport, metabolize, or degrade proteins (e.g., glycosylation and ubiquitination) within the cell. Posttranslational modifications can be extensive, exponentially expanding the "functional proteome" into more than 100,000 different protein entities within the cell. To quantify such a diverse class of molecules, several innovative approaches using liquid chromatography and mass spectrometry exist to facilitate new discoveries in proteomics. Initially, these methods were limited to the human medical sciences, but over the past two decades, such methods have been applied to study the toxicology of chemicals in fishes (Martyniuk et al. 2012; Martyniuk and Denslow 2009; Sanchez et al. 2011).

Recent reviews on proteomics in fish toxicology have been published, describing advances in the field and types of chemicals investigated at the level of the proteome (Liang et al. 2020). Proteomic studies in fishes include those investigating the effects of triclosan, bisphenol A (BPA), di-*n*-butyl phthalate (DBP), 2,3,7,8-tetrachlorodibenzo-*p*-dioxin (TCDD), cadmium ($Cd^{2+}$), diluted bitumen and oil, benzotriazole (BT), and pesticides (e.g., organochlorine pesticides like dieldrin) which are summarized by others (Liang et al. 2020). There are several advantages of proteomics over other omics methods. Proteins are typically longer lived compared to transcripts which can be more transient in expression. For example, vitellogenin (Vtg) mRNA precedes the synthesis of Vtg protein, but it is degraded within a short time. Vtg protein, on the other hand, is relatively stable and still remains present in the blood of males a month after a short exposure to 17β-estradiol (Denslow et al. 2005). Proteins can exhibit stability over longer periods, which may be more useful for assessing chronic impacts of a chemical exposure. For example, increased enzyme abundance and activity for proteins involved in biotransformation of chemicals are useful indicators of adverse effect. Proteins are also the functional units of the cell; a transcript can increase or decrease in steady-state levels; however, this does not necessarily translate into a functional consequence for the individual. Posttranscriptional modification and regulation by, for example, miRNAs and alternative splicing are key processes that determine whether the corresponding protein is translated from the message. As such, it is important to point out that each of the omics has inherent strengths and weaknesses.

## PROTEOMICS METHODS IN FISH TOXICOLOGY

Quantitative proteomics (i.e., determining differences in protein abundance among treatment groups) in environmental risk assessment has not been as widely utilized compared to transcriptomics or metabolomics. However, the use of proteomics in fish toxicology has been relatively consistent over the last decade (Figure 11.1). Some of the early fish models used in proteomics experiments included zebrafish, fathead minnow, Japanese medaka, Atlantic cod (*Gadus morhua*), and largemouth bass (*Micropterus salmoides*) to name but a few species. To simplify, there are two major approaches in quantitative proteomics that are typically employed in fish toxicology experiments. Label-based methods utilize a "tag" that is chemically attached to a specific amino acid or peptide. In the mass spectrometer, the tag is ultimately cleaved from the peptide and acts as a reporter ion to ultimately quantify relative protein abundance. Labeling methods in proteomics include 2-D Difference Gel Electrophoresis (DiGE), which utilizes fluorescence labeling, stable isotope labeling (SILAC), isotope-coded affinity tags (ICAT), isobaric tagging for relative and absolute quantitation (iTRAQ), and tandem mass tags (or TMT). Conversely, label-free proteomics includes spectral counting and absolute quantitation. Below, we provide an overview of the different methodologies available for proteomics investigations.

### Labeling Methods: Two-Dimensional Differential Gel Electrophoresis (2D-DIGE)

Many of the first studies using proteomics in environmental toxicology and fishes utilized 2D-polyacrylamide gel electrophoresis (PAGE) and western blotting, methods that are still relied upon today. In 2D-PAGE, proteins are first separated in one-dimensional isoelectric focusing gels using an electrical current. Isoelectric focusing refers to the process of the protein moving along a pH gradient in the first dimension under electrical current until it rests at a point in which the overall charge of the protein is 0 (isoelectric point or point in which the protein carries a neutral charge). More specifically, the gel has a pH gradient that typically spans from pH 3 to pH 11. At any one pH, a protein has either a positive or negative charge due to the mix of amino acids, and the charge each carries. The charge on the amino acids varies with pH, so as the protein moves in the pH gradient due to its charge, it reaches a pH at which all the negative charges equal the positive charges (i.e., overall neutral charge). At this isoelectric point, the protein stops moving in the electric field. Following separation in the first dimension, proteins are then further separated on denaturing PAGE gels, which separates the proteins by their mass. After proteins have been separated, Coomassie dye, silver stains, or fluorescent stains such as SYPRO™ RUBY are used to visual the proteins in the gel. Software can be used to detect unique protein spots which are subsequently compared across gels and quantified (Marouga et al. 2005).

Building upon these principles, DiGE is a method used to separate complex mixtures of proteins using the isoelectric point of the protein and the size of the protein, as above, but the proteins are first labeled with one of two different fluorescent dyes (cyanine 3 or 5). For example, proteins derived from control tissue are labeled with a dye called cyanine 3 (Cy3), and a second set of proteins derived from treatment tissue are labeled with a second dye called cyanine 5 (Cy5). A common standardized pool of proteins, generated from both the control and treatment samples, is used as an internal standard, and is labeled with a third fluorescent dye called cyanine 2 (Cy2). The internal standard is added to every sample and acts as a calibrator or "spike" to correct for any interspecific variability among the gels.

Two-dimensional differential gel electrophoresis was used in some of the earliest experiments involving proteomics in fish to study chemical toxicants as previously reviewed (Martyniuk and Denslow 2009). Examples of these first studies included experiments to understand the effect of perfluorinated chemicals, methylmercury, and pesticides. Perfluorooctane sulfonate (PFOS), for example, is a ubiquitous environmental contaminant that is persistent and detectable in diverse animal species (Dorts et al. 2011). The impacts of PFOS exposure on the gill proteome of the European bullhead (*Cottus gobio)* were measured using DIGE after a 96-h exposure. Labeled CyDye gels were scanned with a Typhoon 9400 scanner (GE Healthcare) to reveal proteins that were increased or decreased in abundance in the gill. In addition, enzyme activity was assessed for citrate synthase and cytochrome c oxidase (Dorts et al. 2011). Proteome responses in the gill, as well as suppressed enzyme activity, suggested that acute exposure to PFOS could affect proteins and enzymes related to glycolysis, Krebs cycle, and oxidative phosphorylation. Effects on energy-related pathways can have significant implications for the gill and respiration. Another interesting finding from the study by Dorts et al. (2011) included the observation of nonmonotonic responses in the gill to low and high levels of PFOS at the level of the proteome, highlighting the fact that chemicals do not necessarily cause a dose-dependent effect on all proteins.

In another study, the effects of methylmercury in marine medaka were determined in the liver (Wang et al. 2013). Increasing hepatic levels of total Hg were related to hepatocyte damage, liver parenchyma, and increased numbers of lipid vesicles as well as apoptotic bodies. The proteomics data confirmed much of what was being observed with microstructural changes in hepatocytes, and proteins related to cytoskeletal assembly and oxidative stress were determined to be different in relative abundance compared to fish from the unexposed control. Noteworthy from the study by Wang et al. (2013) was the recognition that the toxic mechanisms of low and high Hg concentrations were different, and very few proteins were identified to be in common between the two doses. This highlights an important observation often observed in proteomics studies, that being the dose-dependent effect on the proteome.

There are other examples of 2D gel-based proteomics in fish. The toxicity of Dechlorane Plus (DP), a polychlorinated flame retardant and alternative to decabromodiphenyl ether (BDE-209), was investigated in Chinese sturgeon (*Acipenser sinensis*) following 14 days after a single intraperitoneal injection of the chemical (Liang et al. 2014). The liver of Chinese sturgeon was excised and processed for 2D gel electrophoresis. Proteins related to metabolism, calcium ion binding, protein folding, and cell structure were altered in abundance, and HSP70 was significantly upregulated which was subsequently validated by Western blot (Liang et al. 2014). The study demonstrates how two complimentary techniques (2D gel electrophoresis and Western blotting) can be used to confirm proteomic responses in tissues.

Lastly, the issue of mixtures has garnered significant attention in fish toxicology. In one study, the effects of complex environmental mixtures of herbicides and fungicides and stressors (i.e., temperature) were examined in goldfish (*Carassius auratus*). The response of the hepatic proteome was quantified using 2-D gel electrophoresis after exposure to pesticide mixtures at three concentrations at 22°C or 32°C (Gandar et al. 2017). The liver proteome responded to the treatments, and several proteins were identified as different in abundance. These proteins included those involved in glycolysis and gluconeogenesis, such as fructose-bisphosphate aldolase, glyceraldehyde-3-phosphatedehydrogenase, and triosephosphate isomerase 1b. Proteins associated with methionine degradation and abnormal DNA methylation were also altered by pesticide mixtures and temperature. These examples nicely highlight the use of 2D gel electrophoresis in fish toxicology studies to elucidate mechanisms and biomarkers of toxicity.

Recently, researchers have more frequently turned to gel-free approaches using liquid chromatography tandem mass spectrometry (LC MS/MS) to circumvent some of the limitations with 2D gel electrophoresis. This is also driven by the fact that sufficiently more information is gathered using mass spectrometry (100s versus 1,000s of proteins quantified). Limitations to 2D gel electrophoresis include variability from excision of the protein spots on the gel and lower overall sensitivity compared to MS/MS. Often in 2D gel experiments, the most abundant proteins are repeatedly detected due to post-translational modifications, limiting the scope and depth of the proteome analysis. It is also a laborious task to excise multiple spots from the gel for matrix assisted laser desorption ionization mass spectrometry (MALDI MS) analysis to identify specific proteins. However, one significant advantage for 2D gel electrophoresis over some techniques is that post-translation modifications of proteins can be identified and visualized in the gel, providing novel insight into the biology underlying the chemical exposure. Even though 2D gel electrophoresis is less frequently used in proteomics compared to LC MS/MS labeling and nonlabeling proteomics methods, important observations were made using the approach that is often recapitulated and supported using the newer methods in fish proteomics.

## Labeling Methods: iTRAQ and Tandem Mass Tags

Isobaric tags for relative and absolute quantitation or iTRAQ™ is a method that uses relative quantitation of tags to inform on the abundance of peptides in a pooled sample (Figure 11.3). A peptide reactive group acts as a reporter ion (isobaric tags

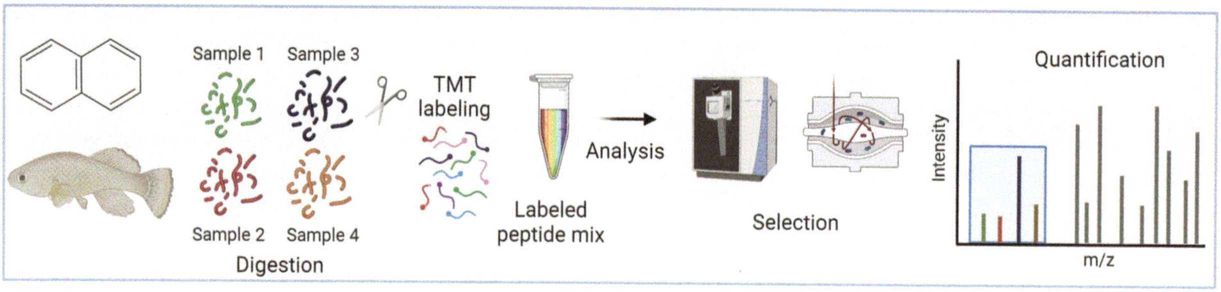


**FIGURE 11.3** General workflow for relative quantification of proteins using either isobaric tags for relative and absolute quantitation (iTRAQ™) or Tandem Mass Tag™ (TMT) systems. Fish are exposed to contaminants and their protein extracted from a tissue. Proteins are enzymatically digested into smaller peptides which are labeled through chemical modification (i.e., tags). Labeled peptides are pooled and analyzed using LC/MS/MS. During the analytical analysis, peptides are fragmented, and tags released and observed in MS/MS spectra (colored bars in blue box). These tags, which differ slightly in mass, are compared on a relative scale to indicate abundance of peptides in the initial pool, which subsequently corresponds to the initial amount of protein in the tissue.

containing unique reporter ions with masses of 113, 114, 115, 116, 117, 118, 119, 121), which is covalently bound to a neutral balance group. The total mass of the reactive group and the balancer is constructed to be equal across each of the eight tags (e.g., isobaric and chromatographically identical), ensuring they elute and are detected in a single spectrum. The important component of the iTRAQ label is the reactive group, which covalently attaches at amino groups that are present at the N-termini and at lysine side chains of peptides. Each protein sample is digested first with trypsin, which cleaves proteins at arginine or lysine amino acids to generate peptide fragments that are subsequently labeled with a unique iTRAQ reagent tag (113 up to 121). Up to eight samples can be labeled individually and pooled into a multiplex reaction, each sample in the pool having been labeled with one of the unique tags. The pooled sample is then mixed and processed using high-resolution LC/MS/MS. Collision-induced dissociation (CID) using a gas (e.g., helium, nitrogen, or argon) and high-energy collisions result in fragmentation of selected ions. The fragmentation spectra also contain the reporters that have been lost from the peptide (Figure 11.3). The relative abundance of the reporter tags informs the relative abundance of peptides originating in the pooled sample. As the tags are unique to a specific protein sample, one can statistically determine whether there are differences in relative abundance of a protein between a control and a treatment group (e.g., $n=4$ biological replicates in the pool for control and treatment group). Multiple peptides with reporter ions are typically assigned to a single protein and can be used in a weight of evidence approach to make conclusions on relative protein abundance.

There are additional approaches for peptide labeling comparable to the iTRAQ method. The TMT Systems (Thermo Scientific) can also label protein samples and, most recently, can accommodate up to 16 different samples in a single pool (16-plex). The method utilizes amine-, cysteine-, and carbonyl-reactive chemistries to label peptide functional groups, and each mass reporter contains different numbers and combinations of $^{13}C$ and $^{15}N$ isotopes to improve accuracy and specificity. While chemistries differ, the general principles of the workflow are similar in many ways to iTRAQ. Protein extracts are reduced, alkylated, and enzymatically digested. To improve specificity and detection, an anti-TMT immobilized resin can be used to enrich the labeled peptides from each sample; however, in many cases, the TMT is efficient at labeling N-termini of peptides after digestion and enrichment is not required. Labeled peptides are processed using high-resolution LC-MS/MS, and reporter ions are used to quantify relative abundance of peptides among samples.

One of the first studies to use iTRAQ to address toxicity of chemicals to fishes investigated endocrine disruptors. The potent androgen 17β-trenbolone and the antiandrogen flutamide were studied in fathead minnow, a popular fish model in ecotoxicology, to elucidate the proteome responses to compounds that interfere with androgen receptor signaling (Martyniuk et al. 2009). Flutamide is a pharmaceutical that antagonizes androgen receptors, while 17β-trenbolone is a growth promoting androgen used in the beef cattle industry to build muscle mass. The use of 17β-trenbolone in the beef industry has led to the phenomenon of female masculinization of fish inhabiting aquatic environments surrounding agricultural regions that support cattle. These two model compounds were used to modulate androgen receptor signaling in female fish. After a short-term exposure, the hepatic proteome of female fathead minnows was quantified using iTRAQ to identify protein biomarkers sensitive to chemicals that act via androgen receptors (i.e., by comparing protein responses following activation and inhibition of the receptor). The study revealed that proteins involved in cellular metabolism (e.g., glyceraldehyde 3-phosphate dehydrogenase, phosphoglycerate mutase), oxidative stress (e.g., heat shock proteins), and ribosomal proteins were reciprocally regulated in the liver of fathead minnows by 17β-trenbolone and flutamide. Protein networks were then constructed to discern molecular interactomes mediated by androgen receptors.

The iTRAQ and TMT methods have been used in fish toxicology to evaluate proteome responses to endocrine disruptors, organochlorine pesticides, and several other types of chemicals (Table 11.1). Toxicity mechanisms of pesticides have successfully been revealed with iTRAQ experiments. Diquat dibromide is an herbicide that is added directly into water bodies or applied on agricultural fields. The proteome of both early life stage swim-up fry and juvenile rainbow trout was measured using iTRAQ following pulse exposure experiments to diquat formulations to mimic exposures in the natural

**TABLE 11.1**
**Fish Proteome Studies Using Mass Tags to Study Environmental Chemicals**

| Fish Species | Method | Chemical/Toxicant | Tissue | Number of Proteins Detected | References |
|---|---|---|---|---|---|
| Common carp (*Cyprinus carpio*) | TMT | Nanosilver and silver nitrate | Gill | 9,783 | Xiang et al. (2020) |
| Fathead minnow (*Pimephales promelas*) | iTRAQ | 17beta-trenbolone | Liver | 293 | Martyniuk et al. (2009) |
| Fathead minnow (*Pimephales promelas*) | iTRAQ | 17alpha-ethinylestradiol | Telencephalon | 192 | Martyniuk et al. (2010) |
| Fathead minnow (*Pimephales promelas*) | iTRAQ | Trenbolone | Plasma | 148 | Schultz et al. (2013) |
| Largemouth Bass (*Micropterus salmoides*) | iTRAQ | Dieldrin | Hypothalamus | 1,523 | Martyniuk et al. (2010) |
| Marine medaka (*Oryzias melastigma*) | iTRAQ | BDE-47 | Testes/Ovary | 133/144 | Fong et al. (2014) |
| Marine medaka (*Oryzias melastigma*) | TMT | Butenolide and 4,5-dichloro-2-*n*-octyl-4-isothiazolin-3-one (DCOIT) | Whole brain | 2,228 | Chen et al. (2014) |
| Olive flounder (*Paralichthys olivaceus*) | iTRAQ | Cadmium | Mitochondria from gills | 4,690 | Lu et al. (2020) |
| Rainbow trout (*Oncorhynchus mykiss*), common carp (*Cyprinus carpio*) and gibel carp (*Carassius auratus gibelio*) | iTRAQ | Copper | Gills | 1,358/979/401 | Eyckmans et al. (2012) |
| Rainbow trout (*Oncorhynchus mykiss*), common carp (*Cyprinus carpio*) and gibel carp (*Carassius auratus gibelio*) | iTRAQ | Diquat dibromide | Liver | 1,207 | McCuaig et al. (2020) |
| Rockfish (*Sebastes schlegelii*) | iTRAQ | Arsenic | Whole fish | 2,053 | Xu et al. (2019) |
| Sockeye (*Oncorhynchus nerka*) | iTRAQ | Bitumen | Serum | 513 | Alderman et al. (2017) |
| Zebrafish (*Danio rerio*) | iTRAQ | Dieldrin | Liver | 3,991 | Simmons et al. (2019) |
| Zebrafish (*Danio rerio*) | iTRAQ | Dieldrin | Hypothalamus | 3,941 | Cowie et al. (2017) |
| Zebrafish (*Danio rerio*) | iTRAQ | Mercury | Larval fish | 819 | Cuello et al. (2012) |
| Zebrafish (*Danio rerio*) | iTRAQ | Tris(1,3-dichloro-2-propyl) phosphate (TDCPP) | Larval fish | 1,362 | Fu et al. (2013) |
| Zebrafish (*Danio rerio*) | iTRAQ | Bisphenol A and di-*n*-butyl phthalate | Larval fish | 4,466 | Dong et al. (2018) |

environment (McCuaig et al. 2020). The experiments revealed several hepatic proteins differentially expressed in the early developmental stage and in the liver of juveniles, including proteins related to oxidative stress, energy homeostasis, and apoptosis. Most intriguing was that there were significantly more proteins affected in swim-up fry (total 315) compared to later developmental stages (total 84), suggesting chemical exposures early in development can significantly impact the proteome. In this study, there was some evidence for mitochondrial function, and it was hypothesized that the activation of the Atk/mTOR pathway may be related to impairment of the electron transport chain (McCuaig et al. 2020).

In one of the first studies to use iTRAQ in fish, Eyckmans et al. (2012) studied responses of rainbow trout (*O. mykiss*), common carp (*C. carpio*), and gibel carp (*C. gibelio*) in the gill following exposure to copper using both 2D-DiGE and iTRAQ. A weighted gene coexpression network analysis was used to determine protein correlates to copper toxicity, and processes such as transcription and translation, histones, energy homeostasis, cell structure, iron homeostasis, and blood function were represented by the altered proteome. This study effectively compares two methodologies (one gel and one nongel) in proteomics to study chemical exposures in fish (Eyckmans et al. 2012).

A final example of the use of iTRAQ proteomics to study toxicants in fish is one involving common carp (*Cyprinus carpio*). Silver nanoparticles (AgNPs) are widely used commercialized nanomaterials, and in the study, carps were exposed for 96 hours and gill tissues were examined for damage. TMT tags were used to quantify proteome responses in the gill to the nanosilver in parallel with histopathology (Xiang et al. 2020). A total of 9,783 proteins were identified, and differentially expressed proteins were involved in the vascular endothelial growth factor (VEGF) signaling pathway, arachidonic acid metabolism, vascular smooth muscle contraction, and oxidative phosphorylation. The study also validated altered proteins with parallel reaction monitoring (PRM) to confirm protein expression. Based upon these data, a model highlighting mechanisms of gill damage was developed for nanosilver toxicity.

These studies are examples of how proteomics can be used to determine toxicity mechanisms of aquatic contaminants in fishes. As one can observe in Table 11.1, there has been a steady increase in the number of proteins identified in iTRAQ and TMT experiments over the past decade (~10-fold increase), which bodes well for future experiments.

A final point to make is that proteomics can be used to improve the understanding of environmental exposures, which can be associated with several wildlife and human diseases. Such studies improve linkages between environmental chemical exposure and long-term health effects. There are examples of this from fish toxicology studies. Proteomics has been used to study molecular responses in the whole brain of fish to elucidate neurotoxicity mechanisms, as well as in distinct regions of the brain important for behavior and neuroendocrine function (e.g., hypothalamus and telencephalon). The organochlorine pesticide dieldrin is a persistent legacy contaminant, which was heavily used in agriculture in the 1960s and 1970s. However, it was subsequently banned due to reports of negative effects on wildlife and human health. Unfortunately, this pesticide is resistant to degradation in the natural environment and can bioaccumulate in lipid rich tissues including the brain, thus there are long-term consequences of exposure for wildlife and humans. Several studies revealed that dieldrin acted as an endocrine disruptor in fish (Martyniuk et al. 2020). Dietary exposures conducted in largemouth bass (*Micropterus salmoides*) and zebrafish have uncovered effects on proteins related to the immune system, mitochondrial oxidative phosphorylation, and structural components of the cytoskeleton among others (Cowie et al. 2017; Martyniuk et al. 2010; Simmons et al. 2019). Perhaps most significant is that dietary intake of dieldrin is associated with higher body burden levels and disruption of proteins that are often altered in abundance in neurodegenerative diseases such as Parkinson's and Alzheimer's disease. Pesticide exposure has long been identified as a risk factor for neurodegeneration, and these studies identify biomarkers in the central nervous system that may relate to these human diseases.

## Label-Free Methods

Proteins in a complex mixture can also be quantified without the use of mass tags. Label-free methods such as spectral counting rely on differences in spectral intensity across samples, typically using total ion current (e.g., sum of the areas under all the peaks contained in a MS/MS spectrum) to quantify protein abundance. Label-free protein quantification assumes a linear relationship between peptide mass peak signal intensity and the concentration of a specific peptide.

In addition to quantifying mass peak for several thousands of peptides, targeted approaches using synthetically produced peptide standards can also be conducted to quantify peptide abundance with extremely high accuracy and precision. Absolute quantitation (AQUA) utilizes synthesized isotopic peptides that are spiked into samples at known amounts, which eventually coelute during LC MS/MS; the peptide quantitation is then based on the ratio of a known spike amount and a standard curve of the synthetic peptide.

Label-free proteomics has been used to study chemical exposures in several species of fishes, including studies on the effects of dietary zinc oxide (ZnO) nanoparticles (NPs) in common carp (Chupani et al. 2017), commercial herbicide (e.g., Reward®) effects on early life stages and adult fathead minnows (Moreton et al. 2020), and polycyclic aromatic carbon exposure on the plasma proteome of Atlantic cod (Skogland Enerstvedt et al. 2017).

Label-free proteomics has been used to determine the effects of aromatase inhibition in fish. Aromatase is the enzyme that converts testosterone into 17β-estradiol, and it is an important regulatory checkpoint for masculinization and feminization of an individual. Antagonism of this enzyme can lead to masculinization of fishes, and there are pharmaceuticals detected in aquatic environments that can alter the expression and function of this crucial protein (e.g., 17β-trenbolone in water systems next to cattle feed lots). Ralston-Hooper and colleagues used a Waters Synapt G2 quadrupole time-of-flight (QToF) mass spectrometer operating in data-independent mode of acquisition to measure the hepatic proteome of female fathead minnows in response to aromatase inhibition (Ralston-Hooper et al. 2013). The liver was investigated because it produces vitellogenin and other egg yolk precursor proteins that are critical for female reproduction; these lipoproteins are synthesized in the liver, released into the bloodstream, and are subsequently incorporated into eggs to provide nutrients to developing embryos and larvae. Fadrozole is a pharmaceutical used to antagonize aromatase, and it is a model compound for understanding biological consequences of enzyme inhibition. In the study, fathead minnows were exposed to fadrozole-contaminated water as a continuous flow at a range of concentrations (0.04 up to 5.0 μg of fadrozole). The label-free proteomics experiments identified 782 unique proteins and noted that egg yolk precursor proteins like vitellogenin and zona radiata proteins were reduced in abundance, proposed to be the result of lower 17β-estradiol production from the ovary due to aromatase inhibition. Such events can have population level consequences for fish because fewer viable eggs are produced by the female. Intriguingly, in the study, protein networks altered by fadrozole were those related to molecular transport, endocrine system disorder, and gastrointestinal disease. Pathway analysis also uncovered differentially regulated proteins that were preferentially associated with peroxisome proliferator-activated receptor (PPAR)-alpha, lending support to the hypothesis that lipid regulation is subsequently disrupted by aromatase inhibitors (Ralston-Hooper et al. 2013).

Label-free proteomics has yielded intriguing data regarding biological phenomena in toxicology. The Developmental Origins of Health and Disease (DOHaD) is a hypothesis that states the environment encountered *in utero* or during early development can have lasting and long-term consequences for an organism later in life. As such, exposures early in development can shape susceptibility and the likelihood of developing tissue pathology and disease as adults. Delayed effects of pharmaceuticals have been observed in the Mangrove killifish, *Kryptolebias marmoratus*, exposed to one concentration of 4 or 120 ng/L 17-α-ethinylestradiol (EE2) over 28 days exposure (Voisin et al. 2016). The pharmaceutical EE2 is widely used in birth control pills as the active ingredient. It is also a ubiquitous contaminant in aquatic ecosystems, responsible for the feminization of individuals from several populations of fishes in the natural environment. In another study in adult mangrove killifish, the brain, liver, and gonad were investigated for persistent protein responses due to early EE2 exposure (Voisin et al. 2019). Using an impact™ HD UHR-QTOF (high definition, ultrahigh resolution QtOF) System, the researchers identified 327 proteins that were common to all three organs investigated, and it was determined that early-life EE2 exposure triggered persistent changes in the abundance of proteins related to reproduction, lipid metabolism, and the innate immune system. In addition, the study revealed that insulin growth factor could also play a predominant role in mediating transgenerational effects, influencing longevity, apoptosis, and reproduction.

In another study, Atlantic cods were exposed to various concentrations of polychlorinated biphenyl (PCB) 153 (0.5, 2, and 8 mg/kg per body weight). This chemical class is highly persistent in the environment, and these chemicals were once used in appliances as coolants, paints and sealants, wood floor finishing, and waterproofing material; however, their use in products is now more limited and in decline. In the study, the fish were injected with PCB153 and sacrificed 14 days later. An LTQ-Orbitrap Velos Pro mass spectrometer identified 1,272 unique cod proteins, of which 78 were differentially altered by PCB153. The study revealed that lipid metabolism-related pathways (such as PPAR gamma signaling), and the sphingosine-1-phosphate (S1P) receptor signaling pathway, were putative targets of PCB153 exposure. Protein responses were also associated with increasing levels of plasma triglycerides in the cod, lending support to the hypothesis that chemical exposures affect metabolism and lipid stores leading to an obese phenotype (Yadetie et al. 2017).

Other environmental contaminants such as metals have also been investigated with the label-free proteomics in fishes. Yadetie et al. (2016) exposed juvenile Atlantic cod to either 0.5, 2, or 8 mg/kg body methylmercury for 14 days. An LTQ-Orbitrap Velos Pro identified 1,143 unique cod proteins, again many of which were related to energy pathways, specifically mitochondrial metabolic pathways (e.g., electron transport chain and fatty acid beta-oxidation) suggesting MeHg may induce oxidative stress. In a follow-up, select proteins (Tryptophan 2,3-dioxygenase (T23O), glutamine synthetase (GLNA), epidermal growth factor receptor kinase substrate 8-like protein 2 (EPS8L2), Apolipoprotein A4 (APOA4), and Ras-Related Protein 1B, (RAP1B), Cathepsin Z (CATZ)) were further investigated using stable isotope-labeled synthetic peptides in selected reacting monitoring (SRM) to validate the label-free proteomics data. As a result, five of the six proteins were verified as differentially expressed in cod following MeHg exposure.

Small fish and embryo toxicity testing play a critical role in prioritizing chemicals and discerning mechanisms of action for compounds. The Organisation for Economic Cooperation and Development (OECD) has developed validated protocols for such testing which includes the Fish Embryo Acute Toxicity or FET test (OECD Test #236). This validated assay is amendable for studying diverse classes of chemicals. Application of fish embryo toxicity testing yields valuable mechanistic toxicity data for chemicals. Elucidating protein responses in embryos can also generate clues about molecular initiating events and key events for adverse outcome pathways. There are several examples in literature, but only a few are highlighted below.

Using embryos from medaka, Sun and colleagues (Sun et al. 2016) investigated exposure effects of 1,2,5,6-tetrabromocyclooctane (TBCO) (10 up to 1,000 μg/L) using a Linear ion trap (LTQ) Orbitrap Velos hybrid mass spectrometer and a label-free proteomics approach. The study revealed that 189 proteins were altered in abundance in embryos (68 upregulated and 121 downregulated) following exposure to TBCO. Many of these proteins were related to ribosomal function suggesting that TBCO may inhibit biosynthesis of proteins. Other proteins that were altered by TBCO exposure included those related to general metabolism and cardiac-muscle contraction. Furthermore, the study conducted transcriptomics and reported that there was a significant correlation between differentially expressed genes and proteins ($r=0.26$, $p<0.001$). However, we point out that mRNA-protein relationships can be confounded by mRNA turnover rates and maternal transfer of transcripts to the embryo. Nevertheless, the study supports the hypothesis that proteomics can discern toxicity pathways in embryos that can be associated with apical morphological endpoints like deformities.

Zebrafish are also widely used in fish embryo toxicity testing (FET). Vieira and colleagues (Vieira et al. 2020) used the FET to study 3,4-dichloroaniline, a hazardous environmental chemical and one that is commonly used as a teratogenic positive control in fish embryo testing. The study quantified 1,409 proteins, and differentially expressed proteins included those in lipid-related pathways (e.g., lipid homeostasis, triglyceride catabolic process, and regulation of steroid metabolic process). These proteins were predicted to be a subset of those associated with malformations in embryos during the FET test. Moreover, the fish short-term reproduction assay (FSTRA) is often used in environmental toxicology for *in vivo* screening of hormonally active agents. After 21 days, exposed adults are bred, and endpoints related to cumulative fecundity and fertilization rates are recorded. Ayobahan et al. (2019) exposed zebrafish to the aromatase inhibitor fadrozole in the FSTRA to quantify proteomic changes in reproductive tissues in relation to egg production. Across tissues, 787, 1,130, 464, and 676 proteins were measured in zebrafish male liver, testis, female liver, and ovary, respectively, characterizing disruptions in lipid regulation and tissue-specific dysfunction at the protein level. The study identifed a subset of proteins that can act as biomarkers for reproductive dysfunction.

An exciting new advance in label-free proteomics has been the application of SWATH/DIA-MS or Sequential

Window Acquisition of All Theoretical Spectra, a relatively new method that combines aspects of both shotgun and targeted proteomics (Monroe et al. 2020). This method facilitates the discovery of more proteins in a sample, increasing the coverage of the proteome. Data-independent acquisition (DIA) comprehensively detects all proteins in a sample by generating a permanent recording of all fragment ions of peptide precursors. While this approach has yet to be used in fish toxicology, it has recently been used to study environmental stressors, specifically heat stress in relation to climate change. Quan et al. (2021) used SWATH/DIA-MS to investigate the hepatic proteome of rainbow trout (*Oncorhynchus mykiss*) in relation to increasing temperatures. Using a spectral library, SWATH/DIA-MS recorded a significant number of precursors (90,827), peptides (67,028), and proteins (6,770) by including every ion fragment in the sample. In fish toxicology experiments, a typical proteomics experiment with other methods may yield 2,000 proteins on average (Table 11.1 for example), and SWATH/DIA-MS is able to triple the amount of information gleaned from an experiment (Quan et al. 2021).

### Phosphoproteomics

Phosphoproteomics aims to identify and characterize proteins containing a phosphate group via analysis of a complete set of phosphorylation sites using enrichment and mass spectrometry. Phosphorylation kinetics is highly relevant to biological systems as many proteins require the addition of phosphate groups for their activities such as enzymes and cell signaling molecules. Predominantly, these phosphorylation events occur on amino acids that include threonine, tyrosine, and serine but can happen to a lesser extent on other amino acids. Prior to analysis, proteins must be purified or enriched using methods such as $TiO_2$ phosphopeptide enrichment with columns (metal oxide affinity chromatography (MOAC)). Other enrichment methods include phosphopeptide enrichment by chemical modification (e.g., oxidation–reduction condensation or β-elimination coupled with a Michael addition) and enrichment by phosphopeptide precipitation, each with the objective of increasing detection of unique and rare phosphorylation sites on proteins.

While this method has been used to study dietary responses of fish to supplements (Qin et al. 2020), there remain few examples in fish toxicology. This is a significant knowledge gap given the fact that measuring protein abundance does not infer functional consequences of a chemical exposure. Rather it becomes more important to understand whether a protein is activated or deactivated with a chemical.

In one recent study, phosphoproteomics was employed to investigate the effects of two constituents of the birth control pill in the brain of fathead minnow, a widely used freshwater fish model to quantify organismal impacts of environmental chemical exposure (Smith et al. 2018). The authors set out to identify phosphorylation of proteins within 30 minutes of exposure, reasoned to reflect nongenomic signaling pathways activated by exposure to the synthetic estrogen, EE2, and the progestin, levonorgestrel (LNG). A water-borne exposure to EE2 and LNG at environmentally relevant concentrations for surface water and sewage treatment effluents (5 ng/L or 100 ng/L, respectively) revealed that each pharmaceutical induced unique phosphorylation profiles in the FHM brain. The study identified 186 unique proteins within the control and treatment groups, and select proteins showed differential responses in phosphorylation patterns to pharmaceuticals, for example, dihydropyrimidinase-like 2 (DPSYL2) and regulating synaptic membrane exocytosis 2 (RIMS2), indicating that a synthetic estrogen and progestin can have opposing effects on protein phosphorylation. A pathway analysis revealed that proteins involved in calcium ion homeostasis, neurogenesis, cell proliferation, migration, and apoptosis were preferentially targeted for phosphorylation. Activation of these pathways by synthetic steroids is consistent with the mammalian literature, and it is well established that neuro-estrogens and other neurosteroids mediate brain development and neuronal migration, and neural cell differentiation. This study is significant because it highlights the need for deeper understanding into post-translational mechanisms that alter the proteome in fish tissues. This information is crucial for the construction of adverse outcome pathways, as phosphorylation events can be considered key events (or molecular initiating events) leading to adverse outcomes in individuals in a population.

## PROTEOMICS: A ROLE IN ADVERSE OUTCOME PATHWAYS

As noted above, several studies have used proteomics to study chemicals of concern in fishes. Such data can be used to elucidate mechanisms of action of chemicals and prioritize compounds for further investigation. Protein interaction with a ligand can be considered either a molecular initiating event or key event in the adverse outcome pathway (AOP) framework. There are several examples for the use of proteomics in AOPs. Molecular responses at the level of the transcriptome, proteome, or metabolome can be integrated to reveal regulatory networks most predictive of an adverse outcome. Here we present two examples of integration of proteomics into the AOP framework. One example considers proteomics in fish embryo toxicity testing (Test No. 236: FET), and the second example highlights the role of proteomics data in discerning toxicity pathways in non-model species, specifically the white sturgeon (*Acipenser transmontanus*).

The zebrafish is a widely used toxicity model due to its rapid development, transparency of embryos, high fecundity, and flexibility in experimental designs. Proteomics has been integrated into high-throughput assays with zebrafish embryos and larvae to study chemical toxicity. Thyroid hormone mediates developmental patterning and early embryogenesis in vertebrates. Thyroid hormone is also critical for normal growth in fishes. Several aquatic contaminants can negatively impact thyroid hormone receptor signaling, leading to altered development of the skeletal system and organogenesis. These key events downstream of thyroid hormone receptor signaling can result in deformed larvae and ultimately, reduced probability of survival (adverse outcome at the population level). To better annotate an adverse outcome pathway for thyroid hormone disruption in fishes, Reinwald et al. (2021) treated zebrafish with either triiodothyronine (T3) (an agonist for the thyroid hormone receptor) or propyl-2-thiouracil (PTU) (blocks thyroid hormone synthesis). In the study, gene expression analysis was coupled to proteomics, and the authors identified several transcripts and proteins that showed reciprocal expression based upon the treatment. Reciprocal molecular responses to increased/decreased thyroid hormone activation is strong evidence that these downstream molecular entities are mediated by thyroid hormone receptor. In the study, several key events were proposed in the AOP based upon the transcriptome and proteome data, including muscle function and contraction, demonstrating that proteomics data can be used to identify molecular biomarkers and refine adverse outcome pathways (Reinwald et al. 2021).

Such approaches are not limited to model species with well-annotated genomes. Dioxin-like compounds and coplanar PCBs can cause any number of adverse effects in fishes, including immune and hematopoietic system dysfunction, metabolic derangements, impaired reproduction, reduced growth, and mortality. Direct effects of dioxin-like compounds are mediated through aryl-hydrocarbon receptor signaling, and prolonged activation can lead to hepatic necrosis, vacuolization of the liver, and irreversible liver damage. Doering et al. (2016) studied three model compounds that activate aryl hydrocarbon receptor (AHR) signaling (TCDD, PCB 77, and benzo[a]pyrene) in juvenile white sturgeon. Proteomics revealed signatures for each chemical in the liver, and Venn diagrams were used to identify conserved proteins that were increased or decreased in relative abundance. Network analysis revealed clusters of proteins related to amino acid metabolism, nicotinamide adenine dinucleotide phosphate (NADP) metabolic processes, response to hypoxia, and lipid homeostasis as those likely altered by AHR-activating chemicals in the liver of sturgeon. Elucidating such protein biomarkers can yield valuable insights into key events that lead to adverse outcome, in this case liver damage and subsequent mortality. It was noted in the study that AOP development for AHR is challenging because it involves several molecular-initiating events that can cross-talk with other signaling pathways (PPAR, estrogen receptor (ER), and androgen receptor (AR) signaling), and AHR-mediated responses can link to adverse outcomes (Doering et al. 2016). Nevertheless, there are clear opportunities for proteomics in AOPs, whether it be clues to uncharacterized molecular initiating events or linkages between specific cell signaling cascades and apical factors in fishes.

## CHALLENGES AND OPPORTUNITIES FOR PROTEOMICS IN FISH TOXICOLOGY

Careful consideration is required when using omics in toxicology, and it is important to acknowledge that there are inherent advantages and disadvantages with each technology. Quantitative ecotoxicoproteomics studies have undoubtedly proved useful in elucidating toxicity pathways in several fishes (Lee et al. 2018; Martyniuk et al. 2009; Ralston-Hooper et al. 2013; Simmons et al. 2019), and the approaches outlined above are feasible across diverse tissues. The technology has greatly improved over a short period, and mass spectrometers are now highly sensitive and include Q-TOF, Orbitrap, and Fourier transform ion cyclotron resonance mass spectrometers, all of which allow for higher resolution and mass accuracy. Despite higher rates of detection and improved sensitivity, one significant challenge that remains is that protein databases are incomplete for many nonmodel species, making peptide-protein assignments more difficult compared to genome models like zebrafish and Japanese medaka. However, coverage of fish proteomes is expected to improve as genome databases grow with massive parallel sequencing, facilitating a greater number of peptide-protein assignments.

There are other knowledge gaps that can be addressed in future studies. Little is known about the stability and turnover rates of proteins under various environmental conditions in fish, and posttranslational mechanisms in nonmodel species are almost nonexistent in fish toxicology literature. Two additional areas that need attention in proteomics include the quantitation of posttranslational modifications (PTMs) and target mass spectrometry or multiple reaction monitoring (MRM). The MRM method is ideal for risk assessment, zeroing in on key protein molecular initiating events that can be targeted following chemical treatment (Trapp et al., 2014).

One question that has been addressed is the capacity of different groups to detect the same protein signature in fish exposed to a chemical. Such investigations are necessary for establishing the technology as robust and reliable for monitoring and risk assessment. Researchers typically must decide to employ either label-free or a labeling approach to quantify relative protein abundance in tissues; however, there are little data derived from fishes about the ability of different approaches to capture the same proteomics data. In a laboratory experiment, Simmons et al. (2019) set out to quantify protein signatures using both label-free and iTRAQ in zebrafish liver to compare and contrast the information gained about the toxicant exposure. The researchers quantified the liver proteome in zebrafish following a dietary exposure to the organochlorine pesticide dieldrin. This legacy environmental contaminant contains multiple chlorine moieties in a cyclic chemical structure and is problematic for wildlife health because of its high propensity to bioaccumulation in lipid rich tissues, as well as its resistance to microbial degradation, and its multiple adverse effects on endocrine systems (Martyniuk et al. 2020). Fish were fed the pesticide mixed in feed at three different doses for 21 days, and over the exposure, zebrafish significantly accumulated dieldrin into their body. At the end of the experiment, the same liver was excised and processed for two different proteomics methods conducted at two different collaborating institutions. Label-free proteomics quantified 1,563 proteins, while the iTRAQ method quantified more than 3,500 proteins. Most intriguingly, the iTRAQ and label-free proteomics revealed unique perspectives on hepatotoxicity, and there was surprisingly low congruence between the two methods and only about 1% of the measured proteome identified as different following dieldrin exposure. The authors proposed reasons for the discrepancy and pointed out that the label-free proteins were digested with formic acid (cleaves peptides at either the carboxyl or amino side of aspartyl (Asp) residues), while the iTRAQ labeling approach uses the enzyme trypsin (cleaves peptides at the carboxyl side of the amino acids lysine (Lys) or arginine (Arg)). Thus, differences in enzyme digestion can lead to technical bias toward the detection of specific peptides and proteins. Moreover, enzymes show specificity for the proteins they digest. As an example, trypsin does not digest membrane and hydrophobic proteins as efficiently as formic acid which can lead to further differences in the proteins quantified. The authors also pointed out that using different peptide-protein search engines at the back end of the proteomics pipeline was another source of variation between the two methods. When the same search engine was used, there was a slight increase in the number of common proteins detected. Lastly, data were deemed more congruent when higher-level analyses were used to describe proteome responses, for example, the use of pathway analysis homogenized data and centered around major processes impacted by dieldrin in the liver such as the immune system, glucose/energy metabolism, and processes related to liver function. Such experiments are important for quantifying sources of variability among proteomics methods and reveal important information to improve experimental designs in fish toxicology.

For those interested in utilizing proteomics in fish toxicology, there are useful resources available for methodology and workflow. As one example, Lee et al. (2018) prepared a methodology paper that provides a comprehensive overview of the steps involved in measuring proteins in zebrafish using label-free and iTRAQ approaches. The methodology paper uses the teratogen TCDD as an example, describing methods of exposure in embryos and adults. Detailed methods and strategies for protein extraction, shotgun proteomics, labeling protein with 8-plex ITRAQ reagents, mass spectrometry analysis, and two-dimensional liquid chromatography separation of peptides are also outlined, as well as helpful tips.

## METABOLOMICS AND LIPIDOMICS

Metabolomics/metabonomics (hereafter both referred to as metabolomics) is the holistic study of the metabolites that are changed after an exposure to a chemical or due to disease. Metabolites that are usually measured are small molecules (<1,500 da) and include sugars, amino acids, nucleotides, vitamins, hormones, phospholipids, fatty acids, and reaction intermediates. These molecules are the end products of complex metabolic pathways that are altered in fish by exposures to chemicals or by disease. Interestingly, these small molecules are highly conserved across most species from invertebrates to mammals, thus enabling one to understand how metabolism is affected by exposure or disease. Alterations in the levels of these molecules can be used as biomarkers to indicate pathological processes due to contaminant exposure. There are several technological improvements in instrumentation and bioinformatics that have enabled the quantification and identification of important metabolites. Lipidomics is a subsection of metabolomics but with thousands of lipid structures to analyze; it has become its own discipline with methods that are directed specifically for lipids and thus will be discussed separately here.

Adequate experimental design is important to obtain interpretable data. For metabolomics, it has been suggested that well-controlled cell-based experiments include an $n=6$ and that this number be increased to an $n=10$ for model organisms tested in a laboratory where interindividual variance might be a little higher (Southam et al. 2017). It is also recommended to include technical replicates to make sure the data are reproducible. Critical to all of the methods is the proper collection and preservation of samples. Tissue or fluid samples from fish should be collected as quickly as possible and frozen at minus 80°C until extraction. The recommended method is to use a polytron homogenizer or similar type method to disrupt cellular integrity in the presence of organic solvents. The method of Bligh and Dyer (Bligh and Dyer 1959), which uses a final ratio of methanol/chloroform/water of 2:2:1.8, respectively, gives an excellent yield of metabolites, with polar metabolites in the aqueous portion and the lipids in the organic layer. Other extraction methods that can be optimized for different analytes of interest are also available. A step-by-step protocol can be found in the publication of Southam et al. (2017).

### METABOLOMICS

There are two main technological developments that have spurred the study of metabolomics, including nuclear magnetic resonance (NMR) and LC MS/MS. Each analytical technique presents distinct advantages and disadvantages, and to sample the entire metabolome, it is best to use a combination of techniques and platforms (Emwas et al. 2019). The two methods are used frequently for metabolomics, but LC MS/MS appears to be more sensitive and has gained in popularity. While most researchers think of the metabolic pathway charts, for example, the products generated during glycolysis and the Krebs cycle, as constituting the important metabolites, there are many other small molecules that have not yet been identified that are important for vertebrates, including fish. This is especially true when thinking of the myriad of bacterial products made in the microbiome of vertebrate gastrointestinal system that we now recognize as important to human and animal health.

## NMR

NMR technology, as other analytical methods, has advanced tremendously over the past 20 years, making routine analyses of metabolites possible. Analytes can be measured directly without chemical derivatization or labeling and even without additional separation by chromatography. The method is highly reproducible, and it may be able to detect some metabolites that are not easily analyzed by LC MS/MS such as sugars, organic acids, alcohols, or highly polar metabolites (Emwas et al. 2019). Metabolites do not need to be extracted from tissues or biofluids, as tissues or biofluids can be placed directly into the instrument. In fact, intact tissues and even small animals can be measured directly. One can record NMR spectra for different nuclei including $^{1}H$, $^{13}C$, $^{15}N$, $^{19}F$, $^{31}P$ either separately or at the same time, enabling multiplexing of metabolite classes. There are multidimensional NMR methods that can also be employed. Each of the nuclei has a characteristic frequency range over which they can be measured with $^{15}N$ centered around 61 MHz, $^{13}C$ around 151 MHz, $^{31}P$ around 243 MHz, $^{19}F$ around 565 MHz, and $^{1}H$ around 600 MHz (Emwas et al. 2019).

An interesting application of NMR to fathead minnows was done to evaluate the alterations in the liver metabolome upon exposure of fathead minnows to endocrine disrupting chemicals (Ekman et al. 2007; Ekman et al. 2008; Ekman et al. 2011). Fathead minnows (FHM) are an excellent model for ecotoxicology, and significant ecotoxicological databases exist for this model. It is also the preferred fish model by the US Environmental Protection Agency (US EPA). The liver metabolomes of FHM exposed to either EE2 or 17β-trenbolone (TB) were investigated in separate experiments but using a similar experimental approach (Ekman et al. 2008; Ekman et al. 2011). In each case, FHMs were exposed for 8 days to two concentrations of the endocrine disruptor and then depurated for 8 days. Sampling occurred at days 1, 2, 4, and 8 of the exposure regimes, for EE2 after 8 days of depuration, and for TB, after 1, 2, 4, and 8 days of depuration. Alterations in the polar fraction of the liver metabolome of males exposed to EE2 and females exposed to TB were monitored by NMR.

The two concentrations of EE2 were 10 ng/L and 100 ng/L, concentrations that have significant effects on the liver for the synthesis of vitellogenin, the egg yolk precursor protein that is directly controlled by exposure to estradiol. The main increases noted in the livers of male FHMs were for the amino acids, alanine and glutamic acid, and lactic acid that were related by the authors to increased demand for the synthesis of vitellogenin. Decreases in glycogen, glucose, and betaine were also observed. Since the metabolome was sampled at different time points, it was possible to chart the changes in the metabolome over the 8-day exposure and return to normal using a principal component analysis. The alterations of the metabolome were much greater with the higher concentration of EE2 (Ekman et al. 2008).

The analysis of effects in females exposed to TB was more complicated. The two concentrations used were 33 ng/L and 472 ng/L, again over the 8-day time frame. While the spectra during the time course were quite variable, some metabolite alterations were noted at different time points and concentrations. An increase in taurine was seen in fish exposed to both concentrations but on different days; otherwise during the exposure phase most of the liver, metabolite differences were in reduction of metabolites, for example, in glucose, glutamine, and bile acids. During the depuration phase, on the other hand, there was an almost immediate increase in glycogen, a decrease in taurine, glutamic acid, and lactic acid for the high dose and an increase in bile acids in the low dose after 2 days. The highly variable response was suggested to be due to compensatory changes in the fish, as indicated by changes in the transcriptome. However, livers of reproductive females are normally engaged in high metabolic activity geared toward the synthesis of vitellogenin, which is disrupted by the exposure to androgens, and possibly, the high variability in response is due to interindividual differences in coping with the change, when the natural response in reproductive females is to produce eggs (Ekman et al. 2011).

NMR has also been applied by the same group to fathead minnow urine (Collette et al. 2010; Ekman et al. 2007). Urine is an excellent tissue for metabolomics research as it contains many small molecular weight metabolites that are under 1 KDa and that represent the metabolome of the organism, so that any changes in the metabolome due to exposure to contaminants are reflected in the metabolome of the urine. In the case of the fathead minnow, the investigators were able to obtain 5–160 ul of urine at the end of an experiment by gently pressing on the fish abdomen and using a microcapillary tube applied directly to the cloaca (Ekman et al. 2007). They were able to measure changes in the fish urine metabolome with endocrine disruptors. These novel approaches reflect alterations of the metabolome of fishes, as they are exposed to various changes in their environments.

## LC/GC Tandem Mass Spectrometry

As for proteomics, discussed above, metabolites can be measured either in a holistic, nontargeted approach, usually using high-resolution liquid chromatography tandem mass spectrometry, or with a targeted approach. Of the metabolome universe, each method focuses on a slightly different fraction, and while there is significant overlap in what the different methods can analyze, no one method can analyze the entire metabolome (Gika et al. 2019). Nontargeted approaches tend to use high-resolution mass spectrometers such as the Fourier transform ion cyclotron resonance (FT-ICR) MS, orbitrap, QTOF, or

Time-of-Flight×Time-of-Flight (TOF TOF) approaches, while targeted methods are more likely to use triple quadrupole instruments, but there is significant commonality in the instruments used for targeted and nontargeted approaches. While LC or GC is the more common separation approach, there are also innovative methods that use hydrophilic interaction chromatography (HILIC) or capillary electrophoresis to capture polar metabolites that may not otherwise bind to reverse phase columns, for example, highly charged molecules, sugars, and nucleotides (Garcia et al. 2017; Gika et al. 2019).

Another innovative method is the use of supercritical fluid chromatography for hydrophobic molecules (van de Velde et al. 2020) as the initial separation method. This method uses a supercritical fluid, normally $CO_2$, in combination with an organic modifier for the mobile phase of the chromatography. The supercritical fluid is achieved by increasing both the temperature and the pressure to their respective critical point. These fluids have density close to that of a liquid, but viscosity and diffusivity properties close to that of gases. This separation can be carried out with both normal and reverse phase matrices and is amenable to high-throughput analyses. The method requires high-performance liquid chromatography (HPLC) that is adapted to running the supercritical fluids.

An issue that is never quite overcome by any method is the large range of metabolite concentrations in body fluids. For example, in the human blood metabolome, analytes can extend from the pM range (e.g., diacylglycerides) to >100 mM range (e.g., sodium and chlorine), often displaying a difference in nine or more significant figures (Psychogios et al. 2011). The estimate is for over 4,000 metabolites in human blood (Psychogios et al. 2011). The metabolome of fish blood is likely similar in complexity. The other issue is that metabolites can vary in their polarity from quite polar to very hydrophobic molecules, requiring different approaches for column chromatography. Hydrophobic molecules are best separated by reverse phase chromatography, whereas hydrophilic molecules are best separated by HILIC. There are also many column resins that have been developed with functional groups that can interact with a wide range of metabolite physical properties, changing their retention time, peak shape, and elution profiles (Gika et al. 2019). Of interest is the use of ultra-high-performance liquid (uHPLC) chromatography that provides significant improvement in the resolution of metabolites by making sharper peaks with less overlap. Thus, to be able to analyze the less abundant metabolites, it is necessary to perform prefractionation and enrichment using special solid phase extraction resins and a variety of metabolomics approaches.

Nontargeted approaches generally rely on analyzing as many molecules as possible in the fractions of interest. Normally, samples are separated by column chromatography in tandem with a high-end mass spectrometer to identify as many masses and fragments as possible. Identification of the mass spectral information (i.e., mass-to-charge ratios (m/z) of precursor and fragment ions) for thousands of ion features is difficult and requires automated software and metabolite libraries. To generate the features, the mass spectrometers can be set up in data-dependent mode, where the most intense ions are analyzed first. Depending on the duty cycle of the instrument, the mass spectrometer can analyze many masses and daughter ions. It is possible to then place the identified ions on an exclusion list and perform the analysis a second time to analyze ions that are less abundant or that are ionized to lower levels. There are software programs from many vendors that contain thousands of features that can be used to putatively identify the metabolites. Because the approach targets as many metabolites as possible, many of the features cannot be annotated, with normally 10% or less with a validated annotation. But, as libraries of metabolites keep increasing, the fraction that can be positively identified will also increase.

One approach to nontargeted analyses is the use of direct infusion mass spectrometry, which as the name suggests consists of direct infusion of the metabolite fraction directly into the mass spectrometer. One method uses a chip-based nanoelectrospray ionization sample delivery system with an Orbitrap mass spectrometer (Southam et al. 2017). This method is rapid and reduces the time to analyze each sample, giving a higher throughput for metabolomics.

For targeted approaches, generally, there is a list of metabolites that will be targeted for analysis. Because the analytes are known, the mass spectrometer can be set up to analyze specific masses that come off the column at specific retention times and further to identify known daughter ions that are generated in the collision chamber. Generally, two daughter ions are monitored, one for quantitation and a second for qualification. This approach known as multiple reaction monitoring (MRM) is powerful and extremely sensitive and is generally used in hypothesis-driven studies. The targeted approach is more specific and can provide better quantitation, especially if internal standards are added to the samples prior to analysis.

There are many case studies that show the power of LC-based approaches for metabolomics. In one study, zebrafish were treated with dydrogesterone, a widely used progestin that also binds and activates androgen receptors (Jiang et al. 2019) and can masculinize fish. Three concentrations of dydrogesterone were used (4.4, 44, and 440 ng/L) for 140 days to encompass gonadal differentiation in zebrafish, which occurs normally around 35 days posthatch. At the highest concentration, and after 140 days of exposure, it was determined that all the fish were male. Using both an LC QTOF and a gas chromatography (GC) QTOF, these investigators measured alterations in the metabolome at 35 days postfertilization. They saw significant changes in at least 60 metabolites, with both up- and downregulation, compared to controls. Among the amino acids that were changed, there was upregulation of taurine and glycine, among others, and downregulation of glutamic acid, as seen previously by Eckman et al (Ekman et al. 2011) for masculinizing fathead minnows. In addition, there was a universal increase in lactic acid, maleic acid, and fumaric acid.

There was also a significant decrease in six phospholipids and an increase in one. The alterations in metabolites fit nicely into a proposed pathway for masculinization of the gonad that is consistent with other studies suggesting a reduction of NF-κB and Wnt/β-catenin pathways, and an increase in p53 pathway (Jiang et al. 2019). Important for this study was the actual measurement of the treatment exposures, and showing the masculinization of the fish, in a dose-dependent manner.

## LIPIDOMICS

Lipids are only a fraction of the metabolome, but because there are so many possible phospholipids (estimate of 100,000 to 200,000 different molecules (Yetukuri et al. 2008)), lipidomics has become its own separate omics area of interest. The lipids are found in the bottom organic layer of the Bligh and Dyer method (Bligh and Dyer 1959). Lipids that are most easily analyzed include free fatty acids, cholesterol, cholesterol esters, ceramides, sphingomyelins, triglycerides, diglycerides, monoglycerides, phosphatidylethanolamines, phosphatidylcholines, phosphatidylinositols, phosphatidylserines, phosphatidylglycerols and their lyso-derivatives, among others. There are specific methods for different classes of oxidized lipids (Li et al. 2019) that take advantage of the molecular structure and charge for getting them to fly in an LC MS/MS system.

Lipids can be analyzed via GC MS/MS or LC MS/MS. GC MS is the method of choice for free fatty acids and can also analyze more hydrophobic lipids. However, LC MS/MS has become very versatile for phospholipids with new methods now able to quantify over 1,500 phospholipids using a targeted approach with scheduled MRM (Breitkopf et al. 2017). As for metabolomics in general, lipids can also be analyzed by non-targeted methods, but identifications rely on matching profiles to databases. As with metabolomics, many different column matrices have been used including C18 which is excellent for the more hydrophobic lipids, including triglycerides, diglycerides, monoglycerides, and the ceramide class and HILIC or amide columns, that work for the majority of the more polar phospholipids.

Historically, lipids were thought to be only involved in forming lipid bilayers. However, over the past 30 years, it has become apparent that they are involved in many processes and in fact may regulate most cellular processes. While the main components of membranes are glycerophospholipids, sphingolipids, and sterols, many other lipid classes, which are made on site when needed, are also important for controlling cellular processes (Yang et al. 2018). The glycerophospholipids are configured with glycerol as a backbone and have two fatty acids attached to positions sn1 and sn2 and a phosphoric acid at position sn3 via ester bonds. The phosphoric acid can be further modified by several different head groups to make the main classes of glycerophospholipids, including phosphatidylcholine, phosphatidylethanolamine, phosphatidylserine, and phosphatidylinositol. Plasmalogens are phospholipids in which the fatty acid in position sn1 is linked via an ether bond rather than an ester bond, which gives the lipid special characteristics. Sphingolipids are related lipids, but instead of a glycerol backbone, they have a serine backbone. They are found mostly on the exofacial side of the membrane, and they can have many different sized head groups ranging from the size of a hydroxyl group up to a complex sugar group. Sterols such as cholesterol are also found in membranes.

In addition to their function in membranes, phospholipids also have roles in many cellular processes, and some are potent signal molecules. Their varied degree of saturation in addition to the characteristics of head groups and the distribution of phospholipids give membranes their curvature, fluidity, and specific characteristics to bind with membrane proteins. They are thought to be repositories for signaling molecules that can be made from the lipids via specific enzymes when they are needed. They give rise to prostaglandins, leukotrienes, eicosanoids, diacylglycerols, lyso-glycerophospholipids, among others that are potent signals within cells. A family of specific phospholipases is involved in these transformations. Lipids have active roles in metabolism and cellular function, including the immune system.

The most abundant phospholipid class is phosphatidylcholine. The carbon chains of the fatty acids can vary in length, and they can contain double bonds at different locations, properties which are integral to membrane fluidity at 37 °C. Phosphatidylethanolamine is the next most abundant glycerophospholipid, and it too is highly abundant in membranes and allows for membrane curvature. Phosphatidylserine and phosphatidyl inositol are found in lower abundance in membranes and provide additional characteristics. Phosphatidylserine gives the membrane an overall negative charge due to its polarity, and phosphatidyl inositol can be phosphorylated at three positions in the inositol ring, giving rise to bioactive molecules.

### Lipidomics Experiments in Fish

Specific focus on lipidomics in fish has resulted in several studies of importance. Lipid changes with respect to exposure to contaminants have important implications because of the changes in bioactive lipids and the pathways that they may interact with or control. In one study, alterations in the zebrafish lipidome during development either by a mitochondrial toxicant, 2,4-dinitrophenol (2,4-DNP), or heat stress, or a combination of the two stressors was measured (Dreier et al. 2021). During embryonic development, lipids are important to provide critical energy supplies, membrane support for differentiating cells and tissues and for their important roles in cell signaling. The mitochondrial membrane is important for respiration and it plays a role in apoptosis, both important aspects of development. Dreier et al. (2021) treated zebrafish-fertilized eggs based upon the FET, starting with eggs that were 3 hours postfertilization. Lipid profiles were analyzed in the hatched fish after the egg yolk was completely consumed. The abundance of many lipids was altered by higher temperature and 2,4-DNP, but what stood out was the upregulation of ceramides. Ceramides are bioactive lipids involved in many areas of metabolism and specifically as regulators of cell proliferation, growth, and apoptosis (Pettus et al. 2002). Furthermore, ceramides are linked to important signaling pathways involving protein phosphatases and kinases (Hannun and Obeid 2018). Other lipid classes that are considered bioactive were also altered including diacylglycerides (DAGs) and lysophospholipids. These alterations may give insight into the toxicity of 2,4-DNP.

Bisphenol A (BPA) is known as an obesogen and likely involved in disruption of metabolism. To test this hypothesis during development, zebrafish embryos were exposed to none or to three concentrations of BPA (0, 4, 6, and 8 mg/L) during development (Martinez et al. 2020). There was a clear dose-dependent alteration of lipid classes at each of the development stages, including at 4-, 5-, and 6-days post-fertilization. High-resolution LC MS/MS showed increases in triglycerides, diglycerides phosphatidylcholines, and phosphatidylinositols, compared to the control group. Also, there were major differences in degree of unsaturation and chain length. BPA is known to agonize and antagonize several nuclear receptors including ER, AR, glucocorticoid receptor (GR), retinoid X receptor (RXR), thyroid receptor (TR), and PPAR (Mustieles et al. 2020), making its effect relatively unpredictable due to the modulation of several signaling pathways controlled by each receptor. From this work, BPA was shown to alter lipid metabolism and energy availability for development, which probably results in alterations of various developmental processes.

In another study with chlorpyrifos, lipidomic analyses of juvenile rainbow trout (*Oncorhynchus mykiss*) showed that this pesticide (at relatively low concentrations and time of exposure, 10–40 ug/L for 7 days) could alter neurodevelopment in fish (Greer et al. 2019). Chlorpyrifos is known as a potent acetylcholinesterase inhibitor, but alternative effects on brain development had been hypothesized without a clear mechanism of action. The hypothesis tested was that in addition to cholinesterase inhibition, chlorpyrifos could alter serine lipase activity in fish brains and potentially alter the lipidome. Direct effects on two isolated serine lipases, monoacylglycerol lipase and fatty acid amide hydrolase, were confirmed in vitro. Importantly, in the treated fish, there was a decrease in the levels of lysophospholipids, with concomitant increases in their respective parent phospholipids, showing the direct effect *in vivo*. Lysophospholipids are bioactive lipids that are generated from their parent phospholipids by lipases on site when they are needed and may be involved in neurodevelopment (van Meeteren et al. 2006; Yung et al. 2015). While more work needs to be performed to confirm the direct effect of chlorpyrifos on neurodevelopment, this study suggested a possible mechanism.

A universal present-day contaminant is found in the family of per- and polyfluoroalkyl substances (PFAS) that are located widely in the environment and linked to immunotoxicity (DeWitt et al. 2012), altered lipid metabolism (Wang et al. 2014), and other toxicological endpoints (Fuentes et al. 2007; Harada et al. 2006). Thus, examining changes in the lipidome of fish exposed to these contaminants may provide interesting insight into their modes of action. In a study with Atlantic cod (*Gadus morhua*) exposed by intraperitoneal injections on 2 days 1 week apart to a mixture of PFAS (PFOS, perfluorooctanoic acid (PFOA), perfluorononanoic acid (PFNA), and pentacosafluorotridecanoic acid (PFTrDA), chosen based on their relative concentrations in wild fish), a group led by Goksoyr identified lipid degradation as a main pathway affected by the mixture (Dale et al. 2020). One important pathway, fatty acid β-oxidation, appeared to be increased in a proteomics experiment performed on the same fish. This was matched in a lipidomic experiment with decreases in triacylglycerols, suggesting that PFAS increased the level of catabolism in fish. Other lipids that were changed comprised several phospholipids including some phosphatidylinositols, which are important for their signaling potential. Approaches such as this one add some clarity to the serious effects of exposure to PFAS.

Finally, lipidomics can also be applied to the fish collected from the natural environment. In a study by Porte's group (Marqueño et al. 2019), two fish species (*Barbus meridionalis* and *Squalius laietanus*) were collected at polluted sites in the Ripoll River receiving urban and industrial wastewater. The lipidomes of the fillets of these fish were analyzed by an untargeted method, and they were distinct from fish collected at a reference site. The differentially quantified features were assigned to glycerophospholipid and sphingolipid classes. This group also used a targeted approach to identify with absolute mass some of the varied lipids. Interestingly, there were more altered lipids in the *B. meridionalis* fish than the *S. laietanus*, suggesting the species may be more sensitive to the pollutants found in the wastewater contaminated sites. This finding was supported by laboratory-based exposures.

While lipidomics studies are still in their infancy with respect to fish, the few studies summarized as examples here suggest a wealth of information that can be obtained by studying fish lipidomes. Lipid pathways are still underdeveloped, with only a handful available through "Lipid Maps," a Welcome Trust supported consortium (Fahy et al. 2011). However, as more information is determined for the activity of lipids and how they relate back to alterations seen in other omics technologies, more information about their signaling abilities will be generated.

## INTEGRATION OF PROTEOMICS AND METABOLOMICS

In the past, omics methods were more commonly applied in isolated experiments in the laboratory or in the field in environmental toxicology. However, computational advances, as well as reduced cost and increasing accessibility to technology, have fueled a new area of research in multiomics toxicology. Multiomics is an approach whereby data sets of multiple "omes" are integrated to reveal biological responses in a more holistic manner in the organism. Figure 11.4 shows how environmental contaminants disrupt lipid related processes in fish. The advantage of such an approach is greater understanding of the molecular events that underscore physiological adaptation and compensation to chemical stressors. In the case of proteomics and metabolomics, for example, enzyme-metabolite relationships can yield information on metabolic flux and regulatory checkpoints in key pathways, as altered abundance of enzymes can be associated with specific precursor metabolites (i.e., substrates) and their product. Given the fact that exposure to several chemical classes can alter the metabolism of organisms, either through specific mechanisms like peroxisome proliferator-activated receptors (PPARs), or through more general effects on metabolism, a multiomics approach measuring both proteins and metabolites is warranted.

Next-generation multiomics studies have included both proteomics and metabolomics in fish toxicology. Despite clear advantages compared to single omics studies, there remain few examples in fish toxicology that consider both proteomic and metabolite responses together. As one example, juvenile rockfish (*S. schlegelii*) exposed to either 5 or 50 mg/L arsenic (As) for 2 weeks were assessed for metabolome and proteome changes in the whole fish (Xu et al. 2019). Rockfish accumulated significant levels of arsenic over the 2-week exposure, and NMR spectroscopy-based metabolomic analysis and iTRAQ-based quantitative proteomic analysis uncovered several molecular entities linked to metabolism and cell signaling. In the study, 34 metabolites were identified as affected by As, including several amino acids, organic acids (e.g., lactate and acetate), and metabolites related to ATP production, glycolysis, and intermediates in the Krebs cycle. In addition, a total of 2,053 proteins were detected with iTRAQ, of which 163 proteins were significantly altered in abundance by As. Based upon these data, the authors were able to generate a schematic diagram of molecular responses, highlighting the impact of As exposure on mitochondrial energy metabolism and glycolysis. While a limitation of the study was that entire fish were used rather than assessing metabolites and proteins in specific tissues, it was clear that a combined proteomic and metabolomic analysis offers a global overview of how certain metals of environmental concern affect fish.

The major challenges with multiomics data sets are, not unexpectedly, the organization, integration, and interpretation of these different data sets. Molecular connections in multilayered data are not always clear, considering there exist temporal and spatial specificity for genes, proteins, and metabolites. Following exposure, chemical biotransformation by the liver and rates of distribution to target tissues can vary over time which adds complexity to biological responses. To assist with such realistic exposure scenarios, there are exciting new opportunities on the horizon

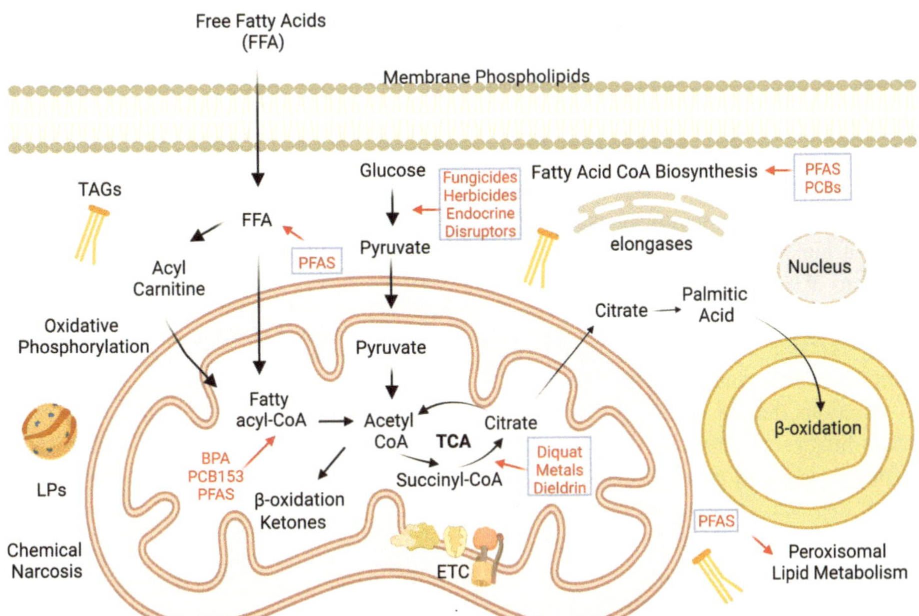


**FIGURE 11.4** Environmental contaminants disrupt lipid related processes in teleost fishes. Fungicides, herbicides, metals, and endocrine disruptors can alter glycolysis and the Krebs cycle, compromising energy production and cellular metabolism. Consequently, free fatty acid biosynthesis and β-oxidation may be adjusted to compensate for loss in energy homeostasis. The regulation of lipid stores can subsequently become unhinged with chemical exposures, impacting triacylglycerides, lipoproteins, and phospholipids in membranes. These actions can compromise cell integrity and lead to adverse outcomes that include lipid insufficiency for reproduction or hepatic steatosis. Boxes provide examples of chemicals that impact metabolism. BPA, bisphenol A; FFA, free fatty acids; LPs, lipoproteins; PCB, polychlorinated biphenyl; PFAS, per- and polyfluoroalkyl substances; TAGs, triacyl glycerides. Figure produced in Biorender (biorender.com).

in machine learning and artificial intelligence. Computer-based learning approaches can be supervised or unsupervised, each generating new models to predict toxicity and characterize patterns in molecular data associated with chemical exposure. In addition, deep learning uses neural networks that can utilize different layered data to uncover novel findings previously undiscovered in omics data. Leveraging computational expertise and principles will result in greater understanding of the complexity of biological response, leading to stronger predictors and validated biomarkers of toxicity.

## FUTURE OPPORTUNITIES

Over the past decade, there has been a significant increase in the number of studies in fish that use both proteomic and metabolite tools to characterize the toxicity of environmental chemicals. Suggested recommendations for future investigations using proteomics and metabolomics include the following:

1. Application of targeted proteomics and metabolomics for metabolism studies in fishes is warranted, for example, targeted assays for regulatory enzymes, precursor metabolites and products in glycolysis, Krebs cycle, and the electron transport chain. Disruptions in energy related processes are a frequent outcome in many fish toxicology studies; however, untargeted mass spectrometry may not detect important metabolic intermediates in energy related pathways. Proteomics will often measure the most abundant proteins in the cell or tissue and in-depth data on mechanisms of metabolic disruption can be lacking. This is especially important as many environmental contaminants are labeled as metabolic disruptors or obesogens.
2. A confounding variable in many studies to date is that subpopulations of cells in a tissue may respond in unique ways to a chemical. Investigations using proteomics and metabolomics in fish typically spans whole embryos or larvae, or entire tissues to circumvent limitations due to availability of biological material. Single-cell proteomics or metabolomics is expected to yield novel information at the cellular level for chemical toxicants.
3. Use of matrix-assisted laser desorption ionization (MALDI) imaging technology to identify metabolites, lipids, and proteins, as this method will provide spatial distributions of important biomolecules in tissues. This will help us understand the integrated function of the biomolecules, akin to in situ hybridization for mRNAs and immunohistochemistry for proteins.
4. Integration of omics technologies through bioinformatic approaches to get a complete picture of how contaminants alter the proteome and metabolome.

## REFERENCES


Alderman SL, Dindia LA, Kennedy CJ, Farrell AP, Gillis TE. 2017. Proteomic analysis of sockeye salmon serum as a tool for biomarker discovery and new insight into the sublethal toxicity of diluted bitumen. *Comparative Biochemistry and Physiology Part D: Genomics and Proteomics*. 22:157–66.

Ayobahan SU, Eilebrecht E, Kotthoff M, Baumann L, Eilebrecht S, Teigeler M, Hollert H, Kalkhof S, Schäfers C. 2019. A combined FSTRA-shotgun proteomics approach to identify molecular changes in zebrafish upon chemical exposure. *Scientific Reports*. 9(1):6599.

Bligh EG, Dyer WJ. 1959. A rapid method of total lipid extraction and purification. *Canadian Journal of Biochemistry and Physiology*. 37(8):911–917.

Breitkopf SB, Ricoult SJH, Yuan M, Xu Y, Peake DA, Manning BD, Asara JM. 2017. A relative quantitative positive/negative ion switching method for untargeted lipidomics via high resolution LC-MS/MS from any biological source. *Metabolomics*. 13(3):1–21.

Chen L, Zhang H, Sun J, Wong YH, Han Z, Au DW, Bajic VB, Qian PY. 2014. Proteomic changes in brain tissues of marine medaka (Oryzias melastigma) after chronic exposure to two antifouling compounds: Butenolide and 4, 5-dichloro-2-n-octyl-4-isothiazolin-3-one (DCOIT). *Aquatic Toxicology*. 157:47–56.

Chupani L, Zusková E, Niksirat H, Panáček A, Lünsmann V, Haange S-B, von Bergen M, Jehmlich N. 2017. Effects of chronic dietary exposure of zinc oxide nanoparticles on the serum protein profile of juvenile common carp (*Cyprinus carpio* l.). *Science of the Total Environment*. 579:1504–1511.

Collette TW, Teng Q, Jensen KM, Kahl MD, Makynen EA, Durhan EJ, Villeneuve DL, Martinovic-Weigelt D, Ankley GT, Ekman DR. 2010. Impacts of an anti-androgen and an androgen/anti-androgen mixture on the metabolite profile of male fathead minnow urine. *Environmental Science & Technology*. 44(17):6881–6886.

Cowie AM, Sarty KI, Mercer A, Koh J, Kidd KA, Martyniuk CJ. 2017. Molecular networks related to the immune system and mitochondria are targets for the pesticide dieldrin in the zebrafish (*Danio rerio*) central nervous system. *Journal of Proteomics*. 157:71–82.

Cuello S, Ximénez-Embún P, Ruppen I, Schonthaler HB, Ashman K, Madrid Y, Luque-Garcia JL, Cámara C. 2012. Analysis of protein expression in developmental toxicity induced by MeHg in zebrafish. *Analyst*. 137(22):5302–5311.

Dale K, Yadetie F, Muller MB, Pampanin DM, Gilabert A, Zhang X, Tairova Z, Haarr A, Lille-Langoy R, Lyche JL et al. 2020. Proteomics and lipidomics analyses reveal modulation of lipid metabolism by perfluoroalkyl substances in liver of atlantic cod (*Gadus morhua*). *Aquatic Toxicology*. 227:105590.

Denslow ND, Knoebl I, Larkin P. 2005. Approaches in proteomics and genomics for eco-toxicology. In: Mommsen TP, Moon TW, editors. *Biochemistry and Molecular Biology of Fishes*. Elsevier. pp. 85–116.

DeWitt JC, Peden-Adams MM, Keller JM, Germolec DR. 2012. Immunotoxicity of perfluorinated compounds: Recent developments. *Toxicologic Pathology*. 40(2):300–311.

Doering JA, Tang S, Peng H, Eisner BK, Sun J, Giesy JP, Wiseman S, Hecker M. 2016. High conservation in transcriptomic and proteomic response of white sturgeon to equipotent concentrations of 2, 3, 7, 8-tcdd, pcb 77, and benzo [a] pyrene. *Environmental Science & Technology*. 50(9):4826–4835.

Dong X, Qiu X, Meng S, Xu H, Wu X, Yang M. 2018. Proteomic profile and toxicity pathway analysis in zebrafish embryos exposed to bisphenol A and di-n-butyl phthalate at environmentally relevant levels. *Chemosphere*. 193:313–320.

Dorts J, Kestemont P, Marchand P-A, D'Hollander W, Thézenas M-L, Raes M, Silvestre F. 2011. Ecotoxicoproteomics in gills of the sentinel fish species, *Cottus gobio*, exposed to perfluorooctane sulfonate (PFOS). *Aquatic Toxicology*. 103(1–2):1–8.

Dreier DA, Nouri MZ, Denslow ND, Martyniuk CJ. 2021. Lipidomics reveals multiple stressor effects (temperature x mitochondrial toxicant) in the zebrafish embryo toxicity test. *Chemosphere*. 264:128472.

Ekman DR, Teng Q, Jensen KM, Martinovic D, Villeneuve DL, Ankley GT, Collette TW. 2007. NMR analysis of male fathead minnow urinary metabolites: A potential approach for studying impacts of chemical exposures. *Aquatic Toxicology*. 85(2):104–112.

Ekman DR, Teng Q, Villeneuve DL, Kahl MD, Jensen KM, Durhan EJ, Ankley GT, Collette TW. 2008. Investigating compensation and recovery of fathead minnow (*Pimephales promelas*) exposed to 17alpha-ethynylestradiol with metabolite profiling. *Environmental Science & Technology*. 42(11):4188–4194.

Ekman DR, Villeneuve DL, Teng Q, Ralston-Hooper KJ, Martinovic-Weigelt D, Kahl MD, Jensen KM, Durhan EJ, Makynen EA, Ankley GT et al. 2011. Use of gene expression, biochemical and metabolite profiles to enhance exposure and effects assessment of the model androgen 17beta-trenbolone in fish. *Environmental Toxicology and Chemistry*. 30(2):319–329.

Emwas AH, Roy R, McKay RT, Tenori L, Saccenti E, Gowda GAN, Raftery D, Alahmari F, Jaremko L, Jaremko M et al. 2019. *NMR Spectroscopy for Metabolomics Research. Metabolites*. 9(7):123.

Eyckmans M, Benoot D, Van Raemdonck GA, Zegels G, Van Ostade XW, Witters E, Blust R, De Boeck G. 2012. Comparative proteomics of copper exposure and toxicity in rainbow trout, common carp and gibel carp. *Comparative Biochemistry and Physiology, Part D. Genomics Proteomics*. 7(2):220–232.

Fahy E, Cotter D, Sud M, Subramaniam S. 2011. Lipid classification, structures and tools. *Biochimical Biophysical Acta*. 1811(11):637–647.

Lipid maps online tools. 2007. https://www.lipidmaps.org; [accessed 2021].

Fong CC, Shi YF, Yu WK, Wei F, van De Merwe JP, Chan AK, Ye R, Au DW, Wu RS, Yang MS. 2014. iTRAQ-based proteomic profiling of the marine medaka (*Oryzias melastigma*) gonad exposed to BDE-47. *Marine Pollution Bulletin*. 85(2):471–478.

Fu J, Han J, Zhou B, Gong Z, Santos EM, Huo X, Zheng W, Liu H, Yu H, Liu C. 2013. Toxicogenomic responses of zebrafish embryos/larvae to tris (1, 3-dichloro-2-propyl) phosphate (TDCPP) reveal possible molecular mechanisms of developmental toxicity. *Environmental Science & Technology*. 47(18):10574–10582.

Fuentes S, Vicens P, Colomina MT, Domingo JL. 2007. Behavioral effects in adult mice exposed to perfluorooctane sulfonate (PFOS). *Toxicology*. 242(1–3):123–129.

Gandar A, Laffaille P, Marty-Gasset N, Viala D, Molette C, Jean S. 2017. Proteome response of fish under multiple stress exposure: effects of pesticide mixtures and temperature increase. *Aquatic Toxicology*. 184:61–77.

Garcia A, Godzien J, Lopez-Gonzalvez A, Barbas C. 2017. Capillary electrophoresis mass spectrometry as a tool for untargeted metabolomics. *Bioanalysis*. 9(1):99–130.

Gika H, Virgiliou C, Theodoridis G, Plumb RS, Wilson ID. 2019. Untargeted LC/MS-based metabolic phenotyping (metabonomics/metabolomics): The state of the art. *Journal Chromatography B* 1117:136–147.

Greer JB, Magnuson JT, Hester K, Giroux M, Pope C, Anderson T, Liu J, Dang V, Denslow ND, Schlenk D. 2019. Effects of chlorpyrifos on cholinesterase and serine lipase activities and lipid metabolism in brains of rainbow trout (*Oncorhynchus mykiss*). *Toxicological Sciences*. 172(1):146–154.

Hannun YA, Obeid LM. 2018. Sphingolipids and their metabolism in physiology and disease. *Nature Reviews Molecular Cell Biology*. 19(3):175–191.

Harada KH, Ishii TM, Takatsuka K, Koizumi A, Ohmori H. 2006. Effects of perfluorooctane sulfonate on action potentials and currents in cultured rat cerebellar purkinje cells. *Biochemical and Biophysical Research Communications*. 351(1):240–245.

Jiang YX, Shi WJ, Ma DD, Zhang JN, Ying GG, Zhang H, Ong CN. 2019. Male-biased zebrafish sex differentiation and metabolomics profile changes caused by dydrogesterone. *Aquatic Toxicology*. 214:105242.

Lee YM, Li C, Lam SH, Gong Z, Lin Q. 2018. Proteomic analysis of zebrafish (*Danio rerio*) after chemical exposure. *Methods in Molecular Biology*. 1797:443–459.

Li L, Zhong S, Shen X, Li Q, Xu W, Tao Y, Yin H. 2019. Recent development on liquid chromatography-mass spectrometry analysis of oxidized lipids. *Free Radical Biology and Medicine*. 144:16–34.

Liang X, Li W, Martyniuk CJ, Zha J, Wang Z, Cheng G, Giesy JP. 2014. Effects of dechlorane plus on the hepatic proteome of juvenile chinese sturgeon (*Acipenser sinensis*). *Aquatic Toxicology*. 148:83–91.

Liang X, Martyniuk CJ, Simmons DB. 2020. Are we forgetting the "proteomics" in multi-omics ecotoxicology? *Comparative Biochemistry and Physiology Part D: Genomics and Proteomics*. 36:100751.

Lu Z, Wang S, Ji C, Li F, Cong M, Shan X, Wu H. 2020. iTRAQ-based proteomic analysis on the mitochondrial responses in gill tissues of juvenile olive flounder Paralichthys olivaceus exposed to cadmium. *Environmental Pollution*. 257:113591.

Marouga R, David S, Hawkins E. 2005. The development of the dige system: 2D fluorescence difference gel analysis technology. *Analytical and Bioanalytical Chemistry*. 382(3):669–678.

Marqueño A, Blanco M, Maceda-Veiga A, Porte C. 2019. Skeletal muscle lipidomics as a new tool to determine altered lipid homeostasis in fish exposed to urban and industrial wastewaters. *Environmental Science & Technology*. 53(14):8416–8425.

Martinez R, Navarro-Martin L, van Antro M, Fuertes I, Casado M, Barata C, Pina B. 2020. Changes in lipid profiles induced by bisphenol A (BPA) in zebrafish eleutheroembryos during the yolk sac absorption stage. *Chemosphere*. 246:125704.

Martyniuk CJ, Alvarez S, Denslow ND. 2012. DiGE and iTRAQ as biomarker discovery tools in aquatic toxicology. *Ecotoxicology Environmental Safety*. 76(2):3–10.

Martyniuk CJ, Alvarez S, McClung S, Villeneuve DL, Ankley GT, Denslow ND. 2009. Quantitative proteomic profiles of androgen receptor signaling in the liver of fathead minnows (*Pimephales promelas*). *Journal Proteome Research*. 8(5):2186–2200.

Martyniuk CJ, Denslow ND. 2009. Towards functional genomics in fish using quantitative proteomics. *General and Comparative Endocrinology*. 164(2–3):135–141.

Martyniuk CJ, Kroll KJ, Doperalski NJ, Barber DS, Denslow ND. 2010. Genomic and proteomic responses to environmentally relevant exposures to dieldrin: Indicators of neurodegeneration? *Toxicological Sciences*. 117(1):190–199.

Martyniuk CJ, Mehinto AC, Denslow ND. 2020. Organochlorine pesticides: agrochemicals with potent endocrine-disrupting properties in fish. *Molecular and Cellular Endocrinology*. 507:110764.

McCuaig LM, Martyniuk CJ, Marlatt VL. 2020. Morphometric and proteomic responses of early-life stage rainbow trout (*Oncorhynchus mykiss*) to the aquatic herbicide diquat dibromide. *Aquatic Toxicology*. 222:105446.

Monroe AA, Zhang H, Schunter C, Ravasi T. 2020. Probing SWATH-MS as a tool for proteome level quantification in a nonmodel fish. *Molecular Ecology Resources*. 20(6):1647–1657.

Moreton ML, Lo BP, Simmons DB, Marlatt VL. 2020. Toxicity of the aquatic herbicide, Reward(r), on the fathead minnow with pulsed-exposure proteomic profile. *Comparative Biochemistry and Physiology Part D: Genomics and Proteomics*. 33:100635.

Mustieles V, D'Cruz SC, Couderq S, Rodriguez-Carrillo A, Fini JB, Hofer T, Steffensen IL, Dirven H, Barouki R, Olea N et al. 2020. Bisphenol A and its analogues: a comprehensive review to identify and prioritize effect biomarkers for human biomonitoring. *Environment International*. 144:105811.

Pettus BJ, Chalfant CE, Hannun YA. 2002. Ceramide in apoptosis: an overview and current perspectives. *Biochimica et Biophysica Acta (BBA)-Molecular and Cell Biology of Lipids*. 1585(2-3):114–125.

Psychogios N, Hau DD, Peng J, Guo AC, Mandal R, Bouatra S, Sinelnikov I, Krishnamurthy R, Eisner R, Gautam B et al. 2011. The human serum metabolome. *PLoS One*. 6(2):e16957.

Qin G, Xu D, Lou B, Chen R, Wang L, Tan P. 2020. Itraq-based quantitative phosphoproteomics provides insights into the metabolic and physiological responses of a carnivorous marine fish (*Nibea albiflora*) fed a linseed oil-rich diet. *Journal of Proteomics*. 228:103917.

Quan J, Kang Y, Li L, Zhao G, Sun J, Liu Z. 2021. Proteome analysis of rainbow trout (*Oncorhynchus mykiss*) liver responses to chronic heat stress using DIA/SWATH. *Journal of Proteomics*. 233:104079.

Ralston-Hooper KJ, Turner ME, Soderblom EJ, Villeneuve D, Ankley GT, Moseley MA, Hoke RA, Ferguson PL. 2013. Application of a label-free, gel-free quantitative proteomics method for ecotoxicological studies of small fish species. *Environmental Science & Technology*. 47(2):1091–1100.

Reinwald H, König A, Ayobahan SU, Alvincz J, Sipos L, Göckener B, Böhle G, Shomroni O, Hollert H, Salinas G et al. 2021. Toxicogenomic fin(ger)prints for thyroid disruption AOP refinement and biomarker identification in zebrafish embryos. *Science of the Total Environment*. 760:143914.

Sanchez BC, Ralston-Hooper K, Sepúlveda MS. 2011. Review of recent proteomic applications in aquatic toxicology. *Environmental Toxicology and Chemistry*. 30(2):274–282.

Schultz IR, Nagler JJ, Swanson P, Wunschel D, Skillman AD, Burnett V, Smith D, Barry R. 2013. Toxicokinetic, toxicodynamic, and toxicoproteomic aspects of short-term exposure to trenbolone in female fish. *Toxicological Sciences*. 136(2):413–429.

Simmons DBD, Cowie AM, Koh J, Sherry JP, Martyniuk CJ. 2019. Label-free and iTRAQ proteomics analysis in the liver of zebrafish (*Danio rerio*) following dietary exposure to the organochlorine pesticide dieldrin. *Journal of Proteomics*. 202:103362.

Skogland Enerstvedt K, Sydnes MO, Pampanin DM. 2017. Study of the plasma proteome of atlantic cod (*Gadus morhua*): Effect of exposure to two PAHs and their corresponding diols. *Chemosphere*. 183:294–304.

Smith LC, Lavelle CM, Silva-Sanchez C, Denslow ND, Sabo-Attwood T. 2018. Early phosphoproteomic changes for adverse outcome pathway development in the fathead minnow (*Pimephales promelas*) brain. *Scientific Reports*. 8(1):10212.

Southam AD, Weber RJM, Engel J, Jones MR, Viant MR. 2017. A complete workflow for high-resolution spectral-stitching nanoelectrospray direct-infusion mass-spectrometry-based metabolomics and lipidomics. *Nature Protocols*. 12(2):310–328.

Sun J, Tang S, Peng H, Saunders DM, Doering JA, Hecker M, Jones PD, Giesy JP, Wiseman S. 2016. Combined transcriptomic and proteomic approach to identify toxicity pathways in early life stages of japanese medaka (*Oryzias latipes) exposed to* 1, 2, 5, 6–tetrabromocyclooctane (TBCO). *Environmental Science & Technology*. 50(14):7781–7790.

Trapp J, Armengaud J, Salvador A, Chaumot A, Geffard O. 2014. Next-generation proteomics: toward customized biomarkers for environmental biomonitoring. *Environmental Science & Technology*. 2;48(23):13560–13572. doi: 10.1021/es501673s. Epub 2014 Nov 17. PMID: 25345346.

van de Velde B, Guillarme D, Kohler I. 2020. Supercritical fluid chromatography - mass spectrometry in metabolomics: Past, present, and future perspectives. *Journal of Chromatography B Analyst Technology Biomedical Life Science*. 1161:122444.

van Meeteren LA, Ruurs P, Stortelers C, Bouwman P, van Rooijen MA, Pradere JP, Pettit TR, Wakelam MJ, Saulnier-Blache JS, Mummery CL et al. 2006. Autotaxin, a secreted lysophospholipase d, is essential for blood vessel formation during development. *Molecular and Cellular Biology*. 26(13):5015–5022.

Vieira LR, Hissa DC, de Souza TM, Sá CA, Evaristo JA, Nogueira FC, Carvalho AF, Farias DF. 2020. Proteomics analysis of zebrafish larvae exposed to 3, 4-dichloroaniline using the fish embryo acute toxicity test. *Environmental Toxicology*. 35(8):849–860.

Voisin A-S, Fellous A, Earley RL, Silvestre F. 2016. Delayed impacts of developmental exposure to 17-α-ethinylestradiol in the self-fertilizing fish *Kryptolebias marmoratus*. *Aquatic Toxicology*. 180:247–257.

Voisin A-S, Kültz D, Silvestre F. 2019. Early-life exposure to the endocrine disruptor 17-α-ethinylestradiol induces delayed effects in adult brain, liver and ovotestis proteomes of a self-fertilizing fish. *Journal of Proteomics*. 194:112–124.

Wang L, Wang Y, Liang Y, Li J, Liu Y, Zhang J, Zhang A, Fu J, Jiang G. 2014. PFOS induced lipid metabolism disturbances in balb/c mice through inhibition of low density lipoproteins excretion. *Scientific Reports* 4:4582.

Wang M, Wang Y, Zhang L, Wang J, Hong H, Wang D. 2013. Quantitative proteomic analysis reveals the mode-of-action for chronic mercury hepatotoxicity to marine medaka (*Oryzias melastigma*). *Aquatic Toxicology*. 130:123–131.

Xiang Q-Q, Gao Y, Li Q-Q, Ling J, Chen L-Q. 2020. Proteomic profiling reveals the differential toxic responses of gills of common carp exposed to nanosilver and silver nitrate. *Journal of Hazardous Materials*. 394:122562.

Xu L, Lu Z, Ji C, Cong M, Li F, Shan X, Wu H. 2019. Toxicological effects of as (v) in juvenile rockfish *Sebastes schlegelii* by a combined metabolomic and proteomic approach. *Environmental Pollution*. 255(Pt 2):113333.

Yadetie F, Bjørneklett S, Garberg HK, Oveland E, Berven F, Goksøyr A, Karlsen OA. 2016. Quantitative analyses of the hepatic proteome of methylmercury-exposed atlantic cod (*Gadus morhua*) suggest oxidative stress-mediated effects on cellular energy metabolism. *BMC Genomics*. 17(1):554.

Yadetie F, Oveland E, Døskeland A, Berven F, Goksøyr A, Karlsen OA. 2017. Quantitative proteomics analysis reveals perturbation of lipid metabolic pathways in the liver of atlantic cod (*Gadus morhua*) treated with pcb 153. *Aquatic Toxicology*. 185:19–28.

Yang Y, Lee M, Fairn GD. 2018. Phospholipid subcellular localization and dynamics. *Journal Biological Chemistry*. 293(17):6230–6240.

Yetukuri L, Ekroos K, Vidal-Puig A, Oresic M. 2008. Informatics and computational strategies for the study of lipids. *Molecular BioSystems*. 4(2):121–127.

Yung YC, Stoddard NC, Mirendil H, Chun J. 2015. Lysophosphatidic acid signaling in the nervous system. *Neuron*. 85(4):669–682.

# 12 Adverse Outcome Pathways

*Daniel L. Villeneuve*
Great Lakes Toxicology and Ecology Division

*Gerald T. Ankley*
US EPA Center for Computational Toxicology and Exposure

**Abstract**

Adverse outcome pathways (AOPs) organize scientific information concerning the linkage between biological perturbations measured at low levels of biological organization (e.g., molecular, biochemical, cellular) and adverse effects of concern for risk assessment, regulatory decision-making, and/or resource management. AOP development is (1) stressor agnostic, (2) modular in structure, (3) pragmatically simple, but (4) capable of representing biological complexity as networks, and (5) continuously evolving as scientific knowledge expands. The foundation of AOP development is the assembly and evaluation weight of evidence that establishes causal relationships between measurable key events that serve as markers in the progression of toxicity from the initial perturbation of a biological target to an adverse outcome of concern. As of fall 2021, there were 33 fish-relevant AOPs out of over 360 described in an open-access AOP-Wiki. There are many opportunities for experts in fish toxicology, physiology, aquaculture, and other related fields to expand this complement of AOPs. Concerted efforts in fish AOP development support expanded use of data from new approach methodologies for chemical safety assessment in aquatic toxicology and ecological risk assessment.

## INTRODUCTION TO ADVERSE OUTCOME PATHWAYS

New information in toxicology and other biological disciplines is being generated at a rapid and exponentially increasing rate. Collectively, the global scientific community is publishing over 2.5 million articles per year, with somewhere between 7 and 9 million researchers contributing (Ware and Mabe 2015). In 2014, there were estimated to be over 28,000 active English-language peer-reviewed scientific, technical, and medical journals with another 6,000 or more non-English language journals (Ware and Mabe 2015). All these publications translate into a global scientific output that doubles roughly every 9 years (Van Noorden 2014). Given this, it is reasonable to ask how regulators, resource managers, and a wide range of other stakeholders and decision-makers can keep pace with emerging scientific data and knowledge, let alone be able to synthesize, evaluate, and assimilate this incredible scope of information in a way that supports decision-making. While computational solutions like search engines, databases, and systematic review software can aid these processes, integration and assembly of scientific information into knowledge that can support decisions is a time-consuming and challenging process.

The adverse outcome pathway (AOP) framework is an epistemic approach that aims to provide a systematic and transparent method for assembling biological information into actionable knowledge that can support decision-making concerning the effects of contaminants or other stressors (Ankley et al. 2010; Fitzgerald 2020; Villeneuve et al. 2014a, 2014b). The AOP framework draws its origins from long-standing efforts in toxicology to make more effective use of molecular, biochemical, cellular, and behavioral responses, as well as in silico approaches like quantitative structure-activity relationships (QSARs), as meaningful inputs in risk assessment and decision-making (Huggett 1992; Bradbury et al. 2004; Mekenyan et al. 2003). However, the formalization of a framework for describing, evaluating, and disseminating those relationships in a scientifically robust and transparent manner first started to take shape around 2010 (Ankley et al. 2010) in conjunction with a vision articulated by the U.S. National Research Council (NRC) for a strategic approach to toxicity testing in the 21st century (Krewski et al. 2010). The NRC was charged with addressing the question of how the toxicology community could keep pace with regard to generating the depth and breadth of toxicological information needed for evaluation of 10,000s of chemicals currently in commerce as well as the new chemicals being produced each year. The vision advocated a move away from the direct observation of apical toxicity, i.e., direct measures of adverse effect in intact laboratory animals, toward an approach that utilized high throughput, mechanistically oriented assays to screen chemicals for biological activities that might result in toxicity. A critical need for implementing such a strategy was development of a core understanding of the biological pathways whose perturbation can be expected to result in adverse outcomes meaningful to risk assessment (Krewski et al. 2010). Furthermore, it was important to understand the magnitude and duration of perturbation that could plausibly result in toxicity and account for factors related to sex, life stage, taxonomic lineage, or individual variability that define and/or modulate susceptibility (Krewski et al. 2010). The AOP concept was proposed to address these needs (Ankley et. al 2010).

An adverse outcome pathway was defined as "a conceptual construct that portrays existing knowledge concerning the linkage between a direct molecular initiating event and an adverse outcome at a biological level of organization relevant to risk assessment" (Ankley et al. 2010; Figure 12.1). One key



DOI: 10.1201/9781003160694-12

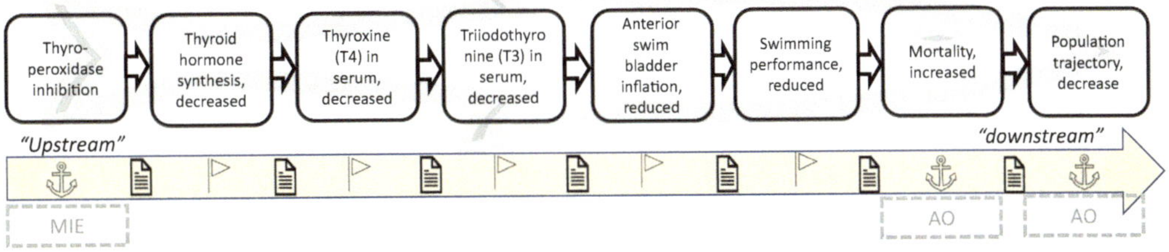


**FIGURE 12.1** Example of an adverse outcome pathway (https://aopwiki.org/aops/159), illustrating key features including a molecular initiating event (MIE); one or more adverse outcomes (AOs) at a level of organization relevant to risk assessment; intermediate key events (KEs; ) that identify key milestones or checkpoints in the sequential progression from the MIE to AO and bridge across levels of biological organization; and key event relationships (KERs; ) that describe the scientific support for, and quantitative understanding of (where known), a causal linkage between the "upstream" event and its "downstream" partner.

feature of an AOP is the identification of an initiating event. This initiating event is the point of direct interaction between the stressor and the biological system that causes its perturbation (Ankley et al. 2010). Because this interaction typically occurs at the molecular level, it is called the "molecular initiating event" (MIE; Figure 12.1). For example, inhibition of the enzyme thyroid peroxidase (TPO) is the MIE in an AOP linking TPO inhibition to increased mortality in early-life-stage fish due to impaired swim bladder inflation (https://aopwiki.org/aops/159; Figure 12.1). Contaminants can act directly on the function of this enzyme to elicit a perturbation that in-turn impacts downstream biological processes, so inhibition of thyroid peroxidase was identified as the MIE. However, in some cases, this first point of direct interaction may be unknown. In cases where a specific MIE cannot be identified, AOP developers are guided to identify an effect as early in the pathway as possible as the "upstream" anchor for the pathway description (OECD 2018). The second key feature, and an equally critical one, was to identify one or more "downstream" effects that occurred at a level of organization considered relevant to risk assessment (Ankley et al. 2010; Figure 12.1). For example, increased mortality and potential decreases in population trajectory are identified as adverse outcomes (AOs) in AOP 159 (Figure 12.1). In ecotoxicology, the AO is typically at the level of an individual or a population and directly reflects endpoints of accepted demographic significance, such as impacts on survival, growth, or reproduction. In human health toxicology, the AO of regulatory concern may be recognized at the lower levels of the organization such as the organ or organ system. Regardless, the aim of an AOP was to connect from the MIE to AO via a sequence of measurable key events (Figure 12.1) that represent salient milestones or signposts in the progression from the initial perturbation of the system to an ultimate adverse effect (Hutchinson et al. 2006).

In its initial incarnation (Ankley et al. 2010), the AOP served as a type of graphical short-hand for the accumulated knowledge of an expert in the field. However, the knowledge supporting the graphical overview was still not entirely accessible/transparent. While one could probe the details of a given AOP through discussion with the subject matter experts, there wasn't necessarily documentation of the underlying evidence which other scientists or decision-makers could use to evaluate the veracity and quality of the proposed relationships. Thus, as the demand for knowledge to bridge from increasingly available mechanistic, pathway-based data to relevant AOs increased, so, too, did the need to document and present the knowledge in a manner that is both accessible and transparent.

To take this crucial next step, those involved in developing the AOP framework drew upon the experiences of experts in the human toxicology community. Specifically, as part of the international programme on chemical safety assessment (IPCS), a structured approach to assess the overall weight of evidence for a proposed mode of action (termed the mode of action [MoA] framework) was developed (Sonich-Mullin et al. 2001; Boobis et al. 2006; Boobis et al. 2008; Meek et al. 2011; Meek et al. 2014). The framework included both a description of the postulated MoA and the identification of key events (KEs) that are both measurable and expected to be reliably observed across a body of experiments probing the consequences of a given biological perturbation (Sonich-Mullin et al. 2001). Importantly, the IPCS framework also incorporated formal consideration and documentation of various lines of evidence supporting causal associations between events (Sonich-Mullin et al. 2001; Boobis et al. 2006). These were adapted from a set of criteria proposed by Bradford-Hill in 1965 to evaluate whether observed connections between environmental factors and disease were causal or simply correlative (Hill 1965). Elements of documentation and weight of evidence consideration were taken on board as the AOP framework evolved from a concept to a knowledge assembly framework, which was formalized with the launch of the organization for economic cooperation and development (OECD)'s AOP program in 2014 (https://www.oecd.org/chemicalsafety/testing/adverse-outcome-pathways-molecular-screening-and-toxicogenomics.htm; Becker et al. 2015).

## PRINCIPLES OF AOP DEVELOPMENT

While the AOP framework drew upon the IPCS MoA framework in terms of formal documentation of the weight of evidence supporting a proposed pathway, there also were notable conceptual and practical differences between the AOP and MoA frameworks. These differences become apparent when considering the five principles of AOP development formulated during an international expert workshop held in 2014 (Villeneuve et al. 2014a). The principles reflect both the basic aim of identifying and integrating generalizable biological information and updating that information as new evidence emerges.

### AOPs Are not Chemical Specific

The first principle of AOP development is that **AOPs are not chemical specific** (Villeneuve et al. 2014a). The IPCS MoA framework was chemical-specific. Modes of action and associated KEs were described for specific chemicals, often including understanding of that specific compound's potency, adsorption, distribution, metabolism, and elimination (ADME; Sonich-Mullin et al. 2001; Boobis et al. 2008). This had the advantage that toxicity thresholds at the various KEs along the MoA could be precisely defined. In this sense, the MoA framework was very much rooted in traditional approaches to toxicological testing. In most cases, the amount of information required for a thorough chemical-specific MoA description would facilitate chemical risk assessment, without much need for alternative data. However, the throughput and resource intensity that limits the traditional toxicity testing paradigm also limited broad application of the MoA framework.

In contrast, the AOP framework was explicitly intended to support extrapolation from high throughput measurements and other mechanistically oriented endpoints to probable apical hazards, as envisioned for toxicity testing in the 21st century (Krewski et al. 2010). Therefore, the aim of AOP development was to capture the generalizable response of biological systems to specific types of biological perturbations, with the assumption that chemicals, or non-chemical stressors, that interact with the same MIEs and perturb pathways in the same manner would produce similar effects. The basic philosophy was that efficiency would be gained by describing these general patterns of response evoked by a range of stressors acting on the same KEs, rather than describing a MoA for each chemical independently. As a simple example, while the MoA framework would describe a separate pathway for each organophosphate pesticide, the AOP framework would assume that because the organophosphates act on a common target in the same way (inhibition of acetylcholinesterase), biological systems should respond to them in very similar ways once differences in potency and ADME are accounted for (Ankley et al. 2010; Russom et al. 2014). Both potency and ADME remain critical considerations in a chemical-specific assessment, but they are accounted for when the generalizable AOP is subsequently applied to a chemical-specific assessment and draws upon other sources of information captured outside the scope of AOP description, which is biologically focused, not stressor-focused (e.g., Teeguarden et al. 2016; Tan et al. 2018a, 2018b; Hines et al. 2019; Clewell et al. 2020). Overall, the chemically agnostic nature of AOPs better enables consideration of chemicals based on their biological similarity (i.e., ability to interact with an MIE), which can assist in classifying and prioritizing large groups of chemicals relative to assessing potential risk.

### AOPs Are Modular

The second principle of AOP development relates to the way information is organized when describing an AOP. **Adverse outcome pathways are modular** (Villeneuve et al. 2014a). The information defining an AOP is composed of two basic building blocks. The first of these blocks is the KE (Figure 12.1). As in the MoA framework, KEs describe a measurable change in a biological system that we expect to be reliably observed across a body of experiments. Key events are, in essence, observations. As a result, they are described in terms of (1) what is being measured; (2) how those measurements are made; and (3) the biological domain in terms of taxonomic groups, sex, and life stage for which that measurement can be made in a meaningful way (OECD 2018). Key events at the molecular and cellular levels of a biological organization will often be measured in non-intact animal assays, like in vitro or in chemico assays, or modeled in silico, while those measured at higher levels of organization will often be based on in vivo experimentation. Both MIEs and AOs are KEs, but they are considered specialized KEs in that MIEs provide the direct link to chemical-specific properties that interact with specific targets in the biological system, and AOs directly link the pathway to risk assessment/regulatory relevance. Thus, the MIE and AO(s) "anchor" the upstream and downstream ends of the AOP, respectively (Figure 12.1).

The second building block of an AOP is a KE relationship (KER; Figure 12.1). Each KER connects two KEs together as a cause-effect pair. One event ("upstream") causes the other event ("downstream") to occur. For example, the linkage of TPO inhibition to decreased thyroid hormone synthesis (Figure 12.1) is a KER (https://aopwiki.org/relationships/309). Because KERs are focused on establishing causal relationships between specific pairs of KEs, the description of a KER entails assembling weight of evidence that supports that cause-effect association. As such, assembly of the weight of evidence supporting the KERs is considered the most critical aspect of AOP development. While this draws on the modified Bradford–Hill considerations in a manner similar to the MoA framework (Becker et al. 2015), AOPs explicitly represent these connections between KEs as part of the framework and place considerable emphasis on independent documentation of each KER, without reference to other elements of the AOP (OECD 2018; Villeneuve et al. 2014a). The reason for documenting each KE and KER independently, without reference to other elements of the AOP, is a recognition that KEs and KERs reflect fundamental aspects of biology. That biology may respond in a similar manner across a wide range of chemicals or non-chemical stressors. Consequently, individual KEs and KERs can be shared among multiple AOPs (Villeneuve et al. 2014b, 2017; Knapen et al. 2018; Pollesch et al. 2019). For example, the KER linking TPO inhibition to decreased thyroid hormone synthesis is shared by eight different AOPs (https://aopwiki.org/relationships/309; accessed 12-05-2021). Because the description of each KE and KER in an AOP can be a considerable task unto itself, a great deal of redundant effort can be eliminated by describing KEs and KERs as discrete units (building blocks) that can be used in the assembly of several AOPs.

### Individual AOPs Are a Pragmatic Unit of Development and Evaluation

The third principle of AOP development is that **individual AOPs are** simply **a pragmatic unit of development and**

**evaluation** (Villeneuve et al. 2014a). An early criticism of the AOP framework was that it was too linear and simplistic. Biology is complex and cannot be expected to conform to a one-perturbation-to-one-adverse outcome type of representation. However, it is a direct recognition of the mind-spinning complexities of biological systems, and the complex networks of cellular, molecular, and biochemical interactions that underlie them, that informed taking a simplified approach to AOP development. Because a central aim of AOP development is to support extrapolation from responses or endpoints measured at relatively low levels of biological organization (e.g., molecular, cellular, biochemical) to apical outcomes at the level of individuals or even populations, extrapolation from one perturbation to one adverse outcome represents the smallest unit of practical application. While the ideal is to describe the many outcomes that could result from a given perturbation, across a wide range of biological contexts or, conversely, describe the many types of biological perturbations that can cause a given adverse outcome (e.g., reproductive toxicity), describing all those possibilities becomes an overwhelming and incredibly daunting task. In many cases, it would require the contributions of entire disciplines within the toxicology community and allied biological fields, working together, and would likely still not cover every pathway of potential relevance. With that in mind, a connection from one-perturbation-to-one-outcome was viewed as the scale of effort that a few individuals or a collaborative team could reasonably tackle. Likewise, it was viewed as a level of information assembly that a pool of reviewers could critically evaluate to provide an objective second opinion on the quality of the scientific evidence assembled. Thus, while recognized as a dramatic simplification of biological complexity, the individual AOP was defined as the most practical unit for development and evaluation (Villeneuve et al. 2014a; Knapen 2021).

## AOP Networks Are the Functional Unit of Prediction for Most Real-World Scenarios

Real-world exposures to chemicals and other stressors can be expected to cause multi-faceted responses and outcomes which often differ across different types of organisms, exposed during different life stages. A framework that cannot take such complexities into account when trying to apply our knowledge of biological systems is of limited utility. Consequently, the fourth principle of AOP development is that **AOP networks are the functional unit of prediction for most real-world scenarios** (Villeneuve et al. 2014a). When assembled in accordance with the modular AOP framework, an AOP network is defined as a system of two or more AOPs that share one or more key events (Knapen et al. 2018). Through the sharing of KE and KER pages in the AOP-Wiki (aopwiki.org), assembly of AOP descriptions in this online system leads to the de-facto creation of AOP networks (Villeneuve et al. 2014b). This is significant because description of a relatively small number of user-defined pathways can lead to the generation of many so-called "emergent pathways" that may reveal both novel biological relationships and a more comprehensive biological understanding with smaller levels of effort (Knapen et al. 2018; Villeneuve et al. 2018a; Pollesch et al. 2019). This potential was illustrated by a 2019 analysis of the AOP-Wiki that showed that just 187 user-defined AOPs resulted in emergence of over 9400 unique connections from MIEs to AOs in the system (Pollesch et al. 2019). Thus, while individual AOPs are intentional simplifications of biological complexity, that complexity can be captured in the AOP-Wiki (aopwiki.org) and underlying AOP knowledgebase (AOP-KB; aopkb.oecd.org).

Utilizing the network of AOPs described in the AOP-KB, one can begin to address more sophisticated assessment scenarios. The consequence of specific perturbations can be investigated in the context of multiple potential outcomes within different life stages, sex, and taxonomic scenarios by using ontology- or controlled vocabulary-based filters as queries in the AOP-KB (Knapen et al. 2018). Likewise, to consider impacts of multiple stressors as well as individual stressors that may elicit more than one biological perturbation concurrently (LaLone et al. 2017; Knapen et al. 2018; Ankley et al. 2020), there is a potential to overlay various types of quantitative data or information onto a graphical representation of an AOP. Additionally, because AOP networks adhere to the basic node-edge relationships that define diverse types of networks ranging from social networks, transportation networks, and computer networks, to food-webs and financial networks, etc., many of the tools and approaches of network analysis and graph theory (Newman 2003; Pavlopoulos et al. 2011; Trudeau 2013) can be productively applied to analyze AOP networks (Villeneuve et al. 2018a; Pollesch et al. 2019). At the time of this writing, the possibilities, and capabilities of AOP network-based analyses are just starting to be explored (e.g., Corsi et al. 2019; Knapen et al. 2020).

## AOPs Are "Living Documents"

The fifth and final principle of AOP development is rooted in the fact that AOP development is an epistemic concept. The AOP provides a systematic approach to organizing information into knowledge. Because new scientific information continues to be generated, the fifth principle of AOP development is an explicit recognition that **AOPs are "living documents"** (Villeneuve et al. 2014a). In concept, new information could always be added to an AOP in the form of additional weight of evidence, improved quantitative understanding, more in-depth description of biological plausibility, etc. This means that there is no objective "complete AOP". One could always do and add more. Consequently, AOPs should not be judged when they provide a "complete" compilation of all relevant scientific information. Instead, AOPs are evaluated relative to their utility for a given application (i.e., fit-for-purpose; Kleinstreuer et al. 2016; Coady et al. 2019). In that context, it is the intended application that dictates whether the AOP is sufficiently developed. Thus, the decision of when an AOP is "done" is a subjective one that needs to be weighed against the intended purpose and application of the knowledge assembled (Coady et al. 2019). An AOP to be used for hypothesis generation or to identify gaps in existing knowledge may be sparsely supported yet still useful. Likewise, if an AOP is used to design an integrated testing strategy or defined approach to alternative testing (Tollefsen et al. 2014; Patlewicz et al. 2014; Sakuratani et al. 2018; Kolle 2020), less confidence and quantitative understanding may be required as final assessment decisions will be based on empirical results of testing. In contrast, an AOP to be used to support a quantitative and predictive risk assessment of a high-value substance may require extensive amounts of support, quantitative understanding, associated mathematical or computational modeling, and extensive knowledge of modulating factors to be considered fit for purpose.

Because the assembly of information supporting an AOP can be an on-going task, particularly as new applications for the AOP emerge, information is captured in an interactive knowledgebase, rather than static documents. The IPCS MoA framework was focused on the development of documents. However, such hard copies or pdfs are not readily updated as new information becomes available. Consequently, the official source for dissemination of AOPs developed in accordance with the OECD AOP program is via the AOP-KB and AOP-Wiki (aopkb.oecd.org; aopwiki.org). The AOP-KB is the term ascribed to the data model and machine-readable stored content, and the text of the AOP descriptions in their native XML format. The AOP-wiki (aopwiki.org) is the most commonly used interface for adding new content to the AOP-KB, searching and viewing AOP content, etc. However, the AOP-KB is in an open and accessible format that allows for the development of additional applications for querying, accessing, and/or analyzing content. For example, the eAOP portal (aopkb.oecd.org) is an alternative to the AOP-Wiki for searching the AOP-KB. Likewise, as web-based content, most AOPs can be found by their titles or by their unique AOP IDs via conventional web search (e.g., Google search).

While pdf "snapshots" of the content assembled in the AOP-KB can be subjected to peer review and endorsement through the OECD program (OECD 2021) and/or via affiliated scientific journals (e.g., Schmid et al. 2021; Song and Villeneuve 2021), content in the AOP-Wiki can continue to evolve even after initial peer review and publication has been completed. This is a novel approach in scientific knowledge aggregation and dissemination that further distinguishes the AOP framework from both the IPCS MoA framework as well as other fora used for assembling and synthesizing knowledge like review articles, textbooks, etc. Consequently, the principle of AOPs as living documents is fundamental to the framework and how it differs from its origin as a graphical model absent of any formal assembly of scientific support.

## ASSEMBLING EVIDENCE FOR CAUSALITY

Above all else, AOPs are intended to facilitate inference and extrapolation from one biological event to another. Accordingly, the heart of an AOP and its utility rests in the assembly of evidence that supports individual KERs (Villeneuve et al. 2014b). Each KER is intended to address the question "if the upstream event in the relationship is observed, what downstream effect can be expected to result, and what is the scientific support for that expectation?" There are two key components to addressing that question. First, "what" we can expect. This is defined through "qualitative evidence" that establishes a scientifically sound basis for the association of the downstream event with its upstream cause. The second component is "under what conditions" can we expect the upstream event to evoke some defined magnitude or probability of change in the downstream event. This is captured as "quantitative understanding" of the KER (Wittwehr et al. 2017; OECD 2018). When a sequence of KERs is assembled to connect an MIE to an AO, the qualitative evidence across each KER allows for the inference of hazard (i.e., what kind of outcome could occur), while the quantitative understanding across the KERs determines the extent to which one can estimate the probability and/or severity of that outcome for a defined set of conditions (i.e., an assessment scenario). So, in general terms, extrapolation along an AOP is facilitated, to the extent possible, by the assembly of qualitative support and quantitative understanding of its component KERs (Villeneuve et al. 2014b; Wittwehr et al. 2017; Figure 12.2). The remainder of this section explores what constitutes qualitative support for a KER and the elements of quantitative understanding that are viewed as most critical.

### QUALITATIVE SUPPORT

Qualitative support for a KER is assembled from lines of evidence that indicate when the upstream event is observed, the downstream event could be expected (Villeneuve et al. 2014b). That expectation can be based entirely on correlative association (i.e., when A is observed, B is often observed as well). However, without understanding some mechanistic basis for event A to cause event B to happen, a great deal of evidence (many observations) is needed to establish confidence that a correlative association is a sound basis for inference or extrapolation (Becker et al. 2015). Consequently, the most valuable evidence for establishing a KER is establishing or demonstrating a causal relationship between event A (upstream) and event B (downstream). Canonical knowledge of basic biology can often serve as one of the most important lines of evidence (Svingen et al. 2021). For example, it is widely accepted that fins are important to locomotion in fish. Based on that knowledge, it is regarded as highly plausible that severe fin damage would impair locomotion leading to negative outcomes. Likewise, a large body of literature in basic biochemistry and endocrinology has firmly established that the enzyme cytochrome

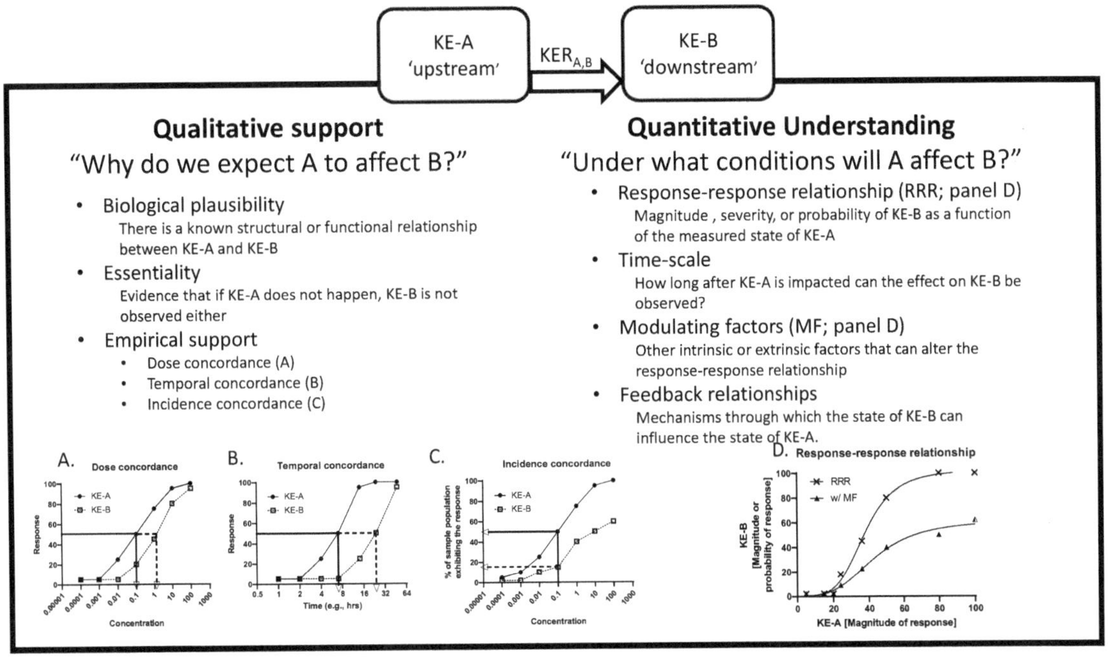


**FIGURE 12.2** Overview of the major lines of evidence used to provide qualitative support for a causal relationship between the key events (KEs) linked by a key event relationship (KER) and the desired information to gather regarding the quantitative understanding of the relationship, when available.

P450 19 (cyp19; aromatase) is rate limiting for the conversion of androgens like testosterone or androstenedione to estrogens like 17β-estradiol or estrone (Miller 1988). Consequently, it is plausible that inhibition of aromatase could result in reduced synthesis of 17β-estradiol (https://aopwiki.org/relationships/45). In both these examples, the biology under "normal" conditions is so well understood to be causally connected that it is not a great leap, scientifically, to infer that if the upstream event is perturbed, the downstream event could also be impacted. Thus, "**biological plausibility**" (Figure 12.2) is viewed as one of the most important lines of evidence for establishing a KER (Becker et al. 2015; OECD 2018).

A second type of evidence that provides compelling scientific support for a KER is termed "**essentiality**" (Figure 12.2). Essentiality addresses the question of whether the upstream event must occur for the downstream event to take place. If A must occur for B to take place, this is compelling support for a causal relationship. As such, evidence for essentiality most often derives from experiments that show that if event A (upstream) is blocked or prevented, event B does not occur. For example, in an AOP-linking activation of the aryl hydrocarbon receptor (AhR) to increased expression of cyclooxygenase 2 (cox2) and subsequent alterations in cardiovascular development and function (AOP 21; https://aopwiki.org/aops/21; Doering et al. 2019), administration of a cox2 inhibitor was shown to block the effect of a strong AhR agonist (2,3,7,8-tetrachlorodibenzo-p-dioxin; TCDD) on precardiac edema in developing zebrafish (Teraoka et al. 2014). This provided a line of evidence for the essentiality of cox2 expression in the pathway leading to AhR-mediated cardiotoxicity (Doering et al. 2019). While that example was based on the administration of a selective inhibitor of a particular enzyme during co-treatment with a stressor that could activate the MIE, other approaches could include for example knock-down or knock-out of a target thought to play a critical role in the pathway. In the case of "essentiality", it is important to keep in mind that the "essential" nature of the upstream KE(s) pertains to the specific sequence of events defined in a single "linear" AOP. While increased cox2 expression may indeed be one of the ways that cardiotoxicity occurs in developing fish, it may not be the only way (e.g., Farhat and Kennedy 2019; https://aopwiki.org/aops/150). If a stressor is present in the experimental system that can cause cardiotoxicity through an alternative mechanism (a different AOP in the AOP network leading to cardiotoxicity), that does not necessarily undermine the causal role cox2 may play in mediating AhR agonism-related cardiotoxicity. Thus, in experiments that would appear to reject the role of a given event as being essential in a particular AOP, it is important to consider whether there are alternative pathways through which the same downstream outcome may be observed.

The third type of evidence that is routinely assembled to support a KER is broadly termed "**empirical support**" (Figure 12.2). Empirical support can be viewed as lines of evidence that are consistent with the hypothesis that Event A (upstream) causes Event B (downstream) to happen. Empirical support typically relies on experimental studies in which both events A and B were evaluated, and one can assess whether the pattern of response for endpoints A and B are consistent with a causal relationship. Specifically, if event A causes event B, one would expect the following: (1) event A would occur at doses of the stressor equal to or lower than those that cause event B to occur (termed **dose concordance**; Figure 12.2A); (2) upon exposure to a stressor, event A would be observed before event B (termed **temporal concordance**; Figure 12.2B); and (3) at a given level of exposure to a stressor, the number of individuals in the sample population for which event A was observed is greater than or equal to the number of individuals in that population for which event B was observed (termed **incidence concordance**; Figure 12.2C). Experimental data in agreement with these patterns provide additional support for the hypothesis that event A causes event B (Becker et al. 2015; OECD 2018). For example, Doering et al. (2019) cited evidence that cox2 expression was increased at the same concentration of TCDD shown to cause increased hemorrhage in developing fish (Dong et al. 2010; Park et al. 2014) as evidence of the dose concordance between increased cox2 expression and altered cardiovascular development and function (https://aopwiki.org/relationships/1351). Yauk et al. (2016; https://aopwiki.org/aops/15) cited evidence that incidence of akyl adducts (upstream event) was significantly increased after treatment of cells with 0.25 µg/ml of the well-known mutagen methylmethanesulfonate, while significant increases in mutations (downstream event) were not detected in treatments less than 1.25 µg/ml (Swenberg et al. 2008; https://aopwiki.org/relationships/25). While, in general, it is viewed as ideal to collect empirical support from studies in which both events were measured in the same study (i.e., same species, comparable time course, same exposure conditions, etc.), this is not always possible. In cases where both events are not measured in the same study, empirical support can still be obtained, but one must be careful to evaluate whether apparent concordance, or lack thereof, is an artifact of study design elements like the relative sensitivity of methods used to observe the events, dose selection across studies, differences in sensitivity of the organism or assay system employed, differences in exposure duration and/or times at which endpoints were measured, etc. Thus, while straight forward in concept, the evaluation of empirical support and overall concordance with a hypothesized cause-effect relationship can be quite challenging in practice, particularly when using studies that were not specifically designed to evaluate the concordance of the relationships between KEs A and B.

The assembly of biological plausibility, evidence for essentiality, empirical support, and any other lines of evidence relevant to establishing causality is the most critical step in AOP development. The overall level of support for the various KERs along the pathway ultimately dictates the AOP's fit for purpose for different types of applications (Coady et al. 2019). Thus, it is the aspect of AOP development on which the most time is typically spent and the developer's assessment of the level of support as being high, medium, or low can help guide application. Guiding questions for evaluating whether each line of evidence assembled for a KER/AOP would be subjectively evaluated as high, medium, or low have been defined (Becker et al. 2015; OECD 2018). However, it is often less clear how to aggregate the evaluations of plausibility, essentiality, and empirical support into an overall ranking for the KER. While it is established that plausibility provides the strongest support, followed by essentiality and empirical support (Becker et al. 2015), it is ultimately some combination of the three that provides the greatest confidence in a KER. Recognizing that there is inherent subjectivity in these evaluations, transparency in the supporting evidence and the approach the developer took to assembling and evaluating that evidence are also viewed as critical to the evaluation of fit for purpose (Coady et al. 2019). Such transparency allows users to evaluate the evidence and draw their own conclusions if they question the "calls" made by the developer.

### Quantitative Understanding

Beyond establishing that the two KEs connected via a KER are causally related, it is also helpful to assemble quantitative information about the nature of the relationship to enhance the utility of AOPs beyond defining what could happen if a perturbation were sufficiently severe (hazard), to being able to define what outcome(s) are expected under a given exposure scenario (Villeneuve et al. 2014a; Conolly et al. 2017). Thus, when assembling the evidence supporting a KER, AOP developers are encouraged to include quantitative understanding of the KER when feasible/available. Important aspects of this are the (a) response-response relationship (RRR) between KEs A and B; (b) time-scale over which KE-A influences KE-B; (c) modulating factors (MFs) known to alter the response-response relationship or time scale; and (d) any known feedback relationships between KE-B or other components of the AOP network or underlying biology that can influence KE-A (Wittwehr et al. 2017; OECD 2018; Foran et al. 2019; Spinu et al. 2020; Figure 12.2).

To assemble this quantitative understanding, it is important to understand what is meant by an RRR. An RRR is a mathematical function that describes the magnitude, probability, or severity of change in the downstream KE (B) as a function of the measured (or predicted) state of the upstream KE (A). Simplistically, an RRR can be envisioned as a two-dimensional plot with the magnitude of response for the upstream KE (A) on the X-axis and the magnitude of response in the downstream KE (B) on the Y-axis (e.g., Figure 12.2D). However, in practice, the RRR may take many forms, from a simple linear or non-linear regression (e.g., Foran et al. 2019) to more complex computational models that incorporate stochasticity, feedback, and compensation, and/or conditional probabilities (e.g., Conolly et al. 2017; Perkins et al. 2019; Burgoon et al. 2020; Moe et al. 2021). However, once defined, an RRR can allow for quantitative inference from one KE to another along the AOP (Conolly et al. 2017; Foran et al. 2019).

A general strategy for defining RRRs is to identify one or more "prototypical stressor(s)" for the AOP of interest. A prototypical stressor is the one known to trigger the MIE (or earliest defined KE in the pathway), for which there is an extensive database with respect to its impacts on the downstream KEs, such that the response of multiple KEs over a range of different exposure scenarios (concentrations, time-course, etc.) have been described. With enough such observations for a given stressor, one can empirically define the relationship between KEs A and B (Foran et al. 2019). Since the magnitude, probability, or incidence of biological response is (conceptually) plotted on each axis, the concentration of the prototypical stressor (usually a chemical) itself drops out of the equation leaving a, potentially, stressor-agnostic relationship that can be applied to a variety of stressors and scenarios given that the response level of the upstream KE (A) is known. The ultimate accuracy, precision, and confidence in the quantitative extrapolation is then based on several assumptions (Safe et al. 2002). The first is that all the stressors for which the RRR is being generalized act through the same biologic pathway. When applied in the context of an AOP, this assumption is, by definition, met. A second assumption is that the concentration–response curves for different stressors acting on the relationship are parallel. That is individual stressors may vary in their potency, but as long as one can adjust for the relative potency compared to some reference stressor (i.e., the prototypical stressor in the AOP), based on evidence available outside the AOP, the basic shape/slope of the response should hold (Putzrath 1997; Oris and Bailer 1997). While this has been established for certain classes of chemicals (e.g., dioxin-like chemicals acting on the AhR [Safe 1990]; estrogenic chemicals acting on the estrogen receptor to a somewhat lesser extent [Safe et al. 2002]), there are many classes of chemicals for which this is demonstrably not the case. In such cases, the degree of uncertainty associated with violating the assumption of parallel dose-response can often be estimated (Villeneuve et al. 2000). Closely tied with the assumption of parallel concentration–response is that the effects of individual stressors, in combination, are essentially additive as sub-maximal levels of exposure. The final assumption is that the fate and distribution of all the stressors acting via a given AOP are the same over the relevant range of concentrations considered. This last consideration falls in the realm of chemical- or stressor-specific ADME and, in that regard, is handled outside of the AOP framework, as noted earlier. Unless these assumptions have been robustly evaluated for a given RRR, there will often be considerable uncertainties associated with its application to new scenarios for which empirical results are not available.

To complement description of the RRR associated with a KER, there are several additional types of information that AOP developers are encouraged to capture when possible. The first of these is description of the time-scale over which the RRR operates. Essentially, the goal is to address the question, "how soon after KE-A is impacted will KE-B change in response?". The aim is to provide a sense of whether the effect takes place in a matter of seconds, minutes, days, weeks, years, etc. This can be critical to both the interpretation of empirical results and evidence. For example, if the effect on KE-B is known to lag a change in KE-A by days, a study in which both A and B are measured concurrently just a few hours after exposure may not provide useful information about whether the perturbation is progressing toward adversity. Likewise, an understanding of the time-scale of transition between KEs A and B can help inform both the implementation of computational models for the relationship and/or experimental approaches to evaluate the support for causality of the KER or derivation of an RRR.

A second complement to overall description of the RRR is the identification of established scientific knowledge about factors that can change the nature (function) of the relationship and/or the time-scale of transition between KEs A and B. Variables that are known to influence or change a given RRR and/or its time-scale are termed "modulating factors". Examples of modulating factors include, but are not limited to, considerations like diet, pre-existing disease or injury, genetic susceptibility or resistance factors, and sex. Where such factors are known to change the shape or mathematical description of an RRR or time-scale, that information should be captured as part of the KER description (Wittwehr et al. 2017; OECD 2018). However, AOP developers should not simply speculate as to what factors may affect the relationship. Only those modulating factors for which the effect on the relationship has been documented and defined should be included as part of the KER description. This is consistent with most conventional toxicity testing and risk assessment practices, where modulating factors are rarely considered except in cases where there is well-documented evidence for their importance/relevance to a given assessment scenario.

The final type of quantitative understanding of KERs that AOP developers are encouraged to identify where possible are any known positive or negative feedback interactions between KEs A and B (Wittwehr et al. 2017; OECD 2018). Adjacent KEs, and AOPs, are graphically represented as sequentially directed whereby KE-A impacts KE-B (Knapen et al. 2018; Villeneuve et al. 2018). However, the biological reality is that there are often cases where KE-B influences KE-A, either directly or indirectly. For example, from the diagram of AOP 159 (Figure 12.1), it is not obvious, but it is known that decreased concentrations of serum thyroxine can elicit a feedback response at the level of the hypothalamus and pituitary gland that can, in turn, alter the expression of thyroperoxidase in the thyroid gland, directly influencing the MIE (Knapen et al. 2018). Another example is highlighted in AOP 25 (Villeneuve 2016; https://aopwiki.org/aops/25) which links aromatase inhibition to reproductive dysfunction in female fish. In this AOP, inhibition of the enzyme aromatase leads to reductions in circulating concentrations of 17β-estradiol. However, experimentally, it has been shown that as plasma concentrations of 17β-estradiol fall, a feedback response along the hypothalamic-pituitary-gonadal axis stimulates the expression of mRNA transcripts coding for the aromatase enzyme (Villeneuve et al. 2009). Thus, at some point, after 17β-estradiol concentrations start to fall, the increased expression of aromatase can begin to off-set the effects on the MIE such that KE-A is not independent of KE-B. Compensatory/adaptive responses such as this are common homeostatic attributes of many biological systems. In the case of AOP 25, this feedback relationship was incorporated into a mathematical model of the RRR, allowing for more accurate prediction of dose-response and time-course behaviors (Conolly et al. 2017; Cheng et al. 2016). Where known, it is helpful to document information on such known feedback relationships as part of the quantitative understanding of the AOP. A feedback symbol (⤳) can also be added to the graphical representation of the pathway to alert users of feedback that may need to be taken into consideration (OECD 2018).

While the aim is to define a generalizable RRR and capture associated information about the time-scale of the relationship, known modulating factors, and any known feedback interactions between KEs, many AOPs, and their associated KERs lack this information. Robust quantitative understanding of a KER often requires extensive experimental characterization over a range of stressor concentrations and time-points to fully describe a range of responses between KEs of interest. Most toxicology studies, focused on documenting concentration–response for selected endpoints are not specifically designed to capture this type of information, and the studies required to do so are not trivial. Consequently, in the literature, there is often a dearth of information that addresses the quantitative aspects of a KER. It, generally, will be dictated by needs of an intended fit-for-purpose as to whether the investment of time and resources needed to derive a quantitative RRR assess the assumptions associated with its generalization across stressors, species, exposure scenarios, etc. is warranted.

## THE STATE OF FISH-ORIENTED AOPS (CA. 2021)

As of fall 2021, over 360 AOP descriptions were under various stages of development in the AOP-Wiki (aopwiki.org). Based on a search of AOP titles and AOP page content using the keyword "fish", 33 of those AOPs, representing just over 9% of all AOPs under development, were identified as being specifically pertinent to fish (Figure 12.3A). A few of these (e.g., AOP 12; AOP 43; AOP 218; aopwiki.org) were primarily focused on other species, but specifically

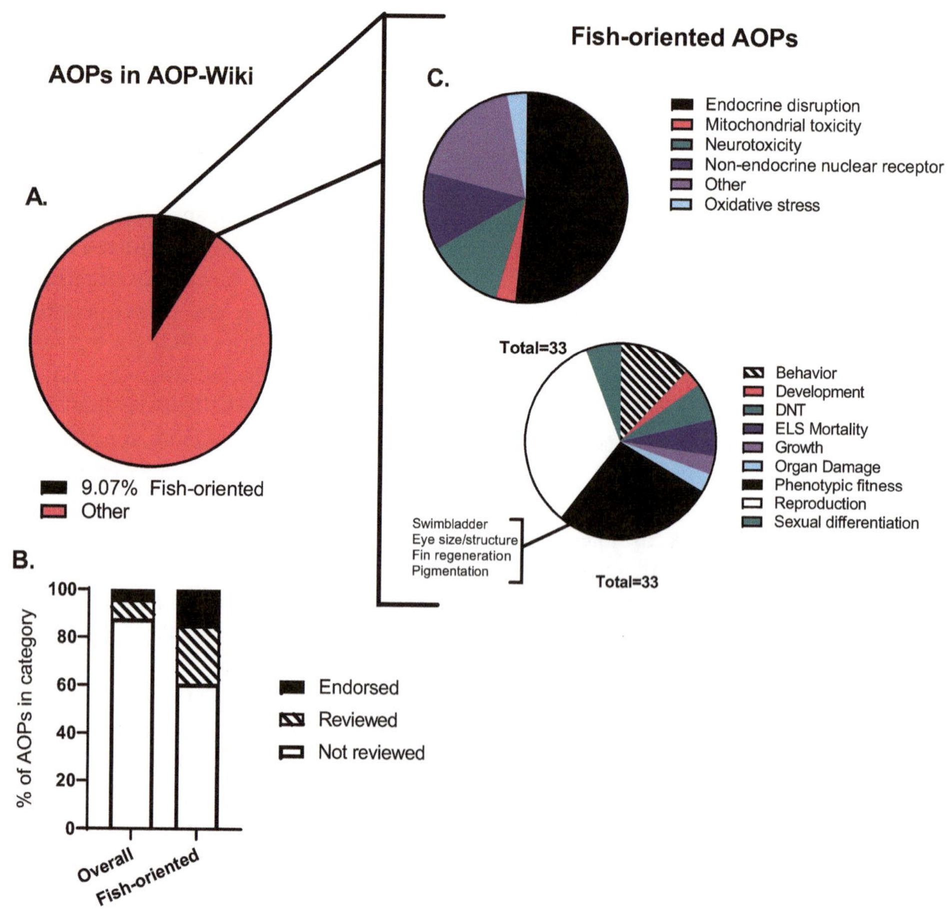


**FIGURE 12.3** The state of fish-oriented AOPs in the AOP-Wiki, circa September 2021. (A) Proportion of fish-oriented AOPs relative to all AOPs in the AOP-Wiki. (B) Proportion of AOPs that have been reviewed, endorsed, or not reviewed. (C) Mode(s) of action (C.1) and apical endpoints (C.2) covered by the complement of fish-oriented AOPs.

noted fish as being among their plausible domain of applicability. However, 30 of the 33 AOPs (91%) were specifically targeted at describing effects in fish and drew heavily from evidence gathered from studies in fish.

While fish-oriented AOPs represent only around 10% of AOPs under development in the AOP-Wiki, at the time of this writing, they constitute a disproportionate share of the highly developed and vetted AOPs. In fall 2021, just 16 out of 360 (4.4%) AOPs under development in the AOP-Wiki had been reviewed and endorsed by one or more Working parties of the organization for economic cooperation and development (OECD; https://www.oecd.org/chemicalsafety/testing/projects-adverse-outcome-pathways.htm). An additional 28 (7.7%) had been subjected to technical peer review organized either by the OECD Extended advisory group on molecular screening and toxicogenomics or an affiliated scientific journal. However, among the 33 fish-oriented AOPs, eight (24%) had undergone technical peer review, and five (15%) had already been endorsed by OECD Working parties (Figure 12.3B). Thus, fish-oriented AOPs have provided some of the strongest examples of quality AOP development and various applications of AOPs as they are integrated into the practice of 21$^{st}$- century toxicology. A specific example of this was development of one of the first quantitative AOPs published in the open literature, which linked altered steroid synthesis to egg production/population status in the fathead minnow and other fish species (Conolly et al. 2017).

Endocrine disruption has been a significant focus for fish AOP development, with just over half of the fish-oriented AOPs in the AOP-Wiki focused on various modes of endocrine disruption (Figure 12.3C). Multiple AOPs provide evidence linking the inhibition of enzymes involved in steroid biosynthesis to reproductive toxicity or altered sexual differentiation (e.g., https://aopwiki.org/aops/25; https://aopwiki.org/aops/289; https://aopwiki.org/aops/346). Several AOPs also trace hazards associated with agonism or antagonism of steroid hormone receptors (e.g., estrogen receptor, androgen receptor), once again with a focus on adverse effects on reproduction and/or sexual differentiation (https://aopwiki.org/aops/23; https://aopwiki.org/aops/30; https://aopwiki.org/aops/29; https://aopwiki.org/aops/346). While thyroid axis disruption has generally been regarded as one of the understudied aspects of endocrine disruption in fish, concerted efforts in thyroid AOP development starting around 2014 have yielded a significant cluster of thyroid-related AOPs focused on fish (reviewed by Knapen et al. 2020). Hazards associated with thyroid disruption in fish include impaired inflation of the posterior and/or anterior chamber of the swim bladder, altered eye development, and impaired fertility (aopwiki.org). With complementary thyroid-related AOP development in both mammalian and amphibian models, the overall network of thyroid AOPs in the AOP-Wiki provides a compelling example of a cross-species AOP network that bridges both human health and ecological toxicity (Knapen et al. 2018; Noyes et al. 2019; Holbech et al. 2020; Knapen et al. 2020).

While endocrine disruption has been a major focus for fish-oriented AOP development, other notable biological pathways/processes also have been explored (Figure 12.3C). Given the extensive research conducted on the effects of dioxin and dioxin-like chemicals over decades in the field of aquatic toxicology, several well-developed AOPs linking AhR activation to fish early life stage mortality have been described and endorsed by the OECD (Doering et al. 2019; Farhat and Kennedy 2019). Hazards associated with other MIEs involving nuclear receptors have also been documented, including AOPs linking peroxisome proliferator-activated receptor gamma to impaired reproduction (https://aopwiki.org/aops/323) and glucocorticoid receptor activation to impaired fin regeneration (https://aopwiki.org/aops/334). Neurotoxicity has also been the subject of fish-oriented AOP development. AOPs outlining effects of stressors that alter neurotransmitter signaling (e.g., serotonin [https://aopwiki.org/aops/98] or N-methyl-D-aspartate [https://aopwiki.org/aops/12]) or disrupt action potentials through effects on ion channels (https://aopwiki.org/aops/91) have all been initiated. Adverse effects in fish due to inhibition of a variety of enzymes including tyrosinase, cystathione beta synthase, retinaldehyde dehydrogenase, Src family tyrosine kinase A (fyna), cytochrome P450 7b, and lysyl oxidase have also been described (aopwiki.org). A few AOPs have also provided evidence linking fairly general mechanisms of toxicity such as oxidative stress and mitochondrial toxicity to target organ effects (e.g., https://aopwiki.org/aops/413) or growth impacts (Song and Villeneuve 2021; https://aopwiki.org/aops/263). The complement of AOPs found in the AOP-Wiki ca. fall 2021 serve as useful examples of how information regarding the effects of biological perturbation in fish can be organized, using the AOP framework, to facilitate translation and application of mechanistic data.

## OPPORTUNITIES FOR FISH AOP DEVELOPMENT

There is a strong foundation of well-developed fish-oriented AOPs in the AOP-Wiki. However, at the time of this writing, many opportunities for fish AOP development remain. Other chapters in this book (Willett and Aluru 2023) review accumulated scientific knowledge on various target organ toxicities and many of the MIEs and KEs that can lead to adverse outcomes in fish. However, much of this knowledge has not been assembled and organized in an AOP format, making it more difficult to apply in the context of the emerging vision for 21st-century toxicology (Krewski et al. 2010) or the use of "new approach methodologies" (NAMs) for chemical risk assessment (Kavlock et al. 2018). For example, despite extensive research on immune function in fish, as of fall 2021, there were no AOPs related to adverse effects of stressors on immune function in fish described in the AOP-Wiki (aopwiki.org). Also, there are many AOPs that might be applicable to fish, but their full domain of biological applicability is not well defined. For example, there are several mammalian-focused AOPs related to liver toxicity, carcinogenesis, and various target organ toxicities with MIEs related to oxidative stress, but few of these include specific evidence demonstrating their relevance to fish. However, given the general nature of many of the basic mechanisms (e.g., oxidative stress; inflammation; immunosuppression) underlying target organ toxicities and/or carcinogenesis (Villeneuve et al. 2018; Smith et al. 2020), it is likely that many fish-relevant AOPs could be developed via a modest effort by supplementing existing mammalian AOPs with evidence from fish and/or linking various fish-specific KEs and KERs to existing AOP elements describing these universal mechanisms.

Although there is the opportunity to leverage elements of AOP developed for mammals and other vertebrates to rapidly and efficiently expand AOP coverage for fish, it is also critical to consider unique aspects of fish physiology that may require a more explicitly fish-centric approach to AOP development. For example, while sharing several functional similarities with the mammalian lung (Progatzky et al. 2016), the gill is a rather unique tissue in fish that plays a critical role in respiration, osmoregulation, and excretion (Evans 1987) while being in intimate contact with the aqueous environment and both dissolved and particle-bound contaminants therein (Sollid and Nilsson 2006). As such, there is potential for adverse effects in fish related to impacts of stressors on gill development, morphology, and function (Evans 1987) that could be described using the AOP framework. Osmoregulation also plays a critical role in fish survival, making fish potentially sensitive to ion-mediated toxicities and indicating that interference with various mechanisms of osmoregulation can be expected to be detrimental to fish (Griffith 2017). Given the different aims of osmoregulation in marine versus freshwater fish (Hammerschlag 2006), it can be expected that the relevant AOPs may differ for these two environments. Another set of pathways novel to fish involves those related to olfaction/homing behavior. For certain groups of anadromous fish like salmonids (*Salmonidae*) and lampreys (*Pretromyzonidae*), olfaction plays a critical role in homing behaviors and navigation (Bett and Hinch 2016). Physical or contaminant-induced damage to olfactory structures, which are in direct contact with the aqueous environment, can adversely affect migration or other important olfaction-directed behaviors in fish (Tierney et al. 2010; Williams et al. 2016). The lateral line represents yet another sensory structure in fish that is vulnerable to damage by toxicants and plays a critical role in fish survival and fitness (Mogdans 2019; Coffin and Ramcharitar 2016). Likewise, there is evidence to suggest electrosensory systems critical to survival in some groups of fish (e.g., many Chonrichthyes) can be impaired by exposure to environmental contaminants (Cave and Kajiura 2020). While the examples here are not comprehensive, it is safe to assume that there are many aspects of fish physiology that can be altered or disrupted by chemical contaminants and other types of stressors. That information is often highly dispersed among various journals related to zoology, fish physiology, endocrinology, aquaculture, marine biology, aquatic and environmental toxicology, fisheries management, etc. Nevertheless, in many cases, there is extensive biological understanding and evidence available to support the initial development and description of an AOP.

## APPLICATION OF AOPS

A range of publications have discussed and provided examples of the practical application of AOPs to different facets of chemical risk assessment, ranging from providing a platform for knowledge assembly to supporting predictive safety assessments in a regulatory setting (Ankley and Edwards 2017; Coady et al. 2019; Fitzgerald 2020). The most basic function of an AOP is to provide a systematic approach for synthesizing and integrating different knowledge streams concerning generalized responses to various biological perturbations. A good example of this type of application for risk assessment was provided by LaLone et al. (2017), who employed the AOP framework to integrate molecular, physiological, and behavioral data from more than 200 studies focused on the effects of neonicotinoid pesticides on honeybee colonies (AOPs 77–80, 87–90, 178; aopwiki.org). They were able to demonstrate the plausibility of a linkage between activation of the nicotinic acetylcholine receptor and adverse effects and, just as importantly, highlight knowledge gaps and research needs for risk assessors concerned with the effects of chemical stressors on pollinators, an issue of global concern. On the other end of the spectrum are applications in which the AOP framework has been effectively used to move from employing whole animal testing to computational and in vitro assays to support regulatory decisions concerning chemical safety. The most advanced example of this involves skin sensitization associated with personal care products in humans. Specifically, an approach using a combination of in vitro assays for multiple KEs for a skin sensitization AOP (https://aopwiki.org/aops/40) in conjunction with Bayesian modeling has been advanced as an animal-free alternative for safety assessments in the European Union (Schultz et al. 2016).

The most complete examples of the application of AOPs to chemical assessments in fish involve perturbations in endocrine function associated with the hypothalamic-pituitary-gonad and -thyroid (HPG/T) axes. For decades, there has been concern for the potential effects of endocrine-disrupting chemicals on human health and the environment, and regulatory strategies have been developed throughout the world to identify specific compounds with the potential to elicit these effects (World Health Organization 2002; Hotchkiss et al. 2008). A core theme in all these strategies has been the need to link measurable molecular/biochemical changes in HPG/T pathways to adverse effects in individuals (and in some cases populations), a major challenge relative to establishing causality (Matthiessen et al. 2017). One approach to addressing the regulatory need of associating perturbation of specific endocrine pathways with effects meaningful to risk assessment has been the development of novel fish toxicity tests that included endpoints measured at different biological levels of organization, ranging from basic measures of endocrine function to apical effects on development and/or reproduction (Ankley and Tyler 2013). However, even with these types of multi-endpoint assays, establishing causal relationships between changes in endocrine function and adverse effects on apical responses can be difficult. Making this linkage is, of course, the goal of the AOP framework. Given this focus, development of AOPs involve perturbations of the HPG/T axes and associated effects in fish was a logical extension of extensive research on endocrine active chemicals.

Different fish AOPs (as well as those developed for endocrine pathways in other non-mammalian species) directly contribute to meeting the challenge of linking pathway-specific molecular and biochemical alterations (i.e., early KEs) to adverse effects of ecological significance (Coady et al. 2017). A practical example of application of these types of AOPs involves the relationship of changes in levels of vitellogenin (VTG; egg yolk precursor protein) in reproductively active female fish with their ability to produce offspring. Changes in VTG in females can be indicative of chemicals that impact several endocrine pathways of regulatory concern, including production of steroids, activation of the androgen receptor, and antagonism of the estrogen receptor (USEPA, 2019). Consequently, many fish assays developed for detecting endocrine disruption routinely measure levels of VTG in females (Ankley and Johnson 2004). To support interpretation of this endpoint in the context of regulatory decision-making, fish-specific AOPs have been developed and vetted through OECD review that directly relate VTG decreases in females to reproductive success and population status (AOPs 23, 25, 30; aopwiki.org).

Beyond the need to establish a causal connection between disruption of the HPG/T axes and apical adverse effects, AOPs were also used to support development of an alternative approach for endocrine disruptor screening. The USEPA endocrine disruptor screening and testing program has been tasked with assessing the order of 10,000 chemicals (USEPA 2019). Given the sheer number of compounds that needed to be screened for endocrine activity, this was recognized as a logistically and financially improbable task based solely on the collection of effects data from whole animal tests. Consequently, programs involved in the regulatory assessment of endocrine-active chemicals transitioning to greater reliance on NAMs, most prominently high throughput (HTP) in vitro assays, to rapidly screen large numbers of chemicals for endocrine activity. Those identified as active can then be prioritized for in vivo testing using rats, fish, amphibians, etc. (USEPA 2019). In vitro assays used for screening for potential endocrine activity often employ measurements that correspond to MIEs or early KEs in an AOP, such as binding to/activation of receptors or inhibition of specific enzymes. Consequently, the AOP framework has been critical to providing the "phenotypic anchoring" necessary to support the use of NAMs for this type of application (Browne et al. 2017; Noyes et al. 2019; Knapen et al. 2020). Likewise, consideration of additional KEs along the pathway can serve as a basis for tiered testing approaches. Acceptance of the utility of data generated via NAMs by both the regulated and /regulatory community very much depends on demonstration that specific endpoints measured may indeed translate into adverse effects in whole animals.

An ultimate goal in terms of AOP application to chemical risk assessment is quantitative prediction of adverse effects in whole animals based solely on in vitro or in silico data. One relatively advanced quantitative AOP for predicting endocrine outcomes in fish has been developed for this purpose. As mentioned in prior sections, Conolly et al. (2017) developed a quantitative AOP based on AOP 25 (https://aopwiki.org/aops/25), utilizing in vitro inhibition of CYP19 aromatase as a basis to predict egg production and population status in the fathead minnow. While this quantitative AOP captures only one of many HPG pathways of regulatory concern, it represents the first step in the development of a "library" of AOPs capable of translating in vitro perturbations in endocrine function into actionable knowledge for chemical risk assessors/managers.

To successfully realize implementation of this vision, a larger suite of HTP assays representing not only the various endocrine pathways of concern, but other toxicologically relevant perturbations, is required. To date, most HTP assays used for screening chemicals for HPG/T activity have been based on mammalian (primarily rodent) models. While many aspects of biological systems and their function are well conserved across vertebrates, substantial taxonomic differences in structure and function of some biological pathways should not be ignored. In these instances, it may be necessary to develop HTP assays specific to different taxonomic groups to measure bioactivities of concern. Here again, AOPs can be very useful. They can help to identify MIEs and early KEs, relevant to fish, that could be developed into HTP assays. In this context, it is reasonable to view AOP and assay development as a complementary, iterative process critical to the broader use of NAMs to assess chemical effects.

While the primary motivation for formalization of the AOP framework may have been to support prospective evaluations of chemical hazards using NAMs, applications of the AOP framework are not limited to prospective assessments. A role for AOPs and pathway-based bioassays and endpoints in effects-based environmental monitoring has also been recognized. Acknowledging that monitoring of targeted lists of contaminants in the environment may only be capturing the tip of the iceberg relative to the totality of contaminants in the environment (Escher et al. 2013); there has been increased recognition of the role that effects-based monitoring with pathway-based assays could play in environmental management (e.g., Ekman et al. 2013; Villeneuve et al. 2018; Dingemans et al. 2019). For known (measured) chemicals in the environment, comparison to effect concentrations in pathway-based assays has been used to prioritize those contaminants for which there is evidence of connection to specific hazards, using AOPs to translate those pathway perturbations to ecologically relevant impacts (Blackwell et al. 2017; Corsi et al. 2019). Likewise, pathway-based assays anchored to AOPs have been used to directly assess both the known and unknown contaminants contributing to biological activities to evaluate the overall potency/potential of a mixture to elicit adverse effects (Escher et al. 2013; Leusch et al. 2018; Blackwell et al. 2019; Cavallin et al. 2021). Finally, where either pathway-based biological responses or adverse effects in the field have been reported, AOPs can serve to guide selection of assays and endpoints for further monitoring and can help point toward potential causative stressors (e.g., Cavallin et al. 2016; Alvarez et al. 2021).

While a diverse array of applications are described above and documented in the literature, these are still likely just scratching the surface. Utility of AOPs as a knowledge organization framework is expanding beyond its origins in chemical safety assessment (e.g., Draskau et al. 2020; Wittwehr et al. 2021; Nymark et al. 2021; Chauhan et al. 2021). The question moving forward is not one of the utility of the framework or the associated knowledge synthesis. Rather, the question is whether the scientific community can generate sufficient high quality, fit-for-purpose, AOPs to keep pace with the demands for their use. With just 16 AOPs achieving OECD endorsement by fall 2021, there is ample reason to question whether the needed output can be achieved. It is expected that growing demand for knowledge synthesis to support actionable decision-making and growing familiarity with the principles and practice of AOP development will incentivize further participation in the crowd-sourced endeavor of documenting AOPs. Nonetheless, active participation by a broader portion of the scientific community will be critical to ultimate success.

## CONCLUSION

The dispersed nature of our accumulated scientific knowledge and evidence is what makes AOP development an attractive and increasingly important endeavor to support 21st-century toxicology. An overwhelming amount of scientific research is ongoing globally. Even with search engines and ready access to articles from online electronic sources, sorting through all the information available to derive coherent and defensible conceptual models of biological responses to perturbations, caused both by chemical and nonchemical stressors, is both a daunting and time-consuming exercise. The individual risk assessor, regulator, or resource manager generally has neither the time nor the breadth and depth of expertise to assemble and synthesize all this information *de novo*. Thus, there is a critical need for subject matter experts, including fish toxicologists, to take a more active role in the type of knowledge synthesis and science translation that AOPs are intended to support. Adverse outcome pathway descriptions, developed according to internationally harmonized guidance (OECD 2018) and deposited in and/or accessed through the AOP-KB (aopkb.oecd.org) via the AOP-Wiki (aopwiki.org) or other emerging tools, can provide scientific knowledge in an accessible format. It is ultimately that synthetic and transparent access to fundamental scientific knowledge that can support science-based decisions and policy making. Ongoing participation of fish biologists and aquatic toxicologists in AOP development can help facilitate the impact of their specialized knowledge and expertise on decision-making related to aquatic ecosystems and resources.

## ACKNOWLEDGMENTS

The authors thank Prarthana Shankar and Aimen Farraj for comments on an earlier draft of this chapter. We recognize all our colleagues who have contributed to advancing the AOP framework and shaped the content of this chapter. Contents of this chapter neither constitute nor necessarily represent U.S. EPA policy. Mention of trade names or commercial products does not constitute endorsement or recommendation for use.

## REFERENCES

Alvarez, D. A., S. R. Corsi, L. A. De Cicco, D. L. Villeneuve, and A. K. Baldwin. (2021). Identifying chemicals and mixtures of potential biological concern detected in passive samplers from Great Lakes tributaries using high-throughput data and biological pathways. *Environ. Toxicol. Chem.* 40, 2165–2182. doi: 10.1002/etc.5118.

Ankley, G. T., R. S. Bennett, R. J. Erickson, D. J. Hoff, M. W. Hornung, R. D. Johnson, D. R. Mount, J. W. Nichols, C. L. Russom, P. K. Schmieder, J. A. Serrano, J. E. Tietge, and D. L. Villeneuve. (2010). Adverse outcome pathways: a conceptual framework to support ecotoxicology research and risk assessment. *Environ. Toxicol. Chem.* 29(3), 730–741. doi: 10.1002/etc.34.

Ankley, G. T., B. R. Blackwell, J. E. Cavallin, J. A. Doering, D. J. Feifarek, K. M. Jensen, M. D. Kahl, C. A. LaLone, S. T. Poole, E. C. Randolph, T. W. Saari, and D. L. Villeneuve. (2020). Adverse outcome pathway network-based assessment of the interactive effects of an androgen receptor agonist and an aromatase inhibitor on fish endocrine function. *Environ. Toxicol. Chem.* 39(4), 913–922. doi: 10.1002/etc.4668.

Ankley, G. T., and S. W. Edwards. (2017). The adverse outcome pathway: a multifaceted framework supporting 21st century toxicology. *Curr. Opin. Toxicol.* 9, 1–7. doi: 10.1016/j.cotox.2018.03.004. Ankley, G. T., and R. D. Johnson. (2004). Small fish models for identifying and assessing the effects of endocrine-disrupting chemicals. *Inst. Lab. Anim. Res. J.* 45, 469–483. doi: 10.1093/ilar.45.4.469.

Ankley, G. T., and C. R. Tyler. (2013). Development of methods to detect occurrence and effects of endocrine-disrupting chemicals: fueling a fundamental shift in regulatory ecotoxicology. *Environ. Toxicol. Chem.* 32, 1261–1262. doi: 10.1002/etc.2415.

Becker, R. A., G. T. Ankley, S. W. Edwards, S. W. Kennedy, I. Linkov, B. Meek, M. Sachana, H. Segner, B. Van Der Burg, D. L. Villeneuve, H. Watanabe, and T. S. Barton-Maclaren.

(2015). Increasing scientific confidence in adverse outcome pathways: application of tailored bradford-hill considerations for evaluating weight of evidence. *Regul. Toxicol. Pharmacol.* 72(3), 514–537. doi: 10.1016/j.yrtph.2015.04.004.

Bett, N. N., and S. G. Hinch. (2016). Olfactory navigation during spawning migrations: a review and introduction of the hierarchical navigation hypothesis. *Biol. Rev. Camb. Philos. Soc.* 91(3), 728–759. doi: 10.1111/brv.12191.

Blackwell, B. R., G. T. Ankley, P. M. Bradley, K. A. Houck, S. S. Makarov, A. V. Medvedev, J. Swintek, and D. L. Villeneuve. (2019). Potential toxicity of complex mixtures in surface waters from a nationwide survey of United States streams: identifying in vitro bioactivities and causative chemicals. *Environ. Sci. Technol.* 53(2), 973–983. doi: 10.1021/acs.est.8b05304.

Blackwell, B. R., G. T. Ankley, S. R. Corsi, L. A. DeCicco, K. A. Houck, R. S. Judson, S. Li, M. T. Martin, E. Murphy, A. L. Schroeder, E. R. Smith, J. Swintek, and D. L. Villeneuve. (2017). An "EAR" on environmental surveillance and monitoring: a case study on the use of exposure-activity ratios (EARs) to prioritize sites, chemicals, and bioactivities of concern in Great Lakes waters. *Environ. Sci. Technol.* 51(15), 8713–8724. doi: 10.1021/acs.est.7b01613.

Boobis, A. R., S. M. Cohen, V. Dellarco, D. McGregor, M. E. Meek, C. Vickers, D. Willcocks, and W. Farland. (2006). IPCS framework for analyzing the relevance of a cancer mode of action for humans. *Crit. Rev. Toxicol.* 36(10), 781–792. doi: 10.1080/10408440600977677.

Boobis, A. R., J. E. Doe, B. Heinrich-Hirsch, M. E. Meek, S. Munn, M. Ruchirawat, J. Schlatter, J. Seed, and C. Vickers. (2008). IPCS framework for analyzing the relevance of a noncancer mode of action for humans. *Crit. Rev. Toxicol.* 38(2), 87–96. doi: 10.1080/10408440701749421.

Bradbury, S. P., T. C. Feijtel, and C. J. Van Leeuwen. (2004). Meeting the scientific needs of ecological risk assessment in a regulatory context. *Environ. Sci. Technol.* 38(23), 463A–470A. doi: 10.1021/es040675s.

Browne, P., P. D. Noyes, W. Casey, and D. J. Dix. (2017). Application of adverse outcome pathways to US EPA's endocrine disruptor screening program. *Environ. Health Perspect.* 125(9), 096001. doi: 10.1289/EHP1304.

Burgoon, L. D., M. Angrish, N. Garcia-Reyero, N. Pollesch, A. Zupanic, and E. Perkins. (2020). Predicting the probability that a chemical causes steatosis using adverse outcome pathway bayesian networks (AOPBNs). *Risk Anal.* 40(3), 512–523. doi: 10.1111/risa.13423.

Cavallin, J. E., J. Beihoffer, B. R. Blackwell, A. R. Cole, D. R. Ekman, R. Hofer, A. Jastrow, J. Kinsey, K. Keteles, E. M. Maloney, J. Parman, D. L. Winkelman, and D. L. Villeneuve. (2021). Effects-based monitoring of bioactive compounds associated with municipal wastewater treatment plant effluent discharge to the South Platte River, Colorado, USA. *Environ. Pollut.* 289, 117928. doi: 10.1016/j.envpol.2021.117928.

Cavallin, J. E., K. M. Jensen, M. D. Kahl, D. L. Villeneuve, K. E. Lee, A. L. Schroeder, J. Mayasich, E. P. Eid, K. R. Nelson, R. Y. Milsk, B. R. Blackwell, J. P. Berninger, C. A. LaLone, C. Blanksma, T. Jicha, C. Elonen, R. Johnson, and G. T. Ankley. (2016). Pathway-based approaches for assessment of real-time exposure to an estrogenic wastewater treatment plant effluent on fathead minnow reproduction. *Environ. Toxicol. Chem.* 35(3), 702–716. doi: 10.1002/etc.3228.

Cave, E. J., and S. M. Kajiura. (2020). Electrosensory impairment in the Atlantic stingray, Hypanus sabinus, after crude oil exposure. *Zoology (Jena).* 143, 125844. doi: 10.1016/j.zool.2020.125844.

Chauhan, V., D. Stricklin, and D. Cool. (2021). The integration of the adverse outcome pathway framework to radiation risk assessment. *Int. J. Radiat. Biol.* 97(1), 60–67. doi: 10.1080/09553002.2020.1761570.

Cheng, W. Y., Q. Zhang, A. Schroeder, D. L. Villeneuve, G. T. Ankley, and R. Conolly. (2016). Editor's highlight: computational modeling of plasma vitellogenin alterations in response to aromatase inhibition in fathead minnows. *Toxicol. Sci.* 154(1), 78–89. doi: 10.1093/toxsci/kfw142.

Clewell, R. A., J. A. Leonard, C. I. Nicolas, J. L. Campbell, M. Yoon, A. Y. Efremenko, P. D. McMullen, M. E. Andersen, H. J. Clewell 3rd, K. A. Phillips, and Y. M. Tan. (2020). Application of a combined aggregate exposure pathway and adverse outcome pathway (AEP-AOP) approach to inform a cumulative risk assessment: a case study with phthalates. *Toxicol. In Vitro.* 66, 104855. doi: 10.1016/j.tiv.2020.104855.

Coady, K., P. Browne, M. Embry, T. Hill, E. Leinala, T. Steeger, L. Maślankiewicz, and T. Hutchinson. (2019). When are adverse outcome pathways and associated assays "fit for purpose" for regulatory decision-making and management of chemicals? *Integr. Environ. Assess. Manag.* 15(4), 633–647. doi: 10.1002/ieam.4153.

Coady, K. K., R. C. Biever, N. D. Denslow, M. Gross, P. D. Guiney, H. Holbech, N. K. Karouna-Renier, I. Katsiadaki, H. Kruger, S. L. Levine, G. Maack, M. Williams, J. C. Wolf and G. T. Ankley. (2017). Current limitations and recommendations to improve testing for environmental assessment of endocrine active chemicals. *Integ. Environ. Assess. Manag.* 13, 302–316. doi: 10.1002/ieam.1862.

Coffin, A. B., and J. Ramcharitar. (2016). Chemical ototoxicity of the fish inner ear and lateral line. *Adv. Exp. Med. Biol.* 877, 419–437. doi: 10.1007/978-3-319-21059-9_18.

Conolly, R. B., G. T. Ankley, W. Cheng, M. L. Mayo, D. H. Miller, E. J. Perkins, D. L. Villeneuve, and K. H. Watanabe. (2017). Quantitative adverse outcome pathways and their application to predictive toxicology. *Environ. Sci. Technol.* 51(8), 4661–4672. doi: 10.1021/acs.est.6b06230.

Corsi, S. R., L. A. De Cicco, D. L. Villeneuve, B. R. Blackwell, K. A. Fay, G. T. Ankley, and A. K. Baldwin. (2019). Prioritizing chemicals of ecological concern in Great Lakes tributaries using high-throughput screening data and adverse outcome pathways. *Sci. Total Environ.* 686, 995–1009. doi: 10.1016/j.scitotenv.2019.05.457.

Dingemans, M. M., K. A. Baken, R. van der Oost, M. Schriks, and A. P. van Wezel. (2019). Risk-based approach in the revised European Union drinking water legislation: opportunities for bioanalytical tools. *Integr. Environ. Assess. Manage.* 15(1), 126–134. doi: 10.1002/ieam.4096.

Doering, J., et al. (2019), "Adverse outcome pathway on aryl hydrocarbon receptor activation leading to early life stage mortality, via increased COX-2", *OECD Series on Adverse Outcome Pathways*, No. 12, OECD Publishing, Paris, https://doi.org/10.1787/bd46b538-en.

Dong, W., F. Matsumura, and S. W. Kullman. (2010). TCDD induced pericardial edema and relative COX-2 expression in medaka (Oryzias latipes) embryos. *Toxicol. Sci.* 118(1), 213–223. doi: 10.1093/toxsci/kfq254.

Draskau, M. K., C. M. Spiller, J. Boberg, J. Bowles, and T. Svingen. (2020). Developmental biology meets toxicology: contributing reproductive mechanisms to build adverse outcome pathways. *Mol. Hum. Reprod.* 26(2), 111–116. doi: 10.1093/molehr/gaaa001.

Ekman, D. R., G. T. Ankley, V. S. Blazer, T. W. Collette, N. Garcia-Reyero, L. R. Iwanowicz, Z. G. Jorgenson, K. E. Lee, P. M. Mazik, D. H. Miller, and E. J. Perkins.

(2013). Biological effects-based tools for monitoring impacted surface waters in the Great Lakes: a multiagency program in support of the Great Lakes restoration initiative. *Environ. Practice.* 2013 Dec;15(4):409–426. doi: 10.1017/S1466046613000458.

Escher, B. I., C. van Daele, M. Dutt, J. Y. Tang, R. Altenburger. (2013). Most oxidative stress response in water samples comes from unknown chemicals: the need for effect-based water quality trigger values. *Environ. Sci. Technol.* 47(13), 7002–7011. doi: 10.1021/es304793h.

Evans, D. H. (1987). The fish gill: site of action and model for toxic effects of environmental pollutants. *Environ. Health Perspect.* 71, 47–58. doi: 10.1289/ehp.877147.

Farhat, A. and S. Kennedy (2019), "Adverse outcome pathway on aryl hydrocarbon receptor activation leading to early life stage mortality, via reduced VEGF", *OECD Series on Adverse Outcome Pathways*, No. 16, OECD Publishing, Paris, https://doi.org/10.1787/063e1bf4-en.

Fitzgerald, R. E. (2020). Adverse outcome pathway bridge building from research to regulation. *Chem. Res. Toxicol.* 33(4), 849–851. doi: 10.1021/acs.chemrestox.9b00527.

Foran, C. M., T. Rycroft, J. Keisler, E. J. Perkins, I. Linkov, and N. Garcia-Reyero. (2019). A modular approach for assembly of quantitative adverse outcome pathways. *ALTEX.* 36(3), 353–362. doi: 10.14573/altex.1810181.

Griffith, M. B. (2017). Toxicological perspective on the osmoregulation and ionoregulation physiology of major ions by freshwater animals: teleost fish, crustacea, aquatic insects, and Mollusca. *Environ. Toxicol. Chem.* 36(3), 576–600. doi: 10.1002/etc.3676.

Hammerschlag, N. (2006). Osmoregulation in elasmobranchs: a review for fish biologists, behaviourists and ecologists. *Mar. Freshwater Behav. Physiol.* 39(3), 209–228, doi: 10.1080/10236240600815820.

Hill, A. B. (1965). The environment and disease: association or causation? *Proc. Royal Soc. Med.* 58(5), 295–300. doi:10.1177/003591576505800503.

Hines, D. E., R. B. Conolly, and A. M. Jarabek. (2019). A quantitative source-to-outcome case study to demonstrate the integration of human health and ecological end points using the aggregate exposure pathway and adverse outcome pathway frameworks. *Environ. Sci. Technol.* 53(18), 11002–11012. doi: 10.1021/acs.est.9b04639.

Holbech, H., P. Matthiessen, M. Hansen, G. Schüürmann, D. Knapen, M. Reuver, F. Flamant, L. Sachs, W. Kloas, K. Hilscherova, M. Leonard, J. Arning, V. Strauss, T. Iguchi, and L. Baumann. (2020). ERGO: breaking down the wall between human health and environmental testing of endocrine disrupters. *Int. J. Mol. Sci.* 21(8), 2954. doi: 10.3390/ijms21082954.

Hotchkiss, A. K., C. V. Rider, C. R. Blystone, V. S. Wilson, P. C. Hartig, G. T. Ankley, P. M. Foster, C. L. Gray, and L. E. Gray. (2008). Fifteen years after "Wingspread"-environmental endocrine disrupters and human and wildlife health: where we are today and where we need to go. *Toxicol. Sci.* 105, 235–259. doi: 10.1093/toxsci/kfn030.

Huggett, R. J. (1992). *Biomarkers: biochemical, physiological, and histological markers of anthropogenic stress.* Boca Raton: Lewis Publishers.

Hutchinson, T. H., G. T. Ankley, H. Segner, C. R. Tyler. (2006). Screening and testing for endocrine disruption in fish-biomarkers as "signposts," not "traffic lights," in risk assessment. *Environ. Health Perspect.* 114(Suppl 1), 106–114. doi: 10.1289/ehp.8062.

Kavlock, R. J., T. Bahadori, T. S. Barton-Maclaren, M. R. Gwinn, M. Rasenberg, R. S. Thomas. (2018). Accelerating the pace of chemical risk assessment. *Chem. Res. Toxicol.* 31(5), 287–290. doi: 10.1021/acs.chemrestox.7b00339.

Kleinstreuer, N. C., K. Sullivan, D. Allen, S. Edwards, D. L. Mendrick, M. Embry, J. Matheson, J. C. Rowlands, S. Munn, E. Maull, and W. Casey. (2016). Adverse outcome pathways: from research to regulation scientific workshop report. *Regul. Toxicol. Pharmacol.* 76, 39–50. doi: 10.1016/j.yrtph.2016.01.007.

Knapen, D. (2021). Adverse outcome pathways and the paradox of complex simplicity. *Environ. Toxicol. Chem.* doi: 10.1002/etc.5205.

Knapen, D., M. M. Angrish, M. C. Fortin, I. Katsiadaki, M. Leonard, L. Margiotta-Casaluci, S. Munn, J. M. O'Brien, N. Pollesch, L. C. Smith, X. Zhang, and D. L. Villeneuve. (2018). Adverse outcome pathway networks I: development and applications. *Environ. Toxicol. Chem.* 37(6), 1723–1733. doi: 10.1002/etc.4125.

Knapen, D., E. Stinckens, J. E. Cavallin, G. T. Ankley, H. Holbech, D. L. Villeneuve, and L. Vergauwen. (2020). Toward an AOP network-based tiered testing strategy for the assessment of thyroid hormone disruption. *Environ. Sci. Technol.* 54(14), 8491–8499. doi: 10.1021/acs.est.9b07205.

Kolle, S. N., R. Landsiedel, and A. Natsch. (2020). Replacing the refinement for skin sensitization testing: considerations to the implementation of adverse outcome pathway (AOP)-based defined approaches (DA) in OECD guidelines. *Regul. Toxicol. Pharmacol.* 115, 104713. doi: 10.1016/j.yrtph.2020.104713.

Krewski, D., D. Acosta Jr, M. Andersen, H. Anderson, J. C. Bailar 3rd, K. Boekelheide, R. Brent, G. Charnley, V. G. Cheung, S. Green Jr, K. T. Kelsey, N. I. Kerkvliet, A. A. Li, L. McCray, O. Meyer, R. D. Patterson, W. Pennie, R. A. Scala, G. M. Solomon, M. Stephens, J. Yager, L. Zeise. (2010). Toxicity testing in the 21st century: a vision and a strategy. *J. Toxicol. Environ. Health B Crit. Rev.* 13(2-4), 51–138. doi: 10.1080/10937404.2010.483176.

LaLone, C. A., D. L. Villeneuve, J. Wu-Smart, R. Y. Milsk, K. Sappington, K. V. Garber, J. Housenger, and G. T. Ankley. (2017) Weight of evidence evaluation of a network of adverse outcome pathways linking activation of the nicotinic acetylcholine receptor in honey bees to colony death. *Sci. Total Environ.* 584-585, 751–775. doi: 10.1016/j.scitotenv.2017.01.113.

Leusch, F. D. L., P. A. Neale, C. Arnal, N. H. Aneck-Hahn, P. Balaguer, A. Bruchet, B. I. Escher, M. Esperanza, M. Grimaldi, G. Leroy, M. Scheurer, R. Schlichting, M. Schriks, and A. Hebert. (2018) Analysis of endocrine activity in drinking water, surface water and treated wastewater from six countries. *Water Res.* 139, 10–18. doi: 10.1016/j.watres.2018.03.056.

Matthiessen, P., G. T. Ankley, R. Biever, P. Bjerregaard, C. Borgert, K. Brugger, A. Blankenship, J. Chambers, K. Coady, L. Constantine, Z. Dang, N. Denslow, D. A. Dreier, S. Dungey, L. E. Gray, M. Gross, P. D. Guiney, M. Hecker, H. Holbech, T. Iguchi, S. Kadlec, N. Karouna-Renier, I. Katsiadaki, Y. Kawashima, W. Kloas, H. Krueger, L. Lagadic, A. Leopold, S. Levine, G. Maack, S. Marty, J. Meador, E. Mihaich, J. Odum, L. Ortego, J. Parrott, D. Pickford, M. Roberts, C. Schaefers, T. Swartz, K. Solomon, T. Verslycke, L. Weltje, J. Wheeler, M. Williams, J. C. Wolf, and K. Yamazaki, K. (2017). Recommended approaches to the scientific evaluation of ecotoxicological hazards and risks of endocrine-active substances. *Integ. Environ. Assess. Manag.* 13, 267–279. doi: 10.1002/ieam.1885.

Mekenyan O., Dimitrov S., Schmieder P., Veith G. (2003 Oct-Dec). In silico modelling of hazard endpoints: current problems and perspectives. *SAR QSAR Environ Res.* 14(5–6), 361–71. doi: 10.1080/10629360310001623953.

Meek, M. E., A. Boobis, I. Cote, V. Dellarco, G. Fotakis, S. Munn, J. Seed, and C. Vickers. (2014). New developments in the evolution and application of the WHO/IPCS framework on mode of action/species concordance analysis. *J. Appl. Toxicol.* doi: 10.1002/jat.2949.

Meek, M. E., A. R. Boobis, K. M. Crofton, G. Heinemeyer, M. V. Raaij, and C. Vickers. (2011). Risk assessment of combined exposure to multiple chemicals: a WHO/IPCS framework. *Regul. Toxicol. Pharmacol.* 60(2), S1–S14. doi: 10.1016/j.yrtph.2011.03.010.

Miller, W. L. (1988). Molecular biology of steroid hormone synthesis. *Endocr. Rev.* 9(3), 295–318. doi: 10.1210/edrv-9-3-295. PMID: 3061784.

Moe, S. J., R. Wolf, L. Xie, W. G. Landis, N. Kotamäki, and K. E. Tollefsen. (2021). Quantification of an adverse outcome pathway network by Bayesian regression and Bayesian network modeling. *Integr. Environ. Assess. Manag.* 17(1), 147–164. doi: 10.1002/ieam.4348.

Mogdans J. (2019). Sensory ecology of the fish lateral-line system: morphological and physiological adaptations for the perception of hydrodynamic stimuli. *J. Fish Biol.* 95(1), 53–72. doi: 10.1111/jfb.13966.

Newman, M. E. J. (2003). The structure and function of complex networks. *SIAM Rev.* 45: 167–256.

Noyes, P. D., K. P. Friedman, P. Browne, J. T. Haselman, M. E. Gilbert, M. W. Hornung, S. Barone Jr, K. M. Crofton, S. C. Laws, T. E. Stoker, S. O. Simmons, J. E. Tietge, and S. J. Degitz. (2019). Evaluating chemicals for thyroid disruption: opportunities and challenges with in vitro testing and adverse outcome pathway approaches. *Environ. Health Perspect.* 127(9):95001. doi: 10.1289/EHP5297.

Nymark, P., M. Sachana, S. B. Leite, J. Sund, C. E. Krebs, K. Sullivan, S. Edwards, L. Viviani, C. Willett, B. Landesmann, and C. Wittwehr. (2021). Systematic organization of COVID-19 data supported by the adverse outcome pathway framework. *Front. Public Health.* 19(9), 638605. doi: 10.3389/fpubh.2021.638605.

OECD (2018), "Users' handbook supplement to the guidance document for developing and assessing adverse outcome pathways", *OECD Series on Adverse Outcome Pathways*, No. 1, OECD Publishing, Paris, https://doi.org/10.1787/5jlv1m9d1g32-en.

OECD (2021). OECD Series on Adverse Outcome Pathways. doi: 10.1787/2415170X. https://www.oecd-ilibrary.org/environment/oecd-series-on-adverse-outcome-pathways_2415170x.

Oris, J. T., and A. J. Bailer. (1997). Equivalence of concentration-response relationships in aquatic toxicology studies: testing and implications for potency estimation. *Environ. Toxicol. Chem.* 16, 2204- 2209. doi: 10.1002/etc.5620161031.

Park Y. J., Lee M. J., Kim H. R., Chung K. H., Oh S. M. (2014 Sep 1). Developmental toxicity of 2,3,7,8-tetrachlorodibenzo-p-dioxin in artificially fertilized crucian carp (Carassius auratus) embryo. *Sci Total Environ.* 491–492, 271–278. doi: 10.1016/j.scitotenv.2014.03.129

Patlewicz, G., C. Kuseva, A. Kesova, I. Popova, T. Zhechev, T. Pavlov, D. W. Roberts, and O. Mekenyan. (2014). Towards AOP application--implementation of an integrated approach to testing and assessment (IATA) into a pipeline tool for skin sensitization. *Regul. Toxicol. Pharmacol.* 69(3), 529–545. doi: 10.1016/j.yrtph.2014.06.001.

Pavlopoulos, G. A., M. Secrier, C. N. Moschopoulos, T. G. Soldatos, S. Kossida, J. Aerts, R. Schneider, and P. G. Bagos. (2011). Using graph theory to analyze biological networks. *BioData Mining* 4, 10. doi: 10.1186/1756-0381-4-10.

Perkins, E. J., R. Ashauer, L. Burgoon, R. Conolly, B. Landesmann, C. Mackay, C. A. Murphy, N. Pollesch, J. R. Wheeler, A. Zupanic, and S. Scholz. (2019). Building and applying quantitative adverse outcome pathway models for chemical hazard and risk assessment. *Environ. Toxicol. Chem.* 38(9), 1850–1865. doi: 10.1002/etc.4505.

Pollesch, N. L., D. L. Villeneuve, and J. M. O'Brien. (2019). Extracting and benchmarking emerging adverse outcome pathway knowledge. *Toxicol. Sci.* 168(2), 349–364. doi: 10.1093/toxsci/kfz006.

Progatzky, F., H. T. Cook, J. R. Lamb, L. Bugeon, and M. J. Dallman. (2016). Mucosal inflammation at the respiratory interface: a zebrafish model. Am. J. Physiol. Lung Cell. Mol. Physiol. 310(6):L551–L561. doi: 10.1152/ajplung.00323.2015.

Putzrath, R. M. (1997). Estimating relative potency for receptor-mediated toxicity: reevaluating the toxic equivalence factor (TEF) model. *Regul. Toxicol. Pharmacol.* 25, 68- 78. doi: 10.1006/rtph.1996.1063.

Russom, C. L., C. A. LaLone, D. L. Villeneuve, and G. T. Ankley. (2014). Development of an adverse outcome pathway for acetylcholinesterase inhibition leading to acute mortality. *Environ. Toxicol. Chem.* 33(10), 2157–2169. doi: 10.1002/etc.2662.

Safe, S. (1990). Polychlorinated biphenyls (PCBs), dibenzo-*p*-di-oxins (PCDDs), dibenzofurans (PCDFs), and related compounds: environmental and mechanistic considerations which support the development of toxic equivalency factors (TEFs). *Crit. Rev. Toxicol* 21, 51-88. doi: 10.3109/10408449009089873.

Safe, S. H., L. Pallaroni, K. Yoon, K. Gaido, S. Ross, and D. McDonnell. (2002). Problems for risk assessment of endocrine-active estrogenic compounds. *Environ. Health Perspect.* 110(suppl 6), 925–929. doi: 10.1289/ehp.02110s6925.

Sakuratani, Y., M. Horie, and E. Leinala. (2018). Integrated approaches to testing and assessment: OECD activities on the development and use of adverse outcome pathways and case studies. *Basic Clin. Pharmacol. Toxicol.* 123 (Suppl 5), 20–28. doi: 10.1111/bcpt.12955.

Schmid, S., Y. Song, and K. E. Tollefsen. (2021). AOP report: inhibition of chitin synthase 1 leading to increased mortality in arthropods. *Environ. Toxicol. Chem.* 40(8), 2112–2120. doi: 10.1002/etc.5058.

Schultz, T. W., G. Dimitrova, S. Dimitrov, and O. G. Mekenyan. (2016). The adverse outcome pathway for skin sensitization: moving closer to replacing animal testing. *Altern. Lab. Anim.* 44, 453–460. doi: 10.1177/026119291604400515.

Smith, M. T., K. Z. Guyton, N. Kleinstreuer, A. Borrel, A. Cardenas, W. A. Chiu, D. W. Felsher, C. F. Gibbons, W. H. Goodson 3rd, K. A. Houck, A. B. Kane, M. A. La Merrill, H. Lebrec, L. Lowe, C. M. McHale, S. Minocherhomji, L. Rieswijk, M. S. Sandy, H. Sone, A. Wang, L. Zhang, L. Zeise, and M. Fielden. (2020). The key characteristics of carcinogens: relationship to the hallmarks of cancer, relevant biomarkers, and assays to measure them. *Cancer Epidemiol. Biomarkers Prev.* 29(10), 1887–1903. doi: 10.1158/1055-9965.EPI-19-1346.

Sollid, J., and G. E. Nilsson. (2006). Plasticity of respiratory structures--adaptive remodeling of fish gills induced by ambient oxygen and temperature. *Respir. Physiol. Neurobiol.* 154(1-2), 241–251. doi: 10.1016/j.resp.2006.02.006.

Song, Y., and D. L. Villeneuve. (2021). AOP report: uncoupling of oxidative phosphorylation leading to growth inhibition via decreased cell proliferation. *Environ. Toxicol. Chem.* doi: 10.1002/etc.5197.

Sonich-Mullin, C., R. Fielder, J. Wiltse, K. Baetcke, J. Dempsey, P. Fenner-Crisp, D. Grant, M. Hartley, A. Knaap, D. Kroese, I. Mangelsdorf, E. Meek, J. M. Rice, M. Younes. (2001). International Programme on Chemical Safety. IPCS conceptual framework for evaluating a mode of action for chemical carcinogenesis. *Regul. Toxicol. Pharmacol.* 34(2), 146–152. doi: 10.1006/rtph.2001.1493.

Spinu, N., M. T. D. Cronin, S. J. Enoch, J. C. Madden, and A. P. Worth. (2020). Quantitative adverse outcome pathway (qAOP) models for toxicity prediction. *Arch. Toxicol.* 94(5), 1497–1510. doi: 10.1007/s00204-020-02774-7.

Svingen, T., D. L. Villeneuve, D. Knapen, E. M. Panagiotou, M. K. Draskau, P. Damdimopoulou, J. O'Brien. (2021). A pragmatic approach to adverse outcome pathway development and evaluation. *Toxicol. Sci.* (in press). doi: 10.1093/toxsci/kfab113.

Swenberg, J. A., E. Fryar-Tita, Y. Jeong, G. Boysen, T. Starr, V. E. Walker and R. J. Albertini. (2008). Biomarkers in toxicology and risk assessment: informing critical dose-response relationships, *Chem. Res. Toxicol.*, 21(1), 253–265. doi: 10.1021/tx700408t.

Tan, Y. M., J. A. Leonard, S. Edwards, J. Teeguarden, and P. Egeghy. (2018a). Refining the aggregate exposure pathway. *Environ. Sci. Process Impacts.* 20(3), 428–436. doi: 10.1039/c8em00018b.

Tan, Y. M., J. A. Leonard, S. Edwards, J. Teeguarden, A. Paini, and P. Egeghy. (2018b). Aggregate exposure pathways in support of risk assessment. *Curr. Opin. Toxicol.* 9, 8–13. doi: 10.1016/j.cotox.2018.03.006.

Teeguarden, J. G., Y. M. Tan, S. W. Edwards, J. A. Leonard, K. A. Anderson, R. A. Corley, M. L. Kile, S. M. Simonich, D. Stone, R. L. Tanguay, K. M. Waters, S. L. Harper, and D. E. Williams DE. (2016). Completing the link between exposure science and toxicology for improved environmental health decision making: the aggregate exposure pathway framework. *Environ. Sci. Technol.* 50(9), 4579–4586. doi: 10.1021/acs.est.5b05311.

Teraoka, H., Y. Okuno, D. Nijoukubo, A. Yamakoshi, R. E. Peterson, J. J. Stegeman, T. Kitazawa, T. Hiraga, and A. Kubota. (2014). Involvement of COX2-thromboxane pathway in TCDD-induced precardiac edema in developing zebrafish. *Aquat. Toxicol.* 154, 19–26. doi: 10.1016/j.aquatox.2014.04.025.

Tierney, K. B., D. H. Baldwin, T. J. Hara, P. S. Ross, N. L. Scholz, and C. J. Kennedy. (2010). Olfactory toxicity in fishes. *Aquat. Toxicol.* 96(1), 2–6. doi: 10.1016/j.aquatox.2009.09.019.

Tollefsen, K. E., S. Scholz, M. T. Cronin, S. W. Edwards, J. de Knecht, K. Crofton, N. Garcia-Reyero, T. Hartung, A. Worth, and G. Patlewicz. (2014). Applying adverse outcome pathways (AOPs) to support integrated approaches to testing and assessment (IATA). *Regul. Toxicol. Pharmacol.* 70(3), 629–640. doi: 10.1016/j.yrtph.2014.09.009.

Trudeau, R. J. (2013). *Introduction to Graph Theory*. Dover Publications, New York, NY, USA.

US Environmental Protection Agency (USEPA). (2019). *Endocrine disruptor screening program (EDSP) overview*. Washington, DC. https://www.epa.gov/endocrine-disruption/endocrine-disruptor-screening-program-edsp-overview

Van Noorden. 2014. *Nature Newsblog*. https://blogs.nature.com/news/2014/05/global-scientific-output-doubles-every-nine-years.html

Villeneuve, D. (2016), "Adverse outcome pathway on aromatase inhibition leading to reproductive dysfunction (in fish)", *OECD Series on Adverse Outcome Pathways*, No. 4, OECD Publishing, Paris, https://doi.org/10.1787/5jlsv05mx433-en.

Villeneuve, D. L., M. M. Angrish, M. C. Fortin, I. Katsiadaki, M. Leonard, L. Margiotta-Casaluci, S. Munn, J. M. O'Brien, N. L. Pollesch, L. C. Smith, X. Zhang, and D. Knapen. (2018a). Adverse outcome pathway networks II: network analytics. *Environ. Toxicol. Chem.* 37(6), 1734–1748. doi: 10.1002/etc.4124.

Villeneuve, D. L., A. L. Blankenship, and J. P. Giesy. (2000). Derivation and application of relative potency estimates based on in vitro bioassay results. *Environ. Toxicol. Chem.* 19(11), 2835–2843. doi: 10.1002/etc.5620191131.

Villeneuve, D. L., K. Coady, B. I. Escher, E. Mihaich, C. A. Murphy, T. Schlekat, N. Garcia-Reyero. (2019). High-throughput screening and environmental risk assessment: state of the science and emerging applications. *Environ. Toxicol. Chem.* 38(1), 12–26. doi: 10.1002/etc.4315.

Villeneuve, D. L., D. Crump, N. Garcia-Reyero, M. Hecker, T. H. Hutchinson, C. A. LaLone, B. Landesmann, T. Lettieri, S. Munn, M. Nepelska, M. A. Ottinger, L. Vergauwen, and M. Whelan. (2014a). Adverse outcome pathway (AOP) development I: strategies and principles. *Toxicol. Sci.* 142(2), 312–320. doi: 10.1093/toxsci/kfu199.

Villeneuve, D. L., D. Crump, N. Garcia-Reyero, M. Hecker, T. H. Hutchinson, C. A. LaLone, B. Landesmann, T. Lettieri, S. Munn, M. Nepelska, M. A. Ottinger, L. Vergauwen, and M. Whelan. (2014b). Adverse outcome pathway development II: best practices. *Toxicol. Sci.* 142(2), 321–330. doi: 10.1093/toxsci/kfu200.

Villeneuve, D. L., B. Landesmann, P. Allavena, N. Ashley, A. Bal-Price, E. Corsini, S. Halappanavar, T. Hussell, D. Laskin, T. Lawrence, D. Nikolic-Paterson, M. Pallardy, A. Paini, R. Pieters, R. Roth, and F. Tschudi-Monnet. (2018b). Representing the process of inflammation as key events in adverse outcome pathways. *Toxicol. Sci.* 163(2), 346–352. doi: 10.1093/toxsci/kfy047.

Villeneuve, D. L., N. D. Mueller, D. Martinović, E. A. Makynen, M. D. Kahl, K. M. Jensen, E. J. Durhan, J. E. Cavallin, D. Bencic, and G. T. Ankley. (2009). Direct effects, compensation, and recovery in female fathead minnows exposed to a model aromatase inhibitor. *Environ. Health Perspect.* 117(4), 624–631. doi: 10.1289/ehp.11891.

Ware, M. and M. Mabe. (2015). The STM Report, an overview of scientific and scholarly journal publishing. International Association of Scientific, Technical, and Medical Publishers. The Hague, the Netherlands. https://digitalcommons.

unl.edu/scholcom/9?utm_source=digitalcommons.unl.edu%2Fscholcom%2F9&utm_medium=PDF&utm_campaign=PDFCoverPages

Williams, C. R., MacDonald, J. W., Bammler, T. K., Paulsen, M. H., Simpson, C. D, Gallagher EP. (2016 Dec). From the cover: Cadmium exposure differentially alters odorant-driven behaviors and expression of olfactory receptors in juvenile coho salmon (Oncorhynchus kisutch). *Toxicol Sci.* 154(2):267–277. doi: 10.1093/toxsci/kfw172.

Wittwehr, C., H. Aladjov, G. Ankley, H. J. Byrne, J. de Knecht, E. Heinzle, G. Klambauer, B. Landesmann, M. Luijten, C. MacKay, G. Maxwell, M. E. Meek, A. Paini, E. Perkins, T. Sobanski, D. Villeneuve, K. M. Waters, M. Whelan. (2017). How adverse outcome pathways can aid the development and use of computational prediction models for regulatory toxicology. *Toxicol. Sci.* 155(2), 326–336. doi: 10.1093/toxsci/kfw207.

Wittwehr, C., M. J. Amorim, L. A. Clerbaux, C. Krebs, B. Landesmann, D. S. Macmillan, P. Nymark, R. Ram, N. Garcia-Reyero, M. Sachana, K. Sullivan, J. Sund, and C. Willett. (2021). Understanding COVID-19 through adverse outcome pathways -2nd CIAO AOP Design Workshop. *ALTEX*. 38(2), 351–357. doi: 10.14573/altex.2102221.

World Health Organization. 2002. ICPS Global Assessment of the State-of-the-Science of Endocrine Disruptors. International Programme on Chemical Safety, WHO/PCS/EDC/02.2.

Yauk, C., et al. (2016). “Adverse outcome pathway on alkylation of DNA in male pre-meiotic germ cells leading to heritable mutations”, *OECD Series on Adverse Outcome Pathways*, No. 3, OECD Publishing, Paris, doi: 10.1787/5jlsvvxn1zjc-en.

# 13 The Potential for Toxicity to Fishes from Micro- and Nanoplastics, and Their Additives


*Susanne M. Brander*
Oregon State University

*Azora König and Bethanie Carney Almroth*
University of Gothenburg

*Leah Thornton Hampton*
Southern California Coastal Water Research Project



### Abstract

Over the past several decades, the intense production and use of plastics have led to the contamination of all types of ecosystems with microplastics. Microplastics, plastic particles less than 5 mm in size, and nanoplastics which are less than 1 µm, are a diverse contaminant suite comprised of different sizes, shapes, polymer properties, additives and sorbed pollutants, and biofilms. Micro and nanoplastics (MNPs) may pose a potential danger for organisms who are exposed to the particles and the chemicals associated with the plastics. As many marine and freshwater systems are highly contaminated with MNPs, fish are continuously exposed to them, which may lead to a variety of adverse effects.

This chapter describes current knowledge on the relationship between nano- and microplastic occurrence in freshwater and marine ecosystems and the potential toxicological effects in fish. First, the MNP properties, associated additives, pollutants, and biofilms are described. Following, it is illustrated how fish can be exposed to MNPs and which adverse effects exposure can have on the different levels of biological organization of fish. These effects range from altered gene expression, oxidative stress, and immune responses to changes in behavior and impaired reproduction. Finally, new multivariate approaches to address microplastic toxicity and recommended future research directions are summarized.


## INTRODUCTION

While plastics were invented in the late 1800s, large-scale mass production of modern polymers escalated in the mid-20th century (Brander & Betjemann, 2023; Thompson et al., 2004). A handful of early papers documented evidence of microplastics, in the form of pre-production nurdles, in the

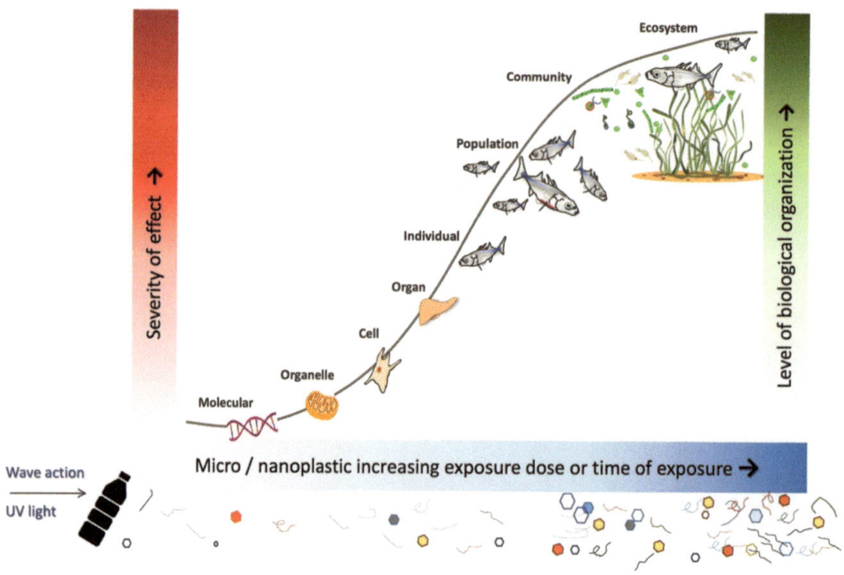


**FIGURE 13.1**


DOI: 10.1201/9781003160694-13

N. Atlantic and in the gut of marine animals (Carpenter et al., 1972). Occurrence of these nurdles was global, and effects of ingestion as well as potential exposure to organic toxicants were studied in seabirds in the 1980s (Ryan et al., 1988; Ryan & Jackson, 1987).

Renewed and more intensive research on microplastic pollution, as well as its smaller counterpart, nanoplastics, has entered its second decade of heavily focused research. While early reports of occurrence primarily focused on larger microplastics in marine ecosystems, research approaches have matured into targeted controlled studies quantifying microplastics in remote marine sites (e.g., marine protected areas, deep ocean organisms), as well as freshwater and terrestrial environments, leading to a better understanding of sources, fate, and transport (Brander et al., 2020; Granek et al., 2020; Miller et al., 2021). Plastics are present in every corner of the planet, and constitute a diverse suite of contaminants rather than a single pollutant (Rochman et al., 2019) (Figure 13.2). A considerable challenge is presented to toxicologists, since findings from research thus far indicate that each physicochemical characteristic of micro- and nanosized plastics (MNPs) should be independently evaluated using yet-to-be standardized laboratory assays (Cowger et al., 2020; Gouin et al., 2019; Thornton Hampton et al., 2022), this is because whether an effect is detected or not appears to vary not only with concentration, but is also dependent on plastic characteristics (e.g., shape, size, polymer), exposure period, extent of plastic weathering, sometimes on the associated chemicals such as additives, and a variety of other factors (e.g., Athey et al., 2020; Bucci et al., 2019, 2021; Choi et al., 2018; de Ruijter et al., 2020; Qiao et al., 2019a; Ziajahromi et al., 2017). An organism's interaction with an MNP is a multiple-stressor exposure in and of itself (Rochman et al., 2019), but should be considered in the context of realistic interactions with wild organisms, where abiotic factors may play an important role in microplastic fate and behavior, and thus further influence organism responses (e.g., McCormick et al., 2020; Shupe et al., 2021). A recent synthesis even suggests that macro- and microplastic ingestion is an evolutionary trap, resulting in overall reduced fitness across teleost (bony fish) families, over 25% of which have been confirmed to ingest plastics (Santos et al., 2021).

As intimated above, toxicological responses to microplastics are complex and can be difficult to reproduce across taxonomic groups and experiments (Bucci et al., 2019). This is not surprising given the complexity of this contaminant suite, but researchers are beginning to arrive at some consensus on the types of responses that are more frequently detected as well as honing in on plastic characteristics with a higher potential to cause toxicity, such as fibers (e.g., Brander et al., 2021; Jacob et al., 2020; Stienbarger et al., 2021, Granek et al., 2022; Siddiqui et al., 2023). Here, we summarize the state of knowledge on MNP toxicity, in terms of their toxicity to finfishes as a taxonomic group. The goals of this work are to first introduce microplastics as a contaminant suite, then to highlight where convergence of responses across fish species occur as well as differences in effects and sensitivity across the biological hierarchy. This includes consideration of the complexities of life stage, habitat usage, and abiotic factors. We will also briefly explore the relative importance of associated chemicals such as additives and investigate why a number of studies show no effect.

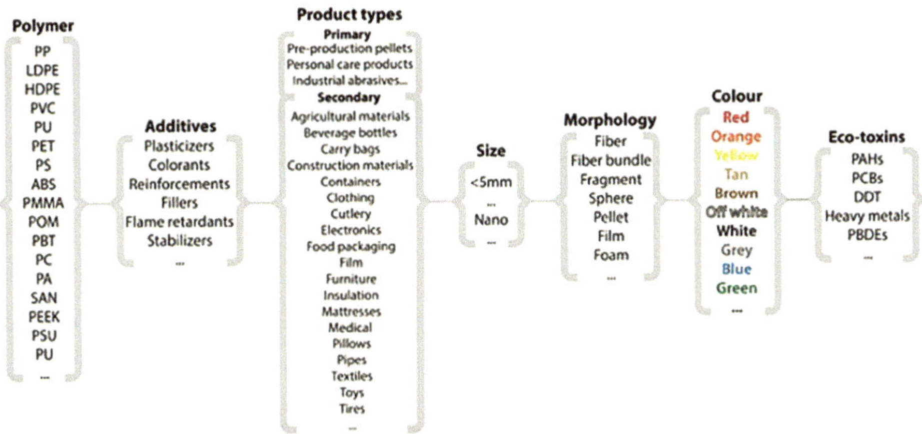


**FIGURE 13.2** Microplastics are made with a variety of polymers, augmented with an array of additives that can be manufactured into a multitude of products. Sources of microplastics can be either primary or secondary, and microplastics may be any size less than 5 mm. Microplastics are described with at least seven morphologies and are found in many different colors. When in the environment, microplastics can sorb numerous chemical contaminants, including heavy metals and persistent organic pollutants. This is not an exhaustive list. PP=polypropylene; LDPE=low-density polyethylene; HDPE=high-density polyethylene; PVC=polyvinyl chloride; PU=polyurethane; PET=polyethylene terephthalate; PS=polystyrene; ABS=acrylonitrile butadiene styrene; PMMA=polymethyl methacrylate; POM=polyoxymethylene; PBT=polybutylene terephthalate; PC=polycarbonate; PA=polyamides; SAN=styrene-acrylonitrile; PEEK=polyether ether ketone; PSU=polyarylsulfone; PAH=polycyclic aromatic hydrocarbon; PCB=polychlorinated biphenyl; DDT=dichlorodiphenyltrichloroethane; PBDE=polybrominated diphenyl ethers. Reprinted from Rochman et al. 2019, Rethinking microplastics as a diverse contaminant suite, with permission from *Environmental Toxicology and Chemistry*.

## PARTICLE CHARACTERISTICS

Microplastics are broadly defined as polymers <5 mm in size, with a lower size range cutoff that varies between 1,000 nm and 100 nm (Brander et al., 2020). The majority of researchers now deem anything smaller than 1 µm a nanoplastic (NP). These size designations are somewhat arbitrary and to an extent are based on current analytical limitations, which only allow for identification of MPs in environmental samples above a size of 1 µm (Primpke et al., 2020). Although micro- and nanoplastics (MNPs) are often referred to as one pollutant type, we now acknowledge MNPs as a diverse suite of contaminants comprised of various polymers, shapes, colors, increasingly smaller sizes, associated chemicals, and the nearly limitless combinations of these categories (Granek et al., 2020; Rochman et al., 2019). Additionally, the ubiquitous presence of semi-synthetic and anthropogenically modified natural fibers is now included under the larger umbrella of 'microparticle' contamination, rather than limiting the challenge to plastics alone (Athey & Erdle, 2021; Granek et al., 2022; Miller et al., 2021). In fact, the State of California now includes any particle with a polymer content of >1% in its recent definition of microplastics in drinking water (CA SWRCB, 2020-0021), which is now being used to describe microplastics in aquatic and marine ecosystems as well. Notably, this definition includes any particle with a polymer content of greater than 1% by mass, and includes nanoplastics.

It has become clear that considering the multidimensionality of MNPs is critical to understanding their toxicity, particularly now that synthesized data strongly suggest that both size and shape both play a role in how hazardous an MNP is (Brander et al., 2021; Jacob et al., 2020; Thornton Hampton et al., 2022). Larger microplastics (MPs) appear to be more likely to cause food dilution, that is, the reduced availability of nutrients paired with reduced food intake in organisms that have ingested MPs (Koelmans et al., 2020). Smaller MNPs (<80 microns in diameter) have a greater tendency to translocate, meaning that these particles may be small enough to cross epithelial barriers, leaving the gastrointestinal tract and becoming embedded in other organs and tissues (Collard et al., 2017; McIlwraith et al., 2021). Microfibers make estimating this translocation cutoff more complicated, since their narrow width (3–15 µm) makes it possible for relatively long MPs to translocate if they are in the right orientation (Granek et al., 2022). Nanoplastics (NP, <1 micron) can become entrapped in or penetrate cellular membranes and even affect specific cellular components such as lysosomes or the mitochondrial membrane (Sendra et al., 2020). Translocation may result in a greater likelihood for reactive oxygen species (ROS) generation, which can cause oxidative damage within the cell as well as to surrounding tissues (Thornton Hampton et al., 2022). ROS generation is posited to be one of the early events in the very preliminary AOPs now published for MNPs (Jeong & Choi, 2019), and may either precede or follow food dilution. In terms of shape, although data are still limited because the majority of studies have been conducted using microspheres due to their wide availability, evidence points to fibers potentially having greater toxicity than other plastic shapes (Jacob et al., 2020; Thornton Hampton et al., 2022; Stienbarger et al., 2021). This is important to seriously consider in the design of future experiments, given that fibers are detected at higher proportions than other plastic types across environmental matrices, including biota (Athey & Erdle, 2021; Baechler et al., 2020; Harris et al., 2021; Ross et al., 2021).

## PARTICLE BEHAVIOR IN THE ENVIRONMENT

As described in detail below, plastics in the environment are associated with diverse mixtures of pollutants and organisms that form biofilms, already present in water and sediment

### DEFINITION OF 'MICROPLASTICS IN DRINKING WATER'*

'Microplastics in Drinking Water' are defined as solid[1] polymeric materials[2] to which chemical additives or other substances may have been added, which are particles[2] which have at least three dimensions that are greater than 1 nm and less than 5,000 micrometers (µm).[3] Polymers that are derived in nature that have not been chemically modified (other than by hydrolysis) are excluded.

* The California definition of microplastics in drinking water, adopted in June 2021. '[1]Solid' means a substance or mixture which does not meet the definitions of liquid or gas. 'Liquid' means a substance or mixture which (i) at 50 degrees Celsius (°C) has a vapor pressure less than or equal to 300 kPa; (ii) is not completely gaseous at 20°C and at a standard pressure of 101.3 kPa; and (iii) which has a melting point or initial melting point of 20°C or less at a standard pressure of 101.3 kPa. 'Gas' means a substance which (i) at 50°C has a vapor pressure greater than 300 kPa (absolute); or (ii) is completely gaseous at 20°C at a standard pressure of 101.3 kPa. [2]'Polymeric material' means either (i) a particle of any composition with a continuous polymer surface coating of any thickness, or (ii) a particle of any composition with a polymer content of greater than or equal to 1% by mass. 'Particle' means a minute piece of matter with defined physical boundaries; a defined physical boundary is an interface. 'Polymer' means a substance consisting of molecules characterized by the sequence of one or more types of monomer units. Such molecules must be distributed over a range of molecular weights wherein differences in the molecular weight are primarily attributable to differences in the number of monomer units. A polymer comprises the following: (a) a simple weight majority of molecules containing at least three monomer units which are covalently bound to at least one other monomer unit or other reactant; (b) less than a simple weight majority of molecules of the same molecular weight. 'Monomer unit' means the reacted form of a monomer substance in a polymer. 'Monomer' means a substance which is capable of forming covalent bonds with a sequence of additional like or unlike molecules under the conditions of the relevant polymer-forming reaction used for the particular process. 3Size-based nomenclature within the dimension limits include: 'nanoplastics' (1 nm to <100 nm); 'submicron plastics' (100 nm to <1 µm); 'small microplastics' (1 µm to < 100 µm); 'large microplastics' (100 µm to <5 mm); 'mesoplastics' (5 mm to <2.5 cm); 'macroplastics' (>2.5 cm).

(Bucci et al., 2021). Additionally MNPs weather in response to UV light exposure and wave action (Liu et al., 2020a; Lusher et al., 2020). Weathering can result in changes in density, often causing the MNPs to sink to sediments, where it is estimated that microplastic particles are eventually deposited (Kane & Clare, 2019; Woodall et al., 2014). Weathering can reduce toxicity as additives are leached, depending on the polymer type, but it also increases the presence of cracks and fissures that can provide more surface area for sorbed pollutants or create the potential for more damage if internalized by an organism (Liu et al., 2020a; Simon et al., 2021). These factors are important in determining potential effects in fish. The ultimate fate of MNPs in the environment will dictate concentrations in different niches (e.g., coastal surface waters versus pelagic water column versus sediment), and concentrations will drive potential exposure levels for fish living in those different niches. Current knowledge, built upon best practices in sampling and analytical techniques, shows that MPs are found in all parts of freshwater and marine water columns as well as in sediment, but higher concentrations can be found in coastal regions, fjords, and estuaries (Harris, 2020; Liu et al., 2020b). The relative importance of size, shape, surface structure, and chemical content is still understudied.

## PLASTIC ADDITIVES

Plastics, and therefore MNPs, have very complex chemical profiles. The life cycle of most polymers begins at the extraction phase when raw materials are sourced. In most cases, fossil fuels (oil and gas) are used to produce monomers (CIEL Center for International Environmental Law, 2019) which are processed into polymers, usually with the addition of polymerization catalysis, solvents, or lubricants. Then, manufacturers can add additional chemicals to impart specific qualities to the plastic products. This includes pigments like azodyes or some metals for color, antioxidants to hinder UV degradation, plasticizers such as phthalates to increase flexibility, flame retardants to increase fire resistance (Wiesinger et al., 2021). In addition to these substances, plastics can also contain non-intentionally added substances (NIASs) which include contaminants, breakdown products, and by-products from production processes (Groh et al., 2019).

While the chemical profile of different plastic products is often not publicly available, a recent study of the global market concluded that 10,000 substances may be used in the production of, and potentially released from, plastic products including plastic monomers, additives, and processing aids (Wiesinger et al., 2021). These authors identified over 2,400 substances of potential concern based on the toxicity criteria in the EU; persistence, bioaccumulation, and toxicity. So plastics that reach the environment can contain potentially thousands of hazardous substances. In beach litter surveys, some of the most commonly identified types of plastic products are packaging materials (i.e., food packaging) and food contact materials (i.e., cutlery, straws, bottles, and lids) (Hanke, 2016). Groh and co-authors (2019) identified 3,377 chemicals as possibly associated with plastic packaging and 148 of these were identified as hazardous (see Figure 13.3). In addition to this, microplastic fibers from synthetic textiles are a major source of MPs (Acharya et al., 2021; Xu et al., 2020). Over 8,000 different chemical compounds, with more added each year, are used across the entire production chain of textile fibers and garments, from agricultural production of natural and synthetic fibers, garment production, shipment, storage, sales, and consumer use (Lacasse & Baumann, 2004). Textile garments can contain up to 30% chemicals by weight, and

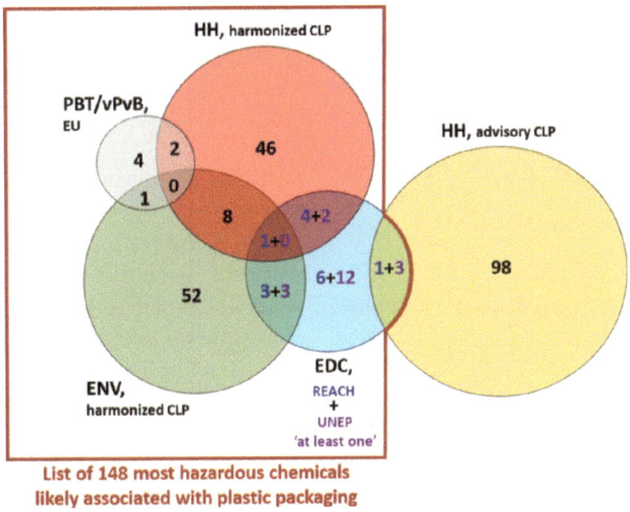


**FIGURE 13.3** Showing overlap between the groups of the most hazardous substances likely associated with plastic packaging. These were identified according to (i) environmental hazards (ENV) based on Classification, Labeling, and Packaging (CLP); (ii) human health hazards (HH) based on CLP; (iii) persistent, bioaccumulative, toxic (PBT) or very persistent, very bioaccumulative (vPvB) substances, as defined by the European Union (EU); (iv) endocrine disrupting chemicals (EDC) as identified by the Registration, Evaluation, Authorization, and restriction of CHemicals (REACH) legislation in the EU, the 2018 United Nations Environmental Programme (UNEP), or the World Health Organisation (WHO)/International Proteam on Chemical Safety (IPCS). This totals 148 substances. An additional 98 substances identified via CLP as human health hazards (HH) were also noted. From Groh et al. (2019).

many of these chemicals are hazardous to human and environmental health (KEMI, Swedish Chemicals Agency, 2016). Numerous different mechanisms of toxicity have been attributed to plastic-associated chemicals, including endocrine disruption, carcinogenicity, neurotoxicity, and developmental toxicity, leading to multiple calls to identify plastic waste as hazardous (Brander et al., 2012; CIEL Center for International Environmental Law, 2019; Rochman et al., 2019).

However, plastics and MNP particles in the environment are chemically different from original plastic products. Plastics leach many of these additives and NIASs during their use phases (Schellenberger et al., 2019; Muncke, 2021) and, once they reach the environment, will continue to leach additives to the surrounding water, sediment, or other particles, and biota (Jang et al., 2017). The chemical load of plastics will change further as they sorb contaminants present in the environment, owing to the hydrophobic nature of plastic materials. A recent inventory placed the total number of chemicals registered for production or use in global markets over 350,000, a number which far exceeds earlier estimates (Wang et al., 2020a). While not all of these chemicals have been studied with regard to environmental concentrations or hazards, especially not in marine organisms, environmental thresholds have been reported for thousands of them, including biocides, priority pollutants, pesticides, pharmaceuticals, and industrial chemicals (listed here in order of decreasing toxicity) (Gustavsson et al., 2017). The sorption of a number of priority compounds, including persistent organic pollutants (POPs) like PAHs, PCBs, pesticides, and metals, on to plastics in the environment been studied (Ogata et al., 2009; Turner & Holmes, 2015; Yamashita et al., 2019), showing that sorption does occur and can result in concentrations of chemicals that are significantly higher than in the surrounding water. The sorption rates and concentrations of chemicals binding to plastic pellets differ significantly among plastic types and are dependent on the chemical contaminants present at specific localities, aging and weather of the plastics, and biofilm formation (Rochman, 2013; Turner & Holmes, 2015).

## MICROPLASTICS AS CHEMICAL VECTORS

The role of plastics in the fate of chemicals in the environment is widely discussed, and the importance of microplastics as a vector for chemicals into biota has been investigated using empirical data (Ašmonaitė et al., 2020; Bour, et al., 2020a; Rochman et al., 2013), correlation studies (Fossi et al., 2012; Herzke et al., 2016; Ryan et al., 1988), and modelling based on physicochemical properties of the particles and chemicals (Bakir et al., 2016; Koelmans et al., 2016). In order to understand the propensity of MNPs to act as chemical vectors, sorption and desorption mechanisms, carefully planned studies are necessary. Many factors will influence release of chemicals from MNPs and uptake in fish, including polymer types, surface area, gut retention time, pH, surfactants, presence of other food items and persistent chemicals in the food itself, and concentrations of chemical in the plastics and in the fish (Bour et al., 2020a; Koelmans et al., 2016; Velez et al., 2018; Ziccardi et al., 2016). Indeed, some work shows that the transfer of chemicals from plastic to biota is biphasic and reversible (Mohamed et al., 2019).

The choice of plastics used in these experiments is important, as it will determine the chemicals present in the particles prior to spiking and the physicochemical interactions between the polymers and chemicals. Virgin or unplasticized materials, for example, may have fewer plastic additives, but can still contain a suite of other non-intentionally added substances (Ašmonaitė et al., 2018). Particles originating from products, or weathered, might have very different chemical profiles. In addition, rubbery versus glassy polymers will interact differently with chemical substances. Sorption takes place in the amorphous regions of the plastic polymer, which, depending on their glass transition temperature, can be glassy (e.g., PS) or rubbery (e.g., PE). The polymer chains in rubbery polymers have a higher mobility and faster successive diffusion of molecules, compared to glassy polymers (Hartmann et al., 2017; Teuten et al., 2009). The choice of chemical substances is also important, and consideration must be given to the physicochemical characteristics, including molecular size and hydrophobicity.

In laboratory exposure experiments, researchers have used a number of different approaches to address chemical transfer. In some cases, they have artificially spiked plastic pellets with specific compounds (Bour et al., 2020a; Coffin et al., 2020; Wardrop et al., 2016). This can allow researchers to use artificially high levels of chemicals (higher than concentrations measured in environmental matrices), ensuring that concentrations will be above detection levels and thereby avoiding methodological limitations in chemical detection. In other studies, researchers have exposed plastics in situ in marine waters. While many studies have measured hydrophobic chemicals in plastics either collected in aquatic systems, or exposed in situ, fewer have also used these particles in exposure experiments. Rochman et al. (2013) did this and found evidence of vector mediated transfer of chemicals from particles exposed in a harbor, to the fish fed with these particles. Ašmonaitė et al. (2018), on the other hand, measured chemicals in plastic particles incubated in situ in an industrial harbor and in sewage effluent, but relied on biomarker responses in the exposed fish, rather than chemical analyses. When spiking particles or incubating in situ in contaminated environments, it is important to have a good understanding of whether compounds adsorbed to the surface of the particles, or diffused into the polymer matrix, as this will affect both chemical sorption and desorption kinetics (Koelmans et al., 2021). These may differ for glassy versus rubber polymers and in new versus weathered particles. Weathered particles may have accumulated changes to polymer backbones, or changes in surface charge via, for example, oxidation reactions, which will change interactions with chemicals (Hartmann et al., 2017). The presence of a biofilm on weathered particles can also affect chemical binding.

There are also a number of studies that find evidence of transfer of chemicals from MNPs to biota by correlating evidence of

MP ingestion (e.g., gut content analyses) with chemical analyses in tissues of the studied animals. Early work in birds demonstrated that this occurred in wild populations, though the relationship between MPs and POPs is not entirely clear. Ryan et al. (1988) determined the concentrations of organochlorines in the fat and eggs of great shearwaters and found that the mass of ingested plastics was positively correlated to PCB concentrations in the birds, concluding that assimilation of PCBs from plastics was one of several probable exposure routes. On the other hand, a study in the northern fulmar, correlating plastic ingestion with concentrations of several POPs in the tissues of the birds did not find that the MPs were acting as vectors but rather as passive samplers reflecting ingestion of POPs by prey animals as well (Herzke et al., 2016). While few studies address this relationship in wild fish, evidence from a laboratory study in fish showed that MPs can increase accumulation of organochlorine pesticides and (OCPs) and polychlorinated biphenyls (PCBs) over time, but that contamination in food is also an important source of exposure (Rios-Fuster et al., 2021). This is in agreement with other works showing that PCBs can be taken up via exposure to contaminated MPs in food, though assimilation efficiencies from PCB spike food matrices are higher (Grigorakis and Drouillard, 2018). In a feeding study in stickleback, authors used three different types of particles and three different toxic chemicals and found that less than 1% of chemicals spiked onto particles were detected in fish tissues (Ašmonaitė et al., 2020). Similarly, larval herring tube-fed plastic particles spiked with PCB-153 found that there was no measurable transfer of the chemical into fish tissues (Norland et al., 2021). Chemical uptake likely depends on a number of factors, including digestive period, and the amount as well as chemical properties of the sorbed pollutants.

Several groups have used modelling to predict the vector properties of MPs. These are based on polymer properties, particle size, surface area and mass, chemical molecular properties, and fugacity (Koelmans et al., 2014, 2016). These studies have proven useful in advancing our understanding of these mechanisms but should also be tested with experimentation and empirical data in order to understand how well they predict actual uptake. In addition, risk assessment comparing exposure pathways via MPs versus other routes has helped us to understand the current, relatively small importance of MPs as a route of exposure (Everaert et al., 2020; Gouin et al., 2011).

The general consensus of all of this research is that chemical exposure and uptake via ingested MNPs will likely be an insignificant exposure pathway when compared to other exposure routes (Bour et al., 2020a; Gouin et al., 2011; Ziccardi et al., 2016). These exposure pathways include natural prey or uptake from the natural environment (e.g., from sediment or water). Importantly, the uptake of chemicals in fish will be a function of uptake directly from water (across dermal membranes, the gills, or gut), uptake via food, and uptake via the ingested plastic particles, and chemical concentrations measured in fish will also be affected by the amount of chemicals lost to metabolism, depuration, egestion, and defecation (Koelmans et al., 2021). See Figure 13.4. Current environmental concentrations of MPs are relatively low, compared to other particles and matrices in the environment, so the total chemical exposure via MPs is also low, save in some hot spots with locally high MP concentrations (Everaert et al., 2020), and risk could increase in the future as plastic concentrations increase in aquatic environments. What this does indicate, however, is the ubiquitous nature of plastic-associated chemicals, their pervasiveness in the environment, and their uptake into biota via multiple routes. These are still substances that are problematic for human and environmental health, even if exposure via MNPs is not currently dominant.

## BIOFILMS

Biofilms play important roles in the fate and behavior of MNPs in ecosystems, in sorption of chemical contaminants, in the interactions between organisms in MPs, and in their

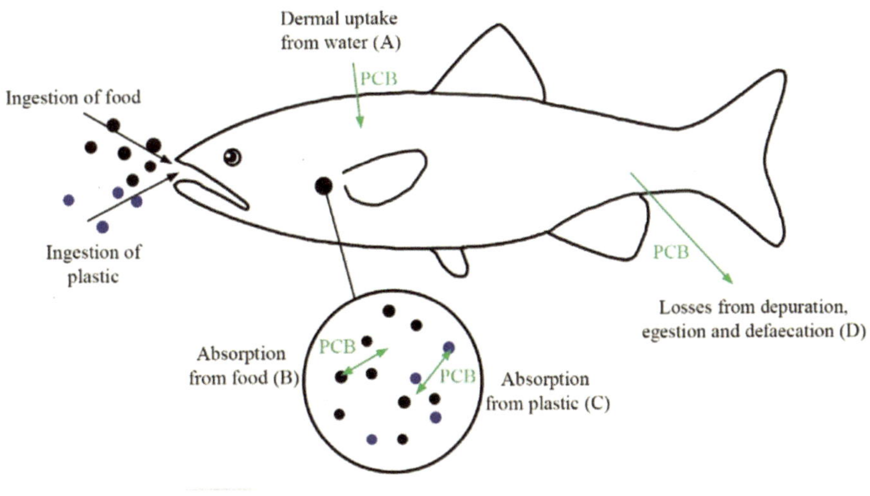


**FIGURE 13.4** Diagram illustrating the uptake and elimination pathways for chemicals by and from fish. PCB is used to represent chemicals in general. Processes A, B, C, and D represent the concomitant terms in equations described in Koelmans et al. (2021).

potential toxicity. Biofilms form as biomolecules (polysaccharides, proteins, lipids) adhere to surfaces, followed by planktonic microbes that colonize surfaces, excreting a matrix of extracellular polymeric substances (EPS) that acts as a glue. This corona formation is followed by a succession of diverse microbes (bacteria, algae, fungus) and then larger organisms (larvae, zooplankton, etc.). These microorganisms form distinct communities, organized within three-dimensional structures. They can develop phenotypic behaviors or functions that differ from the planktonic microbes, as the organisms succeed one another and specialize to fulfill necessary functions (motility, matrix production, metabolism) of the biofilm community (An & Parsek, 2007; Momeni, 2018). Biofilms can impact the fate and potential effects of MPs in the environment, see Figure 13.5.

## EFFECT OF BIOFILMS ON MICROPLASTIC FATE AND EXPOSURE TO FISH

The formation of the biofilm can change the physicochemical characteristics of the particles, thereby affecting their fate. The physical and chemical characteristics of the microplastic affect microbial colonization of the particle (Amaral-Zettler et al., 2020). As the biofilm grows, the density of microplastics can change, although these density changes have not yet been fully quantified (Skalska et al., 2020). Early work indicates that biofouling and de-fouling can result in yo-yo like behavior of MPs in the water column (Ye & Andrady, 1991). (Note, changes in MPs density are also affected by other mechanisms, including fragmentation and degradation). Biofilm growth on plastic particles can therefore impact their buoyancy, and, consequently, their fate in aquatic systems. Plastics were originally predicted to remain in surface waters due to their low density, though evidence refuted this idea as early as 1975 when plastic films were identified in the depths of Skagerrak in the Northeast Atlantic (Holmström, 1975; Shupe et al., 2021). More recent evidence demonstrates that as organic particles and microorganisms attach on the surface of microplastics, that is, the biofilm, changes do occur with either increases or decreases in the density of microplastic particles (Rummel et al., 2017; Ye & Andrady, 1991).

In addition, when microplastics enter the water column, they can form aggregates with other molecules and particles, such as suspended sediments and microorganisms (Yan et al., 2021). For smaller MNPs, in particular those in the NP size range, environmental conditions such as salinity and the presence of organic matter can increase the agglomeration of these nanoscopic particles, potentially increasing their size to the extent that they are more likely to be mistaken for a prey item or altering sinking rates (Shupe et al., 2021). Heteroaggregation with inorganic particles (i.e., suspended sediments and metal oxides) can also increase the density of microplastics (Procter et al., 2019a; Savoca et al., 2016a; Wu et al., 2020; Leiser et al., 2020). Both of these mechanisms (biofilming and aggregate formation) can change the presence of MPs in different environmental niches, from surface waters to the water column to sediments and therefore availability to different fish species. Planktivores, for example, may ingest particles that remain in the water column (Boerger et al., 2010).

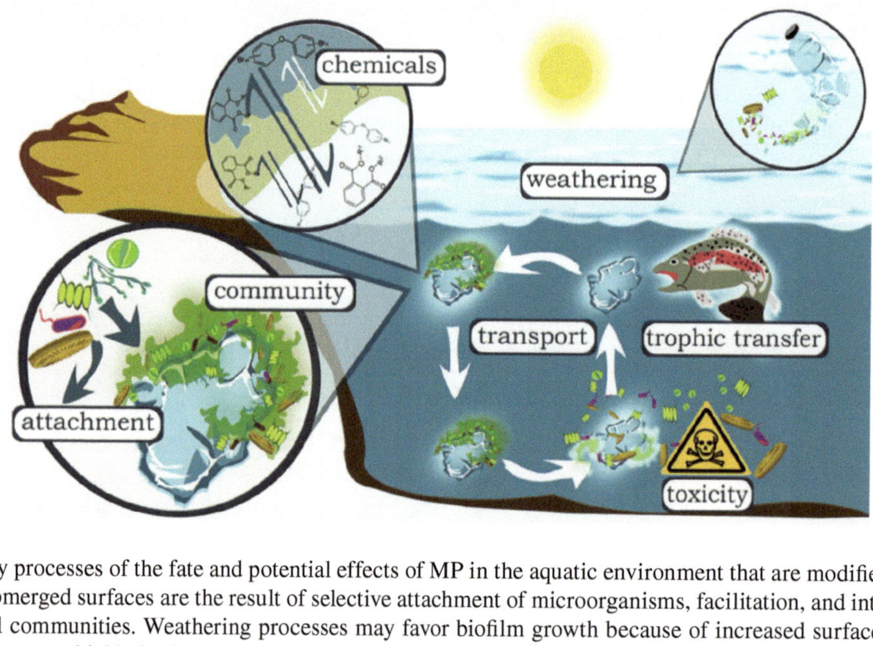


**FIGURE 13.5** 'Key processes of the fate and potential effects of MP in the aquatic environment that are modified by biofilm formation. Biofilms on submerged surfaces are the result of selective attachment of microorganisms, facilitation, and inter-specific competition in the microbial communities. Weathering processes may favor biofilm growth because of increased surface areas available for settling, which in turn may shield plastic debris from ultraviolet light. However, biofilms have the capacity to biodegrade the polymer. In addition, vertical transport and the uptake and release of plastic-associated chemicals are influenced by biofilm formation on plastic debris. Biological implications of biofilm formation include effects on trophic transfer of MP and associated contaminants, community structure of microbial assemblages, and potential toxicity to grazers'. The biofilm is represented throughout the figure in green. Reprinted from Rummel et al. (2017).

A number of studies have shown that olfactory cues, including dimethyl sulfide from microalgae and diatoms, can enhance grazing and ingestion by a wide number of species ranging from copepods to fish to marine birds (Procter et al., 2019b; Savoca et al., 2016b). Once MPs are in the sediment, which is considered to be a major sink for MPs (Woodall et al., 2014), they will become available to fishes mainly via trophic transfer from benthic invertebrates, though there are indications that benthic invertebrate foragers are less susceptible to microplastic ingestion than species which exhibit more general preferences and methods of prey capture (Peters et al., 2017). In addition, there is evidence that microplastics can adhere to algal surfaces and transfer up through the food chain via grazing of common periwinkles (Gutow et al., 2016).

While biofilms are an important community niche in aquatic ecosystems, the impacts of MPs on the structure and function of periphyton communities (material growing on submerged surfaces in freshwater ecosystems) have not been fully studied, nor have the role of the periphyton in the fate of micro- and nanoparticles been fully investigated. There are some studies showing that exposure to nano- and microplastics can induce oxidative stress and have the potential to induce changes in the function of biofilm communities (Miao et al., 2019). This is of concern considering that changes in function at the base of an ecological system have the potential to drive indirect effects at higher levels in the system.

## EFFECT OF BIOFILMS ON THE SORPTION AND DESORPTION OF CHEMICALS

In addition to affecting distribution and therefore bioavailability, biofilms can also affect interactions between MNPs and toxic chemicals. The presence of biofilms on the surface of particles can affect sorption/desorption mechanisms and interactions between chemicals and the plastic polymers (Mincer et al., 2019; Verla et al., 2019). Biofilms can change the polarity of MNPs, rendering them increasingly hydrophobic. In addition, microbes can create pits in the surface of MPs, thereby increasing the surface area (Zettler et al., 2013).

Accumulation of contaminants in biofilms is well known, and biofilms have previously been identified as a sink for metals (van Hullebusch et al., 2003; Wu et al., 2014). Microbial biofilms are useful in remediation efforts to remove metals from aquatic environments (Wu et al., 2012). Other types of environmental pollutants, including pesticides, pharmaceuticals, PAHs, and PCBs, can also bioaccumulate in biofilms (Bonnineau et al., 2021), as can steroidal hormones (Writer et al., 2011). The microorganisms in biofilms may metabolize or sequester the pollutants that are taken up, thereby changing their form and distribution in the environment, and fate within food webs. In some cases, biofilms can facilitate accumulation of pollutants onto MPs. A recent study showed that biofilms can promote accumulation of metal on MPs (Richard et al., 2019) and these authors suggest that these small increases in sorption capacity could potentially affect the propensity for MPs to act as vectors of exposure, especially in the open ocean gyres where plastic is the dominant type of debris. Others have shown that biofilms can enhance the adsorption of tetracycline, a commonly used antibiotic, onto MPs (Wang et al., 2020b). Biofilms can also increase solubility of hydrophobic organic compounds in water, and biodegrade some compounds, but sorption and/or degradation of pollutants in MP biofilms will be highly variable and dependent upon both biotic (microbial community diversity and composition) and abiotic factors (temperature, pH) (Jin et al., 2020).

Several studies have addressed the impacts of weathering (including fragmentation, polymer degradation, and biofilm formation) on microplastics and how these changes can impact the propensity for MNPs to act as vectors for contaminants. Weathering can impact the leaching of various plastic additives into surroundings (Liu et al., 2020a). However, most researchers conclude that additives will leach out of plastics relatively quickly in aquatic environments (De Frond et al., 2019; Suhrhoff & Scholz-Böttcher, 2016), and that those chemicals transferred to fish upon ingestion will be environmental pollutants that have sorbed to the MNPs. This is supported by research into the chemical fingerprints of virgin versus weathered particles, and the differences in toxicity that result when fish are exposed (Bucci et al., 2021; Mato et al., 2001; Rochman et al., 2013).

## THE PLASTISPHERE AND IMPORTANCE OF MICROPLASTIC MICROBIAL COMMUNITIES

The formation of new, diverse communities of microorganisms on the surfaces of MPs in aquatic environments has been the subject of study, building on knowledge concerning corona formation and microbial communities inhabiting the new niche, now designated as the plastisphere. The plastisphere is a diverse microbial community of heterotrophs, autotrophs, predators, and symbionts that inhabit the surface of plastic particles (Zettler et al., 2013). Biofilms are important benthic communities that fill important functions in aquatic ecosystems, including carbon and nutrient cycling (Battin et al., 2003). There is evidence that the microbial communities of the plastisphere are divergent in composition from the planktonic organisms in the surrounding water (Dussud et al., 2018) as well as microbes on other types of particles (Kirstein et al., 2018). However, there is currently no consensus on the degree of biodiversity in the plastisphere; some studies indicate that biofilm communities on microplastic particles are more diverse than those on other types of particles, while others report lower levels of diversity (Amaral-Zettler et al., 2020). The exchange of genetic material among microbes on MPs is higher than in other environments, occurring via plasmid transfers (Arias-Andres et al., 2018). This includes genes for antibiotic resistance and has potential to affect genetic diversity in the plastisphere, as well as contribute to the spread of antibiotic resistant bacteria. Human pathogens and multidrug-resistant bacteria have been identified on plastic litter (Rasool et al., 2021).

Plastics in aquatic systems have been identified as a conduit for spread of invasive species via hitchhiking (GESAMP, 2020) and this includes spread of pathogens to new areas (Bowley et al., 2021). Microplastics can also harbor fish pathogens including *Aeromonas* sp. Water samples collected along the Norwegian coastline, in an area where aquaculture of salmon is a major activity, were found to contain MPs carrying *A. salmonicida*, in addition to several opportunistic human pathogens (Radisic et al., 2020; Viršek et al., 2017). *A. salmonicida* is a pathogen that can severely impact salmon health, and the microbes isolated in the study were found to be both potentially virulent and carrying several antibiotic resistance genes.

## EXPOSURE

### Oral Ingestion

Fish may be exposed to microplastic particles via a variety of routes, though to date the majority of studies have focused on the oral ingestion of microplastics (Jabeen et al., 2018; Jovanović et al., 2018; Lusher et al., 2013). MNPs may be actively ingested during feeding when they are confused with food items or unintentionally ingested when drinking or via trophic transfer (Figure 13.6). Once ingested, most particles pass through the gastrointestinal tract of fishes relatively quickly (~24–96 hours) and are excreted (Athey et al., 2020; Elizalde-Velázquez et al., 2020; Xiong et al., 2018), though others may be retained for longer in the gut (Ory et al., 2018) or translocate to other tissues where they may remain for extended periods of time (Collard et al., 2017; Wang et al., 2019b). Microplastic ingestion, egestion, and possible translocation are a function of both particle characteristics (e.g., size, color, morphology) and life history traits (e.g., foraging strategy, age, behavior) (Roch et al., 2020), and may be influenced by a variety of abiotic and biotic factors.

Oral ingestion of plastics has been repeatedly observed in laboratory settings (Jabeen et al., 2018; Jovanović et al., 2018), and microplastics are frequently detected in the gastrointestinal tracts of wild-caught fish (Lasdin et al., 2023; McIlwraith et al., 2021; Munno et al., 2021; Zhu et al., 2021). Microplastic ingestibility is first dictated by particle size as fish may only ingest particles that are smaller than their maximum mouth opening in at least one dimension. For instance, if a larval fish has a mouth opening of approximately 200 μm, it may theoretically consume particles 200 μm or smaller. It is also important to note that some particles may only be ingested in specific orientations. The length of a microplastic fiber may exceed the mouth opening size, but if the fiber is oriented longitudinally or folded over, it may be ingested (similar to slurping a spaghetti noodle) (Koelmans et al., 2020).

### Consumption during Feeding

Microplastics are much more likely to be ingested in the presence of food than when presented alone (Kim et al., 2019; Ory et al., 2018), and specific types of MNPs are more readily consumed, particularly if they resemble familiar food items. Several laboratory studies have demonstrated species-specific preferences for certain types of microplastics (Ory et al., 2018; Xiong et al., 2019). For instance, a study by Xiong et al. (2019) found that goldfish (*Carassius auratus*) preferentially consumed black and green microplastics as opposed to those that were red, blue, or white. Fish were also more likely to consume films over fragments or fibers, possibly due to the resemblance with flake fish feeds typically used in laboratory settings. However, microplastics greater than 2 mm were generally avoided or rejected, suggesting larger plastics may be more easily recognized as non-food items. Other studies have also observed fish actively rejecting microplastics via 'spitting behavior' (Kim et al., 2019). The behavior of particles in the aquatic environment may also influence exposure as some species of fish prefer sinking over floating microplastics

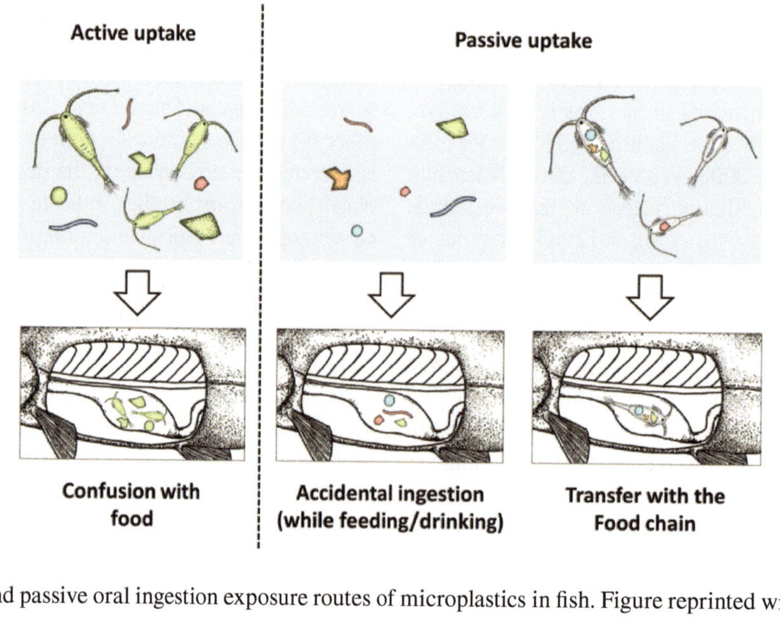


**FIGURE 13.6** Active and passive oral ingestion exposure routes of microplastics in fish. Figure reprinted with permission from Roch et al. (2020).

(Roch et al., 2020), depending on feeding strategy and preferred prey type. Though these studies provide evidence that fish may be more likely to feed upon certain types of microplastics, it is important to note that these preferences are largely driven by foraging strategies and life history traits, which may vary widely across species or even individuals. In a recent study by Roch et al. (2020), the effects of domestication (i.e., wild-caught or laboratory-reared) and foraging style (i.e., visual or chemosensory) on MP uptake were explored. Here, chemosensory foragers and wild-caught fish ingested significantly fewer microplastics than visual foragers and laboratory-reared fish. These results suggest that taste and odor as well as foraging experience are also important factors for differentiating plastics from non-plastics.

Fish may also unknowingly ingest microplastics through trophic transfer. Laboratory studies have found that MNP uptake is much greater via trophic transfer than direct ingestion when particles are presented in the water column (Athey et al., 2020; Hasegawa & Nakaoka, 2021; Stienbarger et al., 2021). For example, inland silverside larvae (*Menidia beryllina*) had almost twice the number of microspheres in their gut when they were presented with single-celled ciliates laden with microplastics than when they were exposed directly (Athey et al., 2020), suggesting that trophic transfer may be a significant exposure pathway for microplastics in fish. Given these and other similar findings, some researchers have hypothesized that microplastics may bioaccumulate in the environment (Miller et al., 2020). However, most studies have found no evidence of bioaccumulation in the field. For instance, a recent survey of Lake Simcoe in Ontario Canada found no relationship between trophic level and the concentrations of microplastics in the livers or fillet of seven different species of sportfish (McIlwraith et al., 2021), and these findings have been corroborated by meta-analyses (Gouin, 2020) and modelling exercises (Diepens & Koelmans, 2018) which also find no evidence of microplastic bioaccumulation. The lack of bioaccumulation is most often attributed to the relatively rapid egestion rate of microplastics in most organisms, including fish (Diepens & Koelmans, 2018; Gouin, 2020). However, it should be noted that most field studies, including those focused on potential bioaccumulation, may be undercounting smaller particles (<10 µm) due to analytical limitations (Brander et al., 2020).

### Consumption during Drinking or Respiring

Small microplastics and nanoplastics may also be consumed by fish when drinking. Researchers have suggested that marine species, which must drink more frequently to maintain osmoregularity in high salinities, are more susceptible to this exposure route than freshwater species (Assas et al., 2020; Bour, et al., 2020a; Roch et al., 2020). For instance, when Japanese medaka (*Oryzias latipes*), a freshwater species, and Java medaka (*Oryzias javanicus*), a marine species, were exposed to similar concentrations of polystyrene microspheres, the estimated bioconcentration factor was four times greater in Java than Japanese medaka (Assas et al., 2020). Similarly, modelling exercises that incorporate ingestion, egestion, and drinking rates as well as body mass suggest that microplastic accumulation may occur via drinking in large, marine species at high environmental concentrations (Roch et al., 2020). However, this phenomenon has not been studied extensively, and more research is required to fully understand the relative contribution of drinking relative to other exposure routes.

Fish may also ingest smaller microplastics and nanoplastics during respiration as water is pulled in through the mouth and over the gill tissue. Microplastics are commonly detected in gill tissues of wild-caught fish (Hurt et al., 2020; Koongolla et al., 2020). Laboratory studies in other taxa suggest that microplastics may become lodged in the gill tissues and cause adverse effects on respiratory function (Watts et al., 2014, 2016). Though current understanding of respiration as an uptake route in fish is somewhat limited, existing evidence suggests that microplastics may damage gill tissue as they pass through and ultimately cause adverse effects. Microplastic fibers may be particularly problematic as studies have found structural damage to the gill lamellae of Japanese medaka (Hu et al., 2020), including denuding of the epithelium on gill arches and membrane swelling, as well as increased mucus production, and increased respiratory stress in juvenile black sea bass (*Centropristis striata*) following exposure to microplastic fibers (Stienbarger et al., 2021).

## EGESTION

The egestion of microplastics from the gut following oral ingestion has been extensively studied under laboratory conditions. In general, most species of fish are capable of eliminating microplastics via feces within 24–96 hours following ingestion (Athey et al., 2020; Elizalde-Velázquez et al., 2020; Xiong et al., 2019). Egestion rates are predominantly determined by the characteristics of a given microplastic particle, where more irregularly shaped particles typically take longer to egest. For instance, in laboratory studies, zebrafish (*Danio rerio*) exposed to similarly sized (~20 µm) microplastic fragments, fibers, and spheres demonstrated that fibers had the greatest propensity to accumulate following 7 days of exposure (8.0 µg/mg) followed by fragments (1.7 µg/mg) and spheres (0.5 µg/mg) (Qiao et al., 2019a). The presence of food may also impact egestion rates. Some studies have found that egestion rates do not change in the presence of food (Elizalde-Velázquez et al., 2020), whereas others have found egestion rates increased (Bour et al., 2020b). These conflicting results may be due to differences in species, exposure scenarios, or even the particles themselves.

## TRANSLOCATION

Translocation, or the movement of MNPs from the gastrointestinal tract to other organs, is hypothesized to occur by one of four mechanisms (Figure 13.7, Hirt & Body-Malapel, 2020; Powell et al. (2010)): (1) Endocytosis: Epithelial cells in the gastrointestinal tract lining uptake particles via invagination of the cell membrane. (2) M-cell Uptake: In mammals, M-cells are specialized cells found in gut-associated

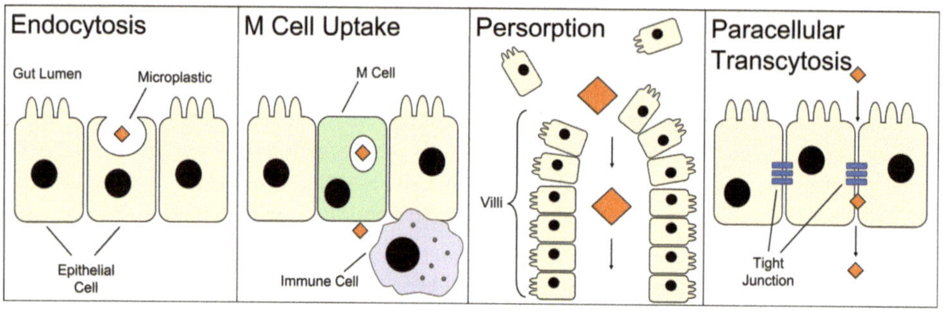


**FIGURE 13.7** Hypothesized mechanisms of translocation for microplastic particles.

lymphoid tissue, namely Peyer's patches, where they play a major role in the activation of the immune system by transporting antigens from the gut to other immune cells. This is considered the most common particle uptake route. In fish, M-type cells have been identified and partially described in teleosts (Kato et al., 2018). (3) Persorption: Particles pass in-between epithelial cells of the gut lining when they are shed from the tip of the intestinal villi. This may be a potential mechanism for the translocation of large particles, but is considered to be inefficient as compared to M-cell uptake. (4) Paracellular Transcytosis: Under specific conditions, very small nanoparticles may slip through the tight junctions between cells, but this phenomenon is often to considered to be rare at best.

In the laboratory, translocation of MNPs from the gut to other tissues has been observed in a wide variety of species. For instance, 10-µm polystyrene particles were detected in the livers of marine medaka (*Oryzias melastigma*) following 60 days of exposure (Wang et al., 2019a). Similarly, 5-µm spheres were also found in the livers of zebrafish but only after 7 days of exposure (Lu et al., 2016). Particles on the nanoscale may also be more likely to infiltrate certain tissues as 0.1-µm and 180-nm particles were found in the brain tissue of red tilapia (*Oreochromis niloticus*) and Crucian carp (*Carassius carassius*), respectively (Ding et al., 2018; Mattsson et al., 2017). Nanoplastics may also be able to penetrate the chorion as 51-nm particles were observed in the gut, gallbladder, liver, pancreas, heart, and brain of embryonic zebrafish (Pitt et al., 2018). Some studies have even reported that polystyrene nanoparticles (~50 nm) may be maternally transferred to larvae due to their affinity for lipophilic proteins such as vitellogenin (Pitt et al., 2018; Rossi et al., 2014; Zhao et al., 2021). Though the aforementioned studies provide evidence of particle translocation, it is important to note that methods for confirming microplastic quantification in tissues are still developing, and some results should be interpreted with caution, particularly those which may rely on techniques using fluorescent particles as fluorophores may leach from particles and result in a false-positive (Schür et al., 2019). For example, translocation of large microplastics have been reported in laboratory studies, but results are often conflicting (Jovanović et al., 2018; Kim et al., 2020). The use of tissue-clearing approaches to determine and better visualize particle presence may be preferable, particularly for very small specimens, and reduce the possibility of false-positives (Siddiqui et al., 2022).

In the field, there is also evidence of translocation as microplastics have been found in the fillet (Abbasi et al., 2018; McIlwraith et al., 2021) and livers (Abbasi et al., 2018; Collard et al., 2017, 2018; McIlwraith et al., 2021) of wild-caught fish. Interestingly, translocated particles in the field are often much larger than those reported in laboratory studies. For example, particles in the livers of European anchovies (*Engraulis encrasicolus*) ranged from 124 to 438 µm (Collard et al., 2017), and particles from the livers and fillet of a variety of sport fish species sampled from Lake Simcoe ranged from 12 µm to over 5 mm, though it is important to note that larger particles were most often fibers (McIlwraith et al., 2021). In addition to differences in size, field studies also provide evidence that particles with irregular morphologies may also translocate. Most laboratory studies focused on translocation use spheres (Ding et al., 2018; Lu et al., 2016; Mattsson et al., 2017; Wang et al., 2019a), whereas field studies provide evidence of the translocation of fibers, fragments, and even films (Collard et al., 2017; McIlwraith et al., 2021). Though the reasons behind these differences are partly attributable to particle selection (i.e., using spheres for translocation studies rather than fragments or fibers), some have hypothesized that laboratory studies do not observe the translocation of larger particles due to relatively short exposure times as compared to the lifelong exposures of wild-caught fish (McIlwraith et al., 2021).

Current knowledge suggests that the likelihood of a microplastic particle to translocate is predominately dictated by particle size as the frequency of translocated particles often increases with decreasing particle size (Wright et al., 2013), but other particle characteristics may also influence the possibility of translocation. For instance, McIlwraith et al. found that while most tissues were dominated by fibers. However, approximately a third of the particles in the fillet were fragments, and foams were only observed in the liver and fillet but not in the gut (McIlwraith et al., 2021). This suggests that particle morphology heavily influences translocation. The weathering of particles may also play a role as microplastics incubated in fresh or saltwater were internalized by cells at ten times the rate of microplastics that were incubated in sterile water (Ramsperger et al., 2020).

## MOLECULAR-LEVEL EFFECTS ON FISHES

The study of molecular responses to MNP exposure remains in its infancy, as many of the early studies in fishes focused primarily on apical endpoints such as growth and reproduction (Brander et al., 2004). To date, studies evaluating gene expression and other biochemical markers like reactive oxygen species (ROS) production suggest particle-mediated toxicity as the dominant mechanism, which is likely a result of the size-dependent interaction of these three-dimensional foreign objects with tissues and cells (Jacob et al., 2020; Thornton Hampton et al., 2022). This contrasts with the specific receptor-mediated toxicity observed following exposure to soluble chemicals such as plasticizers, which are capable of either binding to membrane receptors or crossing lipid membranes to interact with transcription factors such as nuclear receptors, or the active sites of enzymes (Brander, 2013; Kar et al., 2021). Advanced tools enabling the derivation of QSARs (quantitative structure activity relationships) and read-across approaches used to predict molecular interactions with soluble compounds (e.g., (McArdle et al., 2020; Patlewicz et al., 2017) simply are not yet available for MNPs.

### Potential Mechanisms of Toxicity

As a field, our current understanding of these less targeted interactions is limited and makes it challenging to pinpoint biomarkers of exposure and effect for MNPs. Evidence supports the assertion that as MNPs become smaller, they are more likely to cause oxidative stress and other molecular-level perturbations (Mehinto et al., 2022), because they are more readily taken up by phagocytes and transported across epithelial membranes into tissues and at the nanoscale, into cells. In fact, for MNPs, the mechanism of greatest concern may simply be the process of translocation and entrapment of particles internally causing oxidative damage via the generation of reactive oxygen species (Jacob et al., 2020; Lu et al., 2016; Yu, 1994; P. Yu et al., 2018). The magnitude of the effect is likely based on the total surface area of the particle(s), which increases as overall size decreases. The markers of exposure indicated across studies thus far include the generation of ROS, with hydroxide radicals or ions (OH–), hydrogen peroxide ($H_2O_2$), and superoxide anions (O–), among others, commonly measured in MNP studies on fishes (Jacob et al., 2020) as well as in other taxonomic groups (not covered in this chapter). ROS production is often a non-specific reaction to cellular stress and in healthy organisms is followed by the transcription and translation of antioxidant scavenging enzymes like superoxide dismutase (SOD) or peroxidases, which serve to neutralize free radicals and/or ions to protect cellular structures and organelles (Yu, 1994). Alternatively, ROS production can be an early sign of toxicity and may either precede or result from alterations in diet leading to reduced caloric intake and alterations in metabolic state, and linked to food dilution when MNPs are present in the digestive tract (Jacob et al., 2020; Koelmans et al., 2020) discussed in detail above. In fishes and other vertebrates, ROS production is associated with apoptosis, inflammation, and changes in lipid metabolism (Barboza et al., 2018; Cedervall et al., 2012; Jacob et al., 2020; Qiao et al., 2019a; Wan et al., 2019).

### Molecular and Cellular Alterations

As with other toxicants, molecular- and cellular-level changes following MNP exposure often provide the first indication of a potentially adverse response. For example, impacts on gene expression, hormone levels, and lipid or protein levels occur across studies in two commonly tested species, Japanese medaka and marine medaka – *Oryzias latipes, melastigma* (Assas et al., 2020; Chen et al., 2020; Pannetier et al., 2020; Wan et al., 2019; Yan et al., 2020). In marine medaka, following exposure to PS particles, males showed significant differences from controls in estradiol (E2) and testosterone (T) at the highest concentrations (200 mg/L) and females showed changes in E2, T, and E2/T (Wang et al., 2019a). This study notably produced some of the lowest NOEC (no effect concentration) levels in a recent risk assessment of aquatic organisms that used species sensitivity distributions to determine thresholds for MNPs in ambient waters (Mehinto et al., 2022). In *D. labrax* (European sea bass), NP exposure caused altered gene expression in two different peroxisome proliferator receptor isoforms: PPARa when exposed to concentrations of 2 mg/L, PPARy when exposed to 0.2 and 2 mg/L, and NADH dehydrogenase subunit 5 which increased when exposed to circular and non-uniform PPMA at concentrations of 0.02, 0.2, 2, and 20 mg/L, while oxidative stress significantly decreased at 0.02 and 0.2mg/L (Brandts et al., 2018). Changes in genes such as PPAR indicate alterations to lipid metabolism, and NADH dehydrogenase represents a critical step in the electron transport chain that facilitates mitochondrial respiration. Together their expression points to altered metabolic capacity in *D. labrax*. In another study with the same species that used plastics collected from the marine environment ground to <100 μm, oxidative stress-related enzymes catalase (CAT), glutathione-S-transferase (GST), SOD, and thiobarbituric acid-reactive substances (TBARS), a by-product of lipid peroxidation, all significantly increased following MP exposure (Zitouni et al., 2021).

Other genes related to oxidative stress are commonly upregulated in response to MNP exposure. Expression of the oxidative stress gene (nrf2) increased in gilthead seabream (*S. aurata*) in the presence of 40–150 μm particles of PE and PVC at 100 mg/mL (Espinosa et al., 2018). Nrf2 was also upregulated in mammals exposed to PS MNPs, in combination with other responses encompassing inflammation, apoptosis, and generalized oxidative stress response (Li et al., 2021). A large array of oxidative stress markers including SOD, malondialdehyde (MDA), ethoxyresorufin-O-deethylase (EROD) – used to measure Cyp1a activity – benzyloxy-4-trifluoromethylcoumarin-O-debenzyloxylase (BFCOD) – used to measure CYP3A activity – and acetylcholinesterase (AChE), were altered by MNP exposure in *Oreochromis niloticus* in a size- and time-dependent manner relative to the control (Ding et al., 2020). Alterations in liver function, kidney function, and biochemical changes including glucose, albumin, total protein, globulin, albumin:globulin ratio, and cholesterol, which was dependent on exposure/recovery period and concentration, were also altered following MP exposure in the same species (Hamed et al., 2019), and point to changes in digestive processes potentially caused by food dilution.

More traditionally assessed detoxification pathways were also affected by MNP exposure in fishes. Mazurais et al. (2015) found that expression of the detoxifying enzyme Cyp1a1 had increased in *D. labrax* following exposure to 10–45 μm PE beads at 20 days post-hatch (dph); however, the cytokine il-1 beta, indicative of immune response, was not significantly changed at either 27 or 43 dph. Cyp1a1 often provides the first line of defense against xenobiotics, although in terms of MNPs it is important to assess the toxicity of additives or sorbed pollutants and the particle itself separately, which may not have been done in this study, since responses may be from chemicals associated with the MNPs, the particles themselves, or a combination of the two. However, activation of detoxification pathways also occurs in freshwater fishes, with EROD activity increasing and CYP1A1 differentially expressed at 0.33 g/L and 1.0 g/L concentrations of 38–355 μm micronized crumb rubber in mummichogs (*Fundulus heteroclitus*) as well as being elevated in fathead minnows (*Pimephales promelas*) exposed to 0.3, 1.9, and 6.0 g/L of the same particles compared to the controls (LaPlaca & van den Hurk, 2020). Assessing changes in gene expression helps to elucidate the underlying biological pathways causing toxic responses to MNPs, and collectively, similar responses across fish species contribute important information that can eventually be used to determine the underlying mechanisms causing adverse outcomes in fishes across different polymer types, shapes, and sizes.

## Putative Adverse Outcome Pathways

At the level of gene expression, discerning the difference between protective responses and those signaling toxicity that may impact organisms at higher levels of organization is challenging. However, understanding the downstream effects of damage-associated molecular patterns (DAMPs, Villeneuve et al., 2018), which appear to be triggered by MNP exposure and generally lead to an inflammatory response, allows for more accurate documentation of the events that can precede cell damage and even dysfunction at the tissue or organ levels. The challenge is that these types of responses are pleiotropic and can vary depending on the tissue or cell-type being impacted, and are also dependent on the duration of the stress-inducing event (Schaefer, 2014; Villeneuve et al., 2018), and likely the physical characteristics and concentrations of the MNPs too (described above). Although the lack of specific cellular or molecular mechanisms currently stymies the development of an AOP describing MNP toxicity, researchers have begun to lay down a roadmap (Jeong & Choi, 2019, Figure 13.8).

Adverse outcome pathways are used to conceptually describe the events leading to a toxicity outcome, beginning with a molecular initiating event (MIE) and followed stepwise by key events occurring at each tier of the biological hierarchy (cell, tissue, organ, organism, population; (Ankley et al., 2010). A recently published particle-focused putative AOP, incorporating studies on the fish species *Mullus surmuletus*, zebrafish, gilthead seabream (*Sparus aurata*), sheepshead minnow (*Cyprinodon variegatus*) European sea bass, common goby (*Pomatoschistus microps*), put forth evidence for several potential key events that underlie the adverse responses documented in MNP-exposed organisms, these include metabolic and inflammation-based biomarkers. The authors stop short of suggesting any molecular initiating events (MIEs), which falls in line with the current understanding that MNP toxicity is induced by non-specific interactions with tissues and cells. Possible key events include oxidative stress (KE1392), ROS generation (KE1278), increased inflammation (KE149), and lipid peroxidation (KE1511) (Jeong & Choi, 2019). Notably, responses described in the collection of mostly fishes, with some invertebrates included as well, correspond with existing or drafted AOPs describing adverse effects induced by particles or other non-specific stressor exposures. This includes AOP 173 which describes the events underlying an outcome of lung fibrosis, or proposed pathways triggered by oxidative stress (e.g., AOP 151) resulting immune or cytoprotective responses. Adverse outcomes and biomarkers associated with microplastic exposure also align to some extent with what is known about nanoparticles (e.g., $TiO_2$), with KEs of oxidative stress induction and inflammation in the liver and digestive tract (Brand et al., 2020). Given all of the uncertainties and data gaps, it is not possible to suggest MIEs at this juncture, and due to the nature of MNPs and the non-specificity of their interactions, it may not be advisable. However, consideration of how MNP-induced toxicity aligns with our existing understanding of the roles inflammation, oxidative stress, and metabolic alteration interplay to cause particle toxicity is critical, as is understanding the difference between localized and systemic effects.

## Microbiome Effects

Given that metabolic alterations are a common response to MNP exposure (Jacob et al., 2020), interactions between MNPs and the microbiome of fishes are now being explored. Zebrafish are the most extensively studied fish in terms of gut microbiota. Studies using polystyrene and polyethylene MPs observed microbiome dysbiosis and other alterations to the metabolic profile of exposed fishes (Qiao et al., 2019a; Wan et al., 2019; Zhao et al., 2021). Similar to studies on molecular responses to MNPs, two of the most commonly studied marine fishes are European sea bass and marine medaka. A 2018 study looking at the gut microbiota of *D. labrax* in response to MNPs found minimal impacts – there was a 94–96% similarity in the microbiome depending on exposure time and whether the plastics were weathered or not, after 90 days of exposure (Caruso et al., 2018).

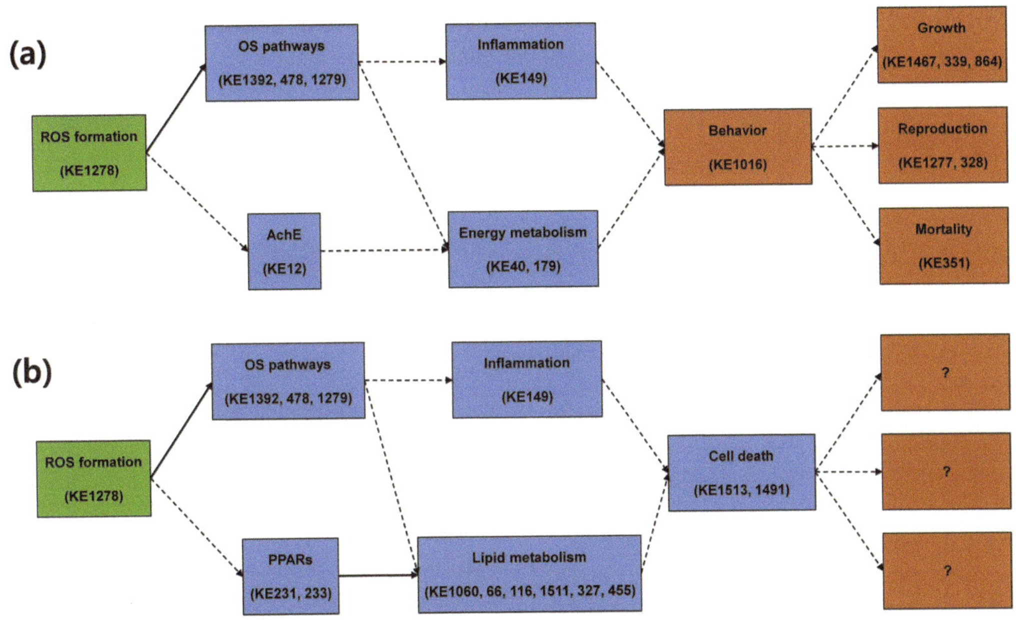


**FIGURE 13.8** Schematic diagram of the putative AOPs for microplastics. The AOPs for ecotoxicity (a) and human health implications (b). Green boxes: molecular initiating events, blue boxes: key events, orange boxes: adverse outcomes, solid lines: strong evidence, dashed lines: weak evidence (ROS: reactive oxygen species; OS: oxidative stress; AchE: acetylcholinesterase; PPARs: peroxisome proliferator-activated receptors). Reprinted from Jeong & Choi. 2019.. Adverse outcome pathways potentially related to hazard identification of microplastics based on toxicity mechanisms, with permission from *Chemosphere*.

However, there was a significant change in heterotrophic bacteria with time in the microbiome. In marine medaka, the diversity and abundance of intestinal microbiota were reduced following exposure to PS MPs (2.5 μm) (Yan et al., 2020). These microbiome-specific impacts may also be size-dependent, as demonstrated recently also in marine medaka. Larger 200-μm polystyrene (PS) MPs caused gut microbiota dysbiosis, which led to disordered lipid metabolism, while smaller 2- and 10-μm PS MPs triggered hepatic inflammation (Zhang et al., 2021). In zebrafish, gut-based toxicity was shape dependent, with fibers accumulating more than fragments or beads, and resulting in more severe intestinal toxicity (Qiao et al., 2019a). These changes can be problematic as different microbiota have specific functions that assist with digestion and nutrient absorption, and affect processes such as hormone circulation and even neural circuitry, and thus, impacts on the microbiome can have systemic and complex implications for fishes, including an increased likelihood for impacts on growth and other apical endpoints.

## Epigenetics, Genetic Variation, and Genotoxicity

As shown with other classes of pollutants (Bacchetta et al., 2017; Beedanagari et al., 2014; Bhandari, 2016; Brander et al., 2017; DeCourten et al., 2020), MNPs can potentially cause multigenerational effects, influence genetic variation, and cause genotoxicity. Thus far possible epigenetic effects are only demonstrated in invertebrates, presumably because the investment in and length of experiments are reduced in comparison with comparable work in fishes. A handful of publications suggests there could be cause for concern. For example, marine copepods were adversely impacted in the F1 generation from parental MP exposure, with energy storage capabilities being reduced, but recovery being observed in the F2 (Zhang et al., 2019). Although no true multi-generational experiments (to the F2) are yet published on fishes, there is evidence that fish may not necessarily adapt in response to selective pressures placed on their parents. Huuskonen et al. (2020) conducted a breeding experiment in European whitefish to test whether parental exposure to PS spheres affected the feeding behavior of their offspring. PS ingestion by parents showed no significant effect on additive genetic variation, implying that the offspring cannot adaptively avoid ingesting MPs. There is evidence published by the same group that sperm exposure to NPs prior to fertilization can affect both offspring size and swimming performance (Yaripour et al., 2021). Finally, genotoxicity, or physical damage to DNA (e.g., strand breaks) that can cause errors in replication and increased likelihood of mutation, can increase in the presence of ROS, shown with nanoparticles (Yu et al., 2020), and already demonstrated in many studies to occur in response to MNP exposure (Jacob et al., 2020; Thornton Hampton et al., 2022). Notably, Singh et al. (2021) saw indications of genotoxicity in zebrafish, such as increased TDNA% (percentage of DNA in the tail) and raised LPO levels, following exposure to NPs (0.055, 0.1 μm) at 1 mg/L for just 5 days. A synthesis by Sun et al. (2021) suggests that these responses hold across aquatic taxa and that the potential genotoxicity of MNPs to fishes was higher compared to aquatic invertebrates. Further research is needed to discern whether these observed responses are adaptive or if they indicate adverse effects that could influence fitness and population resilience.

## ORGANISMAL-LEVEL EFFECTS ON FISHES

### Particle Toxicity

Although the study of molecular impacts from MNP exposure is relatively nascent, it is well established that acute or chronic exposure to microplastics can lead to a variety of adverse effects at the organismal level in fish. A recent review by Jacob et al. (2020) analyzed data of 46 published studies where they found a total of 782 investigated endpoints in fish after virgin micro- and nanoplastics exposure experiments. Thirty-two percent of all studied endpoints were significantly affected by micro- or nanoplastics exposure. These include changes in natural behavior, fecundity, reproductive success, body condition, growth, and development on the organismal level of fish in all life stages.

Adverse effects on the organismal level of fish have been observed in several studies, but it must be pointed out that the concentrations of microplastic particles, as well as the size distribution and polymer composition, often are not environmentally relevant in the applied exposure scenarios (Bucci et al., 2019; Gouin, 2020). However, application of higher concentrations of MNPs can be used to clarify whether microplastics pose a threat to fish at predicted future concentrations, to aid in understanding potential mechanisms driving toxicity, and then to find the respective limits or thresholds (e.g., Adam et al., 2021; Mehinto et al., 2022).

There are currently a number of studies indicating alterations in behavior by microplastic exposure, and these changes can pose a danger to the survival of individuals, and consequently whole populations of fish. For example, exposure to MNPs can affect swimming behavior in fish. Swimming, or locomotion, has been indicated as an important endpoint in ecotoxicological studies (Ford et al., 2021), and is an endpoint that the Organisation for Economic Co-operation and Development (OECD) recommends observing in the fish acute toxicity test (OECD, 1992). Limonta et al. (2019) food-exposed adult zebrafish to a 50:50 mix of high-density polyethylene (HDPE) and polystyrene (PS) microplastics in a size distribution of 90% <90 µm; 50% <50 µm; 10% <25 µm in two concentrations (100 and 1,000 µg/L) for 20 days, which led to an altered daily activity rhythm. The control fish synchronized their daily rhythm with locomotor activity to the 12:12 hours light:dark cycle in the test setup with a higher activity during the light phase. The fish exposed to MPs, however, showed a 10–20% higher nocturnal activity.

Changes in swimming behaviors and activity can also be connected to feeding and foraging behaviors, both of which were affected by exposure to MNPs in various fish species. Common goby juveniles exposed to MPs were found to confuse the plastic particles with their natural prey and displayed reduced predatory performance and efficiency (de Sá et al., 2015). Additionally, the authors found a potential influence on early larval developmental and sensitivity to MP toxicity. Fish which were raised in a polluted river demonstrated stronger adverse effects than those in a cleaner environment in the MP exposure experiments (de Sá et al., 2015). Juvenile marine jacopevers (*Sebastes schlegelii*) were exposed to PS MP beads with a 15 µm diameter, in a concentration of $10^6$ microspheres/L. After 14 days, exposed fish spent twice as much time feeding compared to the control fish, indicating decreased feeding efficiency. Additionally, foraging time was reduced. Fish exposed to PS swam shorter distances and used less space in their tank while they decreased the distance within their school. This was interpreted as an impaired hunting behavior of the MP-treated fish (Yin et al., 2018). Siddiqui and colleagues exposed inland silversides to tire wear particles in micro- (1–20 µm) and nanosize (<1 µm) fractions, at mass concentrations comparable to those found in the environment. Thigmotaxis, or area of fish habitat occupied, and the amount of random movement (e.g., meander, freezing, turn angle) were impacted by TWP, suggesting a stress response to the presence of rinsed particles. Leachate from the same tires' particles were created from did not impact fish behavior as significantly as TWP did (Siddiqui et al., 2022), suggesting that particle-based toxicity was more important in this species, in contrast to studies on leachate in salmonids showing lethality at environmentally relevant levels (Tian et al., 2021). Similar effects on behavior have also been observed with microfibers (Siddiqui et al. 2023).

The impact of microplastics on a population of juvenile ambon damselfish (*Pomacentrus amboinensis*) was studied in interaction with the degradation of their natural habitat, coral reefs in the Indo-Pacific (McCormick et al., 2020). The fish were exposed to 200- to 300-µm transparent PS beads in a concentration of 167 particles/L, which is similar to a relevant concentration in an area near the study region, for 4 days in presence of *Artemia naupli*i, under laboratory conditions. One hour after the last feeding trial, each tagged fish was introduced to a natural reef area, which contained either live, healthy corals (*Pocillopora damicornis*) or dead-degraded corals. The behavior of the individual fish was observed over 3 minutes, regarding four different behavioral endpoints: bite rate, total distance moved in the reef, maximum distance ventured from the habitat reef spot, and boldness after being approached by a pencil tip. Additionally, survival of the fish was monitored over 3 days, and it was assumed that fish that moved far from the home reef were caught by predators. The results demonstrated an interaction between effects of a plastic exposure and the behavior of the fish in a healthy versus a degraded coral reef. Fish which were fed with MPs later showed a higher activity and boldness in the coral reefs. Additionally, both exposed and unexposed fish on dead corals had an even higher activity level than fish on healthy corals. Fish on healthy corals were less bold, so they stayed closer to their home coral, which increased their survival chances significantly. The fish exposed to MPs and placed on dead corals showed the highest boldness and distance moved away from the corals. This behavior led to a mortality of this group of 100%, even before the end of the 72-h observation period was reached. The survival of non-exposed fish was six times higher in a healthy coral reef. This study demonstrates that MP-exposure can lead to behavioral changes, including increases in risky behavior (McCormick et al., 2020).

Other studies have shown that exposure to plastic particles can affect growth parameters in fish. Juvenile planktivorous reef fish (*Acanthochromis polyacanthus*) were exposed to polyethylene terephthalate (PET) particles in the size of 1–2 mm. After 1 week, the fish exposed to plastic particles showed decreased growth compared to the control fish. In a second part of the study, the plastic particles were mixed with commercial fish food, and then the spiked food was exposed to salt water to induce a growth of a microbial biofilm on the surface of the microplastics for an environmentally relevant exposure. The fish fed with plastic spiked food caught up to the control fish and no longer displayed the decrease seen after the acute exposure. In general, there was a decrease in body condition after chronic exposure to a high concentration of MPs, as evident in the hepatosomatic index. There was also an observable trend toward a decline in aggression with increasing plastic concentration (Critchell & Hoogenboom, 2018). Polyvinylchloride (PVC) MPs, 100–200 µm in diameter, were used in feeding experiments with (*Cyprinus carpio var.*) larvae, resulting in both an inhibition of weight gain and decreased length growth in carp, after a chronic 30- to 60-day food exposure at three concentrations (46 µg, 91 µg, and 137 µg of MPs/L water (Xia et al., 2020). Similar results were obtained in a study on juvenile Glassfish (*Ambassis dussumieri*) in chronic exposure over 95 days to environmentally relevant concentrations of different virgin and harbor-water-exposed plastic pellets. Authors reported a decrease in body length and body depth growth, in plastic-exposed fish compared to the control group. Notably, exposure to virgin pellets also induced a decrease of fish mass. Ultimately, both microplastic exposures resulted in an increased mortality of the juvenile fish over the exposure duration (Naidoo & Glassom, 2019). The study on behavioral impacts of TWP with inland silversides described above observed that growth in larval fishes was reduced after a brief exposure (Siddiqui et al., 2022), as did a study by Athey et al. (2020) which found that a 2-hour exposure to high levels of PE microspheres caused reduced mass in larval fish (also silversides) reared for 2 weeks following MP ingestion and excretion. These impacts on growth may be related to some of the changes in gene expression reported above, particularly for studies demonstrating changes in metabolic markers or in the microbiome (Jacob et al., 2020).

Whereas these studies all demonstrated possible effects of an MNP exposure, there are also indications that MNPs may have no influence on body parameters of fish. Juvenile gilthead seabream were exposed to low-density polyethylene (LDPE), which had no effect on growth or health parameters in long-term study over 90 days with 30-day depuration phase, though oxidative stress was evident in the fish after 90 days (Solomando et al., 2020). This may be an indication that sometimes oxidative stress and the ensuing antioxidant response can be adaptive. In addition, fathead minnow (*Pimephales promelas*) larvae were not negatively impacted by MP exposure with regard to their growth or feeding behavior after 30 days. The plastic particles used there were 180- to 500-µm PE spheres, a size range that is equivalent to the natural prey size for these larvae. The applied experimental concentrations were higher than MP densities that larvae would be exposed to in the environment (0, 50, 100, and 500 particles/L). Both larvae age groups of 7 and 14 dph (days post-hatching) showed no significant difference in the tested end points (*Artemia* nauplii consumption and body growth) between the control and exposure groups (Malinich et al., 2018). There are many parameters that will impact the exposure and ingestion of MPs by different fish species, some more or less sensitive than others, and the subsequent impacts of behavior and growth, so more research is required to determine the variables responsible for the observed effects, or lack thereof, and to standardize assay design so responses between studies are comparable (e.g., Cowger et al., 2020; Gouin, 2020; Thornton Hampton et al., 2022).

## Plastic Additive and Sorbed Pollutant Toxicity

It is known that some of the health risks from microplastics do not necessarily come from the plastic polymers themselves. As described above, plastic particles in the environment can contain additives that were used in the manufacturing processes, as well as chemical or biological (e.g., bacteria or fungi) pollutants from the environment that can adhere to the particles. While the full potential of MPs to act as vectors delivering these chemicals into fish is yet to be determined and the contribution of an individual particle's pollutant load minimal (e.g., Koelmans, 2015), others cited earlier in chapter), plastic-associated chemicals are ubiquitous in the environment in measurable concentrations in water, sediment, and prey (see review by Hermabessiere et al., 2017) and there are a number of studies addressing impacts of exposure.

An elaborate study, Rochman et al. (2013) examined the effects of virgin LDPE particles on fish (Japanese medaka) health, in comparison with the same particles which had previously been exposed in situ to the mixture of chemicals present in harbor water for 3 months. The occurrence of the components PAHs, PCBs, and PBDEs was investigated in the virgin and contaminated pellets. In total, six PAHs, ten PCBs, and seven PBDEs were detected in the pellets, but with different concentrations in both conditions (virgin and exposed) of pellets. The concentration of PAHs, PCBs, and PDBEs was 4, 15, and 1.4 times higher in the harbor water-exposed pellets than in the virgin pellets. After the dietary exposure, these chemicals were also transferred to the fish tissue. The exposure was long term, resulting in low doses of chemical compounds found in the fish. Measured effects on the fish were sub-lethal, though the mortality in the contaminated pellet group was 6%, while it was 4% in the virgin pellet and control diets. After being fed with both virgin and pollutant-sorbed pellets, Medaka showed impaired liver health including glycogen depletion, fatty vacuolation, and single-cell necrosis. There were differences in the severity of the liver effects in the treatments. After exposure to pollutant-sorbed pellets, 74% of fish showed severe glycogen depletion, 47% had fatty vacuolation and 11% single-cell necrosis in their liver. A diet with virgin pellets on the other hand caused severe glycogen depletion in 46% of fish, fatty vacuolation in 29%, and single-cell necrosis in 0%. In the control group, none of the fish showed signs of severe glycogen depletion or necrosis, but 21% had fatty vacuolation in the liver. The effects on the liver health of medaka were therefore stronger in the higher polluted particle exposure, clearly showing that pollutants adhering to microplastics can be transferred from the environment to fish and can then lead to health risks (Rochman et al., 2013).

In addition to the common persistent organic pollutants studied there, a large number of the additives that are, or have been, used in the production of plastic products are identified as toxic. Often associated with plastics are perfluoroalkylated substances (PFAS) (Glüge et al., 2020; Timshina et al., 2021). The potential toxicity of plasticizers, flame retardants has been described for a wide variety of organisms, including fish as well as humans (Lyche et al., 2015; Voelker et al., 2021). Possible toxicological effects of substances associated with plastic pollution in environmentally relevant concentrations were studied in various fish species which showed abnormal morphological development, impaired reproduction, mortality, and behavioral changes (Corradetti et al., 2013; Jantzen et al., 2016; Kinch et al, 2016; Van Essen et al., 2021; Usenko et al., 2016). Adverse effects can be additive, when fish were exposed to a mixture of pollutants, such as it would occur in their natural habitats (Kinch et al., 2016).

While it is clear that additives in plastic can have adverse effects on fish health, the question arises as to whether fish can perceive the presence of these chemicals and even avoid ingesting the polluted microplastic particles. Numerous studies have addressed chemical avoidance or attraction in fish, though the sheer numbers of chemical contaminants, which exceeds 1000s of substances, and the complexity of potential responses and underlying modes of actions, makes it difficult to draw general conclusions (see review by Tierney, 2016).

The odor (chemicals) extracted from virgin plastic and marine plastic debris was tested in comparison on the foraging behavior of the adult anchovy (*Engraulis* sp.). A filtered seawater solution was used as a control, and food odor was made of homogenized pacific krill (*Eu. Pacifica*). For the marine plastic debris odor solution, polypropylene (PP) plastic beads in the size range of 4–6 mm in diameter were left to biofoul for 3 weeks in the sea. The same plastic beads without the fouling step were used for the virgin plastic extracts. Odor extracts were injected into the tanks of adult anchovy schools. The anchovy schools showed a significantly increased aggregation and clustering behavior in response to the injection of the biofouled plastic debris odor in their tanks, compared to both the control and virgin plastic odor. In fact, anchovy schools reacted to the marine plastic debris odor in a similar manner to the food odor. The anchovy schools showed a positive rheotaxis (orientation toward oncoming water flow), prior to any introduction of odors, after introduction of the control seawater solution and the virgin plastic odor. An injection of the marine plastic debris odor induced a similar negative rheotaxis response as the food odor (Savoca et al., 2017). This study demonstrated how fish could confuse the odor of biofouled, weathered and pollutant exposed plastic, possibly by accumulation of biota and/or sorbed pollutants, as a food source and potentially ingest plastic particles which have been floating in the water, exposed to a variety of pollutants.

## POPULATION-LEVEL EFFECTS ON FISHES

### Particle Toxicity

At the population level, the negative effects of microplastics are directly apparent as a threat to the survival of the whole fish population, and can manifest in decreased survival, fitness, nutritional status, and reproductive output in addition to ecosystem impacts changing predator prey interactions. The ingestion of MNPs by fish is indisputable and it is also proven that fish are the main consumer of microplastic among all animal groups with the highest total number of species ingesting macro- and microplastic in the literature from 1980 to 2021 (Santos et al., 2021). This leads to the assumption that fish can be affected by microplastic exposure even on the population level. Santos et al. (2021) describe the exposure to plastic as an 'evolutionary trap', as it triggers behavioral changes in feeding strategies and a confusion of the plastic to food. However, few studies have demonstrated concrete population effects of MP exposures in fish (Gove et al., 2019; McCormick et al., 2020), though several studies have investigated this in invertebrates, algae, and plants (see review by Bucci et al., 2019).

The occurrence of MP particles in the habitat of larval fish nurseries, and the risk of reduced survival of the larval fish, was studied in ambient and slick water in the coastal ocean in Hawaii (Gove et al., 2019). Larval fish feed on phyto- and zooplankton in surface waters, so their occurrence in relation to MP particles, and the ratio of larval fish to MPs, can greatly impact interactions between the fish and particles. This study revealed that slick water contained a higher density of phyto- and zooplankton than close-by ambient waters, which is in line with the hypothesis that slick waters function as larval fish nurseries, and evident in the significantly higher number of larvae compared to ambient waters (2.7–28 times, depending on the fish family) (Gove et al., 2019). As the slick waters are important habitats for fish larvae, but also prone to accumulation of plastic garbage due to runoff (median plastic density in this study was 126 times higher in slick than in ambient water), larvae risk exposure to MPs in the size range of their prey which may be ingested when feeding on phyto- and zooplankton. The average larval fish was 6.1 mm in length in surface slicks, and the majority of sampled plastic here were <5 mm fragments. The ratio of plastic particles to larval fish density was 7:1, while the ratio of prey-size plastics to prey for larvae in slicks was 1:55. Most abundant types of polymers found, listed in order of occurrence, were LDPE, unknown PE, HDPE, PP, PP/PE, and others. It was shown that the amount of MPs in larval fish nurseries is higher than the larvae themselves and their prey. This results in a high chance of larvae ingesting microplastic particles. Ingestion could ultimately impair the survival of populations of fish (Gove et al., 2019).

There are several studies investigating the potential reproductive effects of MNP exposures, which would endanger fish populations. The following three studies focused on polystyrene particles and their effects on the reproductive system of fish. Adult zebrafish were exposed to a concentration of 0, 10, 100, or 1,000 μg/L of polystyrene particles with a diameter of 1 mm in their tanks for 21 days. The exposure of the two higher concentrations of microplastic particles led to increased ROS levels in gonads of both male and female zebrafish. Additionally in the testes of male fish, increased apoptosis was observed after higher polystyrene microplastics exposure levels (Qiang & Cheng, 2021; Wang et al., 2019a) demonstrated similar effects on the gonads of marine medaka after polystyrene particle exposure for 60 days. Concentrations of 2, 20, and 200 mg/L particles delayed the gonad maturation and decreased the fecundity and disruption in steroidogenesis in female medaka. This study investigated the transgenerational effects of this exposure. The exposure of 20 mg/L plastic particles to parental fish caused a decreased hatching rate, heart rate and body length in larvae of the next generation (Wang et al., 2019a). In the study by Zhu et al. (2020), Japanese medaka were fed food spiked with 500, 1,000, or 2,000 μg/g 10 μm diameter polystyrene particles for 10 weeks during their transition phase from juveniles to mature adult fish. After the exposure to PS microplastics, female medaka showed a decreased egg number, in a dose-dependent manner (Zhu et al., 2020).

The study by McCormick et al. (2020) which is described in detail in the previous section on behavioral effects on organismal levels revealed that ambon damselfish which were exposed to PS beads showed a higher activity and boldness in coral reefs. This risky behavior ultimately led to a 100% mortality of this group. It demonstrates that MP exposure can lead to behavioral changes, including increases in risky behavior, and that this type of microplastic pollution, together with coral degradation, becomes a serious hazard to the fish population survival (McCormick et al., 2020).

### Plastic Additive and Sorbed Pollutant Toxicity

Exposure to additives and pollutants that are associated with MNPs has the potential to have an impact on the population level of fish. It is important to remember here that the potential for MPs to act as a vector for these compounds, and therefore as a drive of these impacts, is thought to be limited. While exposure routes to plastic-associated chemicals are many, evidence does indicate that exposure to many substances can result in changes that can affect population health in fish, including impaired spermatogenesis, reproduction, and embryo development.

Most fish in the environment will endure chronic exposures to diverse pollutants and stressors during their life cycle, rather than acute, short-term exposures, and this is especially important to study when addressing population effects. A chronic exposure was previously defined as an exposure lasting minimum 10% of the lifespan of the test species (Sprangue, 2005), whereas most chronic studies would not meet this definition. To study the chronic toxicity effects of pollutants sorbed to MP particles, adult Japanese medaka were exposed to with virgin PE pellets, or to pellets that had been deployed *in situ* in harbor water for 3 months (marine-plastic pellets) in the size of <0.5 mm in diameter (Rochman et al., 2014). For a duration of 2 months, the fish received either a control food without plastic, food with 10% virgin polyethylene particles, or food with 10% marine PE particles. After the exposure, selected gene expression was measured in liver samples. Exposure to pollutant-sorbed plastic led to endocrine disruption in both male and female fish, with biomarkers indicating both the presence of anti-estrogenic substances or antagonists to the estrogen receptor. Male medaka fed with marine pollutant-sorbed particles showed a significantly decreased expression of choriogenin H (Chg H), and female fish exposed to the sorbed plastic particles had significantly decreased expression of Vtg I, ChgH, and ERα than the control-food or virgin particles-food groups. After 2 months, also the virgin particle diet induced a decrease of Chg H expression in female fish. Histopathology of the gonads revealed an abnormal proliferation of germ cells of one individual male fish in the marine plastic treatment, which might be related to the sorbed plastic exposure. The results of this study suggest that there could be a mixture of pollutants in the harbor water, which can impair reproductive success and harm whole fish populations. Those pollutants could sorb to the microplastic particles and could be transferred to the organism when ingested. Therefore, microplastic, already with potential endocrine disrupting potential on its own, acts as a mediator to transfer chemicals to the fish (Rochman et al., 2014).

Adverse effects of plastic additives and pollutants have been demonstrated, but it is also important to understand how these chemicals accumulate in fish tissue. This could cause a risk for tropic transfer through food chains, and potentially to humans. The presence of the common plastic additives group of organophosphates (OPPAs) was investigated in seven freshwater fish species in a river in China in a large sampling. The dorsal muscle, belly fat, liver, and swimming bladder of silver carp (*Hypophthalmichthys molitrix*), common carp (*Cyprinus carpio*), grass carp (*Ctenopharyngodon idella*), Guangdong black bream (*Megalobrama hoffmanni*), barbel chub (*Squaliobarbus curriculus*), Redbelly tilapia (*Oreochromis mossambicus*), and catfish (*Silurus asotus*) were scanned for 12 OPPAs. Of these 12 OPPAs, 8 were detected in over 50% of the tissue samples, as well as in the water and sediment of the river. The concentration of present OPPAs in tissues depended on the respective fish species, body size, season, and compound characteristics, such as lipophilicity. In general, the fish tissue with the highest concentration of OPPAs was the liver, pointing out the factor of lipophilicity in bioaccumulation. Regarding inter-specific differences of accumulation, the redbelly tilapia's tissues contained significantly less OPPAs than all other fish species. This could be explained by its differing feeding habits (detritivorous). In addition, female fish in this study showed higher amounts of OPPAs in their tissues than male fish. This leads to the conclusion that female fish could accumulate more additives in their tissue based on a different metabolism or behavioral differences. The organophosphates were also detected in both eggs/ovaries of the females and testes of the males. Again, the concentration was higher in the female than in the male gonads. It could be predicted that the additives are transferred to next generation fish embryos, at least through the eggs/ovaries in the female fish. A paternal transfer through spermatozoa could be possible as well (Peng et al., 2021). Plastic additives pose a risk through different exposure routes and cause a variety of adverse effects on fish.

## MULTIVARIATE APPROACHES FOR ADDRESSING MICROPLASTIC TOXICITY

The diversity of microplastics presents a challenge for identifying which particle characteristics may be most likely to contribute to toxicity as well as which organisms and biological endpoints may be most sensitive to microplastic exposure. To address this, some researchers have explored approaches that used large, curated toxicity datasets to gain a better, more comprehensive understanding of microplastic toxicity. Recent meta-analyses have highlighted interesting trends in microplastic toxicity. For instance, Bucci et al. (Bucci et al., 2019) provides evidence that plastic toxicity in aquatic organisms varies widely though it is influenced by a combination of test concentrations, particle characteristics (e.g., size, morphology, polymer type), levels of biological organization, taxa, exposure duration, and the type of endpoint studied (e.g., mortality, reproduction, growth, etc.). These findings, though perhaps not surprising, were supported by a separate meta-analysis specifically focused on the effects of virgin nano- and microplastics in fish (Jacob et al., 2020). Here, the occurrence of biological responses to microplastic exposure was also categorized by the physiological system. It was found that responses associated with metabolism are well studied, and there is substantial evidence that microplastic particles cause oxidative stress in fish (Jacob et al., 2020). The analysis also revealed important gaps in data as there were few studies focused on the effects of microplastics on behavior. Yet, despite the lack of data coverage, fish behavior may be a particularly sensitive endpoint as 57% of the studied endpoints reported an effect following microplastic exposure (Jacob et al., 2020). Finally, while the approaches used in each of these meta-analyses have revealed a unique perspective of microplastic toxicity, it is important to note that meta-analyses in aquatic organisms have consistently found that smaller-sized particles are more likely to induce adverse effects (Bucci et al., 2019; Jacob et al., 2020; Sun et al., 2021), suggesting that particle size is a major determinant of toxicity.

In addition to meta-analyses, species sensitivity distributions (SSDs) have also been a popular approach to better understand microplastic toxicity in aquatic organisms. Species sensitivity distributions (SSDs) are particularly useful for understanding inter-species variability and community-level effects. Initial attempts to construct SSDs for microplastics used traditional approaches by populating their model using no and lowest observed effect concentrations from published toxicity studies (i.e., NOECs – no effect concentrations /LOECs – lowest effect concentrations) (Burns & Boxall, 2018; Everaert et al., 2018; Jung et al., 2021). These studies generate hazard concentrations at which a specific proportion of species might be negatively affected (e.g., HC5 – hazardous concentration for 5% of organisms), which is then compared to environmental concentrations to determine risk. However, given the variability in both toxicity and occurrence data, others have suggested that a probabilistic approach may be more appropriate (Adam et al., 2019, 2021). Here, probability distributions are generated from toxicity data to derive a predicted no-effect concentration. This is then compared to a probability distribution for environmental concentrations of microplastics. If these distributions overlap, there is risk. Regardless of the SSD approach used (i.e., traditional or probabilistic), these studies suggest that there is currently no substantial risk to aquatic organisms due to microplastic toxicity. However, it should be noted that current methods are limited in their ability to detect and characterize some types of microplastics (Brander et al., 2020), and microplastic concentrations are expected to continuously increase in the absence of significant changes to plastic production and use (Everaert et al., 2020; Lau et al., 2020). Thus, it is necessary to continuously refine risk assessment methods and frameworks for microplastics as research progresses.

While the previously described studies provide a foundation for determining risks posed by MNPs, they do not address the disparity in particle types often found across toxicity tests. First, effect concentrations are most often reported as particle counts or mass. Secondly, toxicity studies typically use distinct types of microplastic particles (e.g., 10-μm polystyrene spheres), which are neither comparable to particle types used in other toxicity studies nor particle types typically found in the aquatic environment. Some studies have attempted to curate datasets to be more representative of environmental microplastics or selected data pertaining to only certain types of particles (e.g., Jung et al., 2021). However, this can substantially reduce the amount of data available for analysis and may cause biases in the results. Recent modelling approaches have been proposed to address each of these challenges (Koelmans et al., 2020; Kooi et al., 2021; Kooi & Koelmans, 2019). Here, toxicity test concentrations are converted between mass and count using simulated probability distributions of particles (Koelmans et al., 2020). Once toxicity concentrations are presented in the same units, the data may be 'aligned' to environmental distributions of microplastics based on an ecologically relevant effect metric (ERM) predetermined to be of specific biological interest. For example, Koelmans et al. (2020) demonstrated this concept by using volume as the ERM for effects pertaining to food dilution. Together, these methods provide powerful tools for comparing microplastic toxicity data across studies and to concentrations reported in the field.

## CONCLUSION AND FUTURE RESEARCH DIRECTIONS

The effects of MNPs across polymer types, shapes, and sizes, in different fish species, and along all tiers of the biological hierarchy are evidenced in the studies summarized herein. However, it is important to acknowledge that some studies show little to no impact of MNPs on fishes (e.g., Jacob et al., 2020), emphasizing the importance of continued efforts to standardize assays used to evaluate MNP toxicity, so observations of effect vs. no effect can be compared more directly and accurately to better assess risk (Gouin et al., 2019). Additionally, more studies should be designed with environmental relevance in mind, including consideration of concentrations and polymer shapes and types detected in the field (e.g., microfibers), weathering, biofilms, and with an emphasis on mechanistically linking responses at the molecular level with fitness-relevant endpoints such as growth and reproduction. Given that anthropogenic mass already exceeds that of all natural biomass globally (Elhacham et al., 2020) and that plastic pollution is predicted to increase exponentially through the remainder of the century (Borrelle et al., 2020; Everaert et al., 2018; Lau et al., 2020), the issue of MNP toxicity in fishes and other aquatic organisms is unlikely to be quickly resolved.

Fishes spanning different species, life history strategies, and developmental stages can be impacted by MNP exposure. At the cellular and molecular levels, ROS production and concomitant expression of antioxidant enzymes (e.g., catalase) are observed across studies (Jacob et al., 2020), but it can often be unclear as to whether the presence of these markers indicates stress and the potential for higher-level impacts (e.g., on tissues, organs), or rather a protective/adaptive response. At the organism level, effects on growth (due to food dilution), behavior, and at times reproduction or impacts on offspring are evident; however, often these responses are only observed at very high concentrations. It is difficult to make conclusions using studies on single species often exposed to monodispersed (one plastic type) solutions; thus, multivariate approaches (e.g., Koelmans et al., 2020) as described above are key to discerning larger patterns. The role of additives, dyes, and adsorbed pollutants continues to be debated, and there is evidence on both sides attesting to their level of importance, and how the toxicity of these compounds should be evaluated in concert with the particles themselves. Whole ecosystem studies conducted over multiple years, such as the one currently in progress in Canada's Experimental Lake System which uses a polydispersed (multiple plastic types) dosing approach across trophic levels (including fish), are likely to unveil impacts that as a relatively nascent field we have yet to imagine.

It is also important to emphasize that the majority of studies thus far have used PS or PE spheres; thus, our understanding of how more realistic plastic shapes (fibers) impact organisms is also in its infancy. There are indications that microfibers are more toxic to fishes compared to other microplastic morphologies (Qiao et al., 2019a; Stienbarger et al., n.d.), and also that the challenge of fiber contamination extends beyond polymeric materials and that it includes fibers from anthropogenically modified textiles such as wool and cotton (Athey & Erdle, 2021; Granek et al., 2022). Though much uncertainty remains on exposure and the creation of realistic dosing in laboratory studies, there are at least some common responses across fish species, such as oxidative stress and food dilution (de Ruijter et al., 2020). And although MNPs are generally not lethally toxic, their impacts appear to occur through more subtle mechanisms such as damage to the gut wall, changes to the microbiome, or impacts on overall energy budget (e.g., Cormier et al., 2021; Qiao et al., 2019b; Wan et al., 2019; Zhao et al., 2021). Ultimately impacts on physiological processes that are crucial to successful reproduction and long-term survival could cumulatively reduce population size and have higher level effects on communities or ecosystems (Salerno et al., 2021).

Finally, one of the biggest obstacles to truly understanding the impact of MNPs is the lack of standardization, and proper controls, across assays used in the laboratory (Bour et al., 2021; Cowger et al., 2020), as well as the need for better documentation and measurement of smaller particles in the wild on which to base experimental design (Brander et al., 2020). More targeted dose or concentration–response studies are needed across at least three or four concentrations, so the assessment of risk is made possible. Having more chronic studies on which to base the assessment of risk is also imperative, not only in water but also in sediment (where the majority of MNPs are deposited), which necessitates studies to be run for a month or longer. This can be challenging in terms of resources, but acute exposures run over a matter of days do not accurately represent exposures in the wild. Shifting the focus to longer-term studies that closely mimic environmental conditions in terms of exposure levels, MNP morphologies, habitat type (emphasis on benthos), and considering this in the context of changes in climatic conditions due to global change, will ultimately lead to more accurate estimates of hazard from MNP exposure and will allow for better assessment of risk.

## REFERENCES


Abbasi, S., Soltani, N., Keshavarzi, B., Moore, F., Turner, A., & Hassanaghaei, M. (2018). Microplastics in different tissues of fish and prawn from the Musa Estuary, Persian Gulf. *Chemosphere*, *205*, 80–87. https://doi.org/10.1016/j.chemosphere.2018.04.076

Acharya, S., Rumi, S. S., Hu, Y., & Abidi, N. (2021). Microfibers from synthetic textiles as a major source of microplastics in the environment: A review. *Textile Research Journal*, 0040517521991244. https://doi.org/10.1177/0040517521991244

Adam, V., Yang, T., & Nowack, B. (2019). Toward an ecotoxicological risk assessment of microplastics: Comparison of available hazard and exposure data in freshwaters. *Environmental Toxicology and Chemistry*, *38*(2), 436–447. https://doi.org/10.1002/etc.4323

Adam, V., von Wyl, A., & Nowack, B. (2021). Probabilistic environmental risk assessment of microplastics in marine habitats. *Aquatic Toxicology*, *230*, 105689. https://doi.org/10.1016/j.aquatox.2020.105689

Amaral-Zettler, L. A., Zettler, E. R., & Mincer, T. J. (2020). Ecology of the plastisphere. *Nature Reviews Microbiology*, *18*(3), 139–151. https://doi.org/10.1038/s41579-019-0308-0

An, D., & Parsek, M. R. (2007). The promise and peril of transcriptional profiling in biofilm communities. *Current Opinion in Microbiology*, *10*(3), 292–296. https://doi.org/10.1016/j.mib.2007.05.011

Ankley, G. T., Bennett, R. S., Erickson, R. J., Hoff, D. J., Hornung, M. W., Johnson, R. D., Mount, D. R., Nichols, J. W., Russom, C. L., Schmieder, P. K., Serrrano, J. A., Tietge, J. E., & Villeneuve, D. L. (2010). Adverse outcome pathways: A conceptual framework to support ecotoxicology research and risk assessment. *Environmental Toxicology and Chemistry*, *29*(3), 730–741. https://doi.org/10.1002/etc.34

Arias-Andres, M., Klümper, U., Rojas-Jimenez, K., & Grossart, H.-P. (2018). Microplastic pollution increases gene exchange in aquatic ecosystems. *Environmental Pollution*, *237*, 253–261. https://doi.org/10.1016/j.envpol.2018.02.058

Ašmonaitė, G., Larsson, K., Undeland, I., Sturve, J., & Carney Almroth, B. (2018). Size matters: Ingestion of relatively large microplastics contaminated with environmental pollutants posed little risk for fish health and fillet quality. *Environmental Science & Technology*, *52*(24), 14381–14391. https://doi.org/10.1021/acs.est.8b04849

Ašmonaitė, G., Tivefälth, M., Westberg, E., Magnér, J., Backhaus, T., & Carney Almroth, B. (2020). Microplastics as a vector for exposure to hydrophobic organic chemicals in fish: A comparison of two polymers and silica particles spiked with three model compounds. *Frontiers in Environmental Science*, *8*, 87. https://doi.org/10.3389/fenvs.2020.00087

Assas, M., Qiu, X., Chen, K., Ogawa, H., Xu, H., Shimasaki, Y., & Oshima, Y. (2020). Bioaccumulation and reproductive effects of fluorescent microplastics in medaka fish. *Marine Pollution Bulletin*, *158*, 111446. https://doi.org/10.1016/j.marpolbul.2020.111446

Athey, S. N., Albotra, S. D., Gordon, C. A., Monteleone, B., Seaton, P., Andrady, A. L., Taylor, A. R., & Brander, S. M. (2020). Trophic transfer of microplastics in an estuarine food chain and the effects of a sorbed legacy pollutant. *Limnology and Oceanography Letters*, *5*(1), 154–162. https://doi.org/10.1002/lol2.10130

Athey, S. N., & Erdle, L. M. (2021). Are we underestimating anthropogenic microfiber pollution? a critical review of occurrence, methods and reporting. *Environmental Toxicology and Chemistry*. https://doi.org/10.1002/etc.5173

Bacchetta, C., Ale, A., Simoniello, M. F., Gervasio, S., Davico, C., Rossi, A. S., Desimone, M. F., Poletta, G., López, G., & Monserrat, J. M. (2017). Genotoxicity and oxidative stress in fish after a short-term exposure to silver nanoparticles. *Ecological Indicators*, *76*, 230–239. https://doi.org/10.1016/j.ecolind.2017.01.018

Baechler, B. R., Granek, E. F., Mazzone, S. J., Nielsen-Pincus, M., & Brander, S. M. (2020). Microplastic exposure by Razor clam recreational harvester-consumers along a sparsely populated coastline. *Frontiers in Marine Science*, *7*, 980. https://doi.org/10.3389/fmars.2020.588481

Bakir, A., O'Connor, I. A., Rowland, S. J., Hendriks, A. J., & Thompson, R. C. (2016). Relative importance of microplastics as a pathway for the transfer of hydrophobic organic chemicals to marine life. *Environmental Pollution*, *219*, 56–65. https://doi.org/10.1016/j.envpol.2016.09.046

Barboza, L. G. A., Vieira, L. R., & Guilhermino, L. (2018). Single and combined effects of microplastics and mercury on juveniles of the European seabass (*Dicentrarchus labrax*): Changes in behavioural responses and reduction of swimming velocity and resistance time. *Environmental Pollution*, *236*, 1014–1019. https://doi.org/10.1016/j.envpol.2017.12.082

Battin, T. J., Kaplan, L. A., Denis Newbold, J., & Hansen, C. M. E. (2003). Contributions of microbial biofilms to ecosystem processes in stream mesocosms. *Nature*, *426*(6965), 439–442. https://doi.org/10.1038/nature02152

Beedanagari, S., Vulimiri, S. V., Bhatia, S., & Mahadevan, B. (2014). Chapter 43-Genotoxicity biomarkers: Molecular basis of genetic variability and susceptibility. In R. C. Gupta (Ed.), *Biomarkers in Toxicology* (pp. 729–742). Academic Press, Boston. https://doi.org/10.1016/B978-0-12-404630-6.00043-9

Bhandari, R. K. (2016). Medaka as a model for studying environmentally induced epigenetic transgenerational inheritance of phenotypes. *Environmental Epigenetics*, *2*(1), dvv010–dvv010. https://doi.org/10.1093/eep/dvv010

Boerger, C.M., Lattin, G.L., Moore, S.L., Moore, C.J. (2010). Plastic ingestion by planktivorous fishes in the North Pacific Central Gyre. *Marine Pollution Bulletin 60*, 2275–2278. https://doi.org/10.1016/j.marpolbul.2010.08.007

Bonnineau, C., Artigas, J., Chaumet, B., Dabrin, A., Faburé, J., Ferrari, B. J. D., Lebrun, J. D., Margoum, C., Mazzella, N., Miège, C., Morin, S., Uher, E., Babut, M., & Pesce, S. (2021). Role of biofilms in contaminant bioaccumulation and trophic transfer in aquatic ecosystems: Current state of knowledge and future challenges. In P. de Voogt (Ed.), *Reviews of Environmental Contamination and Toxicology Volume 253* (pp. 115–153). Springer International Publishing, New York. https://doi.org/10.1007/398_2019_39

Borrelle, S. B., Ringma, J., Law, K. L., Monnahan, C. C., Lebreton, L., McGivern, A., Murphy, E., Jambeck, J., Leonard, G. H., Hilleary, M. A., Eriksen, M., Possingham, H. P., De Frond, H., Gerber, L. R., Polidoro, B., Tahir, A., Bernard, M., Mallos, N., Barnes, M., & Rochman, C. M. (2020). Predicted growth in plastic waste exceeds efforts to mitigate plastic pollution. *Science*, *369*(6510), 1515. https://doi.org/10.1126/science.aba3656

Bour, A., Sturve, J., Höjesjö, J., & Carney Almroth, B. (2020a). Microplastic vector effects: Are fish at risk when exposed via the trophic chain? *Frontiers in Environmental Science*, *8*, 90. https://doi.org/10.3389/fenvs.2020.00090

Bour, A., Hossain, S., Taylor, M., Sumner, M., & Carney Almroth, B. (2020b). Synthetic microfiber and microbead exposure and retention time in model aquatic species under different exposure scenarios. *Frontiers in Environmental Science*, *8*, 83. https://doi.org/10.3389/fenvs.2020.00083

Bour, A., Hamann Sandgaard, M., Syberg, K., Palmqvist, A., & Carney Almroth, B. (2021). Comprehending the complexity of microplastic organismal exposures and effects, to improve testing frameworks. *Journal of Hazardous Materials*, *415*, 125652. https://doi.org/10.1016/j.jhazmat.2021.125652

Bowley, J., Baker-Austin, C., Porter, A., Hartnell, R., & Lewis, C. (2021). Oceanic Hitchhikers - assessing pathogen risks from marine microplastic. *Trends in Microbiology*, *29*(2), 107–116. https://doi.org/10.1016/j.tim.2020.06.011

Brand, W., Peters, R. J., Braakhuis, H. M., Maślankiewicz, L., & Oomen, A. G. (2020). Possible effects of titanium dioxide particles on human liver, intestinal tissue, spleen and kidney after oral exposure. *Nanotoxicology*, *14*(7), 985–1007. https://doi.org/10.1080/17435390.2020.1778809

Brander, S. M., Fontana, R. E., Mata, T. M., Gravem, S. A., Hettinger, A., Bean, J. R., Szoboszlai, A. I., Keiper, C. A., & Marrero, M. E. (2012). The ecotoxicology of plastic marine debris. *The American Biology Teacher*, *73*(8), 474–478. https://doi.org/10.1525/abt.2011.73.8.9

Brander, S. M. (2013). Thinking outside the box: Assessing endocrine disruption in aquatic life. In Satinder Ahuja (Ed.), *Monitoring Water Quality: Pollution Assessment, Analysis, and Remediation*. Elsevier, Waltham, MA, 103–147. https://doi.org/10.1016/B978-0-444-59395-5.00005-4

Brander, S. M., Biales, A. D., & Connon, R. E. (2017). The role of epigenomics in aquatic toxicology. *Environmental Toxicology and Chemistry*, *36*(10), 2565–2573. https://doi.org/10.1002/etc.3930

Brander, S. M., Renick, V. C., Foley, M. M., Steele, C., Woo, M., Lusher, A., Carr, S., Helm, P., Box, C., & Cherniak, S. (2020). Sampling and quality assurance and quality control: A guide for scientists investigating the occurrence of microplastics across matrices. *Applied Spectroscopy*, *74*(9), 1099–1125. https://doi.org/10.1177/0003702820945713

Brander, S. M., Hoh, E., Unice, K. M., Cook, A., Holleman, R. C., Kone, D. V., Rochman, C. M., & Thayer, J. A. (2021). *Microplastic Pollution in California: A Precautionary Framework and Scientific Guidance to Assess and Address Risk to the Marine Environment*. California Ocean Science Trust, Sacramento, CA.

Brander, S., Cunningham, B., Dickens, J., Lasdin, K., Plafcan, M., Van Brocklin, J., Harper, S., Granek, E. *In Press. An Unexpected Snack: Microplastic Toxicity across Marine Taxa, in: Microplastics in the Sea: Occurrence and Impacts*, Sandra E. Shumway and J. Evan Ward (Eds.). *Academic Press*, Cambridge, Massachusetts.

Brander, S. M., & Betjemann, P. 2023. Marine pollution: Science, art, and the accumulations of time. In Spalding, A., & Suman, D. (Eds.), *Oceans and Society: An Introduction to Marine Studies* (1st ed.). Routledge. https://doi.org/10.4324/9781003058151.

Brandts, I., Teles, M., Tvarijonaviciute, A., Pereira, M. L., Martins, M. A., Tort, L., & Oliveira, M. (2018). Effects of polymethylmethacrylate nanoplastics on *Dicentrarchus labrax*. *Genomics*, *110*(6), 435–441. https://doi.org/10.1016/j.ygeno.2018.10.006

Bucci, K., Tulio, M., & Rochman, C. M. (2019). What is known and unknown about the effects of plastic pollution: A meta-analysis and systematic review. *Ecological Applications*, *30*(2), e02044. https://doi.org/10.1002/eap.2044

Bucci, K., Bikker, J., Stevack, K., Watson-Leung, T., & Rochman, C. (2021). Impacts to larval fathead minnows vary between preconsumer and environmental microplastics. *Environmental Toxicology and Chemistry*, *n/a*(n/a). https://doi.org/10.1002/etc.5036

Burns, E. E., & Boxall, A. B. A. (2018). Microplastics in the aquatic environment: Evidence for or against adverse impacts and major knowledge gaps. *Environmental Toxicology and Chemistry*, *37*(11), 2776–2796. https://doi.org/10.1002/etc.4268

California State Water Resources Control Board. Adoption of definition of "microplastics in drinking water." Resolution No. 2020-0021.

Carpenter, E. J., Anderson, S. J., Harvey, G. R., Miklas, H. P., & Peck, B. B. (1972). Polystyrene spherules in coastal waters. *Science*, *178*(4062), 749–750. https://doi.org/10.1126/science.178.4062.749

Caruso, G., Pedà, C., Cappello, S., Leonardi, M., La Ferla, R., Lo Giudice, A., Maricchiolo, G., Rizzo, C., Maimone, G., Rappazzo, A. C., Genovese, L., & Romeo, T. (2018). Effects of microplastics on trophic parameters, abundance and metabolic activities of seawater and fish gut bacteria in mesocosm conditions. *Environmental Science and Pollution Research International*, *25*(30), 30067–30083. https://doi.org/10.1007/s11356-018-2926-x

Cedervall, T., Hansson, L.-A., Lard, M., Frohm, B., & Linse, S. (2012). Food chain transport of nanoparticles affects behaviour and fat metabolism in fish. *PloS One*, *7*(2), e32254. https://doi.org/10.1371/journal.pone.0032254

Chen, J.-C., Chen, M.-Y., Fang, C., Zheng, R.-H., Jiang, Y.-L., Zhang, Y.-S., Wang, K.-J., Bailey, C., Segner, H., & Bo, J. (2020). Microplastics negatively impact embryogenesis and modulate the immune response of the marine medaka *Oryzias melastigma*. *Marine Pollution Bulletin*, *158*, 111349. https://doi.org/10.1016/j.marpolbul.2020.111349

Choi, J. S., Jung, Y.-J., Hong, N.-H., Hong, S. H., & Park, J.-W. (2018). Toxicological effects of irregularly shaped and spherical microplastics in a marine teleost, the sheepshead minnow (Cyprinodon variegatus). *Marine Pollution Bulletin*, *129*(1), 231–240. https://doi.org/10.1016/j.marpolbul.2018.02.039

CIEL (2019). Plastics and Climate: The Hidden Costs of a Plastic Planet. Kistler, A., Muffett, C. (editors). Available online at www.ciel.org/plasticandclimate.

Coffin, S., Magnuson, J. T., Vliet, S. M. F., Volz, D. C., & Schlenk, D. (2020). Effects of short-term exposure to environmentally-relevant concentrations of benzo(a) pyrene-sorbed polystyrene to White seabass (*Atractoscion nobilis*). *Environmental Pollution*, *263*, 114617. https://doi.org/10.1016/j.envpol.2020.114617

Collard, F., Gilbert, B., Compère, P., Eppe, G., Das, K., Jauniaux, T., & Parmentier, E. (2017). Microplastics in livers of European anchovies (*Engraulis encrasicolus*, L.). *Environmental Pollution*, *229*, 1000–1005. https://doi.org/10.1016/j.envpol.2017.07.089

Collard, F., Gasperi, J., Gilbert, B., Eppe, G., Azimi, S., Rocher, V., & Tassin, B. (2018). Anthropogenic particles in the stomach contents and liver of the freshwater fish Squalius cephalus. *Science of The Total Environment*, *643*, 1257–1264. https://doi.org/10.1016/j.scitotenv.2018.06.313

Cormier, B., Le Bihanic, F., Cabar, M., Crebassa, J.-C., Blanc, M., Larsson, M., Dubocq, F., Yeung, L., Clérandeau, C., Keiter, S. H., Cachot, J., Bégout, M.-L., & Cousin, X. (2021). Chronic feeding exposure to virgin and spiked microplastics disrupts essential biological functions in teleost fish. *Journal of Hazardous Materials*, *415*, 125626. https://doi.org/10.1016/j.jhazmat.2021.125626

Corradetti, B., Stronati, A., Tosti, L., Manicardi, G., Carnevali, O., & Bizzaro, D. (2013). Bis-(2-ethylexhyl) phthalate impairs spermatogenesis in zebrafish (*Danio rerio*). *Reproductive Biology*, *13*(3), 195–202. https://doi.org/10.1016/j.repbio.2013.07.003

Cowger, W., Booth, A. M., Hamilton, B. M., Thaysen, C., Primpke, S., Munno, K., Lusher, A. L., Dehaut, A., Vaz, V. P., Liboiron, M., Devriese, L. I., Hermabessiere, L., Rochman, C., Athey, S. N., Lynch, J. M., De Frond, H., Gray, A., Jones, O. A. H., Brander, S., ... Nel, H. (2020). Reporting guidelines to increase the reproducibility and comparability of research on microplastics. *Applied Spectroscopy*, 0003702820930292. https://doi.org/10.1177/0003702820930292

Critchell, K., & Hoogenboom, M. O. (2018). Effects of microplastic exposure on the body condition and behaviour of planktivorous reef fish (*Acanthochromis polyacanthus*). *PloS One*, *13*(3), e0193308. https://doi.org/10.1371/journal.pone.0193308

De Frond, H. L., van Sebille, E., Parnis, J. M., Diamond, M. L., Mallos, N., Kingsbury, T., & Rochman, C. M. (2019). Estimating the mass of chemicals associated with ocean plastic pollution to inform mitigation efforts. *Integrated Environmental Assessment and Management*, *15*(4), 596–606. https://doi.org/10.1002/ieam.4147

de Ruijter, V. N., Redondo-Hasselerharm, P. E., Gouin, T., & Koelmans, A. A. (2020). Quality criteria for microplastic effect studies in the context of risk assessment: A critical review. *Environmental Science & Technology*, *54*(19), 11692–11705. https://doi.org/10.1021/acs.est.0c03057

de Sá, L. C., Luís, L. G., & Guilhermino, L. (2015). Effects of microplastics on juveniles of the common goby (*Pomatoschistus microps*): Confusion with prey, reduction of the predatory performance and efficiency, and possible influence of developmental conditions. *Environmental Pollution*, *196*, 359–362. https://doi.org/10.1016/j.envpol.2014.10.026

DeCourten, B. M., Forbes, J. P., Roark, H. K., Burns, N. P., Major, K. M., White, J. W., Li, J., Mehinto, A. C., Connon, R. E., & Brander, S. M. (2020). Multigenerational and transgenerational effects of environmentally relevant concentrations of endocrine disruptors in an estuarine fish model. *Environmental Science & Technology*, *54*(21), 13849–13860. https://doi.org/10.1021/acs.est.0c02892

Diepens, N. J., & Koelmans, A. A. (2018). Accumulation of plastic debris and associated contaminants in aquatic food webs. *Environmental Science & Technology*, *52*(15), 8510–8520. https://doi.org/10.1021/acs.est.8b02515

Ding, J., Zhang, S., Razanajatovo, R. M., Zou, H., & Zhu, W. (2018). Accumulation, tissue distribution, and biochemical effects of polystyrene microplastics in the freshwater fish red tilapia (*Oreochromis niloticus*). *Environmental Pollution*, *238*, 1–9. https://doi.org/10.1016/j.envpol.2018.03.001

Ding, J., Huang, Y., Liu, S., Zhang, S., Zou, H., Wang, Z., Zhu, W., & Geng, J. (2020). Toxicological effects of nano- and micropolystyrene plastics on red tilapia: Are larger plastic particles more harmless? *Journal of Hazardous Materials*, *396*, 122693. https://doi.org/10.1016/j.jhazmat.2020.122693

Dussud, C., Meistertzheim, A. L., Conan, P., Pujo-Pay, M., George, M., Fabre, P., Coudane, J., Higgs, P., Elineau, A., Pedrotti, M. L., Gorsky, G., & Ghiglione, J. F. (2018). Evidence of niche partitioning among bacteria living on plastics, organic particles and surrounding seawaters. *Environmental Pollution*, *236*, 807–816. https://doi.org/10.1016/j.envpol.2017.12.027

Elhacham, E., Ben-Uri, L., Grozovski, J., Bar-On, Y. M., & Milo, R. (2020). Global human-made mass exceeds all living biomass. *Nature*, *588*(7838), 442–444. https://doi.org/10.1038/s41586-020-3010-5

Elizalde-Velázquez, A., Carcano, A. M., Crago, J., Green, M. J., Shah, S. A., & Cañas-Carrell, J. E. (2020). Translocation, trophic transfer, accumulation and depuration of polystyrene microplastics in *Daphnia magna and Pimephales promelas*. *Environmental Pollution*, *259*, 113937. https://doi.org/10.1016/j.envpol.2020.113937

Espinosa, C., García Beltrán, J. M., Esteban, M. A., & Cuesta, A. (2018). In vitro effects of virgin microplastics on fish head-kidney leucocyte activities. *Environmental Pollution (Barking, Essex: 1987)*, *235*, 30–38. https://doi.org/10.1016/j.envpol.2017.12.054

Everaert, G., Van Cauwenberghe, L., De Rijcke, M., Koelmans, A. A., Mees, J., Vandegehuchte, M., & Janssen, C. R. (2018). Risk assessment of microplastics in the ocean: Modelling approach and first conclusions. *Environmental Pollution*, *242*, 1930–1938. https://doi.org/10.1016/j.envpol.2018.07.069

Everaert, G., De Rijcke, M., Lonneville, B., Janssen, C. R., Backhaus, T., Mees, J., van Sebille, E., Koelmans, A. A., Catarino, A. I., & Vandegehuchte, M. B. (2020). Risks of floating microplastic in the global ocean. *Environmental Pollution*, *267*, 115499. https://doi.org/10.1016/j.envpol.2020.115499

Ford, A. T., Ågerstrand, M., Brooks, B. W., Allen, J., Bertram, M. G., Brodin, T., Dang, Z., Duquesne, S., Sahm, R., Hoffmann, F., Hollert, H., Jacob, S., Klüver, N., Lazorchak, J. M., Ledesma, M., Melvin, S. D., Mohr, S., Padilla, S., Pyle, G. G., ... Maack, G. (2021). The role of behavioral ecotoxicology in environmental protection. *Environmental Science & Technology*, *55*(9), 5620–5628. https://doi.org/10.1021/acs.est.0c06493

Fossi, M. C., Panti, C., Guerranti, C., Coppola, D., Giannetti, M., Marsili, L., & Minutoli, R. (2012). Are baleen whales exposed to the threat of microplastics? A case study of the Mediterranean fin whale (*Balaenoptera physalus*). *Marine Pollution Bulletin*, *64*(11), 2374–2379. https://doi.org/10.1016/j.marpolbul.2012.08.013

Fossi, M. C., Coppola, D., Baini, M., Giannetti, M., Guerranti, C., Marsili, L., Panti, C., de Sabata, E., & Clò, S. (2014). Large filter feeding marine organisms as indicators of microplastic in the pelagic environment: The case studies of the Mediterranean basking shark (*Cetorhinus maximus*) and fin whale (*Balaenoptera physalus*). *Marine Environmental Research*, *100*, 17–24. https://doi.org/10.1016/j.marenvres.2014.02.002

GESAMP. (2020). *Proceedings of the GESAMP International Workshop on assessing the risks associated with plastics and microplastics in the marine environment* (Vol. 103, p. 68). IMO/FAO/UNESCO-IOC/UNIDO/WMO/IAEA/UN/UNEP/UNDP/ISA Joint Group of Experts on the Scientific Aspects of Marine Environmental Protection.

Glüge, J., Scheringer, M., Cousins, I. T., DeWitt, J. C., Goldenman, G., Herzke, D., Lohmann, R., Ng, C. A., Trier, X., & Wang, Z. (2020). An overview of the uses of per-and polyfluoroalkyl substances (PFAS). *Environmental Science: Processes & Impacts*, *22*(12), 2345–2373. https://doi.org/10.1039/D0EM00291G

Gouin, T., Roche, N., Lohmann, R., & Hodges, G. (2011). A thermodynamic approach for assessing the environmental exposure of chemicals absorbed to microplastic. *Environmental Science & Technology*, *45*(4), 1466–1472. https://doi.org/10.1021/es1032025

Gouin, T., Becker, R. A., Collot, A.-G., Davis, J. W., Howard, B., Inawaka, K., Lampi, M., Ramon, B. S., Shi, J., & Hopp, P. W. (2019). Toward the development and application of an environmental risk assessment framework for microplastic. *Environmental Toxicology and Chemistry*, *38*(10), 2087–2100. https://doi.org/10.1002/etc.4529

Gouin, T. (2020). Toward an improved understanding of the ingestion and trophic transfer of microplastic particles: Critical review and implications for future research. *Environmental Toxicology and Chemistry*, *39*(6), 1119–1137. https://doi.org/10.1002/etc.4718

Gove, J. M., Whitney, J. L., McManus, M. A., Lecky, J., Carvalho, F. C., Lynch, J. M., Li, J., Neubauer, P., Smith, K. A., Phipps, J. E., Kobayashi, D. R., Balagso, K. B., Contreras, E. A., Manuel, M. E., Merrifield, M. A., Polovina, J. J., Asner, G. P., Maynard, J. A., & Williams, G. J. (2019). Prey-size plastics are invading larval fish nurseries. *Proceedings of the National Academy of Sciences*, *116*(48), 24143–24149. https://doi.org/10.1073/pnas.1907496116

Granek, E. F., Brander, S. M., & Holland, E. B. (2020). Microplastics in aquatic organisms: Improving understanding and identifying research directions for the next decade. *Limnology and Oceanography Letters*, *5*(1), 1–4. https://doi.org/10.1002/lol2.10145

Granek, E.F., Traylor, S.D., Tissot, A.G., Hurst, P.T., Wood, R.S., Brander, S.M., 2022. Clothes encounters of the microfibre kind: The effects of natural and synthetic textiles on organisms. In Judith S. Weis, Francesca De Falco, Maria Cristina Cocca (Eds.), *Polluting Textiles* (pp. 63–99). Routledge, Oxfordshire.

Grigorakis, S., Drouillard, K.G., 2018. Effect of microplastic amendment to food on diet assimilation efficiencies of PCBs by fish. *Environmental Science & Technology*, *52*, 10796–10802. https://doi.org/10.1021/acs.est.8b02497

Groh, K. J., Backhaus, T., Carney-Almroth, B., Geueke, B., Inostroza, P. A., Lennquist, A., Leslie, H. A., Maffini, M., Slunge, D., & Trasande, L. (2019). Overview of known plastic packaging-associated chemicals and their hazards. *Science of the Total Environment*, *651*, 3253–3268. https://doi.org/10.1016/j.scitotenv.2018.10.015

Gustavsson, M. B., Hellohf, A., & Backhaus, T. (2017). Evaluating the environmental hazard of industrial chemicals from data collected during the REACH registration process. *Science of the Total Environment*, *586*, 658–665. https://doi.org/10.1016/j.scitotenv.2017.02.039

Gutow, L., Eckerlebe, A., Giménez, L., & Saborowski, R. (2016). Experimental evaluation of seaweeds as a vector for microplastics into marine food webs. *Environmental Science & Technology*, *50*(2), 915–923. https://doi.org/10.1021/acs.est.5b02431

Hamed, M., Soliman, H. A. M., Osman, A. G. M., & Sayed, A. E.-D. H. (2019). Assessment the effect of exposure to microplastics in Nile Tilapia (Oreochromis niloticus) early juvenile: I. blood biomarkers. *Chemosphere*, *228*, 345–350. https://doi.org/10.1016/j.chemosphere.2019.04.153

Hanke, G. (2016). *Marine Beach Litter in Europe - Top Items* (JRC Technical Reports, p. 29). https://mcc.jrc.ec.europa.eu/documents/Marine_Litter/MarineLitterTOPitems_final_24.1.2017.pdf

Harris, P. T. (2020). The fate of microplastic in marine sedimentary environments: A review and synthesis. *Marine Pollution Bulletin*, *158*, 111398. https://doi.org/10.1016/j.marpolbul.2020.111398

Harris, L. S. T., La Beur, L., Olsen, A. Y., Smith, A., Eggers, L., Pedersen, E., Van Brocklin, J., Brander, S. M., & Larson, S. (2021). Temporal variability of microparticles under the Seattle Aquarium, WA: Documenting the global covid-19 pandemic. *Environmental Toxicology and Chemistry*, *41*, 917–930. https://doi.org/10.1002/etc.5190

Hartmann, N. B., Rist, S., Bodin, J., Jensen, L. H., Schmidt, S. N., Mayer, P., Meibom, A., & Baum, A. (2017). Microplastics as vectors for environmental contaminants: Exploring sorption, desorption, and transfer to biota. *Integrated Environmental Assessment and Management*, *13*(3), 488–493. https://doi.org/10.1002/ieam.1904.

Hasegawa, T., & Nakaoka, M. (2021). Trophic transfer of microplastics from mysids to fish greatly exceeds direct ingestion from the water column. *Environmental Pollution*, *273*, 116468. https://doi.org/10.1016/j.envpol.2021.116468

Hermabessiere, L., Dehaut, A., Paul-Pont, I., Lacroix, C., Jezequel, R., Soudant, P., & Duflos, G. (2017). Occurrence and effects of plastic additives on marine environments and organisms: A review. *Chemosphere*, *182*, 781–793. https://doi.org/10.1016/j.chemosphere.2017.05.096

Herzke, D., Anker-Nilssen, T., Nøst, T. H., Götsch, A., Christensen-Dalsgaard, S., Langset, M., Fangel, K., & Koelmans, A. A. (2016). Negligible impact of ingested microplastics on tissue concentrations of persistent organic pollutants in Northern Fulmars off Coastal Norway. *Environmental Science & Technology*, *50*(4), 1924–1933. https://doi.org/10.1021/acs.est.5b04663

Hirt, N., & Body-Malapel, M. (2020). Immunotoxicity and intestinal effects of nano- and microplastics: A review of the literature. *Particle and Fibre Toxicology*, *17*(1), 57. https://doi.org/10.1186/s12989-020-00387-7

Holmström, A., 1975. Plastic films on the bottom of the Skagerack. *Nature*, *255*, 622–623. https://doi.org/10.1038/255622a0

Hu, L., Chernick, M., Lewis, A. M., Ferguson, P. L., & Hinton, D. E. (2020). Chronic microfiber exposure in adult Japanese medaka (*Oryzias latipes*). *PLoS One*, *15*(3), e0229962. https://doi.org/10.1371/journal.pone.0229962

Hurt, R., O'Reilly, C. M., & Perry, W. L. (2020). Microplastic prevalence in two fish species in two U.S. reservoirs. *Limnology and Oceanography Letters*, *5*(1), 147–153. https://doi.org/10.1002/lol2.10140

Huuskonen, H., Subiron i Folguera, J., Kortet, R., Akkanen, J., Vainikka, A., Janhunen, M., & Kekäläinen, J. (2020). Do whitefish (*Coregonus lavaretus*) larvae show adaptive variation in the avoidance of microplastic ingestion? *Environmental Pollution*, *262*, 114353. https://doi.org/10.1016/j.envpol.2020.114353

Jabeen, K., Li, B., Chen, Q., Su, L., Wu, C., Hollert, H., & Shi, H. (2018). Effects of virgin microplastics on goldfish (*Carassius auratus*). *Chemosphere*, *213*, 323–332. https://doi.org/10.1016/j.chemosphere.2018.09.031

Jacob, H., Besson, M., Swarzenski, P. W., Lecchini, D., & Metian, M. (2020). Effects of virgin micro-and nanoplastics on fish: Trends, meta-analysis, and perspectives. *Environmental Science & Technology*, *54*(8), 4733–4745. https://doi.org/10.1021/acs.est.9b05995

Jang, M., Shim, W. J., Han, G. M., Rani, M., Song, Y. K., & Hong, S. H. (2017). Widespread detection of a brominated flame retardant, hexabromocyclododecane, in expanded polystyrene marine debris and microplastics from South Korea and the Asia-Pacific coastal region. *Environmental Pollution*, *231*, 785–794. https://doi.org/10.1016/j.envpol.2017.08.066

Jantzen, C. E., Annunziato, K. M., & Cooper, K. R. (2016). Behavioral, morphometric, and gene expression effects in adult zebrafish (*Danio rerio*) embryonically exposed to PFOA, PFOS, and PFNA. *Aquatic Toxicology*, *180*, 123–130. https://doi.org/10.1016/j.aquatox.2016.09.011

Jeong, J., & Choi, J. (2019). Adverse outcome pathways potentially related to hazard identification of microplastics based on toxicity mechanisms. *Chemosphere*, *231*, 249–255. https://doi.org/10.1016/j.chemosphere.2019.05.003

Jin, M., Yu, X., Yao, Z., Tao, P., Li, G., Yu, X., Zhao, J.-L., & Peng, J. (2020). How biofilms affect the uptake and fate of hydrophobic organic compounds (HOCs) In microplastic: Insights from an In situ study of Xiangshan Bay, China. *Water Research, 184*, 116118. https://doi.org/10.1016/j.watres.2020.116118

Jovanović, B., Gökdağ, K., Güven, O., Emre, Y., Whitley, E. M., & Kideys, A. E. (2018). Virgin microplastics are not causing imminent harm to fish after dietary exposure. *Marine Pollution Bulletin, 130*, 123–131. https://doi.org/10.1016/j.marpolbul.2018.03.016

Jung, J.-W., Park, J.-W., Eo, S., Choi, J., Song, Y. K., Cho, Y., Hong, S. H., & Shim, W. J. (2021). Ecological risk assessment of microplastics in coastal, shelf, and deep sea waters with a consideration of environmentally relevant size and shape. *Environmental Pollution, 270*, 116217. https://doi.org/10.1016/j.envpol.2020.116217

Kane, I. A., & Clare, M. A. (2019). Dispersion, accumulation, and the ultimate fate of microplastics in deep-marine environments: A review and future directions. *Frontiers in Earth Science, 7*, 80. https://doi.org/10.3389/feart.2019.00080

Kar, S., Sangem, P., Anusha, N., & Senthilkumaran, B. (2021). Endocrine disruptors in teleosts: Evaluating environmental risks and biomarkers. *Aquaculture and Fisheries, 6*(1), 1–26. https://doi.org/10.1016/j.aaf.2020.07.013

Kato, G., Miyazawa, H., Nakayama, Y., Ikari, Y., Kondo, H., Yamaguchi, T., Sano, M., & Fischer, U. (2018). A novel antigen-sampling cell in the teleost gill epithelium with the potential for direct antigen presentation in mucosal tissue. *Frontiers in Immunology, 9*, 2116. https://doi.org/10.3389/fimmu.2018.02116

KEMI, Swedish Chemicals Agency. (2016). *Hazardous chemical substances in textiles - proposals for risk management measures* (No. 8/16; p. 67). https://www.kemi.se/download/18.6df1d3df171c243fb23a98e7/1591454108832/report-8-16-hazardous-chemical-substances-in-textiles.pdf

Kim, S. W., Chae, Y., Kim, D., & An, Y.-J. (2019). Zebrafish can recognize microplastics as inedible materials: Quantitative evidence of ingestion behavior. *Science of The Total Environment, 649*, 156–162. https://doi.org/10.1016/j.scitotenv.2018.08.310

Kim, J., Poirier, D. G., Helm, P. A., Bayoumi, M., & Rochman, C. M. (2020). No evidence of spherical microplastics (10-300 μm) translocation in adult rainbow trout (*Oncorhynchus mykiss*) after a two-week dietary exposure. *PLoS One, 15*(9), e0239128. https://doi.org/10.1371/journal.pone.0239128

Kinch, C. D., Kurrasch, D. M., & Habibi, H. R. (2016). Adverse morphological development in embryonic zebrafish exposed to environmental concentrations of contaminants individually and in mixture. *Aquatic Toxicology, 175*, 286–298. https://doi.org/10.1016/j.aquatox.2016.03.021

Kirstein, I. V., Wichels, A., Krohne, G., & Gerdts, G. (2018). Mature biofilm communities on synthetic polymers in seawater-Specific or general? *Marine Environmental Research, 142*, 147–154. https://doi.org/10.1016/j.marenvres.2018.09.028

Koelmans, A. A., Besseling, E., & Foekema, E. M. (2014). Leaching of plastic additives to marine organisms. *Environmental Pollution, 187*(0), 49–54. https://doi.org/10.1016/j.envpol.2013.12.013

Koelmans, A. A. (2015). Modeling the role of microplastics in bioaccumulation of organic chemicals to marine aquatic organisms. A critical review. *Marine Anthropogenic Litter*, 309–324. https://doi.org/10.1007/978-3-319-16510-3_11

Koelmans, A. A., Bakir, A., Burton, G. A., & Janssen, C. R. (2016). Microplastic as a vector for chemicals in the aquatic environment: critical review and model-supported reinterpretation of empirical studies. *Environmental Science & Technology, 50*(7), 3315–3326. https://doi.org/10.1021/acs.est.5b06069

Koelmans, A. A., Redondo-Hasselerharm, P. E., Mohamed Nor, N. H., & Kooi, M. (2020). Solving the nonalignment of methods and approaches used in microplastic research to consistently characterize risk. *Environmental Science & Technology, 54*(19), 12307–12315. https://doi.org/10.1021/acs.est.0c02982

Koelmans, A. A., Diepens, N. J., & Mohamed Nor, N. (2022). Weight of evidence for the microplastic vector effect in the context of chemical risk assessment. *Plastic in the Environment: Pattern and Process*. 155–197.

Kooi, M., & Koelmans, A. A. (2019). Simplifying microplastic via continuous probability distributions for size, shape, and density. *Environmental Science & Technology Letters, 6*(9), 551–557. https://doi.org/10.1021/acs.estlett.9b00379

Kooi, M., Primpke, S., Mintenig, S. M., Lorenz, C., Gerdts, G., & Koelmans, A. A. (2021). Characterizing the multidimensionality of microplastics across environmental compartments. *Water Research, 202*, 117429. https://doi.org/10.1016/j.watres.2021.117429

Koongolla, J. B., Lin, L., Pan, Y.-F., Yang, C.-P., Sun, D.-R., Liu, S., Xu, X.-R., Maharana, D., Huang, J.-S., & Li, H.-X. (2020). Occurrence of microplastics in gastrointestinal tracts and gills of fish from Beibu Gulf, South China Sea. *Environmental Pollution, 258*, 113734. https://doi.org/10.1016/j.envpol.2019.113734

Lacasse, K., & Baumann, W. (2004). Environmental considerations for textile processes and chemicals. In K. Lacasse & W. Baumann (Eds.), *Textile Chemicals: Environmental Data and Facts* (pp. 484–647). Springer, New York. https://doi.org/10.1007/978-3-642-18898-5_7

LaPlaca, S. B., & van den Hurk, P. (2020). Toxicological effects of micronized tire crumb rubber on mummichog (*Fundulus heteroclitus*) and fathead minnow (*Pimephales promelas*). *Ecotoxicology (London, England), 29*(5), 524–534. https://doi.org/10.1007/s10646-020-02210-7

Lau, W. W. Y., Shiran, Y., Bailey, R. M., Cook, E., Stuchtey, M. R., Koskella, J., Velis, C. A., Godfrey, L., Boucher, J., Murphy, M. B., Thompson, R. C., Jankowska, E., Castillo Castillo, A., Pilditch, T. D., Dixon, B., Koerselman, L., Kosior, E., Favoino, E., Gutberlet, J., ... Palardy, J. E. (2020). Evaluating scenarios toward zero plastic pollution. *Science, 369*(6510), 1455–1461. https://doi.org/10.1126/science.aba9475

Lasdin, Katherine S, Madison Arnold, Anika Agrawal, H William Fennie, Kirsten Grorud-Colvert, Su Sponaugle, Lindsay Aylesworth, Scott Heppell, and Susanne M Brander. "Presence of Microplastics and Microparticles in Oregon Black Rockfish Sampled near Marine Reserve Areas." *PeerJ* 11 (2023): e14564.

Leiser, R., Wu, G.-M., Neu, T.R., Wendt-Potthoff, K., (2020). Biofouling, metal sorption and aggregation are related to sinking of microplastics in a stratified reservoir. *Water Research, 176*, 115748. https://doi.org/10.1016/j.watres.2020.115748

Li, S., Shi, M., Wang, Y., Xiao, Y., Cai, D., & Xiao, F. (2021). Keap1-Nrf2 pathway up-regulation via hydrogen sulfide mitigates polystyrene microplastics induced-hepatotoxic effects. *Journal of Hazardous Materials, 402*, 123933. https://doi.org/10.1016/j.jhazmat.2020.123933

Limonta, G., Mancia, A., Benkhalqui, A., Bertolucci, C., Abelli, L., Fossi, M. C., & Panti, C. (2019). Microplastics induce transcriptional changes, immune response and behavioral alterations in adult zebrafish. *Scientific Reports, 9*(1), 15775. https://doi.org/10.1038/s41598-019-52292-5

Liu, P., Zhan, X., Wu, X., Li, J., Wang, H., & Gao, S. (2020a). Effect of weathering on environmental behavior of microplastics: Properties, sorption and potential risks. *Chemosphere*, *242*, 125193. https://doi.org/10.1016/j.chemosphere.2019.125193

Liu, K., Courtene-Jones, W., Wang, X., Song, Z., Wei, N., & Li, D. (2020b). Elucidating the vertical transport of microplastics in the water column: A review of sampling methodologies and distributions. *Water Research*, *186*, 116403. https://doi.org/10.1016/j.watres.2020.116403

Lu, Y., Zhang, Y., Deng, Y., Jiang, W., Zhao, Y., Geng, J., Ding, L., & Ren, H. (2016). Uptake and accumulation of polystyrene microplastics in Zebrafish (*Danio rerio*) and toxic effects in liver. *Environ Sci Technol*, *50*(7), 4054–4060. https://doi.org/10.1021/acs.est.6b00183

Lusher, A. L., McHugh, M., & Thompson, R. C. (2013). Occurrence of microplastics in the gastrointestinal tract of pelagic and demersal fish from the English Channel. *Marine Pollution Bulletin*, *67*(1–2), 94–99. https://doi.org/10.1016/j.marpolbul.2012.11.028

Lusher, A. L., Munno, K., Hermabessiere, L., & Carr, S. (2020). Isolation and extraction of microplastics from environmental samples: An evaluation of practical approaches and recommendations for further harmonization. *Applied Spectroscopy*, *74*(9), 1049–1065. https://doi.org/10.1177/0003702820938993

Lyche, J. L., Rosseland, C., Berge, G., & Polder, A. (2015). Human health risk associated with brominated flame-retardants (BFRs). *Environment International*, *74*, 170–180. https://doi.org/10.1016/j.envint.2014.09.006

Malinich, T. D., Chou, N., Sepúlveda, M. S., & Höök, T. O. (2018). No evidence of microplastic impacts on consumption or growth of larval *Pimephales promelas*. *Environmental Toxicology and Chemistry*, *37*(11), 2912–2918. https://doi.org/10.1002/etc.4257

Mato, Y., Isobe, T., Takada, H., Kanehiro, H., Ohtake, C., & Kaminuma, T. (2001). Plastic resin pellets as a transport medium for toxic chemicals in the marine environment. *Environmental Science & Technology*, *35*(2), 318–324. https://doi.org/10.1021/es0010498

Mattsson, K., Johnson, E. V., Malmendal, A., Linse, S., Hansson, L.-A., & Cedervall, T. (2017). Brain damage and behavioural disorders in fish induced by plastic nanoparticles delivered through the food chain. *Scientific Reports*, *7*(1), 11452. https://doi.org/10.1038/s41598-017-10813-0

Mazurais, D., Ernande, B., Quazuguel, P., Severe, A., Huelvan, C., Madec, L., Mouchel, O., Soudant, P., Robbens, J., Hüvet, A., & Zambonino-Infante, J. (2015). Evaluation of the impact of polyethylene microbeads ingestion in European sea bass (*Dicentrarchus labrax*) larvae. *Marine Environmental Research*, *112*, 78–85. https://doi.org/10.1016/j.marenvres.2015.09.009

McArdle, M. E., Freeman, E. L., Staveley, J. P., Ortego, L. S., Coady, K. K., Weltje, L., Weyers, A., Wheeler, J. R., & Bone, A. J. (2020). Critical review of read-across potential in testing for endocrine-related effects in vertebrate ecological receptors. *Environmental Toxicology and Chemistry*, *39*(4), 739–753. https://doi.org/10.1002/etc.4682

McCormick, M. I., Chivers, D. P., Ferrari, M. C. O., Blandford, M. I., Nanninga, G. B., Richardson, C., Fakan, E. P., Vamvounis, G., Gulizia, A. M., & Allan, B. J. M. (2020). Microplastic exposure interacts with habitat degradation to affect behaviour and survival of juvenile fish in the field. *Proceedings of the Royal Society B: Biological Sciences*, *287*(1937), 20201947. https://doi.org/10.1098/rspb.2020.1947

McIlwraith, H. K., Kim, J., Helm, P., Bhavsar, S. P., Metzger, J. S., & Rochman, C. M. (2021). Evidence of microplastic translocation in wild-caught fish and implications for microplastic accumulation dynamics in food webs. *Environmental Science & Technology*, *55*(18), 12372–12382. https://doi.org/10.1021/acs.est.1c02922

Mehinto, A.C., Coffin, S., Koelmans, A., Brander, S.M., Wagner, M., Hampton, L., Burton, G.A., Miller, E., Gouin, T., Weisberg, S., Rochman, C. 2022. Risk-based management framework for microplastics in aquatic ecosystems. *Microplastics and Nanoplastics*, *2*(1), 1–10.

Miao, L., Hou, J., You, G., Liu, Z., Liu, S., Li, T., Mo, Y., Guo, S., & Qu, H. (2019). Acute effects of nanoplastics and microplastics on periphytic biofilms depending on particle size, concentration and surface modification. *Environmental Pollution*, *255*, 113300. https://doi.org/10.1016/j.envpol.2019.113300

Miller, M. E., Hamann, M., & Kroon, F. J. (2020). Bioaccumulation and biomagnification of microplastics in marine organisms: A review and meta-analysis of current data. *PLoS One*, *15*(10), e0240792. https://doi.org/10.1371/journal.pone.0240792

Miller, E., Sedlak, M., Lin, D., Box, C., Holleman, C., Rochman, C. M., & Sutton, R. (2021). Recommended best practices for collecting, analyzing, and reporting microplastics in environmental media: Lessons learned from comprehensive monitoring of San Francisco Bay. *Journal of Hazardous Materials*, *409*, 124770. https://doi.org/10.1016/j.jhazmat.2020.124770

Mincer, T. J., Zettler, E. R., & Amaral-Zettler, L. A. (2019). Biofilms on plastic debris and their influence on marine nutrient cycling, productivity, and hazardous chemical mobility. In H. Takada & H. K. Karapanagioti (Eds.), *Hazardous Chemicals Associated with Plastics in the Marine Environment* (pp. 221–233). Springer International Publishing, New York. https://doi.org/10.1007/698_2016_12

Mohamed Nor, N. H., & Koelmans, A. A. (2019). Transfer of PCBs from microplastics under simulated gut fluid conditions is biphasic and reversible. *Environmental Science & Technology*, *53*(4), 1874–1883. https://doi.org/10.1021/acs.est.8b05143

Momeni, B. (2018). Division of labor: How microbes split their responsibility. *Current Biology*, *28*(12), R697–R699. https://doi.org/10.1016/j.cub.2018.05.024

Muncke, J. (2021). Tackling the toxics in plastics packaging. *PLoS Biology*, *19*(3), e3000961.

Munno, K., Helm, P. A., Rochman, C., George, T., & Jackson, D. A. (2021). Microplastic contamination in Great Lakes fish. *Conservation Biology: The Journal of the Society for Conservation Biology*. https://doi.org/10.1111/cobi.13794

Naidoo, T., & Glassom, D. (2019). Decreased growth and survival in small juvenile fish, after chronic exposure to environmentally relevant concentrations of microplastic. *Marine Pollution Bulletin*, *145*, 254–259. https://doi.org/10.1016/j.marpolbul.2019.02.037

Norland, S., Vorkamp, K., Bogevik, A.S., Koelmans, A.A., Diepens, N.J., Burgerhout, E., Hansen, Ø.J., Puvanendran, V., & Rønnestad, I., 2021. Assessing microplastic as a vector for chemical entry into fish larvae using a novel tube-feeding approach. *Chemosphere* 265, 129144. https://doi.org/10.1016/j.chemosphere.2020.129144

OECD. (1992). *203: Fish, acute toxicity test.*

Ogata, Y., Takada, H., Mizukawa, K., Hirai, H., Iwasa, S., Endo, S., Mato, Y., Saha, M., Okuda, K., Nakashima, A., Murakami, M., Zurcher, N., Booyatumanondo, R., Zakaria, M. P., Dung, L. Q., Gordon, M., Miguez, C., Suzuki, S., Moore, C., ... Thompson, R. C. (2009). International Pellet Watch: Global monitoring of persistent organic pollutants (POPs) in coastal waters. 1. Initial phase data on PCBs, DDTs, and HCHs. *Marine Pollution Bulletin*, *58*(10), 1437–1446. https://doi.org/10.1016/j.marpolbul.2009.06.014

Ory, N. C., Gallardo, C., Lenz, M., & Thiel, M. (2018). Capture, swallowing, and egestion of microplastics by a planktivorous juvenile fish. *Environmental Pollution*, *240*, 566–573. https://doi.org/10.1016/j.envpol.2018.04.093

Pannetier, P., Morin, B., Le Bihanic, F., Dubreil, L., Clérandeau, C., Chouvellon, F., Van Arkel, K., Danion, M., & Cachot, J. (2020). Environmental samples of microplastics induce significant toxic effects in fish larvae. *Environment International*, *134*, 105047. https://doi.org/10.1016/j.envint.2019.105047

Patlewicz, G., Helman, G., Pradeep, P., & Shah, I. (2017). Navigating through the minefield of read-across tools: A review of in silico tools for grouping. *Computational Toxicology*, *3*, 1–18. https://doi.org/10.1016/j.comtox.2017.05.003

Peng, X., Chen, G., Fan, Y., Zhu, Z., Guo, S., Zhou, J., & Tan, J. (2021). Lifetime bioaccumulation, gender difference, tissue distribution, and parental transfer of organophosphorus plastic additives in freshwater fish. *Environmental Pollution*, *280*, 116948. https://doi.org/10.1016/j.envpol.2021.116948

Peters, C. A., Thomas, P. A., Rieper, K. B., & Bratton, S. P. (2017). Foraging preferences influence microplastic ingestion by six marine fish species from the Texas Gulf Coast. *Marine Pollution Bulletin*, *124*(1), 82–88. https://doi.org/10.1016/j.marpolbul.2017.06.080

Pitt, J. A., Trevisan, R., Massarsky, A., Kozal, J. S., Levin, E. D., & Di Giulio, R. T. (2018). Maternal transfer of nanoplastics to offspring in zebrafish (*Danio rerio*): A case study with nanopolystyrene. *Science of The Total Environment*, *643*, 324–334. https://doi.org/10.1016/j.scitotenv.2018.06.186

Powell, J. J., Faria, N., Thomas-McKay, E., & Pele, L. C. (2010). Origin and fate of dietary nanoparticles and microparticles in the gastrointestinal tract. *Journal of Autoimmunity*, *34*(3), J226–J233. https://doi.org/10.1016/j.jaut.2009.11.006

Primpke, S., Christiansen, S. H., Cowger, W., De Frond, H., Deshpande, A., Fischer, M., Holland, E., Meyns, M., O'Donnell, B. A., Ossmann, B., Pittroff, M., Sarau, G., Scholz-Böttcher, B. M., & Wiggin, K. (2020). EXPRESS: Critical assessment of analytical methods for the harmonized and cost efficient analysis of microplastics. *Applied Spectroscopy*, 0003702820921465. https://doi.org/10.1177/0003702820921465

Procter, J., Hopkins, F. E., Fileman, E. S., & Lindeque, P. K. (2019a). Smells good enough to eat: Dimethyl sulfide (DMS) enhances copepod ingestion of microplastics. *Marine Pollution Bulletin*, *138*, 1–6. https://doi.org/10.1016/j.marpolbul.2018.11.014

Procter, J., Hopkins, F. E., Fileman, E. S., & Lindeque, P. K. (2019b). Smells good enough to eat: Dimethyl sulfide (DMS) enhances copepod ingestion of microplastics. *Marine Pollution Bulletin*, *138*, 1–6. https://doi.org/10.1016/j.marpolbul.2018.11.014

Qiang, L., & Cheng, J. (2021). Exposure to polystyrene microplastics impairs gonads of zebrafish (*Danio rerio*). *Chemosphere*, *263*, 128161. https://doi.org/10.1016/j.chemosphere.2020.128161

Qiao, R., Deng, Y., Zhang, S., Wolosker, M. B., Zhu, Q., Ren, H., & Zhang, Y. (2019a). Accumulation of different shapes of microplastics initiates intestinal injury and gut microbiota dysbiosis in the gut of zebrafish. *Chemosphere*, *236*, 124334. https://doi.org/10.1016/j.chemosphere.2019.07.065

Qiao, R., Sheng, C., Lu, Y., Zhang, Y., Ren, H., & Lemos, B. (2019b). Microplastics induce intestinal inflammation, oxidative stress, and disorders of metabolome and microbiome in zebrafish. *Science of The Total Environment*, *662*, 246–253. https://doi.org/10.1016/j.scitotenv.2019.01.245

Radisic, V., Nimje, P. S., Bienfait, A. M., & Marathe, N. P. (2020). Marine plastics from norwegian west coast carry potentially virulent fish pathogens and opportunistic human pathogens harboring new variants of antibiotic resistance genes. *Microorganisms*, *8*(8), 1200. https://doi.org/10.3390/microorganisms8081200

Ramsperger, A. F. R. M., Narayana, V. K. B., Gross, W., Mohanraj, J., Thelakkat, M., Greiner, A., Schmalz, H., Kress, H., & Laforsch, C. (2020). Environmental exposure enhances the internalization of microplastic particles into cells. *Science Advances*, *6*(50), eabd1211. https://doi.org/10.1126/sciadv.abd1211

Rasool, F. N., Saavedra, M. A., Pamba, S., Perold, V., Mmochi, A. J., Maalim, M., Simonsen, L., Buur, L., Pedersen, R. H., Syberg, K., & Jelsbak, L. (2021). Isolation and characterization of human pathogenic multidrug resistant bacteria associated with plastic litter collected in Zanzibar. *Journal of Hazardous Materials*, *405*, 124591. https://doi.org/10.1016/j.jhazmat.2020.124591

Richard, H., Carpenter, E. J., Komada, T., Palmer, P. T., & Rochman, C. M. (2019). Biofilm facilitates metal accumulation onto microplastics in estuarine waters. *Science of The Total Environment*, *683*, 600–608. https://doi.org/10.1016/j.scitotenv.2019.04.331

Rios-Fuster, B., Alomar, C., Viñas, L., Campillo, J.A., Pérez-Fernández, B., Álvarez, E., Compa, M., Deudero, S. (2021). Organochlorine pesticides (OCPs) and polychlorinated biphenyls (PCBs) occurrence in Sparus aurata exposed to microplastic enriched diets in aquaculture facilities. Marine Pollution Bulletin 173, 113030. https://doi.org/10.1016/j.marpolbul.2021.113030

Roch, S., Friedrich, C., & Brinker, A. (2020). Uptake routes of microplastics in fishes: Practical and theoretical approaches to test existing theories. *Scientific Reports*, *10*(1), 3896. https://doi.org/10.1038/s41598-020-60630-1

Rochman, C. M. (2013). Plastics and priority pollutants: A multiple stressor in aquatic habitats. *Environmental Science & Technology*. https://dx.doi.org/10.1021/es400748b

Rochman, C. M., Hoh, E., Kurobe, T., & Teh, S. J. (2013). Ingested plastic transfers hazardous chemicals to fish and induces hepatic stress. *Scientific Reports*, *3*. https://dx.doi.org/10.1038/srep03263

Rochman, C. M., Kurobe, T., Flores, I., & Teh, S. J. (2014). Early warning signs of endocrine disruption in adult fish from the ingestion of polyethylene with and without sorbed chemical pollutants from the marine environment. *Science of The Total Environment*, *493*, 656–661. https://doi.org/10.1016/j.scitotenv.2014.06.051

Rochman, C. M., Brookson, C., Bikker, J., Djuric, N., Earn, A., Bucci, K., Athey, S., Huntington, A., McIlwraith, H., Munno, K., De Frond, H., Kolomijeca, A., Erdle, L., Grbic, J., Bayoumi, M., Borrelle, S. B., Wu, T., Santoro, S., Werbowski, L. M., ... Hung, C. (2019). Rethinking microplastics as a diverse contaminant suite. *Environmental Toxicology and Chemistry*, *38*(4), 703–711. https://doi.org/10.1002/etc.4371

Ross, P. S., Chastain, S., Vassilenko, E., Etemadifar, A., Zimmermann, S., Quesnel, S.-A., Eert, J., Solomon, E., Patankar, S., Posacka, A. M., & Williams, B. (2021). Pervasive distribution of polyester fibres in the Arctic Ocean is driven by Atlantic inputs. *Nature Communications*, *12*(1), 106. https://doi.org/10.1038/s41467-020-20347-1

Rossi, G., Barnoud, J., & Monticelli, L. (2014). Polystyrene nanoparticles perturb lipid membranes. *The Journal of Physical Chemistry Letters*, *5*(1), 241–246. https://doi.org/10.1021/jz402234c

Rummel, C. D., Jahnke, A., Gorokhova, E., Kühnel, D., & Schmitt-Jansen, M. (2017). Impacts of biofilm formation on the fate and potential effects of microplastic in the aquatic environment. *Environmental Science & Technology Letters*, *4*(7), 258–267. https://doi.org/10.1021/acs.estlett.7b00164

Ryan, P. G., & Jackson, S. (1987). The lifespan of ingested plastic particles in seabirds and their effect on digestive efficiency. *Marine Pollution Bulletin*, *18*(5), 217–219.

Ryan, P. G., Connell, A. D., & Gardner, B. D. (1988). Plastic ingestion and PCBs in seabirds: Is there a relationship? *Marine Pollution Bulletin*, *19*(4), 174–176. https://doi.org/10.1016/0025-326X(88)90674-1

Salerno, M., Berlino, M., Mangano, M. C., & Sarà, G. (2021). Microplastics and the functional traits of fishes: A global meta-analysis. *Global Change Biology*, *27*(12), 2645–2655. https://doi.org/10.1111/gcb.15570

Santos, R. G., Machovsky-Capuska, G. E., & Andrades, R. (2021). Plastic ingestion as an evolutionary trap: Toward a holistic understanding. *Science*, *373*(6550), 56–60. https://doi.org/10.1126/science.abh0945

Savoca, M. S., Wohlfeil, M. E., Ebeler, S. E., & Nevitt, G. A. (2016a). Marine plastic debris emits a keystone infochemical for olfactory foraging seabirds. *Science Advances*, *2*(11), e1600395. https://doi.org/10.1126/sciadv.1600395

Savoca, M. S., Wohlfeil, M. E., Ebeler, S. E., & Nevitt, G. A. (2016b). Marine plastic debris emits a keystone infochemical for olfactory foraging seabirds. *Science Advances*, *2*(11), e1600395. https://doi.org/10.1126/sciadv.1600395

Savoca, M. S., Tyson, C. W., McGill, M., & Slager, C. J. (2017). Odours from marine plastic debris induce food search behaviours in a forage fish. *Proceedings of the Royal Society B*, *284*(1860), 20171000.

Schaefer, L. (2014). Complexity of danger: The diverse nature of damage-associated molecular patterns. *Journal of Biological Chemistry*, *289*(51), 35237–35245. https://doi.org/10.1074/jbc.R114.619304

Schellenberger, S., Jönsson, C., Mellin, P., Levenstam, O. A., Liagkouridis, I., Ribbenstedt, A., Hanning, A.-C., Schultes, L., Plassmann, M. M., Persson, C., Cousins, I. T., & Benskin, J. P. (2019). Release of side-chain fluorinated polymer-containing microplastic fibers from functional textiles during washing and first estimates of perfluoroalkyl acid emissions. *Environmental Science & Technology*, *53*(24), 14329–14338. https://doi.org/10.1021/acs.est.9b04165

Schür, C., Rist, S., Baun, A., Mayer, P., Hartmann, N. B., & Wagner, M. (2019). When fluorescence is not a particle: The tissue translocation of microplastics in daphnia magna seems an artifact. *Environmental Toxicology and Chemistry*, *38*(7), 1495–1503. https://doi.org/10.1002/etc.4436

Sendra, L., Miguel, A., Navarro-Plaza, M.C., Herrero, M.J., de la Higuera, J., Cháfer-Pericás, C., Aznar, E., Marcos, M.D., Martínez-Máñez, R., Rojas, L.A., Alemany, R., Aliño, S.F., 2020. Gold nanoparticle-assisted virus formation by means of the delivery of an oncolytic adenovirus genome. *Nanomaterials* 10, 1183. https://doi.org/10.3390/nano10061183

Shupe, H. J., Boenisch, K. M., Harper, B. J., Brander, S. M., & Harper, S. L. (2021). Effect of nanoplastic type and surface chemistry on particle agglomeration over a salinity gradient. *Environmental Toxicology and Chemistry*, *40*(7), 1820–1826. https://doi.org/10.1002/etc.5030

Siddiqui, S., Dickens, J.M., Cunningham, B.E., Hutton, S.J., Pedersen, E.I., Harper, B., Harper, S., Brander, S.M. (2022). Internalization, reduced growth, and behavioral effects following exposure to micro and nano tire particles in two estuarine indicator species. *Chemosphere* 296, 133934. https://doi.org/10.1016/j.chemosphere.2022.133934

Siddiqui, S., S. J. Hutton, J. M. Dickens, E. I. Pedersen, S. L. Harper, and S. M. Brander. "Natural and Synthetic Microfibers Alter Growth and Behavior in Early Life Stages of Estuarine Organisms." *Frontiers in Marine Science* 9 (2023). https://www.frontiersin.org/articles/10.3389/fmars.2022.991650..

Simon, M., Hartmann, N. B., & Vollertsen, J. (2021). Accelerated weathering increases the release of toxic leachates from microplastic particles as demonstrated through altered toxicity to the green algae *Raphidocelis subcapitata*. *Toxics*, *9*(8), 185. https://doi.org/10.3390/toxics9080185

Singh, N., Bhagat, J., Tiwari, E., Khandelwal, N., Darbha, G. K., & Shyama, S. (2021). Metal oxide nanoparticles and polycyclic aromatic hydrocarbons alter nanoplastic's stability and toxicity to zebrafish. *Journal of Hazardous Materials*, *407*, 124382. https://doi.org/10.1016/j.jhazmat.2020.124382

Skalska, K., Ockelford, A., Ebdon, J. E., & Cundy, A. B. (2020). Riverine microplastics: Behaviour, spatio-temporal variability, and recommendations for standardised sampling and monitoring. *Journal of Water Process Engineering*, *38*, 101600. https://doi.org/10.1016/j.jwpe.2020.101600

Solomando, A., Capó, X., Alomar, C., Álvarez, E., Compa, M., Valencia, J. M., Pinya, S., Deudero, S., & Sureda, A. (2020). Long-term exposure to microplastics induces oxidative stress and a pro-inflammatory response in the gut of *Sparus aurata Linnaeus*, 1758. *Environmental Pollution (Barking, Essex: 1987)*, *266*(Pt 1), 115295. https://doi.org/10.1016/j.envpol.2020.115295

Sprangue, J. B. (2005). *Guidance Document on Statistical Methods for Environmental Toxicity Tests*. Environment Ottawa, Ontario, Canada.

Stienbarger, C. D., Joseph, J., Athey, S. N., Monteleone, B., Andrady, A. L., Watanabe, W. O., Seaton, P., Taylor, A. R., & Brander, S. M. (2021). Direct ingestion, trophic transfer, and physiological effects of microplastics in the early life stages of Centropristis striata, a commercially and recreationally valuable fishery species. *Environmental Pollution*, *285*, 117653. https://doi.org/10.1016/j.envpol.2021.117653

Suhrhoff, T. J., & Scholz-Böttcher, B. M. (2016). Qualitative impact of salinity, UV radiation and turbulence on leaching of organic plastic additives from four common plastics-A lab experiment. *Marine Pollution Bulletin*, *102*(1), 84–94. https://doi.org/10.1016/j.marpolbul.2015.11.054

Sun, T., Zhan, J., Li, F., Ji, C., & Wu, H. (2021). Evidence-based meta-analysis of the genotoxicity induced by microplastics in aquatic organisms at environmentally relevant concentrations. *Science of The Total Environment*, *783*, 147076. https://doi.org/10.1016/j.scitotenv.2021.147076

Teuten, E. L., Saquing, J. M., Knappe, D. R. U., Barlaz, M. A., Jonsson, S., Björn, A., Rowland, S. J., Thompson, R. C., Galloway, T. S., Yamashita, R., Ochi, D., Watanuki, Y., Moore, C., Viet, P. H., Tana, T. S., Prudente, M., Boonyatumanond, R., Zakaria, M. P., Akkhavong, K., ... Takada, H. (2009). Transport and release of chemicals from plastics to the environment and to wildlife. *Philosophical Transactions of the Royal Society of London*. Series B, Biological Sciences, *364*(1526), 2027–2045. https://doi.org/10.1098/rstb.2008.0284

Thompson, R. C., Olsen, Y., Mitchell, R. P., Davis, A., Rowland, S. J., John, A. W. G., McGonigle, D., & Russell, A. E. (2004). Lost at sea: Where is all the plastic? *Science*, *304*(5672), 838. https://doi.org/10.1126/science.1094559

Thornton Hampton, L.M., Bouwmeester, H., Brander, S., Cole, M., Hermabessiere, L., Mehinto, A.C., Miller, E., Rochman, C. M., & Weisberg, S. B. (2022). Research recommendations to better understand the potential health impacts of microplastics to humans and aquatic ecosystems. *Microplastics and Nanoplastics, 2*(18).

Thornton Hampton, L.M., Brander, S.M., Coffin S., Cole, M., Hermabessiere L., Koelmans, A.A., Rochman, C.M. (August 6, 2022). Characterizing microplastic hazards: Which concentration metrics and particle characteristics are most informative for understanding toxicity in aquatic organisms? *Microplastics and Nanoplastics, 2*(1), 20. https://doi.org/10.1186/s43591-022-00040-4

Tian, Z., Zhao, H., Peter, K. T., Gonzalez, M., Wetzel, J., Wu, C., Hu, X., Prat, J., Mudrock, E., Hettinger, R., Cortina, A. E., Biswas, R. G., Kock, F. V. C., Soong, R., Jenne, A., Du, B., Hou, F., He, H., Lundeen, R., ... Kolodziej, E. P. (2021). A ubiquitous tire rubber-derived chemical induces acute mortality in coho salmon. *Science, 371*(6525), 185. https://doi.org/10.1126/science.abd6951

Tierney, K. B. (2016). Chemical avoidance responses of fishes. *Aquatic Toxicology, 174*, 228–241. https://doi.org/10.1016/j.aquatox.2016.02.021

Timshina, A., Aristizabal-Henao, J. J., Da Silva, B. F., & Bowden, J. A. (2021). The last straw: Characterization of per-and polyfluoroalkyl substances in commercially-available plant-based drinking straws. *Chemosphere, 277*, 130238.

Turner, A., & Holmes, L. A. (2015). Adsorption of trace metals by microplastic pellets in fresh water. *Environmental Chemistry, 12*(5), 600–610. https://doi.org/10.1071/EN14143

Usenko, C., Abel, E., Hopkins, A., Martinez, G., Tijerina, J., Kudela, M., Norris, N., Joudeh, L., & Bruce, E. (2016). Evaluation of common use Brominated Flame Retardant (BFR) toxicity using a zebrafish embryo model. *Toxics, 4*(3), 21. https://doi.org/10.3390/toxics4030021

Van Essen, D., Devoy, C., Miller, J., Jones, P. D., & Wiseman, S. (2021). Effects of the brominated flame retardant, TBCO, on development of zebrafish (*Danio rerio*) embryos. *Chemosphere, 266*, 129195. https://doi.org/10.1016/j.chemosphere.2020.129195

van Hullebusch, E. D., Zandvoort, M. H., & Lens, P. N. L. (2003). Metal immobilisation by biofilms: Mechanisms and analytical tools. *Reviews in Environmental Science and Biotechnology, 2*(1), 9–33. https://doi.org/10.1023/B:RESB.0000022995.48330.55

Velez, J. F. M., Shashoua, Y., Syberg, K., & Khan, F. R. (2018). Considerations on the use of equilibrium models for the characterisation of HOC-microplastic interactions in vector studies. *Chemosphere, 210*, 359–365. https://doi.org/10.1016/j.chemosphere.2018.07.020

Verla, A. W., Enyoh, C. E., Verla, E. N., & Nwarnorh, K. O. (2019). Microplastic-toxic chemical interaction: A review study on quantified levels, mechanism and implication. *SN Applied Sciences, 1*(11), 1400. https://doi.org/10.1007/s42452-019-1352-0

Villeneuve, D. L., Landesmann, B., Allavena, P., Ashley, N., Bal-Price, A., Corsini, E., Halappanavar, S., Hussell, T., Laskin, D., Lawrence, T., Nikolic-Paterson, D., Pallardy, M., Paini, A., Pieters, R., Roth, R., & Tschudi-Monnet, F. (2018). Representing the process of inflammation as key events in adverse outcome pathways. *Toxicological Sciences, 163*(2), 346–352. https://doi.org/10.1093/toxsci/kfy047

Viršek, M. K., Lovšin, M. N., Koren, Š., Kržan, A., & Peterlin, M. (2017). Microplastics as a vector for the transport of the bacterial fish pathogen species Aeromonas salmonicida. *Marine Pollution Bulletin, 125*(1), 301–309. https://doi.org/10.1016/j.marpolbul.2017.08.024

Voelker, J., Ashcroft, F., Vedoy, A., Zimmermann, L., & Wagner, M. (2021). Adipogenic activity of chemicals used in plastic consumer products. *Pharmacology and Toxicology*. https://doi.org/10.1101/2021.07.29.454199

Wan, Z., Wang, C., Zhou, J., Shen, M., Wang, X., Fu, Z., & Jin, Y. (2019). Effects of polystyrene microplastics on the composition of the microbiome and metabolism in larval zebrafish. *Chemosphere, 217*, 646–658. https://doi.org/10.1016/j.chemosphere.2018.11.070

Wang, J., Li, Y., Lu, L., Zheng, M., Zhang, X., Tian, H., Wang, W., & Ru, S. (2019a). Polystyrene microplastics cause tissue damages, sex-specific reproductive disruption and transgenerational effects in marine medaka (*Oryzias melastigma*). *Environmental Pollution, 254*, 113024. https://doi.org/10.1016/j.envpol.2019.113024

Wang, W., Gao, H., Jin, S., Li, R., & Na, G. (2019b). The ecotoxicological effects of microplastics on aquatic food web, from primary producer to human: A review. *Ecotoxicology and Environmental Safety, 173*, 110–117. https://doi.org/10.1016/j.ecoenv.2019.01.113

Wang, W., Ge, J., Yu, X., & Li, H. (2020a). Environmental fate and impacts of microplastics in soil ecosystems: Progress and perspective. *Science of The Total Environment, 708*, 134841. https://doi.org/10.1016/j.scitotenv.2019.134841

Wang, Y., Wang, X., Li, Y., Li, J., Wang, F., Xia, S., & Zhao, J. (2020b). Biofilm alters tetracycline and copper adsorption behaviors onto polyethylene microplastics. *Chemical Engineering Journal, 392*, 123808. https://doi.org/10.1016/j.cej.2019.123808

Wardrop, P., Shimeta, J., Nugegoda, D., Morrison, P. D., Miranda, A., Tang, M., & Clarke, B. O. (2016). Chemical pollutants sorbed to ingested microbeads from personal care products accumulate in fish. *Environmental Science & Technology, 50*(7), 4037–4044. https://doi.org/10.1021/acs.est.5b06280

Watts, A. J., Lewis, C., Goodhead, R. M., Beckett, S. J., Moger, J., Tyler, C. R., & Galloway, T. S. (2014). Uptake and retention of microplastics by the shore crab *Carcinus maenas*. *Environmental Science & Technology, 48*(15), 8823–8830. https://doi.org/10.1021/es501090e

Watts, A. J. R., Urbina, M. A., Goodhead, R., Moger, J., Lewis, C., & Galloway, T. S. (2016). Effect of microplastic on the gills of the shore crab *Carcinus maenas*. *Environmental Science & Technology, 50*(10), 5364–5369. https://doi.org/10.1021/acs.est.6b01187

Wiesinger, H., Wang, Z., & Hellweg, S. (2021). Deep dive into plastic monomers, additives, and processing aids. *Environmental Science & Technology, 55*(13), 9339–9351. https://doi.org/10.1021/acs.est.1c00976

Woodall, L. C., Sanchez-Vidal, A., Canals, M., Paterson, G. L., Coppock, R., Sleight, V., Calafat, A., Rogers, A. D., Narayanaswamy, B. E., & Thompson, R. C. (2014). The deep sea is a major sink for microplastic debris. *Royal Society Open Science, 1*(4), 140317. https://doi.org/10.1098/rsos.140317

Wright, S. L., Thompson, R. C., & Galloway, T. S. (2013). The physical impacts of microplastics on marine organisms: A review. *Environmental Pollution, 178*(0), 483–492. https://doi.org/10.1016/j.envpol.2013.02.031

Writer, J. H., Ryan, J. N., & Barber, L. B. (2011). Role of biofilms in sorptive removal of steroidal hormones and 4-nonylphenol compounds from streams. *Environmental Science & Technology, 45*(17), 7275–7283. https://doi.org/10.1021/es2008038

Wu, Y., Li, T., & Yang, L. (2012). Mechanisms of removing pollutants from aqueous solutions by microorganisms and their aggregates: A review. *Bioresource Technology, 107*, 10–18. https://doi.org/10.1016/j.biortech.2011.12.088

Wu, Y., Xia, L., Yu, Z., Shabbir, S., & Kerr, P. G. (2014). In situ bioremediation of surface waters by periphytons. *Bioresource Technology, 151*, 367–372. https://doi.org/10.1016/j.biortech.2013.10.088

Wu, N., Zhang, Y., Li, W., Wang, J., Zhang, X., He, J., Li, J., Ma, Y., Niu, Z., (2020). Co-effects of biofouling and inorganic matters increased the density of environmental microplastics in the sediments of Bohai Bay coast. *Science of The Total Environment 717*, 134431. https://doi.org/10.1016/j.scitotenv.2019.134431

Xia, X., Sun, M., Zhou, M., Chang, Z., & Li, L. (2020). Polyvinyl chloride microplastics induce growth inhibition and oxidative stress in *Cyprinus carpio* var. Larvae. *Science of The Total Environment, 716*, 136479. https://doi.org/10.1016/j.scitotenv.2019.136479

Xiong, X., Chen, X., Zhang, K., Mei, Z., Hao, Y., Zheng, J., Wu, C., Wang, K., Ruan, Y., Lam, P. K. S., & Wang, D. (2018). Microplastics in the intestinal tracts of East Asian finless porpoises (*Neophocaena asiaeorientalis sunameri*) from Yellow Sea and Bohai Sea of China. *Marine Pollution Bulletin, 136*, 55–60. https://doi.org/10.1016/j.marpolbul.2018.09.006

Xiong, X., Tu, Y., Chen, X., Jiang, X., Shi, H., Wu, C., & Elser, J. J. (2019). Ingestion and egestion of polyethylene microplastics by goldfish (*Carassius auratus*): Influence of color and morphological features. *Heliyon, 5*(12), e03063. https://doi.org/10.1016/j.heliyon.2019.e03063

Xu, C., Zhang, B., Gu, C., Shen, C., Yin, S., Aamir, M., & Li, F. (2020). Are we underestimating the sources of microplastic pollution in terrestrial environment? *Journal of Hazardous Materials, 400*, 123228. https://doi.org/10.1016/j.jhazmat.2020.123228

Yamashita, R., Tanaka, K., Yeo, B. G., Takada, H., van Franeker, J. A., Dalton, M., & Dale, E. (2019). Hazardous chemicals in plastics in marine environments: International pellet watch. In H. Takada & H. K. Karapanagioti (Eds.), *Hazardous ChemicalsAssociatedwithPlasticsintheMarineEnvironment* (pp. 163–183). Springer International Publishing, Switzerland. https://doi.org/10.1007/698_2018_299

Yan, W., Hamid, N., Deng, S., Jia, P.-P., & Pei, D. S. (2020). Individual and combined toxicogenetic effects of microplastics and heavy metals (Cd, Pb, and Zn) perturb gut microbiota homeostasis and gonadal development in marine medaka (*Oryzias melastigma*). *Journal of Hazardous Materials, 397*, 122795. https://doi.org/10.1016/j.jhazmat.2020.122795

Yan, M., Wang, L., Dai, Y., Sun, H., & Liu, C. (2021). Behavior of microplastics in inland waters: Aggregation, settlement, and transport. *Bulletin of Environmental Contamination and Toxicology*. https://doi.org/10.1007/s00128-020-03087-2

Yaripour, S., Huuskonen, H., Rahman, T., Kekäläinen, J., Akkanen, J., Magris, M., Kipriianov, P. V., & Kortet, R. (2021). Pre-fertilization exposure of sperm to nano-sized plastic particles decreases offspring size and swimming performance in the European whitefish (*Coregonus lavaretus*). *Environmental Pollution, 291*, 118196. https://doi.org/10.1016/j.envpol.2021.118196

Ye, S., & Andrady, A. L. (1991). Fouling of floating plastic debris under Biscayne Bay exposure conditions. *Marine Pollution Bulletin, 22*(12), 608–613. https://doi.org/10.1016/0025-326X(91)90249-R

Yin, L., Chen, B., Xia, B., Shi, X., & Qu, K. (2018). Polystyrene microplastics alter the behavior, energy reserve and nutritional composition of marine jacopever (*Sebastes schlegelii*). *Journal of Hazardous Materials, 360*, 97–105. https://doi.org/10.1016/j.jhazmat.2018.07.110

Yu, B. P. (1994). Cellular defenses against damage from reactive oxygen species. *Physiological Reviews, 74*(1), 139–162. https://doi.org/10.1152/physrev.1994.74.1.139

Yu, P., Liu, Z., Wu, D., Chen, M., Lv, W., & Zhao, Y. (2018). Accumulation of polystyrene microplastics in juvenile *Eriocheir sinensis* and oxidative stress effects in the liver. *Aquatic Toxicology, 200*, 28–36. https://doi.org/10.1016/j.aquatox.2018.04.015

Yu, Z., Li, Q., Wang, J., Yu, Y., Wang, Y., Zhou, Q., & Li, P. (2020). Reactive oxygen species-related nanoparticle toxicity in the biomedical field. *Nanoscale Research Letters, 15*(1), 1–14. https://doi.org/10.1186/s11671-020-03344-7

Zettler, E. R., Mincer, T. J., & Amaral-Zettler, L. A. (2013). Life in the plastisphere: Microbial communities on plastic marine debris. *Environmental Science & Technology, 47*(13), 7137–7146. https://doi.org/10.1021/es401288x

Zhang, C., Jeong, C.-B., Lee, J.-S., Wang, D., & Wang, M. (2019). Transgenerational proteome plasticity in resilience of a marine copepod in response to environmentally relevant concentrations of microplastics. *Environmental Science & Technology, 53*(14), 8426–8436. https://doi.org/10.1021/acs.est.9b02525

Zhang, X., Wen, K., Ding, D., Liu, J., Lei, Z., Chen, X., Ye, G., Zhang, J., Shen, H., Yan, C., Dong, S., Huang, Q., & Lin, Y. (2021). Size-dependent adverse effects of microplastics on intestinal microbiota and metabolic homeostasis in the marine medaka (*Oryzias melastigma*). *Environment International, 151*, 106452. https://doi.org/10.1016/j.envint.2021.106452

Zhao, Y., Qin, Z., Huang, Z., Bao, Z., Luo, T., & Jin, Y. (2021). Effects of polyethylene microplastics on the microbiome and metabolism in larval zebrafish. *Environmental Pollution, 282*, 117039. https://doi.org/10.1016/j.envpol.2021.117039

Zhu, M., Chernick, M., Rittschof, D., & Hinton, D. E. (2020). Chronic dietary exposure to polystyrene microplastics in maturing Japanese medaka (*Oryzias latipes*). *Aquatic Toxicology, 220*, 105396. https://doi.org/10.1016/j.aquatox.2019.105396

Zhu, X., Munno, K., Grbic, J., Werbowski, L.M., Bikker, J., Ho, A., Guo, E., Sedlak, M., Sutton, R., Box, C., Lin, D., Gilbreath, A., Holleman, R.C., Fortin, M.-J., Rochman, C. (2021). Holistic assessment of microplastics and other anthropogenic microdebris in an urban bay sheds light on their sources and fate. *ACS EST Water, 1*, 1401–1410. https://doi.org/10.1021/acsestwater.0c00292

Ziajahromi, S., Kumar, A., Neale, P. A., & Leusch, F. D. (2017). Impact of microplastic beads and fibers on waterflea (*Ceriodaphnia dubia*) survival, growth, and reproduction: Implications of single and mixture exposures. *Environmental Science & Technology, 51*(22), 13397–13406. https://doi.org/10.1021/acs.est.7b03574

Ziccardi, L. M., Edgington, A., Hentz, K., Kulacki, K. J., & Driscoll, S. K. (2016). Microplastics as vectors for bioaccumulation of hydrophobic organic chemicals in the marine environment: A state-of-the-science review. *Environmental Toxicology and Chemistry, 35*(7), 1667–1676. https://doi.org/10.1002/etc.3461

Zitouni, N., Bousserrhine, N., Missawi, O., Boughattas, I., Chèvre, N., Santos, R., Belbekhouche, S., Alphonse, V., Tisserand, F., Balmassiere, L., Dos Santos, S. P., Mokni, M., Guerbej, H., & Banni, M. (2021). Uptake, tissue distribution and toxicological effects of environmental microplastics in early juvenile fish *Dicentrarchus labrax*. *Journal of Hazardous Materials, 403*, 124055. https://doi.org/10.1016/j.jhazmat.2020.124055

# 14 Effects of Pharmaceuticals

*Karl Fent*
ETH Zurich, Switzerland

*Nadja R. Brun*
University of Bergen, Norway

*Kun Zhang and Yanbin Zhao*
Shanghai Jiao Tong University, China

**Abstract**

Pharmaceutical residues are very widespread in the aquatic environment. Several hundred compounds and metabolites are documented in wastewaters, fresh and marine waters, groundwaters, soils, and biota. The current perspective is that some pharmaceuticals are of concern to fish and aquatic life. Environmental concentrations need to be reduced in aquatic systems to protect aquatic life, particularly the highly active compounds (i.e., steroids). In Chapter 14, information on different categories of pharmaceuticals with a focus on human pharmaceuticals is compiled and critically reviewed. Specific drug categories are discussed in light of their environmental occurrence and fate in the aquatic environment, modes of actions, and acute and chronic effects on fish. Thereby, we focus on environmentally important classes and compounds. We answer the question: What is the exposure and what are the effects and risks for fish health?

## INTRODUCTION

Pharmaceuticals are extensively and increasingly used in human and veterinary medicine. While antibiotics play the dominant role in veterinary medicine, many different classes are applied in human medicine. Over the last decade, not only sales figures increased but also the variety of compounds that are produced in substantial amounts (Fent *et al.*, 2006). The pharmaceuticals and their metabolites are released into the environment via excretion, and thus, untreated and treated municipal and hospital wastewaters are the major sources for human pharmaceuticals, while veterinary pharmaceuticals enter the environment via manure and runoff from livestock. Some of them may reach concentrations that may induce ecotoxicological effects including reproductive and developmental effects.

Pharmaceutical residues are very widespread in the aquatic environment. Several hundred compounds and metabolites are documented in wastewaters, fresh and marine waters, groundwaters, soils, and biota. The concentrations correlate positively with consumption figures. Generally, they range from below ng/L to more than µg/L (Ternes, 1998). In treated wastewater, they reach up to low µg/L (Verlicchi *et al.*, 2012), while in freshwater, they rarely exceed 100 ng/L (Kolpin *et al.*, 2002; Zhang *et al.*, 2020), but can also be found at low ng/L in marine waters, groundwater, or even in drinking water (Fent *et al.*, 2006). Extremely high concentrations of up to 100 µg/L or even mg/L occur in some countries below manufacturing plants (Larsson *et al.*, 2007; Phillips *et al.*, 2010; Sim *et al.*, 2011).

Due to the often polar and non-volatile properties, most pharmaceuticals are dissolved in water, degraded to a variable extent and distributed by aqueous transport in the environment. Food-chain accumulation and dispersal play only a very minor role. In wastewater treatment plants (WWTPs), pharmaceuticals and their metabolites are not completely eliminated although they undergo biodegradation in activated sludge treatment (aerobic degradation). Hydrophobic and ionic compounds are adsorbed to sludge and undergo anaerobic degradation. The degradation rates vary considerably depending on the type and performance of WWTP but also on the pharmaceutical. Degradation rates range about 40–99% in WWTPs with secondary and tertiary treatment (Verlicchi *et al.*, 2012; Gros *et al.*, 2010). Consequently, residues occur in effluents of WWTPs in the range from ng/L to µg/L (Fent *et al.*, 2006). However, there are compounds such as iodinated contrast agents and the antiepileptic compound carbamazepine that is not or minimally degraded (Heberer, 2002).

Historically, adverse effects of pharmaceuticals were unexpected, as it was believed that they were non-toxic and without risks. Since the 1990s, there is evidence that pharmaceuticals pose a particular risk to fish because of similar mechanisms of actions across vertebrates and their widespread exposure to chronic levels often for their whole life (Fent *et al.*, 2006; Corcoran *et al.*, 2010; Zheng et al., 2021), but also birds are at threat. The catastrophic decline of endemic vulture populations in the Indian subcontinent by over 90% (Balmford, 2013) originates from diclofenac, a widely used analgesic and non-steroidal anti-inflammatory drug (NSAID) in livestock. Diclofenac residues in carcasses of treated livestock lead to kidney failure in vultures that feed on them (Oaks *et al.*, 2004). Despite the ban of diclofenac in veterinary formulations, the population decline is not yet stopped due to the illegal use and application of other NSAIDs (aceclofenac, ketoprofen), bringing these vultures to the brink of extinction (Cornwall, 2018). This is the first reported case, where a common pharmaceutical has


DOI: 10.1201/9781003160694-14

such a devastating effect at the population level on wildlife. Diclofenac is still used in livestock in other regions, such as across Africa, resulting in similar effects in African vultures.

Chronic exposure of aquatic animals to low concentrations of pharmaceuticals is common, but there are also high concentration effects downstream from manufacture discharge, for instance of antibiotics in India and China (Fick *et al.*, 2009) or of endocrine active pharmaceuticals in France, leading to decreased population density and lack of sensitive fish species (Sanchez *et al.*, 2011). Hospital wastewaters and excretions from treated livestock also promote antibiotic resistance, which is a serious global problem.

The highest risks stem from the highly active steroid hormones. The estrogenic steroid in contraceptives, 17α-ethinylestradiol (EE2), impairs fish reproduction at very low concentrations (Länge *et al.*, 2001). Additional steroid hormones, particularly progestins (Fent, 2015), equine estrogens (Tyler *et al.*, 2009), and trenbolone (Garcia-Reyero *et al.*, 2009), affect reproduction at very low ng/L levels. Many of those compounds are found at effective concentrations in surface water and therefore represent the most critical pharmaceuticals in the environment posing a high risk for fish reproduction. As steroids occur in mixtures, they may have an additive joint effect, as do other pharmaceuticals.

Although pharmaceuticals have a diverse spectrum of activities and modes of action (MoA), many of them are designed to interact with specific targets including receptors, enzymes, and ion channels in humans. These biomolecules and their associated function are evolutionary conserved. Thus, fish are similarly affected in their physiological function as humans, although the receptors differ in specific structures. Consequently, the biological activity in fish is similar to mammals but the potency may differ. As with diclofenac, some species react even stronger than humans, suggesting important species differences. Pharmaceuticals of a given class are also not homogenous in their potency and effects may differ by several orders of magnitude (Zhang *et al.*, 2020).

Generally, the assessment of potential ecotoxicological effects for regulatory purposes is based on acute toxicity experiments in laboratory organisms including fish according to OECD guidelines or *in vitro* assessments. However, there is very little information about chronic toxicity, and *in vitro* systems are limited in this respect. In reaction to ecotoxicological effects, novel pharmaceuticals now need a risk assessment according to regulations implemented in several countries (Europe, USA, and Japan) in order to identify and reduce their environmental risks.

The current perspective is that some pharmaceuticals are of concern to fish and aquatic life. Environmental concentrations need to be reduced in aquatic systems to protect aquatic life, particularly the highly active compounds (i.e., steroids). What can be done to manage and mitigate the risks besides the assessment of novel pharmaceuticals? Direct discharge of untreated wastewater (particularly from production and hospitals) must be circumvented. In wastewater treatment, removal efficiencies should be increased by advanced techniques. Mitigation processes such as advanced wastewater treatment processes need to be implemented to reduce the pharmaceutical load in wastewater. Furthermore, hospital wastewaters need specific pretreatment prior to release into municipal treatment plants.

In this chapter, information on different categories of pharmaceuticals with a focus on human pharmaceuticals is compiled and critically reviewed. Specific drug categories are discussed in light of their environmental occurrence and fate in the aquatic environment, MoAs, and acute and chronic effects on fish. Thereby, we focus on environmentally important classes and compounds. We answer the question: What is the exposure and what are the effects and risks for fish health?

## ANALGESICS

NSAIDs and opioids are two dominant classes of analgesic drugs that have pain-relieving properties. NSAIDs are among the most frequently used pharmaceuticals worldwide in human and veterinary medicine, and also listed among the top ten persistent pollutants (Green, 2001). NSAIDs and opioids are commonly detected in the environment with concentrations ranging from several ng/L to dozens of µg/L (Verlicchi *et al.*, 2012; Petrie *et al.*, 2015; Campos-Mañas *et al.*, 2018). Most frequently detected are diclofenac, ibuprofen, and morphine. Diclofenac and ibuprofen may pose a potential hazard to fish due to the effects on reproduction and offspring development at relatively low concentrations, and their predicted no-effect concentrations (PNECs) are in the range of measured concentrations in hospital and WWTP effluents.

### Concentrations and Fate in the Aquatic Environment

NSAIDs are commonly detected in wastewater and surface water. Acetaminophen (paracetamol), ibuprofen, diclofenac, fenoprofen, and salicylic acid (the metabolite of aspirin) are the most frequently detected NSAIDs in raw influents and treated effluents of WWTPs with total concentrations of up to hundreds of µg/L, depending on the types of compounds and sampling time. Acetaminophen, ibuprofen, and naproxen concentrations in raw urban wastewater are of up to 246, 373, and 53 µg/L, respectively (Verlicchi *et al.*, 2012). In a Korean water reclamation facility, ibuprofen concentrations in influents vary from 641 to 1,768 ng/L (average 1,077 ng/L) and ibuprofen removal ranges from 90% to 96% (Park and Lee, 2018). High concentrations of ibuprofen, ketoprofen, and diclofenac also occur in treated wastewater in three Slovenian and Croatian WWTPs at concentrations of 4,330–6,130 ng/L, 53.8–2460 ng/L, and 113–812 ng/L, respectively (Česen *et al.*, 2019). Even higher levels occur in Portugal for acetaminophen (Togola and Budzinski, 2008) Currently, there are limited studies on the occurrence of opioids, but morphine and codeine concentrations can reach up to several thousands of ng/L in raw wastewaters (Campos-Mañas *et al.*, 2018).

In general, analgesics are fairly well treated in WWTPs. Thus, they are detected with lower concentrations in downstream surface waters. The most well-investigated NSAIDs are ibuprofen, naproxen, and ketoprofen. They exhibit moderate-to-high removal rates. The average removal efficiencies are about 90%, 75%, and 50%, respectively. In particular, the elimination of ibuprofen is relatively constant and commonly higher than 70%. On the other hand, the removal of diclofenac is inefficient (on average 36%) and variable (Verlicchi *et al.*, 2012; Luo *et al.*, 2014).

The concentrations of NSAIDs and opioids are lower in surface waters being generally at several to hundreds of ng/L. This is the result of dilution, dispersion, biotransformation, photolysis, sorption, and volatilization. For instance, ibuprofen is detected in surface waters at median concentrations of 0.98 ng/L (Canada), <8 ng/L (France), 1–67 ng/L (Greece), and <15–414 ng/L (Korea). The average concentration of ibuprofen in some European groundwaters is 3 ng/L, with a maximal concentration of 395 ng/L (Luo *et al.*, 2014). Similar concentrations occur for naproxen, ketoprofen, diclofenac, as well as mefenamic acid. Diclofenac, naproxen, ketoprofen, acetaminophen, and codeine are also among the most detected analgesics in European seawaters (Rodríguez-Navas *et al.*, 2013; Gros *et al.*, 2012). Ketoprofen occurs of up to 90 ng/L and the ibuprofen metabolite carboxyibuprofen up to 1,227 ng/L among others (Lolić *et al.*, 2015).

## Modes of Action

The primary mechanism by which NSAIDs exert their effects is the inhibition of the synthesis of prostaglandin, a group of lipids made at sites of tissue damage or infection, through blockade of the enzyme cyclooxygenase (COX) (Díaz-González and Sánchez-Madrid, 2015). Two types of cyclooxygenases exist, COX-1 and COX-2. COX-1 is constitutively present in virtually all tissues and is the dominant source of prostanoids that serve housekeeping functions. The location of the COX-1 enzyme dictates the function of the prostaglandin it produces. COX-2 is induced by inflammatory stimuli, hormones, and growth factors, and stimulates the production of prostaglandins that lead to symptoms of pain, heat, and swelling (Dubois *et al.*, 1998). Therefore, the benefits of NSAIDs in terms of their analgesic and anti-inflammatory properties mainly come from COX-2 inhibition.

Opioids are analgesics that act at several levels of the nervous system. In particular, they inhibit neurotransmitter release from the primary afferent terminals in the spinal cord. Opioids mimic the actions of endogenous opioid peptides by interacting with three types of opioid receptors located on neuronal cell membranes (Bovill, 1997). These receptors are coupled to guanine nucleotide-binding proteins known as G-proteins to produce effects. Opioids reduce neuronal excitability by closing N-type voltage-operated calcium channels and opening calcium-dependent inwardly rectifying potassium channels that result in cellular hyperpolarization (Bovill, 1997).

## Effects on Fish

NSAIDs and opioids act similarly in fish. Due to the persistence in surface waters, acetaminophen, morphine, and codeine reach plasma concentrations in fish that are 10 to 1,000 times higher than their concentrations in river water. Based on the fish plasma model (FPM), that includes effect ratios and the potential to activate evolutionary conserved biological targets, analgesics are regarded as of concern (Malev *et al.*, 2020).

### NSAIDs

Numerous studies have indicated that NSAIDs such as diclofenac cause histopathological effects in fish. Key organs are kidneys (degeneration of tubular epithelia, interstitial nephritis) and the gills (necrosis of pillar cells) where alterations occur in rainbow trout after 28 days of 5 μg/L diclofenac exposure (Schwaiger *et al.*, 2004). Subtle subcellular effects are further observed at a diclofenac concentration of 1 μg/L (Triebskorn *et al.*, 2004).

Oxidative stress is another effect of NSAIDs. Diclofenac increases superoxide dismutase (SOD), glutathione peroxidase (GPx) activities, and lipoperoxidation in traira fish *Hoplias malabaricus* (Guiloski *et al.*, 2015). A 28-day exposure of zebrafish to ibuprofen increases the activity of GPx at 500 μg/L and above (Bartoskova *et al.*, 2013). Besides increases in cellular oxidation and antioxidant biomarkers (SOD, CAT, and GPx enzymes), ibuprofen causes embryonic development disorders including pericardial edema, yolk deformation, and hypopigmentation in common carp (Gutiérrez-Noya *et al.*, 2020).

Prostaglandins play an essential role in fish development and reproductive processes. Thus, some NSAIDs such as ibuprofen and diclofenac may also act as endocrine disrupters and impact fish fertility. Exposure of zebrafish to ibuprofen for 21 days results in a decrease in egg production and adult growth at 266 μg/L, and reduces the time to hatch and total hatching success in the F1 generation (Constantine *et al.*, 2020). Exposure of Japanese medaka to ibuprofen for several weeks displays similar effects, including changes in egg production at 10 μg/L and above, and delayed hatching of eggs in the F1 generation (Han *et al.*, 2010). Ibuprofen also increases the concentrations of 17β-estradiol and testosterone in female zebrafish, while it decreases those of testosterone in males. Upregulations of *fshβ*, *lhβ*, *fshr*, and *lhr* mRNA are observed in females, whereas those transcripts are downregulated in males (Ji *et al.*, 2013). Similar alterations in reproduction, hatching success, steroid levels, and sperm quality occur for diclofenac and acetaminophen (Kim *et al.*, 2012; Guiloski *et al.*, 2015).

### Opioids

The opioid system in fish interacts with other neurotransmitter systems in modulating the behavior (Thörnqvist *et al.*, 2019). The presynaptic action of opioids to inhibit neurotransmitter release is a major effect on the nervous system. Morphine administered to rainbow trout ameliorates the effect of a potentially painful stimulus. It reduces ventilatory responses and anomalous behaviors, which are exhibited by noxiously treated fish (Baker *et al.*, 2013).

Morphine also ameliorates adverse changes in behavior after temperature threshold testing (Lopez-Luna *et al.*, 2017). Injection of salvinorin A, a potent κ opioid agonist, evokes dose-dependent alterations in adult zebrafish swimming, which accelerates at low (0.1 or 0.2 μg/kg) and slowed down at high doses (5 μg/kg and above) (Braida *et al.*, 2007). In the most sensitive test of zebrafish anxiety (novel tank test), acute morphine exposure increases the exploratory behavior of the top part of the tank and lowers locomotor behavior (Wong *et al.*, 2010).

There are only a few studies in which the reproductive and developmental effects of opioids on fish are explored. Reduced release of the luteinizing hormone is measured after carp pituitary tissue is exposed *in vitro* to morphine (Sokolowska-Mikolajczyk *et al.*, 2005). Treatment of Japanese medaka with codeine reduces egg production and the number of mature oocytes in ovaries at 100 ng/L and above, which matches with its concentrations in surface waters (as high as 815 ng/L) (Petrie *et al.*, 2015).

### Risk Assessment

Current knowledge indicates that levels of NSAIDs and opioids in aquatic systems do not pose a risk for acute toxicity. Environmental concentrations are generally several orders of magnitude lower than known $LC_{50}$ values. In terms of chronic effects, the environmental concentrations of most investigated compounds are too low to induce developmental dysfunctions in fish (Figure 14.1). However, the LOEC of diclofenac for histological effects is in the range of wastewater concentrations in some places (Schwaiger *et al.*, 2004; Triebskorn *et al.*, 2004). This suggests some risks of diclofenac to fish at highly contaminated sites. The environmental risk assessment for some frequently detected analgesics based on their hazard quotient (HQ) values for marine and freshwater fish calculated according to EU guidelines (Lolić *et al.*, 2015) suggests that diclofenac shows an HQ value above one, and thus, represents a potential risk for aquatic organisms. Ibuprofen also shows elevated HQs for fish, while the threshold limit is not exceeded (Lolić *et al.*, 2015). The situation is different for hospital and wastewater effluents, where ibuprofen shows HQ values higher than one for fish and other aquatic organisms. Therefore, it seems that diclofenac, and to a lesser extent, ibuprofen are the most critical NSAIDs currently known, which might be problematic at some highly contaminated sites (Santos *et al.*, 2013).

## CARDIOVASCULAR DRUGS

Cardiovascular diseases, including stroke, hypertension, and coronary artery disease, are the leading causes of death globally, taking an estimated 17.9 million lives each year (World Health Organization, https://www.who.int/). Cardiovascular drugs, therefore, are among the most prescribed medications. Among the top 200 most prescribed drugs in the USA, 49 (24.5%) are cardiovascular drugs (Fuentes *et al.*, 2018), and this holds similarly for Europe and many other highly developed countries (Gore *et al.*, 2008). Among the topmost prescribed drugs are lisinopril, amlodipine, metoprolol, and hydrochlorothiazide (Fuentes *et al.*, 2018).

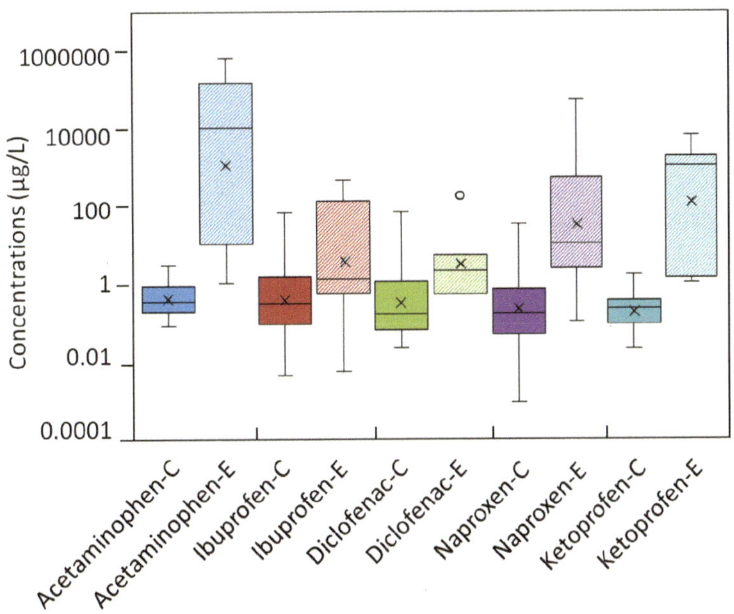


**FIGURE 14.1** Tentative risk assessment of analgesic drugs. Shown are the highest residues levels reported in surface waters and the chronic LOECs in aquatic organisms according to published literature. Data are presented as box and whisker plots, where boxes encompass values between the 25th and 75th percentiles, horizontal lines represent median values, and the whiskers give the minima and maxima, thus the variability outside the upper and lower quartile. Plots (x) represent mean values. C: The highest concentrations determined in surface waters; E: LOECs for fish.

Cardiovascular drugs comprise an array of compounds that act on the heart and blood vessels and include seven different drug classes, including beta-blockers, calcium channel blockers and angiotensin-converting enzyme inhibitors (ACEs), α adrenergic receptor modulators, angiotensin II receptor antagonist, antiarrhythmics, as well as diuretics (Zhang *et al.*, 2020; Fuentes *et al.*, 2018). These drugs are designed to have a long half-life in the body and therefore persist when excreted. They cannot be completely removed in WWTPs, and residues in treated wastewater reach up to several µg/L. In surface waters, their concentrations are at dozens to hundreds of ng/L (Hughes *et al.*, 2013; Stankiewicz *et al.*, 2015; Zhang *et al.*, 2020). Despite their frequent detection, these drugs have received not enough attention for their potential ecotoxicological implications in chronic terms.

## Concentrations and Fate in the Aquatic Environment

The environmental occurrence of cardiovascular drugs is well documented (Godoy *et al.*, 2015; Stankiewicz *et al.*, 2015; Zhang *et al.*, 2020). As with other drugs, municipal wastewater is the principal source, in which beta-blockers, diuretics, and angiotensin II receptor antagonists are dominant. Their total concentrations in influents and effluents of WWTPs can reach up to dozens of µg/L. For instance, valsartan and furosemide have been found in WWTPs effluents at concentrations of up to 15.7 µg/L (Huerta-Fontela *et al.*, 2010; Gurke *et al.*, 2015). Atenolol occurs at up to 33.1 µg/L and 7.6 µg/L in municipal WWTPs influents and effluents, respectively, in South Wales (UK) (Kasprzyk-Hordern *et al.*, 2009). Most studies stem from Europe, Asia, and North America. Average concentrations are between 0.89, 1.57, 0.78, and 2.23 µg/L in WWTP influents, and 0.26, 0.37, 0.39, and 0.78 µg/L in effluents in the USA, the UK, Germany, and Spain, respectively (Godoy *et al.*, 2015; Zhang *et al.*, 2020).

The removal efficiency of cardiovascular drugs in WWTPs is not known to a full extent. When tested, it is mainly estimated by comparing concentrations in influents and effluents. Among beta-blockers, metoprolol and propranolol have low removal rates, which are generally of up to 42% only. In comparison, atenolol seems to have also high degradation rates of up to 97% in WWTPs (Zhang *et al.*, 2020). Additional removal of cardiovascular drugs in WWTPs may be achieved by advanced treatment technologies such as advanced oxidation. However, related investigations are still rare.

In surface water, concentrations are not much lower in many places as in treated wastewater being in the range of low ng/L to several µg/L. Of these drugs, diuretics (furosemide and hydrochlorothiazide), angiotensin II receptor antagonists (olmesartan and telmisartan), and the beta-blocker atenolol are dominant, which goes parallel to their high concentrations in wastewater (Stankiewicz *et al.*, 2015; Zhang *et al.*, 2020). Examples of high concentrations are the principal rivers of the Madrid Region (Spain), where the highest concentrations corresponded to hydrochlorothiazide (17.6 µg/L), atenolol (6.0 µg/L), and furosemide (3.2 µg/L), which in most cases exceeds concentrations in other locations (Muñoz *et al.*, 2009). However, atenolol also occurs at high concentrations in the Umgeni River of South Africa (0.96–39.10 µg/L; median concentration 11.09 µg/L) (Agunbiade and Moodley, 2014). As other pharmaceuticals diuretics and beta-blockers occur in surface waters as mixtures with other drugs (Muñoz *et al.*, 2009; Huerta-Fontela *et al.*, 2010). However, typical concentrations are in the ng/L level.

## Modes of Action

There are multiple classes of antihypertensive medications that can lower blood pressure by different means. Of them, the most recommended classes used as first-line for treatment are as follows:

1. Thiazide diuretics have been the cornerstone of hypertension treatment for more than five decades and are recommended as first-line drugs for patients with hypertension. They achieve their diuretic action via inhibition of the $Na^+/Cl^-$ cotransporter in the renal distal convoluted tubule (Gamba, 2009).
2. Calcium channel blockers act by blocking the influx of calcium ions into vascular smooth muscle and cardiac muscle cells during membrane depolarization. Because muscle contraction is largely dependent upon the influx of calcium, its inhibition causes relaxation of vascular and arterial smooth muscle cells resulting in arterial vasodilation (Godfraind, 2017).
3. Beta-blockers bind to beta-adrenoreceptors and work by blocking the action of certain hormones in the nervous system, such as adrenaline. Thereby, they help prevent the activation of the "fight-or-flight" stress response and sympathetic stimulation of the heart (Hainer *et al.*, 2005).
4. Angiotensin-converting enzyme (ACE) inhibitors and angiotensin II receptors antagonists, as the name implies, interfere with the different steps of the renin–angiotensin–aldosterone system, which is responsible for regulating the blood pressure and electrolyte balance (Chrysant, 2010). ACE inhibitors block an angiotensin-converting enzyme that converts angiotensin I to angiotensin II. Decreased production of angiotensin II enhances natriuresis and lowers blood pressure. Angiotensin II receptors antagonists work by blocking receptors that the hormone acts on, specifically AT1 receptors. Blocking the action of angiotensin II helps to lower blood pressure and prevent damage to the heart (Celle-Jeanton *et al.*, 2014).

Antiarrhythmics are used to suppress abnormal rhythms of the heart (cardiac arrhythmias), such as atrial fibrillation and ventricular tachycardia. They comprise various antiarrhythmic agents that act in multiple MoAs. Generally, they can be classified into five main classes based on the targets to block, that is (I) interfere with the sodium ($Na^+$) channel; (II) affect the anti-sympathetic nervous system; (III) affect potassium ($K^+$) efflux; (IV) affect calcium channels; or (V) work by unknown mechanisms (Lei *et al.*, 2018).

## Effects on Fish

Currently, there are many studies on the impact of cardiovascular drugs on fish, but there is still a lack of chronic and mechanism-based studies. Most studies focus on beta-blockers but the most commonly detected diuretics (including furosemide and hydrochlorothiazide) and ACE inhibitors and angiotensin II receptor antagonists (including valsartan and telmisartan), have not received enough attention. Only a dozen drugs are tested for their adverse effects on fish, especially on zebrafish and fathead minnows. Beta-blockers, including propranolol and metoprolol are identified as of potential concern (Zhang *et al.*, 2020).

### Reproductive Effect

Effects on fish reproduction and the endocrine system are evaluated for beta-blockers including atenolol, metoprolol, and propranolol. Collectively, the data do not suggest that they act as potent reproductive toxicants although they act as endocrine disruptors. A 21-day reproductive study with propranolol in fathead minnow suggests that the most sensitive endpoints are the decrease in hatchability and increase in gonadal somatic index. No effects on egg production are observed at concentrations of up to 100 µg/L (Giltrow *et al.*, 2009). A long-term exposure to metoprolol, from the fertilized egg until 80 days post-hatch, also indicates low reproductive activity and a reduction in growth in Nile tilapia (Gröner *et al.*, 2017). In medaka and zebrafish, propranolol leads to significant changes in the levels of fish plasma steroids, including estradiol and testosterone, at concentrations of several to dozens µg/L (Huggett *et al.*, 2002; Mitchell and Moon, 2016).

### Behavior

Behavior analyses are mainly conducted for propranolol. For example, the swimming pattern of the neotropical fish *Phalloceros harpagos* is altered during a 96-h acute exposure to 1 µg/L (Matus *et al.*, 2018). Reproductive behavior is altered by propranolol as indicated by fathead minnow males visiting the nest less (Lorenzi *et al.*, 2012). Propranolol also reduces anxiety-like behaviors in adult zebrafish, which coincided with reduced testosterone and estradiol levels (Mitchell and Moon, 2016). Similarly, verapamil modulates the aggressive behavior of Siamese fighting fish. It alters the latencies to the first chase and also the attack frequency and duration at 8 µg/L and above (Kania *et al.*, 2015). Diltiazem leads to changes in swimming behavior of larval fathead minnow, including reduced light:dark bursting distance traveled and the number of movements (Saari *et al.*, 2018).

### Effects on Cardiac Development

Mainly zebrafish and Japanese medaka are studied and targets of cardiovascular drugs such as cardiac developmental morphology and functions including heartbeat, cardiac output, and heart muscle contraction are affected. Propranolol leads to a significant reduction in the heart rate of zebrafish and medaka embryos at concentrations of 0.08 µg/L and 0.09 µg/L, respectively, along with an increase in the distance between the sinus venosus and bulbus arteriosus in zebrafish but not medaka (Finn *et al.*, 2012). At high concentrations, propranolol alters multiple developmental parameters including the heart rate in zebrafish. It also weakened pigmentation, induced tail curvatures, and pericardial edema. Among them, the heart rate is the most sensitive target (Fraysse *et al.*, 2006; Sun *et al.*, 2014). Metoprolol also causes a reduction in heart rate and hatching rate in zebrafish embryos but at extremely high concentrations of 4 mg/L and above. Changes in the transcription of the β-AR genes such as *adrb1* and *adrb2b* are observed at a metoprolol concentration of 10 µg/L and above (Sun *et al.*, 2014). Furthermore, metoprolol increases blood flows of zebrafish by up to 14% at 0.04 µM, while showing no effects on fish cardiac outputs (Meng *et al.*, 2020).

## Risk Assessment

Based on risk quotient (RQ) values, in which the surface water concentrations are compared with their LOECs in aquatic organisms (Figure 14.2), a tentative risk assessment is conducted. The LOECs are generally in the range of several µg/L to dozens of mg/L. However, the concentrations of cardiovascular drugs in surface waters are much lower (dozens to hundreds ng/L). Only in a few cases, observed effects occur at environmentally relevant concentrations (Figure 14.2). Generally, they include transcriptional and/or biochemical effects (Godoy *et al.*, 2015; Zhang *et al.*, 2020). Based on the RQ values, low environmental risks are suggested for the beta-blockers atenolol, metoprolol, and propranolol, where effect concentrations and surface water concentrations partially overlap (RQ > 1). Indeed, propranolol concentrations in surface waters are regarded as with no risk for fish health after a thorough analysis of available data (Sumpter *et al.*, 2021).

Low risks exist for furosemide and verapamil but it should be noted that they occur in mixtures with other compounds and some of the adverse effects occur near to their maximal environmental concentrations (Zhang *et al.*, 2020). Further research is needed paying attention to some newer generations of cardiovascular drugs, in particular regarding their environmental fate and occurrence, and their potential chronic effects on aquatic biota. Based on the MoA of these compounds, more emphasis on

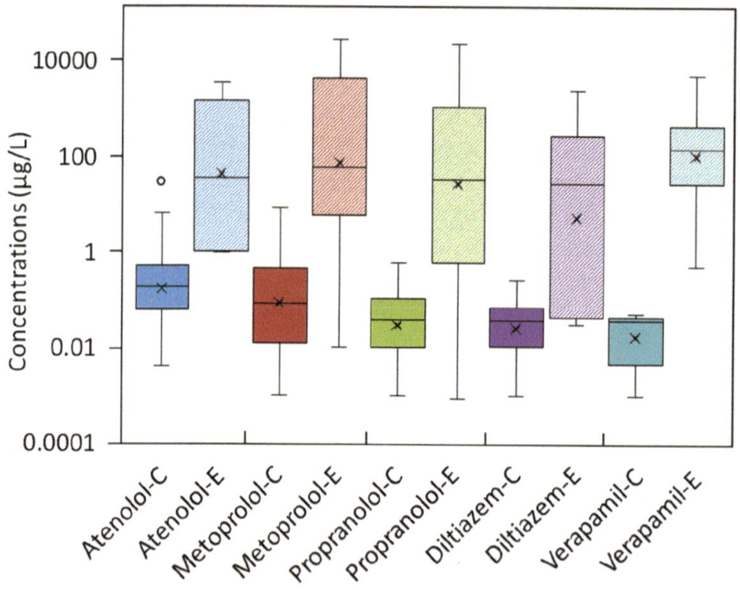


**FIGURE 14.2** Tentative risk assessment of cardiovascular drugs. Shown are the highest residues levels reported in surface waters and the chronic LOECs in aquatic organisms according to published literature. Data are presented as box and whisker plots, where boxes encompass values between the 25th and 75th percentiles, horizontal lines represent median values, and the whiskers give the minima and maxima, thus the variability outside the upper and lower quartile. Plots (x) represent mean values. C: Compiled data representing highest concentrations determined in surface waters; E: Compiled data representing LOECs for fish.

cardiovascular function should be placed to identify potential risks for fish. Presently, these drugs seem to pose no or very little risk for fish, as indicated in Figure 14.2.

## LIPID-REGULATING AGENTS

Lipid-regulating agents, also called lipid-lowering drugs, are a diverse group of medications that are used in the treatment of high levels of lipids, such as cholesterol, in the blood. Lipid-regulating agents are among the most prescribed medications worldwide. Their average usage (daily dose) has at least tripled in OECD countries between 2000 and 2017 (OECD, https://www.oecd-ilibrary.org/). Lipid-regulating agents comprise several classes of hypolipidemic drugs that differ in mechanisms of cholesterol regulation and their adverse effects. Of them, statins and fibrates are two major groups. In raw and treated wastewater, their concentrations are up to several µg/L. In surface waters, several to hundreds of ng/L concentrations are generally detected (Ternes, 1998; López-Serna *et al.*, 2012; Muñoz *et al.*, 2009). As in humans, lipid-regulating agents alter lipid metabolism in fish too. Additional evidence further suggested that lipid-regulating agents, such as bezafibrate and gemfibrozil, affect reproductive success in fish (Galus *et al.*, 2013; Zhang *et al.*, 2020).

### Concentrations and Fate in the Aquatic Environment

Gemfibrozil, bezafibrate, and clofibric acid are most frequently detected. For the newer generation of lipid-regulating agents, such as ezetimibe and fluvastatin, only limited data exist. The concentrations in influents and effluents of WWTPs could reach up to several µg/L. For instance, in WWTPs of Rome (Italy), gemfibrozil concentrations are up to 1.77 and 1.17 µg/L in influents and effluents, respectively, and removal rates vary from 19% to 78% (Patrolecco *et al.*, 2015). In influents of a WWTP in Madrid (Spain), concentrations of gemfibrozil, bezafibrate, and clofibric acid are up to 17.06, 0.36, and 0.12 µg/L, respectively. In effluents, they reach up to 5.23, 0.28, and 0.09 µg/L, respectively (Rosal *et al.*, 2010).

In general, removal rates of fibrates vary with the type of WWTP and its performance. For instance, the removal rate of bezafibrate is 9% in Madrid (Spain), while it reaches up to 87–99% in Guangdong (China) and 98% in Dublin (Ireland) (Lacey *et al.*, 2012; Tang *et al.*, 2019). Similarly, gemfibrozil shows no or very poor removal in the WWTPs of Castellon (Spain), low in Rome (Italy), while very high removal rates in Lausanne (Switzerland) (≈97%) (Zhang *et al.*, 2020). Statins seem to have more stable removal efficiencies compared to fibrates. For instance, the removal rates of losartan and valsartan range from 82% to 93% and 85% to 95%, respectively (Zhang *et al.*, 2020).

In surface water, concentrations are generally lower with typical levels in the ng/L range. Gemfibrozil, bezafibrate, and clofibric acid levels vary from several to up to hundreds of ng/L but are sometimes higher. In surface waters of Madrid, 26–1,014 ng/L gemfibrozil are reported and up to 630 ng/L clofibric acid in the drinking water system of Southern California (USA) (Loraine and Pettigrove, 2006; López-Serna *et al.*, 2012). Gemfibrozil, bezafibrate, clofibric acid, and fenofibric acid occur in German rivers, with concentrations of up to 510, 3,100, 550, and 280 ng/L, respectively (Ternes, 1998). In South Wales (UK), bezafibrate and clofibric acid are detected at up to 66 and 164 ng/L (Kasprzyk-Hordern *et al.*, 2009). These concentrations are rather high and above typical levels, which are in the ng/L range.

## Modes of Action

Lipid-lowering pharmaceuticals are mainly classified into two groups: statins and fibrates. Both differ in mechanisms of lipid metabolism regulation and also in their adverse effects (Al-Qirim *et al.*, 2015). Statins, including atorvastatin, rosuvastatin, and simvastatin, inhibit HMG-CoA reductase activity, a key rate-limiting enzyme of the cholesterol biosynthetic pathway in vertebrates. Statins are similar in structure to HMG-CoA on a molecular level and thereby can inhibit HMG-CoA reductase activity by competitively binding with an affinity >1,000 times greater than that of the natural substrate (Barter and Rye, 2016; Pahan, 2006). In addition to direct inhibition, statins also lower plasma cholesterol levels by upregulation of the low-density lipoprotein receptors (Vaziri and Liang, 2004).

Fibrates, including gemfibrozil, bezafibrate, and clofibrate, and its metabolite clofibric acid, activate peroxisome proliferator-activated receptor α (PPARα) (Al-Qirim *et al.*, 2015). As a transcription factor, PPARα regulates the expressions of a series of genes involved in multiple metabolic pathways including lipid metabolism. Thereby, fibrates stimulate cellular fatty acid uptake, conversion to acyl-CoA derivatives, and catabolism by β-oxidation pathways leading to the reduction in fatty acid and triglyceride synthesis (Barter and Rye, 2016).

Besides, statins and fibrates display other pleiotropic effects including the improvement of endothelial function, enhancement of the stability of atherosclerotic plaques, and decrease of oxidative stress and inflammation (Liao and Laufs, 2005). These pleiotropic effects may reduce cardiovascular morbidity. Due to the conserved evolutionary nature of the targets, statins and fibrates may also act on fish and even in invertebrates that lack an active blood-circulating system (Zhang *et al.*, 2020).

## Effects on Fish

### Lipid Metabolism

Exposure of rainbow trout to the fibrate gemfibrozil for 15 days results in decreased plasma lipoprotein levels and reduced size of all lipoprotein classes. The lipoprotein composition is also modified, characterized by the reduced abundance of long-chain n−3 fatty acids, thereby potentially reducing the nutritional quality of exposed fish (Prindiville *et al.*, 2011). In fathead minnows, gemfibrozil alters plasma triglycerides and cholesterol levels in addition to the changes of transcriptional levels of representative genes involved in lipid metabolism such as *apoa1* and *apoeb* (Skolness *et al.*, 2012). A similar phenomenon is noted in zebrafish upon gemfibrozil exposure (Al-Habsi *et al.*, 2016). Bezafibrate also decreases plasma cholesterol concentrations in zebrafish (Velasco-Santamaría *et al.*, 2011).

Statins show similar effects. Feeding adult zebrafish with atorvastatin over 30 days results in the reduction of whole-body cholesterol content. Concomitant changes in gene expressions related to lipid homeostasis including *srebp-1*, *srebp-2*, *pparα*, and *pparγ* occur (Al-Habsi *et al.*, 2016). Simvastatin promotes the formation of lipid droplets in the hepatic cytoplasm in mosquitofish (Bao *et al.*, 2018).

### Reproductive Effects

Some lipid-regulating agents cause endocrine disorders and compromise reproduction in fish. Long-term exposure of Japanese medaka to gemfibrozil decreases fecundity and reduces hatchability at a high concentration of 3.7 mg/L (Lee *et al.*, 2019). Chronic exposure of zebrafish to 0.5–17.5 μg/L gemfibrozil for weeks also decreases fecundity. Atretic oocytes and altered ovarian histology suggest a direct effect on oocyte development in females (Galus *et al.*, 2013). Furthermore, gemfibrozil decreases testosterone and estradiol levels in fish blood and leads to downregulation of the transcription of genes involved in steroidogenesis, such as *er*, *StAR*, and *cyp17* in adults (Mimeault *et al.*, 2005).

Clofibric acid shows a lower potency: A decline in fecundity is noted at a very high concentration of 108.9 mg/L in fathead minnows (Weston *et al.*, 2009). However, effects on male gonadal development are more potent. Clofibric acid impairs spermatogenesis resulting in a reduction of sperm count, altered sperm motility, and a slight reduction of plasma androgen levels in fathead minnows at 1 mg/L (Runnalls *et al.*, 2007).

Exposure of zebrafish to bezafibrate (70 mg/g food) leads to disruption in gonadal steroidogenesis and spermatogenesis. Bezafibrate reduces plasma 11-ketotestosterone levels after a 21-day exposure and increases transcriptional expression of *star* and *cyp1*. Gonadal histology reveals a testicular degeneration (Velasco-Santamaría *et al.*, 2011). Statins can also have endocrine disruptive effects, but there are only limited data. Exposure of adult zebrafish for 30 days to atorvastatin reduces testosterone and estradiol plasma levels at 0.53 μg/g food (Al-Habsi *et al.*, 2016).

### Fish Early-Life Stages

Effects on fish development are mainly observed for simvastatin, clofibrate, and gemfibrozil. Chronic exposure (90 days) of zebrafish embryos to simvastatin leads to impacts including decreased body weight and morphologic tail abnormalities (Barros *et al.*, 2018). Other morphological abnormalities occur in the eyes and yolk sac, as well as the occurrence of pericardial edema and impairment of primordial germ cell migration (Campos *et al.*, 2016; Ribeiro *et al.*, 2015). Clofibrate or gemfibrozil induces an embryonic malabsorption syndrome with very little yolk consumption, resulting in small-sized fish larvae (Raldúa *et al.*, 2008). Clofibrate further delays hatching time and induces higher density, round-shaped neuromuscular junctions, pericardial edema, and impairs thyroid gland morphogenesis (Raldúa *et al.*, 2008).

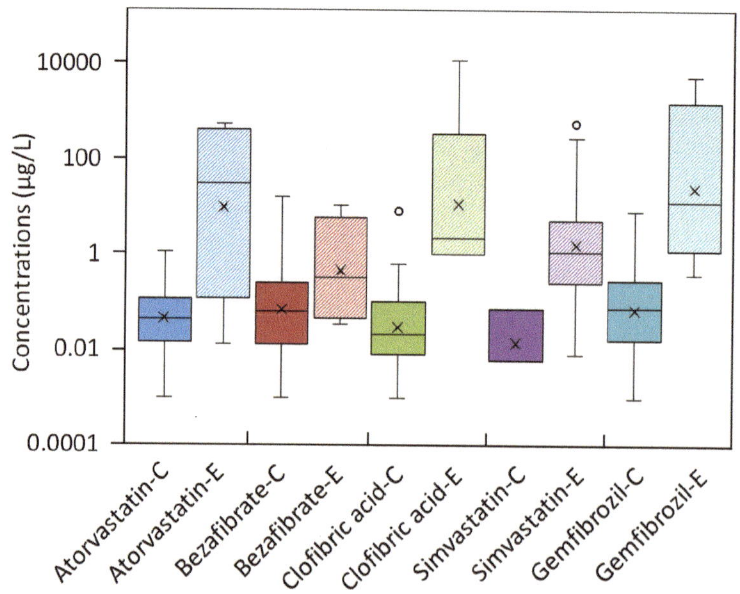


**FIGURE 14.3** Tentative risk assessment of lipid-regulating agents. Shown are the highest residues levels reported in surface waters and the chronic LOECs in aquatic organisms according to published literature. Data are presented as box and whisker plots, where boxes encompass values between the 25th and 75th percentiles, horizontal lines represent median values, and the whiskers give the minima and maxima, thus the variability outside the upper and lower quartile. Plots (x) represent mean values. C: Compilation of data on highest concentrations determined in surface waters; E: Compilation of LOECs for fish.

### Risk Assessment

Bezafibrate is suggested to be more critical than other drugs such as gemfibrozil, clofibric acid, and simvastatin (Figure 14.3). There are examples of high bezafibrate concentrations of up to 15.1 μg/L in the Llobregat river (Spain) (Muñoz *et al.*, 2009) and 3.1 μg/L in the Frankfurt region (Germany) (Ternes, 1998). Such concentrations may induce most sensitive effects, such as genotoxic effects, similar to atorvastatin that reaches up to 1.1 μg/L in the Aranyhegyi stream (Hungary) and 0.86 μg/L in southwest Nova Scotia, Canada (Varga *et al.*, 2013; Ghoshdastidar *et al.*, 2015). Short-term exposure to atorvastatin and bezafibrate leads to DNA damage in zebrafish at ng/L concentrations (Rocco *et al.*, 2012) and alters transcriptional levels of genes involved in oxidative stress, apoptosis, and biotransformation in rainbow trout at a concentration of 1.03 μg/L (Ellesat *et al.*, 2012). Taken together, the risk for physiological effects of lipid-lowering drugs to fish is suggested to be low in surface water.

## NEUROACTIVE COMPOUNDS

Neuroactive pharmaceuticals are prescribed widely to treat diseases where neuronal signaling is dysbalanced including depression, attention-deficit hyperactivity disorder, epilepsy, and bipolar disorder. In addition to that, illicit drugs are consumed to alter moods, perceptions, and behaviors. Both neuroactive pharmaceuticals and illicit drugs are on the rise in production volume, and all of these drugs find their entry into surface waters primarily through untreated as well as treated wastewater. For example, selective serotonin reuptake inhibitors (SSRIs) used to treat depression readily pass through WWTPs and accumulate in wild-caught fish species. Often, the mechanism of action of the prescribed drugs is not fully elucidated in humans, let alone in fish. Predominantly, neuroactive compounds and illicit drugs interfere with synapse functioning such as binding to neurotransmitter receptors or dysregulation of calcium homeostasis altering activity, feeding behavior, and boldness in fish. There are many direct functional interactions between the nervous system and the endocrine system. Therefore, neuroactive compounds are often found to impair reproduction while reversely endocrine active compounds may alter behavior. Laboratory experiments have shown that for a few examples, environmentally relevant concentrations can alter both activity and reproductive output of fish. Such effects can have consequences for population and community structure, yet due to the scarce data and complexity of multi-level interactions, the impact on fish and ecosystems remains largely unknown.

### Concentrations and Fate in the Environment

Neuroactive pharmaceuticals in the environment have two main sources: sewage from households and psychiatric hospitals. With elimination in WWTPs being far from complete and in many cases wastewater being untreated, neuroactive compounds are measured as complex mixtures

**TABLE 14.1**
**Representative Concentrations of Selected Neuroactive Compounds in Surface Water**

| Therapeutic Class | Neuroactive Compounds | Concentration (ng/L) | References |
|---|---|---|---|
| SSRI | Fluoxetine | 3–141[a] | Metcalfe *et al.* (2010) |
| | | 1–9 | Schultz *et al.* (2010) |
| | | 5.8–13.5[b] | Baker and Kasprzyk-Hordern (2013) |
| | | 25.8–26.8[b] | Subedi *et al.* (2021) |
| | Citalopram | 620 | Vasskog *et al.* (2008) |
| | | 78.7–170[b] | Subedi *et al.* (2021) |
| | Venlafaxine | 212–364[b] | Subedi *et al.* (2021) |
| Sodium channel blocker | Carbamazepine | 520±130 | Zind *et al.* (2021) |
| | | 370±14 | Du *et al.* (2014) |
| | | 830±436 | Voloshenko-Rossin *et al.* (2015) |
| | | 129–168[b] | Subedi *et al.* (2021) |
| Benzodiazepine ($GABA_A$ receptor binding) | Diazepam | 0.7–0.9[b] | Baker and Kasprzyk-Hordern (2013) |
| | | 3.16–3.25[b] | Subedi *et al.* (2021) |
| | | 15.26–79.33[b] | Xiang *et al.* (2018) |
| | Oxazepam | 5.66–8.97[b] | Subedi *et al.* (2021) |
| Illicit drugs | Cocaine | 0.3–3.7[b] | Baker and Kasprzyk-Hordern (2013) |
| | | 15–44[b] | Zuccato *et al.* (2008) |
| | | 11.5–14.9[b] | Subedi *et al.* (2021) |
| | Amphetamine | 0.7–3.8[b] | Baker and Kasprzyk-Hordern (2013) |
| | | <0.65 | Zuccato *et al.* (2008) |
| | | 2.32–3.06[b] | Subedi *et al.* (2021) |
| | Methamphetamine | 0.1–0.3[b] | Baker and Kasprzyk-Hordern (2013) |
| | | 1.7–2.1[b] | Zuccato *et al.* (2008) |
| | | 1.9–130 | Koba *et al.* (2018) |
| | | 15.2–18.4[b] | Subedi *et al.* (2021) |
| | MDMA | 0.5–5.8[b] | Baker and Kasprzyk-Hordern (2013) |
| | | 1.1–1.7[b] | Zuccato *et al.* (2008) |
| | Morphine | 1.3–9.5[b] | Baker and Kasprzyk-Hordern (2013) |
| | | 3.5–38[b] | Zuccato *et al.* (2008) |
| | | 10.7–14.1[b] | Subedi *et al.* (2021) |
| | Codeine | 11.7–83.7[b] | Baker and Kasprzyk-Hordern (2013) |
| | | 12–51[b] | Zuccato *et al.* (2008) |
| | | 13.1–15.1[b] | Subedi *et al.* (2021) |
| | Cannabis (THC-COOH) | <0.48–3.7[b] | Zuccato *et al.* (2008) |

[a] Range of mean values from different sampling sites.
[b] Range of mean values of different samplings from the same site.

in surface waters (see Table 14.1). For example, a creek that runs through an urban area of Kentucky (USA) and receives treated wastewater had levels of benzodiazepine tranquilizers (oxazepam and temazepam), anxiolytics (carbamazepine, diazepam), antidepressants (fluoxetine, sertraline, venlafaxine, and citalopram), illicit stimulant (cocaine and one metabolite), opioids (heroin, codeine, morphine, methadone, methadone metabolite, and oxycodone), and a hallucinogen (MDA) (Subedi *et al.*, 2021). Swimming in this drug mixture, fish can accumulate these compounds in various tissues, with the brain being a particular target organ for antidepressants. During an environmental monitoring in Germany, the UK, and the Czech Republic, 94 pharmaceuticals were analyzed in blood plasma from wild bream, chub, and roach (Cerveny *et al.*, 2021). The study identified three neuroactive compounds, flupentixol, haloperidol, and risperidone, at levels that were close to or exceeded the therapeutic plasma concentrations in humans. Pharmaceuticals are designed to be ultimately metabolized and excreted by the patients; thus, it can be expected that most other vertebrates do so too and therefore bioaccumulation factors are rather low. Illicit drugs, on the other hand, have been reported to have bioaccumulation factors of multiple thousands in fish (Yin *et al.*, 2019) which may be related to high accumulation and slow elimination as shown, for example, for cocaine in zebrafish (Kirla *et al.*, 2016).

## Selective Serotonin Reuptake Inhibitors (SSRIs)

One of the most prevalent neuroactive pharmaceuticals in the surface waters of the western world is the antidepressant fluoxetine, which is an SSRI and the active ingredient of Prozac. Fluoxetine and its active metabolite norfluoxetine are regularly detected in freshwater systems below WWTPs, where it readily absorbs to sediments (Mennigen *et al.*, 2011). Typical concentrations of fluoxetine and other SSRI compounds in water bodies downstream of wastewater effluents are several ng/L (Schultz *et al.*, 2010) up to several hundred ng/L (Subedi *et al.*, 2021). Despite a low $k_{ow}$ value, fluoxetine is found to accumulate in lipid-rich brain tissue of wild-caught fish at concentrations of 0.1 ng/g (Schultz *et al.*, 2010), which translates to a bioconcentration factor of 10 or higher. The bioconcentration factor appears to be not solely related to the $k_{ow}$ value but to the ionization capacity of fluoxetine at ambient pH values or a lower biotransformation capacity in the liver of fish compared to humans. Fluoxetine is only one of several SSRIs. The combined concentrations of SSRIs and metabolites in close proximity to wastewater effluents are up to 840 ng/L (Vasskog *et al.*, 2008).

### Carbamazepine

The antiepileptic drug carbamazepine is one of the most prevalent pharmaceuticals detected in the environment, and its metabolites are measured at concentrations of multiple hundred ng/L in surface waters across the entire globe (Zind *et al.*, 2021; Du *et al.*, 2014; Voloshenko-Rossin *et al.*, 2015). Carbamazepine is highly resistant to photodegradation and has a long half-life time of 82±11 days as measured in mesocosms (Lam *et al.*, 2004). On top of that, removal efficiencies of WWTPs are no more than 10% as it is resistant to biodegradation at low concentrations and hardly adsorbs to sludge (Zhang *et al.*, 2008), although improved removal can be obtained by advanced activated carbon treatment. In a neotropical fish, *Jenynsia multidentate*, carbamazepine accumulated in all studied organs with the highest concentrations in the brain and liver (Valdés *et al.*, 2016). The bioconcentration factor is generally reported to be below 10 for fish which may be related to active biotransformation (Valdés *et al.*, 2016).

### Benzodiazepines

Diazepam (Valium™) is the most widely prescribed benzodiazepine, which is used to treat a variety of medical conditions from anxiety disorders to muscle spasms. Due to the worldwide consumption, incomplete metabolism in the human body, and only partial elimination in WWTPs, diazepam is frequently detected in effluents. However, the presence of diazepam is relatively low in surface waters (0.7–79.33 ng/L) (Baker and Kasprzyk-Hordern, 2013; Xiang *et al.*, 2018; Subedi *et al.*, 2021). Benzodiazepines can cross the blood–brain barrier relatively rapidly, and therefore, the plasma concentrations are proportionately related to concentrations in the brain (Huerta *et al.*, 2016). In a laboratory experiment with fathead minnow exposed to oxazepam, another benzodiazepine, the highest concentrations were measured in the brain, followed by liver and plasma, whereas the lowest values were detected in the muscle (Huerta *et al.*, 2016). Bioconcentration factors across the sampled organs and the plasma were in the range of 0.9 (muscle) and 51 (brain).

### Illicit Drugs

Illicit drugs are non-prescribed drugs or psychotropic substances. They are recognized as environmental contaminants of emerging concern to both human and environmental health due to the increased global production and consumption—owing only in part to global population growth—as well as largely unknown impacts on environmental health (UNODC, 2021). Illicit stimulants and opioids partially pass through WWTPs and find their way into receiving water bodies, particularly in urban areas (Zuccato *et al.*, 2008). In WWTPs, the activated sludge treatment is most efficient in eliminating illicit drugs but efficiency depends on the compound: While the removal of amphetamine and cocaine can be more than 90%, for 3,4-methylenedioxymethamphetamine (MDMA) or methamphetamine, it is often below 50% (Yadav *et al.*, 2017). Bioaccumulation studies with illicit drugs in fish are still scarce. In European eels, cocaine accumulated at the highest levels in the brain, muscle, and digestive tract, while the lowest levels were measured in the spleen, and gills and gonads appeared to be cocaine-free (Capaldo *et al.*, 2012). Notably, in zebrafish larvae, cocaine accumulated to the largest extent in the eyes (Kirla *et al.*, 2016). In wild carp caught in urban rivers in Beijing, amphetamine, methylenedioxyamphetamine (MDA), codeine, and heroin were detected with high frequencies (Yin *et al.*, 2019). Amphetamine and MDA were measured at up to multiple ng/g in muscle, gastrointestinal tract, and liver. The concentrations of codeine and heroin were an order of magnitude lower. The same study reports on high bioaccumulation factor of up to 19,993 for amphetamine and 15,260 for codeine in whole fish. Further research will be required to get a better understanding of the fate and bioaccumulation of illicit drugs in aquatic environments.

## Modes of Action

The gross neurological processes such as neurotransmitter systems are conserved even across phyla, as shown, for example, by the serotonin transporter (SERT) acting as MDMA binding site in octopuses (Edsinger and Dölen, 2018). Therefore, it can be expected that the MoAs of neuroactive pharmaceuticals are similar between humans and other aquatic animals such as fish.

One of the main MoAs of neuroactive compounds is the interference with electric and chemical signal transduction, mainly through binding to transporters and receptors. For example, SSRI antidepressants block the serotonin transporter, antiepileptic drugs such as carbamazepine inhibit sodium channel firing—although its mechanism of action is not fully elucidated—and benzodiazepine drugs such as diazepam enhance GABAergic transmission at benzodiazepine-sensitive $GABA_A$ receptors. These pharmaceuticals have been studied for their activity in fish, finding parallels with their activities in mammals. Besides the SSRIs, there are other antidepressants (mianserin, bupropion, and clomipramine) that interact with histamine, κ-opioid, nicotinic, acetylcholine, and adrenergic receptors. These pharmaceuticals have been measured in wild fish (Schultz *et al.*, 2010), but have received considerably less attention in terms of their MoAs and effects in fish.

The endocrine and the neuronal system are tightly coupled, together building the neuroendocrine system. Thus, besides pharmaceuticals that directly interfere with the synthesis and/or metabolism of neuropeptides and neurotransmitters, pharmaceuticals that have endocrine activity (see Chapter 8) can interfere with the synthesis of neurohormones and thereby impair neurological function. The neuroendocrine system regulates the neuronal development, thyroid, metabolic, and stress axes, and its deregulation can have profound biological consequences including the reproduction, development, growth, and response to stress and other challenges. For example, fluoxetine can affect multiple neuroendocrine pathways and behavioral circuits, altering reproduction and feeding in fish.

## Effects

Adverse physiological effects of neuroactive compounds encompass reproductive and behavioral alterations. The latter is particularly interesting as fish are extremely diverse, displaying a wide variety of hunting, mating, and social behaviors. It is thus not surprising that neuroactive pharmaceuticals and illicit drugs have been found to alter locomotion, feeding behavior, schooling, boldness, and addiction. For example, diazepam (Oggier *et al.*, 2010) or the antipsychotic drug risperidone (Kalichak *et al.*, 2017) induces hyperactivity in exposed zebrafish larvae. Of particular concern are transgenerational effects as shown for fluoxetine and carbamazepine, and the irreversible effects on the developing nervous system that is more sensitive than the one of mature fish. Table 14.2 summarizes the important effects of some neuroactive pharmaceuticals and illicit drugs in laboratory experiments.

For most neuroactive pharmaceuticals, ecotoxicity data and underlying toxic mechanisms in fish are still lacking. In the following, we highlight SSRIs, carbamazepine, diazepam, risperidone, and illicit drugs for their effects in fish (for opioids, see the "Analgesics" subchapter). Yet, there are many more neuroactive compounds measured in surface waters and wild fish, warranting more studies to clarify the potency and actions in fish.

### Selective Serotonin Reuptake Inhibitors (SSRIs)

The molecular target of SSRIs, the serotonin transporter (Slc6a4), is evolutionarily conserved in vertebrates, raising the concern for ecotoxicological effects in fish. SSRIs (and their metabolites) inhibit synaptic serotonin (5-HT) reuptake leading to an increase in extracellular 5-HT concentrations. While this is the short-term consequence of SSRIs, the long-term effects in fish can go in both directions: Some studies report an increase in 5-HT after several weeks of exposure while other studies report a decrease or no change in 5-HT, suggesting complex pharmacokinetics of chronic SSRI exposure and an adjustment of compensatory mechanisms of at least 21 days (McDonald, 2017).

5-HT plays a prominent role in the stress axis in both mammals and fish, and the associated changes in cortisol levels are responsible for many of the reported behavioral changes. In fish, fluoxetine can alter anxiety-related behavior, which includes increased boldness in predator–prey interactions, reduced swimming activity, and increased shoal cohesion. There is some ambiguity in responses reported which may be due to different exposure duration and the associated varying stage of compensatory mechanisms (McDonald, 2017) or intraindividual variation in animal behavior (Grzesiuk and Pawelec, 2021). Yet, any of these behavioral changes can have consequences for social and predator–prey interactions. 5-HT also regulates the hypothalamic–pituitary–gonadal (HPG) axis, which is responsible for the control of reproduction. Therefore, fish can respond to SSRIs with reduced spermatogenesis in males (Citalopram; Prasad *et al.*, 2015) and altered estradiol levels (Fluoxetine; Mennigen *et al.*, 2010) which can affect the quality of eggs produced (Parrott and Metcalfe, 2019).

**TABLE 14.2**
**Neuroactive Compounds, Effects, and Lowest Effects Concentration (LOEC) in Selected Fish**

| Pharmaceutical/ Illicit Drug | Effect | LOEC (ng/L) | Fish Species | References |
|---|---|---|---|---|
| Fluoxetine | Reduced aggression | 3,000 | Arabian killifish (*Aphanius dispar*) | Barry (2013) |
| | Reduced swimming speed | 30 | | |
| | schooling behavior in response to a predator alarm | 300 | | |
| | Low-cortisol phenotype for two generations | 540 | Zebrafish (*Danio rerio*) | Vera-Chang *et al.* (2018) |
| | Shorter reaction time<br>Increased boldness | 360 | Monkey goby (*Neogobius fluviatilis*)<br>Gudgeon (*Gobio gobio*) | Grzesiuk and Pawelec (2021) |
| | Decreased pheromone-stimulated milt volume<br>Increased estradiol levels | 540 | Goldfish (*Carassius auratus*) | Mennigen *et al.* (2010) |

(*Continued*)

**TABLE 14.2 (*Continued*)**
**Neuroactive Compounds, Effects, and Lowest Effects Concentration (LOEC) in Selected Fish**

| Pharmaceutical/ Illicit Drug | Effect | LOEC (ng/L) | Fish Species | References |
|---|---|---|---|---|
| | Earlier hatch<br>Lower deformities in fry<br>Transient decrease in length of F1 larvae | 1,000 | Fathead minnow (*Pimephales promelas*) | Parrott and Metcalfe (2019) |
| | Reduced feeding behavior | 540 | Goldfish (*Carassius auratus*) | Mennigen, Sassine, *et al.* (2010) |
| Venlafaxine | Increased growth of the head area, eye area, and body length, hyperactivity, increased cortisol | 1,000 | Zebrafish (*Danio rerio*) | Tang *et al.* (2021) |
| Carbamazepine | Decreased steroid production | 10,000 | Zebrafish (*Danio rerio*) | Fraz *et al.* (2018) |
| | Increased hatching and body length<br>Decreased rest period | 1,000 | Zebrafish (*Danio rerio*) | Qiang *et al.* (2016) |
| | Dyslipidemia and hyperlipidemia<br>Lipid droplet accumulation in liver<br>Increased body weight | 100,000<br>1,000<br>1,000 | Chinese rare minnow (*Gobiocypris rarus*) | Xin *et al.* (2021) |
| | Transgenerational effect on reproductive output | 10,000 | Zebrafish (*Danio rerio*) | Fraz *et al.* (2019) |
| Diazepam | Alteration of genes involved in circadian rhythm<br>Hyperactivity | 273<br>273,000 | Zebrafish (*Danio rerio*) | Oggier *et al.* (2010) |
| | Hypoactivity<br>Hyperactivity | 1,200,000<br>12,000–120,000 | Zebrafish (*Danio rerio*) | Wu *et al.* (2020) |
| | Anxiety-like behavior | 1,250,000 | Zebrafish (*Danio rerio*) | Bencan *et al.* (2009) |
| | Reduced locomotion | 1,200,000 | Eastern mosquitofish (*Gambusia holbrooki*) | Sousa and Nunes (2020) |
| Risperidone | Hyperactivity | 3 | Zebrafish (*Danio rerio*) | Kalichak *et al.* (2017) |
| Methamphetamine | Addiction<br>Hypoactivity during withdrawal | 1,000 | Brown trout (*Salmo trutta*) | Horký *et al.* (2021) |
| Cocaine | Increased brain dopamine, plasma catecholamine, cortisol, and thyroid-stimulating hormone levels, lowered adrenocorticotropic hormone, corticosterone, and triiodothyronine levels | 6,000 | European eel (*Anguilla anguilla*) | Gay *et al.* (2013) |
| Codeine | Reduced egg production<br>Reduced number of mature oocytes<br>Reduced 17β-estradiol | 100 | Japanese medaka (*Oryzias latipes*) | Fischer *et al.* (2021) |
| Cannabis (THC) | Hypoactivity | 24,000 | Zebrafish (*Danio rerio*) | Carty *et al.* (2019) |
| Ketamine | Oxidative stress<br>p53-regulated apoptosis<br>Hyperactivity | 951 | Medaka (*Oryzias latipes*) | Liao *et al.* (2018) |

Another prominent effect of serotonin is reduced feeding behavior, which can be partly attributed to suppressed appetite and partly to increased lethargy, which has been shown in a variety of exposed fish species and may consequently affect overall fitness and reproductive output (McDonald, 2017; Mennigen, Sassine, *et al.*, 2010). The majority of behavioral effects in laboratory settings are reported at concentrations ranging from 30 to 100 μg/L fluoxetine, but concentrations as low as 0.03 μg/L were reported to affect swimming speed, schooling behavior, and response to a predator alarm (Sumpter *et al.*, 2014; Barry, 2013). For venlafaxine, 1 μg/L can increase growth, activity, and cortisol levels in zebrafish larvae (Tang *et al.*, 2021). An environmentally realistic concentration of 0.09 μg/L fluoxetine (as part of a mixture with other antidepressant drugs) exposed to fathead minnow for a full life cycle shows no effect on nest defense behaviors or any other health endpoints measured while a ten times higher concentration caused subtle effects on egg/fry health and weight gain in adult male minnows (Parrott and Metcalfe, 2019). A concentration of 0.54 μg/L fluoxetine can induce a low-cortisol phenotype for two generations in zebrafish (Vera-Chang *et al.*, 2018).

## Carbamazepine

Carbamazepine is an anticonvulsant drug and mood-stabilizing drug used to control seizures resulting from epilepsy and bipolar disorder among other neurological diseases. Despite being described worldwide for the treatment of such diseases, its mechanism of action has not been fully understood. It is thought that it exerts its therapeutic action by binding to the voltage-dependent sodium channels of excitatory neurons and thereby reduces the excitability of nerve cells (Thorn *et al.*, 2011). Carbamazepine is also thought to increase dopamine turnover and increase GABA transmission. While it effectively manages seizures, the numerous side effects including epidermal necrolysis and hepatoxicity hint at a more complex mechanism of action. Carbamazepine is mainly eliminated by the cytochrome P450 system, and its major metabolite, carbamazepine epoxide, is an active anticonvulsant with the same mechanism of action.

In fish, carbamazepine disrupts lipid metabolism leading to increased body weight (Xin *et al.*, 2021) similar to metabolic disorders in epilepsy patients. In male fish, carbamazepine decreases steroid production (Fraz *et al.*, 2018). The effects on larvae are the promotion of early hatching and growth as well as increased sensitivity to touch and light (Qiang *et al.*, 2016). Similar to fluoxetine, carbamazepine can have long-term impacts, causing lower reproductive output up to the F4 generation (Fraz *et al.*, 2019).

### Diazepam

Diazepam (Valium™) binds to a specific site on the $GABA_A$ receptor and thereby increases the frequency by which the chlorine channel opens leading to an increase in the inflow of $Cl^-$ ions which in turn reduces the excitability of resting neurons. Regulating this mechanism is desired to treat seizure disorders, anxiety disorders, insomnia, and muscle spasm. In fish, similar benzodiazepine–GABA receptors binding characteristics occur, and the resulting reduced excitability of neurons is likely mediating aberrant fish behavior. For example, adult mosquitofish treated with diazepam show reduced locomotion (Sousa and Nunes, 2020), and adult zebrafish treated with diazepam show reduced anxiety-like behaviors, in which untreated fish usually exhibit when placed in a novel environment (Bencan *et al.*, 2009). However, such behavioral effects were measured at high concentrations of more than 1,200,000 ng/L. At lower concentrations (12,000–273,000 ng/L), larval and juvenile zebrafish exhibit hypoactivity after treatment with diazepam (Oggier *et al.*, 2010; Wu *et al.*, 2020), while exposing fathead minnows to more environmentally relevant concentrations of 100–10,000 ng/L did not alter reproductive behavior or reproductive output (Lorenzi *et al.*, 2016).

### Illicit Drugs

Illicit drugs have potent pharmacological activities, altering behavior, and perception in various forms depending on the drug's mechanism of action. The effects in fish can be expected to be as diverse, yet only a few studies have begun to explore such consequences. An important study shows that the opioid analgesic codeine exhibits effects at an environmentally relevant concentration (100 ng/L) in female medaka, reducing the number of eggs produced as well as the number of mature oocytes in ovaries while reducing the levels of 17β-estradiol (Fischer *et al.*, 2021). This suggests that opioids are contributing to the endocrine disruptive effects observed in fish exposed to wastewater discharges. Cocaine too impairs the endocrine system, changing brain dopamine and plasma catecholamine levels and the activity of pituitary–adrenal/thyroid axes in European eel (Gay *et al.*, 2013). Besides the potential alteration of reproductive processes, illicit drugs are thought to alter fish behavior. Ketamine induces hyperactivity in larval medaka which coincided with several apoptosis and oxidative stress markers at concentrations of 951 ng/L (Liao *et al.*, 2018). On the contrary, THC exposure in zebrafish larvae leads to hypoactivity (Carty *et al.*, 2019). Similarly, methamphetamine exposure in brown trout leads to hypoactivity and preference for areas in the tank with methamphetamine during withdrawal, which was considered an indicator of addiction (Horký *et al.*, 2021).

## Risk Assessment

Neuroactive pharmaceuticals and illicit drugs have the potential to affect ecologically important behaviors and reproduction in exposed wildlife. Not only are neuroactive pharmaceuticals measured in multitudes downstream of WWTPs, but moreover they can occur together with unknown and known neurotoxic compounds such as heavy metals and pesticides. As single compounds, most neuroactive pharmaceuticals exhibit an effect at much higher concentrations than what is measured in surface waters (Tables 14.1 and 14.2). However, there are a few exceptions that warrant careful observation.

Fluoxetine and codeine were among the measured psychotic and illicit drug residues in a wastewater receiving creek in Kentucky (USA), and both were detected at concentrations that can alter behavior as a single compound (Subedi *et al.*, 2021). An environmentally relevant mixture of those residues altered transcripts related to the nervous system in exposed zebrafish while several other genes were associated with the immune system (Subedi *et al.*, 2021).

The antipsychotic drugs risperidone and flupentixol as well as the neuroleptic haloperidol were measured in chub (*Squalius cephalus*) from a stream in the Czech Republic at levels higher than 10% of the therapeutic effect in humans (Cerveny *et al.*, 2021). This emphasizes the potential to alter fish behavior in polluted environments. While risperidone was found to be present in wild fish (Cerveny *et al.*, 2021) and very potent in inducing behavioral changes in laboratory fish (Kalichak *et al.*, 2017), environmental concentrations in surface waters seem unknown.

We have only begun to unravel the behavioral consequences in fish of such drugs in well-controlled laboratory settings. Translating these behavioral effects to ecological fitness in complex ecosystems with multiple species interactions remains a challenge to be solved.

## STEROID HORMONES

Mammalian steroid hormones regulate a range of important physiological processes, including development, metabolism, osmoregulation, reproduction, and behavior. For instance, the female sex hormone 17β-estradiol (E2) regulates female characteristics and controls together with progesterone (P4) the menstrual cycle in women. The male sex hormone testosterone controls the development and maintenance of male characteristics and regulates sperm development. Besides reproductive function in adults, these sex hormones control the development of gonads in embryogenesis. P4 is also important in maintaining pregnancy. Gluco- and mineralocorticoids have different functions, including osmoregulation and regulatory function in metabolism and embryogenesis.

Hormones are normally excreted as parent compounds and metabolites by humans and mammals. They reach surface waters from wastewater but also from solid and liquid wastes from animal farms. In addition, there are numerous synthetic hormones used as pharmaceuticals in contraception and diverse medical applications in both human and veterinary medicine. Some hormone pharmaceuticals, including glucocorticoids and sex hormones in contraception, are very frequently prescribed. Municipal wastewater but also wastes from farms and manure contain numerous hormones (Zhang *et al.*, 2017, 2021; Fent, 2015). They act at very low concentrations in the human body and in mammals. When occurring in the environment, they are equally very potent acting at extremely low concentrations at ng/L or even less. Due to the high homology of hormone receptors between humans and fish, these pharmaceuticals share a similar effect pattern (Gunnarsson *et al.*, 2008; Christen *et al.*, 2010). Some of them represent the most important endocrine disrupters in fish having the highest risk for fish health as they elicit adverse effects at environmental concentrations. While sex steroids negatively affect fertility and reproduction, glucocorticoids alter metabolism in adult fish. In embryos, there is a critical window when environmental steroids are most effective. During sexual differentiation, estrogens, androgens, and progestins adversely affect sex development, while glucocorticoids affect the development of organs.

### Concentrations and Fate in the Environment

The different classes of steroid hormones and their metabolites are found in surface waters at ng/L concentrations and rarely up to µg/L (Kolpin *et al.*, 2002; Zhang *et al.*, 2017). Higher concentrations occur in untreated municipal and hospital wastewaters (Chang *et al.*, 2009; Fent, 2015; Zhang and Fent, 2018), but livestock runoff can also contain considerable concentrations (Liu *et al.*, 2012). The ubiquitous presence of natural and pharmaceutical estrogens in the aquatic environment is well documented, while less is known about other steroids (Weizel *et al.*, 2018). Synthetic steroids tend to be more persistent than natural steroids as they are designed to have a longer half-life in the body and therefore also in the environment. Steroids accumulate in biota to only a minor extent.

17α-Ethinylestradiol (EE2) and other estrogens, including E2 and degradation products estriol and estrone, are commonly detected in wastewater contaminated surface waters (Runnalls *et al.*, 2010). EE2 is more persistent than E2 and its metabolites, and thus, bears a higher ecotoxicological risk. Other synthetic estrogens used in medical applications include equine estrogens and estrone sulfate, which are also detected in WWTP effluents. All of these estrogens occur below or at low ng/L concentrations in treated wastewater and in surface water (Table 14.3).

Androgens and some androgenic progestins are used in livestock in the USA, China, and other countries to promote animal growth, but this practice is forbidden in Europe. As growth promotor, trenbolone acetate is widely applied. It is also used as an illicit drug in bodybuilding. This steroid is detected along with several metabolites at low ng/L concentrations in surface waters associated with cattle feedlots (Bartelt-Hunt *et al.*, 2012). Natural androgens such as testosterone and androstenedione not only occur in animal farm effluents with treatment systems (Zhang *et al.*, 2021) but also in surface water at similar concentrations (Chang *et al.*, 2009; Zhang and Fent, 2018). In WWTPs, androgens are reduced as other steroids in the range of about 70–90% (Zhang and Fent, 2018). Trenbolone acetate metabolites (17α- and 17β-trenbolone) can occur in livestock runoff (Liu *et al.*, 2012) and be washed off from agricultural fields treated with manure into streams and rivers, where they can affect fish.

Together with synthetic estrogens, progestins (also called progestogens or gestagens) are among the most important group of environmental pharmaceuticals that may exhibit adverse effects in fish. They originate from excretion by humans but also livestock (contaminating feedlots and manure). P4 and progestins are ubiquitous in the aquatic environment. They are consumed in larger amounts than estrogens or androgens. About 20 different synthetic progestins are applied in contraception and medical applications together with P4, but only about half of these compounds have been analyzed in environmental media. P4 is very widespread at low ng/L concentrations in surface water and occurs at higher levels of up to µg/L in untreated municipal wastewater and hospital wastewater (Zhang *et al.*, 2017). P4 is almost always detected, followed by levonorgestrel, norethindrone, norgestrel, megestrol acetate, medroxyprogesterone acetate, and some other progestins. When detected, they occur typically in the range of some ng/L or below but may reach over 100 ng/L in treated wastewater and surface water. In WWTP treatment, the elimination rates of P4 and progestins are in the range of 90% (Fan *et al.*, 2011) or higher (Zhang *et al.*, 2017). In waste and runoff of livestock, up to several µg/L occur.

**TABLE 14.3**
**Representative Concentrations of Selected Steroid Hormones Used in Surface Water**

| Steroid Class | Steroid | Concentration (ng/L) | References |
|---|---|---|---|
| **Estrogens** | 17β-Estradiol (E2) | 0.34–1.2 | Zhang *et al.* (2017) |
| | Estrone (E1) | 0.08–3.7 | |
| | 17α-Ethinylestradiol (EE2) | 0.4 | Kuch and Ballschmiter (2001) |
| | | 7.53–27.40 | Chen *et al.* (2007) |
| | Equine estrogens | 0.1–0.43 | Tyler *et al.* (2009) |
| **Androgen** | Trenbolone | 0.002–50 | Ankley *et al.* (2018) |
| | | 27.6 | Liu *et al.* (2012) |
| **Progestins** | Progesterone (P4) | 0.1–5.4 | Zhang *et al.* (2017) |
| | | 30.5 | Liu *et al.* (2012) |
| | Norethindrone | 5.9–14 | Zhang *et al.* (2017) |
| | Megestrol acetate | 0.98–4.6 | Zhang *et al.* (2017) |
| | Dienogest | 0.05–2.3 | Weizel *et al.* (2018) |
| | Cyproterone acetate | 0.2–2.6 | Weizel *et al.* (2018) |
| | Medroxyprogesterone acetate | 14.4 | Liu *et al.* (2012) |
| | Norgestrel | 465 | Liu *et al.* (2012) |
| **Glucocorticoids** | Fludrocortisone acetate | 5–14 | Ammann *et al.* (2014) |
| | Prednisone | <0.02–1.3 | Shen *et al.* (2020) |
| | Prednisolone | 2.8–94 | Shen *et al.* (2020) |
| | Clobetasol propionate (CLO) | 0.05–3.4 | Weizel *et al.* (2018) |
| | | 1.7–7.3 | Shen *et al.* (2020) |
| | Triamcinolone acetonide | 1.3–14 | Shen *et al.* (2020) |
| | Dexamethasone | 0.52–12 | Shen *et al.* (2020) |
| | | 37.8 | Liu *et al.* (2012) |
| **Mineralocorticoid** | Spironolactone | 1–4 | Ammann *et al.* (2014) |

Glucocorticoids are important pharmaceuticals used in a variety of medical applications due to their anti-inflammatory, anti-allergic, and immune-suppressive properties. Many different compounds are in use, and they are ubiquitous in wastewater-contaminated surface waters. They enter surface waters via hospital and municipal wastewater and from their use in veterinary medicine, which results in contamination of manure and livestock runoff. The environmental concentrations are similar to those of sex steroids: They are in the range of 0.1–50 ng/L in surface water and up to hundreds ng/L in untreated wastewater. In WWTPs, elimination is also similar to that of sex steroids and generally about 90% (Fan *et al.*, 2011). Mineralocorticoids, such as canrenone and spironolactone, occur in river water in a similar range of below to a few ng/L (Weizel *et al.*, 2018).

## Modes of Action

In target cells of mammals and fish, steroids either bind to specific hormone receptors in the cytosol or membrane (e.g., progesterone receptor). In fish, there are two forms of nuclear estrogen, androgen, and progesterone receptors that may have different ligand specificities. Nuclear receptors are ligand-activated transcription factors that induce genomic effects. They are activated upon binding of hormones and translocated to the nucleus, where they bind to response elements, which in turn activate gene expression leading to the synthesis of proteins. Synthetic steroids used as pharmaceuticals act similarly to natural steroids, while the binding affinity and potency of action may differ. Some synthetic hormones have even a higher activity in fish compared to natural steroids, such as the synthetic glucocorticoid clobetasol propionate (CLO) compared to cortisol (Faltermann *et al.*, 2020). By this mechanism, steroids primarily act on the brain (hypothalamus, pituitary) and reproductive organs. During embryogenesis, sex steroids play a major role in sex determination, brain and sex development, and sexual maturation. Gluco- and mineralocorticoids, on the other hand, regulate stress reactions, metabolism, and the immune system. Besides genomic effects through nuclear receptor binding, steroids are also thought to have non-genomic effects that occur more rapidly and activate signal transduction pathways.

Progestins interact with progesterone receptors but also with other steroid receptors in addition. P4 is a ligand to the mammalian progesterone receptor but has only a low affinity to the fish receptor, due to structural differences. Fish specific progestins are 17α,20β-dihydroxy-4-pregnen-3-one (DHP) and 17α,20β,21-trihydroxy-4-pregnen-3-one (20β-S). As do estrogens and androgens, progestins act via the HPG axis. Naturally, they lead to oocyte maturation in female gonads and act on sperm maturation and motility in males. Progestins affect reproduction through both progesterone- and androgen-mediated signal pathways. Via membrane progesterone receptors, they also mediate non-genomic actions. Progesterone receptors have strong structural similarities to mineralo- and glucocorticoid receptors, but elicit distinct physiological responses. Some progestins act also as antiandrogens, such as cyproterone acetate, and others display estrogenic activity.

Synthetic glucocorticoids and mineralocorticoids act via the hypothalamus–pituitary–interrenal (HPI) axis and are involved in the stress reaction, carbohydrate, protein and lipid metabolism, and immune response. In fish, they have also an osmoregulatory function. The potency and targets of the different pharmaceuticals vary depending on their binding affinity to the nuclear receptors. Together, steroid pharmaceuticals exhibit adverse endocrine effects in fish. Some of them represent the most active endocrine disrupters, such as EE2 and levonorgestrel, acting via the HPG axis, and adversely interfering with fertility and reproduction of fish.

## Effects

Effects of pharmaceutical steroids in fish were first described with estrogens (Sumpter and Jobling, 2013) and only subsequently, androgens and progestins, and recently, glucocorticoids came into the focus, while mineralocorticoids are less studied. Sex steroids have distinctly different effects than glucocorticoids in developing embryos, but both induce irreversible effects in embryos and reversible effects in adults.

EE2 used as estrogen in contraception is probably the most studied pharmaceutical (Caldwell *et al.*, 2008). Table 14.4 summarizes important effects of some steroid pharmaceuticals that are active at low concentrations in laboratory experiments. It should be noted that steroids

**TABLE 14.4**
**Steroid Hormones, Effects and Lowest Effects Concentration (LOEC) in Selected Fish**

| Pharmaceutical | Effect | LOEC (ng/L) | Fish Species | References |
|---|---|---|---|---|
| Estrogens | | | | |
| 17β-Estradiol (E2) | Feminization of males<br>VTG induction, feminization of males | 4<br>25 | Roach (*Rutilus rutilus*) | Jobling *et al.* (2006)<br>Brion *et al.* (2004) |
| 17α-Ethinylestradiol (EE2) | Ovo-testis (intersex gonads)<br>Reduced spawning<br>Feminization of male fish, sex reversal<br>Sex reversed males<br>Reduced fecundity, population failure<br>Population collapse | 0.1<br>0.4<br>4<br>4<br>5<br>5–6 | Japanese medaka (*Oryzias latipes*)<br>Fathead minnow (*Pimephales promelas*)<br>Fathead minnows (*Pimephales promelas*)<br>Roach (*Rutilus rutilus*)<br>Zebrafish (*Danio rerio*)<br>Fathead minnows (*Pimephales promelas*) | Metcalfe *et al.* (2001)<br>Runnalls *et al.* (2015)<br>Länge *et al.* (2001)<br>Lange *et al.* (2009)<br>Nash *et al.* (2004)<br>Kidd *et al.* (2007) |
| 17β-Dihydroequilenin | VTG induction | 0.6 | Fathead minnows (*Pimephales promelas*) | Tyler *et al.* (2009) |
| Androgens | | | | |
| 17α-trenbolone | Masculinization, reduced VTG and steroids, fecundity reduction | 7–16 | Fathead minnow (*Pimephales promelas*) | Jensen *et al.* (2006) |
| 17β-Trenbolone | Sex reversal to males<br>Population decrease | 9.7<br>27 | Zebrafish (*Danio rerio*) embryos<br>Fathead minnow (*Pimephales promelas*) | Holbech *et al.* (2006), Ankley *et al.* (2018)<br>Miller and Ankley (2004) |
| 17-Methyltestosterone | Fecundity decrease, masculinization of females | 200 | Fathead minnow (*Pimephales promelas*) | Ankley *et al.* (2001) |
| Progestins | | | | |
| Progesterone (P4) | Transcriptional alterations, reduced reproduction, altered circadian rhythm<br>Sex ratio changes | 2–3.5, 306<br>742<br>7<br>63 | Zebrafish (*Danio rerio*) embryos and adults | Zhao *et al.* (2015a), Zucchi *et al.* (2012, 2013, 2014)<br>Liang *et al.* (2015) |
| Levonorgestrel | Reduced reproduction<br>Masculinization of females<br>Transcriptional alterations | 0.8–100<br>2 | Fathead minnow (*Pimephales promelas*)<br>Zebrafish embryos | Zeilinger *et al.* (2009), Runnalls *et al.* (2013)<br>Zucchi *et al.* (2012) |
| Norethindrone | Reduced fecundity and reproduction | 1–10<br>25 | Fathead minnow (*Pimephales promelas*)<br>Medaka | Paulos *et al.* (2010) |

*(Continued)*

**TABLE 14.4 (*Continued*)**
**Steroid Hormones, Effects and Lowest Effects Concentration (LOEC) in Selected Fish**

| Pharmaceutical | Effect | LOEC (ng/L) | Fish Species | References |
|---|---|---|---|---|
| Norgestrel | Sex ratio change | 34 | Zebrafish (*Danio rerio*) | Liang *et al.* (2015, 2018) |
| Medroxyprogesterone acetate | Transcriptional alteration<br>Reduced fecundity | 4.6<br>342 | Zebrafish (*Danio rerio*) embryos<br>Adult Zebrafish | Zhao *et al.* (2015b) |
| Mifepristone (RU486) | Transcriptional alteration | 3–39 | Fathead minnow (*Pimephales promelas*) | Blüthgen *et al.* (2013) |
| Drospirenone | Transcriptional alteration<br>Reduced fecundity | 55, 99<br>2,763 | Zebrafish (*Danio rerio*) | Zucchi *et al.* (2014), Zhao *et al.* (2015a) |
| Dydrogesterone | Transcriptional alteration<br>Reduced fecundity | 5<br>89<br>1,263 | Zebrafish (*Danio rerio*) embryos<br>Adult zebrafish | Zhao *et al.* (2015b) |
| Gestodene | Reduced fecundity | 1 | Fathead minnow (*Pimephales promelas*) | Runnalls *et al.* (2013) |
| Desogestrel | Reduced fecundity | 1,000 | Fathead minnow (*Pimephales promelas*) | Paulos *et al.* (2010) |
| Glucocorticoids | | | | |
| Beclomethasone dipropionate | Elevated blood glucose, induced VTG in females | 100 | Fathead minnow (*Pimephales promelas*) | Kugathas *et al.* (2013) |
| CLO | Increased heart rate, accelerated hatching, decreased muscle contractions, transcriptional alterations | 91–9,380 | Zebrafish (*Danio rerio*) embryos | Willi *et al.* (2018) |
| Fluticasone propionate | Increased heart rate, decreased muscle contractions, transcriptional alterations | 976 | Zebrafish (*Danio rerio*) embryos | Willi, Salgueiro-González, Carcaiso, *et al.* (2019) |
| Triamcinolone acetonide | Increased heart rate, decreased muscle contractions, transcriptional alterations | 1,050 | Zebrafish (*Danio rerio*) embryos | Willi, Salgueiro-González, Carcaiso, *et al.* (2019) |
| Betamethasone | Increased heart rate, decreased muscle contractions | 110 | Zebrafish (*Danio rerio*) embryos | Willi, Salgueiro-González, Faltermann, *et al.* (2019) |

always occur as mixtures in surface water, and thus, the combinatory action of different steroid classes is of importance as they often sum up to a joint additive activity (see below).

### Estrogens

EE2 is extremely potent and exhibits adverse effects in fish at environmental concentrations. Induction of vitellogenin (VTG), the yolk protein precursor, is commonly found to have an adverse effect in juvenile and male fish (it occurs normally in females only). Generally, VTG induction serves as a biomarker for estrogenic activity of chemicals. EE2 induces VTG at 1 ng/L and even below in a wide range of fish species, but some species are less sensitive (Sumpter and Jobling, 1995). Additional effects are alteration of gene expression in different pathways (mitochondrial function, energy metabolism, cell cycle control) and of steroid hormone receptors (Rossier *et al.*, 2016), alteration of sex steroid levels in the blood, and changes in gonad histology and function (Table 14.4).

EE2 leads to feminization of male fish during sex development in the embryonic stage. Concentrations as low as 1 ng/L result in the formation of oocytes in testes. These intersex gonads are called ovo-testis (Jobling *et al.*, 2006). Responsible for ovo-testis formation are also other estrogens such as E2 and estrogenic chemicals present in wastewater. Reproduction of these fish is dependent on the degree of ovo-testis formation; only a moderate to high extent of ovo-testis induces a compromised reproduction (Lange *et al.*, 2011). Ovo-testis is widespread in wild populations of roaches in waste-water polluted rivers. Exposure of roach for 3 years to 4 ng/L EE2 resulted in feminization of the entire population (Lange *et al.*, 2009, 2011) while dosing of a whole lake in Canada with 5–6 ng/L EE2 resulted in complete reproduction failure of fathead minnows followed by population collapse (Kidd *et al.*, 2007).

E2 is used in estrogen replacement therapies and has an identical effects profile as EE2 in fish, although the potency is slightly lower. Demonstrated effects include transcriptional alterations, induction of VTG in juvenile and male fish, development of ovo-testis, compromised fertility and fecundity, and reduced reproduction (Table 14.4). Estrogens used in medical estrogen replacement therapy, estrone sulfate, and equine estrogens are similarly active in fish as EE2 and E2 with effective concentrations between 0.6 and 4.2 ng/L (Tyler *et al.*, 2009). Furthermore, the non-steroidal antiestrogen, tamoxifen used in cancer therapy, is also present in the environment but exhibits adverse effects in fish at higher than environmental concentrations.

### Androgens

In contrast to mammals, where testosterone is the principal androgenic hormone, in fish, it is 11-ketotestosterone (11-KT). 11-KT shows the strongest agonistic binding to the fish androgen receptor (AR), while testosterone and androstenedione have a lower affinity, and therefore, lower potency. However, the androgenic pharmaceutical trenbolone acetate and its metabolites, 17α-trenbolone and 17β-trenbolone, have also a high affinity to the AR. Trenbolone that occurs in livestock-affected aquatic environments in the USA can cause sex reversal or a skewed sex ratio in developing fish and leads to masculinization in females (Table 14.4).

Trenbolone acetate and its metabolites have adverse impacts on fertility and fecundity in fish. Adverse effects may occur at environmentally realistic concentrations of 10 ng/L but most effects are found at 30–100 ng/L (Ankley *et al.*, 2018). Trenbolone changes sex steroid metabolism and androgen signaling with transcriptomic and proteomic alterations. Furthermore, trenbolone induces secondary sex characteristics in female fish, and alters sex steroid levels in the blood and the histology of gonads. Transcriptional changes are noted in fathead minnows exposed to 17β-trenbolone (Garcia-Reyero *et al.*, 2009).

### Progestins

Progestins have high endocrine activity in fish and are used in contraception in combination with EE2 in humans Levonorgestrel is equally active as EE2 and inhibits reproduction in fish at very low concentrations (Zeilinger *et al.*, 2009). Progestins adversely affect the development of sex gonads and the brain in embryos and fertility and reproduction in adult fish. P4 and progestins not only alter gene expression (Zucchi *et al.*, 2012, 2013, 2014; Zhao *et al.*, 2015a) but also hormone levels, lead to masculinization of females (Runnalls *et al.*, 2013), and alter the circadian rhythm (Zheng *et al.*, 2021), gonad histology (Zhao *et al.*, 2015b), and mating behavior (Steinbach *et al.*, 2019) (Table 14.4). Alterations in sex development have been demonstrated in zebrafish at ng/L levels of P4 and norgestrel (Liang *et al.*, 2015). The effective concentrations of levonorgestrel, norethindrone, gestodene, and medroxyprogesterone acetate are comparable to those of EE2. They all occur at environmental concentrations (Fent, 2015).

Progestins have additional hormonal activities besides progestogenic activity depending on their structure, as they additionally bind to estrogen, androgen, and glucocorticoid receptors. They are structurally related to either testosterone or P4, or are derived from spironolactone. Older progestins used in contraception (levonorgestrel, norethindrone, gestodene, medroxyprogesterone acetate) bear an androgenic activity, while newer compounds (drospirenone, dienogest) bind more specifically to the progesterone receptor.

Physiological effects in adult fish occur at higher concentrations than transcriptional effects and have so far been studied for P4 and about ten synthetic progestins. Progestins reduce VTG levels in the blood and fecundity of female fish, which leads to adverse reproductive effects, as shown for many fish species. Progestins vary in their potency. Some of them induce androgenic effects and lead to masculinization. Levonorgestrel is very potent and inhibits egg production in adult females and leads to masculinization at 1–3 ng/L (Zeilinger *et al.*, 2009). Gestodene and norethindrone show a similar potency (Runnalls *et al.*, 2013; Paulos *et al.*, 2010). P4 reduces fecundity in female zebrafish at 742 ng/L and drospirenone at 2,763 ng/L (Zhao *et al.*, 2015b). Besides alteration in gene expression, P4 and other progestins induce histological changes in ovaries of female (Zucchi *et al.*, 2013, 2014, Zhao *et al.*, 2015b) and in testes of male fish (Zhao *et al.*, 2015a; Zeilinger *et al.*, 2009).

Mifepristone (RU486) is a potent antiprogesterone (a derivative of norethindrone), which is used to induce medical abortion and applied in therapy. It competitively binds to the progesterone receptor as an antagonist and shows transcriptional alterations and reproductive effects in zebrafish at 5 and 77 ng/L (Blüthgen *et al.*, 2013).

### Glucocorticoids

Exposure to glucocorticoid pharmaceuticals can lead to physiological adverse effects in fish. In zebrafish embryos, they lead to increased heart rate, decreased spontaneous muscle contractions, behavioral changes, and accelerated hatching (Table 14.4). At high concentrations, they induce edema, deformations, and paralysis (Willi *et al.*, 2018). Transcriptional alterations of target genes are involved in glucose metabolism, immune system regulation, and steroidogenesis, or are hormone receptors (Willi et al. 2018, Willi et al. 2019a, 2019b, Zhao et al. 2016). In adult fish, glucocorticoids cause alteration of glucose metabolism, reproductive effects, and immune suppression (Kugathas *et al.*, 2013; Zhao *et al.*, 2016). As do progestins, glucocorticoids deregulate circadian rhythm and alter swimming activity (Zhao *et al.*, 2016; Zheng *et al.*, 2021). Fludrocortisone acetate displays transcriptional effects in zebrafish at environmentally relevant concentrations of 6 and 42 ng/L and alters blood glucose levels and leukocyte counts at 42 ng/L. Embryos of the F1 generation show alterations in heartbeat, hatching success, and swimming behavior (Zhao *et al.*, 2016). CLO shows embryonic effects at 90 ng/L (Willi *et al.*, 2018).

Endogenous glucocorticoids regulate embryonic development, and consequently, exogenous glucocorticoid exposure is disruptive in the development of fish embryos. In adult fish, glucocorticoids induce stress symptoms, characterized by alterations in gene transcription and metabolism (blood glucose, amino and fatty acids) levels, immunosuppression (white blood cells), and other reactions. Thus, the adverse effects are distinct from those of sex steroids. Although some pharmaceuticals are effective at environmental concentrations, they are rather effective at higher concentrations, often in the μg/L range. However, as with other steroids, they occur as mixtures with joint activity.

## Mixture Effects

### Concepts

Pharmaceuticals always occur as mixtures in the environment, but their joint effects are poorly known. Do compounds act together and result in additive effects or do they act synergistically or antagonistically? There are two distinct interactions: pharmaceuticals that act on the same receptor(s) and thus exhibit similar actions, and pharmaceuticals that have dissimilar actions as they interact with different receptors or have different modes of actions (MoAs). Compounds with similar MoAs act additively. There are two established concepts for assessing their activity. The concentration addition (CA) concept assumes that one compound can be replaced by an equal concentration of another compound of the same potency. Thus, each component in the mixture contributes to the combined effect in proportion to its concentration and potency. This seems a valid concept for steroids of a given class. On the other hand, when compounds have dissimilar actions in a mixture and act on different receptors, the concept of independent action (IA) is valid. In this case, the joint effects can be estimated by multiplying the response probabilities. Both concepts rely on compound additivity (Kortenkamp *et al.*, 2009). There also exist more than additive (synergistic) and less than additive (antagonistic) interactions.

Generally, there is only little empirical data on the effects of pharmaceutical mixtures in fish. Most studies are performed with two or three pharmaceuticals. Such experiments are challenging and laborious. Ideally, the toxicity profile of individual components is determined prior to the assessment of mixtures. Furthermore, the compounds are mixed at equally effective concentrations to determine their interactions. There are only a few published studies that follow this procedure (Brian *et al.*, 2007; Thrupp *et al.*, 2018; Rossier *et al.*, 2016; Siegenthaler *et al.*, 2017). Conclusions about the type of compound interactions can only be made by such systematic approaches (Schmid, Willi, and Fent, 2020; Willi *et al.*, 2020; Thrupp *et al.*, 2018; Brian *et al.*, 2007; Silva *et al.*, 2002). Often compounds are mixed at equal concentrations, independent of their potencies. When pharmaceutical mixtures are tested on the basis of a specific contamination situation (David *et al.*, 2020; Schmitz *et al.*, 2018), they do not allow conclusions about the type of compound interactions.

### Steroid Mixtures

Additive action is first demonstrated for estrogens (Brian *et al.*, 2007; Silva *et al.*, 2002). Studies with steroids that act by the same MoA are more frequently performed than those with different class steroids (Schmid, Willi, and Fent, 2020; Willi, Salgueiro-González, Faltermann, *et al.*, 2019; Willi, Salgueiro-González, Carcaiso, *et al.*, 2019). Steroids of the same class act mainly additive and can be predicted by the CA model, as demonstrated with estrogens (Thorpe *et al.*, 2003), progestins (Zhao *et al.*, 2015b; Schmid, Willi, Salgueiro-González, *et al.*, 2020; Zucchi *et al.*, 2014), and glucocorticoids (Willi, Salgueiro-González, Carcaiso, *et al.*, 2019; Willi, Salgueiro-González, Faltermann, *et al.*, 2019). Consequently, a mixture can have a measurable effect even when the individual compounds show no effects, which is shown for estrogenic chemicals including steroids (Brian *et al.*, 2007; Silva *et al.*, 2002; Thrupp *et al.*, 2018).

How is the mixture activity with multiple class steroids? Often, two steroids of different classes are mixed, such as progestin and EE2 mimicking the composition in contraceptive pills. Equi-effective mixtures of two progestins (drospirenone, progesterone) and of drospirenone and EE2 show antagonistic effects on the abundance of some transcripts in zebrafish embryos (Rossier *et al.*, 2016). The interaction is different with the type of progestin. When the progestin cyproterone acetate is mixed with EE2, the interaction is additive (Siegenthaler *et al.*, 2017). Furthermore, inhibition of egg production in fathead minnows by levonorgestrel and EE2 is additive and can be best interpreted by the model of IA in mixtures of levonorgestrel and EE2 (Runnalls *et al.*, 2015).

A study with five different steroids (EE2, trenbolone, beclomethasone dipropionate, desogestrel, and levonorgestrel) based on individual concentration–response data revealed an additive effect on egg production in the mixture. This effect can be predicted better by the model of IA than CA (Thrupp *et al.*, 2018). Our transcriptomic study in zebrafish embryos with a mixture of E2, the glucocorticoid CLO, and three progestins (dienogest, medroxyprogesterone acetate, drospirenone) shows that each steroid class activates its own target gene transcripts, independently from the other steroids. Thus, each steroid class basically activates its own receptor, and the mixture activity is the sum of the individual steroid effects. The transcriptional signature of each steroid class becomes visible. There are only a very few genes that are regulated by both, E2 and CLO, suggesting that these steroids interact with the estrogen and glucocorticoid receptors (Schmid, Willi, and Fent, 2020). The interactive additive effects indicate that these genes have response elements in their promoter region for both steroid receptors. E2 and CLO show prominent activity, while the progestins are less active (Schmid, Willi, and Fent, 2020; Willi *et al.*, 2020).

Hence, physiological and transcriptional effects of a mixture of environmental estrogens, androgens, progestins, and glucocorticoids in zebrafish embryos are steroid-specific with the exception of some genes that are regulated by more than one steroid class (Schmid, Willi, and Fent, 2020; Willi *et al.*, 2020). Thus, the transcriptional activity in a mixture represents merely the additive sum of individual steroid classes. Consequently, steroids in mixtures that mimic river water composition act additively and affect zebrafish embryo development and the abundance of steroidogenesis gene transcripts (Zhang *et al.*, 2017).

Taken together, steroids of the same class act merely additive in a mixture. In mixtures with steroids of different classes, each steroid exhibits its own effect profile which contributes to an overall sum of effects. Depending on the specific steroids, there are no interactions, or they can be additive or more complex, including antagonistic interactions.

### Risk Assessment

Presently, risk assessment is generally performed with single compounds only, but as shown above, mixtures of steroids are relevant; individual compounds sum up additively to an overall joint effect. There is a need in the future to implement mixtures and their combined activity in risk considerations. As shown for steroids, mixtures exhibit effects even if single steroids are below the effect level (Thrupp *et al.*, 2018; Schmid, Willi, and Fent, 2020). The risk of steroids for adverse effects in fish presumably increases with the homology of target receptors. In the case of steroid pharmaceuticals, they interact with highly homologous fish receptors, and thus, the potential for adverse activities in fish is high (Christen *et al.*, 2010).

Effects in wild fish are indeed documented in field studies. Downstream of a pharmaceutical manufacturing plant, wild gudgeon exhibits signs of endocrine disruption (induced VTG concentration, intersex, male-biased sex ratio) probably related to the presence of synthetic hormones in river water (Sanchez *et al.*, 2011). Clear evidence for an association between exposure and adverse effects in wild fish comes from an experimental field study (Kidd *et al.*, 2007) and from wild roach in UK rivers, where EE2 contributes together with E2 to the feminization of wild roach populations (Jobling *et al.*, 2006). The severity of intersex manifestation in these roaches is associated with reduced fertility (Lange *et al.*, 2011). Thus, for EE2, there is evidence of adverse effects in polluted environments (Corcoran *et al.*, 2010).

The risk of steroids in fish can effectively be estimated by comparing environmental concentrations and the lowest effect concentrations of individual steroids obtained in laboratory experiments. Generally, transcriptional alterations occur at lower concentrations than physiological effects, but they may translate to physiological outcomes. EE2 and some progestins exhibit transcriptional effects at environmental concentrations and pose a high risk for reproductive effects in fish, particularly EE2 which also exhibits reproductive effects (Jobling *et al.*, 2006; Fent, 2015). The "no observed effect concentration" of EE2 was suggested to be 0.35 ng/L and by considering a safety factor of 10, even lower (Caldwell *et al.*, 2008). Such concentrations occur in contaminated rivers. Therefore, the application of EE2 and highly active progestins in contraception poses the highest risks of all steroids to fish, and pharmaceuticals in general. Mixtures of both steroids may add to an overall adverse effect on reproduction by additive action of the individual steroids.

Among the progestins, levonorgestrel, norgestrel, and norethindrone pose the highest risks for fish reproduction, as effects occur at environmental concentrations (Fent, 2015). Levels of levonorgestrel in the blood of wild fish can reach human therapeutic levels (Fick *et al.*, 2010). Furthermore, effects of dydrogesterone and medroxyprogesterone acetate in zebrafish occur at environmental levels (Zhao *et al.*, 2015b). Among other steroids, trenbolone poses a risk in aquatic systems adjacent to livestock runoff in the USA as concentrations may reach effective levels. Risks of equine estrogens seem of less concern but environmental levels are largely unknown. Taken together, there is a risk for reproductive effects and sex development in fish in wastewater-contaminated surface waters, mainly originating from EE2 and progestins. Forthcoming studies should investigate the risks of glucocorticoids on fish development and the effects of steroid mixtures.

## ANTIBIOTICS

Antibiotics inhibit the growth of or kill microorganisms such as bacteria. They are extensively used in human and veterinary medicine and also applied in aquaculture and orchards. Today, about equal amounts are prescribed for humans and animals. While in human medicine, antibiotics are specifically used to treat diseases, livestock producers also use antibiotics as growth promoter. This is now prohibited in Europe but still a questionable praxis in other countries, as unfortunately, this promotes antibiotics resistance.

Antibiotics are a diverse group of chemicals that can be divided according to their chemical structures and mechanisms of action into different classes. The most frequently detected classes in aquatic environments are quinolones, macrolides, sulfonamides, and tetracyclines (Kümmerer, 2009). The latter two occur at higher levels in regions with livestock and aquaculture industries (Liu *et al.*, 2021). Antibiotics in rivers of highly polluted areas of China can reach multiple thousand nanograms per liter (Xu *et al.*, 2021), but concentrations are usually one or two orders of magnitude lower in European rivers (Loos *et al.*, 2013). Fish exposed to such concentrations exhibit increased heart rate, decreased nutrient digestibility, suppression of innate immunity, and altered behaviors including hyperactivity and increased feeding time besides the inherent dysbiosis of gut microbiota. The mechanistic link between microbial changes and adverse health effects remains unknown. The data on chronic effects in fish are scarce, but indicate that risks arising from ecosystems polluted with high levels of antibiotics cannot be ignored. The most urgent problem of antibiotics in the environment is the promotion of antibiotic-resistant bacteria, including highly pathogenic ones.

### Concentrations and Fate in the Environment

The main sources of antibiotics in surface waters are livestock, domestic, and hospital wastewater. Aquaculture is also important, while use in orchards is not very widespread. Ingested antibiotics are metabolized to a wide range of degrees, with the non-metabolized fraction being excreted as a still-active compound (Kümmerer, 2009). Both parent antibiotic molecules and their metabolites are usually soluble. In conventional WWTPs, antibiotics are only partially eliminated, as anaerobic fermentation and anaerobic–aerobic combined process have limited removal capacity and physical adsorption cannot degrade antibiotics. In addition to that, chemical oxidation (chlorination) likely generates secondary pollution (Liu *et al.*, 2021). Thus, advanced treatments have recently been suggested to increase antibiotic removal efficiency. These include UV and fungal treatment (Lucas *et al.*, 2016), full-scale ozonation, adsorption onto activated carbon (Östman *et al.*, 2019), advanced oxidation methods (Xu *et al.*, 2021), and addition of synthesized antibiotic-degrading enzymes (Liu *et al.*, 2021).

Besides the treatment technique, the removal efficiencies also depend on the antibiotic. For instance, relatively high removal efficiencies are achieved for ciprofloxacin (48–88%) and sulfamethoxazole (51–100%) while tetracycline (36–39%) and erythromycin (26–43%) can leave the WWTP in larger amounts (Burch *et al.*, 2019). This results in concentrations in surface waters in the lower ng-per-liter to lower µg-per-liter range (Table 14.5). In aquatic ecosystems, antibiotics can be transformed by environmental factors such as ambient light, pH, and temperature (Kümmerer, 2009), and sediments can be an important sink but also a potential source of antibiotics, particularly for quinolones and tetracyclines (Zhou *et al.*, 2016; Liu *et al.*, 2021). As a place of accumulation, benthic sediments are a likely origin for microbial antibiotic resistance. In fish, antibiotics accumulate primarily in the liver and muscle tissue but are also detected in the brain and gills where concentrations of several to several hundred µg/kg are measured (Yang *et al.*, 2020). Bioaccumulation patterns across tissues are thought to vary depending on the class of antibiotic (Liu *et al.*, 2021).

### Modes of Action

Antibiotics specifically target microorganisms and have two main mechanisms of action: bactericidal (kill the bacteria) or bacteriostatic (inhibit bacterial growth). There are various mechanisms through which antibiotics can achieve that, including interfering with the synthesis of proteins (tetracyclines, macrolides) and inhibiting the bacteria's control of nucleic acid synthesis, resulting in impaired DNA replication and cell death (quinolones, sulfonamides). For fish, this means that antibiotics interfere with the bacterial community colonizing their outer surface that includes skin and gills the gastrointestinal tract, and the chorion for embryonic stages. Currently, the associated molecular mechanisms of action in fish tissues are unknown.

### Effects

Antibiotics have a strong inhibitory effect on the gut microbial community, which, in humans, is associated with various diseases, especially when newborns are exposed. These microbiome–host interactions leading to disease pathology are complex and not fully understood. In fish, observed effects include cardiovascular damage, delayed and abnormal development, changes in hatching time, alteration of nutrient metabolism, inflammatory response, and altered behavior (Yang *et al.*, 2020). However, these effects commonly occur at high doses. Studies investigating chronic and low-dose effects are scarce. For sulfonamides and tetracyclines, such effects are summarized below. The effects of quinolones and macrolides at environmental concentrations have only been assessed by a handful of studies. Ciprofloxacin (quinolones) increases mortality in juvenile sunfish (*Lepomis gibbosus*) at 10,000 ng/L when combined with other pharmaceuticals such as ibuprofen or fluoxetine (Richards *et al.*, 2004). Erythromycin (macrolides) induces genotoxic damage and oxidative stress at 800 ng/L and 50 ng/L, respectively, after chronic exposure to rainbow trout (*Oncorhynchus mykiss*) (Rodrigues *et al.*, 2016). The effects of different antibiotics at relatively low concentrations are summarized in Table 14.6. The widespread use and presence of sulfonamides, tetracyclines, quinolones, and macrolides antibiotics merit further investigation into chronic effects at low concentrations to establish regulatory concentrations.

In addition to the pathological consequences of microbial dysbiosis, the presence of antibiotics increases the potential that antibiotic-resistant genes are spread and novel antibiotic-resistant genes are generated. Such genes are present in the environment, and low concentrations of

**TABLE 14.5**
**Representative Concentrations of Selected Antibiotics in Surface Water**

| Class | Antibiotic | Concentration (ng/L) | References |
|---|---|---|---|
| Quinolones | Ciproflaxacin | 23–14,331 | Reviewed in Danner *et al.* (2019) |
| Sulfonamides | Sulfamethoxazole | 2.29–234[a] | Zhou *et al.* (2016) |
| Tetracyclines | Oxytetracycline | nd–361,107[b] | Reviewed in Xu *et al.* (2021) |
| Macrolides | Erythromycin | 14–2,500[c] | Reviewed in Rodrigues *et al.* (2016) |

[a] Minimum and maximum value of one sampling site.
[b] Minimum and maximum value of different references reviewed in Xu *et al.* (2021).
[c] Mean value of different sampling sites reviewed in Rodrigues *et al.* (2016).

**TABLE 14.6**
**Antibiotics, Effects, and Lowest Effects Concentration (LOEC) in Selected Fish**

| Antibiotic | Effect | LOEC (ng/L) | Fish Species | References |
|---|---|---|---|---|
| Ciprofloxacin | Mortality in combination with other pharmaceuticals | 10,000 | *Lepomis gibbosus* | Richards *et al.* (2004) |
| Sulfamethazine | Increased heart rate, malformations, oxidative stress response | 200 | *Danio rerio* | Yan *et al.* (2018) |
| | Increased heart rate, oxidative stress response | 1,000 | *Danio rerio* | Lin *et al.* (2014) |
| Sulfamethoxazole | Weight loss, reduced hepatosomatic index, decreased nutrient digestibility, microbiota dysbiosis, suppression of innate immunity, stimulation of inflammatory and detoxification responses | 260 | *Oreochromis niloticus* | Limbu *et al.* (2018) |
| Oxytetracycline | Weight loss, reduced hepatosomatic index, decreased nutrient digestibility, microbiota dysbiosis, suppression of innate immunity, stimulation of inflammatory and detoxification responses | 420 | *Oreochromis niloticus* | Limbu *et al.* (2018) |
| | Hyperactivity and increased feeding time, decreased energetic reserves | 100 | *Danio rerio* | Almeida *et al.* (2019) |
| Erythromycin | Oxidative stress | 50 | *Oncorhynchus mykiss* | Rodrigues *et al.* (2016) |
| | Genotoxicity | 800 | | |

antibiotics from ng to µg per liter can put selective pressure on the microbial community that can provoke the spread of drug-resistant bacteria (Yang *et al.*, 2020).

### Sulfamethazine (Sulfonamides)

Sulfamethazine is a broad-spectrum antibiotic belonging to the sulfonamides class and prescribed for both humans and animals. Like other antibiotics, sulfamethazine can induce an oxidative stress response as shown for zebrafish at a concentration of 200–1,000 ng/L (Lin *et al.*, 2014; Yan *et al.*, 2018). At the physiological level, these doses of sulfamethazine increase the heart rate and spontaneous movement of zebrafish embryos (Lin *et al.*, 2014; Yan *et al.*, 2018). However, the mechanisms underlying these effects of sulfamethazine are not understood.

### Oxytetracycline (Tetracyclines)

Oxytetracycline is another broad-spectrum antibiotic belonging to the tetracycline class. It is widely used in livestock production being effective both against Gram-positive and Gram-negative bacteria. Runoffs from agriculture can lead to exposure of fish where long-term exposure of an environmentally relevant concentration (100 ng/L) to adult zebrafish can reduce lipid levels which coincides with behavioral changes including increased feeding time and locomotion (Almeida *et al.*, 2019). Similarly, Nile tilapia (*Oreochromis niloticus*) exposed to 420 ng/L oxytetracycline show reduced nutrient digestibility, lipid body composition, and feed efficiency (Limbu *et al.*, 2018). While the underlying mechanism of these alterations remains unknown, oxytetracycline changes the water and gut bacterial communities and the latter may be related to increased energy demand and consumption (Almeida *et al.*, 2019; Limbu *et al.*, 2018).

## Risk Assessment

Aquatic and oceanic ecosystems prone to chronic antibiotic contamination are catchment areas of wastewater, agricultural runoff, or aquaculture, where concentrations of multiple hundred nanograms per liter are often reported (Table 14.5). However, the assessment of chronic effects in fish and other aquatic organisms at environmentally relevant concentrations is limited (Table 14.6). Effects summarized here indicated that low doses applied over weeks or months can affect fish health and behavior, albeit abiotic factors specific to each ecosystem may influence the potency of the antibiotic. Therefore, chronic or environmentally relevant exposure scenarios are much needed for environmental risk assessments.

Some studies have evaluated the environmental risk of antibiotics in surface water including mainly acute toxicity data. A comprehensive study at the European level investigating WWTP effluents predicted the impact of antibiotics on the environment as moderate with all RQ values below 0.9 (Rodriguez-Mozaz *et al.*, 2020). When evaluating hospital wastewaters in Portugal, ciprofloxacin and ofloxacin are determined to pose a risk at all trophic levels and have a HQ > 1. Consequently, these antibiotics are suggested to be prioritized in environmental monitoring (Santos *et al.*, 2013). Comparing antibiotic concentrations in surface waters worldwide reveals that the highest concentrations are measured in China (Danner *et al.*, 2019). A study focusing on different tetracyclines from China calculated RQ values ranging from <0.001 to 36 (Xu *et al.*, 2021). These assessments indicate that in heavily polluted areas, antibiotics such as quinolones and tetracyclines may pose a risk to the aquatic environment. It should be noted that the highest risks of antibiotics in the environment are the promotion of antibiotic resistance in bacteria and their adverse effects on aquatic biota other than fish, particularly on bacteria and algae.

## AZOLE FUNGICIDES: IMIDAZOLES AND TRIAZOLES

Azole compounds, classified into imidazoles (two nitrogen atoms) and triazoles (three nitrogen atoms), are widely used as fungicides in agriculture and in veterinary and human medicine to treat a broad spectrum of fungal diseases. Some azoles are also employed in household products such as hair shampoos and dermal creams (Chen and Ying, 2015; Bhagat *et al.*, 2020). Azoles constitute a substantial portion of fungicides used worldwide amounting to around 300,000 tons per year. In the USA alone, 2,940 metric tons of azoles have been used in the year 2015, which is more than a 500% increase since 2000 (Battaglin *et al.*, 2011).

Azoles are fairly stable; thus, their excessive use in agriculture and pharmaceuticals results in contamination of water, soil, and aquatic organisms (Chen and Ying, 2015). Azoles and their metabolites are found in surface waters at ng/L to several μg/L (Wick *et al.*, 2010; Ccanccapa *et al.*, 2016). Higher concentrations occur in untreated municipal wastewaters (Peng *et al.*, 2012). By exhibiting endocrine-disruptive and neurotoxic action, azoles may pose a potential risk to aquatic organisms, as well as human health (Chen and Ying, 2015; Bhagat *et al.*, 2020).

### Concentrations and Fate in the Aquatic Environment

Domestic sewage and hospital wastewater are the essential sources for the azole contamination of the aquatic environment. Azoles are recorded with high detection frequencies in WWTPs in the majority of European, North American, and Asian countries (Lacey *et al.*, 2012; Parajulee *et al.*, 2017; Wick *et al.*, 2010; Peng *et al.*, 2012). In WWTPs, azoles are removed inefficiently.

High concentrations of climbazole, clotrimazole, fluconazole, and ketoconazole are reported in influents and effluents of WWTPs. The total concentration could reach up to several μg/L. For instance, in Koblenz (Germany), concentrations of climbazole range from 0.48 to 1.35 μg/L in WWTP influents and 0.31–0.44 μg/L in effluents (Wick *et al.*, 2010). In Guangdong (China), concentrations reach up to 0.28 and 0.07 μg/L in WWTP influents and effluents, respectively (Chen *et al.*, 2012). The reported concentrations of clotrimazole range from the limit of quantification to 0.65 and 8.65 μg/L, respectively, in WWTP influents and effluents (Dublin, Ireland) (Lacey *et al.*, 2012). In Guangdong (China), the maximum concentrations of fluconazole are 0.17 and 0.14 μg/L in WWTP influents and effluents, respectively (Peng *et al.*, 2012).

Azoles differ in their hydrophilicity and biodegradation rates in WWTPs. The removal rates of clotrimazole are 88% and 92% in two WWTPs in Switzerland and 18% to 56% in WWTP Howdon (UK) (Lacey *et al.*, 2012). Biotransformation and absorption to solids are two important elimination processes. The decrease of dissolved azole concentrations occurs in biological processes but neither in the primary sedimentation nor in UV irradiation/chlorination. In contrast, the primary sedimentation decreases the total concentration of azole pharmaceuticals on average by 40–80%, suggesting that this portion is adsorbed to solids (Peng *et al.*, 2012).

Generally, residues of various azoles in surface waters are in the low ng/L. Climbazole, clotrimazole, and fluconazole are dominant, which is consistent with their increased levels in wastewaters. For instance, in Guangdong (China), concentrations of climbazole and fluconazole in surface waters are up to 264 and 153 ng/L, respectively (Chen *et al.*, 2014). In the Main River (Germany), concentrations of climbazole are up to 530 ng/L (Wick *et al.*, 2010). Interestingly, benzotriazoles, which have high commercial use due to their effectiveness in inhibiting metal corrosion, are ubiquitous in wastewater and surface water with concentrations up to 2,390 ng/L in Toronto, Canada (Parajulee *et al.*, 2017). Similar concentrations are also reported in many other countries, including Switzerland and Spain (Kahle *et al.*, 2008; Ccanccapa *et al.*, 2016). Azoles occur as mixtures in surface waters. Several imidazoles and triazoles are detected simultaneously in most of the water samples (Chen *et al.*, 2014). Concentrations are lower in the wet season than in the dry season due to dilution by precipitation (Chen *et al.*, 2014).

### Modes of Action

Azole fungicides act as antifungal agents by inhibiting the biosynthesis of ergosterol, which is an essential sterol constituent in cell membranes in fungi. They block the conversion of lanosterol to ergosterol via inhibition of the sterol 14α-demethylase (CYP51 from cytochrome P450 family) that catalyzes an oxidative removal of lanosterol, 14-methyl group, and leads to a 14–15 double bond, resulting in decreased ergosterol synthesis and loss of fungal cell membrane integrity (Synnott *et al.*, 2010). CYP51 is not only found in fungi but also in many other species including aquatic organisms. Thus, azoles can also act as inhibitors of CYP51 in non-target species (Zarn *et al.*, 2003; Brown *et al.*, 2011). For instance, high concentrations of clotrimazole result in downregulation of the *cyp51* transcript in male zebrafish (Brown *et al.*, 2011).

Azoles also inhibit other CYP450 enzymes, which in turn may affect metabolism. 7-ethoxyresorufin O-deethylase (EROD, encoded by the *cyp1a* gene) and

7-benzyloxy-4-trifluoromethylcoumarin O-debenzylase (BFCOD, encoded by the *cyp3a* gene) are essential hepatic detoxification enzymes that are involved in endogenous and exogenous metabolism. Azoles, including clotrimazole, prochloraz, and ketoconazole, impact the expression and/or enzyme activity of *cyp1a* and *cyp3a* of various fish species, such as rainbow trout (Burkina *et al.*, 2013) and clearfin livebearer (Lacey *et al.*, 2012). Azoles also affect the expression of phase II conjugating enzyme genes such as α-glutathione S-transferase (*gstα*) and glutathione S-transferase p (*gstp*), and other drug transporter genes such as canalicular multispecific organic anion transporter 1 (*cmoat*) in teleost fish (Corcoran *et al.*, 2014).

Aromatase (encoded by the *cyp19* gene) is another target of azoles (Beijer *et al.*, 2018; Zarn *et al.*, 2003; Brown *et al.*, 2011). In steroidogenesis, aromatase converts androgens into corresponding estrogens. Consequently, azoles are implicated in endocrine disruptive effects (Zarn *et al.*, 2003; Brown *et al.*, 2011). For instance, chronic exposure to clotrimazole lowers the transcriptional expression of *cyp19a1a* and disrupts steroidogenesis in male zebrafish (Brown *et al.*, 2011). Besides, inhibition of acetylcholinesterase (AChE) activity can also be observed when fish are exposed to azoles, including tebuconazole and propiconazole, suggesting that azoles may exhibit neurotoxic activity in fish (Liu *et al.*, 2016; Altenhofen *et al.*, 2017; Souders II *et al.*, 2019).

## Effects on Fish

Effects in fish are demonstrated for several frequently detected compounds including clotrimazole, ketoconazole, and fluconazole. Acute values of azoles are generally at mg/L levels in several fish species (Zhu *et al.*, 2014; Beijer *et al.*, 2018). Targets of chronic exposures are the oxidative stress, the reproductive and endocrine system, and the nervous system.

### Oxidative Stress

Azoles alter levels of antioxidant enzymes and stimulate the production of reactive oxygen species (ROS) in fish. Propiconazole can induce ROS levels and alter the homeostasis of antioxidant SOD in rainbow trout (Li *et al.*, 2010). Significant alteration of SOD and catalase (CAT) activities are further observed in rainbow trout (Li *et al.*, 2010) and medaka (Chu *et al.*, 2016). A similar phenomenon was observed for clotrimazole and ketoconazole (Liu *et al.*, 2017). Exposure of African catfish to sublethal concentrations of clotrimazole increases oxidative stress parameters including CAT, SOD, lipid peroxidation, glutathione reductase, and glutathione peroxide in both liver and gills (Melefa and Nwani, 2021).

### Reproductive Effects and Endocrine Disruption

Azole compounds may cause reproductive and endocrine effects in fish. Early-life exposure of medaka fish to triadimenol at environmentally relevant concentrations (3 μg/L and above) interrupts hormone balance, alters ovary development, decreases reproduction in females, and results in a skewed sex ratio in progeny (Chu *et al.*, 2016). In zebrafish, decreased fecundity along with decreasing estradiol synthesis has been reported when fish are exposed to prochloraz (De Oliveira *et al.*, 2020) and propiconazole (Teng *et al.*, 2019). Prochloraz also reduces the fecundity of fathead minnows at 100 μg/L, followed by alterations of plasma steroid (17β-estradiol, testosterone, 11-ketotestosterone) levels and gonadal histology. This is consistent with inhibitory effects on steroidogenesis (Ankley *et al.*, 2012). In fathead minnows, propiconazole decreases plasma concentrations of 17β-estradiol, testosterone, and VTG in females, and reduces weight and egg production (Skolness *et al.*, 2013). Fadrozole has similar effects in zebrafish with a decrease in plasma VTG concentrations, oocyte atresia, and a decrease in fecundity (Ayobahan *et al.*, 2019).

At the transcription level, inhibitory effects on the expression of genes involved in steroidogenesis, especially aromatase that is responsible for the production of estrogen, are described in various fish species such as fathead minnow (Ankley *et al.*, 2005), rainbow trout (Beijer *et al.*, 2018), and zebrafish (Ayobahan *et al.*, 2019). A comparative study with six azoles shows that clotrimazole was the most active compound in rainbow trout inhibiting aromatase activity (IC50 at 20 nM), followed by letrozole, imazalil, miconazole, bifonazole, and prochloraz (Beijer *et al.*, 2018). In most teleost fish, two forms of the *cyp19* gene exist, and they may be differentially affected among tissues (Ankley *et al.*, 2005). Besides, reduced transcription of genes such as *cyp11a, cyp11c, cyp17a, hsd3b1, hsd11b3,* as well as steroidogenic acute regulatory protein (*star*) and follicle-stimulating hormone receptor (*fshr*), all encoding enzymes or receptors involved in steroidogenesis, are reported (Ankley *et al.*, 2012; Skolness *et al.*, 2013).

### Neurotoxicity

AChE is a primary target of azoles. AChE plays a critical role in modulating cholinergic neurotransmission and serves as an important biomarker for neurotoxicity in animals including fish. Benzotriazole, tebuconazole, and cyproconazole can inhibit AChE activity, leading to reduced traveled distance and locomotor activity in early-staged zebrafish (Altenhofen *et al.*, 2017; Souders II *et al.*, 2019). Inhibition of AChE activity and lower swimming activity occurs in several other species including goldfish and catfish when exposed to clotrimazole and ketoconazole (Zhu *et al.*, 2014; Liu *et al.*, 2017). A comparative study with five widely used triazoles conducted in early-staged rainbow trout shows that fluconazole and myclobutanil are the most active compounds in inhibiting AChE activity with effects at 200 μg/L (Zhu *et al.*, 2014).

## Risk Assessment

A risk assessment for azoles is still incomplete due to the lack of sufficient environmental residue and effects data. Current data suggest that potential risks are associated with climbazole and propiconazole. This is based on the

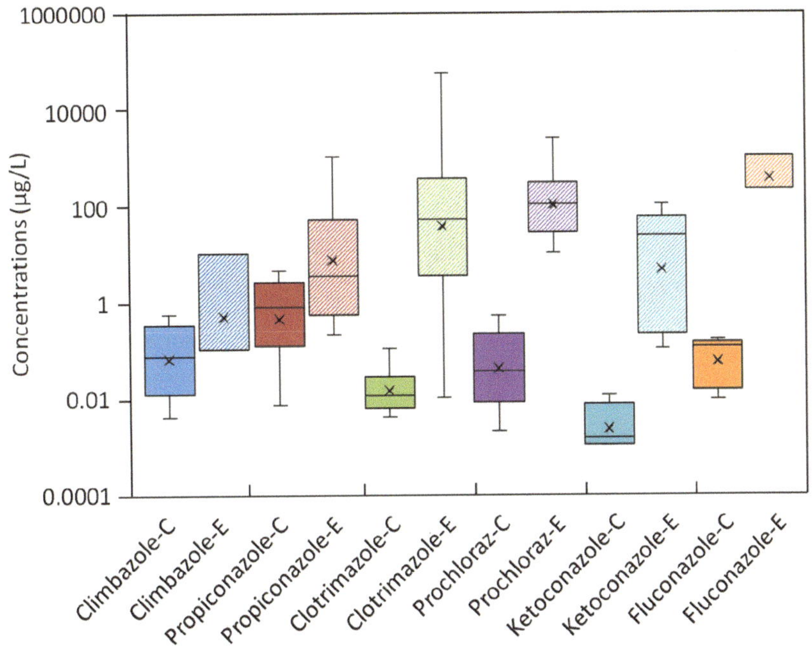


**FIGURE 14.4** Tentative risk assessment of azole fungicides. Shown are the highest residues levels reported in surface waters and the chronic LOECs in aquatic organisms according to published literature. Data are presented as box and whisker plots, where boxes encompass values between the 25th and 75th percentiles, horizontal lines represent median values, and the whiskers give the minima and maxima, thus the variability outside the upper and lower quartile. Plots (x) represent mean values. C: Compiled data on highest concentrations determined in surface waters; E: Compiled LOECs for fish.

ratio between the measured environmental concentration (MEC) and their LOECs (Figure 14.4). Generally, the concentrations of azoles in surface waters are lower than the related LOECs. In the case of propiconazole, the highest environmental concentrations are sometimes in the range of LOECs. Ketoconazole, miconazole, and fluconazole pose low risks with RQ of <0.1. However, clotrimazole may suggest a risk in 1.9% of water samples (RQ≥1), especially in WWTP effluents, and consequently, at some discharge points (Chen and Ying, 2015). Although the toxic profile of azoles is important and critical, the levels in the environment seem too low to induce adverse effects in fish. Based on the current knowledge, the risks of azoles for fish health are low with the exception of climbazole and propiconazole in some sites.

## CYTOSTATICS

Cytostatics, also known as anticancer drugs or anti-neoplastic drugs, are of potential concern due to their high cellular activity. Here, we also regard cancer chemotherapeutics that interfere with the sex hormone system (antiestrogens, antiandrogens). They enter the aquatic ecosystem predominantly via hospital and municipal wastewater effluents and are measured at concentrations of several ng/L to several hundred ng/L in surface waters (Li *et al.*, 2021; Gouveia *et al.*, 2019).

Cytostatics are classified as cytotoxic drugs according to their chemical nature and function. Cytotoxic drugs can be further divided into alkylating drugs, antimetabolites, antibiotics, plant alkaloids, platinum complexes, protein kinase inhibitors, and immunosuppressants due to their different MoAs (Li *et al.*, 2021). Given the lacking selectivity to kill cancer cells, these drugs may cause genotoxic, mutagenic, teratogenic, and endocrine disruptor effects in healthy cells of living organisms including fish (Kümmerer *et al.*, 2016). Despite such potential risks, studies addressing the occurrence of cytostatics in different water matrices remain scarce, which may be due to their generally low concentrations and lack of accurate quantitative analysis method (Pieczyńska *et al.*, 2017; Gouveia *et al.*, 2019).

### Concentrations and Fate in the Aquatic Environment

In general, cytostatics are detected in hospital and municipal wastewaters with concentrations ranging from several ng/L to hundreds of μg/L. In surface waters, they occur in the ng/L range (Li *et al.*, 2021; Gouveia *et al.*, 2019). Here, we focus on the four most important categories.

#### Alkylating Drugs

Cyclophosphamide and ifosfamide, the most commonly used anticancer alkylating drugs, are extensively studied. Their highest residue levels are found in hospital wastewaters, with concentrations of up to dozens μg/L. In WWTPs effluents, they are up to several μg/L (Li *et al.*, 2021). Cyclophosphamide has been frequently reported as a difficult compound to degrade in WWTPs, withstanding primary, secondary, and even tertiary treatments. This is also observed for ifosfamide (Gouveia *et al.*, 2019). In surface waters, concentrations of cyclophosphamide range from <0.2 to 0.096 µg/L (Buerge *et al.*, 2006). Up to 0.5 µg/L cyclophosphamide occurs in the Jingmei River receiving hospital discharges (Lin *et al.*, 2013). Similar surface water concentrations occur for ifosfamide that may reach up to 0.041 µg/L in the Guadarrama River (Spain) (Valcárcel *et al.*, 2011).

### Antimetabolites

Among antimetabolites, methotrexate and 5-fluorouracil are the most prescribed drugs (Besse *et al.*, 2012). They are widely detected in wastewaters and surface waters. Generally, they occur in µg/L concentrations in hospital and WWTP wastewater influents (Besse *et al.*, 2012). Detection frequencies range from 23% to 100% (Li *et al.*, 2021). The highest reported concentrations of methotrexate and 5-fluorouracil are 4.8 and 124 µg/L in hospital wastewaters in Austria and Spain, respectively (Mahnik *et al.*, 2007; Olalla *et al.*, 2018). The degradation of antimetabolites in WWTPs is not well known. Methotrexate has been reported to be resistant to chlorine treatment and UV disinfection, while high removal rates could be achieved via biodegradation. Moreover, membrane bioreactors and activated sludge have been reportedly effective in the degradation of 5-fluorouracil (Gouveia *et al.*, 2019).

In surface waters, concentrations are considerably lower. Methotrexate occurs at levels lower than 0.005 µg/L (Franquet-Griell *et al.*, 2017) and 5-fluorouracil up to 0.578 µg/L in the Chao Phraya River (Thailand) (Usawanuwat *et al.*, 2014). Several other compounds such as cytarabine and tegafur are in the same concentration range (Li *et al.*, 2021).

### Cytotoxic Antibiotics

Only a couple of studies explore the occurrence of cytotoxic antibiotics in the aquatic environment. In hospital effluents in Brazil, daunorubicin, epirubicin, and doxorubicin concentrations range from hundreds of ng/L to several µg/L. Doxorubicin has concentrations of up to 10.4 µg/L with a detection frequency of 57% (Souza *et al.*, 2018). Bleomycin occurs at dozens of ng/L in domestic wastewater effluents, receiving streams as well as drinking waters in the southeast of England (Aherne *et al.*, 1990). Limited data are available on the removal efficiency of cytotoxic antibiotics. Some studies suggest that they may be poorly biodegradable and resistant to both biological and physical removal processes during wastewater treatment (Souza *et al.*, 2018).

### Endocrine Therapy Drugs

Tamoxifen, megestrol acetate, and flutamide are highly administered with up to thousands of kg per year in Spain, France, and other EU countries. Their occurrence in the aquatic environment and their activity profile are of concern (Franquet-Griell *et al.*, 2017). The antiestrogen tamoxifen is among the most frequently detected endocrine therapy drugs. Detection frequencies are 37–100% in hospital effluents, 4–100% in WWTP wastewaters, and 19–100% in surface waters. Its concentration is up to 170, 740, and 212 ng/L in hospital effluents, WWTP wastewaters, and surface waters, respectively (Gouveia *et al.*, 2019; Besse *et al.*, 2012; Li *et al.*, 2021). In general, tamoxifen undergoes limited removal throughout the WWTPs with activated sludge treatment, trickling biofilters, membrane bioreactors, and oxidation ditches. Moreover, poor removal has also been reported for sand filtration and ticking filters. Ozonation and UV radiation are relatively efficient. However, their implementation in WWTPs is still scarce (Gouveia *et al.*, 2019; Li *et al.*, 2021). Very little information is available for other drugs, but it is expected that they occur in similar concentration ranges. Megestrol is detected at concentrations of up to 1.3 µg/L in hospital wastewaters and up to 20 ng/L in WWTPs effluents (Gómez-Canela *et al.*, 2014). Letrozole and anastrozole are frequently detected but at slightly lower concentrations of up to 2.4 and 3.7 ng/L, respectively, in hospital effluent and up to 0.3 and 0.4 ng/L, respectively, in WWTP effluents (Liu *et al.*, 2010).

## Modes of Action

Alkylating agents are among the first classes of drugs used in cancer treatment. They act by inhibiting the transcription of DNA into RNA and thereby stopping protein synthesis. Though the alkyl carbon groups of alkylating drugs typically bind to a broad range of biological molecules, aromatic nitrogen and extra-cyclic oxygen of DNA bases are the primary targets. They generate cross-links within the DNA chain and thereby result in cytotoxic, mutagenic, and carcinogenic effects (Drabløs *et al.*, 2004). This action occurs in all cells, but cancer cells are among the most affected due to their rapid division.

Antimetabolites are highly prescribed anti-cancer drugs. They act as a substitute for the actual metabolites that would be used in the normal metabolism and thereby halt cell growth and cell division. Some of the important drugs from this class are 5-fluorouracil, methotrexate, fludarabine, and cytarabine. These molecules interfere with one or more enzymes or their reactions that are necessary for DNA synthesis, especially, DNA polymerases, including thymidylate synthetase and ribonucleotide reductase (Valenzuela *et al.*, 2015) 5-Fluorouracil acts also as a RNA base analogue.

Cytotoxic antibiotics are a large group of drugs that act through various mechanisms of action. In general, they interrupt cell division and are thus employed in chemotherapy (Gerber, 2008). The most important subgroups are anthracyclines and bleomycin. Anthracyclines, including doxorubicin and daunorubicin, act as DNA intercalators (molecules inserted between the two strands of DNA) that generate highly reactive free radicals, leading to damage of intercellular molecules and inhibition of topoisomerase (Martins-Teixeira and Carvalho, 2020).

Endocrine therapy drugs are designed to inhibit the production or activity of hormones, leading to a reduction in cellular proliferation. There are two major groups of endocrine therapy drugs: non-steroidal antiestrogens and antiandrogens. Antiestrogens act either by inhibiting the estrogen biosynthesis (aromatase inhibitors) or by direct estrogen receptor antagonism. Antiandrogens act as AR antagonists (Besse *et al.*, 2012). For instance, tamoxifen is a selective estrogen receptor modulator used to prevent estrogen receptor-positive tumor cells from growing, including breast cancer cells. It acts by competitively inhibiting estrogen binding to its receptors to block the promotional role of estradiol. Tamoxifen also leads to a decrease in tumor growth factor α and an increase in sex hormone-binding globulin, leading to an inhibition of the amount of freely available estradiol. Thus, it reduces levels of factors that stimulate tumor growth (Yu and Bender, 2001).

## Effects on Fish

### Oxidative Damage

Given the specific MoAs for anti-cancer drugs for cytotoxicity, cytotoxic action and oxidative damage are expected to be generated in various aquatic organisms. Cyclophosphamide leads to an increased frequency of sister chromatid exchanges in the gills of eels (Santos and Pacheco, 1995). It also inhibits the enzymatic antioxidant system (SOD, GPx) and GST activity at 0.5 and 1 μg/L, and leads to DNA damage (Fonseca *et al.*, 2018).

Some antimetabolites (such as 5-fluorouracil and capecitabine) and antibiotics (doxorubicin) are evaluated for their genotoxicity in aquatic biota. In a two-generation study in zebrafish, 5-fluorouracil increases DNA damage and the frequency of micronuclei in the liver and blood cells at 0.01 μg/L and above (Kovács *et al.*, 2015).

### Reproductive Effects and Endocrine Disruption

Drugs used in endocrine therapy may be of concern for their reproductive and endocrine disruptive effects due to their specific hormonal or anti-hormonal properties. In Japanese medaka exposed to tamoxifen for 21 days, males show significantly increased plasma VTG levels, as well as reduced fecundity and fertility at 625 μg/L (L. Sun *et al.*, 2007). In zebrafish, tamoxifen alters VTG gene expression, 17β-estradiol levels, and disturbs spermatogenesis (Maradonna *et al.*, 2009). Very limited studies are performed to explore the potential reproductive effects of other types of cytotoxic and cytostatic drugs.

### Embryonic Development

Cytostatic drugs target rapidly dividing cells during cell division, and thereby, they may lead to developmental deformation in fish. Antimetabolites and several cytotoxic antibiotics seem to be of concern. In zebrafish embryos, methotrexate leads to disruption in cardiovascular development, including abnormal cardiac twist, reduced heart rate, and ventricular shortening fraction, as well as disrupted vascular development (Sun *et al.*, 2009). Mycophenolic acid results in missing tails, severe pericardial edema, bradycardia and enlarged yolk sacs in zebrafish embryos, and thalidomide and gambogic acid disturb pectoral fin development (Gao *et al.*, 2014).

Tamoxifen decreases heart rate in zebrafish embryos and induces mild defects in the caudal fin and skin at concentrations of 50 and 500 μg/L. Expression of genes involved in steroidogenesis is altered similarly as in adults at 0.5 μg/L and above (Xia *et al.*, 2016). Tamoxifen also affects hatchability and alters the sex ratios of Japanese medaka embryos at 5 μg/L and above (D. Sun *et al.*, 2007).

## Risk Assessment

Figure 14.5 shows the comparison between the detected highest surface water concentrations and the lowest values of LOECs in fish for several cytostatic drugs. Generally, the LOECs range from several μg/L to hundreds of mg/L, which are several orders of magnitude higher than the highest concentrations in surface waters. However, in some cases, observed effects occur at environmentally relevant concentrations. This is the case for bicalutamide. Thus, environmental risks are suggested for this compound, and in some cases, for tamoxifen, where effect concentrations and surface water concentrations partially overlap (Figure 14.5). However, this is the worst-case situation.

Similar estimates are reported in other studies, where the HQs for 31 cytostatic drugs are estimated based on the LOECs for different endpoints in aquatic organisms. Cyclophosphamide (HQ=22000), 5-fluorouracil (HQ=150), and bicalutamide (HQ=125) suggest high HQ values and thereby display hazards concerning chronic exposure of aquatic organisms including fish. Tamoxifen has a moderate hazard for aquatic organisms with a HQ value of 2.1. The widely detected cytostatics methotrexate and daunorubicin exhibit low risks, with HQ values of 0.08 and 0.07, respectively (Li *et al.*, 2021). In another study, moderate-to-high risks are suggested for tamoxifen, 5-fluorouracil, imatinib, methotrexate, and mycophenolic acid. However, cyclophosphamide, bicalutamide, and daunorubicin were considered to have low risks. Such inconsistent results are mainly due to some reports with abnormally high cytostatic concentrations, which are not always considered for risk assessment (Gouveia *et al.*, 2019). Overall, though current data on the occurrence of cytostatic drugs and their effects in aquatic organisms are scarce, risks posed by cytostatics seem to be restricted to bicalutamide, tamoxifen, and 5-fluorouracil.

# ANTIDIABETICS

High blood sugar (hyperglycemia) is a symptom of diabetes mellitus. Type 2 diabetes mellitus is among the most common metabolic disorders, currently affecting over 6% of the world's population (WHO, 2016), and more than 90% of the patients are treated orally with hypoglycemic agents (e.g., metformin). Some antidiabetic drugs like metformin are also used to treat obesity, cancer, and polycystic ovary syndrome. Due to the high doses of metformin required for

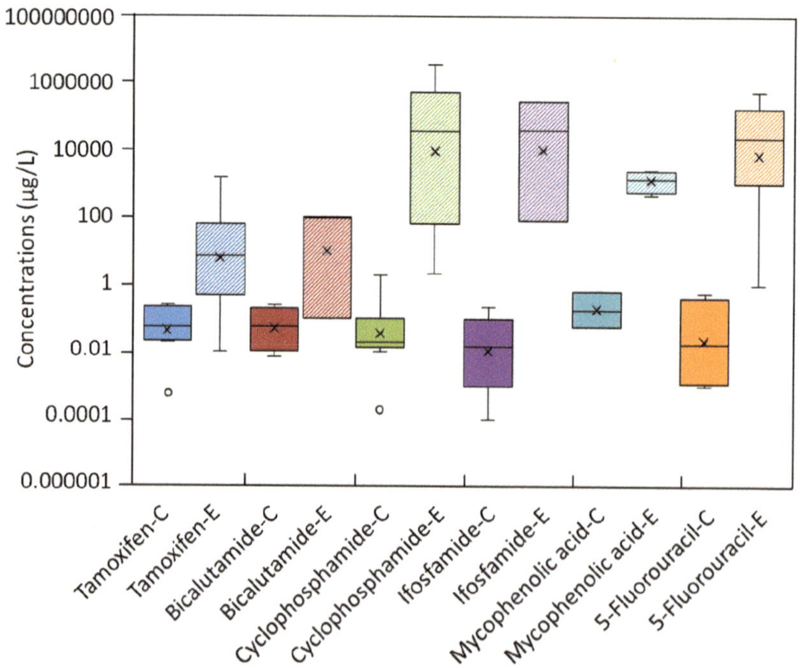


**FIGURE 14.5** Tentative risk assessment of cytostatics. Shown are the highest residues levels reported in surface waters and the chronic LOECs in aquatic organisms according to published literature. Data are presented as box and whisker plots, where boxes encompass values between the 25th and 75th percentiles, horizontal lines represent median values, and the whiskers give the minima and maxima, thus the variability outside the upper and lower quartile. Plots (x) represent mean values. C: Compiled data on highest concentrations determined in surface waters; E: Compiled LOECs for fish.

therapeutic effects (daily doses range from 500 to 2,500 mg/L (Trautwein and Kümmerer, 2011)), the chronic nature of diabetes, as well as its increasing diagnosis, large quantities of antidiabetic drugs are produced and consumed. In the case of metformin, the drug is not metabolized in humans, and the transformation product generated in WWTPs, if transformed at all, retains its pharmacological activity. Therefore, metformin and its primary degradation product are frequently detected in surface waters at concentrations of hundreds to thousands of ng/L. In larval fish, these concentrations lead to reduced growth metrics, and in adult fish, alter endocrine function including upregulation of vitellogenin (VTG), in males. Information on the environmental concentration and effects of other blood sugar-lowering pharmaceuticals with molecular mechanisms that differ from metformin is rare.

### Concentrations and Fate in the Environment

Antidiabetics are only partially metabolized by the human body or as in the case of the most prominent antidiabetic metformin, not at all (Markiewicz *et al.*, 2017). Thus, these drugs are distributed primarily via aqueous transport of sewage but landfills with expired medicines also present a point source. In WWTPs, flocculation, activated carbon filtration, ozonation, and photodegradation are ineffective in removing the antidiabetic metformin and its degradation product guanylurea (Scheurer *et al.*, 2012; Elizalde-Velázquez and Gómez-Oliván, 2020). However, most antidiabetics can be biologically transformed by bacteria in activated sewage sludge (Markiewicz *et al.*, 2017). For metformin, this results in the formation of guanylurea which is thought to be a dead-end product as it is stable against further photo- and biodegradation (Trautwein and Kümmerer, 2011). However, some studies suggest that guanylurea can be fully mineralized (Straub *et al.*, 2019). Chlorination, adsorption in graphene oxide, and phytoremediation are wastewater treatments that have shown high rates of removal (Elizalde-Velázquez and Gómez-Oliván, 2020; Scheurer *et al.*, 2012). For example, phytoremediation using *Typha latifolia* or *Helianthus annuus* results in high removal efficiency of metformin. Yet, these results are achieved under laboratory conditions, and methods such as phytoremediation or graphene oxide are not widely applied in WWTPs. Taking into account several studies across multiple countries, metformin concentrations in WWTP influent range from several ng/L to 94,600 ng/L and in effluents from several ng/L to 82,700 ng/L (Meador *et al.*, 2016; Elizalde-Velázquez and Gómez-Oliván, 2020).

Both metformin and its degradation product guanylurea are highly water-soluble and therefore measured at relatively high concentrations in surface water worldwide (Table 14.7). In fish, metformin is accumulating at concentrations of up to 28 ng/g as measured for sculpin (*Leptocottus armatus*) (Meador *et al.*, 2016), and in fish organs, metformin is measured in the highest concentrations in gut>liver>gills according to a laboratory accumulation study performed in zebrafish (*Danio rerio*) (Lin *et al.*, 2021). The antidiabetic

**TABLE 14.7**
**Representative Concentrations of Selected Antidiabetics in Surface Water**

| Antidiabetics and Metabolites | Concentrations (ng/L) | References |
|---|---|---|
| **Metformin** | 120–9,200[a] | Blair *et al.* (2013) |
| | 60–3,100[a] | Scheurer *et al.* (2012) |
| | 1–2,635[a] | Bradley *et al.* (2016) |
| | 0.2–33,600[a] | Reviewed in Elizalde-Velázquez and Gómez-Oliván (2020) and Elliott *et al.* (2017) |
| | 105–832[a] | Meador *et al.* (2016) |
| **Guanylurea** | 100–2,200[a] | Scheurer *et al.* (2012) |
| | 1.2–222,000[a] | Reviewed in Elizalde-Velázquez and Gómez-Oliván (2020) |
| **Glibenclamide** | 1.4[c] | Al-Odaini *et al.* (2012) |
| | 13[d] | Ternes *et al.* (2001) |

[a] Range of values from different sampling sites.
[b] Range of mean values of different samplings.
[c] Mean value of different sampling sites.
[d] Maximum concentration.

drug repaglinide is only detected in 3 out of 40 wild-caught fish in the Czech Republic (0.2±0.561 ng/mL plasma) and the UK (0.09±0.291 ng/mL plasma) (Cerveny *et al.*, 2021). The depuration rate of Japanese medaka (*Oryzias latipes*) larvae exposed for 24h to metformin is rather rapid, with the time required for larvae to depurate 50% of metformin being 4.88 h (Ussery *et al.*, 2018). This is supported by other studies, stating bioconcentration factors of metformin between 0.02 and 21.72 (reviewed in Elizalde-Velázquez and Gómez-Oliván, 2020) suggesting that metformin is unlikely to bioconcentrate in aquatic organisms. However, exposure to metformin for 30 days results in measurable metformin levels in zebrafish even after a 30-day recovery period (Lin *et al.*, 2021).

## Modes of Action

All anti-diabetic drugs are designed to lower blood glucose levels, and therefore most of their MoA is intimately tied to glucose metabolism. Yet, there are seven classes of anti-diabetic drugs, each with different mechanisms of action (Pappachan *et al.*, 2019).

1. Inhibitors of α-glucosidases modulate the rate of digestion of complex carbohydrates and disaccharides by competitively and reversibly inhibiting α-glucosidases in the intestine. Thereby, inhibitors of α-glucosidases reduce intestinal glucose absorption.
2. Sulfonylureas close adenosine triphosphate (ATP)-sensitive potassium channels in the pancreatic β-cells, which results in membrane depolarization and subsequent calcium influx. The calcium influx then stimulates insulin secretion.
3. Similarly, meglitinides (e.g., repaglinide) promote insulin secretion by acting on ATP-dependent potassium channels in the membrane of pancreatic β-cells. In contrast to sulfonylureas, meglitinides have a rapid onset of action as well as a short duration of activity.
4. GLP-1 receptor agonists directly stimulate the GLP-1 receptor, while DPP-4 inhibitors inhibit GLP-1 degradation, promoting glucose-dependent insulin secretion.
5. Thiazolidinediones (rosiglitazone and pioglitazone) are potent activators of the nuclear receptor PPARγ, promoting transactivation of a large number of target genes involved in hepatic gluconeogenesis and insulin-dependent uptake of glucose in muscle and fat, among others. Thereby, thiazolidinediones increase insulin sensitivity in tissues.
6. Sodium-dependent glucose cotransporter 2 (SGLT2) inhibitors promote urinary loss of glucose through the inhibition of glucose uptake (SGLT2) in the kidney.
7. Biguanides include the most often prescribed metformin, and they exert their effect in the liver by blocking gluconeogenesis, thereby improving the insulin sensitivity of peripheral tissue. However, despite almost 60 years of research, the exact mechanisms of metformin action are not fully understood. Biguanides are thought to act as an inhibitor of the mitochondrial respiratory-chain complex 1, activating AMP-activated protein kinase signaling.

## Effects

Following the concept of evolutionary conserved molecular drug targets, antidiabetic drugs may lower glucose levels in fish. However, healthy fish already maintain low blood glucose concentrations in comparison to other vertebrates. There thus may be physiological consequences

**TABLE 14.8**
**Effects, and LOEC of Metformin in Selected Fish**

| Antidiabetic Drug (Class) | Effect | LOEC (ng/L) | Fish Species | References |
|---|---|---|---|---|
| Metformin (Biguanide) | Liver glycogen increased; weight decreased, altered gut microbiome | 1,000 | *Salmo trutta f. fario* | Jacob *et al.* (2018) |
| | Hardening of chorion | 10,000 | *Oryzias latipes*, embryo | Ussery *et al.* (2018) |
| | Weight and length decreased | 1,000 | *Oryzias latipes*, larva | Ussery *et al.* (2018) |
| | 11-Ketotestosterone increased in ♂ | 3,200 | *Oryzias latipes*, full life-cycle | Ussery *et al.* (2018) |
| | *VTG* mRNA, cyp3a126 mRNA, and gnrh3 mRNA increased | 1,000 | *Pimephales promelas*, males | Crago *et al.* (2016) |
| | *VTG* mRNA increased | 40,000 | *Pimephales promelas*, males | Niemuth *et al.* (2015) |
| | Upregulation of ar, 3β-hsd, 17β-hsd, cyp19a1, sult2a1in ♂ gonads | 40,000 | *Pimephales promelas* | Niemuth and Klaper (2018) |
| | DNA methylation | 100 | *Danio rerioww* | Lin *et al.* (2021) |

of exposure that can include reduced growth (Table 14.8). Adverse effects on fish have only been studied for metformin, the most consumed antidiabetic worldwide, and there is a clear lack of knowledge on the adverse effects of non-biguanide antidiabetic drugs with different mechanisms of action.

### Metformin

Metformin has been investigated for effects in fish at all life stages, including whole life-cycle exposures. Data indicate that while in developing fish, metformin can affect growth and metabolism, in adults, metformin may have endocrine-disrupting effects.

In the earliest stage of fish development, while the embryo is still in its eggshell, metformin can harden the chorion, as shown for Japanese medaka (*Oryzias latipes*) (Ussery *et al.*, 2018). The same study shows that metformin reduces growth metrics, alters metabolomes, and changes the expression of genes associated with cell growth. As metformin reduces glucose levels, reduced growth in developing vertebrates seems an expected outcome. In contrast, no reduced growth occurs in brown trout (*Salmo trutta f. fario*) larvae exposed to 10,000 ng/L, yet weight is reduced (Jacob *et al.*, 2018).

In adult fish, evidence has emerged that metformin can alter endocrine function, for example, through interactions between insulin signaling and steroidogenesis. The classic biomarker for environmental estrogens, the egg protein VTG, as well as genes involved in E2 synthesis (*gnrh3*) and metabolism (*cyp3a126*), are upregulated at the transcriptional level in adult male fathead minnows (*Pimephales promelas*) (Niemuth and Klaper, 2015; Crago *et al.*, 2016; Niemuth *et al.*, 2015). Likewise, full life-cycle exposure of metformin to Japanese medaka stimulates the production of the steroid hormone 11-ketotestosterone, among others, in adult female medaka (Ussery *et al.*, 2018). Similar long-term exposure to metformin from early-life stage to adulthood in fathead minnows results in intersex and reduced size in males (Niemuth and Klaper, 2015, 2018). Juvenile fathead minnows seem more susceptible to the estrogenic effects of metformin during a 7-d exposure than older, sexually mature male fish (Crago *et al.*, 2016). Beyond the potential endocrine-disrupting effects, metformin can affect global DNA methylation and the associated DNA methyltransferase (*dnmt1*) gene in the liver of zebrafish at a concentration of 100 ng/L, suggesting epigenetic effects of metformin (Lin *et al.*, 2021).

#### Risk Assessment

Metformin is one of the most prevalent pharmaceuticals in surface waters owing to its incomplete metabolism in the human body and WWTPs. The median MEC of metformin is 193 ng/L (Caldwell *et al.*, 2019), with the maximum concentration reported from the tributaries of the Laurentian Great Lakes (USA) being 33,600 ng/L (Elliott *et al.*, 2017). While the median concentration is above the value that induces epigenetic changes in zebrafish (Lin *et al.*, 2021), two studies performing an environmental risk assessment conclude that there is no environmental concern for metformin. A PEC/PNEC and MEC/PNEC risk characterization ratio of <1 for metformin and guanylurea is estimated (Caldwell *et al.*, 2019), which includes data from standard OECD tests with *Pseudokirchneriella subcapitata* growth, *Daphnia magna* reproduction, *Danio rerio,* and *Pimephales promelas* early-life stage tests. Similarly, a RQ of lower than 1 is estimated (Godoy *et al.*, 2018), dividing the highest $MEC_{\text{surface water}}$ by the PNEC based on long-term toxicity data (including *Lemna minor, Hydra attenuate, Daphnia similis, and D. rerio).* Thus, ecological risks are not expected for the chronic exposure of pelagic freshwater species to metformin. The influences of other classes of antidiabetic drugs on aquatic biota as well as potential interactive effects with other pharmaceuticals such as lipid-lowering agents and steroid hormones merit further investigation.

## OVERALL HAZARD AND RISK ASSESSMENT AND CONCLUSIONS

Pharmaceuticals are ubiquitous in wastewater-influenced environments at sub-ng/L to low ng/L concentrations or even higher. Do these concentrations pose a risk to fish? For the different pharmaceuticals, environmental concentrations, their potencies and risks vary significantly. Low environmental concentrations intuitively suggest little concern, which is not the case for highly potent pharmaceuticals such as synthetic steroid hormones. These pose the highest risk among pharmaceuticals for fish development and reproduction despite their low environmental concentrations. To evaluate the risk of pharmaceuticals to fish, several concepts are applied. In our opinion, the best way is to compare the chronic effects, which focus on the MoA of the pharmaceutical, with environmental concentrations in surface waters. We have taken this approach and represent the outcome in Figures 14.1–14.5 and their interpretation in the chapters above.

### Concepts for Risk Evaluation and Their Limitations

There are a few concepts to identify and rate the environmental risks of pharmaceuticals. Intuitively, the comparison of chronic effect concentrations with surface water concentrations gives a very good indication of whether or not a compound poses a risk. If environmental concentrations are in the effects range, a risk is given. In 2006, The European Medicines Agency (EMEA) issued a guideline (EMEA/CHMP/SWP/4447/00, 2006) for estimating the potential environmental risk of newly marketed medicinal products by a stepwise approach. First, the predicted environmental concentration (PEC) is estimated in surface water considering different parameters. If the PEC is below 10 ng/L, no environmental concern seems apparent and the compound is assumed as unlikely to represent a risk unless the compound is toxic, bioaccumulative, persistent, and deserves special attention (such as synthetic steroid hormones). Pharmaceuticals with a PEC equal to above 10 ng/L undergo an environmental fate and effect analysis. Tests on growth inhibition in algae, reproduction in *Daphnia*, and developmental toxicity in early-life stages of fish are performed according to OECD guidelines to derive and predict the no-effect concentration. To account for species differences, a safety or application factor is applied. When the ratio between PEC and PNEC is above 1, further evaluation of the compound is necessary.

The EMEA approach is pragmatic and straightforward, but has limitations, and thus needs improvement. For instance, the tests assess acute toxicity, and the MoA of the drug is not specifically considered. Thus, health-relevant physiological, biochemical, and molecular effects are not regarded. Consequently, a valuable risk assessment needs to include such important endpoints for comparison of chronic effect levels with environmental levels of the compound.

Some additional methods for drug prioritization and risk assessment approaches exist. One is the FPM used to estimate potentially hazardous compounds (Huggett *et al.*, 2003). Based on the conservation of drug targets (enzymes, receptors) across fish species, a theoretical model is developed to estimate the fish plasma concentration of a pharmaceutical based on physico-chemical properties, environmental concentration, and information from mammalian pharmacology and toxicology. Considering the environmental concentration of a drug, the estimated fish plasma concentration is compared to the human therapeutic plasma concentration. The closer the fish plasma and human therapeutic concentrations are, the higher the risk is that the compound elicits an effect in fish. This hypothesis assumes that fish receptors and physiology are comparable to humans.

Two examples illustrate this approach. Diclofenac exposure resulting in fish plasma levels close to human therapeutic levels results in differentially regulated hepatic gene transcripts, which is consistent with the MoA of diclofenac (Cuklev *et al.*, 2011). When the fish plasma concentration of SSRI sertraline in fathead minnow exceeds human therapeutic doses, behavioral changes occur (Valenti *et al.*, 2012). Ibuprofen exposure to 0.37 and 0.47 mg/L results in fish plasma concentrations that are 2–6-fold below the human therapeutic dose. This concentration decreases prostaglandin E metabolite levels as expected from human therapy (Patel *et al.*, 2016). There are also critical aspects to this model. There is no sufficient evidence for the assumption of equal sensitivity in fish and mammals. Only very few studies (Malev *et al.*, 2020; Cerveny *et al.*, 2021) have actually measured fish plasma concentrations and tested this hypothesis. One reason is the analytical challenge in determining the fish plasma concentration. Theoretical estimates are also not sufficient for bioaccumulative drugs; the levels may diverge and can be much higher as shown for levonorgestrel (Fick *et al.*, 2010).

Other concepts focus on the MoAs of the pharmaceutical and the homology between human drug targets and potential targets in fish (Gunnarsson *et al.*, 2008; Christen *et al.*, 2010). Thereby, it is assumed that the risk for adverse effects in fish increases with the degree of homology between the human and fish target receptors. High receptor homologies between mammals and fish occur for steroid hormone receptors, thyroid hormone receptors, retinoic acid receptors, peroxisome proliferator-activated receptors, and GABA receptors. By applying this MoA and target conservation concept, steroid hormones (estrogens, progestins, glucocorticoids), tamoxifen, and partly diazepam are identified as high-risk compounds (Christen *et al.*, 2010). The conservation of molecular targets is fundamental to this hypothesis, but many drugs have multiple mechanisms, and therefore, a range of effects.

There also exists a so-called read-across hypothesis (Rand-Weaver *et al.*, 2013). Thereby, the MoAs and receptor homology hypothesis are combined with the fish plasma concentration hypothesis. Although many aspects are similar to mammals, currently, no study considers and measures fish plasma concentrations and mode-of-action-based effects at the same time. Among the limitations of this hypothesis are the species specificity and the occurrence of multiple mechanisms of drugs and analytical challenges in assessing plasma levels.

### Risk Assessment: Comparison of Chronic Effects with Environmental Concentrations

All concepts indicate that the highest risks for fish health are posed by synthetic steroids (Fent *et al.*, 2006; Fent, 2015; Christen *et al.*, 2010; Gunnarsson *et al.*, 2019; Rand-Weaver *et al.*, 2013). In Table 14.9, we summarize the potential risks of the drug classes discussed in the chapters above. The best method for assessing risks is the comparison of environmental concentrations with the lowest chronic effect concentrations as performed with progestins (Fent, 2015) and many other pharmaceuticals (Fent *et al.*, 2006; Zhang *et al.*, 2020).

In the case of steroids, the highest environmental risks are associated with synthetic hormones employed in contraception and hormone therapies, including EE2, levonorgestrel, norethindrone, and norgestrel. Possibly, progesterone may also pose a risk, as there is an overlap between concentrations at highly polluted environments and the lowest observed effect concentrations (LOECs). Table 14.9 includes physiological and biochemical effects that are indicative of physiological changes. Transcriptional effects are very sensitive, and for many pharmaceuticals, they occur at environmental concentrations, for instance for the progestins dydrogesterone and medroxyprogesterone acetate. However, their translation to physiological outcomes and significance is sometimes unclear, unless they are combined with physiological effects (Fent and Sumpter, 2011).

Apart from these high-risk compounds, for most pharmaceuticals, the levels in surface water are at least an order of magnitude lower than the levels that cause an effect, and thus, they can be regarded as of no or low risk. However, pharmaceuticals occur as mixtures and effects may be additive. Effects of the mixture can occur even when individual compounds have no effects as shown with steroids. Most current data rely on relatively short-term

**TABLE 14.9**
**Pharmaceuticals Posing the Highest Risk to Fish**

| Class | Pharmaceutical | Representative Freshwater Concentrations (ng/L) | Lowest Chronic Effect Concentrations (ng/L) | Endpoints | References |
|---|---|---|---|---|---|
| **Steroids** | Ethinylestradiol (EE2) | 0.1–1<br>0.35[a] | 0.1<br>0.4–4<br>5–6 | Intersex<br>Fertility, reproduction<br>Population collapse | Metcalfe *et al.* (2001)<br>Länge *et al.* (2001), Runnalls *et al.* (2015), Kidd *et al.* (2007) |
| | Levonorgestrel | 4–11 | 0.8–100<br>2 | Fecundity, reproduction, masculinization<br>Transcripts | Zeilinger *et al.* (2009), Runnalls *et al.* (2015)<br>Zucchi *et al.* (2012) |
| | Norgestrel | 4–22 | 5<br>34 | Transcripts, sex determination | Liang *et al.* (2015) |
| | Norethindrone | 1–2 | 1–10 | Masculinization, reproduction | Paulos *et al.* (2010) |
| | Progesterone | 0.5–10 | 2<br>10–742<br>63 | Transcripts<br>Fecundity, reproduction<br>Increased females in sex determination | Zucchi *et al.* (2012), Liang *et al.* (2015)<br>Liang *et al.* (2015) |
| | Mifepristone | <0.4 | 3–39 | Fecundity increase, gonad histology | Blüthgen *et al.* (2013) |
| | Gestodene | nd | 1–100 | Fecundity, steroid level | Runnalls *et al.* (2013) |
| | 17β-Trenbolone | 0.02–50 | 10<br>27 | Masculinization<br>Population decrease | Holbech *et al.* (2006), Ankley *et al.* (2018), Miller and Ankley (2004) |
| **NSAID** | Diclofenac | 6–1,800 | 1,000–5,000 | Damage to gill, liver, kidney<br>Kidney histology | Schwaiger *et al.* (2004; Triebskorn *et al.* (2004) |
| **SSRI** | Fluoxetine | 1–141 | 540<br>30–3,000 | Increased estradiol level, reduced feeding<br>Behavioral alterations | Mennigen, Lado *et al.* (2010)<br>Barry (2013) |

[a] Median concentration.

exposures in a few species of laboratory fish and do not consider the vast variety of fish in the field. There may be more sensitive fish species or effects may occur only after exposure for a long time, which is not considered in laboratory experiments.

## CONCLUSIONS

Pharmaceuticals are ubiquitous contaminants in marine and freshwaters that are influenced by municipal and hospital wastewaters. The development of high-sensitivity chemical analytical techniques has allowed us to accurately determine the contamination situation. In contrast, effect studies often lack high sensitivity unless focused on chronic exposures and considering sensitive physiological endpoints, preferably considering the MoAs of the compound. This is the way to go in forthcoming studies.

Current knowledge suggests that among the variety of pharmaceuticals, synthetic steroids of all classes pose the highest risk to fish with respect to development and reproduction. This is particularly true for mixtures of steroids that show additive effects. Additionally, there are a few compounds that occur at levels in polluted surface waters, which may negatively affect fish (Table 14.9). Among them are the NSAIDs diclofenac (Figure 14.1) and the neuroactive compound fluoxetine. Only in heavily polluted environments, quinolone and tetracycline antibiotics, the NSAID ibuprofen, and the lipid-regulating agent bezafibrate (Figure 14.3) may pose a risk to aquatic organisms including fish.

Given the increasing amount of pharmaceutical consumption, a further decrease in surface water concentration is important, particularly as pharmaceuticals act as mixtures. The mixture activity of common pharmaceuticals needs further investigation, as they may act additively. Of course, the treatment of wastewater from production plants is a must. Specific treatment of hospital wastewater prior to its release to municipal wastewater plants should be implemented as it lowers concentrations there. Additional and advanced treatment steps should be implemented in municipal WWTPs, either activated carbon treatment, UV irradiation, or ozonation. All these treatments lower the concentrations of pharmaceuticals in the effluent (Zhang *et al.*, 2017). These techniques result in a lower pharmaceutical load and contribute to lowering the risk of pharmaceuticals to fish health.

## REFERENCES


Agunbiade, F.O. and Moodley, B. (2014) Pharmaceuticals as emerging organic contaminants in Umgeni River water system, KwaZulu-Natal, South Africa. *Environ. Monit. Assess.*, **186**, 7273–7291.

Aherne, G.W., Hardcastle, A., and Nield, A.H. (1990) Cytotoxic drugs and the aquatic environment: Estimation of bleomycin in river and water samples. *J. Pharm. Pharmacol.*, **42**, 741–742.

Al-Habsi, A.A., Massarsky, A., and Moon, T.W. (2016) Exposure to gemfibrozil and atorvastatin affects cholesterol metabolism and steroid production in zebrafish (Danio rerio). *Comp. Biochem. Physiol. Part B Biochem. Mol. Biol.*, **199**, 87–96.

Al-Odaini, N.A., Zakaria, M.P., Zali, M.A., Juahir, H., Yaziz, M.I., and Surif, S. (2012) Application of chemometrics in understanding the spatial distribution of human pharmaceuticals in surface water. *Environ. Monit. Assess.*, **184**, 6735–6748.

Al-Qirim, T., Shattat, G., Sheikha, G.A., Sweidan, K., Al-Hiari, Y., and Jarab, A. (2015) Synthesis of novel N-(4-benzoylphenyl)-2-furamide derivatives and their pharmacological evaluation as potent antihyperlipidemic agents in rats. *Drug Res. (Stuttg).*, **65**, 158–163.

Almeida, A.R., Tacão, M., Machado, A.L., Golovko, O., Zlabek, V., Domingues, I., and Henriques, I. (2019) Long-term effects of oxytetracycline exposure in zebrafish: A multilevel perspective. *Chemosphere*, **222**, 333–344.

Altenhofen, S., Nabinger, D.D., Wiprich, M.T., Pereira, T.C.B., Bogo, M.R., and Bonan, C.D. (2017) Tebuconazole alters morphological, behavioral and neurochemical parameters in larvae and adult zebrafish (Danio rerio). *Chemosphere*, **180**, 483–490.

Ammann, A.A., Macikova, P., Groh, K.J., Schirmer, K., and Suter, M.J.F. (2014) LC-MS/MS determination of potential endocrine disruptors of cortico signalling in rivers and wastewaters. *Anal. Bioanal. Chem.*, **406**, 7653–7665.

Ankley, G.T., Coady, K.K., Gross, M., Holbech, H., Levine, S.L., Maack, G., and Williams, M. (2018) A critical review of the environmental occurrence and potential effects in aquatic vertebrates of the potent androgen receptor agonist 17β-trenbolone. *Environ. Toxicol. Chem.*, **37**, 2064–2078.

Ankley, G.T., Cavallin, J.E., Durhan, E.J., Jensen, K.M., Kahl, M.D., Makynen, E.A., Thomas, L.M., Wehmas, L.C., and Villeneuve, D.L. (2012) A time-course analysis of effects of the steroidogenesis inhibitor ketoconazole on components of the hypothalamic-pituitary-gonadal axis of fathead minnows. *Aquat. Toxicol.*, **114**, 88–95.

Ankley, G.T., Jensen, K.M., Kahl, M.D., Korte, J.J., and Makynen, E.A. (2001) Description and evaluation of a short-term reproduction test with the fathead minnow (Pimephales promelas). *Environ. Toxicol. Chem.*, **20**, 1276–1290.

Ankley, G.T., Jensen, K.M., Durhan, E.J., Makynen, E.A., Butterworth, B.C., Kahl, M.D., Villeneuve, D.L., Linnum, A., Gray, L.E., and Cardon, M. (2005) Effects of two fungicides with multiple modes of action on reproductive endocrine function in the fathead minnow (Pimephales promelas). *Toxicol. Sci.*, **86**, 300–308.

Ayobahan, S.U., Eilebrecht, E., Kotthoff, M., Baumann, L., Eilebrecht, S., Teigeler, M., Hollert, H., Kalkhof, S., and Schäfers, C. (2019) A combined FSTRA-shotgun proteomics approach to identify molecular changes in zebrafish upon chemical exposure. *Sci. Rep.*, **9**, 1–12.

Baker, D.R. and Kasprzyk-Hordern, B. (2013) Spatial and temporal occurrence of pharmaceuticals and illicit drugs in the aqueous environment and during wastewater treatment: New developments. *Sci. Total Environ.*, **454–455**, 442–456.

Baker, T.R., Baker, B.B., Johnson, S.M., and Sladky, K.K. (2013) Comparative analgesic efficacy of morphine sulfate and butorphanol tartrate in koi (Cyprinus carpio) undergoing unilateral gonadectomy. *J. Am. Vet. Med. Assoc.*, **243**, 882–890.

Balmford, A. (2013) Pollution, politics, and vultures. *Science*, **339**, 653–654.

Bao, S., Nie, X., Liu, Y., Wang, C., and Liu, S. (2018) Response of PXR signaling pathway to simvastatin exposure in mosquitofish (Gambusia affinis) and its histological changes. *Ecotoxicol. Environ. Saf.*, **154**, 228–236.

Barros, S., Montes, R., Quintana, J.B., Rodil, R., André, A., Capitão, A., Soares, J., Santos, M.M., and Neuparth, T. (2018) Chronic environmentally relevant levels of simvastatin disrupt embryonic development, biochemical and molecular responses in zebrafish (Danio rerio). *Aquat. Toxicol.*, **201**, 47–57.

Barry, M.J. (2013) Effects of fluoxetine on the swimming and behavioural responses of the Arabian killifish. *Ecotoxicology*, **22**, 425–432.

Bartelt-Hunt, Shannon, L., Snow, Daniel, D., Kranz, William, L., Mader, Terry, L., Shapiro, Charles, A., van Donk, Simon, J., Shelton, David, P., Tarkalson, David, D., and Zhang, Tian, C. (2012) Effect of growth promotants on the occurrence of endogenous and synthetic steroid hormones on feedlot soils and in runoff from beef cattle feeding operations. *Environ. Sci. Technol.*, **46**, 1352–1360.

Barter, P.J. and Rye, K.-A. (2016) New era of lipid-lowering drugs. *Pharmacol. Rev.*, **68**, 458–475.

Bartoskova, M., Dobsikova, R., Stancova, V., Zivna, D., Blahova, J., Marsalek, P., Zelnickova, L., Bartos, M., Di Tocco, F.C., and Faggio, C. (2013) Evaluation of ibuprofen toxicity for zebrafish (Danio rerio) targeting on selected biomarkers of oxidative stress. *Neuroendocrinol. Lett.*, **34**, 102–108.

Battaglin, W.A., Sandstrom, M.W., Kuivila, K.M., Kolpin, D.W., and Meyer, M.T. (2011) Occurrence of azoxystrobin, propiconazole, and selected other fungicides in US streams, 2005–2006. *Water, Air, Soil Pollut.*, **218**, 307–322.

Beijer, K., Jönsson, M., Shaik, S., Behrens, D., Brunström, B., and Brandt, I. (2018) Azoles additively inhibit cytochrome P450 1 (EROD) and 19 (aromatase) in rainbow trout (Oncorhynchus mykiss). *Aquat. Toxicol.*, **198**, 73–81.

Bencan, Z., Sledge, D., and Levin, E.D. (2009) Buspirone, chlordiazepoxide and diazepam effects in a zebrafish model of anxiety. *Pharmacol. Biochem. Behav.*, **94**, 75–80.

Besse, J.-P., Latour, J.-F., and Garric, J. (2012) Anticancer drugs in surface waters: What can we say about the occurrence and environmental significance of cytotoxic, cytostatic and endocrine therapy drugs? *Environ. Int.*, **39**, 73–86.

Bhagat, J., Singh, N., Nishimura, N., and Shimada, Y. (2020) A comprehensive review on environmental toxicity of azole compounds to fish. *Chemosphere*, **262**, 128335.

Blair, B.D., Crago, J.P., Hedman, C.J., and Klaper, R.D. (2013) Pharmaceuticals and personal care products found in the Great Lakes above concentrations of environmental concern. *Chemosphere*, **93**, 2116–2123.

Blüthgen, N., Castiglioni, S., Sumpter, J.P., and Fent, K. (2013) Effects of low concentrations of the antiprogestin mifepristone (RU486) in adults and embryos of zebrafish (Danio rerio): 1. Reproductive and early developmental effects. *Aquat. Toxicol.*, **144–145**, 83–95.

Bovill, J.G. (1997) Adverse drug interactions in anesthesia. *J. Clin. Anesth.*, **9**, 3S–13S.

Bradley, P.M., Journey, C.A., Button, D.T., Carlisle, D.M., Clark, J.M., Mahler, B.J., Nakagaki, N., Qi, S.L., Waite, I.R., and VanMetre, P.C. (2016) Metformin and other pharmaceuticals widespread in wadeable streams of the southeastern United States. *Environ. Sci. Technol. Lett.*, **3**, 243–249.

Braida, D., Limonta, V., Pegorini, S., Zani, A., Guerini-Rocco, C., Gori, E., and Sala, M. (2007) Hallucinatory and rewarding effect of salvinorin A in zebrafish: κ-opioid and CB 1-cannabinoid receptor involvement. *Psychopharmacology (Berl).*, **190**, 441–448.

Brian, J. V, Harris, C.A., Scholze, M., Kortenkamp, A., Booy, P., Lamoree, M., Pojana, G., Jonkers, N., Marcomini, A., and Sumpter, J.P. (2007) Evidence of estrogenic mixture effects on the reproductive performance of fish. *Environ. Sci. Technol.*, **41**, 337–344.

Brion, F., Tyler, C.R., Palazzi, X., Laillet, B., Porcher, J.M., Garric, J., and Flammarion, P. (2004) Impacts of 17β-estradiol, including environmentally relevant concentrations, on reproduction after exposure during embryo-larval-, juvenile- and adult-life stages in zebrafish (*Danio rerio*). *Aquat. Toxicol.*, **68**, 193–217.

Brown, A.R., Bickley, L.K., Le Page, G., Hosken, D.J., Paull, G.C., Hamilton, P.B., Owen, S.F., Robinson, J., Sharpe, A.D., and Tyler, C.R. (2011) Are toxicological responses in laboratory (inbred) zebrafish representative of those in outbred (wild) populations?– A case study with an endocrine disrupting chemical. *Environ. Sci. Technol.*, **45**, 4166–4172.

Buerge, I.J., Buser, H.-R., Poiger, T., and Müller, M.D. (2006) Occurrence and fate of the cytostatic drugs cyclophosphamide and ifosfamide in wastewater and surface waters. *Environ. Sci. Technol.*, **40**, 7242–7250.

Burch, K.D., Han, B., Pichtel, J., and Zubkov, T. (2019) Removal efficiency of commonly prescribed antibiotics via tertiary wastewater treatment. *Environ. Sci. Pollut. Res.*, **26**, 6301–6310.

Burkina, V., Zlabek, V., and Zamaratskaia, G. (2013) Clotrimazole, but not dexamethasone, is a potent in vitro inhibitor of cytochrome P450 isoforms CYP1A and CYP3A in rainbow trout. *Chemosphere*, **92**, 1099–1104.

Caldwell, D., H. Hutchinson, T., Heijerick, D., D. Anderson, P., and P. Sumpter, J. (2008) Derivation of an aquatic predicted no-effect concentration for the synthetic hormone, 17α-ethinyl estradiol. *Environ. Sci. Technol.*, **42**, 7046–7054.

Caldwell, D.J., D'Aco, V., Davidson, T., Kappler, K., Murray-Smith, R.J., Owen, S.F., Robinson, P.F., Simon-Hettich, B., Straub, J.O., and Tell, J. (2019) Environmental risk assessment of metformin and its transformation product guanylurea: II. Occurrence in surface waters of Europe and the United States and derivation of predicted no-effect concentrations. *Chemosphere*, **216**, 855–865.

Campos, L.M., Rios, E.A., Guapyassu, L., Midlej, V., Atella, G.C., Herculano-Houzel, S., Benchimol, M., Mermelstein, C., and Costa, M.L. (2016) Alterations in zebrafish development induced by simvastatin: Comprehensive morphological and physiological study, focusing on muscle. *Exp. Biol. Med.*, **241**, 1950–1960.

Campos-Mañas, M.C., Ferrer, I., Thurman, E.M., and Agüera, A. (2018) Opioid occurrence in environmental water samples-a review. *Trends Environ. Anal. Chem.*, **20**, e00059.

Capaldo, A., Gay, F., Maddaloni, M., Valiante, S., De Falco, M., Lenzi, M., and Laforgia, V. (2012) Presence of cocaine in the tissues of the European Eel, Anguilla anguilla, exposed to environmental cocaine concentrations. *Water. Air. Soil Pollut.*, **223**, 2137–2143.

Carty, D.R., Miller, Z.S., Thornton, C., Pandelides, Z., Kutchma, M.L., and Willett, K.L. (2019) Multigenerational consequences of early-life cannabinoid exposure in zebrafish. *Toxicol. Appl. Pharmacol.*, **364**, 133–143.

Ccanccapa, A., Masiá, A., Navarro-Ortega, A., Picó, Y., and Barceló, D. (2016) Pesticides in the Ebro River basin: Occurrence and risk assessment. *Environ. Pollut.*, **211**, 414–424.

Celle-Jeanton, H., Schemberg, D., Mohammed, N., Huneau, F., Bertrand, G., Lavastre, V., and Le Coustumer, P. (2014) Evaluation of pharmaceuticals in surface water: Reliability of PECs compared to MECs. *Environ. Int.*, **73**, 10–21.

Cerveny, D., Grabic, R., Grabicová, K., Randák, T., Larsson, D.G.J., Johnson, A.C., Jürgens, M.D., Tysklind, M., Lindberg, R.H., and Fick, J. (2021) Neuroactive drugs and other pharmaceuticals found in blood plasma of wild European fish. *Environ. Int.*, **146**: 106188.

Česen, M., Ahel, M., Terzić, S., Heath, D.J., and Heath, E. (2019) The occurrence of contaminants of emerging concern in Slovenian and Croatian wastewaters and receiving Sava river. *Sci. Total Environ.*, **650**, 2446–2453.

Chang, H., Wan, Y., and Hu, J. (2009) Determination and source apportionment of five classes of steroid hormones in urban rivers. *Environ. Sci. Technol.*, **43**, 7691–7698.

Chen, C.-Y., Wen, T.-Y., Wang, G.-S., Cheng, H.-W., Lin, Y.-H., and Lien, G.-W. (2007) Determining estrogenic steroids in Taipei waters and removal in drinking water treatment using high-flow solid-phase extraction and liquid chromatography/tandem mass spectrometry. *Sci. Total Environ.*, **378**, 352–365.

Chen, Z.-F., Ying, G.-G., Lai, H.-J., Chen, F., Su, H.-C., Liu, Y.-S., Peng, F.-Q., and Zhao, J.-L. (2012) Determination of biocides in different environmental matrices by use of ultra-high-performance liquid chromatography-tandem mass spectrometry. *Anal. Bioanal. Chem.*, **404**, 3175–3188.

Chen, Z.-F. and Ying, G.-G. (2015) Occurrence, fate and ecological risk of five typical azole fungicides as therapeutic and personal care products in the environment: A review. *Environ. Int.*, **84**, 142–153.

Chen, Z.-F., Ying, G.-G., Liu, Y.-S., Zhang, Q.-Q., Zhao, J.-L., Liu, S.-S., Chen, J., Peng, F.-J., Lai, H.-J., and Pan, C.-G. (2014) Triclosan as a surrogate for household biocides: An investigation into biocides in aquatic environments of a highly urbanized region. *Water Res.*, **58**, 269–279.

Christen, V., Hickmann, S., Rechenberg, B., and Fent, K. (2010) Highly active human pharmaceuticals in aquatic systems: A concept for their identification based on their mode of action. *Aquat. Toxicol.*, **96**, 167–181.

Chrysant, S.G. (2010) Current status of dual renin angiotensin aldosterone system blockade for the treatment of cardiovascular diseases. *Am. J. Cardiol.*, **105**, 849–852.

Chu, S.-H., Liao, P.-H., and Chen, P.-J. (2016) Developmental exposures to an azole fungicide triadimenol at environmentally relevant concentrations cause reproductive dysfunction in females of medaka fish. *Chemosphere*, **152**, 181–189.

Constantine, L.A., Green, J.W., and Schneider, S.Z. (2020) Ibuprofen: Fish short-term reproduction assay with zebrafish (Danio rerio) based on an extended OECD 229 Protocol. *Environ. Toxicol. Chem.*, **39**, 1534–1545.

Corcoran, J., Lange, A., Cumming, R.I., Owen, S.F., Ball, J.S., Tyler, C.R., and Winter, M.J. (2014) Bioavailability of the imidazole antifungal agent clotrimazole and its effects on key biotransformation genes in the common carp (Cyprinus carpio). *Aquat. Toxicol.*, **152**, 57–65.

Corcoran, J., Winter, M.J., and Tyler, C.R. (2010) Pharmaceuticals in the aquatic environment: A critical review of the evidence for health effects in fish. *Crit. Rev. Toxicol.*, **40**, 287–304.

Cornwall, W. (2018) Looking for love. *Science*, **359**, 1086–1089.

Crago, J., Bui, C., Grewal, S., and Schlenk, D. (2016) Age-dependent effects in fathead minnows from the anti-diabetic drug metformin. *Gen. Comp. Endocrinol.*, **232**, 185–190.

Cuklev, F., Kristiansson, E., Fick, J., Asker, N., Förlin, L., and Larsson, D.G.J. (2011) Diclofenac in fish: Blood plasma levels similar to human therapeutic levels affect global hepatic gene expression. *Environ. Toxicol. Chem.*, **30**, 2126–2134.

Danner, M.C., Robertson, A., Behrends, V., and Reiss, J. (2019) Antibiotic pollution in surface fresh waters: Occurrence and effects. *Sci. Total Environ.*, **664**, 793–804.

David, V., Joachim, S., Catteau, A., Nott, K., Ronkart, S., Robert, C., Gillard, N., Bado-Nilles, A., Chadili, E., Palluel, O., *et al.* (2020) Effects of chronic exposure to a pharmaceutical mixture on the three-spined stickleback (Gasterosteus aculeatus) population dynamics in lotic mesocosms. *Aquat. Toxicol.*, **224**, 105499.

Díaz-González, F. and Sánchez-Madrid, F. (2015) NSAIDs: Learning new tricks from old drugs. *Eur. J. Immunol.*, **45**, 679–686.

Drabløs, F., Feyzi, E., Aas, P.A., Vaagbø, C.B., Kavli, B., Bratlie, M.S., Peña-Diaz, J., Otterlei, M., Slupphaug, G., and Krokan, H.E. (2004) Alkylation damage in DNA and RNA-repair mechanisms and medical significance. *DNA Repair (Amst).*, **3**, 1389–1407.

Du, B., Haddad, S.P., Luek, A., Scott, W.C., Saari, G.N., Kristofco, L.A., Connors, K.A., Rash, C., Rasmussen, J.B., Chambliss, C.K., *et al.* (2014) Bioaccumulation and trophic dilution of human pharmaceuticals across trophic positions of an effluent-dependent wadeable stream. *Philos. Trans. R. Soc. B Biol. Sci.*, **369**, 20140058.

Dubois, R.N., Abramson, S.B., Crofford, L., Gupta, R.A., Simon, L.S., Van De Putte, L.B.A., and Lipsky, P.E. (1998) Cyclooxygenase in biology and disease. *FASEB J.*, **12**, 1063–1073.

Edsinger, E. and Dölen, G. (2018) A conserved role for serotonergic neurotransmission in mediating social behavior in octopus. *Curr. Biol.*, **28**, 3136–3142.e4.

Elizalde-Velázquez, G.A. and Gómez-Oliván, L.M. (2020) Occurrence, toxic effects and removal of metformin in the aquatic environments in the world: Recent trends and perspectives. *Sci. Total Environ.*, **702**, 134924.

Ellesat, K.S., Holth, T.F., Wojewodzic, M.W., and Hylland, K. (2012) Atorvastatin up-regulate toxicologically relevant genes in rainbow trout gills. *Ecotoxicology*, **21**, 1841–1856.

Elliott, S.M., Brigham, M.E., Lee, K.E., Banda, J.A., Choy, S.J., Gefell, D.J., Minarik, T.A., Moore, J.N., and Jorgenson, Z.G. (2017) Contaminants of emerging concern in tributaries to the Laurentian Great Lakes: I. Patterns of occurrence. *PLoS One*, **12**, 1–21.

EMEA /CHMP/SWP/4447/00 (2006) *Guideline on the Environmental Risk Assessment of Medicinal Products for Human Use.* London, United Kingdom.

Faltermann, S., Hettich, T., Küng, N., and Fent, K. (2020) Effects of the glucocorticoid clobetasol propionate and its mixture with cortisol and different class steroids in adult female zebrafish. *Aquat. Toxicol.*, **218**, 105372.

Fan, Z., Wu, S., Chang, H., and Hu, J. (2011) Behaviors of glucocorticoids, androgens and progestogens in a municipal sewage treatment plant: Comparison to estrogens. *Environ. Sci. Technol.*, **45**, 2725–2733.

Fent, K., Weston, A.A., and Caminada, D. (2006) Ecotoxicology of human pharmaceuticals. *Aquat. Toxicol.*, **76**, 122–159.

Fent, K. (2015) Progestins as endocrine disrupters in aquatic ecosystems: Concentrations, effects and risk assessment. *Environ. Int.*, **84**, 115–130.

Fent, K. and Sumpter, J.P. (2011) Progress and promises in toxicogenomics in aquatic toxicology: Is technical innovation driving scientific innovation? *Aquat. Toxicol.*, **105**, 25–39.

Fick, J., Söderström, H., Lindberg, R.H., Phan, C., Tysklind, M., and Larsson, D.G.J. (2009) Contamination of surface, ground, and drinking water from pharmaceutical production. *Environ. Toxicol. Chem.*, **28**, 2522–2527.

Fick, J., H. Lindberg, R., Parkkonen, J., Arvidsson, B., Tysklind, M., and G. Joakim Larsson, D. (2010) Therapeutic levels of levonorgestrel detected in blood plasma of fish: Results from screening rainbow trout exposed to treated sewage effluents. *Environ. Sci. Technol.*, **44**, 2661–2666.

Finn, J., Hui, M., Li, V., Lorenzi, V., de la Paz, N., Cheng, S.H., Lai-Chan, L., and Schlenk, D. (2012) Effects of propranolol on heart rate and development in Japanese medaka (Oryzias latipes) and zebrafish (Danio rerio). *Aquat. Toxicol.*, **122**, 214–221.

Fischer, A.J., Kerr, L., Sultana, T., and Metcalfe, C.D. (2021) Effects of opioids on reproduction in Japanese medaka, Oryzias latipes. *Aquat. Toxicol.*, **236**, 105873.

Fonseca, D., Baquero, J., Murphy, M.R., Aruggoda, G., Varriano, S., Sapienza, C., Mashadova, O., Rahman, S., and Kleiman, F.E. (2018) mRNA processing factor CstF-50 and ubiquitin escort factor p97 are BRCA1/BARD1 cofactors involved in chromatin remodeling during the DNA damage response. *Mol. Cell. Biol.*, **38**, e00364-17.

Franquet-Griell, H., Cornadó, D., Caixach, J., Ventura, F., and Lacorte, S. (2017) Determination of cytostatic drugs in Besòs River (NE Spain) and comparison with predicted environmental concentrations. *Environ. Sci. Pollut. Res.*, **24**, 6492–6503.

Fraysse, B., Mons, R., and Garric, J. (2006) Development of a zebrafish 4-day embryo-larval bioassay to assess toxicity of chemicals. *Ecotoxicol. Environ. Saf.*, **63**, 253–267.

Fraz, S., Lee, A.H., and Wilson, J.Y. (2018) Gemfibrozil and carbamazepine decrease steroid production in zebrafish testes (Danio rerio). *Aquat. Toxicol.*, **198**, 1–9.

Fraz, S., Lee, A.H., Pollard, S., Srinivasan, K., Vermani, A., David, E., and Wilson, J.Y. (2019) Paternal exposure to carbamazepine impacts zebrafish offspring reproduction over multiple generations. *Environ. Sci. Technol.*, **53**, 12734–12743.

Fuentes, A. V, Pineda, M.D., and Venkata, K.C.N. (2018) Comprehension of top 200 prescribed drugs in the US as a resource for pharmacy teaching, training and practice. *Pharmacy*, **6**, 43.

Galus, M., Kirischian, N., Higgins, S., Purdy, J., Chow, J., Rangaranjan, S., Li, H., Metcalfe, C., and Wilson, J.Y. (2013) Chronic, low concentration exposure to pharmaceuticals impacts multiple organ systems in zebrafish. *Aquat. Toxicol.*, **132**, 200–211.

Gamba, G. (2009) The thiazide-sensitive Na+-Cl– cotransporter: Molecular biology, functional properties, and regulation by WNKs. *Am. J. Physiol. Physiol.*, **297**, F838–F848.

Gao, J., Zhang, W., Dang, W., Mou, Y., Gao, Y., Sun, B.-J., and Du, W.-G. (2014) Heat shock protein expression enhances heat tolerance of reptile embryos. *Proc. R. Soc. B Biol. Sci.*, **281**, 20141135.

Garcia-Reyero, N., Villeneuve, D.L., Kroll, K.J., Liu, L., Orlando, E.F., Watanabe, K.H., Sepúlveda, M.S., Ankley, G.T., and Denslow, N.D. (2009) Expression signatures for a model androgen and antiandrogen in the fathead minnow (Pimephales promelas) ovary. *Environ. Sci. Technol.*, **43**, 2614.

Gay, F., Maddaloni, M., Valiante, S., Laforgia, V., and Capaldo, A. (2013) Endocrine disruption in the European eel, Anguilla anguilla, exposed to an environmental cocaine concentration. *Water. Air. Soil Pollut.*, **224**, 1579.

Gerber, D.E. (2008) Targeted therapies: A new generation of cancer treatments. *Am. Fam. Physician*, **77**, 311–319.

Ghoshdastidar, A.J., Fox, S., and Tong, A.Z. (2015) The presence of the top prescribed pharmaceuticals in treated sewage effluents and receiving waters in Southwest Nova Scotia, Canada. *Environ. Sci. Pollut. Res.*, **22**, 689–700.

Giltrow, E., Eccles, P.D., Winter, M.J., McCormack, P.J., Rand-Weaver, M., Hutchinson, T.H., and Sumpter, J.P. (2009) Chronic effects assessment and plasma concentrations of the β-blocker propranolol in fathead minnows (Pimephales promelas). *Aquat. Toxicol.*, **95**, 195–202.

Godfraind, T. (2017) Discovery and development of calcium channel blockers. *Front. Pharmacol.*, **8**, 286.

Godoy, A.A., Domingues, I., Arsénia Nogueira, A.J., and Kummrow, F. (2018) Ecotoxicological effects, water quality standards and risk assessment for the anti-diabetic metformin. *Environ. Pollut.*, **243**, 534–542.

Godoy, A.A., Kummrow, F., and Pamplin, P.A.Z. (2015) Occurrence, ecotoxicological effects and risk assessment of antihypertensive pharmaceutical residues in the aquatic environment-A review. *Chemosphere*, **138**, 281–291.

Gómez-Canela, C., Ventura, F., Caixach, J., and Lacorte, S. (2014) Occurrence of cytostatic compounds in hospital effluents and wastewaters, determined by liquid chromatography coupled to high-resolution mass spectrometry. *Anal. Bioanal. Chem.*, **406**, 3801–3814.

Gore, M., Sadosky, A., Leslie, D., and Sheehan, A.H. (2008) Selecting an appropriate medication for treating neuropathic pain in patients with diabetes: A study using the UK and Germany Mediplus databases. *Pain Pract.*, **8**, 253–262.

Gouveia, T.I.A., Alves, A., and Santos, M.S.F. (2019) New insights on cytostatic drug risk assessment in aquatic environments based on measured concentrations in surface waters. *Environ. Int.*, **133**, 105236.

Green, G.A. (2001) Understanding NSAIDs: From aspirin to COX-2. *Clin. Cornerstone*, **3**, 50–59.

Gröner, F., Höhne, C., Kleiner, W., and Kloas, W. (2017) Chronic exposure to the ß-blocker metoprolol reduces growth and alters gene expression of gonadotropins and vitellogenin in Nile tilapia (Oreochromis niloticus). *Ecotoxicol. Environ. Saf.*, **141**, 271–279.

Gros, M., Rodríguez-Mozaz, S., and Barceló, D. (2012) Fast and comprehensive multi-residue analysis of a broad range of human and veterinary pharmaceuticals and some of their metabolites in surface and treated waters by ultra-high-performance liquid chromatography coupled to quadrupole-linear ion trap tandem. *J. Chromatogr. A*, **1248**, 104–121.

Gros, M., Petrović, M., Ginebreda, A., and Barceló, D. (2010) Removal of pharmaceuticals during wastewater treatment and environmental risk assessment using hazard indexes. *Environ. Int.*, **36**, 15–26.

Grzesiuk, M. and Pawelec, A. (2021) Fluoxetine results in misleading conclusions on fish behavior. *Ecol. Evol.*, **11**(14), 9707–9714.

Guiloski, I.C., Ribas, J.L.C., da Silva Pereira, L., Neves, A.P.P., and de Assis, H.C.S. (2015) Effects of trophic exposure to dexamethasone and diclofenac in freshwater fish. *Ecotoxicol. Environ. Saf.*, **114**, 204–211.

Gunnarsson, L., Jauhiainen, A., Kristiansson, E., Nerman, O., and Larsson, D.G.J. (2008) Evolutionary conservation of human drug targets in organisms used for environmental risk assessments. *Environ. Sci. Technol.*, **42**, 5807–5813.

Gunnarsson, L., Snape, J.R., Verbruggen, B., Owen, S.F., Kristiansson, E., Margiotta-Casaluci, L., Österlund, T., Hutchinson, K., Leverett, D., Marks, B., *et al.* (2019) Pharmacology beyond the patient - The environmental risks of human drugs. *Environ. Int.*, **129**, 320–332.

Gurke, R., Rößler, M., Marx, C., Diamond, S., Schubert, S., Oertel, R., and Fauler, J. (2015) Occurrence and removal of frequently prescribed pharmaceuticals and corresponding metabolites in wastewater of a sewage treatment plant. *Sci. Total Environ.*, **532**, 762–770.

Gutiérrez-Noya, V.M., Gómez-Oliván, L.M., del Carmen Ramírez-Montero, M., Islas-Flores, H., Galar-Martínez, M., Dublán-García, O., and Romero, R. (2020) Ibuprofen at environmentally relevant concentrations alters embryonic development, induces teratogenesis and oxidative stress in Cyprinus carpio. *Sci. Total Environ.*, **710**, 136327.

Hainer, J., Papademetriou, V., Frishman, W., Sugg, J., and Sorof, J. (2005) P-228: Metoprolol succinate extended release: Antihypertensive dose response characteristics. *Am. J. Hypertens.*, **18**, 88A.

Han, S., Choi, Kyungho, Kim, J., Ji, K., Kim, S., Ahn, B., Yun, J., Choi, Kyunghee, Khim, J.S., and Zhang, X. (2010) Endocrine disruption and consequences of chronic exposure to ibuprofen in Japanese medaka (Oryzias latipes) and freshwater cladocerans Daphnia magna and Moina macrocopa. *Aquat. Toxicol.*, **98**, 256–264.

Heberer, T. (2002) Occurrence, fate, and removal of pharmaceutical residues in the aquatic environment: A review of recent research data. *Toxicol. Lett.*, **131**, 5–17.

Holbech, H., Kinnberg, K., Petersen, G.I., Jackson, P., Hylland, K., Norrgren, L., and Bjerregaard, P. (2006) Detection of endocrine disrupters: Evaluation of a Fish Sexual Development Test (FSDT). *Comp. Biochem. Physiol. Part C Toxicol. Pharmacol.*, **144**, 57–66.

Horký, P., Grabic, R., Grabicova, K., Brooks, B.W., Douda, K., Hubena, P., and Santos, E.M.S. (2021) Methamphetamine pollution elicits addiction in wild fish. *J. Exp. Biol.*, **224**, jeb242145.

Huerta, B., Margiotta-Casaluci, L., Rodríguez-Mozaz, S., Scholze, M., Winter, M.J., Barceló, D., and Sumpter, J.P. (2016) Anti-anxiety drugs and fish behavior: Establishing the link between internal concentrations of oxazepam and behavioral effects. *Environ. Toxicol. Chem.*, **35**, 2782–2790.

Huerta-Fontela, M., Galceran, M.T., and Ventura, F. (2010) Fast liquid chromatography-quadrupole-linear ion trap mass spectrometry for the analysis of pharmaceuticals and hormones in water resources. *J. Chromatogr. A*, **1217**, 4212–4222.

Huggett, D.B., Cook, J.C., Ericson, J.F., and Williams, R.T. (2003) A theoretical model for utilizing mammalian pharmacology and safety data to prioritize potential impacts of human pharmaceuticals to fish. *Hum. Ecol. Risk Assess. An Int. J.*, **9**, 1789–1799.

Huggett, D.B., Brooks, B.W., Peterson, B., Foran, C.M., and Schlenk, D. (2002) Toxicity of select beta adrenergic receptor-blocking pharmaceuticals (B-blockers) on aquatic organisms. *Arch. Environ. Contam. Toxicol.*, **43**, 229–235.

Hughes, S.R., Kay, P., and Brown, L.E. (2013) Global synthesis and critical evaluation of pharmaceutical data sets collected from river systems. *Environ. Sci. Technol.*, **47**, 661–677.

Jacob, S., Dötsch, A., Knoll, S., Köhler, H.R., Rogall, E., Stoll, D., Tisler, S., Huhn, C., Schwartz, T., Zwiener, C., *et al.* (2018) Does the antidiabetic drug metformin affect embryo development and the health of brown trout (Salmo trutta f. fario)? *Environ. Sci. Eur.*, **30**, 48.

Jensen, K.M., Makynen, E.A., Kahl, M.D., and Ankley, G.T. (2006) Effects of the feedlot contaminant 17α-trenbolone on reproductive endocrinology of the fathead minnow. *Environ. Sci. Technol.*, **40**, 3112–3117.

Ji, K., Liu, X., Lee, S., Kang, S., Kho, Y., Giesy, J.P., and Choi, K. (2013) Effects of non-steroidal anti-inflammatory drugs on hormones and genes of the hypothalamic-pituitary-gonad axis, and reproduction of zebrafish. *J. Hazard. Mater.*, **254**, 242–251.

Jobling, S., Williams, R., Johnson, A., Taylor, A., Gross-Sorokin, M., Nolan, M., Tyler, C.R., van Aerle, R., Santos, E., and Brighty, G. (2006) Predicted exposures to steroid estrogens in U.K. rivers correlate with widespread sexual disruption in wild fish populations. *Environ. Health Perspect.*, **114 Suppl**, 32–39.

Kahle, M., Buerge, I.J., Hauser, A., Müller, M.D., and Poiger, T. (2008) Azole fungicides: Occurrence and fate in wastewater and surface waters. *Environ. Sci. Technol.*, **42**, 7193–7200.

Kalichak, F., Idalencio, R., Da Rosa, J.G.S., De Alcântara Barcellos, H.H., Fagundes, M., Piato, A., and Barcellos, L.J.G. (2017) Psychotropic in the environment: Risperidone residues affect the behavior of fish larvae. *Sci. Rep.*, **7**, 1–7.

Kania, B.F., Debski, B., Wronska, D., and Zawadzka, E. (2015) Verapamil-L type voltage gated calcium channel inhibitor diminishes aggressive behavior in male Siamese fighting fish. *Pol. J. Vet. Sci.*, **18**, 401–406.

Kasprzyk-Hordern, B., Dinsdale, R.M., and Guwy, A.J. (2009) The removal of pharmaceuticals, personal care products, endocrine disruptors and illicit drugs during wastewater treatment and its impact on the quality of receiving waters. *Water Res.*, **43**, 363–380.

Kidd, K.A., Blanchfield, P.J., Mills, K.H., Palace, V.P., Evans, R.E., Lazorchak, J.M., and Flick, R.W. (2007) Collapse of a fish population after exposure to a synthetic estrogen. *Proc. Natl. Acad. Sci. U. S. A.*, **104**, 8897–8901.

Kim, P., Park, Y., Ji, K., Seo, J., Lee, S., Choi, Kyunghee, Kho, Y., Park, J., and Choi, Kyungho (2012) Effect of chronic exposure to acetaminophen and lincomycin on Japanese medaka (Oryzias latipes) and freshwater cladocerans Daphnia magna and Moina macrocopa, and potential mechanisms of endocrine disruption. *Chemosphere*, **89**, 10–18.

Kirla, K.T., Groh, K.J., Steuer, A.E., Poetzsch, M., Banote, R.K., Stadnicka-Michalak, J., Eggen, R.I.L., Schirmer, K., and Kraemer, T. (2016) Zebrafish larvae are insensitive to stimulation by cocaine: Importance of exposure route and toxicokinetics. *Toxicol. Sci.*, **154**, 183–193.

Koba, O., Grabicova, K., Cerveny, D., Turek, J., Kolarova, J., Randak, T., Zlabek, V., and Grabic, R. (2018) Transport of pharmaceuticals and their metabolites between water and sediments as a further potential exposure for aquatic organisms. *J. Hazard. Mater.*, **342**, 401–407.

Kolpin, D.W., Furlong, E.T., Meyer, M.T., Thurman, E.M., Zaugg, S.D., Barber, L.B., and Buxton, H.T. (2002) Pharmaceuticals, hormones, and other organic wastewater contaminants in U.S. streams, 1999–2000: A national reconnaissance. *Environ. Sci. Technol.*, **36**, 1202–1211.

Kortenkamp, A., Backhaus, T., and Faust, M. (2009) State of the Art report on Mixture Toxicity. Directorate General for the Environment: Report to the EU Commission.

Kovács, R., Csenki, Z., Bakos, K., Urbányi, B., Horváth, Á., Garaj-Vrhovac, V., Gajski, G., Gerić, M., Negreira, N., and de Alda, M.L. (2015) Assessment of toxicity and genotoxicity of low doses of 5-fluorouracil in zebrafish (Danio rerio) two-generation study. *Water Res.*, **77**, 201–212.

Kuch, H.M. and Ballschmiter, K. (2001) Determination of endocrine-disrupting phenolic compounds and estrogens in surface and drinking water by HRGC–(NCI)–MS in the picogram per liter range. *Environ. Sci. Technol.*, **35**, 3201–3206.

Kugathas, S., J. Runnalls, T., and P. Sumpter, J. (2013) Metabolic and reproductive effects of relatively low concentrations of beclomethasone dipropionate, a synthetic glucocorticoid, on fathead minnows. *Environ. Sci. Technol.*, **47**, 9487–9495.

Kümmerer, K. (2009) Antibiotics in the aquatic environment - A review - Part I. *Chemosphere*, **75**, 417–434.

Kümmerer, K., Haiß, A., Schuster, A., Hein, A., and Ebert, I. (2016) Antineoplastic compounds in the environment-substances of special concern. *Environ. Sci. Pollut. Res.*, **23**, 14791–14804.

Lacey, C., Basha, S., Morrissey, A., and Tobin, J.M. (2012) Occurrence of pharmaceutical compounds in wastewater process streams in Dublin, Ireland. *Environ. Monit. Assess.*, **184**, 1049–1062.

Lam, M.W., Young, C.J., Brain, R.A., Johnson, D.J., Hanson, M.L., Wilson, C.J., Richards, S.M., Solomon, K.R., and Mabury, S.A. (2004) Aquatic persistence of eight pharmaceuticals in a microcosm study. *Environ. Toxicol. Chem.*, **23**, 1431–1440.

Lange, A., C. Paull, G., B. Hamilton, P., Iguchi, T., and R. Tyler, C. (2011) Implications of persistent exposure to treated wastewater effluent for breeding in wild roach (Rutilus rutilus) populations. *Environ. Sci. Technol.*, **45**, 1673–1679.

Lange, A., Paull, G.C., Coe, T.S., Katsu, Y., Urushitani, H., Iguchi, T., and Tyler, C.R. (2009) Sexual reprogramming and estrogenic sensitization in wild fish exposed to ethinylestradiol. *Environ. Sci. Technol.*, **43**, 1219–1225.

Länge, R., Hutchinson, T.H., Croudace, C.P., Siegmund, F., Schweinfurth, H., Hampe, P., Panter, G.H., and Sumpter, J.P. (2001) Effects of the synthetic estrogen 17α-ethinylestradiol on the life-cycle of the fathead minnow (*Pimephales promelas*). *Environ. Toxicol. Chem.*, **20**, 1216–1227.

Larsson, D.G.J., de Pedro, C., and Paxeus, N. (2007) Effluent from drug manufactures contains extremely high levels of pharmaceuticals. *J. Hazard. Mater.*, **148**, 751–755.

Lee, G., Lee, S., Ha, N., Kho, Y., Park, K., Kim, P., Ahn, B., Kim, S., and Choi, K. (2019) Effects of gemfibrozil on sex hormones and reproduction related performances of Oryzias latipes following long-term (155 d) and short-term (21 d) exposure. *Ecotoxicol. Environ. Saf.*, **173**, 174–181.

Lei, M., Wu, L., Terrar, D.A., and Huang, C.L.-H. (2018) Modernized classification of cardiac antiarrhythmic drugs. *Circulation*, **138**, 1879–1896.

Li, D., Chen, H., Liu, H., Schlenk, D., Mu, J., Lacorte, S., Ying, G.-G., and Xie, L. (2021) Anticancer drugs in the aquatic ecosystem: Environmental occurrence, ecotoxicological effect and risk assessment. *Environ. Int.*, **153**, 106543.

Li, Z.-H., Zlabek, V., Li, P., Grabic, R., Velisek, J., Machova, J., and Randak, T. (2010) Biochemical and physiological responses in liver and muscle of rainbow trout after long-term exposure to propiconazole. *Ecotoxicol. Environ. Saf.*, **73**, 1391–1396.

Liang, Y.Q., Huang, G.Y., Liu, S.S., Zhao, J.L., Yang, Y.Y., Chen, X.W., Tian, F., Jiang, Y.X., and Ying, G.G. (2015) Long-term exposure to environmentally relevant concentrations of progesterone and norgestrel affects sex differentiation in zebrafish (Danio rerio). *Aquat. Toxicol.*, **160**, 172–179.

Liang, Y.Q., Huang, G.Y., Lin, Z., Li, J., Yang, J.W., Zhong, L.Y., and Ying, G.G. (2018) Reproductive effects of synthetic progestin norgestrel in zebrafish (Danio rerio). *Chemosphere*, **190**, 17–24.

Liao, J.K. and Laufs, U. (2005) Pleiotropic effects of statins. *Annu. Rev. Pharmacol. Toxicol.*, **45**, 89–118.

Liao, P.H., Yang, W.K., Yang, C.H., Lin, C.H., Hwang, C.C., and Chen, P.J. (2018) Illicit drug ketamine induces adverse effects from behavioral alterations and oxidative stress to p53-regulated apoptosis in medaka fish under environmentally relevant exposures. *Environ. Pollut.*, **237**, 1062–1071.

Limbu, S.M., Zhou, L., Sun, S.X., Zhang, M.L., and Du, Z.Y. (2018) Chronic exposure to low environmental concentrations and legal aquaculture doses of antibiotics cause systemic adverse effects in Nile tilapia and provoke differential human health risk. *Environ. Int.*, **115**, 205–219.

Lin, A.Y.-C., Wang, X.-H., and Lee, W.-N. (2013) Phototransformation determines the fate of 5-fluorouracil and cyclophosphamide in natural surface waters. *Environ. Sci. Technol.*, **47**, 4104–4112.

Lin, T., Yu, S., Chen, Y., and Chen, W. (2014) Integrated biomarker responses in zebrafish exposed to sulfonamides. *Environ. Toxicol. Pharmacol.*, **38**, 444–452.

Lin, W., Yan, Y., Ping, S., Li, P., Li, D., Hu, J., Liu, W., Wen, X., and Ren, Y. (2021) Metformin-induced epigenetic toxicity in zebrafish: Experimental and molecular dynamics simulation studies. *Environ. Sci. Technol.*, **55**, 1672–1681.

Liu, C., Tan, L., Zhang, L., Tian, W., and Ma, L. (2021) A review of the distribution of antibiotics in water in different regions of China and current antibiotic degradation pathways. *Front. Environ. Sci.*, **9**, 1–24.

Liu, J., Lu, G., Yang, H., Yan, Z., Wang, Y., and Wang, P. (2016) Bioconcentration and metabolism of ketoconazole and effects on multi-biomarkers in crucian carp (Carassius auratus). *Chemosphere*, **150**, 145–151.

Liu, J., Lu, G., Cai, Y., Wu, D., Yan, Z., and Wang, Y. (2017) Modulation of erythromycin-induced biochemical responses in crucian carp by ketoconazole. *Environ. Sci. Pollut. Res.*, **24**, 5285–5292.

Liu, S., Ying, G.-G., Zhou, L.-J., Zhang, R.-Q., Chen, Z.-F., and Lai, H.-J. (2012) Steroids in a typical swine farm and their release into the environment. *Water Res.*, **46**, 3754–3768.

Liu, X., Zhang, J., Yin, J., Duan, H., Wu, Y., and Shao, B. (2010) Analysis of hormone antagonists in clinical and municipal wastewater by isotopic dilution liquid chromatography tandem mass spectrometry. *Anal. Bioanal. Chem.*, **396**, 2977–2985.

Lolić, A., Paíga, P., Santos, L.H., Ramos, S., Correia, M., and Delerue-Matos, C. (2015) Assessment of non-steroidal anti-inflammatory and analgesic pharmaceuticals in seawaters of North of Portugal: Occurrence and environmental risk. *Sci. Total Environ.*, **508**, 240–250.

Loos, R., Carvalho, R., António, D.C., Comero, S., Locoro, G., Tavazzi, S., Paracchini, B., Ghiani, M., Lettieri, T., Blaha, L., *et al.* (2013) EU-wide monitoring survey on emerging polar organic contaminants in wastewater treatment plant effluents. *Water Res.*, **47**, 6475–6487.

Lopez-Luna, J., Al-Jubouri, Q., Al-Nuaimy, W., and Sneddon, L.U. (2017) Reduction in activity by noxious chemical stimulation is ameliorated by immersion in analgesic drugs in zebrafish. *J. Exp. Biol.*, **220**, 1451–1458.

López-Serna, R., Petrović, M., and Barceló, D. (2012) Occurrence and distribution of multi-class pharmaceuticals and their active metabolites and transformation products in the Ebro River basin (NE Spain). *Sci. Total Environ.*, **440**, 280–289.

Loraine, G.A. and Pettigrove, M.E. (2006) Seasonal variations in concentrations of pharmaceuticals and personal care products in drinking water and reclaimed wastewater in southern California. *Environ. Sci. Technol.*, **40**, 687–695.

Lorenzi, V., Choe, R., and Schlenk, D. (2016) Effects of environmental exposure to diazepam on the reproductive behavior of fathead minnow, Pimephales promelas. *Environ. Toxicol.*, **31**, 561–568.

Lorenzi, V., Mehinto, A.C., Denslow, N.D., and Schlenk, D. (2012) Effects of exposure to the β-blocker propranolol on the reproductive behavior and gene expression of the fathead minnow, Pimephales promelas. *Aquat. Toxicol.*, **116**, 8–15.

Lucas, D., Badia-Fabregat, M., Vicent, T., Caminal, G., Rodríguez-Mozaz, S., Balcázar, J.L., and Barceló, D. (2016) Fungal treatment for the removal of antibiotics and antibiotic resistance genes in veterinary hospital wastewater. *Chemosphere*, **152**, 301–308.

Luo, Y., Guo, W., Ngo, H.H., Nghiem, L.D., Hai, F.I., Zhang, J., Liang, S., and Wang, X.C. (2014) A review on the occurrence of micropollutants in the aquatic environment and their fate and removal during wastewater treatment. *Sci. Total Environ.*, **473**, 619–641.

Mahnik, S.N., Lenz, K., Weissenbacher, N., Mader, R.M., and Fuerhacker, M. (2007) Fate of 5-fluorouracil, doxorubicin, epirubicin, and daunorubicin in hospital wastewater and their elimination by activated sludge and treatment in a membrane-bio-reactor system. *Chemosphere*, **66**, 30–37.

Malev, O., Lovrić, M., Stipaničev, D., Repec, S., Martinović-Weigelt, D., Zanella, D., Ivanković, T., Sindičić Đuretec, V., Barišić, J., Li, M., *et al.* (2020) Toxicity prediction and effect characterization of 90 pharmaceuticals and illicit drugs measured in plasma of fish from a major European river (Sava, Croatia). *Environ. Pollut.*, **266**, 115162.

Maradonna, F., Batti, S., Marino, M., Mita, D.G., and Carnevali, O. (2009) Tamoxifen as an emerging endocrine disruptor: Effects on fish reproduction and detoxification target genes. *Ann. N. Y. Acad. Sci.*, **1163**, 457–459.

Markiewicz, M., Jungnickel, C., Stolte, S., Białk-Bielińska, A., Kumirska, J., and Mrozik, W. (2017) Primary degradation of antidiabetic drugs. *J. Hazard. Mater.*, **324**, 428–435.

Martins-Teixeira, M.B. and Carvalho, I. (2020) Antitumour anthracyclines: Progress and perspectives. *ChemMedChem*, **15**, 933–948.

Matus, G.N., Pereira, B.V.R., Silva-Zacarin, E.C.M., Costa, M.J., Dos Santos, A.C.A., and Nunes, B. (2018) Behavior and histopathology as biomarkers for evaluation of the effects of paracetamol and propranolol in the neotropical fish species Phalloceros harpagos. *Environ. Sci. Pollut. Res.*, **25**, 28601–28618.

McDonald, M.D. (2017) An AOP analysis of selective serotonin reuptake inhibitors (SSRIs) for fish. *Comp. Biochem. Physiol. Part - C Toxicol. Pharmacol.*, **197**, 19–31.

Meador, J.P., Yeh, A., Young, G., and Gallagher, E.P. (2016) Contaminants of emerging concern in a large temperate estuary. *Environ. Pollut.*, **213**, 254–267.

Melefa, T.D. and Nwani, C.D. (2021) Imidazole antifungal drug clotrimazole alters the behavior, brain acetylcholinesterase and oxidative stress biomarkers in African catfish Clarias gariepinus (Burchell 1822). *Comp. Biochem. Physiol. Part C Toxicol. Pharmacol.*, **248**, 109108.

Meng, H., Liang, J., Zheng, X., Zhang, K., and Zhao, Y. (2020) Using a high-throughput zebrafish embryo screening approach to support environmental hazard ranking for cardiovascular agents. *Sci. Total Environ.*, **702**, 134703.

Mennigen, J.A., Stroud, P., Zamora, J.M., Moon, T.W., and Trudeau, V.L. (2011) Pharmaceuticals as neuroendocrine disruptors: Lessons learned from fish on prozac. *J. Toxicol. Environ. Heal. - Part B Crit. Rev.*, **14**, 387–412.

Mennigen, J.A., Sassine, J., Trudeau, V.L., and Moon, T.W. (2010) Waterborne fluoxetine disrupts feeding and energy metabolism in the goldfish Carassius auratus. *Aquat. Toxicol.*, **100**, 128–137.

Mennigen, J.A., Lado, W.E., Zamoraa, J.M., Duarte-Guterman, P., Langlois, V.S., Metcalfe, C.D., Chang, J.P., Moon, T.W., and Trudeau, V.L. (2010) Waterborne fluoxetine disrupts the reproductive axis in sexually mature male goldfish, Carassius auratus. *Aquat. Toxicol.*, **100**, 354–364.

Metcalfe, C.D., Chu, S., Judt, C., Li, H., Oakes, K.D., Servos, M.R., and Andrews, D.M. (2010) Antidepressants and their metabolites in municipal wastewater, and downstream exposure in an urban watershed. *Environ. Toxicol. Chem.*, **29**, 79–89.

Metcalfe, C.D., Metcalfe, T.L., Kiparissis, Y., Koenig, B.G., Khan, C., Hughes, R.J., Croley, T.R., March, R.E., and Potter, T. (2001) Estrogenic potency of chemicals detected in sewage treatment plant effluents as determined by in vivo assays with Japanese medaka (Oryzias latipes). *Environ. Toxicol. Chem.*, **20**, 297–308.

Miller, D.H. and Ankley, G.T. (2004) Modeling impacts on populations: Fathead minnow (Pimephales promelas) exposure to the endocrine disruptor 17beta-trenbolone as a case study. *Ecotoxicol. Environ. Saf.*, **59**, 1–9.

Mimeault, C., Woodhouse, A.J., Miao, X.-S., Metcalfe, C.D., Moon, T.W., and Trudeau, V.L. (2005) The human lipid regulator, gemfibrozil bioconcentrates and reduces testosterone in the goldfish, Carassius auratus. *Aquat. Toxicol.*, **73**, 44–54.

Mitchell, K.M. and Moon, T.W. (2016) Behavioral and biochemical adjustments of the zebrafish Danio rerio exposed to the β-blocker propranolol. *Comp. Biochem. Physiol. Part B Biochem. Mol. Biol.*, **199**, 105–114.

Muñoz, I., López-Doval, J.C., Ricart, M., Villagrasa, M., Brix, R., Geiszinger, A., Ginebreda, A., Guasch, H., de Alda, M.J.L., and Romaní, A.M. (2009) Bridging levels of pharmaceuticals in river water with biological community structure in the Llobregat river basin (northeast Spain). *Environ. Toxicol. Chem.*, **28**, 2706–2714.

Nash, J.P., Kime, D.E., Van der Ven, L.T.M., Wester, P.W., Brion, F., Maack, G., Stahlschmidt-Allner, P., and Tyler, C.R. (2004) Long-term exposure to environmental concentrations of the pharmaceutical ethynylestradiol causes reproductive failure in fish. *Environ. Health Perspect.*, **112**, 1725–1733.

Niemuth, N.J. and Klaper, R.D. (2015) Emerging wastewater contaminant metformin causes intersex and reduced fecundity in fish. *Chemosphere*, **135**, 38–45.

Niemuth, N.J. and Klaper, R.D. (2018) Low-dose metformin exposure causes changes in expression of endocrine disruption-associated genes. *Aquat. Toxicol.*, **195**, 33–40.

Niemuth, N.J., Jordan, R., Crago, J., Blanksma, C., Johnson, R., and Klaper, R.D. (2015) Metformin exposure at environmentally relevant concentrations causes potential endocrine disruption in adult male fish. *Environ. Toxicol. Chem.*, **34**, 291–296.

Oaks, J.L., Gilbert, M., Virani, M.Z., Watson, R.T., Meteyer, C.U., Rideout, B.A., Shivaprasad, H.L., Ahmed, S., Iqbal Chaudhry, M.J., Arshad, M., *et al.* (2004) Diclofenac residues as the cause of vulture population decline in Pakistan. *Nature*, **427**, 630–633.

Oggier, D.M., Weisbrod, C.J., Stoller, A.M., Zenker, A.K., and Fent, K. (2010) Effects of diazepam on gene expression and link to physiological effects in different life stages in zebrafish Danio rerio. *Environ. Sci. Technol.*, **44**, 7685–7691.

Olalla, A., Negreira, N., de Alda, M.L., Barceló, D., and Valcárcel, Y. (2018) A case study to identify priority cytostatic contaminants in hospital effluents. *Chemosphere*, **190**, 417–430.

De Oliveira, J., Chadili, E., Piccini, B., Turies, C., Maillot-Maréchal, E., Palluel, O., Pardon, P., Budzinski, H., Cousin, X., and Brion, F. (2020) Refinement of an OECD test guideline for evaluating the effects of endocrine disrupting chemicals on aromatase gene expression and reproduction using novel transgenic cyp19a1a-eGFP zebrafish. *Aquat. Toxicol.*, **220**, 105403.

Östman, M., Björlenius, B., Fick, J., and Tysklind, M. (2019) Effect of full-scale ozonation and pilot-scale granular activated carbon on the removal of biocides, antimycotics and antibiotics in a sewage treatment plant. *Sci. Total Environ.*, **649**, 1117–1123.

Pahan, K. (2006) Lipid-lowering drugs. *Cell. Mol. life Sci. C.*, **63**, 1165–1178.

Pappachan, J.M., Fernandez, C.J., and Chacko, E.C. (2019) Diabesity and antidiabetic drugs. *Mol. Aspects Med.*, **66**, 3–12.

Parajulee, A., Lei, Y.D., De Silva, A.O., Cao, X., Mitchell, C.P.J., and Wania, F. (2017) Assessing the source-to-stream transport of benzotriazoles during rainfall and snowmelt in urban and agricultural watersheds. *Environ. Sci. Technol.*, **51**, 4191–4198.

Park, S. and Lee, W. (2018) Removal of selected pharmaceuticals and personal care products in reclaimed water during simulated managed aquifer recharge. *Sci. Total Environ.*, **640**, 671–677.

Parrott, J.L. and Metcalfe, C.D. (2019) Assessing the effects of environmentally relevant concentrations of antidepressant mixtures to fathead minnows exposed over a full life cycle. *Sci. Total Environ.*, **648**, 1227–1236.

Patel, A., Panter, G.H., Trollope, H.T., Glennon, Y.C., Owen, S.F., Sumpter, J.P., and Rand-Weaver, M. (2016) Testing the "read-across hypothesis" by investigating the effects of ibuprofen on fish. *Chemosphere*, **163**, 592–600.

Patrolecco, L., Capri, S., and Ademollo, N. (2015) Occurrence of selected pharmaceuticals in the principal sewage treatment plants in Rome (Italy) and in the receiving surface waters. *Environ. Sci. Pollut. Res.*, **22**, 5864–5876.

Paulos, P., Runnalls, T.J., Nallani, G., La Point, T., Scott, A.P., Sumpter, J.P., and Huggett, D.B. (2010) Reproductive responses in fathead minnow and Japanese medaka following exposure to a synthetic progestin, Norethindrone. *Aquat. Toxicol.*, **99**, 256–262.

Peng, X., Huang, Q., Zhang, K., Yu, Y., Wang, Z., and Wang, C. (2012) Distribution, behavior and fate of azole antifungals during mechanical, biological, and chemical treatments in sewage treatment plants in China. *Sci. Total Environ.*, **426**, 311–317.

Petrie, B., Barden, R., and Kasprzyk-Hordern, B. (2015) A review on emerging contaminants in wastewaters and the environment: Current knowledge, understudied areas and recommendations for future monitoring. *Water Res.*, **72**, 3–27.

Phillips, P.J., Smith, S.G., Kolpin, D.W., Zaugg, S.D., Buxton, H.T., Furlong, E.T., Esposito, K., and Stinson, B. (2010) Pharmaceutical formulation facilities as sources of opioids and other pharmaceuticals to wastewater treatment plant effluents. *Environ. Sci. Technol.*, **44**, 4910–4916.

Pieczyńska, A., Fiszka Borzyszkowska, A., Ofiarska, A., and Siedlecka, E.M. (2017) Removal of cytostatic drugs by AOPs: A review of applied processes in the context of green technology. *Crit. Rev. Environ. Sci. Technol.*, **47**, 1282–1335.

Prasad, P., Ogawa, S., and Parhar, I.S. (2015) Serotonin reuptake inhibitor citalopram inhibits GnRH synthesis and spermatogenesis in the male zebrafish. *Biol. Reprod.*, **93**, 1–10.

Prindiville, J.S., Mennigen, J.A., Zamora, J.M., Moon, T.W., and Weber, J.-M. (2011) The fibrate drug gemfibrozil disrupts lipoprotein metabolism in rainbow trout. *Toxicol. Appl. Pharmacol.*, **251**, 201–208.

Qiang, L., Cheng, J., Yi, J., Rotchell, J.M., Zhu, X., and Zhou, J. (2016) Environmental concentration of carbamazepine accelerates fish embryonic development and disturbs larvae behavior. *Ecotoxicology*, **25**, 1426–1437.

Raldúa, D., André, M., and Babin, P.J. (2008) Clofibrate and gemfibrozil induce an embryonic malabsorption syndrome in zebrafish. *Toxicol. Appl. Pharmacol.*, **228**, 301–314.

Rand-Weaver, M., Margiotta-Casaluci, L., Patel, A., Panter, G.H., Owen, S.F., and Sumpter, J.P. (2013) The read-across hypothesis and environmental risk assessment of pharmaceuticals. *Environ. Sci. Technol.*, **47**, 11384–11395.

Ribeiro, S., Torres, T., Martins, R., and Santos, M.M. (2015) Toxicity screening of diclofenac, propranolol, sertraline and simvastatin using Danio rerio and Paracentrotus lividus embryo bioassays. *Ecotoxicol. Environ. Saf.*, **114**, 67–74.

Richards, S.M., Wilson, C.J., Johnson, D.J., Castle, D.M., Lam, M., Mabury, S.A., Sibley, P.K., and Solomon, K.R. (2004) Effects of pharmaceutical mixtures in aquatic microcosms. *Environ. Toxicol. Chem.*, **23**, 1035–1042.

Rocco, L., Frenzilli, G., Zito, G., Archimandritis, A., Peluso, C., and Stingo, V. (2012) Genotoxic effects in fish induced by pharmacological agents present in the sewage of some Italian water-treatment plants. *Environ. Toxicol.*, **27**, 18–25.

Rodrigues, S., Antunes, S.C., Correia, A.T., and Nunes, B. (2016) Acute and chronic effects of erythromycin exposure on oxidative stress and genotoxicity parameters of Oncorhynchus mykiss. *Sci. Total Environ.*, **545–546**, 591–600.

Rodriguez-Mozaz, S., Vaz-Moreira, I., Varela Della Giustina, S., Llorca, M., Barceló, D., Schubert, S., Berendonk, T.U., Michael-Kordatou, I., Fatta-Kassinos, D., Martinez, J.L., *et al.* (2020) Antibiotic residues in final effluents of European wastewater treatment plants and their impact on the aquatic environment. *Environ. Int.*, **140**, 105733.

Rodríguez-Navas, C., Björklund, E., Bak, S.A., Hansen, M., Krogh, K.A., Maya, F., Forteza, R., and Cerdà, V. (2013) Pollution pathways of pharmaceutical residues in the aquatic environment on the island of Mallorca, Spain. *Arch. Environ. Contam. Toxicol.*, **65**, 56–66.

Rosal, R., Rodríguez, A., Perdigón-Melón, J.A., Petre, A., García-Calvo, E., Gómez, M.J., Agüera, A., and Fernández-Alba, A.R. (2010) Occurrence of emerging pollutants in urban wastewater and their removal through biological treatment followed by ozonation. *Water Res.*, **44**, 578–588.

Rossier, N.M., Chew, G., Zhang, K., Riva, F., and Fent, K. (2016) Activity of binary mixtures of drospirenone with progesterone and 17α-ethinylestradiol in vitro and in vivo. *Aquat. Toxicol.*, **174**, 109–122.

Runnalls, T.J., Beresford, N., Kugathas, S., Margiotta-Casaluci, L., Scholze, M., Scott, A.P., and Sumpter, J.P. (2015) From single chemicals to mixtures-reproductive effects of levonorgestrel and ethinylestradiol on the fathead minnow. *Aquat. Toxicol.*, **169**, 152–167.

Runnalls, T.J., Margiotta-Casaluci, L., Kugathas, S., and Sumpter, J.P. (2010) Pharmaceuticals in the aquatic environment: Steroids and anti-steroids as high priorities for research. *Hum. Ecol. Risk Assess.*, **16**, 1318–1338.

Runnalls, T.J., Hala, D.N., and Sumpter, J.P. (2007) Preliminary studies into the effects of the human pharmaceutical Clofibric acid on sperm parameters in adult fathead minnow. *Aquat. Toxicol.*, **84**, 111–118.

Runnalls, T.J., Beresford, N., Losty, E., Scott, A.P., and Sumpter, J.P. (2013) Several synthetic progestins with different potencies adversely affect reproduction of fish. *Environ. Sci. Technol.*, **47**, 2077–2084.

Saari, G.N., Corrales, J., Haddad, S.P., Chambliss, C.K., and Brooks, B.W. (2018) Influence of diltiazem on fathead minnows across dissolved oxygen gradients. *Environ. Toxicol. Chem.*, **37**, 2835–2850.

Sanchez, W., Sremski, W., Piccini, B., Palluel, O., Maillot-Maréchal, E., Betoulle, S., Jaffal, A., Aït-Aïssa, S., Brion, F., Thybaud, E., *et al.* (2011) Adverse effects in wild fish living downstream from pharmaceutical manufacture discharges. *Environ. Int.*, **37**, 1342–1348.

Santos, L.H., Gros, M., Rodriguez-Mozaz, S., Delerue-Matos, C., Pena, A., Barceló, D., and Montenegro, M.C.B.S.M. (2013) Contribution of hospital effluents to the load of pharmaceuticals in urban wastewaters: Identification of ecologically relevant pharmaceuticals. *Sci. Total Environ.*, **461**, 302–316.

Santos, M.A. and Pacheco, M. (1995) Mutagenicity of cyclophosphamide and kraft mill effluent and sediment on the eel Anguilla anguilla L. *Sci. Total Environ.*, **171**, 127–130.

Scheurer, M., Michel, A., Brauch, H.J., Ruck, W., and Sacher, F. (2012) Occurrence and fate of the antidiabetic drug metformin and its metabolite guanylurea in the environment and during drinking water treatment. *Water Res.*, **46**, 4790–4802.

Schmid, S., Willi, R.A., and Fent, K. (2020) Effects of environmental steroid mixtures are regulated by individual steroid receptor signaling. *Aquat. Toxicol.*, **226**, 105562.

Schmid, S., Willi, R.A., Salgueiro-González, N., and Fent, K. (2020) Effects of new generation progestins, including as mixtures and in combination with other classes of steroid hormones, on zebrafish early life stages. *Sci. Total Environ.*, **709**, 136262.

Schmitz, M., Beghin, M., Mandiki, S.N.M., Nott, K., Gillet, M., Ronkart, S., Robert, C., Baekelandt, S., and Kestemont, P. (2018) Environmentally-relevant mixture of pharmaceutical drugs stimulates sex-steroid hormone production and modulates the expression of candidate genes in the ovary of juvenile female rainbow trout. *Aquat. Toxicol.*, **205**, 89–99.

Schultz, M.M., Furlong, E.T., Kolpin, D.W., Werner, S.L., Schoenfuss, H.L., Barber, L.B., Blazer, V.S., Norris, D.O., and Vajda, A.M. (2010) Antidepressant pharmaceuticals in two U.S. effluent-impacted streams: Occurrence and fate in water and sediment and selective uptake in fish neural tissue. *Environ. Sci. Technol.*, **44**, 1918–1925.

Schwaiger, J., Ferling, H., Mallow, U., Wintermayr, H., and Negele, R.D. (2004) Toxic effects of the non-steroidal anti-inflammatory drug diclofenac: Part I: histopathological alterations and bioaccumulation in rainbow trout. *Aquat. Toxicol.*, **68**, 141–150.

Shen, X., Chang, H., Sun, Y., and Wan, Y. (2020) Determination and occurrence of natural and synthetic glucocorticoids in surface waters. *Environ. Int.*, **134**, 105278.

Siegenthaler, P.F., Bain, P., Riva, F., and Fent, K. (2017) Effects of antiandrogenic progestins, chlormadinone and cyproterone acetate, and the estrogen 17α-ethinylestradiol (EE2), and their mixtures: Transactivation with human and rainbowfish hormone receptors and transcriptional effects in zebrafish (Danio rerio). *Aquat. Toxicol.*, **182**, 142–162.

Silva, E., Rajapakse, N., and Kortenkamp, A. (2002) Something from "Nothing"–Eight weak estrogenic chemicals combined at concentrations below NOECs produce significant mixture effects. *Environ. Sci. Technol.*, **36**, 1751–1756.

Sim, W.-J., Lee, J.-W., Lee, E.-S., Shin, S.-K., Hwang, S.-R., and Oh, J.-E. (2011) Occurrence and distribution of pharmaceuticals in wastewater from households, livestock farms, hospitals and pharmaceutical manufactures. *Chemosphere*, **82**, 179–186.

Skolness, S.Y., Durhan, E.J., Jensen, K.M., Kahl, M.D., Makynen, E.A., Villeneuve, D.L., and Ankley, G.T. (2012) Effects of gemfibrozil on lipid metabolism, steroidogenesis, and reproduction in the fathead minnow (Pimephales promelas). *Environ. Toxicol. Chem.*, **31**, 2615–2624.

Skolness, S.Y., Blanksma, C.A., Cavallin, J.E., Churchill, J.J., Durhan, E.J., Jensen, K.M., Johnson, R.D., Kahl, M.D., Makynen, E.A., and Villeneuve, D.L. (2013) Propiconazole inhibits steroidogenesis and reproduction in the fathead minnow (Pimephales promelas). *Toxicol. Sci.*, **132**, 284–297.

Sokolowska-Mikolajczyk, M., Socha, M., Mikolajczyk, T., Chyb, J., Szymacha, J., and Epler, P. (2005) Differential effects of morphine and naltrexone on the in vitro LH secretion from male and female carp pituitary gland. *Comp. Biochem. Physiol. Part C Toxicol. Pharmacol.*, **141**, 325–331.

Souders II, C.L., Xavier, P., Perez-Rodriguez, V., Ector, N., Zhang, J.-L., and Martyniuk, C.J. (2019) Sub-lethal effects of the triazole fungicide propiconazole on zebrafish (Danio rerio) development, oxidative respiration, and larval locomotor activity. *Neurotoxicol. Teratol.*, **74**, 106809.

Sousa, B. and Nunes, B. (2020) Reliability of behavioral test with fish: How neurotransmitters may exert neuromodulatory effects and alter the biological responses to neuroactive agents. *Sci. Total Environ.*, **734**, 139372.

Souza, D.M., Reichert, J.F., and Martins, A.F. (2018) A simultaneous determination of anti-cancer drugs in hospital effluent by DLLME HPLC-FLD, together with a risk assessment. *Chemosphere*, **201**, 178–188.

Stankiewicz, A., Giebułtowicz, J., Stankiewicz, U., Wroczyński, P., and Nałęcz-Jawecki, G. (2015) Determination of selected cardiovascular active compounds in environmental aquatic samples-methods and results, a review of global publications from the last 10 years. *Chemosphere*, **138**, 642–656.

Steinbach, C., Císař, P., Šauer, P., Klicnarová, J., Schmidt - Posthaus, H., Golovko, O., and Kocour Kroupová, H. (2019) Synthetic progestin etonogestrel negatively affects mating behavior and reproduction in Endler's guppies (Poecilia wingei). *Sci. Total Environ.*, **663**, 206–2015.

Straub, J.O., Caldwell, D.J., Davidson, T., D'Aco, V., Kappler, K., Robinson, P.F., Simon-Hettich, B., and Tell, J. (2019) Environmental risk assessment of metformin and its transformation product guanylurea. I. Environmental fate. *Chemosphere*, **216**, 844–854.

Subedi, B., Anderson, S., Croft, T.L., Rouchka, E.C., Zhang, M., and Hammond-Weinberger, D.R. (2021) Gene alteration in zebrafish exposed to a mixture of substances of abuse. *Environ. Pollut.*, **278**, 116777.

Sumpter, J.P., Runnalls, T.J., Donnachie, R.L., and Owen, S.F. (2021) A comprehensive aquatic risk assessment of the beta-blocker propranolol, based on the results of over 600 research papers. *Sci. Total Environ.*, **793**, 148617.

Sumpter, J.P., Donnachie, R.L., and Johnson, A.C. (2014) The apparently very variable potency of the anti-depressant fluoxetine. *Aquat. Toxicol.*, **151**, 57–60.

Sumpter, J.P. and Jobling, S. (2013) The occurrence, causes, and consequences of estrogens in the aquatic environment. *Environ. Toxicol. Chem.*, **32**, 249–251.

Sumpter, J.P. and Jobling, S. (1995) Vitellogenesis as a biomarker for estrogenic contamination of the aquatic environment. *Environ. Health Perspect.*, **103**, 173–178.

Sun, C., Wang, J.J., Mackey, D.A., and Wong, T.Y. (2009) Retinal vascular caliber: Systemic, environmental, and genetic associations. *Surv. Ophthalmol.*, **54**, 74–95.

Sun, D., Sharma, A.K., Dellinger, R.W., Blevins-Primeau, A.S., Balliet, R.M., Chen, G., Boyiri, T., Amin, S., and Lazarus, P. (2007) Glucuronidation of active tamoxifen metabolites by the human UDP glucuronosyltransferases. *Drug Metab. Dispos.*, **35**, 2006–2014.

Sun, L., Xin, L., Peng, Z., Jin, R., Jin, Y., Qian, H., and Fu, Z. (2014) Toxicity and enantiospecific differences of two β-blockers, propranolol and metoprolol, in the embryos and larvae of zebrafish (Danio rerio). *Environ. Toxicol.*, **29**, 1367–1378.

Sun, L., Zha, J., Spear, P.A., and Wang, Z. (2007) Toxicity of the aromatase inhibitor letrozole to Japanese medaka (Oryzias latipes) eggs, larvae and breeding adults. *Comp. Biochem. Physiol. Part C Toxicol. Pharmacol.*, **145**, 533–541.

Synnott, J.M., Guida, A., Mulhern-Haughey, S., Higgins, D.G., and Butler, G. (2010) Regulation of the hypoxic response in Candida albicans. *Eukaryot. Cell*, **9**, 1734–1746.

Tang, Y., Mi, P., Li, M., Zhang, S., Li, J., and Feng, X. (2021) Environmental level of the antidepressant venlafaxine induces behavioral disorders through cortisol in zebrafish larvae (Danio rerio). *Neurotoxicol. Teratol.*, **83**, 106942.

Tang, Y., Guo, L., Hong, C., Bing, Y., and Xu, Z. (2019) Seasonal occurrence, removal and risk assessment of 10 pharmaceuticals in 2 sewage treatment plants of Guangdong, China. *Environ. Technol.*, **40**, 458–469.

Teng, M., Zhao, F., Zhou, Y., Yan, S., Tian, S., Yan, J., Meng, Z., Bi, S., and Wang, C. (2019) Effect of propiconazole on the lipid metabolism of zebrafish embryos (Danio rerio). *J. Agric. Food Chem.*, **67**, 4623–4631.

Ternes, T., Bonerz, M., and Schmidt, T. (2001) Determination of neutral pharmaceuticals in wastewater and rivers by liquid chromatography-electrospray tandem mass spectrometry. *J. Chromatogr. A*, **938**, 175–185.

Ternes, T.A. (1998) Occurrence of drugs in German sewage treatment plants and rivers. *Water Res.*, **32**, 3245–3260.

Thorn, C.F., Leckband, S.G., Kelsoe, J., Steven Leeder, J., Müller, D.J., Klein, T.E., and Altman, R.B. (2011) PharmGKB summary: Carbamazepine pathway. *Pharmacogenet. Genomics*, **21**, 906–910.

Thörnqvist, P.-O., McCarrick, S., Ericsson, M., Roman, E., and Winberg, S. (2019) Bold zebrafish (Danio rerio) express higher levels of delta opioid and dopamine D2 receptors in the brain compared to shy fish. *Behav. Brain Res.*, **359**, 927–934.

Thorpe, K.L., Cummings, R.I., Hutchinson, T.H., Scholze, M., Brighty, G., Sumpter, J.P., and Tyler, C.R. (2003) Relative potencies and combination effects of steroidal estrogens in fish. *Environ. Sci. Technol.*, **37**, 1142–1149.

Thrupp, T.J., Runnalls, T.J., Scholze, M., Kugathas, S., Kortenkamp, A., and Sumpter, J.P. (2018) The consequences of exposure to mixtures of chemicals: Something from 'nothing' and 'a lot from a little' when fish are exposed to steroid hormones. *Sci. Total Environ.*, **619–620**, 1482–1492.

Togola, A. and Budzinski, H. (2008) Multi-residue analysis of pharmaceutical compounds in aqueous samples. *J. Chromatogr. A*, **1177**, 150–158.

Trautwein, C. and Kümmerer, K. (2011) Incomplete aerobic degradation of the antidiabetic drug Metformin and identification of the bacterial dead-end transformation product Guanylurea. *Chemosphere*, **85**, 765–773.

Triebskorn, R., Casper, H., Heyd, A., Eikemper, R., Köhler, H.-R., and Schwaiger, J. (2004) Toxic effects of the non-steroidal anti-inflammatory drug diclofenac: Part II. Cytological effects in liver, kidney, gills and intestine of rainbow trout (Oncorhynchus mykiss). *Aquat. Toxicol.*, **68**, 151–166.

Tyler, C.R., Filby, A.L., Bickley, L.K., Cumming, R.I., Gibson, R., Labadie, P., Katsu, Y., Liney, K.E., Shears, J.A., Silva-Castro, V., *et al.* (2009) Environmental health impacts of equine estrogens derived from hormone replacement therapy. *Environ. Sci. Technol.*, **43**, 3897–3904.

UNODC (2021) United Nations Office on Drugs and Crime, World Drug Report 2017.

Usawanuwat, J., Boontanon, N., and Boontanon, S.K. (2014) Analysis of three anticancer drugs (5-fluorouracil, cyclophosphamide and hydroxyurea) in water samples by HPLC-MS/MS. *Int. J. Adv. Agr. Environ. Eng*, **1**, 72–76.

Ussery, E., Bridges, K.N., Pandelides, Z., Kirkwood, A.E., Bonetta, D., Venables, B.J., Guchardi, J., and Holdway, D. (2018) Effects of environmentally relevant metformin exposure on Japanese medaka (Oryzias latipes). *Aquat. Toxicol.*, **205**, 58–65.

Valcárcel, Y., Alonso, S.G., Rodríguez-Gil, J.L., Gil, A., and Catalá, M. (2011) Detection of pharmaceutically active compounds in the rivers and tap water of the Madrid Region (Spain) and potential ecotoxicological risk. *Chemosphere*, **84**, 1336–1348.

Valdés, M.E., Huerta, B., Wunderlin, D.A., Bistoni, M.A., Barceló, D., and Rodriguez-Mozaz, S. (2016) Bioaccumulation and bioconcentration of carbamazepine and other pharmaceuticals in fish under field and controlled laboratory experiments. Evidences of carbamazepine metabolization by fish. *Sci. Total Environ.*, **557–558**, 58–67.

Valenti, T.W., Gould, G.G., Berninger, J.P., Connors, K.A., Keele, N.B., Prosser, K.N., and Brooks, B.W. (2012) Human therapeutic plasma levels of the Selective Serotonin Reuptake Inhibitor (SSRI) sertraline decrease serotonin reuptake transporter binding and shelter-seeking behavior in adult male fathead minnows. *Environ. Sci. Technol.*, **46**, 2427–2435.

Valenzuela, M.M.A., Castro, I., Gonda, A., Osterman, C.J.D., Jutzy, J.M., Aspe, J.R., Khan, S., Neidigh, J.W., and Wall, N.R. (2015) Cell death in response to antimetabolites directed at ribonucleotide reductase and thymidylate synthase. *Onco. Targets. Ther.*, **8**, 495.

Varga, R., Somogyvári, I., Eke, Z., and Torkos, K. (2013) Seasonal monitoring of cardiovascular and antiulcer agents' concentrations in stream waters encompassing a capital city. *J. Pharm.*, **2013**, 753928.

Vasskog, T., Anderssen, T., Pedersen-Bjergaard, S., Kallenborn, R., and Jensen, E. (2008) Occurrence of selective serotonin reuptake inhibitors in sewage and receiving waters at Spitsbergen and in Norway. *J. Chromatogr. A*, **1185**, 194–205.

Vaziri, N.D. and Liang, K.H. (2004) Acyl-coenzyme A: Cholesterol acyltransferase inhibition ameliorates proteinuria, hyperlipidemia, lecithin-cholesterol acyltransferase, SRB-1, and low-density lipoprotein receptor deficiencies in nephrotic syndrome. *Circulation*, **110**, 419–425.

Velasco-Santamaría, Y.M., Korsgaard, B., Madsen, S.S., and Bjerregaard, P. (2011) Bezafibrate, a lipid-lowering pharmaceutical, as a potential endocrine disruptor in male zebrafish (Danio rerio). *Aquat. Toxicol.*, **105**, 107–118.

Vera-Chang, M.N., St-Jacques, A.D., Gagné, R., Martyniuk, C.J., Yauk, C.L., Moon, T.W., and Trudeau, V.L. (2018) Transgenerational hypocortisolism and behavioral disruption are induced by the antidepressant fluoxetine in male zebrafish Danio rerio. *Proc. Natl. Acad. Sci.*, **115**, E12435–E12442.

Verlicchi, P., Al Aukidy, M., and Zambello, E. (2012) Occurrence of pharmaceutical compounds in urban wastewater: Removal, mass load and environmental risk after a secondary treatment-A review. *Sci. Total Environ.*, **429**, 123–155.

Voloshenko-Rossin, A., Gasser, G., Cohen, K., Gun, J., Cumbal-Flores, L., Parra-Morales, W., Sarabia, F., Ojeda, F., and Lev, O. (2015) Emerging pollutants in the Esmeraldas watershed in Ecuador: Discharge and attenuation of emerging organic pollutants along the San Pedro-Guayllabamba-Esmeraldas rivers. *Environ. Sci. Process. Impacts*, **17**, 41–53.

Weizel, A., P. Schlüsener, M., Dierkes, G., and A. Ternes, T. (2018) Occurrence of glucocorticoids, mineralocorticoids, and progestogens in various treated wastewater, rivers, and streams. *Environ. Sci. & Technol.*, **52**, 5296–5307.

Weston, A., Caminada, D., Galicia, H., and Fent, K. (2009) Effects of lipid-lowering pharmaceuticals bezafibrate and clofibric acid on lipid metabolism in fathead minnow (Pimephales promelas). *Environ. Toxicol. Chem.*, **28**, 2648–2655.

WHO (2016) Global Report on Diabetes. Executive summary (No. WHO/NMH/NVI/16.3). World Health Organization.

Wick, A., Fink, G., and Ternes, T.A. (2010) Comparison of electrospray ionization and atmospheric pressure chemical ionization for multi-residue analysis of biocides, UV-filters and benzothiazoles in aqueous matrices and activated sludge by liquid chromatography-tandem mass spectrometry. *J. Chromatogr. A*, **1217**, 2088–2103.

Willi, R.A., Faltermann, S., Hettich, T., and Fent, K. (2018) Active glucocorticoids have a range of important adverse developmental and physiological effects on developing zebrafish embryos. *Environ. Sci. Technol.*, **52**, 877–885.

Willi, R.A., Salgueiro-González, N., Faltermann, S., Hettich, T., and Fent, K. (2019) Environmental glucocorticoids corticosterone, betamethasone and flumethasone induce more potent physiological than transcriptional effects in zebrafish embryos. *Sci. Total Environ.*, **672**, 183–191.

Willi, R.A., Salgueiro-González, N., Carcaiso, G., and Fent, K. (2019) Glucocorticoid mixtures of fluticasone propionate, triamcinolone acetonide and clobetasol propionate induce additive effects in zebrafish embryos. *J. Hazard. Mater.*, **374**, 101–109.

Willi, R.A., Castiglioni, S., Salgueiro-González, N., Furia, N., Mastroianni, S., Faltermann, S., and Fent, K. (2020) Physiological and transcriptional effects of mixtures of environmental estrogens, androgens, progestins, and glucocorticoids in zebrafish. *Environ. Sci. Technol.*, **54**, 1092–1101.

Wong, K., Elegante, M., Bartels, B., Elkhayat, S., Tien, D., Roy, S., Goodspeed, J., Suciu, C., Tan, J., and Grimes, C. (2010) Analyzing habituation responses to novelty in zebrafish (Danio rerio). *Behav. Brain Res.*, **208**, 450–457.

Wu, M., Qiu, X., Chen, C., Chen, K., Li, M., Xu, H., Wu, X., Shimasaki, Y., and Oshima, Y. (2020) Short-term and persistent impacts of sublethal exposure to diazepam on behavioral traits and brain GABA levels in juvenile zebrafish (Danio rerio). *Sci. Total Environ.*, **740**, 140392.

Xia, L., Zheng, L., and Zhou, J.L. (2016) Transcriptional and morphological effects of tamoxifen on the early development of zebrafish (Danio rerio). *J. Appl. Toxicol.*, **36**, 853–862.

Xiang, J., Wu, M., Lei, J., Fu, C., Gu, J., and Xu, G. (2018) The fate and risk assessment of psychiatric pharmaceuticals from psychiatric hospital effluent. *Ecotoxicol. Environ. Saf.*, **150**, 289–296.

Xin, J., Yan, S., Hong, X., Zhang, H., and Zha, J. (2021) Environmentally relevant concentrations of carbamazepine induced lipid metabolism disorder of Chinese rare minnow (Gobiocypris rarus) in a gender-specific pattern. *Chemosphere*, **265**, 129080.

Xu, L., Zhang, H., Xiong, P., Zhu, Q., Liao, C., and Jiang, G. (2021) Occurrence, fate, and risk assessment of typical tetracycline antibiotics in the aquatic environment: A review. *Sci. Total Environ.*, **753**, 141975.

Yadav, M.K., Short, M.D., Aryal, R., Gerber, C., van den Akker, B., and Saint, C.P. (2017) Occurrence of illicit drugs in water and wastewater and their removal during wastewater treatment. *Water Res.*, **124**, 713–727.

Yan, Z., Yang, Q., Jiang, W., Lu, J., Xiang, Z., Guo, R., and Chen, J. (2018) Integrated toxic evaluation of sulfamethazine on zebrafish: Including two lifespan stages (embryo-larval and adult) and three exposure periods (exposure, post-exposure and re-exposure). *Chemosphere*, **195**, 784–792.

Yang, C., Song, G., and Lim, W. (2020) A review of the toxicity in fish exposed to antibiotics. *Comp. Biochem. Physiol. Part - C Toxicol. Pharmacol.*, **237**, 108840.

Yin, X., Guo, C., Teng, Y., and Xu, J. (2019) Development and application of the analytical method for illicit drugs and metabolites in fish tissues. *Chemosphere*, **233**, 532–541.

Yu, F. and Bender, W. (2001) The mechanism of tamoxifen in breast cancer prevention. In: *Breast Cancer Research.* Springer, p. 1.

Zarn, J.A., Brüschweiler, B.J., and Schlatter, J.R. (2003) Azole fungicides affect mammalian steroidogenesis by inhibiting sterol 14 alpha-demethylase and aromatase. *Environ. Health Perspect.*, **111**, 255–261.

Zeilinger, J., Steger-Hartmann, T., Maser, E., Goller, S., Vonk, R., and Länge, R. (2009) Effects of synthetic gestagens on fish reproduction. *Environ. Toxicol. Chem.*, **28**, 2663–2670.

Zhang, J.-N., Chen, J., Yang, L., Zhang, M., Yao, L., Liu, Y.-S., Zhao, J.-L., Zhang, B., and Ying, G.-G. (2021) Occurrence and fate of androgens, progestogens and glucocorticoids in two swine farms with integrated wastewater treatment systems. *Water Res.*, **192**, 116836.

Zhang, K., Zhao, Y., and Fent, K. (2020) Cardiovascular drugs and lipid regulating agents in surface waters at global scale: Occurrence, ecotoxicity and risk assessment. *Sci. Total Environ.*, **729**, 138770.

Zhang, K. and Fent, K. (2018) Determination of two progestin metabolites (17α-hydroxypregnenolone and pregnanediol) and different classes of steroids (androgens, estrogens, corticosteroids, progestins) in rivers and wastewaters by high-performance liquid chromatography-tandem mass spectrometry. *Sci. Total Environ.*, **610–611**, 1164–1172.

Zhang, K., Zhao, Y., and Fent, K. (2017) Occurrence and ecotoxicological effects of free, conjugated, and halogenated steroids including 17α-hydroxypregnenolone and pregnanediol in Swiss wastewater and surface water. *Environ. Sci. Technol.*, **51**, 6498–6506.

Zhang, Y., Geißen, S.U., and Gal, C. (2008) Carbamazepine and diclofenac: Removal in wastewater treatment plants and occurrence in water bodies. *Chemosphere*, **73**, 1151–1161.

Zhao, Y., Zhang, K., and Fent, K. (2016) Corticosteroid fludrocortisone acetate targets multiple end points in zebrafish (Danio rerio) at low concentrations. *Environ. Sci. Technol.*, **50**, 10245–10254.

Zhao, Y., Castiglioni, S., and Fent, K. (2015a) Environmental progestins progesterone and drospirenone alter the circadian rhythm network in zebrafish (Danio rerio). *Environ. Sci. Technol.*, **49**, 10155–10164.

Zhao, Y., Castiglioni, S., and Fent, K. (2015b) Synthetic progestins medroxyprogesterone acetate and dydrogesterone and their binary mixtures adversely affect reproduction and lead to histological and transcriptional alterations in zebrafish (Danio rerio). *Environ. Sci. Technol.*, **49**, 4636–4645.

Zheng, X., Zhang, K., Zhao, Y., and Fent, K. (2021) Environmental chemicals affect circadian rhythms: An underexplored effect influencing health and fitness in animals and humans. *Environ. Int.*, **149**, 106159.

Zhou, L.J., Wu, Q.L., Zhang, B.B., Zhao, Y.G., and Zhao, B.Y. (2016) Occurrence, spatiotemporal distribution, mass balance and ecological risks of antibiotics in subtropical shallow Lake Taihu, China. *Environ. Sci. Process. Impacts*, **18**, 500–513.

Zhu, B., Liu, L., Gong, Y.-X., Ling, F., and Wang, G.-X. (2014) Triazole-induced toxicity in developing rare minnow (Gobiocypris rarus) embryos. *Environ. Sci. Pollut. Res.*, **21**, 13625–13635.

Zind, H., Mondamert, L., Remaury, Q.B., Cleon, A., Leitner, N.K.V., and Labanowski, J. (2021) Occurrence of carbamazepine, diclofenac, and their related metabolites and transformation products in a French aquatic environment and preliminary risk assessment. *Water Res.*, **196**, 117052.

Zuccato, E., Castiglioni, S., Bagnati, R., Chiabrando, C., Grassi, P., and Fanelli, R. (2008) Illicit drugs, a novel group of environmental contaminants. *Water Res.*, **42**, 961–968.

Zucchi, S., Castiglioni, S., and Fent, K. (2013) Progesterone alters global transcription profiles at environmental concentrations in brain and ovary of female zebrafish (Danio rerio). *Environ. Sci. Technol.*, **47**, 12548–12556.

Zucchi, S., Castiglioni, S., and Fent, K. (2012) Progestins and antiprogestins affect gene expression in early development in zebrafish (Danio rerio) at environmental concentrations. *Environ. Sci. Technol.*, **46**(9), 5183–5192.

Zucchi, S., Mirbahai, L., Castiglioni, S., and Fent, K. (2014) Transcriptional and physiological responses induced by binary mixtures of drospirenone and progesterone in zebrafish (Danio rerio). *Environ. Sci. Technol.*, **48**, 3523–3531.

# 15 Oil Spills

*Fernando Galvez*
Louisiana State University

**Abstract**

Crude oils are naturally occurring mixtures of thousands of known chemicals and potentially hundreds of thousands of unknown chemicals that are still not analytically resolved using standard techniques. Despite the large suite of potentially toxic chemicals within oil, only a small subset of PAHs and their alkylated products is usually reported during routine chemical analyses and considered as potentially toxic to aquatic animals in laboratory and field-based toxicity testing. Chapter 15 provides an overview of the chemical behavior of crude oil in water and describes how the chemical and physical diversity of these complex mixtures influence the results of laboratory toxicity testing. The concept of narcosis and how it is applied in the target lipid model to predict the toxic effects of a mixture based on the underlying lipid–water partitioning of the chemical constituents. Finally, the chapter reviews recent research highlighting the sublethal effects of crude oil in developing fish embryos. Most of the discussion highlights the specific actions of tricyclic and tetracyclic PAHs on various targets impacting cardiovascular development in fish embryos and describes the role that the aryl hydrocarbon receptor pathway plays in mediating the sublethal responses to tricyclic and tetracyclic PAHs.

## INTRODUCTION

Crude oils are naturally occurring mixtures of thousands of known chemicals (Marshall and Rodgers 2004) and potentially hundreds of thousands of unknown chemicals that are still not analytically resolved using standard techniques (Sutton et al. 2005). Despite the large suite of potentially toxic chemicals within oil, only a small subset of PAHs and their alkylated products is usually reported during routine chemical analyses and considered as potentially toxic to aquatic animals in laboratory and field-based toxicity testing. This chapter will review the chemical behavior of crude oil in water and describe how the chemical and physical diversity of these complex mixtures influence the results of laboratory toxicity testing. It will then discuss the concept of narcosis and how it is applied in the target lipid model to predict the toxic effects of a mixture based on the underlying lipid–water partitioning of the chemical constituents. Finally, the chapter will review the recent research highlighting the sublethal effects of crude oil in developing embryos. Most of the discussion will highlight the specific actions of tricyclic and tetracyclic PAHs on various targets impacting cardiovascular development in fish embryos. It will also describe the role the aryl hydrocarbon receptor pathway plays in mediating the sublethal responses to tricyclic and tetracyclic PAHs.

## GLOBAL OIL PRODUCTION AND OIL CONSUMPTION

Despite attempts to diversify the sources of energy production, oil remains a critical commodity due to its use for transportation, heating, and as a raw material for plastics production. Aside from brief periods of economic downturn, as occurred during the financial crisis in the late 2000s and due to the COVID pandemic in 2020, there has been a progressive increase in the annual production of oil over the past two decades. In 2019, approximately 88.5 million barrels per day of crude oil, shale oil, oil sands, or natural gas liquids were extracted worldwide (BP Statistical Review of World Energy 2021). Although the Middle East continues to provide 31.7% of the world's supply of crude oil and condensates, the relative proportion produced in North America and the Commonwealth of Independent States has increased in the last ten years. Collectively, these two regions now produce 37.7% of the global supply (BP Statistical Review of World Energy 2021). Meanwhile, there has been a sharp decline in oil production in most parts of Europe over the same time frame. Despite these changing trends, oil production has increased approximately 1.4% per year from 2010 to 2019. There has also been a shift in the location of petroleum extraction from nearshore toward deep-sea, offshore facilities. For example, approximately 62% of the 1,332 offshore rigs worldwide are located within the North Sea, the U.S. waters of the Gulf of Mexico, the Persian Gulf, and the eastern and southeastern waters of Asia. This transition from nearshore to offshore oil extraction brings inherent risks associated with exploring and extracting from locations of greater depths and distance from the shore. No clearer example of this risk was the 2010 Deepwater Horizon oil spill (DHOS). It is expected that oil production will continue to increase until other sources of energy production, such as renewables, can offset the increased consumption needs (Figure 15.1).

Concurrent with the increase in total oil production has been a change in the regional oil consumption patterns. Interestingly, although oil consumption throughout much of the world has remained relatively constant since 2010, there has been an increase of 3.2% consumption in Asia Pacific (BP Statistical Review of World Energy 2021) (Figure 15.2). Most of this regional increase has been driven by the increased oil consumption in China and India. As the significant regions of oil production and oil consumption change, it is expected that the primary oil input sources to the aquatic environment are also likely to change. Emerging from these global trends of oil production and oil consumption

DOI: 10.1201/9781003160694-15

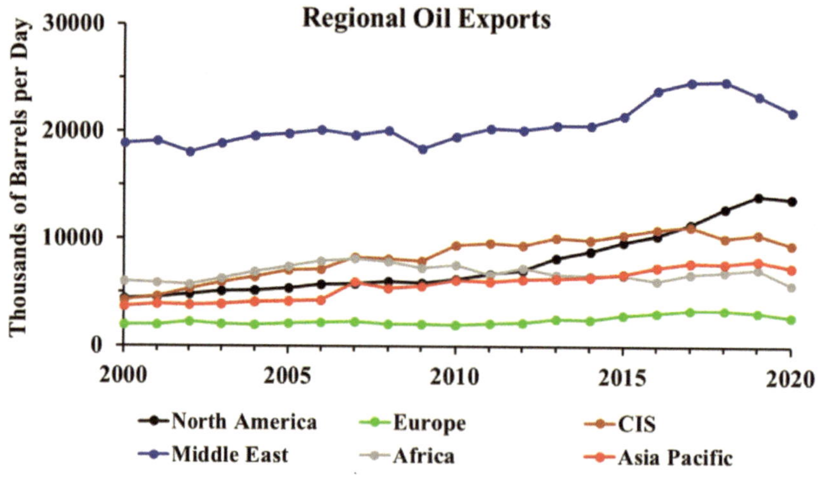


**FIGURE 15.1** Oil exports from different regions of the world. Data are expressed in thousands of barrels per day expressed on an annual basis from the years 2000 to 2020. Adapted from BP Statistical Review of World Energy (2021).

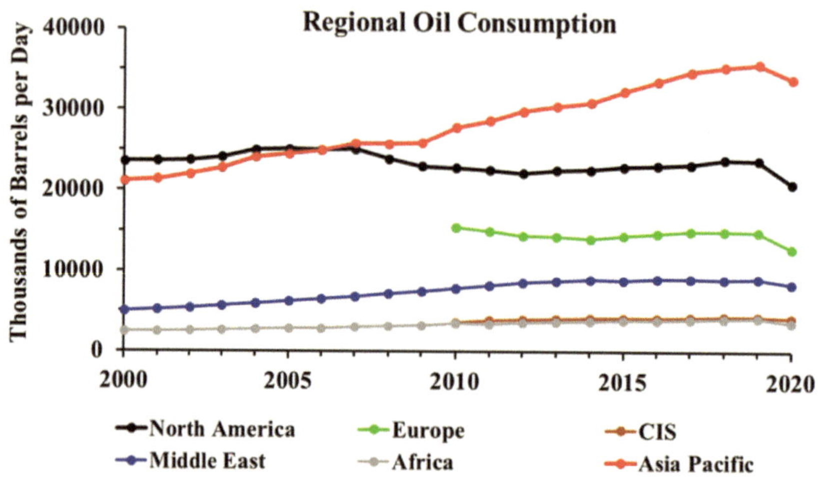


**FIGURE 15.2** Oil consumption in different regions of the world. Data are expressed in thousands of barrels per day expressed on an annual basis from the years 2000 to 2020. Adapted from BP Statistical Review of World Energy (2021).

are patterns that differ considerably from those reported by the National Academies of Sciences in its report on the inputs, fates, and effects of marine oil spills (NRC 2003). These changing patterns are expected to influence the major sources of input of oil into the aquatic environment.

## SOURCES OF INPUT OF CRUDE OIL INTO THE AQUATIC ENVIRONMENT

Oil can enter the aquatic environment from diverse sources, but these sources can be categorized into four major groups: natural seeps, petroleum extraction, petroleum transportation, and petroleum consumption. Natural seeps, which originate from geological formations rich in crude oil, contribute approximately 47% of the total petroleum entering marine waters worldwide (NRC 2003). Although seeps are the largest source of oil in the aquatic environment (Kvenvolden and Cooper 2003), regions where they occur have rich indigenous microbial communities that biodegrade oil (Head et al. 2006). As such, it is unlikely that these seeps are acutely toxic to fish, even though they may produce chronic effects or alter the sensitivity of animals to large-scale spills. However, regional differences between the rate of natural seepage and the rate hydrocarbons are removed by biodegradation, and other sources of weathering are likely important (Lee et al. 2015).

Petroleum extraction in oceanic regions coincides with areas of large natural seeps of oil (MacDonald et al. 2015). In contrast to the magnitude of natural seeps, petroleum extraction represents less than 3% of the world's oil input into the marine environment, making it the smallest contributor of oil to oceans (NRC 2003). However, this source, localizing exclusively to active oil and gas exploration areas, is episodic and variable in size. Stemming from regions of oil production is petroleum transportation, which contributes approximately 166 million liters each year, or roughly 12% of the total worldwide input (NRC 2003). Oil spills during petroleum transport can occur anywhere along transportation routes but are concentrated in populated coastal centers. Finally, petroleum consumption represents approximately 37% of the total environmental input (NRC 2003). One of the major sources

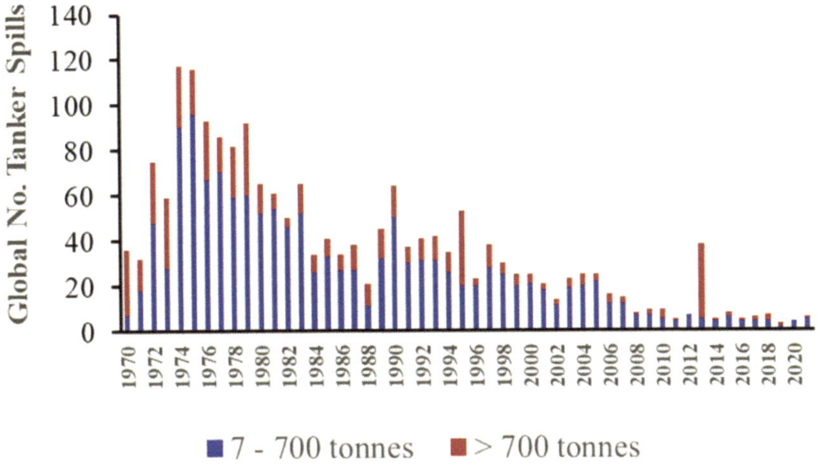


**FIGURE 15.3** The number of tanker spills globally from 1971 to 2021. Each bar represents the cumulative total of spills releasing 7–700 tons of crude oil versus those releasing greater than 700 tons per spill. Adapted from ITOPF (2022).

of consumption-based input is runoff from a large suite of land-based sources, which can eventually enter marine environments by riverine or storm-water transport. Unlike other sources, this diffuse input tends to be widespread and chronic. However, oil entering the oceans via this route is difficult to accurately quantify given the many assumptions and uncertainties in its calculation (NRC 2003). Aside from natural seeps, other sources of oil input into the environment are anthropogenic, with the diverse suite of activities categorized as petroleum consumption representing nearly 70% of the total anthropogenic source of oil in the world's oceans (NRC 2003). Furthermore, whereas oil from natural seeps and petroleum extraction is restricted to areas rich in oil reserves, oil spills due to transportation or consumption can occur ubiquitously, although they tend to concentrate more heavily in high population centers (Lee et al. 2015; NRC 2003).

Petroleum extraction accounts minimally for the input of oil in marine environments; however, oil-spill-related accidents have the potential to be large when they occur. Fortunately, undersea blowouts during petroleum exploration and production are rare due to technological advances. In the last 50 years, there have been two production-related spills greater than 94 million liters each, both occurring in the Gulf of Mexico (Soto et al. 2014). The Ixtoc I oil spill in 1979 led to the release of approximately 475 million liters at 100 kilometers from Mexico's coast and at a depth of 50 meters (Soto et al. 2014). Before being capped ten months after the blowout, the oil spill covered an area of 2,800 kilometers squared and oiled 261 kilometers of coastline (Soto et al. 2014). Meanwhile, the DHOS in 2010 released about 530 million liters at 80 kilometers from the Louisiana coast at a depth of 1,500 meters (Soto et al. 2014). The DWH spill, which was the largest marine spill in recorded history, lasted 87 days and formed surface slicks over approximately 112,000 square kilometers and oiled at least 2,100 kilometers of coastline. Although both the Ixtoc I spill and the DHOS were very large spill events, considerably more scientific research followed in the aftermath of the DHOS. According to a Web of Science search in June 2020, the search term "Ixtoc" produced 59 publications. In comparison, the search terms "Deepwater Horizon" produced almost 2,000 publications, which have contributed dramatically to our understanding of oil spills.

Oil tankers are the primary method for the long-distance transport of oil across oceans and large seas. Given the large volume of oil transported by this method, it also has the potential to be a major source of oil input into marine waters. Since 1970, the International Tanker Owners Pollution Federation (ITOPF) has maintained a database documenting over 10,000 oil-tanker-related spills ranging in size from <50 barrels to >5,000 barrels (ITOPF 2022) (Figure 15.3). These data show that there has been a progressive downward trend in tanker-related spills in the last five decades. In the 1970s, the average number of medium (50–5,000 barrels) and large (>5,000 barrels) spills was 78.8 per year (ITOPF 2022). This number decreased to 6.3 spills of equivalent size per year by the 2010s. The mandating of double-hulled oil tankers by the International Maritime Organization in 1990 has contributed much to this decline (ITOPF 2022). However, despite the added precautions, there have been seven tanker spills, each releasing over 150 million liters for each spill due to accidental groundings or allisions. Interestingly, none of these spills have sparked much oil-spill-related research. In fact, it has been smaller spills such as the Exxon Valdez oil spill (EVOS) of 1989 in Prince William Sound near Cordova, Alaska, and Hebei Spirit off the coast of South Korea that received the media attention that drove oil spill research at the time (Barron et al. 2020).

## COMPOSITION OF CRUDE OIL

Crude oils contain hydrocarbons consisting of carbon and hydrogen atoms organized into diverse structures but also molecules containing heteroatoms such as sulfur, nitrogen, and oxygen, and trace elements such as metals (Marshall and Rodgers 2004). The chemistry of oil compounds ranges from small and volatile molecules to complex and nonvolatile macromolecules that exist in colloidal dispersions. Most constituents in crude oil can be classified as saturates, aromatic hydrocarbons, resins, and asphaltenes based on solubility in liquid-solid chromatography (Lee et al. 2015). These categories are collectively given the acronym SARA. The relative abundance of these four different classes of molecules in a crude oil influences the oil's density, viscosity, water solubility, and tendency to adhere to surfaces. Whereas saturates are hydrocarbons with a maximum number of hydrogen bonds, olefins (i.e., unsaturated compounds) have at least one double bond between adjacent carbon atoms and are thus not fully saturated with hydrogen atoms. Aromatics, resins, and asphaltenes are characterized as olefins.

Saturates are defined as hydrocarbon molecules with single bonds between carbon atoms and the maximum number of hydrogen atoms bound to each carbon. The molecules can exist either as straight (i.e., paraffins or strait-chained alkanes) or branched chains (i.e., isoparaffins) with a general chemical formula of $C_nH_{2n+2}$. Collectively, these molecules comprise the largest series of hydrocarbons found in petroleum. A third type of saturate is the cycloparaffins with one or more saturated ring structures but with a general formula of $C_nH_{2n}$. A fourth type of saturate consists of large and complex molecules with paraffinic and cycloparaffinic features. Collectively, these saturates can exist either as a gas (i.e., low molecular weight molecules with five or less carbon atoms), as a liquid (i.e., medium molecular weight molecules with six to sixteen carbon atoms), or as waxy solids (i.e., high molecular weight molecules with greater than twenty carbon atoms) (reviewed in Lee et al. 2015).

Aromatic hydrocarbons are cyclic, planar structures composed of a varying number of ring structures of hydrocarbons, ranging from one ring for the monoaromatics to five or more aromatic rings fused together (Sahoo et al. 2020). Although crude oil can contain aromatics of varying size, those with smaller ring number are more common than those with more fused ring backbones. For typical crude oils, approximately 90% of the aromatics have only 1–3 fused rings (reviewed in Lee et al. 2015). Furthermore, the proportion of larger aromatics decreases with increasing molecular weight. The most common monoaromatics in petroleum hydrocarbons are benzene, toluene, ethylbenzene, and the xylenes, collectively given the acronym, BTEX. Aromatics with two ringed structures are termed diaromatics, and those with three or more ringed structures are termed polycyclic aromatic hydrocarbons (PAHs). Sometimes diaromatics are simply classified as PAHs. The monoaromatics, diaromatics, and polyaromatics only have carbon in their aromatic ring structures. However, adding to the complexity of aromatics is the fact that PAHs encompass a large suite of diverse molecules with alkyl chains of varying lengths and numbers branching from different carbons on the aromatic backbone. The bulk of the PAHs consist of these alkylated derivatives rather than the parent PAHs. In general, the water solubility decreases with increasing ring number (Sahoo et al. 2020). There are also examples of aromatics having heteroatoms termed aromatic heterocycles. Although they are not PAHs since they contain one or more sulfur, nitrogen, or oxygen atoms within the cyclic ring structure (i.e., heteroatoms), they are often classified within the aromatic fraction. In the literature, PAHs and these aromatic heterocycles are collectively termed polyaromatic compounds (PACs) (Hodson et al. 2020).

Resins are a mixture of complex molecules classified based on their tendency to dissolve in n-heptane and n-pentane rather than any unique molecular structure (Wen et al. 2016). However, resins are not considered hydrocarbons due to the presence of heteroatoms within molecules. As such, they tend to be more polar than the aromatics from which they are often derived (reviewed in Lee et al. 2015). Many of the molecules that constitute resins are water soluble due to their polarity. This characteristic may increase their toxicity to aquatic animals (Lee et al. 2015). Resins are an important constituent in heavier crude oils, but they are often chemically unresolved by conventional analytical techniques such as gas chromatography. As with resins, asphaltene is a complex mixture of chemicals that has neither been completely characterized nor has a specific molecular formula (OECD 2019). Although they appear to be related to resins, asphaltenes tend to be larger in size and have a higher aromatic content (Wen et al. 2016). Due to their aromaticity, asphaltenes have a low water solubility despite their slight polarity. However, their polarity causes them to form stable aggregates that stabilize crude oil droplets in water (Mousavi et al. 2016). In doing so, these compounds contribute to the viscosity of heavy oils such as bitumens.

## FATE OF CRUDE OIL IN THE AQUATIC ENVIRONMENT

There has been extensive research characterizing the fate of crude oil in the aquatic environment. Although much of this research is beyond the scope of this chapter, a brief review of critical aspects relevant to the aquatic toxicity of oil will be given. When crude oils are spilt in water, their physical and chemical composition changes over time by processes collectively known as weathering (NRC 2003). Weathering processes may include evaporation, emulsification, dispersion, dissolution, microbial degradation, photooxidation, and others (reviewed in Lee et al. 2015). Some of these processes such as evaporation and dissolution begin immediately and decrease in importance over time. Other processes such as dispersion, emulsification, and biodegradation increase in relative importance later following a spill (NRC 2014). In turn, photooxidation, which was once thought to be only of limited importance and only affected spills in the long term (NRC 2003), is now known to begin immediately and considered an important determinant of floating surface oil (Ward and Overton 2020). Weathering has an important influence on the fate and behavior of oil in the environment and, as such, will influence the toxicity of oils to aquatic animals. As weathering progressively leaves behind heavier components, effectively increasing their mole fraction, the dissolved concentration for these compounds increases due to Raoult's law (Forth et al. 2017; Redman and Parkerton 2015). Raoult's law can estimate the dissolved concentrations of crude oil components for a saturated solution at equilibrium. The relative importance of each of these weathering processes is dependent on the crude oil type and on various factors at the location and time of a specific spill.

Evaporation is the process of loss of hydrocarbons to the atmosphere by volatilization. Following the mixing of oil in water, volatilization will begin immediately and remove many of the low- and medium molecular weight compounds within the first 24- to 48-hours following a spill (Gros et al. 2014). The highly volatile compounds include many of the aliphatic saturates of less than 15 carbons, soluble monoaromatic hydrocarbons (i.e., BTEX), and LMW PAHs. Condensates and ultralight oils may lose up to 75% of their mass within hours. In comparison, heavy crude oils may lose less than 10% of their mass by evaporation due to their much lower proportion of LMW compounds (NRC 2003). As the LMW components are lost by volatilization, the density of the residual oil increases as the HMW compounds become a larger proportion of the remaining oil. For example, ultralight condensates used as diluents to transport bitumen are quick to evaporate following a spill of dilbit (i.e., the name given to the diluent and bitumen) (Lee et al. 2015). Dilbit has a binary nature between fresh and weathered oil. As dilbits weather, the lighter fraction volatilizes, leaving behind a much heavier fraction that has the potential to sink in the water column (reviewed in Lee et al. 2015). For most conventional oils, the residual oils may not have a high enough density to sink, but they will become more viscous as the oil mass decreases. On one hand, large, alkylated PAHs will concentrate in the residual oil, where they may exert chronic effects on animals. On the other hand, highly volatile compounds such as the monoaromatics, which are very acutely toxic to aquatic animals, will be removed from the water column by evaporation.

Unlike evaporation, dissolution is the process by which petroleum compounds dissolve into the water column. In general, most hydrocarbons have low solubility in water. As such, the rate of dissolution for most hydrocarbons is considerably less than the rate of evaporation (Gros et al. 2014). The rate of dissolution may be limited even for compounds with high water solubility since the rate of evaporation would typically outpace the rate of dissolution when a molecule is both volatile and water soluble (NRC 2003). The aqueous solubility of oil compounds is governed by factors such as the molecular size and structure of the compounds (McCarty et al. 1992) but also by the mole fraction a specific hydrocarbon represents to the crude oil according to Raoult's law (Redman et al. 2012; Redman et al. 2017b). Within any class of petroleum compounds, water solubility decreases rapidly with increasing molecular size. Monoaromatics such as BTEX are very water soluble, whereas PAHs of greater than 16 carbons (i.e., 4 or more aromatic rings) have low water solubility. In comparison, PAHs with less than 16 carbons (i.e., 2–3 aromatic rings) have moderate water solubility. The solubilities of standard aromatics vary over 6-orders of magnitude (Lee et al. 2015). Benzene has a solubility of approximately 1,700 mg/L, whereas chrysene has a solubility limit of only 0.002 mg/L. The same relationship also applies to most other classes of hydrocarbons. The solubility of petroleum compounds is also strongly influenced by the molecular structure independent of molecular size. When comparing hydrocarbons with similar carbon number, the saturated alkanes are extremely insoluble in water compared with the aromatics. As an example, the ten-carbon PAH, naphthalene, is approximately 600-fold more water soluble than is the ten-carbon saturated alkane, n-decane (NRC 2003).

The mechanical mixing of oil and water will lead to an almost instantaneous dissolution of the lighter and more water-soluble components (Payne and Driskell 2018). Meanwhile, the less water-soluble components may become incorporated within oil droplets (Redman 2015). Modeling of the fate of marine oil spills suggests that large oil droplets (>1 mm) tend to ascend through the water column and collect on the water surface (French-McCay et al. 2021; Payne and Driskell 2018). Most of the compounds in oil droplets reaching the surface eventually evaporate or undergo photooxidation (French-McCay et al. 2021; Lee et al. 2015). In comparison, small droplets can become suspended stably within the water column and transported long distances from the site of a spill (French-McCay et al. 2021). This issue was pertinent during the DHOS due to the depth of the blowout.

The oxidation of crude oil either by photooxidation or microbial oxidation involves the chemical transformation of hydrocarbons into alcohols, ketones, and organic acids. The incorporation of oxygen molecules into the hydrocarbons makes the overall oil more polar in nature, thus increasing its water solubility (Aeppli 2022). In general, smaller hydrocarbons with less than 20 carbons are oxidized preferentially. Also, *n*-alkanes, followed by branched and cyclic alkanes, are oxidized before the less susceptible PAHs. Recent research suggests that photooxidation was the single most relevant process for oxidation of surface oil in the early phase of the DWH oil spill (Ward et al. 2018). This process transformed over 50% of the estimated mass of surface slicks into photoproducts within days to weeks (Aeppli et al. 2012; Lewan et al. 2014), making the fraction of oil photoproducts at least 50 times larger than the PAH fraction. These oil photoproducts were found to persist in the environment for several years (Aeppli et al. 2014; White et al. 2016), while over 99% of many of the PAHs were found to degrade within days to weeks (Brakstad et al. 2004; Brakstad et al. 2015; Campo et al. 2013; Howard et al. 2005; Mackay et al. 2006; Prince et al. 2008; Prince et al. 2013; Prince et al. 2007; Venosa and Holder 2007). An emerging view suggests that sunlight substantially alters the toxicity of oil in the early phase of a marine oil spill through the partial photooxidation of oil compounds (Nicodem et al. 2001).

## STANDARDS FOR THE TESTING OF CRUDE OIL TOXICITY

Oil toxicity tests are controlled experiments to assess the relationship between known concentrations of oil and the biological effects produced by these exposures (Hodson et al. 2019). Animals are exposed to a graded concentration series, and the biological responses to these oil concentrations are measured. These data are used to establish a concentration-effect relationship to calculate an estimate of the potency of the crude oil to produce a toxic response. The most common point estimate of potency is the median lethal concentration (i.e., LC50), which is the concentration of oil needed to kill 50% of a group of animals. However, median effective concentration (i.e., EC50) based on the concentration needed to produce an effect in 50% of animals is also routinely calculated.

The authors of a 2015 Report of the Royal Society of Canada (Lee et al. 2015) proposed guidelines to improve the quality of oil toxicity testing data for environmental risk assessments. First, toxicity tests must ensure that only the concentrations of oil vary between treatments and that these concentrations are proportional to the nominal loadings. All other variables other than oil concentration should remain constant between treatments. To calculate potency estimates such as LC50 or EC50 values, the oil concentrations used in testing should be within the range needed to produce differential responses in test subjects. Second, tests should include suitable chemical analyses of the total oil concentrations and the chemical compositions of the test treatments. Third, analyses should be performed with enough frequency to ensure the concentration and composition of oil within a treatment remain constant over time. Finally, the method used to mix oil and water should be adequate to ensure that oil constituents are freely dissolved within the test solutions (Hodson et al. 2019).

Testing the toxicity of crude oils on aquatic animals is notoriously difficult to perform. Crude oils are complex mixtures of thousands of compounds with unique physicochemical characteristics (Redman and Parkerton 2015). No two oils are alike, and each oil will respond uniquely to specific set of experimental conditions. Furthermore, the composition of oils will vary considerably over time due to weathering (reviewed in Lee et al. 2015). Unlike with hydrophilic chemicals, crude oil contains numerous constituents that are negligibly soluble in water. Not surprisingly, it is often difficult to interpret data derived from toxicity tests for oil unless specific guidelines are followed carefully (Adams et al. 2017; Hodson et al. 2019; OECD 2018; Redman and Parkerton 2015). Given the methodological differences between two studies, it is difficult to know for certain whether variations in the reported toxicity of crude oil and its constituents represent differences in the sensitivity of species or on how a toxicity test was completed. It is likely that the lack of standardization in oil toxicity testing contributes to much of the variation in test results. The uncertainty on the sources of variation and the substandard reporting of testing methods precludes the use of much of the toxicity data for environmental risk assessment (Hodson et al. 2020). It is therefore critical to review the role of different methodological procedures on the behavior of oil in toxicity testing. Box 15.1 highlights important principles of oil toxicity testing. Modified from Redman and Parkerton (2015).

### BOX 15.1 PRINCIPLES OF OIL TOXICITY TESTING

**PRINCIPLES**

- The concentration and composition of dissolved hydrocarbons generally determine the aquatic toxicity of physically and chemically dispersed oils.
- Dissolved hydrocarbon exposures depend on WAF preparation method.
- Dissolved hydrocarbon exposures depend on the amount of headspace in WAF test system.
- Dissolved hydrocarbon exposures depend on substance composition.
- Soluble hydrocarbons contribute more to toxicity under time-variable exposures.
- Passive samplers allow direct measurement of dissolved hydrocarbon exposures.

*Redman and Parkerton (2015).*

## DESCRIPTION OF TOXICITY TEST CONDITIONS

### Preparation of Oil-in-Water Solutions for Toxicity Testing

Researchers have attempted to understand the influence of testing methods on oil toxicity. Adams et al. (2017) completed a review on the toxicity of 185 oil types for 74 fish species published between 1973 and 2016. The authors described the large variation in testing methodology and in the terminology used to classify the toxicity testing solutions. In the 144 studies reviewed, over 54 distinct terms were given to classify the testing solutions used in experiments. These solutions were produced by approximately 223 distinct methods. Some of this diversity represented actual differences in methodology. However, some of the variation simply reflected the choice of nomenclature used by researchers. Despite the considerable variation between studies, most of these test solutions were based on derivations of one of four main types of oil-in-water solutions referred to as the water accommodated fraction (WAF), the chemically enhanced WAF (CEWAF), dispersed oil, and the WSF (Adams et al. 2017). The WAF is produced by the mechanical agitation of an oil-on-water mixture. After mixing, the WAF is characterized as a multicomponent mixture consisting of partially dissolved compounds and other components in a nondissolved phase such as a stable dispersion or emulsion (OECD 2019). In comparison, CEWAF is prepared much like WAF solutions except for the addition of a chemical dispersant prior to mixing. As a result of the dispersant, CEWAFs contain considerably more microdroplets than WAFs, although these droplets are typically smaller in diameter. Meanwhile, the dissolved compounds in CEWAFs are similar to those found in WAFs. Dispersed oil specifically refers to the particulate oil generated during WAF and CEWAF production, respectively. Finally, WSF represents oil compounds exclusively within the dissolved phase. Considering that even a small number of droplets are formed by the other methods, WSF require the removal of suspended undissolved or emulsified components from the WAF or CEWAF (OECD 2019). The most common separatory methods are filtration or centrifugation.

The oil toxicity literature highlights many differences in the production and classification of WAF, CEWAF, dispersed oil, and dissolved oil. Arguably, the largest source of differences is in the methods used to produce WAF. The most common method was promulgated by the Chemical Response to Oil Spills: Ecological Effects Research Forum (CROSERF) (Aurand and Coelho 2005). CROSERF consisted of members from government and industry who established several standard protocols for WAF and CEWAF production of crude oils. CROSERF recommended making WAF by mixing an oil-on-water mixture using a magnetic stirrer at a low enough mixing energy to prevent the formation of a vortex on the water surface. Under these conditions, few oil compounds distribute to the nonaqueous phase since mixing energy is insufficient to disrupt the surface oil. However, subsequent studies suggested using a mixing energy that produced a 20–25% vortex in total fluid column height within 1- to 20-liter carboys (Barron and Ka'aihue 2003). This newly adopted procedure, modified from the CROSERF method, is often referred to as low energy WAF (LEWAF). The basis for the increase in mixing energy was to better mimic the action of waves on slick oil. It also made the mixing procedure more consistent with the CEWAF production method developed by CROSERF.

Other WAF preparations termed medium-energy WAFs (MEWAFs) and high-energy WAFs (HEWAFs) are commonly seen in the literature (Allmon et al. 2021; Alloy et al. 2017; Duffy et al. 2016; Incardona et al. 2013; Morris et al. 2018; Reddam et al. 2017). These are all relative terms since there is no reliable method to expeditiously evaluate the mixing energy during each of these distinct WAF preparations. As a result, the descriptive terms used to classify WAFs are largely subjective. Some of these higher-energy WAFs were used routinely in the Natural Resource Damage Assessment (NRDA) research following the DHOS (Incardona et al. 2013). HEWAF involves mixing crude oil briefly with a commercial blender at very high speeds. Unlike with the original CROSERF method, the procedures using higher mixing energies produce WAFs with disproportionately more oil within the nondissolved phase. Furthermore, these nonsoluble components within the droplets make the composition more similar to the parent oil. A study suggested that LEWAF and HEWAF solutions differ considerably in droplet abundance but have similar dissolved concentrations of oil compounds (Morris et al. 2018). The authors found that both methods produced similar toxicological effects in developing fish embryos, supporting the assumption that dissolved components are most biologically active. However, as may be expected, the presence of the "less toxic" oil in the oil droplets obscures the analytical chemical analysis of the more toxic oil in the dissolved phase.

Another issue found commonly in the oil toxicity literature is the inconsistent or imprecise use of the term WSF. WSF is used despite no evidence for the absence of particulate oil in a solution. Although the original CROSERF method of WAF likely contains a small number of oil droplets, some researchers classify these testing solutions as WSFs. Other studies have often referred to the oiled effluent produced from the water passing across oiled gravel (Carls et al. 1999; Incardona et al. 2009; Marty et al. 1997) or through oiled ceramic beads (Kennedy and Farrell 2006) as WSF. However, these oiled effluents have rarely been tested for oil droplets. There are important implications of this imprecise use of the term WSF. As will be described later in more detail, oil droplets contain disproportionately more water-insoluble compounds than found in the dissolved phase. As a result, the analytical chemistry of testing solutions containing oil droplets will not reflect the concentration of compounds in the dissolved phase. Also, compounds within the oil phase have the potential to dissolute into the dissolved phase at different rates. This will affect the chemistry of the dissolved phase during the time course of an exposure in ways that are difficult to predict (Redman et al. 2012). Finally, oil droplets differ in their residence time within the WAF based on their diameter. Large droplets often dissipate quickly whereas small droplets may persist in the water column for a longer time (Forth et al. 2017). This in part explains why oil droplets in HEWAFs are lost quickly, whereas the small microdroplets in CEWAF persist for a long time (Sandoval et al. 2017).

### Exposure Regimes

Oil compounds tend to degrade over time making it difficult to maintain a constant concentration and composition in solutions during toxicity testing. The magnitude of this problem is affected appreciably by the nature of the oil and the manner used to create a dissolved oil mixture (Redman and Parkerton 2015). HEWAFs lose hydrocarbons quickly when large oil droplets suspended in solution coalesce and rise to the water surface (Forth et al. 2017). Meanwhile, water-soluble volatiles in light crude oils are lost to the atmosphere due to evaporation; whereas, water-insoluble components, like those found in heavier crude oils or weathered oil, tend to partition to surfaces or are bioaccumulated into testing subjects. These oils will experience losses in oil concentration over time but for different reasons. These concerns should be considered when adopting the most appropriate exposure regime for toxicity tests.

The three most common exposure regimes include static, static renewal, and continuous flow-through exposures (Hodson et al. 2019). Static exposures involve the one-time addition of oiled solutions to test chambers. Under static conditions, no exchange of the test solution is made, and there is no attempt to regulate the concentration of oil. Sandoval et al. (2017) found that oil concentrations decreased by approximately 90% from starting concentrations after only 24 hours. These losses were exacerbated in tests using HEWAF solutions. To counteract this loss of oil during static exposures, many studies will partially or fully replace testing solutions periodically. Much like with the static exposures, the oil concentration of the test solution in static replacement tests is highest at the start of the exposure and drops exponentially over time (Hodson et al. 2019). However, the replacement of test solutions, usually at 24- or 48-hour intervals, leads to a transient spike in oil concentration followed by a secondary exponential loss of oil (Hodson et al. 2019). Static exposures may be appropriate if the duration of a test is short or if the goal is to mimic the exposure profile typical of an open water spill. However, one of the inherent problems with the static and the static replacement exposure regimes is the difficulty in assessing the concentration of test solutions. As mentioned previously, toxicity testing requires establishing a concentration–response relationship. How this relationship is best established when concentrations are changing over time is unclear. Some studies use the starting concentrations or an average of oil concentrations at the start and at the end of a specified amount of time. Guidelines proposed by the Organisation for Economic Co-operation and Development (OECD) propose utilizing the integrated concentration calculated from the area under the concentration versus time relationship (OECD 2018, 2019). The one caveat of this approach is that establishing a time-weight average of concentration requires extensive chemical sampling. These static exposures are potentially suitable if the oil concentration of the mixture is expected to remain reasonably stable over time (i.e., CEWAF exposures) or if tests are short in duration (i.e., a few hours).

Other studies have developed flow-through exposure regimes to improve the consistency of oil chemistry in testing solutions. These methods are optimal for oil mixtures that have chemistries that tend to change considerably over time. One technique involves the use of substrate (i.e., gravel or ceramic bead) desorption columns (Carls et al. 1999; Incardona et al. 2005; Kennedy and Farrell 2006). Prior to tests, substrate is preloaded with oil and placed in glass columns fitted with intake and outflow ports. As water is passed through oil-loaded columns, it picks up the oil desorbed from the substrate. The effluent emerging from these columns is then delivered continuously to treatment vessels. Although the oil concentration of the effluent still decreases over time, the rate of loss is substantially lower than that produced by static exposures. More recently, a pressurized fluid injection (PFI) droplet generator system has been used to generate dispersions with oil droplets of predictable size distribution (Hansen et al. 2019a; Nordtug et al. 2011a; Olsvik et al. 2011). Oil and water are introduced into a series of chambers connected sequentially via narrow nozzles. As the mixture passes through these nozzles, small oil droplets of uniform diameter are formed. The oil-in-water mixture can be filtered to generate a WSF or left unfiltered to produce a mixture characteristic of a WAF. Effluent emerging from these chambers is introduced to test chambers either filtered or unfiltered. Using this system, Nordtug et al. (2011b) were able to perform side-by-side comparisons between WSFs at equilibrium and WAFs with oil droplets of varying size. The one benefit of this system is that the oil droplets generated by the system help to maintain a testing solution with very stable concentrations regardless of whether the droplets are later filtered.

Another method of generating WSF with reasonable consistency in oil concentration is known as passive dosing. Passive dosing involves the use of a polymer that is preloaded with crude oil and then allowed to equilibrate with the aqueous phase at target concentrations of compounds at or below saturation (Bera et al. 2018; Butler et al. 2013; Butler et al. 2016; OECD 2019; Redman et al. 2017a). A recent study found that the passive loading of silicone tubing or O-rings loaded to equilibration for 72 hours with

Macondo-252 oil could subsequently be used to produce WSFs with a constant concentration and chemical composition (Bera et al. 2018). Moreover, this method of passive dosing could deliver WSFs at multiple oil loadings that did not exceed the solubility limits of the individual or the total PAHs in the mixture. The method combines the simplicity of static exposures with the benefits of consistency in oil concentrations produced more reliably with flow-through exposure regimes.

### Preparation of WAF with Different Head Spaces

During the preparation of WAF stocks, there is evidence that the volume of the headspace used to prepare the WAF has a strong influence on the toxicity of the resulting oiled preparations. In the absence of a headspace, LMW volatiles such as BTEX and monoaromatic hydrocarbons dissolve in water due to their high water solubility and their inability to evaporate into the air phase (Esbaugh et al. 2016; Greer et al. 2019a). Under this condition, the high dissolved concentrations of these volatiles in water would be acutely toxic to animals. However, even a small headspace during WAF preparation would allow these compounds to evaporate, thus preventing these compounds from dissolving in the water phase to the same extent. Similarly, whether a toxicity test is performed in chambers that are open or closed to the atmosphere will have considerable effects on the acute toxicity of oil. In open systems, volatile hydrocarbons would be quickly lost to the atmosphere due to evaporation. This evaporation would cause large shifts in the chemistry of the test solution, making it difficult to interpret results. The one benefit of an open system is that it might be more predictive of the effects of spilt oil in the environment. However, the closed systems provide a better assessment of the toxicity of an oil under its native state making them more amenable to regulatory hazard assessments. Furthermore, the chemistry of the test solutions will be less likely to change during a study.

One concern evaluating the oil toxicity literature is the lack of sufficient documentation regarding the head space in the containers used for WAF preparation and toxicity testing. For instance, Hodson et al. (2019) documented that approximately 90% of studies lacked any information on head space. A recent review suggested that toxicity studies on oil need to report whether the containers used for WAF preparations and testing are closed or open to the atmosphere, and if closed, the volume of the head space (Redman and Parkerton 2015). This is particularly true when conducting tests on lighter oils with a large proportion of volatiles subject to evaporation.

### Influence of Different Oil-to-Water Ratios

Oil toxicity testing requires assessing the effects of oil across a concentration range. Two approaches are predominantly used to achieve a concentration series of distinct oil-to-water ratios (OWR). These methods include the variable loading method and the variable dilution method. In the variable loading method, distinct WAFs are created for each test concentration. An example of the variable loading method was proposed by CROSERF (Aurand and Coelho 2005). In these guidelines, WAFs were created by adding between 0.01 and 25 grams of oil per liter of water, and then creating multiples WAFs at different OWRs (Aurand and Coelho 2005). Following mixing and settling, the individual WAFs would be used directly as testing solutions. The variable dilution method is also commonly used to create a concentration series for testing. This approach involves creating a single WAF stock at a high OWR. Following mixing and settling, the single WAF solution would be diluted serially with varying amounts of water to create the serial dilutions of the toxicity test solutions (Barron and Ka'aihue 2003). In a review of different methods used to generate aqueous mixtures for oil toxicity tests, Girling (1989) noted that OWRs varied from 1:18 to 1:1000 with mixing times from 4 to 240 hours. Adams et al. (2017) found an even larger disparity in OWRs, ranging from 1:106,667 to 1:1. However, the most common OWR reported was 1:9 based on Anderson et al. (1974). The choice between these two methods can have important repercussions on the toxicity of most oils, particularly at high mixing energies.

The OWR used in WAF preparations and in the generation of serially diluted test solutions can have a large effect on the composition and concentration of oil compounds. Although each procedure produces a series of decreasing concentrations of solutions, these methods differ in their compositional makeup. In a recent review of acute toxicity methods, it was noted that WAFs prepared with high OWRs (i.e., high relative amounts of oil to water) of unweathered oil produced water-soluble fractions dominated by BTEX (~80% of the total) (Adams et al. 2017). In comparison, an OWR with 20,000-times less OWR had only 15–50% BTEX. In the former case, the acute toxicity would be higher, but toxicity of the test solution would decrease quickly as BTEX was lost to the atmosphere (assuming the test is performed under open conditions). As a result, Barron and Ka'aihue (2003) suggested using the variable dilution approach rather than a variable oil loading approach as suggested in the CROSERF guidelines.

As oil weathers and loses its volatile components, the OWR continues to influence the size of oil droplets in a nonlinear fashion, thus complicating the interpretation of toxicity results. Sandoval et al. (2017) investigated the mean diameter of oil droplets in HEWAFs prepared from mixtures of weathered oil with different OWRs. They found that at low OWR ranging from 20 ppm to 100 ppm, there was two-fold decrease in the mean oil droplet diameter (i.e., 2.53–1.07 μm) in the HEWAF solutions. However, at OWRs above 100 ppm to as high as 1,000 ppm, there was no further change in oil droplet diameter. Given the importance of oil droplet diameter on the stability of HEWAFs, care should be taken when deciding on an appropriate OWR.

The authors also commented that at higher OWRs, more of the oil is likely to adsorb to the container walls, reducing its ability to contribute to the dispersion. Reddam et al. (2017) and Redman and Parkerton (2015) also described a nonlinear relationship at high OWR due to the multicomponent dissolution behavior of the constituents from oil droplets. In their analysis, they found that constituents dissoluted from the oil droplets differentially based on the OWR leading to distinct toxic potentials for each WAF.

### Influence of Oil Droplets on Toxicity

It is well known that the dissolved fraction in WAFs is the most toxic to fish. Although the prevailing view is that oil droplets exert little direct toxicity (Carls et al. 2008; Di Toro et al. 2007; Hansen et al. 2019a; Redman et al. 2017a), some studies suggest droplets may serve as a reservoir of low-solubility hydrocarbons (Payne and Driskell 2018), thus helping to buffer soluble PAH concentrations in the dissolved phase (Landry et al. 2019). Under such a mechanism, PAHs could diffuse from the droplet phase to the dissolved phase as oil weathers. Also, it is possible that PAHs within oil droplets could bind to biota and accumulate following contact without directly dissolving in the testing solution (Hansen et al. 2018; Petersen and Kristensen 1998; Redman et al. 2012; Sorensen et al. 2017). As such, it is important to standardize conditions that are known to produce oil droplets. Factors such as the oil-loading rate, the mixing energy and procedure used during WAF preparation, and the materials that come in contact with the WAF during mixing may affect the oil droplet formation (Adams et al. 2017). Furthermore, the size and density of oil droplets in WAFs are dependent on the density, viscosity, and composition of the source oil.

## OCTANOL-WATER PARTITION COEFFICIENT OF CRUDE OIL CONSTITUENTS

The persistence of a chemical in the environment and how it partitions between the various compartments in the water column (e.g., dissolved phase, droplet phase, suspended particles, sediment, biota) are influenced by its structure and physical and chemical properties. One parameter used to characterize the portioning of a chemical between the various compartments is its lipophilicity, which represents its tendency to partition to lipids. Lipophilic compounds bind to particles and sediments or accumulate in biota. The opposite is the case for hydrophilic compounds with a greater tendency to sequester in the aqueous phase. A common method used to quantify a chemical's relative lipophilicity is the octanol-water partition coefficient ($K_{ow}$). This coefficient, which is normally expressed as a log $K_{ow}$, is calculated from the ratio of the chemical concentrations partitioning to the *n*-octanol phase relative to the water phase at equilibrium. The lower the value of the log $K_{ow}$ the greater is its polar characteristics. The increased polarity of a chemical would reduce its portioning to lipophilic compartments. Conversely, chemicals with high log $K_{ow}$ values can partition to lipids but at the expense of limited solubility in water (McCarty et al. 1992). There is evidence that the relationship between $K_{ow}$ and accumulation breaks apart for compounds with $K_{ow}>6$. Compounds with this magnitude of lipophilicity dissolve negligibly in water and require a long time to reach equilibrium between the lipid and water phases (McCarty and Mackay 1993). Furthermore, these molecules can often be large, creating a scenario where their ability to pass through biological membranes is impeded compared to the much higher permeability of small molecules. Also, the low concentration of highly hydrophobic chemicals in water attenuates their accumulation potential despite their propensity to partition into cell membranes.

The study of lipid-water partition coefficients (such as the log $K_{ow}$) began as far back as the pioneering work of Meyers (1906) and Overton (1901) on the nonspecific actions of general anesthetics. They independently proposed that lipophilic anesthetics, which they measured using the olive oil-water coefficient (i.e., olive oil was used previously as a lipid solvent rather than *n*-octanol), could intercalate into the lipids of cell membranes, disrupt cellular functions, and produce narcotic effects. They found a strong positive correlation between narcosis and the lipophilicity of anesthetics. These findings drove interest on assessing the effects of other lipophilic compounds on biological membrane functions. This narcosis model has been used extensively since the early 1980s to describe the action of lipophilic compounds, such as hydrocarbons, on fish (Mackay 1982; McGrath et al. 2021; McGrath et al. 2005). Early reports proposed that hydrocarbons exert their acute toxicity through a narcotic mechanism similar to that proposed by Meyers and Overton (Di Toro et al. 2000; Mackay 1982). However, the exact mechanism of narcotic action is not known (McGrath et al. 2021). The criterion for this model is that hydrocarbons, due to their lipophilicity, produce nonselective toxicity due to general disruption of cellular functions stemming from interaction with lipid membranes.

The narcosis model of action is the basis for several computational models that predict the acute toxicity of hydrocarbons based on their lipophilicity. In the report by Lee et al. (2015), multiple assumptions associated with the implementation of predictive models integrating the concepts of lipophilicity ($K_{ow}$) and narcosis were summarized as follows:

a. Compounds must exert their effects by narcosis.
b. Compounds within a mixture must exert additive effects; there should be no synergistic or antagonistic modes of interaction between compounds.
c. Compounds should be in the dissolved phase, or the dissolved concentration must be calculatable, and any compound in nonaqueous phases must not elicit effects.

d. Compounds must not interact with other compounds in water, natural, or anthropogenic.
e. Compound concentrations in water must be constant with time.
f. Compound concentrations in tissue lipids must be at equilibrium with water concentrations.
g. Compound concentrations must be well characterized and remain constant with time (i.e., must not be affected by metabolism or by changing rates of accumulation or depuration).
h. Metabolites produced from parent compounds must have the same toxicity as their parent compounds.

In addition, application of these models should consider the following aspects of hydrocarbon chemistry, such that it

i. predicts the toxicity of hydrocarbons across a broad range of log $K_{ow}$, including those with log $K_{ow}$ >6;
j. operates for both unsubstituted PAHs and alkyl PAHs;

## NARCOSIS AND THE MODELS BASED ON THE TARGET LIPID MODEL

One of these models is the target lipid model (TLM). The TLM is a quantitative structure-activity model originally developed to predict the acute toxicity of Type 1 narcotic chemicals such as aliphatics, and monocyclic and polycyclic aromatic hydrocarbons (Di Toro et al. 2000). As such, the TLM has been applied to predict the toxicity of complex mixtures such as crude oils. The original model for the TLM was developed from an acute lethality data set for 33 animal species, including fish, and the physicochemical data, including log $K_{ow}$ values, for 156 chemicals (Di Toro et al. 2000). The TLM database now contains lethality data from 79 species from eight taxonomic groups and chemical data for 228 chemicals (McGrath et al. 2018). At the core of the TLM is the assumption that lipophilic chemicals exert their toxic effects nonspecifically by narcosis (McCarty and Mackay 1993). Acute lethality is predicted to occur when the total accumulation of chemicals to the target lipid reaches a critical body residue of 5 mM (McCarty et al. 1992; McCarty and Mackay 1993). In essence, the TLM relies on the empirical relationships between log $K_{ow}$ and accumulation, and the relationship between chemical accumulation and mortality. The proponents of the TLM state that it is possible to predict acute toxicity of a mixture of dissolved hydrocarbons using physical chemistry data for the components within the mixture. Narcosis, which is the underlying foundation of the TLM, has been used by many to predict the acute toxicity of low molecular weight components such as monoaromatics (BTEX), diaromatics (naphthalene and alkyl-naphathalenes), and short chain alkanes (<C12) (Lee et al. 2015; McGrath et al. 2005; McGrath et al. 2004).

PetroTox, a tool developed by Conservation of Clean Air and Water in Europe (CONCAWE), has also been used to predict the acute and chronic toxicity of complex mixtures of petroleum chemicals. One of the features of this model is that it calculates the mass balance of individual components distributing to the air, water (i.e., dissolved), and oil (i.e., nonaqueous phase liquid) phases using Henry's law to calculate the interactions with air and Raoult's law to calculate components dissolved in water (Forth et al. 2017; Redman et al. 2012). The output is a set of dissolved concentrations for oil components, which are grouped together in blocks of compounds with similar physicochemical properties. The cumulative concentrations of chemicals in a block are measured, and the ratio of the dissolved concentrations to the $LC_{50}$ value for each of the hydrocarbon blocks is calculated. This ratio is termed a toxic unit (TU). Based on the assumption of additivity in the mode of action for each of the compounds, the TU for each of the hydrocarbon blocks is summed to provide a cumulative TU for the entire crude oil. A TU of 1 represents a dissolved exposure that would lead to a sufficient accumulation in the target lipid to elicit a 50% response (i.e., mortality, $LC_{50}$). In other words, this is the dissolved concentration needed to achieve the median lethal loading to the target lipid (Redman et al. 2017b). A TU>1 occurs with a dissolved exposure greater than needed to produce the $LC_{50}$, and a TU<1 suggests that the dissolved exposure is less than required to reach the $LC_{50}$. Redman et al. (2017b) was able to use this model to predict acute toxicity of a mixture of petrochemicals within a factor of 3. Redman and Parkerton (2015) also used the output of this model to assess the hydrocarbon blocks contributing most to the toxicity of a complex mixture such as crude oil. Despite the use of the TLM as a tool in environmental risk assessment, several authors have argued against a nonselective mode of action for petroleum hydrocarbons (Grosell and Pasparakis 2021; Incardona et al. 2006). The implications of these studies supporting specific modes of toxic action refuting the narcotic mode of baseline toxicity will be discussed below.

## ARYL HYDROCARBON RECEPTOR PATHWAY

The aryl hydrocarbon receptor (AHR) pathway consists of a suite of genes, gene transcripts, and enzyme products involved in the metabolism and excretion of organics such as planar aromatic compounds regulated by the transcription factor, AHR (Marris et al. 2020). The AHR is localized initially in the cytosol where it is protected against degradation by heat-shock protein 90 (hsp90) and AHR-interacting protein (AIP). When ligands bind to the AHR the complex translocates to the nucleus, where it then dissociates from hsp90 and AIP. The newly formed AHR-ligand complex heterodimerizes with the aryl hydrocarbon nuclear translocator (ARNT) (Billiard et al. 2006; Cherr et al. 2017). This complex binds to xenobiotic response elements (XREs), also known as dioxin response elements or AHR response elements. Binding of the AHR complex to the XREs stimulates the widespread transcription of several genes associated with phase I and phase II metabolism. Phase 1 enzymes mediate oxidative, reductive, or hydrolytic reactions that convert highly lipophilic molecules to more water-soluble products through the formation of polar functional groups on hydrophobic compounds. Phase 2 enzymes are involved in conjugative reactions that further increase the water solubility of compounds by the addition of polar moieties using glucuronidation, acetylation, and sulfation reactions. Collectively, these metabolic reactions serve to increase the water solubility and depuration of hydrophobic compounds (Denison and Nagy 2003; Hahn 1998).

Several components of the AHR pathway have been commonly used as biomarkers of exposure to crude oil (Bak et al. 2019; Dubansky et al. 2017; Ramon Leggieri et al. 2019). A common phase I product is the cytochrome P4501A (CYP1A), involved in a hydroxylation reaction associated with PAH metabolism (Denison and Nagy 2003; Hahn 1998). AHR activation upregulates CYP1A and many other genes. Microarrays and RNA sequencing have gone on to corroborate the link between crude oil exposure and AHR activation in "model" species such as zebrafish (Bui et al. 2012; Li et al. 2021) and in "nonmodel" fish species such as Gulf killifish (Dubansky et al. 2013; Pilcher et al. 2014; Whitehead et al. 2012a; Whitehead et al. 2012b), mahi mahi (Greer et al. 2019a; Greer et al. 2019b; Xu et al. 2017b; Xu et al. 2016; Xu et al. 2017c; Xu et al. 2018), red drum (Xu et al. 2019; Xu et al. 2017a), and flounder species (Alejandro Zamora-Briseno et al. 2021; Jung et al. 2017; Rodgers et al. 2021; Zhu et al. 2016). Collectively, they provide a comprehensive picture of the transcriptomic responses to crude oil. For example, fish collected from a field-site oiled by the DHOS showed the differential regulation of a battery of genes, many of which are known to be regulated upon AHR activation. These gene-expression patterns were consistent with an AHR-mediated activation of phase I and phase II metabolic pathways but a pronounced downregulation of the estrogen-regulated genes, choriogenin, vitellogenin, and zona pellucida. The nature of the gene expression changes suggested that oiled fish may have been less reproductively fit, producing embryos with reduced provisioning of yolk reserves and disrupted integrity. Several related studies found CYP1A induction in the tissues of Gulf killifish exposed to weathered Macondo-252 oil in the laboratory (Dubansky et al. 2013; Pilcher et al. 2014) or in fish collected from oil-contaminated sites in the northern Gulf of Mexico (Dubansky et al. 2017; Whitehead et al. 2012a). CYP1a activity can also be assessed indirectly by measuring ethoxyresorufin-*O*-deethylase (EROD) activity (Bak et al. 2019; Collier et al. 1992) or by analyzing the concentration of metabolites in tissues and bile (Dearnley et al. 2020).

Although most PAHs are metabolized by metabolic enzymes activated by AHR induction, the link between AHR activation and PAH metabolism is not always straightforward. Organic constituents in petroleum serve as ligands to the AHR (Aranguren-Abadia et al. 2020; Cunha et al. 2020); however, PAHs and other organic chemicals bind to AHR with different ligand binding affinities (Zhang et al. 2022). Molecules also vary in the efficacy by which they are metabolized by phase I and phase II enzymes. For instance, low molecular weight compounds of crude oil may activate the AHR pathway with relatively low ligand-binding affinity but are differentially metabolized (reviewed by Incardona 2017). Conversely, other molecules such as dioxins and dioxin-like polychlorinated biphenyls are strong agonists for AHR but are themselves poorly metabolized (Cherr et al. 2017). In addition, the metabolic products of some PAHs are more toxic than the parent molecules creating a situation where AHR activation stimulates the toxicity of some but not all PAHs. A study in pink salmon embryos exposed to low ppb levels of crude oil suggested that CYP1A induction is related to a variety of lethal and sublethal effects at the cellular, organism, and population levels, and can be used to predict these responses (Carls et al. 2005). In the latter case, this creates a situation where PAHs may exert their toxic effects on fish either in an AHR-dependent or an AHR-independent mode of action. For instance, there appear to be at least two distinct pathways mediating cardiotoxicity (Incardona 2017); one involving AHR-dependent activation by four-ringed PAHs (Incardona 2017; Incardona et al. 2006) and the other an AHR-independent pathway primarily activated by three-ringed PAHs (Incardona et al. 2009; Incardona et al. 2005; Incardona et al. 2004). Other compounds such as 5-ringed PAHs can produce cardiotoxic effects either in an AHR-dependent or AHR-independent mode of action (Andreasen et al. 2002; Aranguren-Abadia et al. 2020; Cunha et al. 2020). Finally, AHR has undergone multiple gene duplications in teleost fish. Depending on species, fish have either three or four paralogs of AHR (AHR1a, AHR1b, AHR2a, AHR2b) (Cherr et al. 2017). There are species differences in the AHR paralog mediating hydrocarbon toxicity. Recent morpholino knockdowns of the various AHR paralogs suggest that activation of AHR or some of the downstream genes can stimulate, reduce, or have no effect on PAH toxicity (Cherr et al. 2017). For instance, knockdown of AHR1 and AHR was shown to enhance the toxicity of the four-ringed PAH, pyrene (Incardona et al. 2005; Incardona et al. 2006), but had no effect on phenanthrene or dibenzothiophene on toxicity (Incardona et al. 2005). Given the complexity of the AHR pathway, the use of genome engineering technologies (i.e., morpholinos or CRISPR–Cas9) coupled with genome-wide sequencing technologies (Shankar et al. 2019) should prove valuable.

## CURRENT UNDERSTANDING OF CRUDE OIL TOXICITY IN FISH EMBRYOS

### Cardiovascular Effects

For over 40 years, researchers have documented a suite of phenotypes associated with early-life toxicity to crude oil including early mortality, edema, hemorrhaging, disruption of cardiac form and function, and craniofacial and spinal cord deformities (Carls et al. 2002; Carls et al. 1999; Incardona and Scholz 2016). There has been particular interest in studying the heart and cardiovascular-related disturbances due to the sensitivity of fish embryo heart development to oil (Carls et al. 1999; Heintz et al. 1999). In one of the earliest studies describing heart effects, Baltic herring embryos exposed to Venezuelan light crude oil were shown to have slow heart beats relative to controls and to accumulate fluids in the area anterior to the yolk sac (i.e., yolk sac edema) (Lindén 1976, 1978). Exposure of pink salmon (Heintz et al. 1999) and pacific herring (Carls et al. 1999) to weathered Alaska North Slope oil (ANSO) at concentrations analogous to those found in Prince William Sound after the EVOS produced a large suite of cardiovascular effects leading to a high incidence of pericardial edema. Edema was the developmental abnormality linked most often with larval mortality in Pacific herring exposed to ANSO. Others have suggested that pericardial edema is the most robust phenotypic indicator of cardiac injury in mahi mahi (Edmunds et al. 2015; Esbaugh et al. 2016; Mager et al. 2014; Xu et al. 2016). The edema produced by crude oil is similar to the generalized response produced in developing fish to dioxins and PCBs (Andreasen et al. 2002). Attempts have been made to characterize the oil-induced alterations to heart structure in greater detail. Middaugh et al. (1996) found a suite of cardiovascular phenotypes in Inland silverside embryos exposed to ANSO, ranging in severity from edematous heart (i.e., tube hearts) to beating tissues with no heart structure or no discernable heart. These phenotypes are still measured routinely in studies; however, there is increasing interest in characterizing the mechanistic underpinnings of these cardiovascular effects. Also, research has tried to associate toxicological effects with distinct components of crude oil.

Three methodological approaches have been used extensively to characterize the mechanistic basis of oil-induced cardiotoxicity in fish. Morpholino oligomers have been used in zebrafish embryos to genetically knockdown specific AHR paralogs and other gene products to assess their role in cardiotoxicity (Brown et al. 2015; Clark et al. 2010; Scott et al. 2011). Furthermore, isolated cardiomyocyte preparations derived from atrial (Abramochkin et al. 2021) and ventricular (Ainerua et al. 2020; Brette et al. 2014; Heuer et al. 2019) cardiomyocytes have been developed for multiple fish species including mahi mahi (Greer et al. 2021; Heuer et al. 2019), zebrafish (Kompella et al. 2021), and rainbow trout (Vehniainen et al. 2019). These cardiomyocytes have been amenable for use with the whole cell patch clamp technique to study the biophysical properties of surface membrane ion channels of the heart (Kompella et al. 2021; Vehniainen et al. 2019). Finally, the advent of molecular approaches such as microarrays and RNA-sequencing has been used to provide unbiased assessment of genome-wide transcriptomic responses in fish including those associated with cardiovascular effects (Allmon et al. 2021; Greer et al. 2019a; Incardona 2017; Li et al. 2021; Sorhus et al. 2021a; Sorhus et al. 2021b; Xu et al. 2019).

Crude oil exposure can slow the rate of heart contraction, produce arrhythmias involving the uncoordinated contraction of the atrium and ventricle of the heart, and reduce the force of ventricle contraction in fish embryos. Zebrafish exposed to isolated PAH solutions were susceptible to tricyclic PAHs, particularly, the fluorene, dibenzothiophene, and phenanthrene (Incardona et al. 2004). The authors noted that the effects elicited by these high concentrations of tricyclic compounds recapitulated the cardiovascular defects produced by exposure to ANSO (Incardona et al. 2009; Incardona et al. 2005; Incardona et al. 2006). These results support earlier studies suggesting a putative relationship between specific PAHs in weathered crude oil and cardiovascular deformities. Tricyclic PAHs appear to reduce heart rate by inhibiting the outward movement of potassium ($K^+$ ions) through delayed rectifier $K^+$ channels (i.e., $I_{Kr}$) in both chambers of the heart (Brette et al. 2014; Brette et al. 2017). Reducing the potassium conductance through these ion channels would decrease the rate of repolarization of contractile cardiomyocytes. Furthermore, blockade of this channel in cells of the atrioventricular canal would partially or completely impede the conductance of depolarizing current from the atrium to the ventricles of the heart leading to cardiac arrythmias. Complete atrioventricular nodal blockade would electrically uncouple the atrium from the ventricle leading to a complete or partial stoppage of ventricle excitation-contraction coupling (Incardona et al. 2009).

Specific tricyclics, such as fluorene, dibenzothiophene, and phenanthrene, were shown to reduce calcium ($Ca^{2+}$ ion) permeability through the L-type $Ca^{2+}$ channel (LTCC) on the cell membrane of the transverse tubules. There is indirect evidence that tricyclic compounds may interfere with the $Ca^{2+}$-mediated opening of the ryanodine receptor (i.e., RyR; a $Ca^{2+}$-gated $Ca^{2+}$ channel on the intracellular, sarcoplasmic reticulum (Incardona 2017; Marris et al. 2020). Both processes will lead to a reduction in cytosolic $Ca^{2+}$ during the plateau phase of cardiac action potentials. The net effect of this response would be a decrease in the strength of cardiomyocyte contraction. Furthermore, there is evidence that tricyclics interfere with the activity of $Ca^{2+}$ ATPase on the sarcoplasmic reticulum (SERCA) restoring cytosolic $Ca^{2+}$ to resting levels at the end of the excitation–contraction coupling (Zhang et al. 2013). These effects would further decrease heart rate by increasing the time it takes for contractile cells of the heart to relax following an action potential.

Genetic knockdown experiments using morpholinos targeting the AHR found that these tricyclic effects on $K^+$ and $Ca^{2+}$ occurred in the absence of AHR activation (Incardona et al. 2005; Jung et al. 2013). Specifically, morpholino knockdown of the AHR2 paralogs failed to show any change in the magnitude of tricyclic-mediated cardiotoxicity (Brown et al. 2015; Incardona et al. 2005). These mechanisms are termed as AHR-independent. However, unlike the actions of tricyclics, larger PAHs were shown to exert their cardiotoxic effects in an AHR-dependent manner. These tetracyclic PAHs appeared to elicit similar cardiac responses as those produced by dioxins and PCBs. Briefly, the tricyclics acted as strong AHR agonists leading to a decrease in cell proliferation and an impairment of cardiac looping during the latter stages of embryonic development (Edmunds et al. 2015; Jung et al. 2013). Also, toward the end of heart development, activation of the AHR-dependent pathway was also associated with impaired cytosolic $Ca^{2+}$ homeostasis and cardiac contractility as produced by the tricyclic action via an AHR-independent pathway (Incardona 2017).

Several studies found that five-ringed PAHs produced developmental abnormalities similar to those produced by exposure to potent AHR ligands such as dioxin (Incardona 2017). Exposure to HMW PAHs can elicit gross morphological abnormalities in early life-stage fish by disrupting AHR-mediated migration of neural crest cells in mummichog (Bozinovic et al. 2013), rockfish (He et al. 2011), and zebrafish (de Soysa et al. 2012), and through disruptions in skeletal ossification in medaka (Seemann et al. 2015) and zebrafish (Xiong et al. 2008). Not surprisingly, these PAH-induced dysmorphogenetic effects produce morphological abnormalities such as jaw malformations and spinal deformities (i.e., curvature) in Atlantic cod (Hansen et al. 2019b).

## Other Developmental Effects

Crude oil has long been shown to impair craniofacial development, leading to alterations in embryonic development of the ocular region of the cranium (Carls et al. 1999; Hose et al. 1996; Incardona et al. 2004; Middaugh and Whiting 1995). Previous work has also shown that the five-ringed PAH, benzo[*a*]pyrene, produced ocular defects in rainbow trout alevin (Hose et al. 1984). Other studies also suggest that phenanthrene (Huang et al. 2013; Sorhus et al. 2021a; Sorhus et al. 2021b) and oxygenated chrysene (Diamante et al. 2017) can inhibit eye development. Moreover, recent studies have begun specifically investigating the biological effects of embryonic and larval crude oil exposure on eye structure and function, and the underlying molecular and biochemical determinants of eye development (Colavecchia et al. 2007; Lie et al. 2019; Magnuson et al. 2020; Magnuson et al. 2018).

The role of AHR activation in oil-induced visual impairment has also been investigated in earlier studies. Fathead minnows and white sucker exposed as embryos to Athabasca oil sands exhibited a suite of structural, degenerative, inflammatory, and cytoplasmic alterations to the eye (Colavecchia et al. 2007). The most pronounced effects included structural defects in eye development such as poor retinal differentiation, microphthalmia, optic fissures, and distorted lenses. In both species, the severity of eye pathology correlated with CYP1A expression suggesting that activation of the AHR pathway may mediate this response in both species. Furthermore, the severity of eye pathology correlated with CYP1A expression in the eye in a concentration-dependent manner with TPAH concentration in the soils. Worth noting is the recent work suggesting that AHR1A expression may be critical for eye development in developing Atlantic cod embryos and larvae (Aranguren-Abadia et al. 2020).

Several studies suggest that cardiovascular development in embryos is the primary target of oil exposure, and that other deformities are secondary due to decreased blood flow owing to impaired heart function (Incardona et al. 2004). Sorhus et al. (2021b) found that Atlantic haddock embryos exposed transiently to weathered crude oil had impaired small and deformed eyes with low concentrations of total fatty acid, cholesterol, and attenuated polyunsaturated fatty levels (Sorhus et al. 2021b). Although cardiac function in oil-exposed embryos was not measured, the study proposed that impaired cardiovascular function, as seen during a concomitant phenanthrene exposure, could have led to some of the ocular dysfunctions. The authors suggested that attenuated cardiac function could reduce the transport of yolk lipids to the developing eye, leading to the changes in fatty acid metabolism associated with oil exposure. However, Grosell and Pasparakis (2021) noted that in mahi mahi, ocular effects occurred at PAH concentrations below those reported to have elicited cardiovascular effects. They used this as evidence that ocular effects in this species are unlikely secondary to cardiovascular impairment. The eyes as a primary target of crude oil is also supported by Li et al. (2021) who suggest that the visual system may be even a more sensitive target to oil than it is to the cardiovascular system.

New molecular approaches such as RNA-sequencing studies have been used to track genome-wide transcriptomic responses in fish exposed to crude oil. Xu et al. (2016) were first to use this technique to assess the transcriptomic responses in larval mahi mahi to weathered slick oil from the DHOS. At 96 hours postfertilization, larvae exposed to slick oil had multiple differentially expressed genes associated with altered visual perception, phototransduction, and detection and response to light stimuli. Bioinformatic analysis suggested that the top diseases characteristic of these transcriptomic responses was associated with retinitis pigmentosa, leber congenital amaurosis, and cataract formation. Similar results were seen in red drum embryos exposed to DWH oil. Another study in zebrafish embryos suggests that exposure to environmental concentrations of phenanthrene alone is sufficient to induce visual system developmental defects and dysfunction similar to those described in mahi mahi (Huang et al. 2013). Interestingly, phenanthrene is the primary chemical constituent in the weathered DWH oil used by Xu et al. (2016). Huang et al. (2013) proposes that AHR activation by phenanthrene specifically downregulates the zinc finger E-box homeobox 1 (ZEB1), leading to the downstream upregulation of microphthalmia-associated transcription factor (Mtif). Upregulation of Mtif has been implicated in a decrease in cell proliferation and an increase in apoptosis within the retina. Magnuson et al. (2020) has gone on to show that apoptosis in the zebrafish eye in embryos due to DWH oil exposure leads to retinal degeneration sufficient to impair optomotor behavioral responses in live fish.

Regardless of the mechanism of action, impaired eye development is likely impaired visual acuity and impact animal fitness. Various assays to assess visual function have been developed including the optomotor and optokinetic responses. Crude oil was shown to impair the optokinetic behavioral response in zebrafish larvae following embryonic exposure to weathered oil at 89.8 μg/L PAH (Magnuson et al. 2020).

### Fitness Level Effects

The effects of embryonic oil exposure on fitness-level effects later in life have been assessed in multiple species of fish. This work stems from earlier reports that exposure of pink salmon and Pacific herring embryos to low level concentrations of crude oil produced embryonic heart defects that persisted until later in life (Incardona et al. 2015). It has also been shown that even when embryos were exposed to an oil concentration insufficient to cause observable developmental abnormalities, cardiac performance could be impacted in exposed fish later in life (Hicken et al. 2011).

Fitness in adults was accessed using aerobic ($U_{crit}$) and anaerobic ($U_{burst}$) swim trials, or analyses of oxygen consumption under standard and maximum metabolic conditions to assess aerobic scope. Several studies have investigated that the effects of acute oiling on swim performance in adult mahi mahi (Mager et al. 2014), red drum (Johansen and Esbaugh 2017), and cobia (Nelson et al. 2017) decreased fitness in adult fish. Studies found that embryonic crude oil exposure produced effects that persisted to late in life. Specifically, adult fish exposed to oil as embryos had reduced aerobic scope due to a reduction in the maximum metabolic rate. However, there was no change in standard metabolic rate in fish. This impaired aerobic scope was used to explain the reduced aerobic swim capacity in oiled fish. Similarly, adult mahi mahi exposed to a low concentration of 50 selected PAHs had impaired aerobic scope due to a reduced metabolic activity (Stieglitz et al. 2016). Although most studies focused on the effects of embryonic exposures on adult fitness, a few studies also assessed the impacts of oil exposure in adults on swim performance concomitant with hypoxia exposure. Ackerly and Esbaugh (2020) found that a combined exposure to hypoxia and crude oil led to an additive reduction in aerobic scope in adult red drum.

## CONCLUDING REMARKS

- Crude oil is made up of a highly complex mixture of chemicals with unique physical and chemical properties. Differences in the water solubility of these petroleum compounds influence the behavior of the mixture. In particular, the low solubility of many of the hydrocarbons affects the composition of the oil-in-water mixtures used in toxicity testing.
- There is considerable diversity in the published literature on the methods used to test the toxicity of crude oil. The inconsistent use of methods and the imprecise terminology to describe certain procedures complicate the interpretation of data and reduce the likelihood they can be used in the environmental risk assessment of crude oils.
- Many of the procedures used to produce testing solutions promote the formation of oil droplets. However, oil droplets likely contribute little to the toxicity of the crude oil mixture but obscure the chemical analyses. Furthermore, oil droplets in water-accommodated fractions are likely to increase the variability of testing solutions over time during toxicity testing.
- Special care needs to be taken when conducting toxicity tests for crude oil. The methods used to create oil-in-water mixtures should be well documented, and the specific procedures should consider the characteristics of the oil and the goals of the study.
- Narcosis is the underlying framework of the target lipid model used in the environmental risk assessments of oil. The target lipid model assumes that hydrocarbons are toxic due to a nonspecific accumulation on target lipids. Unfortunately, the target lipid is not congruent with much of the research suggesting certain PAHs exert toxicity through their action on specific targets.
- Many of the fish studies since the aftermath of the 1989 Exxon Valdez oil spill suggest oil produces sublethal effects in fish embryos at total dissolved PAH concentrations at the high ng/L to low μg/L range. Despite the heterogenous nature of oil, most studies only implicate a small number of 3- to 5-ringed PAHs in the suite of sublethal effects targeting early development in fish embryos. Most studies suggest that the embryo heart is the most sensitive organ and that tricyclic and tetracyclic PAHs explain much of the sublethal toxicity seen in embryos exposed to crude oil.
- Several studies have shown that tricyclic compounds act predominantly by a mechanism independent of aryl hydrocarbon receptor activation. The most proximate effects of tricyclic PAHs appear to be impairment of ion channels on the heart regulating the rate and strength of cardiomyocyte contraction. Several studies have proposed adverse outcome pathways promoting impairment of potassium and calcium ion channel conductivities as the molecular initiating events causally linked to the suite of developmental and functional abnormalities seen in the heart. Meanwhile, tetracyclic PAHs act in an AHR-dependent manner to produce a distinct set of developmental abnormalities to the heart. However, there is no direct evidence that any of

these PAHs directly interact with these ion channels. Additional research is needed to create causal links to the suite of cardiovascular effects impacted by oil.

- Despite the large suite of potentially toxic chemicals within oil, only a small subset of PAHs and their alkylated products is usually reported during routine chemical analyses and considered as potentially toxic to aquatic animals in laboratory- and field-based toxicity testing. Curiously, much of the research assessing the "specific" action of tricyclic and tetracyclic PAHs was performed at concentrations much higher than those reported in the dissolved phase of crude oil. The concentrations are also above the solubility limits of the individual PAHs. Research should continue to explore the effects of tricyclic and tetracyclic PAHs but should also consider other constituents in oil given that PAHs represent only a minor fraction of the total mixture.

## REFERENCES

Abramochkin DV, Kompella SN, Shiels HA. 2021. Phenanthrene alters the electrical activity of atrial and ventricular myocytes of a polar fish, the Navaga cod. *Aquat Toxicol.* 235:105823.

Ackerly KL, Esbaugh AJ. 2020. The additive effects of oil exposure and hypoxia on aerobic performance in red drum (*Sciaenops ocellatus*). *Sci Total Environ.* 737:140174.

Adams J, Charbonneau K, Tuori D, Brown RS, Hodson PV. 2017. Review of methods for measuring the toxicity to aquatic organisms of the water accommodated fraction (WAF) and chemically-enhanced water accommodated fraction (CEWAF) of petroleum. *DOF Can Sci Advis Sec Res Doc.* 2017/064:xi+110p.

Aeppli C. 2022. Recent advance in understanding photooxidation of hydrocarbons after oil spills. *Curr Opin Chem Eng.* 36:100763.

Aeppli C, Carmichael CA, Nelson RK, Lemkau KL, Graham WM, Redmond MC, Valentine DL, Reddy CM. 2012. Oil weathering after the Deepwater Horizon disaster led to the formation of oxygenated residues. *Environ Sci Technol.* 46(16):8799–8807.

Aeppli C, Nelson RK, Radović JR, Carmichael CA, Valentine DL, Reddy CM. 2014. Recalcitrance and degradation of petroleum biomarkers upon abiotic and biotic natural weathering of Deepwater Horizon oil. *Environ Sci Technol.* 48(12):6726–6734.

Ainerua MO, Tinwell J, Kompella SN, Sorhus E, White KN, van Dongen BE, Shiels HA. 2020. Understanding the cardiac toxicity of the anthropogenic pollutant phenanthrene on the freshwater indicator species, the brown trout (*Salmo trutta*): From whole heart to cardiomyocytes. *Chemosphere.* 239:124608.

Alejandro Zamora-Briseno J, Amendola-Pimenta M, Antonio Ortega-Rosas D, Pereira-Santana A, Margarita Hernandez-Velazquez I, Eduardo Gonzalez-Penagos C, Antonio Perez-Vega J, del Rio-Garcia M, Arcega-Cabrera F, Rodriguez-Canul R. 2021. Gill and liver transcriptomic responses of *Achirus lineatus* (Neopterygii: Achiridae) exposed to water-accommodated fraction (WAF) of light crude oil reveal an onset of hypoxia-like condition. *Environ Sci Pollut.* 28(26):1–19.

Allmon E, Serafin J, Chen S, Rodgers ML, Griffitt RJ, Bosker T, de Guise S, Sepulveda MS. 2021. Effects of polycyclic aromatic hydrocarbons and abiotic stressors on *Fundulus grandis* cardiac transcriptomics. *Sci Total Environ.* 752:142156.

Alloy M, Garner TR, Bridges K, Mansfield C, Carney M, Forth HP, Krasnec MO, Lay CR, Takeshita R, Morris JM et al. 2017. Co-exposure to sunlight enhances the toxicity of naturally weathered Deepwater Horizon oil to early lifestage red drum (Sciaenops ocellatus) and speckled seatrout (*Cynoscion nebulosus*). *Environ Technol Sci.* 36(3):780–785.

Anderson JW, Neff JM, Cox BA, Tatem HE, Hightower GM. 1974. Characteristics of dispersions and water-soluble extracts of crude and refined oils and their toxicity to estuarine crustaceans and fish. *Mar Biol.* 27(1):75–88.

Andreasen EA, Spitsbergen JM, Tanguay RL, Stegeman JJ, Heideman W, Peterson RE. 2002. Tissue-specific expression of AHR2, ARNT2, and CYP1A in zebrafish embryos and larvae: Effects of developmental stage and 2,3,7,8-tetrachlorodibenzo-p-dioxin exposure. *Toxicol Sci.* 68(2):403–419.

National Research Council. 2014. *Responding to Oil Spills in the U.S. Arctic Marine Environment.* Washington, DC: The National Academies Press. https://doi.org/10.17226/18625.

Aranguren-Abadia L, Donald CE, Eilertsen M, Gharbi N, Tronci V, Sorhus E, Mayer P, Nilsen TO, Meier S, Goksoyr A et al. 2020. Expression and localization of the aryl hydrocarbon receptors and cytochrome p450 1a during early development of atlantic cod (*Gadus morhua*). *Aquat Toxicol.* 226:105558.

Aurand D, Coelho G. 2005. Cooperative aquatic toxicity testing of dispersed oil and the chemical response to oil spills: Ecological effects research forum (CROSERF). *Ecosystem Management and Associates, Inc Technical Report.* 105 p.

Bak S-M, Nakata H, Koh D-H, Yoo J, Iwata H, Kim E-Y. 2019. *In vitro* and *in silico* AHR assays for assessing the risk of heavy oil-derived polycyclic aromatic hydrocarbons in fish. *Ecotoxicol Environ Saf.* 181:214–223.

Barron MG, Ka'aihue L. 2003. Critical evaluation of CROSERF test methods for oil dispersant toxicity testing under subarctic conditions. *Mar Pollut Bull.* 46(9):1191–1199.

Barron MG, Vivian DN, Heintz RA, Yim UH. 2020. Long-term ecological impacts from oil spills: Comparison of Exxon Valdez, Hebei Spirit, and Deepwater Horizon. *Environ Sci Technol.* 54(11):6456–6467.

Bera G, Parkerton T, Redman A, Turner NR, Renegar DA, Sericano JL, Knap AH. 2018. Passive dosing yields dissolved aqueous exposures of crude oil comparable to the CROSERF (chemical response to oil spill: Ecological effects research forum) water accommodated fraction method. *Environ Toxicol Chem.* 37(11):2810–2819.

Billiard SM, Timme-Laragy AR, Wassenberg DM, Cockman C, Di Giulio RT. 2006. The role of the aryl hydrocarbon receptor pathway in mediating synergistic developmental toxicity of polycyclic aromatic hydrocarbons to zebrafish. *Toxicol Sci.* 92(2):526–536.

Bozinovic G, Sit TL, Di Giulio R, Wills LF, Oleksiak MF. 2013. Genomic and physiological responses to strong selective pressure during late organogenesis: Few gene expression changes found despite striking morphological differences. *BMC Genom.* 14(1):779.

BP Statistical Review of World Energy. 2021. London. https://www.bp.com/content/dam/bp/business-sites/en/global/corporate/pdfs/energy-economics/statistical-review/bp-stats-review-2021-full-report.pdf

Brakstad OG, Bonaunet K, Nordtug T, Johansen Ø. 2004. Biotransformation and dissolution of petroleum hydrocarbons in natural flowing seawater at low temperature. *Biodegradation.* 15(5):337–346.

Brakstad OG, Nordtug T, Throne-Holst M. 2015. Biodegradation of dispersed Macondo oil in seawater at low temperature and different oil droplet sizes. *Mar Pollut Bull.* 93(1–2):144–152.

Brette F, Machado B, Cros C, Incardona JP, Scholz NL, Block BA. 2014. Crude oil impairs cardiac excitation-contraction coupling in fish. *Science.* 343(6172):772–776.

Brette F, Shiels HA, Galli GLJ, Cros C, Incardona JP, Scholz NL, Block BA. 2017. A novel cardiotoxic mechanism for a pervasive global pollutant. *Sci Rep.* 7:41476.

Brown DR, Clark BW, Garner LVT, Di Giulio RT. 2015. Zebrafish cardiotoxicity: The effects of CYP1A inhibition and AHR2 knockdown following exposure to weak aryl hydrocarbon receptor agonists. *Environ Sci Pollut.* 22(11):8329–8338.

Bui A, Xiao R, Perveen Z, Kleinow K, Penn A. 2012. Zebrafish embryos sequester and retain petrochemical combustion products: Developmental and transcriptome consequences. *Aquat Toxicol.* 108:23–32.

Butler JD, Parkerton TF, Letinski DJ, Bragin GE, Lampi MA, Cooper KR. 2013. A novel passive dosing system for determining the toxicity of phenanthrene to early life stages of zebrafish. *Sci Total Environ.* 463:952–958.

Butler JD, Parkerton TF, Redman AD, Letinski DJ, Cooper KR. 2016. Assessing aromatic-hydrocarbon toxicity to fish early life stages using passive-dosing methods and target-lipid and chemical-activity models. *Environ Sci Technol.* 50(15):8305–8315.

Campo P, Venosa AD, Suidan MT. 2013. Biodegradability of Corexit 9500 and dispersed South Louisiana crude oil at 5 and 25°C. *Environ Sci Technol.* 47(4):1960–1967.

Carls MG, Heintz RA, Marty GD, Rice SD. 2005. Cytochrome p4501a induction in oil-exposed pink salmon *Oncorhynchus gorbuscha* embryos predicts reduced survival potential. *Mar Ecol Prog Ser.* 301:253–265.

Carls MG, Holland L, Larsen M, Collier TK, Scholz NL, Incardona JP. 2008. Fish embryos are damaged by dissolved PAHs, not oil particles. *Aquat Toxicol.* 88(2):121–127.

Carls MG, Marty GD, Hose JE. 2002. Synthesis of the toxicological impacts of the Exxon Valdez oil spill on Pacific herring (*Clupea pallasi*) in Prince William Sound, Alaska, USA. *Can J Fish Aquat Sci.* 59(1):153–172.

Carls MG, Rice SD, Hose JE. 1999. Sensitivity of fish embryos to weathered crude oil: Part i. Low-level exposure during incubation causes malformations, genetic damage, and mortality in larval pacific herring (*Clupea pallasi*). *Environ Toxicol Chem.* 18(3):481–493.

Cherr GN, Fairbairn E, Whitehead A. 2017. Impacts of petroleum-derived pollutants on fish development. *Annu Rev Anim Biosci.* 5(1):185–203.

Clark BW, Matson CW, Jung D, Di Giulio RT. 2010. AHR2 mediates cardiac teratogenesis of polycyclic aromatic hydrocarbons and PCB-126 in atlantic killifish (*Fundulus heteroclitus*). *Aquat Toxicol.* 99(2):232–240.

Colavecchia MV, Hodson PV, Parrott JL. 2007. The relationships among CYP1A induction, toxicity, and eye pathology in early life stages of fish exposed to oil sands. *J Toxicol Environ.* 70(18):1542–1555.

Collier TK, Connor SD, Eberhart BTL, Anulacion BF, Goksoyr A, Varanasi U. 1992. Using cytochrome-P450 to monitor the aquatic environment - initial results from regional and national surveys. *Mar Environ Res.* 34(1–4):195–199.

Cunha V, Vogs C, Bihanic FL, Dreij K. 2020. Mixture effects of oxygenated PAHs and benzo-a-pyrene on cardiovascular development and function in zebrafish embryos. *Environ Int.* 143(4):105913.

de Soysa TY, Ulrich A, Friedrich T, Pite D, Compton SL, Ok D, Bernardos RL, Downes GB, Hsieh S, Stein R et al. 2012. Macondo crude oil from the Deepwater Horizon spill disrupts specific developmental processes during zebrafish embryogenesis. *BMC Biol.* 10:doi: 10.1186/1741-7007-1110-1140.

Dearnley JM, Killeen C, Davis RL, Palace VP, Tomy GT. 2020. Monitoring polycyclic aromatic compounds exposure in fish using biliary metabolites. *Crit Rev Environ Sci Technol.* 52(18):1–45.

Denison MS, Nagy SR. 2003. Activation of the aryl hydrocarbon receptor by structurally diverse exogenous and endogenous chemicals. *Annu Rev Pharmacol Toxicol.* 43:309–334.

Di Toro DM, McGrath JA, Hansen DJ. 2000. Technical basis for narcotic chemicals and polycyclic aromatic hydrocarbon criteria. I. Water and tissue. *Environ Toxicol Chem.* 19(8):1951–1970.

Di Toro DM, McGrath JA, Stubblefield WA. 2007. Predicting the toxicity of neat and weathered crude oil: Toxic potential and the toxicity of saturated mixtures. *Environ Toxicol Chem: Int J.* 26(1):24–36.

Diamante G, Muller GdAe S, Menjivar-Cervantes N, Xu EG, Volz DC, Dias Bainy AC, Schlenk D. 2017. Developmental toxicity of hydroxylated chrysene metabolites in zebrafish embryos. *Aquat Toxicol.* 189:77–86.

Dubansky B, Rice CD, Barrois LF, Galvez F. 2017. Biomarkers of aryl-hydrocarbon receptor activity in Gulf killifish (*Fundulus grandis*) from Northern Gulf of Mexico marshes following the Deepwater Horizon oil spill. *Arch Environ Contam Toxicol.* 73(1):63–75.

Dubansky B, Whitehead A, Miller JT, Rice CD, Galvez F. 2013. Multitissue molecular, genomic, and developmental effects of the Deepwater Horizon oil spill on resident Gulf killifish (*Fundulus grandis*). *Environ Technol Sci.* 47(10):5074–5082.

Duffy TA, Childress W, Portier R, Chesney EJ. 2016. Responses of bay anchovy (*Anchoa mitchilli*) larvae under lethal and sublethal scenarios of crude oil exposure. *Ecotoxicol Environ Saf.* 134:264–272.

Edmunds RC, Gill JA, Baldwin DH, Linbo TL, French BL, Brown TL, Esbaugh AJ, Mager EM, Stieglitz JD, Hoenig R et al. 2015. Corresponding morphological and molecular indicators of crude oil toxicity to the developing hearts of *Mahi mahi*. *Sci Rep.* 5:17326.

Esbaugh AJ, Mager EM, Stieglitz JD, Hoenig R, Brown TL, French BL, Linbo TL, Lay C, Forth HP, Scholz NL et al. 2016. The effects of weathering and chemical dispersion on Deepwater Horizon crude oil toxicity to mahi-mahi (*Coryphaena hippurus*) early life stages. *Sci Total Environ.* 543:644–651.

Forth HP, Mitchelmore CL, Morris JM, Lay CR, Lipton J. 2017. Characterization of dissolved and particulate phases of water accommodated fractions used to conduct aquatic toxicity testing in support of the Deepwater Horizon natural resource damage assessment. *Environ Toxicol Chem.* 36(6):1460–1472.

French-McCay DP, Robinson HJ, Spaulding ML, Li Z, Horn M, Gloekler MD, Kim YH, Crowley D, Mendelsohn D. 2021. Validation of oil fate and mass balance for the Deepwater Horizon oil spill: Evaluation of water column partitioning. *Mar Pollut Bull.* 173:113064.

Girling AE. 1989. Preparation of aqueous media for aquatic toxicity testing of oils and oil-based products: A review of the published literature. *Chemosphere.* 19(10):1635–1641.

Greer JB, Andrzejczyk NE, Mager EM, Stieglitz JD, Benetti DD, Grosell M, Schlenk D. 2019a. Whole-transcriptome sequencing of epidermal mucus as a novel method for oil exposure assessment in juvenile mahi-mahi (*Coryphaena hippurus*). *Environ Sci Technol Lett.* 6(9):538–544.

Greer JB, Magnuson JT, McGruer V, Qian L, Dasgupta S, Volz DC, Schlenk D. 2021. miR133b microinjection during early development targets transcripts of cardiomyocyte ion channels and induces oil-like cardiotoxicity in zebrafish (*Danio rerio*) embryos. *Chem Res Toxicol*. 34(10):2209–2215.

Greer JB, Pasparakis C, Stieglitz JD, Benetti DD, Grosell M, Schlenk D. 2019b. Effects of Corexit 9500A and Corexit-crude oil mixtures on transcriptomic pathways and developmental toxicity in early life stage mahi-mahi (*Coryphaena hippurus*). *Aquat Toxicol*. 212:233–240.

Gros J, Nabi D, Würz B, Wick LY, Brussaard CPD, Huisman J, van der Meer JR, Reddy CM, Arey JS. 2014. First day of an oil spill on the open sea: Early mass transfers of hydrocarbons to air and water. *Environ Sci Technol*. 48(16):9400–9411.

Grosell M, Pasparakis C. 2021. Physiological responses of fish to oil spills. *Annu Rev Mar Sci*. 13(1):137–160.

Hahn ME. 1998. The aryl hydrocarbon receptor: A comparative perspective. *Comp Biochem Physiol C Pharmacol Toxicol Endocrinol*. 121(1–3):23–53.

Hansen BH, Parkerton T, Nordtug T, Storseth TR, Redman A. 2019a. Modeling the toxicity of dissolved crude oil exposures to characterize the sensitivity of cod (*Gadus morhua*) larvae and role of individual and unresolved hydrocarbons. *Mar Pollut Bull*. 138:286–294.

Hansen BH, Salaberria I, Read KE, Wold PA, Hammer KM, Olsen AJ, Altin D, Overjordet IB, Nordtug T, Bardal T et al. 2019b. Developmental effects in fish embryos exposed to oil dispersions- the impact of crude oil micro-droplets. *Mar Environ Res*. 150: 104753.

Hansen BH, Sorensen L, Carvalho PA, Meier S, Booth AM, Altin D, Farkas J, Nordtug T. 2018. Adhesion of mechanically and chemically dispersed crude oil droplets to eggs of Atlantic cod (*Gadus morhua*) and haddock (*Melanogrammus aeglefinus*). *Sci Total Environ*. 640:138–143.

He C, Zuo Z, Shi X, Li R, Chen D, Huang X, Chen Y, Wang C. 2011. Effects of benzo(a)pyrene on the skeletal development of *Sebastiscus marmoratus* embryos and the molecular mechanism involved. *Aquat Toxicol*. 101(2):335–341.

Head IM, Jones DM, Roling WFM. 2006. Marine microorganisms make a meal of oil. *Nat Rev Microbiol*. 4(3):173–182.

Heintz RA, Short JW, Rice SD. 1999. Sensitivity of fish embryos to weathered crude oil: Part II. Increased mortality of pink salmon (*Oncorhynchus gorbuscha*) embryos incubating downstream from weathered Exxon Valdez crude oil. *Environ Toxicol Chem*. 18(3):494–503.

Heuer RM, Galli GLJ, Shiels HA, Fieber LA, Cox GK, Mager EM, Stieglitz JD, Benetti DD, Grosell M, Crossley DA, II. 2019. Impacts of Deepwater Horizon crude oil on mahi-mahi (*Coryphaena hippurus*) heart cell function. *Environ Sci Technol*. 53(16):9895–9904.

Hicken CE, Linbo TL, Baldwin DH, Willis ML, Myers MS, Holland L, Larsen M, Stekoll MS, Rice SD, Collier TK et al. 2011. Sublethal exposure to crude oil during embryonic development alters cardiac morphology and reduces aerobic capacity in adult fish. *Proc Natl Acad Sci*. 108(17):7086–7090.

Hodson PV, Adams J, Brown RS. 2019. Oil toxicity test methods must be improved. *Environ Toxicol Chem*. 38(2):302–311.

Hodson PV, Wallace SJ, de Solla SR, Head SJ, Hepditch SLJ, Parrott JL, Thomas PJ, Berthiaume A, Langlois VS. 2020. Polycyclic aromatic compounds (PACs) in the Canadian environment: The challenges of ecological risk assessments. *Environ Pollut*. 266:115165.

Hose JE, Hannah JB, Puffer HW, Landolt ML. 1984. Histologic and skeletal abnormalities in benzo(a)pyrene-treated rainbow trout alevins. *Arch Environ Contam Toxicol* 13(6):675–684.

Hose JE, McGurk D, Marty GD, Hinton DE, Brown ED, Baker TT. 1996. Sublethal effects of the Exxon Valdez oil spill on herring embryos and larvae: Morphological, cytogenetic, and histopathological assessments, 1989-1991. *Can J Fish Aquat Sci*. 53(10):2355–2365.

Howard P, Meylan W, Aronson D, Stiteler W, Tunkel J, Comber M, Parkerton TF. 2005. A new biodegradation prediction model specific to petroleum hydrocarbons. *Environ Toxicol Chem*. 24(8):1847–1860.

Huang L, Wang C, Zhang Y, Wu M, Zuo Z. 2013. Phenanthrene causes ocular developmental toxicity in zebrafish embryos and the possible mechanisms involved. *J Hazard Mater*. 261:172–180.

Incardona JP. 2017. Molecular mechanisms of crude oil developmental toxicity in fish. *Arch Environ Contam Toxicol*. 73(1):19–32.

Incardona JP, Carls MG, Day HL, Sloan CA, Bolton JL, Collier TK, Scholz NL. 2009. Cardiac arrhythmia is the primary response of embryonic Pacific herring (*Clupea pallasi*) exposed to crude oil during weathering. *Environ Toxicol Chem*. 43(1):201–207.

Incardona JP, Carls MG, Holland L, Linbo TL, Baldwin DH, Myers MS, Peck KA, Tagal M, Rice SD, Scholz NL. 2015. Very low embryonic crude oil exposures cause lasting cardiac defects in salmon and herring. *Sci Rep*. 5:13499.

Incardona JP, Carls MG, Teraoka H, Sloan CA, Collier TK, Scholz NL. 2005. Aryl hydrocarbon receptor-independent toxicity of weathered crude oil during fish development. *Environ Health Perspect*. 113(12):1755–1762.

Incardona JP, Collier TK, Scholz NL. 2004. Defects in cardiac function precede morphological abnormalities in fish embryos exposed to polycyclic aromatic hydrocarbons. *Toxicol Appl Pharmacol*. 196(2):191–205.

Incardona JP, Day HL, Collier TK, Scholz NL. 2006. Developmental toxicity of 4-ring polycyclic aromatic hydrocarbons in zebrafish is differentially dependent on AH receptor isoforms and hepatic cytochrome P4501A metabolism. *Toxicol Appl Pharmacol*. 217(3):308–321.

Incardona JP, Scholz NL. 2016. The influence of heart developmental anatomy on cardiotoxicity-based adverse outcome pathways in fish. *Aquat Toxicol*. 177:515–525.

Incardona JP, Swarts TL, Edmunds RC, Linbo TL, Aquilina-Beck A, Sloan CA, Gardner LD, Block BA, Scholz NL. 2013. Exxon Valdez to Deepwater Horizon: Comparable toxicity of both crude oils to fish early life stages. *Aquat Toxicol*. 142:303–316.

ITOPF. 2022. Oil tanker spill statistics 2021. ITOPF Ltd, London, UK.

Johansen JL, Esbaugh AJ. 2017. Sustained impairment of respiratory function and swim performance following acute oil exposure in a coastal marine fish. *Aquat Toxicol*. 187:82–89.

Jung J-H, Hicken CE, Boyd D, Anulacion BF, Carls MG, Shim WJ, Incardona JP. 2013. Geologically distinct crude oils cause a common cardiotoxicity syndrome in developing zebrafish. *Chemosphere*. 91(8):1146–1155.

Jung J-H, Ko J, Lee E-H, Choi K-M, Kim M, Yim UH, Lee J-S, Shim WJ. 2017. RNA seq and DEG-based comparison of developmental toxicity in fish embryos of two species exposed to Iranian heavy crude oil. *Comp Biochem Physiol C Pharmacol Toxicol Endocrinol*. 196:1–10.

Kennedy CJ, Farrell AP. 2006. Effects of exposure to the water-soluble fraction of crude oil on the swimming performance and the metabolic and ionic recovery postexercise in Pacific herring (*Clupea pallasi*). *Environ Toxicol Chem.* 25(10):2715–2724.

Kompella SN, Brette F, Hancox JC, Shiels HA. 2021. Phenanthrene impacts zebrafish cardiomyocyte excitability by inhibiting I-Kr and shortening action potential duration. *J Gen Physiol.* 153(2):e202012733.

Kvenvolden KA, Cooper CK. 2003. Natural seepage of crude oil into the marine environment. *GEO-MAR Lett.* 23(3–4):140–146.

Landry ME, Adams E, Bejarano A, Boufadel M, White HK. 2019. *The Use of Dispersants in Marine Oil Spill Response.* Washington, DC: The National Academies Press.

Lee K (chair), Boufadel M, Chen B, Foght J, Hodson PV, Swanson S, Venosa A. 2015. *Expert Panel Report on the Behaviour and Environmental Impacts of Crude Oil Released into Aqueous Environments.* Ottawa, ON: Royal Society of Canada. ISBN: 978-1928140-02-3.

Lewan MD, Warden A, Dias RF, Lowry ZK, Hannah TL, Lillis PG, Kokaly RF, Hoefen TM, Swayze GA, Mills CT et al. 2014. Asphaltene content and composition as a measure of Deepwater Horizon oil spill losses within the first 80 days. *Org Geochem.* 75:54–60.

Li X, Xiong D, Ju Z, Xiong Y, Ding G, Liao G. 2021. Phenotypic and transcriptomic consequences in zebrafish early-life stages following exposure to crude oil and chemical dispersant at sublethal concentrations. *Sci Total Environ.* 763:143053.

Lie KK, Meier S, Sorhus E, Edvardsen RB, Karlsen O, Olsvik PA. 2019. Offshore crude oil disrupts retinoid signaling and eye development in larval Atlantic haddock. *Front Mar Sci.* 6(368). https://doi.org/10.3389/fmars.fmars.2019.00368.

Lindén O. 1976. The influence of crude oil and mixtures of crude oil/dispersants on the ontogenic development of the Baltic herring, *Clupea harengus membras* l. *Ambio.* 5(3):136–140.

Lindén O. 1978. Biological effects of oil on early development of the Baltic herring *Clupea harengus membras. Mar Biol.* 45(3):273–283.

MacDonald IR, Garcia-Pineda O, Beet A, Asl SD, Feng L, Graettinger DG, French-McCay D, Holmes J, Hu C, Huffer F et al. 2015. Natural and unnatural oil slicks in the Gulf of Mexico. *J Geophys Res Oceans.* 120(12):8364–8380.

Mackay D. 1982. Correlation of bioconcentration factors. *Environ Sci Technol.* 16(5):274–278.

Mackay D, Shiu WY, Ma K-C, Lee SC. 2006. *Handbook of Physical-Chemical Properties and Environmental Fate for Organic Chemicals. Vol. I, Introduction and Hydrocarbons* (2nd Edition). Boca Raton: CRC Press, Taylor & Francis Group.

Mager EM, Esbaugh AJ, Stieglitz JD, Hoenig R, Bodinier C, Incardona JP, Scholz NL, Benetti DD, Grosell M. 2014. Acute embryonic or juvenile exposure to Deepwater Horizon crude oil impairs the swimming performance of mahi-mahi (*Coryphaena hippurus*). *Environ Sci Technol.* 48(12):7053–7061.

Magnuson JT, Bautista NM, Lucero J, Lund AK, Xu EG, Schlenk D, Burggren WW, Roberts AP. 2020. Exposure to crude oil induces retinal apoptosis and impairs visual function in fish. *Environ Sci Technol.* 54(5):2843–2850.

Magnuson JT, Khursigara AJ, Allmon EB, Esbaugh AJ, Roberts AP. 2018. Effects of Deepwater Horizon crude oil on ocular development in two estuarine fish species, red drum (*Sciaenops ocellatus*) and sheepshead minnow (*Cyprinodon variegatus*). *Ecotoxicol Environ Saf.* 166:186–191.

Marris CR, Kompella SN, Miller MR, Incardona JP, Brette F, Hancox JC, Sorhus E, Shiels HA. 2020. Polyaromatic hydrocarbons in pollution: A heart-breaking matter. *J Physiol- London.* 598(2):227–247.

Marshall AG, Rodgers RP. 2004. Petroleomics: The next grand challenge for chemical analysis. *Acc Chem Res.* 37(1):53–59.

Marty GD, Short JW, Dambach DM, Willits NH, Heintz RA, Rice SD, Stegeman JJ, Hinton DE. 1997. Ascites, premature emergence, increased gonadal cell apoptosis, and cytochrome P4501A induction in pink salmon larvae continuously exposed to oil-contaminated gravel during development. *Can J Zool.* 75(6):989–1007.

McCarty LS, Dixon DG, MacKay D, Smith AD, Ozburn GW. 1992. Residue-based interpretation of toxicity and bioconcentration QSARs from aquatic bioassays: Neutral narcotic organics. *Environ Toxicol Chem.* 11(7):917–930.

McCarty LS, Mackay D. 1993. Enhancing ecotoxicological modeling and assessment. Body residues and modes of toxic action. *Environ Sci Technol.* 27(9):1718–1728.

McGrath J, Getzinger G, Redman AD, Edwards M, Aparicio AM, Vaiopoulou E. 2021. Application of the target lipid model to assess toxicity of heterocyclic aromatic compounds to aquatic organisms. *Environ Toxicol Chem.* 40(11):3000–3009.

McGrath JA, Fanelli CJ, Di Toro DM, Parkerton TF, Redman AD, Paumen ML, Comber M, Eadsfort CV, den Haan K. 2018. Re-evaluation of target lipid model-derived HC5 predictions for hydrocarbons. *Environ Toxicol Chem.* 37(6):1579–1593.

McGrath JA, Parkerton TE, Hellweger FL, Di Toro DM. 2005. Validation of the narcosis target lipid model for petroleum products: Gasoline as a case study. *Environ Toxicol Chem.* 24(9):2382–2394.

McGrath JA, Parkerton TF, Di Toro DM. 2004. Application of the narcosis target lipid model to algal toxicity and deriving predicted-no-effect concentrations. *Environ Toxicol Chem.* 23(10):2503–2517.

Meyer H. 1906. The theory of narcosis. *J Am Med Assoc.* 46:167–169.

Middaugh DP, Chapman PJ, Shelton ME. 1996. Responses of embryonic and larval inland silversides, *Menidia beryllina*, to a water-soluble fraction formed during biodegradation of artificially weathered Alaska North Slope crude oil. *Arch Environ Contam Toxicol.* 31(3):410–419.

Middaugh DP, Whiting DD. 1995. Responses of embryonic and larval inland silversides, *Menidia beryllina*, to no-2 fuel-oil and oil dispersants in seawater. *Arch Environ Contam Toxicol.* 29(4):535–539.

Morris JM, Gielazyn M, Krasnec MO, Takeshita R, Forth HP, Labenia JS, Linbo TL, French BL, Gill JA, Baldwin DH et al. 2018. Crude oil cardiotoxicity to red drum embryos is independent of oil dispersion energy. *Chemosphere.* 213:205–214.

Mousavi M, Abdollahi T, Pahlavan F, Fini E. 2016. The influence of asphaltene-resin molecular interactions on the colloidal stability of crude oil. *Fuel.* 183:262–271.

Nelson D, Stieglitz JD, Cox GK, Heuer RM, Benetti DD, Grosell M, Crossley DA, II. 2017. Cardio-respiratory function during exercise in the cobia, *Rachycentron canadum*: The impact of crude oil exposure. *Comp Biochem Physiol C Pharmacol Toxicol Endocrinol*. 201:58–65.

Nicodem DE, Guedes CLB, Fernandes MCZ, Severino D, Correa RJ, Coutinho MC, Silva J. 2001. Photochemistry of petroleum. *Prog React Kinet Mech*. 26(2–3):219–238.

Nordtug T, Olsen AJ, Altin D, Meier S, Overrein I, Hansen BH, Johansen Ø. 2011a. Method for generating parameterized ecotoxicity data of dispersed oil for use in environmental modelling. *Mar Pollut Bull*. 62(10):2106–2113.

Nordtug T, Olsen AJ, Altin D, Overrein I, Storoy W, Hansen BH, De Laender F. 2011b. Oil droplets do not affect assimilation and survival probability of first feeding larvae of North-East Arctic cod. *Sci Total Environ*. 412:148–153.

NRC. 2003. *Oil in the Sea III: Inputs, Fates, and Effects*. Washington, D.C: National Academies of Sciences. National Academies Press (US). PMID:25057607.

OECD. 2018. *Guidance Document on Aqueous-phase Aquatic Toxicity Testing of Difficult Test Chemicals*. Paris: OECD Publishing. https://doi.org/10.1787/9789264078406-en.

OECD. 2019. Guidance document on aquatic toxicity testing of difficult substances and mixtures. OECD Series on Testing and Assessment, OECD Publishing, Paris, https://doi.org/10.1787/0ed2f88e-en.

Olsvik PA, Hansen BH, Nordtug T, Moren M, Nolen E, Lie KK. 2011. Transcriptional evidence for low contribution of oil droplets to acute toxicity from dispersed oil in first feeding Atlantic cod (*Gadus morhua*) larvae. *Comp Biochem Physiol C Pharmacol Toxicol Endocrinol*. 154(4):333–345.

Overton CE. 1901. Studien über die narkose: Zugleich ein beitrag zur allgemeinen pharmakologie. *Science*. 14(361):845–846.

Payne JR, Driskell WB. 2018. Macondo oil in Northern Gulf of Mexico waters - part 1: Assessments and forensic methods for Deepwater Horizon offshore water samples. *Mar Pollut Bull*. 129(1):399–411.

Petersen GI, Kristensen P. 1998. Bioaccumulation of lipophilic substances in fish early life stages. *Environ Toxicol Chem*. 17(7):1385–1395.

Pilcher W, Miles S, Tang S, Mayer G, Whitehead A. 2014. Genomic and genotoxic responses to controlled weathered-oil exposures confirm and extend field studies on impacts of the Deepwater Horizon oil spill on native killifish. *PLoS One*. 9(9):e106351.

Prince RC, Haitmanek C, Lee CC. 2008. The primary aerobic biodegradation of biodiesel B20. *Chemosphere*. 71(8):1446–1451.

Prince RC, McFarlin KM, Butler JD, Febbo EJ, Wang FCY, Nedwed TJ. 2013. The primary biodegradation of dispersed crude oil in the sea. *Chemosphere*. 90(2):521–526.

Prince RC, Parkerton TF, Lee C. 2007. The primary aerobic biodegradation of gasoline hydrocarbons. *Environ Sci Technol*. 41(9):3316–3321.

Ramon Leggieri L, De Anna JS, Carcamo JG, Ceron GA, Arias Darraz L, Panebianco A, Luquet CM. 2019. Gills CYP1A of *Oncorhynchus mykiss* as a sensitive biomarker of crude oil pollution in freshwater environments. *Environ Toxicol Pharmacol*. 67:61–65.

Reddam A, Mager EM, Grosell M, McDonald MD. 2017. The impact of acute PAH exposure on the toadfish glucocorticoid stress response. *Aquat Toxicol*. 192:89–96.

Redman AD. 2015. Role of entrained droplet oil on the bioavailability of petroleum substances in aqueous exposures. *Mar Pollut Bull*. 97(1–2):342–348.

Redman AD, Butler JD, Letinski DJ, Parkerton TF. 2017a. Investigating the role of dissolved and droplet oil in aquatic toxicity using dispersed and passive dosing systems. *Environ Toxicol Chem*. 36(4):1020–1028.

Redman AD, McGrath JA, Stubblefield WA, Maki A, Di Toro DM. 2012. Quantifying the concentration of crude oil microdroplets in oil-water preparations. *Environ Toxicol Chem*. 31(8):1814–1822.

Redman AD, Parkerton TF. 2015. Guidance for improving comparability and relevance of oil toxicity tests. *Mar Pollut Bull*. 98(1–2):156–170.

Redman AD, Parkerton TF, Paumen ML, Butler JD, Letinski DJ, den Haan K. 2017b. A re-evaluation of PETROTOX for predicting acute and chronic toxicity of petroleum substances. *Environ Toxicol Chem*. 36(8):2245–2252.

Rodgers ML, Sherwood TA, Tarnecki AM, Griffitt RJ, Wetzel DL. 2021. Characterizing transcriptomic responses of southern flounder (*Paralichthys lethostigma*) chronically exposed to Deepwater Horizon oiled sediments. *Aquat Toxicol*. 230:105716.

Sahoo BM, Ravi Kumar BVV, Banik BK, Borah P. 2020. Polyaromatic hydrocarbons (PAHs): Structures, synthesis and their biological profile. *Curr Org Synth*. 17(8): 625–640.

Sandoval K, Ding Y, Gardinali P. 2017. Characterization and environmental relevance of oil water preparations of fresh and weathered MC-252 Macondo oils used in toxicology testing. *Sci Total Environ*. 576:118–128.

Scott JA, Incardona JP, Pelkki K, Shepardson S, Hodson PV. 2011. AhR2-mediated, CYP1A-independent cardiovascular toxicity in zebrafish (*Danio rerio*) embryos exposed to retene. *Aquat Toxicol*. 101(1):165–174.

Seemann F, Peterson DR, Witten PE, Guo BS, Shanthanagouda AH, Ye RR, Zhang G, Au DWT. 2015. Insight into the transgenerational effect of benzo[a]pyrene on bone formation in a teleost fish (*Oryzias latipes*). *Comp Biochem Physiol C*. 178:60–67.

Shankar P, Geier MC, Truong L, McClure RS, Pande P, Waters KM, Tanguay RL. 2019. Coupling genome-wide transcriptomics and developmental toxicity profiles in zebrafish to characterize polycyclic aromatic hydrocarbon (PAH) hazard. *Int J Mol Sci*. 20(10):2570.

Sorensen L, Sorhus E, Nordtug T, Incardone JP, Linbo TL, Giovanetti L, Karlsen O, Meier S. 2017. Oil droplet fouling and differential toxicokinetics of polycyclic aromatic hydrocarbons in embryos of Atlantic haddock and cod. *PLoS One*. 12(7):e0180048.

Sorhus E, Donald CE, da Silva D, Thorsen A, Karlsen O, Meier S. 2021a. Untangling mechanisms of crude oil toxicity: Linking gene expression, morphology and PAHs at two developmental stages in a cold-water fish. *Sci Total Environ*. 757:143896.

Sorhus E, Meier S, Donald CE, Furmanek T, Edvardsen RB, Lie KK. 2021b. Cardiac dysfunction affects eye development and vision by reducing supply of lipids in fish. *Sci Total Environ*. 800:149460.

Soto LA, Botello AV, Licea-Duran S, Lizarraga-Partida ML, Yanez-Arancibia A. 2014. The environmental legacy of the Ixtoc-I oil spill in Campeche Sound, southwestern Gulf of Mexico. *Front Mar Sci*. 1(57):1–9.

Stieglitz JD, Mager EM, Hoenig RH, Benetti DD, Grosell M. 2016. Impacts of Deepwater Horizon crude oil exposure on adult mahi-mahi (*Coryphaena hippurus*) swim performance. *Environ Toxicol Chem*. 35(10):2613–2622.

Sutton PA, Lewis CA, Rowland SJ. 2005. Isolation of individual hydrocarbons from the unresolved complex hydrocarbon mixture of a biodegraded crude oil using preparative capillary gas chromatography. *Org Geochem*. 36(6):963–970.

Vehniainen E-R, Haverinen J, Vornanen M. 2019. Polycyclic aromatic hydrocarbons phenanthrene and retene modify the action potential via multiple ion currents in rainbow trout *Oncorhynchus mykiss* cardiac myocytes. *Environ Toxicol Chem*. 38(10):2145–2153.

Venosa AD, Holder EL. 2007. Biodegradability of dispersed crude oil at two different temperatures. *Mar Pollut Bull*. 54(5):545–553.

Ward CP, Overton EB. 2020. How the 2010 Deepwater Horizon spill reshaped our understanding of crude oil photochemical weathering at sea: A past, present, and future perspective. *Environ Sci Process Impacts*. 22(5):1125–1138.

Ward CP, Sharpless CM, Valentine DL, French McCay DP, Aeppli C, White HK, Rodgers RP, Gosselin KM, Nelson RK, Reddy CM. 2018. Partial photochemical oxidation was a dominant, previously understated fate of Deepwater Horizon surface oil. *Proc Natl Acad Sci USA*. Feb 20;52(4):1797–1805.

Wen JB, Zhang JJ, Wang ZH, Zhang YX. 2016. Correlations between emulsification behaviors of crude oil-water systems and crude oil compositions. *JPSE*. 146:1–9.

White HK, Wang CH, Williams PL, Findley DM, Thurston AM, Simister RL, Aeppli C, Nelson RK, Reddy CM. 2016. Long-term weathering and continued oxidation of oil residues from the Deepwater Horizon spill. *Mar Pollut Bull*. 113(1–2):380–386.

Whitehead A, Dubansky B, Bodinier C, Garcia TI, Miles S, Pilley C, Raghunathan V, Roach JL, Walker N, Walter RB et al. 2012a. Genomic and physiological footprint of the Deepwater Horizon oil spill on resident marsh fishes. *PNAS*. 109(50):20298–20302.

Whitehead A, Dubansky B, Bodinier C, Garcia TI, Miles S, Pilley C, Raghunathan V, Roach JL, Walker N, Walter RB et al. 2012b. Genomic and physiological footprint of the Deepwater Horizon oil spill on resident marsh fishes. *PNAS*. 109(50):20775–20775.

Xiong KM, Peterson RE, Heideman W. 2008. Aryl hydrocarbon receptor-mediated down-regulation of sox9b causes jaw malformation in zebrafish embryos. *Mol Pharmacol*. 74(6):1544–1553.

Xu EG, Khursigara AJ, Li S, Esbaugh AJ, Dasgupta S, Volz DC, Schlenk D. 2019. mRNA-miRNA-seq reveals neuro-cardio mechanisms of crude oil toxicity in red drum (*Sciaenops ocellatus*). *Environ Sci Technol*. 53(6):3296–3305.

Xu EG, Khursigara AJ, Magnuson JT, Hazard ES, Hardiman G, Esbaugh AJ, Roberts AP, Schlenk D. 2017a. Larval red drum (*Sciaenops ocellatus*) sublethal exposure to weathered Deepwater Horizon crude oil: Developmental and transcriptomic consequences. *Environ Technol Sci*. 51(17):10162–10172.

Xu EG, Mager EM, Grosell M, Hazard ES, Hardiman G, Schlenk D. 2017b. Novel transcriptome assembly and comparative toxicity pathway analysis in mahi-mahi (*Coryphaena hippurus*) embryos and larvae exposed to Deepwater Horizon oil. *Sci Rep*. 7: 44546.

Xu EG, Mager EM, Grosell M, Pasparakis C, Schlenker LS, Stieglitz JD, Benetti D, Hazard ES, Courtney SM, Diamante G et al. 2016. Time- and oil-dependent transcriptomic and physiological responses to Deepwater Horizon oil in mahi-mahi (*Coryphaena hippurus*) embryos and larvae. *Environ Technol Sci*. 50(14):7842–7851.

Xu EG, Mager EM, Grosell M, Stieglitz JD, Hazard ES, Hardiman G, Schlenk D. 2017c. Developmental transcriptomic analyses for mechanistic insights into critical pathways involved in embryogenesis of pelagic mahi mahi (*Coryphaena hippurus*). *PLoS One*. 12(7): e0180454.

Xu EG, Magnuson JT, Diarnante G, Mager E, Pasparakis C, Grosell M, Roberts AP, Schlenk D. 2018. Changes in microRNA-mRNA signatures agree with morphological, physiological, and behavioral changes in larval mahi-mahi treated with Deepwater Horizon oil. *Environ Sci Technol*. 52(22):13501–13510.

Zhang W, Xie HQ, Li Y, Zhou M, Zhou Z, Wang R, Hahn ME, Zhao B. 2022. The aryl hydrocarbon receptor: A predominant mediator for the toxicity of emerging dioxin-like compounds. *J Hazard Mater*. 426:128084.

Zhang Y, Huang L, Zuo Z, Chen Y, Wang C. 2013. Phenanthrene exposure causes cardiac arrhythmia in embryonic zebrafish via perturbing calcium handling. *Aquat Toxicol*. 142:26–32.

Zhu L, Qu K, Xia B, Sun X, Chen B. 2016. Transcriptomic response to water accommodated fraction of crude oil exposure in the gill of Japanese flounder, *Paralichthys olivaceus*. *Mar Pollut Bull*. 106(1–2):283–291.

# Index

**Note:** **Bold** page numbers refer to tables; *italic* page numbers refer to figures.